Encyclopedia of Optimization
Second Edition

C. A. Floudas and P. M. Pardalos (Eds.)

Encyclopedia of Optimization

Second Edition

With 613 Figures and 247 Tables

Volume 1
A–C

CHRISTODOULOS A. FLOUDAS
Department of Chemical Engineering
Princeton University
Princeton, NJ 08544-5263
USA
floudas@titan.princeton.edu

PANOS M. PARDALOS
Center for Applied Optimization
Department of Industrial and Systems Engineering
University of Florida
Gainesville, FL 32611-6595
USA
pardalos@cao.ise.ufl.edu

Library of Congress Control Number: 2008927531

ISBN: 978-0-387-74758-3

The electronic publication is available under ISBN: 978-0-387-74759-0
The print and electronic bundle is available under ISBN: 978-0-387-74760-6

© 2009 Springer Science+Buisiness Media, LLC.

All rights reserved. This work may not be translated or copied in whole or in part without the written permission of the publisher (Springer Science+Business Media, LLC., 233 Spring Street, New York, NY 10013, USA), except for brief excerpts in connection with reviews or scholarly analysis. Use in connection with any form of information storage and retrieval, electronic adaptation, computer software, or by similar or dissimilar methodology now known or hereafter developed is forbidden.
The use in this publication of trade names, trademarks, service marks, and similar terms, even if they are not identified as such, is not to be taken as an expression of opinion as to whether or not they are subject to proprietary rights.

springer.com

Printed on acid free paper   SPIN: 11680840 2109letex – 5 4 3 2 1 0

Preface to the Second Edition

Optimization may be regarded as the cornerstone of many areas of applied mathematics, computer science, engineering, and a number of other scientific disciplines. Among other things, optimization plays a key role in finding feasible solutions to real-life problems, from mathematical programming to operations research, economics, management science, business, medicine, life science, and artificial intelligence, to mention only several.

Optimization entails engaging in an action to find the best solution. As a flourishing research activity, it has led to theoretical and computational advances, new technologies and new methods in developing more optimal designs of different systems, efficiency, and robustness, in minimizing the costs of operations in a process, and maximizing the profits of a company.

The first edition of the encyclopedia of optimization was well received by the scientific community and has been an invaluable source of scientific information for researchers, practitioners, and students.

Given the enormous yearly increases in this field since the appearance of the first edition, additional optimization knowledge has been added to this second edition. As before, entries are arranged in alphabetical order; the style of the entries has been retained to emphasize the expository and survey-type nature of the articles. Also many older entries have been updated and revised in light of new developments. Finally, several improvements have been made in the format to allow for more links to appropriate internet cites and electronic availability.

Acknowledgments

We wish to thank all invited contributors for their excellent efforts in writing their article in an expository way so that it is accessible to most scientists. The editors want to also take this opportunity to express gratitude to all the contributors, their research groups, their families for their support and encouragement, their research sponsors, and to Princeton University, and the University of Florida.

Many thanks go especially to Kerstin Kindler at Springer for her efficiency, terrific organization skills, and friendly spirit through the entire project. We would like also to thank the advisory board for their support and suggestions. Finally, we would like to thank the editors of Springer, Ann Kostant and Elizabeth Loew, for their thoughtful guidance, assistance and support during the planning and preparation of the second edition.

C.A. Floudas and P.M. Pardalos
Editors

About the Editors

 Christodoulos A. Floudas

Christodoulos A. Floudas is the Stephen C. Macaleer '63 Professor in Engineering and Applied Science and Professor of Chemical Engineering at Princeton University. Concurrent Faculty positions include Center for Quantitative Biology at the Lewis-Sigler Institute, Program in Applied and Computational Mathematics, and Department of Operations Research and Financial Engineering. He has held Visiting Professor positions at Imperial College, Swiss Federal Institute of Technology, ETH, University of Vienna, and the Chemical Process Engineering Research Institute (CPERI), Thessaloniki, Greece.

Dr. Floudas obtained his Ph. D. from Carnegie Mellon University, and today is a world authority in mathematical modeling and optimization of complex systems. His research interests lie at the interface of chemical engineering, applied mathematics, and operations research with principal areas of focus including chemical process synthesis and design, process control and operations, discrete-continuous nonlinear optimization, local and global optimization, and computational chemistry and molecular biology.

He has received numerous awards for teaching and research including the NSF Presidential Young Investigator Award, the Engineering Council Teaching Award, the Bodossaki Foundation Award in Applied Sciences, the Best Paper Award in Computers and Chemical Engineering, Aspen Tech Excellence in Teaching Award, the 2001 AIChE Professional Progress Award for Outstanding Progress in Chemical Engineering, the 2006 AIChE Computing in Chemical Engineering Award, and the 2007 Graduate Mentoring Award. Dr. Floudas has served on the editorial boards of *Industrial Engineering Chemistry Research*, *Journal of Global Optimization*, *Computers and Chemical Engineering*, and various book series. He has authored 2 graduate textbooks, has co-edited several volumes, has published over 200 articles, and has delivered over 300 invited lectures and seminars.

Panos M. Pardalos

Panos M. Pardalos is Distinguished Professor of Industrial and Systems Engineering at the University of Florida and the director of the Center for Applied Optimization. He is also an affiliated faculty member of the Computer Science Department, the Hellenic Studies Center, and the Biomedical Engineering Program. Pardalos has held visiting appointments at Princeton University, DIMACS Center, Institute of Mathematics and Applications, Fields Institute, AT & T Labs Research, Trier University, Linkoping Institute of Technology, and universities in Greece.

Dr. Pardalos obtained his Ph. D. from the University of Minnesota, and today is a world leading expert in global and combinatorial optimization. His primary research interests include network design problems, optimization in telecommunications, e-commerce, data mining, biomedical applications, and massive computing. He has been an invited lecturer at several universities and research institutes around the world and has organized several international conferences.

He has received numerous awards including University of Florida Research Foundation Professor, UF Doctoral Dissertation Advisor/Mentoring Award, Foreign Member of the Royal Academy of Doctors (Spain), Foreign Member Lithuanian Academy of Sciences, Foreign Member of the Ukrainian Academy of Sciences, Foreign Member of the Petrovskaya Academy of Sciences and Arts (Russia), and Honorary Member of the Mongolian Academy of Sciences. He has also received an Honorary Doctorate degree from Lobachevski University, he is a fellow of AAAS, a fellow of INFORMS, and in 2001 he was awarded the Greek National Award and Gold Medal for Operations Research. Panos Pardalos is the editor-in-chief of the *Journal of Global Optimization, Journal of Optimization Letters*, and *Computational Management Science*. Dr. Pardalos is also the managing editor of several book series, and a member of the editorial board of various international journals. He has authored 8 books, has edited several volumes, has written numerous articles, and has also developed several well known software packages.

Editorial Board Members

DIMITRI P. BERTSEKAS
McAfee Professor of Engineering
Massachusetts Institute of Technology
Cambridge, MA, USA

JOHN R. BIRGE
Jerry W. and Carol Lee Levin
Professor of Operations Management
The University of Chicago Graduate School
of Business
Chicago, IL, USA

JONATHAN M. BORWEIN
FRSC, Canada Research Chair
Computer Science Department
Dalhousie University
Halifax, NS, Canada

VLADIMIR F. DEMYANOV
Applied Mathematics Department
St. Petersburg State University
St. Petersburg, Russia

FRED GLOVER
University of Colorado
Boulder, CO, USA

OLVI L. MANGASARIAN
Computer Sciences Department
University of Wisconsin
Madison, WI, USA

ROBERT R. MEYER
Computer Sciences Department
University of Wisconsin
Madison, WI, USA

BORIS MORDUKHOVICH
Department of Mathematics
Wayne State University
Detroit, MI, USA

WALTER MURRAY
Department of Management Science and Engineering
Stanford University
Stanford, CA, USA

GEORGE L. NEMHAUSER
Chandler Chaired Professor
School of Industrial and Systems Engineering
Georgia Institute of Technology
Atlanta, GA, USA

JAMES B. ORLIN
Edward Pennell Brooks Professor
of Operations Research
MIT Sloan School of Management
Cambridge, MA, USA

J. BEN ROSEN
Computer Science Department
UCSD and University of Minnesota
La Jolla, CA, and Minneapolis, MN, USA

ROBERT B. SCHNABEL
Computer Science Department
University of Colorado
Boulder, CO, USA

HANIF D. SHERALI
W. Thomas Rice Chaired Professor of Engineering
Virginia Polytechnic Institute and State University
Blacksburg, VA, USA

RALPH E. STEUER
Department of Banking & Finance
Terry College of Business
University of Georgia
Athens, GA, USA

TAMÁS TERLAKY
Canada Research Chair in Optimization
Director, McMaster School of Computational
Engineering and Science
McMaster University
Hamilton, ON, Canada

HOANG TUY
Department of Optimization and Control
Institute of Mathematics, VAST
Hanoi, Vietnam

J.-PH. VIAL
Operations Management
University of Geneva
Geneva, Switzerland

HENRY WOLKOWICZ
Faculty of Mathematics
University of Waterloo
Waterloo, ON, Canada

List of Contributors

ACKERMANN, JUERGEN
DLR Oberpfaffenhofen
Wessling, Germany

ADJIMAN, CLAIRE S.
Imperial College
London, UK

AGGOUN, ABDERRAHMANE
KLS-OPTIM
Villebon sur Yvette, France

AHMED, SHABBIR
University of Illinois
Urbana-Champaign, IL, USA

AHUJA, RAVINDRA K.
University of Florida
Gainesville, FL, USA

AIT SAHLIA, FARID
University of Florida
Gainesville, FL, USA

ALEVRAS, DIMITRIS
IBM Corporation
West Chester, PA, USA

ALEXANDROV, NATALIA M.
NASA Langley Res. Center
Hampton, VA, USA

ALIZAMIR, SAED
University of Florida
Gainesville, FL, USA

ALKAYA, DILEK
Carnegie Mellon University
Pittsburgh, PA, USA

ALVES, MARIA JOÃO
University Coimbra and INESC
Coimbra, Portugal

ANASTASSIOU, GEORGE A.
University of Memphis
Memphis, TN, USA

ANDERSEN, ERLING D.
Odense University
Odense M, Denmark

ANDREI, NECULAI
Center for Advanced Modeling and Optimization
and
Academy of Romanian Scientists
Bucharest, Romania

ANDRICIOAEI, IOAN
Boston University
Boston, MA, USA

ANDROULAKIS, IOANNIS P.
Rutgers University
Piscataway, NJ, USA

ANSTREICHER, KURT M.
University of Iowa
Iowa City, IA, USA

ARABIE, P.
Rutgers University
Newark, DE, USA

AURENHAMMER, FRANZ
Graz University of Technology
Graz, Austria

AUSTIN-RODRIGUEZ, JENNIFER
Louisiana State University
Baton Rouge, LA, USA

BAGIROV, ADIL
University of Ballarat
Victoria, Australia

BALAS, EGON
Carnegie Mellon University
Pittsburgh, PA, USA

BALASUNDARAM, B.
Texas A&M University
College Station, TX, USA

BANERJEE, IPSITA
Rutgers University
Piscataway, NJ, USA

BAO, GANG
Michigan State University
East Lansing, MI, USA

BARD, JONATHAN F.
University of Texas
Austin, TX, USA

BARDOW, ANDRÉ
RWTH Aachen University
Aachen, Germany

BARNETTE, GREGORY
University of Florida
Gainsville, FL, USA

BARNHART, CYNTHIA
Massachusetts Institute of Technology
Cambridge, MA, USA

BATTITI, ROBERTO
Universitá Trento
Povo (Trento), Italy

BEASLEY, JOHN E.
The Management School, Imperial College
London, England

BECKER, OTWIN
University of Heidelberg
Heidelberg, Germany

BEHNKE, HENNING
Institute Math. TU Clausthal
Clausthal, Germany

BELIAKOV, GLEB
Deakin University
Victoria, Australia

BENDSØE, MARTIN P.
Technical University of Denmark
Lyngby, Denmark

BENHAMOU, FRÉDÉRIC
Université de Nantes
Nantes, France

BENSON, HAROLD P.
University of Florida
Gainesville, FL, USA

BEN-TAL, AHARON
Technion – Israel Institute of Technology
Haifa, Israel

BERTSEKAS, DIMITRI P.
Massachusetts Institute of Technology
Cambridge, MA, USA

BIEGLER, L. T.
Carnegie Mellon University
Pittsburgh, PA, USA

BILLUPS, STEPHEN C.
University of Colorado, Denver
Denver, CO, USA

BIRGE, JOHN R.
Northwestern University
Evanston, IL, USA

BIRGIN, ERNESTO G.
University of São Paulo
São Paulo, Brazil

BISCHOF, CHRISTIAN H.
RWTH Aachen University
Aachen, Germany

BJÖRCK, ÅKE
Linköping University
Linköping, Sweden

BOARD, JOHN L.G.
London School of Economics and Political Sci.
London, UK

BOMZE, IMMANUEL M.
University of Vienna
Wien, Austria

BORCHERS, BRIAN
New Mexico Technology
Socorro, NM, USA

BORGWARDT, KARL HEINZ
University of Augsburg
Augsburg, Germany

BOUYSSOU, DENIS
LAMSADE University Paris Dauphine
Paris, France

BOYD, STEPHEN
Stanford University
Stanford, CA, USA

BRANDEAU, MARGARET
Stanford University
Stanford, CA, USA

BRÄNNLUND, ULF
Kungliga Tekniska Högskolan
Stockholm, Sweden

BREZINSKI, CLAUDE
Université Sci. et Techn. Lille Flandres–Artois
Lille, France

BRIMBERG, JACK
Royal Military College of Canada
Kingston, ON, Canada

BRUALDI, RICHARD A.
University of Wisconsin
Madison, WI, USA

BRUCKER, PETER
Universität Osnabrück
Osnabrück, Germany

BRUGLIERI, MAURIZIO
University of Camerino
Camerino, Italy

BURKARD, RAINER E.
Technical University of Graz
Graz, Austria

BUSYGIN, STANISLAV
University of Florida
Gainesville, FL, USA

BUTENKO, S.
Texas A&M University
College Station, TX, USA

CALVETE, HERMINIA I.
Universidad de Zaragoza
Zaragoza, Spain

CALVIN, J. M.
New Jersey Institute of Technology
Newark, DE, USA

CAMBINI, ALBERTO
University of Pisa
Pisa, Italy

CAPAR, ISMAIL
Texas A&M University
College Station, TX, USA

CARLSON, DEAN A.
University of Toledo
Toledo, OH, USA

CARON, RICHARD J.
University of Windsor
Windsor, ON, Canada

CERVANTES, ARTURO
Mellon University
Pittsburgh, PA, USA

CHA, MEEYOUNG
Korean Advanced Institute of Science and Technology
Daejeon, Korea

CHANG, BYUNGMAN
Seoul National University of Technology
Seoul, Korea

CHAOVALITWONGSE, PAVEENA
University of Florida
Gainesville, FL, USA

CHAOVALITWONGSE, W. ART
Rutgers University
Piscataway, NJ, USA

CHARDAIRE, PIERRE
University of East Anglia
Norwich, UK

CHEN, ANTHONY
Utah State University
Logan, UT, USA

CHEN, JIANER
Texas A&M University
College Station, TX, USA

CHEN, QING
Louisiana State University
Baton Rouge, LA, USA

CHRISTIANSEN, MARIELLE
Norwegian University of Science and Technology
Trondheim, Norway

CIESLIK, DIETMAR
University of Greifswald
Greifswald, Germany

CIFARELLI, C.
Università di Roma "La Sapienza"
Roma, Italy

CIRIC, AMY
University of Cincinnati
Cincinnati, OH, USA

CLÍMACO, JOÃO
University Coimbra and INESC
Coimbra, Portugal

COMBETTES, PATRICK L.
University Pierre et Marie Curie
Paris, France
and
New York University
New York, NY, USA

COMMANDER, CLAYTON W.
University of Florida
Gainesville, FL, USA

CONEJEROS, RAÚL
Cambridge University
Cambridge, UK

CONSTANTIN, ZOPOUNIDIS
Technical University of Crete
Chania, Greece

CORLISS, GEORGE F.
Marquette University
Milwaukee, WI, USA

COTTLE, RICHARD W.
Stanford University
Stanford, CA, USA

CRAVEN, B. D.
University of Melbourne
Melbourne, VIC, Australia

CSENDES, TIBOR
University of Szeged
Szeged, Hungary

CUCINIELLO, SALVATORE
Italian Research Council
Napoli, Italy

DADUNA, JOACHIM R.
Fachhochschule für Wirtschaft (FHW) Berlin
Berlin, Germany

DANNINGER-UCHIDA, GABRIELE E.
University of Vienna
Vienna, Austria

DASCI, ABDULLAH
York University
Toronto, ON, Canada

DE ANGELIS, PASQUALE L.
Naval Institute
Naples, Italy

DE LEONE, R.
University degli Studi di Camerino
Camerino, Italy

DEMPE, STEPHAN
Freiberg University of Mining and Technology
Freiberg, Germany

DEMYANOV, VLADIMIR F.
St. Petersburg State University
St. Petersburg, Russia

DENG, NAIYANG
China Agricultural University
Beijing, China

DENG, XIAOTIE
City University of Hong Kong
Kowloon, China

DIDERICH, CLAUDE G.
Swiss Federal Institute of Technology
Lausanne, Switzerland

DI GIACOMO, LAURA
Università di Roma "La Sapienza"
Roma, Italy

DIMAGGIO JR., PETER A.
Princeton University
Princeton, NJ, USA

DIXON, LAURENCE
University of Hertfordshire
Hatfield, England

DOUMPOS, MICHAEL
Financial Engineering Lab. Techn. University Crete
Chania, Greece

DREZNER, TAMMY
California State University
Fullerton, CA, USA

DUA, PINKY
Imperial College
London, UK
and
GlaxoSmithKline Research & Development Limited
Harlow, UK

DUA, VIVEK
Imperial College
London, UK

DU, DING-ZHU
University of Texas at Dallas
Richardson, TX, USA

DUNN, JOSEPH C.
North Carolina State University
Raleigh, NC, USA

DUPAČOVÁ, JITKA
Charles University
Prague, Czech Republic

DYE, SHANE
University of Canterbury
Christchurch, New Zealand

EDIRISINGHE, CHANAKA
University of Tennessee
Knoxville, TN, USA

EDMONSON, WILLIAM
Hampton University
Hampton, VA, USA

EGLESE, RICHARD
Lancaster University
Lancaster, UK

GOUNARIS, CHRYSANTHOS E.
Princeton University
Princeton, NJ, USA

EKSIOGLU, BURAK
Mississippi State University
Mississippi State, MS, USA

EKSIOGLU, SANDRA DUNI
University of Florida
Gainesville, FL, USA

ELHEDHLI, SAMIR
McGill University
Montréal, QC, Canada

EMMONS, HAMILTON
Case Western Reserve University
Cleveland, OH, USA

ENGE, ANDREAS
University of Augsburg
Augsburg, Germany

ERLEBACH, THOMAS
University of Leicester
Leicester, UK

ERMOLIEV, YURI
International Institute for Applied Systems Analysis
Laxenburg, Austria

ESCUDERO, LAUREANO F.
M. Hernández University
Elche, Spain

ESPOSITO, WILLIAM R.
Princeton University
Princeton, NJ, USA

EVSTIGNEEV, IGOR
Russian Acadamy of Science
Moscow, Russia

FÁBIÁN, CSABA I.
Eötvös Loránd University
Budapest, Hungary

FAGERHOLT, KJETIL
Norwegian University of Science and Technology
and
Norwegian Marine Technology Research Institute (MARINTEK)
Trondheim, Norway

FAÍSCA, NUNO P.
Imperial College
London, UK

FANG, SHU-CHERNG
North Carolina State University
Raleigh, NC, USA

FANG, YUGUANG
University of Florida
Gainesville, FL, USA

FAN, YA-JU
Rutgers University
Piscataway, NJ, USA

FEMINIANO, DAVIDE
Italian Research Council
Napoli, Italy

FENG, F. J.
University of Sussex
Sussex, England

FERREIRA, AFONSO
INRIA Sophia Antipolis
Sophia–Antipolis, France

FESTA, PAOLA
Universitá Salerno
Baronissi, Italy

FIACCO, A. V.
George Washington University
Washington, DC, USA

FISCHER, HERBERT
Technical University of Munich
München, Germany

FLÅM, SJUR DIDRIK
University of Bergen
Bergen, Norway

FLOUDAS, CHRISTODOULOS A.
Princeton University
Princeton, NJ, USA

FORSGREN, ANDERS
Royal Institute of Technology (KTH)
Stockholm, Sweden

FOULDS, L. R.
University Waikato
Waikato, New Zealand

FRAUENDORFER, KARL
University of St. Gallen
St. Gallen, Switzerland

FRENK, HANS
Erasmus University
Rotterdam, The Netherlands

FU, CHANG-JUI
National Tsing Hua University
Hsinchu, Taiwan

FUNG, HO KI
Princeton University
Princeton, NJ, USA

FÜRER, MARTIN
Pennsylvania State University
University Park, PA, USA

GAGANIS, CHRYSOVALANTIS
Technical University of Crete
Chania, Greece

GALÉ, CARMEN
Universidad de Zaragoza
Zaragoza, Spain

GAO, DAVID YANG
Virginia Polytechnic Institute and State University
Blacksburg, VA, USA

GARCÍA-PALOMARES, UBALDO M.
University of Simón Bolívar
Caracas, Venezuela

GEHRLEIN, WILLIAM V.
University of Delaware
Newark, DE, USA

GENGLER, MARC
Université Méditerrannée
Marseille, France

GEROGIORGIS, DIMITRIOS I.
Imperial College
London, UK

GEUNES, JOSEPH
University of Florida
Gainesville, FL, USA

GIACOMO, LAURA DI
Università di Roma "La Sapienza"
Roma, Italy

GIANNESSI, FRANCO
University of Pisa
Pisa, Italy

GOELEVEN, DANIEL
I.R.E.M.I.A. University de la Réunion
Saint-Denis, France

GOFFIN, JEAN-LOUIS
McGill University
Montréal, QC, Canada

GOUNARIS, CHRYSANTHOS E.
Princeton University
Princeton, NJ, USA

GRAMA, ANANTH
Purdue University
West Lafayette, IN, USA

GRIEWANK, ANDREAS
Technical University of Dresden
Dresden, Germany

GRIPPO, LUIGI
Università di Roma "La Sapienza"
Roma, Italy

GROSSMANN, IGNACIO E.
Carnegie Mellon University
Pittsburgh, PA, USA

GRUNDEL, DON
671 ARSS/SYEA
Eglin AFB, FL, USA

GUARRACINO, MARIO R.
Italian Research Council
Napoli, Italy

GUDDAT, JÜRGEN
Humboldt University
Berlin, Germany

GUERRA, FRANCISCO
University of the Americas
Cholula, Mexico

GÜMÜŞ, ZEYNEP H.
University of Cincinnati
Cincinnati, OH, USA
and
Cornell University
New York, NY, USA

GUPTA, KAPIL
Georgia Institute of Technology
Atlanta, GA, USA

GÜRSOY, KORHAN
University of Cincinnati
Cincinnati, OH, USA

GUSTAFSON, SVEN-ÅKE
Stavanger University
Stavanger, Norway

GUTIN, GREGORY
University of London
Egham, UK

HADDAD, CAROLINE N.
State University of New York
Geneseo, NY, USA

HADJISAVVAS, NICOLAS
University of the Aegean
Hermoupolis, Greece

HAFTKA, RAPHAEL T.
University of Florida
Gainesville, FL, USA

HAMACHER, HORST W.
Universität Kaiserslautern
Kaiserslautern, Germany

HAN, CHI-GEUN
Kyung Hee University
Seoul, Korea

HANSEN, PIERRE
GERAD and HEC Montréal
Montréal, QC, Canada

HANSMANN, ULRICH H.E.
Michigan Technological University
Houghton, MI, USA

HARDING, S. T.
Princeton University
Princeton, NJ, USA

HARHAMMER, PETER G.
Technical University of Vienna
Vienna, Austria

HARJUNKOSKI, IIRO
Åbo Akademi University
Turku, Finland

HASLINGER, JAROSLAV
Charles University
Prague, Czech Republic

HATZINAKOS, DIMITRIOS
University of Toronto
Toronto, ON, Canada

HAUPTMAN, HERBERT A.
Hauptman–Woodward Medical Research Institute Inc.
Buffalo, NY, USA

HAURIE, ALAIN B.
University of Geneva
Geneva, Switzerland

HEARN, DONALD W.
University of Florida
Gainesville, FL, USA

HEINKENSCHLOSS, MATTHIAS
Rice University
Houston, TX, USA

HERTZ, DAVID
RAFAEL Department 82
Haifa, Israel

HETTICH, RAINER
University of Trier
Trier, Germany

HE, XIAOZHENG
University of Minnesota
Minneapolis, MN, USA

HICKS, ILLYA V.
Rice University
Houston, TX, USA

HIGLE, JULIA L.
University of Arizona
Tucson, AZ, USA

HOFFMAN, KARLA
George Mason University
Fairfax, VA, USA

HOLDER, ALLEN
University Colorado
Denver, CO, USA

HOLMBERG, KAJ
Linköping Institute of Technology
Linköping, Sweden

HOOKER, J. N.
Carnegie Mellon University
Pittsburgh, PA, USA

HOVLAND, PAUL D.
Argonne National Lab.
Argonne, IL, USA

HUANG, HONG-XUAN
Tsinghua University
Beijing, P. R. China

HUANG, XIAOXIA
University of Florida
Gainesville, FL, USA

HUBERT, L. J.
University of Illinois
Champaign, IL, USA

HUNT III, H. B.
University of Albany
New York, NY, USA

HÜRLIMANN, TONY
University of Fribourg
Fribourg, Switzerland

HURSON, CHRISTIAN
University Rouen, CREGO
Mont Saint Aignan, France

IERAPETRITOU, MARIANTHI
Rutgers University
Piscataway, NJ, USA

IRI, MASAO
Chuo University
Tokyo, Japan

ISAC, GEORGE
Royal Military College of Canada
Kingston, ON, Canada

İZBIRAK, GÖ KHAN
Eastern Mediterranean University
Mersin-10, Turkey

JACOBSEN, STEPHEN E.
University of California
Los Angeles, CA, USA

JAHN, JOHANNES
University of Erlangen–Nürnberg
Erlangen, Germany

JANAK, STACY L.
Princeton University
Princeton, NJ, USA

JANSEN, KLAUS
Univerisität Kiel
Kiel, Germany

JANSSON, C.
Techn. Universität Hamburg-Harburg
Hamburg, Germany

JEFFCOAT, DAVID
AFRL/RWGN
Eglin AFB, FL, USA

JEYAKUMAR, V.
University of New South Wales
Sydney, NSW, Australia

JHA, KRISHNA C.
GTEC
Gainesville, FL, USA

JHONES, ALINA RUIZ
University of Havana San Lázaro y L
Ciudad Habana, Cuba

JIA, ZHENYA
Rutgers University
Piscataway, NJ, USA

JONES, DONALD R.
General Motors Corp.
Warren, MI, USA

JONGEN, HUBERTUS T.
RWTH Aachen University
Aachen, Germany

JUDSON, RICHARD S.
Genaissance Pharmaceuticals
New Haven, CT, USA

KAKLAMANIS, CHRISTOS
University of Patras
Patras, Greece

KALLRATH, JOSEF
BASF Aktiengesellschaft
Ludwigshafen, Germany
and
University of Florida
Gainesville, FL, USA

KALMAN, DAN
American University
Washington, DC, USA

KAMMERDINER, ALLA R.
University of Florida
Gainesville, FL, USA

KAPLAN, ALEXANDER
University of Trier
Trier, Germany

KAPLAN, UĞUR
Koç University
Istanbul, Turkey

KASAP, SUAT
University of Oklahoma
Norman, OK, USA

KAS, PÉTER
Eastern Mediterranean University
Mersin-10, Turkey

KATOH, NAOKI
Kyoto University
Kyoto, Japan

KEARFOTT, R. BAKER
University of Louisiana at Lafayette
Lafayette, LA, USA

KELLEY, C. T.
North Carolina State University
Raleigh, NC, USA

KENNINGTON, JEFFERY L.
Southern Methodist University
Dallas, TX, USA

KESAVAN, H. K.
University of Waterloo
Waterloo, ON, Canada

KHACHIYAN, LEONID
Rutgers University
Piscataway, NJ, USA

KIM, DUKWON
University of Florida
Gainesville, FL, USA

KISIALIOU, MIKALAI
University of Minnesota
Minneapolis, MN, USA

KLAFSZKY, EMIL
Technical University
Budapest, Hungary

KLATTE, DIETHARD
University of Zurich
Zurich, Switzerland

KLEPEIS, JOHN L.
Princeton University
Princeton, NJ, USA

KNIGHT, DOYLE
Rutgers University
New Brunswick, NJ, USA

KOBLER, DANIEL
Swiss Federal Institute of Technology
Lausanne, Switzerland

KOHOUT, LADISLAV J.
Florida State University
Tallahassee, FL, USA

KOMÁROMI, ÉVA
Budapest University of Economic Sciences
Budapest, Hungary

KONNOV, IGOR V.
Kazan University
Kazan, Russia

KORHONEN, PEKKA
Internat. Institute Applied Systems Analysis
Laxenburg, Austria
and
Helsinki School Economics and Business Adm.
Helsinki, Finland

KOROTKICH, VICTOR
Central Queensland University
Mackay, QLD, Australia

KORTANEK, K. O.
University of Iowa
Iowa City, IA, USA

KOSMIDIS, VASSILEIOS D.
Imperial College
London, UK

KOSTREVA, MICHAEL M.
Clemson University
Clemson, SC, USA

KOURAMAS, K.I.
Imperial College
London, UK

KRABS, W.
University Darmstadt
Darmstadt, Germany

KRARUP, JAKOB
DIKU Universitetsparken 1
Copenhagen, Denmark

KRISHNAN, NIRANJAN
Massachusetts Institute of Technology
Cambridge, MA, USA

KROKHMAL, PAVLO A.
University of Iowa
Iowa City, IA, USA

KRUGER, ALEXANDER Y.
University of Ballarat
Ballarat, VIC, Australia

KUBOTA, KOICHI
Chuo University
Tokyo, Japan

KUEHRER, MARTIN
Siemens AG (NYSE: SI)
Wien, Austria

KUMAR, ARVIND
GTEC
Gainesville, FL, USA

KUMAR, VIPIN
University of Minnesota
Minneapolis, MN, USA

KUNDAKCIOGLU, O. ERHUN
University of Florida
Gainesville, FL, USA

KUNO, TAKAHITO
University of Tsukuba
Ibaraki, Japan

KYPARISIS, GEORGE J.
Florida International University
Miami, FL, USA

KYRIAKI, KOSMIDOU
Athens University Economics and Business
Athens, Greek

LAMAR, BRUCE W.
The MITRE Corp.
Bedford, MA, USA

LANCASTER, LAURA C.
Clemson University
Clemson, SC, USA

LAPORTE, GILBERT
HEC Montréal
Montréal, QC, Canada

LAURENT, MONIQUE
CWI
Amsterdam, The Netherlands

LAVOR, CARLILE
State University of Campinas (IMECC-UNICAMP)
Campinas, Brazil

LAWPHONGPANICH, SIRIPHONG
Naval Postgraduate School
Monterey, CA, USA

LECLERC, ANTHONY P.
College of Charleston
Charleston, SC, USA

LEDZEWICZ, URSZULA
Southern Illinois University at Edwardsville
Edwardsville, IL, USA

LEE, EVA K.
Georgia Institute of Technology
Atlanta, GA, USA

LEE, WEN
University of Florida
Gainesville, FL, USA

LEOPOLD-WILDBURGER, ULRIKE
Karl-Franzens University of Graz
Graz, Austria

LEPP, RIHO
Tallinn Technical University
Tallinn, Estonia

LETCHFORD, ADAM
Lancaster University
Lancaster, UK

LEWIS, KAREN R.
Southern Methodist University
Dallas, TX, USA

LEYFFER, S.
University of Dundee
Dundee, UK

LIANG, ZHE
Rutgers University
Piscataway, NJ, USA

LIANG, ZHIAN
Shanghai University of Finance and Economics
Shanghai, P.R. China

LIBERTI, LEO
LIX
Palaiseau, France

LI, GUANGYE
Silicon Graphics, Inc.
Houston, TX, USA

LI, HAN-LIN
National Chiao Tung University
Hsinchu, Taiwan

LIM, GINO J.
University of Houston
Houston, TX, USA

LINDBERG, P. O.
Linköping University
Linköping, Sweden

LIN, YOUDONG
University of Notre Dame
Notre Dame, IN, USA

LISSER, ABDEL
France Telecom
Issy les Moulineaux, France

LIU, WENBIN
University of Kent
Canterbury, England

LIWO, ADAM
Cornell University
Ithaca, NY, USA

L. JANAK, STACY
Princeton University
Princeton, NJ, USA

LOCKHART BOGLE, IAN DAVID
University College London
London, UK

LOUVEAUX, FRANCOIS
University of Namur
Namur, Belgium

LOWE, TIMOTHY J.
University of Iowa
Iowa City, IA, USA

LU, BING
University of Minnesota
Minneapolis, MN, USA

LUCIA, ANGELO
University of Rhode Island
Kingston, RI, USA

LUO, ZHI-QUAN
University of Minnesota
Minneapolis, MN, USA

LUUS, REIN
University of Toronto
Toronto, ON, Canada

MAAREN, HANS VAN
Delft University of Technology
Delft, The Netherlands

MACULAN, NELSON
Federal University of Rio de Janeiro (COPPE-UFRJ)
Rio de Janeiro, Brazil

MAGNANTI, THOMAS L.
Massachusetts Institute of Technology
Cambridge, MA, USA

MAIER, HELMUT
University of Ulm
Ulm, Germany

MÄKELÄ, MARKO M.
University of Jyväskylä
Jyväskylä, Finland

MÁLYUSZ, LEVENTE
Technical University
Budapest, Hungary

MAMMADOV (MAMEDOV), MUSA
University of Ballarat
Ballarat, VIC, Australia

MAPONI, PIERLUIGI
University of Camerino
Camerino, Italy

MARANAS, COSTAS D.
Pennsylvania State University
University Park, PA, USA

MARAVELIAS, CHRISTOS
University of Wisconsin – Madison
Madison, WI, USA

MARCOTTE, PATRICE
University of Montréal
Montréal, QC, Canada

MARINAKI, MAGDALENE
Technical University of Crete
Chania, Greece

MARINAKIS, YANNIS
Technical University of Crete
Chania, Greece

MARINO, MARINA
University of Naples 'Federico II' and CPS
Naples, Italy

MARQUARDT, WOLFGANG
RWTH Aachen University
Aachen, Germany

MARTEIN, LAURA
University of Pisa
Pisa, Italy

MARTI, KURT
University of Munich
Neubiberg, Germany

MARTÍNEZ, J. M.
University of Campinas
Campinas, Brazil

MATOS, ANA C.
Université Sci. et Techn. Lille Flandres–Artois
Lille, France

MAVRIDOU, THELMA D.
University of Florida
Gainesville, FL, USA

MAVROCORDATOS, P.
Algotheque and Université Paris 6
Paris, France

MCALLISTER, S. R.
Princeton University
Princeton, NJ, USA

MCDONALD, CONOR M.
E.I. DuPont de Nemours & Co.
Wilmington, DE, USA

MEDVEDEV, VLADIMIR G.
Byelorussian State University
Minsk, Republic Belarus

MEULMAN, J.
Leiden University
Leiden, The Netherlands

MIETTINEN, MARKKU
University of Jyväskylä
Jyväskylä, Finland

MINOUX, MICHEL
University of Paris
Paris, France

MISSEN, RONALD W.
University of Toronto
Toronto, ON, Canada

MISTAKIDIS, EURIPIDIS
University of Thessaly
Volos, Greece

MITCHELL, JOHN E.
Math. Sci. Rensselaer Polytechnic Institute
Troy, NY, USA

MLADENOVIĆ, NENAD
Brunel University
Uxbridge, UK

MOCKUS, JONAS
Institute Math. and Informatics
Vilnius, Lithuania

MOHEBI, HOSSEIN
University of Kerman
Kerman, Iran

MONDAINI, RUBEM P.
Federal University of Rio de Janeiro, Centre
of Technology/COPPE
Rio de Janeiro, Brazil

MONGEAU, MARCEL
University of Paul Sabatier
Toulouse, France

MÖNNIGMANN, MARTIN
Technische Universität Braunschweig
Braunschweig, Germany

MOON, SUE B.
Korean Advanced Institute of Science and Technology
Daejeon, Korea

MOORE, RAMON E.
Worthington, OH, USA

MORTON, DAVID P.
University of Texas at Austin
Austin, TX, USA

MOTREANU, DUMITRU
University of Alexandru Ioan Cuza
Iasi, Romania

MULVEY, JOHN M.
Princeton University
Princeton, NJ, USA

MURLI, ALMERICO
University of Naples Federico II
and
Center for Research on Parallel Computing
and Supercomputers of the CNR (CPS-CNR)
Napoli, Italy

MUROTA, KAZUO
Res. Institute Math. Sci. Kyoto University
Kyoto, Japan

MURPHEY, ROBERT
US Air Force Research Labor.
Eglin AFB, FL, USA

MURRAY, WALTER
Stanford University
Stanford, CA, USA

MURTY, KATTA G.
University of Michigan
Ann Arbor, MI, USA

MUTZEL, PETRA
University of Wien
Wien, Austria

NAGURNEY, ANNA
University of Massachusetts
Amherst, MA, USA

NAHAPETYAN, ARTYOM G.
University of Florida
Gainesville, FL, USA

NAKAYAMA, HIROTAKA
Konan University
Kobe, Japan

NAZARETH, J. L.
Washington State University
Pullman, WA, USA
and
University of Washington
Seattle, WA, USA

NEMIROVSKI, ARKADI
Technion: Israel Institute Technology
Technion-City, Haifa, Israel

NGO, HUANG
State University of New York at Buffalo
Buffalo, NY, USA

NICKEL, STEFAN
Universität Kaiserslautern
Kaiserslautern, Germany

NIELSEN, SØREN S.
University of Copenhagen
Copenhagen, Denmark

NIÑO-MORA, JOSÉ
Universidad Carlos III de Madrid
Getafe, Spain

NOOR, MUHAMMAD ASLAM
Dalhousie University in Halifax
Halifax, NS, Canada

NOWACK, DIETER
Humboldt University
Berlin, Germany

OKAMOTO, YUKO
Graduate University of Adv. Studies
Okazaki, Japan

OLAFSSON, SIGURDUR
Iowa State University
Ames, IA, USA

OLSON, DAVID L.
University of Nebraska
Lincoln, NE, USA

ONN, SHMUEL
Technion – Israel Institute of Technology
Haifa, Israel

ORLIK, PETER
University of Wisconsin
Madison, WI, USA

ORLIN, JAMES B.
Massachusetts Institute of Technology
Cambridge, MA, USA

OULTON, R.F.
University of California at Berkeley
Berkeley, CA, USA

PACHTER, RUTH
Air Force Research Laboratory Materials &
Manufacturing Directorate
Wright-Patterson AFB, USA

PADBERG, MANFRED
New York University
New York, NY, USA

PALAGI, LAURA
Universitá di Roma "La Sapienza"
Roma, Italy

PANAGOULI, OLYMPIA
Aristotle University
Thessaloniki, Greece

PANICUCCI, BARBARA
University of Pisa
Pisa, Italy

PAPAGEORGIOU, LAZAROS G.
UCL (University College London)
London, UK

PAPAJORGJI, PETRAQ
University of Florida
Gainesville, FL, USA

PAPALEXANDRI, KATERINA P.
bp Upstream Technology
Middlesex, UK

PAPARRIZOS, KONSTANTINOS
University of Macedonia
Thessaloniki, Greece

PAPPALARDO, MASSIMO
University of Pisa
Pisa, Italy

PARDALOS, PANOS M.
University of Florida
Gainesville, FL, USA

PARPAS, PANOS
Imperial College
London, UK

PASIOURAS, FOTIOS
University of Bath
Bath, UK

PATRIKSSON, MICHAEL
Chalmers University of Technology
Göteborg, Sweden

PATRIZI, GIACOMO
Università di Roma "La Sapienza"
Roma, Italy

PELILLO, MARCELLO
Università Ca' Foscari di Venezia
Venice, Italy

PERSIANO, GIUSEPPE
Università di Salerno
Fisciano, Italy

PFLUG, GEORG
University of Vienna
Vienna, Austria

PHILLIPS, ANDREW T.
University of Wisconsin–Eau Claire
Eau Claire, WI, USA

PIAO TAN, MENG
Princeton University
Princeton, NJ, USA

PICKENHAIN, SABINE
Brandenburg Technical University Cottbus
Cottbus, Germany

PĬNAR, MUSTAFA Ç.
Bilkent University
Ankara, Turkey

PINTÉR, JÁNOS D.
Pintér Consulting Services, Inc.,
and Dalhousie University
Halifax, NS, Canada

PISTIKOPOULOS, EFSTRATIOS N.
Imperial College
London, UK

PITSOULIS, LEONIDAS
Princeton University
Princeton, NJ, USA

POLYAKOVA, LYUDMILA N.
St. Petersburg State University
St. Petersburg, Russia

POPOVA, ELMIRA
University of Texas at Austin
Austin, TX, USA

PÖRN, RAY
Åbo Akademi University
Turku, Finland

POTVIN, JEAN-YVES
University of Montréal
Montréal, QC, Canada

POURBAIX, DIMITRI
Royal Observatory of Belgium
Brussels, Belgium

PRÉKOPA, ANDRÁS
RUTCOR, Rutgers Center for Operations Research
Piscataway, NJ, USA

PROKOPYEV, OLEG
University of Pittsburgh
Pittsburgh, PA, USA

QI, LIQUN
University of New South Wales
Sydney, NSW, Australia

QUERIDO, TANIA
University of Florida
Gainesville, FL, USA

QUEYRANNE, MAURICE
University of British Columbia
Vancouver, BC, Canada

RADZIK, TOMASZ
King's College London
London, UK

RAGLE, MICHELLE A.
University of Florida
Gainesville, FL, USA

RAI, SANATAN
Case Western Reserve University
Cleveland, OH, USA

RAJGARIA, R.
Princeton University
Princeton, NJ, USA

RALL, L. B.
University of Wisconsin–Madison
Madison, WI, USA

RALPH, DANIEL
University of Melbourne
Melbourne, VIC, Australia

RAPCSÁK, TAMÁS
Hungarian Academy of Sciences
Budapest, Hungary

RASSIAS, THEMISTOCLES M.
University Athens Zografou Campus
Athens, Greece

RATSCHEK, H.
Universität Düsseldorf
Düsseldorf, Germany

RAYDAN, MARCOS
Universidad Central de Venezuela
Caracas, Venezuela

REBENNACK, STEFFEN
University of Florida
Gainesville, FL, USA

RECCHIONI, MARIA CRISTINA
University of Ancona
Ancona, Italy

REEMTSEN, REMBERT
Brandenburg Technical University Cottbus
Cottbus, Germany

REINEFELD, ALEXANDER
ZIB Berlin
Berlin, Germany

RESENDE, MAURICIO G.C.
AT&T Labs Res.
Florham Park, NJ, USA

RIBEIRO, CELSO C.
Catholic University Rio de Janeiro
Rio de Janeiro, Brazil

RIPOLL, DANIEL R.
Cornell University
Ithaca, NY, USA

R. KAMMERDINER, ALLA
University of Florida
Gainesville, FL, USA

ROKNE, J.
University of Calgary
Calgary, AB, Canada

ROMA, MASSIMO
Universitá di Roma "La Sapienza"
Roma, Italy

ROMEIJN, H. EDWIN
University of Florida
Gainsville, FL, USA

ROOS, KEES
Delft University of Technology
AJ Delft, The Netherlands

RUBINOV, A. M.
School Inform. Techn. and Math. Sci. University Ballarat
Ballarat, VIC, Australia

RUBIN, PAUL A.
Michigan State University
East Lansing, MI, USA

RUBIO, J. E.
University of Leeds
Leeds, UK

RÜCKMANN, JAN-J.
Ilmenau University of Technology
Ilmenau, Germany

RUSTEM, BERÇ
Imperial College
London, UK

SAFONOV, MICHAEL G.
University Southern California
Los Angeles, CA, USA

SAGASTIZÁBAL, CLAUDIA
IMPA
Jardim Botânico, Brazil

SAHINIDIS, NIKOLAOS V.
University of Illinois
Urbana-Champaign, IL, USA

SAHIN, KEMAL
University of Cincinnati
Cincinnati, OH, USA

SAKIZLIS, V.
Bechtel Co. Ltd.
London, UK

SAMARAS, NIKOLAOS
University of Macedonia
Thessaloniki, Greece

SANCHEZ, SALVADOR NIETO
Louisiana State University
Baton Rouge, LA, USA

SARAIVA, PEDRO M.
Imperial College
London, UK

SAVARD, GILLES
École Polytechnique
Montréal, QC, Canada

SAVELSBERGH, MARTIN W.P.
Georgia Institute of Technology
Atlanta, GA, USA

SAYIN, SERPIL
Koç University
İstanbul, Turkey

SAYYADY, FATEMEH
North Carolina State University
Raleigh, NC, USA

SCHAIBLE, SIEGFRIED
University of California
Riverside, CA, USA

SCHÄTTLER, HEINZ
Washington University
St. Louis, MO, USA

SCHERAGA, HAROLD A.
Cornell University
Ithaca, NY, USA

SCHOEN, FABIO
Universita Firenze
Firenze, Italy

SCHULTZ, RÜDIGER
Gerhard-Mercator University
Duisburg, Germany

SCHÜRLE, MICHAEL
University of St. Gallen
St. Gallen, Switzerland

SCHWEIGER, CARL A.
Princeton University
Princeton, NJ, USA

SEN, SUVRAJEET
University of Arizona
Tucson, AZ, USA

SHAIKH, AMAN
AT&T Labs – Research
Florham Park, NJ, USA

SHAIK, MUNAWAR A.
Princeton University
Princeton, NJ, USA

SHALLOWAY, DAVID
Cornell University
Ithaca, NY, USA

SHAPIRO, ALEXANDER
Georgia Institute of Technology Atlanta
Atlanta, GA, USA

SHERALI, HANIF D.
Virginia Polytechnic Institute and State University
Blacksburg, VA, USA

SHETTY, BALA
Texas A&M University
College Station, TX, USA

SHI, LEYUAN
University of Wisconsin
Madison, WI, USA

SIM, MELVYN
NUS Business School Office
Singapore

SIMONE, VALENTINA DE
University Naples 'Federico II' and CPS
Naples, Italy

SIMONS, STEPHEN
University of California
Santa Barbara, CA, USA

SIRLANTZIS, K.
University of Kent
Canterbury, England

SISKOS, YANNIS
Technical University of Crete
Chania, Greece

SIVAZLIAN, B. D.
University of Florida
Gainesville, FL, USA

SLOWINSKI, ROMAN
Poznań University Technology
Poznań, Poland

SMITH, ALEXANDER BARTON
Carnegie Mellon University
Pittsburgh, PA, USA

SMITH, J. MACGREGOR
University of Massachusetts, Amherst
Amherst, MA, USA

SMITH, WILLIAM R.
University of Guelph
Guelph, ON, Canada

SO, ANTHONY MAN-CHO
The Chinese University of Hong Kong
Hong Kong, China

SOBIESZCZANSKI-SOBIESKI, JAROSLAW
NASA Langley Research Center
Hampton, VA, USA

SOKOLOWSKI, JAN
Université Henri Poincaré
Nancy, France

SOLODOV, MICHAEL V.
Institute Mat. Pura e Apl.
Rio de Janeiro, Brazil

SPEDICATO, EMILIO
University of Bergamo
Bergamo, Italy

SPIEKSMA, FRITS
Maastricht University
Maastricht, The Netherlands

STADTHERR, MARK A.
University of Notre Dame
Notre Dame, IN, USA

STAVROULAKIS, GEORGIOS E.
Carolo Wilhelmina Technical University
Braunschweig, Germany

STEIN, OLIVER
University of Karlsruhe
Karlsruhe, Germany

STILL, CLAUS
Åbo Akademi University
Åbo, Finland

STILL, GEORG
University of Twente
Enschede, The Netherlands

STRAUB, JOHN
Boston University
Boston, MA, USA

STRONGIN, ROMAN G.
Nizhni Novgorod State University
Nizhni Novgorod, Russia

SUGIHARA, KOKICHI
Graduate School of Engineering, University of Tokyo
Tokyo, Japan

SUN, DEFENG
University of New South Wales
Sydney, NSW, Australia

SUTCLIFFE, CHARLES M.S.
University of Southampton
Southampton, UK

SVENSSON, LARS
Royal Institute of Technology (KTH)
Stockholm, Sweden

SZÁNTAI, TAMÁS
Technical University
Budapest, Hungary

TAN, MENG PIAO
Princeton University
Princeton, NJ, USA

TAWARMALANI, MOHIT
University of Illinois
Urbana-Champaign, IL, USA

TEBOULLE, MARC
Tel-Aviv University
Ramat-Aviv, Tel-Aviv, Israel

TEGHEM, JACQUES
Polytechnique Mons
Mons, Belgium

TERLAKY, TAMÁS
McMaster University
Hamilton, ON, Canada

TESFATSION, LEIGH
Iowa State University
Ames, IA, USA

THENGVALL, BENJAMIN
CALEB Technologies Corp.
Austin, TX, USA

THOAI, NGUYEN V.
University of Trier
Trier, Germany

THOMAS, REKHA R.
University of Washington
Seattle, WA, USA

TICHATSCHKE, RAINER
University of Trier
Trier, Germany

TIND, JØRGEN
University of Copenhagen
Copenhagen, Denmark

TITS, ANDRÉ L.
University of Maryland
College Park, MD, USA

TORALDO, GERARDO
University of Naples 'Federico II' and CPS
Naples, Italy

TORVIK, VETLE I.
Louisiana State University
Baton Rouge, LA, USA

TRAFALIS, THEODORE B.
University of Oklahoma
Norman, OK, USA

TRAUB, J. F.
Columbia University
New York, NY, USA

TRIANTAPHYLLOU, EVANGELOS
Louisiana State University
Baton Rouge, LA, USA

TRLIFAJOVÁ, KATEŘINA
Charles University
Prague, Czech Republic

TSAI, JUNG-FA
National Taipei University of Technology
Taipei, Taiwan

TSAO, JACOB H.-S.
San José State University
San José, CA, USA

TSENG, PAUL
University of Washington
Seattle, WA, USA

TSIPLIDIS, KONSTANTINOS
University of Macedonia
Thessaloniki, Greece

TUNÇEL, LEVENT
University of Waterloo
Waterloo, ON, Canada

TÜRKAY, METIN
Koç University
Istanbul, Turkey

TUY, HOANG
Vietnamese Academy of Science and Technology
Hanoi, Vietnam

UBHAYA, VASANT A.
North Dakota State University
Fargo, ND, USA

URYASEV, S.
University of Florida
Gainesville, FL, USA

VAIDYANATHAN, BALACHANDRAN
University of Florida
Gainesville, FL, USA

VAIRAKTARAKIS, GEORGE
Case Western Reserve University
Cleveland, OH, USA

VANCE, PAMELA H.
Emory University
Atlanta, GA, USA

VANDENBERGHE, LIEVEN
University of California
Los Angeles, CA, USA

VAN DEN HEEVER, SUSARA
Carnegie Mellon University
Pittsburgh, PA, USA

VASANTHARAJAN, SRIRAM
Mobil Technology Company
Dallas, TX, USA

VASSILIADIS, VASSILIOS S.
Cambridge University
Cambridge, UK

VAVASIS, STEPHEN A.
Cornell University
Ithaca, NY, USA

VAZACOPOULOS, ALKIS
Dash Optimization
Englewood Cliffs, NJ, USA

VEMULAPATI, UDAYA BHASKAR
University of Central Florida
Orlando, FL, USA

VERTER, VEDAT
McGill University
Montréal, QC, Canada

VIAL, JEAN-PHILIPPE
University of Genève
Geneva, Switzerland

VICENTE, LUIS N.
University of Coimbra
Coimbra, Portugal

VINCKE, PH.
Université Libre de Bruxelles, Gestion
Brussels, Belgium

VISWESWARAN, VISWANATHAN
SCA Technologies, LLC
Pittsburgh, PA, USA

VLADIMIROU, HERCULES
University of Cyprus
Nicosia, Cyprus

VOPĚNKA, PETR
Charles University
Prague, Czech Republic

VOSS, STEFAN
University of Hamburg
Hamburg, Germany

VURAL, ARIF VOLKAN
Mississippi State University
Mississippi State, MS, USA

WALLACE, STEIN W.
Norwegian University Sci. and Techn.
Trondheim, Norway

WALTERS, JAMES B.
Marquette University
Milwaukee, WI, USA

WANG, YANJUN
Shanghai University of Finance and Economics
Shanghai, China

WANG, ZHIQIANG
Air Force Research Laboratory Materials &
Manufacturing Directorate
Wright-Patterson AFB, OH, USA

WATSON, LAYNE T.
Virginia Polytechnic Institute and State University
Blacksburg, VA, USA

WEI, JAMES
Princeton University
Princeton, NJ, USA

WERSCHULZ, A. G.
Fordham University
and
Columbia University
New York, NY, USA

WESOLOWSKY, GEORGE O.
McMaster University
Hamilton, ON, Canada

WESTERLUND, TAPIO
Åbo Akademi University
Åbo, Finland

WOLKOWICZ, HENRY
University of Waterloo
Waterloo, ON, Canada

WOOD, GRAHAM
Massey University
Palmerston North, New Zealand

WU, SHAO-PO
Stanford University
Stanford, CA, USA

WU, TSUNG-LIN
Georgia Institute of Technology
Atlanta, GA, USA

WU, WEILI
University of Minnesota
Minneapolis, MN, USA

WYNTER, LAURA
University of Versailles
Versailles-Cedex, France

XANTHOPOULOS, PETROS
University of Florida
Gainesville, FL, USA

XIA, ZUNQUAN
Dalian University of Technology
Dalian, China

XIE, WEI
American Airlines Operations Research and Decision Support Group
Fort Worth, TX, USA

XI, SHAOLIN
Beijing Polytechnic University
Beijing, China

XU, CHENGXIAN
Xian Jiaotong University
Xian, China

XUE, JUE
City University of Hong Kong
Kowloon, Hong Kong

XU, YINFENG
Xian Jiaotong University
Xian, China

YAJIMA, YASUTOSHI
Tokyo Institute of Technology
Tokyo, Japan

YANG, ERIC
Rutgers University
Piscataway, NJ, USA

YATES, JENNIFER
AT&T Labs – Research
Florham Park, NJ, USA

YAVUZ, MESUT
University of Florida
Gainesvile, FL, USA

YE, YINYU
University of Iowa
Iowa City, IA, USA

YU, GANG
University of Texas at Austin
Austin, TX, USA

YÜKSEKTEPE, FADIME ÜNEY
Koç University
Istanbul, Turkey

ZABINSKY, ZELDA B.
University of Washington
Seattle, WA, USA

ZAMORA, JUAN M.
Universidad Autónoma Metropolitana-Iztapalapa
Mexico City, Mexico

ZELIKOVSKY, ALEXANDER
Georgia State University
Atlanta, GA, USA

ZENIOS, STAVROS A.
University of Cyprus
Nicosia, Cyprus

ZHANG, JIANZHONG
City University of Hong Kong
Kowloon Tong, Hong Kong

ZHANG, LIWEI
Dalian University of Technology
Dalian, China

ZHANG, QINGHONG
University of Iowa
Iowa City, IA, USA

ZHU, YUSHAN
Tsinghua University
Beijing, China

ZIEMBA, WILLIAM T.
University of British Columbia
Vancouver, BC, Canada

ŽILINSKAS, ANTANAS
Institute of Mathematics and Informatics
Vilnius, Lithuania

ŽILINSKAS, JULIUS
Institute of Mathematics and Informatics
Vilnius, Lithuania

ZIRILLI, FRANCESCO
Università di Roma "La Sapienza"
Roma, Italy

ZISSOPOULOS, DIMITRIOS
Technical Institute of West Macedonia
Kozani, Greece

ZLOBEC, SANJO
McGill University
West Montréal, QC, Canada

ZOCHOWSKI, ANTONI
Systems Research Institute of the Polish Academy of Sciences
Warsaw, Poland

ZOPOUNIDIS, CONSTANTIN
Technical University of Crete
Chania, Greece

ZOWE, JOCHEM
University of Nürnberg–Erlangen
Erlangen, Germany

ABS Algorithms for Linear Equations and Linear Least Squares

EMILIO SPEDICATO
Department Math., University Bergamo,
Bergamo, Italy

MSC2000: 65K05, 65K10

Article Outline

Keywords
Synonyms
The Scaled ABS Class: General Properties
Subclasses of the ABS Class
The Implicit LU Algorithm
  and the Huang Algorithm
Other ABS Linear Solvers
ABS Methods for Linear Least Squares
See also
References

Keywords

Linear algebraic equations; Linear least squares; ABS methods; Abaffian matrices; Huang algorithm; Implicit LU algorithm; Implicit LX algorithm

Synonyms

Abaffi–Broyden–Spedicato algorithms for linear equations and linear least squares

The Scaled ABS Class: General Properties

ABS methods were introduced by [1], in a paper dealing originally only with solving linear equations via what is now called the *basic* or *unscaled ABS class*. The basic ABS class was later generalized to the so-called *scaled ABS class* and subsequently applied to linear least squares, nonlinear equations and optimization problems, see [2]. Preliminary work has also been initiated concerning *Diophantine equations*, with possible extensions to combinatorial optimization, and the eigenvalue problem. There are presently (1998) over 350 papers in the ABS field, see [11]. In this contribution we will review the basic properties and results of ABS methods for solving linear determined or underdetermined systems and overdetermined linear systems in the least squares sense.

Let us consider the linear determined or underdetermined system, where rank(A) is arbitrary

$$Ax = b, \quad x \in \mathbb{R}^n, b \in \mathbb{R}^m, \quad m \leq n, \quad (1)$$

or

$$a_i^\top x - b_i = 0, \quad i = 1, \dots, m, \quad (2)$$

where

$$A = \begin{pmatrix} a_1^\top \\ \vdots \\ a_m^\top \end{pmatrix}. \quad (3)$$

The steps of the scaled ABS class algorithms are as follows:

A) Let $x_1 \in \mathbf{R}^n$ be arbitrary, $H_1 \in \mathbf{R}^{n,n}$ be nonsingular arbitrary, v_1 be an arbitrary nonzero vector in $\mathbf{R}^m$; set $i = 1$.

B) Compute the residual $r_i = Ax_i - b$. If $r_i = 0$, stop (x_i solves the problem); else compute $s_i = H_i A^\top v_i$. If $s_i \neq 0$, then go to C). If $s_i = 0$ and $\tau = v_i^\top r_i = 0$, then set $x_{i+1} = x_i$, $H_{i+1} = H_i$ and go to F), else stop (the system has no solution).

C) Compute the search vector p_i by

$$p_i = H_i^\top z_i, \quad (4)$$

where $z_i \in \mathbf{R}^n$ is arbitrary save for the condition

$$v_i^\top A H_i^\top z_i \neq 0. \quad (5)$$

D) Update the estimate of the solution by

$$x_{i+1} = x_i - \alpha_i p_i, \quad (6)$$

where the stepsize α_i is given by

$$\alpha_i = \frac{v_i^\top r_i}{r_i^\top A p_i}. \quad (7)$$

E) Update the matrix H_i by

$$H_{i+1} = H_i - \frac{H_i A^\top v_i w_i^\top H_i}{w_i^\top H_i A^\top v_i}, \quad (8)$$

where $w_i \in \mathbf{R}^n$ is arbitrary save for the condition

$$w_i^\top H_i A^\top v_i \neq 0. \quad (9)$$

F) If $i = m$, then stop (x_{m+1} solves the system), else define v_{i+1} as an arbitrary vector in $\mathbf{R}^m$ but linearly independent from $v_1, \ldots, v_i$, increment i by one and go to B).

The matrices H_i appearing in step E) are generalizations of (oblique) projection matrices. They probably first appeared in [16]. They have been named *Abaffians* since the first international conference on ABS methods (Luoyang, China, 1991) and this name will be used here.

The above recursion defines a class of algorithms, each particular method being determined by the choice of the parameters H_1, v_i, z_i, w_i. The *basic ABS class* is obtained by taking $v_i = e_i$, e_i being the ith unitary vector in $\mathbf{R}^m$. The parameters w_i, z_i, H_1 have been introduced respectively by J. Abaffy, C.G. Broyden and E. Spedicato, whose initials are referred to in the name of the class. It is possible to show that the scaled ABS class is a complete realization of the so-called *Petrov–Galerkin iteration* for solving a linear system (but the principle can be applied to more general problems), where the iteration has the form $x_{i+1} = x_i - \alpha_i p_i$ with α_i, p_i chosen so that the orthogonality relation $r_{i+1}^\top v_j = 0, j = 1,$ $\ldots, i$, holds, the vectors v_j being arbitrary linearly independent. It appears that all deterministic algorithms in the literature having finite termination on a linear system are members of the scaled ABS class (this statement has been recently shown to be true also for the *quasi-Newton methods*, which are known to have under some conditions termination in at most $2n$ steps: the iterate of index $2i - 1$ generated by Broyden's iteration corresponds to the ith iterate of a certain algorithm in the ABS class).

Referring [2] for proofs, we give some of the general properties of methods of the scaled ABS class, assuming, for simplicity, that A has full rank.

- Define $V_i = (v_1, \ldots, v_i)$, $W_i = (w_1, \ldots, w_i)$. Then $H_{i+1} A^\top V_i = 0$, $H_{i+1}^\top W_i = 0$, meaning that vectors $A^\top v_j, w_j, j = 1, \ldots, i$, span the null spaces of H_{i+1} and its transpose, respectively.

- The vectors $H_i A^\top v_i, H_i^\top w_i$ are nonzero if and only if a_i, w_i are linearly independent from $a_1, \ldots, a_{i-1}$, $w_1, \ldots, w_{i-1}$, respectively.

- Define $P_i = (p_1, \ldots, p_i)$. Then the implicit factorization $V_i^\top A_i^\top P_i = L_i$ holds, where L_i is nonsingular lower triangular. From this relation, if $m = n$, one obtains the following semi-explicit factorization of the inverse, with $P = P_n, V = V_n, L = L_n$

$$A^{-1} = P L^{-1} V^\top. \quad (10)$$

For several choices of the matrix V the matrix L is diagonal, hence formula (10) gives a fully explicit factorization of the inverse as a byproduct of the ABS solution of a linear system, a property that does not hold for the classical solvers. It can also be shown that all possible factorizations of the form (10) can be obtained by proper parameter choices in the scaled ABS class, another completeness result.

- Define S_i and R_i by $S_i = (s_1, \ldots, s_i)$, $R_i = (r_1, \ldots, r_i)$, where $s_i = H_i A^\top v_i, r_i = H_i^\top w_i$. Then the Abaffian can be written in the form $H_{i+1} = H_1 - S_i R_i^\top$ and the vectors s_i, r_i can be built via a *Gram–Schmidt type iterations* involving the previous vectors (the search vector p_i can be built in a similar way). This representation of the Abaffian in terms of $2i$ vectors is computationally convenient when the number of equations is much less than the number of variables. Notice that there is also a representation in terms of $n - i$ vectors.

- A compact formula of the Abaffian in terms of the parameter matrices is the following

$$H_{i+1} = H_1 - H_1 A^\top V_i (W_i^\top H_1 A^\top V_i)^{-1} W_i^\top H_1. \quad (11)$$

Letting $V = V_m$, $W = W_m$, one can show that the parameter matrices H_1, V, W are admissible (i.e. are such that condition (9) is satisfied) if and only if the matrix $Q = V^\top A H_1^\top W$ is *strongly nonsingular* (i.e. is LU factorizable). Notice that this condition can always be satisfied by suitable exchanges of the columns of V or W, equivalent to a row or a column pivoting on the matrix Q. If Q is strongly nonsingular and we take, as is done in all algorithms insofar considered, $z_i = w_i$, then condition (5) is also satisfied.

It can be shown that the *scaled ABS class* corresponds to applying (implicitly) the unscaled ABS algorithm to the scaled (or preconditioned) system $V^\top A x = V^\top b$, where V is an arbitrary nonsingular matrix of order m. Therefore we see that the scaled ABS class is also complete with respect to all possible left preconditioning matrices, which in the ABS context are defined implicitly and dynamically (only the ith column of V is needed at the ith iteration, and it can also be a function of the previous column choices).

Subclasses of the ABS Class

In [1], nine subclasses are considered of the scaled ABS class. Here we quote three important subclasses.

- The *conjugate direction subclass*. This class is well defined under the condition (sufficient but not necessary) that A is symmetric and positive definite. It contains the *implicit Choleski algorithm*, the *Hestenes–Stiefel* and the *Lanczos algorithms*. This class generates all possible algorithms whose search directions are A-conjugate. The vector x_{i+1} minimizes the *energy* or A-weighted Euclidean norm of the error over $x_1 + \mathrm{Span}(p_1, \ldots, p_i)$. If $x_1 = 0$, then the solution is approached monotonically from below in the energy norm.
- The *orthogonally scaled subclass*. This class is well defined if A has full column rank and remains well defined even if m is greater than n. It contains the ABS formulation of the QR algorithm (the so-called *implicit QR algorithm*), of the *GMRES* and of the *conjugate residual algorithms*. The scaling vectors are orthogonal and the search vectors are $AA^\top$-conjugate. The vector x_{i+1} minimizes the Euclidean norm of the residual over $x_1 + \mathrm{Span}(p_1, \ldots, p_i)$. In general, the methods in this class can be applied to overdetermined systems to obtain the solution in the least squares sense.
- The *optimally scaled subclass*. This class is obtained by the choice $v_i = A^{-\top} p_i$. The inverse of $A^\top$ disappears in the actual formulas, if we make the change of variables $z_i = A^\top u_i$, u_i being now the parameter that defines the search vector. For $u_i = e_i$ the Huang method is obtained and for $u_i = r_i$ a method equivalent to *Craig's conjugate gradient type algorithm*. From the general implicit factorization relation one obtains $P^\top P = D$ or $V^\top A A^\top V = D$, a relation which was shown in [5] to characterize the optimal choice of the parameters in the general Petrov–Galerkin process in terms of minimizing the effect of a single error in x_i on the final computed solution. Such a property is therefore satisfied by the Huang (and the Craig) algorithm, but not, for instance, by the implicit LU or the implicit QR algorithms. A. Galantai [8] has shown that the condition characterizing the optimal choice of the scaling parameters in terms of minimizing the final residual Euclidean norm is $V^\top V = D$, a condition satisfied by the implicit QR algorithm, the GMRES method, the implicit LU algorithm and again by the Huang algorithm, which therefore satisfies both conditions). The methods in the optimally stable subclass have the property that x_{i+1} minimizes the Euclidean norm of the error over $x_1 + \mathrm{Span}(p_1, \ldots, p_i)$.

The Implicit LU Algorithm and the Huang Algorithm

Specific algorithms of the scaled ABS class are obtained by choosing the available parameters. The *implicit LU algorithm* is given by the choices $H_1 = I$, $z_i = w_i = v_i = e_i$. We quote the following properties of the implicit LU algorithm.

a) The algorithm is well defined if and only if A is *regular* (i.e. all principal submatrices are nonsingular). Otherwise column pivoting has to be performed (or, if $m = n$, equations pivoting).

b) The Abaffian H_{i+1} has the following structure, with $K_i \in \mathbf{R}^{n-i,i}$:

$$H_{i+1} = \begin{pmatrix} 0 & 0 \\ \vdots & \vdots \\ 0 & 0 \\ K_i & I_{n-i} \end{pmatrix}. \quad (12)$$

c) Only the first i components of p_i can be nonzero and the ith component is one. Hence the matrix P_i is unit upper triangular, so that the implicit factorization $A = LP^{-1}$ is of the LU type, with units on the diagonal, justifying the name.

d) Only K_i has to be updated. The algorithm requires $nm^2 - 2m^3/3$ multiplications plus lower order terms, hence, for $m = n$, $n^3/3$ multiplications plus lower order terms. This is the same overhead required by the *classical LU factorization* or *Gaussian elimination* (which are two essentially equivalent processes).

e) The main storage requirement is the storage of K_i, whose maximum value is $n^2/4$. This is two times less than the storage needed by Gaussian elimination and four times less than the storage needed by the LU factorization algorithm (assuming that A is not overwritten). Hence the implicit LU algorithm is computationally better than the classical Gaussian elimination or LU algorithm, having the same overhead but less memory cost.

The implicit LU algorithm, implemented in the case $m = n$ with row pivoting, has been shown in experiments of M. Bertocchi and Spedicato [3] to be numerically stable and in experiments of E. Bodon [4] on the vector processor Alliant FX 80 with 8 processors to be about twice faster than the *LAPACK* implementation of the classical LU algorithm.

The *Huang algorithm* is obtained by the parameter choices $H_1 = I$, $z_i = w_i = a_i$, $v_i = e_i$. A mathematically equivalent, but numerically more stable, formulation of this algorithm is the so-called *modified Huang algorithm* where the search vectors and the Abaffians are given by formulas $p_i = H_i(H_i a_i)$ and $H_{i+1} = H_i - p_i p_i^\top / p_i^\top p_i$. Some properties of this algorithm follow.

- The search vectors are orthogonal and are the same vectors obtained by applying the classical Gram–Schmidt orthogonalization procedure to the rows of A. The modified Huang algorithm is related, but is not numerically identical, with the *Daniel-Gragg-Kaufmann-Stewart reorthogonalized Gram-Schmidt algorithm* [6].

- If x_1 is the zero vector, then the vector x_{i+1} is the solution with least Euclidean norm of the first i equations and the solution x^+ of least Euclidean norm of the whole system is approached monotonically and from below by the sequence x_i. L. Zhang [17] has shown that the Huang algorithm can be applied, via the *Goldfarb–Idnani active set strategy* [9], to systems of linear inequalities. The process in a finite number of steps either finds the solution with least Euclidean norm or determines that the system has no solution.

- While the error growth in the Huang algorithm is governed by the square of the number $\eta_i = \| a_i \| / \| H_i a_i \|$, which is certainly large for some i if A is ill conditioned, the error growth depends only on η_i if p_i or H_i are defined as in the modified Huang algorithm and, at first order, there is no error growth for the modified Huang algorithm.

- Numerical experiments, see [15], have shown that the modified Huang algorithm is very stable, giving usually better accuracy in the computed solution than both the implicit LU algorithm and the classical LU factorization method.

The *implicit LX algorithm* is defined by the choices $H_1 = I$, $v_i = e_i$, $z_i = w_i = e_{k_i}$, where k_i is an integer, $1 \leq k_i \leq n$, such that

$$e_{k_i}^\top H_i a_i \neq 0. \quad (13)$$

Notice that by a general property of the ABS class for A with full rank there is at least one index k_i such that (13) is satisfied. For stability reasons it may be recommended to select k_i such that $\eta_i = |e_{k_i}^\top H_i a_i|$ is maximized.

The following properties are valid for the implicit LX algorithm. Let N be the set of integers from 1 to n, $N = (1, \ldots, n)$. Let B_i be the set of indexes $k_1, \ldots, k_i$ chosen for the parameters of the implicit LX algorithm up to the step i. Let N_i be the set $N \setminus B_i$. Then:

- The index k_i is selected in the set N_{i-1}.
- The rows of H_{i+1} of index $k \in B_i$ are null rows.
- The vector p_i has $n - i$ zero components; its k_ith component is equal to one.
- If $x_1 = 0$, then x_{i+1} is a basic type solution of the first i equations, whose nonzero components may lie

only in the positions corresponding to the indices $k \in B_i$.
- The columns of H_{i+1} of index $k \in N_i$ are the unit vectors e_k, while the columns of H_{i+1} of index $k \in B_i$ have zero components in the jth position, with $j \in B_i$, implying that only $i(n-i)$ elements of such columns have to be computed.
- At the ith step $i(n-i)$ multiplications are needed to compute $H_i a_i$ and $i(n-i)$ to update the nontrivial part of H_i. Hence the total number of multiplications is the same as for the implicit LU algorithm (i. e. $n^3/3$), but no pivoting is necessary, reflecting the fact that no condition is required on the matrix A.
- The storage requirement is the same as for the implicit LU algorithm, i. e. at most $n^2/4$. Hence the implicit LX algorithm shares the same storage advantage of the implicit LU algorithm over the classical LU algorithm, with the additional advantage of not requiring pivoting.
- Numerical experiments by K. Mirnia [10] have shown that the implicit LX method gives usually better accuracy, in terms of error in the computed solution, than the implicit LU algorithm and often even than the modified Huang algorithm. In terms of size of the final residual, its accuracy is comparable to that of the LU algorithm as implemented (with row pivoting) in the MATLAB or LAPACK libraries, but it is better again in terms of error in the solution.

Other ABS Linear Solvers

ABS reformulations have been obtained for most algorithms proposed in the literature. The availability of several formulations of the linear algebra of the ABS process allows alternative formulations of each method, with possibly different values of overhead, storage and different properties of numerical stability, vectorization and parallelization. The reprojection technique, already seen in the case of the modified Huang algorithm and based upon the identities $H_i q = H_i(H_i q)$, $H_i^\top = H_i^\top(H_i^\top q)$, valid for any vector q if $H_1 = I$, remarkably improves the stability of the algorithm. The ABS versions of the *Hestenes–Stiefel* and the *Craig algorithms* for instance are very stable under the above reprojection. The *implicit QR algorithm*, defined by the choices $H_1 = I$, $v_i = A p_i$, $z_i = w_i = e_i$ can be implemented in a very stable way using the reprojection in both the definition of the search vector and the scaling vector. It should also be noticed that the classical iterative refinement procedure, which amounts to a Newton iteration on the system $Ax - b = 0$ using the approximate factors of A, can be reformulated in the ABS context using the previously defined search vectors p_i. Experiments of Mirnia [11] have shown that ABS refinement works excellently.

For problems with special structure ABS methods can often be implemented taking into account the effect of the structure on the Abaffian matrix, which often tends to reflect the structure of the matrix A. For instance, if A has a banded structure, the same is true for the Abaffian matrix generated by the implicit LU, the implicit QR and the Huang algorithm, albeit the band size is increased. If A is SPD and has a ND structure, the same is true for the Abaffian matrix. In this case the implementation of the implicit LU algorithm has much less storage cost, for large n, than the cost required by an implementation of the Choleski algorithm. For matrices having the Kuhn–Tucker structure (KT structure) large classes of ABS methods have been devised, see ▶ ABS algorithms for optimization. For matrices with general sparsity patterns little is presently known about minimizing the fill-in in the Abaffian matrix. Careful use of BLAS4 routines can however substantially reduce the number of operations and make the ABS implementation competitive with a sparse implementation of say the LU factorization (e. g. by the code MA28) for values of n not too big.

It is possible to implement the ABS process also in block form, where several equations, instead of just one, are dealt with at each step. The block formulation does not deteriorate the numerical accuracy and can lead to reduction of overhead on special problems or to faster implementations on vector or parallel computers.

Finally infinite iterative methods can be obtained by the finite ABS methods via two approaches. The first one consists in restarting the iteration after $k < m$ steps, so that the storage will be of order $2kn$ if the representation of the Abaffian in terms of $2i$ vectors is used. The second approach consists in using only a limited number of terms in the Gram–Schmidt type processes that are alternative formulations of the ABS procedure. For both cases convergence at a linear rate has been established using the technique developed in [7]. The infinite iteration methods obtained by these approaches define a very large class of methods, that contains not only all *Krylov space type methods* of the literature, but also non-Krylov type methods as the *Gauss–Seidel*, the *De La Garza* and the *Kackmartz methods*, with their generalizations.

ABS Methods for Linear Least Squares

There are several ways of using ABS methods for solving in the least squares sense an overdetermined linear system without forming the normal equations of Gauss, which are usually avoided on the account of their higher conditioning. One possibility is to compute explicitly the factors associated with the implicit factorization and then use them in the standard way. From results of [14] the obtained methods work well, giving usually better results than the methods using the QR factorization computed in the standard way. A second possibility is to use the representation of the *Moore–Penrose pseudo-inverse* that is provided explicitly by the ABS technique described in [13]. Again this approach has given very good numerical results. A third possibility is based upon the equivalence of the normal system $A^\mathsf{T} A x = A^\mathsf{T} b$ with the extended system in the variables $x \in \mathbf{R}^n, y \in \mathbf{R}^m$, given by the two subsystems $Ax = y$, $A^\mathsf{T} y = A^\mathsf{T} b$. The first of the subsystems is overdetermined but must be solvable. Hence y must lie in the range of A^T, which means that y must be the solution of least Euclidean norm of the second underdetermined subsystem. Such a solution is computed by the Huang algorithm. Then the ABS algorithm, applied to the first subsystem, in step B) recognizes and eliminates the $m - k$ dependent equations, where k is the rank of A. If $k < n$ there are infinite solutions and the one of least Euclidean norm is obtained by using again the Huang algorithm on the first subsystem.

Finally a large class of ABS methods can be applied directly to an overdetermined system stopping after n iterations in a least squares solution. The class is obtained by defining $V = AU$, where U is an arbitrary nonsingular matrix in $\mathbf{R}^n$. Indeed at the point x_{n+1} the satisfied Petrov–Galerkin condition is just equivalent to the normal equations of Gauss. If $U = P$ then the orthogonally scaled class is obtained, implying, as already stated in section 2, that the methods of this class can be applied to solve linear least squares (but a suitable modification has to be made for the *GMRES* method). A version of the implicit QR algorithm, with reprojection on both the search vector and the scaling vector, tested in [12], has outperformed other ABS algorithms for linear least squares methods as well as methods in the *LINPACK* and *NAG library* based upon the classical QR factorization via the Householder matrices.

See also

- ▶ ABS Algorithms for Optimization
- ▶ Cholesky Factorization
- ▶ Gauss–Newton Method: Least Squares, Relation to Newton's Method
- ▶ Generalized Total Least Squares
- ▶ Interval Linear Systems
- ▶ Large Scale Trust Region Problems
- ▶ Large Scale Unconstrained Optimization
- ▶ Least Squares Orthogonal Polynomials
- ▶ Least Squares Problems
- ▶ Linear Programming
- ▶ Nonlinear Least Squares: Newton-type Methods
- ▶ Nonlinear Least Squares Problems
- ▶ Nonlinear Least Squares: Trust Region Methods
- ▶ Orthogonal Triangularization
- ▶ Overdetermined Systems of Linear Equations
- ▶ QR Factorization
- ▶ Solving Large Scale and Sparse Semidefinite Programs
- ▶ Symmetric Systems of Linear Equations

References

1. Abaffy J, Broyden CG, Spedicato E (1984) A class of direct methods for linear systems. Numerische Math, 45:361–376
2. Abaffy J, Spedicato E (1989) ABS projection algorithms: Mathematical techniques for linear and nonlinear equations. Horwood, Westergate
3. Bertocchi M, Spedicato E (1989) Performance of the implicit Gauss–Choleski algorithm of the ABS class on the IBM 3090 VF. In: Proc. 10th Symp. Algorithms, Strbske Pleso, pp 30–40
4. Bodon E (1993) Numerical experiments on the ABS algorithms for linear systems of equations. Report DMSIA Univ Bergamo 93(17)
5. Broyden CG (1985) On the numerical stability of Huang's and related methods. JOTA 47:401–412
6. Daniel J, Gragg WB, Kaufman L, Stewart GW (1976) Reorthogonalized and stable algorithms for updating

the Gram–Schmidt QR factorization. Math Comput 30: 772–795
7. Dennis J, Turner K (1987) Generalized conjugate directions. Linear Alg & Its Appl 88/89:187–209
8. Galantai A (1991) Analysis of error propagation in the ABS class. Ann Inst Statist Math 43:597–603
9. Goldfarb D, Idnani A (1983) A numerically stable dual method for solving strictly convex quadratic programming. Math Program 27:1–33
10. Mirnia K (1996) Numerical experiments with iterative refinement of solutions of linear equations by ABS methods. Report DMSIA Univ Bergamo 32/96
11. Nicolai S, Spedicato E (1997) A bibliography of the ABS methods. OMS 8:171–183
12. Spedicato E, Bodon E (1989) Solving linear least squares by orthogonal factorization and pseudoinverse computation via the modified Huang algorithm in the ABS class. Computing 42:195–205
13. Spedicato E, Bodon E (1992) Numerical behaviour of the implicit QR algorithm in the ABS class for linear least squares. Ricerca Oper 22:43–55
14. Spedicato E, Bodon E (1993) Solution of linear least squares via the ABS algorithm. Math Program 58:111–136
15. Spedicato E, Vespucci MT (1993) Variations on the Gram-Schmidt and the Huang algorithms for linear systems: A numerical study. Appl Math 2:81–100
16. Wedderburn JHM (1934) Lectures on matrices. Colloq Publ Amer Math Soc
17. Zhang L (1995) An algorithm for the least Euclidean norm solution of a linear system of inequalities via the Huang ABS algorithm and the Goldfarb–Idnani strategy. Report DMSIA Univ Bergamo 95/2

ABS Algorithms for Optimization

EMILIO SPEDICATO[1], ZUNQUAN XIA[2], LIWEI ZHANG[2]
[1] Department Math., University Bergamo, Bergamo, Italy
[2] Department Applied Math., Dalian University Technol., Dalian, China

MSC2000: 65K05, 65K10

Article Outline

Keywords
A Class of ABS Projection Methods
   for Unconstrained Optimization
Applications to Quasi-Newton Methods
ABS Methods for Kuhn–Tucker Equations
Reformulation of the Simplex Method
   via the Implicit LX Algorithm
ABS Unification of Feasible Direction Methods
   for Minimization with Linear Constraints
See also
References

Keywords

Linear equations; Optimization; ABS methods; Quasi-Newton methods; Linear programming; Feasible direction methods; KT equations; Interior point methods

The *scaled ABS* (Abaffy–Broyden–Spedicato) class of algorithms, see [1] and ▶ ABS algorithms for linear equations and linear least squares, is a very general process for solving linear equations, realizing the so-called *Petrov–Galerkin approach*. In addition to solving general determined or underdetermined linear systems $Ax = b$, $x \in \mathbb{R}^n$, $b \in \mathbb{R}^m$, $m \leq n$, rank$(A) \leq m$, $A = [a_1, \ldots a_m]^\mathsf{T}$, ABS methods can also solve linear least squares problems and nonlinear algebraic equations. In this article we will consider applications of ABS methods to optimization problems. We will consider only the so-called *basic ABS class*, defined by the following procedure for solving $Ax = b$:

A) Let $x_1 \in \mathbb{R}^n$ be arbitrary, $H_1 \in \mathbb{R}^{n,n}$ be nonsingular arbitrary, set $i = 1$.
B) Compute $s_i = H_i a_i$. IF $s_i \neq 0$, go to C).
   IF $s_i = 0$ and $\tau = a_i^\mathsf{T} x_i - b_i = 0$, THEN set $x_{i+1} = x_i$, $H_{i+1} = H_i$ and go to F), ELSE stop, the system has no solution.
C) Compute the search vector p_i by $p_i = H_i^\mathsf{T} z_i$, where $z_i \in \mathbb{R}^n$ is arbitrary save for the condition $a_i^\mathsf{T} H_i^\mathsf{T} z_i \neq 0$.
D) Update the estimate of the solution by $x_{i+1} = x_i - \alpha_i p_i$, where the stepsize α_i is given by $\alpha_i = (a_i^\mathsf{T} p_i - b_i)/a_i^\mathsf{T} p_i$.
E) Update the matrix H_i by $H_{i+1} = H_i - H_i a_i w_i^\mathsf{T} H_i / w_i^\mathsf{T} H_i a_i$, where $w_i \in \mathbb{R}^n$ is arbitrary save for the condition $w_i^\mathsf{T} H_i a_i \neq 0$.
F) IF $i = m$, THEN stop; x_{m+1} solves the system, ELSE increment i by one and go to B).

Among the properties of the ABS class the following is fundamental in the applications to optimization. Let

$m < n$ and, for simplicity, assume that rank(A) = m. Then the linear variety containing all solutions of the underdetermined system $Ax = b$ is represented by the vectors x of the form

$$x = x_{m+1} + H_{m+1}^\top q, \qquad (1)$$

where $q \in \mathbb{R}^n$ is arbitrary. In the following the matrices generated by the ABS process will be called *Abaffians*. It is recalled that the matrix H_{i+1} can be represented in terms of either $2i$ vectors or of $n - i$ vectors, which is also true for the representation of the search vector p_i. The first representation is computationally convenient for systems where the number of equations is small (less than $n/2$), while the second one is suitable for problems where m is close to n. In the applications to optimization, the first case corresponds to problems with few constraints (many degrees of freedom), the second case to problems with many constraints (few degrees of freedom).

Among the algorithms of the basic ABS class, the following are particularly important.

a) The *implicit LU algorithm* is given by the choices $H_1 = I$, $z_i = w_i = e_i$, where e_i is the ith unit vector in $\mathbb{R}^n$. This algorithm is well defined if and only if A is regular (otherwise pivoting of the columns has to be performed, or of the equations, if $m = n$). Due to the special structure of the Abaffian induced by the parameter choices (the first i rows of H_{i+1} are identically zero, while the last $n - i$ columns are unit vectors) the maximum storage is $n^2/4$, hence 4 times less than for the classical LU factorization or twice less than for *Gaussian elimination*; the number of multiplications is $nm^2 - 2m^3/3$, hence, for $m = n$, $n^3/3$, i. e. the same as for Gaussian elimination or the LU factorization algorithm.

b) The *Huang algorithm* is obtained by the parameter choices $H_1 = I$, $z_i = w_i = a_i$. A mathematically equivalent, but numerically more stable, formulation of this algorithm is the so-called *modified Huang algorithm* where the search vectors and the Abaffians are given by formulas $p_i = H_i(H_i a_i)$ and $H_{i+1} = H_i - p_i p_i^\top / p_i^\top p_i$. The search vectors are orthogonal and are equal to the vectors obtained by applying the classical *Gram-Schmidt orthogonalization* procedure to the rows of A. If x_1 is the zero vector, then the vector x_{i+1} is the solution of least Euclidean norm of the first i equations and the solution x^+ of least Euclidean norm of the whole system is approached monotonically and from below by the sequence x_i.

c) The *implicit LX algorithm*, where 'L' refers to the lower triangular left factor while 'X' refers to the right factor, which is a matrix obtainable after row permutation of an upper triangular matrix, considered by Z. Xia, is defined by the choices $H_1 = I$, $z_i = w_i = e_{k_i}$ where k_i is an integer, $1 \leq k_i \leq n$, such that

$$e_{k_i}^\top H_i a_i \neq 0. \qquad (2)$$

If A has full rank, from a property of the basic ABS class the vector $H_i a_i$ is nonzero, hence there is at least one index k_i such that (2) is satisfied. The implicit LX algorithm has the same overhead as the implicit LU algorithm, hence the same as Gaussian elimination, and the same storage requirement, i. e. less than Gaussian elimination or the LU factorization algorithm. It has the additional advantage of not requiring any condition on the matrix A, hence pivoting is not necessary. The structure of the Abaffian matrix is somewhat more complicated than for the implicit LU algorithm, the zero rows of H_{i+1} being now in the positions $k_1, \ldots, k_i$ and the columns that are unit vectors being in the positions that do not correspond to the already chosen indices k_i.

The vector p_i has $n - i$ zero components and its k_ith component is equal to one. It follows that if $x_1 = 0$, then x_{i+1} is a basic type solution of the first i equations, whose nonzero components correspond to the chosen indices k_i.

In this paper we will present the following applications of ABS methods to optimization problems. In Section 2 we describe a class of ABS related methods for the unconstrained optimization problem. In Section 3 we show how ABS methods provide the general solution of the quasi-Newton equation, also with sparsity and symmetry and we discuss how SPD solutions can be obtained. In Section 4 we present several special ABS methods for solving the Kuhn–Tucker equations. In Section 5 we consider the application of the implicit LX algorithm to the linear programming (LP) problem. In Section 6 we present ABS approaches to the general linearly constrained optimization problem, which unify linear and nonlinear problems.

A Class of ABS Projection Methods for Unconstrained Optimization

ABS methods can be applied directly to solve *unconstrained optimization* problems via the iteration $x_{i+1} = x_i - \alpha_i H_i^\top z_i$, where H_i is reset after n or less steps and z_i is chosen so that the descent condition holds, i. e. $g_i^\top H_i^\top z_i > 0$, with g_i the gradient of the function at x_i. If the function to be minimized is quadratic, one can identify the matrix A in the Abaffian update formula with the Hessian of the quadratic function. Defining a perturbed point x' by $x' = x_i - \beta v_i$ one has on quadratic functions $g' = g - \beta A v_i$, hence the update of the Abaffian takes the form $H_{i+1} = H_i - H_i y_i w_i^\top H_i / w_i^\top H_i y_i$, where $y_i = g' - g_i$. The above defined class has termination on quadratic functions and local superlinear (n-step Q-quadratic) rate of convergence on general functions. It is a special case of a class of projection methods developed in [7]. Almost no numerical results are available about the performance of the methods in this class.

Applications to Quasi-Newton Methods

ABS methods have been used to provide the general solution of the quasi-Newton equation, also with the additional conditions of symmetry, sparsity and positive definiteness. While the general solution of only the quasi-Newton equation was already known from [2], the explicit formulas obtained for the sparse symmetric case are new, and so is the way of constructing sparse SPD updates.

Let us consider the quasi-Newton equation defining the new approximation to a Jacobian or a Hessian, in the transpose form

$$d^\top B' = y^\top, \quad (3)$$

where $d = x' - x$, $y = g' - g$. We observe that (3) can be seen as a set of n linear underdetermined systems, each one having just one equation and differing only in the right-hand side. Hence the general solution can be obtained by one step of the ABS method. It can be written in the following way

$$B' = B - \frac{s(B^\top d - y)^\top}{d^\top s} + \left(I - \frac{s d^\top}{d^\top s}\right) Q, \quad (4)$$

where $Q \in \mathbb{R}^{n,n}$ is arbitrary and $s \in \mathbb{R}^n$ is arbitrary subject to $s^\top d \neq 0$. Formula (4), derived in [9], is equivalent to the formula in [2].

Now the conditions that some elements of B' should be zero, or have constant value or that B' should be symmetric can be written as the additional linear constraints, where b'_i is the ith column of B'

$$(b'_i)^\top e_k = \eta_{ij}, \quad (5)$$

where $\eta_{ij} = 0$ implies sparsity, η_{ij} = const implies that some elements do not change their value and $\eta_{ij} = \eta_{ji}$ implies symmetry. The ABS algorithm can deal with these extra conditions, see [11], giving the solution in explicit form, columnwise in presence of symmetry. By adding the additional condition that the diagonal elements be sufficiently large, it is possible to obtain formulas where B' is quasi positive definite or quasi diagonally dominant, in the sense that the principal submatrix of order $n - 1$ is positive definite or diagonally dominant. It is not possible in general to force B' to be SPD, since SPD solutions may not exist, which is reflected in the fact that no additional conditions can be put on the last diagonal element, since the last column is fully determined by the $n - 1$ symmetry conditions and the quasi-Newton equation. This result can however be exploited to provide SPD approximations by imbedding the original minimization problem of n variables in a problem of $n + 1$ variables, whose solution with respect to the first n variables is the original solution (just set, for instance, $f(x') = f(x) + x_{n+1}^2$). This imbedding modifies the quasi-Newton equation so that SPD solutions exist.

ABS Methods for Kuhn–Tucker Equations

The *Kuhn–Tucker equations* (KT equations), which should more appropriately be named *Kantorovich–Karush–Kuhn–Tucker equations* (KKKT equations), are a special linear system, obtained by writing the optimality conditions of the problem of minimizing a quadratic function with Hessian G subject to the linear equality constraint $Cx = b$. They are the system $Ax = b$, where A is a symmetric indefinite matrix of the following form, with $G \in \mathbb{R}^{n,n}$, $C \in \mathbb{R}^{m,n}$

$$A = \begin{pmatrix} G & C^\top \\ C & 0 \end{pmatrix}. \quad (6)$$

If G is nonsingular, then A is nonsingular if and only if $CG^{-1}C^T$ is nonsingular. Usually G is nonsingular, symmetric and positive definite, but this assumption, required by several classical solvers, is not necessary for the ABS solvers.

ABS classes for solving the KT problem can be derived in several ways. Observe that system (6) is equivalent to the two subsystems

$$Gp + C^T z = g, \quad (7)$$
$$Cp = c, \quad (8)$$

where $x = (p^T, z^T)^T$ and $b = (g^T, C^T)^T$. The general solution of subsystem (8) has the form, see (1)

$$p = p_{m+1} + H_{m+1}^T q, \quad (9)$$

with q arbitrary. The parameter choices made to construct p_{m+1} and H_{m+1} are arbitrary and define therefore a class of algorithms.

Since the KT equations have a unique solution, there must be a choice of q in (9) which makes p be the unique n-dimensional subvector defined by the first n components of the solution x. Notice that since H_{m+1} is singular, q is not uniquely defined (but would be uniquely defined if one takes the representation of the Abaffian in terms of $n - m$ vectors).

By multiplying equation (7) on the left by H_{m+1} and using the ABS property $H_{m+1} C^T = 0$, we obtain the equation

$$H_{m+1} G p = H_{m+1} g, \quad (10)$$

which does not contain z. Now there are two possibilities to determine p:

A1) Consider the system formed by equations (8) and (10). Such a system is solvable but overdetermined. Since rank(H_{m+1}) = $n - m$, m equations are recognized as dependent and are eliminated in step B) of any ABS algorithm applied to this system.

A2) In equation (10) substitute p with the expression of the general solution (9) obtaining

$$H_{m+1} G H_{m+1}^T q = H_{m+1} g - H_{m+1} G p_{m+1}. \quad (11)$$

The above system can be solved by any ABS method for a particular solution q, m equations being again removed at step B) of the ABS algorithm as linearly dependent.

Once p is determined, there are two approaches to determine z, namely:

B1) Solve by any ABS method the overdetermined compatible system

$$C^T z = g - Gp \quad (12)$$

by removing at step B) of the ABS algorithm the $n - m$ dependent equations.

B2) Let $P = (p_1, \ldots p_m)$ be the matrix whose columns are the search vectors generated on the system $Cp = c$. Now $CP = L$, with L nonsingular lower diagonal. Multiplying equation (12) on the left by P^T we obtain a triangular system, defining z uniquely

$$L^T z = P^T g - P^T G p. \quad (13)$$

Extensive numerical testing has evaluated the accuracy of the above considered ABS algorithms for KT equations for certain choices of the ABS parameters (corresponding to the implicit LU algorithm with row pivoting and the modified Huang algorithm). The methods have been tested against classical methods, in particular the method of Aasen and methods using the QR factorization. The experiments have shown that some ABS methods are the most accurate, in both residual and solution error; moreover some ABS algorithms are cheaper in storage and in overhead, up to one order, especially for the case when m is close to n.

In many interior point methods the main computational cost is to compute the solution for a sequence of KT problems where only G, which is diagonal, changes. In such a case the ABS methods, which initially work on the matrix C, which is unchanged, are advantaged, particularly when m is large, where the dominant cubic term decreases with m and disappears for $m = n$, so that the overhead is dominated by second order terms. Again numerical experiments show that some ABS methods are more accurate than the classical ones. For details see [8].

Reformulation of the Simplex Method via the Implicit LX Algorithm

The implicit LX algorithm has a natural application to a reformulation of the simplex method for the LP problem in standard form, i. e. the problem

$$\begin{cases} \min & c^\mathsf{T} x \\ \text{s.t.} & Ax = b \\ & x \geq 0 \, . \end{cases}$$

The applicability of the implicit LX method is a consequence of the fact that the iterate x_{i+1} generated by the method, started from the zero vector, is a basic type vector, with a unit component in the position k_i, non identically zero components corresponding to indices $j \in B_i$, where B_i is the set of indices of the unit vectors chosen as the z_i, w_i parameters, i. e. the set $B_i = (k_1, \ldots, k_i)$, while the components of x_{i+1} of indices in the set $N_i = N/B_i$ are identically zero, where $N = (1, \ldots n)$. Therefore, if the nonzero components are nonnegative, the point defines a vertex of the polytope containing the feasible points defined by the constraints of the LP problem.

In the simplex method one moves from a vertex to another one, according to some rules and usually reducing at each step the value of the function $c^\mathsf{T} x$. The direction along which one moves from a vertex to another one is an edge direction of the polytope and is determined by solving a linear system, whose coefficient matrix A_B, the *basic matrix*, is defined by m linearly independent columns of the matrix A, called the *basic columns*. Usually such a system is solved by the LU factorization method or occasionally by the *QR method*, see [5]. The new vertex is associated to a new basic matrix $A_{B'}$, which is obtained by substituting one of the columns in A_B by a column of the matrix A_N, which comprises the columns of A that do not belong to A_B. The most efficient algorithm for solving the modified system, after the column interchange, is the *Forrest–Goldfarb method* [6], requiring m^2 multiplications. Notice that the classical simplex method requires m^2 storage for the matrix A_B plus mn storage for the matrix A, which must be kept in general to provide the columns for the exchange.

The application of the implicit LX method to the simplex method, developed in [4,10,13,17] exploits the fact that in the implicit LX algorithm the interchange of a jth column in A_B with a kth column in A_N corresponds to the interchange of a previously chosen parameter vector $z_j = w_j = e_j$ with a new parameter $z_k = w_k$ = e_k. This operation is a special case of the perturbation of the Abaffian after a change in the parameters and can be done using a general formula of [15], without explicit use of the kth column in A_N. Moreover since all quantities which are needed for the construction of the search direction (the edge direction) and for the interchange criteria can as well be implemented without explicit use of the columns of A, it follows that the ABS approach needs only the storage of the matrix H_{m+1}, which, in the case of the implicit LX algorithm, has a cost of at most $n^2/4$. Therefore for values of m close to n the storage required by the ABS formulation is about 8 times less than for the classical simplex method.

Here we give the basic formulas of the simplex method in the classical and in the ABS formulation. The column in A_N substituting an old column in A_B is often taken as the column with minimal relative cost. In terms of the ABS formulation this is equivalent to minimize with respect to $i \in N_m$ the scalar $\eta_i = c^\mathsf{T} H^\mathsf{T} e_i$. Let N^* be the index chosen in this way. The column in A_B to be exchanged is usually chosen with the criterion of the maximum displacement along an edge which keeps the basic variables nonnegative. Define $\omega_i = x^\mathsf{T} e_i / e_i^\mathsf{T} H^\mathsf{T} e_{N^*}$, where x is the current basic feasible solution. Then the above criterion is equivalent to minimize ω_i with respect the set of indices $i \in B_m$ such that

$$e_i^\mathsf{T} H^\mathsf{T} e_{N^*} > 0. \tag{14}$$

Notice that $H^\mathsf{T} e_{N^*} \neq 0$ and that an index i such that (14) is satisfied always exists, unless x is a solution of the LP problem.

The update of the Abaffian after the interchange of the unit vectors, which corresponds to the update of the LU factors after the interchange of the basic with the nonbasic column, is given by the following formula

$$H' = H - (He_{B^*} - e_{B^*})\frac{e_{N^*}^\mathsf{T} H}{e_{N^*}^\mathsf{T} He_{B^*}}. \tag{15}$$

The search direction d, which in the classical formulation is obtained by solving the system $A_B d = -Ae_{N^*}$, is given by $d = H_{m+1}^\mathsf{T} e_{N^*}$, hence at no cost. Finally, the relative cost vector r, classically given by $r = c - A^\mathsf{T} A_B^{-1} c_B$, where c_B consists of the components of c with indices corresponding to those of the basic columns, is simply given by $r = H_{m+1} c$.

Let us now consider the computational cost of update (15). Since He_{B*} has at most $n - m$ nonzero components, while $H^\top e_{N*}$ has at most m, no more than $m(n - m)$ multiplications are required. The update is most expensive for $m = n/2$ and gets cheaper the smaller m is or the closer it is to n. In the dual steepest edge Forrest–Goldfarb method [6] the overhead for replacing a column is m^2, hence formula (15) is faster for $m > n/2$ and is recommended on overhead considerations for m sufficiently large. However we notice that ABS updates having a $O(m^2)$ cost can also be obtained by using the representation of the Abaffian in terms of $2m$ vectors. No computational experience has been obtained till now on the new ABS formulation of the simplex method.

Finally, a generalization of the *simplex method*, based upon the use of the Huang algorithm started with a suitable singular matrix, has been developed in [16]. In this formulation the solution is approached by points lying on a face of the polytope. Whenever the point hits a vertex the remaining iterates move among vertices and the method is reduced to the simplex method.

ABS Unification of Feasible Direction Methods for Minimization with Linear Constraints

ABS algorithms can be used to provide a unification of feasible point methods for nonlinear minimization with linear constraints, including as a special case the LP problem. Let us first consider the problem with only linear equality constraints:

$$\begin{cases} \min_{x \in \mathbb{R}^n} & f(x) \\ \text{s.t.} & Ax = b \\ & A \in \mathbb{R}^{m,n}, \quad m \leq n, \\ & \text{rank}(A) = m. \end{cases}$$

Let x_1 be a feasible starting point; then for an iteration procedure of the form $x_{i+1} = x_i - \alpha_i d_i$, the search direction will generate feasible points if and only if

$$Ad_i = 0. \tag{16}$$

Solving the underdetermined system (16) for d_i by the ABS algorithm, the solution can be written in the following form, taking, without loss of generality, the zero vector as a special solution

$$d_i = H_{m+1}^\top q, \tag{17}$$

where the matrix H_{m+1} depends on the arbitrary choice of the parameters H_1, w_i and v_i used in solving (16) and $q \in \mathbb{R}^n$ is arbitrary. Hence the general feasible direction iteration has the form

$$x_{i+1} = x_i - \alpha_i H_{m+1}^\top q. \tag{18}$$

The search direction is a descent direction if and only if $d^\top \nabla f(x) = q^\top H_{m+1} \nabla f(x) > 0$. Such a condition can always be satisfied by choice of q unless $H_{m+1} \nabla f(x) = 0$, which implies, from the null space structure of H_{m+1}, that $\nabla f(x) = A^\top \lambda$ for some λ, hence that x_{i+1} is a KT point and λ is the vector of the Lagrange multipliers. When x_{i+1} is not a KT point, it is immediate to see that the search direction is a descent directions if we select q as $q = W H_{m+1} \nabla f(x)$, where W is a symmetric and positive definite matrix.

Particular well-known algorithms from the literature are obtained by the following choices of q, with $W = I$:

- The *Wolfe reduced gradient method*. Here, H_{m+1} is constructed by the implicit LU (or the implicit LX) algorithm.
- The *Rosen gradient projection method*. Here, H_{m+1} is built using the Huang algorithm.
- The *Goldfarb–Idnani method*. Here, H_{m+1} is built via the modification of the Huang algorithm where H_1 is a symmetric positive definite matrix approximating the inverse Hessian of $f(x)$.

If there are inequalities two approaches are possible:

A) The *active set* approach. In this approach the set of linear equality constraints is modified at every iteration by adding and/or dropping some of the linear inequality constraints. Adding or deleting a single constraint can be done, for every ABS algorithm, in order two operations, see [15]. In the ABS reformulation of the Goldfarb–Idnani method, the initial matrix is related to a quasi-Newton approximation of the Hessian and an efficient update of the Abaffian after a change in the initial matrix is discussed in [14].

B) The *standard form* approach. In this approach, by introducing slack variables, the problem with both types of linear constraints is written in the equivalent form

$$\begin{cases} \min & f(x) \\ \text{s.t.} & Ax = b \\ & x \geq 0. \end{cases}$$

The following general iteration, started with x_1 a feasible point, generates a sequence of feasible points for the problem in standard form

$$x_{i+1} = x_i - \alpha_i \beta_i H_{m+1} \nabla f(x), \tag{19}$$

where the parameter α_i can be chosen by a line search along the vector $H_{m+1} \nabla f(x)$, while the relaxation parameter $\beta_i > 0$ is selected to avoid that the new point has some negative components.

If $f(x)$ is nonlinear, then H_{m+1} can be determined once and for all at the first step, since $\nabla f(x)$ generally changes from iteration to iteration, therefore modifying the search direction. If, however, $f(x) = c^\mathsf{T} x$ is linear (we have then the LP problem) to modify the search direction we need to change H_{m+1}. As observed before, the simplex method is obtained by constructing H_{m+1} with the implicit LX algorithm, every step of the method corresponding to a change of the parameters e_{k_i}. It can be shown, see [13], that the *method of Karmarkar* (equivalent to an earlier *method of Evtushenko* [3]), corresponds to using the generalized Huang algorithm, with initial matrix $H_1 = \text{Diag}(x_i)$ changing from iteration to iteration. Another method, faster than Karmarkar's and having superlinear against linear rate of convergence and $O(\sqrt{n})$ against $O(n)$ complexity, again first proposed by Y. Evtushenko, is obtained by the generalized Huang algorithm with initial matrix $H_1 = \text{Diag}(x_i^2)$.

See also

- ABS Algorithms for Linear Equations and Linear Least Squares
- Gauss–Newton Method: Least Squares, Relation to Newton's Method
- Generalized Total Least Squares
- Least Squares Orthogonal Polynomials
- Least Squares Problems
- Nonlinear Least Squares: Newton-type Methods
- Nonlinear Least Squares Problems
- Nonlinear Least Squares: Trust Region Methods

References

1. Abaffy J, Spedicato E (1989) ABS projection algorithms: Mathematical techniques for linear and nonlinear equations. Horwood, Westergate
2. Adachi N (1971) On variable metric algorithms. JOTA 7:391–409
3. Evtushenko Y (1974) Two numerical methods of solving nonlinear programming problems. Soviet Dokl Akad Nauk 251:420–423
4. Feng E, Wang XM, Wang XL (1997) On the application of the ABS algorithm to linear programming and linear complementarity. Optim Methods Softw 8:133–142
5. Fletcher R (1997) Dense factors of sparse matrices. In: Buhmann MD, Iserles A (eds) Approximation Theory and Optimization. Cambridge Univ. Press, Cambridge, pp 145–166
6. Forrest JH, Goldfarb D (1992) Steepest edge simplex algorithms for linear programming. Math Program 57:341–374
7. Psenichny BN, Danilin YM (1978) Numerical methods in extremal problems. MIR, Moscow
8. Spedicato E, Chen Z, Bodon E (1996) ABS methods for KT equations. In: Di Pillo G, Giannessi F (eds) Nonlinear Optimization and Applications. Plenum, New York, pp 345–359
9. Spedicato E, Xia Z (1992) Finding general solutions of the quasi-Newton equation in the ABS approach. Optim Methods Softw 1:273–281
10. Spedicato E, Xia Z, Zhang L (1995) Reformulation of the simplex algorithm via the ABS algorithm. Preprint Univ Bergamo
11. Spedicato E, Zhao J (1992) Explicit general solution of the quasi-Newton equation with sparsity and symmetry. Optim Methods Softw 2:311–319
12. Xia Z (1995) ABS generalization and formulation of the interior point method. Preprint Univ Bergamo
13. Xia Z (1995) ABS reformulation of some versions of the simplex method for linear programming. Report DMSIA Univ Bergamo 10/95
14. Xia Z, Liu Y, Zhang L (1992) Application of a representation of ABS updating matrices to linearly constrained optimization. Northeast Oper Res 7:1–9
15. Zhang L (1995) Updating of Abaffian matrices under perturbation in W and A. Report DMSIA Univ Bergamo 95/16
16. Zhang L (1997) On the ABS algorithm with singular initial matrix and its application to linear programming. Optim Methods Softw 8:143–156
17. Zhang L, Xia ZH (1995) Application of the implicit LX algorithm to the simplex method. Report DMSIA Univ Bergamo 9/95

Adaptive Convexification in Semi-Infinite Optimization

OLIVER STEIN
School of Economics and Business Engineering,
University of Karlsruhe, Karlsruhe, Germany

MSC2000: 90C34, 90C33, 90C26, 65K05

Article Outline

Synonyms
Introduction
    Feasibility in Semi-Infinite Optimization
    Convex Lower Level Problems
    The αBB Method
Formulation
    αBB for the Lower Level
    The MPCC Reformulation
Method
    Refinement Step
    The Algorithm
    A Consistent Initial Approximation
    A Certificate for Global Optimality
Conclusions
See also
References

Synonyms

ACA

Introduction

The adaptive convexification algorithm is a method to solve semi-infinite optimization problems via a sequence of *feasible iterates*. Its main idea [6] is to adaptively construct convex relaxations of the lower level problem, replace the relaxed lower level problems equivalently by their Karush–Kuhn–Tucker conditions, and solve the resulting mathematical programs with complementarity constraints. The convex relaxations are constructed with ideas from the αBB method of global optimization.

Feasibility in Semi-Infinite Optimization

In a (standard) semi-infinite optimization problem a finite-dimensional decision variable is subject to infinitely many inequality constraints. For adaptive convexification one assumes the form

$$SIP:\ \min_{x \in X} f(x) \quad \text{subject to} \quad g(x, y) \le 0,$$
$$\text{for all } y \in [0, 1]$$

with objective function $f \in C^2(\mathbb{R}^n, \mathbb{R})$, constraint function $g \in C^2(\mathbb{R}^n \times \mathbb{R}, \mathbb{R})$, a box constraint set $X = [x^\ell, x^u] \subset \mathbb{R}^n$ with $x^\ell < x^u \in \mathbb{R}^n$, and the set of infinitely many indices $Y = [0, 1]$. Adaptive convexification easily generalizes to problems with additional inequality and equality constraints, a finite number of semi-infinite constraints as well as higher-dimensional box index sets [6]. Reviews on semi-infinite programming are given in [8,13], and [9,14,15] overview the existing numerical methods.

Classical numerical methods for *SIP* suffer from the drawback that their approximations of the feasible set $X \cap M$ with

$$M = \{x \in \mathbb{R}^n \mid g(x, y) \le 0 \text{ for all } y \in [0, 1]\}$$

may contain infeasible points. In fact, discretization and exchange methods approximate M by finitely many inequalities corresponding to finitely many indices in $Y = [0, 1]$, yielding an outer approximation of M, and reduction based methods solve the Karush–Kuhn–Tucker system of *SIP* by a Newton-SQP approach. As a consequence, the iterates of these methods are not necessarily feasible for *SIP*, but only their limit might be. On the other hand, a first method producing feasible iterates for *SIP* was presented in the articles [3,4], where a branch-and-bound framework for the global solution of *SIP* generates convergent sequences of lower and upper bounds for the globally optimal value.

In fact, checking feasibility of a given point $\bar{x} \in \mathbb{R}^n$ is the crucial problem in semi-infinite optimization. Clearly we have $\bar{x} \in M$ if and only if $\varphi(\bar{x}) \le 0$ holds with the function

$$\varphi: \mathbb{R}^n \to \mathbb{R}, \quad x \mapsto \max_{y \in [0,1]} g(x, y).$$

The latter function is the optimal value function of the so-called lower level problem of *SIP*,

$$Q(x): \max_{y \in \mathbb{R}} g(x, y) \quad \text{subject to} \quad 0 \le y \le 1.$$

The difficulty lies in the fact that $\varphi(\bar{x})$ is the *globally* optimal value of $Q(\bar{x})$ which might be hard to determine numerically. In fact, standard NLP solvers can only be expected to produce a *local* maximizer y_{loc} of $Q(\bar{x})$ which is not necessarily a global maximizer y_{glob}. Even if $g(\bar{x}, y_{\text{loc}}) \leq 0$ is satisfied, $\bar{x}$ might be infeasible since $g(\bar{x}, y_{\text{loc}}) \leq 0 < \varphi(\bar{x}) = g(\bar{x}, y_{\text{glob}})$ may hold.

Convex Lower Level Problems

Assume for a moment that $Q(x)$ is a convex optimization problem for all $x \in X$, that is, $g(x, \cdot)$ is concave on $Y = [0, 1]$ for these x. An approach developed for so-called generalized semi-infinite programs from [18,19] then takes advantage of the fact that the solution set of a differentiable convex lower level problem satisfying a constraint qualification is characterized by its first order optimality condition. In fact, SIP and the Stackelberg game

$$SG: \min_{x,y} f(x) \quad \text{subject to} \quad g(x, y) \leq 0,$$

$$\text{and } y \text{ solves } Q(x)$$

are equivalent problems, and the restriction 'y solves $Q(x)$' in SG can be equivalently replaced by its Karush–Kuhn–Tucker condition. For this reformulation we use that the Lagrange function of $Q(x)$,

$$\mathcal{L}(x, y, \gamma_\ell, \gamma_u) = g(x, y) + \gamma_\ell y + \gamma_u(1 - y),$$

satisfies

$$\nabla_y \mathcal{L}(x, y, \gamma_\ell, \gamma_u) = \nabla_y g(x, y) + \gamma_\ell - \gamma_u$$

and obtain that the Stackelberg game is equivalent to the following mathematical program with complementarity constraints:

$$MPCC: \min_{x,y,\gamma_\ell,\gamma_u} f(x) \text{ subject to } g(x, y) \leq 0$$

$$\nabla_y g(x, y) + \gamma_\ell - \gamma_u = 0$$

$$\gamma_\ell y = 0$$

$$\gamma_u(1 - y) = 0$$

$$\gamma_\ell, \gamma_u \geq 0$$

$$y, 1 - y \geq 0.$$

Overviews of solution methods for MPCC are given in [10,11,17]. One approach to solve MPCC is the reformulation of the complementarity constraints by a so-called NCP function, that is, a function $\phi: \mathbb{R}^2 \to \mathbb{R}$ with

$$\phi(a, b) = 0$$

$$\text{if and only if} \quad a \geq 0, \quad b \geq 0, \quad ab = 0.$$

For numerical purposes one can regularize these non-differentiable NCP functions. Although MPCC does not necessarily have to be solved via the NCP function formulation, in the following we will use NCP functions to keep the notation concise. In fact, MPCC can be equivalently rewritten as the nonsmooth problem

$$P: \min_{x,y,\gamma_\ell,\gamma_u}$$

$$f(x) \quad \text{subject to} \quad g(x, y) \leq 0$$

$$\nabla_y g(x, y) + \gamma_\ell - \gamma_u = 0$$

$$\phi(\gamma_\ell, y) = 0$$

$$\phi(\gamma_u, 1 - y) = 0.$$

The αBB Method

In αBB, a convex underestimator of a nonconvex function is constructed by decomposing it into a sum of nonconvex terms of special type (e.g., linear, bilinear, trilinear, fractional, fractional trilinear, convex, univariate concave) and nonconvex terms of arbitrary type. The first type is then replaced by its convex envelope or very tight convex underestimators which are already known. A complete list of the tight convex underestimators of the above special type nonconvex terms is provided in [5].

For the ease of presentation, here we will treat all terms as arbitrarily nonconvex. For these terms, αBB constructs convex underestimators by adding a quadratic relaxation function ψ. With the obvious modification we use this approach to construct a concave overestimator for a nonconcave function $g: [y^\ell, y^u] \to \mathbb{R}$ being C^2 on an open neighborhood of $[y^\ell, y^u]$. With

$$\psi(y; \alpha, y^\ell, y^u) = \frac{\alpha}{2}(y - y^\ell)(y^u - y) \tag{1}$$

we put

$$\tilde{g}(y; \alpha, y^\ell, y^u) = g(y) + \psi(y; \alpha, y^\ell, y^u).$$

In the sequel we will suppress the dependence of $\tilde{g}$ on y^ℓ, y^u. For $\alpha \geq 0$ the function $\tilde{g}$ clearly is an overestimator of g on $[y^\ell, y^u]$, and it coincides with g at the endpoints y^ℓ, y^u of the domain. Moreover, $\tilde{g}$ is twice continuously differentiable with second derivative

$$\nabla_y^2 \tilde{g}(y; \alpha) = \nabla^2 g(y) - \alpha$$

on $[y^\ell, y^u]$. Consequently $\tilde{g}$ is concave on $[y^\ell, y^u]$ for

$$\alpha \geq \max_{y \in [y^\ell, y^u]} \nabla^2 g(y) \quad (2)$$

(cf. also [1,2]). The computation of α thus involves a global optimization problem itself. Note, however, that one may use *any* upper bound for the right-hand side in (2). Such upper bounds can be provided by interval methods (see, e. g., [5,7,12]). An α satisfying (2) is called *convexification parameter*.

Combining these facts shows that for

$$\alpha \geq \max\left(0, \max_{y \in [y^\ell, y^u]} \nabla^2 g(y)\right)$$

the function $\tilde{g}(y; \alpha)$ is a concave overestimator of g on $[y^\ell, y^u]$.

Formulation

For $N \in \mathcal{N}$ let $0 = \eta^0 < \eta^1 < \ldots < \eta^{N-1} < \eta^N = 1$ define a subdivision of $Y = [0, 1]$, that is, with $K = \{1, \ldots, N\}$ and

$$Y^k = [\eta^{k-1}, \eta^k], \quad k \in K,$$

we have

$$Y = \bigcup_{k \in K} Y_k.$$

A trivial but very useful observation is that the single semi-infinite constraint

$$g(x, y) \leq 0 \quad \text{for all} \quad y \in Y$$

is equivalent to the finitely many semi-infinite constraints

$$g(x, y) \leq 0 \quad \text{for all} \quad y \in Y^k, k \in K.$$

Given a subdivision, one can construct concave overestimators for each of these finitely many semi-infinite constraints, solve the corresponding optimization problem, and adaptively refine the subdivision.

The following lemma formulates the obvious fact that replacing g by overestimators on each subdivision node Y^k results in an approximation of M by *feasible* points.

Lemma 1 *For each $k \in K$ let $g^k: X \times Y^k \to \mathbb{R}$, and let $\tilde{x} \in X$ be given such that for all $k \in K$ and all $y \in Y^k$ we have $g(\tilde{x}, y) \leq g^k(\tilde{x}, y)$. Then the constraints*

$$g^k(\tilde{x}, y) \leq 0 \quad \text{for all} \quad y \in Y^k, k \in K,$$

entail $\tilde{x} \in M$.

αBB for the Lower Level

For the construction of these overestimators one uses ideas of the αBB method. In fact, for each $k \in K$ we put

$$g^k: X \times Y^k \to \mathbb{R}, (x, y) \mapsto g(x, y) + \psi(y; \alpha_k, \eta^{k-1}, \eta^k) \quad (3)$$

with the quadratic relaxation function ψ from (1) and

$$\alpha_k > \max\left(0, \max_{(x,y) \in X \times Y^k} \nabla^2_y g(x, y)\right). \quad (4)$$

Note that the latter condition on α_k is uniform in x. We emphasize that with the single bound

$$\bar{\alpha} > \max\left(0, \max_{(x,y) \in X \times Y} \nabla^2_y g(x, y)\right) \quad (5)$$

the choices $\alpha_k := \bar{\alpha}$ satisfy (4) for all $k \in K$. Moreover, the α_k can always be chosen such that $\alpha_k \leq \bar{\alpha}, k \in K$.

The following properties of g^k are easily verified.

Lemma 2 ([6]) *For each $k \in K$ let g^k be given by (3). Then the following holds:*
(i) For all $(x, y) \in X \times Y^k$ we have $g(x, y) \leq g^k(x, y)$.
(ii) For all $x \in X$, the function $g^k(x, \cdot)$ is concave on Y^k.

Now consider the following approximation of the feasible set M, where $E = \{\eta^k | k \in K\}$ denotes the set of subdivision points, and α the vector of convexification parameters:

$$M_{\alpha BB}(E, \alpha) = \{x \in \mathbb{R}^n | g^k(x, y) \leq 0,$$
$$\text{for all } y \in Y^k, k \in K\}.$$

By Lemma 1 and Lemma 2(i) we have $M_{\alpha BB}(E, \alpha) \subset M$. This means that any solution concept for

$$SIP_{\alpha BB}(E, \alpha): \min_{x \in X} f(x) \quad \text{subject to}$$
$$x \in M_{\alpha BB}(E, \alpha),$$

be it global solutions, local solutions or stationary points, will at least lead to feasible points of SIP (provided that $SIP_{\alpha BB}(E,\alpha)$ is consistent).

The problem $SIP_{\alpha BB}(E,\alpha)$ has finitely many lower level problems $Q^k(x)$, $k \in K$, with

$$Q^k(x): \max_{y \in \mathbb{R}} g^k(x,y) \quad \text{subject to} \quad \eta^{k-1} \leq y \leq \eta^k.$$

Since the inequality (4) is strict, the convex problem $Q^k(x)$ has a unique solution $y^k(x)$ for each $k \in K$ and $x \in X$. Recall that $y \in Y^k$ is called active for the constraint $\max_{y \in Y^k} g^k(x,y) \leq 0$ at $\bar{x}$ if $g^k(\bar{x},y) = 0$ holds. By the uniqueness of the global solution of $Q^k(\bar{x})$ there exists at most one active index for each $k \in K$, namely $y^k(\bar{x})$. Thus, one can consider the finite active index sets

$$K_0(\bar{x}) = \{ k \in K | \ g^k(\bar{x}, y^k(\bar{x})) = 0 \},$$
$$Y_0^{\alpha BB}(\bar{x}) = \{ y^k(\bar{x}) | \ k \in K_0(\bar{x}) \}.$$

The MPCC Reformulation

Following the ideas to treat convex lower level problems, y^k solves $Q^k(x)$ if and only if $(x, y^k, \gamma_\ell^k, \gamma_u^k)$ solves the system

$$\nabla_y g^k(x,y) + \gamma_\ell - \gamma_u = 0$$
$$\phi(\gamma_\ell, y - \eta^{k-1}) = 0$$
$$\phi(\gamma_u, \eta^k - y) = 0$$

with some $\gamma_\ell^k, \gamma_u^k$, and ϕ denoting some NCP function. With

$$w := (x, y^k, \gamma_\ell^k, \gamma_u^k, k \in K)$$
$$F(w) := f(x)$$
$$G^k(w; E, \alpha) := g(x, y^k) + \frac{\alpha_k}{2}(y^k - \eta^{k-1})(\eta^k - y^k)$$
$$H^k(w; E, \alpha) :=$$
$$\begin{pmatrix} \nabla_y g(x, y^k) + \alpha_k \left(\frac{\eta^{k-1}+\eta^k}{2} - y^k \right) + \gamma_\ell^k - \gamma_u^k \\ \phi(\gamma_\ell^k, y^k - \eta^{k-1}) \\ \phi(\gamma_u^k, \eta^k - y^k) \end{pmatrix}$$

one can thus replace $SIP_{\alpha BB}(E,\alpha)$ equivalently by the nonsmooth problem

$$P(E,\alpha): \min_w F(w) \quad \text{subject to}$$
$$G^k(w; E, \alpha) \leq 0,$$
$$H^k(w; E, \alpha) = 0, \quad k \in K.$$

The latter problem can be solved to local optimality by MPCC algorithms [10,11,17]. For a local minimizer $\bar{w}$ of $P(E,\alpha)$ the subvector $\bar{x}$ of $\bar{w}$ is a local minimizer and, hence, a stationary point of $SIP_{\alpha BB}(E,\alpha)$.

Method

The main idea of the adaptive convexification algorithm is to compute a stationary point $\bar{x}$ of $SIP_{\alpha BB}(E,\alpha)$ by the approach from the previous section, and terminate if $\bar{x}$ is also stationary for SIP within given tolerances. If $\bar{x}$ is not stationary it refines the subdivision E in the spirit of exchange methods [8,15] by adding the active indices $Y_0^{\alpha BB}(\bar{x})$ to E, and constructs a refined problem $SIP_{\alpha BB}(E \cup Y_0^{\alpha BB}(\bar{x}), \tilde{\alpha})$ by the following procedure. Note that, in view of Carathéodory's theorem, the number of elements of $Y_0^{\alpha BB}(\bar{x})$ may be bounded by $n + 1$.

Refinement Step

For any $\tilde{\eta} \in Y_0^{\alpha BB}(\bar{x})$, let $k \in K$ be the index with $\tilde{\eta} \in [\eta^{k-1}, \eta^k]$. Put $Y^{k,1} = [\eta^{k-1}, \tilde{\eta}]$, $Y^{k,2} = [\tilde{\eta}, \eta^k]$, let $\alpha_{k,1}$ and $\alpha_{k,2}$ be the corresponding convexification parameters, put

$$g^{k,1}(x,y) = g(x,y) + \frac{\alpha_{k,1}}{2}(y - \eta^{k-1})(\tilde{\eta} - y),$$
$$g^{k,2}(x,y) = g(x,y) + \frac{\alpha_{k,2}}{2}(y - \tilde{\eta})(\eta^k - y),$$

and define $M_{\alpha BB}(E \cup \{\tilde{\eta}\}, \tilde{\alpha})$ by replacing the constraint

$$g^k(x,y) \leq 0, \quad \text{for all } y \in Y^k$$

in $M_{\alpha BB}(E,\alpha)$ by the two new constraints

$$g^{k,i}(x,y) \leq 0, \quad \text{for all } y \in Y^{k,i}, \quad i = 1,2,$$

and by replacing the entry α_k of α by the two new entries $\alpha_{k,i}$, $i = 1,2$.

The Algorithm

The point $\bar{x}$ is stationary for $SIP_{\alpha BB}(E,\alpha)$ (in the sense of Fritz John) if $\bar{x} \in M_{\alpha BB}(E,\alpha)$ and if there exist $y^k \in Y_0^{\alpha BB}(\bar{x})$, $1 \leq k \leq n+1$, and $(\kappa, \lambda) \in S^{n+1}$ (the $(n+1)$-dimensional standard simplex) with

$$\kappa \nabla f(\bar{x}) + \sum_{k=1}^{n+1} \lambda_k \nabla_x g(\bar{x}, y^k) = 0$$
$$\lambda_k \cdot g^k(\bar{x}, y^k) = 0, \ 1 \leq k \leq n+1.$$

For the adaptive convexification algorithm the notions of *active index*, *stationarity*, and *set unification* are relaxed by certain tolerances.

Definition 1 For $\varepsilon_{\text{act}}, \varepsilon_{\text{stat}}, \varepsilon_\cup > 0$ we say that
(i) y^k is ε_{act}-active for g^k at $\bar{x}$ if $g^k(\bar{x}, y^k) \in [-\varepsilon_{\text{act}}, 0]$,
(ii) $\bar{x}$ is $\varepsilon_{\text{stat}}$-stationary for *SIP* with ε_{act}-active indices if $\bar{x} \in M$ and if there exist $y^k \in Y, 1 \le k \le n+1$, and $(\kappa, \lambda) \in S^{n+1}$ such that

$$\left\| \kappa \nabla f(\bar{x}) + \sum_{k=1}^{n+1} \lambda_k \nabla_x g(\bar{x}, y^k) \right\| \le \varepsilon_{\text{stat}}$$

$$\lambda_k \cdot g(\bar{x}, y^k) \in [-\lambda_k \cdot \varepsilon_{\text{act}}, 0], \quad 1 \le k \le n+1,$$

hold, and
(iii) the $\varepsilon_\cup$-union of E and $\tilde{\eta}$ is $E \cup \{\tilde{\eta}\}$ if

$$\min\{\tilde{\eta} - \eta^{k-1}, \eta^k - \tilde{\eta}\} > \varepsilon_\cup \cdot (\eta^k - \eta^{k-1})$$

holds for the $k \in K$ with $\tilde{\eta} \in [\eta^{k-1}, \eta^k]$, and E otherwise (i. e., $\tilde{\eta}$ is not unified with E if its distance from E is too small).

In [6] it is shown that Algorithm 1 is well-defined, convergent and finitely terminating. Furthermore, the following feasibility result holds.

Theorem 2 ([6]) *Let $(x^\nu)_\nu$ be a sequence of points generated by Algorithm 1. Then all x^ν, $\nu \in \mathbb{N}$, are feasible for SIP, the sequence $(x^\nu)_\nu$ has an accumulation point, each such accumulation point x^* is feasible for SIP, and $f(x^*)$ provides an upper bound for the optimal value of SIP.*

Numerical examples for the performance of the method from Chebyshev approximation and design centering are given in [6].

A Consistent Initial Approximation

Even if the feasible set M of *SIP* is consistent, there is no guarantee that its approximations $M_{\alpha BB}(E, \alpha)$ are also consistent. For Step 1 of Algorithm 1 [6] suggests the following phase I approach: use Algorithm 1 to construct adaptive convexifications of

$$SIP^{ph.I}: \min_{(x,z) \in X \times \mathbb{R}} z \quad \text{subject to} \quad g(x, y) \le z$$

$$\text{for all } y \in [0, 1]$$

Algorithm 1
(Adaptive convexification algorithm)

Step 1: Determine a uniform convexification parameter $\bar{\alpha}$ with (5), choose $N \in \mathbb{N}$, $\eta^k \in Y$ and $\alpha_k \le \bar{\alpha}$, $k \in K = \{1, \ldots, N\}$, such that $SIP_{\alpha BB}(E, \alpha)$ is consistent, as well as tolerances $\varepsilon_{\text{act}}, \varepsilon_{\text{stat}}, \varepsilon_\cup > 0$ with $\varepsilon_\cup \le 2\varepsilon_{\text{act}}/\bar{\alpha}$.
Step 2: By solving $P(E, \alpha)$, compute a stationary point x of $SIP_{\alpha BB}(E, \alpha)$ with ε_{act}-active indices $y^k, 1 \le k \le n+1$, and multipliers (κ, λ).
Step 3: Terminate if x is $\varepsilon_{\text{stat}}$-stationary for *SIP* with $(2\varepsilon_{\text{act}})$-active indices $y^k, 1 \le k \le n+1$, from Step 2 and multipliers (κ, λ) from Step 2. Otherwise construct a new set $\tilde{E}$ of subdivision points as the $\varepsilon_\cup$-union of E and $\{y^k | 1 \le k \le n+1\}$, and perform a refinement step for the elements in $\tilde{E} \setminus E$ to construct a new feasible set $M_{\alpha BB}(\tilde{E}, \tilde{\alpha})$.
Step 4: Put $E = \tilde{E}, \alpha = \tilde{\alpha}$, and go to Step 2.

Adaptive Convexification in Semi-Infinite Optimization, Algorithm 1

until a feasible point $(\bar{x}, \bar{z})$ with $\bar{z} \le 0$ of $SIP_{\alpha BB}^{ph.I}(E, \alpha)$ is found with some subdivision E and convexification parameters α. The point $\bar{x}$ is then obviously also feasible for $SIP_{\alpha BB}(E, \alpha)$ and can be used as an initial point to solve the latter problem. Due to the possible nonconvexity of the upper level problem of *SIP*, this phase I approach is not necessarily successful, but possible remedies for this situation are given in [6].

To initialize Algorithm 1 for phase I, select some point $\bar{x}$ in the box X and put $E^1 = \{0, 1\}$, that is, $Y^1 = Y = [0, 1]$. Compute α_1 according to (4) and solve the convex optimization problem $Q^1(\bar{x})$ with standard software. With its optimal value $\bar{z}$, the point $(\bar{x}, \bar{z})$ is feasible for $SIP_{\alpha BB}^{ph.I}(E^1, \alpha_1)$.

A Certificate for Global Optimality

After termination of Algorithm 1 one can exploit that the set $E \subset [0, 1]$ contains indices that should also yield a good outer approximation of M. The optimal value of the problem

$$P_{\text{outer}}: \min_{x \in X} f(x) \quad \text{subject to} \quad g(x, \eta) \le 0, \eta \in E,$$

yields a rigorous *lower* bound for the optimal value of *SIP*. If P_{outer} can actually be solved to global optimality (e. g., if a standard NLP solver is used, due to convexity with respect to *x*), then a comparison of this lower bound for the optimal value of *SIP* with the upper bound from Algorithm 1 can yield a certificate of global optimality for *SIP* up to some tolerance.

Conclusions

The adaptive convexification algorithm provides an easily implementable way to solve semi-infinite optimization problems with feasible iterates. To explain its basic ideas, in [6] the algorithm is presented in its simplest form. It can be improved in a number of ways, for example in the magnitude of the convexification parameters and in their adaptive refinement, or by using other convexification techniques. Although the numerical results from [6] are very promising, further work is needed on error estimates on the numerical solution of the auxiliary problem $P(E, \alpha)$, which is assumed to be solved to exact local optimality by the present adaptive convexification algorithm.

See also

- ▶ αBB Algorithm
- ▶ Bilevel Optimization: Feasibility Test and Flexibility Index
- ▶ Convex Discrete Optimization
- ▶ Generalized Semi-infinite Programming: Optimality Conditions

References

1. Adjiman CS, Androulakis IP, Floudas CA (1998) A global optimization method, αBB, for general twice-differentiable constrained NLPs – I: theoretical advances. Comput Chem Eng 22:1137–1158
2. Adjiman CS, Androulakis IP, Floudas CA (1998) A global optimization method, αBB, for general twice-differentiable constrained NLPs – II: implementation and computational results. Comput Chem Eng 22:1159–1179
3. Bhattacharjee B, Green WH Jr, Barton PI (2005) Interval methods for semi-infinite programs. Comput Optim Appl 30:63–93
4. Bhattacharjee B, Lemonidis P, Green WH Jr, Barton PI (2005) Global solution of semi-infinite programs. Math Program 103:283–307
5. Floudas CA (2000) Deterministic global optimization, theory, methods and applications. Kluwer, Dordrecht
6. Floudas CA, Stein O (2007) The adaptive convexification algorithm: a feasible point method for semi-infinite programming. SIAM J Optim 18:1187–1208
7. Hansen E (1992) Global optimization using interval analysis. Dekker, New York
8. Hettich R, Kortanek KO (1993) Semi-infinite programming: theory, methods, and applications. SIAM Rev 35:380–429
9. Hettich R, Zencke P (1982) Numerische Methoden der Approximation und semi-infiniten Optimierung. Teubner, Stuttgart
10. Kočvara M, Outrata J, Zowe J (1998) Nonsmooth approach to optimization problems with equilibrium constraints: theory, applications and numerical results. Kluwer, Dordrecht
11. Luo Z, Pang J, Ralph D (1996) Mathematical programs with equilibrium constraints. Cambridge University Press, Cambridge
12. Neumaier A (1990) Interval methods for systems of equations. Cambridge University Press, Cambridge
13. Polak E (1987) On the mathematical foundation of nondifferentiable optimization in engineering design. SIAM Rev 29:21–89
14. Polak E (1997) Optimization, algorithms and consistent approximations. Springer, Berlin
15. Reemtsen R, Görner S (1998) Numerical methods for semi-infinite programming: a survey. In: Reemtsen R, Rückmann J-J (eds) Semi-infinite programming. Kluwer, Boston, pp 195–275
16. Reemtsen R, Rückmann J-J (eds) (1998) Semi-infinite programming. Kluwer, Boston
17. Scholtes S, Stöhr M (1999) Exact penalization of mathematical programs with equilibrium constraints. SIAM J Control Optim 37:617–652
18. Stein O (2003) Bi-level strategies in semi-infinite programming. Kluwer, Boston
19. Stein O, Still G (2003) Solving semi-infinite optimization problems with interior point techniques. SIAM J Control Optim 42:769–788

Adaptive Global Search

J. M. CALVIN
Department Computer and Information Sci., New Jersey Institute Techn., Newark, USA

MSC2000: 60J65, 68Q25

Article Outline

Keywords
See also
References

Keywords

Average case complexity; Adaptive algorithm; Wiener process; Randomized algorithms

This article contains a survey of some well known facts about the complexity of *global optimization*, and also describes some results concerning the *average-case complexity*.

Consider the following optimization problem. Given a class F of objective functions f defined on a compact subset of d-dimensional Euclidean space, the goal is to approximate the global minimum of f based on evaluation of the function at sequentially selected points. The focus will be on the error after n observations

$$\Delta_n = \Delta_n(f) = f_n - f^*,$$

where f_n is the smallest of the first n observed function values (other approximations besides f_n are often considered).

Complexity of optimization is usually studied in the worst- or average-case setting. In order for a *worst-case analysis* to be useful the class of objective functions F must be quite restricted. Consider the case where F is a subset of the continuous functions on a compact set. It is convenient to consider the class $F = C^r([0, 1]^d)$ of real-valued functions on $[0, 1]^d$ with continuous derivatives up to order $r \geq 0$. Suppose that $r > 0$ and f^r is bounded. In this case $\Theta(\epsilon^{-d/r})$ function evaluations are needed to ensure that the error is at most ϵ for any $f \in F$; see [8].

An *adaptive algorithm* is one for which the $(n + 1)$st observation point is determined as a function of the previous observations, while a nonadaptive algorithm chooses each point independently of the function values. In the worst-case setting, adaptation does not help much under quite general assumptions. If F is convex and symmetric (in the sense that $-F = F$), then the maximum error under an adaptive algorithm with n observations is not smaller than the maximum error of a nonadaptive method with $n + 1$ observations; see [4].

Virtually all global optimization methods in practical use are adaptive. For a survey of such methods see [6,9]. The fact that the worst-case performance can not be significantly improved with adaptation leads to consideration of alternative settings that may be more appropriate. One such setting is the average-case setting, in which a probability measure P on F is chosen. The object of study is then the sequence of random variables $\Delta_n(f)$, and the questions include under what conditions (for what algorithms) the error converges to zero and for convergent algorithms the speed of convergence. While the average-case error is often defined as the mathematical expectation of the error, it is useful to take a broader view, and consider for example convergence in probability of $a_n \Delta_n$ for some normalizing sequence $\{a_n\}$.

With the average-case setting one can consider less restricted classes F than in the worst-case setting. As F gets larger, the worst-case deviates more and more from the average case, but may occur on only a small portion of the set F. Even for continuous functions the worst-case is arbitrarily bad.

Most of what is known about the average-case complexity of optimization is in the one-dimensional setting under the *Wiener probability measure* on $C([0, 1])$. Under the Wiener measure, the increments $f(t) - f(s)$ have a normal distribution with mean zero and variance $t - s$, and are independent for disjoint intervals. Almost every f is nowhere differentiable, and the set of local minima is dense in the unit interval. One can thus think of the Wiener measure as corresponding to assuming 'only' continuity; i. e., a worst-case probabilistic assumption.

K. Ritter proved [5] that the best nonadaptive algorithms have error of order $n^{-1/2}$ after n function evaluations; the optimal order is achieved by observing at equally spaced points. Since the choice of each new observation point does not depend on any of the previous observations, the computation can be carried out in parallel. Thus under the Wiener measure, the optimal nonadaptive order of convergence can be accomplished with an algorithm that has computational cost that grows linearly with the number of observations and uses constant storage. This gives the base on which to compare adaptive algorithms.

Recent studies (as of 2000) have formally established the improved power of adaptive methods in the average-case setting by analyzing the convergence rates of certain adaptive algorithms. A *randomized algorithm* is described in [1] with the property that for any $0 < \delta < 1$, a version can be constructed so that under the Wiener measure, the error converges to zero at rate $n^{-1+\delta}$. This algorithm maintains a memory of two past observation values, and the computational cost grows linearly with the number of iterations. Therefore, the convergence rate of this adaptive algorithm improves from the non-adaptive $n^{-1/2}$ rate to $n^{-1+\delta}$ with only a constant increase in storage.

Algorithms based on a random model for the objective function are well-suited to average-case analysis. H. Kushner proposed [3] a global optimization method based on modeling the objective function as a Wiener process. Let $\{z_n\}$ be a sequence of positive numbers, and let the $(n + 1)$st point be chosen to maximize the probability that the new function value is less than the previously observed minimum minus z_n. This class of algorithms, often called *P-algorithms*, was given a formal justification by A. Žilinskas [7].

By allowing the $\{z_n\}$ to depend on the past observations instead of being a fixed deterministic sequence, it is possible to establish a much better convergence rate than that of the randomized algorithm described above. In [2] an algorithm was constructed with the property that the error converges to zero for any continuous function and furthermore, the error is of order e^{-nc_n}, where $\{c_n\}$ (a parameter of the algorithm) is a deterministic sequence that can be chosen to approach zero at an arbitrarily slow rate. Notice that the convergence rate is now almost exponential in the number of observations n. The computational cost of the algorithm grows quadratically, and the storage increases linearly, since all past observations must be stored.

See also

- ▶ Adaptive Simulated Annealing and its Application to Protein Folding
- ▶ Global Optimization Based on Statistical Models

References

1. Calvin J (1997) Average performance of a class of adaptive algorithms for global optimization. Ann Appl Probab 7:711–730
2. Calvin J (2001) A one-dimensional optimization algorithm and its convergence rate under the Wiener measure. J Complexity
3. Kushner H (1962) A versatile stochastic model of a function of unknown and time varying form. J Math Anal Appl 5:150–167
4. Novak E (1988) Deterministic and stochastic error bounds in numerical analysis. Lecture Notes in Mathematics, vol 1349. Springer, Berlin
5. Ritter K (1990) Approximation and optimization on the Wiener space. J Complexity 6:337–364
6. Törn A, Žilinskas A (1989) Global optimization. Springer, Berlin
7. Žilinskas A (1985) Axiomatic characterization of global optimization algorithm and investigation of its search strategy. OR Lett 4:35–39
8. Wasilkowski G (1992) On average complexity of global optimization problems. Math Program 57:313–324
9. Zhigljavsky A (1991) Theory of global random search. Kluwer, Dordrecht

Adaptive Simulated Annealing and its Application to Protein Folding
ASA

Ruth Pachter, Zhiqiang Wang
Air Force Research Laboratory Materials & Manufacturing Directorate, Wright–Patterson AFB, Wright–Patterson AFB, USA

MSC2000: 92C05

Article Outline

Keywords
The ASA Method
    Monte-Carlo Configurations
    Annealing Schedule
    Re-Annealing
Application to Protein Folding
    Computational Details
    Met-Enkephalin
    Poly(L-Alanine)
Conclusion
Recent Studies and Future Directions
See also
References

Keywords

Optimization; Adaptive simulated annealing; Protein folding; Met-Enkephalin; Poly(L-Alanine)

The adaptive simulated annealing (ASA) algorithm [3] has been shown to be faster and more efficient than simulated annealing and genetic algorithms [4]. In this article we first outline some of the aspects of the method and specific computational details, and then review the application of the ASA method to biomolecular structure determination [15], specifically for Met-Enkephalin and a model of the poly(L-Alanine) system.

The ASA Method

For a system described by a cost function $E(\{p^i\})$, where all p^i ($i = 1, \ldots, D$) are parameters (variables) having ranges $[A_i, B_i]$, the ASA procedure to find the global optimum of 'E' contains the following elements.

Monte-Carlo Configurations

As the kth point is saved in a D-dimensional configuration space, the new point p^i_{k+1} is generated by:

$$p^i_{k+1} = p^i_k + y^i(B_i - A_i), \quad (1)$$

where the random variables y^i in $[-1, 1]$ (non-uniform) are generated from a random number u^i uniformly distributed in $[0, 1]$, and the temperature T_i associated with parameter p^i, as follows:

$$y^i = \mathrm{sgn}(u^i - 0.5)T_i\left[\left(1 + \frac{1}{T_i}\right)^{|2u^i - 1|} - 1\right]. \quad (2)$$

Note that if p^i_{k+1} is outside the range of $[A_i, B_i]$ it will be disregarded, with the process being repeated until it falls within the range. The choice of y^i is made so that the probability density distribution of the D parameters will satisfy the distribution of each parameter:

$$g^i(y^i; T_i) = \frac{1}{2(|y^i| + T_i)(1 + \frac{1}{T_i})}, \quad (3)$$

which is chosen to ensure that any point in configuration space can be sampled infinitely often in annealing time with a cooling schedule outlined below. Thus, at any annealing time k_0, the probability of not generating a global optimum, given infinite time, is zero:

$$\prod_{k=k_0}^{\infty} (1 - g_k) = 0, \quad (4)$$

where g_k is the distribution function at time step k. Note that all atoms move at each Monte-Carlo step in ASA. A Boltzmann acceptance criterion is then applied to the difference in the cost function.

Annealing Schedule

The annealing schedule for each parameter temperature from a starting temperature T_{0i}, and similarly for the cost temperature, is given by:

$$T_i(k_i) = T_{0i} \exp\left(-c_i k_i^{\frac{1}{D}}\right), \quad (5)$$

where c_i and k_i are the annealing scale and ASA step of parameter p^i. The index for re-annealing the cost function is determined by the number of accepted points instead of the number of generated points as is being used for the parameters. This choice was made since the Boltzmann acceptance criterion uses an exponential distribution which is not as 'fat-tailed' as the ASA distribution used for the parameters.

Re-Annealing

The temperatures may be periodically re-annealed or re-scaled according to the sensitivity of the cost function. At any given annealing time, the temperature range is 'stretched out' over the relatively insensitive parameters, thus guiding the search 'fairly' among the parameters. The sensitivity of the energy to each parameter is calculated by:

$$S_i = \frac{\partial E}{\partial p^i}, \quad (6)$$

while the re-annealing temperature is determined by:

$$T_i(k') = T_i(k)\frac{S_i}{S_{\max}}. \quad (7)$$

In this way, less sensitive parameters anneal faster. This is done approximately every 100 accepted events.

For comparison, within conventional simulated annealing [6] the cooling schedule is given by:

$$ST_k = T_0 e^{-(1-c)k} \quad (0 < c < 1), \quad (8)$$

where trial and error are applied to determine the annealing rate $c-1$ as well as the starting temperature T_0. A Monte-Carlo simulation is carried out at each temperature step k with temperature T_k. This cooling schedule is equivalent to $T_{k+1} = T_k c$.

The ASA algorithm is mostly suited to problems for which less is known about the system, and has proven to be more robust than other simulated annealing techniques for complex problems with multiple local minima, e.g., as compared to Cauchy annealing where T_i

= T_0/k, and Boltzmann annealing where $T_i = T_0/\ln k$. The annealing schedule in (8), faster than ASA for a large dimension of D, does not pass the infinitely often annealing-time test in (4), and is therefore referred to as *simulated quench* in the terminology of ASA.

Application to Protein Folding

Computational Details

A protein can be defined as a biopolymer of hundreds of amino acids bonded by peptide bonds, while the test models in this article contain less amino acids, namely oligopeptides. The Met-Enkephalin model was constructed as (H-Tyr-Gly-Gly-Phe-Met-OH). For 14(L-Alanine), the neutral —NH_2 and —COOH end groups were substituted at the termini. The conformation of a protein is described by the dihedral angles of the backbone (ϕ_i, ψ_i), side-chains (χ_i^j), and peptide bond (ω_i, often very close to 180°). Therefore, the conformation determination of the most stable protein is to find the set of $\{\phi, \psi, \chi, \omega\}$ which give the global minimal potential energy $E(\phi, \psi, \chi, \omega)$. Within the ASA nomenclature, the 'cost function' is the potential energy, while a 'parameter' is a dihedral angle variable.

Conformational analyses using conventional simulated annealing were carried out previously [9,11]. The modifications in these works include moving a number of dihedral angles in a Monte-Carlo step; adjusting the maximum deviation of the variables as the temperature decreases to insure that the acceptance ratio is more than 25%; and treating the variables differently according to their importance in the folding process, e. g., by increasing sampling for the backbone dihedral angles as compared to those of the side-chains. It is interesting to point out that within ASA these modifications are implicitly included.

Each ASA run in our work was started from a random initial configuration $\{\phi, \psi, \chi\}$. The dihedral angle ω was fixed to 180° in all of the ASA runs. The initial temperature was determined by the average energy of 5 or 10 random samplings, and a full search range of the dihedral angles ($-\pi, \pi$) was set. The typical maximum number of calls to the energy function was 30000. An ASA run was terminated if it repeated the best energy value for 3 or 5 re-annealing cycles (each cycle generates 100 configurations). Further refinement of the final ASA optimized configuration was carried out by using the local minimizer SUMSL [1], or the conjugate gradient method. The combination of the ASA application and a local minimizer improved the efficiency of the search.

The ASA calculation is governed by various control parameters [3], for which the most important setting is the annealing rate for the temperatures of 'cost' and 'parameters', determined by the so-called 'temperature-ratio-scale' (the ratio of the final to the initial temperature after certain annealing steps) and the 'cost-parameter-scale'. The control parameters were varied to improve the search efficiency. Adequate control parameters used for obtaining the results reported in this study were: 'temperature-ratio-scale' = 10^{-4}; 'cost-parameter-scale' = 0.5. These parameter settings correspond to an annealing rate for energy of $c_{\text{cost}} = 3.6$, and for all dihedral angles of $c_{\text{parameter}} = 7.2$. Note that the annealing rate for all dihedral angles was chosen to be the same.

Met-Enkephalin

Met-Enkephalin has a complicated energy surface [11,16]. The lowest energy for Met-Enkephalin was found to be −12.9 kcal/mol with the force field being ECEPP/2 (Empirical Conformation Energy Program for Peptides) [8]. With all ω fixed, the lowest energy was found to be −10.7 kcal/mol by MCM [14]. Using different initial conformations and control parameter settings of the cooling schedule as described above, 55 independent ASA runs were carried out. Table 1 summarizes the energy distribution of these calculations. Most of the ASA calculations result in energies in the range of −8 to −3 kcal/mol, with 7 of the results determining conformations having energies that are only 3 kcal/mol above the known lowest energy, thus exhibiting the effectiveness of the approach. Moreover, as the range of search was somewhat narrowed, almost all of the ASA runs reach the global energy minimum.

Adaptive Simulated Annealing and its Application to Protein Folding, Table 1
The energy (in kcal/mol) distribution of ASA runs for Met-Enkephalin using a full search range

Energy	< − 8	(−8, −5)	(−5, −3)	> − 3
No. of runs	7	19	19	10

Adaptive Simulated Annealing and its Application to Protein Folding, Table 2
Energy and dihedral angles of the lowest energy conformations of Met-Enkephalin calculated by ASA. RMSD1 is the root-mean-square deviation (in Å) for backbone atoms, while RMSD2 is for all atoms

	A0	A	1	2	3	4
E	−12.9	−10.7	−10.6	−10.4	−10.1	−8.5
ϕ_1	−86	−87	−87	−87	−87	−87
ψ_1	156	154	153	153	156	153
ϕ_2	−155	−162	−161	−162	−166	−166
ψ_2	84	71	72	75	87	72
ϕ_3	84	64	64	63	68	63
ψ_3	−74	−93	−94	−95	−91	−97
ϕ_4	−137	−82	−83	−81	−103	−74
ψ_4	19	−29	−26	−30	−13	−30
ϕ_5	−164	−81	−79	−76	−76	−82
ψ_5	160	144	133	132	137	143
χ_1^1	−173	−180	180	179	−166	−180
χ_1^2	79	−111	−110	71	88	73
χ_1^3	−166	145	145	−35	−148	−179
χ_4^1	59	180	72	−179	71	179
χ_4^2	−86	−100	84	−100	−93	−100
χ_5^1	53	−65	−171	−173	−65	−65
χ_5^2	175	−179	176	176	−178	−179
χ_5^3	−180	−179	180	179	−178	−179
χ_5^4	−58	−180	−60	60	−178	−179
RMSD1		0	0.04	0.07	0.51	0.26
RMSD2		0	2.52	1.92	2.08	1.29

Adaptive Simulated Annealing and its Application to Protein Folding, Table 3
The conformation of a model 14(L-Alanine) peptide as calculated by ASA

	2	3	4	5	6
ϕ	−99.4	−68.2	−68.0	−69.3	−66.9
ψ	158.1	−34.3	−38.8	−38.5	−38.6
	7	8	9	10	11
ϕ	−68.3	−66.7	−68.8	−67.1	−69.4
ψ	−39.2	−38.0	−38.7	−37.7	−39.6
	12	13	14	15	
ϕ	−65.0	−67.2	−87.7	−75.9	
ψ	−40.0	−44.6	65.8	−40.1	

For the full range search, we identified three conformations with energies of −10.6, −10.4, and −10.1 kcal/mol, that exhibit the configuration of the known lowest geometry of −10.7 kcal/mol. Table 2 lists the conformations of these lowest energy configurations, as well as an additional low energy structure. Conformations A0 and A are the lowest-energy conformations with ω nonfixed and fixed, respectively, taken from [11,14]. The first two conformations, #1 and #2, have almost the same backbone configuration as that of A (−10.7 kcal/mol), with a backbone root-mean-square deviation (RMSD) of only 0.04 and 0.07Å, respectively. The all-atom RMSD of the listed conformations with energies ranging from −8.5 to −10.6 kcal/mol are about 2Å. For conformations #1 and #2, the noted differences are in the side-chains, corresponding to a 0.1 and 0.3 kcal/mol difference in energy, respectively.

Poly(L-Alanine)

The ASA algorithm was applied to a model of (L-Alanine) that is known to assume a dominant right-handed α-helical structure [13]. For a search range of dihedral angles that include both the right-handed (RH) α-helix and the β-sheet region in the Ramachandran's diagram, ψ: (−115°, −180°) and ϕ: (−115°, 0°), it was significant to find RH α-helices with $\phi \approx -68°$ and $\psi \approx -38°$ in all backbones except those near the end-groups, as shown in Table 3. The energy of such a geometry is typically −10.2 kcal/mol after a local minimization. The energy surfaces of the RH α-helical regions were found to be less complex than those of Met-Enkephalin. These results are consistent with a previous study [16].

Conclusion

The adaptive simulated annealing as a global optimization method intrinsically includes some of the modifications of conventional simulated annealing used for biomolecular structure determination. As applied to Met-Enkephalin, the performance of ASA is comparable to the simulated annealing study reported in [12], while better than the one reported in [11], although some differences other than the algorithms are noted. Utilizing a partial search range improves the efficiency significantly, showing that ASA may be useful for refinement of a molecular structure predicted or measured by other methods. A dominant right-handed α-helical conformation was found for the 14 residue (L-Alanine) model, with deviations observed only near the end groups.

Recent Studies and Future Directions

Recent studies have shown improved efficiency in the conformational search of Met-Enkephalin, *e.g.*, the so-called conformation space annealing (CSA), which combines the ideas of genetic algorithms, simulated annealing, a build up procedure, and local minimization [7]. The use of the multicanonical ensemble algorithm (ME) (one of the generalized-ensemble algorithms [2]), allows free random walks in energy space, escaping from any energy barrier. Both the ME and CSA algorithms outperform genetic algorithms (GA), simulated annealing (SA), GA with minimization (GAM) and Monte-Carlo with minimization (MCM). Our own work (unpublished) and the work in ref. [5] both show that simple GA alone underperforms simulated annealing for the Met-Enkephalin conformational search problem. Table 4 compares these algorithms for efficiency (the number of evaluations of energy and energy gradient, or the number of local minimizations) and effectiveness (the number of runs reaching the ground state conformation (hits) versus the number of total independent runs). Caution should be exercised since some differences exist between these studies, such as the version of the ECEPP potential used, the treatment of the peptide dihedral angle ω, etc. Ground state conformations are those having energy within approximately 1eV from the known global minimum energy. Note that the generalized-ensemble method can be carried out with both Monte-Carlo and molecular dynamics.

In comparison to the studies summarized in Table 4, ASA seems to be using too small a number of function evaluations. Optimizing control parameters such as the annealing schedule and increasing the number of energy evaluations may improve the effectiveness. Search efficiency could also be improved by adopting parallellization to achieve scalable simulation for various algorithms. Extensive research on the protein conformational search using various hybrids of genetic algorithms and parallelization is in progress (as of 1999).

Adaptive Simulated Annealing and its Application to Protein Folding, Table 4
Comparison of the conformation search efficiency and effectiveness of Met-Enkephalin using different algorithms. N_E, $N_{\nabla E}$, and N_{minz} are the number of the evaluations of energy, energy gradient, and number of local minimizations of each run, in the unit of 10^3

	hits/total	N_E	$N_{\nabla E}$	N_{minz}
ME [2]	10/10	< 1900	0	0
MCM [11]	24/24	*	*	15
GAM [10]	5/5	*	*	50
ME [2]	18/20	950	0	0
CSA [7]	99/100	300	250	5
ME [2]	21/50	400	0	0
CSA [7]	50/100	170	130	2.6
SA [2]	8/20	1000	0	0
GA [5]	< 1/27	100	0	0.001

*: The total number of $E, \nabla E$ evaluations are not given, but can be estimated based on roughly 100 evaluations for each minimization.

See also

- Adaptive Global Search
- Bayesian Global Optimization
- Genetic Algorithms
- Genetic Algorithms for Protein Structure Prediction
- Global Optimization Based on Statistical Models
- Global Optimization in Lennard–Jones and Morse Clusters
- Global Optimization in Protein Folding
- Molecular Structure Determination: Convex Global Underestimation
- Monte-Carlo Simulated Annealing in Protein Folding
- Multiple Minima Problem in Protein Folding: αBB Global Optimization Approach
- Packet Annealing
- Phase Problem in X-ray Crystallography: Shake and Bake Approach
- Protein Folding: Generalized-ensemble Algorithms
- Random Search Methods
- Simulated Annealing
- Simulated Annealing Methods in Protein Folding
- Stochastic Global Optimization: Stopping Rules
- Stochastic Global Optimization: Two-phase Methods

References

1. Gay DM (1983) Subroutines for unconstrained minimization using a model/trust-region approach. ACM Trans Math Softw 9:503

2. Hansmann UHE (1998) Generalized ensembles: A new way of simulating proteins. Phys A 254:15
3. Ingber L (1989) Very fast simulated re-annealing. Math Comput Modelling 12:967 ASA code is available from: ftp.alumni.caltech.edu:pub/ingber
4. Ingber L, Rosen B (1992) Genetic algorithm and very fast simulated re-annealing: A comparison. Math Comput Modelling 16:87
5. Jin AY, Leung FY, Weaver DF (1997) Development of a novel genetic algorithm search method (GAP1.0) for exploring peptide conformational space. J Comput Chem 18:1971
6. Kirkpatrick S, Gelatt CD, Vecchi MP Jr (1983) Optimization by simulated annealing. Science 220:671
7. Lee J, Scheraga HA, Rackovsky S (1997) New optimization method for conformational energy calculations on polypeptides: conformational space annealing. J Comput Chem 18:222
8. Li Z, Scheraga HA (1987) Monte Carlo-minimization approach to the multiminima problem in protein folding. Proc Nat Aca Sci USA 84:6611
9. Li Z, Scheraga HA (1988) Structure and free energy of complex thermodynamic systems. J Mol Struct (Theochem) 179:333
10. Merkle LD, Lamont GB, Gates GH, Pachter R (May, 1996) Hybrid genetic algorithms for minimization of polypeptide specific energy model. Proc. IEEE Int. Conf. Evolutionary Computation, p 192
11. Nayeem A, Vila J, Scheraga HA (1991) A comparative study of the simulated-annealing and Monte Carlo-with minimization approaches to the minimum-energy structures of polypeptides: Met-Enkephalin. J Comput Chem 12:594
12. Okamoto Y, Kikuchi T, Kawai H (1992) Prediction of low-energy structure of Met-Enkephalin by Monte Carlo simulated annealing. Chem Lett (Chem Soc Japan):1275
13. Piela L, Scheraga HA (1987) On the multiple-minima problem in the conformational analysis of polypeptides: I. backbone degrees of freedom for a perturbed α-helix. Biopolymers 26:S33
14. Vasquez M (1999) Private communication
15. Wang Z, Pachter R (1997) Prediction of polypeptide conformation by the adaptive simulated annealing approach. J Comput Chem 18:323
16. Wilson SR, Cui W (1990) Applications of simulated annealing to peptides. Biopolymers 29:225

Affine Sets and Functions

LEONIDAS PITSOULIS
Princeton University, Princeton, USA

MSC2000: 51E15, 32B15, 51N20

Article Outline

Keywords
See also
References

Keywords

Linear algebra; Convex analysis

A subset S of $\mathbf{R}^n$ is an *affine set* if

$$(1-\lambda)x + \lambda y \in S,$$

for any $x, y \in S$ and $\lambda \in \mathbf{R}$. A function $f: \mathbf{R}^n \to \mathbf{R}$ is an *affine function* if f is finite, convex and concave (cf. ▶ Convex max-functions).

See also

▶ Linear Programming
▶ Linear Space

References

1. Rockafellar RT (1970) Convex analysis. Princeton Univ. Press, Princeton

Airline Optimization

GANG YU[1], BENJAMIN THENGVALL[2]
[1] Department Management Sci. and Information Systems Red McCombs School of Business, University Texas at Austin, Austin, USA
[2] CALEB Technologies Corp., Austin, USA

MSC2000: 90B06, 90C06, 90C08, 90C35, 90C90

Article Outline

Keywords
See also
References

Keywords

Network design and schedule construction; Fleet assignment; Aircraft routing; Crew scheduling; Revenue management; Irregular operations; Air traffic control and ground delay programs

The airline industry was one of the first to apply operations research methodology and techniques on a large scale. As early as the late 1950s, operations researchers were beginning to study how the developing fields of mathematical programming could be used to address a number of very difficult problems faced by the airline industry. Since that time many airline related problems have been the topics of active research [26]. Most optimization-related research in the airline industry can be placed in one of the following areas:
- network design and schedule construction;
- fleet assignment;
- aircraft routing;
- crew scheduling;
- revenue management;
- irregular operations;
- air traffic control and ground delay programs.

In the following, each of these problem areas will be defined along with a brief discussion of some of the operations research techniques that have been applied to solve them. The majority of applications utilize network-based models. Solution of these models range from traditional mathematical programming approaches to a variety of novel heuristic approaches. A very brief selection of references is also provided.

Construction of flight schedules is the starting point for all other airline optimization problems and is a critical operational planning task faced by an airline. The *flight schedule* defines a set of flight segments that an airline will service along with corresponding origin and destination points and departure and arrival times for each flight segment. An airline's decision to offer certain flights will depend in large part on market demand forecasts, available aircraft operating characteristics, available manpower, and the behavior of competing airlines [11,12].

Of course, prior to the construction of flight schedules, an airline must decide which markets it will serve. Before the 1978 'Airline Deregulation Act', airlines had to fly routes as assigned by the Civil Aeronautics Board regardless of the demand for service. During this period, most airlines emphasized long point-to-point routes. Since deregulation, airlines have gained the freedom to choose which markets to serve and how often to serve them. This change led to a fundamental shift in most airlines routing strategies from point-to-point flight networks to hub-and-spoke oriented flight networks. This, in turn, led to new research activities for finding optimal hub [3,18] and maintenance base [13] locations.

Following network design and schedule construction, an aircraft type must be assigned to each flight segment in the schedule. This is called the *fleet assignment problem*. Airlines generally operate a number of different fleet types, each having different characteristics and costs such as seating capacity, landing weights, and crew and fuel costs. The majority of fleet assignment methods represent the flight schedule via some variant of a time-space network with flight arcs between stations and inventory arcs at each station. A *multicommodity network flow problem* can then be formulated with arcs and nodes duplicated as appropriate for all fleets that can take a particular flight. Side constraints must be implemented to ensure each flight segment is assigned to only one fleet. In domestic fleet assignment problems, a common simplifying assumption is that every flight is flown every day of the week. Under this assumption, the network model need only account for one day's flights and a looping arc connects the end of the day with the beginning. The resulting models are mixed integer programs [1,16,27,30].

Aircraft routing is a fleet by fleet process of assigning individual aircraft to fly each flight segment assigned to a particular fleet. A primary consideration at this stage is *maintenance* requirements mandated by the Federal Aviation Administration. There are different types of maintenance activities that must be performed after a given number of flight hours. The majority of these maintenance activities can be performed overnight; however, not all stations are equipped with proper maintenance facilities for all fleets. During the aircraft routing process, individual aircraft from each fleet must be assigned to fly all flight segments assigned to that fleet in a manner that provides maintenance opportunities for all aircraft at appropriate stations within the required time intervals. This problem has been formulated and solved in a number of ways including as a general integer programming problem solved by Lagrangian relaxation [9] and as a set partitioning problem solved with a branch and bound algorithm [10].

As described above, the problems of fleet assignment and aircraft routing have been historically solved in a sequential manner. Recently, work has been done to solve these problems simultaneously using a string-based model and a branch and price solution approach [5].

Crew scheduling, like aircraft routing, is done following fleet assignment. The first of two sequentially solved crew scheduling problems is the crew pairing problem. A crew pairing is a sequence of flight legs beginning and ending at a crew base that satisfies all governmental and contractual restrictions (some times called legalities). These crew pairings generally cover a period of 2–5 days. The problem is to find a minimum cost set of such crew pairings such that all flight segments are covered. This problem has generally been modeled as a set partitioning problem in which pairings are enumerated or generated dynamically [15,17]. Other attempts to solve this problem have employed a decomposition approach based on graph partitioning [4] and a linear programming relaxation of a set covering problem [21]. Often a practice called *deadheading* is used to reposition flight crews in which a crew will fly a flight segment as passengers. Therefore, in solving the crew-pairing problem, all flight segments must be covered, but they may be covered by more than one crew.

The second problem to be solved relating to crew scheduling is the monthly crew rostering problem. This is the problem of assigning individual crew members to crew pairings to create their monthly schedules. These schedules must incorporate time off, training periods, and other contractual obligations. Generally, a *preferential bidding system* is used to make the assignments in which each personalized schedule takes into account an employee's pre-assigned activities and weighted bids representing their preferences. While the crew pairing problem has been widely studied, a limited number of publications have dealt with the monthly crew rostering problem. Approaches include an integer programming scheme [14] and a network model [24].

Revenue management is the problem of determining fare classes for each flight in the flight schedule as well as the allocation of available seats to each fare class. Not only are seats on an airplane partitioned physically into sections such as first class and coach, but also seats in the same section are generally priced at many different levels. The goal is to maximize the expected revenue from a particular flight segment by finding the proper balance between gaining additional revenue by selling more inexpensive seats and losing revenue by turning away higher fare customers. A standard assumption is that fare classes are filled sequential from the lowest to the highest. This is often the case where discounted fares are offered in advance, while last minute tickets are sold at a premium. Recent research includes a probabilistic decision model [6], a dynamic programming formulation [31] and some calculus-based booking policies [8].

When faced with a lack of resources, airlines often are not able to fly their published flight schedule. This is frequently the result of aircraft mechanical difficulties, inclement weather, or crew shortages. As situations like these arise, decisions must be made to deal with the shortage of resources in a manner that returns the airline to the originally planned flight schedule in a timely fashion while attempting to reduce operational cost and keep passengers satisfied. This general situation is called the *airline irregular operations problem* and it involves aircraft, crew, gates, and passenger recovery.

The aircraft schedule recovery problem deals with re-routing aircraft during irregular operations. This problem has received significant attention among irregular operations topics; papers dealing with crew scheduling during irregular operations have only recently started to appear [28,35]. Most approaches for dealing with aircraft schedule recovery have been based on network models. Some early models were pure networks [19]. Recently, more comprehensive models have been developed that better represent the problem, but are more difficult to solve as side constraints have been added to the otherwise network structure of these problems [2,33,36]. In practice, many airlines use heuristic methods to solve these problems as their real-time nature does not allow for lengthy optimization run times.

Closely related to the irregular operations problem is the *ground delay problem* in air traffic control. Ground delay is a program implemented by the Federal Aviation Administration in cases of station congestion. During ground delay, aircraft departing for a congested station are held on the ground before departure. The rational for this behavior is that ground delays are less expensive and safer than airborne delays. Several optimization models have been formulated to decrease the total minutes of delay experienced throughout the system during a ground delay program. These problems have generally been modeled as integer programs ([22,23]), but the problem has also been solved using stochastic linear programming [25] and by heuristic methods [34].

Optimization based methods have also been applied to a myriad of other airline related topics such as gate assignment [7], fuel management [29], short term fleet assignment swapping [32], demand modeling [20], and others. Airline industry is an exciting arena for the interplay between optimization theory and practice. Many more optimization applications in the airline industry will evolve in the future.

See also

▶ Integer Programming
▶ Vehicle Scheduling

References

1. Abara J (1989) Applying integer linear programming to the fleet assignment problem. Interfaces 19(4):20–28
2. Argüello MF, Bard JF, Yu G (1997) Models and methods for managing airline irregular operations aircraft routing. In: Yu G (ed) Operations Research in Airline Industry. Kluwer, Dordrecht
3. Aykin T (1994) Lagrangian relaxation based approaches to capacitated hub-and-spoke network design problem. Europ J Oper Res 79(3):501–523
4. Ball M, Roberts A (1985) A graph partitioning approach to airline crew scheduling. Transport Sci 19(2):107–126
5. Barnhart C, Boland NL, Clarke LW, Johnson EL, Nemhauser G, Shenoi RG (1998) Flight string models for aircraft fleeting and routing. Transport Sci 32(3):208–220
6. Belobaba PP (1989) Application of a probabilistic decision model to airline seat inventory control. Oper Res 37:183–197
7. Brazile RP, Swigger KM, Wyatt DL (1994) Selecting a modelling technique for the gate assignment problems: Integer programming, simulation, or expert system. Internat J Modelling and Simulation 14(1):1–5
8. Brumelle SL, McGill JI (1993) Airline seat allocation with multiple nested fare classes. Oper Res 41(1):127–137
9. Daskin MS, Panagiotopoulos ND (1989) A Lagrangian relaxation approach to assigning aircraft to routes in hub and spoke networks. Transport Sci 23(2):91–99
10. Desaulniers G, Desrosiers J, Dumas Y, Solomon MM, Soumis F (1997) Daily aircraft routing and scheduling. Managem Sci 43(6):841–855
11. Dobson G, Lederer PJ (1993) Airline scheduling and routing in a hub-and-spoke system. Transport Sci 27(3):281–297
12. Etschamaier MM, Mathaisel DFX (1985) Airline scheduling: An overview. Transport Sci 9(2):127–138
13. Feo TA, Bard JF (1989) Flight scheduling and maintenance base planning. Managem Sci 35(12):1415–1432
14. Gamache M, Soumis F, Villeneuve D, Desrosiers J (1998) The preferential bidding system at Air Canada. Transport Sci 32(3):246–255
15. Graves GW, McBride RD, Gershkoff I, Anderson D, Mahidhara D (1993) Flight crew scheduling. Managem Sci 39(6):736–745
16. Hane CA, Barnhart C, Johnson EL, Marsten RE, Nemhauser GL, Sigismondi G (1995) The fleet assignment problem: Solving a large-scale integer program. Math Program 70(2):211–232
17. Hoffman KL, Padberg M (1993) Solving airline crew scheduling problems by branch-and-cut. Managem Sci 39(6):657–680
18. Jaillet P, Song G, Yu G (1997) Airline network design and hub location problems. Location Sci 4(3):195–212
19. Jarrah AIZ, Yu G, Krishnamurthy N, Rakshit A (1993) A decision support framework for airline flight cancellations and delays. Transport Sci 27(3):266–280
20. Jorge-Calderon JD (1997) A demand model for scheduled airline services on international European routes. J Air Transport Management 3(1):23–35
21. Lavoie S, Minoux M, Odier E (1988) A new approach for crew pairing problems by column generation with an application to air transportation. Europ J Oper Res 35:45–58
22. Luo S, Yu G (1997) On the airline schedule perturbation problem caused by the ground delay program. Transport Sci 31(4):298–311
23. Navazio L, Romanin-Jacur G (1998) The multiple connections multi-airport ground holding problem: Models and algorithms. Transport Sci 32(3):268–276
24. Nicoletti B (1975) Automatic crew rostering. Transport Sci 9(1):33–48
25. Richetta O, Odoni AR (1993) Solving optimally the static ground-holding policy problem in air traffic control. Transport Sci 27(3):228–238
26. Richter H (1989) Thirty years of airline operations research. Interfaces 19(4):3–9
27. Rushmeier RA, Kontogiorgis SA (1997) Advances in the optimization of airline fleet assignment. Transport Sci 31(2):159–169
28. Stojkovic M, Soumis F, Desrosiers J (1998) The operational airline crew scheduling problem. Transport Sci 32(3):232–245
29. Stroup JS, Wollmer RD (1992) A fuel management model for the airline industry. Oper Res 40(2):229–237
30. Subramanian R, Scheff RP Jr, Quillinan JD, Wiper DS, Marsten RE (1994) Coldstart: Fleet assignment at delta air lines. Interfaces 24(1):104–120
31. Tak TC, Hersh M (1993) A model for dynamic airline seat inventory control with multiple seat bookings. Transport Sci 27(3):252–265
32. Talluri KT (1993) Swapping applications in a daily airline fleet assignment. Transport Sci 30(3):237–248

33. Thengvall BT, Bard JF, Yu G (2000) Balancing user preferences for aircraft schedule recovery during airline irregular operations. IEE Trans Oper Eng 32(3):181–193
34. Vranas PB, Bertsemas DJ, Odoni AR (1994) The multiairport ground-holding problem in air traffic control. Oper Res 42(2):249–261
35. Wei G, Song G, Yu G (1997) Model and algorithm for crew management during airline irregular operations. J Combin Optim 1(3):80–97
36. Yan S, Tu Y (1997) Multifleet routing and multistop flight scheduling for schedule perturbation. Europ J Oper Res 103(1):155–169

Algorithmic Improvements Using a Heuristic Parameter, Reject Index for Interval Optimization

TIBOR CSENDES
University of Szeged, Szeged, Hungary

MSC2000: 65K05, 90C30

Article Outline

Keywords and Phrases
Introduction
Subinterval Selection
Multisection
Heuristic Rejection
References

Keywords and Phrases

Branch-and-bound; Interval arithmetic; Optimization; Heuristic parameter

Introduction

Interval optimization methods (▶ interval analysis: unconstrained and constrained optimization) have the guarantee not to loose global optimizer points. To achieve this, a deterministic branch-and-bound framework is applied. Still, heuristic algorithmic improvements may increase the convergence speed while keeping the guaranteed reliability.

The indicator parameter called RejectIndex

$$pf^*(X) = \frac{f^* - \underline{F}(X)}{\overline{F}(X) - \underline{F}(X)}$$

was suggested by L.G. Casado as a measure of the closeness of the interval X to a global minimizer point [1]. It was first applied to improve the work load balance of global optimization algorithms.

A subinterval X of the search space with the minimal value of the inclusion function $\underline{F}(X)$ is usually considered as the best candidate to contain a global minimum. However, the larger the interval X, the larger the overestimation of the range $f(X)$ on X compared to $F(X)$. Therefore a box could be considered as a good candidate to contain a global minimum just because it is larger than the others. To compare subintervals of different sizes we normalize the distance between the global minimum value f^* and $\underline{F}(X)$.

The idea behind pf^* is that in general we expect the overestimation to be symmetric, i. e., the overestimation above $f(X)$ is closely equal to the overestimation below $f(X)$ for small subintervals containing a global minimizer point. Hence, for such intervals X the relative place of the global optimum value inside the $F(X)$ interval should be high, while for intervals far from global minimizer points pf^* must be small. Obviously, there are exceptions, and there exists no theoretical proof that pf^* would be a reliable indicator of nearby global minimizer points.

The value of the global minimum is not available in most cases. A generalized expression for a wider class of indicators is

$$p(\hat{f}, X) = \frac{\hat{f} - \underline{F}(X)}{\overline{F}(X) - \underline{F}(X)},$$

where the $\hat{f}$ value is a kind of approximation of the global minimum. We assume that $\hat{f} \in F(X)$, i. e., this estimation is realistic in the sense that $\hat{f}$ is within the known bounds of the objective function on the search region. According to the numerical experience collected, we need a good approximation of the f^* value to improve the efficiency of the algorithm.

Subinterval Selection

I. Among the possible applications of these indicators the most promising and straightforward is in the *subinterval selection*. The theoretical and computational properties of the interval branch-and-bound optimization has been investigated extensively [6,7,8,9]. The most important statements proved are the following for algorithms with balanced subdivision direction selection:

1. Assume that the inclusion function of the objective function is isotone, it has the zero convergence property, and the $p(f_k,Y)$ parameters are calculated with the f_k parameters converging to $\hat{f} > f^*$, for which there exists a point $\hat{x} \in X$ with $f(\hat{x}) = \hat{f}$. Then the branch-and-bound algorithm that selects that interval Y from the working list which has the maximal $p(f_i,Z)$ value can converge to a point $\hat{x} \in X$ for which $f(\hat{x}) > f^*$, i.e., to a point which is not a global minimizer point of the given problem.

2. Assume that the inclusion function of the objective function has the zero convergence property and f_k converges to $\hat{f} < f^*$. Then the optimization branch-and-bound algorithm will produce an everywhere dense sequence of subintervals converging to each point of the search region X regardless of the objective function value.

3. Assume that the inclusion function of the objective function is isotone and has the zero convergence property. Consider the interval branch-and-bound optimization algorithm that uses the cutoff test, the monotonicity test, the ▶ interval Newton step, and the concavity test as accelerating devices, and that selects as the next leading interval that interval Y from the working list which has the maximal $p(f_i,Z)$ value. A necessary and sufficient condition for the convergence of this algorithm to a set of global minimizer points is that the sequence $\{f_i\}$ converges to the global minimum value f^*, and there exist at most a finite number of f_i values below f^*.

4. If our algorithm applies the interval selection rule of maximizing the $p(f^*, X) = pf^*(X)$ values for the members of the list L (i.e., if we can use the known exact global minimum value), then the algorithm converges exclusively to global minimizer points.

5. If our algorithm applies the interval selection rule of maximizing the $p(\tilde{f}, X)$ values for the members of the list L, where $\tilde{f}$ is the best available upper bound for the global minimum, *and its convergence to f^* can be ensured*, then the algorithm converges exclusively to global minimizer points.

6. Assume that for an optimization problem $\min_{x \in X} f(x)$ the inclusion function $F(X)$ of $f(x)$ is isotone and α-convergent with given positive constants α and C. Assume further that the pf^* parameter is less than 1 for all the subintervals of X. Then an arbitrary large number $N(>0)$ of consecutive leading intervals of the basic B&B algorithm that selects the subinterval with the smallest lower bound as the next leading interval may have the following properties:

   i. None of these processed intervals contains a stationary point.

   ii. During this phase of the search the pf^* values are maximal for these intervals.

7. Assume that the inclusion function of the objective function is isotone and it has the zero convergence property. Consider the interval branch-and-bound optimization algorithm that uses the cutoff test, the monotonicity test, the interval Newton step, and the concavity test as accelerating devices and that selects as the next leading interval that interval Y from the working list which has the maximal $pf(f_k,Z)$ value.

   i. The algorithm converges exclusively to global minimizer points if

   $$\underline{f}_k \leq f_k < \delta(\overline{f}_k - \underline{f}_k) + \underline{f}_k$$

   holds for each iteration number k, where $0 < \delta < 1$.

   ii. The above condition is sharp in the sense that $\delta = 1$ allows convergence to not optimal points. Here $\underline{f}_k = \min\{\underline{F}(Y^l), l = 1, \ldots, |L_k|\} \leq f_k < \tilde{f}_k = \overline{f}_k$, where $|L|$ stands for the cardinality of the elements of the list L.

II. These theoretical results are in part promising (e.g., 7), in part disappointing (5 and 6). The conclusions of the detailed numerical comparisons were that if the global minimum value is known, then the use of the pf^* parameter in the described way can accelerate the interval optimization method by orders of magnitude, and this improvement is especially strong for hard problems.

In case the global minimum value is not available, then its estimation, f_k, which fulfills the conditions of 7, can be utilized with similar efficacy, and again the best results were achieved on difficult problems.

Multisection

I. The multisection technique is a way to accelerate branch-and-bound methods by subdividing the actual interval into several subintervals in a single algorithm step. In the extreme case half of the function evaluations can be saved [5,10]. On the basis of the RejectIndex value of a given interval it is decided whether simple bisection or two higher-degree multisections are to be applied [2,11]. Two threshold values, $0 < P_1 < P_2 < 1$, are used for selecting the proper multisection type.

This algorithm improvement can also be cheated in the sense that there exist global optimization problems for which the new method will follow for an arbitrary long number of iterations an embedded interval sequence that contains no global minimizer point, or that intervals in which there is a global minimizer have misleading indicator values.

According to the numerical tests, the new multisection strategies result in a substantial decrease both in the number of function evaluations and in the memory complexity.

II. The multisection strategy can also be applied to constrained global optimization problems [11]. The feasibility degree index for constraint $g_j(x) \leq 0$ can be formulated as

$$pu_{G_j}(X) = \min\left\{\frac{-G_j(X)}{w(G_j(X))}, 1\right\}.$$

Notice that if $pu_{G_j}(X) < 0$, then the box is certainly infeasible, and if $pu_{G_j}(X) = 1$ then X certainly satisfies the constraint. Otherwise, the box is undetermined for that constraint. For boxes that are not certainly infeasible, i.e., for which $pu_{G_j}(X) \geq 0$ for all $j = 1, \ldots, r$ holds, the total infeasibility index is given by

$$pu(X) = \prod_{j=1}^{r} pu_{G_j}(X).$$

We must only define the index for such boxes since certainly infeasible boxes are immediately removed by the algorithm from further consideration. With this definition,
- $pu(X) = 1 \Leftrightarrow X$ is certainly feasible and
- $pu(X) \in [0, 1) \Leftrightarrow X$ is undetermined.

Using the $pu(X)$ index, we now propose the following modification of the RejectIndex for constrained problems:

$$pup(\hat{f}, X) = pu(X) \cdot p(\hat{f}, X),$$

where $\hat{f}$ is a parameter of this indicator, which is usually an approximation of f^*. This new index works like $p(\hat{f}, X)$ if X is certainly feasible, but if the box is undetermined, then it takes the feasibility degree of the box into account: the less feasible the box is, the lower the value of $pu(X)$ is.

A careful theoretical analysis proved that the new interval selection and multisection rules enable the branch-and-bound interval optimization algorithm to converge to a set of global optimizer points assuming we have a proper sequence of $\{\hat{f}_k\}$ parameter values. The convergence properties obtained were very similar to those proven for the unconstrained case, and they give a firm basis for computational implementation.

A comprehensive numerical study on standard global optimization test problems and on facility location problems indicated [11] that the constrained version interval selection rules and, to a lesser extent, also the new adaptive multisection rules have several advantageous features that can contribute to the efficiency of the interval optimization techniques.

Heuristic Rejection

RejectIndex can also be used to improve the efficiency of interval global optimization algorithms on very hard to solve problems by applying a rejection strategy to get rid of subintervals not containing global minimizer points. This heuristic rejection technique selects those subintervals on the basis of a typical pattern of changes in the pf^* values [3,4].

The RejectIndex is not always reliable: assume that the inclusion function $F(X)$ of $f(x)$ is isotone and α-convergent. Assume further that the RejectIndex parameter pf^* is less than 1 for all the subintervals of X. Then an arbitrary large number $N(> 0)$ of consecutive leading intervals may have the following properties:
i. Neither of these processed intervals contains a stationary point, and
ii. During this phase of the search the pf^* values are maximal for these intervals as compared with the subintervals of the current working list.

Also, when a global optimization problem has a unique global minimizer point x^*, there always exists an isotone and α-convergent inclusion function $F(X)$ of $f(x)$ such that the new algorithm does not converge to x^*.

In spite of the possibility of losing the global minimum, obviously there exist such implementations that allow a safe way to use heuristic rejection. For example, the selected subintervals can be saved on a hard disk for further possible processing if necessary.

Although the above theoretical results were not encouraging, the computational tests on very hard global optimization problems were convincing: when the whole list of subintervals produced by the B&B algorithm is too large for the given computer memory, then the use of the suggested heuristic rejection technique decreases the number of working list elements without missing the global minimum. The new rejection test may also make it possible to solve hard-to-solve problems that are otherwise unsolvable with the usual techniques.

References

1. Casado LG, García I (1998) New Load Balancing Criterion for Parallel Interval Global Optimization Algorithms. In: Proceedings of the 16th IASTED International Conference on Applied Informatics, Garmisch-Partenkirchen, pp 321–323
2. Casado LG, García I, Csendes T (2000) A new multisection technique in interval methods for global optimization. Computing 65:263–269
3. Casado LG, García I, Csendes T (2001) A heuristic rejection criterion in interval global optimization algorithms. BIT 41:683–692
4. Casado LG, García I, Csendes T, Ruiz VG (2003) Heuristic Rejection in Interval Global Optimization. JOTA 118:27–43
5. Csallner AE, Csendes T, Markót MC (2000) Multisection in Interval Branch-and-Bound Methods for Global Optimization I. Theoretical Results. J Global Optim 16:371–392
6. Csendes T (2001) New subinterval selection criteria for interval global optimization. J Global Optim 19:307–327
7. Csendes T (2003) Numerical experiences with a new generalized subinterval selection criterion for interval global optimization. Reliab Comput 9:109–125
8. Csendes T (2004) Generalized subinterval selection criteria for interval global optimization. Numer Algorithms 37:93–100
9. Kreinovich V, Csendes T (2001) Theoretical Justification of a Heuristic Subbox Selection Criterion for Interval Global Optimization. CEJOR 9:255–265
10. Markót MC, Csendes T, Csallner AE (2000) Multisection in Interval Branch-and-Bound Methods for Global Optimization II. Numerical Tests. J Global Optim 16:219–228
11. Markót MC, Fernandez J, Casado LG, Csendes T (2006) New interval methods for constrained global optimization. Math Programm 106:287–318

Algorithms for Genomic Analysis

Eva K. Lee, Kapil Gupta
Center for Operations Research in Medicine and HealthCare,
School of Industrial and Systems Engineering,
Georgia Institute of Technology, Atlanta, USA

MSC2000: 90C27, 90C35, 90C11, 65K05, 90-08, 90-00

Article Outline

Abstract
Introduction
Phylogenetic Analysis
   Methods Based on Pairwise Distance
   Parsimony Methods
   Maximum Likelihood Methods
Multiple Sequence Alignment
   Scoring Alignment
   Alignment Approaches
   Progressive Algorithms
   Graph-Based Algorithms
   Iterative Algorithms
Novel Graph-Theoretical Genomic Models
   Definitions
   Construction of a Conflict Graph from Paths
      of Multiple Sequences
   Complexity Theory
   Special Cases of MWCMS
   Computational Models:
      Integer Programming Formulation
Summary
Acknowledgement
References

Abstract

The genome of an organism not only serves as its blueprint that holds the key for diagnosing and curing diseases, but also plays a pivotal role in obtaining a holistic view of its ancestry. Recent years have witnessed a large number of innovations in this field, as exemplified by the Human Genome Project. This chapter provides an overview of popular algorithms used in genome analysis and in particular explores two important and deeply interconnected problems: phylogenetic analysis and multiple sequence alignment. We also describe our novel graph-theoretical approach that encompasses a wide variety of genome sequence analysis problems within a single model.

Introduction

Genomics encompasses the study of the genome in human and other organisms. The rate of innovation in this field has been breathtaking over the last decade, especially with the completion of Human Genome Project. The purpose of this chapter is to review some well-known algorithms that facilitate genome analysis. The material is presented in a way that is interesting to both the specialists working in this area and others. Thus, this review includes a brief sketch of the algorithms to facilitate a deeper understanding of the concepts involved. The list of problems related to genomics is very extensive; hence, the scope of this chapter is restricted to the following two related important problems: (1) phylogenetic analysis and (2) multiple sequence alignment. Readers interested in algorithms used in other fields of computational biology are recommended to refer to reviews by Abbas and Holmes [1] and Blazewicz et al. [7].

Genome refers to the complete DNA sequence contained in the cell. The DNA sequence consists of the four nucleotides adenine (A), thymine (T), cytosine (C), and guanine (G). Associated with each DNA strand (sequence) is a complementary DNA strand of the same length. The strands are complementary in that each nucleotide in one strand uniquely defines an associated nucleotide in the other: A and T are always paired, and C and G are always paired. Each pairing is referred to as a base pair; and bound complementary strands make up a DNA molecule. Typically, the number of base pairs in a DNA molecule is between thousands and billions, depending on the complexity of a given organism. For example, a bacterium contains about 600,000 base pairs, while human and mouse have some three billion base pairs. Among humans, 99.9% of base pairs are the same between any two unrelated persons. But that leaves millions of single-letter differences, which provide genetic variation between people.

Understanding the DNA sequence is extremely important. It is considered as the blueprint for an organism's structure and function. The sequence order underlies all of life's diversity, even dictating whether an organism is human or another species such as yeast or a fruit fly. It helps in understanding the evolution of mankind, identifying genetic diseases, and creating new approaches for treating and controlling those diseases. In order to achieve these goals, research in genome analysis has progressed rapidly over the last decade.

The rest of this chapter is organized as follows. Section "Phylogenetic Analysis" discusses techniques used to infer the evolutionary history of species and Sect. "Multiple Sequence Alignment" presents the multiple sequence alignment problem and recent advances. In Sect. "Novel Graph-Theoretical Genomic Models", we describe our research effort for advancing genomic analysis through the design of a novel graph-theoretical approach for representing a wide variety of genomic sequence analysis problems within a single model. We summarize our theoretical findings, and present computational models based on two integer programming formulations. Finally, Sect. "Summary" summarizes the interdependence and the pivotal role played by the abovementioned two problems in computational biology.

Phylogenetic Analysis

Phylogenetic analysis is a major aspect of genome research. It refers to the study of evolutionary relationships of a group of organisms. These hierarchical relationships among organisms arising through evolution are usually represented by a phylogenetic tree (Fig. 1). The idea of using trees to represent evolution dates back to Darwin. Both rooted and unrooted tree representations have been used in practice [17]. The branches of a tree represent the time of divergence and the root represents the ancestral sequence (Fig. 2).

The study of phylogenies and processes of evolution by the analysis of DNA or amino acid sequence data is

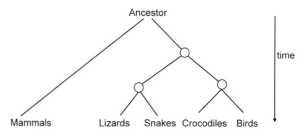

Algorithms for Genomic Analysis, Figure 1
An example of an evolutionary tree

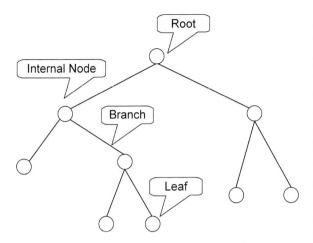

Algorithms for Genomic Analysis, Figure 2
Tree terminology

called molecular phylogenetics. In this study, we will focus on methods that use DNA sequence data. There are two processes involved in inferring both rooted and unrooted trees. The first is estimating the branching structure or topology of the tree. The second is estimating the branch lengths for a given tree. Currently, there are wide varieties of methods available to conduct this analysis [16,19,55,79]. These available approaches can be classified into three broad groups: (1) distance methods; (2) parsimony methods; and (3) maximum likelihood methods. Below, we will discuss each of them in detail.

Methods Based on Pairwise Distance

In distance methods, an evolutionary distance d_{ij} is computed between each pair i, j of sequences, and a phylogenetic tree is constructed from these pairwise distances. There are many different ways of defining pairwise evolutionary distance used for this purpose. Most of the approaches estimate the number of nucleotide substitutions per site, but other measures have also been used [70,71]. The most popular one is the Jukes–Cantor distance [37], which defines d_{ij} as $-\frac{3}{4}\log(1-\frac{4f}{3})$, where f is the fraction of sites where nucleotides differ in the pairwise alignment [37].

There are a large number of distance methods for constructing evolutionary trees [78]. In this article, we discuss methods based on *cluster analysis* and *neighbor joining*.

Cluster Analysis: Unweighted Pair Group Method Using Arithmetic Averages The conceptually simplest and most known distance method is the unweighted pair group method using arithmetic averages (UPGMA) developed by Sokal and Michener [66]. Given a matrix of pairwise distances between each pair of sequences, it starts with assigning each sequence to its own cluster. The distances between the clusters are defined as $d_{ij} = \frac{1}{|C_i||C_j|} \sum_{p \in C_i, q \in C_j} d(p,q)$, where C_i and C_j denote sequences in clusters i and j, respectively. At each stage in the process, the least distant pair of clusters are merged to create a new cluster. This process continues until only one cluster is left. Given n sequences, the general schema of UPGMA is shown in Algorithm 1.

Algorithm 1 (UPGMA)
1. Input: Distance matrix $d_{ij}, 1 \leq i, j \leq n$
2. **For** $i = 1$ to n **do**
3.   Define singleton cluster C_i comprising of sequence i
4.   Place cluster C_i as a tree leaf at height zero
5. **End for**
6. **Repeat**
7.   Determine two clusters i, j such that d_{ij} is minimal.
8.   Merge these two clusters to form a new cluster k having a distance from other clusters defined as the weighted average of the comprising two clusters. If C_k is the union of two clusters C_i and C_j, and if C_l is any other cluster, then $d_{kl} = \frac{d_{il}|C_i|+d_{jl}|C_j|}{|C_i|+|C_j|}$.
9.   Define a node k at height $\frac{d_{ij}}{2}$ with daughter nodes i and j.
10. Until just a single cluster remains

The time and space complexity of UPGMA is $O(n^2)$, since there are $n-1$ iterations of complexity $O(n)$. A number of approaches have been developed which are motivated by UPGMA. Li [52] developed a similar approach which also makes corrections for unequal rates of evolution among lineages. Klotz and Blanken [43] presented a method where a present-day sequence serves as an ancestor in order to determine the tree regardless of the rates of evolution of the sequences involved.

Neighbor Joining Neighbor joining is another very popular algorithm based on pairwise distances [63]. This approach yields an unrooted tree and overcomes the assumption of the UPGMA method that the same rate of evolution applies to each branch.

Given a matrix of pairwise distances between each pair of sequences d_{ij}, it first defines the modified distance matrix $\bar{d}_{ij}$. This matrix is calculated by subtracting average distances to all other sequences from the d_{ij}, thus compensating for long edges. In each stage, the two nearest nodes (minimal $\bar{d}_{ij}$) of the tree are chosen and defined as neighbors in the tree. This is done recursively until all of the nodes are paired together.

Given n sequences, the general schema of neighbor joining is shown in Algorithm 2.

Algorithm 2 (Neighbor joining)
1. Input: Distance matrix d_{ij}, $1 \leq i, j \leq n$
2. **For** $i = 1$ to n
3.   Assign sequence i to the set of leaf nodes of the tree (T)
4. **End for**
5. Set list of active nodes (L) = T
6. **Repeat**
7.   Calculate the modified distance matrix $\bar{d}_{ij} = d_{ij} - (r_i + r_j)$, where $r_i = \frac{1}{|L|-2} \sum_{k \in L} d_{ik}$
8.   Find the pair i, j in L having the minimal value of $\bar{d}_{ij}$
9.   Define a new node u and set $d_{uk} = \frac{1}{2}(d_{ik} + d_{jk} - d_{ij})$, for all k in L
10.   Add u to T joining nodes i, j with edges of length given by: $d_{iu} = \frac{1}{2}(d_{ij} + r_i - r_j), d_{ju} = d_{ij} - d_{iu}$
11.   Remove i and j from L and add u
12. **Until** only two nodes remain in L
13. Connect remaining two nodes i and j by a branch of length d_{ij}

Neighbor joining has a execution time of $O(n^2)$, like UPGMA. It has given extremely good results in practice and is computationally efficient [63,72]. Many practitioners have developed algorithms based on this approach. Gascuel [24] improved the neighbor-joining approach by using a simple first-order model of the variances and covariances of evolutionary distance estimates. Bruno et al. [10] developed a weighted neighbor joining using a likelihood-based approach. Goeffon et al. [25] investigated a local search algorithm under the maximum parsimony criterion by introducing a new subtree swapping neighborhood with an effective array-based tree representation.

Parsimony Methods

In science, notion of parsimony refers to the preference of simpler hypotheses over complicated ones. In the parsimony approach for tree building, the goal is to identify the phylogeny that requires the fewest necessary changes to explain the differences among the observed sequences. Of the existing numerical approaches for reconstructing ancestral relationships directly from sequence data, this approach is the most popular one. Unlike distance-based methods which build trees, it evaluates all possible trees and gives each a score based on the number of evolutionary changes that are needed to explain the observed sequences. The most parsimonious tree is the one that requires the fewest evolutionary changes for all sequences to derive from a common ancestor [69]. As an example, consider the trees in Fig. 3 and Fig. 4. The tree in Fig. 3 requires only one evolutionary change (marked by the star) compared with the tree in Fig. 4, which requires two changes. Thus, Fig. 3 shows the more parsimonious tree.

There are two distinct components in parsimony methods: given a labeled tree, determine the score; determine global minimum score by evaluating all possible trees, as discussed below.

Score Computation Given a set of nucleotide sequences, parsimony methods treat each site (position) independently. The algorithm evaluates the score at each position and then sums them up over all the positions. As an example, suppose we have the following three aligned nucleotide sequences:

CCC

GGC

CGC

Then, for a given tree topology, we would calculate the minimal number of changes required at each of the three sites and then sum them up. Here, we investigate a traditional parsimony algorithm developed by Fitch [21], where the number of substitutions required is taken as a score. For a particular topology, this ap-

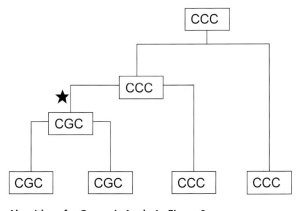

Algorithms for Genomic Analysis, Figure 3
Parsimony tree 1

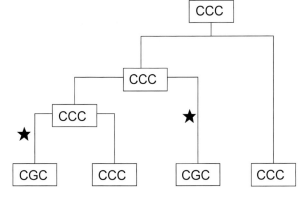

Algorithms for Genomic Analysis, Figure 4
Parsimony tree 2

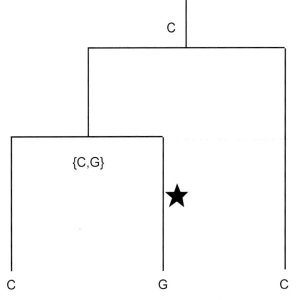

Algorithms for Genomic Analysis, Figure 5
The sets R_k for the first site of given three sequences

5.　　if $R_k = \emptyset$ then
6.　　　　$R_k = R_i \bigcup R_j$
7.　　　　$S = S + 1$
8.　　end if
9.　End for
10.　Minimal score $= S$

proach starts by placing nucleotides at the leaves and traverses toward the root of the tree. At each node, the nucleotides common to all of the descendant nodes are placed. If this set is empty then the union set is placed at this node. This continues until the root of the tree is reached. The number of union sets { equals} the number of substitutions required.

The general scheme for every position is shown in Algorithm 3.

Algorithm 3 (Parsimony: score computation)
1. Each leaf l is labeled with set R_l having observed nucleotide at that position.
2. Score $S = 0$
3. **For all** internal nodes k with children i and j having labels R_i and R_j **do**
4.　　$R_k = R_i \bigcap R_j$

Figure 5 shows the set R_k obtained by Algorithm 3. The computation is done for the first site of the three sequences shown above. The minimal score given by the algorithm is 1.

A wide variety of approaches have been developed by modifying Fitch's algorithm [68]. Sankoff and Cedergren [64] presented a generalized parsimony method which does not just count the number of substitutions, but also assigns a weighted cost for each substitution.

Ronquist [62] improved the computational time by including strategies for rapid evaluation of tree lengths and increasing the exhaustiveness of branch swapping while searching topologies.

Search of Possible Tree Topologies　The number of possible tree topologies dramatically increases with the number of sequences. Consequently, in practice usually only a subset of them are examined using efficient search strategies. The most commonly used strategy is branch and bound methods to select branching patterns [60]. For large-scale problems, heuristic methods are typically used [69]. These exact and heuristic tree search strategies are implemented in various programs like PHYLIP (phylogeny inference package) and MEGA (molecular evolutionary genetic analysis) [20,47].

Maximum Likelihood Methods

The method of maximum likelihood is one of the most popular statistical tools used in practice. In molecular phylogenetics, maximum likelihood methods find the tree that has the highest probability of generating observed sequences, given an explicit model of evolution. The method was first introduced by Felenstein [18]. We discuss herein both the evolution models and the calculation of tree likelihood.

Model of Evolution A model of evolution refers to various events like mutation, which changes one sequence to another over a period of time. It is required to determine the probability of a sequence S_2 arising from an ancestral sequence S_1 over a period of time t. Various sophisticated models of evolution have been suggested, but simple models like the Jukes–Cantor model are preferred in maximum likelihood methods.

The Jukes–Cantor [37] model assumes that all nucleotides (A, C, T, G) undergo mutation with equal probability, and change to all of the other three possible nucleotides with the same probability. If the mutation rate is 3α per unit time per site, the mutation matrix P_{ij} (probability that nucleotide i changes to nucleotide j in unit time) takes the form

$$\begin{pmatrix} 1-3\alpha & \alpha & \alpha & \alpha \\ \alpha & 1-3\alpha & \alpha & \alpha \\ \alpha & \alpha & 1-3\alpha & \alpha \\ \alpha & \alpha & \alpha & 1-3\alpha \end{pmatrix}.$$

The above matrix is integrated to evaluate mutation rates over time t and is then used to calculate $P(nt_2|nt_1, t)$, defined as the probability of nucleotide nt_1 being substituted by nucleotide nt_2 over time t.

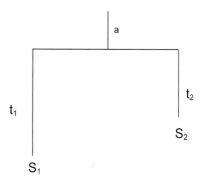

Algorithms for Genomic Analysis, Figure 6
A simple tree

Various other evolution models like the Kimura model have also been mentioned in the literature [9,42].

Likelihood of a Tree The likelihood of a tree is calculated as the probability of observing a set of sequences given the tree.

$$L(\text{tree}) = \text{probability}[\text{sequences}|\text{tree}]$$

We begin with the simple case of two sequences S^1 and S^2 of length n having a common ancestor **a** as shown in Fig. 6. It is assumed that all different sites (positions) evolve independently, and thus the total likelihood is calculated as the product of the likelihood of all sites [15]. Here, the likelihood of each site is obtained using substitution probabilities based on an evolution model.

Given q_a is the equilibrium distribution of nucleotide a, the likelihood for the simple tree in Fig. 6 is calculated as $L(\text{tree}) = P(S^1, S^2) = \prod_{i=1}^{n} P(S_i^1, S_i^2)$, where $P(S_i^1, S_i^2) = \sum_a q_a P(S_i^1|a) P(S_i^2|a)$. To generalize this approach for m sequences, it is assumed that diverged sequences evolve independently after diverging. Hence, the likelihood for every node in a tree depends only on its immediate ancestral node and a recursive procedure is used to evaluate the likelihood of the tree. The conditional likelihood $L_{k,a}$ is defined as the likelihood of the subtree rooted at node k, given that the nucleotide at node k is a. The general schema for every site is shown in Algorithm 4. The likelihood is then maximized over all possible tree topologies and branch lengths.

Algorithm 4 (Likelihood: computation at given site)
1. **For all** leaf *l* **do**
2.   **if** leaf has nucleotide *a* at that site **then**
3. $L_{l,a} = 1$
4.   **else**
5. $L_{l,a} = 0$
6.   **end if**
7. **End for**
8. **For all** internal nodes *k* with children *i* and *j*
9.   define the conditional likelihood
$L_{k,a} = \sum_{b,c} [P(b|a)L_{i,b}][P(c|a)L_{j,c}]$
10. **End for**
11. Likelihood at given site = $\sum_a q_a L_{root,a}$

Recent Improvements The maximum likelihood approach has received great attention owing to the existence of powerful statistical tools. It has been made more sophisticated using advance tree search algorithms, sequence evolution models, and statistical approaches. Yang [80] extended it to the case where the rate of nucleotide substitutions differ over sites. Huelsenbeck and Crandall [34] incorporated the improvements in substitution models. Piontkivska [59] evaluated the use of various substitution models in the maximum likelihood approach and inferred that simple models are comparable in terms of both efficiency and reliability with complex models.

The enormously large number of possible tree topologies, especially while working with a large number of sequences, makes this approach computationally intensive [72]. It has been proved that reconstructing the maximum likelihood tree is nondeterministic polynomial time hard (NP) hard even for certain approximations [14]. In order to reduce computational time, Guindon and Gascuel [31] developed a simple hill-climbing algorithm based on the maximum-likelihood principle that adjusts tree topology and branch lengths simultaneously. Recently, parallel computation has been used to address huge computational requirement. Stamatakis et al. [67] have used OpenMP–parallelization for symmetric multiprocessing machines and Keane et al. [39] developed a distributed platform for phylogeny reconstruction by maximum likelihood.

Multiple Sequence Alignment

Multiple sequence alignment is arguably among the most studied and difficult problems in computational biology. It is a vital tool because it compactly represents conserved or variable features among the family members. Alignment also allows character-based analysis compared to distance-based analysis and thus helps to elucidate evolutionary relationships better. Consequently, it plays a pivotal role in a wide range of sequence analysis problems like identifying conserved motifs among given sequences, predicting secondary and tertiary structures of protein sequences, and molecular phylogenetic analysis. It is also used for sequence comparison to find the similarity of a new sequence with pre-existing ones. This helps in gathering information about the function and structure of newly found sequences from existing ones in databases like GenBank in the USA and EMBL in Europe.

The multiple sequence alignment problem can be stated formally as follows. Let $\sum$ be the alphabet and let $\hat{\sum} = \sum \bigcup \{-\}$, where "-" is a symbol to represent "gaps" in sequences. For DNA sequences, alphabet $\sum = \{A,T,C,G,-\}$.

An alignment for N sequences $S_1, \ldots, S_N$ is given by a set $\hat{S} = \{\hat{S}_1, \ldots, \hat{S}_N\}$ over the alphabet $\hat{\sum}$ which satisfy the following two properties: (1) the strings in $\hat{S}$ are of the same length; (2) S_i can be obtained from $\hat{S}_i$ by removing the gaps. Thus, an alignment in which each string $\hat{S}_i$ has length K can be interpreted as an alignment matrix of N rows and K columns, where row i corresponds to sequence S_i. Alphabets that are placed into the same column of the alignment matrix are said to be aligned with each other.

Figure 7 shows two possible alignments for given three sequences: $S_1 = $ CCC, $S_2 = $ CGGC, and $S_3 = $ CGC.

For two sequences, the optimal multiple sequence alignment is easily obtained using dynamic program-

C	C	C	—		C	—	C	C
C	G	G	C		C	G	G	C
C	G	—	C		C	G	C	—

Algorithms for Genomic Analysis, Figure 7
Two possible alignments for given three sequences

ming (Needleman–Wunsch algorithm). Unfortunately, the problem becomes much harder for more than two sequences, and the optimal solution can be found only for a limited number of sequences of moderate length (approximately 100) [8]. Researchers have tried to solve it by generalizing the dynamic programming approach to a multidimensional space. However, this approach has huge time and memory requirements and thus cannot be used in practice even for small problems of five sequences of length 100 each. This algorithm has been improved by identifying the portion of hyperspace which does not contribute to the solution and excluding it from the computation [11]. But even this approach of Carrillo and Lipman implemented in the multiple sequence alignment program can only align up to ten sequences [53]. Although, Gupta et al. [32] improved the space and time usage of this approach, it cannot align large data sets. To reduce the huge time and memory expenses, a wide variety of heuristic approaches for multiple sequence alignment have been developed [56].

There are two components for finding the multiple sequence alignment: (1) searching over all the possible multiple alignments; (2) scoring each of them to find the best one.

The problem becomes more complex for remotely related homologous sequences, i. e., sequences which are not derived from a common ancestor [28]. Numerous approaches have been proposed, but the quest for an approach which is accurate and fast is continuing. It must be remembered that even the choice of sequences and calculating the score of alignment is a nontrivial task and is an active research field in itself.

Scoring Alignment

There is no unanimous way of characterizing an alignment as the correct one and the strategy depends on the biological context. Different alignments are possible and we never know for sure which alignment is correct. Thus, one scores every alignment according to an appropriate objective function and alignments with higher scores are deemed to be better. A typical alignment scoring scheme consists of the following steps.

Independent Columns The score of alignment is calculated in terms of columns of alignments. The individual columns are assumed to be independent and thus the total score of an alignment is a simple summation over column scores. Thus, the score for an alignment $score(A) = \sum_j score(A_j)$, where A_j is column j of the multiple alignment A. Now, the score for every column j is calculated as the "sum-of-pairs" function using the scoring matrices described below. The sum-of-pairs score for column A_j is obtained as $score(A_j) = \sum_{k<l} score(A_j^k, A_j^l)$, where A_j^k and A_j^l are nucleotides in column j of the alignment corresponding to sequences k and l, respectively. If the gap costs are linear, score(nucleotide, –) and score(–, nucleotide) will be the insertion cost. But, this approach would not differentiate between opening a gap and extending it. So, affine gap penalties are often used where gap opening and extension penalty are treated as two different parameters. The correct value of both of these parameters is a major concern since their values can be set only empirically [75]. Also most schemes used in practice score columns as the weighted sum of pairwise substitutions instead of just addition as described before. The weights are decided in accordance with the amount of independent information each sequence possesses [4].

Both the assumption of treating every column independently and using the sum-of-pairs score for the column have limitations. The problem increases as the number of sequences increases.

Scoring Matrices Any alignment can be obtained by performing three evolution operations: insertion, deletion, and substitution. It is assumed that all the different operations occur independently and thus the complete score is evaluated as the sum of scores from every operation. Insertion and deletion scores are calculated as either linear or affine gap penalty. Substitutions scores are stored as a substitution score matrix, which contains the score for every pair of nucleotides. Thus, these scores $S(A,B)$ can be treated as the score of aligning nucleotide A with nucleotide B.

These substitution score matrices can be obtained in various ways. One could adopt an ad hoc approach of setting up a score matrix which produces good alignments for a given set of sequences. The second approach would be more fundamental and look into the physical and chemical properties of nucleotides. If two nucleotides have similar properties, they would be more likely to be substituted by one another. The third and the most prominent one is a statistical approach where the maximum likelihood principle is used in conjunction with probabilistic models of evolution [3].

Alignment Approaches

The number of different approaches for the multiple sequence alignment problem has steadily increased over the last decade and thus being exhaustive will not be possible. In this chapter, we will emphasize the most widely used class of algorithms and the new emerging and most promising approaches:

1. Progressive alignment algorithms: The most widely used type of algorithm based on using pairwise alignment information of input sequences. It assumes that input sequences are phylogenetically related, and uses these relationships to guide the alignment [13].
2. Graph-based algorithms: A new trend where graph-based models are used to approach this problem.
3. Iterative alignment algorithms: Typically an alignment is produced and is then refined through a series of iterations until no more improvement can be made.

Progressive Algorithms

Progressive alignment constitutes one of the simplest and most effective ways for multiple alignment. This strategy was introduced by various researchers, like Waterman and Perlwitz [77]. Among all the progressive algorithms, ClustalW is the most famous one. It is a noniterative, deterministic algorithm that attempts to optimize the weighted sums-of-pairs with affine gap penalties [73].

The typical progressive algorithm scheme is as follows:

- Compute the distance between all pairs of given sequences by aligning them. The distances represent the divergence of each pair of sequences. These distances could be calculated by fast approximation methods or by slower but more precise methods like complete dynamic programming. Since for given N sequences $\frac{N(N-1)}{2}$ pairwise scores have to be calculated and the scores are used just for construction of a guide tree and not the alignment itself, it is desirable to use approximation methods like k tuple matches.
- Find a guide tree from the distance matrix. This is typically achieved using the clustering algorithms discussed in the construction of an evolutionary tree. Once again, since the aim is to get the alignment and not the tree itself, approximation methods are used to construct the evolution trees.
- Align sequences progressively according to the branching order in the guide tree. The basic idea is to start from the leaves of the guide tree and move toward its root and to use a series of pairwise alignments to align larger and larger groups of sequences. Some algorithms have only a single growing alignment to which every remaining sequence is aligned, whereas other approaches align a subgroup of sequences and then merge the alignments.

There are three main shortcomings of the progressive algorithms.

1. There does not exist an undisputable "best" way of ordering the given sequences.
2. Once a sequence has been aligned, that alignment will not be modified even if it conflicts with sequences added later in the process. Hence, the order in which sequences are added becomes crucial, and since there is no undisputed best way to order the sequences, this approach returns suboptimal solutions.
3. For a given set of n sequences, $\binom{n}{2}$ pairwise alignments are generated; but while computing the final multiple alignment, most of these algorithms use fewer than n pairwise alignments. Thus, the resulting multiple alignment agrees with only a small amount of information available in the data.

Therefore, there is a growing need for an algorithm to align extremely divergent sequences whose pairwise alignments are likely to be incorrect. In order to address all these issues, some techniques have been developed; while they are innovative, it is understandable that they have their own assumptions and drawbacks.

Graph-Based Algorithms

Over the last few years, the field of genomics has undergone evolutionary changes with a rapid increase in new solution strategies. The use of graph-based models is easily seen as one of the most emerging and far-reaching trends. Just and Vedova [38] used a relation between the facility location problem and sequence alignment to prove the NP-hardness of multiple sequence alignment. In this section, we review the most prominent integer programming approaches for finding multiple sequence alignment.

Maximum-Weight Trace Kececioglu et al. [40] used a solution of the maximum trace problem to construct alignment. The algorithm starts by calculating all pairwise alignments and using them to find a trace. To achieve this, given n sequences, an input alignment graph $G = (V, E)$ is constructed. It is an n-partite graph whose vertex set V represents the characters of the given sequences and whos edge set E represents the pairs of characters matched in the pairwise alignments. The subset of matching in E realized by an alignment is called a trace.

Alignment graph $G = (V, E)$ is extended to a mixed graph $G' = (V, E, A)$ by adding arc set A which connects the characters of every sequence to the next character in the same sequence. The objective of the algorithm is to find the maximum weight trace by finding cycles termed as "critical mixed cycles" in graph G' such that they satisfy sequence alignment properties [61].

The integer programming model for this problem is formulated as

$$\text{Maximize} \sum_{e \in E} w_e x_e \qquad (1)$$

$$\text{subject to} \sum_{e \in P \cap E} x_e = |E \cap P| - 1 \ \forall \ \text{critical mixed}$$

cycles P in G', $x_e \in \{0, 1\}$ for all $e \in E$.

$$(2)$$

An implementation of a branch-and-cut algorithm is used to solve the above problem. Various valid inequalities for the polytope are added as cuts, some of which are facet-defining. The algorithm is capable of giving an exact solution under the sum-of-pairs objective function with linear gap costs. Kececioglu et al. [40] have made a significant contribution by introducing a polyhedral approach capable of obtaining exact solutions for a subclass of multiple sequence alignment. However, this method has its own drawbacks like not being able to capture the order of insertions and deletions between two matchings and affine gap costs. Recently, Althaus et al. [2] proposed a general model using this approach in which arbitrary gap costs are allowed.

Minimum-Spanning Tree and Traveling Salesman Problem Shyu et al. [65] explored the use of minimum spanning trees to determine the order of sequences. The idea of the approach is to preserve the most informative distances among the set of given sequences. The criterion used is meaningful and capable of working better than the traditional criteria like those in sum-of-pairs. The algorithm itself is very efficient for practical usage, and can be easily implemented. However, it fails to address the issue of using all the information in pairwise alignments, since it only uses the score and not the pairwise alignments themselves. Moreover, this approach has all the drawbacks of the progressive strategy.

A similar approach was also developed by Korostensky and Gonnet [44] using the traveling salesman problem. In this technique, a circular sum measure is used instead of a sum-of-pairs score. The cities in the traveling salesman problem correspond to the sequences and the scores of pairwise alignment are taken as the distances. The problem is to find the longest tour where each sequence is visited exactly once [45].

Eulerian Path Approach Zhang and Waterman [81] proposed a new approach motivated by the Eulerian method for fragment assembly in DNA sequencing. In their work, a consensus sequence is found and later pairwise alignments are obtained between each input sequence and consensus sequence. Finally, multiple sequence alignment is obtained according to these pairwise alignments. The most significant advantage of this method is the linear time and memory cost for finding the consensus sequence. And, if the consensus sequence is the one closest to all given sequences, good quality alignment can be obtained in a reasonable amount of time. Once again, this approach suffers from the prominent drawback of the progressive strategy and issues in graph formation while finding the consensus sequence.

Iterative Algorithms

The main shortcoming of the progressive strategy is the failure to remove errors in the alignment, which are introduced early. The iterative algorithms are developed precisely to overcome this flaw. They are based on the idea of reconsidering and realigning previously aligned sequences with the goal of improving the overall alignment score. Each modification step is an iteration to improve the quality of the alignment.

These available approaches can be classified into two broad categories: probabilistic iterative algorithms, and deterministic iterative algorithms. We will briefly discuss them below.

Probabilistic Algorithms We will discuss both the traditional probabilistic optimization approaches like the genetic algorithm and relatively recent approaches based on a Bayesian idea.

- *Simulated annealing and genetic algorithm.* Simulated annealing and the genetic algorithm are very popular stochastic methods for solving complex optimization problems. While they are often viewed as separate and competing paradigms, both of them are iterative algorithms which search for new solutions "near" to already known good solutions. The fundamental difference between simulated annealing and the genetic algorithm is that simulated annealing performs a local move only on one solution to create a new solution, whereas the genetic algorithm also creates solutions by combining information from two different solutions. The performance of simulated annealing and the genetic algorithm varies with the problem and representation used.
  The algorithms starts with an initial alignment and the alignment score is taken to be the objective function [57]. Various operations like mutation, insertion, and substitution constitute the local move which is used to a get new solution from existing ones. Flexibility in the scoring systems and the ability to correct for errors introduced during the early phase makes these approaches desirable [41].
- *Hidden Markov model and Gibbs sampler.* The hidden Markov model and the Gibbs sampler are relatively recent approaches which view multiple sequence alignment in a statistical context. Both of them use the central Bayesian idea of simultaneously maximizing the data and the model. The Gibbs sampler find motifs using local alignment techniques [49]. It is essentially similar to the hidden Markov model with no insert and delete states.
  The hidden Markov model is a statistical model based on the Markov process, which has gained importance in various fields related to pattern recognition. It determines the hidden parameters of the system on the basis of the observable parameters of the model. For multiple sequence alignment, the hidden Markov model consists of three types of states: match states, insert states, and delete states [46]. Each state has its own emission probability of nucleotides and transition probability to other states. The standard expectation-maximization algorithm or gradient descent algorithms are used to train the model and evaluate the parameters.
  Although the hidden Markov model has been successfully used in other areas, it faces a lot of challenges. There need to be some minimum number of sequences (approximately 50) required to train the model and the hidden Markov model can be easily trapped in local optima like other hill-climbing approaches [35].

Deterministic Algorithms A deterministic iterative algorithm starts with an initial alignment and then attempts to improve it. This helps in overcoming the drawback of a progressive alignment strategy where partial alignments are "frozen" [6]. A typical scheme is as follows:

- Given N sequences $S_1, S_2, \ldots, S_N$, find alignment A.
- Remove sequence S_1 from alignment A and realign it to the profile of other aligned sequences $S_2, \ldots, S_N$ to get new alignment A'.
- Calculate the score of the new alignment A' and if it is better replace A by A'.
- Remove sequence S_2 from A' and realign it. Continue this procedure for $S_3, \ldots, S_N$.
- Repeat the realignment steps until the alignment score converges or the number of iterations reaches the user-specified limit.

Many iteration strategies which enable very accurate alignments have been developed [76]. The aim is to reduce the greedy nature of the algorithm and avoid getting trapped in a local optimum. One approach is to remove and realign every sequence to the rest in each iteration. Then, the alignment with the best score is taken to be the input for the next iteration. The other famous approach is to randomly split a set of sequences into two sets, which are then realigned.

Some researchers have incorporated the iterative strategy in the progressive alignment procedure itself. For instance, a double iteration loop has been used to make the alignment, guide tree, and sequence weights mutually consistent [27]. Recently, Chakrabarti et al. [12] developed an approach which provides a fast and accurate method for refining existing block-based alignments.

Novel Graph-Theoretical Genomic Models

In this section, we present our research effort for a novel graph-theoretical approach for representing a wide variety of genomic sequence analysis problems within a single model [50]. The model allows incorporation of the operations "insertion," "deletion," and "substitution," and various parameters such as relative distances and weights. Conceptually, we refer the problem as the *minimum weight common mutated sequence* (MWCMS) problem. The MWCMS model has many applications, including the multiple sequence alignment problem, phylogenetic analysis, the DNA sequencing problem, and the sequence comparison problem, which encompass a core set of very difficult problems in computational biology. Thus, the model presented in this section lays out a mathematical modeling framework that allows one to investigate theoretical and computational issues, and to forge new advances for these distinct, but related problems.

DNA sequencing refers to determining the exact order of nucleotide sequences in a segment of DNA. This was the greatest technical challenge in the Human Genome Project. Achieving this goal has helped reveal the estimated 30,000 human genes that are the basic physical and functional units of heredity. The resulting DNA sequence maps are being used by scientists to explore human biology and other complex phenomena.

The structure of a DNA strand (sequence) is determined by experimentation. Typically, short sequences are determined to be in the strand, and the short sequences identified are then "connected" to form a long sequence. Recent advances attempting to identify DNA strand structure involve sequencing by hybridization [5,36]. Sequencing by hybridization is the process where every possible sequence of length n (4^n possibilities) is compared with a full DNA strand. Practical values for n are 8–12. Each short string either binds or does not bind to the full strand. Biologists can thus determine exactly which short strings are contained in the DNA strand and which are not.

However, the experiment does not identify the exact location of each short string in the full strand. Hence, an important issue involves how these short strings are connected together to form the complete strand. This problem can be viewed as a shortest common superstring problem and has been studied extensively [22,23,54]. Unfortunately, errors may arise during sequencing experiments. Three types of errors are deletions (a letter appears in an input string that should not be in the final sequence), insertions (a letter is missing from an input string), and substitutions (a letter in an input string should be substituted with another letter). *The MWCMS problem can be used to model and solve this shortest common superstring problem while addressing the issue of possible errors.*

Sequence comparison is one of the most crucial problems faced by researchers in the area of bioinformatics. The sequence patterns are conserved during evolution. Given a new sequence, it will be of interest to understand how much similarity it has with pre-existing sequences. Significant similarity between two sequences implies similarities in their structures and/or functions. There are lots of DNA databases containing DNA sequences and their functions. The major ones are GenBank in the USA and the EMBL data library in Europe. If one finds a new sequence similar to existing ones in these databases, one can transfer information about the function and structure [78]. Hence, an algorithm for sequence comparison which is efficient for a large number of sequences will play a pivotal role in rapid sequence analysis. The MWCMS problem can be used to address this issue.

Definitions

Our motivation for first defining the problem arose from the desire to help quantify the concept of the "best" representative sequence in the evolutionary distance problem. The evolutionary distance problem involves finding the DNA sequence of the most likely ancestor associated with a given set of DNA sequences from distinct but similar organisms. In other words, find the DNA strand that best represents a possible ancestor, if each of the organisms evolved from the same ancestor. Changes that contribute to differences between the given sequences and the ancestor are referred to as insertions, deletions, and substitutions.

These operations account for both evolutionary mutations and experimental errors in sequencing. Mathematically, given two sequences S and B, let $\text{ord}(S, B)$ be an ordered collection of insertions, deletions, and substitutions to convert sequence S to sequence B. (For any two sequences S and B, there are an infinite number of collections $\text{ord}(S, B)$.) Let $w(\text{ord}(S, B))$ be the weight of the conversion from S to B, where the weight is the sum of an expression involving values η, δ, and $\psi \in \Re^+$ which represent the weights associated with a single insertion, deletion, and substitution, respectively. Let $\text{ord}^*(S, B)$ be such that $w(\text{ord}^*(S, B)) \leq w(\text{ord}(S, B))$ for all $\text{ord}(S, B)$. Define $d(S, B) = w(\text{ord}^*(S, B))$. Formally, the MWCMS problem can be stated as follows: Given positive weights η, δ, and ψ corresponding to a single insertion, deletion, and substitution respectively, a positive threshold κ, and finite sequences $S_1, \ldots, S_m$ from a finite alphabet, does there exist a sequence B such that $\sum_{i=1}^{m} d(S_i, B) \leq \kappa$?

We have defined the MWCMS problem—which incorporates the notions of insertion, deletion, and substitution—to help quantify the concept of the "best" representative sequence in the evolutionary distance problem. We now define precisely the operations of *insertion*, *deletion*, and *substitution*. Let $S = \{s_1, \ldots, s_n\}$ be a finite sequence of letters from a finite alphabet:

1. An *insertion* of an element x in position i of the sequence S is characterized by the addition of x between elements s_i and s_{i+1}. An insertion carries an associated penalty cost of η.
2. A *deletion* of an element in position i of S amounts to deleting s_i from the sequence S. The penalty for deletion is represented by δ.
3. A *substitution* of an element in position i of S amounts to replacing s_i with another letter from the alphabet. The penalty for substitution is represented by ψ.

We remark that a penalty cost for an operation could, more generally, depend on the position where the operation is performed and/or the element to be inserted/deleted/substituted.

Let $S_1 = \{s_{11}, \ldots, s_{1m}\}$ and $S_2 = \{s_{21}, \ldots, s_{2n}\}$ be two finite sequences of letters from a finite alphabet $\sum$. We say that the *relative distance* between elements s_{1i} and s_{2j} is k if $|i - j| = k$. We define a k-restrictive bipartite graph as a graph $G_k = (V_1, V_2, E_k)$ such that the nodes in V_1 and V_2 correspond, respectively, to each of the elements from the first and the second sequences. We assume the nodes in V_i are ordered in the same order as they appear in the sequence S_i. There is an edge between nodes $u \in V_1$ and $v \in V_2$ if u and v are identical (i.e., the same letter of the alphabet $\sum$) and if the relative distance between these two elements is less than or equal to k. The problem of identifying the "greatest similarity" between these two sequences can then be approached as the problem of finding a maximum cardinality matching between the associated node sets, subject to restrictions on which matchings are allowed. In particular, one must take into consideration the ordering of nodes so as to preserve the relative occurrence of the elements in the matching. In addition, matchings that have edge crossings must be prevented. When $k = \max\{|S_1|, |S_2|\} - 1$, we denote the graph by $G = (V_1, V_2, E)$, and the problem is equivalent to the well-studied longest common subsequence problem for two sequences, which is polynomial time solvable [23].

Construction of a Conflict Graph from Paths of Multiple Sequences

Let S_i, $i = 1, \ldots, m$, be a collection of finite sequences, each of length n, over a common alphabet $\sum$. Let $G_k = (V_1, \ldots V_m, E_1, E_2, \ldots, E_{m-1})$ be the k-restrictive *multilayer* graph in which each element in S_i forms a distinct node in V_i. Assume the nodes in V_i are ordered in the same order as they appear in the sequence S_i. E_i denotes the set of edges between nodes in V_i and V_{i+1}. There is an edge between nodes $u \in V_i$ and $v \in V_{i+1}$ if and only if u and v are the same letter in the alphabet $\sum$, and the relative distance between them is less than or equal to k. The multiple sequence comparison problem involves finding the longest common subsequence within the sequences S_i, $i = 1, \ldots, m$. We call a path $P = p_1, p_2, \ldots, p_m$ a *complete* path in G_k if $p_i \in V_i$ and $p_i p_{i+1} \in E_i$. Two complete paths are said to be *parallel* if their node sets are disjoint and the edges do not cross. Hence, a set of parallel complete paths in G_k corresponds to a feasible solution to longest common subsequence problem on the collection of sequences S_i, $i = 1, \ldots, m$. We say that two complete paths P_1 and P_2 *cross* if they are not parallel. We remark that the longest common subsequence problem with the number of sequences bounded, is polynomial time solvable using dynamic programming [23]. In general, the problem remains NP-complete.

We can incorporate insertions by generating new paths which include inserted nodes on various layers. The weight for such a new path will be affected by the total number of insertions in the path. In particular, if L is a common subsequence for S_i and $|S_i| = n$ for all $i = 1, \ldots, m$, then the total number of unmatched elements remaining will be $m(n - |L|)$. These elements can be deleted completely, or for a given unmatched element, one can increase the size of L by 1 by appropriately inserting this element into various sequences. By doing so, one decreases the number of unmatched elements. Let l be the number of insertions needed to generate a new complete path. Then the number of unmatched elements will decrease by $m - l$. If we assume that at the end of the sequencing process all unmatched elements will be deleted, then the penalty for generating this new complete path will be given by $l\eta - (m - l)\delta$.

We next define the concept of a conflict graph relative to the complete paths in G_k.

Definition 1 Let $\mathcal{P} = \{P_1, \ldots, P_s\}$ be a finite collection of complete paths in G_k. The *conflict graph* $C_\mathcal{P} = (V_\mathcal{P}, E_\mathcal{P})$ associated with $\mathcal{P}$ is constructed as follows:
- $V_\mathcal{P} = \{P_1, \ldots, P_s\}$;
- there is an edge between two nodes P_i and P_j in $V_\mathcal{P}$ if and only if P_i and P_j cross each other.

This definition applies to any multilayer graph in general. Note that any stable set of nodes in $C_\mathcal{P}$ corresponds to a set of parallel complete paths for G_k, and thereby to a feasible solution to the longest common subsequence problem on the collection of sequences S_i, $i = 1, \ldots, m$.

We remark that when $m = 2$, the resulting conflict graph is weakly triangulated, and thus is perfect. For $m > 2$, the conflict graph can contain an antihole of size 6. However, these complete paths can be viewed as continuous functions on the interval from 0 to 1; thus, by construction, $C_\mathcal{P}$ is perfect [26].

Complexity Theory

Recall that the notation $\mathrm{ord}(S, B)$, $w(\mathrm{ord}(S, B))$, $\mathrm{ord}^*(S, B)$, and the formal definition of the MWCMS problem were given in Sect. "Definitions". As an optimization problem, the MWCMS problem can be stated as follows. Given a set of input sequences, the MWCMS problem seeks to mutate every input sequence to the same a priori unknown sequence using the operations of insertion, deletion, and substitution; weights are assigned for each operation, and the total weight associated with all mutations is to be minimized. Levenshtein [51] first considered a special case of this problem by changing a single input sequence to another sequence using insertions, deletions, and substitutions. Our study involves changing multiple input sequences to arrive at an a priori unknown common sequence.

Given positive weights η, δ, and ψ corresponding, respectively, to insertions, deletions, and substitutions and any two sequences S and B, clearly any $\mathrm{ord}^*(S, B)$ will never contain more than $|B|$ insertions or substitutions. Proving that the MWCMS is in NP is not obvious. While one can transform the MWCMS to special applications (as described at beginning of Sect. "Novel Graph-Theoretical Genomic Models") to conclude that it is in NP, here we prove it directly for the general case. One needs to be able to evaluate $d(S, B)$ in polynomial time for any two sequences S and B. We next construct a graph that can be used to establish the existence of a polynomial-time algorithm for obtaining $d(S, B)$. The constructs and arguments used here typify those used to establish many of the results presented in this chapter. It is noteworthy that the notions of both conflict graph and perfect graph come into play.

Let $\sum$ be a finite alphabet, and define $\sum$-cross to be a directed bipartite graph consisting of $|\sum|$ vertices in each bipartition such that each vertex in the bipartition represents a distinct element in $\sum$. There is an arc between two vertices if the vertices correspond to the same element in $\sum$, and the geometric layout is rigidly constructed so that every arc crosses every other arc. This graph will be used as a "supernode" for insertion and substitution operations in our model. Figure 8 shows an example for $\sum$-cross when $\sum = \{A, C, G, T\}$.

We now construct a *three-layer supergraph*, G_L, using the sequences S and B along with the $\sum$-cross graphs. Layers 1 and 2 consist of exactly $|B|(|S| + 1) + |S|$ $\sum$-crosses. The first $|B|$ $\sum$-crosses represent potential insertions before the first letter in S. The next $\sum$-cross represents either the first letter of S or a substitution of this letter. The next $|B|$ $\sum$-crosses represent potential insertions between the first and second letters of S. And this is followed by a $\sum$-cross rep-

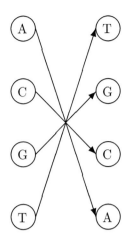

Algorithms for Genomic Analysis, Figure 8
An example of $\sum$-cross when $\sum$ = {A, C, G, T}

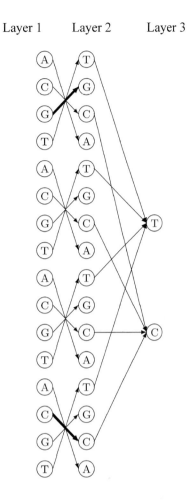

Algorithms for Genomic Analysis, Figure 9
An example of the three-layer supergraph for converting the sequence S = GC to B = TC. *Bold arcs* are used to denote the original letters in S (the weight of these arcs is $-\delta$). For simplicity, we omit the first two insertion supernodes before the first letter G. The first supernode thus represents the letter G from the original sequence, which allows for substitution. The second and third supernodes correspond to insertions, and the fourth supernode corresponds to the letter C and allows substitution as well. There are two more insertion supernodes which are omitted from the graph

resenting either the second letter of S or a substitution of this letter. This continues for each letter in S with the final $|B|$ $\sum$-crosses representing up to $|B|$ insertions after the last letter in S. Each $\sum$-cross is called either an *insertion supernode* or a *substitution supernode*, according to what it represents. The weight of all of the arcs in an insertion supernode is η. An arc in a substitution supernode has weight $-\delta$ if the arc represents the original letter in the sequences, or $\psi - \delta$ if the arc represents a substitution of the original letter. Layer 3 consists of the vertices represented by B. A vertex in layer 2 is connected to a vertex in layer 3 if they have the same letter. The weight of every arc between layers 2 and 3 is $M \leq -(\eta + \delta + \psi)$. A sample of a three-layer supergraph is given in Fig. 9. The bold arcs are used to denote the original letters in S (the weight of these arcs is $-\delta$). For simplicity, we omit the first two insertion supernodes before the first letter G. The first supernode thus represents the letter G from the original sequence, which allows for substitution. The second and third supernodes correspond to insertion supernodes, and the fourth supernode corresponds to the letter C and allows substitution as well. There are two more insertion supernodes which are omitted from the graph.

The main step in proving $d(S, B)$ to be polynomial time solvable for any sequences S and B involves the use of the conflict graph as defined in Definition 1. We state some preliminary theoretical results below. Detailed proofs can be found in Lee et al. [50].

Lemma 1 *The following statements are equivalent:*
1. *There exists a conversion from S to B using no more than a total of $|B|$ insertions or substitutions.*
2. *There exist a set of noncrossing complete paths in the associated three-layer supergraph G_L of size $|B|$.*
3. *There exists a node packing of size $|B|$ in the associated conflict graph C.*

Lemma 2 *Calculating d(S, B) for any sequences S and B can be accomplished in polynomial time.*

The three-layer supergraph can be generalized to a multilayer supergraph when multiple sequences are considered. Clearly, such multilayer supergraphs are much too large for practical purposes, yet polynomiality is preserved in the construction, and it is therefore sufficient. We can now arrive at the result that the MWCMS is in NP.

Theorem 1 *The MWCMS is in NP.*

To prove that the MWCMS is polynomial time solvable when the number of input sequences is bounded by a positive constant, the following lemma is crucial, though trivial.

Lemma 3 *Given η, δ, $\psi \in \Re^+$, an optimal solution B to any MWCMS problem has the following properties. B has no substitutions from letters other than the original letters in S_i, and B will never have an element which is inserted in every sequence (in the same location). Therefore, there are at most $\sum_{i=1}^{m} |S_i|$ insertions in any sequence.*

In addition, we also require the construction of a (directed) $2m$-layer supergraph, G_L^m, similar to the three-layer supergraph, G_L.

Given sequences $S_1, \ldots, S_m$, generate a $2m$-layer (directed) graph $G_L^m = (V, E)$ as follows. Layers $2i-1$ and $2i$ consist of $(\sum_{j=1}^{m} |S_j|)(|S_i| + 1) + |S_i|$ copies of $\sum$-crosses for $i = 1, \ldots, m$, constructed in exactly the same manner as layers 1 and 2 of the three-layer supergraph using the input sequence S_i. The first $\sum_{j=1}^{m} |S_j|$ $\sum$-crosses represent the possibility that $\sum_{j=1}^{m} |S_j|$ different letters can be inserted before the first element in S_i. The next $\sum$-cross corresponds to either the first letter in S_i or a substitution of this letter. This is repeated $|S_i|$ times (for each letter in S_i), and the final $\sum_{j=1}^{m} |S_j|$ $\sum$-crosses represent insertions after the final letter in S_i. Thus, the first $\sum_{j=1}^{m} |S_j|$ $\sum$-crosses represent the insertion supernodes, followed by one $\sum$-cross representing a letter in S_i or a substitution supernode, and so forth. An arc exists from a vertex in layer $2i$ to a vertex in layer $2i + 1$ if the vertices correspond to the same letter. Observe that G_L^m is an acyclic directed graph which is polynomial in the size of the input sequences. Assign every arc between layers $2i$ and $2i + 1$ a weight of 0. There are three different weights for arcs between layers $2i - 1$ and $2i$ each corresponding to an insertion, deletion, or substitution. The assignment of weights on such arcs is analogous to the assignment in G_L: a weight of η is assigned to every arc contained in an insertion supernode; and an arc in a substitution supernode is assigned a weight of $-\delta$ if it corresponds to the original letter, or $\psi - \delta$, otherwise.

Figure 10 shows a sample graph for two sequences: $S_1 = $ GC and $S_2 = $ TG. Observe that at most two insertions are needed in an optimal solution; thus, we can reduce the number of $\sum$-crosses as insertion supernodes from $\sum_{i=1}^{2} |S_i| = 4$ to 2. For simplicity, in the graph shown in Fig. 10, we have not included the two insertion supernodes before the first letter nor those after the last letter of each sequence. Thus, in the figure, the first $\sum$-cross represents the substitution supernode associated with the first letter in S_1. The second and third $\sum$-crosses represent two insertion supernodes. And the last $\sum$-cross represents the substitution supernode associated with the second letter in S_1. For simplicity, we include only arcs connecting vertices associated to the element G between layers 2 and 3. The arcs for other vertices follow similarly.

A conflict graph C associated with G_L^m can be generated by finding all complete paths (paths from layer 1 to layer $2m$) in G_L^m. These complete paths correspond to the set of vertices in C, as in Definition 1. If we assign a weight to each vertex equal to the weight of the associated complete path, then the following result can be established.

Theorem 2 *Every node packing in C represents a candidate solution to the MWCMS if and only if at most $\sum_{i=1}^{m} |S_i|$ letters can be inserted between any two original letters. Furthermore, the weight of the node packing is equal to the weight of the $MWCMS - \sum_{i=1}^{m} |S_i|\delta$.*

The supergraph G_L^m and its associated conflict graph are fundamental to our proof of the following theorem on the polynomial-time solvability of a restricted version of the MWCMS problem.

Theorem 3 *The MWCMS problem restricted to instances for which the number of sequences is bounded by a positive constant is polynomial time solvable.*

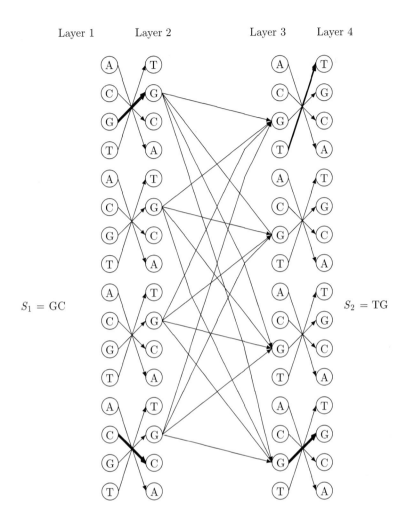

Algorithms for Genomic Analysis, Figure 10
A sample graph G_L^m of MWCMS with $S_1 = GC$ to $S_2 = TG$, where $\sum = \{A,C,G,T\}$

Special Cases of MWCMS

The MWCMS encompasses a very broad class of problems. In computational biology as discussed in this chapter, first and foremost, it represents a model for phylogenetic analysis. The MWCMS as defined is the "most likely ancestor problem," and the concept of the three-layer supergraph as described in Sect. "Complexity Theory" describes the evolutionary distance problem. An optimal solution to a multiple sequence alignment instance can be found using the solution of the MWCMS problem obtained on the $2m$-layer supergraph, G_L^m. The alignment is the character matrix obtained by placing together the given sequences incorporating the insertions into the solution of the MWCMS problem. Furthermore, DNA sequencing can be viewed as the shortest common superstring problem, while sequence comparison of a given sequence B to a collection of N sequences $S_1, \ldots, S_N$ is the MWCMS problem itself.

Broader than the computational biology applications, special cases of the MWCMS include shortest common supersequences, longest common subsequences, and shortest common superstring; these problems are of interest in their own right as combinatorial optimization problems and for their role in complexity theory.

Computational Models: Integer Programming Formulation

The construction of the multilayer supergraphs described in our theoretical study lays the foundation and provides direction for computational models and solution strategies that we will explore in future research. Although the theoretical results obtained are polynomial-time in nature, they present computational challenges. In many cases, calculating the worst-case scenario is not trivial. Furthermore, the polynomial-time result of a node-packing problem for a perfect graph by Grötschel et.al. [29,30] is existential in nature, and relies on the polynomial-time nature of the ellipsoid algorithm. The process itself involves solving an integer program relaxation multiple times. In our case, the variables of the integer program generated are the complete paths in the multilayer supergraph, G_L^m. Formally, the integer program corresponding to our conflict graph can be stated as follows.

Let x_p be the binary variable denoting the use or nonuse of the complete path p with weight w_p. Then the corresponding node-packing problem is

Minimize $\sum w_p x_p$

subject to $x_p + x_q \leq 1$ if complete paths p and q cross

$x_p \in \{0, 1\}$ for all complete paths p in G_L^m.

(MIP1)

We call the inequality $x_p + x_q \leq 1$ an adjacency constraint. A natural approach to improve the solution time for (MIP1) is to decrease the size of the graph G_L^m and thus the number of variables. Reductions in the size of G_L^m can be accomplished for shortest common superstrings, longest common subsequences, and shortest common supersequences. Among these three problems, the graph G_L^m is smallest for longest common subsequences. In longest common subsequences, all insertion and substitution supernodes can be eliminated.

Our theoretical results thus far rely on the creation of *all* complete paths. Clearly, the typical number of complete paths will be on the order of n^m, where $n = \max |S_i|$. In this case, an instance with three sequences and 300 letters in each sequence generates more than one million variables; hence, an exact formulation with all complete paths is impractical in general. A simultaneous column and row generation approach within a parallel implementation may lead to computational advances related to this formulation.

An alternative formulation can be obtained by examining G_L^m from a network perspective using arcs (instead of complete paths) in G_L^m as variables. Namely, let $x_{i,j}$ denote the use or nonuse of arc (i, j) in the final sequence, with $c_{i,j}$ the cost of the arc in G_L^m. The network formulation can be stated as

Minimize $\sum_{(i,j) \in E} c_{i,j} x_{i,j}$

subject to $\sum_{i:(i,j) \in E} x_{i,j} = \sum_{k:(j,k) \in E} x_{j,k}$

for all $j \in V$ in layers $2, \ldots, 2m - 1$

$x_{i,j} + x_{k,l} \leq 1$

for all crossing arcs (i, j) and $(k, l) \in E$

$x_{i,j} \in \{0, 1\}$

for all $(i, j) \in E$.

(MIP2)

The first set of constraints ensures flow in equals flow out in all vertices contained in sequences $2, \ldots, m - 1$ (complete paths). The second set of constraints ensures that no two arcs cross. This model grows linearly in the number of sequences. This alternative integer programming formulation is still large, but is manageable for even fairly large instances.

Utilizing a collection of DNA sequences (each with 40,000 base pairs in length) from a bacterium, and a collection of short sequences associated with genes found in breast cancer patients, computational tests of our graph-theoretical models are under way. We are seeking to develop computational strategies to provide reasonable running times for evolutionary distance problem instances derived from these data. In an initial test, when three sequences each with 100 letters are used, the initial linear program requires more than 10,000 s to provide a solution when tight constraints are employed (in this case, each adjacency constraint is replaced by a maximal clique constraint). Our ongoing computational effort will focus on developing and investigating solution techniques for practical problem instances, including those based on the abovementioned two integer programming formulations, as well as development of fast heuristic procedures.

In [50], we outline a simple yet practical heuristic based on (MIP2) that we developed for solving the multiple sequence alignment problem; and we report on preliminary tests of the algorithm using different sets of sequence data. Motivation for the heuristic is derived from the desire to reduce computational time through various strategies for reducing the number of variables in (MIP2).

Summary

Multiple sequence alignment and phylogenetic analysis are deeply interconnected problems in computational biology. A good multiple alignment is crucial for reliable reconstruction of the phylogenetic tree [58]. On the other hand, most of the multiple alignment methods require a phylogenetic tree as the guide tree for progressive iteration.

Thus, the evolutionary tree construction might be biased by the guide tree used for obtaining the alignment. In order to avoid this pitfall, various algorithms have been developed which simultaneously find alignment and phylogenetic relationship among given sequences. Sankoff and Cedergren [64] developed a parsimony-based algorithm using a character-substitution model of gaps. The algorithm is guaranteed to find the evolutionary tree and alignment which minimizes tree-based parsimony cost. Hein [33] also developed a parsimony-type algorithm but used an affine gap cost, which is more realistic than the character-substitution gap model. This algorithm is also faster than Sankoff and Cedergreen's approach but makes simplifying assumptions in choosing ancestral sequences.

Like parsimony methods for finding a phylogenetic tree, both of the abovementioned approaches require a search over all possible trees to find the global optimum. This makes these algorithms computationally very intensive. Hence, there has been a strong focus on developing an efficient algorithm that considers both alignment and the tree. Vingron and Haeseler [74] have developed an approach based on three-way alignment of prealigned groups of sequences. It also allows change in the alignment made early in the course of computation. Many programs, like MEGA, are trying to develop an efficient integrated computing environment that allows both sequence alignment and evolutionary analysis [48].

We addressed this issue of simultaneously finding alignment and phylogenetic relationships by presenting a novel graph-theoretical approach. Indeed, our model can be easily tailored to find theoretically provable optimum solutions to a wide range of crucial sequence analysis problems. These sequence analysis problems were proven to be NP-hard, and thus understandably present computational challenges. In order to strike a balance between the time and the quality of the solution, a variety of parameters are provided. Ongoing research efforts are exploring the development of efficient computational models and solution strategies in a massive parallel environment.

Acknowledgement

This research was partially supported by grants from the National Science Foundation.

References

1. Abbas A, Holmes S (2004) Bioinformatics and management science: Some common tools and techniques. Oper Res 52(2):165–190
2. Althaus E, Caprara A, Lenhof H, Reinert K (2006) A branch-and-cut algorithm for multiple sequence alignment. Math Program 105(2-3):387–425
3. Altschul S (1991) Amino acid substitution matrices from an information theoretic perspective. J Mol Biol 219(3):555–565
4. Altschul SF, Carroll RJ, Lipman DJ (1989) Weights for data related by a tree. J Mol Biol 207(4):647–653
5. Bains W, Smith G (1988) A novel nethod for DNA sequence determination. J Theor Biol 135:303–307
6. Barton GJ, Sternberg MJE (1987) A strategy for the rapid multiple alignment of protein sequences: confidence levels from tertiary structure comparisons. J Mol Biol 198:327–337
7. Blazewicz J, Formanowicz P, Kasprzak M (2005) Selected combinatorial problems of computational biology. Eur J Oper Res 161:585–597
8. Bonizzoni P, Vedova G (2001) The complexity of multiple sequence alignment with SP-score that is a metric. Theor Comput Sci 259:63–79
9. Bos D, Posada D (2005) Using models of nucleotide evolution to build phylogenetic trees. Dev Comp Immunol 29(3):211–227

10. Bruno WJ, Socci ND, Halpern AL (2000) Weighted neighbor joining: A likelihood-based approach to distance-based phylogeny reconstruction. Mol Biol Evol 17:189–197
11. Carrillo H, Lipman D (1988) The multiple sequence alignment problem in biology. SIAM J Appl Math 48(5):1073–1082
12. Chakrabarti S, Lanczycki CJ, Panchenko AR, Przytycka TM, Thiessen PA, Bryant SH (2006) Refining multiple sequence alignments with conserved core regions. Nucleic Acids Res 34(9):2598–2606
13. Chenna R, Sugawara H, Koike T, Lopez R, Gibson TJ, Higgins DG, Thompson JD (2003) Multiple sequence alignment with the clustal series of programs. Nucleic Acids Res 31(13):3497–3500
14. Chor B, Tuller T (2005) Maximum likelihood of evolutionary trees: hardness and approximation. Bioinf 21(Suppl. 1):I97–I106
15. Clote P, Backofen R (2000) Computational Molecular Biology: An Introduction. Wiley, NY, USA
16. Delsuc F, Brinkmann H, Philippe H (2005) Phylogenomics and the reconstruction of the tree of life. Nature reviews. Genet 6(5):361–375
17. Durbin R, Eddy S, Krogh A, Mitchison G (1998) Biological Sequence Analysis. Cambridge University Press, UK
18. Felsenstein J (1981) Evolutionary trees from DNA sequences: a maximum likelihood approach. J Mol Evol 17(6):368–376
19. Felsenstein J (1988) Phylogenies from molecular sequences: Inference and reliability. Annu Rev Genet 22:521–565
20. Felsenstein J (1989) PHYLIP – phylogeny inference package (version 3.2). Cladistics 5:164–166
21. Fitch WM (1971) Toward defining the course of evolution: Minimum change for a specific tree topology. Syst Zool 20(4):406–416
22. Gallant J, Maider D, Storer J (1980) On finding minimal length superstrings. J Comput Syst Sci 20:50–58
23. Garey M, Johnson D (1979) Computers and Intractability: A Guide to the Theory of NP-Completeness. W.H. Freeman, San Francisco, USA
24. Gascuel O (1997) BIONJ: An improved version of the NJ algorithm based on a simple model of sequence data. Mol Biol Evol 14(7):685–695
25. Goeffon A, Richer J, Hao J (2005) Local search for the maximum parsimony problem. Lect Notes Comput Sci 3612:678–683
26. Golumbic MC, Rotem D, Urrutia J (1983) Comparability graphs and intersection graphs. Discret Math 43:37–46
27. Gotoh O (1996) Significant improvement in accuracy of multiple protein sequence alignments by iterative refinement as assessed by reference to structural alignments. J Mol Biol 264(4):823–838
28. Gotoh O (1999) Multiple sequence alignment: algorithms and applications. Adv Biophys 36:159–206
29. Grötschel M, Lovász L, Schrijver A (1984) Polynomial algorithms for perfect graphs. Annals Discret Math 21:325–356
30. Grötschel M, Lovász L, Schrijver A (1988) Geometric algorithms and combinatorial optimization. Springer, New York
31. Guindon S, Gascuel O (2003) A simple, fast, and accurate algorithm to estimate large phylogenies by maximum likelihood. Syst Biol 52(5):696–704
32. Gupta S, Kececioglu J, Schaeffer A (1995) Improving the practical space and time efficiency of the shortest-paths approach to sum-of-pairs multiple sequence alignment. J Comput Biol 2:459–472
33. Hein J (1989) A new method that simultaneously aligns and reconstructs ancestral sequences for any number of homologous sequences, when the phylogeny is given. Mol Biol Evol 6(6):649–668
34. Huelsenbeck J, Crandall K (1997) Phylogeny estimation and hypothesis testing using maximum likelihood. Annu Rev Ecol Syst 28:437–66
35. Hughey R, Krogh A (1996) Hidden markov models for sequence analysis: extension and analysis of the basic method. Comput Appl Biosci 12(2):95–107
36. Idury RM, Waterman MS (1995) A new algorithm for DNA sequence assembly. J Comput Biol 2(2):291–306
37. Jukes TH, Cantor CR (1969) Evolution of protein molecules. In: Munro HN (ed) Mammalian Protein Metabolism. Academic Press, New York, pp 21–123
38. Just W, Vedova G (2004) Multiple sequence alignment as a facility-location problem. INFORMS J Comput 16(4):430–440
39. Keane T, Naughton T, Travers S, McInerney J, McCormack G (2005) DPRml: distributed phylogeny reconstruction by maximum likelihood. Bioinf 21(7):969–974
40. Kececioglu J, Lenhof H, Mehlhorn K, Mutzel P, Reinert K, Vingron M (2000) A polyhedral approach to sequence alignment problems. Discret Appl Math 104:143–186
41. Kim J, Pramanik S, Chung MJ (1994) Multiple sequence alignment using simulated annealing. Bioinf 10(4):419–426
42. Kimura M (1980) A simple method for estimating evolutionary of base substitution through comparative studies of nucleotide sequences. J Mol Evol 16:111–120
43. Klotz L, Blanken R (1981) A practical method for calculating evolutionary trees from sequence data. J Theor Biol 91(2):261–272
44. Korostensky C, Gonnet GH (1999) Near optimal multiple sequence alignments using a traveling salesman problem approach. In: Proceedings of the String Processing and Information Retrieval Symposium. IEEE, Cancun, pp 105–114
45. Korostensky C, Gonnet GH (2000) Using traveling salesman problem algorithms for evolutionary tree construction. Bioinf 16(7):619–627
46. Krogh A, Brown M, Mian IS, Sjolander K, Haussler D (1994) Hidden markov models in computational biology: Applications to protein modeling. J Mol Biol 235:1501–1531

47. Kumar S, Tamura K, Nei M (1994) MEGA: Molecular evolutionary genetics analysis software for microcomputers. Comput Appl Biosci 10:189–191
48. Kumar S, Tamura K, Nei M (2004) MEGA3: integrated software for molecular evolutionary genetics analysis and sequence alignment. Brief Bioinform 5(2):150–163
49. Lawrence C, Altschul S, Boguski M, Liu J, Neuwald A, Wootton J (1993) Detecting subtle sequence signals: a gibbs sampling strategy for multiple alignment. Science 262:208–214
50. Lee EK, Easton T, Gupta K (2006) Novel evolutionary models and applications to sequence alignment problems. Annals Oper Res 148(1):167–187
51. Levenshtein VL (1966) Binary codes capable of correcting deletions, insertions, and reversals. Cybern Control Theor 10(9):707–710
52. Li W (1981) Simple method for constructing phylogenetic trees from distance matrices. Proc Natl Acad Sci USA 78(2):1085–1089
53. Lipman D, Altschul S, Kececioglu J (1989) A tool for multiple sequence alignment. Proc Natl Acad Sci USA 86(12):4412–4415
54. Maier D, Storer JA (1977) A note on the complexity of the superstring problem. Technical Report 233, Princeton University, USA
55. Nei M (1996) Phylogenetic analysis in molecular evolutionary genetics. Annu Rev Genet 30:371–403
56. Notredame C (2002) Recent progress in multiple sequence alignment: a survey. Pharmacogenomics 3(1):131–144
57. Notredame C, Higgins D (1996) SAGA: sequence alignment by genetic algorithm. Nucleic Acids Res 24(8):1515–1524
58. Phillips A, Janies D, Wheeler W (2000) Multiple sequence alignment in phylogenetic analysis. Mol Phylogenet Evol 16(3):317–330
59. Piontkivska H (2004) Efficiencies of maximum likelihood methods of phylogenetic inferences when different substitution models are used. Mol Phylogenet Evol 31(3):865–873
60. Purdom P, Bradford PG, Tamura K, Kumar S (2000) Single column discrepancy and dynamic max-mini optimizations for quickly finding the most parsimonious evolutionary trees. Bioinformamtics 16:140–151
61. Reinert K, Lenhof H, Mutzel P, Mehlhorn K, Kececioglu J (1997) A branch-and-cut algorithm for multiple sequence alignment. In: Proceedings of the First Annual International Conference on Computational Molecular Biology (RECOMB-97). ACM Press, Santa Fe, pp 241–249
62. Ronquist F (1998) Fast fitch-parsimony algorithms for large data sets. Cladistics 14:387–400
63. Saitou N, Nei M (1987) The neighbor-joining method: a new method for reconstructing phylogenetic trees. Mol Biol Evol 4:406–425
64. Sankoff D, Cedergren RJ (1983) Simultaneous comparison of three or more sequences related by a tree. In: Sankoff D, Kruskal JB (eds) Time Warps, String Edits, and Macromolecules: The Theory and Practice of Sequence Comparison. Addison-Wesley, MA, USA, pp 253–264
65. Shyu SJ, Tsai YT, Lee R (2004) The minimal spanning tree preservation approaches for DNA multiple sequence alignment and evolutionary tree construction. J Comb Optim 8(4):453–468
66. Sokal R, Michener C (1958) A statistical method for evaluating systematic relationships. University of Kansas, Scientific Bull 38:1409–1438
67. Stamatakis A, Ott M, Ludwig T (2005) RAxML-OMP: An efficient program for phylogenetic inference on SMPs. Lect Notes Comput Sci 3606:288–302
68. Swofford DL, Maddison WP (1987) Reconstructing ancestral character states under wagner parsimony. Math Biosci 87:199–229
69. Swofford DL, Olsen GJ (1990) Phylogeny reconstruction. In: Hillis DM, Moritz G (eds) Molecular Systs. Sinauer Associates, MA, USA, pp 411–501
70. Tajima F, Nei M (1984) Estimation of evolutionary distance between nucleotide sequences. Mol Biol Evol 1(3):269–85
71. Tajima F, Takezaki N (1994) Estimation of evolutionary distance for reconstructing molecular phylogenetic trees. Mol Biol Evol 11:278–286
72. Takahashi K, Nei M (2000) Efficiencies of fast algorithms of phylogenetic inference under the criteria of maximum parsimony, minimum evolution, and maximum likelihood when a large number of sequences are used. Mol Biol Evol 17:1251–1258
73. Thompson JD, Higgins DG, Gibson TJ (1994) CLUSTAL W: improving the sensitivity of progressive multiple sequence alignment through sequence weighting, position-specific gap penalties and weight matrix choice. Nucleic Acids Res 22(22):4673–4680
74. Vingron M, Haeseler A (1997) Towards integration of multiple alignment and phylogenetic tree construction. J Comput Biol 4(1):23–34
75. Vingron M, Waterman M (1994) Sequence alignment and penalty choice. review of concepts, case studies and implications. J Mol Biol 235(1):1–12
76. Wallace IM, O'Sullivan O, Higgins DG (2005) Evaluation of iterative alignment algorithms for multiple alignment. Bioinformatics 21(8):1408–14
77. Waterman M, Perlwitz M (1984) Line geometries for sequence comparisons. Bull Math Biol 46(4):567–577
78. Waterman MS (1995) Introduction to Computational Biology: Maps, Sequences and Genomes. Chapman and Hall
79. Whelan S, Lio P, Goldman N (2001) Molecular phylogenetics: state-of-the-art methods for looking into the past. Trends Genet 17(5):262–272
80. Yang Z (1993) Maximum-likelihood estimation of phylogeny from DNA sequences when substitution rates differ over sites. Mol Biol Evol 10(6):1396–401
81. Zhang Y, Waterman M (2003) An eulerian path approach to global multiple alignment for DNA sequences. J Comput Biol 10(6):803–819

Alignment Problem

CLAUDE G. DIDERICH[1], MARC GENGLER[2]
[1] Computer Sci. Department,
  Swiss Federal Institute Technology-Lausanne,
  Lausanne, Switzerland
[2] Ecole Sup. d'Ingénieurs de Luminy,
  University Méditerrannée, Marseille, France

MSC2000: 05-02, 05-04, 15A04, 15A06, 68U99

Article Outline

Keywords
Alignment Problem
  Communication-Free Alignment Problem
  Constant-Degree Parallelism Alignment Problem
Solving the Alignment Problem
  Communication-Free Alignment Approaches
  Alignment Approaches Based
    on Generating HPF like Data Distributions
  Approaches Using a Graph Based Framework
  Approaches Using a Linear Algebra Framework
  Other Approaches
Conclusion
See also
References

Keywords

Alignment problem; Automatic parallelization; Computation and data mapping; Nested loops; Scheduling functions

Since the mid-1990s the need for techniques to parallelize numerical applications has increased. When parallelizing nested loops for distributed memory parallel computers, two major problems have to be solved: the scheduling of the loop iterations and the mapping of the computations and data elements onto the processors. The scheduling functions must satisfy all the data dependences existing in the sequential loop nests. The mapping functions should maximize the degree of parallelism obtained. Furthermore they should minimize the amount of communication overhead due to non local data references.

This survey presents the *alignment problem*, that is, the problem of mapping computation and data onto the processors. The alignment problem has been studied extensively since the beginning of the nineties, that is, since the beginning of the introduction of massively parallel distributed memory computers. For different sub-problems of the alignment problem, the most interesting results are surveyed.

Alignment Problem

The alignment problem is the problem of finding an alignment of loop iterations with the array elements accessed. This means computing mapping functions of the loop iterations, called computations, and mapping functions of the array elements, called data, to a multidimensional grid of virtual processors. The name of the problem comes from the idea of aligning the processors computing with the ones owning the data. The alignment problem is tightly related to the mapping of the computation and data objects onto a grid of virtual processors.

As input, programs containing nested loops are considered. Each loop nest may contain one or more instructions. For the sake of simplicity, only assignment instructions are considered. The data access functions are described by the functions $F_l: \mathbf{I}_j \rightarrow \mathbf{D}_K$, where $\mathbf{I}_j$ represents the iteration space surrounding instruction S_j and $\mathbf{D}_K$ the domain of the array K.

To solve the alignment problem, computation and data mapping functions C_j and D_K have to be computed such as to minimize the overall execution time of the resulting parallel program.

$$C_j: \mathbf{I}_j \rightarrow \mathbf{P},$$
$$D_K: \mathbf{D}_K \rightarrow \mathbf{P},$$

where $\mathbf{P}$ represents a multidimensional grid of virtual processors.

To minimize the overall execution time a solution to the alignment problem has to address the following needs:
i) maximize the degree of parallelism, that is, use as many dimensions of the virtual grid of processors as possible,
ii) minimize the need for non local data accesses, that is, distribute the array elements such that a minimal amount of communication overhead is required to

access data elements stored on different processors than the ones accessing them,
iii) guarantee the existence of scheduling functions compatible with the computation mapping functions.

Clearly the needs i)–iii) depend on each other. In this survey we only focus on the first two needs.

Need i) can be expressed by maximizing the dimension of the virtual processor grid **P** onto which the computations and data elements are mapped.

The need for a given data access F_l to be local is expressed by the equation (1) being satisfied:

$$C_j(\vec{i}) = D_K(F_l(\vec{i})). \tag{1}$$

Equation (1) is called *alignment constraint* or *locality constraint*. Depending on how the needs i) and ii) are satisfied, various subproblems of the alignment problem can be defined.

Communication-Free Alignment Problem

The *communication-free alignment problem* (CFAP) is the problem of finding computation and data mapping functions for each instruction and for each data array such that no communication is needed and the degree of parallelism obtained is maximal. The CFAP can be formulated as an optimization problem:

$$\begin{cases} \max_{C_j, D_K} & \text{dimension of } \mathbf{P} \\ \text{s.t.} & \forall j, l, K: \ C_j(\vec{i}) = D_K(F_l(\vec{i})). \end{cases}$$

Constant-Degree Parallelism Alignment Problem

Let $\mathcal{F}$ be the set of data access functions from a set of loop nests forming an alignment problem and d a positive constant. Let $c(\mathcal{F}', \mathcal{F})$ be a cost function on a subset $\mathcal{F}' \subseteq \mathcal{F}$ of data access functions. The *constant degree parallelism alignment problem* (CDPAP), denoted by $(\mathcal{F}, d)$, is the problem of finding a subset $\mathcal{F}' \subseteq \mathcal{F}$ of data access functions such that:

1) There exists a solution to the CFAP consisting of all data accesses in the set $\mathcal{F}'$ admitting a degree of parallelism of at least d.
2) The cost function $c(\mathcal{F}', \mathcal{F})$ on the subset $\mathcal{F}'$ is minimized.

As for the CFAP, the CDPAP can be formulated as follows as an optimization problem:

$$\begin{cases} \max_{C_j, D_K} & \sum_{j,l,K} [[C_j(\vec{i}) = D_K(F_l(\vec{i}))]] \\ \text{s.t.} & \text{dimension of } \mathbf{P} \geq d. \end{cases}$$

Example 1 The data accesses in this example are encoded by the three functions $F_1(i,j) = (i\,j+1)$, $F_2(i,j) = (i-1\,j+1)$ and $F_3(i,j) = (i+1\,j+1)$. A possible solution requiring no communication and admitting one degree of parallelism is given by $C(i,j) = j$ and $D_a(i,j) = j-1$, **P** being a one-dimensional processor set.

```
DO i = 2, n − 1
    DO j = 2, n − 1
        a(i, j + 1) = a(i − 1, j + 1) + a(i + 1, j + 1)
    END DO
END DO
```

Solving the Alignment Problem

Communication-Free Alignment Approaches

C.-H. Huang and P. Sadayappan [17], in 1991, were the first to formulate the alignment problem in a linear algebra framework. They focus on a communication-free solution. The data array elements as well as the loop iterations are partitioned in disjoint sets represented by hyperplanes. Each set is mapped onto a different processor. The partitions are sought such that they result in the elimination of communication. A characterization of a necessary and sufficient condition for communication-free hyperplane partitioning is provided. Various results are given characterizing the situation where the iteration and data space can be partitioned along hyperplanes so that no communication is necessary. More precisely, two data elements accessed during a single iteration in a single instruction must be located on a single processor and two iterations in the same instruction accessing a single data element must be executed on the same processor.

In [30], a matrix notation is presented to describe array accesses in fully parallel loop nests. A sufficient condition on the matrices for computing a communication-free mapping of the arrays onto the processors is given. The owner computes rule is assumed for the computation mapping. The presented existence condition for communication-free partitions is based on the connectivity of the data access graph which models the data access patterns. To compute data mapping functions, a set of systems of linear equations is constructed, one system of linear equations per pair of read and write data accesses. If there exists a solution to the set of systems of linear equations, then there exists a communication-free partitioning of the array elements into parallel hyperplanes.

In [2] a linear algebra approach is proposed, based on [17]. The communication-free alignment problem is solved by computing a basis of the null space of the application representing the alignment constraints. The problem of data replication is addressed.

In [6], T.-S. Chen and J.-P. Sheu consider perfect loop nests. They compute iteration and data space partitioning functions requiring no communication. Their work focuses only on uniformly generated data references. Sufficient conditions are given for the existence of a communication-free partition. The method for partitioning the data onto the processors is based on the computation of independent blocks called iteration and data partitions respectively. If no communication-free partitioning exists, data replication is considered.

In [24], an algorithm is presented that extracts all the degrees of communication-free parallelism that can be obtained via loop fission, fusion, interchange, reversal, skewing, scaling, re-indexing and statement reordering. The algorithm first assigns the iterations of the instructions in the program to processors via affine processor mapping functions. Then it generates the correct code by assuring that the semantics of the sequential program are satisfied.

Alignment Approaches Based on Generating HPF like Data Distributions

J. Li and M. Chen [22,23] are interested in the indices of the arrays that have to be aligned with one another to minimize remote data references. The techniques were initially developed for compiling the functional language 'Crystal', but can be applied in the process of compiling imperative languages like 'Fortran'. The parallelism is assumed to be specified explicitly and the single assignment form is used. The goal of their approach is to find alignment functions such that the dimensions of each array are projected onto the same space of a virtual processor grid. They consider four basic alignments:
i)   permutations of the indices,
ii) embeddings,
iii) translations by a constant, and
iv) reflections.

To find a set of data accesses for which valid alignment functions exist, a component affinity graph is constructed. It represents the affinities between cross reference patterns. The nodes of the graph represent the components of the index domains to be aligned. An edge represents an affinity between the two corresponding domain components. The alignment problem then consists in partitioning the set of nodes of the component affinity graph into disjoint subsets with the restriction that no two nodes belonging to the same array are allowed in the same subset. A fast and quite efficient heuristic algorithm is presented.

M. Gupta, in his thesis in 1992 [16], presents a data distribution algorithm that operates in four passes. The first pass serves to compute an alignment of the array dimensions. The algorithm developed is based on the notion of component affinity graph introduced by Li and Chen [22]. In the second phase the arrays are partitioned using either block or cyclic data distributions. In the third pass, the block sizes of the arrays distributed are computed whereas the last pass computes the number of processors on which each array dimension is distributed.

K. Kunchithapadam and B.P. Miller [20], in opposition to other approaches, assume that a user-defined data distribution is given. The data accesses of a program are modeled by a colored proximity graph. Each vertex of the graph represents a part of an array and the color of a vertex represents the current processor to which this array part is assigned. Edges of the graph represent assignments of values arising from part of one or more arrays to part of another array assuming the owner computes rule for the computation mapping. Edges between vertices of different colors are assigned a weight representing the associated communication costs. The problem of improving a given set of data mapping functions is to find a sequence of color exchanges, that is, data redistributions, that minimize the weight of the graph, that is, the communication costs. A possible algorithm for solving this problem is presented.

B. Sinharoy and B.K. Szymanski [32] study the problem of finding computation and data alignment functions for regular iterative algorithms. A loop nest can be represented by a regular iterative algorithm if and only if all the data access functions are constant offset functions and the loop nest's instructions are in single assignment form. The communication cost function used is based on the distance of the processors exchanging data on the virtual processor grid. The authors show that finding computation and data mapping functions is equivalent to minimizing a sum of absolute values composed of sums. An exact enumeration algorithm is presented and a polynomial time algorithm for finding an approximate solution is described.

Approaches Using a Graph Based Framework

K. Knobe, J.D. Lukas and G.L. Steele Jr. [19] study the problem of aligning the array elements accessed amongst each other. They target their approach towards SIMD machines. Two different kinds of preferences are distinguished:
i) identity preferences representing alignment preferences due to different data accesses to the same array, and
ii) conformance preferences relating two different arrays.

To compute what preferences can be satisfied without loosing parallelism, a cyclic preference graph is constructed. Each data access is represented by a vertex and two vertices are related by an undirected weighted edge if there exists a preference between the two data accesses. The weight of each edge is defined by the loop depth at which the data accesses occur. Conflicts between preferences are represented by cycles in the cyclic preference graph. A heuristic, using a greedy approach, is presented to remove annoying cycles or to reduce the parallelism.

In [5] an intermediate representation of a program called the alignment-distribution graph is described. The *alignment-distribution graph* is a directed graph in which nodes represent communication and edges represent the data flow. It exposes the communication requirements of the program. The framework restricts the alignments computed to alignments in which each axis of an array maps to a different axis of an HPF like template and data elements are evenly spaced along the template axis. The alignments computed have three components:
i) the axis,
ii) the stride, and
iii) the offset.

The papers present two separate algorithms called the compact dynamic programming algorithm and the constraint graph method for minimizing a communication cost function.

A. Darte and Y. Robert [8] study the problem of mapping perfectly nested affine loops onto distributed memory parallel computers. The problem is formulated by introducing the communication graph that captures all the required information to align data and computations. Each instruction and each array is represented by a vertex, the directed edges representing read and write data accesses. The problem of message vectorization and the use of global communication operations, like broadcasting, is addressed.

In [11] an algorithm is presented for computing HPF like data distribution functions. A distribution graph is constructed representing the relation between the data access functions and the array accessed. Based on the distribution graph a decision tree, modeling all possible combinations of data distribution functions, is traversed using a branch and bound algorithm. The cost function minimized by the algorithm is based on a communication analysis tool. The computation mapping is done in accordance with the owner computes rule.

M. Wolfe and M. Ikey [33] propose in 1994 an adaption of the techniques introduced by Li and Chen [22,23] for the language 'Crystal' to the imperative language 'Tiny'. The alignment phase is decomposed into four operations:
i) finding reference patterns,
ii) adding implicit dimensions to the arrays when required,
iii) building a component affinity graph, and
iv) partitioning the component affinity graph.

As the partitioning problem is *NP*-hard, a heuristic is used. The authors furthermore describe an algorithm to generate SPMD code based on the alignments computed.

J. Garcia, E. Ayguagé and J. Labarta [15] proposed for an algorithm to compute data distribution functions that can be expressed using HPF distribute statements. This algorithm is based on the construction and traversal of a single data structure, called the computation-parallelism graph. The computation-parallelism graph represents all possible data distributions along the dimensions of the arrays. Parallelism constraints are modeled as hyper-edges. Weights are associated to the edges to represent the associated communication costs. Negative costs are associated with the hyperedges to represent the associated parallelism. It is shown that distributing the data according to one dimension is equivalent to finding a path through the computation-parallelism graph fulfilling some additional constraints. The problem is formulated as a 0–1 integer programming problem. In contrast to other graph based approaches, the computation-parallelism graph models both the possible data distribution, that is, the locality constraints within a single data structure, and the possible parallelism.

W. Kelly and W. Pugh [18] describe a technique to minimize communication while preserving parallelism. The approach is not sensitive to the original program structure. For each array, the possible data mapping functions form a finite set of candidate space mappings. These sets consist of each dimension of the original iteration space being distributed. Next, for each candidate space, that is, for each possible data distribution function, all possible permutations of the surrounding loops are considered and the obtained parallelism measured. In a third step a weighted graph is constructed to model the parallelism as well as the communication cost associated with various data decompositions. One node in this weighted graph represents one candidate space mapping for each statement. The weight associated with a node is its degree of parallelism obtained. The edges represent the communication required and their weight models the communication costs. The alignment problem, as formulated in [18], is the problem of selecting one node per statement such that the sum of the weights of the selected nodes and edges is minimized. An algorithm to find such a set using various pruning strategies to reduce the size of the search space is presented.

Approaches Using a Linear Algebra Framework

Sheu and T.-H. Toi [31] introduced a method for the parallel execution of nested loops with constant loop-carried data dependences by reducing the communication overhead. First the nested loops are partitioned into large blocks which result in little inter-block communication. For a given linear transformation found by the hyperplane method [21], the iterations are partitioned into blocks such that the communication among the blocks is reduced while the execution order defined by the time transformation is not disturbed. The partitioning is based on projection techniques. In a second step these blocks are mapped onto message-passing multiprocessor systems according to specific properties of the target machine.

M. O'Boyle and G.A. Hedayat [26,27] express the alignment problem in a linear algebra framework. In this framework, aligned data can be viewed as forming a subspace in the iteration space. The problem solved is the computation of a transformation of the data access functions relative to one another such as to maximize the number of iteration points in the loop iteration space for which no communication is needed.

P. Feautrier [14] addresses the problem of finding an alignment function that maps the computations on a one-dimensional grid of virtual processors. The data mapping functions are defined by the owner computes rule which is imposed. The alignment constraints between computation and data accesses are derived from the data-flow graph of the program, procedure or loop nest considered. The data-flow graph is a directed graph. Vertices correspond to statements and the arcs to producers and consumers of data. For each statement, the alignment function is assumed to be an affine function of the iteration vectors with unknown parameters. The locality of data accesses is imposed by asking that the producer and the consumer of a data element be the same processor. Feautrier defines distance vectors between all pairs of producers and consumers. To any arc of the data-flow graph corresponds a distance vector that expresses the difference of the indices of the processor that computes the data and the one that uses it. Thus, a computation is local if and only if the corresponding distance vector is zero. The edges are hence transformed into affine equations and the problem consists in determining nontrivial parameters for the computation mappings that zero out as many distance vectors as possible. A heuristic is used to sort the equations in decreasing order of the communication traffic induced. The system of equations, which usually does not have a non trivial solution, is solved by successive

Gauss–Jordan eliminations as long as a feasible solution remains nontrivial. A solution is nontrivial if it has one degree of parallelism.

J.M. Anderson and M.S. Lam [1] describe necessary conditions for the data elements accessed by each processor to be local. They present a greedy algorithm to compute the computation and data mapping functions that can be satisfied. They incrementally add constraints as long as their conditions are satisfied, starting with the most frequently used array access functions. They only consider the linear part of the data access functions, taking care of the constant offsets in a second step. Their heuristic technique is close to the one defined in [9].

A. Platonoff [28,29] develops extensions to Feautrier's [14] automatic data distribution algorithm. A method is presented to extract global broadcast operations as well as translation operations to optimize the data mapping functions. In the data-flow graph, patterns representing broadcast and other global communication patterns are searched for. The data distribution is then chosen such as to maximize the number of global communication operations possible.

M. Dion and Robert [12,13] consider a problem in which all data access functions are of full rank and no smaller than d, the required degree of parallelism. This ensures that the parallelism obtained is indeed as large as wanted. By considering only the linear parts they compute the largest set of alignment constraints that can be satisfied while yielding the given degree of parallelism d. The constant offsets are considered subsequently, using techniques developed by Darte and Robert [8]. They consider a set of candidate solutions and search for an optimal one that verifies the largest number of constraints while effectively yielding the degree of parallelism desired. In their approach, Dion and Robert consider three basic cases depending on the structure of the data access function. Then, they build a directed graph defined as follows. Vertices correspond either to statements or arrays. There is an arc from vertex p to vertex q if and only if a mapping of rank d can be computed for q from a given mapping of rank d for p according to the basic cases enumerated previously. In this graph they search for a tree containing a maximal number of arcs. Obviously, choosing a mapping of rank d for the root of the computed tree implicitly determines mappings of rank d for all other vertices.

C. Mongenet [25] is interested in minimizing communication costs in the presence of systems of affine recurrence equations, that is, single assignment loop nests. The data dependences are subdivided into two classes:
i) auto dependences, and
ii) cross dependences.
Auto-dependences are data dependences between two data accesses to the same array. The domains of these arrays are projected onto hyperplanes such as to minimize the number of remote data accesses. Cross-dependences are dependences between data accesses to different arrays. Unimodular transformations are applied to the projected domains to align the different data array and so minimize the resulting communications. A heuristic based on these two steps is introduced.

C.G. Diderich [9] and Diderich and M. Gengler [10] present and extend the algorithm for solving this problem introduced in [2]. In a second step they introduce the constant degree parallelism alignment problem. It is the problem of finding computation and data mapping functions that minimize the number of remote data accesses for a given degree of parallelism. An exact implicit enumeration algorithm is presented. It proceeds by enumerating all interesting subsets of alignment constraints to be satisfied. To allow large alignment problems to be solved an efficient heuristic is presented and applied to various benchmarks.

Other Approaches

B.M. Chapman, T. Fahringer and H.P. Zima [4] for a software tool to provide automatic support for the mapping of the data onto the processors of the target machine. The computation is mapped by using the owner computes rule. The tool is integrated within the *Vienna Fortran Compilation System*, a compiler for Vienna Fortran, an HPF like Fortran dialect. The tool makes use of performance analysis methods and uses, via heuristics, empirical performance data. Once the performance data has been obtained for a given program, an inter-procedural alignment and pattern matching phase determines a suitable alignment of the arrays within each procedure. The alignments are then propagated through the call graph of the program.

Eventually more versions of a procedure are generated, corresponding to differently distributed actual arguments. Finally code is generated using the selected data distributions.

In [7], P. Crooks and R.H. Perrott present an algorithm for determining data mapping functions by generating HPF like directives. Their approach is based on identifying reference patterns. To each read/write pair is associated an ideal data distribution that minimized inter-processor communication. Once the preferences for the individual accesses are determined, a performance estimator is used to select the combination of preferences that gives the best performance estimate.

R. Bixby, K. Kennedy and U. Kremer [3] present an automatic data layout algorithm based on 0–1 integer programming techniques. The data mapping functions, following the HPF alignment structure, are optimized for a target distributed memory machine, a specific problem size and the number of available processors. The distribution analysis uses the alignment search space, that is, the space of all possible HPF like alignments, to build candidate data layout search spaces of reasonable data mapping functions for each loop nest. In a second step the inter-phase or inter-loop nests data layout problem is addressed. By using an integer programming formulation, a data mapping function is selected for each loop nest such that a single global cost function, modeling the communication costs, is minimized.

Conclusion

This article presents major advancements made in solving the alignment problem. Different subproblems are defined and described. One major open problem is how to incorporate scheduling information into the algorithms computing efficient alignment functions. See [9] for a first approach towards computing scheduling functions compatible with computation and data mapping functions. The question of which cost function to use when computing alignment functions has to be addressed with more details.

See also

▶ Integer Programming

References

1. Anderson JM, Lam MS (1993) Global optimizations for parallelism and locality on scalable parallel machines. In: ACM SIGPLAN Conf. Programming Language Design and Implementation (PLDI '93). ACM, New York, pp 112–125
2. Bau D, Kodukula I, Kotylar V, Pingali K, Stodghill P (1994) Solving alignment using elementary linear algebra. In: 7th Internat. Workshop Languages and Compilers for Parallel Computing (LCPC '94). In: Lecture Notes Computer Sci, vol 892. Springer, Berlin, pp 46–60
3. Bixby R, Kennedy K, Kremer U (1994) Automatic data layout using 0–1 integer programming. Internat. Conf. Parallel Architectures and Compilation Techniques (PACT '94). pp 111–122
4. Chapman BM, Fahringer T, Zima HP (1993) Automatic support for data distribution on distributed memory multiprocessor systems. In: 6th Internat. Workshop Languages and Compilers for Parallel Computing (LCPC '93). In: Lecture Notes Computer Sci, vol 768. Springer, Berlin, pp 184–199
5. Chatterjee S, Gilbert JR, Schreiber R, Sheffler TJ (1994) Array distribution in data-parallel programs. In: 7th Internat. Workshop Languages and Compilers for Parallel Computing (LCPC '94). In: Lecture Notes Computer Sci, vol 892. Springer, Berlin, pp 78–91
6. Chen T-S, Sheu J-P (1994) Communication-free data allocation techniques for parallelizing compilers on multicomputers. IEEE Trans Parallel and Distributed Systems 5(9):921–938
7. Crooks P, Perrott RH (1993) An automatic data distribution generator for distributed memory MIMD machines. In: 4th Internat. Workshop Compilers for Parallel Computers, pp 33–44
8. Darte A, Robert Y (1994) On the alignment problem. Parallel Proc Lett 4(3):259–270
9. Diderich CG (1998) Automatic data distribution for massively parallel distributed memory computers. PhD Thesis. Computer Sci. Dept. Swiss Federal Inst. Tech., Lausanne
10. Diderich CG, Gengler M (1997) The alignment problem in a linear algebra framework. In: Proc. Hawaii Internat. Conf. System Sci. (HICSS-30); Software Techn. Track. IEEE Computer Soc Press, New York, pp 586–595
11. Dierstein A, Hayer R, Rauber T (1994) The ADDAP system on the iPSC/860: Automatic data distribution and parallelization. J Parallel Distributed Comput 32(9):1–10
12. Dion M (1996) Alignement et distribution en parallélisation automatique. PhD Thesis. Ecole Normale Sup. Lyon (In French)
13. Dion M, Robert Y (1996) Mapping affine loop nests. Parallel Comput 22:1373–1397
14. Feautrier P (1992) Towards automatic distribution. Parallel Proc Lett 4(3):233–244
15. Garcia J, Ayguadé E, Labarta J (1995) A novel approach towards automatic data distribution. In: Supercomputing '95 Conf

16. Gupta M (1992) Automatic data partitioning on distributed memory multicomputers. PhD Thesis. Univ. Illinois at Urbana-Champaign, Urbana, IL
17. Huang C-H, Sadayappan P (1991) Communication-free hyperplane partitioning of nested loops. In: 4th Internat Workshop Languages and Compilers for Parallel Computing (LCPC '91), vol 589. In: Lecture Notes Computer Sci, vol 589. Springer, Berlin, pp 186–200
18. Kelly W, Pugh W (1996) Minimizing communication while preserving parallelism. In: 1996 ACM Internat. Conf. Supercomputing (ICS '96). ACM, New York, pp 52–60
19. Knobe K, Lukas JD, Steele GL Jr (1990) Data optimization: Allocation of arrays to reduce communication on SIMD machines. J Parallel Distributed Comput 8(2):102–118
20. Kunchithapadam K, Miller BP (1994) Optimizing array distributions in data-parallel programs. In: 7th Internat. Workshop Languages and Compilers for Parallel Computing (LCPC '94). In: Lecture Notes Computer Sci, vol 892. Springer, Berlin, pp 470–484
21. Lamport L (1974) The parallel execution of DO loops. Comm ACM 17(2):83–93
22. Li J, Chen M (1990) Index domain alignment: Minimizing cost of cross-referencing between distributed arrays. In: 3rd Symp. Frontiers of Massively Parallel Computation (Frontiers '90). IEEE Computer Soc Press, New York, pp 424–433
23. Li J, Chen M (1991) The data alignment phase in compiling programs for distributed-memory machines. J Parallel Distributed Comput 13:213–221
24. Lim AW, Lam MS (1994) Communication-free parallelization via affine transformations. In: 7th Internat. Workshop Languages and Compilers for Parallel Computing (LCPC '94). In: Lecture Notes Computer Sci, vol 892. Springer, Berlin, pp 92–106
25. Mongenet C (1995) Mappings for communications minimization using distribution and alignment. In: Internat. Conf. Parallel Architectures and Compilation Techniques (PACT '95). pp 185–193
26. O'Boyle M (1993) A data partitioning algorithm for distributed memory compilation. Techn Report Ser Univ Manchester, England UMCS-93-7-1
27. O'Boyle M, Hedayat GA (1992) Data alignment: Transformation to reduce communication on distributed memory architectures. In: Scalable High Performance Computing Conf. (SHPCC '92). IEEE Computer Soc Press, New York, pp 366–371
28. Platonoff A (1995) Automatic data distribution for massively parallel computers. In: Int. Workshop Compilers for Parallel Computers, pp 555–570
29. Platonoff A (1995) Contribution à la distribution automatique des données pour machines massivement parallèles. PhD Thesis. Ecole Normale Sup. Mines de Paris (In French)
30. Ramanujam J, Sadayappan P (1991) Compile-time techniques for data distribution in distributed memory machines. IEEE Trans Parallel and Distributed Systems 2(4):472–482
31. Sheu J-P, Tai T-H (1991) Partitioning and mapping nested loops on multiprocessor systems. IEEE Trans Parallel and Distributed Systems 2(4):430–439
32. Sinharoy B, Szymanski BK (1994) Data and task alignment in distributed memory architectures. J Parallel Distributed Comput 21:61–74
33. Wolfe M, Ikei M (1994) Automatic array alignment for distributed memory multicomputers. 27th Annual Hawaii Internat. Conf. System Sci., vol II. IEEE Computer Soc. Press, New York, pp 23–32

αBB Algorithm

CLAIRE S. ADJIMAN,
CHRISTODOULOS A. FLOUDAS
Department Chemical Engineering,
Princeton University, Princeton, USA

MSC2000: 49M37, 65K10, 90C26, 90C30

Article Outline

Keywords
General Framework
Convexification and Underestimation Strategy
    Function Decomposition
    Linear and Convex Terms
    Bilinear Terms
    Trilinear, Fractional and Fractional Trilinear Terms
    Univariate Concave Terms
    General Nonconvex Terms
    Overall Convexification/Relaxation Strategy
    Equality Constraints
Branching Variable Selection
    Least Reduced Axis Rule
    Term Measure
    Variable Measure
Variable Bound Updates
    Optimization-Based Approach
    Interval-Based Approach
Algorithmic Procedure
Computational Experience
Conclusions
See also
References

Keywords

Global optimization; Interval arithmetic; Twice-differentiable NLPs; Branch and bound; αBB algorithm

Deterministic global optimization techniques for nonconvex NLPs have been the subject of growing interest because they can potentially provide a very complete characterization of the problem being considered. In addition to guaranteeing identification of the global solution within arbitrary accuracy, they enable the location of all local and global solutions of the problem. As a result, they can be used to determine the feasibility of a given problem with certainty [1,2,3,4], or to find all solutions of a nonlinear system of equations [13]. They are especially valuable in the study of systems in which the global optimum solution is the only physically meaningful solution, as is the case of the phase equilibrium of non ideal mixtures [16,17,18,19,20]. Traditionally, a major theoretical limitation of these approaches has been their inability to tackle problems with arbitrary nonconvexities. However, the recent development of rigorous convex relaxation techniques for general twice continuously differentiable functions [2,3,4] has greatly expanded the class of problems that can be addressed through deterministic global optimization. These approaches have been incorporated within a branch and bound framework to create the αBB *global optimization algorithm* for twice continuously differentiable problems [3,6,12]. The theoretical basis of the algorithm as well as the efficient search strategies it uses are discussed in this article.

General Framework

The αBB algorithm guarantees finite ϵ-convergence to the global solution of nonlinear programming problems (NLPs) belonging to the general class

$$\begin{cases} \min_\mathbf{x} & f(\mathbf{x}) \\ \text{s.t.} & \mathbf{g}(\mathbf{x}) \leq 0 \\ & \mathbf{h}(\mathbf{x}) = 0 \\ & \mathbf{x} \in [\mathbf{x}^L, \mathbf{x}^U], \end{cases} \quad (1)$$

where $f(\mathbf{x})$, $\mathbf{g}(\mathbf{x})$ and $\mathbf{h}(\mathbf{x})$ are continuous twice-differentiable functions.

The solution scheme is based on the generation of a nonincreasing sequence of upper bounds and a nondecreasing sequence of lower bounds on the global solution. The monotonicity of these sequences is ensured through successive partitioning of the search space which enables the construction of increasingly tight relaxations of the problem. The validity of the bounds obtained is of crucial importance in a rigorous global optimization approach. The *upper bounding* step does not present any theoretical difficulties and consists of a local optimization of the nonconvex problem. The *lower bounding* step is a more challenging operation in which the nonconvex problem must be convexified and underestimated in the current subdomain. The strategy adopted dictates the applicability of the algorithm and plays a pivotal role in its performance as it determines the tightness of the lower bounds obtained. The procedure followed in the αBB algorithm is discussed in the next section. Finally, the *branching* step involves the partition of the solution domain with the smallest lower bound on the global optimum solution into a covering set of subdomains. Although this is a simple task, the choice of partition has implications for the rate of convergence of the algorithm and efficient branching rules must be used.

Convexification and Underestimation Strategy

A convex relaxation of problem (1) is obtained by constructing convex underestimators for the nonconvex objective function and inequality constraints and by relaxing the nonlinear equality constraints, replacing them with less stringent linear equality constraints or a set of two convex inequalities. The general convexification/relaxation procedure used is first discussed for the objective function and nonconvex inequalities.

Function Decomposition

A convex underestimator for a twice continuously differentiable function is constructed by following a two-stage procedure. In the first stage, the function is decomposed into a summation of terms of special structure, such as linear, convex, bilinear, trilinear, fractional, fractional trilinear, concave in one variable and general nonconvex terms. Then, based on the fact that the summation of convex functions results in a convex function, a tailored convex underestimator is used for each different term type. Thus, a twice-differentiable function $F(\mathbf{x})$ defined over the domain $[\mathbf{x}^L, \mathbf{x}^U]$ is written as

$$F(\mathbf{x}) = c^T \mathbf{x} + F_C(\mathbf{x}) + \sum_{i=1}^{bt} b_i x_{B_i,1} x_{B_i,2}$$
$$+ \sum_{i=1}^{tt} t_i x_{T_i,1} x_{T_i,2} x_{T_i,3} + \sum_{i=1}^{ft} f_i \frac{x_{F_i,1}}{x_{F_i,2}} \quad (2)$$
$$+ \sum_{i=1}^{ftt} ft_i \frac{x_{FT_i,1} x_{FT_i,2}}{x_{FT_i,3}} + \sum_{i=1}^{uct} F_{UC_i}(x_{UC_i})$$
$$+ \sum_{i=1}^{nct} F_{NC_i}(\mathbf{x}),$$

where c is a scalar vector; $F_C(\mathbf{x})$ is a convex function; bt is the number of bilinear terms, b_i is the coefficient of the ith bilinear term and $x_{B_i,1}$ and $x_{B_i,2}$ are the two variables participating in the bilinear term; tt is the number of trilinear terms, t_i is the coefficient of the ith trilinear term and $x_{T_i,1}$ $x_{T_i,2}$ and $x_{T_i,3}$ are the three variables participating in the trilinear term; ft is the number of fractional terms, f_i is the coefficient of the ith fractional term and $x_{F_i,1}$ and $x_{F_i,2}$ are the two variables participating in the fractional term; ftt is the number of fractional trilinear terms, ft_i is the coefficient of the ith fractional trilinear term and $x_{FT_i,1}$, $x_{FT_i,2}$ and $x_{FT_i,3}$ are the three variables participating in the fractional trilinear term; uct is the number of univariate concave terms, F_{UC_i} is the ith univariate concave term and x_{UC_i} is the variable participating in the univariate concave term; nct is the number of general nonconvex terms and $F_{NC_i}(\mathbf{x})$ is the ith general nonconvex term.

The decomposition phase serves two purposes: it can lead to the construction of a tight underestimator by taking advantage of the special structure of the function and it may reduce the complexity of the underestimation strategy by permitting the treatment of terms which involve a smaller number of variables than the overall nonconvex function. As will become apparent, this is especially important for general nonconvex terms.

Linear and Convex Terms

Any term that has been identified as linear or convex does not need to be modified during the convexification/underestimation procedure.

Bilinear Terms

The bilinear terms can be replaced by their convex envelope [5,15]. A new variable w_B substitutes a bilinear term $x_1 x_2$ and is bounded by a set of four inequality constraints which depend on the variable bounds.

$$\begin{cases} w_B \geq x_1^L x_2 + x_2^L x_1 - x_1^L x_2^L, \\ w_B \geq x_1^U x_2 + x_2^U x_1 - x_1^U x_2^U, \\ w_B \leq x_1^U x_2 + x_2^L x_1 - x_1^U x_2^L, \\ w_B \leq x_1^L x_2 + x_2^U x_1 - x_1^L x_2^U. \end{cases} \quad (3)$$

Trilinear, Fractional and Fractional Trilinear Terms

For trilinear, fractional and fractional trilinear terms, the convex underestimators proposed in [13] can be used. They are constructed in a fashion similar to the bilinear term underestimators: a new variable replaces the term and a set of inequality constraints provides bounds on this variable. For a trilinear term $x_1 x_2 x_3$, for instance, the substitution variable w_T is subject to

$$\begin{cases} w_T \geq x_1 x_2^L x_3^L + x_1^L x_2 x_3^L \\ \quad + x_1^L x_2^L x_3 - 2 x_1^L x_2^L x_3^L, \\ w_T \geq x_1 x_2^U x_3^U + x_1^U x_2 x_3^U \\ \quad + x_1^U x_2^U x_3 - x_1^U x_2^U x_3^U - x_1^U x_2^U x_3^U, \\ w_T \geq x_1 x_2^L x_3^L + x_1^L x_2 x_3^U \\ \quad + x_1^L x_2^U x_3 - x_1^L x_2^U x_3^U - x_1^L x_2^L x_3^L, \\ w_T \geq x_1 x_2^U x_3^L + x_1^U x_2 x_3^U \\ \quad + x_1^L x_2^U x_3 - x_1^U x_2^U x_3^U - x_1^U x_2^U x_3^U, \\ w_T \geq x_1 x_2^L x_3^U + x_1^L x_2 x_3^L \\ \quad + x_1^U x_2^L x_3 - x_1^U x_2^L x_3^U - x_1^L x_2^L x_3^L, \\ w_T \geq x_1 x_2^L x_3^U + x_1^L x_2 x_3^U \\ \quad + x_1^U x_2^U x_3 - x_1^L x_2^U x_3^U - x_1^U x_2^U x_3^U, \\ w_T \geq x_1 x_2^U x_3^L + x_1^U x_2 x_3^L \\ \quad + x_1^L x_2^L x_3 - x_1^U x_2^L x_3^L - x_1^L x_2^L x_3^L, \\ w_T \geq x_1 x_2^U x_3^U + x_1^U x_2 x_3^U \\ \quad + x_1^U x_2^U x_3 - 2 x_1^U x_2^U x_3^U. \end{cases} \quad (4)$$

For a fractional term x_1/x_2 with $x_2^L > 0$, the new variable w_F is bounded by

$$w_F \geq \begin{cases} \frac{x_1^L}{x_2} + \frac{x_1}{x_2^U} - \frac{x_1^L}{x_2^U} & \text{if } x_1^L \geq 0, \\ \frac{x_1}{x_2^U} - \frac{x_1^L x_2}{x_2^L x_2^U} + \frac{x_1^L}{x_2^L} & \text{if } x_1^L < 0, \end{cases}$$

$$w_F \geq \begin{cases} \frac{x_1^U}{x_2} + \frac{x_1}{x_2^L} - \frac{x_1^U}{x_2^L} & \text{if } x_1^U \geq 0, \\ \frac{x_1}{x_2^L} - \frac{x_1^U x_2}{x_2^L x_2^U} + \frac{x_1^U}{x_2^U} & \text{if } x_1^U < 0. \end{cases} \quad (5)$$

Finally, for a fractional trilinear term $x_1 x_2/x_3$ with $x_1^L, x_2^L \geq 0$ and $x_3^L > 0$, the substitution variable w_{FT} is subject to

$$\begin{cases} w_{FT} \geq \frac{x_1 x_2^L}{x_3^U} + \frac{x_1^L x_2}{x_3^U} \\ \qquad + \frac{x_1^L x_2^L}{x_3} - \frac{2 x_1^L x_2^L}{x_3^U}, \\ w_{FT} \geq \frac{x_1 x_2^L}{x_3^U} + \frac{x_1^L x_2}{x_3^L} \\ \qquad + \frac{x_1^L x_2^U}{x_3} - \frac{x_1^L x_2^U}{x_3^L} - \frac{x_1^L x_2^L}{x_3^U}, \\ w_{FT} \geq \frac{x_1 x_2^U}{x_3^L} + \frac{x_1^U x_2}{x_3^U} \\ \qquad + \frac{x_1^U x_2^L}{x_3} - \frac{x_1^U x_2^L}{x_3^U} - \frac{x_1^U x_2^L}{x_3^L}, \\ w_{FT} \geq \frac{x_1 x_2^U}{x_3^U} + \frac{x_1^U x_2}{x_3^L} \\ \qquad + \frac{x_1^L x_2^U}{x_3} - \frac{x_1^L x_2^U}{x_3^U} - \frac{x_1^U x_2^U}{x_3^L}, \\ w_{FT} \geq \frac{x_1 x_2^L}{x_3^U} + \frac{x_1^U x_2}{x_3^L} \\ \qquad + \frac{x_1^U x_2^L}{x_3} - \frac{x_1^U x_2^L}{x_3^L} - \frac{x_1^L x_2^L}{x_3^U}, \\ w_{FT} \geq \frac{x_1 x_2^U}{x_3^U} + \frac{x_1^U x_2}{x_3^L} \\ \qquad + \frac{x_1^L x_2}{x_3} - \frac{x_1^L x_2^U}{x_3^L} - \frac{x_1^L x_2^U}{x_3^U}, \\ w_{FT} \geq \frac{x_1 x_2^L}{x_3^U} + \frac{x_1^L x_2}{x_3^L} \\ \qquad + \frac{x_1^U x_2^L}{x_3} - \frac{x_1^U x_2^L}{x_3^L} - \frac{x_1^L x_2^L}{x_3^U}, \\ w_{FT} \geq \frac{x_1 x_2^U}{x_3^L} + \frac{x_1^U x_2}{x_3^L} \\ \qquad + \frac{x_1^U x_2^U}{x_3} - \frac{2 x_1^U x_2^U}{x_3^L}. \end{cases} \quad (6)$$

Univariate Concave Terms

For univariate concave terms, the convexification/underestimation procedure does not require the introduction of new variables or constraints: a simple linearization of the term suffices. Thus, a univariate concave term $F_{UC}(x)$ is replaced by the linear term

$$F_{UC}(x^L) + \frac{F_{UC}(x^U) - F_{UC}(x^L)}{x^U - x^L}(x - x^L). \quad (7)$$

General Nonconvex Terms

For a general nonconvex term $F_{NC}(\mathbf{x})$, a convex underestimator $\check{F}_{NC}(\mathbf{x})$ over $[\mathbf{x}^L, \mathbf{x}^U]$ is constructed by subtracting a positive separable quadratic term from $F_{NC}(\mathbf{x})$ [12]:

$$\check{F}_{NC}(\mathbf{x}) = F_{NC}(\mathbf{x}) - \sum_{j=1}^{n} \alpha_j (x_j - x_j^L)(x_j^U - x_j), \quad (8)$$

where n is the number of variables and the α parameters are positive scalars.

The magnitude of the α parameters determines both the quality of the convex underestimator, that is, its tightness, and its convexity. It was shown in [12] that the maximum separation distance, $d_{\max}$, between the nonconvex term $F_{NC}(\mathbf{x})$ and its convex underestimator $\check{F}_{NC}(\mathbf{x})$ is given by

$$d_{\max} = \max_{\mathbf{x}} \left(F_{NC}(\mathbf{x}) - \check{F}_{NC}(\mathbf{x}) \right)$$

$$= \frac{1}{4} \sum_{j=1}^{n} \alpha_j (x_j^U - x_j^L)^2 . \quad (9)$$

Thus, small α values are needed to construct a tight underestimator. The dependence of the maximum separation distance on the square of the variable ranges is especially important for the convergence proof of the algorithm [12]. Provided that the α values do not increase from a parent node to a child node, relation (9) guarantees that the convex relaxations become increasingly tight as the branch and bound iterations progress and smaller subdomains are generated. In the limit, the convex underestimators match the original functions. As a result, the monotonicity of the lower bound sequence can be ensured.

To meet the convexity requirement of $\check{F}_{NC}(\mathbf{x})$, the positive quadratic term needs to be sufficiently large to overcome the nonconvexity of $F_{NC}(\mathbf{x})$. This is achieved by manipulating the value of the α parameters. Based on the properties of convex functions, a necessary and sufficient condition for the convexity of $\check{F}_{NC}(\mathbf{x})$ is the positive semidefiniteness of the matrix $H_{F_{NC}}(\mathbf{x}) + 2\,\mathrm{diag}(\alpha_j)$ for all $\mathbf{x} \in [\mathbf{x}^L, \mathbf{x}^U]$, where $H_{F_{NC}}(\mathbf{x})$ is the Hessian matrix of the nonconvex term $F_{NC}(\mathbf{x})$. The diagonal matrix $\Delta = \mathrm{diag}(\alpha_j)$ results in a shift in the diagonal elements of the matrix $H_{F_{NC}}(\mathbf{x})$ and is therefore referred to as the *diagonal shift matrix*. The rigorous derivation of a matrix Δ that satisfies the convexity condition is a difficult matter in the general case, primarily because of the nonlinear dependence of the Hessian matrix on the **x** variables. This problem can be alleviated by using *interval arithmetic* to generate an *interval Hessian matrix* $[H_{F_{NC}}]$ such that $H_{F_{NC}}(\mathbf{x}) \in [H_{F_{NC}}]$ for all $\mathbf{x} \in]\mathbf{x}^L, \mathbf{x}^U[$ [1,3,4]. This process allows the formulation of a sufficient convexity condition for the underestimator: if all real symmetric matrices in $[H_{F_{NC}}] + 2\,\text{diag}(\alpha_j)$ are positive semidefinite, then $\check{F}_{NC}(\mathbf{x})$ is convex over $[\mathbf{x}^L, \mathbf{x}^U]$.

Based on the interval Hessian matrix, a number of methods may be used to automatically and rigorously compute a diagonal shift matrix Δ that guarantees the convexity of $\check{F}_{NC}(\mathbf{x})$. The first class of techniques generates a *uniform* diagonal shift matrix by equating all the diagonal elements of Δ with a single α value. In the second class of techniques, different α values are used and a *nonuniform* diagonal shift matrix is obtained [1,3].

In the first class of methods, the convexity condition is equivalent to the positive semidefiniteness of all real symmetric matrices in $[H_{F_{NC}}] + 2\,\text{diag}(\alpha)$ and is satisfied by any α parameter such that

$$\alpha \geq \max\left\{0, -\frac{1}{2}\lambda_{\min}\left([H_{F_{NC}}]\right)\right\}, \quad (10)$$

where $\lambda_{\min}([H_{F_{NC}}])$ is the minimum eigenvalue of $[H_{F_{NC}}]$ [3,12].

Consider a square symmetric interval Hessian matrix family $[H]$ whose element (ij) is the interval $[\underline{h}_{ij}, \overline{h}_{ij}]$ and whose radius matrix ΔH is defined as $(\Delta H)_{ij} = \frac{(\overline{h}_{ij} - \underline{h}_{ij})}{2}$. A lower bound on the minimum eigenvalue of $[H]$ can be obtained using one of the following methods [1,3,4]:

- Method I.1 — the Gershgorin theorem approach;
- Method I.2a — the E-matrix approach with $E = 0$;
- Method I.2b — the E-matrix approach with $E = \text{diag}(\Delta H)$;
- Method I.3 — Mori–Kokame's approach;
- Method I.4 — the lower bounding Hessian approach;
- Method I.5 — an approach based on the Kharitonov theorem;
- Method I.6 — the Hertz approach.

Method I.1 is an extension of the *Gershgorin theorem* for real matrices to interval matrices. The minimum eigenvalue of $[H]$ is such that

$$\lambda_{\min}([H]) \geq \min_i \left[\underline{h}_{ii} - \sum_{j \neq i} \max\left(\|\underline{h}_{ij}\|, \|\overline{h}_{ij}\|\right)\right].$$

Methods I.2a and I.2b are a generalization of the results presented in [8,23]. It requires the computation of the modified midpoint matrix $\widetilde{H}_M$ such that $(\widetilde{H}_M)_{ij} = \frac{(\overline{h}_{ij} + \underline{h}_{ij})}{2}$ for $i \neq j$ and $(\widetilde{H}_M)_{ii} = 0$, as well as the computation of the modified radius matrix $\widetilde{\Delta H}$ such that $(\widetilde{\Delta H})_{ij} = \frac{(\overline{h}_{ij} - \underline{h}_{ij})}{2}$ for $i \neq j$ and $(\widetilde{\Delta H})_{ii} = \underline{h}_{ij}$. Given an arbitrary real symmetric matrix E, the minimum eigenvalue of the interval Hessian matrix $[H]$ is such that

$$\lambda_{\min}([H]) \geq \lambda_{\min}\left(\widetilde{H}_M + E\right) - \rho\left(\widetilde{\Delta H} + \|E\|\right),$$

where $\rho(M)$ denotes the spectral radius of the real matrix M. In practice, two E-matrices have been used: $E = 0$ (Method I.2a) and $E = \Delta H$ (Method I.2b).

Method I.3 is based on a result presented in [21], which uses the lower vertex matrix $\underline{H}$, such that $(\underline{H})_{ij} = \underline{h}_{ij}$, and the upper vertex matrix $\overline{H}$, such $(\overline{H})_{ij} = \overline{h}_{ij}$. The minimum eigenvalue of $[H]$ is such that

$$\lambda_{\min}([H]) \geq \lambda_{\min}(\underline{H}) - \rho(\overline{H} - \underline{H}).$$

Method I.4 uses a *lower bounding Hessian* of the interval Hessian matrix. Such a matrix is defined in [24] as a real symmetric matrix whose minimum eigenvalue is smaller than the minimum eigenvalue of any real symmetric matrix in the interval Hessian family. It therefore suffices to compute the minimum eigenvalue of this real matrix to obtain the desired lower bound. A lower bounding Hessian $L = (l_{ij})$ can be constructed from the following rule:

$$l_{ij} = \begin{cases} \underline{h}_{ii} + \sum_{k \neq i} \frac{\underline{h}_{ik} - \overline{h}_{ik}}{2}, & i = j, \\ \frac{\underline{h}_{ij} + \overline{h}_{ij}}{2}, & i \neq j. \end{cases}$$

Method I.5 is based on the *Kharitonov theorem* [11] which, by extension, gives a lower bound on the minimum eigenvalue of an interval Hessian matrix family [2]. First, the corresponding characteristic polynomial family must be derived

$$[K] = [\underline{c}_0, \overline{c}_0] + [\underline{c}_1, \overline{c}_1]\lambda + [\underline{c}_2, \overline{c}_2]\lambda^2 + [\underline{c}_3, \overline{c}_3]\lambda^3 + [\underline{c}_4, \overline{c}_4]\lambda^4 + [\underline{c}_5, \overline{c}_5]\lambda^5 + \cdots,$$

where the coefficients of λ depend on the elements of the interval Hessian matrix $[H]$. A lower bound on the roots of this polynomial can then obtained by calculating the minimum roots of only four real polynomials. The appropriate bounding polynomials are the Kharitonov polynomials

$$\begin{cases} K_1 = & \underline{c}_0 + \underline{c}_1\lambda + \overline{c}_2\lambda^2 + \overline{c}_3\lambda^3 \\ & +\underline{c}_4\lambda^4 + \underline{c}_5\lambda^5 + \cdots, \\ K_2 = & \overline{c}_0 + \overline{c}_1\lambda + \underline{c}_2\lambda^2 + \underline{c}_3\lambda^3 \\ & +\overline{c}_4\lambda^4 + \overline{c}_5\lambda^5 + \cdots, \\ K_3 = & \overline{c}_0 + \underline{c}_1\lambda + \underline{c}_2\lambda^2 + \overline{c}_3\lambda^3 \\ & +\overline{c}_4\lambda^4 + \underline{c}_5\lambda^5 + \cdots, \\ K_4 = & \underline{c}_0 + \overline{c}_1\lambda + \overline{c}_2\lambda^2 + \underline{c}_3\lambda^3 \\ & +\underline{c}_4\lambda^4 + \overline{c}_5\lambda^5 + \cdots. \end{cases}$$

Method I.6 allows the computation of the exact minimum eigenvalue of the family of symmetric matrices represented by the interval Hessian matrix. It requires the construction of 2^{n-1} vertex matrices H^k of the interval matrix $[H]$ as defined by

$$(H^k)_{ij} = \begin{cases} \underline{h}_{ii} & \text{if } i = j, \\ \underline{h}_{ij} & \text{if } u_i u_j \geq 0,\ i \neq j, \\ \overline{h}_{ij} & \text{if } u_i u_j < 0,\ i \neq j, \end{cases}$$

where all possible combinations of the signs of the arbitrary scalars u_i and u_j are enumerated. It was shown in [4,10] that the lowest minimum eigenvalue from this set of real matrices is the minimum eigenvalue of the interval matrix.

Three rigorous techniques for the generation of a non uniform shift matrix Δ can be used [1,3]:
- Method II.1a — the scaled Gershgorin theorem approach with scaling vector $\mathbf{d} = \underline{1}$;
- Method II.1b — the scaled Gershgorin theorem approach with scaling vector $\mathbf{d} = \mathbf{x}^U - \mathbf{x}^L$;
- Method II.2 — the H-matrix approach;
- Method II.3 — an approach based on the minimization of the maximum separation distance.

The main advantage of these techniques is that resorting to a different value of the α parameter for each variable may lead to tighter underestimators by taking into account the individual contribution of each variable to the overall nonconvexity of the term being considered. In the case of a uniform diagonal shift, the worst contribution is uniformly assigned to all variables.

Methods II.1a and II.1b bear resemblance with the Gershgorin theorem used for Method I.1. In the present case, however, each row is considered independently and the ith element of the diagonal shift matrix, α_i, is the maximum of zero and

$$-\frac{1}{2}\left(\underline{h}_{ii} - \sum_{j \neq i} \max\left\{\left\|\underline{h}_{ij}\right\|, \left\|\overline{h}_{ij}\right\|\right\}\frac{d_j}{d_i}\right),$$

where $\mathbf{d}$ is an arbitrary positive vector. In practice, $\mathbf{d} = \underline{1}$ (Method II.1a) and $\mathbf{d} = \mathbf{x}^U - \mathbf{x}^L$ (Method II.1b) have been used. The latter choice of scaling often helps to reduce the maximum separation distance between the nonconvex term and its underestimator by assigning smaller α values to variables with a larger range.

Method II.2 is an iterative method based on the properties of H-matrices: a square interval matrix that has the H-matrix property is regular and does not have 0 as an eigenvalue [22]. In order to determine whether a square interval matrix $[H]$ is an H-matrix, its comparison matrix $\langle H \rangle$ must first be defined. For $i \neq j$, the off-diagonal element $(\langle H \rangle)_{ij}$ of the comparison matrix is given by $-\max\{\|\underline{h}_{ij}\|, \|\overline{h}_{ij}\|\}$. A diagonal element $(\langle H \rangle)_{ii}$ of the comparison matrix is given by

$$\begin{cases} 0, & 0 \in [\underline{h}_{ii}, \overline{h}_{ii}], \\ \min\left\{\|\underline{h}_{ii}\|, \|\overline{h}_{ii}\|\right\}, & 0 \notin [\underline{h}_{ii}, \overline{h}_{ii}]. \end{cases}$$

A real matrix such as $\langle H \rangle$ is an M-matrix if all its off-diagonal elements are nonpositive – this is always true for $\langle H \rangle$ – and if there exists a real positive vector $\mathbf{u}$ such that $\langle H \rangle \mathbf{u} > 0$. The interval matrix $[H]$ is an H-matrix if its comparison matrix $\langle H \rangle$ is an M-matrix. Method II.2 follows an iterative procedure to construct a nonuniform diagonal shift matrix Δ such that $[H] + 2\Delta$ is an H-matrix whose modified midpoint matrix is positive definite. If these conditions are met, the diagonal elements of the shift matrix are guaranteed to lead to the construction of a *convex* underestimator for the nonconvex term. The initial guess chosen for Δ is the uniform diagonal shift matrix given by Method I.2.

Method II.3 aims to generate a non uniform diagonal shift matrix which minimizes the maximum separation distance between the nonconvex term and its underestimator. For this purpose, the following semidefinite programming problem is solved using an interior point method [25]:

$$\begin{cases} \min_{\alpha_i} & (\mathbf{x}^U - \mathbf{x}^L)^\top \Delta (\mathbf{x}^U - \mathbf{x}^L) \\ \text{s.t.} & L + 2\,\text{diag}(\alpha_i) \geq 0 \\ & \alpha_i \geq 0, \quad \forall i, \end{cases}$$

where L is the lower bounding Hessian matrix defined in Method I.4. Because this approach is based on the lower bounding Hessian matrix rather than the exact $\mathbf{x}$-dependent Hessian matrix, the solution found does not correspond to the smallest achievable maximum separation distance, but can be expected to be smaller than when Method I.4 is used.

A comparative study [1,3] of all the methods available for the generation of a diagonal shift matrix found that Methods II.1a, II.1b and II.3 usually give the tightest underestimators. However, Method II.3 is computationally intensive and therefore results in poorer convergence rates than Methods II.1a and II.1b. Since the least computationally expensive techniques for the generation of the diagonal shift matrix, Methods I.1, II.1a and II.1b, are of order $O(n^2)$, the decomposition of the nonconvex terms into a summation of terms involving a smaller number of variables may have a significant impact on the performance of the algorithm.

Overall Convexification/Relaxation Strategy

Based on the rigorous convexification/underestimation schemes for bilinear, trilinear, fractional, fractional trilinear, univariate concave and general nonconvex terms, the overall convex underestimator $\check{F}(\mathbf{x}, \mathbf{w})$ for a twice continuously differentiable function $F(\mathbf{x})$ decomposed according to (2) is

$$\check{F}(\mathbf{x}, \mathbf{w}) = c^\top \mathbf{x} + F_C(\mathbf{x}) + \sum_{i=1}^{bt} b_i w_{B_i}$$

$$+ \sum_{i=1}^{tt} t_i w_{T_i} + \sum_{i=1}^{ft} f_i w_{F_i} + \sum_{i=1}^{ftt} f t_i w_{FT_i}$$

$$+ \sum_{i=1}^{uct} \left(F_{UC_i}(x_{UC_i}^L) \right.$$

$$+ \frac{F_{UC_i}(x_{UC_i}^U) - F_{UC_i}(x_{UC_i}^L)}{x_{UC_i}^U - x_{UC_i}^L}(x_{UC_i} - x_{UC_i}^L) \right)$$

$$+ \sum_{i=1}^{nct} \left(F_{NC_i}(\mathbf{x}) - \sum_{j=1}^{n} \alpha_{ij}(x_j - x_j^L)(x_j^U - x_j) \right), \quad (11)$$

where the notation is as defined for (2). The introduction of the new variables w_{B_i}, w_{T_i}, w_{F_i} and w_{FT_i} is accompanied by the addition of convex inequalities of the type given in (3), (4), (5) and (6). For the trilinear, fractional and fractional trilinear terms, the specific form of these equations depends on the sign of the term coefficients and variable bounds.

The form given by (11) can be used to construct convex underestimators for the objective function and inequality constraints.

Equality Constraints

For nonlinear equality constraints, two different convexification/relaxation schemes are used, depending on the mathematical structure of the function. If the equality $h(\mathbf{x}) = 0$ involves only linear, bilinear, trilinear, fractional and fractional trilinear terms, it is first decomposed into the equivalent equality constraint

$$c^\top \mathbf{x} + \sum_{i=1}^{bt} b_i x_{B_i,1} x_{B_i,2} + \sum_{i=1}^{tt} t_i x_{T_i,1} x_{T_i,2} x_{T_i,3}$$

$$+ \sum_{i=1}^{ft} f_i \frac{x_{F_i,1}}{x_{F_i,2}} + \sum_{i=1}^{ftt} f t_i \frac{x_{FT_i,1} x_{FT_i,2}}{x_{FT_i,3}} = 0, \quad (12)$$

where the notation is as previously defined. (12) is then replaced by

$$c^\top \mathbf{x} + \sum_{i=1}^{bt} b_i w_{B_i} + \sum_{i=1}^{tt} t_i w_{T_i}$$

$$+ \sum_{i=1}^{ft} f_i w_{F_i} + \sum_{i=1}^{ftt} f t_i w_{FT_i} = 0, \quad (13)$$

with the addition of convex inequalities of the type given by (3), (4), (5) and (6). If the nonlinear equality contains at least one convex, univariate concave or general nonconvex term, the convexification/relaxation strategy must first transform the equality constraint $h(\mathbf{x})$ into a set of two equivalent inequality constraints

$$\begin{cases} h(\mathbf{x}) \leq 0 \\ -h(\mathbf{x}) \leq 0, \end{cases} \quad (14)$$

which can then be convexified and underestimated independently using (11).

The transformation of a nonconvex twice-differentiable problem into a convex lower bounding problem described in this section allows the generation of valid and increasingly tight lower bounds on the global optimum solution.

Branching Variable Selection

Once upper and lower bounds have been obtained for all the existing nodes of the branch and bound tree, the region with the smallest lower bound is selected for branching. The partitioning of the solution space can have a significant effect on the quality of the lower bounds obtained because of the strong dependence of the convex underestimators described by (3)–(8) on the variable bounds. It is therefore important to identify the variables which most contribute to the separation between the original problem and the convex lower bounding problem at the current node. Several branching variable selection criteria have been designed for this purpose [1].

Least Reduced Axis Rule

The first strategy leads to the selection of the variable that has least been branched on to arrive at the current node. It is characterized by the largest ratio

$$\frac{x_i^U - x_i^L}{x_{i,0}^U - x_{i,0}^L},$$

where $x_{i,0}^L$ and $x_{i,0}^U$ are the lower and upper bounds on variable x_i at the first node of the branch and bound tree and x_i^L and x_i^U are the current lower and upper bounds on variable x_i.

The main disadvantage of this simple rule is that it does not account for the specificities of the participation of each variable in the problem and therefore cannot accurately identify the critical variables that determine the quality of the underestimators.

Term Measure

A more sophisticated rule is based on the computation of a term measure μ_j^t for term t_j defined as

$$\mu_j^t = t_j(\mathbf{x}^*) - \check{t}_j(\mathbf{x}^*, \mathbf{w}^*), \quad (15)$$

where $t_j(\mathbf{x})$ is a bilinear, trilinear, fractional, fractional trilinear, univariate concave or general nonconvex term, $\check{t}_j(\mathbf{x}, \mathbf{w})$ is the corresponding convex underestimator, $\mathbf{x}^*$ is the solution vector corresponding to the minimum of the convex lower bounding problem, and $\mathbf{w}^*$ is the solution vector for the new variables at the minimum of the convex lower bounding problem. One of the variables participating in the term with the largest measure μ_j^t is selected for branching.

Variable Measure

A third strategy is based on a variable measure μ_i^v which is computed from the term measures μ_j^t. For variable x_i, this measure is

$$\mu_i^v = \sum_{j \in T_i} \mu_j^t, \quad (16)$$

where T_i is the set of terms in which x_i participates. The variable with the largest measure μ_i^v is branched on.

Variable Bound Updates

The effect of the variable bounds on the convexification/relaxation procedure motivates the tightening of the variable bounds. However, the trade-off between tight underestimators generated at a large computational cost and looser underestimators obtained more rapidly must be taken into account when designing a variable bound update strategy. For this reason, one of several approaches can be adopted, depending on the degree of nonconvexity of the problem [1,3]:

- variable bound updates
  - at the beginning of the algorithmic procedure only; or
  - at each iteration;
- bound updates
  - for all variables in the problem; or
  - bound updates for those variables that most affect the quality of the lower bounds as measured by the variable measure μ_i^v.

Two different techniques can be used to tighten the variable bounds. The first is based on the generation and solution of a series of convex optimization problems while the second is an iterative procedure relying on the interval evaluation of the functions in the nonconvex NLP.

Optimization-Based Approach

In the optimization approach, a new lower or upper bound for variable x_i is obtained by solving the convex problem

$$\begin{cases} \min_{\mathbf{x},\mathbf{w}} \text{ or } \max_{\mathbf{x},\mathbf{w}} & x_i \\ \text{s.t.} & \check{f}(\mathbf{x},\mathbf{w}) \leq \overline{f}^* \\ & \check{g}(\mathbf{x},\mathbf{w}) \leq 0 \\ & \check{\mathbf{h}}_N^+(\mathbf{x},\mathbf{w}) \leq 0 \\ & \check{\mathbf{h}}_N^-(\mathbf{x},\mathbf{w}) \leq 0 \\ & \check{\mathbf{h}}_L(\mathbf{x},\mathbf{w}) = 0 \\ & \mathbf{n}(\mathbf{x},\mathbf{w}) \leq 0 \\ & \mathbf{x} \in [\mathbf{x}^L, \mathbf{x}^U], \\ & \mathbf{w} \in [\mathbf{w}^L, \mathbf{w}^U], \end{cases} \quad (17)$$

where $\check{p}(\mathbf{x},\mathbf{w})$ denotes the convex underestimator of function $p(\mathbf{x})$ as defined in (11), $\overline{f}^*$ denotes the current best upper bound on the global optimum solution, $\mathbf{h}_L(\mathbf{x})$ denotes the set of equality constraints which involve only linear, bilinear, trilinear, fractional and fractional trilinear terms, $\mathbf{h}_N^+(\mathbf{x})$ denotes the set of equality constraints that involve other term types and $\mathbf{h}_N^-(\mathbf{x})$ denotes the negative of that set, $\mathbf{n}(\mathbf{x},\mathbf{w})$ denotes the set of additional constraints that arise from the underestimation of bilinear, trilinear, fractional and fractional trilinear terms, and $\mathbf{w}$ is the corresponding set of new variables.

Interval-Based Approach

In the interval-based approach, an iterative procedure is followed for each variable whose bounds are to be updated. The original functions in the problem are used without any transformations. An inequality constraint $g(\mathbf{x}) \leq 0$ is infeasible in the domain $[\mathbf{x}^L, \mathbf{x}^U]$ if its range $[g^L, g^U]$, computed so that $g(\mathbf{x}) \in [g^L, g^U] \; \forall \, \mathbf{x} \in [\mathbf{x}^L, \mathbf{x}^U]$, is such that $g^L > 0$. Similarly, an equality constraint $h(\mathbf{x}) = 0$ is infeasible in this domain if its range $[h^L, h^U]$, computed so that $h(\mathbf{x}) \in [h^L, h^U], \; \forall \, \mathbf{x} \in [\mathbf{x}^L, \mathbf{x}^U]$, is such that $0 \notin [h^L, h^U]$. The variable bounds are updated based on the feasibility of the constraints in the original problem and the additional constraint that the objective function should be less than or equal to the current best upper bound $\overline{f}^*$. The feasible region is therefore defined as

$$F = \left\{ \mathbf{x}: \begin{array}{l} g(\mathbf{x}) \leq 0, \mathbf{h}(\mathbf{x}) = 0, \\ f(\mathbf{x}) \leq \overline{f}^*, \mathbf{x} \in [\mathbf{x}^L, \mathbf{x}^U] \end{array} \right\}.$$

The lower (upper) bound on variable $x_i \in [x_i^L, x_i^U]$ is updated as follows:

PROCEDURE interval-based bound update()
    Set initial bounds $L = x_i^L$ and $U = x_i^U$;
    Set iteration counter $k = 0$;
    Set maximum number of iterations K;
    DO $k < K$
        Compute midpoint $M = (U + L)/2$;
        Set left region $\{\mathbf{x} \in F : x_i \in [L, M]\}$;
        Set right region $\{\mathbf{x} \in F : x_i \in [L, M]\}$;
        Test interval feasibility of left (right region);
        IF feasible,
            Set $U = M$ $(L = M)$;
        ELSE,
            Test interval feasibility of right (left) region;
            IF feasible,
                Set $L = M$ $(U = M)$;
            ELSE,
                Set $L = U$ $(U = L)$;
                Set $U = x_i^U$ $(U = x_i^L)$
        IF $k = 0$ and $L = x_i^U$ $(U = x_i^L)$,
            RETURN(infeasible node);
        Set $k = k + 1$;
    OD;
    RETURN($x_i^L = L$ $(x_i^U = U)$);
END interval-based bound update;

Interval-based bound update procedure

In general, the interval-based bound update strategy is less computationally expensive than the optimization-based approach. However, at the beginning of the branch and bound search, when the bound updates are most critical and the variable ranges are widest, the overestimations inherent in interval computations often lead to looser updated bounds in the interval-based approach than in the optimization-based technique.

Algorithmic Procedure

Based on the developments presented in previous sections, the procedure for the αBB algorithm can be summarized by the following pseudocode:

```
PROCEDURE αBB algorithm()
    Decompose functions in problem;
    Set tolerance ε;
    Set f* = f⁰ = −∞ and f̄* = f̄⁰ = +∞;
    Initialize list of lower bounds {f⁰};
    DO f̄* − f* > ε
        Select node k with smallest lower bound, fᵏ,
        from list of lower bounds;
        Set f* = fᵏ;
        (Optional) Update variable bounds for cur-
        rent node using optimization or interval
        approach;
        Select branching variable;
        Partition to create new nodes;
        DO for each new node i
            Generate convex lower bounding NLP
                Introduce new variables, constraints;
                Linearize univariate concave terms;
                Compute interval Hessian matrices;
                Compute α values;
            Find solution fⁱ of convex lower bound-
            ing NLP;
            IF infeasible or fⁱ > f̄* + ε
                Fathom node;
            ELSE
                Add fⁱ to list of lower bounds;
                Find a solution f̄ⁱ of nonconvex NLP;
                IF f̄ⁱ < f̄*
                    Set f̄* = f̄ⁱ;
        OD;
    OD;
    RETURN(f̄* and variables values at correspond-
    ing node);
END αBB algorithm;
```

A pseudocode for the αBB algorithm

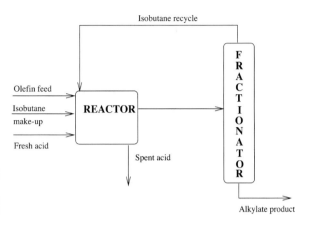

αBB Algorithm, Figure 1
Simplified alkylation process flowsheet

Computational Experience

Significant computational experience with the αBB algorithm has been acquired through the solution of a wide variety of problems involving different types of nonconvexities and up to 16000 variables [1,2,3,4,6,9,12]. These include problems such as pooling/blending, design of reactor networks, design of batch plants under uncertainty [9], stability studies belonging to the class of generalized geometric programming problems, characterization of phase-equilibrium using activity coefficient models, identification of stable molecular conformations and the determination of all solutions of systems of nonlinear equations.

In order to illustrate the performance of the algorithm and the importance of variable bound updates, a medium-size example is presented. The objective is to maximize the profit for the simplified alkylation process presented in [7] and shown in Fig. 1.

An olefin feed (100% butene), a pure isobutane recycle and a 100% isobutane make up stream are introduced in a reactor together with an acid catalyst. The reactor product stream is then passed through a fractionator where the isobutane and the alkylate product are separated. The spent acid is also removed from the reactor. The formulation used here includes 7 variables and 16 constraints, 12 of which are nonlinear. The variables are defined as follows: x_1 is the olefin feed rate in barrels per day; x_2 is the acid addition rate in thousands of pounds per day; x_3 is the alkylate yield in barrels per day; x_4 is the acid strength (weight percent); x_5 is the motor octane number; x_6 is the external isobutane-to-olefin ratio; x_7 is the F-4 performance number. The profit maximization problem is then expressed as:

$$\text{Profit} = -\min(1.715x_1 + 0.035x_1x_6 \\ + 4.0565x_3 + 10.0x_2 - 0.063x_3x_5)$$

subject to:

$$0.0059553571 x_6^2 x_1 + 0.88392857 x_3$$
$$- 0.1175625 x_6 x_1 - x_1 \leq 0,$$
$$1.1088 x_1 + 0.1303533 x_1 x_6$$
$$- 0.0066033 x_1 x_6^2 - x_3 \leq 0,$$
$$6.66173269 x_6^2 + 172.39878 x_5$$
$$- 56.596669 x_4 - 191.20592 x_6 \leq 10000,$$
$$1.08702 x_6 + 0.32175 x_4 - 0.03762 x_6^2$$
$$- x_5 \leq -56.85075,$$
$$0.006198 x_7 x_4 x_3 + 2462.3121 x_2$$
$$- 25.125634 x_2 x_4 - x_3 x_4 \leq 0,$$
$$161.18996 x_3 x_4 + 5000.0 x_2 x_4$$
$$- 489510.0 x_2 - x_3 x_4 x_7 \leq 0,$$
$$0.33 x_7 - x_5 + 44.333333 \leq 0,$$
$$0.022556 x_5 - 0.007595 x_7 \leq 1,$$
$$0.00061 x_3 - 0.0005 x_1 \leq 1,$$
$$0.819672 x_1 - x_3 + 0.819672 \leq 0,$$
$$24500.0 x_2 - 250.0 x_2 x_4 - x_3 x_4 \leq 0,$$
$$1020.4082 x_4 x_2 + 1.2244898 x_3 x_4$$
$$- 100000 x_2 \leq 0,$$
$$6.25 x_1 x_6 + 6.25 x_1 - 7.625 x_3 \leq 100000,$$
$$1.22 x_3 - x_6 x_1 - x_1 + 1 \leq 0,$$
$$1500 \leq x_1 \leq 2000,$$
$$1 \leq x_2 \leq 120,$$
$$3000 \leq x_3 \leq 3500,$$
$$85 \leq x_4 \leq 93,$$
$$90 \leq x_5 \leq 95,$$
$$3 \leq x_6 \leq 12,$$
$$145 \leq x_7 \leq 162.$$

The maximum profit is $1772.77 per day, and the optimal variable values are $x_1^* = 1698.18$, $x_2^* = 53.66$, $x_3^* = 3031.30$, $x_4^* = 90.11$, $x_5^* = 95.00$, $x_6^* = 10.50$, $x_7^* = 153.53$. In this example, variable bound tightening is performed using the optimization-based approach. An update of all the variable bounds therefore involves the solution of 14 convex NLPs. The computational cost is significant and may not always be justified by the corresponding decrease in number of iterations. Two extreme tightening strategies were used to illustrate this trade-off: an update of all variable bounds at the onset of the algorithm only ('Single Up'), or an update of all bounds at each iteration of the αBB algorithm ('One Up/Iter'). An intermediate strategy might involve bound updates for those variables that affect the underestimators most significantly or bound updates at only a few levels of the branch and bound tree. The results of runs performed on an HP9000/730 are summarized in the table below. t_U denotes the percentage of CPU time devoted to the construction of the convex underestimating problem.

Although the approach relying most heavily on variable bound updates results in tighter underestimators, and hence a smaller number of iterations, the time requirements for each iteration are significantly larger than when no bounds updates are performed. Thus, the overall CPU requirements often increase when all variable bounds are updated at each iteration.

Meth	Single up			One Up/Iter		
	Iter.	CPU sec.	t_U (%)	Iter.	CPU sec.	t_U (%)
I.1	74	37.5	0.5	31	41.6	0.0
I.2a	61	30.6	1.6	25	37.2	0.2
I.2b	61	29.2	1.0	25	35.4	0.1
I.3	69	32.8	1.9	25	31.5	0.2
I.4	61	31.6	1.4	25	33.1	0.2
I.5	61	32.8	12.3	25	36.7	1.7
I.6	59	32.9	1.4	25	32.8	0.5
II.1a	56	24.9	0.3	30	36.5	0.3
II.1b	38	13.6	1.7	17	19.9	0.5
II.2	62	32.7	0.6	25	34.5	0.3
II.3	54	21.8	16.7	23	30.4	5.0

Alkylation process design results

In order to determine the best technique for the construction of convex underestimators, the percentage of computational effort dedicated to this purpose, t_U, is tracked. As can be seen in the above table, the generation of the convex lower bounding does not consume a large share of the computational cost, regardless of the method. It is, however, significantly larger for Methods I.5 and II.3 as they require the solution of a polynomial and a semidefinite programming problem respectively. t_U decreases when bound updates are performed at each iteration as a large amount of time is spent solving the bound updates problems. In this example, the scaled Gershgorin approach with $d_i = (x_i^U - x_i^L)$ (Method II.1b) gives the best results both in terms of number of iterations and CPU time.

Conclusions

The αBB algorithm is guaranteed to identify the global optimum solution of problems belonging to the broad class of twice continuously differentiable NLPs. It is a branch and bound approach based on a rigorous convex relaxation strategy, which involves the decomposition of the functions into a sum of terms with special mathematical structure and the construction of different convex underestimators for each class of term. In particular, the treatment of general nonconvex terms requires the analysis of their Hessian matrix through interval arithmetic. Efficient branching and variable bound update strategies can be used to enhance the performance of the algorithm.

See also

- ▶ Bisection Global Optimization Methods
- ▶ Continuous Global Optimization: Applications
- ▶ Continuous Global Optimization: Models, Algorithms and Software
- ▶ Convex Envelopes in Optimization Problems
- ▶ D.C. Programming
- ▶ Differential Equations and Global Optimization
- ▶ DIRECT Global Optimization Algorithm
- ▶ Eigenvalue Enclosures for Ordinary Differential Equations
- ▶ Generalized Primal-relaxed Dual Approach
- ▶ Global Optimization Based on Statistical Models
- ▶ Global Optimization in Batch Design Under Uncertainty
- ▶ Global Optimization in Binary Star Astronomy
- ▶ Global Optimization in Generalized Geometric Programming
- ▶ Global Optimization Methods for Systems of Nonlinear Equations
- ▶ Global Optimization in Phase and Chemical Reaction Equilibrium
- ▶ Global Optimization Using Space Filling
- ▶ Hemivariational Inequalities: Eigenvalue Problems
- ▶ Interval Analysis: Eigenvalue Bounds of Interval Matrices
- ▶ Interval Global Optimization
- ▶ MINLP: Branch and Bound Global Optimization Algorithm
- ▶ MINLP: Global Optimization with αBB
- ▶ Reformulation-linearization Methods for Global Optimization
- ▶ Reverse Convex Optimization
- ▶ Semidefinite Programming and Determinant Maximization
- ▶ Smooth Nonlinear Nonconvex Optimization
- ▶ Topology of Global Optimization

References

1. Adjiman CS, Androulakis IP, Floudas CA (1998) A global optimization method, αBB, for general twice-differentiable constrained NLPs – II. Implementation and computational results. Comput Chem Eng 22:1159
2. Adjiman CS, Androulakis IP, Maranas CD, Floudas CA (1996) A global optimization method, αBB, for process design. Comput Chem Eng 20:S419–S424
3. Adjiman CS, Dallwig S, Floudas CA, Neumaier A (1998) A global optimization method, αBB, for general twice-differentiable constrained NLPs – I. Theoretical advances. Comput Chem Eng 22:1137
4. Adjiman CS, Floudas CA (1996) Rigorous convex underestimators for twice-differentiable problems. J Global Optim 9:23–40
5. Al-Khayyal FA, Falk JE (1983) Jointly constrained biconvex programming. Math Oper Res 8:273–286
6. Androulakis IP, Maranas CD, Floudas CA (1995) αBB: A global optimization method for general constrained nonconvex problems. J Global Optim 7:337–363
7. Bracken J, McCormick GP (1968) Selected applications of nonlinear programming. Wiley, New York
8. Deif AS (1991) The interval eigenvalue problem. Z Angew Math Mechanics 71:61–64
9. Harding ST, Floudas CA (1997) Global optimization in multiproduct and multipurpose batch design under uncertainty. Industr Eng Chem Res 36:1644–1664
10. Hertz D (1992) The extreme eigenvalues and stability of real symmetric interval matrices. IEEE Trans Autom Control 37:532–535
11. Kharitonov VL (1979) Asymptotic stability of an equilibrium position of a family of systems of linear differential equations. Differential Eq:1483–1485
12. Maranas CD, Floudas CA (1994) Global minimum potential energy conformations of small molecules. J Global Optim 4:135–170
13. Maranas CD, Floudas CA (1995) Finding all solutions of nonlinearly constrained systems of equations. J Global Optim 7:143–182

14. Maranas CD, Floudas CA (1997) Global optimization in generalized geometric programming. Comput Chem Eng 21:351–370
15. McCormick GP (1976) Computability of global solutions to factorable nonconvex programs: part I – Convex underestimating problems. Math Program 10:147–175
16. McDonald CM, Floudas CA (1994) Decomposition based and branch and bound global optimization approaches for the phase equilibrium problem. J Global Optim 5:205–251
17. McDonald CM, Floudas CA (1995) Global optimization and analysis for the Gibbs free energy function for the UNIFAC, Wilson, and ASOG equations. Industr Eng Chem Res 34:1674–1687
18. McDonald CM, Floudas CA (1995) Global optimization for the phase and chemical equilibrium problem: Application to the NRTL equation. Comput Chem Eng 19:1111–1141
19. McDonald CM, Floudas CA (1995) Global optimization for the phase stability problem. AIChE J 41:1798–1814
20. McDonald CM, Floudas CA (1997) GLOPEQ: A new computational tool for the phase and chemical equilibrium problem. Comput Chem Eng 21:1–23
21. Mori T, Kokame H (1994) Eigenvalue bounds for a certain class of interval matrices. IEICE Trans Fundam E77-A:1707–1709
22. Neumaier A (1992) An optimality criterion for global quadratic optimization. J Global Optim 2:201–208
23. Rohn J (1996) Bounds on eigenvalues of interval matrices. Techn Report Inst Computer Sci Acad Sci Prague 688
24. Stephens C (1997) Interval and bounding Hessians. In: Bonze IM et al (eds) Developments in Global Optimization. Kluwer, Dordrecht, pp 109–199
25. Vandenberghe L, Boyd S (1996) Semidefinite programming. SIAM Rev 38:49–95

Alternative Set Theory

AST

PETR VOPĚNKA, KATEŘINA TRLIFAJOVÁ
Charles University, Prague, Czech Republic

MSC2000: 03E70, 03H05, 91B16

Article Outline

Keywords
Classes, Sets and Semisets
Infinity
Axiomatic System of AST
Rational and Real Numbers
Infinitesimal Calculus
Topology
  Basic Definitions
Motion
Utility Theory
Conclusion
See also
References

Keywords

Sets; Semisets; Infinity; Countability; Continuum; Topology; Indiscernibility; Motion; Utility theory

Alternative set theory has been created and, together with his colleagues at Charles University, developed by P. Vopěnka since the 1970s. In agreement with Husserl's phenomenology, he based his theory on the natural world and the human view thereof.

The most important for any set theory is the way it treats infinity. A different approach to infinity forms the key difference between AST and classical set theories based on the *Cantor set theory* (CST). Cantor's approach led to the creation of a rigid, abstract world with an enormous scale of infinite cardinalities while Vopěnka's infinity, based on the notion of horizon, is more natural and acceptable.

Another source of inspiration were nonstandard models of Peano arithmetics with infinitely large (nonstandard) numbers. The way to build them in AST is easy and natural.

The basic references are [9,10,11].

Classes, Sets and Semisets

AST, as well as CST, builds on notions of 'set', 'class', 'element of a set' and, in addition, introduces the notion of 'semiset'. A *class* is the most general notion used for any collection of distinct objects. *Sets* are such classes that are so clearly defined and clean-cut that their elements could be, if necessary, included in a list. *Semisets* are classes which are not sets, because their borders are vague, however, they are parts of sets. For example, all living people in the world form a class—some are being born, some are dying, we do not know where all of them are. The citizens of Prague, registered at the given moment in the register, form a set. However, all the beautiful women in Prague or brave men in Prague form a semiset, since it is not clear who belongs to this collection and who not.

In the real world, we may find many other semisets. Almost each property defines a semiset of objects, e. g., people who are big, happy or sick. Many properties are naturally connected with a vagueness. Also, what we see and perceive can be vague and limited by a horizon. Objects described in this way may form a semiset, e. g. flowers I can see in the blooming meadow, all my friends, sounds I can hear.

Infinity

This interpretation differs from the normal one and corresponds more to the etymological origin of the word infinity. We will call *finite* those classes any part of which is surveyable and forms a set. Any finite class is a set.

$$\mathrm{Fin}(X) \Leftrightarrow (\forall Y)(Y \subseteq X \Longrightarrow \mathrm{Set}(Y)).$$

On the other side, *infinite classes* include ungrasped parts, semisets. This phenomenon may occur also when watching large sets in the case when it is not possible to capture them clearly as a whole.

There are two different forms of infinity traditionally called denumerability and continuum.

A countable (denumerable) class, in a way, represents a road towards the horizon. Its beginning is clear and definite but it comes less and less clear and its end loses in a vagueness. A *countable class* is defined as an infinite class with a linear ordering such that each initial part (segment) is finite. For instance, a railway track with cross-ties leading straight to the horizon, days of our life we are to live or ever smaller and smaller reflections in two mirrors facing each other. The most important example is a class of natural numbers that will be discussed later.

The phenomenon of denumerability corresponds to a road towards the horizon. Though we get to the last point we can see, we can still go a bit further, the road will not disappear immediately. People have always tried to look a bit behind the horizon, to gain understanding and to overcome it in their thinking. This experience is expressed here by the important *axiom of prolongation* (see Axiom A6).

The other type of infinity, *continuum*, is based on the following experience. If we watch an object, however, are not able to distinguish individual elements which form it since they lie beyond the horizon of our perception. For example, the class of all geometric points in the plane, class of all atoms forming a table or grains of sand which together form a heap.

In fact the classical infinite mathematics, when applied to the real world, then solely to the above two types of infinity.

The intention of AST is to built on the natural world and human intuition. There is no reason for other types of infinity which are enforced in CST by its assumption that natural numbers form a set and that a power set is a set. That is why there are only two infinite cardinalities in AST: denumerability and continuum (see Axiom A8).

All examples from mathematical and real worlds are intentionally set out here together. They serve the purpose of inspiration to see where the idea of infinity comes from, they should be kept in mind when one deals with infinity.

The mathematical world is an ideal one, it is a perfect world of objective truths abstracted from all that is external. There is only little space for subjectivity of perception in it. That is why not all semisets from the real world may be interpreted directly.

The axiomatic system bellow describes that part of the AST which can be expressed in a strictly formal way. This basis provides space for extending AST by semisets which are parts of big, however, classically finite sets and thus make a lot of applications possible.

Axiomatic System of AST

[3] The language of AST uses symbols $\in$ and $=$, symbols $X, Y, Z, \ldots$ for class variables and symbols $x, y, z, \ldots$ for set variables. *Sets* are created by iteration from the empty set by Axiom A3. *Classes* are defined by formulas by Axiom A2. Every set is a class. Formally, a set is a class that is a member of another class:

$$\mathrm{Set}(X) \Leftrightarrow (\exists Y)(X \in Y).$$

AST is a theory with the following axioms:
- A1 (*extensionality*). $(X = Y) \Leftrightarrow (\forall Z)(Z \in X) \Leftrightarrow (Z \in Y)$;
- A2 (*existence of classes*). If ψ is a formula, then

$$(\exists Y)(\forall x)(x \in Y \Leftrightarrow \psi(x, X_1, \ldots, X_n));$$

- A3 (*existence of sets*).

$$\text{Set}(\emptyset) \wedge (\forall x, y) \text{Set}(x \cup \{y\}).$$

A *set-formula* is a formula in which only set variables and constants occur.
- A4 (*induction*). If ψ is a set-formula, then $(\psi(\emptyset) \wedge (\forall x, y)(\psi(x) \Rightarrow \psi(x \cup \{y\}))) \Rightarrow (\forall x) \psi(x)$.
- A5 (*regularity*). If ψ is a set-formula, then $(\exists x)\psi(x) \Rightarrow (\exists x)(\psi(x) \wedge (\forall y \in x)\neg\psi(y))$.

As usual, the class of *natural numbers* N is defined in the von Neumann way

$$N = \left\{x: \begin{array}{c}(\forall y \in x)(y \subseteq x) \\ \wedge (\forall y, z \in x)(y \in z \vee y = z \vee z \in y)\end{array}\right\}$$

The class of *finite natural numbers* (FN) consists of the numbers represented by a finite set. They are accessible, easy to overlook and lie before the horizon:

$$FN = \{x \in N: \text{ Fin}(x)\}$$

FN forms a countable class in the sense described above. The class FN correspond to classical natural numbers and the class N to their nonstandard model. Both N and FN satisfy the axioms of Peano arithmetic.

Two classes X, Y are *equivalent* if there is a one-one mapping of X onto Y, i.e. $X \approx Y$.
- A6 (*prolongation*). Every countable function can be prolonged to a function which is a set, i.e. $(\forall F)((\text{Fnc}(F) \wedge (F \approx FN)) \Rightarrow (\exists f)(\text{Fnc}(f) \wedge F \subseteq f))$.

An easy corollary is that a countable class is a semiset. Also FN is a semiset and it can be prolonged to a set which is an element of N and which is greater than all finite natural numbers and so it represents an infinitely large natural number. Consequently, the class N is not countable.

The *universal class* V includes all sets created by iteration from the empty set.
- A7 (*choice*). The universal class V can be well ordered.
- A8 (*two cardinalities*). Every two infinite classes that are not countable are equivalent.

Thus, any infinite class is either equivalent to FN or N.

Using ultrapowers, the relative consistency of AST can be proved.

Rational and Real Numbers

Rational numbers Q are constructed in the usual way from N as the quotient field of the class $N \cup \{-n; n \in N\}$. Because N includes infinitely large numbers, Q includes infinitely small numbers.

Finite rational numbers FQ are similarly constructed from finite natural numbers FN. They include quantities that are before the horizon with respect to distance and depth. Surely $FQ \subseteq Q$.

We define that $x, y \in Q$ are *infinitely near* by

$$x \doteq y \Leftrightarrow (\forall n \in FN) \begin{cases} |x - y| < \frac{1}{n} \\ \vee (x > n \wedge y > n) \\ \vee (x < -n \wedge y < -n). \end{cases}$$

This relation is an equivalence. The corresponding partition classes are called *monads*. For $x \in Q$

$$\text{Mon}(x) = \{y: y \doteq x\}.$$

Rational numbers x that are elements of $\text{Mon}(0)$, i.e. ($x \doteq 0$), are *infinitely small*. All monads are of the same nature except for the two limit ones. These consists of *infinitely large positive* and *negative* numbers. The class of bounded rational numbers is

$$BQ = \{x \in Q: (\exists n)((n \in FN) \wedge (|x| < n))\}$$

Now, it is easy and natural to construct *real numbers*:

$$\mathbb{R} = \{\text{Mon}(x): x \in BQ\}.$$

Real numbers built in this way display the same characteristics as real numbers in CST.

This motivation for expressing real numbers as monads of rational numbers corresponds rather to etymology than to the traditional interpretation. Rational numbers are constructed by reason, perfectly exact; their existence is purely abstract. On the other hand, real numbers are more similar to those that are used in the real world. If we say: one eighth of a cake, we surely do not expect it to be the ideal eighth, it is rather a portion which differs from the ideal one by a difference which is beyond the horizon of our perception. A similar situation occurs in the case of a pint of milk or twenty miles.

Infinitesimal Calculus

[12] Infinitesimal calculus in AST is based on the same point of view and intuition as that of its founders, I. Newton and G.W. Leibniz. It is so because infinitely small or infinitesimal quantities are naturally available in AST. For example, the limit of a function and the continuity in $a \in Q$ are defined, respectively, by:

$$\lim_{x \to a} f(x) = b$$
$$\Leftrightarrow (\forall x)((x \doteq a \wedge x \neq a) \Rightarrow f(x) \doteq b));$$
$$(\forall x)(x \doteq a \Rightarrow f(x) \doteq f(a)).$$

This topic is discussed in detail in [9]. As a method, these definitions were successfully used for teaching students.

Topology

Classes described by arbitrary formulas can be complex and difficult to capture. The easiest are sets, also classes described by using set-formulas, so-called *set-definable classes* (Sd-classes) can be described well. Semisets which are defined by a positive property (big, blue or happy and also distinguishable or to be a finite natural number) can be described as a countable union of Sd-classes, the so-called σ-classes. On the other hand, classes whose definition is based on negation (not big, not happy, indistinguishable), are the so-called π-classes—countable intersections of Sd-classes. A class which is at the same time π and σ is an Sd-class. Using combinations of π and σ, a set hierarchy can be described.

One of the most important tasks of mathematics is to handle the notion of the continuum. AST is based on the assumption that this phenomenon is caused by that of the indiscernibility of elements of the observed class. That is why, for the study of topology, the basic notion is a certain *relation of indiscernibility* ($\equiv$). Two elements are indiscernible if, when observed, available criteria that might distinguish them fail. It is a negative feature, therefore it must be a π-class. The relation of indiscernibility is naturally reflexive and symmetric. In pure mathematics, it is in addition transitive (because FN is closed under addition), thus it is an equivalence. This relation must also be compact, i. e. for each infinite set $u \subseteq \text{dom}(\equiv)$ there are $x, y \in u$ such that $x \neq y \wedge x \equiv y$. The corresponding topological space is a compact metric space.

The relation of infinite nearness in rational numbers represents a special case of equivalence of indiscernibility.

Monads and *figures* correspond to phenomena of points and shapes, respectively:

$$\text{Mon}(x) = \{y : y \equiv x\},$$
$$\text{Fig}(X) = \{y : (\exists x \in X)(y \equiv x)\}.$$

Basic Definitions

Two classes X, Y are *separable*, $\text{Sep}(X, Y) \Leftrightarrow (\exists Z)(\text{Sd}(Z) \wedge \text{Fig}(X) \subseteq Z \wedge \text{Fig}(Y) \cap Z = \emptyset)$.

A *closure* $\overline{X}$ of a class X is defined as $\overline{X} = \{x : \neg \text{Sep}(\{x\}, X)\}$.

A class X is *closed* if $X = \overline{X}$.

A set u is *connected* if $(\forall w)(\emptyset \neq = u \Rightarrow \text{Fig}(w) \cap (u-w) \neq \emptyset)$.

It is quite easy to prove basic topological theorems. Also proofs of some classical theorems are much simpler here. For instance the *Sierpinski theorem*: If v is a connected set then $\text{Fig}(v)$ cannot be expressed as a countable union of disjoint closed sets.

The *fundamental indiscernibility* $\stackrel{\circ}{=}_c$ is defined as follows. If c is a set then $x \stackrel{\circ}{=}_c y$ if for any set-formula ψ with the constants from c and for any x, it is $\psi(x) \Leftrightarrow \psi(y)$.

This relation has a special position. For any relation of indiscernibility $\equiv$ there is a set c such that $\stackrel{\circ}{=}_{\{c\}}$ is finer than $\equiv$ i. e. $\stackrel{\circ}{=}_{\{c\}} \subseteq \equiv$.

Motion

Unlike classical mathematics, the motion is captured in AST by any relation of indiscernibility $\equiv$.

Everybody knows the way films work. Pictures coming one after another are almost indiscernible from each other, however, when shown in a rapid sequence, the pictures start to move. The continuous motion may be viewed like this, as a sequence of indiscernible stages in certain time intervals.

A function d is a *motion of a point* in the time $\delta \in N$ if $\text{dom}(f) = \delta \wedge (\forall \alpha < \delta)(d(\alpha) \equiv d(\alpha+1))$.

If $\delta \in FN$ then the point does not move, it can move only in an infinitely big time interval.

A sequence $\{d(\alpha): \alpha \in \text{dom}(d)\}$ is a sequence of states. The number $\delta = \text{dom}(d)$ is the number of moments and $\text{rng}(d)$ is the trace of a moving point.

A trace is a connected set and for each nonempty connected set u there is a motion of a point such that u is the trace of d.

A *motion of a set* is defined similarly, only the last condition is different: $(\forall \alpha < \delta)(\text{Fig}(d(\alpha)) = \text{Fig}(d(\alpha+1)))$.

The following theorem is proved in [10,11]: Each motion of a set may be divided into motions of points. This does not involve only the mechanical motion, but any motion describing a continuous change. Thus, for example, even the growth of a tree from a planted seed may be divided into movements of individual points while all of their initial stages are already contained in the seed. In addition, it is possible to describe conditions under which such a change is still continuous.

Utility Theory

[7] The utility theory is one of nice examples of applying AST. Its aim is to find a valuation of elements of a class S. There is a preference relation $\succ$ on linear combinations of elements of S with finite rational coefficients, i. e. on the class

$$\left\{ \sum_{i=1}^{n} \alpha_i u_i : \begin{array}{c} (n \in FN) \\ \wedge (\forall i)((i \leq n)(u_i \in S) \wedge (\alpha_i \in FQ)) \\ \wedge \sum_{i=1}^{n} \alpha_i = 1 \end{array} \right\}.$$

An interpretation of a combination is a game in which every u_i can be won with the probability α_i. The *preference relation* $\succ$ declares which of the two games is preferred.

The *valuation* is a function F from the class S to Q for which

$$\sum_{i=1}^{n} \alpha_i u_i \succ \sum_{j=1}^{m} \beta_j u_j \Leftrightarrow \sum_{i=1}^{n} \alpha_i F(u_i) > \sum_{j=1}^{m} \beta_j F(u_j).$$

It is not necessary to require the so-called Archimedes property on the relation of preference thanks to the possibility of using infinitely small and infinitely large rational numbers. It is possible to capture finer and more complex relations than in classic mathematics, e. g. the fact that the value of one element is incomparably higher than that of another element or it is possible to compare infinitely small differences of values.

For each class S with a preference relation a valuation may be found. Such a valuation is not uniquely defined, it is possible to construct it so that $\text{rng}(F) \subseteq N$.

Conclusion

The aim of this short survey is to demonstrate the basic ideas of AST. Yet, there are other areas of mathematics which were studied in it, for instance measurability [8], ultrafilters [6], endomorphic universes [5] and automorphisms of natural numbers [2], representability [1] metamathematics [3] and models of AST [4].

See also

- ▶ Boolean and Fuzzy Relations
- ▶ Checklist Paradigm Semantics for Fuzzy Logics
- ▶ Finite Complete Systems of Many-valued Logic Algebras
- ▶ Inference of Monotone Boolean Functions
- ▶ Optimization in Boolean Classification Problems
- ▶ Optimization in Classifying Text Documents

References

1. Mlček J (1979) Valuation of structures. Comment Math Univ Carolinae 20:681–695
2. Mlček J (1985) Some automorphisms of natural numbers in AST. Comment Math Univ Carolinae 26:467–475
3. Sochor A (1992) Metamathematics of AST. From the logical point of view 1:61–75
4. Sochor A, Pudlák P (1984) Models of AST. J Symbolic Logic 49:570–585
5. Sochor A, Vopěnka P (1979) Endomorfic universes and their standard extensions. Comm Math Univ Carolinae 20:605–629
6. Sochor A, Vopěnka P (1981) Ultrafilters of sets. Comment Math Univ Carolinae 22:698–699
7. Trlifajová K, Vopěnka P (1985) Utility theory in AST. Comment Math Univ Carolinae 26:699–711
8. Čuda K (1986) The consistency of measurability of projective semisets. Comment Math Univ Carolinae 27:103–121
9. Čuda K, Sochor A, Vopěnka P, Zlatoš P (1989) Guide to AST. Proc. First Symp. Mathematics in AST, Assoc. Slovak Mathematicians and Physicists, Bratislava
10. Vopěnka P (1979) Mathematics in AST. Teubner, Leipzig
11. Vopěnka P (1989) Introduction to mathematics in AST. Alfa Bratislava, Bratislava
12. Vopěnka P (1996) Calculus infinitesimalis-pars prima. Práh Praha, Praha

Approximation of Extremum Problems with Probability Functionals

APF

RIHO LEPP

Tallinn Technical University, Tallinn, Estonia

MSC2000: 90C15

Article Outline

Keywords
See also
References

Keywords

Discrete approximation; Probability functionals

To ensure a certain level of reliability for the solution of an extremum problem under uncertainty it has become a spread approach to introduce probabilistic (chance) cost and/or constraints into the model. The stability analysis of chance constraint problems is rather complicated due to complicated properties of the *probability function* $v_t(x)$, defined as

$$v_t(x) = \mathsf{P}\{s: f(x,s) \le t\}. \tag{1}$$

Here $f(x, s)$ is a real valued function, defined on $\mathbf{R}^r \times \mathbf{R}^r$, t is a fixed level of reliability, s is a random vector and P denotes probability. The function $v_t(x)$ is never convex, only in some cases (e. g., $f(x, s)$ linear in s and distribution of the random parameter s normal), it is quasiconvex. Note that for a fixed x function $v_t(x)$, as a function of t, is the distribution function of the random variable $f(x, s)$.

The 'inverse', the *quantile function* $w_\alpha(x)$, to the probability function $v_t(x)$ is defined in such a way that the probability level α, $0 < \alpha < 1$, is fixed earlier, and the purpose is to minimize the reliability level t:

$$w_\alpha(x) = \min_t \{t: \mathsf{P}\{s: f(x,s) \le t\} \ge \alpha\}. \tag{2}$$

Varied examples of extremum problems with probability and quantile functions are presented in [7] and in [8]. Some of these models have such a complicated structure, see [8, Chap. 1.8], about correction of a satellite orbit, that we are forced to look for a solution x from a certain class of strategies, that means, the solution x itself depends on the random parameter s, $x = x(s)$.

This class of probability functions was introduced to *stochastic programming* by E. Raik, and lower semicontinuity and continuity properties of $v_t(x)$ and $w_\alpha(x)$ in Lebesgue L^p-spaces, $1 \le p < \infty$, were studied in [12]. Simultaneously, in [4] problems with various classes of solutions $x(s)$ (measurable, continuous, linear, etc) were considered. Since the paper [4] solutions $x(s)$ are called *decision rules*, and we will follow also this terminology.

Differently from [4], here we will consider approximation of a decision rule $x(s)$ by sequences of vectors $\{x_n\}$, $x_n = (x_{1n}, \ldots, x_{nn})$, $n = 1, 2, \ldots$, with increasing dimension in order to maximize the value of the probability functional $v(x)$ under certain set C of decision rules. It will be assumed that the set C will be bounded in the space $L^1(S, \Sigma, \sigma) = L^1(\sigma)$ of integrable functions $x(s)$, $x \in L^1(\sigma)$:

$$\max_{x \in C} v_t(x) = \max_{x \in C} \mathsf{P}\{s: f(x(s), s) \le t\}. \tag{3}$$

Here S is the support of random variable s with distribution (probability measure) $\sigma(\cdot)$ and Σ denotes the sigma-algebra of Borel measurable sets from $\mathbf{R}^r$.

Due to technical reasons we are forced to assume that the random parameter s has bounded support $S \subset \mathbf{R}^r$, diam $S < \infty$, and its distribution σ is atomless,

$$\sigma\{s: |s - s_0| = \text{const}\} = 0, \quad \forall s_0 \in \mathbb{R}^r. \tag{4}$$

Since the problem (3) is formulated in the function space $L^1(\sigma)$ of σ-integrable functions, the first step in its solution is the approximation step where we will replace the initial problem (3) by a sequence of finite-dimensional optimization problems with increasing dimension. Second step, solution methods were considered in a series of papers of the author (see, e. g., [9]), where the gradient projection method was suggested together with simultaneous Parzen–Rosenblatt kernel-type smooth approximation of the discontinuous integrand from (1).

There are several ways to divide the support S of the probability measure σ into smaller parts in discretization, e. g., taking disjoint subsets S_j, $j = 1, \ldots, k$, of S

from the initial sigma-algebra Σ as in [11], or using in the partition of S only convex sets from Σ, as in [5].

We will divide the support S into smaller parts by using only sets A_{in}, $i = 1, \ldots, n$, $n \in N = \{1, 2, \ldots\}$, with σ-measure zero of their boundary, i.e., σ (intA_{in}) = σ (A_{in}) = σ (clA_{in}), where int A and cl A denote topological interior and closure of a set A, respectively. Such division is equivalent to *weak convergence* of a sequence of *discrete measures* $\{(m_n, s_n)\}$ to the initial probability measure σ, see, e.g. [14]:

$$\sum_{i=1}^{n} h(s_{in}) m_{in} \to \int_{S} h(s) \sigma(ds), \quad n \in \mathbb{N}, \quad (5)$$

for any continuous on S function $h(s)$, $h \in C(S)$.

The usage of the weak convergence of discrete measures in stochastic programming has its disadvantages and advantages. An example in [13] shows that, in general, the stability of a probability function with respect to weak convergence cannot be expected without additional smoothness assumptions on the measure σ. This is one of the reasons, why we should use only continuous measures with the property (4). An advantage of the usage of the weak convergence is that it allows us to apply in the approximation process instead of conditional means [11] the more simple, grid point approximation scheme.

Since the functional $v_t(x)$ is not convex, we are not able to exploit in the stability analysis of discrete approximation of the problem (3) the more convenient, weak topology, but only the strong (norm) topology. As the first step we will approximate $v_t(x)$ so, that the discrete analogue of continuous convergence of a sequence of approximate functionals will be guaranteed.

Schemes of *stability analysis* (e.g., finite-dimensional approximations) of extremum problems in Banach spaces require from the sequence of solutions of 'approximate' problems certain kind of compactness. Assuming that the constraint set C is compact in $L^1(\sigma)$, we, as the second step, will approximate the set C by a sequence of finite-dimensional sets $\{C_n\}$ with increasing dimension so, that the sequence of solutions of approximate problems is compact in a certain (discrete convergence) sense in $L^1(\sigma)$. Then the approximation scheme for the discrete approximation of (3) will follow formed schemes of approximation of extremum problems in Banach spaces, see e.g. [2,3,15].

Redefine the functional $v_t(x)$ by using the Heaviside zero-one function χ:

$$v_t(x) = \int_{S} \chi(t - f(x(s), s)) \sigma(ds), \quad (6)$$

where

$$\chi(t - f(x(s), s)) = \begin{cases} 1 & \text{if } f(x(s), s) \leq t, \\ 0 & \text{if } f(x(s), s) > t. \end{cases}$$

Since the integrand $\chi(\cdot)$ itself, as a zero-one function, is discontinuous, we will assume that the function $f(x, s)$ is continuous both in (x, s) and satisfies following growth and 'platform' conditions:

$$|f(x, s)| \leq a(s) + \alpha |x|, \quad (7)$$
$$a \in L^1(\sigma), \quad \alpha > 0,$$
$$\sigma \{s: f(x, s) = \text{const}\} = 0, \quad (8)$$
$$\forall (x, s) \in \mathbb{R}^r \times S.$$

The continuity assumption is technical in order to simplify the description of the approximation scheme below. The growth condition (7) is essential: without it the superposition operator $f(x) = f(x(s), s)$ will not map an element from L^1 to L^1 (is even not defined). Condition (8) means that the function $f(x, s)$ should not have horizontal platforms with positive measure.

Constraint set C is assumed to be a set of integrable functions $x(s)$, $x \in L^1(\sigma)$, with properties

$$\int_{S} |x(s)| \sigma(ds) \leq M < \infty, \quad \forall x \in C \quad (9)$$

for some $M > 0$ (C is bounded in $L^1(\sigma)$);

$$\int_{D} |x(s)| \leq K\sigma(D), \quad \forall x \in C, \quad D \in \Sigma \quad (10)$$

for some $K > 0$;

$$(x(s) - x(t), s - t) \geq 0 \quad \text{for a.a. } s, t \in S \quad (11)$$

(functions $x \in C$ are monotone almost everywhere and a.a. denotes abbreviation of 'almost all').

Conditions (9), (10) guarantee that the set C is weakly compact (i.e., compact in the (L^1, L^∞)-topology, see, e.g., [6, Chap. 9.1.2]). Condition (11) guarantees now, following [1, Lemma 3], that the set C is strongly compact in $L^1(\sigma)$. Then, following [11], we can conclude that assumptions (7)–(11) together with atomless assumption (4) for the measure σ guarantee the existence of a solution of problem (3) in the Banach space $L^1(\sigma)$ of σ-integrable functions (the cost functional $v_t(x)$ is continuous in x and the constraint set C is compact in $L^1(\sigma)$).

Since approximate problems will be defined in $\mathbf{R}^m$, we should define a system of connection operators $\mathcal{P} = \{p_n\}$ between spaces $L^1(\sigma)$ and $\mathbf{R}^m$, $n \in \mathbf{N}$. In L^p-spaces, $1 \leq p \leq \infty$, systems of connection operators should be defined in a piecewise integral form (as conditional means):

$$(p_n x)_{in} = \sigma(A_{in})^{-1} \int_{A_{in}} x(s)\, \sigma(ds), \qquad (12)$$

where $i = 1, \ldots, n$, and sets A_{in}, $i = 1, \ldots, n$, $n \in \mathbf{N}$, that define connection operators (12), satisfy following conditions A1)–A7):

A1) $\sigma(A_{in}) > 0$;
A2) $A_{in} \cap A_{jn} = \emptyset$, $i \neq j$;
A3) $\cup_{i=1}^n A_{in} = S$;
A4) $\sum_{i=1}^n |m_{in} - \sigma(A_{in})| \to 0$, $n \in \mathbf{N}$;
A5) $\max_i \operatorname{diam} A_{in} \to 0$, $n \in \mathbf{N}$;
A6) $s_{in} \in A_{in}$;
A7) $\sigma(\operatorname{int} A_{in}) = \sigma(A_{in}) = \sigma(\operatorname{cl} A_{in})$.

Remark 1 Weak convergence (5) is equivalent to the partition $\{\mathcal{A}_n\}$ of S, $\mathcal{A}_n = \{A_{1n}, \ldots, A_{nn}\}$, with properties A1)–A7), see [14].

Remark 2 Collection of sets $\{A_{in}\}$ with the property A7) constitutes an algebra $\Sigma_0 \subset \Sigma$, and if $S = [0, 1]$ and if σ is Lebesgue measure on $[0, 1]$, then integrability relative to $\sigma|_{\Sigma_0}$ means Riemann integrability.

Define now the *discrete convergence* for the space $L^1(\sigma)$ of σ-integrable functions.

Definition 3 A sequence of vectors $\{x_n\}$, $x_n \in \mathbf{R}^m$, $\mathcal{P}$-*converges* (or *converges discretely*) to an integrable function $x(s)$, if

$$\sum_{i=1}^n |x_{in} - (p_n x)_{in}|\, m_{in} \to 0, \quad n \in \mathbb{N}. \qquad (13)$$

Remark 4 Note that in the space $L^1(\sigma)$ of σ-integrable functions we are also able to use the projection methods approach, defining convergence of $\{x_n\}$ to $x(s)$ as follows:

$$\int_S \left| x(s) - \sum_{i=1}^n x_{in} \chi_{A_{in}}(s) \right|\, \sigma(ds) \to 0, \quad n \in \mathbb{N}.$$

Remark 5 Projection methods approach does not work in the space $L^\infty(\sigma)$ of essentially bounded measurable functions with vraisup-norm topology ($L^\infty(\sigma)$ is a nonseparable Banach space and the space $C(S)$ of continuous functions is not dense there).

We need the space $L^\infty(\sigma)$, which is the topological dual to the space $L^1(\sigma)$ of σ-integrable functions, in order to define also the discrete analogue of the weak convergence in $L^1(\sigma)$.

Definition 6 A Sequence of vectors $\{x_n\}$, $x_n \in \mathbf{R}^m$, $n \in \mathbf{N}$, $w\mathcal{P}$-*converges* (or *converges weakly discretely*) to an integrable function $x(s)$, $x \in L^1(\sigma)$, if

$$\sum_{i=1}^n (z_{in}, x_{in}) m_{in} \to \int_S (z(s), x(s))\, \sigma(ds), \qquad (14)$$

$n \in \mathbb{N}$,

for any sequence $\{z_n\}$ of vectors, $z_n \in \mathbf{R}^m$, $n \in \mathbf{N}$, and function $z(s)$, $z \in L^\infty(\sigma)$, such that

$$\max_{1 \leq i \leq n} |z_{in} - (p_n z)_{in}| \to 0, \quad n \in \mathbb{N}. \qquad (15)$$

In order to formulate the discretized problem and to simplify the presentation, we will assume that in partition $\{\mathcal{A}_n\}$ of S, where $\mathcal{A}_n = \{A_{1n}, \ldots, A_{nn}\}$, with properties A1)–A7), in property A4) we will identify m_{in} and $\sigma(A_{in})$, i. e. $m_{in} = \sigma(A_{in})$ (e. g. squares with decreasing diagonal in $\mathbf{R}^2$).

Discretize now the probability functional $v_t(x)$:

$$v_{tn}(x_n) = \sum_{i=1}^n \chi(t - f(x_{in}, s_{in})) m_{in}, \qquad (16)$$

and formulate the discretized problem:

$$\max_{x_n \in C_n} v_{tn}(x_n)$$
$$= \max_{x_n \in C_n} \sum_{i=1}^n \chi(t - f(x_{in}, s_{in})) m_{in}, \qquad (17)$$

where constraint set C_n will satisfy discrete analogues of conditions (9)–(11), covered to the set C:

$$\sum_{i=1}^n |x_{in}|\, m_{in} \leq M \quad \forall x_n \in C_n, \qquad (18)$$

$$\sum_{i \in I_n} |x_{in}|\, m_{in} \leq K \sum_{i \in I_n} m_{in},$$

$$\forall x_n \in C_n, \quad \forall I_n \subset \{1, \ldots, n\}, \qquad (19)$$

$$\sum_{k=1}^{r}(x_{i_k n}^k - x_{j_k n}^k)(i_k - j_k) \geq 0, \quad \forall i_k, j_k : \quad i_k < j_k, \tag{20}$$

and such that $0 \leq i_k, j_k \leq n, \forall n \in \mathbf{N}$.

Definition 7 A sequence of sets $\{C_n\}$, $C_n \subset \mathbf{R}^m$, $n \in \mathbf{N}$, converges to the set $C \subset L^1(\sigma)$ in the *discrete Mosco sense* if

1) for any subsequence $\{x_n\}$, $n \in \mathbf{N}' \subset \mathbf{N}$, such that $x_n \in C_n$, from convergence $w\mathcal{P}$-lim $x_n = x$, $n \in \mathbf{N}$, it follows that $x \in C$;
2) for any $x \in C$ there exists a sequence $\{x_n\}$, $x_n \in C_n$, which $\mathcal{P}$-converges to x, $\mathcal{P}$-lim $x_n = x$, $n \in \mathbf{N}$.

Remark 8 If in the above definition also 'for any' part 1) is defined for $\mathcal{P}$-convergence of vectors, then it is said that sequence of sets $\{C_n\}$ converges to the set C in the *discrete Painlevé–Kuratowski sense*.

Denote optimal values and optimal solutions of problems (3) and (17) by v^*, x^* and v_n^*, x_n^*, respectively.

Let function $f(x, s)$ be continuous in both variables (x, s) and satisfy growth and platform conditions (7) and (8). Then from convergence $\mathcal{P}$-lim $x_n = x$, $n \in \mathbf{N}$, for any monotone a.e. function $x(s)$, it follows convergence $v_n(x_n) \to v(x)$, $n \in \mathbf{N}$.

Verification of this statement is quite lengthy and technically complicated: we should first approximate discontinuous function $\chi(t - f(x, s))$ by continuous function $\chi_c(t - f(x, s))$ in the following way:

$$\chi_c(t - f(x, s))$$
$$= \begin{cases} 1 & \text{if } f(x, s) \leq t, \\ 1 - \delta^{-1}[f(x, s) - t] & \text{if } t < f(x, s) \leq t + \delta, \\ 0 & \text{if } f(x, s) > t + \delta \end{cases}$$

for some (small) δ, and then a discontinuous solution $x(s)$, $x \in L^1(\sigma)$, by continuous function $x_c(s)$ (in L^1-norm topology).

Let constraint sets C and C_n satisfy conditions (9)–(11) and (18)–(20), respectively. Let discrete measures $\{(m_n, s_n)\}$ converge weakly to the measure σ. Then the sequence of sets $\{C_n\}$ converges to the set C in the discrete Painlevé–Kuratowski sense.

Verification of this statement relies on the two following convergences:

1) sequence of sets, determined by inequalities (18), (19) converges, assuming weak convergence of discrete measures (5), in discrete Mosco sense to the weakly compact in $L^1(\sigma)$ set, determined by inequalities (9), (10);
2) adding to both, approximate and initial sets of admissible solutions monotonicity conditions (20) and (11), respectively, we can guarantee the discrete convergence of sequence $\{C_n\}$ to C in Painlevé–Kuratowski sense.

Now we can formulate the discrete approximation conditions for a stochastic programming problem with probability cost function in the class of integrable decision rules.

Let function $f(x, s)$ be continuous in both variables (x, s) and satisfy growth and platform conditions (7) and (8), constraint set C satisfy conditions (9)–(11) and let discrete measures $\{(m_n, s_n)\}$ converge weakly to the atomless measure σ. Then $v_n^* \to v^*$, $n \in \mathbf{N}$, and sequence of solutions $\{x_n^*\}$ of approximate problems (17) has a subsequence, which converges discretely to a solution of the initial problem (3).

Remark 9 The usage of the space $L^1(\sigma)$ of integrable functions is essential. In reflexive L^p-spaces, $1 < p < \infty$, serious difficulties arise with application of the strong (norm) compactness criterion for a maximizing sequence.

As a rule, problems with probability cost function are maximized, whereas stochastic programs with quantile cost are minimized, see, e. g., [8,10].

Consider at last discrete approximation of the quantile minimization problem (2):

$$\min_{x \in C} w_\alpha(x)$$
$$= \min_{x \in C} \min_t \{P(f(x(s), s) \leq t) \geq \alpha\}, \tag{21}$$

It was verified in [10] that under certain (quasi)-convexity-concavity assumptions the quantile minimization problem (21) is equivalent to the following Nash game:

$$\max_{x \in C} v_t(x) = J_1^*, \tag{22}$$

$$\min_t (v_t(x) - \alpha)^2 = J_2^*. \tag{23}$$

Discretizing $v_t(x)$ as in (16) and $w_\alpha(x)$ as

$$w_{\alpha n}(x_n) = \min_t \left\{ \sum_{i=1}^n \chi(t - f(x_{in}, s_{in})) m_{in} \geq \alpha \right\},$$

we can, analogously to the probability functional approximation, approximate the quantile minimization problem (21) too. In other words, to replace the Nash game (22), (23) with the following finite-dimensional game:

$$\max_{x_n \in C_n} v_{tn}(x_n) = J_{1n}^*, \qquad (24)$$

$$\min_t (v_{tn}(x_n) - \alpha)^2 = J_{2n}^*. \qquad (25)$$

Verification of convergences $J_{1n}^* \to J_1^*$ and $J_{2n}^* \to J_2^*$, $n \in \mathbf{N}$, is a little bit more labor-consuming compared with approximate maximization of probability functional $v_t(x)$, since we should guarantee also convergence of the sequence of optimal quantiles $\{t_n^*\}$ of minimization problems (25).

See also

- ▶ Approximation of Multivariate Probability Integrals
- ▶ Discretely Distributed Stochastic Programs: Descent Directions and Efficient Points
- ▶ Extremum Problems with Probability Functions: Kernel Type Solution Methods
- ▶ General Moment Optimization Problems
- ▶ Logconcave Measures, Logconvexity
- ▶ Logconcavity of Discrete Distributions
- ▶ L-Shaped Method for Two-Stage Stochastic Programs with Recourse
- ▶ Multistage Stochastic Programming: Barycentric Approximation
- ▶ Preprocessing in Stochastic Programming
- ▶ Probabilistic Constrained Linear Programming: Duality Theory
- ▶ Probabilistic Constrained Problems: Convexity Theory
- ▶ Simple Recourse Problem: Dual Method
- ▶ Simple Recourse Problem: Primal Method
- ▶ Stabilization of Cutting Plane Algorithms for Stochastic Linear Programming Problems
- ▶ Static Stochastic Programming Models
- ▶ Static Stochastic Programming Models: Conditional Expectations
- ▶ Stochastic Integer Programming: Continuity, Stability, Rates of Convergence
- ▶ Stochastic Integer Programs
- ▶ Stochastic Linear Programming: Decomposition and Cutting Planes
- ▶ Stochastic Linear Programs with Recourse and Arbitrary Multivariate Distributions
- ▶ Stochastic Network Problems: Massively Parallel Solution
- ▶ Stochastic Programming: Minimax Approach
- ▶ Stochastic Programming Models: Random Objective
- ▶ Stochastic Programming: Nonanticipativity and Lagrange Multipliers
- ▶ Stochastic Programming with Simple Integer Recourse
- ▶ Stochastic Programs with Recourse: Upper Bounds
- ▶ Stochastic Quasigradient Methods in Minimax Problems
- ▶ Stochastic Vehicle Routing Problems
- ▶ Two-stage Stochastic Programming: Quasigradient Method
- ▶ Two-Stage Stochastic Programs with Recourse

References

1. Banaš J (1989) Integrable solutions of Hammerstein and Urysohn integral equations. J Austral Math Soc (Ser A) 46:61–68
2. Daniel JW (1971) The approximate minimization of functionals. Prentice-Hall, Englewood Cliffs
3. Esser H (1973) Zur Diskretisierung von Extremalproblemen. Lecture Notes Math, vol 333. Springer, Berlin, pp 69–88
4. Garstka J, Wets RJ-B (1974) On decision rules in stochastic programming. Math Program 7:117–143
5. Hernandez-Lerma O, Runggaldier W (1994) Monotone approximations for convex stochastic control problems. J Math Syst, Estimation and Control 4:99–140
6. Ioffe AD, Tikhomirov VM (1979) Theory of extremal problems. North-Holland, Amsterdam
7. Kall P, Wallace SW (1994) Stochastic programming. Wiley, New York
8. Kibzun AI, Kan YS (1995) Stochastic programming problems with probability and quantile functions. Wiley, New York
9. Lepp R (1983) Stochastic approximation type algorithm for the maximization of the probability function. Proc Acad Sci Estonian SSR Phys Math 32:150–156

10. Malyshev VV, Kibzun AI (1987) Analysis and synthesis of high precision aircraft control. Mashinostroenie, Moscow, Moscow
11. Olsen P (1976) Discretization of multistage stochastic programming problems. Math Program Stud 6:111–124
12. Raik E (1972) On stochastic programming problem with probability and quantile functionals. Proc Acad Sci Estonian SSR Phys Math 21:142–148
13. Römisch W, Schultz R (1988) On distribution sensitivity in chance constrained programming. Math Res 45:161–168. Advances in Mathematical Optimization, In: Guddat J et al (eds)
14. Vainikko GM (1971) On convergence of the method of mechanical curvatures for integral equations with discontinuous kernels. Sibirsk Mat Zh 12:40–53
15. Vasin VV (1982) Discrete approximation and stability in extremal problems. USSR Comput Math Math Phys 22:57–74

Approximation of Multivariate Probability Integrals

TAMÁS SZÁNTAI
Technical University, Budapest, Hungary

MSC2000: 65C05, 65D30, 65Cxx, 65C30, 65C40, 65C50, 65C60, 90C15

Article Outline

Keywords
Lower and Upper Bounds
Monte-Carlo Simulation Algorithm
One- and Two-Dimensional Marginal Distribution Functions
Examples
Remarks
See also
References

Keywords

Boole–Bonferroni bounds; Hunter–Worsley bounds; Approximation; Probability integrals; Variance reduction; Probabilistic constrained stochastic programming

Approximation of *multivariate probability integrals* is a hard problem in general. However, if the domain of the probability integral is multidimensional interval, then the problem reduces to the approximation of *multivariate probability distribution function* values.

Lower and Upper Bounds

Let $\xi^T = (\xi_1, \ldots, \xi_n)$ be a random vector with given multivariate probability distribution. Introducing the events

$$A_1 = \{\xi_1 < x_1\}, \ldots, A_n = \{\xi_n < x_n\},$$

where $x_1, \ldots, x_n$ are arbitrary real values the multivariate probability distribution function of the random vector ξ can be expressed in the following way:

$$\begin{aligned}
F(x_1, \ldots, x_n) &= P(\xi_1 < x_1, \ldots, \xi_n < x_n) \\
&= P(A_1 \cap \cdots \cap A_n) \\
&= 1 - P(\overline{A}_1 \cup \cdots \cup \overline{A}_n) \\
&= 1 - \overline{S}_1 + \overline{S}_2 - \cdots + (-1)^n \overline{S}^n,
\end{aligned}$$

where

$$\overline{A}_i = \{\xi_i \geq x_i\}, \quad i = 1, \ldots, n,$$

and

$$\overline{S}_k = \sum_{1 \leq i_1 < \cdots < i_k \leq n} P(\overline{A}_{i_1} \cap \cdots \cap \overline{A}_{i_k}), k = 1, \ldots, n.$$

First one shows that $\overline{S}_1, \overline{S}_2$ and so the individual probabilities $P(\overline{A}_i), i = 1, \ldots, n$, $P(\overline{A}_i \cap \overline{A}_j), i = 1, \ldots, n-1$, $j = i+1, \ldots, n$, involved in them can be expressed by $F_i(x_i), i = 1, \ldots, n$, and $F_{ij}(x_i, x_j), i = 1, \ldots, n-1, j = i+1, \ldots, n$, the *one-* and *two-dimensional marginal probability distribution functions* of the random vector ξ. One has

$$\overline{S}_1 = \sum_{i=1}^n P(\overline{A}_i) = \sum_{i=1}^n P(\xi_i \geq x_i)$$
$$= n - \sum_{i=1}^n P(\xi_i < x_i) = n - \sum_{i=1}^n F_i(x_i)$$

and

$$\bar{S}_2 = \sum_{1 \leq i < j \leq n} P(\bar{A}_i \cap \bar{A}_j)$$

$$= \sum_{1 \leq i < j \leq n} P(\xi_i \geq x_i, \xi_j \geq x_j)$$

$$= \sum_{1 \leq i < j \leq n} \{1 - P(\xi_i < x_i)$$

$$-P(\xi_j < x_j) + P(\xi_i < x_i, \xi_j < x_j)\}$$

$$= \frac{n(n-1)}{2} - (n-1)\sum_{i=1}^{n} F_i(x_i)$$

$$+ \sum_{1 \leq i < j \leq n} F_{ij}(x_i, x_j).$$

So if one can calculate the one- and two-dimensional marginal probability distribution functions of the random vector ξ then one can bound the multivariate probability distribution function by the very simple bounds given by C.E. Bonferroni [1]:

$$1 - \bar{S}_1 \leq F(x_1, \ldots, x_n) \leq 1 - \bar{S}_1 + \bar{S}_2,$$

or by the sharp bounds, called *Boole–Bonferroni bounds* discovered independently by many authors (see [11] for a summary):

$$1 - \bar{S}_1 + \frac{2}{n}\bar{S}_2$$
$$\leq F(x_1, \ldots, x_n)$$
$$\leq 1 - \frac{2}{k^* + 1}\bar{S}_1 + \frac{2}{k^*(k^* + 1)}\bar{S}_2,$$

where

$$k^* = 1 + \left\lfloor \frac{2\bar{S}_2}{\bar{S}_1} \right\rfloor.$$

When applying the above bounds usually the upper bound proves to be sharper. However one can improve the lower bound by the application of the bound discovered independently by D. Hunter [5] and K.J. Worsley [18]. This bound is an upper bound for $P(\bar{A}_1 \cup \cdots \cup \bar{A}_n)$ by the use of $\bar{S}_1$ and the individual probabilities $P(\bar{A}_i \cap \bar{A}_j), 1 \leq i < j \leq n$. It is constructed in the following way. Construct a nonoriented complete graph with n nodes and assign to node i the event $\bar{A}_i$ (or the probability $P(\bar{A}_i)$) and to arc (i, j) the weight $P(\bar{A}_i \cap \bar{A}_j)$. Let T^* be

a maximum weight spanning tree in this nonoriented complete graph then one has

$$P(\bar{A}_1 \cup \cdots \cup \bar{A}_n) \leq \bar{S}_1 - \sum_{(i,j) \in T^*} P(\bar{A}_i \cap \bar{A}_j),$$

which is called the *Hunter–Worsley upper bound*. This results the following lower bound on the multivariate probability distribution function:

$$1 - \bar{S}_1 + \sum_{(i,j) \in T^*} P(\bar{A}_i \cap \bar{A}_j) \leq F(x_1, \ldots, x_n).$$

The individual probabilities $P(\bar{A}_i \cap \bar{A}_j), 1 \leq i < j \leq n$, can be stored when one calculates the value of $\bar{S}_2$ and the maximum weight spanning tree can be found by several fast algorithms, for example by *Kruskal's algorithm*, see [9]. Now one has three lower and two upper bounds on the multivariate probability distribution function and all of them are computable if the one- and two-dimensional marginal probability distribution functions are known. Let us denote these bounds in the following way:

$$L_1 = 1 - \bar{S}_1,$$
$$L_2 = 1 - \bar{S}_1 + \frac{2}{n}\bar{S}_2,$$
$$L_3 = 1 - \bar{S}_1 + \sum_{(i,j) \in T^*} P(\bar{A}_i \cap \bar{A}_j),$$
$$U_1 = 1 - \bar{S}_1 + \bar{S}_2,$$
$$U_2 = 1 - \frac{2}{k^* + 1}\bar{S}_1 + \frac{2}{k^*(k^* + 1)}\bar{S}_2.$$

As one has $L_1 \leq L_2 \leq L_3$ and $U_2 \leq U_1$, the best lower bound is L_3 and the best upper bound is U_2.

Monte-Carlo Simulation Algorithm

One can take the differences between the multivariate probability distribution function and its lower and upper bounds introduced before:

$$F(x_1, \ldots, x_n) - L_1 = \bar{S}_2 - \bar{S}_3 + \cdots + (-1)^n \bar{S}_n,$$
$$F(x_1, \ldots, x_n) - L_2$$
$$= \left(1 - \frac{2}{n}\right)\bar{S}_2 - \bar{S}_3 + \cdots + (-1)^n \bar{S}_n,$$

$$F(x_1, \ldots, x_n) - L_3$$
$$= - \sum_{(i,j) \in T^*} P(\overline{A}_i \cap \overline{A}_j) + \overline{S}_2 - \overline{S}_3 + \cdots + (-1)^n \overline{S}_n,$$

$$F(x_1, \ldots, x_n) - U_1 = \overline{S}_3 + \cdots + (-1)^n \overline{S}_n,$$

$$F(x_1, \ldots, x_n) - U_2$$
$$= \left(\frac{2}{k^* + 1} - 1\right) \overline{S}_1 + \left(1 - \frac{2}{k^*(k^* + 1)}\right) \overline{S}_2$$
$$- \overline{S}_3 + \cdots + (-1)^n \overline{S}_n.$$

A *Monte-Carlo simulation procedure* of the multivariate probability distribution function value based on the estimation of the differences above will be given. First however the so called crude Monte-Carlo simulation procedure will be described. Let the random vectors $(\xi_1^s, \ldots, \xi_n^s), s = 1, \ldots, S$, be distributed according to the multivariate probability distribution function to be approximated. One must check the inequalities $\xi_1^s < x_1, \ldots, \xi_n^s < x_n$ for all sample elements, $s = 1, \ldots, S$. For this purpose let be defined the random values

$$v_0^s = \begin{cases} 1 & \text{if } \xi_1^s < x_1, \ldots, \xi_n^s < x_n, \\ 0 & \text{otherwise}, \end{cases} \quad s = 1, \ldots, S.$$

These random values are identically distributed and stochastically independent. All of them take on the value 1 with probability equal to the approximated multivariate probability distribution function value. The sum of them has binomial probability distribution with parameters S and $F(x_1, \ldots, x_n)$. So the random variable

$$v_0 = \frac{1}{S}(v_0^1 + \cdots + v_0^S)$$

has expected value $P = F(x_1, \ldots, x_n)$ and variance $\frac{P(1-P)}{S}$. This is why v_0 can be regarded as an estimate, the so called crude Monte-Carlo estimate of $F(x_1, \ldots, x_n)$. If one introduces κ^s as the number of those $\xi_1^s < x_1, \ldots, \xi_n^s < x_n$ inequalities which are not fulfilled, i.e. the number of those $\xi_1^s \geq x_1, \ldots, \xi_n^s \geq x_n$ inequalities which are fulfilled, or the number of those $\overline{A}_1^s, \ldots, \overline{A}_n^s$ events which occur, $s = 1, \ldots, S$, the v_0^s random values can be expressed as

$$v_0^s = \begin{cases} 1 & \text{if } \kappa^s = 0, \\ 0 & \text{otherwise} \end{cases} \quad s = 1, \ldots, S,$$

and on the other hand for the *binomial moments* of κ^s one has

$$E\left[\binom{\kappa^s}{k}\right] = \overline{S}_k, k = 0, \ldots, n, \quad s = 1, \ldots, S.$$

The simplest proof of these equalities was given by L. Takács [17] and it was reproduced by A. Prékopa in [11]. If the random numbers $\lambda^s, s = 1, \ldots, S$, are also introduced as the number of those $\overline{A}_i \cap \overline{A}_j = \{\xi_i^s \geq x_i, \xi_j^s \geq x_j\}, (i, j) \in T^*$, events which occur then for the expected value of λ^s one has

$$E(\lambda^s) = \sum_{(i,j) \in T^*} P(\overline{A}_i \cap \overline{A}_j), \quad s = 1, \ldots, S.$$

Using these equalities one easily can see that the following random values have expected values equal to the differences between the multivariate probability distribution function and its bounds:

$$v_{L_1}^s = \binom{\kappa^s}{2} - \binom{\kappa^s}{3} + \cdots + (-1)^n \binom{\kappa^s}{n},$$

$$v_{L_2}^s = \left(1 - \frac{2}{n}\right) \binom{\kappa^s}{2} - \binom{\kappa^s}{3} + \cdots$$
$$+ (-1)^n \binom{\kappa^s}{n},$$

$$v_{L_3}^s = -\lambda^s + \binom{\kappa^s}{2} - \binom{\kappa^s}{3} + \cdots$$
$$+ (-1)^n \binom{\kappa^s}{n},$$

$$v_{U_1}^s = -\binom{\kappa^s}{3} + \cdots + (-1)^n \binom{\kappa^s}{n},$$

$$v_{U_2}^s = \left(\frac{2}{k^* + 1} - 1\right) \binom{\kappa^s}{1}$$
$$+ \left(1 - \frac{2}{k^*(k^* + 1)}\right) \binom{\kappa^s}{2}$$
$$- \binom{\kappa^s}{3} + \cdots + (-1)^n \binom{\kappa^s}{n}.$$

By the binomial theorem one has

$$\binom{\kappa^s}{0} - \binom{\kappa^s}{1} + \cdots + (-1)^n \binom{\kappa^s}{n} = 0$$

and the above random values can be expressed as

$$v^s_{L_1} = \begin{cases} \kappa^s - 1 & \text{if } \kappa^s \geq 2, \\ 0 & \text{otherwise}, \end{cases}$$

$$v^s_{L_2} = \begin{cases} \frac{1}{n}(\kappa^s - 1)(n - \kappa^s) & \text{if } \kappa^s \geq 2, \\ 0 & \text{otherwise}, \end{cases}$$

$$v^s_{L_3} = \begin{cases} \kappa^s - 1 - \lambda^s & \text{if } \kappa^s \geq 2, \\ 0 & \text{otherwise}, \end{cases}$$

$$v^s_{U_1} = \begin{cases} \frac{1}{2}(\kappa^s - 1)(2 - \kappa^s) & \text{if } \kappa^s \geq 3, \\ 0 & \text{otherwise}, \end{cases}$$

$$v^s_{U_2} = \begin{cases} \frac{(k^* - \kappa^s)(\kappa^s - k^* - 1)}{k^*(k^* + 1)} & \text{if } \kappa^s \geq 1, \\ 0 & \text{otherwise}. \end{cases}$$

Taking the new random values $v_{L_1}, v_{L_2}, v_{L_3}, v_{U_1}, v_{U_2}$ and the estimate v_0 introduced before:

$$v_0 = \frac{1}{S}(v^1_0 + \cdots + v^S_0),$$

$$v_{L_1} = L_1 + \frac{1}{S}(v^1_{L_1} + \cdots + v^S_{L_1}),$$

$$v_{L_2} = L_2 + \frac{1}{S}(v^1_{L_2} + \cdots + v^S_{L_2}),$$

$$v_{L_3} = L_3 + \frac{1}{S}(v^1_{L_3} + \cdots + v^S_{L_3}),$$

$$v_{U_1} = U_1 + \frac{1}{S}(v^1_{U_1} + \cdots + v^S_{U_1}),$$

$$v_{U_2} = U_2 + \frac{1}{S}(v^1_{U_2} + \cdots + v^S_{U_2}),$$

one gets altogether six estimates of the multivariate probability distribution function. These estimates obviously are not stochastically independent so one can mix them to get a new estimate with minimal possible variance. This technique is called *regression method* and it means forming the estimate

$$v = w_0 v_0 + w_{L_1} v_{L_1} + w_{L_2} v_{L_2} + w_{L_3} v_{L_3} + w_{U_1} v_{U_1} + w_{U_2} v_{U_2}$$

with $w_0 + w_{L_1} + w_{L_2} + w_{L_3} + w_{U_1} + w_{U_2} = 1$, where $w_0, w_{L_1}, w_{L_2}, w_{L_3}, w_{U_1}, w_{U_2}$ are chosen so that the variance of v be minimized. Let

$$\begin{pmatrix} c_{00} & c_{0L_1} & c_{0L_2} & c_{0L_3} & c_{0U_1} & c_{0U_2} \\ c_{L_10} & c_{L_1L_1} & c_{L_1L_2} & c_{L_1L_3} & c_{L_1U_1} & c_{L_1U_2} \\ c_{L_20} & c_{L_2L_1} & c_{L_2L_2} & c_{L_2L_3} & c_{L_2U_1} & c_{L_2U_2} \\ c_{L_30} & c_{L_3L_1} & c_{L_3L_2} & c_{L_3L_3} & c_{L_3U_1} & c_{L_3U_2} \\ c_{U_10} & c_{U_1L_1} & c_{U_1L_2} & c_{U_1L_3} & c_{U_1U_1} & c_{U_1U_2} \\ c_{U_20} & c_{U_2L_1} & c_{U_2L_2} & c_{U_2L_3} & c_{U_2U_1} & c_{U_2U_2} \end{pmatrix}$$

be the covariance matrix C of the six estimates, where C is a symmetrical matrix. Then the variance of v is $w^\mathsf{T} C w$, where $w = (w_0, w_{L_1}, w_{L_2}, w_{L_3}, w_{U_1}, w_{U_2})^\mathsf{T}$. The Lagrangian problem:

$$\begin{cases} \min & w^\mathsf{T} C w \\ \text{s.t.} & w_0 + w_{L_1} + w_{L_2} + w_{L_3} + w_{U_1} + w_{U_2} = 1 \end{cases}$$

can easily be solved. In fact, the gradient of $w^\mathsf{T} C w$ equals $2 w^\mathsf{T} C$, hence one has to solve the system of linear equations

$$c_{00} w_0 + c_{0L_1} w_{L_1} + c_{0L_2} w_{L_2} + c_{0L_3} w_{L_3}$$
$$+ c_{0U_1} w_{U_1} + c_{0U_2} w_{U_2} - \lambda = 0,$$
$$c_{L_10} w_0 + c_{L_1L_1} w_{L_1} + c_{L_1L_2} w_{L_2} + c_{L_1L_3} w_{L_3}$$
$$+ c_{L_1U_1} w_{U_1} + c_{L_1U_2} w_{U_2} - \lambda = 0,$$
$$c_{L_20} w_0 + c_{L_2L_1} w_{L_1} + c_{L_2L_2} w_{L_2} + c_{L_2L_3} w_{L_3}$$
$$+ c_{L_2U_1} w_{U_1} + c_{L_2U_2} w_{U_2} - \lambda = 0,$$
$$c_{L_30} w_0 + c_{L_3L_1} w_{L_1} + c_{L_3L_2} w_{L_2} + c_{L_3L_3} w_{L_3}$$
$$+ c_{L_3U_1} w_{U_1} + c_{L_3U_2} w_{U_2} - \lambda = 0,$$
$$c_{U_10} w_0 + c_{U_1L_1} w_{L_1} + c_{U_1L_2} w_{L_2} + c_{U_1L_3} w_{L_3}$$
$$+ c_{U_1U_1} w_{U_1} + c_{U_1U_2} w_{U_2} - \lambda = 0,$$
$$c_{U_20} w_0 + c_{U_2L_1} w_{L_1} + c_{U_2L_2} w_{L_2} + c_{U_2L_3} w_{L_3}$$
$$+ c_{U_2U_1} w_{U_1} + c_{U_2U_2} w_{U_2} - \lambda = 0,$$
$$w_0 + w_{L_1} + w_{L_2} + w_{L_3} + w_{U_1} + w_{U_2} - \lambda = 1.$$

for the unknowns $w_0, w_{L_1}, w_{L_2}, w_{L_3}, w_{U_1}, w_{U_2}, \lambda$. As the covariance matrix C is not known in advance, so one must estimate its elements from the random sample during the Monte-Carlo simulation procedure. This means that one must sum up not only the individual random values $v^s_0, v^s_{L_1}, v^s_{L_2}, v^s_{L_3}, v^s_{U_1}, v^s_{U_2}$ but their crossproducts, too. The crossproducts are many times trivial, so their calculation is not necessary. For example v^s_0 equals $v^s_0 v^s_0$, further when v^s_0 equals nonzero ($\kappa^s = 0$) then all other random values $v^s_{L_1}, v^s_{L_2}, v^s_{L_3}, v^s_{U_1}, v^s_{U_2}$ are equal zero, so the corresponding crossproducts are all zero. One should also notice that the random values $v^s_{L_1}, v^s_{L_2}, v^s_{L_3}$ are always nonnegative while the random values $v^s_{U_1}, v^s_{U_2}$ are always nonpositive. So the corresponding crossproducts cannot be positive even they are many times negative yielding real variance reduction in the final estimate.

One- and Two-Dimensional Marginal Distribution Functions

For the applicability of the Monte-Carlo simulation algorithm of the previous section one has to show that the one- and two-dimensional marginal distribution function values can be evaluated efficiently. As in the cases of the *multivariate normal distribution*, (one parameter) gamma and Dirichlet distributions the marginal distributions are also normal, gamma and Dirichlet and the one-dimensional Dirichlet distribution is the beta distribution, the one-dimensional marginal probability distribution functions can be evaluated by known algorithms. For example in the *IMSL subroutine library* [6] the subroutines MDNOR, MDGAM and MDBETA provide these calculations. In the case of the normal distribution the two-dimensional marginal probability distribution function also can be evaluated by a standard IMSL subroutine called MDBNOR. Some details of the calculations provided by these subroutines can be found in [8].

In the case of the *multivariate gamma distribution*, introduced by Prékopa and T. Szántai in [12], only the evaluation of the joint probability distribution function of the random variables

$$\xi_1 = \eta_1 + \eta_2,$$
$$\xi_2 = \eta_1 + \eta_3$$

is necessary. Here the random variables η_1, η_2 and η_3 are independent and gamma distributed with parameters ϑ_1, ϑ_2 and ϑ_3. Taking the joint characteristic function of ξ_1 and ξ_2 and applying the inversion formula one easily gets the joint probability density function of them. This is in the form of series expansion involving Laguerre polynomials. Using some integral formulae of these orthogonal polynomials one can integrate the joint probability density function to get the final formula for the evaluation of the joint probability distribution function in the following form

$$F(z_1, z_2) = F_{\vartheta_1+\vartheta_2}(z_1) F_{\vartheta_1+\vartheta_3}(z_2)$$
$$+ \sum_{k=1}^{\infty} C(\vartheta_1, \vartheta_2, \vartheta_3, k)$$
$$\times f_{\vartheta_1+\vartheta_2+1}(z_1) L_{k-1}^{\vartheta_1+\vartheta_2}(z_1)$$
$$\times f_{\vartheta_1+\vartheta_3+1}(z_2) L_{k-1}^{\vartheta_1+\vartheta_3}(z_2),$$

where

$$C(\vartheta_1, \vartheta_2, \vartheta_3, k) = \frac{(k-1)!}{k} \frac{\Gamma(\vartheta_1 + k)}{\Gamma(\vartheta_1)}$$
$$\times \frac{\Gamma(\vartheta_1 + \vartheta_2 + 1)}{\Gamma(\vartheta_1 + \vartheta_2 + k)} \frac{\Gamma(\vartheta_1 + \vartheta_3 + 1)}{\Gamma(\vartheta_1 + \vartheta_3 + k)}$$

and $f_\vartheta(z)$ and $F_\vartheta(z)$ are the one-dimensional gamma probability density, respectively distribution, functions. For the calculation of the Laguerre polynomial the following recursion formula can be used

$$(k+1) L_{k+1}^\vartheta(z)$$
$$= (2k + \vartheta + 1 - z) L_k^\vartheta(z) - (k + \vartheta) L_{k-1}^\vartheta(z),$$
$$k = 0, 1, \ldots,$$

where $L_0^\vartheta(z) = 1$ and $L_1^\vartheta(z) = \vartheta + 1 - z$. The convergence of the series for calculation of $F(z_1, z_2)$ has been established by Szántai in [14].

In the case of *Dirichlet distribution* the two-dimensional marginal probability density function of the components ξ_i, ξ_j is given by

$$f(z_1, z_2) = \frac{\Gamma(a)\Gamma(b)\Gamma(c)}{\Gamma(a+b+c)} \cdot z_1^{a-1} z_2^{b-1} (1-z_1-z_2)^{c-1},$$
$$\text{if } z_1 + z_2 \leq 1, \ z_1 \geq 0, \ z_2 \geq 0,$$

where $a = \vartheta_i$, $b = \vartheta_j$ and $c = \sum_{k=1}^{n+1} \vartheta_k - \vartheta_i - \vartheta_j$. One obtains by direct calculation for the two-dimensional probability distribution function

$$F(z_1, z_2) = \frac{\Gamma(a+b+c)}{\Gamma(a)\Gamma(b)\Gamma(c)}$$
$$\times \int_0^{z_1} \int_0^{z_2} t_1^{a-1} t_2^{b-1} (1 - t_1 - t_2)^{c-1} dt_2 \, dt_1$$
$$= \frac{z_1^a}{a} \frac{z_2^b}{b} + \sum_{m=1}^{\infty} (1-c) \cdots (m-c)$$
$$\times \left[\frac{z_1^a}{a} \frac{z_2^{b+m}}{(b+m) m!} \right.$$
$$\left. + \sum_{k=0}^{m} \frac{z_1^{a+k}}{(a+k) k!} \frac{z_2^{b+m-k}}{(b+m-k)(m-k)!} \right].$$

The above formula is valid only if $z_1 + z_2 \leq 1, z_1 \geq 0, z_2 \geq 0$; otherwise the statement a) of the following more general theorem can be applied.

Theorem 1 *Let $z_1^* \leq \cdots \leq z_n^*$ be the ordered sequence of $z_1, \ldots, z_n$, the arguments of the n-dimensional Dirichlet distribution function.*

a) If $z_1^* + z_2^* > 1$ then one has

$$F(z_1, \ldots, z_n) = 1 - n + \sum_{i=1}^{n} F_i(z_i).$$

b) If $z_1^* + z_2^* + z_3^* > 1$ then one has

$$F(z_1, \ldots, z_n)$$
$$= \frac{n-1}{2} - (n-2) \sum_{i=1}^{n} F_i(z_i) + \sum_{1 \leq i < j \leq n} F_{ij}(z_i, z_j).$$

Here $F_i(z_i)$ and $F_{ij}(z_i, z_j)$ are the one- and two-dimensional marginal probability distribution functions.

This theorem was formulated and proved by Szántai in [13]. It also can be found in [11].

Examples

For illustrating the lower and upper bounds on the multivariate normal probability distribution function value and the efficiency of the *variance reduction technique* described before one can regard the following examples.

Example 2

$n = 10$,

$x_1 = 1.7$, $x_2 = 0.8$, $x_3 = 5.1$,
$x_4 = 3.2$, $x_5 = 2.4$, $x_6 = 1.8$,
$x_7 = 2.7$, $x_8 = 1.5$, $x_9 = 1.2$,
$x_{10} = 2.6$,
$r_{ij} = 0.0$, $i = 2, \ldots, 10$, $j = 1, \ldots, i-1$,

except $r_{21} = -0.6$, $r_{43} = 0.9$, $r_{65} = 0.4$, $r_{87} = 0.2$, $r_{10,9} = -0.8$.

Number of trials: 10000.

Lower bound by S1, S2	0.524736
Lower bound by Hunter	0.563719
Upper bound by S1, S2	0.588646
Estimated value	0.582743
Standard deviation	0.000608
Time in seconds (PC-586)	0.77
Efficiency	65.73

Example 3

$n = 15$,
$x_1 = 2.9$, $x_2 = 2.9$, $x_3 = 2.9$,
$x_4 = 2.9$, $x_5 = 2.9$, $x_6 = 2.9$,
$x_7 = 2.9$, $x_8 = 2.9$, $x_9 = 2.9$,
$x_{10} = 2.9$, $x_{11} = 2.9$, $x_{12} = 2.7$,
$x_{13} = 1.6$, $x_{14} = 1.2$, $x_{15} = 2.1$,
$r_{ij} = 0.2$, $i = 2, \ldots, 10$, $j = 1, \ldots, i-1$,
$r_{ij} = 0.0$, $i = 11, \ldots, 15$, $j = 1, \ldots, i-1$

except $r_{13,12} = 0.3$, $r_{15,14} = -0.95$.
Number of trials = 10000.

Lower bound by S1, S2	0.790073
Lower bound by Hunter	0.798730
Upper bound by S1, S2	0.801745
Estimated value	0.801304
Standard deviation	0.000193
Time in seconds (PC-586)	1.38
Efficiency	417.84

Both of the above examples are taken from [2, Exam. 4; 6] and they are according to standard multivariate normal probability distributions, i.e. all components of the normally distributed random vector have expected value zero and variance one. The efficiency of the Monte-Carlo simulation algorithm was calculated according to the crude Monte-Carlo algorithm in the usual way, i.e. it equals to the fraction $(t_0 \sigma_0^2)/(t_1 \sigma_1^2)$ where t_0, t_1 are the calculation times and σ_0^2, σ_1^2 are the variances of the crude and the compared simulation algorithms.

Remarks

In many applications one may need finding the *gradient of multivariate distribution functions*, too. As one has the general formula

$$\frac{\partial F(z_1, \ldots, z_n)}{\partial z_i}$$
$$= F(z_1, \ldots, z_{i-1}, z_{i+1}, \ldots, z_n | z_i) \cdot f_i(z_i),$$

where $F(z_1, \ldots, z_{i-1}, z_{i+1}, \ldots, z_n | z_i)$ is the conditional probability distribution function of the random variables $\xi_1, \ldots, \xi_{i-1}, \xi_{i+1}, \ldots, \xi_n$, given that $\xi_i = z_i$, and

$f_i(z)$ is the probability density function of the random variable ξ_i, finding the gradient of a multivariate probability distribution function can be reduced to finding conditional distribution functions. In the cases of multivariate normal and Dirichlet distributions the conditional distributions are also multivariate normal and Dirichlet, and in the case of multivariate gamma distribution they are different and more complicated as it was obtained by Prékopa and Szántai [12].

In the case of multivariate normal probability distribution I. Deák [2] proposed another simulation technique which proved to be as efficient as the method described here. The main advantage of Deák's method is that it easily can be generalized for calculation the probability content of more general sets in the multidimensional space, like convex polyhedrons, hyperellipsoids, circular cones, etc. Its main drawback is that it works only for the multivariate normal probability distribution. The methods of Szántai and Deák have been combined by H. Gassmann to compute the probability of an n-dimensional rectangle in the case of multivariate normal distribution (see [3]). Also in the case of multivariate normal probability distribution A. Genz proposed the transformation of the original integration region to the unit hypercube $[0, 1]^n$ and then the application of a crude Monte-Carlo method or some lattice rules for the numerical integration of the resulting multidimensional integral. A comparison of methods for the computation of multivariate normal probabilities can be found in [4]. When the three-dimensional marginal probability distribution function values are also calculated by numerical integration there exist some new, sharper bounds. See [16] for these bounds and their effect on the efficiency of the Monte-Carlo simulation algorithm.

Approximation of multivariate probability integrals has a central role in probabilistic constrained stochastic programming when the probabilistic constraints are joint. The *computer code PCSP* (probabilistic constrained stochastic programming) originally was developed for handling the multivariate normal probability distributions in this framework (see [15]). A new version of the code now can handle multivariate gamma and Dirichlet distributions as well. The calculation procedures of this paper also has been applied by J. Mayer in his code solving this type of stochastic programming problems by reduced gradient algorithm (see [10]).

These codes have been integrated by P. Kall and Mayer into a more advanced computer system for modeling in *stochastic linear programming* (see [7]).

See also

- ▶ Approximation of Extremum Problems with Probability Functionals
- ▶ Discretely Distributed Stochastic Programs: Descent Directions and Efficient Points
- ▶ Extremum Problems with Probability Functions: Kernel Type Solution Methods
- ▶ General Moment Optimization Problems
- ▶ Logconcave Measures, Logconvexity
- ▶ Logconcavity of Discrete Distributions
- ▶ L-shaped Method for Two-stage Stochastic Programs with Recourse
- ▶ Multistage Stochastic Programming: Barycentric Approximation
- ▶ Preprocessing in Stochastic Programming
- ▶ Probabilistic Constrained Linear Programming: Duality Theory
- ▶ Probabilistic Constrained Problems: Convexity Theory
- ▶ Simple Recourse Problem: Dual Method
- ▶ Simple Recourse Problem: Primal Method
- ▶ Stabilization of Cutting Plane Algorithms for Stochastic Linear Programming Problems
- ▶ Static Stochastic Programming Models
- ▶ Static Stochastic Programming Models: Conditional Expectations
- ▶ Stochastic Integer Programming: Continuity, Stability, Rates of Convergence
- ▶ Stochastic Integer Programs
- ▶ Stochastic Linear Programming: Decomposition and Cutting Planes
- ▶ Stochastic Linear Programs with Recourse and Arbitrary Multivariate Distributions
- ▶ Stochastic Network Problems: Massively Parallel Solution
- ▶ Stochastic Programming: Minimax Approach
- ▶ Stochastic Programming Models: Random Objective
- ▶ Stochastic Programming: Nonanticipativity and Lagrange Multipliers
- ▶ Stochastic Programming with Simple Integer Recourse
- ▶ Stochastic Programs with Recourse: Upper Bounds

- Stochastic Quasigradient Methods in Minimax Problems
- Stochastic Vehicle Routing Problems
- Two-stage Stochastic Programming: Quasigradient Method
- Two-stage Stochastic Programs with Recourse

References

1. Bonferroni CE (1937) Teoria statistica delle classi e calcolo delle probabilita. Volume in onordi Riccardo Dalla Volta: 1–62
2. Deák I (1980) Three digit accurate multiple normal probabilities. Numerische Math 35:369–380
3. Gassmann H (1988) Conditional probability and conditional expectation of a random vector. In: Ermoliev Y, Wets RJ-B (eds) Numerical Techniques for Stochastic Optimization. Springer, Berlin, pp 237–254
4. Genz A (1993) Comparison of methods for the computation of multivariate normal probabilities. Computing Sci and Statist 25:400–405
5. Hunter D (1976) Bounds for the probability of a union. J Appl Probab 13:597–603
6. IMSL (1977) Library 1 reference manual. Internat. Math. Statist. Library
7. Kall P, Mayer J (1995) Computer support for modeling in stochastic linear programming. In: Marti K, Kall P (eds) Stochastic Programming: Numerical Methods and Techn. Applications. Springer, Berlin, pp 54–70
8. Kennedy WJ Jr, Gentle JE (1980) Statistical computing. M. Dekker, New York
9. Kruskal JB (1956) On the shortest spanning subtree of a graph and the travelling salesman problem. Proc Amer Math Soc 7:48–50
10. Mayer J (1988) Probabilistic constrained programming: A reduced gradient algorithm implemented on PC. Working Papers IIASA WP-88-39
11. Prékopa A (1995) Stochastic programming. Akad. Kiadó and Kluwer, Budapest–Dordrecht
12. Prékopa A, Szántai T (1978) A new multivariate gamma distribution and its fitting to empirical streamflow data. Water Resources Res 14:19–24
13. Szántai T (1985) Numerical evaluation of probabilities concerning multivariate probability distributions. Thesis Candidate Degree Hungarian Acad Sci (in Hungarian)
14. Szántai T (1986) Evaluation of a special multivariate gamma distribution function. Math Program Stud 27:1–16
15. Szántai T (1988) A computer code for solution of probabilistic-constrained stochastic programming problems. In: Ermoliev Y, Wets RJ-B (eds) Numerical Techniques for Stochastic Optimization. Springer, Berlin, pp 229–235
16. Szántai T: Improved bounds and simulation procedures on the value of multivariate normal probability distribution functions. Ann Oper Res (to appear) Special Issue: Research in Stochastic Programming (Selected refereed papers from the VII Internat. Conf. Stochastic Programming, Aug. 10–14, Univ. British Columbia, Vancouver, Canada).
17. Takács L (1955) On the general probability theorem. Comm Dept Math Physics Hungarian Acad Sci 5:467–476 (In Hungarian.)
18. Worsley KJ (1982) An improved Bonferroni inequality and applications. Biometrika 69:297–302

Approximations to Robust Conic Optimization Problems

MELVYN SIM
NUS Business School, National University of Singapore, Singapore, Republic of Singapore

Article Outline

Introduction
Formulation
   Affine Data Dependency
   Tractable Approximations
      of a Conic Chance Constrained Problem
References

Introduction

We consider a general conic optimization problem under parameter uncertainty is as follows:

$$\begin{aligned}
\max \quad & c'x \\
\text{s.t.} \quad & \sum_{j=1}^{n} \tilde{A}_j x_j - \tilde{B} \in \mathcal{K} \\
& x \in X,
\end{aligned} \quad (1)$$

where the cone $\mathcal{K}$ is a regular cone, i.e., a closed, convex and pointed cone. The space of the data $(\tilde{A}_1, \ldots, \tilde{A}_n, \tilde{B})$ depends on the cone, $\mathcal{K}$. The most common cone is the cone of non-negative orthant, $\Re^m_+$ in which the conic constraint in Problem (1) becomes a set of m linear constraints. Two important cones, which have many applications, include the second-order cone,

$$\mathcal{L}^{m+1} = \{(y_0, y) : \|y\|_2 \leq y_0, y \in \Re^m\}$$

and the cone of symmetric positive semidefinite matrix,

$$S^m = \{Y : Y \text{ is a symmetric postive semidefinite matrix}\}.$$

The interested reader may refer to the references of Ben-Tal and Nemirovski [3] and Pardalos and Wolkowicz [13].

In the uncertain conic optimization problem (1), the data $(\tilde{A}_1, \ldots, \tilde{A}_n, \tilde{B})$ are uncertain. It is therefore conceivable that as the data take values different than the nominal ones, the conic constraint may be violated, and the optimal solution found using the nominal data may no longer be feasible at the conic constraint. To control the feasibility level of the conic constraint, one may consider a conic chance constrained model as follows:

$$\begin{aligned} \max \quad & c'x \\ \text{s.t.} \quad & P(\sum_{j=1}^{n} \tilde{A}_j x_j - \tilde{B} \in \mathcal{K}) \geq 1 - \epsilon \\ & x \in X, \end{aligned} \quad (2)$$

in which the level of constraint violation is controlled probabilistically. Unfortunately, the chance constrained conic optimization problem (2) destroys the convexity of the problem and hence its computational tractability.

Formulation

In modern robust optimization, we represent data uncertainty using uncertainty sets instead of probability distributions. We allow the data $(\tilde{A}_1, \ldots, \tilde{A}_n, \tilde{B})$ to vary within an uncertainty set $\mathcal{U}$ without having to violate the conic constraint. We call the following problem the *robust counterpart* of Problem (1)

$$\begin{aligned} \max \quad & c'x \\ \text{s.t.} \quad & \sum_{j=1}^{n} A_j x_j - B \in \mathcal{K} \\ & \forall (A_1, \ldots, A_n, B) \in \mathcal{U} \\ & x \in X. \end{aligned} \quad (3)$$

The robust counterpart is introduced by Ben-Tal and Nemirovski [1] and independently by El-Ghoui et al. [9]. An immediate consequence of the robust counterpart is the preservation of the convexity. Unfortunately, due to the possibly infinite number of scenarios corresponding to the extreme points of the uncertainty set $\mathcal{U}$, optimizing the robust counterpart for general conic optimization problems is intractable.

It is noteworthy that in robust optimization, the ellipsoidal uncertainty set is a popular choice because of the motivation from the laws of large numbers and normal distributions. Under the assumption of normality, we could design an ellipsoidal set that is large enough so that the robust model will remain feasible with high probability. However, it turns out this approach can grossly over estimate the size of ellipsoid necessary to ensure the same level of robustness. To illustrate this issue, consider a linear constraint $\tilde{a}'x \geq b$ such that $\tilde{a}$ is a multivariate normal with mean $\bar{a}$ and covariance $\sum$. It is natural to design an ellipsoidal uncertainty set of the form $\mathcal{U} = \{a : (a - \bar{a})\Sigma^{-1}(a - \bar{a}) \leq \alpha^2\}$ so that the problem remains feasible if $\tilde{a} \in \mathcal{U}$, which has a probability of $\chi_n^2(\alpha^2)$. However, when solving the equivalent robust counterpart, $\bar{a}'x - \alpha\sqrt{x'\Sigma x} \geq b$, the robust solution has a feasibility probability of at least $\Phi(\alpha)$, where $\Phi(\alpha)$ is the standard normal function. Clearly, the value $\chi_n^2(\alpha^2)$ would be a gross over estimate of the robustness of the uncertain linear constraint compared to the value $\Phi(\alpha)$. The reason for this disparity is the fact that the uncertainty set chosen does not take into account the structure of cone.

We focus on the robust optimization framework proposed by Bertsimas and Sim [5], which offers a simple and tractable approximation of uncertain conic optimization problems. Moreover, under reasonable probabilistic assumptions on data variation, the framework approximates the conic chance constraint problem (2) by relating its feasibility probability with the size of the uncertainty set and the structure of the cone. Note that more refined approximations of chance constrained problem are available for the case of linear cones, $\mathcal{K} = \mathfrak{R}_+^m$. Interested readers can refer to Ben-Tal and Nemirovski [2], Bertsimas and Sim [4], Chen, Sim and Sun [8], Chen and Sim [6], Chen et al.[7], Lin et al. [10] and Janak et al. [11].

Affine Data Dependency

We first assume that uncertain data $(\tilde{A}_1, \ldots, \tilde{A}_n, \tilde{B})$ are affinely dependent on some primitive uncertainty vector, $\tilde{z} \in \Re^N$, as follows

$$\tilde{A}_i = A_i(\tilde{z}) \triangleq A_i^0 + \sum_{j=1}^{N} A_i^j \tilde{z}_j \quad i = 1, \ldots, n$$

$$\tilde{B} = B(\tilde{z}) \triangleq B^0 + \sum_{j=1}^{N} B^j \tilde{z}_j.$$

Note that we can always define a bijection mapping from a vector space of $\tilde{z}$ to the data space of $(\tilde{A}_1, \ldots, \tilde{A}_n, \tilde{B})$. Therefore, under the affine data dependency, it is always possible to map all the data uncertainties affecting the conic constraint to the primitive uncertainty vector, $\tilde{z}$. It is more convenient to define the following linear function mapping with respect to (z_0, z),

$$Y((z_0, z)) = \sum_{j=0}^{N} Y_j z_j,$$

in which the variables x are affinely mapped to the variables $(Y_0, \ldots, Y_N)$ as follows

$$Y_j = \sum_{i=1}^{n} A_i^j x_i - B^j \quad \forall j = 0, \ldots, N.$$

For instance, under such transformation, Problem (2) is equivalent to

$$\begin{array}{ll} \max & c'x \\ \text{s.t.} & Y_j = \sum_{i=1}^{n} A_i^j x_i - B^j \quad \forall j = 0, \ldots, N \\ & P(Y((1, \tilde{z})) \in \mathcal{K}) \geq 1 - \epsilon \\ & x \in X, \end{array} \qquad (4)$$

and Problem (3) is the same as

$$\begin{array}{ll} \max & c'x \\ \text{s.t.} & Y_j = \sum_{i=1}^{n} A_i^j x_i - B^j \quad \forall j = 0, \ldots, N \\ & Y((1, \tilde{z})) \in \mathcal{K} \quad \forall z \in \mathcal{V} \\ & x \in X, \end{array} \qquad (5)$$

in which the uncertainty set $\mathcal{U}$ is mapped accordingly to the uncertainty set $\mathcal{V}$.

Example: Quadratic Chance Constraint Consider the following quadratic chance constraint,

$$P(\|A(\tilde{z})x\|_2^2 + b(\tilde{z})'x + c(\tilde{z}) \leq 0) \geq 1 - \epsilon,$$

where $x \in \Re^n$ is the decision variable and $(A(\tilde{z}), b(\tilde{z}), c(\tilde{z})) \in \Re^{m \times n} \times \Re^n \times \Re$ are the input data, which are affinely dependent on its primitive uncertainties as follows:

$$\begin{array}{rcl} A(\tilde{z}) & \triangleq & A^0 + \sum_{j=1}^{N} A^j \tilde{z}_j \\ b(\tilde{z}) & \triangleq & b^0 + \sum_{j=1}^{N} b^j \tilde{z}_j \\ c(\tilde{z}) & \triangleq & c^0 + \sum_{j=1}^{N} c^j \tilde{z}_j. \end{array}$$

Note that a quadratic constraint

$$\|A(\tilde{z})x\|_2^2 + b(\tilde{z})'x + c(\tilde{z}) \leq 0$$

is second-order cone representable as follows

$$\begin{bmatrix} \frac{1-b(\tilde{z})'x-c(\tilde{z})}{2} \\ A(\tilde{z})x \\ \frac{1+b(\tilde{z})'x+c(\tilde{z})}{2} \end{bmatrix} \in \mathcal{L}^{m+2}.$$

Therefore, under the affine relation,

$$y_0 = \begin{bmatrix} A^0 x \\ \frac{1+b^{0'}x+c^0}{2} \\ \frac{1-b^{0'}x-c^0}{2} \end{bmatrix},$$

and

$$y_j = \begin{bmatrix} A^j x \\ \frac{b^{j'}x+c^j}{2} \\ \frac{-b^{j'}x-c^j}{2} \end{bmatrix} \quad \forall j = 1, \ldots, N$$

we transform the quadratic chance constraint problem into the following conic chance constraint

$$P\left(y_0 + \sum_{j=1}^{n} y_j \tilde{z}_j \in \mathcal{L}^{m+2} \right).$$

Hence, we treat the quadratic constraint as a special case of second-order cone constraint.

Tractable Approximations of a Conic Chance Constrained Problem

We focus on deriving a tractable approximation on the following conic chance constraint:

$$P(Y((1, \tilde{z})) \in \mathcal{K}) \geq 1 - \epsilon. \qquad (6)$$

For notational convenience, we define

$$X \triangleq (Y_0, \ldots, Y_N).$$

For a given a reference vector (or matrix), $V \in \text{int}(\mathcal{K})$, where $\text{int}(\mathcal{K})$ denotes the interior of the cone $\mathcal{K}$, we can define the function

$$f(X, (z_0, z)) \triangleq \max\{\theta : Y((z_0, z)) - \theta V \in \mathcal{K}\},$$

which has the following properties:

Proposition 1 *For any $V \in \text{int}(\mathcal{K})$, the function $f(X, (z_0, z))$ satisfies the properties:*
(a) *$f(X, (z_0, z))$ is bounded and concave in X and (z_0, z).*
(b) *$f(X, k(z_0, z)) = k f(X, (z_0, z))$, $\forall k \geq 0$.*
(c) *$f(X, (z_0, z)) \geq s$ if and only if $Y((z_0, z)) - s V \in \mathcal{K}$.*
(d) *$f(X, (z_0, z)) > s$ if and only if $Y((z_0, z)) - s V \in \text{int}(\mathcal{K})$.*

Hence, the conic chance constraint of (6) is equivalent to the following chance constraint

$$P(f(X, (1, \tilde{z})) \geq 0) \geq 1 - \epsilon. \tag{7}$$

In order to build a tractable framework that approximates the conic chance constraint problem, we first analyze the robust counterpart approach to uncertainty. Given an ellipsoidal uncertainty set

$$\mathcal{E}(\rho) = \{z : \|z\|_2 \leq \rho\},$$

the robust counterpart

$$f(X, (1, z)) \geq 0 \quad \forall z \in \mathcal{E}(\rho), \tag{8}$$

despite its convexity, is generally intractable. Instead we consider the following robust counterpart:

$$f(X, (1, 0)) + \sum_{j=1}^{N} \{f(X, (0, e_j))v_j \\ + f(X, (0, -e_j))w_j\} \geq 0, \\ \forall (v, w) \in \mathcal{V}(\rho) \tag{9}$$

where $e_j \in \Re^N$ is a unit vector with one at the jth entry and the uncertainty set

$$\mathcal{V}(\rho) = \{(v, w) \in \Re_+^N \times \Re_+^N \mid \|v + w\|_2 \leq \rho\}. \tag{10}$$

Proposition 2 *The robust counterpart (9) is tractable relaxation of the robust counterpart, (8).*

Theorem 1
(a) *The constraint (9) is equivalent to*

$$f(X, (1, 0)) \geq \rho \|s\|_2, \tag{11}$$

where

$$s_j = \max\{-f(X, (0, e_j)), -f(X, (0, -e_j))\}, \\ \forall j = 1, \ldots, N.$$

(b) *Eq. (11) can be written as:*

$$f(X, (1, 0)) \geq \rho y \\ f(X, (0, e_j)) + t_j \geq 0, \quad \forall j \in N \\ f(X, (0, -e_j)) + t_j \geq 0, \quad \forall j \in N \quad (12) \\ \|t\|_2 \leq y \\ \text{for some } y \in \Re, \; t \in \Re^N.$$

From Proposition 1 and noting that

$$Y((1, 0)) = Y_0$$

and

$$Y((0, \pm e_j)) = \pm Y_j,$$

we can also represent the formulation (12) explicitly in conic constraints as follows:

$$Y_0 - \rho y V \in \mathcal{K} \\ Y_j + t_j V \in \mathcal{K}, \quad \forall j \in N \\ -Y_j + t_j V \in \mathcal{K}, \quad \forall j \in N \quad (13) \\ \|t\|_2 \leq y \\ \text{for some } y \in \Re, \; t \in \Re^N,$$

for a given reference vector, V in the interior of the cone, $\mathcal{K}$. The formulation (12) becomes a cartesian product of $2N + 1$ cones of the nominal problem plus an additional second-order cone, which is a computationally tractable cone. Hence, in theory the formulation (12) is not much harder to solve compared with its nominal problem.

One natural question is whether the simple approximation is overly conservative with respect to Problem (8). While there is lack of theoretical evidence on the closeness of the approximation, the framework does lead to an approximation of the conic chance constraint problem. An important component of the analysis is the relation among different norms, which we will subsequently present.

Recall that a norm satisfies $\|A\| \geq 0$, $\|kA\| = |k| \cdot \|A\|$, $\|A+B\| \leq \|A\|+\|B\|$, and $\|A\| = 0$, implies that $A = 0$. For a given regular cone, $\mathcal{K}$, and its interior, V, we define the following cone induced norm

$$\|Y\|_{\mathcal{K},V} \triangleq \min\{y, \, yV - Y \in \mathcal{K}, \, Y + yV \in \mathcal{K}\}. \quad (14)$$

Proposition 3

$$\max\{-f(X, (z_0, z)), -f(X, -(z_0, z))\} = \|Y((z_0, z))\|_{\mathcal{K},V}.$$

We consider the common cones and the respective norms.

(a) Second-order cone:

Let $e_1 \in \text{int}(\mathcal{L}^{n+1})$ be the reference vector, we have for any vector $(y_0, y) \in \Re^{n+1}$

$$\|(y_0, y)\|_{\mathcal{L}^{n+1}, e_{n+1}}$$
$$= \min\{\theta: \|y\|_2 \leq \theta - y_0, \|y\|_2 \leq \theta + y_0\}$$
$$= \|y\|_2 + |y_0|$$

(b) Cone of symmetric positive definite matrix:

Let the identity matrix I be the reference matrix, then for any $m \times m$ symmetric matrix, Y,

$$\|Y\|_{S_+^m, I} = \min\{y, \, yI - Y \in S_+^m, \, Y - yI \in S_+^m\}$$
$$= \|Y\|_2.$$

Proposition 4 *Suppose X is feasible in Problem (12) then*

$$P(f(X, (1, \tilde{z})) < 0)$$
$$\leq P\left(\left\|\sum_{j=1}^N Y_j \tilde{z}_j\right\|_{\mathcal{K},V} > \rho \sqrt{\sum_{j \in N} \|Y_j\|_{\mathcal{K},V}^2}\right).$$

To obtain explicit bounds, we focus on primitive uncertainties, $\tilde{z}$ that are normally and independently distributed with mean zero and variance one. For a sum of random scalers, we have

$$P\left(\left|\sum_{j=1}^N y_j \tilde{z}_j\right| > \rho \sqrt{\sum_{j=1}^N y_j^2}\right) \leq 1 - 2\Phi(\rho).$$

To derive a similar large deviation result for the sum of random vectors used in Proposition 4, we consider the following generalization:

$$P\left(\left\|\sum_{j=1}^N Y_j \tilde{z}_j\right\|_{\mathcal{K},V} > \rho \sqrt{\sum_{j=1}^N \|Y_j\|_{\mathcal{K},V}^2}\right) \leq \phi(\rho),$$

where $\phi(\rho)$ is a non-trivial probability bound that depends on the choice of cone, $\mathcal{K}$, and possibly the dimension and the reference vector, V.

An important component of the analysis is the relation among different norms. We denote by $\langle \, , \, \rangle$ the inner product on a vector space, $\Re^m$, or the space of m by m symmetric matrices. The inner product induces a norm $\|X\| \triangleq \sqrt{\langle X, X \rangle}$. For a vector space, the natural inner product is the Euclidian inner product, $\langle x, y \rangle = x'y$, and the induced norm is the Euclidian norm $\|x\|_2$. For the space of symmetric matrices, the natural inner product is the trace product or $\langle X, Y \rangle = \text{trace}(XY)$ and the corresponding induced norm is the Frobenius norm, $\|Y\|_F$.

We analyze the relation of the inner product norm $\sqrt{\langle X, X \rangle}$ with the norm $\|X\|_{\mathcal{K},V}$ for the conic optimization problems we consider. Since $\|X\|_{\mathcal{K},V}$ and the inner product norm $\|X\|$ are valid norms in a finite dimensional space, there exist finite $\alpha_1, \alpha_2 > 0$ such that

$$\frac{1}{\alpha_1}\|X\|_{\mathcal{K},V} \leq \|X\| \leq \alpha_2 \|X\|_{\mathcal{K},V},$$

for all X in the relevant space. Hence, we define the parameter

$$\alpha_{\mathcal{K},V} = \underbrace{\left(\max_{\|X\|=1} \|X\|_{\mathcal{K},V}\right)}_{=\alpha_1} \underbrace{\left(\max_{\|X\|_{\mathcal{K},V}=1} \|X\|\right)}_{=\alpha_2} \quad (15)$$

which measures the disparity between the norm $\|\cdot\|_{\mathcal{K},V}$ and the inner product norm $\|\cdot\|$.

Parameter $\alpha_{\mathcal{K},V}$ of Common Cones

(a) Second-order cone:

Let e_{n+1} be the reference vector, then

$$\|(y, y_{n+1})\|_{\mathcal{L}^{n+1}, V} = \|y\|_2 + |y_{n+1}|.$$

Therefore,

$$\frac{1}{\sqrt{2}}\|(y, y_{n+1})\|_{\mathcal{L}^{n+1}, e_{n+1}} \leq \|(y, y_{n+1})\|_2$$
$$\leq \|(y, y_{n+1})\|_{\mathcal{L}^{n+1}, e_{n+1}}$$

Approximations to Robust Conic Optimization Problems, Table 1
Probability bounds of $P(f(X, (1, \tilde{z})) < 0)$ for $\tilde{z} \sim \mathcal{N}(0, I)$.

Type	Probability bound of infeasibility
$\mathcal{L}^{m+1}$	$\sqrt{\frac{e}{2}}\Omega \exp(-\frac{\Omega^2}{4})$
$\mathcal{S}^m_+$	$\sqrt{\frac{e}{m}}\Omega \exp(-\frac{\Omega^2}{2m})$

and hence,

$$\alpha_{\mathcal{L}^{n+1}, v} = \sqrt{2}.$$

(b) Cone of symmetric positive definite matrix:
Let I be the reference matrix, then for any $m \times m$ symmetric matrix Y

$$\|Y\|_{\mathcal{S}^m_+, I} = \|Y\|_2.$$

Let λ_j, $j = 1, \ldots, m$ be the eigenvalues of the matrix Y. Since $\|Y\|_F = \sqrt{\text{trace}(Y^2)} = \sqrt{\sum_j \lambda_j^2}$ and $\|Y\|_2 = \max_j |\lambda_j|$, we have

$$\|Y\|_2 \leq \|A\|_F \leq \sqrt{m} \|Y\|_2.$$

Hence,

$$\alpha_{\mathcal{S}^m_+, I} = \sqrt{m}.$$

Theorem 2 *Given an inner product norm $\|\cdot\|$ and under the assumption that $\tilde{z}_j$ are normally and independently distributed with mean zero and variance one, i.e., $\tilde{z} \sim \mathcal{N}(0, I)$, then*

$$P\left(\left\|\sum_{j=1}^N Y_j \tilde{z}_j\right\|_{\mathcal{K}, V} > \rho \sqrt{\sum_{j \in N} \|Y_j\|^2_{\mathcal{K}, V}}\right)$$
$$\leq \frac{\sqrt{e}\rho}{\alpha_{\mathcal{K}, V}} \exp\left(-\frac{\rho^2}{2\alpha^2_{\mathcal{K}, V}}\right), \quad (16)$$

for all $\rho > \alpha_{\mathcal{K}, V}$.

In order to have the smallest budget of uncertainty, ρ, it is reasonable to select V that minimizes $\alpha_{\mathcal{K}, V}$, i.e.,

$$\alpha_{\mathcal{K}} = \min_{V \in \text{int}(\mathcal{K})} \alpha_{\mathcal{K}, V}.$$

For general conic optimization, we have shown that the probability bound depends on the the choice of $V \in \text{int}(\mathcal{K})$. A cone, $\mathcal{K} \subseteq \Re^n$ is *homogenous* if for any pair of points $A, B \in \text{int}(\mathcal{K})$ there exists an invertible linear map $M: \Re^n \to \Re^n$ such that $M(A) = B$ and $M(\mathcal{K}) = \mathcal{K}$. It turns out that for *homogenous cones*, of which semidefinite and second-order cones are special cases, the probability bound does not depend on $V \in \text{int}(\mathcal{K})$.

Theorem 3 *Suppose the cone $\mathcal{K}$ is homogenous. For any $V \in \text{int}(\mathcal{K})$, the probability bound of conic infeasibility satisfies*

$$P(Y((1, z)) \notin \mathcal{K}) \leq \frac{\sqrt{e}\rho}{\alpha_{\mathcal{K}}} \exp\left(-\frac{\rho^2}{2\alpha^2_{\mathcal{K}}}\right).$$

For the second-order cone, $\alpha_{\mathcal{L}^{n+1}} = \sqrt{2}$ and for the symmetric positive semidefinite cone, $\alpha_{\mathcal{S}^m_+} = \sqrt{m}$.

While different V lead to the same probability bounds, some choices of V may lead to better objectives. The following theorem suggests an iterative improvement strategy.

Theorem 4 *For any $V \in \text{int}(\mathcal{K})$, if X, t and $y > 0$ are feasible in (13), then they are also feasible in the same problem in which V is replaced by*

$$W = Y_0/(\rho y).$$

While we focus on the primitive uncertainty vector $\tilde{z}$ being normally distributed, using the large deviation bounds of Nemirovski [12], we can also apply the same framework to other distributions. The interested reader may refer to Bertsimas and Sim [5].

References

1. Ben-Tal A, Nemirovski A (1998) Robust convex optimization. Math Oper Res 23:769–805
2. Ben-Tal A, Nemirovski A (2000) Robust solutions of Linear Programming problems contaminated with uncertain data. Math Program 88:411–424
3. Ben-Tal A, Nemirovski A (2001) Lectures on Modern Convex Optimization: Analysis, Algorithms, and Engineering Applications. MPR-SIAM Series on Optimization. SIAM, Philadelphia
4. Bertsimas D, Sim M (2004) Price of robustness. Oper Res 52:35–53
5. Bertsimas D, Sim M (2006) Tractable Approximations to Robust Conic Optimization Problems. Math Program 107(1):5–36

6. Chen W, Sim M (2007) Goal Driven Optimization. Working Paper, NUS Business School
7. Chen W, Sim M, Sun J, Teo C-P (2007) From CVaR to Uncertainty Set: Implications in Joint Chance Constrained Optimization. Working Paper, NUS Business School
8. Chen X, Sim M, Sun P (2006) A robust optimization perspective on stochastic programming. to appear in: Oper Res 55(6)
9. El Ghaoui L, Oustry F, Lebret H (1998) Robust Solutions to Uncertain Semidefinite Programs. SIAM J Optim 9(1):33–52
10. Lin X, Janak SL, Floudas CA (2004) A New Robust Optimization Approach for Scheduling under Uncertainty: I Bounded Uncertainty. Comput Chem Eng 28:1069–1085
11. Janak SL, Lin X, Floudas CA (2007) A New Robust Optimization Approach for Scheduling under Uncertainty: II Uncertainty with Known Probability Distribution. Comput Chem Eng 31:171–195
12. Nemirovski A (2003) On tractable approximations of ramdomly perturbed convex constraints. Proceedings of the 42nd IEEE Conference on Decision and Control, Maui, Hawaii, USA, December, pp 2419–2422
13. Pardalos PM, Wolkowicz H (1998) Topics in Semidefinite and Interior-Point Methods. Fields Institute Communications Series vol 18, American Mathematical Society

Archimedes and the Foundations of Industrial Engineering

PETROS XANTHOPOULOS
Industrial and Systems Engineering, University of Florida, Gainesville, USA

MSC2000: 01A20

Article Outline

Biographical Sketch
  Archimedes' Work
Conclusion
References

Biographical Sketch

Archimedes (287–212 B.C.) was a famous Greek mathematician, engineer and philosopher. Born in the city of Syracuse on the Island of Sicily to an astronomer and mathematician named Phidias, Archimedes spent the first years of his life in his home city and went to Alexandria in Egypt to study mathematics. He soon became friends with Konon of Samos and Eratosthenes. After spending a considerable amount of time in Alexandria, he returned to Syracuse, where he remained for the rest of his life conducting mathematical research. He had a good relationships with king Hieron of Syracuse and his son Gelon. We know that he assisted king Hieron numerous times either with his inventions during the Second Punic War or by solving problems like the well-known case (the one that Archimedes jumped out of his bathtub crying out eureka) with the crown of king Hieron during peacetime.

In this article we will concentrate on the work of Archimedes, which is closely related to what we call today industrial engineering (including the mathematical theory of optimization, operations research, theory of algorithms, etc.). In particular, we will present Archimedes' definition of convex sets, his method of exhaustion for computing finite integrals, his contribution to recursive algorithms, and his approach to solving real-life operations research problems during the Second Punic War.

Archimedes' Work

One very important concept for optimization is the definition of convex sets. The first such definition was given by Euclid in his books *Elements*, but Archimedes elaborated this definition and gave us his definition, which was used until the first decades of the 20th century. In his work *On the sphere and the cylinder* he gives the following definition of the convex arc:

Definition 1 I call convex in one and the same directions the surfaces for which the straight line joining two arbitrary points lies on the same side of the surface.

On his work *On the equilibrium of planes* he gives a definition of the convex set using the center-of-gravity concept:

Definition 2 In any figure whose perimeter is convex the center of gravity must be within the figure.

It is worth mentioning that Archimedes' definitions of convex arcs and convex sets were those used until 1913, when E. Steinitz introduced the modern definitions of convexity.

Archimedes had invented a geometrical method called the method of exhaustion (or method of infinitesimals) in order to be able to compute areas under convex curves. This was one of the first geometrical methods devised to compute what we call today definite

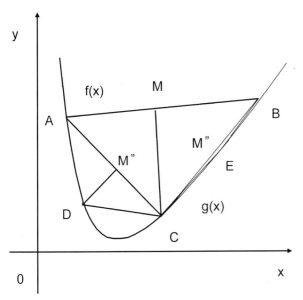

Archimedes and the Foundations of Industrial Engineering, Figure 1
Illustration of Archimedes' exhaustion method

integrals. In modern notation Archimedes was able to compute

$$\int_a^b [f(x) - g(x)]dx, \qquad (1)$$

where $f(x)$ is a line segment and $g(x)$ a convex function (usually parabola). An illustration of this method can be found in Fig. 1.

Suppose that we want to compute the area over a curve and below the line segment AB. Archimedes considered the triangle $\widehat{ABC}$, where C is the point below the midpoint M of the line segment AB (MC is the middle vertical of AB). If we iteratively repeat this process, we can see that the next two parabolic triangles have an area that is $\frac{1}{4}$ of the initial triangle. Therefore, the area of the curve was the infinite sum of $1 + \frac{1}{4} + \frac{1}{8} + \ldots$, where 1 corresponds to the area of the initial triangle $\widehat{ABC}$. In this way Archimedes was able to geometrically approximate the area of a convex parabolic curve.

According to [7] Archimedes was the first (in around 220 B.C.) to use a double recursive algorithm to solve the problem of the sand reckoner (Psammitis). In this book he tries to come up with of a number that is much larger than the number of grains of sand in the world and therefore prove that the number of grains of sand in the world is not infinite. For this he fixes a number α and defines the number $p_k(x)$ as follows (using a double recursion scheme):

$$p_0(x) = 1,$$
$$p_{k+1}(0) = p_k(\alpha), \qquad (2)$$
$$p_{n+1}(x+1) = ap_{k+1}(x).$$

Therefore, $p_k(x) = \alpha^{xk}$. Then he considers $p_\alpha(\alpha)$ for $\alpha = 10^8$, which was the largest number known at that time, and he comes up with the number $10^{10}17$, which was the largest number used in mathematics until 1933.

Apart from Archimedes' exceptional skills in theoretical research, he also became famous for his ability to deal with everyday life problems. Although operations research was developed during World War II, when mathematicians were looking for ways to make better decisions in utilizing certain materials subject to some constraint, some consider Archimedes the father of operations research as he helped his home city defend itself against the Romans during the Second Punic War.

Before King Hieron died, he asked Archimedes to organize the complete defense of Syracuse against Roman general Marcelus. Archimedes is said to have invented many mechanical war machines like the claw of Archimedes, a new version of catapult, an array of mirrors that was able to burn enemy ships, etc.

Archimedes was also responsible for organizing the defense of Syracuse and the redecoration of Fort Euryalus [6]. Due to Archimedes' clever defense plans, Syracuse managed to survive the Roman siege for 2 years.

Conclusion

Archimedes was a perfect example of a scientist who managed to combine theoretical research with practical problem solving. He managed to distinguish between the two by referring to his mechanical inventions as *parergon*. This shows that Archimedes was capable of performing both basic and applied research, but he regarded basic research as more important. In this sense he can be considered the father of the modern industrial engineer who utilizes theoretical methods to solve problems that arise in everyday life.

References

1. Brunschwig J, Lloyd GER (eds) (2000) Greek thought: a guide to classical knowledge. Belknap, Cambridge, Massachusetts
2. Dijksterhuis EJ (1987) Archimedes. Princeton University Press, Princeton, NJ
3. Gow M (2005) Archimedes: Mathematical Genius of the Ancient World. Great Minds of Science series. Enslow, Berkeley Heights, NJ
4. Heath TL (1897) The works of Archimedes. Cambridge University Press, Cambridge
5. Heath TL (ed) (2002) The Works of Archimedes. Dover, New York
6. Lawrence AW (1946) Archimedes and the Design of Euryalus Fort. J Hell Stud – JSTOR
7. Odifreddi P. Recursive Functions. In: Edward Zalta N (ed) The Stanford Encyclopedia of Philosophy (Summer 2005 edn). http://plato.stanford.edu/archives/sum2005/entries/recursive-functions. Accessed 21 Mar 2008
8. Simms DL (1995) Archimedes the Engineer. History of Technology, vol 17. Continuum International, London, pp 45–111
9. Stein S (1999) Archimedes: What Did He Do Besides Cry Eureka? Mathematical Association of America, Washington, DC

Asset Liability Management Decision Support System

KOSMIDOU KYRIAKI[1], ZOPOUNIDIS CONSTANTIN[2]

[1] Department of International European Economic Studies, Athens University Economics and Business, Athens, Greek

[2] Department of Production Engineering and Management, Technical University of Crete, Chania, Greece

MSC2000: 90C29, 65K99

Article Outline

Introduction
Background
Model
Conclusions
References

Introduction

Asset Liability Management (ALM) is an important dimension of risk management, where the exposure to various risks is minimized while maintaining the appropriate combination of asset and liability, in order to satisfy the goals of the firm or the financial institution (Kosmidou and Zopounidis [18]).

Up to the 1960's, liability management was aimless. In their majority, the banking institutions considered liabilities as exogenous factors contributing to the limitation of asset management. Indeed, for a long period the greater part of capital resources originated from savings deposits and deposits with agreed maturity.

Nevertheless, the financial system has radically changed. Competition among the banks for obtaining capital has become intense. Liability management is the main component of each bank strategy in order to ensure the cheapest possible financing. At the same time, the importance of decisions regarding the amount of capital adequacy is enforced. Indeed, the adequacy of the bank as far as equity, contributes to the elimination of bankruptcy risk, a situation in which the bank cannot satisfy its debts towards clients who make deposits or others who take out loans. Moreover, the capital adequacy of banks is influenced by the changes of stock prices in relation to the amount of the capital stock portfolio. Finally, the existence of a minimum amount of equity is an obligation of commercial banks to the Central Bank for supervisory reasons. It is worth mentioning that based on the last published data (31/12/2001) the Bank of Greece assigns the coefficient for the Tier 1 capital at 8%, while the corresponding European average is equal to 6%. This results in the configuration of the capital adequacy of the Greek banking system at higher levels than the European average rate. The high capital adequacy index denotes large margins of profitability amelioration, which reduces the risk of a systematic crisis.

Asset management in a contemporary bank cannot be distinct from liability management. The simultaneous management of assets and liabilities, in order to maximize the profits and minimize the risk, demands the analysis of a series of issues.

Firstly, there is the substantive issue of strategic planning and expansion. That is, the evaluation of the total size of deposits that the bank wishes to attract and the total number of loans that it wishes to provide.

Secondly, there is the issue of determination of the "best temporal structure" of the asset liability management, in order to maximize the profits and to ensure the robustness of the bank. Deposits cannot all be liquidated in the same way. From the point of view of assets, the loans and various placements to securities constitute commitments of the bank's funds with a different duration time. The coordination of the temporal structure of the asset liability management is of major importance in order to avoid the problems of temporary liquidity reduction, which might be very injurious.

Thirdly, there is the issue of risk management of assets and liabilities. The main focus is placed on the assets, where the evaluation of the quality of the loans portfolio (credit risk) and the securities portfolio (market risk) is more easily measurable.

Fourthly, there is the issue of configuration of an integrated invoice, which refers to the entire range of bank operations. It refers mainly to the determination of interest rates for the total of loans and deposits as well as for the various commissions which the bank charges for specific mediating operations. It is obvious that in a bank market which operates in a competitive environment, there is no issue of pricing. This is true even in the case where all interest rates and commissions are set by monetary authorities, as was the situation in Greece before the liberalization of the banking system.

In reality, bank markets have the basic characteristics of monopolistic competition. Thus, the issue of planning a system of discrete pricing and product diversification is of major importance. The problem of discrete pricing, as far as the assets are concerned, is connected to the issue of risk management. It is a common fact that the banks determine the borrowing interest rate on the basis of the interest rates which increase in proportion to the risk as they assess it in each case. The product diversification policy includes all the loan and deposit products and is based on thorough research which ensures the best possible knowledge of market conditions.

Lastly, the management of operating cost and technology constitutes an important issue. The collaboration of a well-selected and fully skilled personnel, as well as contemporary computerization systems and other technological applications, constitutes an important element in creating a low-cost bank. This results in the acquisition of a significant competitive advantage against other banks, which could finally be expressed through a more aggressive policy of attracting loans and deposits with low loan interest rates and high deposit interest rates. The result of this policy is the increase of the market stake. However, the ability of a bank to absorb the input of the best strategic technological innovations depends on the human resources management.

The present research focuses on the study of bank asset liability management. Many are the reasons that lead us to study bank asset liability management, as an application of ALM. Firstly, bank asset/liability management has always been of concern to bank managers, but in the last years and especially today its importance has grown more and more. The development of information technology has led to such an increasing public awareness that the bank's performance, its politics and its management are closely monitored by the press and the bank's competitors, shareholders and customers and thereby highly affect the bank's public standing.

The increasing competition in the national and international banking markets, the changeover towards the monetary union and the new technological innovations herald major changes in the banking environment and challenge all banks to make timely preparations in order to enter into the new competitive monetary and financial environment.

All the above drove banks to seek out greater efficiency in the management of their assets and liabilities. Thus, the central problem of ALM revolves around the bank's balance sheet and the main question that arises is: What should be the composition of a bank's assets and liabilities on average given the corresponding returns and costs, in order to achieve certain goals, such as maximization of the bank's gross revenues?

It is well known that finding an appropriate balance between profitability, risk and liquidity considerations is one of the main problems in ALM. The optimal balance between these factors cannot be found without considering important interactions that exist between the structure of a bank's liabilities and capital and the composition of its assets.

Bank asset/liability management is defined as the simultaneous planning of all asset and liability positions on the bank's balance sheet under consideration of the different banking and bank management objectives and legal, managerial and market constraints. Banks are looking to maximize profit and minimize risk.

Taking into consideration all the above, the purpose of this paper is to develop a goal programming system into a stochastic environment, focusing, mainly, on the change of the interest rate risk. This system provides the possibility to the administrative board and the managers of the bank to proceed to various scenarios related to their future economic process, aiming mainly to the management of the risks, emerged from the changes of the market parameters.

The rest of the paper is organized as follows. The next section includes a brief overview of bank ALM techniques. Section "Model" outlines the methodology used and describes the development of the ALM decision support system. Finally, the conclusions of the paper as well as future research perspectives are discussed in the last section.

Background

Looking to the past, we find the first mathematical models in the field of bank management. Asset and liability management models can be deterministic or stochastic (Kosmidou and Zopounidis [17]).

Deterministic models use linear programming, assume particular realizations for random events, and are computationally tractable for large problems. The deterministic linear programming model of Chambers and Charnes [6] is the pioneer in ALM. Chambers and Charnes were concerned with formulating, exploring and interpreting the use and construction which may be derived from a mathematical programming model which expresses more realistically than past efforts the actual conditions of current operations. Their model corresponds to the problem of determining an optimal portfolio for an individual bank over several time periods in accordance with requirements laid down by bank examiners which are interpreted as defining limits within which the level of risk associated with the return on the portfolio is an acceptable one.

Cohen and Hammer [9], Robertson [31], Lifson and Blackman [23], Fielitz and Loeffler [14] have realized successful applications of Chambers and Charnes' model. Even though these models have differed in their treatment of disaggregation, uncertainty and dynamic considerations, they all have in common the fact that they are specified to optimize a single objective profit function subject to the relevant linear constraints.

Eatman and Sealey [12] developed a multiobjective linear programming model for commercial bank balance sheet management considering profitability and solvency objectives subject to policy and managerial constraints.

Giokas and Vassiloglou [15] developed a goal-programming model for bank asset and liability management. They supported the idea that apart from attempting to maximize revenues, management tries to minimize risks involved in the allocation of the bank's capital, as well as to fulfill other goals of the bank, such as retaining its market share, increasing the size of its deposits and loans, etc. Conventional linear programming is unable to deal with this kind of problem, as it can only handle a single goal in the objective function. Goal programming is the most widely used approach that solves large-scale multi-criteria decision making problems.

Apart from the deterministic models, several stochastic models have been proposed since the 1970s. These models, including the use of chance-constrained programming [7,8,29], dynamic programming [13,25,26,32], sequential decision theory [3,35] and stochastic linear programming under uncertainty [2,10,11,16], presented computational difficulties. The stochastic models, in their majority, originate from the portfolio selection theory of Markowitz [24] and they are known as static mean-variance methods. Pyle [30] and Brodt [4] adapted Markowitz's theory and presented an efficient dynamic balance sheet management plan that considers only the risk of the portfolio and not other possible uncertainties or maximizes profits for a given amount of risk over a multi-period planning horizon respectively.

Wolf [35] proposed the sequential decision theoretic approach that employs sequential decision analysis to find an optimal solution through the use of implicit enumeration.

An alternative approach in considering stochastic models, is the stochastic linear programming with simple recourse. Kusy and Ziemba [19] employed a multi-period stochastic linear program with simple recourse to model the management of assets and liabilities in banking while maintaining computational feasibility. Their results indicate that the proposed ALM model is theoretically and operationally superior to a corresponding deterministic linear programming model and that the computational effort required for its implementation is comparable to that of the deterministic model. Another application of the multistage stochastic programming is the Russell-Yasuda Kasai model [5], which aims at maximizing the long term wealth of the firm while producing high income returns.

Mulvey and Vladimirou [27] used dynamic generalized network programs for financial planning problems under uncertainty and they developed a model in the framework of multi-scenario generalized network that captures essential features of various discrete time financial decision problems.

Finally, Mulvey and Ziemba [28] present a more detailed overview of various asset and liability modeling techniques, including models for individuals and financial institutions such as banks and insurance companies.

Moreover, over the years, many models have been developed in the area of financial analysis and financial planning techniques. Kvanli [20], Lee and Lerro [22], Lee and Chesser [21], Baston [1], Sharma et al. [34], among others have applied goal programming to investment planning. Giokas and Vassiloglou [15], Seshadri et al. [33] presented bank models using goal programming. These studies focus on the areas of banking and financial institutions and they use data from the bank financial statements.

Model

Kosmidou and Zopounidis [18] developed an asset liability management (ALM) methodology into a stochastic environment of interest rates in order to select the best direction strategies to the banking financial planning. The ALM model was developed through goal programming in terms of a one-year time horizon. The model used balance sheet and income statement information for the previous year of the year t to produce a future course of ALM strategy for the year $t + 1$. As far as model variables are concerned, we used variables familiar to management and facilitated the specification of the constraints and goals. For example, goals concerning measurements such as liquidity, return and risk have to be expressed in terms of utilized variables.

More precisely, the asset liability management model that was developed can be expressed as follows:

$$\min z = \sum_{P} p_k (d_k^- + d_k^+) \quad (1)$$

subject to constraints:

$$K\Phi_{X'} \leq X' \leq A\Phi_{X'} \quad (2)$$

$$K\Phi_{Y'} \leq Y' \leq A\Phi_{Y'} \quad (3)$$

$$\sum_{i=1}^{n} X_i = \sum_{j=1}^{m} Y_j \quad \forall i = 1, \ldots, n, \quad \forall j = 1, \ldots, m \quad (4)$$

$$\sum_{j \in \Pi_{Y''}} Y_j - a \sum_{i \in E_{X''}} X_i = 0 \quad (5)$$

$$\sum_{j \in \Pi_1} Y_j - \sum_{i \in E} w_i X_i - d_s^+ + d_s^- = k_1 \quad (6)$$

$$\sum_{i \in E_x} X_i - k_2 \sum_{j \in \Pi_k} Y_j + d_l^- - d_l^+ = 0 \quad (7)$$

$$\sum_{i=1}^{n} R_i^X X_i - \sum_{j=1}^{m} R_j^Y Y_j - d_r^+ + d_r^- = k_3 \quad (8)$$

$$\sum_{i \in E_p} X_i + d_p^- - d_p^+ = l_p, \quad \forall p \quad (9)$$

$$\sum_{j \in \Pi_p} Y_j + d_p^- - d_p^+ = l_p, \quad \forall p \quad (10)$$

$$X_i \geq 0, Y_j \geq 0, d_k^+ \geq 0, d_k^- \geq 0,$$
$$\text{for all} \quad i = 1, \ldots, n, j = 1, \ldots, m, k \in P \quad (11)$$

where

X_i: the element i of asset, $\forall i = 1, \ldots, n$, n is the number of asset variables

Y_j: the element j of liability, $\forall j = 1, \ldots, m$, m is the number of liability variables

$K\Phi_{X'}$ ($K\Phi_{Y'}$): is the low bound of specific asset accounts X' (liability Y')

$A\Phi_{X'}$ ($A\Phi_{Y'}$): is the upper bound of specific asset accounts X' (liability Y')

$E_{X''}$: specific categories of asset accounts

$\Pi_{Y''}$: specific categories of liability accounts

α: the desirable value of specific asset and liability data

Π_1: the liability set, which includes the equity

E: the set of assets

w_i: the degree of riskness of the asset data

k_1: the solvency ratio, as it is defined from the European Central Bank.

k_2: the liquidity ratio, as it is defined from the bank policy

E_χ: the set of asset data, which includes the loans

Π_κ: the set of liability data, which includes the deposits

R_i^X: the expected return of the asset i, $\forall i = 1, \ldots, n$

R_j^Y: the expected return of the liability j, $\forall j = 1, \ldots, m$

k_3: the expected value for the goal of asset and liability return

P: the goal imposed from the bank

L_p: the desirable value goal for the goal constraint p defined by the bank

d_k^+: the over-achievement of the goal k, $\forall k \in P$

d_k^-: the under-achievement of the goal k, $\forall k \in P$

p_k: the priority degree (weight) of the goal k

Certain constraints are imposed by the banking regulation on particular categories of accounts. Specific categories of asset accounts (X') and liability accounts (Y') are detected and the minimum and maximum allowed limit for these categories are defined based on the strategy and policy that the bank intends to follow (constraints 2–3).

The structural constraints (4–5) include those that contribute to the structure of the balance sheet and especially to the performance of the equation Assets = Liabilities + Net Capital.

The bank management should determine specific goals, such as the desirable structure of each financial institution's assets and liabilities for the units of surplus and deficit, balancing the low cost and the high return. The structure of assets and liabilities is significant, since it affects swiftly the income and profits of the bank.

Referring to the goals of the model, the solvency goal (6) is used as a risk measure and is defined as the ratio of the bank's equity capital to its total weighted assets. The weighting of the assets reflects their respective risk, greater weights corresponding to a higher degree of risk. This hierarchy takes place according to the determination of several degrees of significance for the variables of assets and liabilities. That is, the variables with the largest degrees of significance correspond to categories of the balance sheet accounts with the highest risk stages.

Moreover, a basic policy of the commercial banks is the management of their liquidity and specifically the measurement of their needs that is relative to the progress of deposits and loans. The liquidity goal (7) is defined as the ratio of liquid assets to current liabilities and indicates the liquidity risk, that indicates the possibility of the bank to respond to its current liabilities with a security margin, which allows the probable reduction of the value of some current data.

Furthermore, the bank aims at the maximization of its efficiency that is the accomplishment of the largest possible profit from the best placement of its funds. Its aim is the maximization of its profitability and therefore precise and consistent decisions should be taken into account during the bank management. These decisions will guarantee the combined effect of all the variables that are included on the calculation of the profits. This decision taking gives emphasis to several selected variables that are related to the bank management, such as to the management of the difference between the asset return and the liability cost, the expenses, the liquidity management and the capital management. The goal (8) determines the total expected return based on the expected returns for all the assets R^X and liabilities R^Y.

Beside the goals of solvency, liquidity and return of assets and liabilities, the bank could determine other goals that concern specific categories of assets and liabilities, in proportion to the demands and preferences of the bank managers. These goals are the deposit goal, the loan goal and the goal of asset and liability return.

The drawing of capital, especially from the deposits constitutes a major part of commercial bank management. All sorts of deposits constitute the major source of capital for the commercial banks, in order to proceed to the financing of the economy, through the financing of firms. Thus, it is given special significance to the deposits goal.

The goal of asset and liability return defines the goal for the overall expected return of the selected asset-liability strategy over the year of the analysis.

Finally, there are goals reflecting that variables such as cash, cheques receivables, deposits to the Bank of Greece and fixed assets, should remain at the levels of previous years. More analytically, it is known that the fixed assets are the permanent assets, which have a natural existence, such as buildings, machines, locations and equipment, etc. Intangible assets are the fixed assets, which have no natural existence but constitute rights and benefits. They have significant economic value, which sometimes is larger than the value of the tangible fixed assets. These data have stable character and are used productively by the bank for the regular operation and performance of its objectives. Since the fixed assets, tangible or intangible, are presented at the balance sheet at their book value that is the initial value of cost minus the depreciation till today, it is assumed that their value does not change during the development of the present methodology.

At this point, Kosmidou and Zopounidis [18] took into account that the banks should manage the interest rate risk, the operating risk, the credit risk, the market risk, the foreign exchange risk, the liquidity risk and the country risk.

More specifically, the interest rate risk indicates the effect of the changes to the net profit margin between the deposit and borrowing values, which are evolved as a consequence of the deviations to the dominant interest rates of assets and liabilities. When the interest rates diminish, the banks accomplish high profits since they can refresh their liabilities to lower borrowing values. The reverse stands to high borrowing values. It is obvious, that the changes of the inflation have a relevant impact on the above sorts of risk.

Considering the interest rate risk as the basic uncertainty parameter to the determination of a bank asset liability management strategy, the crucial question that arises concerns the determination of the way through which this factor of uncertainty affects the profitability of the pre-specified strategy. The estimation of the expected return of the pre-specified strategy and of its variance can render a satisfactory response to the above question.

The use of Monte Carlo techniques constitutes a particular widespread approach for the estimation of the above information (expected return – variance of bank asset liability management strategies). Monte Carlo simulation consists in the development of various random scenarios for the uncertain variable (interest rates) and the estimation of the essential statistical measures (expected return and variance), which describe the effect of the interest rate risk to the selected strategy. The general procedure of implementation of Monte Carlo simulation based on the above is presented in Fig. 1.

During the first stage of the procedure the various categories of the interest rate risks are identified. The risk and the return of the various data of bank asset and liability are determined from the different forms of interest rates. For example, the investments of a bank to government or corporate bonds are determined from the interest rates that prevail in the bond market, which are affected so by the general economic environment as by the rules of demand and supply. Similarly, the deposits and loans of the bank are determined from the corresponding interest rates of deposits and loans, which are assigned by the bank according to the conditions that prevail to the bank market. At this stage, the categories of the interest rates, which constitute crucial uncertain variables for the analysis, are detected. The determined interest rates categories depend on the type of the bank. For example, for a decisive commercial bank, the deposit and loan interest rates have a role, whereas for an investment bank more emphasis is given

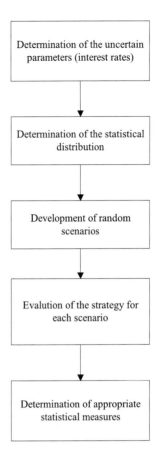

Asset Liability Management Decision Support System, Figure 1
General Monte Carlo simulation procedure for the evaluation of the asset liability management strategies

to the interest rates and the returns of investment products (repos, bonds, interest-bearing notes, etc.).

After the determination of the various categories of interest rates, which determine the total interest rate risk, at the second stage of the analysis the statistical distribution that follows each of the pre-specified categories should be determined.

Having determined the statistical distribution that describes the uncertain variables of the analysis (interest rates), a series of random independent scenarios is developed, through a random number generator. Generally, the largest the number of scenarios that are developed, the more reliable conclusions can be derived. However, the computational effort increases significantly, since for each scenario the optimal asset liability strategy should be determined and moreover its evaluation for each other scenario should take place. Thus, the determination of the number volume N of simulations (scenarios), which will take place should be determined, taking into account both the reliability of the results and the available computational resources.

For each scenario s_i ($i = 1, 2, \ldots, N$) over the interest rates the optimal asset liability management strategy Υ_i is determined through the solution of the goal programming problem. It is obvious that this strategy is not expected to be optimal for each of the other scenarios s_j ($j \neq i$). Therefore the results obtained from the implementation of the strategy Y_i under the rest $N-1$ possible scenarios s_j should be evaluated. The evaluation of the results can be implemented from various directions. The most usual is the one that uses the return. Representing as r_{ij} the outcome (return) of the strategy Υ_i under the scenario s_j, the expected return $\bar{r}_i$ of the strategy can be easily determined based on all the other $N-1$ scenarios s_j ($j \neq i$), as follows:

$$\bar{r}_i = \frac{1}{N-1} \sum_{j=1, j \neq i}^{N} r_{ij} \quad (12)$$

At the same time, the variance σ_i^2 of the expected return can be determined as a risk measure of the strategy Y_i, as follows:

$$\sigma_i^2 = \frac{1}{N-1} \sum_{j=1, j \neq i}^{N} \left(r_{ij} - \bar{r}_i\right)^2 \quad (13)$$

These two statistical measures (average and variance) contribute to the extraction of useful conclusions concerning the expected efficiency of the asset liability management strategy, as well as the risks that it carries. Moreover, these two basic statistical measures can be used for the expansion of the analysis of the determination of other useful statistical information, such as the determination of the confidence interval for the expected return, the quantiles, etc.

Conclusions

The banking business has recently become more sophisticated due to technological expansion, economic development, creation of financial institutions and increased competition. Moreover, the mergers and acquisitions that have taken place the last years create large groups of banking institutions. The success of a bank depends mainly on the quality of its asset and liability management, since the latter deals with the efficient management of sources and uses of bank funds concentrating on profitability, liquidity, capital adequacy and risk factors.

It is obvious that in the last two decades modern finance has developed into a complex mathematically challenging field. Various and complicated risks exist in financial markets. For banks, interest rate risk is at the core of their business and managing it successfully is crucial to whether or not they remain profitable. Therefore, it has been essential the creation of the department of financial risk management within the banks. Asset liability management is associated with the changes of the interest rate risk. Although several models exist regarding asset liability management, most of them are focused on the general aspects and methodologies of this field and do not refer extensively to the hedging of bank interest rate risk through asset liability management. Thus, the main purpose of the present paper was to describe the development of a bank ALM decision support system, which gives the possibility to the decision maker to proceed to various scenarios of the economic process of the bank in order to monitor its financial situation and to determine the optimal strategic implementation of the composition of assets and liabilities. Moreover, we believe that the development of a bank asset liability management model that takes into account the exogenous factors and the economic parameters of the market as well as the uncertainty of variations of the financial risks become essential.

Finally, despite the approaches described in this paper, little academic work has been done so far to develop a model for the management of assets and liabilities in the European banking industry. Based on the above we conclude that the quality of asset liability management in the European banking system has become significant as a resource of competitive advantage. Therefore, the development of new technological approaches in bank asset liability management in Europe is worth further research.

References

1. Baston RG (1989) Financial planning using goal programming. Long Range Plan 22(17):112–120
2. Booth GG (1972) Programming Bank Portfolios under Uncertainty: An Extension. J Bank Res 2:28–40
3. Bradley SP, Crane DB (1972) A Dynamic Model for Bond Portfolio Management. Manage Sci 19:139–151
4. Brodt AI (1978) Dynamic Balance Sheet Management Model for a Canadian Chartered Bank. J Bank Financ 2(3):221–241
5. Carino DR, Kent T, Muyers DH, Stacy C, Sylvanus M, Turner AL, Watanabe K, Ziemba WT (1994) The Russell-Yasuda Kasai Model: An Asset/Liability Model for a Japanese Insurance Company Using Multistage Stochastic Programming. Interfaces 24:29–49
6. Chambers D, Charnes A (1961) Inter-Temporal Analysis and Optimization of Bank Portfolios. Manage Sci 7:393–410
7. Charnes A, Littlechild SC (1968) Intertemporal Bank Asset Choice with Stochastic Dependence. Systems Research Memorandum no.188, The Technological Institute, Nortwestern University, Evanston, Illinois
8. Charnes A, Thore S (1966) Planning for Liquidity in Financial Institution: The Chance Constrained Method. J Finance 21(4):649–674
9. Cohen KJ, Hammer FS (1967) Linear Programming and Optimal Bank Asset Management Decisions. J Financ 22:42–61
10. Cohen KJ, Thore S (1970) Programming Bank Portfolios under Uncertainty. J Bank Res 2:28–40
11. Crane B (1971) A Stochastic Programming Model for Commercial Bank Bond Portfolio Management. J Finance Quant Anal 6:955–976
12. Eatman L, Sealey W (1979) A Multi-objective Linear Programming Model for Commercial bank Balance Sheet Management. J Bank Res 9:227–236
13. Eppen GD, Fama EF (1971) Three Asset Cash Balance and Dynamic Portfolio Problems. Manage Sci 17:311–319
14. Fielitz D, Loeffler A (1979) A Linear Programming Model for Commercial Bank Liquidity Management. Finance Manage 8(3):44–50
15. Giokas D, Vassiloglou M (1991) A Goal Programming Model for Bank Assets and Liabilities. Eur J Oper Res 50:48–60
16. Kallberg JG, White RW, Ziemba WT (1982) Short Term Financial Planning under Uncertainty. Manage Sci 28:670–682
17. Kosmidou K, Zopounidis C (2001) Bank Asset Liability Management Techniques: An Overview. In: Zopounidis C, Pardalos PM, Baourakis G (eds) Fuzzy Set Systems in Management and Economy. World Scientific Publishers, pp 255–268
18. Kosmidou K, Zopounidis C (2004) Combining Goal Programming Model with Simulation Analysis for Bank Asset Liability Management. Inf Syst Oper Res J 42(3):175–187
19. Kusy IM, Ziemba TW (1986) A Bank Asset and Liability Management model. Oper Res 34(3):356–376
20. Kvanli AH (1980) Financial planning using goal programming-OMEGA. Int J Manag Sci 8(2):207–218
21. Lee SM, Chesser DL (1980) Goal programming for portfolio selection. J Portf Manag 6:22–26
22. Lee SM, Lerro AJ (1973) Optimizing the portfolio selection for mutual funds. J Financ 28:1086–1101
23. Lifson KA, Blackman BR (1973) Simulation and Optimization Models for Asset Deployment and Funds Sources Balancing Profit Liquidity and Growth. J Bank Res 4(3):239–255
24. Markowitz HM (1959) Portfolio Selection. Efficient Diversification of Investments. Wiley, New York
25. Merton RC (1969) Lifetime portfolio selection under certainty: the continuous time case. Rev Eco Stat 3:373–413
26. Merton RC (1990) Continuous-Time Finance. Blackwell Publishers, Merton, UK
27. Mulvey JM, Vladimirou H (1989) Stochastic Network Optimization Models of Investment Planning. Annal Oper Res 20:187–217
28. Mulvey JM, Ziemba WT (1998) Asset and liability management systems for long-term investors: discussion of the issues. In: Ziemba W, Mulvey J (eds) Worldwide Asset and Liability Modelling. Cambridge University Press, Mulvey, UK, pp 3–38
29. Pogue GA, Bussard RN (1972) A Linear Programming Model for Short Term Financial Planning under Uncertainty. Sloan Manag Rev 13:69–98
30. Pyle DH (1971) On the Theory of Financial Intermediation. J Financ 26:737–746
31. Robertson M (1972) A Bank Asset Management Model. In: Eilon S, Fowkes TR (eds) Applications of Management Science in Banking and Finance. Gower Press, Epping, Essex, pp 149–158
32. Samuelson P (1969) Lifetime portfolio selection by dynamic stochastic programming. Rev Eco Stat (August), 239–246

33. Seshadri S, Khanna A, Harche F, Wyle R (1999) A method for strategic asset-liability management with an application to the federal home loan bank of New York. Oper Res 47(3):345–360
34. Sharma JK, Sharma DK, Adeyeye JO (1995) Optimal portfolio selection: A goal programming approach. Indian J Financ Res 7(2):67–76
35. Wolf CR (1969) A Model for Selecting Commercial Bank Government Security Portfolios. Rev Econ Stat 1:40–52

Assignment and Matching

AM

DIMITRIS ALEVRAS
IBM Corporation, West Chester, USA

MSC2000: 90C35, 90C27, 90C10, 90C05

Article Outline

Keywords
Maximum Cardinality Bipartite Matching Problem
Weighted Bipartite Matching Problem
Weighted Matching Problem
Maximum Cardinality Matching Problem
See also
References

Keywords

Optimization; Integer programming; Graph theory; Marriage problem

Matching problems comprise an important set of problems that link the areas of graph theory and combinatorial optimization. The maximum cardinality matching problem (see below) is one of the first integer programming problems that was solved in polynomial time. Matchings are of great importance in graph theory (see [9]) as well as in combinatorial optimization (see e. g. [15]).

The matching problem and its variations arise in cases when we want to find an 'optimal' pairing of the members of two (not necessarily disjoint) sets. In particular, if we are given two sets of 'objects' and a 'weight' for each pair of objects, we want to match the objects into pairs in such a way that the total weight is maximal. In graph theory, the problem is defined on a graph $G = (V, E)$ where V is the node set of the graph, corresponding to the union of the two sets of objects, and E is the edge set of the graph corresponding to the possible pairs. A pair is possible if there exists an edge between the corresponding nodes. A *matching* M is a subset of the edges E with the property that each node in V is incident to at most one edge in M. If each node in V is met by exactly one edge in M, then M is called a *perfect matching*. There exist several versions of the matching problem, depending on whether the graph G is bipartite or not (i. e., the two sets of objects are disjoint or not), and on whether we want to find the maximum size (cardinality) or the maximum weight of the matching. The book [1] gives several applications of the matching problem.

Maximum Cardinality Bipartite Matching Problem

The graph G is *bipartite* if the node set V can be partitioned into two disjoint sets V_1 and V_2 such that no edge in E connects nodes from the same set. Finding a maximum cardinality matching on a bipartite graph can be solved by several efficient algorithms with a worst-case bound of $O(\sqrt{n}m)$, where n is the number of nodes and m the number of edges of the graph. See [1] for details.

Weighted Bipartite Matching Problem

This problem is known as the *assignment* or the *marriage* problem. In the traditional definition it is required that the sets V_1 and V_2 are of equal size, but even if not, one can add 'dummy' nodes to the smaller set to satisfy this condition. This problem can be formulated as a zero-one linear programming problem as follows:

$$\begin{cases} \min & \sum_{(u,v) \in E} f(u,v) x_{uv} \\ \text{s.t.} & \sum_{(u,v) \in E} x_{uv} = 1 \quad \text{for all } u \in V_1, \\ & \sum_{(u,v) \in E} x_{uv} = 1 \quad \text{for all } v \in V_2, \\ & x_{uv} \in \{0,1\} \quad \text{for all } u \in V_1, v \in V_2. \end{cases}$$

The *assignment problem* has the property that if solved as a linear programming problem in nonnegative x_{uv} it yields an integer solution, i. e., the zero-one integrality condition in the formulation is not necessary. This is so because the constraint matrix of the equations is *totally unimodular*, i. e., the determinant of every square submatrix of it is 0 or ± 1. This means that if the right-hand sides of the equations are integer numbers, as is the case in the assignment problem, then the solution will be integer.

Linear programming algorithms are not as efficient as specialized algorithms for solving the assignment problem. The assignment problem is a special case of the minimum cost flow problem, and adaptations of algorithms for that problem that take into account the special structure of the assignment problem yield the most efficient algorithms. Probably the best known algorithm is the so called *Hungarian algorithm*, see [8], which is a primal-dual algorithm for the minimum cost flow problem. See [1] for details and other algorithms.

Variations of the bipartite matching include among others the order preserving assignment problem and the stable marriage problem. In the *order preserving assignment problem* the assignment must be such that a prespecified order among the objects of one of the node partitions is preserved. Although the linear programming formulation of this problem is more complicated than that of the assignment problem, the problem itself is easier to solve than the assignment problem and can be solved in $O(m)$ time where m is the number of edges in the graph; see [2,12]. In the *stable marriage problem* each object of one partition has a ranking (or preference) for each of the objects of the other partition, and the assignment must be such that there is no nonmatched pair of objects that its members prefer each other to the ones they are matched against. This problem can be solved in $O(n^2)$ time using a greedy algorithm (n is the number of nodes in one partition). See [1].

Weighted Matching Problem

The weighted matching problem can be formulated as a 0–1 programming problem as follows:

$$\begin{cases} \max & \sum_{(u,v)\in E} f(u,v) x_{uv} \\ \text{s.t.} & \sum_{(u,v)\in E} x_{uv} \leq 1 \quad \text{for all } u \in V, \\ & x_{uv} \in \{0,1\} \quad \text{for all } (u,v) \in E. \end{cases}$$

Unlike the case of the assignment problem, relaxing the integrality constraints yields, in general, a fractional solution.

Maximum Cardinality Matching Problem

J. Edmonds showed in [5] that one more set of inequalities—the *odd-set constraints*—is needed in order to get a linear programming formulation of the matching problem. The odd-set or *blossom* inequalities are

$$\sum_{(u,v)\in E(U)} x_{uv} \leq \left\lfloor \frac{|U|}{2} \right\rfloor, \quad \forall \text{odd } U \subseteq V, |U| \geq 3,$$

where $E(U)$ is the set of all edges in E with both end nodes in U. An odd set is a set of odd cardinality. See also [11].

Solving the matching problem on nonbipartite graphs is considerably more difficult than on bipartite ones. This is so because the path augmenting algorithms used in the case of bipartite matchings, may fail when a structure called blossom is encountered. Edmonds provided an $O(n^4)$ algorithm that would find an integer solution to the linear programming relaxation of the formulation (including the odd-set constraints) for any objective function, proving this way the completeness of the formulation. Several implementations that improved the performance of the algorithm have been proposed (see [1,10], among others) as well as data structures for the efficient implementation of such algorithms (see [3]). M. Grötschel and O. Holland [6] gave a cutting plane algorithm for the weighted matching problem, where they used an efficient separation algorithm to identify violated *blossom inequalities*, based on the algorithm of M.W. Padberg and M.R. Rao [14] for the *b-matching problem*.

The b-matching problem is an important generalization of the matching problem. In the b-matching problem each node $v \in V$ is met by no more than b_v edges; thus, in this context, the previous definition of matching corresponds to an 1-matching. A *perfect b-matching* is one in which each node $v \in V$ is met by exactly b_v edges. If it is permitted to chose an edge more than one times then the problem becomes a general integer program instead of a 0–1 program. The b-matching problem can be reduced to 1-matching problem on an appropriately constructed graph. Although this procedure is not polynomial in general—and thus, Edmonds' algorithm can not be readily applied—the b-matching problem is polynomially solvable; see [14] and [7]. A linear inequality description for the integer b-matching problem is given in [15]. See also [11]. The perfect 0–1 2-matching problem is a relaxation of the *traveling salesman problem* (TSP). Solving the 0–1 2-matching problem yields a heuristic solution to the TSP which is an *NP*-hard problem; see [13].

See also

- ▶ Assignment Methods in Clustering
- ▶ Bi-Objective Assignment Problem
- ▶ Communication Network Assignment Problem
- ▶ Frequency Assignment Problem
- ▶ Maximum Partition Matching
- ▶ Quadratic Assignment Problem

References

1. Ahuja R, Magnanti T, Orlin J (1994) Network flows. Wiley, New York
2. Alevras D (1997) Order preserving assignments without contiguity. Discret Math 163:1–11
3. Ball MO, Derigs U (1983) An analysis of alternate strategies for implementing matching algorithms. Networks 13:517–549
4. Edmonds J (1965) Maximum matching and a polyhedron with (0, 1) vertices. J Res Nat Bureau Standards (B) 69B:125–130
5. Edmonds J (1965) Paths, trees, and flowers. Canad J Math 17:449–467
6. Grötschel M, Holland O (1985) Solving matching problems with linear programming. Math Program 33:243–259
7. Grötschel M, Lovasz L, Schrijver A (1988) Geometric algorithms and combinatorial optimization. Springer, Berlin
8. Kuhn HW (1955) The Hungarian method for the assignment problem. Naval Res Logist Quart 2:83–97
9. Lovasz L, Plummer M (1986) Matching theory. North-Holland, Amsterdam
10. Micali S, Vazirani VV (1980) An $O(\sqrt{(|v|)}|E|)$ algorithm for finding maximum matching in general graphs. IEEE Symp Found Computer Sci, pp 17–27
11. Nemhauser GL, Wolsey L (1988) Integer and combinatorial optimization. Wiley, New York
12. Padberg M, Alevras D (1994) Order-preserving assignments. Naval Res Logist 41:395–421
13. Padberg MW, Grötschel M (1985) Polyhedral theory. The Traveling Salesman Problem. Wiley, New York, pp 251–305
14. Padberg M, Rao MR (1982) Odd minimum cut-sets and b-matchings. Math Oper Res 7:67–80
15. Pulleyblank WR (1989) Polyhedral combinatorics. Optimization, vol 1 of Handbook Oper Res and Management Sci. North-Holland, Amsterdam, pp 371–446

Assignment Methods in Clustering

L. J. HUBERT[1], P. ARABIE[2]
[1] University Illinois, Champaign, USA
[2] Rutgers University, Newark, USA

MSC2000: 62H30, 90C27

Article Outline

Keywords
Weighting Schemes
 for the Fixed (Target) Matrix Q
  Single Cluster Statistics
  Partition Statistics
  Partition Hierarchy Statistics
Alternative Assignment Indices
Modifications of the Target Matrix Q
See also
References

Keywords

Combinatorial optimization; Quadratic assignment; Clustering

The use of *assignment methods* in the formulation of various optimization problems encountered in *clustering* and *classification*, can be introduced through the well-known *quadratic assignment* (QA) model (see [5] for a comprehensive discussion of most of the topics presented in this entry). In its most basic form the QA optimization task can be stated using two $n \times n$ matrices, say $\mathbf{P} = \{p_{ij}\}$, and $\mathbf{Q} = \{q_{ij}\}$, and the identification of a one-to-one function (or a permutation), $\rho(\cdot)$, on the first n integers, to optimize (either by minimizing or maximizing) the cross-product index

$$\Gamma(\rho) = \sum_{i,j} p_{\rho(i)\rho(j)} q_{ij}. \qquad (1)$$

Typically, the main diagonal entries in **P** and **Q** are considered irrelevant and can be set equal to zero. For arbitrary matrices **P** and **Q**, the cross product index in (1) may be rewritten as

$$\sum_{i,j} \left(\frac{p_{\rho(i)\rho(j)} + p_{\rho(j)\rho(i)}}{2} \right) \left(\frac{q_{ij} + q_{ji}}{2} \right)$$
$$+ \sum_{i,j} \left(\frac{p_{\rho(i)\rho(j)} - p_{\rho(j)\rho(i)}}{2} \right) \left(\frac{q_{ij} - q_{ji}}{2} \right),$$

indicating that the optimization of (1) jointly involves the *symmetric* ($[\mathbf{P}+\mathbf{P}']/2$ versus $[\mathbf{Q}+\mathbf{Q}']/2$) and *skew-symmetric* ($[\mathbf{P}-\mathbf{P}']/2$ versus $[\mathbf{Q}-\mathbf{Q}']/2$) components of both $\mathbf{P}$ and $\mathbf{Q}$. Because of this separation of $\mathbf{P}$ and $\mathbf{Q}$ into symmetric and skew-symmetric components, it is possible in the context of the clustering/classification tasks to be discussed below, to assume that both $\mathbf{P}$ and $\mathbf{Q}$ are symmetric or that both are skew-symmetric.

In applications to clustering, the matrix $\mathbf{P}$ usually contains numerical proximity information between distinct pairs of the n objects from some given set $S = \{O_1, \ldots, O_n\}$ that is of substantive interest. If $\mathbf{P}$ is symmetric, p_{ij} ($=p_{ji}$) denotes the degree to which objects O_i and O_j are similar (and keyed as what is referred to as a *dissimilarity* [or as a *similarity*] measure if smaller [or larger] values reflect greater object similarity). If $\mathbf{P}$ is skew-symmetric, p_{ij} ($=-p_{ji}$) is an index of dominance (or *flow*) between objects O_i and O_j, with the sign reflecting the directionality of *dominance* and the absolute value indicating the degree. The (target) matrix $\mathbf{Q}$, as developed in detail in the next section, will typically be fixed, with the specific pattern of entries characterizing the type of structure to be identified for the set S, e. g., a single object *cluster*, a *partition*, or a *partition hierarchy*. An optimal permutation, say, $\rho^*(\cdot)$, based on the cross-product index in (1) for a specific target matrix $\mathbf{Q}$ will identify the (salient) combinatorial structure sought.

The QA *optimization* task as formulated through (1) has an enormous literature that will not be reviewed here (for an up-to-date and comprehensive source on QA, see [11]). For current purposes, one might consider the optimization of (1) through a simple object interchange heuristic that would begin with some permutation (possibly chosen at random), and then implement local interchanges until no improvement in the index can be made. By repeatedly initializing such a process randomly, a distribution over a set of *local optima* can be achieved. At least within the context of clustering/classification, such a distribution may be highly relevant diagnostically for explaining whatever structure is inherent in the data matrix $\mathbf{P}$, and possibly of even greater interest than the identification of just a single optimal permutation. In a related framework, there are considerable applications for the QA model in a confirmatory context where the distribution of $\Gamma(\rho)$ is constructed over all $n!$ possible permutations considered equally-likely, and the index value associated with some identified permutation is compared to this distribution. Most *nonparametric statistical methods* popular in the literature can be rephrased through the device of defining the matrices $\mathbf{P}$ and $\mathbf{Q}$ appropriately (see [5] for a comprehensive development of these special cases as well as approximation methods based on closed-form expressions for the first three moments of $\Gamma(\rho)$). A few of these applications will be briefly noted below.

Weighting Schemes for the Fixed (Target) Matrix Q

Single Cluster Statistics

To identify a *single* salient cluster of fixed size K (that can be varied by the user), consider $\mathbf{Q}$ to have the partitioned form

$$\mathbf{Q} = \begin{pmatrix} \mathbf{Q}_{11} & \mathbf{Q}_{12} \\ \mathbf{Q}_{21} & \mathbf{Q}_{22} \end{pmatrix},$$

where within each submatrix of the size indicated, the (off-diagonal) entries are constant:

$$\mathbf{Q}_{11} = \begin{pmatrix} 0 & \cdots & q_{11} \\ \vdots & \ddots & \vdots \\ q_{11} & \cdots & 0 \end{pmatrix}_{K \times K}$$

$$\mathbf{Q}_{12} = \begin{pmatrix} & \vdots & \\ \cdots & q_{12} & \cdots \\ & \vdots & \end{pmatrix}_{K \times (n-K)}$$

$$\mathbf{Q}_{21} = \begin{pmatrix} & \vdots & \\ \cdots & q_{21} & \cdots \\ & \vdots & \end{pmatrix}_{(n-K) \times K}$$

$$\mathbf{Q}_{22} = \begin{pmatrix} 0 & \cdots & 0 \\ \vdots & \ddots & \vdots \\ 0 & \cdots & 0 \end{pmatrix}_{(n-K) \times (n-K)}$$

Depending on how the values for q_{11}, q_{12}, and q_{21} are defined, different indices can be generated that measure the salience of the subset constructed by any permutation $\rho(\cdot)$, i.e., for the identified cluster $S_\rho \equiv \{O_{\rho(1)}, \ldots, O_{\rho(K)}\}$.

For symmetric **P**:

A) letting

$$q_{11} = \frac{1}{K(K-1)}, \quad q_{12} = q_{21} = 0,$$

the index $\Gamma(\rho)$ is the average proximity within the subset S_ρ and defines a measure of *cluster 'compactness'*;

B) letting

$$q_{11} = 0, \quad q_{12} = q_{21} = \frac{1}{2K(n-K)},$$

$\Gamma(\rho)$ is the average proximity between the subset S_ρ and its complement, and defines a measure of *cluster 'isolation'* for either S_ρ or $S - S_\rho$; alternatively, it can be considered a measure of *'separation'* between S_ρ or $S - S_\rho$;

C) by contrasting A) and B) as

$$q_{11} = \frac{1}{K(K-1)},$$
$$q_{12} = q_{21} = -\frac{1}{2K(n-k)},$$

$\Gamma(\rho)$ characterizes the salience of the subset S_ρ by a trade-off between compactness and isolation. The optimization of $\Gamma(\rho)$ based on these latter weights identifies a cluster that would be both relatively compact and isolated, whereas the emphasis in A) and B) are on clusters that may be either compact or isolated but not necessarily both.

For skew-symmetric **P**:

D) letting

$$q_{11} = 0, \quad q_{12} = \frac{1}{2K(n-K)}, \quad q_{21} = -q_{12},$$

the index $\Gamma(\rho)$ is the average dominance (or flow) from the subset S_ρ to its complement, minus the average dominance (or flow) from the complement to the subset. Thus, its optimization (e.g., maximization) identifies a subset of S whose members tend to dominate those in its complement (or where aggregate outflow exceeds aggregate inflow).

In a confirmatory comparison context, the single-cluster statistic $\Gamma(\rho)$ can be used to generate a number of nonparametric test statistics for comparing the difference between two independent groups. For example, suppose observations are available on n objects, $x_1, \ldots, x_n$, where the first K belong to group I and the last $n - K$ to group II. If the (now asymmetric) proximity matrix is defined as $\mathbf{P} = \{p_{ij}\}$, where $p_{ij} = 1$ if $x_j < x_i$ and $= 0$ if $x_j \geq x_i$ then the weighting scheme in B) gives (a simple linear transform of) the well-known *Mann–Whitney statistic* for comparing two-independent groups, i.e., if two observations are drawn at random from groups I and II, then $\Gamma(\rho_o)$, for ρ_o the identity permutation, is the probability that the group I observation is the larger. The distribution of $\Gamma(\rho)$ over all $n!$ permutations generates the null distribution against which the observed index $\Gamma(\rho_o)$ can be compared. Because of the structure of $\mathbf{Q}$, this null distribution is based on all $n!/(K!(n-K)!)$ distinct subsets considered equally-likely to be formed from the collection of size n. (See [3, Chap. 7], for a more complete discussion of the two-independent sample problem in this type of nonparametric framework.)

Although single-cluster statistics that depend on the comparison of mean proximities may be the most obvious to consider, a number of possible alternatives can be constructed by varying the definition for the weight matrices in $\mathbf{Q}$. For example, for symmetric **P**, if $\mathbf{Q}_{11}$ is (re)defined to have the form

$$\begin{pmatrix} 0 & 1 & 0 & \cdots & 0 & 0 & 0 \\ 1 & 0 & 1 & \cdots & 0 & 0 & 0 \\ \vdots & \vdots & \vdots & \ddots & \vdots & \vdots & \vdots \\ 0 & 0 & 0 & \cdots & 1 & 0 & 1 \\ 0 & 0 & 0 & \cdots & 0 & 1 & 0 \end{pmatrix},$$

with entries of all ones immediately above and below the main diagonal, and $q_{12} = q_{21} = 0$, the salience of S_ρ is now based on (twice) the sum of adjacent proximities along a path of *length K* considered in the object order $O_{\rho(1)} \leftrightarrow \cdots \leftrightarrow O_{\rho(K)}$. Or, if $\mathbf{Q}_{11}$ is (re)defined to have the form

$$\begin{pmatrix} 0 & 1 & 1 & \cdots & 1 & 1 \\ 1 & 0 & 0 & \cdots & 0 & 0 \\ \vdots & \vdots & \vdots & \ddots & \vdots & \vdots \\ 1 & 0 & 0 & \cdots & 0 & 0 \end{pmatrix},$$

and $q_{12} = q_{21} = 0$, the salience of S_ρ is now based on (twice) the sum of proximities between $O_{\rho(1)}$ and the remaining objects $O_{\rho(2)}, \ldots, O_{\rho(K)}$ (this is called a 'star' cluster of size K with object $O_{\rho(1)}$ as its center; see [10, Sect. 4.5.2] for a further discussion of clustering based on stars).

Partition Statistics

To identify a salient partition of S into M subsets, S_1, ..., S_M, of fixed sizes $n_1, \ldots, n_M$, respectively, consider $\mathbf{Q}$ to have the partitioned form

$$\mathbf{Q} = \begin{pmatrix} \mathbf{Q}_{11} & \mathbf{Q}_{12} & \cdots & \mathbf{Q}_{1M} \\ \vdots & \vdots & \ddots & \vdots \\ \mathbf{Q}_{M1} & \mathbf{Q}_{M2} & \cdots & \mathbf{Q}_{MM} \end{pmatrix},$$

where the (off-diagonal) entries in each submatrix $\mathbf{Q}_{mm'}$ of size $n_m \times n_{m'}$, are all equal to a constant $q_{mm'}$, $1 \leq m$, $m' \leq M$. Again, depending on how these latter values are defined, a variety of different indices can be generated that now measure the salience of the partition generated by a permutation $\rho(\cdot)$. For a symmetric $\mathbf{P}$, three of the most popular alternatives are noted below that differ only in how the weights q_{mm}, $1 \leq m \leq M$, are defined, and which all assume $q_{mm'} = 0$ for $m \neq m'$:

a) $q_{mm} = 1$: each subset in a partition contributes in direct proportion to the number of object pairs it contains;
b) $q_{mm} = 1/(n_m(n_m - 1))$: each subset contributes equally irrespective of the number of objects (or object pairs) it contains;
c) $q_{mm'} = 1/n_m$: each subset contributes in direct proportion to the number of objects it contains.

In a confirmatory comparison context, the partition statistic $\Gamma(\rho)$ with weighting option c) can be used to construct a test-statistic equivalent to the common F-ratio in a *one-way analysis of variance* for assessing whether mean differences exist over K independent groups. Explicitly, suppose observations are available on n objects, $x_1, \ldots, x_n$, with the first n_1 belonging to group 1, the second n_2 belonging to group 2, and so on. If proximity is defined as $\mathbf{P} = \{p_{ij}\}$, where $p_{ij} = (x_i - x_j)^2$, then the weights in c) produce $\Gamma(\rho_o)$, for ρ_o the identity permutation, equal to twice the within group sum of squares. The distribution of $\Gamma(\rho)$ over all $n!$ permutations generates a distribution over all $n!/(n_1! \ldots n_M!)$ equally-likely ways the n observations can be grouped into subsets of sizes $n_1, \ldots, n_M$, and against which the observed index $\Gamma(\rho_o)$ can be compared. (See [9] for a more thorough discussion of thus evaluating a priori classifications.)

For a skew-symmetric $\mathbf{P}$, the partitioning of S would now be into M ordered subsets, $S_1 \prec \ldots \prec S_M$ of fixed sizes $n_1, \ldots, n_M$, with the most natural weights being $q_{mm} = 0$ for $1 \leq m \leq M$, $q_{mm'} = +1$ if $m < m'$, and $= -1$ if $m > m'$. Maximizing $\Gamma(\rho)$ is this case would be a search for an *ordered partition* in which objects in S_m tend to dominate those in $S_{m'}$ if $m < m'$, i. e., there are generally positive dominance values from a lower-placed subset to one that is higher.

There are several special cases of interest for the partition statistic:
i) for symmetric $\mathbf{P}$ and if for convenience it is assumed n is even and $n_m = 2$ for $1 \leq m \leq M$ (so, $n = 2M$), the weights in a) make $\Gamma(\rho)$ the index for a *matching* of the objects in S induced by $\rho(\cdot)$;
ii) if the proximity matrix $\mathbf{P}$ is itself constructed from a partition of S, then the index $\Gamma(\rho)$ can be interpreted as a measure of association for a *contingency table* defined by the n objects cross-classified using $\rho(\cdot)$ and the two partitions underlying $\mathbf{P}$ and $\mathbf{Q}$.

Depending on the choice of weights for $\mathbf{Q}$, and how proximity is defined in $\mathbf{P}$ based on its underlying partition, a number of well-known indices of association can be obtained: *Pearson's chi-square statistic, Goodman–Kruskal's τ_b*, and *Rand's index*. For a more complete discussion of these special cases, including the necessary definitions for $\mathbf{P}$, consult [5].

Partition Hierarchy Statistics

One straightforward strategy for extending QA to identify salient partition hierarchies having a specific form, begins with a given collection of T partitions of S, $\mathcal{P}_1$, ..., $\mathcal{P}_T$, that are hierarchically related. Here, $\mathcal{P}_1$ contains all n objects in n separate classes, $\mathcal{P}_T$ contains all n objects in one class, and $\mathcal{P}_{t+1}$ is formed from $\mathcal{P}_t$ for $t \geq 1$ by uniting one or more of the classes in the latter. If $\mathbf{Q} = \{q_{ij}\}$ is defined by $q_{ij} = \min\{t - 1: O_i, O_j \in$ common object class in $\mathcal{P}_t\}$, then these latter entries satisfy the defining property of being an *ultrametric*, i. e., $q_{ij} \leq \max\{q_{ik}, q_{kj}\}$ for all $O_i, O_j, O_k \in S$ (see [2,10, Chap. 7] for an extensive discussion of ultrametrics). For symmetric $\mathbf{P}$, the optimization of $\Gamma(\rho)$ in (1) would be the search for a salient partition hierarchy having the generic form defined by $\mathcal{P}_1, \ldots, \mathcal{P}_T$, and which optimizes the cross-product between the proximity information in **P** and the levels at which the object pairs are first placed into common classes in the hierarchy. It might be noted that both single clusters and partitions could be considered special cases of a partition hierarchy when $T = 3$ and the only nontrivial partition is $\mathcal{P}_2$, i.e., to obtain a single cluster, $\mathcal{P}_2$ can be defined by one subset of size K and $n - K$ subsets each of size one; to obtain a single partition, $\mathcal{P}_2$ merely has to be that partition with the desired number of classes and class sizes.

Alternative Assignment Indices

There are a variety of alternatives for replacing the cross-product in the QA index in (1) by a different function between the entries in **P** and **Q**. Depending on how the proximity information in **P** and the target given by **Q** are specified, one might adopt, for example, the sum of absolute differences, $\sum_{i,j} |p_{\rho(i)\rho(j)} - q_{ij}|$, or the sum of dichotomous indicators for equality, $\sum_{i,j} g(p_{\rho(i)\rho(j)}, q_{ij})$, where $g(x, y) = 1$ if $x = y$ and 0 otherwise, or even use 'bottleneck' measures such as $\min_{i,j} p_{\rho(i)\rho(j)} q_{ij}$ or $\max_{i,j} p_{\rho(i)\rho(j)} q_{ij}$. Somewhat more well-developed in the literature than these possibilities (e.g., see [5, Chap. 5]) are generalizations of (1) that would maintain the basic cross-product structure but which would rely on higher-order functions of the entries in **P** and **Q** before the cross-products were taken. Again, variations would be possible, but two of the more obvious forms of extension are given below that depend solely on the order of the entries within **P** and within **Q**:

- *Three-argument functions*: Given **P** and **Q**, and letting $\text{sign}(x) = +1$ if $x > 0$, $= 0$ if $x = 0$, and $= -1$ if $x < 0$, define

$$\mathcal{A}(\rho) = \sum_{\substack{i \neq j \\ i \neq k}} \text{sign}(p_{\rho(i)\rho(j)} - p_{\rho(i)\rho(k)}) \, \text{sign}(q_{ij} - q_{ik}).$$

The index $\mathcal{A}(\rho)$ can be interpreted as the difference between two counts, say $\mathcal{A}^+(\rho)$ and $\mathcal{A}^-(\rho)$, where $\mathcal{A}^+(\rho)$ (respectively, $\mathcal{A}^-(\rho)$) is the number of consistencies (inconsistencies) in the ordering of pairs of off-diagonal entries in $\{p_{\rho(i)\rho(j)}\}$ and their counterparts in $\{q_{ij}\}$, where the former pairs share a common (row) object $O_{\rho(i)}$.

- *Four-argument functions*: Define

$$\mathcal{B}(\rho) = \sum_{\substack{i \neq j \\ k \neq l}} \text{sign}(p_{\rho(i)\rho(j)} - p_{\rho(k)\rho(l)}) \, \text{sign}(q_{ij} - q_{kl}).$$

Again, the index $\mathcal{B}(\rho)$ can be viewed as the difference between $\mathcal{B}^+(\rho)$ and $\mathcal{B}^-(\rho)$, where $\mathcal{B}^+(\rho)$ (respectively, $\mathcal{B}^-(\rho)$) is the number of consistencies (inconsistencies) in the ordering of pairs of off-diagonal entries in $\{p_{\rho(i)\rho(j)}\}$ and their counterparts in $\{q_{ij}\}$. In contrast to $\mathcal{A}(\rho)$, however, no common object need be present in the pairs of off-diagonal entries. The distinction between $\mathcal{A}(\rho)$ and $\mathcal{B}(\rho)$ in measuring the correspondence between **P** and **Q** rests on whether the proximity entries in **P** are strictly comparable only within rows (i.e., to what are called *row conditional proximity data*, e.g., see [1, p. 192]) or whether such comparisons make sense when performed across rows.

To illustrate the interpretation of $\mathcal{A}(\rho)$ and $\mathcal{B}(\rho)$ in the single cluster statistic context, suppose **Q** has the weight structure in A) that generated through (1) the measure of cluster compactness as the average within group proximity in $S_\rho = \{O_{\rho(1)}, \ldots, O_{\rho(K)}\}$. In using this specific target **Q** for $\mathcal{A}(\rho)$, the index is, in words, twice the difference between the number of instances in which a proximity for two objects both within S_ρ is greater than the proximity from one of these two objects to another in $S - S_\rho$, and the number of instances in which it is less. Depending on whether proximity is keyed as a similarity or a dissimilarity, a compact subset would be one for which $\mathcal{A}(\rho)$ is maximized or minimized, respectively. If instead, the weight structure for **Q** given in B) that defined the measure of cluster isolation, the index $\mathcal{A}(\rho)$ would now be twice the difference between the number of instances in which a proximity between two objects that span S_ρ and $S - S_\rho$ is greater than the proximity between two objects within S_ρ or within $S - S_\rho$ (where the latter have one member in common with the two that span S_ρ and $S - S_\rho$), and the number of instances in which it is less. Now, an isolated subset would be identified by maximizing or minimizing $\mathcal{A}(\rho)$ depending on the keying of proximity as a dissimilarity or similarity, respectively. For $\mathcal{B}(\rho)$, and the weight structure in A), the index is, in words, twice the difference between the number of instances in which a proximity for two objects both within S_ρ is greater than the proximity between *any* two objects that span S_ρ and $S - S_\rho$ and the number of instances in which it is less. The index $\mathcal{B}(\rho)$ for the weight matrix in B) would be twice the difference between the number of instances in which a proximity between two objects that span S_ρ and $S - S_\rho$ is greater than the proximity between *any* two objects within S_ρ or within $S - S_\rho$.

In the partition context, a similar interpretation to the use of the single subset compactness measure would be present for $\mathcal{A}(\rho)$ and $\mathcal{B}(\rho)$ and for all of the three weighting options mentioned, but now all aggregated over the M subsets of the partition. In the partition hierarchy framework, the correspondence between $\{p_{\rho(i)\rho(j)}\}$ and $\mathbf{Q}$ is measured by the degree of consistency in the ordering of the object pairs by proximity and the ordering of the object pairs by the levels in which the objects are first placed into a common class.

In addition to replacing the QA index in (1) by the higher order functions adopted in $\mathcal{A}(\rho)$ and $\mathcal{B}(\rho)$ to effect a reliance only on the order properties of the entries within $\mathbf{P}$ and $\mathbf{Q}$, there are several other uses in a clustering/classification context for the definition of three- or four-argument functions. One alternative will be mentioned here that deals with what can be called the *generalized single cluster statistic*. Explicitly, suppose three- and four-argument function of the entries in $\mathbf{P}$ are denoted by $u(\cdot, \cdot, \cdot)$ and $r(\cdot, \cdot, \cdot, \cdot)$, respectively, and those in $\mathbf{Q}$ by $v(\cdot, \cdot, \cdot)$ and $s(\cdot, \cdot, \cdot, \cdot)$, and consider the general cross-product forms of

$$C(\rho) = \sum_{i,j,k} u(\rho(i), \rho(j), \rho(k)) v(i, j, k),$$

$$D(\rho) = \sum_{i,j,k,l} r(\rho(i), \rho(j), \rho(k), \rho(l)) s(i, j, k, l).$$

It will be assumed here that both $v(\cdot, \cdot, \cdot)$ and $s(\cdot, \cdot, \cdot, \cdot)$ are merely indicator functions for a subset of size K, so $v(i, j, k) = 1$ if $1 \leq i, j, k \leq K$, and $= 0$ otherwise; $s(i, j, k, l) = 1$ if $1 \leq i, j, k, l \leq K$, and $= 0$ otherwise. Thus, the optimization of $C(\rho)$ or $D(\rho)$ can be viewed as the search for a subset of size K with extreme values for the indices $\sum_{1 \leq i,j,k \leq K} u(\rho(i), \rho(j), \rho(k))$ or $\sum_{1 \leq i,j,k,l \leq K} r(\rho(i), \rho(j), \rho(k), \rho(l))$, and depending on how the functions $u(\cdot, \cdot, \cdot)$ and $r(\cdot, \cdot, \cdot, \cdot)$ are defined, a subset that is very salient with respect to the property that characterizes the latter.

A number of properties that may be desirable to optimize in a subset of size K have been considered (see [4] for a more complete discussion), of which the two listed below are directly relevant to the clustering/classification context:

i) a proximity matrix (with a dissimilarity keying) represents a perfect partition hierarchy if it satisfies the property of being an ultrametric: for all $1 \leq i, j, k \leq n$, $p_{ij} \leq \max\{p_{ik}, p_{kj}\}$, or equivalently, the two largest values among p_{ij}, p_{ik}, and p_{kj} are equal. Thus, if $u(\rho(i), \rho(j), \rho(k))$ equals the absolute difference between the two largest values among $p_{\rho(i)\rho(j)}$, $p_{\rho(i)\rho(k)}$, and $p_{\rho(j)\rho(k)}$, the minimization of $C(\rho)$ seeks a subset of size K that is as close to being an ultrametric as possible (as measured by $C(\rho)$);

ii) a proximity matrix (again, with a dissimilarity keying) represents a perfect *additive tree* where proximities can be reconstructed by minimum path lengths in a tree if they satisfy the four-point property: for all $1 \leq i, j, k, l \leq n$, $p_{ij} + p_{kl} \leq \max\{p_{ik} + p_{jl}, p_{il} + p_{jk}\}$, or equivalently, the largest two sums among $p_{ij} + p_{kl}$, $p_{ik} + p_{jl}$, and $p_{il} + p_{jk}$ are equal. Thus, if $r(\rho(i), \rho(j), \rho(k), \rho(l))$ equals the absolute difference between the two largest values among $p_{\rho(i)\rho(j)} + p_{\rho(k)\rho(l)}$, $p_{\rho(i)\rho(k)} + p_{\rho(j)\rho(l)}$, and $p_{\rho(i)\rho(l)} + p_{\rho(j)\rho(k)}$, the minimization of $D(\rho)$ seeks a subset of size K that is as close to satisfying the four-point condition as possible (as measured by $D(\rho)$).

Modifications of the Target Matrix Q

The optimization of an assignment index such as (1) assumes that the target matrix $\mathbf{Q}$ is fixed and given a priori. Based on this invariance, maximizing (1), for example, could be equivalently stated as the minimization of

$$\sum_{i,j} (p_{\rho(i)\rho(j)} - q_{ij})^2. \tag{2}$$

There has been a substantial recent literature (e. g., [6,7,8]) where not only is an optimal permutation, say $\rho^*(\cdot)$, sought that would minimize (2), but in which a specific target matrix $\mathbf{Q}$ is also constructed based on a collection of (linear inequality) constraints that would characterize some type of classificatory structure fitted to $\{p_{\rho(i)\rho(j)}\}$. The constraints imposed on $\mathbf{Q}$ are possibly based on the (sought for) permutation $\rho^*(\cdot)$.

In minimizing (2) but allowing the target matrix $\mathbf{Q}$ to itself be estimated, a typical iterative process would proceed as follows: on the basis of an initial target matrix $\mathbf{Q}^{(0)}$, find a permutation, say $\rho^{(1)}(\cdot)$, to maximize the cross-product in (1). Using $\rho^{(1)}(\cdot)$, fit a target matrix $\mathbf{Q}^{(1)}$ to $\{p_{\rho^{(1)}(i)\rho^{(1)}(j)}\}$ minimizing (2). Continue the process for $\rho^{(t)}$ and $\mathbf{Q}^{(t)}$ for $t > 1$ until convergence. A variety of constraints for $\mathbf{Q}$ have been considered. Among these, there are

i) a sum of matrices each having what are called anti-Robinson forms (i.e., a matrix is *anti-Robinson* if within each row and column, the entries never decrease moving in any direction away from the main diagonal [6]);
ii) a sum of ultrametric matrices (characterized by the ultrametric condition given earlier [7]);
iii) a sum of additive tree matrices (again, as characterized by the four-point condition given earlier [7]);
iv) *unidimensional scales* (i.e., a matrix is a *linear unidimensional scale* if its entries can be given by $\{|x_j - x_i| + c\}$, where the estimated coordinates are $x_1 \leq \cdots \leq x_n$ and c is an estimated constant [8]); and
v) *circular unidimensional scales* (i.e., a matrix is so characterized if it can be represented as $\{\min\{|x_j - x_i|, x_0 - |x_j - x_i|\} + c\}$, where $x_1 \leq \cdots \leq x_n$, x_0 is the circumference of the circular structure, and c is an estimated constant [8]).

See also

- ▶ Assignment and Matching
- ▶ Bi-Objective Assignment Problem
- ▶ Communication Network Assignment Problem
- ▶ Frequency Assignment Problem
- ▶ Maximum Partition Matching
- ▶ Quadratic Assignment Problem

References

1. Carroll JD, Arabie P (1998) Multidimensional scaling. In: Birnbaum MH (ed) Measurement, judgement, and decision making. Handbook Perception and Cognition. Acad Press, New York, pp 179–250
2. De Soete G, Carroll JD (1996) Tree and other network models for representing proximity data. In: Arabie P, Hubert LJ, De Soete G (eds) Clustering and classification. World Sci, Singapore, pp 157–198
3. Gibbons JD (1971) Nonparametric statistical inference. McGraw-Hill, New York
4. Hubert LJ (1980) Analyzing proximity matrices: The assessment of internal variation in combinatorial structure. J Math Psych 21:247–264
5. Hubert LJ (1987) Assignment methods in combinatorial data analysis. M. Dekker, New York
6. Hubert LJ, Arabie P (1994) The analysis of proximity matrices through sums of matrices having (anti-)Robinson forms. British J Math Statist Psych 47:1–40
7. Hubert LJ, Arabie P (1995) Iterative projection strategies for the least-squares fitting of tree structures to proximity data. British J Math Statist Psych 48:281–317
8. Hubert LJ, Arabie P, Meulman J (1997) Linear and circular unidimensional scaling for symmetric proximity matrices. British J Math Statist Psych 50:253–284
9. Mielke PW, Berry KJ, Johnson ES (1976) Multi-response permutation procedures for a priori classifications. Comm Statist A5:1409–1424
10. Mirkin B (1996) Mathematical classification and clustering. Kluwer, Dordrecht
11. Pardalos PM, Wolkowicz H (eds) (1994) Quadratic assignment and related problems. DIMACS, Amer. Math. Soc., Providence, RI

Asymptotic Properties of Random Multidimensional Assignment Problem

PAVLO A. KROKHMAL
Department of Mechanical and Industrial Engineering, The University of Iowa, Iowa City, USA

MSC2000: 90C27, 34E05

Article Outline

Keywords and Phrases
Introduction
Expected Optimal Value of Random MAP
Expected Number of Local Minima in Random MAP
   Local Minima and p-exchange Neighborhoods in MAP
   Expected Number of Local Minima in MAP with $n = 2$
   Expected Number of Local Minima in a Random MAP with Normally Distributed Costs
Conclusions
References

Keywords and Phrases

Multidimensional assignment problem; Random assignment problem; Expected optimal value; Asymptotical analysis; Convergence bounds

Introduction

The Multidimensional Assignment Problem (MAP) is a higher dimensional version of the two-dimensional, or Linear Assignment Problem (LAP) [24]. If a classical textbook formulation of the Linear Assignment Problem is to find an optimal assignment of "N jobs to M workers", then, for example, the 3-dimensional Assignment Problem can be interpreted as finding an optimal assignment of "N jobs to M workers in K time slots", etc. In general, the objective of the MAP is to find tuples of elements from given sets, such that the total cost of the tuples is minimized. The MAP was first introduced by Pierskalla [26], and since then has found numerous applications in the areas of data association [4], image recognition [31], multisensor multitarget tracking [18,27], tracking of elementary particles [28], etc. For a discussion of the MAP and its applications see, for example, [7] and references therein.

Without loss of generality, a d-dimensional axial MAP can be written in a form where each dimension has the same number n of elements, i. e.,

$$\min_{x \in \{0,1\}^{n^d}} \left\{ \sum_{\substack{i_k \in \{1,\ldots,n\} \\ k \in \{1,\ldots,d\}}} c_{i_1 \cdots i_d} x_{i_1 \cdots i_d} \,\middle|\, \sum_{\substack{i_k \in \{1,\ldots,n\} \\ k \in \{1,\ldots,d\} \setminus j}} x_{i_1 \cdots i_d} = 1, \right.$$
$$\left. i_j = 1, \ldots, n, \ j = 1, \ldots, d \right\}. \quad (1)$$

An instance of the MAP with different numbers of elements in each dimension, $n_1 \geq n_2 \geq \cdots \geq n_d$, is reducible to form (1) by introduction of dummy variables.

Problem (1) admits the following geometric interpretation: given a d-dimensional cubic matrix, find such a permutation of its rows and columns that the sum of the diagonal elements is minimized (which explains the term "axial"). This rendition leads to an alternative formulation of the MAP (1) in terms of permutations $\pi_1, \ldots, \pi_{d-1}$ of numbers 1 to n, i. e., one-to-one mappings $\pi_i : \{1, \ldots, n\} \mapsto \{1, \ldots, n\}$,

$$\min_{\pi_1,\ldots,\pi_{d-1} \in \Pi^n} \sum_{i=1}^n c_{i,\pi_1(i),\ldots,\pi_{d-1}(i)},$$

where Π^n is the set of all permutations of the set $\{1, \ldots, n\}$. A feasible solution to the MAP (1) can be conveniently described by specifying its cost,

$$z = c_{i_1^{(1)} \cdots i_d^{(1)}} + c_{i_1^{(2)} \cdots i_d^{(2)}} + \cdots + c_{i_1^{(n)} \cdots i_d^{(n)}}, \quad (2)$$

where $\left(i_j^{(1)}, i_j^{(2)}, \ldots, i_j^{(n)}\right)$ is a permutation of the set $\{1, 2, \ldots, n\}$ for every $j = 1, \ldots, d$. In contrast to the LAP that represents a $d = 2$ special case of the MAP (1) and is polynomially solvable [7], the MAP with $d \geq 3$ is generally NP-hard, a fact that follows from reduction of the 3-dimensional matching problem (3DM) [8].

Despite its inherent difficulty, several exact and heuristic algorithms [1,6,11,25] have been proposed to this problem. Most of these algorithms rely, at least partly, on repeated local searches in neighborhoods of feasible solutions, which brings about the question of how the number of local minima in a MAP impact these solution algorithms. Intuitively, if the number of local minima is small then one may expect better performance from meta-heuristic algorithms that rely on local neighborhood searches. A solution landscape is considered to be rugged if the number of local minima is exponential with respect to the dimensions of the problem [21]. Evidence in [5] showed that ruggedness of the solution landscape has a direct impact on the effectiveness of the simulated annealing heuristic in solving at least one other hard problem, the quadratic assignment problem. Thus, one of the issues that we address below is estimation of the expected number $E[M]$ of local minima in random MAPs with respect to different local neighborhoods.

Another problem that we discuss is the behavior of the expected optimal value $Z_{d,n}^*$ of random large-scale MAPs, whose assignment costs are assumed to be independent identically distributed (iid) random variables from a given continuous distribution.

During the last two decades, expected optimal values of random assignment problems have been studied intensively in the context of random LAP. Perhaps, the most widely known result in this area is the conjecture by Mézard and Parisi [17] that the expected optimal value $E[L_n] := Z_{2,n}^*$ of a LAP of size n with iid uniform or exponential with mean 1 cost coefficients satisfies $\lim_{n \to \infty} E[L_n] = \frac{\pi^2}{6}$. In fact, this conjecture was preceded by an upper bound on the expected optimal value of the LAP with uniform (0,1) costs: $\limsup_{n \to \infty} L_n \leq 3$ due to

Walkup [32], which was soon improved by Karp [12]: $\limsup_{n\to\infty} L_n \leq 2$. A lower bound on the limiting value of L_n was first provided by Lazarus [14]: $\liminf_{n\to\infty} L_n \geq 1 + e^{-1} \approx 1.37$, and then has been improved to 1.44 by Goemans and Kodialam [9] and 1.51 by Olin [20]. Experimental evidence in support of the Mézard-Parisi conjecture was provided by Pardalos and Ramakrishnan [22]. Recently, Aldous [2] has shown that indeed $\lim_{n\to\infty} \mathsf{E}[L_n] = \frac{\pi^2}{6}$, thereby proving the conjecture. Another conjecture due to Parisi [23] stating that the expected optimal value of a random LAP of finite size n with exponentially distributed iid costs is equal to $\mathsf{E}[L_n] = Z^*_{2,n} = \sum_{i=1}^{n} i^{-2}$ has been proven independently in [16] and [19].

Our work contributes to the existing literature on random assignment problems by establishing the limiting value and asymptotic behavior of the expected optimal cost $Z^*_{d,n}$ of random MAP with iid cost coefficients for a broad class of continuous distributions. The presented approach is constructive in the sense that it allows for deriving converging asymptotical lower and upper bounds for $Z^*_{d,n}$, as well as for estimating the rate of convergence for $Z^*_{d,n}$ in special cases.

Expected Optimal Value of Random MAP

Our approach to determining the asymptotic behavior of the expected optimal cost $Z^*_{d,n}$ of an MAP (1) with random cost coefficients involves analysis of the so-called *index tree*, a graph structure that represents the set of feasible solutions of the MAP. First introduced by Pierskalla [26], the index tree graph $G = (V, E)$ of the MAP (1) has a set of vertices V which is partitioned into n levels[1] and a distinct *root node*. A node at level j of the graph represents an assignment $(i_1, \ldots, i_d)$ with $i_1 = j$ and cost $c_{ji_2\cdots i_d}$, whereby each level contains $\kappa = n^{d-1}$ nodes. The set E of arcs in the index tree graph is constructed in such a way that any feasible solution of the MAP (1) can be represented as a path connecting the root node to a leaf node at level n (such a path is called a *feasible* path); evidently, the index tree contains $n!^{d-1}$ feasible paths, by the number of feasible solutions of the MAP (1).

The index tree representation of MAP aids in construction of lower and upper bounds for the expected

---

[1]In the general case of MAP with n_i elements in dimension $i = 1, \ldots, d$, the index graph would contain n_1 levels.

optimal cost of MAP (1) with random iid costs via the following lemmata [10].

Lemma 1. *Given the index tree graph $G = (V, E)$ of $d \geq 3$, $n \geq 3$ MAP whose assignment costs are iid random variables from an absolutely continuous distribution, construct set $\mathcal{A} \subset V$ by randomly selecting α different nodes from each level of the index tree. Then, $\mathcal{A}$ is expected to contain a feasible solution of the MAP if*

$$\alpha = \left\lceil \frac{n^{d-1}}{n!^{\frac{d-1}{n}}} \right\rceil. \tag{3}$$

Lemma 2. *For a $d \geq 3$, $n \geq 3$ MAP whose cost coefficients are iid random variables from an absolutely continuous distribution F with existing first moment, define*

$$\underline{Z}^*_{d,n} := n\mathsf{E}_F\left[X_{(1|\kappa)}\right] \quad \text{and} \quad \overline{Z}^*_{d,n} := n\mathsf{E}_F\left[X_{(\alpha|\kappa)}\right], \tag{4}$$

where $X_{(i|\kappa)}$ is the ith order statistic of $\kappa = n^{d-1}$ iid random variables with distribution F, and parameter α is determined as in (3). Then, $\underline{Z}^*_{d,n}$ and $\overline{Z}^*_{d,n}$ constitute lower and upper bounds for the expected optimal cost $Z^*_{d,n}$ of the MAP, respectively: $\underline{Z}^*_{d,n} \leq Z^*_{d,n} \leq \overline{Z}^*_{d,n}$.

Proofs of the lemmas are based on the probabilistic method [3] and can be found in [10]. In particular, the proof of Lemma 2 considers a set $\mathcal{A}_{\min}$ that is constructed by selecting from each level of the index tree α nodes with the smallest costs among the κ nodes at that level. The continuity of distribution F ensures that assignment costs in the MAP (1) are all different almost surely, hence locations of the nodes that comprise the set $\mathcal{A}_{\min}$ are random with respect to the array of nodes in each level of $G(V, E)$. In the remainder of the paper, we always refer to α and κ as defined above.

By definition, the parameter $\kappa = n^{d-1}$ approaches infinity whenever n or d does; this allows us to denote the corresponding cases by $\kappa \xrightarrow{n} \infty$ and $\kappa \xrightarrow{d} \infty$, respectively. If certain statement holds for both cases of $n \to \infty$ and $d \to \infty$, we indicate this by $\kappa \xrightarrow{n,d} \infty$. The behavior of quantity α (3) when n or d increases is more contrasting. In the case $n \to \infty$ it approaches a finite limiting value,

$$\alpha \to \alpha^* := \lceil e^{d-1} \rceil, \quad \kappa \xrightarrow{n} \infty, \tag{5}$$

while in the case of fixed n and unbounded d it increases exponentially:

$$\alpha \sim \kappa^{\gamma_n}, \quad \kappa \xrightarrow{d} \infty, \quad \text{where} \quad \gamma_n = 1 - \frac{\ln n!}{n \ln n}, \quad (6)$$

and it is important to observe that $0 < \gamma_n < \frac{1}{2}$ for $n \geq 3$ [13].

The presented lemmata addresses MAPs with $d \geq 3, n \geq 3$. The case $d = 2$ represents, as noted earlier, the Linear Assignment Problem, whose asymptotic behavior is distinctly different from that of MAPs with $d \geq 3$. It can be shown that in the case of $d = 2$ Lemmas 1 and 2 produce only trivial bounds that are rather inefficient in determining the asymptotic behavior of the expected optimal value of the LAP within the presented approach. In the case $n = 2$ the costs of feasible solutions to the MAP (1) have the form

$$z = c_{i_1^{(1)} \ldots i_d^{(1)}} + c_{i_1^{(2)} \ldots i_d^{(2)}},$$
$$\text{where} \quad i_j^{(1)}, i_j^{(2)} \in \{1, 2\}, i_j^{(1)} \neq i_j^{(2)},$$

and consequently are iid random variables with distribution F_2, which is the convolution of F with itself: $F_2 = F * F$ [11]. This fact allows for computing the expected optimal value of $n = 2$ MAP exactly, without resorting to bounds (4):

$$Z_{d,2}^* = \mathsf{E}_{F*F}\left[X_{(1|2!^{d-1})}\right]. \quad (7)$$

In the general case $d \geq 3, n \geq 3$ the main challenge is constituted by computation of the upper bound $\overline{Z}_{d,n}^* = n\mathsf{E}_F\left[X_{(\alpha|\kappa)}\right]$, where $X_{(\alpha|\kappa)}$ is the α-th order statistic among κ independent F-distributed random variables. The subsequent analysis relies on representation of $\overline{Z}_{d,n}^*$ in the form

$$\overline{Z}_{d,n}^* = \frac{n\Gamma(\kappa+1)}{\Gamma(\alpha)\Gamma(\kappa-\alpha+1)}$$
$$\cdot \int_0^1 F^{-1}(u) u^{\alpha-1}(1-u)^{\kappa-\alpha} du, \quad (8)$$

where F^{-1} denotes the inverse of the c.d.f. F of the the distribution of assignment costs in MAP (1). While it is practically impossible to evaluate the integral in (8) exactly in the general case, its asymptotic behavior for large n and d can be determined for a wide range of distributions F. For instance, in the case when distribution F has a finite left endpoint of its support set, the asymptotic behavior of the integral in (8) is obtained by means of the following

Lemma 3. *Let function $h(u)$ have the following asymptotic expansion at $0+$,*

$$h(u) \sim \sum_{s=0}^{\infty} a_s u^{(s+\lambda-\mu)/\mu}, \quad u \to 0+, \quad (9)$$

where $\lambda, \mu > 0$. Then for any positive integer m one has

$$\int_0^1 h(u) u^{\alpha-1}(1-u)^{\kappa-\alpha} du$$
$$= \sum_{s=0}^{m-1} a_s \phi_s(\kappa) + \mathcal{O}(\phi_m(\kappa)), \quad \kappa \xrightarrow{n,d} \infty, \quad (10)$$

where $\phi_s(\kappa) = B\left(\frac{s+\lambda}{\mu}+\alpha-1, \kappa-\alpha+1\right), s = 0, 1, \ldots,$ provided that the integral is absolutely convergent for $\kappa = \alpha = 1$.

Above, $B(x, y)$ is the Beta function. Using similar results for the cases when the support set of distribution F is unbounded from below, we obtain that the limiting behavior of the expected optimal value $Z_{d,n}^*$ of random MAP is determined by the location of the left endpoint of the support of F [13].

Theorem 1. Expected Optimal Value of Random MAP *Consider a $d \geq 3, n \geq 2$ MAP (1) with cost coefficients that are iid random variables from an absolutely continuous distribution F with existing first moment. If the distribution F satisfies either of the following conditions,*
1. $F^{-1}(u) = F^{-1}(0+) + \mathcal{O}(u^\beta), u \to 0+, \beta > 0$
2. $F^{-1}(u) \sim -\nu u^{-\beta_1}\left(\ln \frac{1}{u}\right)^{\beta_2}, u \to 0+, 0 \leq \beta_1 < 1, \beta_2 \geq 0, \beta_1 + \beta_2 > 0, \nu > 0$

where $F^{-1}(0+) = \lim_{u \to 0+} F^{-1}(u)$, the expected optimal value of the MAP satisfies

$$\lim Z_{d,n}^* = \lim nF^{-1}(0+),$$

where both limits are taken at either $n \to \infty$ or $d \to \infty$.

The obtained results can be readily employed to construct upper and lower asymptotical bounds for the expected optimal value of MAP when one of the parameters n or d is large but finite. The following statement follows directly from Lemma 3 and Theorem 1.

Corollary 1. *Consider a $d \geq 3, n \geq 3$ MAP (1) with cost coefficients that are iid random variables from an absolutely continuous distribution with existing first moment. Let $a \in \mathbb{R}$ be the left endpoint of the support set of this distribution, $a = F^{-1}(0+)$, and assume that the inverse $F^{-1}(u)$ of the c.d.f. $F(u)$ of the distribution is such that*

$$F^{-1}(u) \sim a + \sum_{s=1}^{\infty} a_s u^{s/\mu}, \quad u \to 0+, \mu > 0. \quad (11)$$

Then, for any integer $m \geq 1$, lower and upper bounds $\underline{Z}^*_{d,n}, \overline{Z}^*_{d,n}$ (4) on the expected optimal cost $Z^*_{d,n}$ of the MAP can be asymptotically evaluated as

$$\underline{Z}^*_{d,n} = an + \sum_{s=1}^{m-1} a_s \frac{n\Gamma(\kappa+1)\Gamma\left(\frac{s}{\mu}+1\right)}{\Gamma\left(\kappa+\frac{s}{\mu}+1\right)}$$
$$+ \mathcal{O}\left(n\frac{\Gamma(\kappa+1)\Gamma\left(\frac{m}{\mu}+1\right)}{\Gamma\left(\kappa+\frac{m}{\mu}+1\right)}\right), \kappa \xrightarrow{n,d} \infty, \quad (12a)$$

$$\overline{Z}^*_{d,n} = an + \sum_{s=1}^{m-1} a_s \frac{n\Gamma(\kappa+1)\Gamma\left(\frac{s}{\mu}+\alpha\right)}{\Gamma(\alpha)\Gamma\left(\kappa+\frac{s}{\mu}+1\right)}$$
$$+ \mathcal{O}\left(n\frac{\Gamma(\kappa+1)\Gamma\left(\frac{m}{\mu}+\alpha\right)}{\Gamma(\alpha)\Gamma\left(\kappa+\frac{m}{\mu}+1\right)}\right), \kappa \xrightarrow{n,d} \infty. \quad (12b)$$

It can be shown that the lower and upper bounds defined by (12a, 12b) are convergent, i.e., $|\overline{Z}^*_{d,n} - \underline{Z}^*_{d,n}| \to 0$, $\kappa \xrightarrow{n,d} \infty$, whereas the corresponding asymptotical bounds for the case of distributions with support unbounded from below may be divergent in the sense that $|\overline{Z}^*_{d,n} - \underline{Z}^*_{d,n}| \not\to 0$ when $\kappa \xrightarrow{n,d} \infty$.

The asymptotical representations (12a, 12b) for the bounds $\underline{Z}^*_{d,n}$ and $\overline{Z}^*_{d,n}$ are simplified when the inverse F^{-1} of the c.d.f. of the distribution has a regular power series expansion in the vicinity of zero. Assume, for example, that function F^{-1} can be written as

$$F^{-1}(u) = a_1 u + \mathcal{O}(u^2), \quad u \to 0+. \quad (13)$$

It is then easy to see that for $n \gg 1$ and d fixed the expected optimal value of the MAP is asymptotically bounded as

$$\frac{a_1}{n^{d-2}} + \mathcal{O}\left(\frac{1}{n^{d-1}}\right) \leq Z^*_{d,n}$$
$$\leq \frac{a_1 \lceil e^{d-1} \rceil}{n^{d-2}} + \mathcal{O}\left(\frac{1}{n^{d-1}}\right), \quad n \to \infty, \quad (14)$$

which immediately yields the rate of convergence to zero for $Z^*_{d,n}$ as n approaches infinity:

Corollary 2. *Consider a $d \geq 3, n \geq 3$ MAP (1) with cost coefficients that are iid random variables from an absolutely continuous distribution with existing first moment. Let the inverse F^{-1} of the c.d.f. of the distribution satisfy (13). Then, for a fixed d and $n \to \infty$ the expected optimal value $Z^*_{d,n}$ of the MAP converges to zero as $\mathcal{O}\left(n^{-(d-2)}\right)$.*

For example, the expected optimal value of 3-dimensional ($d = 3$) MAP with uniform $U(0, 1)$ or exponential distributions converges to zero as $\mathcal{O}(n^{-1})$ when $n \to \infty$.

We illustrate the tightness of the developed bounds (12a, 12b) by comparing them to the computed expected optimal values of MAPs with coefficients $c_{i_1 \cdots i_d}$ drawn from the uniform $U(0, 1)$ distribution and exponential distribution with mean 1. It is elementary that the inverse functions $F^{-1}(\cdot)$ of the c.d.f.'s for both these distributions are representable in form (13) with $a_1 = 1$.

The numerical experiments involved solving multiple instances of randomly generated MAPs with the number of dimensions d ranging from 3 to 10, and the number n of elements in each dimension running from 3 to 20. The number of instances generated for estimation of the expected optimal value of the MAP with a given distribution of cost coefficients varied from 1000 (for smaller values of d and n) to 50 (for problems with largest n and d).

To solve the problems to optimality, we used a branch-and-bound algorithm that navigated through the index tree representation of the MAP. Figures 1 and 2 display the obtained expected optimal values of MAP with uniform and exponential iid cost coefficients when d is fixed at $d = 3$ or 5 and $n = 3, \ldots, 20$, and when $n = 3$ or 5 and d runs from 3 to 10. This "asymmetry" in reporting of the results is explained by

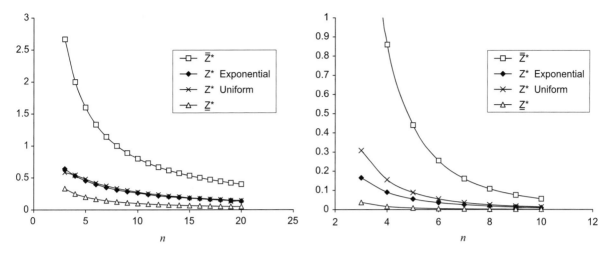

Asymptotic Properties of Random Multidimensional Assignment Problem, Figure 1
Expected optimal value $Z^*_{d,n}$, lower and upper bounds $\underline{Z}^*_{d,n}, \overline{Z}^*_{d,n}$ of an MAP with fixed $d = 3$ (left) and $d = 5$ (right) for uniform $U(0, 1)$ and exponential (1) distributions

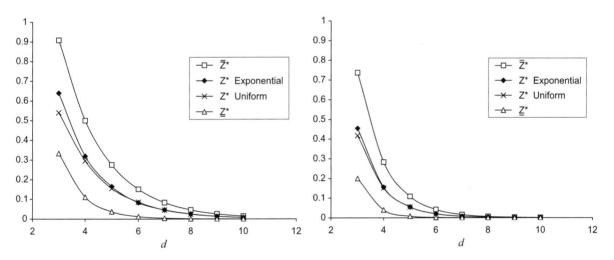

Asymptotic Properties of Random Multidimensional Assignment Problem, Figure 2
Expected optimal value $Z^*_{d,n}$, lower and upper bounds $\underline{Z}^*_{d,n}, \overline{Z}^*_{d,n}$ of an MAP with fixed $n = 3$ (left) and $n = 5$ (right) for uniform $U(0, 1)$ and exponential(1) distributions

the fact that the implemented branch-and-bound algorithm based on index tree is more efficient in solving "shallow" MAPs, i.e., instances that have larger n and smaller d. The solution times varied from several seconds to 20 hours on a 2GHz PC.

The conducted numerical experiments suggest that the constructed lower and upper bounds for the expected optimal cost of random MAPs are quite tight, with the upper bound $\overline{Z}^*_{d,n}$ being tighter for the case of fixed n and large d (see Figs. 1, 2).

Expected Number of Local Minima in Random MAP

Local Minima and p-exchange Neighborhoods in MAP

As it has been mentioned in the Introduction, we consider local minima of a MAP with respect to a local neighborhood, in the sense of [15]. For any $p = 2, \ldots, n$, we define the p-exchange local neighborhood $\mathcal{N}_p(i)$ of the ith feasible solution $\{i_1^{(1)} \cdots i_d^{(1)}, \ldots, i_1^{(n)} \cdots i_d^{(n)}\}$ of the MAP (1) as the set of solutions obtained from i by permuting p or less elements in one of the dimensions $1, \ldots, d$. More formally, $\mathcal{N}_p(i)$ is the set of n-tuples $\{j_1^{(1)} \cdots j_d^{(1)}, \ldots, j_1^{(n)} \cdots j_d^{(n)}\}$ such that $\{j_k^{(1)}, \ldots, j_k^{(n)}\}$ is a permutation of $\{1, \ldots, n\}$ for all $1 \leq k \leq d$, and, furthermore, there exists only one $k_0 \in \{1, \ldots, d\}$ such that

$$2 \leq \sum_{r=1}^{n} \bar{\delta}_{i_{k_0}^{(r)} j_{k_0}^{(r)}} \leq p, \quad \text{while} \quad \sum_{r=1}^{n} \bar{\delta}_{i_k^{(r)} j_k^{(r)}} = 0$$
$$\text{for all} \quad k \in \{1, \ldots, d\} \setminus k_0, \quad (15)$$

where $\bar{\delta}_{ij}$ is the negation of the Kroneker delta, $\bar{\delta}_{ij} = 1 - \delta_{ij}$. As an example, consider the following feasible solution to a $d = 3$, $n = 3$ MAP: $\{111, 222, 333\}$. Then, one of its 2-exchange neighbors is $\{111, 322, 233\}$, another one is $\{131, 222, 313\}$; a 3-exchange neighbor is given by $\{311, 122, 233\}$, etc. Evidently, one has $\mathcal{N}_p \subset \mathcal{N}_{p+1}$ for $p = 2, \ldots, n - 1$.

Proposition 1. *For any* $p = 2, \ldots, n$, *the size* $|\mathcal{N}_p|$ *of the p-exchange local neighborhood of a feasible solution of a MAP (1) is equal to*

$$|\mathcal{N}_p| = d \sum_{k=2}^{p} D(k) \binom{n}{k},$$
$$\text{where} \quad D(k) = \sum_{j=0}^{k} (-1)^{k-j} \binom{k}{j} j! . \quad (16)$$

The quantity $D(k)$ in (16) is known as the number of *derangements* of a k-element set [29], i. e., the number of permutations $\{1, 2, \ldots, k\} \mapsto \{i^{(1)}, i^{(2)}, \ldots, i^{(k)}\}$ such that $i^{(1)} \neq 1, \ldots, i^{(k)} \neq k$, and can be easily calculated by means of the recurrent relation (see [29])

$$D(k) = kD(k-1) + (-1)^k, \quad D(1) = 0,$$

so that, for example, $D(2) = 1$, $D(3) = 2$, $D(4) = 9$, and so on. Then, according to Proposition 1, the size of a 2-exchange neighborhood is $|\mathcal{N}_2| = d\binom{n}{2}$, the size of a 3-exchange neighborhood is $|\mathcal{N}_3| = d\left[\binom{n}{2} + 2\binom{n}{3}\right]$, etc.

Note also that size of the p-exchange neighborhood is *linear* in the number of dimensions d. Depending on p, $|\mathcal{N}_p|$ is either *polynomial* or *exponential* in the number of elements n per dimension, as follows from the representation

$$D(n) = n! \left(1 - \frac{1}{1!} + \frac{1}{2!} - \frac{1}{3!} + \cdots + \frac{(-1)^n}{n!}\right) \approx \frac{n!}{e},$$
$$n \gg 1.$$

The definition of a local minimum with respect to the p-exchange neighborhood is then straightforward. The kth feasible solution with cost z_k is a p-exchange local minimum iff $z_k \leq z_j$ for all $j \in \mathcal{N}_p(k)$. Continuing the example above, the solution $\{111, 222, 333\}$ is a 2-exchange local minimum iff its cost $z_1 = c_{111} + c_{222} + c_{333}$ is less than or equal to costs of all of its 2-exchange neighbors.

The number M_p of local minima of the MAP is obtained by counting the feasible solutions that are local minima with respect to neighborhoods $\mathcal{N}_p$. In a random MAP, where the assignment costs are random variables, M_p becomes a random quantity itself. In this paper we are interested in determining the expected number $\mathsf{E}[M_p]$ of local minima in random MAPs that have iid assignment costs with continuous distribution.

Expected Number of Local Minima in MAP with $n = 2$

As it was noted above, in the special case of random MAP with $n = 2$, $d \geq 3$, the costs of feasible solutions are iid random variables with distribution $F * F$, where F is the distribution of the assignment costs. This special structure of the feasible set allows for a closed-form expression for the expected number of local minima $\mathsf{E}[M]$ (note that in a $n = 2$ MAP the largest local neighborhood is $\mathcal{N}_2$, thus $M = M_2$), as established in [11].

Theorem 2. *In a* $n = 2$, $d \geq 3$ *MAP with cost coefficients that are iid continuous random variables, the expected number of local minima is given by*

$$\mathsf{E}[M] = \frac{2^{d-1}}{d+1} . \quad (17)$$

Equality (17) implies that in a $n = 2$, $d \geq 3$ MAP the number of local minima $\mathsf{E}[M]$ is *exponential* in d, when the cost coefficients are independently drawn from *any* continuous distribution.

Expected Number of Local Minima in a Random MAP with Normally Distributed Costs

Our ability to derive a closed-form expression (17) for the expected number of local minima $\mathsf{E}[M]$ in the previous section has relied on the independence of feasible solution costs (2) in a $n = 2$ MAP. As it is easy to verify directly, in the case $n \geq 3$ the costs of feasible solutions are generally not independent. This complicates analysis significantly if an arbitrary continuous distribution for assignment costs $c_{i_1 \cdots i_d}$ in (1) is assumed. However, as we show below, one can derive upper and lower bounds for $\mathsf{E}[M]$ in the case when the costs coefficients of (1) are independent normally distributed random variables. First, we develop bounds for the number of local minima $\mathsf{E}[M_2]$ defined with respect to 2-exchange neighborhoods $\mathcal{N}_2$ that are most widely used in practice.

2-exchange Local Neighborhoods Noting that in the general case the number N of the feasible solutions to MAP (1) is equal to $N = (n!)^{d-1}$, the expected number of local minima $\mathsf{E}[M_2]$ with respect to local 2-exchange neighborhoods can be written in the form

$$\mathsf{E}[M_2] = \sum_{k=1}^{N} \mathsf{P}\left[\bigcap_{j \in \mathcal{N}_2(k)} z_k - z_j \leq 0\right], \quad (18)$$

where $\mathcal{N}_2(k)$ is the 2-exchange neighborhood of the kth feasible solution, and z_i is the cost of the ith feasible solution, $i = 1, \ldots, N$. If we allow the n^d cost coefficients $c_{i_1 \cdots i_d}$ of the MAP to be independent standard normal $N(\mu, \sigma^2)$ random variables, then the probability term in (18) can be expressed as

$$\mathsf{P}\left[\bigcap_{j \in \mathcal{N}_2(k)} z_k - z_j \leq 0\right] = F_\Sigma(\mathbf{0}), \quad (19)$$

where F_Σ is the c.d.f. of the $|\mathcal{N}_2|$-dimensional random vector

$$\mathbf{Z} = (Z_{121}, \ldots, Z_{12d}, Z_{131}, \ldots, Z_{13d}, \cdots$$
$$\cdots, Z_{rs1}, \ldots, Z_{rsd}, \cdots, Z_{n-1,n,1}, \ldots, Z_{n-1,n,d}),$$
$$r < s. \quad (20)$$

Vector $\mathbf{Z}$ has a normal distribution $N(\mathbf{0}, \Sigma)$ with the covariance matrix Σ defined as

$$\mathrm{Cov}(Z_{rsq}, Z_{ijk}) = \begin{cases} 4\sigma^2, & \text{if } i = r, j = s, q = k, \\ 2\sigma^2, & \text{if } i = r, j = s, q \neq k, \\ \sigma^2, & \text{if } (i = r, j \neq s) \text{ or } (i \neq r, j = s), \\ 0, & \text{if } i \neq r, j \neq s. \end{cases} \quad (21)$$

While the value of $F_\Sigma(\mathbf{0})$ in (19) is difficult to compute exactly for large d and n, lower and upper bounds can be constructed using Slepian's inequality [30]. To this end, we introduce covariance matrices $\underline{\Sigma} = (\underline{\sigma}_{ij})$ and $\overline{\Sigma} = (\bar{\sigma}_{ij})$ as

$$\underline{\sigma}_{ij} = \begin{cases} 4\sigma^2, & \text{if } i = j, \\ 2\sigma^2, & \text{if } i \neq j \text{ and} \\ & (i-1) \operatorname{div} d = (j-1) \operatorname{div} d \\ 0, & \text{otherwise} \end{cases}, \quad (22a)$$

$$\bar{\sigma}_{ij} = \begin{cases} 4\sigma^2, & \text{if } i = j, \\ 2\sigma^2, & \text{otherwise} \end{cases}, \quad (22b)$$

so that $\underline{\sigma}_{ij} \leq \sigma_{ij} \leq \bar{\sigma}_{ij}$ holds for all $1 \leq i, j \leq |\mathcal{N}_2|$, with σ_{ij} being the components of the covariance matrix Σ (21). Then, Slepian's inequality claims that

$$F_{\underline{\Sigma}}(\mathbf{0}) \leq F_\Sigma(\mathbf{0}) \leq F_{\overline{\Sigma}}(\mathbf{0}), \quad (23)$$

where $F_{\underline{\Sigma}}(\mathbf{0})$ and $F_{\overline{\Sigma}}(\mathbf{0})$ are c.d.f.'s of random variables $\mathbf{X}_{\underline{\Sigma}} \sim N(\mathbf{0}, \underline{\Sigma})$ and $\mathbf{X}_{\overline{\Sigma}} \sim N(\mathbf{0}, \overline{\Sigma})$ respectively. The structure of matrices $\underline{\Sigma}$ and $\overline{\Sigma}$ allows the corresponding values $F_{\underline{\Sigma}}(\mathbf{0})$ and $F_{\overline{\Sigma}}(\mathbf{0})$ to be computed in a closed form, which leads to the following bounds for the expected number of local minima in random MAP with iid normal coefficients:

Theorem 3. *In a $n \geq 3, d \geq 3$ MAP with iid normal cost coefficients, the expected number of 2-exchange local minima is bounded as*

$$\frac{(n!)^{d-1}}{(d+1)^{n(n-1)/2}} \leq \mathsf{E}[M_2] \leq \frac{2(n!)^{d-1}}{n(n-1)d+2}. \quad (24)$$

Note that both the lower and upper bounds in (24) coincide with the exact expression (17) for $\mathsf{E}[M_2]$ in the case $n = 2$. Also, from (24) it follows that for fixed $n \geq 3$, the expected number of local minima is *exponential* in the number of dimensions d for a fixed n.

Higher-Order Neighborhoods ($p \geq 3$) The outlined approach is applicable to general p-exchange neighborhoods. For convenience, here we consider the neighborhoods $\mathcal{N}_p^*$ as defined in Sect. "Local Minima and p-exchange Neighborhoods in MAP", i.e., the neighborhoods obtained from a given feasible solution by permuting exactly p elements in one of the d dimensions, so that for any feasible solution $i = \{i_1^{(1)} \cdots i_d^{(1)}, \ldots, i_1^{(n)} \cdots i_d^{(n)}\}$ and its p-exchange neighbor $j = \{j_1^{(1)} \cdots j_d^{(1)}, \ldots, j_1^{(n)} \cdots j_d^{(n)}\} \in \mathcal{N}_p^*(i)$ one has (compare to (15))

$$\sum_{r=1}^{n} \bar{\delta}_{i_{k_0}^{(r)} j_{k_0}^{(r)}} = p, \quad k_0 \in \{1, \ldots, d\}, \quad \text{and}$$

$$\sum_{r=1}^{n} \bar{\delta}_{i_k^{(r)} j_k^{(r)}} = 0 \quad \text{for all} \quad k \in \{1, \ldots, d\} \setminus k_0 \ .$$

(25)

Then, upper and lower bounds for the expected number of local minima $\mathsf{E}[M_p^*]$ defined with respect to p-exchange neighborhoods $\mathcal{N}_p^*$ can be derived in a similar fashion. Namely, the sought probability

$$\mathsf{P}\left[\bigcap_{i \in \mathcal{N}_p^*(k)} z_k - z_i \leq 0\right] = F_{\Sigma_p}(0)$$

can be bounded as $F_{\underline{\Sigma}_p}(0) \leq F_{\Sigma_p}(0) \leq F_{\overline{\Sigma}_p}(0)$, where the matrices $\overline{\Sigma}_p, \underline{\Sigma}_p \in \mathbb{R}^{|\mathcal{N}_p^*| \times |\mathcal{N}_p^*|}$ are such that

$$(\overline{\Sigma}_p)_{ij} = \begin{cases} 2p\sigma^2, & \text{if } i = j, \\ (2p-2)\sigma^2, & \text{if } i \neq j, \end{cases} \quad (26a)$$

$$(\underline{\Sigma}_p)_{ij} = \begin{cases} 2p\sigma^2, & \text{if } i = j, \\ p\sigma^2, & \text{if } i \neq j \text{ and } (i-1) \operatorname{div} (dD(p)) \\ & = (j-1) \operatorname{div} (dD(p)), \\ 0, & \text{otherwise .} \end{cases} \quad (26b)$$

The corresponding bounds for the expected number of local minima $\mathsf{E}[M_p^*]$ are established by the following theorem [11].

Theorem 4. *In a $n \geq 3, d \geq 3$ MAP with iid normal cost coefficients, the expected number of local minima M_p^* with respect to p-exchange local neighborhoods $\mathcal{N}_p^*$ is bounded as*

$$\frac{n!^{d-1}}{[dD(p)+1]^{\binom{n}{p}}} \leq \mathsf{E}[M_p^*] \leq n!^{d-1} \int_{-\infty}^{+\infty} \left[\Phi\left(\sqrt{p-1}z\right)\right]^{d\binom{n}{p}D(p)} d\Phi(z) ,$$

(27)

where $\Phi(z)$ is the c.d.f. of the standard normal $N(0, 1)$ distribution. For 3-exchange neighborhoods $\mathcal{N}_3^*$, an improved upper bound holds:

$$\mathsf{E}[M_3^*] \leq \frac{3n!^{d-1}}{n(n-1)(n-2)d+3} . \quad (28)$$

It is interesting to note that for a fixed p the ratio of number of local minima to the number of feasible solutions becomes infinitely small as the dimensions of the problem increase (see (17), (24), and (27)).

Conclusions

We have discussed asymptotical analysis of the expected optimal value and the expected number of local minima of the Multidimensional Assignment Problem whose assignment costs are iid random variables drawn from a continuous distribution. It has been demonstrated that for a broad class of distributions, the asymptotical behavior of the expected optimal cost of a random MAP in the case when one of the problem's dimension parameters approaches infinity is determined by the location of the left endpoint of the support set of the distribution. The presented analysis is constructive in the sense that it allows for derivation of lower and upper asymptotical bounds for the expected optimal value of the problem for a prescribed probability distribution.

In addition, we have derived a closed-form expression for the expected number of local minima in a $n = 2$ random MAP with arbitrary distribution of assignment costs. In the case $n \geq 3$, bounds for the expected number of local minima have been derived in the assumption that assignment costs are iid normal random variables. It has been demonstrated that the expected number of local minima is exponential in the number of dimensions d of the problem.

References

1. Aiex RM, Resende MGC, Pardalos PM, Toraldo G (2005) GRASP with Path Relinking for Three-Index Assignment. INFORMS J Comput 17(2):224–247
2. Aldous D (2001) The $\zeta(2)$ limit in the random assignment problem. Random Struct Algorithm 18(4):381–418
3. Alon N, Spencer J (2000) The Probabilistic Method, 2nd edn, Interscience Series in Discrete Mathematics and Optimization. Wiley, New York
4. Andrijich SM, Caccetta L (2001) Solving the multi-sensor data association problem. Nonlinear Analysis 47:5525–5536
5. Angel E, Zissimopoulos V (2001) On the landscape ruggedness of the quadratic assignment problem. Theor Comput Sci 263:159–172
6. Balas E, Saltzman MJ (1991) An algorithm for the three-index assignment problem. Oper Res 39:150–161
7. Burkard RE (2002) Selected topics on assignment problems. Discret Appl Math 123:257–302
8. Garey MR, Johnson DS (1979) Computers and Intractability: A Guide to the Theory of NP-completeness. Freeman, San Francisco
9. Goemans MX, Kodialam M (1993) A lower bound on the expected value of an optimal assignment. Math Oper Res 18:267–274
10. Grundel DA, Oliveira CAS, Pardalos PM (2004) Asymptotic properties of random multidimensional assignment problems. J Optim Theory Appl 122(3):487–500
11. Grundel DA, Krokhmal PA, Oliveira CAS, Pardalos PM (2007) Asymptotic properties of random multidimensional assignment problems. J Comb Optim 13(1):1–18
12. Karp RM (1987) An upper bound on the expected cost of an optimal assignment. In: Discret Algorithm Complexity. Academic Press, Boston, pp 1–4
13. Krokhmal PA, Grundel DA, Pardalos P (2007) Asymptotic Behavior of the Expected Optimal Value of the Multidimensional Assignment Problem. Math Program 109(2–3):525–551
14. Lazarus AJ (1993) Certain expected values in the random assignment problem. Oper Res Lett 14:207–214
15. Lin S, Kernighan BW (1973) An effective heuristic algorithm for the traveling salesman problem. Oper Res 21:498–516
16. Linusson S, Wästlund J (2004) A proof of Parisi's conjecture on the random assignment problem. Probab Theory Relat Fields 128(3):419–440
17. Mézard M, Parisi G (1985) Replicas and optimization. J Phys Lett 46(17):771–778
18. Murphey R, Pardalos P, Pitsoulis L (1998) A greedy randomized adaptive search procedure for the multitarget multi-sensor tracking problem. In: DIMACS Series, vol 40, American Mathematical Society, pp 277–302
19. Nair C, Prabhakar B, Sharma M (2005) A Proof of the Conjecture due to Parisi for the Finite Random Assignment Problem. Random Struct Algorithms 27(4):413–444
20. Olin B (1992) Asymptotic properties of the random assignment problem. Ph.D thesis, Royal Institute of Technology, Stockholm, Sweden
21. Palmer R (1991) Optimization on rugged landscapes. In: Perelson A, Kauffman S (eds) Molecular Evolution on Rugged Ladscapes: Proteins, RNA, and the Immune System. Addison Wesley, Redwood City, pp 3–25
22. Pardalos PM, Ramakrishnan KG (1993) On the expected optimal value of random assignment problems: Experimental results and open questions. Comput Optim Appl 2:261–271
23. Parisi G (1998) A conjecture on random bipartite matching. Physics e-Print archive, http://xxx.lang.gov/ps/cond-_mat/9801176
24. Papadimitrou CH, Steiglitz K (1998) Combinatorial Optimization: Algorithms and Complexity. Dover, New York
25. Pasiliao EL (2003) Algorithms for Multidimensional Assignment Problems. PhD thesis, University of Florida
26. Pierskalla W (1968) The multidimensional assignment problem. Oper Res 16:422–431
27. Poore AB (1994) Multidimensional assignment formulation of data association problems arising from multitarget and multisensor tracking. Comput Optim Appl 3:27–54
28. Pusztaszeri J, Rensing PE, Liebling TM (1995) Tracking elementary particles near their primary vertex: a combinatorial approach. J Glob Optim 16:422–431
29. Stanley R (1986) Enumerative Combinatorics. Wadsworth and Brooks, Belmont CA
30. Tong YL (1990) The Multivariate Normal Distribution. Springer, Berlin
31. Veenman CJ, Hendriks EA, Reinders MJT (1998) A fast and robust point tracking algorithm. Proc Fifth IEEE Int Conf Image Processing 653–657, Chicago, USA
32. Walkup DW (1979) On the expected value of a random assignment problem. SIAM J Comput 8:440–442

Asynchronous Distributed Optimization Algorithms

IOANNIS P. ANDROULAKIS
Department of Biomedical Engineering,
Rutgers University, Piscataway, USA

MSC2000: 90C30, 90C30, 90C52, 90C53, 90C55

Article Outline

Keywords
See also
References

Keywords

Asynchronous iterative algorithms; Distributed computing; Optimization

Many iterative algorithms, deterministic or stochastic, admit distributed implementations, whereby the work load for performing computational steps, identified as bottlenecks, is distributed among a variety of computational nodes. Extensive literature regarding distributed implementations of optimization algorithms in particular is available, [19]. In recent years, there has been an extremely fruitful interface between mathematical programming algorithms and computer science. This has resulted in major advances in the development of algorithms and implementation of sophisticated optimization algorithms on high performance parallel and distributed computers, [11,12]. Two major issues are important in designing an efficient distributed implementation, namely, *task allocation*, and *communication protocol*. Task allocation relates to the breakdown of the total work load and this can either be static or dynamic depending. Communication patterns and frequency are important since they can induce substantial overhead in cases where workload irregularities occur. Various important implementational details have been presented, among others, in [10]. The straightforward translation of serial to a distributed algorithm would assume some sort of global synchronization mechanism that would guarantee that information among processing nodes is being exchanged once a computational step has been performed. Processors must then synchronize so as to exchange information and proceed all with the same type of information to their next computational step. *Asynchronous algorithms* relax the assumption of a predetermined synchronization protocol, and allow each processing element to compute and communicate following local rates. The primary motivation for developing algorithms was to address situations in which:

- processors do not need to communicate to each other processor at each time instance;
- processors may keep performing computations without having to wait until they receive the messages that have been transmitted to them;
- processors are allowed to remain idle some of the time;
- some processors may be performing computations faster than others.

Such algorithms can alleviate communication overloads and they are not excessively slowed down by either communication delays nor by differences in the time it takes processors to perform one computation, [18]. Another major motivation is clearly to develop robust algorithms for distributed computation on heterogeneous networks of computers. The ideas of asynchronous, also known as *chaotic*, iterative schemes, can be traced by to [9], in which special schemes for solving linear systems of equations were developed. For discussing the basic principles and conditions of asynchronous iterations, the formalism of [8] will be followed. This work presented the first comprehensive treatment of the recent developments in the theory and practice of asynchronous iterations for a variety of problems, including deterministic and stochastic optimization. In essence, most iterative algorithms can be viewed as the search for a fixed point that corresponds to the solution of the original problem. The basic assumptions of the model of asynchronous (chaotic) iterations for determining fixed point of (non)linear mappings are as follows:

1) Let X be a vector space and $x = (x_1, \ldots, x_n) \in X$ are n-tuples describing any vector from this set. It is also assumed that $X = X_1 \times \cdots \times X_n$, with $x_i \in X_i$, $i = 1, \ldots, n$.
2) Let $f: X \to X$ be a function defined by $f(x) = (f_1(x), \ldots, f_n(x))$, $\forall x \in X$.
3) A point $X^\star \in X$ is a *fixed point* of $f(x)$ if $x^\star = f(x^\star)$ or, equivalently, $x_i^\star = f_i(x^\star)$, $i = 1, \ldots, n$.

For the solution of the aforementioned problem, one can define an iterative method as:

$$x_i := f_i(x), \quad i = 1, \ldots, n,$$

with $x_i(t)$ being the values of the ith component at time (iteration) t. In order to comprehend the concept of asynchronous iterations, we assume that there exists a set of times $T = \{0, 1, \ldots\}$ at which one or more (possibly none) components x_i of x are updated by some processor of a distributed computing system. We defined by T^i the set of times at which x_i is updated. Given that no synchronization protocol dictating the information exchange exists, it is quite conceivable that not all processors have access to the same and most recent values of of the corresponding components of x. It will be therefore assumed that:

$$x_i(t+1) = \begin{cases} f_i(x_1\tau_1^i(t),\ldots,x_n(\tau_n^i(t)), \\ \quad \forall t \in T^i,\ 0 \leq \tau_j^i(t) \leq t, \\ x_i(t), \quad \forall t \notin T^i. \end{cases}$$

In the aforementioned definition of the iterative process, the difference $t - \tau_j^i(t)$ between the current time t and the time $\tau_j^i(t)$ corresponding to the jth component available at the processor updating $x_i(t)$ can be viewed as some form of communication delay. In studying the convergence behavior of algorithms of this type, two cases have to be considered. The operation can either be *totally asynchronous* or *partially asynchronous*. The concept of totally asynchronous algorithms was first introduced in [9], and subsequently analyzed in, among other, [1,5,15]. [5] proposed a general framework that ensembles a variety of instances. The cornerstone of his approach is based on the *asynchronous convergence theorem*, [8]. It defined a general pattern for proving convergence of the asynchronous counterparts of certain sequential algorithms. The asynchronous convergence theorem can be applied to variety of problems including:

- problems involving maximum norm contraction mappings;
- problems involving monotone mappings;
- the shortest path problem;
- linear and nonlinear network flow problems.

Qualitatively speaking, the fundamental difference between a synchronous and an asynchronous iterative mapping, is similar to the differences between a Jacobi and a Gauss–Seidel iteration. Consider the implementation of both these approaches in the minimization of function $F(x)$. The specifics of the minimization algorithm are irrelevant:

- Jacobi:

$$x_i(t+1) = \arg\min_{x_i} F(x_1(t),\ldots,x_n(t));$$

- Gauss–Seidel:

$$x_i(t+1) = \arg\min_{x_i}$$
$$F(x_1(t+1),\ldots,x_i(t),\ldots,x_n(t)).$$

The Gauss–Seidel approach corresponds to the instantaneous communication, in a sequential manner, of the information as it being generated. The Jacobi iteration, forces processors to perform iterations utilizing 'outdated' information. The asynchronous iteration is reminiscent to a Jacobi one. A thorough analysis and comparison of these two extremes is presented in [16]. A major class of iterative schemes that can be shown to be convergent when implemented asynchronously, are defined by mappings which can be shown to be *contraction mappings* with respect to a suitably defined *weighted maximum norm*:

$$\|x\|_\infty^\omega = \max_i \frac{|x_i|}{\omega_i},$$
$$x \in \mathbb{R}^n, \quad \omega \in \mathbb{R}_+^n.$$

Let us consider the minimization of an unconstrained quadratic function F:

$$\begin{cases} \min \quad F(x) = \tfrac{1}{2} x^\top A x - b^\top x \\ \text{s.t.} \quad x \in \mathbb{R}^n, \end{cases}$$

where A is an $n \times n$ positive definite symmetric matrix, and $b \in \mathbf{R}^n$. A gradient iteration of the form

$$x := (I - \gamma A)x + \gamma b$$

will be convergent provided that the maximum row sum of $I - \gamma A$ is less than 1, i.e.:

$$|1 - \gamma \alpha_{ij}| + \sum_{j:j\neq i} \gamma |a_{ij}| < 1, \quad i = 1,\ldots,n,$$

implying the *diagonal dominance condition*:

$$a_{ij} > \sum_{j:j\neq i} |a_{ij}|, \quad \forall i.$$

If we consider the general nonlinear unconstrained optimization problem:

$$\begin{cases} \min \quad g(x) \\ \text{s.t.} \quad x \in \mathbb{R}^n, \end{cases}$$

where $g\colon \mathbf{R}^n \to \mathbf{R}$ is a twice-differentiable convex function, with Hessian matrix $\nabla^2 g(x)$ which is positive definite. If one considers a Newton mapping given by:

$$f(x) = x - [\nabla^2 g(x)]^{-1} \nabla g(x)$$

The norm $\|x\| = \max_i |x_i|$ makes f a contraction mapping in the neighborhood of $x^\star$ (the optimal point). Extensions of the ordinary gradient method

$$f(x) = x - \alpha \nabla g(x)$$

are also discussed in [5]. The shortest path problem is defined in terms of a directed graph consisting of n nodes. We denote by $A(i)$ the set of all nodes j for which there is an outgoing arc (i, j) from node i. The problem is to find a path of minimum length starting at node i and ending at node j. [4] considered the application of the asynchronous convergence theorem to fixed point iterations involving monotone mappings by considering the Bellman–Ford algorithm, [3], applied to the *shortest path problem*. This takes the form:

$$x_i(t+1) = \min_{j \in A(i)} (a_{ij} + x_j(\tau_j^i(t))),$$

$$i = 2, \ldots, n, \quad t \in T^i,$$

$$x_1(t+1) = 0.$$

$A(i)$ is the set of all nodes j for which there exists an arc (i, j). Linear network flow problems are discussed in [8] and asynchronous distributed versions of the auction algorithm are discussed. In the general linear network flow problem we are given a set of N nodes and a set of arcs A, each arc (i, j) has associated with it an integer a_{ij}, referred to as the cot coefficient. The problem is to optimally assign flows, f_{ij} to each one of the arcs, and the problem is represented mathematically as follows:

$$\begin{cases} \min \sum_{(i,j) \in A} a_{ij} f_{ij} \\ \text{s.t.} \sum_{j:(i,j) \in A} f_{ij} - \sum_{j:(j,i)} f_{ji} = s_i, \; \forall i \in N, \\ b_{ij} \leq f_{ij} \leq c_{ij}, \; \forall (i, j) \in A, \end{cases}$$

where a_{ij}, b_{ij}, c_{ij} and s_i are integers. Extensions of the sequential *auction algorithms* are discussed in [6], in which asynchronism manifests itself in the sense that certain processors may be calculating actions bids which other update object prices. [7] extended the analysis to cover certain classes of *nonlinear network flow problems* in which the costs a_{ij} are functions of the flows f_{ij}:

$$\begin{cases} \min \sum_{(i,j) \in A} a_{ij}(f_{ij}) \\ \text{s.t.} \sum_{j:(i,j) \in A} f_{ij} - \sum_{j:(j,i)} f_{ji} = s_i, \; \forall i \in N, \\ b_{ij} \leq f_{ij} \leq c_{ij}, \; \forall (i, j) \in A. \end{cases}$$

Imposing additional reasonable assumptions to the general framework of totally asynchronous iterative algorithms can substantially increase the applicability of the concept. A natural extension is therefore the *partially asynchronous iterative methods*, whereby two major assumptions are be satisfied:

a) each processor performs an update at least once during any time interval of length B;
b) the information used by any processor is outdated by at most B time units.

In other words, the partial asynchronism assumption extends the original model of computation by stating that:

There exists a positive integer B such that:

- For every i and for every $t \geq 0$, at least one of the elements of the set $\{t, \ldots, t + B - 1\}$ belongs to T^i.
- There holds:

$$t - B \leq \tau_j^i(t) \leq t,$$

for all i and j, and all $t \geq 0$ belonging to T^i.
- There holds $\tau_i^i(t) = t$ for all i and $t \in T^i$.

[17] developed a very elegant framework with important implications on the asynchronous minimization of continuous functions. It was established that, while minimize function $F(x)$, the asynchronous implementation of a gradient-based algorithm:

$$x := x - \gamma \nabla F(X)$$

is convergent if and only if the stepsize γ is small compared to the inverse of the asynchronism measure B. Specifically, let $F: \mathbf{R}^n \to \mathbf{R}$ be a cost function to be minimized subject to no constraints. It will be further assumed that:

1) $F(x) > 0, \forall x \in \mathbf{R}^n$;
2) $F(x)$ is *Lipschitz continuous*:

$$\|\nabla F(x) - \nabla F(y)\| \leq K_1 \|x - y\|,$$

$$\forall x, y, \in \mathbb{R}^n.$$

The asynchronous gradient algorithm of the synchronous iteration:

$$x := x - \gamma \nabla F(x)$$

is denoted by:

$$x_i(t+1) := x_i(t) - \gamma s_i(t), \quad i = 1, \ldots, n,$$

where γ is a positive stepsize, and $s_i(t)$ is the update direction. It will be assumed that

$$s_i(t) = 0, \quad \forall t \notin T^i.$$

It is important to realize that processor i at time time t has knowledge of a vector $x^i(t)$ that is a, possibly, outdated version of $x(t)$. In other words: $x^i(t) = ((x_1(\tau_1^i(t)), \ldots, x_n(\tau_n^i(t)))$. It is further assumed that when x_i is being updated, the update direction s_i is a *descent direction*: For every i and t:

$$s_i(t) \nabla_i F(x^i(t)) \leq 0$$

there exists positive constants K_2, K_3 such that

$$K_1 \left| \nabla_i F(x^i(t)) \right| \leq |s_i(t)| \leq K_3 \left| \nabla_i F(x^i(t)) \right|,$$
$$\forall t \in T^i, \quad \forall i.$$

If all of the above is satisfied, then for the asynchronous gradient iteration it can be shown that: There exists some γ_0, depending on n, B, K_1, K_3, such that if $0 < \gamma < \gamma_0$, then $\lim_{t \to \infty} \lambda F(x(t)) = 0$.

It can actually be further shown that the choice

$$\gamma = \frac{1}{K_3 K_1 (1 + B + nB)}$$

can guarantee convergence of the asynchronous algorithm. This results clearly states that one can always, in principle, identify an adequate stepsize for any finite delay.

Furthermore, [14] elaborated on the use of gradient projection algorithm, within the asynchronous iterative framework, for addressing certain classes of constraint nonlinear optimization problems. The constrained optimization problems considered, is that of minimizing a convex function $F: \mathbf{R}^n \to \mathbf{R}$, defined over the space $X = \prod_{i=1}^n X_i$ of lower-dimensional sets $X_i \subset \mathbf{R}^{n_i}$, and $\sum_{i=1}^m n_i = n$. The ith component of the solution vector is now updated by

$$x_i(t+1) = [x_i(t) - \gamma \nabla_i F(x^i(t))]^+$$

where $[\cdot]^+$ denotes the projection on the set X_i. Once again: $x_i(t+1) = x_i(t)$, $t \notin T^i$. Once again, a gradient based algorithm is defined, for which

$$s_i(t) = \begin{cases} \frac{1}{\gamma} \left([x_i(t) - \gamma \nabla_i F(x^i(t))]^+ - x_i(t) \right), \\ \qquad t \in T^i, \\ 0 \quad t \notin T^i. \end{cases}$$

It can actually be shown that for, provided that the partial asynchronism assumption holds, one can always define, in principle, a suitable stepsize γ_0 such that for any $0 < \gamma < 0$ the limit point, x^*, of the sequence generated by the partially asynchronous gradient projection iteration minimizes the Lipschitz continuous, convex function F over the set X. Recently, [2], analyzed asynchronous algorithms for minimizing a function when the communication delays among processors are assumed to be stochastic with Markovian character. The approach is also based on a gradient projection algorithm and was used to address a an optimal routing problem.

A major consideration in asynchronous distributed computing is the fact that since no globally controlling mechanism exists makes the use of any termination criterion which is based on local information obsolete. Clearly, when executing asynchronously a distributed iteration of the form $x_i - f_i(x)$ local error estimates can, and will be, misleading in terms of the global state of the system. Recently [13] made several suggestions as to how the standard model can be supplemented with an additional interprocessor communication protocol so as to address the issue of finite termination of asynchronous iterative algorithms.

See also

- Automatic Differentiation: Parallel Computation
- Heuristic Search
- Interval Analysis: Parallel Methods for Global Optimization
- Load Balancing for Parallel Optimization Techniques
- Parallel Computing: Complexity Classes
- Parallel Computing: Models
- Parallel Heuristic Search
- Stochastic Network Problems: Massively Parallel Solution

References

1. Baudet GM (1978) Asynchronous iterative methods for multiprocessors. J ACM 25:226–244
2. Beidas BF, Papavassilopoulos GP (1995) Distributed asynchronous algorithms with stochastic delays for constrained optimization problems with conditions of time drift. Parallel Comput 21:1431–1450

3. Bellman R (1957) Dynamic programming. Princeton University Press, Princeton
4. Bertsekas DP (1982) Distributed dynamic programming. IEEE Trans Autom Control AC-27:610–616
5. Bertsekas DP (1983) Distributed asynchronous computation of fixed points. Math Program 27:107–120
6. Bertsekas DP, Eckstein J (1987) Distributed asynchronous relaxation methodfs for linear network flow problems. Proc IFAC:39–56
7. Bertsekas DP, El Baz D (1987) Distributed asynchronous relaxation methods for convex network flow problems. SIAM J Control Optim 25:74–85
8. Bertsekas DP, Tsitsiklis JN (1989) Parallel and distributed computation: Numerical methods. Prentice-Hall, Englewood Cliffs, NJ
9. Chazan D, Miranker W (1968) Chaotic reaxation. Linear Alg Appl 2:199–222
10. Ferreira A, Pardalos PM (eds) (1997) Solving combinatorial optimization problems in parallel. Springer, Berlin
11. Pardalos PM, Phillips AT, Rosen JB (eds) (1992) Topics in parallel computing in mathematical programming. Sci Press, Marrickville, Australia
12. Pardalos PM (ed) (1992) Advances in optimization and parallel computing. North-Holland, Amsterdam
13. Savari SA, Bertsekas DP (1996) Finite termination of asynchronous iterative algorithms. Parallel Comput 22:39–56
14. Tseng P (1991) On the rate of convergence of a partially asynchronous gradient projection algorithm. SIAM J Optim 1:603–619
15. Tsitsiklis JN (1987) On the stability of asynchronous iterative processes. Math Syst Theory 20:137–153
16. Tsitsiklis JN (1989) A comparison of Jacobi and Gauss–Seidel parallel iterations. Appl Math Lett 2:167–170
17. Tsitsiklis JN, Bertsekas DP, Athans M (1986) Distributed asynchronous deterministic and stochastic gradient optimization algorithms. IEEE Trans Autom Control ac-31:803–813
18. Tsitsiklis JN, Bertsekas DP, Athnas M (1986) Distributed asynchronous deterministic and stochastic gradient optimization algorithms. IEEE Trans Autom Control AC-31:803–812
19. Zenios AS (1994) Parallel numerical optimization: Current status and annotated bibliography. ORSA J Comput 1:20–42

Auction Algorithms

DIMITRI P. BERTSEKAS
Labor. Information and Decision Systems,
Massachusetts Institute Technol., Cambridge, USA

MSC2000: 90C30, 90C35

Article Outline

Keywords
The Auction Process
Optimality Properties at Termination
Computational Aspects: ϵ-Scaling
Parallel and Asynchronous Implementation
Variations and Extensions
See also
References

Keywords

Linear programming; Optimization; Assignment problem; Transshipment problem

The auction algorithm is an intuitive method for solving the classical assignment problem. It outperforms substantially its main competitors for important types of problems, both in theory and in practice, and is also naturally well suited for parallel computation. In this article, we will sketch the basic principles of the algorithm, we will explain its computational properties, and we will discuss its extensions to more general network flow problems. For a detailed presentation, see the survey paper [3] and the textbooks [2,4]. For an extensive computational study, see [8]. The algorithm was first proposed in the 1979 report [1].

In the classical *assignment problem* there are n persons and n objects that we have to match on a one-to-one basis. There is a benefit a_{ij} for matching person i with object j and we want to assign persons to objects so as to maximize the total benefit. Mathematically, we want to find a one-to-one assignment [a set of person-object pairs $(1, j_1), \ldots, (n, j_n)$, such that the objects $j_1, \ldots, j_n$ are all distinct] that maximizes the total benefit $\sum_{i=1}^{n} a_{ij_i}$.

The assignment problem is important in many practical contexts. The most obvious ones are resource allocation problems, such as assigning personnel to jobs, machines to tasks, and the like. There are also situations where the assignment problem appears as a subproblem in various methods for solving more complex problems.

The assignment problem is also of great theoretical importance because, despite its simplicity, it embodies a fundamental linear programming structure. The most important type of linear programming problems, the linear network flow problem, can be reduced to the assignment problem by means of a simple reformulation. Thus, any method for solving the assignment problem can be generalized to solve the linear network flow problem, and in fact this approach is particularly helpful in understanding the extension of auction algorithms to network flow problems that are more general than assignment.

The classical methods for assignment are based on iterative improvement of some cost function; for example a primal cost (as in primal simplex methods), or a dual cost (as in Hungarian-like methods, dual simplex methods, and relaxation methods). The auction algorithm departs significantly from the cost improvement idea; at any one iteration, it may deteriorate both the primal and the dual cost, although in the end it finds an optimal assignment. It is based on a notion of approximate optimality, called ϵ-complementary slackness, and while it implicitly tries to solve a dual problem, it actually attains a dual solution that is not quite optimal.

The Auction Process

To develop an intuitive understanding of the auction algorithm, it is helpful to introduce an economic equilibrium problem that turns out to be equivalent to the assignment problem. Let us consider the possibility of matching the n objects with the n persons through a market mechanism, viewing each person as an economic agent acting in his own best interest. Suppose that object j has a price p_j and that the person who receives the object must pay the price p_j. Then, the (net) value of object j for person i is $a_{ij} - p_j$ and each person i would logically want to be assigned to an object j_i with maximal value, that is, with

$$a_{ij_i} - p_{j_i} = \max_{j=1,\ldots,n} \{a_{ij} - p_j\}. \tag{1}$$

We will say that a person i is 'happy' if this condition holds and we will say that an assignment and a set of prices are at *equilibrium* when all persons are happy.

Equilibrium assignments and prices are naturally of great interest to economists, but there is also a fundamental relation with the assignment problem; it turns out that an equilibrium assignment offers maximum total benefit (and thus solves the assignment problem), while the corresponding set of prices solves an associated dual optimization problem. This is a consequence of the celebrated duality theorem of linear programming.

Let us consider now a natural process for finding an equilibrium assignment. I will call this process the *naive auction algorithm*, because it has a serious flaw, as will be seen shortly. Nonetheless, this flaw will help motivate a more sophisticated and correct algorithm.

The naive auction algorithm proceeds in 'rounds' (or 'iterations') starting with *any* assignment and *any* set of prices. There is an assignment and a set of prices at the beginning of each round, and if all persons are happy with these, the process terminates. Otherwise some person who is not happy is selected. This person, call him i, finds an object j_i which offers maximal value, that is,

$$j_i \in \arg\max_{j=1,\ldots,n} \{a_{ij} - p_j\}, \tag{2}$$

and then:

a) Exchanges objects with the person assigned to j_i at the beginning of the round;
b) Sets the price of the best object j_i to the level at which he is indifferent between j_i and the second best object, that is, he sets p_{j_i} to

$$p_{j_i} + \gamma_i, \tag{3}$$

where

$$\gamma_i = v_i - w_i, \tag{4}$$

v_i is the best object value,

$$v_i = \max_j \{a_{ij} - p_j\}, \tag{5}$$

and w_i is the second best object value

$$w_i = \max_{j \neq j_i} \{a_{ij} - p_j\}, \tag{6}$$

that is, the best value over objects other than j_i. (Note that γ_i is the largest increment by which the best object price p_{j_i} can be increased, with j_i still being the best object for person i.)

This process is repeated in a sequence of rounds until all persons are happy.

We may view this process as an *auction*, where at each round the bidder i raises the price of his or her preferred object by the *bidding increment* γ_i. Note that γ_i

cannot be negative since $v_i \geq w_i$ (compare (5) and (6)), so the object prices tend to increase. Just as in a real auction, bidding increments and price increases spur competition by making the bidder's own preferred object less attractive to other potential bidders.

Does this auction process work? Unfortunately, not always. The difficulty is that the bidding increment γ_i is zero when more than one object offers maximum value for the bidder i (cf. (4) and (6)). As a result, a situation may be created where several persons contest a smaller number of equally desirable objects without raising their prices, thereby creating a never ending cycle.

To break such cycles, we introduce a perturbation mechanism, motivated by real auctions where each bid for an object must raise its price by a minimum positive increment, and bidders must on occasion take risks to win their preferred objects. In particular, let us fix a positive scalar ϵ and say that a person i is 'almost happy' with an assignment and a set of prices if the value of its assigned object j_i is within ϵ of being maximal, that is,

$$a_{ij_i} - p_{j_i} \geq \max_{j=1,\ldots,n} \{a_{ij} - p_j\} - \epsilon. \tag{7}$$

We will say that an assignment and a set of prices are *almost at equilibrium* when all persons are almost happy. The condition (7), introduced first in 1979 in conjunction with the auction algorithm, is known as ϵ-*complementary slackness* and plays a central role in several optimization contexts. For $\epsilon = 0$ it reduces to ordinary complementary slackness (compare (1)).

We now reformulate the previous auction process so that the bidding increment is always at least equal to ϵ. The resulting method, the *auction algorithm*, is the same as the naive auction algorithm, except that the bidding increment γ_i is

$$\gamma_i = v_i - w_i + \epsilon, \tag{8}$$

(rather than $\gamma_i = v_i - w_i$ as in (4)). With this choice, the bidder of a round is almost happy at the end of the round (rather than happy). The particular increment $\gamma_i = v_i - w_i + \epsilon$ used in the auction algorithm is the maximum amount with this property. Smaller increments γ_i would also work as long as $\gamma_i \geq \epsilon$, but using the largest possible increment accelerates the algorithm. This is consistent with experience from real auctions, which tend to terminate faster when the bidding is aggressive.

We can now show that this reformulated auction process terminates in a finite number of rounds, necessarily with an assignment and a set of prices that are almost at equilibrium. To see this, note that once an object receives a bid for the first time, then the person assigned to the object at every subsequent round is almost happy; the reason is that a person is almost happy just after acquiring an object through a bid, and continues to be almost happy as long as he holds the object (since the other object prices cannot decrease in the course of the algorithm). Therefore, the persons that are not almost happy must be assigned to objects that have never received a bid. In particular, once each object receives at least one bid, the algorithm must terminate. Next note that if an object receives a bid in m rounds, its price must exceed its initial price by at least $m\epsilon$. Thus, for sufficiently large m, the object will become 'expensive' enough to be judged 'inferior' to some object that has not received a bid so far. It follows that only for a limited number of rounds can an object receive a bid while some other object still has not yet received any bid. Therefore, there are two possibilities: either

a) the auction terminates in a finite number of rounds, with all persons almost happy, before every object receives a bid; or

b) the auction continues until, after a finite number of rounds, all objects receive at least one bid, at which time the auction terminates. (This argument assumes that any person can bid for any object, but it can be generalized for the case where the set of feasible person-object pairs is limited, as long as at least one feasible assignment exists.)

Optimality Properties at Termination

When the auction algorithm terminates, we have an assignment that is almost at equilibrium, but does this assignment maximize the total benefit? The answer here depends strongly on the size of ϵ. In a real auction, a prudent bidder would not place an excessively high bid for fear that he might win the object at an unnecessarily high price. Consistent with this intuition, we can show that if ϵ is small, then the final assignment will be 'almost optimal'. In particular, we can show that the total benefit of the final assignment is within $n\epsilon$ of being optimal. To see this, note that an assignment and a set of prices that are almost at equilibrium may be viewed as being at equilibrium for a *slightly different* problem where all benefits a_{ij} are the same as before, except for the n benefits of the assigned pairs which are modified by an amount no more than ϵ.

Suppose now that the benefits a_{ij} are all integer, which is the typical practical case (if a_{ij} are rational numbers, they can be scaled up to integer by multiplication with a suitable common number). Then, the total benefit of any assignment is integer, so if $n\epsilon < 1$, a complete assignment that is within $n\epsilon$ of being optimal must be optimal. It follows, that if

$$\epsilon < \frac{1}{n},$$

and the benefits a_{ij} are all integer, then the assignment obtained upon termination of the auction algorithm is optimal. Let us also note that the final set of prices is within $n\epsilon$ of being an optimal solution of the dual problem

$$\min_{p_j \atop j=1,\ldots,n} \left\{ \sum_{j=1}^{n} p_j + \sum_{i=1}^{n} \max_j \{a_{ij} - p_j\} \right\}. \quad (9)$$

This leads to the interpretation of the auction algorithm as a dual algorithm (in fact an approximate coordinate ascent algorithm; see the cited literature).

Computational Aspects: ϵ-Scaling

The auction algorithm exhibits interesting computational behavior, and it is essential to understand this behavior to implement the algorithm efficiently. First note that the amount of work to solve the problem can depend strongly on the value of ϵ and on the maximum absolute object value

$$C = \max_{i,j} |a_{ij}|.$$

Basically, for many types of problems, the number of bidding rounds up to termination tends to be proportional to C/ϵ. Note also that there is a dependence on the initial prices; if these prices are 'near optimal,' we expect that the number of rounds to solve the problem will be relatively small.

The preceding observations suggest the idea of ϵ-*scaling*, which consists of applying the algorithm several times, starting with a large value of ϵ and successively reducing ϵ up to an ultimate value that is less than some critical value (for example, $1/n$, when the benefits a_{ij} are integer). Each application of the algorithm provides good initial prices for the next application. This is a very common idea in nonlinear programming, encountered for example, in barrier and penalty function methods. An alternative form of scaling, called *cost scaling*, is based on successively representing the benefits a_{ij} with an increasing number of bits, while keeping ϵ at a constant value.

In practice, it is a good idea to at least consider scaling. For sparse assignment problems, that is, problems where the set of feasible assignment pairs is severely restricted, scaling seems almost universally helpful. In theory, scaling leads to auction algorithms with a particularly favorable polynomial complexity (without scaling, the algorithm is pseudopolynomial; see the cited literature).

Parallel and Asynchronous Implementation

Both the bidding and the assignment phases of the auction algorithm are highly parallelizable. In particular, the bidding and the assignment can be carried out for all persons and objects simultaneously. Such an implementation can be termed *synchronous*. There are also *totally asynchronous* implementations of the auction algorithm, which are interesting because they are quite flexible and also tend to result in faster solution in some types of parallel machines. To understand these implementations, it is useful to think of a person as an autonomous decision maker who at unpredictable times obtains information about the prices of the objects. Each person who is not almost happy makes a bid at arbitrary times on the basis of its current object price information (that may be outdated because of communication delays).

See [7] for a careful formulation of the totally asynchronous model, and a proof of its validity, including extensive computational results on a shared memory machine, confirming the advantage of asynchronous over synchronous implementations.

Variations and Extensions

The auction algorithm can be extended to solve a number of variations of the assignment problem, such as the asymmetric assignment problem where the number of objects is larger than the number of persons and there is a requirement that all persons be assigned to some object. Naturally, the notion of an assignment must now be modified appropriately. To solve this problem, the auction algorithm need only be modified in the choice of initial conditions. It is sufficient to require that all initial prices be zero. A similar algorithm can be used for the case where there is no requirement that all persons be assigned. Other variations handle efficiently the cases where there are several groups of 'identical' persons or objects ([5]).

There have been extensions of the auction algorithm for other types of linear network optimization problems. The general approach for constructing auction algorithms for such problems is to convert them to assignment problems, and then to suitably apply the auction algorithm and streamline the computations. In particular, the classical shortest path problem can be solved correctly by the naive auction algorithm described earlier, once the method is streamlined. Similarly, auction algorithms can be constructed for the max-flow problems, and are very efficient. These algorithms bear a close relation to preflow-push algorithms for the max-flow problem, which were developed independently of auction ideas.

The auction algorithm has been extended to solve linear transportation problems ([5]). The basic idea is to convert the transportation problem into an assignment problem by creating multiple copies of persons (or objects) for each source (or sink respectively), and then to modify the auction algorithm to take advantage of the presence of the multiple copies.

There are extensions of the auction algorithm for linear minimum cost flow (*transshipment*) problems, such as the so called ϵ-relaxation method, and the auction/sequential shortest path algorithm algorithm (see the cited literature for a detailed description). These methods have interesting theoretical properties and like the auction algorithm, are well suited for parallelization (see the survey [6], and the textbook [7]).

Let us finally note that there have been proposals of auction algorithms for convex separable network optimization problems with and without gains (but with a single commodity and without side constraints); see [9].

See also

▶ Communication Network Assignment Problem
▶ Directed Tree Networks
▶ Dynamic Traffic Networks
▶ Equilibrium Networks
▶ Evacuation Networks
▶ Generalized Networks
▶ Maximum Flow Problem
▶ Minimum Cost Flow Problem
▶ Network Design Problems
▶ Network Location: Covering Problems
▶ Nonconvex Network Flow Problems
▶ Piecewise Linear Network Flow Problems
▶ Shortest Path Tree Algorithms
▶ Steiner Tree Problems
▶ Stochastic Network Problems: Massively Parallel Solution
▶ Survivable Networks
▶ Traffic Network Equilibrium

References

1. Bertsekas DP (1979) A distributed algorithm for the assignment problem. Working Paper MIT Lab Information & Decision Systems
2. Bertsekas DP (1991) Linear network optimization: Algorithms and codes. MIT, Cambridge, MA
3. Bertsekas DP (1992) Auction algorithms for network flow problems: A tutorial introduction. Comput Optim Appl 1: 7–66
4. Bertsekas DP (1998) Network optimization: Continuous and discrete problems. Athena Sci., Belmont, MA
5. Bertsekas DP, Castañon DA (1989) The auction algorithm for transportation problems. Ann Oper Res 20:67–96
6. Bertsekas DP, Castañon DA, Eckstein J, Zenios S (1995) Parallel computing in network optimization. In: Ball MO, Magnanti TL, Monma CL, Nemhauser GL (eds) Handbooks in OR and MS, vol 7, North-Holland, Amsterdam, pp 331–399
7. Bertsekas DP, Tsitsiklis JN (1989) Parallel and distributed computation: Numerical methods. Prentice-Hall, Englewood Cliffs, NJ
8. Castañon DA (1993) Reverse Auction Algorithms for Assignment Problems. In: Johnson DS, McGeoch CC (eds) Algorithms for network flows and matching. Amer Math Soc, Providence, pp 407–429
9. Tseng P, Bertsekas DP (1996) An epsilon-relaxation method for separable convex cost generalized network flow problems. Math Program (to appear), MIT Lab Information & Decision Systems P-2374

Automatic Differentiation: Calculation of the Hessian

LAURENCE DIXON
Numerical Optim. Centre, University Hertfordshire, Hatfield, England

MSC2000: 90C30, 65K05

Article Outline

Keywords
The Forward Mode
    Illustrative Example 1: Forward Mode
The Mixed Method
    Illustrative Example 2: Reverse Differentiation
Reverse Method
    Illustrative Example 3: Reverse Gradient, Forward Hessian
See also
References

Keywords

Automatic differentiation; Gradient; Hessian; Doublet; Triplet

The *Hessian* of a scalar function $f(x)$ can be computed automatically in at least two ways. The first is a natural extension of the forward method for calculating gradients. The others extend the reverse method.

The Forward Mode

The concept of forward *automatic differentiation* was described by L.B. Rall [14]. When calculating the gradient vector of a function of n variables, a *doublet* data structure is introduced, consisting of $n + 1$ floating point numbers. To calculate the Hessian matrix, this data structure is extended to a *triplet*.

A triplet is a data structure that, in the simplest form, contains $1 + n + n(n+1)/2$ floating point numbers. If X is a variable that occurs in the evaluation of $f(x)$, then the triplet of X consists of

$$\left(X, \frac{\partial X}{\partial x_i}, \frac{\partial^2 X}{\partial x_i \partial x_j}\right)$$

for $i = 1, \ldots, n$ and $j \leq i$.

The doublet consists of the first $n + 1$ elements of the triplet.

At the start of the function evaluation the triplets of the variables x_k must be set and these are simply $(x_k, e_k, 0)$ where e_k is the unit vector with 1 in the kth place, and 0 is the null matrix. If the function evaluation is expanded as a Wengert list [17] consisting of three types of operations,

- addition and subtraction,
- multiplication and division,
- nonlinear scalar functions,

then the arithmetic required to correctly update the triplets is easily deduced.

- If $X_k = X_l + X_m$, $l, m < k$, then to obtain the triplet of X_k, the elements of the triplets of X_l and X_m are simply added together element by element.
- If $X_k = X_l X_m$, $l, m < k$, then the background arithmetic is more complex as

$$\frac{\partial X_k}{\partial x_i} = X_l \frac{\partial X_m}{\partial x_i} + X_m \frac{\partial X_l}{\partial x_i}$$

and

$$\frac{\partial^2 X_k}{\partial x_i \partial x_j} = \frac{\partial X_l}{\partial x_j} \frac{\partial X_m}{\partial x_i} + X_l \frac{\partial^2 X_m}{\partial x_i \partial x_j} + \frac{\partial X_m}{\partial x_j} \frac{\partial X_l}{\partial x_i} + X_m \frac{\partial^2 X_l}{\partial x_i \partial x_j}.$$

As all these terms are stored in the triplets of X_l and X_m, given the triplets of X_l and X_m the triplet of X_k can be computed by a standard routine.

- If $X_k = \phi(X_m)$, $m < k$, then

$$\frac{\partial X_k}{\partial x_i} = \phi'(X_m) \frac{\partial X_m}{\partial x_i}$$

and

$$\frac{\partial^2 X_k}{\partial x_i \partial x_j} = \phi''(X_m) \frac{\partial X_m}{\partial x_i} \frac{\partial X_m}{\partial x_j} + \phi'(X_m) \frac{\partial^2 X_m}{\partial x_i \partial x_j}.$$

To perform this operation the values of $\phi'(X_m)$ and $\phi''(X_m)$ must be calculated with $\phi(X_m)$; all the other data is contained in the triplet of X_m.

Illustrative Example 1: Forward Mode

Consider the simple function

$$f(x) = (x_1 x_2 + \sin x_1 + 4)(3x_2^2 + 6)$$

In this case $n = 2$ and each triplet contains 6 floating point numbers, the value of X, its *gradient*, and the upper half of its Hessian. To evaluate the function, gradient, and Hessian, first expand the function in the Wengert list as shown in column 1 and then evaluate the triplets one by one. The evaluation is performed at the point (0, 1) below.

X_k	triplet(X_k)
$X_1 = x_1$	0, 1, 0, 0, 0, 0
$X_2 = x_2$	1, 0, 1, 0, 0, 0
$X_3 = X_1 X_2$	0, 1, 0, 0, 1, 0
$X_4 = \sin X_1$	0, 1, 0, 0, 0, 0
$X_5 = X_3 + X_4$	0, 2, 0, 0, 1, 0
$X_6 = X_5 + 4$	4, 2, 0, 0, 1, 0
$X_7 = X_2^2$	1, 0, 2, 0, 0, 2
$X_8 = 3X_7$	3, 0, 6, 0, 0, 6
$X_9 = X_8 + 6$	9, 0, 6, 0, 0, 6
$X_{10} = X_6 X_9$	36, 18, 24, 0, 21, 24

The last row contains the values of the function, gradient and Hessian. The values for this simple problem can be easily verified by direct differentiation.

In practice forward automatic differentiation may be implemented in many ways, one possibility in many modern computer languages is to introduce the new data type triplet and over-write the meaning of the standard operators and functions so they perform the arithmetic described above. The code for the function evaluation can then be written normally without recourse to the Wengert list. Details of an implementation in Ada are given in [13]. A single run through a function evaluation code then computes the function, gradient and Hessian. If S is the store required to compute $f(x)$ then this method requires $(1 + n + n(n + 1)/2)S$ store. If M is the number of operations required to compute $f(x)$ then $(1 + 3n + 7n^2)M$ is a pessimistic bound on the operations required to compute the function, gradient and Hessian. Additional overheads are incurred to access the data type and the over-written operator subroutines. The efficiency is often improved by treating the triplet as a vector array and using sparse storage techniques. The number of zeros in the triplets of the above simple example illustrates the strength of the sparse form to calculate full Hessians. Maany reports the following results for the CPU time to differentiate the 50-dimensional Helmholz function (for details see [10]).

	Doublets		triplets	
	full	sparse	full	sparse
f	1.36	0.44	60.29	0.44
$f, \nabla f$	9.24	3.42	68.68	3.52
$f, \nabla f, \nabla^2 f$	N/A	N/A	476.36	20.69

The CPU time for calculating f alone within the full triplet package rises dramatically as although the derivative calculations are switched off the full package still allocates the space for the full triplet. Using the sparse package is also especially helpful if n is large and $f(x)$ is a partially separable function, i. e.

$$f(x) = \sum_k f_k(x)$$

where $f_k(x)$ only depends on a small number V_k of the n variables, as then, throughout the calculation of $f_k(x)$, the *sparse triplet* will only contain at most $1 + V_k + V_k(V_k + 1)/2$ nonzeros, and V_k will replace n in all the operation bounds, to give $\sum_k (1 + 3V_k + 7V_k^2)M_k$ operations.

One of the main purposes for calculating the Hessian matrix is to use it in optimization calculations. The truncated Newton method can be written so that it either requires the user to provide f, ∇f, and $\nabla^2 f$ at each outer iteration or f, ∇f at each outer iteration and $(\nabla^2 f) p$ at each inner iteration. The first method is ideally suited to be combined with sparse triplet differentiation. The algorithm is described in [9] and results given on functions of up to $n = 3000$ in [8]. The calculation of $(\nabla^2 f) p$ can also be undertaken simply by a modification of the triplet method.

In [7] the conclusion was drawn that

the *sparse doublet* and sparse triplet codes in Ada enable normal code to be written for the function f and accurate values of ∇f and $\nabla^2 f$ to be obtained reliably by the computer. The major hope for automatic differentiation is therefore achieved.

Implementations are also available in Pascal.SC, C++, and Fortran90. The NOC Optima Library [1] code, OPFAD, implements the sparse doublet and triplet methods described above in Fortran90.

The Mixed Method

The advent of reverse automatic differentiation, A. Griewank [10], raised the hope that quicker ways could be found. The bound on the operations needed to compute the Hessian by the full forward triplet method contains the term $1/2 n^2 M$; by using a mixed method this is not required. The simplest mixed method is to use reverse automatic differentiation to compute the gradient which, [10], only requires $5M$ operations to compute the function and gradient for any value of n. This can be repeated at appropriate steps h along each axis, i.e. at $x + h e_i, i = 1, \ldots, n$, and simple differences applied to the gradient vectors to calculate the Hessian in less than $5(n + 1)(M + 1)$ operations.

Illustrative Example 2: Reverse Differentiation

To obtain the gradient by *reverse differentiation* we must introduce the adjoint variables X_k^* and reverse back through the list. These rules are discussed in the previous article, but for convenience are repeated. If in the calculation of $f(x)$,

$$X_k = \phi(X_i, X_j), \qquad i, j < k,$$

then in the reverse pass

$$X_i^* = X_i^* + \frac{\partial \phi}{\partial X_i} X_k^*$$

and

$$X_j^* = X_j^* + \frac{\partial \phi}{\partial X_j} X_k^*.$$

For the same example the steps needed to calculate the gradient by reverse differentiation are

X_k^*	X_k^*
$X_{10}^* = 1$	1
$X_9^* = X_{10}^* X_6$	4
$X_6^* = X_{10}^* X_9$	9
$X_8^* = X_9^*$	4
$X_7^* = 3 X_8^*$	12
$X_2^* = 2 X_2 X_7^*$	24
$X_5^* = X_6^*$	9
$X_4^* = X_5^*$	9
$X_3^* = X_5^*$	9
$X_1^* = X_4^* \cos X_1$	9
$X_2^* = X_2^* + X_1 X_3^*$	24
$X_1^* = X_1^* + X_2 X_3^*$	18

giving the gradient as (18, 24) in agreement with the forward calculation. To perform this calculation the values of X_6 and X_9 were required which had been calculated during the function value calculation. The reverse gradient calculation must, therefore, follow a forward function evaluation calculation and the required data must be stored.

The bound $5M$ on the number of operations required to calculate the gradient is often very pessimistic, especially when the function evaluation uses matrix operations, [15], standard subroutines, [5], or when efficient sparse storage is used, [6]. The store required by this simple approach is simply that needed to calculate the gradient by reverse differentiation. The original reverse method required $O(M)$ store, but Griewank [11] describes how the store required can be reduced to $O(S \log M)$ at the cost of increasing the operation bound to $O(M \log M)$.

The accuracy obtained by calculating the Hessian by simple differences will depend on h but will often be sufficient as accurate Hessians are rarely required in optimization. Many software packages for calculating the gradient by reverse differentiation now exist, including the Optima Library Code OPRAD [1].

In 1998 the most widely used code to calculate gradients automatically is probably the *ADIFOR* code, [3], many examples of its use are given in that reference; unfortunately this implements a 'statement level hybrid mode'. In this, each assignment statement

$$Y_i = \Psi(Y_j, j < i, j \in J)$$

is treated in turn and the gradient, $\frac{\partial \Psi}{\partial Y_j}, j \in J$, computed efficiently by RAD but then to obtain the Doublet

$$\frac{\partial Y_i}{\partial x_m} = \sum_j \frac{\partial \Psi}{\partial Y_j} \frac{\partial Y_j}{\partial x_m}$$

many multiplications and additions may be required leading to a high operation count.

Reverse Method

A fully automatic approach could start by obtaining the Wengert list for the function and gradient as calculated by reverse automatic differentiation. This list will contain at most $5M$ steps. Then a forward sparse Doublet pass through this list could be performed that would need less than $(1 + 3n)\, 5M$ operations. The Doublet formed for the same example is illustrated below. In the Wengert list all identical Doublets are merged and composite steps involving more than one operation are split, it will be observed that the last two rows of the Doublet contain the gradient and Hessian, as desired, and that the number of operations, 22, is much less than the bound $5M = 50$. The storage requirement for this approach, when n is large, is considerably greater than that needed by the difference method. An alternative would be to perform a reverse pass through the gradient list. A full discussion is given in [4], who shows the two are identical in arithmetic, storage and operation count. His experience with his Ada implementation showed that the performance was very machine dependent. If the sparse Doublet approach is used with this reverse method on the partially separable function described above then the bound on the operations needed to obtain the Hessian reduces to $\sum_k 5(V_k + 1)(M_k + 1)$, a considerable saving. An early implementation, *PADRE2*, is described in [12]. A more recent code, *ADOL-F*, is described in [16]. Christianson's method is implemented in OPRAD, mentioned above. It should perhaps be mentioned that all the above methods can be hand-coded to solve any important problem without incurring the overheads still associated with most automatic packages, many of the helping hands described in [5] are still not implemented in an automatic package.

Further methods for speeding up the calculation of the Hessian are described in ▶ Automatic Differentiation: Calculation of Newton Steps.

Illustrative Example 3:
Reverse Gradient, Forward Hessian

The variables in the Wengert list of the function and gradient calculation will be denoted by Y.

Y_k	Doublet Y_k
$Y_1 = x_1$	0, 1, 0
$Y_2 = x_2$	1, 0, 1
$Y_3 = Y_1 Y_2$	0, 1, 0
$Y_4 = \sin Y_1$	0, 1, 0
$Y_5 = Y_3 + Y_4$	0, 2, 0
$Y_6 = Y_5 + 4$	4, 2, 0
$Y_7 = Y_2^2$	1, 0, 2
$Y_8 = 3 Y_7$	3, 0, 6
$Y_9 = Y_8 + 6$	9, 0, 6
$Y_{10} = Y_6 Y_9$	36, 18, 24
$Y_{11} = 1$	1, 0, 0
$Y_{12} = Y_{11} Y_6$	4, 2, 0
$Y_{13} = Y_{11} Y_9$	9, 0, 6
$Y_{14} = 3 Y_{12}$	12, 6, 0
$Y_{15} = Y_2 Y_{14}$	12, 6, 12
$Y_{16} = 2 Y_{15}$	24, 12, 24
$Y_{17} = \cos Y_1$	1, 0, 0
$Y_{18} = Y_{17} Y_{13}$	9, 0, 6
$Y_{19} = Y_1 Y_{13}$	0, 9, 0
$Y_{20} = Y_2 Y_{13}$	9, 0, 15
$Y_{21} = Y_{18} + Y_{20}$	18, 0, 21
$Y_{22} = Y_{16} + Y_{19}$	24, 21, 24

See also

- ▶ Automatic Differentiation: Calculation of Newton Steps
- ▶ Automatic Differentiation: Geometry of Satellites and Tracking Stations
- ▶ Automatic Differentiation: Introduction, History and Rounding Error Estimation
- ▶ Automatic Differentiation: Parallel Computation
- ▶ Automatic Differentiation: Point and Interval
- ▶ Automatic Differentiation: Point and Interval Taylor Operators
- ▶ Automatic Differentiation: Root Problem and Branch Problem
- ▶ Nonlocal Sensitivity Analysis with Automatic Differentiation

References

1. Bartholomew-Biggs MC Optima library. Numerical Optim. Centre Univ. Hertfordshire, England
2. Berz M, Bischof Ch, Corliss G, Griewank A (eds) (1996) Computational differentiation: techniques, applications, and tools. SIAM, Philadelphia
3. Bischof Ch, Carle A (1996) Users' experience with ADIFOR 2.0. In: Berz M, Bischof Ch, Corliss G, Griewank A (eds) Computational Differentiation: Techniques, Applications, and Tools. SIAM, Philadelphia, pp 385–392
4. Christianson DB (1992) Automatic Hessians by reverse accumulation. IMA J Numer Anal 12:135–150
5. Christianson DB, Davies AJ, Dixon LCW, Zee P Van der, (1997) Giving reverse differentiation a helping hand. Optim Methods Softw 8:53–67
6. Christianson DB, Dixon LCW, Brown S (1996) Sharing storage using dirty vectors. In: Berz M, Bischof Ch, Corliss G, Griewank A (eds) Computational Differentiation: Techniques, Applications, and Tools. SIAM, Philadelphia, pp 107–115
7. Dixon LCW (1993) On automatic differentiation and continuous optimization. In: Spedicato E (ed) Algorithms for continuous optimization: the state of the art. NATO ASI series. Kluwer, Dordrecht, pp 501–512
8. Dixon LCW, Maany Z, Mohsenina M (1989) Experience using truncated Newton for large scale optimization. In: Bromley K (ed) High Speed Computing, vol 1058. SPIE–The Internat. Soc. for Optical Engineering, Bellingham, WA, pp 94–104
9. Dixon LCW, Maany Z, Mohseninia M (1990) Automatic differentiation of large sparse systems. J Econom Dynam Control 14(2)
10. Griewank A (1989) On automatic differentiation. In: Iri M, Tanabe K (eds) Mathematical programming: recent developments and applications. Kluwer, Dordrecht, pp 83–108
11. Griewank A (1992) Achieving logarithmic growth of temporal and spatial complexity in reverse automatic differentiation. Optim Methods Softw 1:35–54
12. Kubota K (1991) PADRE2, A FORTRAN precompiler yielding error estimates and second derivatives. In: Griewank A and Corliss GF (eds) Automatic differentiation of algorithms: theory, implementation, and application. SIAM, Philadelphia, pp 251–262
13. Maany ZA (1989) Ada automatic differentiation packages. Techn Report Hatfield Polytechnic NOC TR224
14. Rall LB (1981) Automatic differentiation: techniques and applications. Lecture Notes Computer Sci, vol 120. Springer, Berlin
15. Shiriaev D (1993) Fast automatic differentiation for vector processors and reduction of the spatial complexity in a source translation environment. PhD Thesis Inst Angew Math Univ Karlsruhe
16. Shiriaev D, Griewank A (1996) ADOL–F Automatic differentiation of Fortran codes. In: Berz M, Bischof Ch, Corliss G, Griewank A (eds) Computational Differentiation: Techniques, Applications, and Tools. SIAM, Philadelphia, pp 375–384
17. Wengert RE (1964) A simple automatic derivative evaluation program. Comm ACM 7(8):463–464

Automatic Differentiation: Calculation of Newton Steps

LAURENCE DIXON

Numerical Optim. Centre, University Hertfordshire, Hatfield, UK

MSC2000: 90C30, 65K05

Article Outline

Keywords
Jacobian Calculations
The Extended Matrix
Hessian Calculations
The Newton Step
Truncated Methods
See also
References

Keywords

Automatic differentiation; Jacobian matrix; Hessian matrix; Sparsity; Newton step

Many algorithms for solving optimization problems require the minimization of a merit function, which may be the original objective function, or the solution to sets of simultaneous nonlinear equations which may involve the constraints in the problem. To obtain second order convergence near the solution algorithms to solve both rely on the calculation of Newton steps.

When solving a set of nonlinear equations

$$s_j(x) = 0, \qquad j = 1, \ldots, n,$$

the *Newton step* d at $x^{(0)}$, $x \in \mathbf{R}^n$, is obtained by solving the linear set of equations

$$\sum_i \frac{\partial s_j}{\partial x_i} d_i = -s_j, \qquad j = 1, \ldots, n,$$

where both the derivatives $\partial s_j/\partial x_i$ and the vector function s_j are evaluated at a point, which we will denote by $x^{(0)}$.

For convenience we introduce the *Jacobian matrix J* and write the equation as

$$Jd = -s$$

When minimizing a function $f(x)$ the Newton equation becomes

$$\sum_i \frac{\partial^2 f}{\partial x_i \partial x_j} d_j = -\frac{\partial f}{\partial x_j}$$

where all the derivatives are calculated at a point, again denoted by $x^{(0)}$.

In terms of the *Hessian*, H, and the gradient, g, this can be written

$$Hd = -g$$

Automatic differentiation can be used to calculate the gradient, Hessian and Jacobian, but it can also be used to calculate the Newton step directly without calculating the matrices. In this article we will first discuss the calculation of the Jacobian, then extend briefly the calculation of the gradient and Hessian, which was the subject of ▶ Automatic differentiation: Calculation of the Hessian, and finally discuss the direct calculation of the Newton step.

Jacobian Calculations

If the functions s_j were each evaluated as separate entities, requiring M_j operations, then the derivatives could be evaluated by reverse automatic differentiation in 5 M_j operations. For many sets of functions it would, however, be very inefficient to evaluate the set s in this way, as considerable savings could be made by calculating threads of operations common to more than one s_j only once. In such situations the number of operations M required to evaluate the set s may be much less than $\sum_j M_j$. Under these circumstances the decision on how the Jacobian should be evaluated becomes much more complicated.

Before the advent of automatic differentiation the Jacobian was frequently approximated by one-sided differences

$$\frac{\partial s_j}{\partial x_i} = \frac{s_j\left(x^{(0)} + he_i\right) - s_j\left(x^{(0)}\right)}{h}$$

If the vector function s requires M Wengert operations, then the Jacobian would need $(n + 1) M$ operations by this approach. The accuracy of the result depends on a suitable choice of h. If simple forward automatic differentiation using doublets (see ▶ Automatic differentiation: Calculation of the Hessian) is used, an accurate Jacobian is obtained at a cost of $3nM$ operations. If a Newton step is to be calculated then the Jacobian must be square and so the simple reverse mode, which involves a backward pass through the Wengert list for each subfunction, would be bounded by 5 n M operations.

Most large Jacobians are sparse and M.J.D. Powell, A.R. Curtis, and J.R. Reid [5], introduced the idea of combining columns i that had no common nonzeros. Then, provided the sparsity pattern of J is known, the values in those columns can be reconstructed by a reduced number of differences. If the number of such *PCR groups* required to cover all the columns is c then the operations count is reduced to $(c + 1) M$. For example, the columns of the following 5×5 sparse Jacobian could be divided into 3 groups

$$\begin{bmatrix} \blacksquare & \blacktriangle & 0 & 0 & \star \\ \blacksquare & \blacktriangle & 0 & \star & 0 \\ 0 & \blacktriangle & \blacksquare & \star & 0 \\ 0 & 0 & \blacksquare & 0 & \star \\ \blacksquare & \blacktriangle & 0 & 0 & \star \end{bmatrix}$$

indicated by $\blacksquare$, $\blacktriangle$, and $\star$.

This same grouping could be used with forward automatic differentiation to produce an accurate Jacobian in at most $3cM$ operations. If the sparse Doublet is used, the full benefit of *sparsity* within the calculation of the s_j is obtained, as well as the benefit due to sparsity in the Jacobian, without the need to determine the column groupings. Results showing the advantage of calculating large ($n = 5000$) Jacobians this way are given in [15] and summarised in [7].

It is possible for the calculation of some s_j to be independent of other s that do contain a common thread. It would obviously be efficient to calculate these s_j by reverse differentiation, requiring $5M_j$ operations. Reverse differentiation will also be appropriate if the common thread has less outputs than inputs. Then sparse reverse doublets, [2], should be used. These are implemented in OPRAD, see ▶ Automatic differentiation: Calculation of the Hessian.

T.F. Coleman et al. [3,4] demonstrated that calculating some columns using groups in the forward mode and some rows using groups in the reverse mode is considerably more efficient than using either alone. All nonzeros of the Jacobian must be included in a row and/or column computed. Similar results follow if some columns are computed using sparse doublets and some rows using the sparse reverse method. If C is the maximum number of nonzeros in a row within the columns computed forward and R the maximum number of nonzeros in a column within the rows computed in reverse then a crude bound on the number of operations is $(3C + 5R)M$. This bound does not allow for the additional sparsity in the early calculations nor for the fact that for some reverse calculations M_j should replace M. The selection of rows and columns taking account of such considerations is still unresolved.

But the advantages to be obtained can be appreciated by considering the arrow-head Jacobian, where only the diagonal elements and the last row and column contain nonzeros. If the gradient of s_n is computed using sparse reverse doublets this will require at most $5M_n$ operations and if the other gradients are computed using sparse forward doublets, no doublet will contain more than 2 nonzeros, so the operations will be bounded by $6M$. The total operations required in this case is independent of n.

The Extended Matrix

If the calculation of the functions s_j proceeds by a sequence of steps

$$X_k = x_k, \quad k = 1, \ldots, n,$$
$$X_k = \phi_k(X_l, l \in L, l < k),$$
$$k = n+1, \ldots, M+n,$$

with

$$s_j = X_{M+j}, \quad j = 1, \ldots, n,$$

then

$$\frac{\partial X_k}{\partial x_m} = 0, \quad m \neq k, \quad k = 1, \ldots, n,$$
$$\frac{\partial X_k}{\partial x_k} = 1,$$

and

$$\frac{\partial X_k}{\partial x_m} = \sum_l \frac{\partial \phi_k}{\partial X_l} \frac{\partial X_l}{\partial x_m}.$$

If we now denote $\partial X_k / \partial x_m$ by Y_k and $\partial \phi_k / \partial X_l$ by L_{kl}, then this becomes

$$Y_k = \sum_l L_{kl} Y_l$$

i. e. the kth row of the matrix-vector product

$$(I - L)Y,$$

where the elements in the first n rows of L are all zeros, and then

$$\frac{\partial s_j}{\partial x_m} = Y_{M+j}.$$

Obtaining the Jacobian by the forward method may be considered as equivalent to solving

$$(I - L)Y = e_m.$$

Turning now to the reverse method if

$$X_k = \phi_k(X_l),$$

then the adjoint variable X_l^* contains a term

$$X_l^* = X_l^* + \frac{\partial \phi_k}{\partial X_l} X_k^*$$

which is the lth row of the matrix-vector product

$$(I - L^\top)X^*.$$

To obtain the gradient of s_m is therefore equivalent to solving

$$(I - L^\top)X^* = e_{M+m},$$

then

$$\frac{\partial s_m}{\partial x_i} = X_i^*.$$

So both the calculation of the Jacobian by the forward and backward method are equivalent to solving a very sparse set of equations. If the Wengert list is used, each row of L contains at most two nonzeros. It has therefore been suggested that methods for solving linear equations with sparse matrices could be used to calculate J, A. Griewank and S. Reese [14] suggested using the Markowitz rule, while U. Geitner, J. Utke and Griewank [11] applied the *method of Newsam and Ramsdell*.

Hessian Calculations

The calculation of the Hessian, as discussed in ▶ Automatic differentiation: Calculation of the Hessian, can also be formulated as a sparse matrix calculation. Using the notation of ▶ Automatic differentiation: Calculation of the Hessian if the calculation of $f(x)$ consists of

$$X_k = \phi_k(X_m, m < k, m \in M_k),$$

then the reverse gradient calculation consists of

$$X_m^* = X_m^* + X_k^* \frac{\partial \phi_k}{\partial X_m}, \quad m \in M_k.$$

If now we denote

$$Y_k = \frac{\partial X_k}{\partial x_i}, \quad k = 1, \ldots, M,$$

and

$$Y_k = \frac{\partial X_{2M-k+1}^*}{\partial x_i}, \quad k = M+1, \ldots, 2M,$$

then we obtain

$$Y_k = \frac{\partial \phi_k}{\partial X_m} Y_m, \quad k = 1, \ldots, M,$$

and

$$Y_{2M+1-m} = Y_{2M+1-m} + Y_{2M+1-k} \frac{\partial \phi_k}{\partial X_m} + X_k^* \frac{\partial^2 \phi_k}{\partial X_m \partial X_j} Y_j.$$

The second derivatives are 1, if ϕ is a multiplication, 0 if ϕ is an addition, and if ϕ is unary only nonzero if $j = m$. If we denote these second order terms by B, the calculation of $H\, e_i$ is equivalent to solving

$$\begin{bmatrix} I - L & 0 \\ B & I - L^S \end{bmatrix} Y = \begin{pmatrix} e_i \\ 0 \end{pmatrix}.$$

Here the superscript S indicates that L has been transposed through both diagonals. The ith column of the Hessian is then the last n values of Y. For the illustrative example

$$f(x) = (x_1 x_2 + \sin x_1 + 4)(3x_2^2 + 6)$$

used in ▶ Automatic differentiation: Calculation of the Hessian, the off-diagonal nonzeros in the matrix which we will denote by K, are

$$K_{3,1} = K_{20,18} = X_2,$$
$$K_{3,2} = K_{19,18} = X_1,$$
$$K_{4,1} = K_{20,17} = \cos X_1,$$
$$K_{5,3} = K_{18,16} = 1,$$
$$K_{5,4} = K_{17,16} = 1,$$
$$K_{6,5} = K_{16,15} = 1,$$
$$K_{7,2} = K_{19,14} = 2X_2,$$
$$K_{8,7} = K_{14,13} = 3,$$
$$K_{9,8} = K_{13,12} = 1,$$
$$K_{10,6} = K_{15,11} = X_9,$$
$$K_{10,9} = K_{12,11} = X_6,$$
$$K_{12,6} = K_{14,9} = X_{10}^*,$$
$$K_{19,1} = K_{20,2} = X_3^*,$$
$$K_{19,2} = 2X_7^*,$$
$$K_{20,1} = -X_4^* \sin X_1,$$

L contains 11 nonzeros and B contains 6. The matrix is very sparse and the same sparse matrix techniques could be used to solve this system of equations.

The Newton Step

As the notation is easier we will consider the Jacobian case.

We have shown that if we solve $(I - L)\, Y = e_m$, then column m of the Jacobian J is in the last n terms of Y. If we wish to evaluate $J\, p$ we simply have to solve

$$(I - L)Y = p'$$

where p' has its first n terms equal to p and the remaining terms zero. Then the solution is again in the last n terms of Y. To calculate the Newton step we know $J\, d$ as it must be equal to $-s$, but we do not know d. We must therefore add the equations

$$Y_{M+i} = -s_i$$

to the equations, and delete the equations $Y_i = p_i$. For convenience we will partition L, putting the first n columns into A, retaining L for the remainder. So we have to solve

$$\begin{bmatrix} -A & I - L \\ 0 & E \end{bmatrix} \begin{pmatrix} d \\ Y \end{pmatrix} = \begin{pmatrix} 0 \\ -s \end{pmatrix}$$

for d. The matrix E is rectangular and is full of zeros except for the diagonals which are 1. Solving for d gives

$$E(I-L)^{-1}Ad = -s,$$

so

$$J = E(I-L)^{-1}A,$$

which is also the *Schur complement* of the sparse set of equations.

One popular way of solving a sparse set of equations is to form the Schur complement and solve the resulting equations, in this instance this becomes 'form J and solve $J\,d = -s$', which would be the normal indirect method. This also justifies the attention given in this article to the efficient calculation of J.

Griewank [12] observed that it may be possible to calculate the Newton step more cheaply than forming J and then solving the Newton equations. Utke [16] demonstrated that a number of ways of solving the sparse set of equations were indeed quicker. His implementation was compatible with ADOL-C and included many rules for eliminating variables. This approach was motivated by noting that if the Jacobian $J = D + a\,b^\mathsf{T}$, where D is diagonal and a and b vectors, then J is full and so solving $J\,x = -s$ is an $O(n^3)$ operation. However introducing one extra variable $z = b^\mathsf{T}x$ enables the *extended matrix* to be solved very cheaply

$$b^\mathsf{T}x - z = 0,$$
$$Dx + az = -s$$

gives

$$x = -D^{-1}(az + s),$$
$$z = -b^\mathsf{T}D^{-1}(az + s),$$

so

$$z = -(1 + b^\mathsf{T}D^{-1}a)^{-1}b^\mathsf{T}D^{-1}s,$$

and then x may be determined by substitution, which is an $O(n)$ operation. The challenge to find an automatic process that finds such short cuts is still open.

L.C.W. Dixon [6] noted that the extended matrix is an *echelon form*. An echelon matrix of degree k has ones on the k super-diagonal and zeros above it. If the lower part is sparse and contains NNZ nonzeros then the Schur complement can be computed in $kNNZ$ operations and the Newton step obtained by solving the resulting equations in $O(k^3)$ steps. The straight forward sparse system is an echelon form with $k = n$, so he suggested that by re-arranging rows and columns it might be possible to reduce k. This would reduce the operations needed for both parts of the calculation. Many sorting algorithms have been proposed for reducing the echelon index of sparse matrices. J.S. Duff et al. [9] discuss the performance of methods known as P^4 and P^5. R. Fletcher [10] introduced SPK1. Dixon and Z. Maany [8] introduced another which when applied to the extended matrix of the extended Rosenbrock function reduces the echelon index from n to $n/2$ and gives a diagonal Schur complement. It follows that this method, too, has considerable potential.

All these approaches still require further research.

Truncated Methods

Experience using the *truncated Newton* code has led many researchers to doubt the wisdom of calculating accurate Newton steps. Approximate solutions are often preferred in which the *conjugate gradient* method is applied to $H\,d = -g$; this can be implemented by calculating $H\,p$ at each inner iteration. $H\,p$ can be calculated very cheaply by a single forward doublet pass with initial values set at p through list for g obtained by reverse differentiation. The operations required to compute $H\,p$ are therefore bounded by $15M$.

If an iterative method is used to solve $J\,d = -s$, the products $J\,p$ and $J^\mathsf{T}\,v$ can both be obtained cheaply, the first by forward, the second by reverse automatic differentiation.

See also

▶ Automatic Differentiation: Calculation of the Hessian
▶ Automatic Differentiation: Geometry of Satellites and Tracking Stations
▶ Automatic Differentiation: Introduction, History and Rounding Error Estimation
▶ Automatic Differentiation: Parallel Computation
▶ Automatic Differentiation: Point and Interval
▶ Automatic Differentiation: Point and Interval Taylor Operators

▶ Automatic Differentiation: Root Problem and Branch Problem
▶ Dynamic Programming and Newton's Method in Unconstrained Optimal Control
▶ Interval Newton Methods
▶ Nondifferentiable Optimization: Newton Method
▶ Nonlocal Sensitivity Analysis with Automatic Differentiation
▶ Unconstrained Nonlinear Optimization: Newton–Cauchy Framework

References

1. Berz M, Bischof Ch, Corliss G, Griewank A (eds) (1996) Computational differentiation: techniques, applications, and tools. SIAM, Philadelphia
2. Christianson DB, Dixon LCW (1992) Reverse accumulation of Jacobians in optimal control. Techn Report Numer Optim Centre, School Inform Sci Univ Hertfordshire 267
3. Coleman TF, Cai JY (1986) The cyclic coloring problem and estimation of sparse Hessian matrices. SIAM J Alg Discrete Meth 7:221–235
4. Coleman TF, Verma A (1996) Structure and efficient Jacobian calculation. In: Berz M, Bischof Ch, Corliss G, Griewank A (eds) Computational Differentiation: Techniques, Applications, and Tools. SIAM, Philadelphia, pp 149–159
5. Curtis AR, Powell MJD, Reid JK (1974) On the estimation of sparse Jacobian matrices. J Inst Math Appl 13:117–119
6. Dixon LCW (1991) Use of automatic differentiation for calculating Hessians and Newton steps. In: Griewank A, Corliss GF (eds) Automatic Differentiation of Algorithms: Theory, Implementation, and Application. SIAM, Philadelphia, pp 114–125
7. Dixon LCW (1993) On automatic differentiation and continuous optimization. In: Spedicato E (ed) Algorithms for continuous optimisation: the state of the art. NATO ASI series. Kluwer, Dordrecht, pp 501–512
8. Dixon LCW, Maany Z (Feb. 1988) The echelon method for the solution of sparse sets of linear equations. Techn Report Numer Optim Centre, Hatfield Polytechnic NOC TR177
9. Duff IS, Anoli NI, Gould NIM, Reid JK (1987) The practical use of the Hellerman–Ranck P^4 algorithm and the P^5 algorithm of Erisman and others. Techn Report AERE Harwell CSS213
10. Fletcher R, Hall JAJ (1991) Ordering algorithms for irreducible sparse linear systems. Techn Report Dundee Univ NA/131
11. Geitner U, Utke J, Griewank A (1991) Automatic computation of sparse Jacobians by applying the method of Newsam and Ramsdell. In: Griewank A, Corliss GF (eds) Automatic Differentiation of Algorithms: Theory, Implementation, and Application, SIAM, Philadelphia, pp 161–172
12. Griewank A (1991) Direct calculation of Newton steps without accumulating Jacobians. In: Griewank A, Corliss GF (eds) Automatic Differentiation of Algorithms: Theory, Implementation, and Application, SIAM, Philadelphia, pp 126–137
13. Griewank A, Corliss GF (eds) (1991) Automatic differentiation of algorithms: theory, implementation, and application. SIAM, Philadelphia
14. Griewank A, Reese S (1991) On the calculation of Jacobian matrices by the Markowitz rule. In: Griewank A, Corliss GF (eds) Automatic Differentiation of Algorithms: Theory, Implementation, and Application, SIAM, Philadelphia, pp 126–135
15. Parkhurst SC (Dec. 1990) The evaluation of exact numerical Jacobians using automatic differentiation. Techn Report Numer Optim Centre, Hatfield Polytechnic, NOC TR224
16. Utke J (1996) Efficient Newton steps without Jacobians. In: Berz M, Bischof Ch, Corliss G, Griewank A (eds) Computational Differentiation: Techniques, Applications, and Tools, SIAM, Philadelphia, pp 253–264

Automatic Differentiation: Geometry of Satellites and Tracking Stations

DAN KALMAN
American University, Washington, DC, USA

MSC2000: 26A24, 65K99, 85-08

Article Outline

Keywords
Geometric Models
Sample Optimization Problems
   Minimum Range
   Direction Angles and Their Derivatives
   Design Parameter Optimization
Automatic Differentiation
   Scalar Functions and Operations
   Vector Functions and Operations
   Implementation Methods
Summary
See also
References

Keywords

Astrodynamics; Automatic differentiation; Satellite orbit; Vector geometry

Satellites are used in a variety of systems for communication and data collection. Familiar examples of these systems include satellite networks for broadcasting video programming, meteorological and geophysical data observation systems, the global positioning system (GPS) for navigation, and military surveillance systems. Strictly speaking, these are systems in which satellites are just one component, and in which there are other primary subsystems that have no direct involvement with satellites. Nevertheless, they will be referred to as *satellite systems* for ease of reference.

Simple geometric models are often incorporated in simulations of satellite system performance. Important operational aspects of these systems, such as the times when satellites can communicate with each other or with installations on the ground (e. g. *tracking stations*), depend on dynamics of satellite and station motion. The geometric models represent these motions, as well as constraints on communication or data collection. For example, the region of space from which an antenna on the ground can receive a signal might be modeled as a cone, with its vertex centered on the antenna and axis extending vertically upward. The antenna can receive a signal from a satellite only when the satellite is within the cone. Taking into account the motions of the satellite and the earth, the geometric model predicts when the satellite and tracking station can communicate.

Elementary optimization problems often arise in these geometric models. It may be of interest to determine the closest approach of two satellites, or when a satellite reaches a maximum elevation as observed from a tracking station, or the extremes of angular velocity and acceleration for a rotating antenna tracking a satellite. Optimization problems like these are formulated in terms of geometric variables, primarily distances and angles, as well as their derivatives with respect to time. The derivatives appear both in the optimization algorithms, as well as in functions to be optimized. One of the previously mentioned examples illustrates this. When a satellite is being tracked from the ground, the antenna often rotates about one or more axes so as to remain pointed at the satellite. The angular velocity and acceleration necessary for this motion are the first and second derivatives of variables expressed as angles in the geometric configuration of the antenna and satellite. Determining the extreme values of these *derivatives* is one of the optimization problems mentioned earlier.

Automatic differentiation is a feature that can be included in a computer programming language to simplify programs that compute derivatives. In the situation described above, satellite system simulations are developed as computer programs that include computed values for the distance and angle variables of interest. With automatic differentiation, the values of derivatives are an automatic by-product of the computation of variable values. As a result, the computer programmer does not have to develop and implement the computer instructions that go into calculating derivative values. As a specific example of this idea, consider again the rotating antenna tracking a satellite. Imagine that the programmer has worked out the proper equations to describe the angular position of the antenna at any time. The simulation also needs to compute values for the angular velocity and acceleration, the first and second derivatives of angular position. However, the programmer does not need to work out the proper equations for these derivatives. As soon as the equations for angular position are included in the computer program, the programming language provides for the calculation of angular velocity and acceleration *automatically*. That is the effect of automatic differentiation. Because the derivatives of geometric variables such as distances and angles can be quite involved, automatic differentiation results in computer programs that are much easier to develop, debug, and maintain.

The preceding comments have provided a brief overview of geometric models for satellite systems, as well as associated optimization problems and the use of automatic differentiation. The discussion will now turn to a more detailed examination of these topics.

Geometric Models

The geometric models for satellite systems are formulated in the context of three-dimensional real space. A conventional rectangular coordinate system is defined by mutually perpendicular x, y, and z axes. The earth is modeled as a sphere or ellipsoid centered at the origin (0, 0, 0), with the north pole on the positive z axis, and the equator in the xy plane. The coordinate axes are considered to retain a constant orientation relative to the fixed stars, so that the earth rotates about the z axis.

In this setting, tracking station and satellite locations are represented by points moving in space. Each such moving point is specified by a vector valued function $\mathbf{r}(t) = (x(t), y(t), z(t))$ where t represents time. Geometric variables such as angles and distances can be determined using standard vector operations:

$$c(x, y, z) = (cx, cy, cz),$$
$$(x, y, z) \pm (u, v, w) = (x \pm u, y \pm v, z \pm w),$$
$$(x, y, z) \cdot (u, v, w) = xu + yv + zw,$$
$$(x, y, z) \times (u, v, w)$$
$$= (yw - zv, zu - xw, xv - yu),$$
$$\|(x, y, z)\| = \sqrt{x^2 + y^2 + z^2}$$
$$= \sqrt{(x, y, z) \cdot (x, y, z)}.$$

The distance between two points $\mathbf{r}$ and $\mathbf{s}$ is then given by $\| \mathbf{r} - \mathbf{s} \|$. The angle θ defined by rays from point $\mathbf{r}$ through points $\mathbf{p}$ and $\mathbf{q}$ is determined by

$$\cos\theta = \frac{(\mathbf{p} - \mathbf{r}) \cdot (\mathbf{q} - \mathbf{r})}{\|\mathbf{p} - \mathbf{r}\| \cdot \|\mathbf{q} - \mathbf{r}\|}.$$

A more complete discussion of vector operations, their properties, and geometric interpretation can be found in any calculus textbook; [9] is one example.

There are a variety of models for the motions of points representing satellites and tracking stations. The familiar conceptions of a uniformly rotating earth circled by satellites that travel in stable closed orbits is only approximately correct. For qualitative simulations of the performance of satellite systems, particularly at preliminary stages of system design, these models may be adequate. More involved models can take into account such effects as the asphericity of the gravitational field of the earth, periodic wobbling of the earth's axis of rotation, or atmospheric drag, to name a few. Modeling the motions of the earth and satellites with high fidelity is a difficult endeavor, and one that has been studied extensively. Good general references for this subject are [1,2,3,10].

For illustrative purposes, a few of the details will be presented for the simplest models, circular orbits

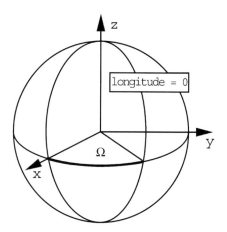

Automatic Differentiation: Geometry of Satellites and Tracking Stations, Figure 1
Earth rotation angle

around a spherical earth, uniformly spinning on a fixed axis. The radius of the earth will be denoted R_e.

As a starting point, the rotation of the earth can be specified by a single function of time, $\Omega(t)$, representing the angular displacement of the prime meridian from a fixed direction, typically the direction specified by the positive x axis (see Fig. 1.). At any time, the positive x axis emerges from the surface of the earth at some point on the equator. Suppose that at a particular time t, the point where the positive x axis emerges happens to be on the prime meridian, located at latitude 0 and longitude 0. Then $\Omega(t) = 0$ for that t. As time progresses, the prime meridian rotates away from the x axis, counter-clockwise as viewed by an observer above the north pole. The function Ω measures the angle of rotation, starting at 0 each time the prime meridian is aligned with the x axis, and increasing toward a maximum of 360° (2 π in radian measure) with each rotation of the earth. With a uniformly spinning earth, Ω increases linearly with t during each rotation.

Once Ω is specified, any terrestrial location given by a latitude ϕ, longitude λ, and altitude a can be transformed into absolute coordinates in space, according to the equations

$$\theta = \lambda + \Omega(t), \tag{1}$$
$$r = R_e + a, \tag{2}$$
$$x = r \cos\theta \cos\phi, \tag{3}$$

$$y = r \sin\theta \cos\phi, \tag{4}$$

$$z = r \sin\phi. \tag{5}$$

Holding latitude, longitude, and altitude constant, these equations express the position in space of a fixed location on the earth for any time, thereby modeling the point's motion. It is also possible to develop models for tracking stations that are moving on the surface of the earth, say on an aircraft or on a ship in the ocean. For example, if it is assumed that the moving craft is traveling at constant speed on a great circle arc or along a line of constant latitude, it is not difficult to express latitude and longitude as functions of time. In this case, the equations above reflect a dependence on t in λ and ϕ, as well as in Ω. A more complicated example would be to model the motion of a missile or rocket launched from the ground. This can be accomplished in a similar way: specify the trajectory in earth relative terms, that is, using latitude, longitude, and altitude, and then compute the absolute spatial coordinates (x, y, z). In each case, the rotation of the earth is accounted for solely by the effect of $\Omega(t)$.

For a satellite in circular orbit, the position at any time is specified by an equation of the following form:

$$\mathbf{r}(t) = r[\cos(\omega t)\mathbf{u} + \sin(\omega t)\mathbf{v}].$$

In this equation, ωt is understood as an angle in radian measure for the sin and cos operations; r, ω, $\mathbf{u}$, and $\mathbf{v}$ are constants. The first, r is the length of the orbit circle's radius. It is equal to the sum of the earth's radius R_e and the satellite's altitude. The constant ω is the angular speed of the satellite. The satellite completes an orbit every $2\pi/\omega$ units of time, thus giving the *orbital period*. Both $\mathbf{u}$ and $\mathbf{v}$ are unit vectors: $\mathbf{u}$ is parallel to the initial position of the satellite; $\mathbf{v}$ is parallel to the initial velocity. See Fig. 2.

Mathematically, the equation above describes some sort of orbit no matter how the constants are selected. But not all of these are accurate descriptions of a free falling satellite in circular orbit. For one thing, $\mathbf{u}$ and $\mathbf{v}$ must be perpendicular to produce a circular orbit. In addition, there is a physical relationship linking r and ω. Assuming that the circular orbit follows Newton's laws of motion and gravitation, r and ω satisfy

$$\omega = K r^{-\frac{3}{2}} \tag{6}$$

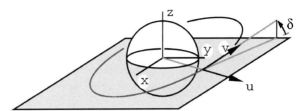

Automatic Differentiation: Geometry of Satellites and Tracking Stations, Figure 2
Circular orbit

where K is a physical constant that depends on both Newton's universal gravitational constant and the mass of the earth. Its numerical value also depends on the units of measurement used for time and distance. For units of hours and kilometers, the value of K is $2.27285 \cdot 10^6$. As this relationship shows, for a given altitude (and hence a given value of r), there is a unique angular speed at which a satellite will maintain a circular orbit. Equivalently, the altitude of a circular orbit determines the constant speed of the satellite, as well as the period of the satellite.

Generally, constants are chosen for a circular orbit based on some geometric description. Here is a typical approach. Assume that the initial position of the satellite is directly above the equator, with latitude 0, a given longitude, and a given altitude. In other words, assume that the initial position is in the plane of the equator, and so has a z coordinate of 0. (This is the situation depicted in Fig. 2.) Moreover, the initial heading of the satellite can be specified in terms of the angle it makes with the xy plane (which is the plane of the equator). Call that angle δ. From these assumptions we can determine values for the constants r, ω, $\mathbf{u}$, and $\mathbf{v}$ in the equation for $\mathbf{r}(t)$. Now the altitude for the orbit is constant, so the initial altitude determines r, as well as ω via equation (6). The initial latitude, longitude, and altitude also provide enough information to determine absolute coordinates (x, y, z) for the initial satellite position using equations (1)–(5). Accordingly, the unit vector $\mathbf{u}$ is given by

$$\mathbf{u} = \frac{(x, y, z)}{\|(x, y, z)\|}.$$

As already observed, the z coordinate of $\mathbf{u}$ will be 0. Finally, the unit vector $\mathbf{v}$ is determined from the initial position and heading. It is known that $\mathbf{v}$ make an angle of δ with the xy plane, and hence makes an angle of $\pi/2 - \delta$ with the z axis. This observation can be expressed as the equation

$$\mathbf{v} \cdot (0, 0, 1) = \sin \delta.$$

It is also known that $\mathbf{v}$ must be perpendicular to $\mathbf{u}$, so

$$\mathbf{v} \cdot \mathbf{u} = 0.$$

Finally, since $\mathbf{v}$ is a unit vector,

$$\mathbf{v} \cdot \mathbf{v} = 1.$$

If $\mathbf{u} = (u_1, u_2, 0)$, then these three equations lead to $\mathbf{v} = (\pm u_2 \cos \delta, \mp u_1 \cos \delta, \sin \delta)$. The ambiguous sign can be resolved by assuming that the direction of orbit is either in agreement with or contrary to the direction of the earth's rotation. Assuming that the orbit is in the same direction as the earth's rotation, $\mathbf{v} = (-u_2 \cos \delta, u_1 \cos \delta, \sin \delta)$. The alternative possibility, that the satellite orbit opposes the rotation of the earth, is generally not practically feasible, so is rarely encountered.

The preceding paragraphs are intended to provide some insight about the mathematics used to describe the movement of satellites and terrestrial observers. Although the models presented here are the simplest ones available, they appear in the same general framework as much more sophisticated models. In particular, in any of these models, it is necessary to be able to compute instantaneous positions for satellites and terrestrial observers at any time during a simulation. Moreover, the use of vector algebra and geometry to set up the simple models is representative of the methods used in more complicated cases.

Sample Optimization Problems

Computer simulations of satellite system performance provide one tool for comparing alternative designs and making cost/benefit trade-offs in the design process. Optimization problems contribute both directly and indirectly. In many cases, system performance is characterized in terms of extreme values of variables: what is the maximum number of users that can be accommodated by a communications system? At a given latitude, what is the longest period of time during which at most three satellites can be detected from some point on the ground? In these examples, the optimization problems are directly connected with the goals of the simulation.

Optimization problems also arise indirectly as part of the logistics of the simulation software. This is particularly the case when a simulation involves events that trigger some kind of system response. Examples of such events include the passage of a satellite into or out of sunlight, reaching a critical level of some resource such as power or data storage, or the initiation or termination of radio contact with a tracking station. The detection of these events typically involves either root location or optimization. These processes are closely related: the root of an equation can usually be characterized as an extreme value of a variable within a suitable domain; conversely, optimization algorithms often generate candidate solutions by solving equations.

In many of these event identification problems, the independent variable is time. The objective functions ultimately depend on the geometric models for satellite and tracking station motion, and so can be formulated in terms of explicit functions of time. In contrast, some of the optimization problems that concern direct estimation of system performance seek to optimize that performance by varying design parameters. A typical approach to this kind of problem is to treat performance measures as functions of the parameters, where the values of the functions are determined through simulation. Both kinds of optimization are illustrated in the following examples.

Minimum Range

As a very simple example of an optimization problem, it is sometimes of interest to determine the closest approach of two orbiting bodies. Assume that a model has been developed, with $\mathbf{r}(t)$ and $\mathbf{s}(t)$ representing the positions at time t for the two bodies. The distance between them is then expressed as $\| \mathbf{r}(t) - \mathbf{s}(t) \|$. This is the objective function to be minimized. Observe that it is simply expressed as a composition of vector operations and the motion models for the two bodies.

A variation of this problem occurs when several satellites are required to stay in radio communication. In that case, an antenna on one satellite (at position A, say) may need to detect signals from two others (at positions B and C). In this setting, the measure of $\angle BAC$ is of interest. If the angle is wide, the antenna requires a correspondingly wide field of view. As the satellites proceed in their orbits, what is the maximum value of the angle? Equivalently, what is the minimum value of the cosine of the angle? As before, the objective function in this minimization problem is easily expressed by applying vector operations to the position models for the satellites. If $\mathbf{a}(t)$, $\mathbf{b}(t)$, and $\mathbf{c}(t)$ are the position functions for the three satellites, then

$$\cos \angle BAC = \frac{(\mathbf{b}-\mathbf{a}) \cdot (\mathbf{c}-\mathbf{a})}{\|\mathbf{b}-\mathbf{a}\| \cdot \|\mathbf{c}-\mathbf{a}\|}.$$

This is a good example of combining vector operations with the models for satellite motion to derive the objective function in an optimization problem. The next example is similar in style, but mathematically more involved.

Direction Angles and Their Derivatives

A common aspect of satellite system simulation is the representation of sensors of various kinds. The images that satellites beam to earth of weather systems and geophysical features are captured by sensors. Sensors are also used to locate prominent astronomical features such as the sun, the earth, and in some cases bright stars, in order to evaluate and control the satellite's attitude. Even the antenna used for communication is a kind of sensor. It is frequently convenient to define a coordinate system that is attached to a sensor, that is, define three mutually perpendicular axes which intersect at the sensor location, and which can be used as an alternate means to assign coordinates to points in space. Such a coordinate system is then used to describe the vectors from the sensor to other objects, and to model sensor sensitivity to signals arriving from various directions. With several different coordinate systems in use, it is necessary to transform information described relative to one system into a form that makes sense in the context of another system. This process also often involves what are called *direction angles*.

As a concrete example, consider an antenna at a fixed location on the earth, tracking a satellite in orbit. The coordinate system attached to the tracking antenna is the natural map coordinate system at that point on the earth: the local x and y axes point east and north, respectively, and the z axis points straight up (Fig. 3). The direction from the station to the satellite is expressed in terms of two angles: the elevation δ of the satellite above the local xy plane, and the compass angle α measured clockwise from north. (See Fig. 4.) To illustrate,

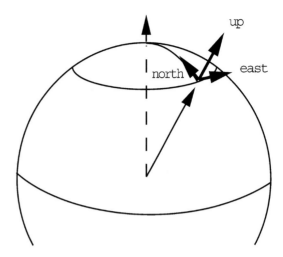

Automatic Differentiation: Geometry of Satellites and Tracking Stations, Figure 3
Local map coordinates

here is the meaning of an elevation of 30 degrees and a compass angle of 270 degrees. Begin by looking due north. Turn clockwise through 270 degrees, maintaining a line of sight that is parallel to the local xy plane. At that point you are looking due west. Now raise the line of sight until it makes a 30 degree angle with the local xy plane. This direction of view, with elevation 30 and compass angle 270 degrees, might thus be described as 30 degrees above a ray 270 degrees clockwise from due north. The elevation and compass angle are examples of direction angles. Looked at another way, if a spherical coordinate system is imposed on the local rectan-

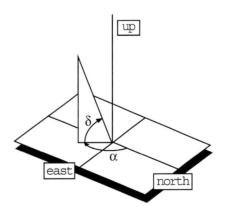

Automatic Differentiation: Geometry of Satellites and Tracking Stations, Figure 4
Compass and elevation angles

gular system at the antenna, then every point in space is described by a distance and two angles. The angles are direction angles. Direction angles can be defined in a similar way for any local coordinate system attached to a sensor.

How are direction angles computed? In general terms, the basic idea is to define the local coordinate system in terms of moving vectors, and then to use vector operations to define the instantaneous value of direction angles. Here is a formulation for the earth based antenna. First, the local z axis points straight up. That means the vector from the center of the earth to the location of the antenna on the surface is parallel to the z axis. Given the latitude, longitude, and altitude of the antenna, its absolute position $\mathbf{r}(t) = (x, y, z)$ is computed using equations (1)–(5), as discussed earlier. The parallel unit vector is then given by $\mathbf{r}/\|\mathbf{r}\|$. To distinguish this from the global z axis, we denote it as the vector **up**. The vector pointing due east must be perpendicular to the up direction. It also must be parallel to the equatorial plane, and hence perpendicular to the global z axis. Using properties of vector cross products, a unit vector pointing east can therefore be expressed as

$$\mathbf{east} = \frac{(0, 0, 1) \times \mathbf{up}}{\|(0, 0, 1) \times \mathbf{up}\|}.$$

Finally, the third perpendicular vector is given by the cross product of the other two: **north** = **up** × **east**. Note that these vectors are defined as functions of time. At each value of t the earth motion model gives an instantaneous value for $\mathbf{r}(t)$, and that, in turn, determines the vectors **up**, **east**, and **north**.

Next, suppose that a satellite is included in the model, with instantaneous position $\mathbf{s}(t)$. The **view** vector from the antenna to the satellite is given by $\mathbf{v}(t) = [\mathbf{s}(t) - \mathbf{r}(t)] / \|\mathbf{s}(t) - \mathbf{r}(t)\|$. The goal is to calculate the direction angles α and δ for $\mathbf{v}$. Since δ measures the angle between $\mathbf{v}$ and the plane of **east** and **north**, the complimentary angle can be measured between $\mathbf{v}$ and **up**. This leads to the equation

$$\sin \delta = \mathbf{up} \cdot \mathbf{v}.$$

The angle α is found from

$$v_n = \mathbf{v} \cdot \mathbf{north}$$
$$v_e = \mathbf{v} \cdot \mathbf{east}$$

according to the equations

$$\cos \alpha = \frac{v_n}{\sqrt{v_n^2 + v_e^2}}$$
$$\sin \alpha = \frac{v_e}{\sqrt{v_n^2 + v_e^2}}.$$

These follow from the fact that the projection of $\mathbf{v}$ into the local xy plane is given by v_e **east** + v_n **north**.

In this example, direction angles play a role in several optimization problems. First, it may be of interest to predict the maximum value of δ as a satellite passes over the tracking station. This maximum value of elevation is an indication of how close the satellite comes to passing directly overhead, and may be used to determine whether communication will be possible between satellite and tracking station.

Additional optimization problems concern the derivatives of α and δ. In many designs, an antenna can turn about horizontal and vertical axes to point the center of the field of view in a particular direction. In order to stay pointed at a passing satellite, the antenna must be rotated on its axes so as to match the motion of the satellite, and α and δ specify exactly how far the antenna must be rotated about each axis at each time. However, there are mechanical limits on how fast the antenna can turn and accelerate. For this reason, during the time that the satellite is in view, the maximum values of the first and second derivatives of α and δ are of interest. If the first derivatives exceed the antenna's maximum turning speed, or if the second derivatives exceed the antenna's maximum acceleration, the antenna will not be able to remain pointed at the satellite.

Design Parameter Optimization

The preceding examples all involve simple kinds of optimization problems with objective functions depending only on time. There are also many situations in which system performance variables are optimized over some domain of design parameters. As one example of this, consider a system with a single satellite traveling in a circular orbit. Assume that the initial point of the orbit falls on the equator, with angle δ between the initial heading and the xy plane, as in Fig. 2. In this example, the object is to choose an optimal value of δ. The optimization problem includes several tracking stations on the ground that are capable of communicating with the satellite. As it orbits, there may be times when the satellite cannot communicate with any of the tracking stations. At other times, one or more stations may be accessible. Over the simulation period, the total amount of time during which at least one tracking station is accessible will depend on the value of δ. It is this total amount of access time (denoted A) that is to be maximized.

In this problem, the objective function A is not given as a mathematical expression involving the variable δ. An appropriate simulation can be created to compute A for any particular δ of interest. This can then be used in conjunction with an optimization algorithm, with the simulation executed each time it is necessary to calculate $A(\delta)$.

The preceding example is a simple one, and the execution time required to compute $A(\delta)$ is small. For more complicated situations, each execution of the simulation can require a significant amount of time. In these cases, it may be more practical to use some sort of interpolation scheme. The idea would be to run the simulation for some values of the parameter(s), and to interpolate between these values as needed during the optimization process.

In some situations, there is a resource allocation problem that can add yet another level of complexity to optimizing system performance. For example, if there are several satellites that must compete for connection time with the various tracking stations, just determining how to assign the tracking stations to the satellites is not a simple matter. In this situation, there may be one kind of optimization problem performed during the simulation to make the resource allocations, and then a secondary optimization that considers the effect of changing system design parameters. An example of this kind of problem is described in detail in [6].

The preceding examples have been provided to illustrate the kinds of optimization problems that arise in simulations of satellite systems. Although there has been very little discussion of methods to solve these optimization problems, it should be clear that standard methods apply, especially in the cases for which the independent variable is time. In that context, the ability to compute derivatives relative to time for the objective function is of interest. In addition, it sometimes occurs that the objective function is, itself, defined as a derivative of some geometric variable, providing another motivation for computing derivatives. The next topic of discussion concerns the use of automatic differentiation for computing the desired derivatives.

Automatic Differentiation

Automatic differentiation refers to a family of techniques for automatically computing derivatives as a byproduct of function evaluation. A survey of different approaches and applications can be found in [5] and in-depth treatment appears in [4]. For the present discussion, attention will be restricted to what is called the *forward mode of automatic differentiation*, and in particular, the approach described in [8]. In this approach, to provide automatic calculation of the first m derivatives of real valued expressions of a single variable x, an algebraic system is defined consisting of real $m+1$ tuples, to which are extended the familiar binary operations and elementary functions generally defined on real variables. For concreteness, m will be assumed to be 3 below, but the discussion can be generalized to other values in an obvious way.

With $m = 3$, the objects manipulated by the automatic differentiation system are 4-tuples. The idea is that each 4-tuple represents the value of a function and its first 3 derivatives, and that the operations on tuples preserve this interpretation. Thus, if $a = (a_0, a_1, a_2, a_3)$ consists of the value of $f(t)$, $f'(t)$, $f''(t)$, and $f'''(t)$ at some t, and if $b = (b_0, b_1, b_2, b_3)$ is similarly defined for function g, then the product ab that is defined for the automatic differentiation system will consist of the value at t of fg and its first 3 derivatives. Similarly, the extension of the squareroot function to 4-tuples is so contrived that $\sqrt{a}$ will consist of the value of $\sqrt{f(t)}$ and its first 3 derivatives.

In the preceding remarks, the functions f and g are assumed to be real valued, but similar ideas work for vector valued functions. The principle difference is this: when $f(t)$ is a vector, then so are its derivatives, and the a_i referred to above are then vectors rather than scalars. In addition, for vector valued functions, there are different operations than for scalar valued functions. For example, vector functions may be combined with a dot product, as opposed to the conventional product of real scalars, and while the squareroot operation is not defined for vector valued functions, the norm operation $\| f(t) \|$ is.

In an automatic differentiation system built along these lines, there must be some functions that are evaluated directly to produce 4-tuples. For example, the constant function with value c can be evaluated directly to produce the tuple $(c, 0, 0, 0)$, and the identity function $I(t) = t$ can be evaluated directly to produce $(t, 1, 0, 0)$. For geometric satellite system simulations, it is also convenient to provide direct evaluation of tuples for the motion models. For example, let $\mathbf{r}(t)$ be the position vector for a tracking station, as developed in equations (1)–(5). It is a simple matter to work out appropriate formulas for the first three derivatives of $\mathbf{r}(t)$, each of which is also a vector. This is included in the automatic differentiation system so that when a particular value of t is given, the motion model computes the 4-tuple $(\mathbf{r}(t), \mathbf{r}'(t), \mathbf{r}''(t), \mathbf{r}'''(t))$. A similar arrangement is made for every moving object represented in the simulation, including satellites, tracking stations, ships, aircraft, and so on.

Here is a simple example of how automatic differentiation is used. In the earlier discussion of optimization problems, there appeared the following equation:

$$\cos \angle BAC = \frac{(\mathbf{b} - \mathbf{a}) \cdot (\mathbf{c} - \mathbf{a})}{\|\mathbf{b} - \mathbf{a}\| \cdot \|\mathbf{c} - \mathbf{a}\|}.$$

Using automatic differentiation, $\mathbf{a}$, $\mathbf{b}$, and $\mathbf{c}$ would be 4-tuples, each consisting of four vectors. These are produced by the motion models for three satellites, as the values of position and its first three derivatives at a specific time. The operations used in the equation, vector difference, dot product, and norm, as well as scalar multiplication and division, are all special modified operations that work directly on 4-tuples. The end result is also a 4-tuple, consisting of the cosine of angle BAC, as well as the first three derivatives of that function, all at the specified value of t. As a result, the programmer can obtain computed values for the derivatives of the function without explicitly coding equations for these derivatives. More generally, after defining appropriate 4-tuples for all of the motion models, the programmer automatically obtains derivatives for any function that is defined by operating on the motion models, just by defining the operations. No explicit representation of the derivatives of the operations is needed. Some details of how the system works follow.

Scalar Functions and Operations

Consider first operations which apply to scalars. There are two basic types: binary operations $(+, -, \times, \div)$ and elementary functions (squareroot, exponential and logarithm, trigonometric functions, etc.). These operations must be defined for the 4-tuples of the automatic differentiation system in such a way that derivatives are correctly propagated.

The definition for multiplication will illustrate the general approach for binary operations. Suppose that (a, b, c, d) and (u, v, w, x) are two 4-tuples of scalars. They represent values of functions and their derivatives, say, $(a, b, c, d) = (f(t), f'(t), f''(t), f'''(t))$ and $(u, v, w, x) = (g(t), g'(t), g''(t), g'''(t))$. The product is supposed to give $((fg)(t), (fg)'(t), (fg)''(t), (fg)''(t))$. Each of these derivatives can be computed using the derivatives of f and g.

$$(fg)(t) = f(t)g(t),$$
$$(fg)'(t) = f'(t)g(t) + f(t)g'(t),$$
$$(fg)''(t) = f''(t)g(t) + 2f'(t)g'(t) + f(t)g''(t),$$
$$(fg)'''(t)) = f'''(t)g(t) + 3f''(t)g'(t)$$
$$+ 3f'(t)g''(t) + f(t)g'''(t)).$$

On the right side of each equation, now substitute the entries of (a, b, c, d) and (u, v, w, x).

$$(fg)(t) = au,$$
$$(fg)'(t) = av + bu,$$
$$(fg)''(t) = aw + 2bv + cu,$$
$$(fg)'''(t)) = ax + 3bw + 3cv + du.$$

This shows that 4-tuples must be multiplied according to the rule

$$(a, b, c, d)(u, v, w, x)$$
$$= (au, av + bu, aw + 2bv + cu,$$
$$ax + 3bw + 3cv + du).$$

For addition, subtraction, and division a similar approach can be used. All that is required is that successive derivatives of the combination of f and g be expressed in terms of the derivatives of f and g separately. Replacing these derivatives with the appropriate components of (a, b, c, d) and (u, v, w, x) produces the desired formula for operating on 4-tuples.

To define the operation on a 4-tuple of an elementary function, a similar approach will work. Consider defining how a function h should apply to a 4-tuple $(a, b, c, d) = (f(t), f'(t), f''(t), f''(t))$. This time, the desired end result should contain derivatives for the composite function $h \circ f$, and so should have the form $((h \circ f)(t), (h \circ f)'(t), (h \circ f)''(t), (h \circ f)'''(t))$ The derivative of $h \circ f$ is given by $h'(f(t))f'(t)$, which becomes $h'(a) b$ after substitution. Similar computations produce expressions for the second and third derivatives:

$$(h \circ f)''(t)$$
$$= h''(f(t))f'(t)^2 + h'(f(t))f''(t)$$
$$= h''(a)b^2 + h'(a)c$$

and

$$(h \circ f)'''(t)$$
$$= h'''(f(t))f'(t)^3 + 3h''(f(t))f'(t)f''(t)$$
$$\quad + h'(f(t))f'''(t)$$
$$= h'''(a)b^3 + 3h''(a)bc + h'(a)d.$$

These results lead to

$$h(a, b, c, d)$$
$$= (h(a), h'(a)b, h''(a)b^2 + h'(a)c,$$
$$\qquad h'''(a)b^3 + 3h''(a)bc + h'(a)d).$$

As an example of how this is applied, let $h(t) = e^t$. Then $h(a) = h'(a) = h''(a) = h'''(a) = e^a$ so

$$e^{(a,b,c,d)}$$
$$= (e^a, e^a b, e^a b^2 + e^a c, e^a b^3 + 3e^a bc + e^a d)$$
$$= e^a(1, b, b^2 + c, b^3 + 3bc + d).$$

Other functions are a little more complicated, but the overall approach is generally correct.

The preceding discussion indicates how operations on 4-tuples would be built into an automatic differentiation system. However, the user of such a system would simply apply the operations. So, if an appropriate definition has been provided for $\Omega(t)$ as discussed earlier, along with the derivatives, the program would compute a 4-tuple for Ω and its derivatives at a particular time. Say that is represented in the program by the variable W. If the program later includes the call $\sin(W)$, the result would be a 4-tuple with values for $\sin(\Omega(t))$, and the first three derivatives.

Vector Functions and Operations

The approach for vector functions is basically the same as for scalar functions. The only modification that is needed is to recognize that the components of 4-tuples are now vectors. Because the rules for computing derivatives of vector operations are so similar to those for scalar operations, there is little difference in the appearance of the definitions. For example, here is the definition for the dot product of two 4-tuples, whose components are vectors:

$$(a, b, c, d) \cdot (u, v, w, x)$$
$$= (a \cdot u, a \cdot v + b \cdot u, a \cdot w + 2b \cdot v + c \cdot u,$$
$$\qquad a \cdot x + 3b \cdot w + 3c \cdot v + d \cdot u).$$

The formulation for vector cross product is virtually identical, as is the product of a scalar 4-tuple with a vector 4-tuple. For the vector norm, simply define

$$\|(a, b, c, d)\| = \sqrt{(a, b, c, d) \cdot (a, b, c, d)}.$$

Since both dot product of vector 4-tuples and square-root of scalar 4-tuples have already been defined in the automatic differentiation system, this equation will propagate derivatives correctly.

With a full complement of scalar and vector operations provided by the automatic differentiation system, all of the geometric variables discussed in previous examples can be included in a computer program, with derivatives generated automatically. As a particular case, reconsider the discussion earlier of computing elevation δ and compass angle α for a satellite as viewed from a tracking station. Assuming that r and s have been defined as 4-tuples for the vector positions of that station and satellite, the following fragment of pseudocode would carry out the computations described earlier:

up	= r/norm(r)
east	= cross(pole, up)
east	= east/norm(east)
north	= cross(up, east)
v	= (s−r)/norm(s−r)
vn	= dot(v, north)
ve	= dot(v, east)
vu	= dot(v, up)
delta	= asin(vu)
alpha	= atan2(ve, vn)

Executed in an automatic differentiation system, this code produces not just the instantaneous values of the angles α and δ, but their first three derivatives, as well. The programmer does not need to derive and code explicit equations for these derivatives, a huge savings in this problem. And all of the derivative information is useful. Recall that the first and second derivatives are of interest for their physical interpretations as angular velocities and accelerations. The third derivatives are used in finding the maximum values of the second derivatives (accelerations).

Implementation Methods

One of the simplest ways to implement automatic differentiation is to use a language like C^{++} that supports the definition of abstract data types and operator overloading. Then the automatic differentiation system would be implemented as a series of data types and operations, and included as part of the code for a simulation. A discussion of one such implementation can be found in [7].

Another approach is to develop a preprocessor that automatically augments code with the steps needed to compute derivatives. With such a system, the programmer develops code in a conventional language such as FORTRAN, with some additional features that control the application of automatic differentiation. Next, this code is operated on by the preprocessor, producing a modified program. That is then compiled and executed in the usual way. Examples of this approach can be found in [5].

Summary

Geometric models are very useful in representing the motions of satellites and terrestrial objects in simulations of satellite systems. These models are defined in terms of vector operations, which permit the convenient formulation of equations for geometric constructs such as distances and angles arising in the satellite system configuration. Equations which specify instantaneous positions in space of moving objects are a fundamental component of the geometric modeling framework.

Optimization problems occur in this framework in two guises. First, there are problems in which the objective functions are directly defined as features of the geometric setting. An example of this would be to find the minimum distance between two satellites. Second, measures of system performance are derived via simulation as a function of design parameters, and these measures are optimized by varying the parameters. An example of this kind of problem would be to seek a particular orbit geometry in order to maximize the total amount of time a satellite has available to communicate with a network of tracking stations.

Automatic differentiation is a feature of an environment for implementing simulations as computer programs. In an automatic differentiation system, the equations which define values of variables automatically produce the values of the derivatives, as well. In the geometric models of satellite systems, derivatives of some variables are of intrinsic interest as velocities and accelerations. Derivatives are also useful in solving optimization problems.

Automatic differentiation can be provided by replacing single operands with tuples, representing the operands and their derivatives. For some tuples, the derivatives must be explicitly provided. This is the case for the motion models. For tuples representing combinations of the motion models, the derivatives are generated automatically. These combinations can be defined using any of the supported operations provided by the automatic differentiation system, typically including the operations of scalar and vector arithmetic, as well as scalar functions such as exponential, logarithmic, and trigonometric functions. Languages which support abstract data types and operator overloading are a convenient setting for implementing an automatic differentiation system.

See also

- ▶ Automatic Differentiation: Calculation of the Hessian
- ▶ Automatic Differentiation: Calculation of Newton Steps
- ▶ Automatic Differentiation: Introduction, History and Rounding Error Estimation
- ▶ Automatic Differentiation: Parallel Computation
- ▶ Automatic Differentiation: Point and Interval
- ▶ Automatic Differentiation: Point and Interval Taylor Operators

▶ Automatic Differentiation: Root Problem and Branch Problem
▶ Nonlocal Sensitivity Analysis with Automatic Differentiation

References

1. Bate RR, Mueller DD, White JE (1971) Fundamentals of astrodynamics. Dover, Mineola, NY
2. Battin RH (1987) An introduction to the mathematics and methods of astrodynamics. AIAA Education Ser. Amer. Inst. Aeronautics and Astronautics, Reston, VA
3. Escobal PR (1965) Methods of orbit determination. R.E. Krieger, Huntington, NY
4. Griewank A (2000) Evaluating derivatives: Principles and techniques of algorithmic differentiation. SIAM, Philadelphia
5. Griewank A, Corliss GF (eds) (1995) Automatic differentiation of algorithms: Theory, implementation, and application. SIAM, Philadelphia
6. Kalman D (1999) Marriages made in the heavens: A practical application of existence. Math Magazine 72(2):94–103
7. Kalman D, Lindell R (1995) Automatic differentiation in astrodynamical modeling. In: Griewank A and Corliss GF (eds) Automatic differentiation of algorithms: Theory, implementation, and application. SIAM, Philadelphia, pp 228–241
8. Rall LB (Dec. 1986) The arithmetic of differentiation. Math Magazine 59(5):275–282
9. Thomas GB Jr, Finney RL (1996) Calculus and analytic geometry, 9th edn. Addison-Wesley, Reading, MA
10. Wertz JR (ed) (1978) Spacecraft attitude determination and control. Reidel, London

Automatic Differentiation: Introduction, History and Rounding Error Estimation

MASAO IRI, KOICHI KUBOTA
Chuo University, Tokyo, Japan

MSC2000: 65D25, 26A24

Article Outline

Keywords
Introduction
   Algorithms
   Complexity
History
Estimates of Rounding Errors
See also
References

Keywords

Differentiation; System analysis; Error analysis

Introduction

Most numerical algorithms for analyzing or optimizing the performance of a nonlinear system require the partial derivatives of functions that describe a mathematical model of the system. The *automatic differentiation* (abbreviated as AD in the following), or its synonym, *computational differentiation*, is an efficient method for computing the numerical values of the derivatives. AD combines advantages of numerical computation and those of symbolic computation [2,4].

Given a vector-valued function $\mathbf{f}: \mathbf{R}^n \to \mathbf{R}^m$:

$$\mathbf{y} = \mathbf{f}(\mathbf{x}) \equiv \begin{pmatrix} f_1(x_1, \ldots, x_n) \\ \vdots \\ f_m(x_1, \ldots, x_n) \end{pmatrix} \qquad (1)$$

of n variables represented by a big program with hundreds or thousands of program statements, one often had encountered (before the advent of AD) some difficulties in computing the partial derivatives $\partial f_i / \partial x_j$ with conventional methods (as will be shown below). Now, one can successfully differentiate them with AD, deriving from the program for $\mathbf{f}$ another program that efficiently computes the numerical values of the partial derivatives.

AD is entirely different from the well-known numerical approximation with quotients of finite differences, or *numerical differentiation*. The quotients of finite differences, such as $(f(x + h) - f(x))/h$ and $(f(x + h) - f(x - h))/2h$, approximate the derivative $f'(x)$, where truncation errors are of $O(h)$ and $O(h^2)$, respectively, but there is an insurmountable difficulty to compute better and better approximation. For, although an appropriately small value of h is chosen, it may fail to compute the values of the function when $x \pm h$ is out of the domain of f, and, furthermore, the effect of rounding errors in computing the values of the functions is of problem.

AD is also different from symbolic differentiation with a symbolic manipulator. The symbolic differentiation derives the expressions of the partial derivatives rather than the values. The mathematical model of a large scale system may be described in thousands of program statements so that it becomes very difficult to handle whole of them with an existing symbolic manipulator. (There are a few manipulators combined with AD, which can handle such large scale programs. They should be AD regarded as a symbolic manipulator.)

Example 1 Program 1 computes an output value y_1 as a composite function f_1 for given input values $x_1 = 2$, $x_2 = 3$, $x_3 = 4$:

$$y_1 = f_1(x_1, x_2, x_3) = \frac{x_1(x_2 - x_3)}{\exp(x_1(x_2 - x_3)) + 1}. \quad (2)$$

```
IF (x₂.le.x₃)
  THEN y₁ = x₁(x₂ − x₃)
  ELSE y₁ = x₁(x₂ + x₃)
ENDIF
y₁ = y₁/(exp(y₁) + 1).
```

Automatic Differentiation: Introduction, History and Rounding Error Estimation, Program 1
Example

The execution of this program is traced by a sequence of assignment statements (Program 2).

```
y₁ ← x₁(x₂ − x₃),
y₁ ← y₁/(exp(y₁) + 1).
```

Automatic Differentiation: Introduction, History and Rounding Error Estimation, Program 2
Program 1 expanded to straight line program for the specified input values

A set of unary or binary arithmetic operators (+, −, ∗, /) and elementary transcendental functions (exp, log, sin, cos, …) that may be used in the programs will be called *basic operations*. (Some special operations such as those generating 'constant' and 'input' are also to be counted among basic operations.) Program 2 can be expanded into a sequence of assignment statements each of whose right side has only one basic operation (Program 3), where $z_1, \ldots, z_s$ are temporary variables ($s = 2$ for this example).

```
1  z₁ ← x₂ − x₃,
2  z₁ ← x₁ ∗ z₁,
3  z₂ ← exp(z₁),
4  z₂ ← z₂ + 1,
5  z₁ ← z₁/z₂.
```

Automatic Differentiation: Introduction, History and Rounding Error Estimation, Program 3
Expanded history of execution with each line having only one basic operation

Moreover, it is useful to rewrite Program 3 into a sequence of *single assignment* statements, in which each variable appears at most once in the left sides (Program 4), hence, '←' can be replaced by ' = '.

```
1  v₁ ← x₂ − x₃,
2  v₂ ← x₁ ∗ v₁,
3  v₃ ← exp(v₂),
4  v₄ ← v₃ + 1,
5  v₅ ← v₂/v₄,
```

Automatic Differentiation: Introduction, History and Rounding Error Estimation, Program 4
Computational process

The sequence is called a *computational process*, where the additional variables $v_1, \ldots, v_5$ are called *intermediate variables* that keep the intermediate results. A graph called a *computational graph*, $G = (V, A)$, may be used to represent the process (see Fig. 1).

Algorithms

There are two modes for AD algorithm, *forward mode* and *reverse mode*. The forward mode is to compute $\partial y_i/\partial x_j$ ($i = 1, \ldots, m$) for a fixed j, whereas the reverse mode is to compute $\partial y_i/\partial x_j$ ($j = 1, \ldots, n$) for a fixed i.

The forward mode corresponds to tracing an expanded program such as Program 3 in the natural order. Assume that execution of the kth assignment in the program is represented as

$$z_c \leftarrow \psi_k(z_a, z_b). \quad (3)$$

When the values of both $\partial z_a/\partial x_j$ and $\partial z_b/\partial x_j$ are known, $\partial z_c/\partial x_j$ can be computed by applying the chain rule of

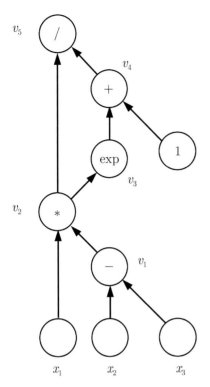

Automatic Differentiation: Introduction, History and Rounding Error Estimation, Figure 1
Computational graph

Automatic Differentiation: Introduction, History and Rounding Error Estimation, Table 1
Elementary partial derivatives

$z_c = \psi(z_a, z_b)$	$\dfrac{\partial \psi}{\partial z_a}$	$\dfrac{\partial \psi}{\partial z_b}$
$z_c = z_a \pm z_b$	1	± 1
$z_c = z_a \cdot z_b$	z_b	z_a
$z_c = z_a/z_b$	$1/z_b$	$-z_a/z_b^2 \ (= -z_c/z_b)$
$z_c = \sqrt{z_a}$	$\tfrac{1}{2}/\sqrt{z_a}(= \tfrac{1}{2}/z_c)$	–
$z_c = \log(z_a)$	$1/z_a$	–
$z_c = \exp(z_a)$	$\exp(z_a)(= z_c)$	–
$z_c = \cos(z_a)$	$-\sin(z_a)$	–
$z_c = \sin(z_a)$	$\cos(z_a)$	–
⋮	⋮	⋮

differentiation to (3):

$$\frac{\partial z_c}{\partial x_j} \leftarrow \frac{\partial \psi_k}{\partial z_a}\frac{\partial z_a}{\partial x_j} + \frac{\partial \psi_k}{\partial z_b}\frac{\partial z_b}{\partial x_j}. \qquad (4)$$

$\partial \psi_k/\partial z_a$ and $\partial \psi_k/\partial z_b$ are called *elementary partial derivatives*, and are computed by Table 1 for various ψ_k.

Introducing new variables $\bar{z}_1, \ldots, \bar{z}_s, \bar{x}_1, \ldots, \bar{x}_n$ corresponding to $\partial z_1/\partial x_j, \ldots, \partial z_s/\partial x_j, \partial x_1/\partial x_j, \ldots, \partial x_n/\partial x_j$, respectively, and initializing $\bar{x}_k \leftarrow 0 \ (1 \leq k \leq n, k \neq j)$ and $\bar{x}_j \leftarrow 1$, we may express (4) as

$$\bar{z}_c \leftarrow \frac{\partial \psi_k}{\partial z_a}\bar{z}_a + \frac{\partial \psi_k}{\partial z_b}\bar{z}_b. \qquad (5)$$

Thus, we can write down the whole program for the forward mode as shown in Program 5.

The reverse mode corresponds to tracing a computational process such as Program 4 backwards. The kth *computational step*, i. e., execution of the kth assignment in the program, can be written in general as

$$v_k = \psi_k(u_{k1}, u_{k2})|_{u_{k1}=v_{\alpha_k}, u_{k2}=v_{\beta_k}}, \qquad (6)$$

Initialization
$\bar{x}_j \leftarrow 1,$
$\bar{x} \leftarrow 0 \ (1 \leq k \leq n, k \neq j),$
Forward algorithm:

1	$z_1 \leftarrow x_2 - x_3,$
1'	$\bar{z}_1 \leftarrow 1 * \bar{x}_2 - 1 * \bar{x}_3,$
2'	$\bar{z}_1 \leftarrow z_1 * \bar{x}_1 + x_1 * \bar{z}_1,$
2	$z_1 \leftarrow x_1 * z_1,$
3	$z_2 \leftarrow \exp(z_1),$
3'	$\bar{z}_2 \leftarrow z_2 * \bar{z}_1,$
4	$z_2 \leftarrow z_2 + 1,$
4'	$\bar{z}_2 \leftarrow 1 * \bar{z}_2,$
5	$z_1 \leftarrow z_1/z_2,$
5'	$\bar{z}_1 \leftarrow (1/z_2) * \bar{z}_1 - (z_1/z_2) * \bar{z}_2$

Automatic Differentiation: Introduction, History and Rounding Error Estimation, Program 5
Forward mode program for differentation

where $u_{k,1}$ and $u_{k,2}$ are formal parameters, v_{α_k} and v_{β_k} are real parameters representing some of $x_1, \ldots, x_n, v_1, \ldots, v_{k-1}$. If ψ_k is unary, $u_{k,2}$ and v_{β_k} are omitted. Let r be the total number of computational steps. In Program 4, we have $r = 5$ and, for $k = 2$, e. g., $\psi_2 = $ '$*$', $v_{\alpha_2} = x_1$ and $v_{\beta_2} = v_1$.

The total differentiation of (6) yields the relations among $dx_1, \ldots, dx_n, dv_1, \ldots, dv_r$ such as follows:

$$dv_k = \frac{\partial \psi_k}{\partial u_{k,1}} dv_{\alpha_k} + \frac{\partial \psi_k}{\partial u_{k,2}} dv_{\beta_k} \quad (k = 1, \ldots, r). \qquad (7)$$

The computation of the partial derivatives of the ith component of the final result $y_i = f_i(x_1, \ldots, x_n)$ in (1) with respect to $x_1, \ldots, x_n$ is that of the coefficients of the relation among $dx_1, \ldots, dx_n$ and dy_i.

Here, new variables $\bar{x}_1, \ldots, \bar{x}_n, \bar{v}_1, \ldots, \bar{v}_r$ are introduced for the computation of those coefficients. Without loss of generality, we may assume that the value of y_i is computed at v_r. After Program 4 is executed in the natural order with all the information on intermediate results preserved, these new variables are initialized as $\bar{x}_j \leftarrow 0$ $(j = 1, \ldots, n)$, $\bar{v}_k \leftarrow 0$ $(k = 1, \ldots, r-1)$ and $\bar{v}_r \leftarrow 1$, then the relation

$$dy = \sum_{j=1}^{n} \bar{x}_j dx_j + \sum_{k=1}^{r} \bar{v}_k dv_k \quad (8)$$

holds. Secondly, $dv_r, dv_{r-1}, \ldots, dv_k$ can be eliminated from (8) in this order by modifying

$$\bar{v}_{\alpha_k} \leftarrow \bar{v}_{\alpha_k} + \bar{v}_k \frac{\partial \psi_k}{\partial v_{\alpha_k}}, \quad (9)$$

$$\bar{v}_{\beta_k} \leftarrow \bar{v}_{\beta_k} + \bar{v}_k \frac{\partial \psi_k}{\partial v_{\beta_k}}. \quad (10)$$

Finally, if we change k in the reverse order, i. e., $k = r$, $r-1, \ldots, 1$, we can successfully eliminate all the dv_k ($k = 1, \ldots, r$) to have

$$dy = \sum_{j=1}^{n} \bar{x}_j dx_j. \quad (11)$$

The final coefficient $\bar{x}_j$ indicates the value of $\partial f_i / \partial x_j$ ($j = 1, \ldots, n$). Program 6 in which modifications (9) and (10) are embedded is the reverse mode program, which is sometimes called the *adjoint program* of Program 4.

It is easy to extend the algorithms for computing a linear combination of the column vectors of the Jacobian matrix J with the forward mode, and a linear combination of the row vectors of J with the reverse mode.

Complexity

It is proved that, for a constant C ($= 4 \sim 6$, varying under different computational models), the total operation count for $\partial y_i / \partial x_j$'s with a fixed j in the forward mode algorithm, as well as that for $\partial y_i / \partial x_j$'s with a fixed i in the reverse mode algorithm, is at most $C \cdot r$, i. e., in $O(r)$. Roughly speaking, r is proportional to the execution time T of the given program, so that the time complexity is in $O(T)$. Furthermore, we have to repeat such computation n times to get all the required partial

Forward sweep:
(insert Program 4 here)
Initialization: ($n = 3, r = 5$)
$\bar{x}_j \leftarrow 0$ $(j = 1, \ldots, n)$,
$\bar{v}_k \leftarrow 0$ $(k = 1, \ldots, r-1)$,
$\bar{v}_r \leftarrow 1$,
Reverse elimination:
5″ $\quad \bar{v}_2 \leftarrow \bar{v}_2 + (1/v_4) * \bar{v}_5$,
$\quad \bar{v}_4 \leftarrow \bar{v}_4 + (-v_5/v_4) * \bar{v}_5$,
4″ $\quad \bar{v}_3 \leftarrow \bar{v}_3 + 1 * \bar{v}_4$,
3″ $\quad \bar{v}_2 \leftarrow \bar{v}_2 + v_3 * \bar{v}_3$,
2″ $\quad \bar{x}_1 \leftarrow \bar{x}_1 + v_1 * \bar{v}_2$,
$\quad \bar{v}_1 \leftarrow \bar{v}_1 + x_1 * \bar{v}_2$,
1″ $\quad \bar{x}_2 \leftarrow \bar{x}_2 + 1 * \bar{v}_1$,
$\quad \bar{x}_3 \leftarrow \bar{x}_3 + (-1) * \bar{v}_1$.

Automatic Differentiation: Introduction, History and Rounding Error Estimation, Program 6
Reverse mode program

derivatives by the forward mode, and m times by the reverse mode. What should be noted here is that the computational time of the forward or reverse mode algorithm for one set of derivatives does not depend on m or n but only on r.

Denoting the spatial complexity of the original program by S, that of the forward mode algorithm is in $O(S)$. However, the spatial complexity of the reverse mode is in $O(T)$, since the reverse mode requires a history of the forward sweep recorded in storage whose size is in $O(T)$.

A rough sketch of the proof is as follows. Without loss of generality, assume that the given program is expanded into a sequence of single assignment statements with a binary or unary basic operation as shown in Program 3 and 4. The operation count for computing the elementary partial derivatives (Table 1) is bounded by a constant. The additional operation count for modifying $\bar{v}_k$'s and $\bar{x}_j$'s in (5), (9) and (10) is also bounded since there are at most two additions and two multiplications. There are r operations in the original program, so that the total operation count in the forward mode algorithm as well as that in the reverse mode algorithm is in $O(r)$.

Note that the computational complexities of the forward mode and the reverse mode may not be optimal, but at least one can compute them in time proportional to that for the computation of the given original program.

One can extend the AD algorithms to compute higher derivatives. In particular, it is well known how to compute a truncated Taylor series to get arbitrarily higher-order derivatives of a function with one variable [14]. One may regard a special function such as a Bessel function or a block of several arithmetic operations, such as the inner product of vectors, as a basic operation if the corresponding elementary partial derivatives are given with computational definitions. An analogy is pointed out in [7] between the algorithms for the partial derivatives and those of the computation of the shortest paths in an acyclic graph.

It has also been pointed out that there may be pitfalls in the derived program with AD. For example, a tricky program

  IF (x.ne.1.0)
  THEN $y = x*x$
  ELSE $y = 1.0 + (x - 1.0) * b$
  ENDIF

can compute the value of a function $f(x) = x^2$ correctly for all x. However, the derived program fails to compute $f'(1.0)$, because the differentiation of the second assignment with respect to x is not 2.0 but b. Thus conditional branches (or equations equivalent to conditional branches) should be carefully dealt with.

History

A brief history of AD is as follows. There were not a few researchers in the world who had more or less independently proposed essentially the same algorithms.

The first publication on the forward mode algorithm was presumably the paper by R.E. Wengert in 1964 [16]. After 15 years, books were published by L.B. Rall [14] and by H. Kagiwada et al. [9] which have been influential on the numerical-computational circle. The practical and famous software system for the forward mode automatic differentiation was Pascal-SC, and its descendants Pascal-XSC and C-XSC are popular now.

The paper [13] might be the first to propose systematically the reverse mode algorithm. But there are many ways through which to approach the reverse mode algorithm. In fact, it is related to Lagrange multipliers, error analysis, generation of adjoint systems, reduction of computational complexity of computing the gradient, neural networks, etc. Of course, the principles of the derived algorithms are the same. Some remarkable works on the reverse mode algorithm had been done by S. Linnainmaa [11] and W. Miller and C. Wrathall [12] from the viewpoint of the error analysis, by W. Baur and V. Strassen [1] from that of complexity, and by P.J. Werbos [17] from that of the optimization of neural networks. A practical program had been developed by B. Speelpenning in 1980 [15] and it was rewritten into Fortran by K.E. Hillstrom in 1985 (now registered in Netlib [5,6]).

Two proceedings of the international workshops held in 1991 and 1996 collect all the theories, techniques, practical programs, current works, and future problems as well as history on automatic differentiation [2,4]. It should be noted that, in 1992, A. Griewank proposed a drastic improvement of the reverse mode algorithm using the so-called checkpointing technique. He succeeded in reducing the order of the size of storage required for the reverse mode algorithm [3]. Several software tools for automatic differentiation have been developed and popular in the world, e. g., ADIC, ADIFOR, ADMIT-1, ADOL-C, ADOL-F, FADBAD, GRESS, Odyssée, PADRE2, TAMC, etc. (See [2,4].)

Estimates of Rounding Errors

In order to solve practical real-world problems, the approximation with floating-point numbers is inevitable so that it is important to analyze and estimate the accumulated rounding errors in a big numerical computation. Moreover, in terms of estimates of the accumulated rounding errors, one can define a normalized (or weighted) norm for a numerically computed vector, that is useful for checking whether the computed vector can be regarded as zero or not from the viewpoint of numerical computation [8].

For the previous example, let us denote as δ_k the rounding error generated at the execution of the basic operation to compute the value of v_k. Then, the rounding errors in the example is explicitly written:

1	$\widetilde{v}_1 = \widetilde{x}_2 - \widetilde{x}_3 + \delta_1,$
2	$\widetilde{v}_2 = \widetilde{x}_1 * \widetilde{v}_1 + \delta_2,$
3	$\widetilde{v}_3 = \exp(\widetilde{v}_2) + \delta_3,$
4	$\widetilde{v}_4 = \widetilde{v}_3 + 1 + \delta_4,$
5	$\widetilde{v}_5 = \widetilde{v}_2/\widetilde{v}_4 + \delta_5.$

Here, $\widetilde{v}_k$ is the value with accumulated rounding errors.

Defining a function $\widetilde{f}$ as

$$\widetilde{f}(x_1, x_2, x_3; \delta_1, \delta_2, \delta_3, \delta_4, \delta_5)$$
$$= \frac{x_1(x_2 - x_3 + \delta_1) + \delta_2}{\exp(x_1(x_2 - x_3 + \delta_1) + \delta_2) + \delta_3 + 1 + \delta_4} + \delta_5,$$

one has

$$\widetilde{v}_5 = \widetilde{f}(x_1, x_2, x_3; \delta_1, \ldots, \delta_5),$$
$$v_5 = \widetilde{f}(x_1, x_2, x_3; 0, \ldots, 0).$$

Here, $\widetilde{v}_5 - v_5$ is the accumulated rounding error in the function value. For $v_5 = v_2/v_4 = \varphi_5(v_2, v_4)$, one has

$$\widetilde{v}_5 - v_5 = \varphi_5(\widetilde{v}_2, \widetilde{v}_4) - \varphi_5(v_2, v_4) + \delta_5$$
$$= \frac{\partial \varphi_5}{\partial v_2}(\xi_2, \xi_4) \cdot (\widetilde{v}_2 - v_2)$$
$$+ \frac{\partial \varphi_5}{\partial v_4}(\xi_2, \xi_4) \cdot (\widetilde{v}_4 - v_4) + \delta_5,$$

where $\xi_2 = \theta' \widetilde{v}_2 + (1-\theta')v_2$ and $\xi_4 = \theta'' \widetilde{v}_4 + (1-\theta'')v_4$ for $0 < \theta', \theta'' < 1$. Expanding $\widetilde{v}_2 - v_2$ and $\widetilde{v}_4 - v_4$ similarly and expanding the other intermediate variables sequentially, the approximation:

$$\widetilde{v}_5 - v_5 \simeq \sum_{k=1}^{5} \frac{\partial \widetilde{f}}{\partial \delta_k} \delta_k \quad (12)$$

is derived [10]. Note that $\frac{\partial \widetilde{f}}{\partial \delta_k}$ are computed as $\overline{v}_k$ in Program 6, which are the final results of (9) and (10).

The locally generated rounding error δ_k for the floating-point number system is bounded by

$$|\delta_k| \leq c \cdot |v_k| \cdot \varepsilon_M, \quad (13)$$

where ε_M indicates so-called 'machine epsilon' and $c = 1$ may be adopted for arithmetic operations according to IEEE754 standard. Then $\Delta[f]_A$, called *absolute estimation*, is defined by

$$\Delta[f]_A \equiv \sum_{k=1}^{r} \left| \frac{\partial \widetilde{f}}{\partial \delta_k} \right| \cdot |v_k| \cdot \varepsilon_M, \quad (14)$$

which is an upper bound on the accumulated rounding error. Regarding the locally generated errors δ_k's as pseudo-probabilistic variables uniformly distributed over $[-|v_k|\varepsilon_M, |v_k|\varepsilon_M]$'s, $\Delta[f]_P$, called *probabilistic estimate*, is defined by

$$\Delta[f]_P \equiv \varepsilon_M \sqrt{\frac{1}{3} \sum_{k=1}^{r} \left(\frac{\partial \widetilde{f}}{\partial \delta_k} \cdot v_k \right)^2}. \quad (15)$$

There are several reports in which these estimates give quite good approximations to the actual accumulated rounding errors [8].

Moreover, one could answer the problem how to choose a norm for measuring the size of numerically computed vector. By means of the estimates of the rounding errors, a weighted norm of a vector $\mathbf{f} = [f_1, \ldots, f_m]$ whose components are numerically computed is defined by

$$\|\mathbf{f}\|_N \equiv \left\| \left[\frac{f_1}{\Delta[f_1]_A}, \ldots, \frac{f_m}{\Delta[f_m]_A} \right] \right\|_p, \quad (16)$$

($p = 1, 2$ or ∞). This weighted norm is called *normalized norm*, because it is normalized with respect to accumulated rounding errors. With this normalized norm, one can determine whether a computed vector approaches to zero or not in reference to the rounding errors accumulated in the components. Note that, since all the components of the vector are divided by the estimates of accumulated rounding errors, they have no physical dimension. The normalized norm may be used effectively as stopping criteria for iterative methods like the Newton–Raphson method.

See also

- Automatic Differentiation: Calculation of the Hessian
- Automatic Differentiation: Calculation of Newton Steps
- Automatic Differentiation: Geometry of Satellites and Tracking Stations
- Automatic Differentiation: Parallel Computation
- Automatic Differentiation: Point and Interval
- Automatic Differentiation: Point and Interval Taylor Operators

▶ Automatic Differentiation: Root Problem and Branch Problem
▶ Nonlocal Sensitivity Analysis with Automatic Differentiation

References

1. Baur W, Strassen V (1983) The complexity of partial derivatives. Theor Comput Sci 22:317–330
2. Berz M, Bischof C, Corliss G, Griewank A (eds) (1996) Computational differentiation: Techniques, applications, and tools. SIAM, Philadelphia
3. Griewank A (1992) Achieving logarithmic growth of temporal and spatial complexity in reverse automatic differentiation. Optim Methods Soft 1:35–54
4. Griewank A, Corliss GF (eds) (1991) Automatic differentiation of algorithms: Theory, implementation, and application. SIAM, Philadelphia
5. Hillstrom KE (1985) Installation guide for JAKEF. Techn Memorandum Math and Computer Sci Div Argonne Nat Lab ANL/MCS-TM-17
6. Hillstrom KE (1985) User guide for JAKEF. Techn Memorandum Math and Computer Sci Div Argonne Nat Lab ANL/MCS-TM-16
7. Iri M (1984) Simultaneous computation of functions, partial derivatives and estimates of rounding errors – Complexity and practicality. Japan J Appl Math 1:223–252
8. Iri M, Tsuchiya T, Hoshi M (1988) Automatic computation of partial derivatives and rounding error estimates with applications to large-scale systems of nonlinear equations. J Comput Appl Math 24:365–392
9. Kagiwada H, Kalaba R, Rasakhoo N, Spingarn K (1986) Numerical derivatives and nonlinear analysis. Math. Concepts and Methods in Sci. and Engin., vol 31. Plenum, New York
10. Kubota K, Iri M (1991) Estimates of rounding errors with fast automatic differentiation and interval analysis. J Inform Process 14:508–515
11. Linnainmaa S (1976) Taylor expansion of the accumulated rounding error. BIT 16:146–160
12. Miller W, Wrathall C (1980) Software for roundoff analysis of matrix algorithms. Acad Press, New York
13. Ostrovskii GM, Wolin JM, Borisov WW (1971) Über die Berechnung von Ableitungen. Wiss Z Techn Hochschule Chemie 13:382–384
14. Rall LB (1981) Automatic differentiation – Techniques and applications. Lecture Notes Computer Science, vol 120. Springer, Berlin
15. Speelpenning B (1980) Compiling fast partial derivatives of functions given by algorithms. Report Dept Computer Sci Univ Illinois UIUCDCS-R-80-1002
16. Wengert RE (1964) A simple automatic derivative evaluation program. Comm ACM 7:463–464
17. Werbos P (1974) Beyond regression: New tools for prediction and analysis in the behavioral sciences. PhD Thesis, Appl. Math. Harvard University

Automatic Differentiation: Parallel Computation

CHRISTIAN H. BISCHOF[1], PAUL D. HOVLAND[2]
[1] Institute Sci. Computing, University Technol., Flachen, Germany
[2] Math. and Computer Sci. Div., Argonne National Lab., Argonne, USA

MSC2000: 65Y05, 68N20, 49-04

Article Outline

Keywords
Background
Implementation Approaches
AD of Parallel Programs
AD-Enabled Parallelism
  Data Parallelism
  Time Parallelism
Parallel AD Tools
Summary
See also
References

Keywords

Automatic differentiation; Parallel computing; MPI

Research in the field of *automatic differentiation* (AD) has blossomed since A. Griewank's paper [15] in 1989 and the Breckenridge conference [17] in 1991. During that same period, the power and availability of parallel machines have increased dramatically. A natural consequence of these developments has been research on the interplay between AD and *parallel computations*. This relationship can take one of two forms. One can examine how AD can be applied to existing parallel programs. Alternatively, one can consider how AD introduces new potential for parallelism into existing sequential programs.

Background

Automatic differentiation relies upon the fact that all programming languages are based on a finite number of elementary functions. By providing rules for the differentiation of these elementary functions, and by combining these elementary derivatives according to the chain rule of differential calculus, an AD system can differentiate arbitrarily complex functions. The chain rule is associative—partial derivatives can be combined in any order. The *forward mode of AD* combines the partial derivatives in the order of evaluation of the elementary functions to which they correspond. The *reverse mode* combines them in the reverse order. For systems with a large ratio of dependent to independent variables, the reverse mode offers lower operation counts, at the cost of increased storage costs [15].

The forward and the reverse mode are the extreme ends of a wide algorithmic spectrum of accumulating derivatives. Recently, hybrid approaches have been developed which combine the forward and the reverse mode [5,10], or apply them in a hierarchical fashion [8,25]. In addition, efficient *checkpointing* schemes have been developed which address the potential storage explosion of the reverse mode by judicious recomputation of intermediate states [16,19]. Viewing the problem of automatic differentiation as an edge elimination problem on the program graph corresponding to a particular code, one can in fact show that the problem of computing derivatives with minimum cost is *NP*-hard [21]. The development of more efficient heuristics is an area of active research (see, for example, several of the papers in [3]).

Implementation Approaches

Automatic differentiation is a particular instantiation of a rule-based semantic transformation process. That is, whenever a floating-point variable changes, an associated derivative object must be updated according to the chain rule of differential calculus. For example, in the forward mode of AD, a derivative object carries the partial derivative(s) of an associated variable with respect to the independent variable(s). In the reverse mode of AD, a derivative object carries the partial derivative(s) of the dependent variable(s) with respect to an associated variable. Thus, any AD tool must provide an instantiation of a 'derivative object', maintain the association between an original variable and its derivative object, and update derivative objects in a timely fashion.

Typically AD is implemented in one of two ways: operator overloading or source transformation. In languages that allow operator overloading, such as C++ and Fortran90, each elementary function can be redefined so that in addition to the normal function, derivatives are computed as well, and either saved for later use or propagated by the chain rule. A simple class definition using the forward mode might be implemented as follows:

```
class adouble{
private:
  double value, grad[GRAD_LENGTH];
public:
  /* constructors omitted */
  friend adouble operator*(const
    adouble &, const adouble &);
  /* similar decs for other ops */
}
adouble operator*(const adouble &g1,
  const adouble &g2){
    int i;
    double newgrad[GRAD_LENGTH];
    for (i=0; i<GRAD_LENGTH;i++){
      newgrad[i] =
        (g1.value)*(g2.grad[i])+
        (g2.value)*(g1.grad[i]);
    }
    return adouble(g1.value*g2.value,
      newgrad);
}
```

An example of how this class could be used is given below.

In languages that do not support operator overloading, it can be faked by manually or automatically replacing operators such as + and * with calls to subroutines.

```
main(){
  double temp[GRAD_LENGTH];
  adouble y;

  /* initialize x1 to (3.0,[1.0  0.0]),
             x2 to (4.0,[0.0 1.0])*/
  temp[0] = 1.0; temp[1] = 0.0;
  adouble  *x1 = new adouble(3.0,temp);
  temp[0] = 0.0; temp[1] = 1.0;
  adouble  *x2 = new adouble(4.0,temp);

  y = (*x1)*(*x2);

  /* output (y,[dy/dx1 dy/dx2])  */
  cout << y;
  /* prints (12.0,[4.0 3.0]) */
}
```

As an alternative to operator overloading, a preprocessor can be used to transform source code for computing the function into source code for computing the function and its derivatives. This approach relies heavily on compiler technology and typically involves a combination of in-lining and subroutine calls to implement the propagation of derivatives. An example of ADIFOR-generated code (edited for clarity) invoking the SparsLinC library for transparent exploitation of *sparsity* [6] follows.

```
c derivation code for f=x*y/z

c preaccumulate partial derivates
      temp1 = x*y/z
      temp2 = 1.0/z
      temp3 = temp2*y
      temp4 = temp2*x
      temp5 = -temp1/z

c propagate derivatives
c (g_x, ... , g_f may by sparse)
      call sspg3q(g_f,temp3,g_x,temp4,
     +            g_y,temp5,g_z)

      f=temp1
```

The advantage of this approach is that it allows the exploitation of computational context in deciding how to propagate derivatives. For example, a recently developed *Hessian* module [1], adaptively determines the best strategy for each assignment statement in the code based on a machine-specific performance model for the implementation kernels employed.

A comparison of these two implementation approaches is provided in [9]. This paper also introduces an implementation design that separates the core issues of automatic differentiation from language-specific issues through the use of an interface layer called AIF (*AD intermediate form*), thus arriving at a system design that allows reuse of differentiation components across front-ends for different languages. Long-term, such a system design also allows the exploitation of the best features of both source transformation and operator overloading.

Current AD tools based on operator overloading include ADOL-C [18] and ADOL-F [29], both of which offer the option of using either the forward or the reverse mode, and to compute derivatives of arbitrary order Source transformation tools that use mostly the forward mode to provide first- and second order derivatives include ADIC [9] and ADIFOR [6]. The Odyssee [28] and TAMC [14] tools use the reverse mode in a source transformation context to provide first order derivatives. A more comprehensive survey of AD tools can be found at the website [31].

AD of Parallel Programs

In 1994, R.L. Hinkins reported on the application of AD to magnetic field calculations implemented in the data parallel languages MPFortran (MasPar Fortran) and CMFortran [22]. In 1997, P. Hovland addressed the larger issue of AD of parallel programs in general, paying close attention to message-passing parallel programs [23], but also considering other parallel programming paradigms, and A. Carle developed ADIFOR-MP, a prototype tool supporting a subset of MPI [30] and PVM [13] constructs. The focus on parallel programs employing a message-passing paradigm can be attributed to the popularity of this parallel programming paradigm and its relevance to all parallel programs targeting nonuniform memory access (NUMA) machines.

Correct AD of message-passing parallel programs requires that we maintain an association between a variable and its derivative object. In particular, when a variable is sent from one processor to another via a message, we must also send the associated derivative object. There are two ways of accomplishing this goal — we can pack the variable and derivative object together in one message or send two separate messages. Packing a variable and its associated derivative object into a single message may incur a copying overhead. On the other hand, sending separate messages requires a mechanism for associating the messages with one another at the receiving end and will increase delivery time on high-latency systems. In general, it is preferable to pack the variable and derivative object together in one message [24], minimizing copying cost through judiciously chosen derivative data structures. Other issues in ensuring correct AD of parallel programs include proper handling of nondeterminism, reduction operations at points of nondifferentiability, and seed matrix initialization [23].

In many instances, only a subset of the program input- and output variables is considered as independent or dependent variables with respect to differentiation. An optimization technique that tries to exploit this fact is activity analysis, which seeks to reduce time and storage costs by identifying variables that do not lie on the computational path from independent to dependent variables. Such variables are termed passive and do not require an associated derivative object. Activity analysis depends on sophisticated compiler technology, namely interprocedural dataflow analysis. In message-passing parallel programs, sends and receives greatly complicate such an analysis. As the analysis needs to guarantee correctness, this fact leads to much more conservative assumptions, and as a result much optimization potential may be lost. Among the available options to circumvent this situation are user annotations, runtime analysis, or the use of a higher-level language such as HPF [26]. These issues are investigated in more detail in [23].

Another issue arising in the parallel setting is the computation of partial derivatives of new elementary functions, such as parallel *reduction operations*. For most of the common reduction operations, such as sum, maximum, and minimum, computing the partial derivatives is trivial. For the product reduction, the situation is more complex. The partial derivative of $y = \prod_{i=1}^{n} x_i$ with respect to x_i is $\partial y / \partial x_i = (\prod_{j=1}^{i-1} x_j)(\prod_{k=i+1}^{n} x_k)$. These partial derivatives can be computed using a parallel prefix and reverse parallel prefix operation. However, propagating the partial derivatives requires an additional sum reduction. We could instead combine the partial derivative computation and propagation into a single reduction. This increases the computational cost, but reduces the communication cost. In [24], Hovland and C. Bischof discuss the conditions under which each approach should be preferred and give experimental results to support the theory.

AD-Enabled Parallelism

As early as 1991, Bischof considered the problem of parallelizing the computation of derivatives computed via AD [4] to distribute the additional work introduced by AD. Applying AD to a program introduces two basic types of parallelism: data parallelism and time parallelism.

Data Parallelism

The potential for data parallelism arises whenever there are multiple independent variables (for the forward mode) or multiple dependent variables (for the reverse mode). Different processes can be employed to propagate partial derivatives with respect to a subset of the independent variables in parallel.

Such an implementation is feasible if one can employ light-weight threads for the parallel derivative computation. A limiting factor is the fact that the derivative computations are interspersed with the function computation. Thus, an alternative approach is to replicate the sequential computation on each processor, thereby virtually eliminating communication costs. This approach has proven effective for computations involving a large number of independent variables [7,32].

Time Parallelism

Time parallelism arises as a consequence of the associativity of the chain rule. By breaking the computation into several phases, we can compute and propagate partial derivatives over each phase simultaneously, then combine the results according to the chain rule. This approach is illustrated in Fig. 1. Before each phase, a derivative computation for that phase is forked off, using as input the results of the previous phase. At the conclusion of the derivative computations, the partial derivatives are combined according to the chain rule.

This illustration assumes the forward mode. If we were using the reverse mode, the derivative computation for phase A would be forked off after phase A had completed. The effectiveness of this approach has been demonstrated for both the forward mode [10] and the reverse mode [2]. The associativity of the chain rule makes it possible to apply this time-parallel approach to arbitrary computational structures, not just the linear schedule illustrated here.

Parallel AD Tools

Research in AD and parallelism is relatively new. Nonetheless, there are several such tools, at varying stages of development.

Hinkins developed special purpose libraries for the AD of programs written in MPFortran or CM-

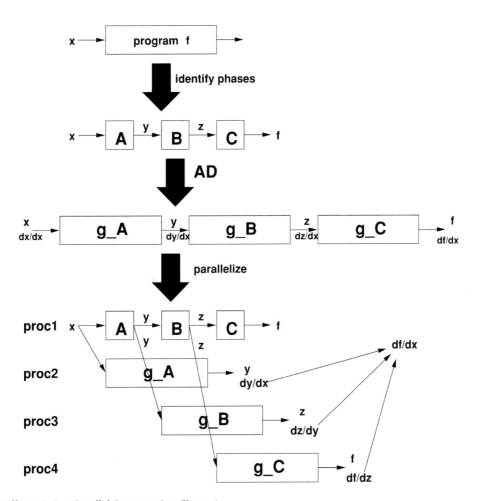

Automatic Differentiation: Parallel Computation, Figure 1

Fortran [22]. Use of these libraries required that each arithmetic operation be manually replaced by a subroutine call. As part of his thesis [23], Hovland developed prototype tools for AD of FortranM [12], Fortran with a subset of MPI [20] message passing, and C with MPI. Carle is developing (1999) a prototype version of ADIFOR [11] supporting MPI and PVM. Roh is developing an extension to ADIC that seeks to automatically exploit the parallelism introduced by AD through the use of threads [27].

Summary

Since 1989, a great deal of progress has been made in the fields of automatic differentiation and parallel computation. Parallel computation and AD interact in two ways. AD can be applied to a parallel program. Alternatively, AD can be used as a source of new parallelism in a computation. Effective strategies exist for exploiting each of the two types of parallelism introduced: time parallelism and data parallelism.

In either case, ensuring that the resulting derivative computation is both correct and efficient requires AD tools that are more sophisticated than in the serial setting. Most of the existing tools are early in their development cycle, but can be expected to mature swiftly as they adopt advanced computational infrastructure developed in other fields of computer science, e.g., parallelizing compilers or parallel runtime systems. Thus, we expect the beginning of 2000 to also provide robust and effective tools for the differentiation of parallel programs and the introduction of parallelism through differentiation.

See also

- ▶ Asynchronous Distributed Optimization Algorithms
- ▶ Automatic Differentiation: Calculation of the Hessian
- ▶ Automatic Differentiation: Calculation of Newton Steps
- ▶ Automatic Differentiation: Geometry of Satellites and Tracking Stations
- ▶ Automatic Differentiation: Introduction, History and Rounding Error Estimation
- ▶ Automatic Differentiation: Point and Interval
- ▶ Automatic Differentiation: Point and Interval Taylor Operators
- ▶ Automatic Differentiation: Root Problem and Branch Problem
- ▶ Heuristic Search
- ▶ Interval Analysis: Parallel Methods for Global Optimization
- ▶ Load Balancing for Parallel Optimization Techniques
- ▶ Nonlocal Sensitivity Analysis with Automatic Differentiation
- ▶ Parallel Computing: Complexity Classes
- ▶ Parallel Computing: Models
- ▶ Parallel Heuristic Search
- ▶ Stochastic Network Problems: Massively Parallel Solution

References

1. Abate J, Bischof Ch, Carle A, Roh L (1997) Algorithms and design for a second-order automatic differentiation module. Proc. Internat. Symp. Symbolic and Algebraic Computing (ISSAC) '97, ACM, New York, pp 149–155
2. Benary J (1996) Parallelism in the reverse mode. In: Berz M, Bischof Ch, Corliss G, Griewank A (eds) Computational Differentiation: Techniques, Applications, and Tools. SIAM, Philadelphia, pp 137–147
3. Berz M, Bischof Ch, Corliss G, Griewank A (1996) Computational differentiation: Techniques, applications, and tools. SIAM, Philadelphia
4. Bischof ChH (1991) Issues in parallel automatic differentiation. In: Griewank A, Corliss G (eds) Automatic Differentiation of Algorithms. SIAM, Philadelphia, pp 100–113
5. Bischof Ch, Carle A, Corliss G, Griewank A, Hovland P (1992) ADIFOR: Generating derivative codes from Fortran programs. Scientif Program 1(1):11–29
6. Bischof Ch, Carle A, Khademi P, Mauer A (1996) ADIFOR 2.0: Automatic differentiation of Fortran 77 programs. IEEE Comput Sci Eng 3(3):18–32
7. Bischof Ch, Green L, Haigler K, Knauff T (1994) Parallel calculation of sensitivity derivatives for aircraft design using automatic differentiation. Proc. 5th AIAA/NASA/USAF/ISSMO Symp. Multidisciplinary Analysis and Optimization, AIAA-94-4261, Amer Inst Aeronautics and Astronautics, Reston, VA, pp 73–84
8. Bischof ChH, Haghighat MR (1996) On hierarchical differentiation. In: Berz M, Bischof Ch, Corliss G, Griewank A (eds) Computational Differentiation: Techniques, Applications, and Tools. SIAM, Philadelphia, pp 83–94
9. Bischof Ch, Roh L, Mauer A (1997) ADIC – An extensible automatic differentiation tool for ANSI-C. Software Practice and Experience 27(12):1427–1456
10. Bischof Ch, Wu Po-Ting (1997) Time-parallel computation of pseudo-adjoints for a leapfrog scheme. Preprint Math and Computer Sci Div Argonne Nat Lab no. ANL/MCS-P639-0197
11. Carle A (1997) ADIFOR-MP – A prototype automatic differentiation tool for Fortran 77 with message-passing extensions. Personal communication
12. Foster IT, Chandy KM (1995) Fortran M: A language for modular parallel programming. J Parallel Distributed Comput 25(1)
13. Geist A, Beguelin A, Dongarra J, Jiang W, Manchek R, Sunderam V (1994) PVM – Parallel virtual machine: A users' guide and tutorial for network parallel computing. MIT, Cambridge, MA
14. Giering R, Kaminski Th (1996) Recipes for adjoint code construction. Max-Planck Inst Meteorologie, Hamburg no. 212
15. Griewank A (1989) On automatic differentiation. In: Iri M, Tanabe K (eds) Mathematical Programming: Recent Developments and Applications. Kluwer, Dordrecht, pp 83–108
16. Griewank A (1992) Achieving logarithmic growth of temporal and spatial complexity in reverse automatic differentiation. Optim Methods Softw 1(1):35–54
17. Griewank A, Corliss G (1991) Automatic differentiation of algorithms. SIAM, Philadelphia
18. Griewank A, Juedes D, Utke J (1996) ADOL-C, a package for the automatic differentiation of algorithms written in C/C++. ACM Trans Math Softw 22(2):131–167
19. Grimm J, Pottier L, Rostaing-Schmidt N (1996) Optimal time and minimum space time product for reversing a certain class of programs. In: Berz M, Bischof Ch, Corliss G, Griewank A (eds) Computational Differentiation, Techniques, Applications, and Tools. SIAM, Philadelphia, pp 95–106

20. Gropp W, Lusk E, Skjellum A (1994) Using MPI – Portable parallel programming with the message passing interface. MIT, Cambridge, MA
21. Herley K (1993) On the NP-completeness of optimum accumulation by vertex elimination. Unpublished Manuscript
22. Hinkins R L (Sept. 1994) Parallel computation of automatic differentiation applied to magnetic field calculations. MSc Thesis Univ Calif
23. Hovland P (1997) Automatic differentiation of parallel programs. PhD Thesis Univ. Illinois at Urbana-Champaign
24. Hovland P, Bischof Ch (1998) Automatic differentiation of message-passing parallel programs. Proc. First Merged Internat. Parallel Processing Symp. and Symp. on Parallel and Distributed Processing, IEEE Computer Soc Press, New York
25. Hovland P, Bischof Ch, Spiegelman D, Casella M (1997) Efficient derivative codes through automatic differentiation and interface contraction: An application in biostatistics. SIAM J Sci Comput 18(4):1056–1066
26. Koelbel C, Loveman D, Schreiber R, Steele G Jr, Zosel M (1994) The high performance Fortran handbook. MIT, Cambridge, MA
27. Roh L (1997) Personal Communication
28. Rostaing N, Dalmas St, Galligo A (Oct. 1993) Automatic differentiation in Odyssee. Tellus 45a(5):558–568
29. Shiriaev D, Griewank A (1996) ADOL-F: Automatic differentiation of Fortran codes. In: Berz M, Bischof Ch, Corliss G, Griewank A (eds) Computational Differentiation: Techniques, Applications, and Tools. SIAM, Philadelphia, pp 375–384
30. Snir M, Otto SW, Huss-Lederman S, Walker DW, Dongarra Jack (1996) MPI: The complete reference. MIT, Cambridge, MA
31. WEB http://www.mcs.anl.gov/autodiff/adtools/
32. Zhang Y, Bischof Ch, Easter R, Wu Po-Ting (1997) Sensitivity analysis of O3 and photochemical indicators using a mixed-phase chemistry box model and automatic differentiation techniques. 90th Air and Waste Management Assoc. Annual Meeting and Exhibition June 8-13, 1997, Toronto. vol 97-WA68A.04, Air and Waste Management Assoc, Pittsburgh, PA, pp 1–16

Automatic Differentiation: Point and Interval

AD

L. B. Rall[1], George F. Corliss[2]
[1] University Wisconsin–Madison, Madison, USA
[2] Marquette University, Milwaukee, USA

MSC2000: 65H99, 65K99

Article Outline

Keywords
See also
References

Keywords

Differentiation; Computational methods

Automatic differentiation (abbreviated AD) is a computational method for evaluating derivatives or Taylor coefficients of algorithmically defined functions. Simply speaking, an *algorithmic definition* of a function is a step-by-step specification of its evaluation by arithmetic operations and library functions. Application of the rules of differentiation to the algorithmic definition of a differentiable function yields values of its derivatives. Examples of algorithmic definitions of functions are code lists, computer subroutines, and even entire computer programs.

Automatic differentiation differs from numerical differentiation based on difference quotients of function values in that automatic differentiation is exact in principle, but of course is subject to roundoff error in practice. In addition to roundoff error, difference quotients entail truncation error. Attempts to reduce this truncation error by decreasing stepsize results in cancellation of significant digits and a catastrophic increase in roundoff error in general. Automatic differentiation also differs significantly from the symbolic differentiation taught in school, the goal of which is the transformation of formulas for functions into formulas for their derivatives. Although automatic differentiation uses the same rules of differentiation as symbolic differentiation, these rules are applied to the algorithmic definition of the function, not to a formula for it, and the results are values of derivatives, not formulas. Furthermore, formulas may not be available for functions of interest defined only algorithmically by computer subroutines or programs to which automatic differentiation can be applied. In summary, automatic differentiation is more accurate than numerical differentiation and requires fewer resources and is more generally applicable than symbolic differentiation.

The simplest type of algorithmic definition of a function is a *code list*, which is similar to the segment of computer code for the evaluation of an expression

(i. e., a formula). For illustration, consider the function defined by the formula

$$f(x, y) = (xy + \sin x + 4)(3y^2 + 6).$$

An equivalent algorithmic definition of this function by a code list is

$$\begin{aligned}
t_1 &= x, & t_6 &= t_5 + 4, \\
t_2 &= y, & t_7 &= t_2^2, \\
t_3 &= t_1 t_2, & t_8 &= 3t_7, \\
t_4 &= \sin t_1, & t_9 &= t_8 + 6, \\
t_5 &= t_3 + t_4, & t_{10} &= t_6 t_9.
\end{aligned}$$

Given the values of x and y, evaluation of the subsequent entries in the code list gives $t_{10} = f(x, y)$. Indeed, the first step in evaluation or symbolic differentiation of a function defined by a formula is to form a corresponding code list, perhaps subconsciously. The conversion of well-formed expressions into code lists is a fundamental process in computer science, sometimes called 'formula translation'. Although both automatic differentiation and symbolic differentiation are applicable in this case, automatic differentiation requires only the code list and produces only values of derivatives for given values of the input variables. To compute the *gradient* ∇f, the rules of differentiation applied to the code list above gives

$$\begin{aligned}
\nabla t_1 &= \nabla x, \\
\nabla t_2 &= \nabla y, \\
\nabla t_3 &= t_1 \nabla t_2 + t_2 \nabla t_1, \\
\nabla t_4 &= (\cos t_1) \nabla t_1, \\
\nabla t_5 &= \nabla t_3 + \nabla t_4, \\
\nabla t_6 &= \nabla t_5, \\
\nabla t_7 &= 2t_2 \nabla t_2, \\
\nabla t_8 &= 3\nabla t_7, \\
\nabla t_9 &= \nabla t_8, \\
\nabla t_{10} &= t_6 \nabla t_9 + t_9 \nabla t_6.
\end{aligned}$$

It is evident from the chain rule that

$$\nabla t_{10} = \nabla f(x, y) = f_x(x, y) \nabla x + f_y(x, y) \nabla y.$$

Thus, once the code list for $f(x, y)$ is given and the 'seed' values of x, ∇x and y, ∇y are known, the values of the function and its gradient can be computed without formulas for either. In case x, y are independent variables, then $\nabla x = [1, 0]$, $\nabla y = [0, 1]$ and

$$\begin{aligned}
\nabla f(x, y) &= [f_x(x, y), f_y(x, y)] \\
&= [t_9(t_2 + \cos t_1), 6t_2 t_6 + t_1 t_9].
\end{aligned}$$

This example illustrates the *forward mode* of automatic differentiation. This process is not restricted to first derivatives as long as the entries t_i of the code list have the desired number of derivatives.

Although the forward mode illustrated above is easy to understand and implement, it is usually more efficient to compute gradients in what is called the *reverse mode*. To explain this process, consider a general code list $t = (t_1, \ldots, t_n)$ which begins with m input variables $t_1, \ldots, t_m$, and ends with p output variables $t_{n-p+1}, \ldots, t_n$. For $i > m$, the entry $t_i = t_j \circ t_k$, where $j, k < i$ and $\circ$ denotes an arithmetic operation, or $t_i = \phi(t_j)$ with $j < i$, where ϕ is a function belonging to a library of standard functions. For convenience, arithmetic operations between constants and entries will be considered library functions in addition to the usual sine, cosine, and so on.

If K_i denotes the set of indices $k < i$ such that the entry t_i of the code list depends explicitly on t_k, then the forward mode of automatic differentiation consists of application of the chain rule in the form

$$\nabla t_i = \sum_{k \in K_i} \frac{\partial t_i}{\partial t_k} \nabla t_k$$

for $i = m + 1, \ldots, n$, to obtain the gradients of the intermediate variables and output. This process works because $\nabla t_1, \ldots, \nabla t_{i-1}$ are known or have been computed before they are needed for the evaluation of ∇t_i. If the seed gradients have dimension at most d, then the forward mode of automatic differentiation requires computational effort proportional to nd, that is, d times the effort required for evaluation of the output t_n. If $d > m$, then it is more efficient to consider the input variables to be independent and then compose ∇f by the standard formula given below. This limits the computational effort for the forward mode to an amount essentially proportional to nm.

The reverse mode is another way to apply the chain rule. Instead of propagating the seed gradients $\nabla t_1, \ldots, \nabla t_m$ throughout the computation, differentiation is applied to the code list in reverse order. In the case of a single output variable t_n, first t_n is differentiated with respect to itself, then with respect to $t_{n-1}, \ldots, t_1$. The resulting *adjoints* $\partial t_n / \partial t_m, \ldots, \partial t_n / \partial t_1$ and the seed gradients then give

$$\nabla t_n = \sum_{i=1}^{m} \frac{\partial t_n}{\partial t_i} \nabla t_i.$$

Formally, the adjoints are given by

$$\frac{\partial t_n}{\partial t_n} = 1, \qquad \frac{\partial t_n}{\partial t_k} = \sum_{i \in I_k} \frac{\partial t_k}{\partial t_i} \frac{\partial t_i}{\partial t_k},$$

$k = n-1, \ldots, 1$, where I_k is the set of indices $i > k$ such that t_i depends explicitly on t_k. It follows that the computational effort to obtain adjoints in the reverse mode is proportional to n, the length of the code list, and is essentially independent of the number of input variables and the dimensionalities of the seed gradients. This can result in significant savings in computational time. In the general case of several output variables, the same technique is applied to each to obtain their gradients.

The reverse mode applied to the example code list gives

$$\frac{\partial t_{10}}{\partial t_{10}} = 1,$$

$$\frac{\partial t_{10}}{\partial t_9} = t_6,$$

$$\frac{\partial t_{10}}{\partial t_8} = \frac{\partial t_{10}}{\partial t_9} \frac{\partial t_9}{\partial t_8} = t_6 \cdot 1,$$

$$\frac{\partial t_{10}}{\partial t_7} = \frac{\partial t_{10}}{\partial t_8} \frac{\partial t_8}{\partial t_7} = t_6 \cdot 3,$$

$$\frac{\partial t_{10}}{\partial t_6} = t_9,$$

$$\frac{\partial t_{10}}{\partial t_5} = \frac{\partial t_{10}}{\partial t_6} \frac{\partial t_6}{\partial t_5} = t_9 \cdot 1,$$

$$\frac{\partial t_{10}}{\partial t_4} = \frac{\partial t_{10}}{\partial t_5} \frac{\partial t_5}{\partial t_4} = t_9 \cdot 1,$$

$$\frac{\partial t_{10}}{\partial t_3} = \frac{\partial t_{10}}{\partial t_5} \frac{\partial t_5}{\partial t_3} = t_9 \cdot 1,$$

$$\frac{\partial t_{10}}{\partial t_2} = \frac{\partial t_{10}}{\partial t_7} \frac{\partial t_7}{\partial t_2} + \frac{\partial t_{10}}{\partial t_3} \frac{\partial t_3}{\partial t_2}$$
$$= (3t_6) \cdot (2t_2) + t_9 \cdot t_1,$$

$$\frac{\partial t_{10}}{\partial t_1} = \frac{\partial t_{10}}{\partial t_4} \frac{\partial t_4}{\partial t_1} + \frac{\partial t_{10}}{\partial t_3} \frac{\partial t_3}{\partial t_1} = t_9 \cdot \cos t_1 + t_9 \cdot t_2.$$

Although this computation appears to be complicated, a comparison of operation counts in the case x, y are independent variables shows that even for this low-dimensional example, the reverse mode requires 13 operations to evaluate ∇f in addition to the operations required to evaluate f itself, while the forward mode requires $22 = 2 + 10 \, m$. In reverse mode, the entire code list has to be evaluated and its values stored before the reverse sweep begins. In forward mode, since the computation of t_i and each component of ∇t_i can be carried out independently, a parallel computer with a sufficient number of processors could compute $t_n, \nabla t_n$ in a single pass through the code list, that is, with effort proportional to n. A more detailed comparison of forward and reverse modes for calculating gradients can be found in the tutorial article [1, pp. 1–18] and the book [3].

Implementation of automatic differentiation can be by *interpretation*, *operator overloading*, or *code transformation*. Early software for automatic differentiation simply interpreted a code list by calling the appropriate subroutines for each arithmetic operation or library function. Although inefficient, this approach is still useful in interactive applications in which functions entered from the keyboard are parsed to form code lists, which are then interpreted to evaluate the functions and their derivatives.

Operator overloading is a familiar concept in mathematics, as the symbol '+' is used to denote addition of such disparate objects as integers, real or complex numbers, vectors, matrices, functions, etc. It follows that a code list as defined above can be evaluated in any mathematical system in which the required arithmetic operations and library function are available, including differentiation arithmetics [14, pp. 73–90]. These arithmetics can be used to compute derivatives or Taylor coefficients of any order of sufficiently smooth functions. In optimization, gradient and Hessian arithmetics are most frequently used. In gradient arithmetic, the basic data type is the ordered pair $(f, \nabla f)$ of a number and a vector representing values of a function and its gradient vector. Arithmetic operations in this system are defined by

$$(f, \nabla f) \pm (g, \nabla g) = (f \pm g, \nabla f \pm \nabla g),$$
$$(f, \nabla f)(g, \nabla g) = (fg, f \nabla g + g \nabla f),$$
$$\frac{(f, \nabla f)}{(g, \nabla g)} = \left(\frac{f}{g}, \frac{g \nabla f - f \nabla g}{g^2} \right),$$

division by 0 excluded. If ϕ is a differentiable library function, then its extension to gradient arithmetic is defined by

$$\phi(f, \nabla f) = (\phi(f), \phi'(f)\nabla f),$$

which is just the chain rule. Hessian arithmetic extends the same idea to triples $(f, \nabla f, Hf)$, where Hf is a matrix representing the value of the *Hessian* of f, $Hf = [\partial^2 f / \partial x_i \, \partial x_j]$.

Programming differentiation arithmetic is convenient in modern computer languages which support operator overloading [9, pp. 291–309]. In this setting, the program is written with expressions or routines for functions in the regular form, and the compiler produces executable code for evaluation of these functions and the desired derivatives. For straightforward implementations such as the one cited above, the differentiation mode will be forward, which has implications for efficiency.

Code transformation essentially consists of analyzing the code for functions to generate code for derivatives. This results in a new computer program which then can be compiled and run as usual. To illustrate this idea, note that in the simple example given above, the expressions

$$f_x(x, y) = t_9(t_2 + \cos t_1),$$
$$f_y(x, y) = 6t_2t_6 + t_1t_9,$$

were obtained for the partial derivatives of the function in either forward or reverse mode. This differs from symbolic differentiation in that values of intermediate entries in the code list for $f(x, y)$ are involved rather than the variables x, y. The corresponding lists for these expressions

$$tx_1 = \cos t_1,$$
$$tx_2 = t_2 + tx_1,$$
$$tx_3 = t_9 tx_2,$$
$$ty_1 = t_2 t_6,$$
$$ty_2 = 6ty_1,$$
$$ty_3 = t_1 t_9,$$
$$ty_4 = ty_2 + ty_3,$$

can then be appended to the code list for the function to obtain a routine with output values $t_{10} = f(x, y)$, $tx_3 = f_x(x, y)$, and $ty_4 = f_y(x, y)$. Further, automatic differentiation can be applied to this list to obtain routines for higher derivatives of f [13]. As a practical matter, duplicate assignments can be removed from such lists before compilation.

Up to this point, the discussion has been of *point AD*, values have been assumed to be real or complex numbers with all operations and library functions evaluated exactly. In reality, the situation is quite different. Expressions, meaning their equivalent code lists, are evaluated in an approximate computer arithmetic known as floating-point arithmetic. This often yields very accurate results, but examples of simple expressions are known for which double and even higher precision calculation gives an answer in which even the sign is wrong for certain input values. Furthermore, such failures can occur without any outward indication of trouble. In addition, values of input variables may not be known exactly, thus increasing the uncertainty in the accuracy of outputs. The use of *interval arithmetic* (abbreviated IA) provides a computational way to attack these problems [11].

The basic quantities in interval arithmetic are finite closed real intervals $X = [x_1, x_2]$, which represent all real numbers x such that $x_1 \leq x \leq x_2$. Arithmetic operations $\circ$ on intervals are defined by

$$X \circ Y = \{x \circ y : x \in X, y \in Y\},$$

again an interval, division by an interval containing zero excluded. Library functions ϕ are similarly extended to interval functions Φ such that $\phi(x) \in \Phi(X)$ for all $x \in X$ with $\Phi(X)$ expected to be an accurate inclusion of the range $\phi(X)$ of ϕ on X. Thus, if $f(x)$ is a function defined by a code list, then assignment of the interval value X to the input variable and evaluation of the entries in interval arithmetic yields the output $F(X)$ such that $f(x) \in F(X)$ for all $x \in X$. The interval function F obtained in this way is called the *united extension* of f [11].

In the floating-point version of interval arithmetic, all endpoints are floating-point numbers and hence exactly representable in the computer. Results of arithmetic operations and calls of library functions are rounded outwardly (upper endpoints up, lower endpoints down) to the closest or very close floating-point numbers to maintain the guarantee of inclusion. Thus, one is still certain that for the interval extension F of f actually computed, $f(x) \in F(X)$ for all $x \in X$. Thus, for example, an output interval $F(X)$ which is very wide for a point input interval $X = [x, x]$ would serve as a warning that the algorithm is inappropriate or ill-conditioned, in contrast to the lack of such information in ordinary floating-point arithmetic.

Automatic differentiation carried out in interval arithmetic is called *interval automatic differentiation*. Interval computation has numerous implications for optimization, with or without automatic differentiation [6]. Maxima and minima of functions can 'slip through' approximate sampling of values at points of the floating-point grid, but have to be contained in the computable interval inclusion $F(X)$ of $f(x)$ over the same interval region X, for example.

Although interval arithmetic properly applied can solve many optimization and other computational problems, a word of warning is in order. The properties of interval arithmetic differ significantly from those of real arithmetic, and simple 'plugging in' of intervals for numbers will not always yield useful results. In particular, interval arithmetic lacks additive and multiplicative inverses, and multiplication is only subdistributive across addition, $X(Y+Z) \subset XY + XZ$ [11]. A real algorithm which uses one or more of these properties of real arithmetic is usually inappropriate for interval computation, and should be replaced by one that is suitable if possible.

To this point, automatic differentiation has been applied only to code lists, which programmers customarily refer to as 'straight-line code'. Automatic differentiation also applies to subroutines and programs, which ordinarily contain loops and branches in addition to expressions. These latter present certain difficulties in many cases. A loop which is traversed a fixed number of times can be 'unrolled,' and thus is equivalent to straight-line code. However, in case the stopping criterion is based on result values, the derivatives may not have achieved the same accuracy as the function values. For example, if the inverse function of a known function is being computed by iterative solution of the equation $f(x) = y$ for $x = f^{-1}(y)$, then automatic differentiation should be applied to f and the derivative of the inverse function obtained from the standard formula $(f^{-1})'(y) = (f'(x))^{-1}$. Branches essentially produce piecewise defined functions, and automatic differentiation then provides the derivative of the function defined by whatever branch is taken. This can create difficulties as described by H. Fischer [4, pp. 43–50], especially since a smooth function can be approximated well in value by highly oscillatory or other nonsmooth functions such as result from table lookups and piecewise rational approximations. For example, one would not expect to obtain an accurate approximation to the cosine function by applying automatic differentiation to the library subroutine for the sine. As with any powerful tool, automatic differentiation should not be expected to provide good results if applied indiscriminately, especially to 'legacy' code. As with interval arithmetic, automatic differentiation will yield the best results if applied to programs written with it in mind.

Current state of the art software for point automatic differentiation of programs are ADOL-C, for programs written in C/C++ [5], and ADIFOR for programs in Fortran 77 [1, pp. 385–392].

Numerous applications of automatic differentiation to optimization and other problems can be found in the conference proceedings [1,4], which also contain extensive bibliographies. An important result with implications for optimization is that automatic differentiation can be used to obtain Newton steps *without* forming Jacobians and solving linear systems, see [1, pp. 253–264].

From a historical standpoint, the principles of automatic differentiation go back to the early days of calculus, but implementation is a product of the computer age, hence the designation 'automatic'. The terminology 'algorithmic differentiation', to which the acronym automatic differentiation also applies, is perhaps better. Since differentiation is widely understood, automatic differentiation literature contains many anticipations and rediscoveries. The 1962 Stanford Ph.D. thesis of R.E. Moore deals with both interval arithmetic and automatic differentiation of code lists to obtain Taylor coefficients of series solution of systems of ordinary differential equations. In 1964, R.E. Wengert [15] published on automatic differentiation of code lists and noted that derivatives could be recovered from Taylor coefficients. Early results in automatic differentiation were applied to code lists in forward mode, as described in [13]. G. Kedem [8] showed that automatic differentiation applies to subroutines and programs, again in forward mode. The reverse mode was anticipated by S. Linnainmaa in 1976 [10], and in the Ph.D. thesis of B. Speelpenning (Illinois, 1980), and published in more complete form by M. Iri in 1984 [7]. automatic differentiation via operator overloading and the concept of differentiation arithmetics, which are commutative rings with identity, were introduced by L.B. Rall [9, pp. 291–309], [14, pp. 73–90], [4, pp. 17–24]. For additional information about the early history of automatic differentiation, see [13] and the article by Iri [4, pp. 3–16] for later developments.

Analysis of algorithms for automatic differentiation has been carried out on the basis of graph theory by Iri [7], A. Griewank [12, pp. 128–161], [3], and equivalent matrix formulation by Rall [2, pp. 233–240].

See also

- ▶ Automatic Differentiation: Calculation of the Hessian
- ▶ Automatic Differentiation: Calculation of Newton Steps
- ▶ Automatic Differentiation: Geometry of Satellites and Tracking Stations
- ▶ Automatic Differentiation: Introduction, History and Rounding Error Estimation
- ▶ Automatic Differentiation: Parallel Computation
- ▶ Automatic Differentiation: Point and Interval Taylor Operators
- ▶ Automatic Differentiation: Root Problem and Branch Problem
- ▶ Bounding Derivative Ranges
- ▶ Global Optimization: Application to Phase Equilibrium Problems
- ▶ Interval Analysis: Application to Chemical Engineering Design Problems
- ▶ Interval Analysis: Differential Equations
- ▶ Interval Analysis: Eigenvalue Bounds of Interval Matrices
- ▶ Interval Analysis: Intermediate Terms
- ▶ Interval Analysis: Nondifferentiable Problems
- ▶ Interval Analysis: Parallel Methods for Global Optimization
- ▶ Interval Analysis: Subdivision Directions in Interval Branch and Bound Methods
- ▶ Interval Analysis: Systems of Nonlinear Equations
- ▶ Interval Analysis: Unconstrained and Constrained Optimization
- ▶ Interval Analysis: Verifying Feasibility
- ▶ Interval Constraints
- ▶ Interval Fixed Point Theory
- ▶ Interval Global Optimization
- ▶ Interval Linear Systems
- ▶ Interval Newton Methods
- ▶ Nonlocal Sensitivity Analysis with Automatic Differentiation

References

1. Berz M, Bischof Ch, Corliss G, Griewank A (eds) (1996) Computational differentiation, techniques, applications, and tools. SIAM, Philadelphia
2. Fischer H, Riedmueller B, Schaeffler S (eds) (1996) Applied mathematics and parallel computing. Physica Verlag, Heidelberg
3. Griewank A (2000) Evaluating derivatives: Principles and techniques of algorithmic differentiation. SIAM, Philadelphia
4. Griewank A, Corliss GF (eds) (1991) Automatic differentiation of algorithms, theory, implementation, and application. SIAM, Philadelphia
5. Griewank A, Juedes D, Utke J (1996) ADOL-C, a package for the automatic differentiation of programs written in C/C++. ACM Trans Math Softw 22:131–167
6. Hansen E (1992) Global optimization using interval analysis. M. Dekker, New York
7. Iri M (1984) Simultaneous computation of functions, partial derivatives, and rounding errors: complexity and practicality. Japan J Appl Math 1:223–252
8. Kedem G (1980) Automatic differentiation of computer programs. ACM Trans Math Softw 6:150–165
9. Kulisch UW, Miranker WL (eds) (1983) A new approach to scientific computation. Acad. Press, New York
10. Linnainmaa S (1976) Taylor expansion of the accumulated rounding error. BIT 16:146–160
11. Moore RE (1979) Methods and applications of interval analysis. SIAM, Philadelphia
12. Pardalos PM (eds) (1993) Complexity in nonlinear optimization. World Sci, Singapore
13. Rall LB (1981) Automatic differentiation: Techniques and applications. Springer, Berlin
14. Ullrich C (eds) (1990) Computer arithmetic and self-validating numerical methods. Acad Press, New York
15. Wengert RE (1964) A simple automatic derivative evaluation program. Comm ACM 7:463–464

Automatic Differentiation: Point and Interval Taylor Operators
AD, Computational Differentiation

James B. Walters, George F. Corliss
Marquette University, Milwaukee, USA

MSC2000: 65K05, 90C30

Article Outline

Keywords
Introduction
   Operator Overloading
   Automatic Differentiation
Taylor Coefficients
Point and Interval Taylor Operators
   Design of Operators
   Use of Interval Operators
One-at-a-Time Coefficient Generation
Trade-Offs
See also
References

Keywords

Automatic differentiation; Code list; Interval arithmetic; Overloaded operator; Taylor series

Frequently of use in optimization problems, automatic differentiation may be used to generate Taylor coefficients. Specialized software tools generate Taylor series approximations, one term at a time, more efficiently than the general AD software used to compute (partial) derivatives. Through the use of operator overloading, these tools provide a relatively easy-to-use interface that minimizes the complications of working with both point and interval operations.

Introduction

First, we briefly survey the tools of automatic differentiation and operator overloading used to compute point- and interval-valued Taylor coefficients. We assume that f is an analytic function $f : \mathbf{R} \to \mathbf{R}$. Automatic differentiation (AD or computational differentiation) is the process of computing the derivatives of a function f at a point $t = t_0$ by applying rules of calculus for differentiation [9,10,17,18]. One way to implement AD uses overloaded operators.

Operator Overloading

An overloaded (or generic) operator invokes a procedure corresponding to the types of its operands. Most programming languages implement this technique for arithmetic operations. The sums of two floating point numbers, two integers, or one floating point number and one integer are computed using three different procedures for addition. Fortran 77 or C denies the programmer the ability to replace or modify the various routines used implicitly for integer, floating point, or mixed-operand arithmetic, but Fortran 95, C++, and Ada support operator overloading for user-defined types. Once we have defined an overloaded operator for each rule of differentiation, AD software performs those operations on program code for f, as shown below. The operators either propagate derivative values or construct a code list for their computation. We give prototypical examples of operators overloaded to propagate Taylor coefficients below.

Automatic Differentiation

The AD process requires that we have f in the form of an algorithm (e. g. computer program) so that we can easily separate and order its operations. For example, given $f(t) = e^t/(2 + t)$, we can express f as an algorithm in Fortran 95 or in C++ (using an assumed AD module or class):

In this section, we use AD to compute first derivatives. In the next section, we extend to point- and interval-valued Taylor series. To understand the AD process, we parse the program above into a sequence of unary and binary operations, called a *code list*, *computational graph*, or 'tape' [9]:

$$x_0 = t_0, \qquad x_2 = 2 + x_0,$$
$$x_1 = \exp(x_0), \qquad x_3 = \frac{x_1}{x_2}.$$

```
program Example1
    use AD_Module
    type(AD_Independent) :: t
        AD_Independent(0)
    type(AD_Dependent) :: f
    f = exp(t)/(2 + t)
end program Example1
#include 'AD_class.h'
void main (void) {
    AD_Independent t(0);
    AD_Dependent f;
    f = exp(t)/(2 + t);
}
```

Automatic Differentiation: Point and Interval Taylor Operators, Figure 1
Fortran and C++ calls to AD operators

Differentiation is a simple mechanical process for propagating derivative values. Let $t = t_0$ represent the value of the independent variable with respect to which we differentiate. We know how to take the derivative of a variable, a constant, and unary and binary operations (i. e. $+$, $-$, $*$, $/$, sin, cos, exp, etc.). Then AD software annotates the code list:

$$x_0 = t_0;$$
$$\nabla x_0 = 1,$$
$$x_1 = \exp(x_0);$$
$$\nabla x_1 = \exp(x_0) * \nabla x_0,$$
$$x_2 = 2 + x_0;$$
$$\nabla x_2 = 0 + \nabla x_0,$$
$$x_3 = \frac{x_1}{x_2};$$
$$\nabla x_3 = \frac{(\nabla x_1 - \nabla x_2 * x_3)}{x_2}.$$

AD propagates values of derivatives, not expressions as symbolic differentiation does. AD values are exact (up to round-off), not approximations of unknown quality as finite differences. For more information regarding AD and its applications, see [2,8,9,10,17,18], or the bibliography [21].

AD software can use overloaded operators in two different ways. Operators can propagate both the value x_i and its derivative ∇x_i, as suggested by the annotated code list above. This approach is easy to understand and to program. We give prototypical Taylor operators of this flavor below.

The second approach has the operators construct and store the code list. Various optimizations and parallel scheduling [1,4,12] may be applied to the code list. Then the code list is interpreted to propagate derivative values. This is the approach of AD tools such as *ADOL-C* [11], *ADOL-F* [20], *AD01* [16], or *INTOPT_90* [13]. The second approach is much more flexible, allowing the code list to be traversed in either the forward or reverse modes of AD (see [9]) or with various arithmetics (e. g. point- or interval-valued series).

AD may be applied to functions of more than one variable, in which partial derivatives with respect to each are computed in turn, and to vector functions, in which the component functions are differentiated in succession. In addition, we can compute higher order derivative values. One application of AD involving higher order derivatives of f is the computation of Taylor (series) coefficients to which we turn in the next section.

Source code transformation is a third approach to AD software used by *ATOMFT* [5] for Taylor series and by *ADIFOR* [3], *PADRE2* [14], or *Odyssée* [19] for partial derivatives. Such tools accept the algorithm for f as data, rather than for execution, and produce code for computing the desired derivatives. The resulting code often executes more rapidly than code using overloaded operators.

Taylor Coefficients

We define the Taylor coefficients of the analytic function f at the point $t = t_0$:

$$(f|t_0)_i := \frac{1}{i!} \frac{d^i f(t_0)}{dt^i},$$

for $i = 0, 1, \ldots$, and let $F := ((f|t_0)_i)$ denote the vector of Taylor coefficients. Then *Taylor's theorem* says that there exists some point τ (usually not practically obtainable) between t and t_0 such that

$$f(t) = \sum_{i=0}^{p} (f|t_0)_i (t-t_0)^i \qquad (1)$$
$$+ \frac{1}{(p+1)!} \frac{d^{p+1} f(\tau)}{dt^{p+1}} (t-t_0)^{p+1}.$$

Computation of Taylor coefficients requires differentiation of f. We generate Taylor coefficients automatically using recursion formulas for unary and binary operations. For example, the recurrences we need for our example $f(t) = e^t/(2+t)$ are

$$x(t) = \exp u(t) \Rightarrow x' = xu',$$
$$(x)_0 = \exp(u)_0,$$
$$(x)_i = \sum_{j=0}^{i-1} (x)_j * (u)_{i-j} * \frac{(i-j)}{i};$$
$$x(t) = u(t) + v(t),$$
$$(x)_i = (u)_i + (v)_i;$$
$$x(t) = \frac{u(t)}{v(t)} \Rightarrow xv = u,$$
$$(x)_i = \frac{\left((u)_i - \sum_{j=0}^{i-1} (x)_j * (v)_{i-j} \right)}{(v)_0}.$$

The recursion relations are described in more detail in [17]. Except for + and −, each recurrence follows from Leibniz' rule for the Taylor coefficients of a product. The relations can be viewed as a lower triangular system. The recurrence represents a solution by forward substitution, but there are sometimes accuracy or stability advantages in an iterative solution to the lower triangular system. The recurrences for each operation can be evaluated in floating-point, complex, interval, or other appropriate arithmetic.

To compute the formal series for $f(t) = e^t/(2+t)$ expanded at $t = 0$,

$$\begin{cases} X_0 := (t_0, 1, 0, \ldots)(0, 1, 0, \ldots), \\ X_1 := \exp X_0 = \left(1, 1, \frac{1}{2!}, \frac{1}{3!}, \ldots\right), \\ X_2 := 2 + X_0 = (2, 1, 0, \ldots), \\ X_3 := \frac{X_2}{X_3} = \left(\frac{1}{2}, \frac{1}{4}, \frac{1}{8}, \frac{1}{48}, \ldots\right). \end{cases} \qquad (2)$$

```
class Taylor {        // Or make a template:
    private:
        cont int Max_Length = 20;
        Value_type coef[Max_Length];
    public:
        Taylor ( Value_type t_0 ) {
            // Constructor for Independents
            coef[0] = t_0; coef[1] = 1;
            for(int i = 2; i ¡ Max_Length; i++)
                { coef[i] = 0; }
        }
        Taylor ( void ) {
            // Constructor for Dependents
            for (int i = 0; i ¡ Max_Length; i++)
                { coef[i] = 0; }
        }
        Taylor ( Taylor &U) {
            // Copy Constructor
            for (int i=0; i ¡ Max_Length; i++)
                { coef[i] = Value_type(U.coef)[i]; }
        }
        friend Taylor operator +
            (int u, Taylor V) {
            V.coef[0] += u; return V;
        }
        friend Taylor operator /
            (Taylor U, Taylor V) {
            Taylor X;
            for (int i = 0; i ¡ Max_Length; i++) {
                Value_type sum = U.coef[i];
                for (int j = 0; j ¡ i; j++)
                    { sum − =X.coef[j] * V.coef[i−j]; }
                X.coef[i] = sum / V.coef[0];
            }
            return X;
        }
        friend Taylor exp (Taylor U)
            { /* Similar to divide */ }
        Value_type getCoef (int i)
            { return coef[i]; }
};    // end class Taylor
```

Point and Interval Taylor Operators

As foreshadowed by this example, we define an abstract data type for Taylor series and use operator overloading to define actions on objects of that type using previously defined floating-point and interval operations.

Design of Operators

In this section, we give prototypical operators for the direct propagation of Taylor coefficients such as might be called from code similar to that shown in Fig. 1. Direct propagation of values works by translating each operation into a call to the appropriate AD routine at compile time. Thus, simply compiling the source code for f and linking it with the overloaded operator routines creates a program that computes the Taylor coefficients of f at $t = t_0$. For illustration, we provide only a stripped-down prototype with operators required for the example $f(t) = e^t/(2 + t)$. We suppress issues of references and the like that are essential to the design of a useful class. See [6] for a description of a set of interval Taylor operators in Ada.

If instead, operators for AD_type record a code list, then an interpreter reads each node from the code list and calls the appropriate operator from class Taylor:

```
Taylor Operand[MemSize];
for (int i = 0; ¡ CodeSize; i++) {
    Node = getNextOperation ();
    switch (Node.OpCode) {
        case PLUS : Operand[Node.Result]
                   = Operand[Node.Left]
                   + Operand[Node.Right];
            break;
        ...
        case EXP : Operand[Node.Result]
                   = exp ( Operand[Node.Left] );
            break;
        ...
    }
}
```

Use of Interval Operators

We have mentioned the possibility of working with interval values but not the significance of doing so. From equation (1) for an interval **t**, and for all $t \in \mathbf{t}$,

$$f(t) \in \sum_{i=0}^{p} (f|t_0)_i (t - t_0)^i$$
$$+ \frac{1}{(p+1)!} \frac{d^{p+1} f(\mathbf{t})}{dt^{p+1}} (t - t_0)^{p+1}. \quad (3)$$

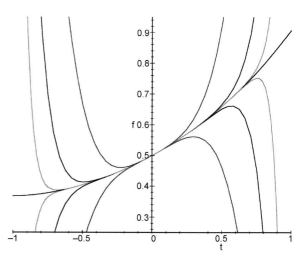

Automatic Differentiation: Point and Interval Taylor Operators, Figure 2
Taylor series enclosures of f

In a computer implementation, the summation is done in interval arithmetic to ensure enclosure. The series Taylor coefficients $(f|t_0)_i$ are narrow intervals whose width comes only from outward rounding. The remainder term is the Taylor coefficient $(f|\mathbf{t})_i$, where the recurrence relations are evaluated in interval arithmetic. The series (3) can be used to bound the range of f, for validated quadrature [7], or for rigorous solution of ODEs [15]. For the example $f(t) = e^t/(2 + t)$, we repeat the sequence of computations of Equation (2) for the interval $\mathbf{t}_0 = [0, 0]$ and for $\mathbf{t} = [-1, 1]$:

$$((f|[0])_i) = \left(\frac{1}{2}, \frac{1}{4}, \frac{1}{8}, \frac{1}{48}, \dots\right),$$

$$((f|[-1, 1])_i)$$
$$= ([0.12, 2.72], [-2.59, 2.68], [-2.64, 4.04], \dots).$$

Assembling these according to (3) yields enclosures for all $t \in [-1, 1]$:

$$f(t) \in (f|[-1, 1])_0 = [0.12, 2.72]$$
$$\in (f|[0])_0 + (f|[-1, 1])_1 (t - 0)$$
$$= \frac{1}{2} + [-2.59, 2.68]t$$
$$\in (f|[0])_0 + (f|[0])_1 (t - 0)$$
$$\quad + (f|[-1, 1])_2 (t - 0)^2$$
$$= \frac{1}{2} + \frac{1}{4}t + [-2.64, 4.04]t^2$$

$$\vdots$$

To demonstrate the true power of this approximation technique, we plot the corresponding 5, 10, and 20 term enclosures in Fig. 2.

One-at-a-Time Coefficient Generation

The Taylor operators described the preceding section accept vectors of p Taylor coefficients for operands u and v and return Taylor coefficients for result x with complexity $O(p^2)$. However, for applications such as ODEs or order-adaptive quadrature, the entire operand series is not known, and we need to compute terms one at a time [6]. For example, for the DE

$$u' = f(t,u) = \frac{\exp(u)}{(2+t)}, \quad u(0) = 1,$$

initial condition $u(0) = 1$ implies

$$(u|0)_0 = 1,$$

and DE $u' = \exp(u)/(2+t)$ implies

$$(u|0)_1 = \frac{\exp(1)}{(2+0)} = \frac{e}{2},$$

$$u'' = \frac{u'\exp(u)}{2+t} - \frac{\exp(u)}{(2+t)^2} \text{ implies}$$

$$(u|0)_2 = \frac{e}{2}\frac{\exp(1)}{(2+0)} - \frac{e}{4} = \frac{e(e-1)}{4},$$

etc.

Successive terms can be computed by interpreting the code list for $f(t,u)$ repeatedly for series of increasing length for u. Each iteration of the automatic generation process yields an additional Taylor coefficient. Unfortunately, a simple implementation of Taylor operators has complexity $O(p^3)$ because already known coefficients of u' are recomputed. However, since the order of operations is the same in each iteration, we can increase the efficiency of the computations by storing intermediate results [6]. Each overloaded operator routine calls a memory allocation procedure that refers it to the next space in an array. If that space is empty, we store Taylor coefficient values for that variable. Otherwise, the space must contain the previously computed Taylor coefficients of that variable, which we can then use to more quickly compute the next coefficient in the set. With clever book-keeping, we compute p floating-point or interval-valued Taylor coefficients one at a time in $O(p^2)$ time.

Trade-Offs

We may strive for three goals when writing software for point and interval Taylor operations: storage space efficiency, time efficiency, and ease of use. These three factors are often at odds with each other.

Carefully implemented operator overloading provides an easy to use interface and provides reasonable time and space efficiency. We may achieve greater time and space efficiency by using source code transformation.

In conclusion, automatic differentiation through Taylor operators shows merit as a technique for computing guaranteed interval enclosures about a function f. Further efforts to refine this technique may provide us with a tool that handles multivariate functions, and runs significantly faster thanks to parallelization and improved optimization techniques.

See also

- ▶ Automatic Differentiation: Calculation of the Hessian
- ▶ Automatic Differentiation: Calculation of Newton Steps
- ▶ Automatic Differentiation: Geometry of Satellites and Tracking Stations
- ▶ Automatic Differentiation: Introduction, History and Rounding Error Estimation
- ▶ Automatic Differentiation: Parallel Computation
- ▶ Automatic Differentiation: Point and Interval
- ▶ Automatic Differentiation: Root Problem and Branch Problem
- ▶ Bounding Derivative Ranges
- ▶ Global Optimization: Application to Phase Equilibrium Problems
- ▶ Interval Analysis: Application to Chemical Engineering Design Problems
- ▶ Interval Analysis: Differential Equations
- ▶ Interval Analysis: Eigenvalue Bounds of Interval Matrices
- ▶ Interval Analysis: Intermediate Terms
- ▶ Interval Analysis: Nondifferentiable Problems
- ▶ Interval Analysis: Parallel Methods for Global Optimization
- ▶ Interval Analysis: Subdivision Directions in Interval Branch and Bound Methods
- ▶ Interval Analysis: Systems of Nonlinear Equations

- ▶ Interval Analysis: Unconstrained and Constrained Optimization
- ▶ Interval Analysis: Verifying Feasibility
- ▶ Interval Constraints
- ▶ Interval Fixed Point Theory
- ▶ Interval Global Optimization
- ▶ Interval Linear Systems
- ▶ Interval Newton Methods
- ▶ Nonlocal Sensitivity Analysis with Automatic Differentiation

References

1. Benary J (1996) Parallelism in the reverse mode. In: Berz M, Bischof Ch, Corliss G, Griewank A (eds) Computational Differentiation: Techniques, Applications, and Tools. SIAM, Philadelphia, pp 137–147
2. Berz M, Bischof Ch, Corliss G, Griewank A (eds) (1996) Computational differentiation: Techniques, applications, and tools. SIAM, Philadelphia
3. Bischof Ch, Carle A, Khademi PM, Mauer A, Hovland P (1994) ADIFOR: 2.0 user's guide. Techn Memorandum Math and Computer Sci Div Argonne Nat Lab ANL/MCS-TM-192
4. Bischof Ch, Haghighat MR (1996) Hierarchical approaches to automatic differentiation. In: Berz M, Bischof Ch, Corliss G, Griewank A (eds) Computational Differentiation: Techniques, Applications, and Tools. SIAM, Philadelphia, pp 83–94
5. Chang YF, Corliss GF (1994) ATOMFT: Solving ODEs and DAEs using Taylor series. Comput Math Appl 28:209–233
6. Corliss GF (1991) Overloading point and interval Taylor operators. In: Griewank A, Corliss GF (eds) Automatic Differentiation of Algorithms: Theory, Implementation, and Application. SIAM, Philadelphia, pp 139–146
7. Corliss GF, Rall LB (1987) Adaptive, self-validating quadrature. SIAM J Sci Statist Comput 8(5):831–847
8. Griewank A (1989) On automatic differentiation. In: Iri M, Tanabe K (eds) Mathematical Programming: Recent Developments and Applications. Kluwer, Dordrecht, pp 83–108
9. Griewank A (1991) The chain rule revisited in scientific computing. SIAM News 24(3):20 Also: Issue 4, page 8.
10. Griewank A, Corliss GF (eds) (1991) Automatic differentiation of algorithms: Theory, implementation, and application. SIAM, Philadelphia
11. Griewank A, Juedes D, Utke J (1996) ADOL-C: A package for the automatic differentiation of algorithms written in C/C++. ACM Trans Math Softw 22(2):131–167
12. Griewank A, Reese S (1991) On the calculation of Jacobian matrices by the Markowitz rule. In: Griewank A, Corliss GF (eds) Automatic Differentiation of Algorithms: Theory, Implementation, and Application. SIAM, Philadelphia, pp 126–135
13. Kearfott RB (1995) A Fortran 90 environment for research and prototyping of enclosure algorithms for nonlinear equations and global optimization. ACM Trans Math Softw 21(1):63–78
14. Kubota K (1996) PADRE2-Fortran precompiler for automatic differentiation and estimates of rounding error. In: Berz M, Bischof Ch, Corliss G, Griewank A (eds) Computational Differentiation: Techniques, Applications, and Tools. SIAM, Philadelphia, 367–374
15. Lohner RJ (1987) Enclosing the solutions of ordinary initial and boundary value problems. In: Kaucher EW, Kulisch UW, Ullrich C (eds) Computer Arithmetic: Scientific Computation and Programming Languages. Wiley and Teubner, Stuttgart, pp 255–286
16. Pryce JD, Reid JK (1997) AD01: A Fortran 90 code for automatic differentiation. Techn Report Rutherford–Appleton Lab RAL-TR-97xxx
17. Rall LB (1981) Automatic differentiation: Techniques and applications. Lecture Notes Computer Sci, vol 120. Springer, Berlin
18. Rall LB, Corliss GF (1996) An introduction to automatic differentiation. In: Berz M, Bischof Ch, Corliss G, Griewank A (eds) Computational Differentiation: Techniques, Applications, and Tools. SIAM, Philadelphia, pp 1–17
19. Rostaing N, Dalmas S, Galligo A (1993) Automatic differentiation in Odyssée. Tellus 45A:558–568
20. Shiriaev D (1996) ADOL-F: Automatic differentiation of Fortran codes. In: Berz M, Bischof Ch, Corliss G, Griewank A (eds) Computational Differentiation: Techniques, Applications, and Tools. SIAM, Philadelphia, pp 375–384
21. Yang W, Corliss G (1996) Bibliography of computational differentiation. In: Berz M, Bischof Ch, Corliss G, Griewank A (eds) Computational Differentiation: Techniques, Applications, and Tools. SIAM, Philadelphia, pp 393–418

Automatic Differentiation: Root Problem and Branch Problem

HERBERT FISCHER

Fakult. Math., Techn. University München, München, Germany

MSC2000: 65K05

Article Outline

Keywords
Root Problem
Branch Problem
See also
References

Keywords

Automatic differentiation; Root problem; Branch problem

Automatic differentiation is a method in which a program for evaluating a function f is transformed into another program that evaluates both the function f and some of its derivatives. The key idea is the repeated use of the chain-rule for composing the derivatives of f from derivatives of parts of f. For more about automatic differentiation (AD), consult [2,3,5].

Proper combinations of differentiable functions produce differentiable functions. Some combinations of nondifferentiable functions also produce differentiable functions. Therefore the mere fact that a program defines a differentiable function is no guarantee that AD will work. Here we investigate two cases, where AD, applied to a program for a differentiable function, fails.

The *root problem* arises when a square-root is combined with other functions so that the resulting function is differentiable but the chain-rule is not applicable for certain arguments.

The *branch problem* arises when a program for evaluating a differentiable function f employs statements of the form $B(x)$ then $S1$ else $S2$, where x is from the domain of f, B is a Boolean function, and $S1$ and $S2$ represent subprograms. This reflects a piece-wise definition of the function f, and the derivative of one or the other piece may be quite different from the derivative of the function f.

Root Problem

An example that is typical of the root problem is shown in Table 1. The program P defines the function

$$f: \mathbb{R}^2 \to \mathbb{R}$$

with

$$f(x) = \sqrt{x_1^4 + x_2^4}.$$

This function is differentiable at any $x \in \mathbb{R}^2$, in particular $f'(0) = [0, 0]$. Standard AD (in the forward mode) transforms P into a program P' by inserting assignment statements for derivatives in proper places (see Table 2).

The program P' is supposed to compute $f(x)$ and $f'(x)$. But for $x = 0$ it does *not* compute the correct value

Automatic Differentiation: Root Problem and Branch Problem, Table 1
Program P for evaluating f at x

input: $x = (x_1, x_2) \in \mathbb{R}^2$		
y_1	$\leftarrow$	x_1
y_2	$\leftarrow$	x_2
y_3	$\leftarrow$	y_1^4
y_4	$\leftarrow$	y_2^4
y_5	$\leftarrow$	$y_3 + y_4$
y_6	$\leftarrow$	$\sqrt{y_5}$
$f(x)$	$\leftarrow$	y_6
output: $f(x)$		

Automatic Differentiation: Root Problem and Branch Problem, Table 2
Program P' for evaluating f and f' at x

input: $x = (x_1, x_2) \in \mathbb{R}^2$					
y_1	$\leftarrow$	x_1	y'_1	$\leftarrow$	$[1,0]$
y_2	$\leftarrow$	x_2	y'_2	$\leftarrow$	$[0,1]$
y_3	$\leftarrow$	y_1^4	y'_3	$\leftarrow$	$4y_1^3 \cdot y'_1$
y_4	$\leftarrow$	y_2^4	y'_4	$\leftarrow$	$4y_2^3 \cdot y'_2$
y_5	$\leftarrow$	$y_3 + y_4$	y'_5	$\leftarrow$	$y'_3 + y'_4$
y_6	$\leftarrow$	$\sqrt{y_5}$	y'_6	$\leftarrow$	$\frac{1}{2\sqrt{y_5}} \cdot y'_5$
$f(x)$	$\leftarrow$	y_6	$f'(x)$	$\leftarrow$	y'_6
output: $f(x)$			output: $f'(x)$		

Automatic Differentiation: Root Problem and Branch Problem, Table 3
Program Q for evaluating f at x

input: $x \in D \subseteq \mathbb{R}^n$		
y_1	$\leftarrow$	$A(x)$
y_2	$\leftarrow$	$\sqrt{y_1}$
y_3	$\leftarrow$	$B(x, y_2)$
$f(x)$	$\leftarrow$	y_3
output: $f(x)$		

$f'(0) = [0, 0]$, but rather it fails because of division by zero.

One can easily see that this failure is not limited to the forward mode, because the reverse mode encounters the same division-by-zero problem. Symbolic manipulation packages such as MAPLE also fail to produce $f'(0)$.

A more general setting for the root problem is shown in Table 3. Here, it is assumed that:
1) D_A is a nonempty open subset of $\mathbf{R}^n$;
2) the function $A: D_A \subseteq \mathbf{R}^n \to \mathbf{R}$ is differentiable;
3) D_B is a nonempty open subset of $\mathbf{R}^{n+1}$;
4) the function $B: D_B \subseteq \mathbf{R}^{n+1} \to \mathbf{R}$ is differentiable;
5) $D := \{x \in D_A: A(x) \geq 0, (x, A(x)) \in D_B\}$;
6) D is nonempty.

The program Q defines the function

$$f: D \subseteq \mathbf{R}^n \to \mathbf{R}$$

with

$$f(x) = B(x, \sqrt{A(x)}).$$

Standard AD (in the forward mode) transforms Q into a program Q'. The steps of Q' in evaluating $f'(x)$ can be seen in the formula

$$f'(x) = B_1(x, y_2) + B_2(x, y_2) \cdot \left(\frac{1}{2\sqrt{y_1}} \cdot y'_1\right),$$

where $[B_1(x, y_2), B_2(x, y_2)]$ is an appropriate partition of $B'(x, y_2)$. For $x \in D$ with $A(x) > 0$, the program Q' will produce $f'(x)$. And for $x \in D$ with $A(x) = 0$, the program Q' fails because of division by zero. The case in which $x^* \in D$ with $A(x^*) = 0$ is ambiguous. It says nothing about the existence of $f'(x^*)$. In this case, we distinguish the following four situations:

A) $f'(x^*)$ does not exist, for instance $n = 2$, $A(x) = x_1^2 + x_2^2$ and $B(x, y) = y$, $x^* = 0$.
B) A alone guarantees existence of $f'(x^*)$, for instance $n = 2$, $A(x) = x_1^4 + x_2^4$, $x^* = 0$.
C) B alone guarantees existence of $f'(x^*)$, for instance $B(x, y) = y^2$.
D) A and B together guarantee existence of $f'(x^*)$, for instance $n = 2$, $A(x) = x_1^2 + x_2^2$ and $B(x, y) = x_1 \cdot x_2 \cdot y$, $x^* = 0$.

What can be done to resolve the root problem?

The use of AD tools for higher derivatives may be helpful. Consider the simple case $n = 1$, $A \in \mathcal{C}^\infty$, $D_B = \mathbf{R}^{n+1}$, $B(x, y) = y$. So we have

$$D := \{x: x \in D_A, A(x) \geq 0\}$$

and $f: D \subseteq \mathbf{R} \to \mathbf{R}$ with $f(x) = \sqrt{A(x)}$.

Assume that for $x \in \mathbf{R}$ it can be decided whether or not $x \in D$, for instance by testing x in a program for evaluating A.

For $x^* \in D$, we require the value of the derivative $f'(x^*)$. Below, we list the relevant implications:
- $A(x^*) > 0 \Rightarrow f'(x^*) = \frac{1}{2\sqrt{A(x^*)}} \cdot A'(x^*)$.
- $A(x^*) = 0 \Rightarrow$ no answer possible.
- $A(x^*) = 0$, $A'(x^*) \neq 0 \Rightarrow f'(x^*)$ does not exist.
- $A(x^*) = 0$, $A'(x^*) = 0 \Rightarrow$ no answer possible.
- $A(x^*) = 0$, $A'(x^*) = 0$, $A''(x^*) \neq 0 \Rightarrow f'(x^*)$ does not exist.
- $A(x^*) = 0$, $A'(x^*) = 0$, $A''(x^*) = 0 \Rightarrow$ no answer possible.
- $A(x^*) = 0$, $A'(x^*) = 0$, $A''(x^*) = 0$, $A'''(x^*) \neq 0 \Rightarrow f'(x^*)$ does not exist.
- $A(x^*) = 0$, $A'(x^*) = 0$, $A''(x^*) = 0$, $A'''(x^*) = 0 \Rightarrow$ no answer possible.
- $A(x^*) = 0$, $A'(x^*) = 0$, $A''(x^*) = 0$, $A'''(x^*) = 0$, $A^{(4)}(x^*) > 0 \Rightarrow f'(x^*) = 0$.
- $A(x^*) = 0$, $A'(x^*) = 0$, $A''(x^*) = 0$, $A''(x^*) = 0$, $A^{(4)}(x^*) < 0 \Rightarrow f'(x^*)$ does not exist.
- $A(x^*) = 0$, $A'(x^*) = 0$, $A''(x^*) = 0$, $A'''(x^*) = 0$, $A^{(4)}(x^*) = 0 \Rightarrow$ no answer possible.

Let $n \in \{1, 2, 3 \ldots\}$ and $A^{(k)}(x^*) = 0$ for $k = 0, \ldots, 2n$.
- $A^{(2n+1)}(x^*) \neq 0 \Rightarrow f'(x^*)$ does not exist.
- $A^{(2n+1)}(x^*) = 0$, $A^{(2n+2)} > 0 \Rightarrow f'(x^*) = 0$.
- $A^{(2n+1)}(x^*) = 0$, $A^{(2n+2)} < 0 \Rightarrow f'(x^*)$ does not exist.
- $A^{(2n+1)}(x^*) = 0$, $A^{(2n+2)} = 0 \Rightarrow$ no answer possible.

For a nonstandard treatment of these implications see [6]. Of course in the general situation given in Table 3, the classification of cases is more problematic.

Branch Problem

A typical example for the branch problem is Gauss-elimination for solving a system of linear equations with parameters. For illustrative purposes, it suffices to consider two equations with a two-dimensional parameter x (see Table 4). Here, it is assumed that:
a) D is a nonempty open subset of $\mathbf{R}^2$;
b) the function $M: D \subseteq \mathbf{R}^2 \to \mathbf{R}^{2,2}$ is differentiable;
c) the function $R: D \subseteq \mathbf{R}^2 \to \mathbf{R}^2$ is differentiable;
d) $x \in D \Rightarrow$ the matrix $M(x)$ is regular.

The program GAUSS defines the function

$$F: D \subseteq \mathbb{R}^2 \to \mathbb{R}^2$$

with

$$M(x) \cdot F(x) = R(x).$$

Automatic Differentiation: Root Problem and Branch Problem, Table 4
Program GAUSS for evaluating f at x

	input: $x \in D$		
	M11	$\leftarrow$	$M_{11}(x)$
	M12	$\leftarrow$	$M_{12}(x)$
	M21	$\leftarrow$	$M_{21}(x)$
	M22	$\leftarrow$	$M_{22}(x)$
	R1	$\leftarrow$	$R_1(x)$
	R2	$\leftarrow$	$R_2(x)$
	IF M11 $\neq$ 0 THEN		
S1:	E	$\leftarrow$	M21 / M11
	M22	$\leftarrow$	M22 − E ∗ M12
	R2	$\leftarrow$	R2 − E ∗ R1
	F2	$\leftarrow$	R2 / M22
	F1	$\leftarrow$	(R1 − M12 ∗ F2) / M11
	ELSE		
S2:	F2	$\leftarrow$	R1 / M12
	F1	$\leftarrow$	(R2 − M22 ∗ F2) / M21
	output: $F(x) = (F1, F2)$		

Since the matrix $M(x)$ is regular for $x \in D$, the program GAUSS and the function f are well-defined. Furthermore, the function f is differentiable.

Standard AD (in the forward mode) transforms GAUSS into a new program by inserting assignment statements for derivatives in proper places. The resulting program GAUSS' is also well-defined, and for $x \in D$ it is supposed to produce $F(x)$ and $F'(x)$.

Now choose

$$D = \{x \in \mathbb{R}^2 : 0 < x_1 < 2,\ 0 < x_2 < 2\}$$

and

$$M(x) = \begin{array}{|c|c|} \hline M_{11}(x) & M_{12}(x) \\ \hline M_{21}(x) & M_{22}(x) \\ \hline \end{array} = \begin{array}{|c|c|} \hline x_1 - x_2 & 1 \\ \hline 10 & x_1 + x_2 \\ \hline \end{array},$$

$$R(x) = \begin{array}{|c|} \hline R_1(x) \\ \hline R_2(x) \\ \hline \end{array} = \begin{array}{|c|} \hline 100(x_1 + 2x_2) \\ \hline 100(x_1 - 2x_2) \\ \hline \end{array}.$$

It is easy to see that D is a nonempty open subset of $\mathbb{R}^2$, that the functions M and R are differentiable, and that $M(x)$ is regular for $x \in D$.

GAUSS' produces

$$F'(1,1) = \begin{array}{|c|c|} \hline -40 & -90 \\ \hline 100 & 200 \\ \hline \end{array},$$

but the correct value is

$$F'(1,1) = \begin{array}{|c|c|} \hline -54 & -76 \\ \hline 170 & 130 \\ \hline \end{array}.$$

One can easily check that the wrong result is not limited to the forward mode, because the reverse mode yields exactly the same wrong result.

To better understand the situation we define

$$D_1 := \{x : x \in D,\ M_{11}(x) \neq 0\},$$
$$D_2 := \{x : x \in D,\ M_{11}(x) = 0\}.$$

The program GAUSS can be considered as a piecewise definition of the function F,

$$F(x) = \begin{cases} F(x) \text{ according to S1}, & \text{for } x \in D_1, \\ F(x) \text{ according to S2}, & \text{for } x \in D_2. \end{cases}$$

Normally, one is not too concerned about the domain of a function. But indeed in this case, we must be concerned.

Let $F|_{D_1}$ denote the restriction of F to D_1 and let $F|_{D_2}$ denote the restriction of F to D_2. Then, of course

$$F(x) = \begin{cases} (F|_{D_1})(x) & \text{for } x \in D_1, \\ (F|_{D_2})(x) & \text{for } x \in D_2. \end{cases}$$

The domain D_1 of the function $F|_{D_1}$ is an open set, $x \in D_1$ is an interior point of D_1, and hence

$$F'(x) = (F|_{D_1})'(x) \quad \text{for } x \in D_1,$$

and this is the value GAUSS' produces.

The domain D_2 of the function $F|_{D_2}$ is too thin, it has no interior points, and hence $F|_{D_2}$ is not differentiable. In other words, the function $F|_{D_2}$ does not provide enough information to obtain $F'(x)$ for $x \in D_2$. Thus GAUSS' *cannot* produce $F'(x)$ for $x \in D_2$. What GAUSS' actually presents for $F'(x)$ is the value for the derivative of another function, which is of no interest here. For more see [1].

In [4] it is claimed that the use of a certain branching function method makes the branch problem vanish.

This is true in certain cases, in our example the branching function method fails because it encounters division by zero. At least this suggests that something went wrong. For a partial solution to the branch problem, see [1] and for a nonstandard treatment of the branch problem, see [6].

A simple example of the branch problem is shown in the informal program

> IF $x \neq 1$ THEN $f(x) \leftarrow x \cdot x$
> ELSE $f(x) \leftarrow 1$.

This program defines the function

$$f: \mathbb{R} \to \mathbb{R} \quad \text{with } f(x) = x^2.$$

Of course, f is differentiable, in particular we have $f'(1) = 2$.

Standard AD software produces the wrong result $f'(1) = 0$. It is not surprising that symbolic manipulation packages produce the same wrong result. Here it is obvious that the else-branch does not carry enough information for computing the correct $f'(1)$.

Sometimes branching is done to *save work*. Consider the function

$$f: D \subseteq \mathbb{R}^n \to \mathbb{R}$$

with

$$f(x) = s(x) + c(x) \cdot E(x),$$

where D is an open set. The real-valued functions s, c, E may be given explicitly or by subroutines. Assume that $f(x)$ has to be evaluated many times for varying x-s, that $c(x) = 0$ for many interesting values of x, and that $E(x)$ is computationally costly. Then it is effective to set up a program for computing $f(x)$ as shown in Table 5.

Assume that the functions s, c, E are differentiable. Then f is differentiable too. For given $x \in D$ we ask for $f'(x)$.

Standard AD (in the forward mode) transforms SW into a new program by inserting assignment statements concerning derivatives. The resulting program SW' is well-defined, and for given $x \in D$ it is supposed to produce $f(x)$ and $f'(x)$.

Define the sets

$$D_1 := \{x: x \in D, \ c(x) \neq 0\},$$
$$D_2 := \{x: x \in D, \ c(x) = 0\}.$$

Automatic Differentiation: Root Problem and Branch Problem, Table 5
Program *SW* for computing f(x)

input: $x \in D$		
$c(x)$	$\leftarrow$	$\ldots$
IF $c(x) \neq 0$ THEN		
S1: $s(x)$	$\leftarrow$	$\ldots$
$E(x)$	$\leftarrow$	$\ldots$
$r(x)$	$\leftarrow$	$s(x) + c(x) \cdot E(x)$
$f(x)$	$\leftarrow$	$r(x)$
ELSE		
S2: $s(x)$	$\leftarrow$	$\ldots$
$f(x)$	$\leftarrow$	$s(x)$
output: $f(x)$		

SW' works correctly to produce

$$f'(x) = r'(x) \quad \text{for } x \in D_1.$$

Looking at SW, it is tempting to assume:

$$f'(x) = s'(x) \quad \text{for } x \in D_2$$

and SW' actually follows this assumption. But it is clear that

$$f'(x) = s'(x) + E(x) \cdot c'(x) + c(x) \cdot E'(x)$$
for $x \in D$,

and in particular

$$f'(x) = s'(x) + E(x) \cdot c'(x)$$

for $x \in D_2$.

If $x \in D_2$, and if either $E(x) = 0$ or $c'(x) = 0$, then SW' produces the correct $F'(x)$, otherwise SW' fails.

See also

▶ Automatic Differentiation: Calculation of the Hessian
▶ Automatic Differentiation: Calculation of Newton Steps
▶ Automatic Differentiation: Geometry of Satellites and Tracking Stations
▶ Automatic Differentiation: Introduction, History and Rounding Error Estimation
▶ Automatic Differentiation: Parallel Computation

- ▶ Automatic Differentiation: Point and Interval
- ▶ Automatic Differentiation: Point and Interval Taylor Operators
- ▶ Nonlocal Sensitivity Analysis with Automatic Differentiation

References

1. Beck T, Fischer H (1994) The if-problem in automatic differentiation. J Comput Appl Math 50:119–131
2. Berz M, Bischof Ch, Corliss GF, Griewank A (eds) (1996) Computational differentiation: Techniques, applications, and tools. SIAM, Philadelphia
3. Griewank A, Corliss GF (eds) (1991) Automatic differentiation of algorithms: Theory, implementation, and application. SIAM, Philadelphia
4. Kearfott RB (1996) Rigorous global search: Continuous problems. Kluwer, Dordrecht
5. Rall LB (1981) Automatic differentiation: Techniques and applications. Lecture Notes Computer Sci, vol 120. Springer, Berlin
6. Shamseddine K, Berz M (1996) Exception handling in derivative computation with nonarchimedean calculus. In: Berz M, Bischof Ch, Corliss GF, Griewank A (eds) Computational Differentiation: Techniques, Applications, and Tools. SIAM, Philadelphia, pp 37–51

Bayesian Global Optimization

BA

JONAS MOCKUS
Institute Math. and Informatics, Vilnius, Lithuania

MSC2000: 90C26, 90C10, 90C15, 65K05, 62C10

Article Outline

Keywords
See also
References

Keywords

Global optimization; Discrete optimization; Bayesian approach; Heuristics

The traditional numerical analysis considers optimization algorithms which guarantee some accuracy for all functions to be optimized. This includes the exact algorithms (that is the worst-case analysis). Limiting the maximal error requires a computational effort that often increases exponentially with the size of the problem. An alternative is *average case analysis* where the average error is made as small as possible. The average is taken over a set of functions to be optimized. The average case analysis is called the *Bayesian approach* (BA) [7,14].

There are several ways of applying the BA in optimization. The direct Bayesian approach (DBA) is defined by fixing a *prior distribution P* on a set of functions $f(x)$ and by minimizing the Bayesian risk function $R(x)$ [6,14]. The risk function describes the average deviation from the global minimum. The distribution P is regarded as a stochastic model of $f(x)$, $x \in \mathbf{R}^m$, where $f(x)$ might be a deterministic or a stochastic function. In the Gaussian case assuming (see [14] that the $(n+1)$th observation is the last one

$$R(x) = \frac{1}{\sqrt{2\pi}\, s_n(x)} \int_{-\infty}^{+\infty} \min(c_n, z) e^{-\frac{1}{2}\left(\frac{y - m_n(x)}{s_n(x)}\right)^2} dz, \quad (1)$$

Here, $c_n = \min_i z_i - \epsilon$, $z_i = f(x_i)$, $m_n(x)$ is the conditional expectation given the values of z_i, $i = 1, \ldots n$, $d_n(x)$ is the conditional variance, and $\epsilon > 0$ is a correction parameter.

The objective of DBA (used mainly in continuous cases) is to provide as small average error as possible while keeping the convergence conditions.

The *Bayesian heuristic approach* (BHA) means fixing a prior distribution P on a set of functions $f_K(x)$ that define the best values obtained using K times some heuristic $h(x)$ to optimize a function $v(y)$ of variables $y \in \mathbf{R}^n$ [15]. As usual, the components of y are discrete variables. The heuristic $h(x)$ defines an expert opinion about the decision priorities. It is assumed that the heuristics or their 'mixture' depend on some continuous parameters $x \in \mathbf{R}^m$, where $m < n$.

The Bayesian stopping rules (BSR) [3] define the best on average stopping rule. In the BSR, the prior distribution is determined regarding only those features of the objective function $f(x)$ which are relevant for the stopping of the algorithm of global optimization.

Now all these ways will be considered in detail starting from the DBA. The Wiener process is common [11,16,19] as a stochastic model applying the DBA in the one-dimensional case $m = 1$.

The *Wiener model* implies that almost all the sample functions $f(x)$ are continuous, that increments $f(x_4) - f(x_3)$ and $f(x_2) - f(x_1)$, $x_1 < x_2 < x_3 < x_4$ are stochastically independent, and that $f(x)$ is Gaussian $(0, \sigma x)$ at any fixed $x > 0$. Note that the Wiener process originally provided a mathematical model of a particle in Brownian motion.

The Wiener model is extended to multidimensional case, too [14]. However, simple approximate stochastic models are preferable, if $m > 1$. These models are designed by replacing the traditional Kolmogorov consistency conditions because they require the inversion of matrices of nth order for computing the conditional expectation $m_n(x)$ and variance $d_n(x)$. The favorable exception is the Markov process, including the Wiener one. Extending the Wiener process to $m > 1$ the Markovian property disappears.

Replacing the regular consistency conditions by:
- continuity of the risk function $R(x)$;
- convergence of x_n to the global minimum;
- simplicity of expressions of $m_n(x)$ and $s_n(x)$,

the following simple expression of $R(x)$ is obtained using the results of [14]:

$$R(x) = \min_{1 \leq i \leq n} z_i - \min_{1 \leq i \leq n} \frac{\|x - x_i\|^2}{z_i - c_n}.$$

The aim of the DBA is to minimize the expected deviation. In addition, DBA has some good asymptotic properties, too. It is shown in [14] that

$$\frac{d^*}{d_a} = \left(\frac{f_a - f^* + \epsilon}{\epsilon}\right)^{1/2}, \quad n \to \infty,$$

where d^* is the density of x_i around the global optimum f^*, d_a and f_a are the average density of x_i and the average value of $f(x)$, and ϵ is the correction parameter in expression (1). That means that DBA provides convergence to the global minimum for any continuous $f(x)$ and greater density of observations x_i around the global optimum, if n is large. Note that the correction parameter ϵ has a similar influence as the temperature in simulated annealing. However, that is a superficial similarity. Using DBA, the good asymptotic behavior should be regarded just as an interesting 'by-product'. The reason is that Bayesian decisions are applied for the small size samples where asymptotic properties are not noticeable.

Choosing the optimal point x_{n+1} for the next iteration by DBA one solves a complicated auxiliary optimization problem minimizing the expected deviation

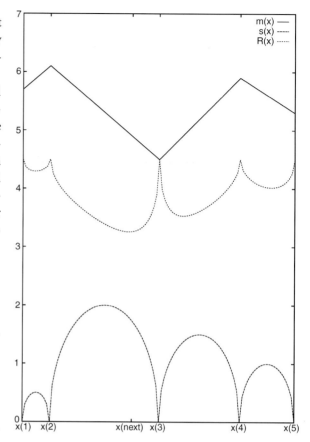

Bayesian Global Optimization, Figure 1
The Wiener model

$R(x)$ from the global optimum (see Fig. 1). That makes the DBA useful mainly for the computationally expensive functions of a few ($m < 20$) continuous variables. This happens in wide variety of problems such as maximization of the yield of differential amplifiers, optimization of mechanical system of shock absorber, optimization of composite laminates, estimation of parameters of immunological model and nonlinear time series, planning of extremal experiments on thermostable polymeric composition [14].

Using DBA the expert knowledge is included by defining the prior distribution. In BHA the expert knowledge is involved by defining the heuristics and optimizing their parameters using DBA.

If the number of variables is large and the objective function is not expensive, the Bayesian heuristic approach is preferable. That is the case in many discrete optimization problems. As usual, these problems are solved using heuristics based on an expert opinion. Heuristics often involve randomization procedures depending on some empirically defined parameters. The examples of such parameters are the initial temperature, if the *simulated annealing* is applied, or the probabilities of different randomization algorithms, if their mixture is used. In these problems, the DBA is a convenient tool for optimization of the continuous parameters of various heuristic techniques. That is the Bayesian heuristic approach [15].

The example of *knapsack problem* illustrates the basic principles of BHA in *discrete optimization*. Given a set of objects $j = 1, \ldots, n$ with values c_j and weights g_j, find the most valuable collection of limited weight:

$$\begin{cases} \max_y \quad v(y) = v(y) = \sum_{j=1}^n c_j y_j \\ \text{s.t.} \quad \sum_{j=1}^n g_j y_j \leq g. \end{cases}$$

Here the objective function $v(y)$ depends on n Boolean variables $y = (y_1, \ldots, y_n)$, where $y_j = 1$ if object j is in the collection, and $y_j = 0$ otherwise. The well-known greedy heuristics $h_j = c_j/g_j$ is the specific value of object j. The greedy heuristic algorithm: 'take the greatest feasible h_j', is very fast but it may get stuck in some nonoptimal decision.

A way to force the heuristic algorithm out of such nonoptimal decisions is to make decision j with probability $r_j = \rho x(h_j)$, where $\rho_x(h_j)$ is an increasing function of h_j and $x = (x_1, \ldots x_m)$ is a parameter vector. The DBA is used to optimize the parameters x by minimizing the best result $f_K(x)$ obtained applying K times the randomized heuristic algorithm $\rho_x(h_j)$. That is the most expensive operation of BHA. Therefore, the parallel computation of $f_K(x)$ should be used when possible reducing the computing time in proportion to a number of parallel processors.

Optimization of x adapts the heuristic algorithm $\rho_x(h_j)$ to a given problem. Let us illustrate the parameterization of $\rho_x(h_j)$ using three randomization functions: $r_i^l = h_i^l / \sum_j h_j^l$, $l = 0, 1, \infty$. Here, the upper index $l = 0$ denotes the uniformly distributed component and $l = 1$ defines the linear component of randomization. The index ∞ denotes the pure heuristics with no randomization where $r_i^\infty = 1$ if $h_i = \max_j h_j$ and $r_i^\infty = 0$, otherwise. Here, parameter $x = (x_0, x_1, x_\infty)$ defines the probabilities of using randomizations $l = 0, 1, \infty$ correspondingly. The optimal x may be applied in different but related problems, too [15]. That is very important in the 'on-line' optimization adapting the BHA algorithms to some unpredicted changes.

Another simple example of BHA application is by trying different permutations of some feasible solution y^0. Then heuristics are defined as the difference $h_i = v(y^i) - v(y^0)$ between the permuted solution y^i and the original one y^0. The well-known simulated annealing algorithm illustrates the parameterization of $\rho_x(h_j)$ related to a single parameter x. Here the probability of accepting a worse solution is equal to $e^{-h_i/x}$, where x is the 'annealing temperature'.

The comparison of BHA with exact branch and bound algorithms solving a set of the flow-show problems is shown by the Table from [15]:

$R = 100, K = 1, J = 10, S = 10, O = 10$					
Technique	f_B	d_B	x_0	x_1	x_∞
BHA	6.18	0.13	0.28	0.45	0.26
CPLEX	12.23	0.00	–	–	–

Here S is the number of tools, J is the number of jobs, O is the number of operations, f_B, x_0, x_1, x_∞ are the mean results, d_B is the variance, and 'CPLEX' denotes the standard MILP technique truncated after 5000 iterations. The table shows that in the randomly generated *flow-shop problems* the average make-span obtained by BHA was almost twice less that obtained by the exact branch and bound procedure truncated at the same time as BHA. The important conclusion is that stopping the exact methods before they reach the exact solution is not a good way to obtain the approximate solution.

The BHA has been used to solve the *batch scheduling* [15] and the clustering (parameter grouping) problems. In the clustering problem the only parameter x was the *initial annealing temperature* [8].

The main objective of BHA is to improve any given heuristic by defining the best parameters and/or the best 'mixtures' of different heuristics. Heuristic decision rules mixed and adapted by BHA often outperform (in terms of speed) even the best individual heuristics as judged by the considered examples. In addition, BHA provides almost sure convergence. However, the final results of BHA depend on the quality of the specific heuristics including the expert knowledge. That means the BHA should be regarded as a tool for *enhancing the heuristics* but not for replacing them.

Many well-known optimization algorithms such as genetic algorithms (GA) [10], *GRASP* [13], and *tabu search* (TS) [14], may be regarded as generalized heuristics that can be improved using BHA. There are many heuristics tailored to fit specific problems. For example, the Gupta heuristic was the best one while applying BHA to the flow-shop problem [15].

Genetic algorithms [10] is an important 'source' of interesting and useful stochastic search heuristics. It is well known [2] that the results of the genetic algorithms depend on the mutation and cross-over parameters. The Bayesian heuristic approach could be used in optimizing those parameters.

In the GRASP system [13] the heuristic is repeated many times. During each iteration a greedy randomized solution is constructed and the neighborhood around that solution is searched for the local optimum. The 'greedy' component constructs a solution, one element at a time until a solution is constructed. A possible application of the BHA in GRASP is in optimizing a random selection of a candidate to be in the solution because different random selection rules could be used and their best parameters should be defined. BHA might be useful as a local component, too, by randomizing the local decisions and optimizing the corresponding parameters.

In tabu search the issues of identifying best combinations of short and long term memory and best balances of intensification and diversification strategies may be obtained using BHA.

Hence the Bayesian heuristics approach may be considered when applying almost any stochastic or heuristic algorithm of discrete optimization. The proven convergence of a discrete search method (see, for example, [1]) is an asset. Otherwise, the convergence conditions are provided by tuning the BHA [15], if needed.

The third way to apply the Bayesian approach is the *Bayesian stopping rules* (BSR) [3]. The first way, the DBA, considers a stochastic model of the whole function to be optimized. In the BSR the stochastic models regard only the features of the objective function which are relevant for the stopping of the multistart algorithm.

In [20] a statistical estimate of the structure of multimodal problems is investigated. The results are applied developing BSR for the multistart global optimization methods [4,5,18].

Besides these three ways, there are other ways to apply the Bayesian approach in global optimization. For example, the Bayes theorem was used to derive the posterior distribution of the values of parameters in the simulated annealing algorithm to make an optimal choice in the trade-off between small steps in the control parameter and short Markov chains and large steps and long Markov chains [12].

In the information approach [17] a prior distribution is considered on the location parameter α of the global optimum of an one-dimensional objective function. Then an estimate of α is obtained maximizing the likelihood function after a number of evaluations of the objective function. This estimate is assumed as the next search point. For the solution of multidimensional problems, it is proposed to transform the problem into a one-dimensional problem by means of *Peano maps*.

See also

- Adaptive Simulated Annealing and its Application to Protein Folding
- Genetic Algorithms for Protein Structure Prediction
- Global Optimization Based on Statistical Models
- Monte-Carlo Simulated Annealing in Protein Folding
- Packet Annealing
- Random Search Methods
- Simulated Annealing
- Simulated Annealing Methods in Protein Folding
- Stochastic Global Optimization: Stopping Rules
- Stochastic Global Optimization: Two-Phase Methods

References

1. Andradottir S (1996) A global serach method for discrete stochastic optimization. SIAM J Optim 6:513–530
2. Androulakis IP, Venkatasubramanian V (1991) A genetic algorithmic framework for process design and optimization. Comput Chem Eng 15:217–228
3. Betro B (1991) Bayesian methods of global optimization. J Global Optim 1:1–14

4. Betro B, Schoen F (1987) Sequential stopping rules for the multistart algorithm in global optimization. Math Program 38:271–286
5. Boender G, Rinnoy-Kan A (1987) Bayesian stopping rules for multi-start global optimization methods. Math Program 37:59–80
6. DeGroot M (1970) Optimal statistical decisions. McGraw-Hill, New York
7. Diaconis P (1988) Bayesian numerical analysis. In: Statistical Decision Theory and Related Topics. Springer, Berlin, pp 163–175
8. Dzemyda G, Senkiene E (1990) Simulated annealing for parameter grouping. Trans Inform Th, Statistical Decision Th, Random Processes, 373–383
9. Glover F (1994) Tabu search: improved solution, alternatives. In: Mathematical Programming. State of the Art. Univ. Michigan, Ann Arbor, MI, pp 64–92
10. Goldberg DE (1989) Genetic algorithms in search, optimization, and machine learning. Addison-Wesley, Reading, MA
11. Kushner HJ (1964) A new method of locating the maximum point of an arbitrary multipeak curve in the presence of noise. J Basic Eng 86:97–100
12. van Laarhoven PJM, Boender CGE, Aarts EHL, Rinnooy-Kan AHG (1989) A Bayesian approach to simulated annealing. Probab Eng Inform Sci 3:453–475
13. Mavridou T, Pardalos PM, Pitsoulis LS, Resende MGC (1997) A GRASP for the biquadratic assignment problem. Europ J Oper Res
14. Mockus J (1989) Bayesian approach to global optimization. Kluwer, Dordrecht
15. Mockus J, Eddy W, Mockus A, Mockus L, Reklaitis G (1997) Bayesian heuristic approach to discrete and global optimization. Kluwer, Dordrecht
16. Saltenis VR (1971) On a metod of multiextremal optimization. Automatics and Computers (Avtomatika i Vychislitelnayya Tekchnika) 3:33–38. (In Russian)
17. Strongin RG (1978) Numerical methods in multi-extremal problems. Nauka, Moscow
18. Timmer GT (1984) Global optimization: A stochastic approach. PhD Thesis, Erasmus Univ. Rotterdam, The Netherlands
19. Törn A, Žilinskas A (1989) Global optimization. Springer, Berlin
20. Zielinski R (1981) A statistical estimate of the sructure of multiextremal problems. Math Program 21:348–356

Bayesian Networks

ALLA R. KAMMERDINER
Department of Industrial and Systems Engineering, University of Florida, Gainesville, USA

Article Outline

Keywords
Synonyms
Introduction
Definitions
   The Chain Rule for Bayesian Network
   Cases/Models
   Methods
   Applications
See also
References

Keywords

Graphical models; Joint probability distribution; Bayesian statistics; Data mining; Optimization

Synonyms

Bayes nets

Introduction

After the initial introduction in 1982, Bayesian networks (BN) have quickly developed into a dynamic area of research. This is largely due to the special structure of Bayesian networks that allows them to be very efficient in modeling domains with inherent uncertainty. In addition, there is a strong connection between Bayesian networks and other adjacent areas of research, including data mining and optimization.

Bayesian networks have their lineage in statistics, and were first formally introduced in the field of artificial intelligence and expert systems by Pearl [17] in 1982 and Spiegelhalter and Knill-Jones [21] in 1984. The first real-life applications of Bayesian networks were Munin [1] in 1989 and Pathfinder [7] in 1992. Since the 1990s, the amount of research in Bayesian networks has increased dramatically, resulting in many modern applications of Bayesian networks to various problems of data mining, pattern recognition, image processing and data fusion, engineering, etc.

Bayesian networks comprise a class of interesting special cases, many of which were in consideration long before the first introduction of Bayesian networks. Among such interesting cases are some frequently used types of the model simplifying assumptions including naïve Bayes, the noisy-OR and noisy-AND models, as well as different models with specialized structure, in particular the time-stamped models, the strictly repetitive models, dynamic Bayesian networks, hidden Markov models, Kalman filter, Markov chains. Artificial neural networks are another subclass of Bayesian networks, which has many applications, in particular in biology and computer science.

Definitions

Based on classical probability calculus, the idea of a Bayesian network has its early origins in Bayesian statistics. On the other hand, it has an added benefit of incorporating the notions of graph theory and networks that allows us to visualize the relationships between the variables represented by the nodes of a Bayesian network. In other words, a Bayesian network is a graphical model providing a compact representation for communicating causal relationships in a knowledge domain. Below we introduce two alternative definitions of the general notion of a Bayesian network, based on the usual concepts of probability and graph theory (e. g. joint probability distribution, conditional probability distribution; nodes and edges of a graph, a parent of a node, a child of a node, etc.).

Roughly speaking, a Bayesian network can be viewed as an application of Bayesian calculus on a causal network. More precisely, one can describe a Bayesian network as a mathematical model for representing the joint distribution of some set of random variables as a graph with the edges characterized by the conditional distributions for each variable given its parents in the graph.

Given a finite collection of random variables $X = \{X_1, X_2, \ldots, X_n\}$, the formal definition of a Bayesian network can be stated as follows:

Definition 1 A *Bayesian network* is an ordered pair (G,D), where
- The first component G represents a directed acyclic graph with nodes, which correspond to the random variables $X_1, X_2, \ldots, X_n$, and directed arcs, which symbolize conditional dependencies between the variables. The set of all the arcs of G satisfies the following assumption: Each random variable in the graph is conditionally independent of its non-descendants in G, given its parents in G.
- The second component D corresponds to the set of parameters that, for each variable $X_i, 1 \leq i \leq n$, define its conditional distribution given its parents in the graph G.

Note that the variables in a Bayesian networks can follow discrete or continuous distributions. Clearly, for continuously distributed variables, there is a correspondent conditional probability density function $f(x_i|Pa(x_i))$ of X_i given its parents $Pa(X_i)$. (From now on we denote by x_i the realization of the correspondent random variable X_i.)

In many real-life applications modeled by Bayesian networks the set of states for each variable (node) in the network is finite. In the special case when all variables have finite sets of mutually exclusive states and follow the discrete distributions, the previous definition of a Bayesian network can be reformulated in the following fashion:

Definition 2 A *Bayesian network* is a structure that consists of the following elements:
- A collection of variables with a finite set of mutually exclusive states;
- A set of directed arcs between the variables symbolizing conditional independence of variables;
- A directed acyclic graph formed by the variables and the arcs between them;
- A potential table $\Pr(X_i|Pa(X_i))$ associated with each variable X_i having a set of parent variables denoted by $Pa(X_i)$.

Observe that we do not require causality in Bayesian networks, i. e. the arcs of a graph do not have to symbolize causal relationship between the variables. However, it is imperative that the so-called *d-separation* rules implied by the structure are satisfied [12,19]. If variables X and Y are *d-separated* in a Bayesian network under the presence of evidence e, then $\Pr(X|Y, e) = \Pr(X|e)$, i. e. the variables are conditionally independent given the evidence.

Furthermore, the *d-separation* rules are applied to prove one of the key laws used in Bayesian networks, a so-called *chain rule for Bayesian networks*.

The joint probability table $\Pr(X) = \Pr(X_1, X_2, \ldots, X_n)$ sufficiently describes the belief structure on the set $X = \{X_1, X_2, \ldots, X_n\}$ of variables in the model. In particular, for each variable X_i, using the joint probability table, one can easily calculate the prior probabilities $\Pr(X_i)$ as well as the conditional probability $\Pr(X_i|e)$ given an evidence e. Nevertheless, with increase in the number of variables, the joint probability table quickly becomes unmanageably large, since the table size grows exponentially fast with the size n of the variable set. Thus, it is necessary to find another representation, which adequately and more efficiently describes the belief structure in the model. A Bayesian network over $X = \{X_1, X_2, \ldots, X_n\}$ provides such a representation. In fact, a graph in a Bayesian network gives a compact representation of conditional dependencies in the network, which allows one to compute the joint probability table from the conditional probabilities specified by the network using the chain rule below.

The Chain Rule for Bayesian Networks [8]

The joint probability distribution $\Pr(X) = \Pr(X_1, X_2, \ldots, X_n)$ of the variables $X = \{X_1, X_2, \ldots, X_n\}$ in a Bayesian network is given by the formula

$$\Pr(X) = \prod_{i=1}^{n} \Pr(X_i|Pa(X_i)), \qquad (1)$$

where $Pa(X_i)$ denotes the set of all parents of variable X_i.

The chain rule for Bayesian networks also provides an efficient way for probability updating when the new information is received about the model. There is a variety of different types of such new information, i. e. evidence. Two most common types of evidence are *finding* and *likelihood evidence*. Finding is evidence that specifies which states are possible for some variables, while likelihood evidence gives a proportion between the probabilities of two given states. Note that some types of evidence including likelihood evidence cannot be given in the form of findings.

Cases/Models

Bayesian networks provide a general framework for a number of specialized models, many of which were identified long before the concept of a Bayesian network was proposed. Such special cases of BN vary in their graph structures as well as the probability distribution.

The probability distributions for a Bayesian network can be defined in several ways. In some situations, it is possible to use theoretically well-defined distributions. In others, the probabilities can be estimated from data as frequencies. In addition, absolutely subjective probability estimates are often used for practical purposes. For instance, when the number of conditional probability distributions to acquire from the data is very large, some simplifying assumptions may be appropriate.

The simplest Bayesian network model is the well-known *naïve Bayes* (or *simple Bayes*) *model* [4], which can be summarized as follows:

- The graph structure of the model consists of one hypothesis variable H, and a finite set of information variables $I = \{I_1, I_2, \ldots, I_n\}$ with the arcs from H to every $I_k, 1 \leq k \leq n$. In other words, the variables form a diverging connection, where the hypothesis variable H is a common parent of variables $I_1, I_2, \ldots, I_n$;
- The probability distributions are given by the values $\Pr(I_k|H)$, for every information variable $I_k, 1 \leq k \leq n$.

The probability updating procedure based on the naïve Bayes model works in the following manner: Given a collection of observations $e_1, e_2, \ldots, e_n$ on the variables $I_1, I_2, \ldots, I_n$ respectively, the *likelihood* of H given $e_1, e_2, \ldots, e_n$ is computed:

$$L(H|e_1, e_2, \ldots, e_n) = \prod_{i=1}^{n} \Pr(e_i|H). \qquad (2)$$

Then the posterior probability of H is obtained from the formula:

$$\Pr(H|e_1, e_2, \ldots, e_n) = C \cdot \Pr(H) \cdot L(H|e_1, e_2, \ldots, e_n), \qquad (3)$$

where C is a normalization constant.

Another special case of BNs is a model underlined by the simplifying assumption called *noisy-OR* [18]. This model can be constructed as follows:

Let $A_1, A_2, \ldots, A_n$ represent some binary variables listing all parents of a binary variable B. Each event $A_i = x, x \in \{0, 1\}$, causes $B = x$ except when an *inhibitor* prevents it, with the probability p_i, i. e. $\Pr(B = 1 - x|A_i = x) = p_i$. Suppose that all inhibitors are independent.

Then the graph of a corresponding Bayesian network is represented by the converging connection with B as the child node of $A_1, A_2, \ldots, A_n$, while the conditional probabilities are given by $\Pr(B = x | A_i = x) = 1 - p_i$. Since the conditional distributions are independent of each other, then

$$\Pr(B = 1 - x | A_1, A_2, \ldots, A_n) = \prod_{i=1}^{n} p_i . \quad (4)$$

The noisy-OR assumption gives a significant advantage for efficient probability updating, since the number of distributions increases linearly with respect to the number of parents.

The construction complementary to noisy-OR is called *noisy-AND*. In the noisy-AND model, the graph is the convergent connection just as in the noisy-OR model, all the causes are required to be on in order to have an effect, and all the causes have mutually independent random inhibitors. Both noisy-OR and noisy-AND are special cases of a general method called *noisy functional dependence*.

Many modeling approaches have been developed which employ introduction of mediating variables in a Bayesian network. One of these methods, called *divorcing*, is the process separating parents $A_1, A_2, \ldots, A_i$ and $A_{i+1}, \ldots, A_n$ of a node B by introducing a mediating variable C as a child of divorced parent nodes $A_1, A_2, \ldots, A_i$ and a parent of the initial child node B. The divorcing of $A_1, A_2, \ldots, A_i$ is possible if the following condition is satisfied:

The set Γ of all configurations of $A_1, A_2, \ldots, A_i$ can be partitioned into the sets $c_1, c_2, \ldots, c_s$ so that for every $1 \leq j \leq m$, any two configurations $\gamma_1, \gamma_2 \in c_j$ have the same conditional probabilities:

$$\Pr(B|\gamma_1, A_{i+1}, \ldots, A_n) = \Pr(B|\gamma_2, A_{i+1}, \ldots, A_n) . \quad (5)$$

Other modeling methods, which engage the mediating variables, involve modeling undirected relations, and situations with expert disagreement. Various types of undirected dependencies, including logical constraints, are represented by adding an artificial child C of the constrained nodes $A_1, A_2, \ldots, A_n$ so that the conditional probability $\Pr(C|A_1, A_2, \ldots, A_n)$ emulates the relation. The situation, where k experts disagree on the conditional probabilities for different variables $B_1, B_2, \ldots, B_n$ in the model can be modeled by introducing a mediating node M with k states $m_1, m_2, \ldots, m_k$ so that the variables $B_1, B_2, \ldots, B_n$ on whose probabilities the experts disagree become the only children of expert node M. Another approach to modeling expert disagreements is by introducing alternative models with weights assigned to each model.

An important type of Bayesian networks are so-called *time-stamped models* [10]. These models reflect the structure which changes over time. By introducing a discrete time stamp in such structures, the time-stamped models are partitioned into submodels for every unit of time. Each local submodel is called a *time slice*. The complete time-stamped model consists of all its time slices connected to each other by *temporal links*.

A *strictly repetitive* model is a special case of a time-stamped model such that all its time slices have the same structure and all the temporal links are alike. The well-studied *hidden Markov models* is a special class of strictly repetitive time-stamped models for which the Markov property holds, i. e. given the present, the past is independent of the future.

A hidden Markov model with only one variable in each time slice connected to the variables outside the time slice is a *Kalman filter*. Furthermore, a *Markov chain* can be represented as a Kalman filter with only one variable in every time slice. It is possible to convert a hidden Markov model into a Markov chain by cross-multiplying all variables in each time slice.

The time-stamped models can have either *finite horizon* or *infinite horizon*. An infinite Markov chain would be an example of a time-stamped model with an infinite horizon. Furthermore, the repetitive time-stamped models with infinite horizon are also known as *dynamical Bayesian networks*. By utilizing the special structure of many repetitive temporal models, they can be compactly represented [2]. Such special representation can often facilitate the design of efficient algorithms in updating procedures.

Artificial neural networks can also be viewed as a special case of Bayesian networks, where the nodes are partitioned into n mutually exclusive *layers*, and the set of arcs represented by the links from the nodes on layer i to the nodes on $i + 1, 1 \leq i \leq n$. Layer 1 is usually called the *input layer*, while layer n is known as the output layer.

Methods

Just as the BNs have their roots in statistics, the approaches for discovering a BN structure utilize statistical methods. That is why a database of cases is instrumental for discovery of the graph configuration of a Bayesian network as well as probability updating. There are three basic types of approaches to extracting BNs from data: *batch learning*, *adaptation*, and *tuning*.

Batch Learning. *Batch Learning* is a process of extracting the information from a database of collected cases in order to establish a graph structure and the probability distributions for a certain Bayesian networks.

Often there are many ways to model a Bayesian network. For example, we may obtain two different probability distributions to model the true distribution of the variable in the network. To make an intelligent choice between two available distributions, it is important to have an appropriate measure of their accuracy. A logical way to approach this subject is by assigning penalties for a wrong forecast on the base of a specified distribution. For example, two widely accepted ways for assigning penalties are the *quadratic (Brier) scoring rule* and the *logarithmic scoring rule*.

Given the true distribution $p = (p_1, p_2, \ldots, p_m)$ of a discrete random variable with m states, and some approximate distribution $q = (q_1, q_2, \ldots, q_m)$, the quadratic scoring rule assigns the expected penalty as:

$$ES_Q(p, q) = \sum_{i=1}^{m} p_i \left((1 - q_i)^2 + \sum_{j \neq i} q_j^2 \right). \qquad (6)$$

The distance between true distribution p and approximation q is given by the formula:

$$d_Q(p, q) = ES_Q(p, q) - ES_Q(p, p). \qquad (7)$$

Hence, from (6) we have:

$$d_Q(p, q) = \sum_{i=1}^{m} (p_i - q_i)^2. \qquad (8)$$

The distance $d_Q(p, q)$ given in (8) is called the *Euclidean distance*.

The logarithmic scoring rule assigns to each outcome i the corresponding penalty $S_L(q, i) = -\log q_i$.

Hence, the expected penalty is calculated as:

$$ES_L(p, q) = -\sum_{i=1}^{m} p_i \log q_i. \qquad (9)$$

From (7), we obtain an expression for the distance between the true distribution p, and the approximation q:

$$d_L(p, q) = \sum_{i=1}^{m} p_i \log \frac{p_i}{q_i}, \qquad (10)$$

which is called the *Kulbach–Leibler divergence*.

Note that both definitions, the Euclidean distance and the Kulbach–Leibler divergence, can be easily extended in the case of continuous random variables. Moreover, both scoring rules, the quadratic and the logarithmic, possess the following useful property: only the true distribution minimizes the score. The scoring rules that exhibit this property are called *strictly proper*. Since the quadratic and the logarithmic scoring rules are strictly proper, then the corresponding distance measures d_Q and d_L both satisfy the following:

$$d(p, q) = 0 \quad \text{if and only if} \quad p = q.$$

Different scoring rules and corresponding distance measures for discrete and continuous random variables have been extensively studied in statistics. A comprehensive review of strictly proper scoring rules is given in [6].

Naturally, among several different Bayesian networks that model the situation equally closely, the one of the smallest "size" would be preferred.

Let M denote a Bayesian network over the variable set $X = \{X_1, X_2, \ldots, X_n\}$. Then the size of M is given by

$$\text{Size}(M) = \sum_{i=1}^{n} s(X_i), \qquad (11)$$

where $s(X_i)$ denotes the number of entries in the conditional probability table $\Pr(X_i | Pa(X_i))$, and $Pa(X_i)$ is the set of parents of X_i.

The following measure accounts for both the size of the model and its accuracy.

Given a Bayesian network M over X with the true probability distribution p, and an approximate Bayesian network model N with distribution q, we define the *acceptance measure* as

$$\alpha(p, N) = \text{Size}(N) + C \cdot d(p, q), \qquad (12)$$

where Size(·) is the network size defined by (11), $d(p,q)$ is a distance measure between probability distributions p and q, and C is a positive real constant.

The general approach to batch learning a Bayesian network from the data set of cases can be summarized as follows:
- Select an appropriate threshold τ for distance measure $d(p,q)$ between two distributions;
- Fix a suitable constant C in a definition of acceptance measure $\alpha(p,N)$;
- Among all Bayesian network models over X and distribution q such that $d(p,q) < \tau$, select the model that minimizes $\alpha(p,N)$.

Although simple, this approach has many practical issues. The data sets in batch learning are usually very large, the model space grows exponentially in the number of variables, there may be missing data in the data set, etc. To extract structure from such data, one often has to employ special heuristics for searching the model space. For instance, causality can be used to cluster the variables according to a causal hierarchy. In other words, we partition the variable set X into subsets $S_1, S_2, \ldots, S_k$, so that the arcs satisfy a partial order relation. If we find the model N having the distance $d(p,q) < \tau$, the search stops; otherwise we consider the submodel of N.

Adaptation It is often desirable to build a system capable of automatically adapting to different settings. *Adaptation* is a process of adjusting a Bayesian network model so that it is better able to accommodate to new accumulated cases.

When building a Bayesian network, usually there is an uncertainty whether the chosen conditional probabilities are correct. This is called the second-order uncertainty.

Suppose that we are not sure which table out of m different conditional probability tables $T_1, T_2, \ldots, T_m$ represents the true distribution for $Pr(X_i|Pa(X_i))$ for some variable X_i in a network. By introducing a so-called *type variable* T with states $t_1, t_2, \ldots, t_m$ into the graph so that T is a parent of X_i, we can model this uncertainty into the network. Then the prior probability $Pr(t_1, t_2, \ldots, t_m)$ represents our belief about the correctness of the tables $T_1, T_2, \ldots, T_m$ respectively. Next, we set $Pr(X_i|Pa(X_i), t_j) = T_j$. Our belief about the correctness of the tables is updated each time we receive new evidence e. In other words, for the next case, we use $Pr(t_1, t_2, \ldots, t_m|e)$ as the new prior probability of tables' accuracy.

Sometimes the second-order uncertainty about the conditional probabilities cannot be modeled by introducing type variables. In such cases, various statistical methods can be applied. Normally such methods exploit various properties of parameters, such as global independence, local independence, etc.

The property of *global independence* states that the second-order uncertainty for the variables is independent, i.e. the probability tables for the variables can be adjusted independently from each other.

The *local independence* property holds if and only if for any two different parent configurations π_1, π_2, the second-order uncertainty on $Pr(A|\pi_1)$ is independent of the second-order uncertainty on $Pr(A|\pi_2)$, and the two distributions can be updated independently from each other. In other words, local independence means the independence of the uncertainties of the distributions for different configurations of parents.

The *fractional updating* scheme [22], is an algorithm for reducing the second-order uncertainty about the distributions based on the received evidence. Suppose that the properties of global and local independence for the second-degree uncertainty hold simultaneously. For every configuration π of parents of variable X_i, the certainty about $Pr(X_i|\pi)$ is given through an artificially selected sample size parameter n_i, and for any state x_i^j of variable X_i we have a corresponding count $n_i^j = n_i \cdot Pr(x_i^j|\pi)$. After receiving an evidence e, we compute probabilities $Pr(x_i^j, \pi|e)$. Then the updated count n_i^j is the sum of $Pr(x_i^j, \pi|e)$ and the old n_i^j. Since $n_i = \sum_j n_i^j$, the old sample size parameter n_i becomes $n_i + Pr(\pi|e)$.

Although efficient in reducing the uncertainty about the distributions, this scheme has some serious drawbacks. In fact, it tends to reduce the second-degree uncertainty too fast, by overestimating the counts. In order to avoid this, one can introduce a so-called *fading factor f*. Then after receiving an evidence, the sample size n_i is changed to $f \cdot n_i + Pr(\pi|e)$, and the counts n_i^j are updated to $f \cdot n_i^j + Pr(x_i^j, \pi|e)$. Therefore, the fading factor f insures that the influence of the past decreases exponentially [16].

After describing some approaches in adapting a Bayesian network to different settings of distribution parameters, it is equally important to discuss the uncertainty in the graph structure. In many cases, we can compensate for the variability in the graph structure of a Bayesian network just by modifying the parameters of distributions in the network. Sometimes it may not be sufficient to adjust the distribution parameters in order to account for the change in the model. In fact, the difference in the graph structure may be so significant that it becomes impossible to accurately reflect the situation by a mere parameter change.

There are two main approaches to graph structure adaptation in Bayesian networks. The first method works by collecting the cases, and re-running the batch learning procedure to update the graph structure. The second method, also known as the expert disagreement approach, works simultaneously with a set of different models, and updates the weight of each model according to the evidence. More precisely, suppose there are m alternative models $M_1, M_2, \ldots, M_m$ with corresponding initial weights $w_1, w_2, \ldots, w_m$ that express our certainty of the models. Let Y be some variable in the network. After receiving an evidence e, we obtain the probabilities $\Pr_i(Y|e) := \Pr(Y|e, M_i)$ and $\Pr_i(e) := \Pr(e|M_i)$ according to each model M_i, for $1 \leq i \leq m$. Then,

$$\Pr(Y|e) := \sum_{i=1}^{m} w_i \cdot \Pr_i(Y|e), \qquad (13)$$

and the updated weights w_i are computed as the probabilities of the corresponding models M_i given the past evidence: $w_i = \Pr(M_i|e)$. Hence, by the well-known Bayes formula:

$$w_i = \frac{\Pr(e|M_i)\Pr(M_i)}{\sum_j w_j \Pr_j(e)}. \qquad (14)$$

Note that the expert disagreement approach to graph structure adaptation can be further extended to include the adaptation of distribution parameters based on the above methods, such as fractional updating.

Tuning Tuning is the process of adjusting the distribution parameters so that some prescribed requests for the model distributions are satisfied. The commonly used approach to tuning is the gradient descent on the parameters similar to training in neural networks.

Let τ represent the set of parameters which are chosen to be altered. Let $p(\tau)$ denote the current model distribution, and q be the target distribution. Suppose $d(p, q)$ represents the distance between two distributions. The following gradient descent tuning algorithm is given in [9]:
- Compute the gradient of $d(p, q)$ with respect to the parameters τ;
- Select a step size $\alpha > 0$, and let $\Delta \tau = -\alpha \cdot \vec{\nabla} d(p, q)(\tau_0)$, i. e. give τ_0 a displacement $\Delta \tau$ in the opposite direction to the gradient of $d(p, q)(\tau_0)$;
- Repeat this procedure until the gradient is sufficiently close to zero.

Evolutionary methods, simulated annealing, expectation-maximization and *non-parametric methods* are among other commonly used methods for tuning or training Bayesian networks.

Applications

The concept of a Bayesian network can be interpreted in different contexts. From a statistical point of view, a Bayesian network can be defined as a compact representation of the joint probability over a given set of variables. From a broader point of view, a Bayesian network is a special type of graphical model capable of reflecting causality, as well as updating its beliefs in view of received evidence. All these features make a Bayesian network a versatile instrument that can be used for various purposes, including facilitating communication between human and computer, extracting hidden information and patterns from data, simplifying decision making, etc.

Due to their special structure, Bayesian networks have found many applications in various areas such as artificial intelligence and expert systems, machine learning and data mining. Bayesian networks are used for modeling knowledge in text analysis, image processing, speech pattern analysis, data fusion, engineering, biomedicine, gene and protein regulatory networks, and even meteorology. Furthermore, it has been expressed that the inductive inference procedures based on Bayesian networks can be used to introduce inductive reasoning in such a previously strictly deductive science as mathematics.

The large scope of different applications of Bayesian networks is especially impressive when taking into account that the theory of Bayesian networks has only been around for about a quarter of a century. Next, several examples of recent real-life applications of Bayesian networks are considered to illustrate this point.

Recent research in the field of automatic speech recognition [13] indicates that dynamic Bayesian networks can effectively model hidden features in speech including articulatory and other phonological features. Both hidden Markov models (HMM), which are a special case of dynamic Bayesian networks (DBN), and more general dynamic Bayesian networks have been applied for modeling audio-visual speech recognition. In particular, a paper by A.V. Nefian et al. [15] describes an application of the *coupled HMM* and the *factorial HMM* as two suitable statistical models for audio-video integration. The factorial HMM is a generalization of HMM, where the hidden state is represented by a collection of variables also called *factors*. These factors, although independent of each other, all impact the observations, and hence become connected indirectly. The coupled HMM is a DBN represented as two regular HMM whose hidden state nodes have links to the hidden state nodes from the next time slice. The coupled HMM has also been applied to model hand gestures, the interaction between speech and hand gestures, etc. In addition, face detection and recognition problems have been studied with the help of Bayesian networks.

Note that different fields of application may call for specialized employment of Bayesian network methods, and conversely, similar approaches can be successfully used in different application areas. For instance, along with the applications to speech recognition above, coupled hidden Markov models have been employed in modeling multi-channel EEG (electroencephalogram) data.

An interesting example of the application of a Bayesian network to expert systems includes developing strategies for troubleshooting complex electro-mechanical systems, presented in [23]. The constructed Bayesian network has the structure of a naïve Bayes model. In the decision tree for the troubleshooting model, the utility function is given by the cost of repair. Hence, the goal is to find a strategy minimizing the expected cost of repair.

An interesting recent study [3] describes some applications of Bayesian networks in meteorology from a data mining point of view. A large database of daily observations of precipitation levels and maximum wind speed is collected. The Bayesian network structure is constructed from meteorological data by using various approaches, including batch learning procedure and simulation techniques. In addition, an important data mining application of Bayesian networks is illustrated by giving an example of missing data values estimation from the evidence received.

Applications of Bayesian Networks to Data Mining; Naïve Bayes Rapid progress in data collection techniques and data storage has enabled an accumulation of huge amounts of experimental, observational and operational data. As the result, massive data sets containing a large amount of information can be found almost everywhere. A well-known example is the data set containing the observed information about the human genome. The need to quickly and correctly analyze or manipulate such enormous data sets facilitated the development of data mining techniques.

Data mining is research aimed at discovery of various types of knowledge from large data warehouses. Data mining can also be seen as an integral part of the more general process of knowledge discovery in databases. Two other parts of this knowledge discovery are preprocessing and postprocessing. As seen above, Bayesian networks can also extract knowledge from data, which is called evidence in the Bayesian framework. In fact, the Bayesian network techniques can be applied to solve data mining problems, in particular, classification.

Many effective techniques in data mining utilize methods from other multidisciplinary research areas such as database systems, pattern recognition, machine learning, and statistics. Many of these areas have a close connection to Bayesian networks. In actuality, data mining utilizes a special case of Bayesian networks, namely, naïve Bayes, to perform effective classification. In a data mining context, classification is the task of assigning objects to their relevant categories. The incentive for performing classification of data is to attain a comprehensive understanding of differences and similarities between the objects in different classes.

In the Bayesian framework, the data mining classification problem translates into finding the class parameter which maximizes the posterior probability of the unknown instance. This statement is called the *maximum a posteriori principle*. As mentioned earlier, the naïve Bayes is an example of a simple Bayesian network model.

Similarly to the naïve Bayes classifier, classification by way of building suitable Bayesian networks is capable of handling the presence of noise in the data as well as the missing values. Artificial neural networks can serve as an example of the Bayesian network classifier designed for a special case.

Application to Global and Combinatorial Optimization In the late 1990s, a number of studies were conducted that described how BN methodology can be applied to solve problems of global and combinatorial optimization. The connection between graphical models (e. g. Bayesian networks) and evolutionary algorithms (applied to optimization problems) was established. In particular, P. Larrañaga et al. combined some techniques from learning BN's structure from data with an evolutionary computation procedure called the Estimation of Distribution Algorithm [11] to devise a procedure for solving combinatorial optimization problems. R. Etxerberria and P. Larrañaga proposed a similar approach for global optimization [5].

Another method based on learning and simulation of BNs that is known as the Bayesian Optimization Algorithm (BOA) was suggested by M. Pelikan et al. [20]. The method works by randomly generating an initial population of solutions and then updating the population by using selection and variation. The operation of selection makes multiple copies of better solutions and removes the worst ones. The operation of variation, at first, constructs a Bayesian network as a model of promising solutions following selection. Then new candidate solutions are obtained by sampling of the constructed Bayesian network. New solutions are incorporated into the population in place of some old candidate solutions, and the next iteration is executed unless a termination criterion is reached.

For additional information on some real-world applications of Bayesian networks to classification, reliability analysis, image processing, data fusion and bioinformatics, see the recent book edited by A. Mittal et al. [14].

See also

▶ Bayesian Global Optimization
▶ Evolutionary Algorithms in Combinatorial Optimization
▶ Neural Networks for Combinatorial Optimization

References

1. Andreassen S (1992) Knowledge representation by extended linear models. In: Keravnou E (ed) Deep Models for Medical Knowledge Engineering. Elsevier, pp 129–145
2. Bangsø O, Wuillemin PH (2000) Top-down Construction and Repetitive Structures Representation in Bayesian Networks, Proceedings of the Thirteenth International FLAIRS Conference. AIII Press, Cambridge, MA
3. Cano R, Sordo C, Gutierrez JM (2004) Applications of Bayesian Networks in Meteorology, Advances in Bayesian Networks. In: Gamez et al (eds) Springer, pp 309–327
4. de Dombal F, Leaper D, Staniland J, McCan A, Harrocks J (1972) Computer-aided diagnostics of acute abdominal pain. Brit Med J 2:9–13
5. Etxerberria R, Larrañaga P (1999) Global optimization with Bayesian networks, II Symposium on Artificial Intelligence, CIMAF-99. Special Session on Distribution and Evolutionary Optimization. ICIMAF, La Habana, Cuba, pp 332–339
6. Gneiting T, Raftery AE (2005) Strictly proper scoring rules, prediction, and estimation, Technical Report no. 463R. Department of Statistics, University of Washington
7. Heckerman D, Horvitz E, Nathwani B (1992) Towards normative expert systems: Part I, the Pathfinder project. Method Inf Med 31:90–105
8. Jensen FV (1996) An Introduction to Bayesian Networks. UCL Press, London
9. Jensen FV (1999) Gradient descent training of Bayesian networks, Proceedings of the Fifth European Conference on Symbolic and Quantitative Approaches to Reasoning with Uncertainty (ECSQARU). Springer, Berlin, pp 190–200
10. Kjærulff U (1995) HUGS: Combining exact inference and Gibbs sampling in junction trees, Proceedings of the Eleventh Conference on Artificial Intelligence. Morgan Kaufmann, San Francisco, CA, pp 368–375
11. Larrañaga P, Etxeberria R, Lozano JA, Peña JM (1999) Optimization by learning and simulation of Bayesian and Gaussian networks, Technical Report EHU-KZAA-IK-4/99. Department of Computer Science and Artificial Intelligence, University of the Basque Country
12. Lauritzen SL (1996) Graphical Models. Oxford University Press, Oxford
13. Livescu K, Glass J, Bilmes J (2003) Hidden feature modeling for speech recognition using dynamic Bayesian networks. Proc. EUROSPEECH, Geneva Switzerland, August–September

14. Mittal A, Kassim A, Tan T (2007) Bayesian Network Technologies: Applications and Graphical Models, Interface Graphics, Inc., Minneapolis, USA
15. Nefian AV, Liang L, Pi X, Liu X, Murphy K (2002) Dynamic Bayesian Networks for Audio-visual Speech Recognition. J Appl Signal Proc 11:1–15
16. Olesen KG, Lauritzen SL, Jensen FV (1992) aHUGIN: A system creating adaptive causal probabilistic networks, Proceedings of the Eighth Conference on Uncertainty in Artificial Intelligence. Morgan Kaufmann, San Francisco, pp 223–229
17. Pearl J (1982) Reverend Bayes on Inference Engines: A Distributed Hierarchical Approach, National Conference on Artificial Intelligence. AAAI Press, Menlo Park, CA, pp 133–136
18. Pearl J (1986) Fusion, propagation, and structuring in belief networks. Artif Intell 29(3):241–288
19. Pearl J (1988) Probabilistic Reasoning in Intelligent Systems: Networks of Plausible Inference Series in Representation and Reasoning. Morgan Kaufmann, San Francisco
20. Pelikan M, Goldberg DE, Cantú-Paz E (1999) BOA: The Bayesian Optimization Algorithm, Proceedings of the Genetic and Evolutionary Computation conference GECCO-99, vol 1. Morgan Kaufmann, San Francisco
21. Spiegelhalter DJ, Knill-Jones RP (1984) Statistical and knowledge-based approaches to clinical decision-support systems. J Royal Stat Soc A147:35–77
22. Spiegelhalter D, Lauritzen SL (1990) Sequential updating of conditional probabilities on directed graphical structures. Networks 20:579–605
23. Vomlel J (2003) Two applications of Bayesian networks, Proceedings of conference Znalosti. Ostrava, Czech Republic, pp 73–82

Beam Selection in Radiotherapy Treatment Design

ALLEN HOLDER
Department of Mathematics and the University of Texas Health Science Center at San Antonio, Department of Radiological Sciences, Trinity University, San Antonio, USA

Article Outline

Synonyms
Introduction
Definitions
Formulation
Models
Conclusions
See also
References

Synonyms

Beam orientation optimization; Beam angle optimization

Introduction

Cancer is typically treated with 3 standard procedures: 1) surgery – the intent of which is to physically rescind the disease, 2) chemotherapy – drug treatment that attacks fast proliferating cells, and 3) radiotherapy – the targeted treatment of cancer with ionizing beams of radiation. About half of all cancer patients receive radiotherapy, which is delivered by focusing high-energy beams of radiation on a patient's tumor(s). Treatment design is traditionally considered in three phases:

Beam Selection The process of deciding the number and trajectory of the beams that will pass through the patient.
Fluence Optimization Calculating the amount of dose to deliver along each of the selected beams so that the patient is treated as well as possible.
Delivery Optimization Deciding how to best deliver the treatment designed in the first two steps.

The fundamental question in optimizing radiotherapy treatments is how to best treat the patient, and such research requires detailed knowledge of medical physics and optimization. Unlike the numerous research pursuits within the field of optimization that require a specific expertise, the goals of this research rely on an overriding understanding of modeling, solving and analyzing optimization problems as well as an understanding of medical physics. The necessary spectrum of knowledge is commonly collected into a research group that is comprised of medical physicists, operations researchers, computer scientists, industrial engineers, and mathematicians.

In a modern clinic, the first phase of selecting beams is accomplished by a treatment planner, and hence, the quality of the resulting treatment depends on the expertise of this person. Fluence optimization is automatically conducted once beams are selected, and the resulting treatment is judged with a variety of metrics and visualization tools. If the treatment is acceptable, the process ends. However, unacceptable treatments are common, and in this scenario the collection of beams is updated and fluence optimization is repeated with the new beams. This trial-and-error approach oscillates between the first two phases of treatment design and often continues for hours until an acceptable treatment is rendered. The third phase of delivery optimization strives to orient the treatment machinery so that the patient is treated as efficiently as possible, where efficiency is interpreted as shortest delivery time, shortest exposure time, etc.

The focus of this entry is Beam Selection, which has a substantial literature in the medical physics community and a growing one in the operations research community. As one would expect, no single phase of treatment design exists in isolation, and although the three phase approach pervades contemporary thinking, readers should be aware that future efforts to optimize the totality of treatment design are being discussed. The presentation below is viewed as part of this bigger goal.

Definitions

An understanding of the technical terms used to describe radiotherapy is needed to understand the scope of Beam Selection. Patient images such as CAT scans or MRI images are used to identify and locate the extent of the disease. Treatment design begins with the tedious task of delineating the target and surrounding tissues on each of the hundreds of images. The resulting 3D structures are individually classified as either a target, a critical structure, or normal tissue. An oncologist prescribes a goal dose for the target and upper bounds on the remaining tissues. This prescription is tailored to the optimization model used in the second phase of treatment design and is far from unique. A discussion of the myriad of models used for fluence optimization exceeds the confines of this article and is fortunately not needed.

The method of treatment depends on the clinic's technology, and we begin with the general concepts common to all modalities. A patient lies on a treatment couch that can be moved vertically and horizontally and rotated in the plane horizontal to the floor. A gantry rotates around the patient in a great circle, the head of which is used to focus the beam on the patient, see Fig. 1. Shaping and modulating the beam is important

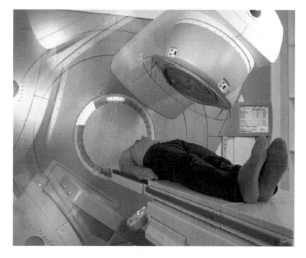

Beam Selection in Radiotherapy Treatment Design, Figure 1
A typical treatment configuration

Beam Selection in Radiotherapy Treatment Design, Figure 2
A multileaf collimator

in all forms of treatment, and although these tasks are accomplished differently depending on the technology, it is common to control smaller divisions of each beam called sub-beams. As an example, the gantry's head often contains a multileaf collimator that is capable of dividing the beam (Fig. 2), a technology that is modeled by replacing the whole beam with a grid of rectangular sub-beams. Previous technology shaped and modulated the beam without a collimator, but the concept of a sub-beam remains appropriate.

The center of the gantry's rotation is called the isocenter, a point that is placed near the center of the target by repositioning the patient via couch adjustments. The beam can essentially be focused on the patient from any point on a sphere with a one meter radius that encompasses the patient, although some positions are not possible due to patient-gantry interference. The beam selection problem is to choose a few of these positions so that the resulting treatment is of high quality. If the selection process is restricted to a single great circle, then the term beam is often replaced with angle (in fact these terms are used synonymously in much of the literature).

The collection of positions on the sphere from which we are allowed to select is denoted by $\mathcal{A}$. This set contains every point of the sphere in the continuum, but in practice $\mathcal{A}$ is a finite set of candidate beams. The problem of selecting beams depends on a judgment function, which is a mapping from the power set of $\mathcal{A}$, denoted $\mathcal{P}(\mathcal{A})$, into the nonnegative extended reals, denoted $\mathbb{R}_+^* = \{x \in \mathbb{R} : x \geq 0\} \cup \{\infty\}$. Assuming that low values correspond with high-quality treatments, we have that a judgment function is a mapping $f : \mathcal{P}(\mathcal{A}) \to \mathbb{R}_+^*$ with the monotonicity property that if $\mathcal{A}'$ and $\mathcal{A}''$ are subsets of $\mathcal{A}$ such that $\mathcal{A}' \supseteq \mathcal{A}''$, then $f(\mathcal{A}') \leq f(\mathcal{A}'')$. The monotonicity condition guarantees that treatment quality can not degrade if beams are added to an existing treatment.

The judgment function is commonly the optimal value from the second phase of treatment design, and for any $\mathcal{A}' \in \mathcal{P}(\mathcal{A})$, we let $X(\mathcal{A}')$ be the feasible region of the optimization problem that decides fluences. An algebraic description of this set relies on the fact that we can accurately model how radiation is deposited as it passes through the anatomy. There are several competing radiobiological models that accomplish this task, each of which produces the rate coefficient $A_{(j, a, i)}$, which is the rate at which sub-beam i in beam a deposits energy into the anatomical position j. These values form a dose matrix A, with rows being indexed by j and columns by (a, i). The term used to measure a sub-beam's energy is fluence, and experimentation validates that anatomical dose, which is measured in Grays (Gy), is linear in fluence. So, if $x_{(a, i)}$ is the fluence of sub-beam i in beam a, then the linear map $x \mapsto Ax$ transforms fluence values into anatomical dose. We partition the rows of the dose matrix into those that correspond with anatomical positions in the target – forming the submatrix A_T, in a critical structure – forming the submatrix A_C, and in normal tissue – forming the subma-

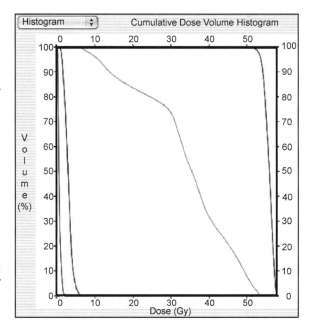

Beam Selection in Radiotherapy Treatment Design, Figure 3
A dose-volume histogram, the horizontal axis is the anatomical dose (measured in Grays) and the vertical axis is the percent of volume

trix A_N. With this notation, $A_T x$, $A_C x$ and $A_N x$ are the delivered doses to the target, the critical structures, and the normal tissues under treatment x.

Treatment planners use visual and numerical methods to evaluate treatments. The two most common visual tools are the dose-volume histogram (DVH) and a collection of isocontours. A DVH is a plot of dose versus volume and allows a treatment planner to quickly gauge the extent to which each structure is irradiated, an example is found in Fig. 3. The curve in the upper right side of the figure corresponds to the target, which is the growth to the left of the brain stem in Fig. 4. The ideal curve for the target would be one that remains at 100% until the desired dose and then falls immediately to zero, and the ideal curves for the remaining structures would be ones that fall immediately to zero. The curve passing through the middle of Fig. 3 corresponds to the brain stem and indicates that approximately 80% of the brain stem is receiving half of the target dose.

What a DVH lacks is spatial detail about the anatomical dose, but this information is provided by the isocontours, which are level curves drawn on each of the patient images. For example, if the target's goal is

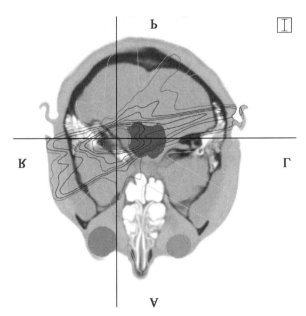

Beam Selection in Radiotherapy Treatment Design, Figure 4
A collection of isocontours on a single patient image

treatment. The prevailing thought is that fewer beams are preferred if all other treatment goals remain satisfactory, and if f adequately measures treatment quality, a model that represents this sentiment is

$$\min\{N : \min\{f(\mathcal{A}') : \mathcal{A}' \in \mathcal{P}(\mathcal{A}), |\mathcal{A}'| = N\} \leq \varepsilon\},$$

where ε defines the quality of an acceptable treatment.

As mentioned in the previous section, the judgment function is typically the objective value from fluence optimization. A common least-squares approach defines $X(\mathcal{A}')$ to be

$$\{x : x \geq 0, \sum_i x_{(a,i)} = 0 \quad \text{for } a \in \mathcal{A}\setminus\mathcal{A}'\}$$

and $f(\mathcal{A}')$ to be

$$\min\{\omega_T \cdot \|A_T x - TG\|_2 + \omega_C \cdot \|A_C x\|_2 \\ + \omega_N \cdot \|A_N x\|_2 : x \in X(\mathcal{A}')\}, \quad (2)$$

where TG is a vector that expresses the target's treatment goal and ω_T, ω_C and ω_N weight the objective terms to express clinical desires. The prescription for this model is TG, but more complicated models with sophisticated prescriptions are common. In particular, dose-volume constraints that restrict the amount of each structure that is permitted to violate a bound are common. Readers interested in fluence optimization are directed to the entry on Cancer Radiation Treatment: Optimization Models.

Models

The N-beam selection problem is often addressed as a mixed integer problem. As an example, for the judgment function in (2) the N-beam selection problem can be expressed as

$$\left.\begin{array}{rl} \min & \omega_T \cdot \|A_T x - TG\|_2 + \omega_C \\ & \cdot \|A_C x\|_2 + \omega_N \cdot \|A_N x\|_2 \\ \text{subject to:} & \sum_i x_{(a,i)} \leq M \cdot y_a, \text{ for } a \in \mathcal{A} \\ & \sum_a y_a \leq N \\ & x \geq 0 \\ & y \in \{0,1\}^{|\mathcal{A}|}, \end{array}\right\} \quad (3)$$

where M is an arbitrarily large value that bounds each beam's fluence. This is one of many possible models, with simple adjustments including the replacement of

80 Gy, then the 90% isocontour contains the anatomical region that receives at least $0.9 \times 80 = 72$ Gy. Figure 4 illustrates the 100%, 90%, ..., 10% isocontours on a single patient image. One would hope that these isocontour would tightly contain the target on each of the patient images, a goal commonly referred to as conformality. Although a DVH is often used to decide if a treatment is unacceptable, both the DVH and the isocontours are used to decide if a treatment is acceptable. Although treatments are commonly evaluated exclusively with a DVH and the isocontours, there are well established numerical scores that are also used. Such scores are called conformality indices and consider the ratios of under and over irradiated tissue, and as such, these values collapse the DVH into a numerical value. We do not discuss these measures here, but the reader should be aware that they exist.

Formulation

The N-beam selection problem for the judgment function f and candidate set of beams $\mathcal{A}$ is

$$\min\{f(\mathcal{A}') : \mathcal{A}' \in \mathcal{P}(\mathcal{A}), |\mathcal{A}'| = N\}. \quad (1)$$

The parameter N is provided by the treatment planner and is intended to control the complexity of the

the 2-norm with the 1 and ∞ norms, both of which result in a linear mixed integer problem.

A modest discretization of the sphere, with 72 great circles through the north and south poles equally spaced at 5 degrees at the equator and each great circle having beams equally spaced at 5 degrees, produces a set of 4902 candidate beams. This means the search tree associated with the mixed integer model above has $\binom{4902}{N}$ terminal nodes, which for the clinically valid $N = 10$ is approximately 2.2×10^{30}. Beyond the immenseness of this search space, branch-and-bound procedures are difficult for two reasons, 1) the number of N element subsets leading to near optimal solutions is substantial, and 2) the evaluation of the judgment function at each node requires the solution to an underlying fluence model, which in itself is time consuming. This inherit difficulty has driven the development of heuristic approaches, which separate into the two steps of: 1) assigning each beam a value that measures its worth to the overall treatment, and 2) using the individual beam values to select a collection of N beams. As a simple example, a scoring technique evaluates each beam and then simply selects the top N beams. The remainder of this section discusses several of the common heuristics.

A selection technique is called *informed* if it requires the evaluation of the underlying judgment function. One example would be to iteratively let $\mathcal{A}'$ be the singleton beam sets and evaluate $f(\mathcal{A}')$ for each. The N beams with the best scores would be selected for the treatment. If a selection method uses the data forming the optimization problem that defines f but fails to evaluate f, then the technique is called *weakly informed*. The preponderance of techniques suggested in the medical physics literature fall into this category. An example based solely on the dose matrix A is to value beam a with

$$\frac{\max_{(i,j)}\{A_{(j,a,i)} : j \in T\}}{\min_{(i,j)}\{A_{(j,a,i)} : j \in C \cup N\}},$$

where we assume the minimums in the denominator are nonzero. This ratio is high if a beam can deliver large amounts of dose to the target without damaging other tissues. A scoring technique based on this would terminate with the collection of N beams with the highest values. Since weakly informed methods do not require the solution of an optimization problem, they tend to be fast.

The concern about the size of the underlying fluence model has lead to a sampling heuristic that reduces the accuracy of the radiobiological model. Clinical relevance mandates that the anatomy be discretized so that dose is measured at distances no greater than 2 mm. For a 20 cm³ portion of the anatomy, roughly the volume of the cranium, this means the coarsest 3D grid permitted in the clinic divides the anatomy into 10^6 sub-regions called voxels, which are indexed by j. Cases in the chest and abdomen are substantially larger and require a significant increase in the number of voxels. A natural question is whether or not all of these regions are needed for beam selection. One approach is to repeatedly sample these regions together with the candidate set of beams and solve (1). Each beam is valued by the number of times it has a high fluence. Beams with high values create $\mathcal{A}$ in (1) with j being indexed over all regions. The goal of this technique is to identify a candidate set of beams whose size is slightly larger than N, which keeps the search space manageable with the full compliment of voxels. The sampling procedure is crucial to the success of the procedure since it is known that beam selection depend on the collection of voxels.

Once beams are valued, there are many ways to use this information to construct a collection of favorable beams. As already discussed, common scoring methods select the best N beams. Another approach is based on set covering, which uses a high-pass filter to decide if a beam adequately treats the target. Allowing ε to be the threshold at which we say beam a treats position j within the target, we let

$$U_{(j,a)} = \begin{cases} 1, & \sum_i A_{(j,a,i)} \geq \varepsilon \\ 0, & \sum_i A_{(j,a,i)} < \varepsilon, \end{cases}$$

for each $j \in T$. If each beam has a value of c_a, where low values are preferred, the set cover heuristic forms a collection of beams by solving

$$\min\left\{\sum_a c_a y_a : \sum_a U_{(j,a)} y_a \geq 1, \right.$$

$$\left. \text{for each } j \in T, y_a \in \{0,1\}\right\}. \quad (4)$$

This in itself is a binary optimization problem, and if ε is small enough to guarantee that every beam treats the target, which is typical, then the size of the search space is the same as the original problem in (1). However, the set cover problem has favorable solution properties, and this problem solves efficiently in practice. The search space decreases in size as ε increases, and designing an appropriate heuristic requires both a judicious selection of ε and an appropriate objective. This method can be informed or weakly informed depending on how the objective coefficients are constructed.

Another approach is to use the beam values as a probability distribution upon normalization. This allows one to address the problem probabilistically, a perspective that has been suggested within column generation and vector quantization. The column generation approach prices beams with respect to the likelihood that they will improve the judgment function, and beams with high probabilities are added to the current collection. The process of adding and deleting beams produces a sequence of beam sets $\mathcal{A}^1, \mathcal{A}^2, \ldots, \mathcal{A}^n$, and problem (1) is solved with $\mathcal{A}$ replaced with $\mathcal{A}^k$, $k = 1, 2, \ldots, n$. Although it is possible for this technique to price all subsets of $\mathcal{A}$ whose cardinality is greater than N, which is significantly greater than the size of the original search space in (1), the pricing scheme tends to limit the number $\mathcal{A}^k$s.

The probabilistic perspective is further incorporated with heuristics based in information science. In particular, a method based on vector quantization, which is a modeling and solution procedure used in data compression, has been suggested. Allowing $\alpha(a)$ to be the probability associated with beam a, this heuristic constructs a collection of beams by solving

$$\min_Q \left\{ \sum_a \alpha(a) \rho(a, Q(a)) : |Q(\mathcal{A})| = N \right\}, \quad (5)$$

where Q is a mapping from $\mathcal{A}$ into itself and ρ is a metric appropriate to the application. A common metric is to let $\rho(a, Q(a))$ be the arc length between a and $Q(a)$.

In the finite case, each N element subset, say $\mathcal{A}'$, of $\mathcal{A}$ uniquely defines Q by setting $Q(\mathcal{A}) = \mathcal{A}'$. Assuming this equality, we complete the definition by setting $Q(a) = a' \in \mathcal{A}'$ if and only if $\rho(a, a') \leq \rho(a, a'')$ for all $a'' \in \mathcal{A}'$, a condition referred to as the nearest neighbor condition. Since the optimization problem in (5) is defined over the collection of these functions, the size of the feasible region is the same as the original beam selection problem in (1). Unlike the set cover approach, which solves (4) to optimality, and the column generation technique, which repeatedly solves (1) to optimality with a restricted beam set, the vector quantization method often solves (5) heuristically. The most common heuristic is the Lloyd algorithm, a technique that begins with an initial collection of N beams and then iterates between

1. defining Q with the nearest neighbor condition, and
2. forming a new collection of beams with the centroids of $Q^{-1}(a)$, where beam a is in the current collection.

This technique guarantees that the objective in (5) decreases with each new collection.

Conclusions

Selecting beams is one of the three sub-problems in the design of radiotherapy treatments, a problem that currently does not have an appropriate solution outside the clinical practice of manually selecting beams through trial-and-error. However, research into automating the selection of beams with optimization is promising. We conclude with a few words on the totality of treatment design.

The overriding goal of treatment design is to remove the threat of cancer while sparing non-cancerous tissues. The status quo is to assume that a patient is static while designing a treatment. Indeed, treatment planners expand targeted regions to address the dynamic patient movement in the static approach, i. e. the target is increased to include the gross volume that contains the estimated movement of the actual target. The primary goal of the third phase of treatment design is to deliver the treatment as efficiently as possible to limit patient movement. This leads to a dilemma. The monotonicity property of the judgment function encourages treatments with many beams, but conventional wisdom dictates that the number of beams and the efficiency of the delivery are inversely proportional. However, in many settings the number of beams is a poor surrogate of efficiency. As an example, the most time demanding maneuver is to rotate the couch since it requires a technician to enter the treatment vault. So, treatments with many beams but fewer couch rotations are preferred to treatments with fewer beams but more couch rotations.

The point to emphasize from the previous paragraph is that the problem of selecting beams is always expressed in terms of the number of beams, which is a byproduct of the three-phase approach. Although the separation of the design process into phases is natural and useful for computation, the division has drawbacks. Fluence models are large and difficult to solve, and every attempt is made to reduce their size. As already discussed, the voxels need to be under 2 mm^3 to reach clinical viability, and hence, the index set for j is necessarily large. The number and complexity of the sub-beams has increased dramatically with advanced technology, similarly making the index set for i large. This leaves the number of beams as the only control, and treatment designers are asked to select beams so that the fluence model is manageable. Years of experience have developed standard collections for many cancers, but asking a designer to select one of the 2.2×10^{30} possible collections for a 10 beam treatment in a non-standard case is daunting. A designer's instinct is to value a beam individually rather than as part of a collection. Several of the weakly informed selection methods from the medical physics literature have the same weakness. Such individual valuation typically identifies all but a few beams of a quality solution, but the last few are often unintuitive. Automating beam selection with an optimization process so that beams are considered within a collection is a step in the right direction.

The future of treatment design is to build global models and solution procedures that simultaneously address all three phases of treatment design. Such models are naturally viewed from the perspective of beam selection. What is missing is a judgment function that includes both fluence and delivery optimization. Learning how to model and solve these holistic models would alleviate the design process from a designer's (lack of) expertise and would provide a uniform level of care available to clinics with comparable technology. Such improvements are the promise of the field.

See also

- ▶ Credit Rating and Optimization Methods
- ▶ Evolutionary Algorithms in Combinatorial Optimization
- ▶ Optimization Based Framework for Radiation Therapy

The literature on beam selection is mature within the medical physics community but is in its infancy within optimization. The five citations below cover the topics discussed in this article and contain bibliographies that adequately cite the work in medical physics.

References

1. Acosta R, Ehrgott M, Holder A, Nevin D, Reese J, Salter B (2007) Comparing Beam Selection Strategies in Radiotherapy Treatment Design: The Influence of Dose Point Resolution. In: Alves C, Pardalos P, Vicente L (eds) Optimization in Medicine, International Center for Mathematics, Springer Optimization and Its Applications. Springer, pp 1–25
2. Aleman D, Romeijn E, Dempsey J (2006) Beam orientation optimization methods in intensity modulated radiation therapy. IIE Conference Proceedings
3. Ehrgott M, Holder A, Reese J (2008) Beam Selection in Radiotherapy Design. In: Linear Algebra and Its Applications, vol 428. pp 1272–1312. doi:10.1016/j.laa.2007.05.039
4. Lim G, Choi J, Mohan R Iterative Solution Methods for Beam Angle and Fluence Map Optimization in Intensity Modulated Radiation Therapy Planning. to appear in OR Spectrum. doi:10.1007/s00291-007-0096-1
5. Lim G, Ferris M, Shepard D, Wright S, Earl M (2007) An Optimization Framework for Conformal Radiation Treatment Planning. INFORMS J Comput 19(3):366–380

Best Approximation in Ordered Normed Linear Spaces

HOSSEIN MOHEBI
Mahani Mathematical Research Center,
and Department of Mathematics,
University of Kerman, Kerman, Iran

MSC2000: 90C46, 46B40, 41A50, 41A65

Article Outline

Keywords and Phrases
Introduction
Metric Projection onto Downward and Upward Sets
Sets Z_+ and Z_-
Downward Hull and Upward Hull
Metric Projection onto a Closed Set
Best Approximation in a Class of Normed Spaces
   with Star-Shaped Cones
Characterization of Best Approximations
Strictly Downward Sets
   and Their Best Approximation Properties
References

Keywords and Phrases

Best approximation; Downward and upward sets; Global minimum; Necessary and sufficient conditions; Star-shaped set; Proximinal set

Introduction

We study the minimization of the distance to an arbitrary closed set in a class of ordered normed spaces (see [8]). This class is broad enough. It contains the space $C(Q)$ of all continuous functions defined on a compact topological space Q and the space $L^\infty(S, \Sigma, \mu)$ of all essentially bounded functions defined on a measure space (S, Σ, μ). It is assumed that these spaces are equipped with the natural order relation and the uniform norm. This class also contains direct products $X = \mathbb{R} \times Y$, where Y is an arbitrary normed space, with the norm $\|(c, y)\| = |c| + \|y\|$. The space X is equipped with the order relation induced by the cone $K = \{(c, y): c \geq \|y\|\}$.

Let U be a closed subset of X, where X is a normed space from the given class, and let $t \in X$. We consider the problem $Pr(U, t)$:

$$\text{minimize} \quad \|u - t\| \quad \text{subject to} \quad u \in U. \qquad (1)$$

It is assumed that there exists a solution of $Pr(U, t)$. This solution is called a metric projection of t onto U, or a best approximation of t by elements of U. We use the structure of the objective function in order to present necessary and sufficient conditions for the global minimum of $Pr(U, t)$ that give a clear understanding of the structure of a metric projection and can be easily verified for some classes of problems under consideration.

We use the so-called *downward and upward* subsets of a space X as a tool for analysis of $Pr(U, t)$. A set $U \subset X$ is called downward if $(u \in U, x \leq u) \implies x \in U$. A set $V \subset X$ is called upward if $(v \in V, x \geq v) \implies x \in V$. Downward and upward sets have a simple structure so the problem $Pr(U, t)$ can be easily analyzed for these sets U. If U is an arbitrary closed subset of X we can consider its downward hull $U_* = U - K$ and upward hull $U^* = U + K$, where $K = \{x \in X: x \geq 0\}$ is the cone of positive elements. These hulls can be used for examination of $Pr(U, t)$. We also suggest an approach based on a division of a normed space under consideration into two homogeneous not necessarily linear subspaces. A combination of this approach with the downward-upward technique allows us to give simple proofs of the proposed necessary and sufficient conditions.

Properties of downward and upward sets play a crucial role in this article. These properties have been studied in [6,13] for $X = \mathbb{R}^n$. We show that some results obtained in [6,13] are valid in a much more general case. In fact, the first necessary and sufficient conditions for metric projection onto closed downward sets in $\mathbb{R}^n$ have been given in [1, p. 132, Theorem 9]. Proposition 1(1) and (2) are extensions of $\mathbb{R}^n$ and $\mathbf{1} = (1, \ldots, 1)$, of [1, Proposition 1(a) and (b)], respectively. Also, Propositions 2 and 3 are extensions of [1, p. 116, Proposition 2]. Furthermore, Corollary 3 is an extension of [1, p. 116, Corollary 2 and p. 117, Remark 2]. In connection with Proposition 6, the downward hull U_* has been introduced in [1, Sect. 1], where the first results on the connection between $d(t, U)$ and $d(t, U_*)$ have been given, for the particular case where U is a normal subset of $\mathbb{R}^n_+$. We use methods of abstract convexity and monotonic analysis (see [11]) in this study.

Let X be a normed space. Let $K \subset X$ be a closed convex and pointed cone. (The latter means that $K \cap (-K) = \{0\}$.) The cone K generates the order relation $\geq$ on X. By definition $x \geq y \iff x - y \in K$. We say that x is greater than y and write $x > y$ if $x - y \in K \setminus \{0\}$. Assume that K is solid, that is, the interior int K of K is nonempty. Let $\mathbf{1} \in \text{int } K$. Using $\mathbf{1}$ we can define the following function:

$$p(x) = \inf\{\lambda \in \mathbb{R}: x \leq \lambda \mathbf{1}\}, \quad (x \in X). \qquad (2)$$

It is easy to check that p is finite. It follows from (2) that

$$x \leq p(x)\mathbf{1}, \quad (x \in X). \qquad (3)$$

It is easy to check (and well known) that p is a sublinear function, that is,

$$p(\lambda x) = \lambda p(x) \quad (\lambda > 0, x \in X),$$
$$p(x + y) \leq p(x) + p(y) \quad (x, y \in X).$$

We need the following definition (see [13] and references therein). A function $s: X \to \mathbb{R}$ is called topical if s is increasing: $x \geq y$ implies $s(x) \geq s(y)$ and $s(x + \lambda \mathbf{1}) = s(x) + \lambda$ for all $x \in X$ and $\lambda \in \mathbb{R}$.

It follows from the definition of p that p is topical. Consider the function

$$\|x\| := \max(p(x), p(-x)). \tag{4}$$

It is easy to check (and well known) that $\|\cdot\|$ is a norm on X. In what follows we assume that the norm (4) coincides with the norm of the space X.

It follows from (3) that

$$x \leq \|x\|\mathbf{1}, \quad -x \leq \|x\|\mathbf{1}, \quad (x \in X). \tag{5}$$

The ball $B(t, r) = \{x \in X : \|x - t\| \leq r\}$ has the form

$$B(t, r) = \{x \in X : t - r\mathbf{1} \leq x \leq t + r\mathbf{1}\}. \tag{6}$$

We now present three examples of spaces under consideration.

Example 1 Let X be a vector lattice with a strong unit $\mathbf{1}$. The latter means that for each $x \in X$ there exists $\lambda \in \mathbb{R}$ such that $|x| \leq \lambda \mathbf{1}$. Then

$$\|x\| = \inf\{\lambda > 0 : |x| \leq \lambda \mathbf{1}\},$$

where norm $\|\cdot\|$ is defined by (4). It is well known (see, for example, [21]) that each vector lattice X with a strong unit is isomorphic as a vector-ordered space to the space $C(Q)$ of all continuous functions defined on a compact topological space Q. For a given strong unit $\mathbf{1}$ the corresponding isomorphism ψ can be chosen in such a way that $\psi(\mathbf{1})(q) = 1$ for all $q \in Q$. The cone $\psi(K)$ coincides with the cone of all nonnegative functions defined on Q. If $X = C(Q)$ and $\mathbf{1}(q) = 1$ for all q, then

$$p(x) = \max_{q \in Q} x(q) \quad \text{and} \quad \|x\| = \max_{q \in Q} |x(q)|.$$

A well-known example of a vector lattice with a strong unit is the space $L^\infty(S, \Sigma, \mu)$ of all essentially bounded functions defined on a measure space (S, Σ, μ). If $\mathbf{1}(s) = 1$ for all $s \in S$, then $p(x) = \operatorname{ess\,sup}_{s \in S} x(s)$ and $\|x\| = \operatorname{ess\,sup}_{s \in S} |x(s)|$.

Example 2 Let $X = \mathbb{R} \times Y$, where Y is a normed space with a norm $\|\cdot\|$, and let $K \subset X$ be the epigraph of the norm $K = \{(\lambda, x) : \lambda \geq \|x\|\}$. The cone K is closed solid convex and pointed. It is easy to check and well known that $\mathbf{1} = (1, 0)$ is an interior point of K. For each $(c, y) \in X$ we have

$$p(c, y) = \inf\{\lambda \in \mathbb{R} : (c, y) \leq \lambda \mathbf{1}\}$$
$$= \inf\{\lambda \in \mathbb{R} : (\lambda, 0) - (c, y) \in K\}$$
$$= \inf\{\lambda \in \mathbb{R} : (\lambda - c, -y) \in K\}$$
$$= \inf\{\lambda \in \mathbb{R} : \lambda - c \geq \|-y\|\} = c + \|y\|.$$

Hence

$$\|(c, y)\| = \max(p(c, y), p(-(c, y)))$$
$$= \max(c + \|y\|, -c + \|y\|) = |c| + \|y\|.$$

Example 3 Consider the space l^1 of all summable sequences with the usual norm. Let $Y = \{x = (x_i) \in l^1 : x_1 = 0\}$. Then we can identify l^1 with the space $\mathbb{R} \times Y$. Let $y \in Y$ and $x = (x_1, y) \in l^1$. Then $\|x\| = |x_1| + \|y\|$. Let $K = \{x = (x_i) \in l^1 : x_1 \geq \sum_{i=2}^\infty |x_i|\}$. Assume that l^1 is equipped with the order relation $\geq$ generated by K: if $x = (x_i)$ and $z = (z_i)$, then

$$x \geq z \iff x_1 - z_1 \geq \sum_{i=2}^\infty |x_i - z_i|.$$

Let $\mathbf{1} = (1, 0, \ldots, 0, \ldots)$. Consider the function p defined on l^1 by

$$p(x) = x_1 + \sum_{i=2}^\infty |x_i|, \quad x = (x_1, x_2, \ldots) \in l^1.$$

Then (see the previous example) $p(x) = \inf\{\lambda \in \mathbb{R} : x \leq \lambda \mathbf{1}\}$ and $\|x\| = \sum_{i=1}^\infty |x_i|$ coincides with $\max(p(x), p(-x))$.

Let X be a normed vector space. For a nonempty subset U of X and $t \in X$, define $d(t, U) = \inf_{u \in U} \|t - u\|$. A point $u_0 \in U$ is called a metric projection of t onto U, or a best approximation of t by elements of U, if $\|t - u_0\| = d(t, U)$.

Let $U \subset X$. For $t \in X$, denote by $P_U(t)$ the set of all metric projections of t onto U:

$$P_U(t) = \{u \in U : \|t - u\| = d(t, U)\}. \tag{7}$$

It is wellknown that $P_U(t)$ is a closed and bounded subset of X. If $t \notin U$, then $P_U(t)$ is located in the boundary of U.

We shall use the following definitions. A pair (U, t) where $U \subset X$ and $t \in X$ is called *proximinal* if there exists a metric projection of t onto U. A pair (U, t) is called

Chebyshev if there exists a unique metric projection of t onto U. A set $U \subset X$ is called proximinal, if the pair (U, t) is proximinal for all $t \in X$. A set $U \subset X$ is called Chebyshev if the pair (U, t) is Chebyshev for all $t \in X$.

A set $U \subset X$ is called boundedly compact if the set $U_r = \{u \in U : \|u\| \leq r\}$ is compact for each $r > 0$. (This is equivalent to the following: the intersection of a closed neighborhood of a point $u \in U$ with U is compact.) Each boundedly compact set is proximinal.

For any subset U of a normed space X we shall denote by int U, cl U, and bd U the interior, the closure, and the boundary of U, respectively.

Metric Projection onto Downward and Upward Sets

Definition 1 A set $U \subset X$ is called downward if $(u \in U, x \leq u) \implies x \in U$.

First we describe some simple properties of downward sets.

Proposition 1 Let U be a downward subset of X and $x \in X$. Then the following assertions are true:
(1) If $x \in U$, then $x - \varepsilon\mathbf{1} \in$ int U for all $\varepsilon > 0$.
(2) int $U = \{x \in X : x + \varepsilon\mathbf{1} \in U \text{ for some } \varepsilon > 0\}$.

Proof
(1) Let $\varepsilon > 0$ be given and $x \in U$. Let $N = \{y \in X : \|y - (x - \varepsilon\mathbf{1})\| < \varepsilon\}$ be an open neighborhood of $(x - \varepsilon\mathbf{1})$. Then, by (6) $N = \{y \in X : x - 2\varepsilon\mathbf{1} < y < x\}$. Since U is a downward set and $x \in U$, it follows that $N \subset U$, and so $x - \varepsilon\mathbf{1} \in$ int U.
(2) Let $x \in$ int U. Then there exists $\varepsilon_0 > 0$ such that the closed ball $B(x, \varepsilon_0) \subset U$. In view of (6), we get $x + \varepsilon_0 \mathbf{1} \in U$.
Conversely, suppose that there exists $\varepsilon > 0$ such that $x + \varepsilon\mathbf{1} \in U$. Then, by (1): $x = (x + \varepsilon\mathbf{1}) - \varepsilon\mathbf{1} \in$ int U, which completes the proof. $\square$

Corollary 1 Let U be a closed downward subset of X and $u \in U$. Then, $u \in$ bd U if and only if $\lambda\mathbf{1} + u \notin U$ for all $\lambda > 0$.

Lemma 1 The closure cl U of a downward set U is downward.

Proof Let $x_k \in U$, $k = 1, 2, \ldots$, and $x_k \to x$ as $k \to +\infty$. Let $\|x_k - x\| = \varepsilon_k (k = 1, 2, \ldots)$. Using (6) we get $x - \varepsilon_k \mathbf{1} \leq x_k$ for all $k \geq 1$. Since U is a downward set and $x_k \in U$ for all $k \geq 1$, we conclude that $x - \varepsilon_k \mathbf{1} \in U$ for all $k \geq 1$. Let $y \leq x$ be arbitrary and $y_k = y - \varepsilon_k \mathbf{1} \leq x - \varepsilon_k \mathbf{1} (k = 1, 2, \ldots)$. Then $y_k \in U (k = 1, \ldots)$. Since $y_k \to y$ as $k \to +\infty$, it follows that $y \in$ cl U. $\square$

Proposition 2 A closed downward subset U of X is proximinal.

Proof Let $t \in X \setminus U$ be arbitrary and $r := d(t, U) = \inf_{u \in U} \|t - u\| > 0$. This implies that for each $\varepsilon > 0$ there exists $u_\varepsilon \in U$ such that $\|t - u_\varepsilon\| < r + \varepsilon$. Then, by (6):

$$-(r + \varepsilon)\mathbf{1} \leq u_\varepsilon - t \leq (r + \varepsilon)\mathbf{1}. \tag{8}$$

Let $u_0 = t - r\mathbf{1}$. Then

$$\|t - u_0\| = \|r\mathbf{1}\| = r = d(t, U).$$

In view of (8), we have $u_0 - \varepsilon\mathbf{1} = t - r\mathbf{1} - \varepsilon\mathbf{1} \leq u_\varepsilon$. Since U is a downward set and $u_\varepsilon \in U$, it follows that $u_0 - \varepsilon\mathbf{1} \in U$ for all $\varepsilon > 0$. The closedness of U implies $u_0 \in U$, and so $u_0 \in P_U(t)$. Thus the result follows. $\square$

Remark 1 We proved that for each $t \in X \setminus U$ the set $P_U(t)$ contains the element $u_0 = t - r\mathbf{1}$ with $r = d(t, U)$. If $t \in U$, then $u_0 = t$ and $P_U(t) = \{u_0\}$.

Proposition 3 Let U be a closed downward subset of X and $t \in X$. Then there exists the least element $u_0 := \min P_U(t)$ of the set $P_U(t)$, namely, $u_0 = t - r\mathbf{1}$, where $r := d(t, U)$.

Proof If $t \in U$, then the result holds. Assume that $t \notin U$ and $u_0 = t - r\mathbf{1}$. Then, by Remark 1, $u_0 \in P_U(t)$. Applying (6) and the equality $\|t - u_0\| = r$ we get

$$x \geq t - r\mathbf{1} = u_0 \quad \forall \, x \in B(t, r).$$

This implies that u_0 is the least element of the closed ball $B(t, r)$.

Now, let $u \in P_U(t)$ be arbitrary. Then $\|t - u\| = r$, and so $u \in B(t, r)$. Therefore, $u \geq u_0$. Hence, u_0 is the least element of the set $P_U(t)$. $\square$

Corollary 2 Let U be a closed downward subset of X, $t \in X$ and $u_0 = \min P_U(t)$. Then, $u_0 \leq t$.

Corollary 3 Let U be a closed downward subset of X and $t \in X$ be arbitrary. Then

$$d(t, U) = \min\{\lambda \geq 0 : t - \lambda\mathbf{1} \in U\}.$$

Proof Let $A = \{\lambda \geq 0: t - \lambda\mathbf{1} \in U\}$. If $t \in U$, then $t - 0 \cdot \mathbf{1} = t \in U$, and so $\min A = 0 = d(t, U)$. Suppose that $t \notin U$; then $r := d(t, U) > 0$. Let $\lambda > 0$ be arbitrary such that $t - \lambda\mathbf{1} \in U$. Thus

$$\lambda = \|\lambda\mathbf{1}\| = \|t - (t - \lambda\mathbf{1})\| \geq d(t, U) = r.$$

Since, by Proposition 3, $t - r\mathbf{1} \in U$, it follows that $r \in A$. Hence, $\min A = r$, which completes the proof. □

The results obtained demonstrate that for the search of a metric projection of an element t onto a downward set U we need to solve the following optimization problem:

$$\text{minimize} \quad \lambda \quad \text{subject to} \quad t - \lambda\mathbf{1} \in U, \ \lambda \geq 0. \quad (9)$$

This is a one-dimensional optimization problem that is much easier than the original problem $Pr(U, t)$. Problem (9) can be solved, for example, by a common bisection procedure: first find numbers ρ_1 and σ_1 such that $t - \rho_1\mathbf{1} \in U$ and $t - \sigma_1\mathbf{1} \notin U$. Let $k \geq 1$. Assume that numbers ρ_k and σ_k are known such that $t - \rho_k\mathbf{1} \in U$ and $t - \sigma_k\mathbf{1} \notin U$. Then consider the number $\pi_k = 1/2(\rho_k + \sigma_k)$. If $t - \pi_k\mathbf{1} \in U$, then put $\rho_{k+1} = \pi_k$, $\sigma_{k+1} = \sigma_k$. If $t - \pi_k\mathbf{1} \notin U$, then put $\rho_{k+1} = \rho_k$, $\sigma_{k+1} = \pi_k$. The number $r = \lim_k \rho_k = \lim_k \sigma_k$ is the optimal value of (9).

The following necessary and sufficient conditions for the global minimum easily follow from the results obtained.

Theorem 1 *Let U be a closed downward set and $t \notin U$. Then $u_0 \in U$ is a solution of the problem $Pr(U, t)$ if and only if*
(i) $u_0 \geq \bar{u} := t - r\mathbf{1}$, where $r = \min\{\lambda \geq 0: t - \lambda\mathbf{1} \in U\}$;
(ii) $p(t - u_0) \geq p(u_0 - t)$.

Proof Let $u_0 \in P_U(t)$. Since $\bar{u} := t - r\mathbf{1}$ is the least element of $P_U(t)$, it follows that $u_0 \geq \bar{u}$, so (i) is proved. We now demonstrate that (ii) is valid. In view of the equality $r = \|t - u_0\| = \max(p(t - u_0), p(u_0 - t))$, we conclude that $p(u_0 - t) \leq r$ and $p(t - u_0) \leq r$. We need to prove that $p(t - u_0) = r$. Assume on the contrary that $p(t - u_0) := \inf\{\lambda: t - u_0 \leq \lambda\mathbf{1}\} < r$. Then there exists $\varepsilon > 0$ such that $t - u_0 \leq (r - \varepsilon)\mathbf{1}$. This implies that $u_0 \geq t - r\mathbf{1} + \varepsilon\mathbf{1} = \bar{u} + \varepsilon\mathbf{1}$. Since $u_0 \in U$ and U is downward, it follows that $\bar{u} + \varepsilon\mathbf{1} \in U$, so $\bar{u}$ is an interior point of U. This contradicts the fact that $\bar{u}$ is a best approximation of t by U.

Assume now that both items (i) and (ii) hold. It follows from (i) that $t - u_0 \leq r\mathbf{1}$. Since p is a topical function, we conclude that $p(t - u_0) \leq r$. Item (ii) implies $\|t - u_0\| = p(t - u_0) \leq r$. Since $r = \min_{u \in U} \|t - u\|$, we conclude that $u \in P_U(t)$. □

We now turn to upward sets.

Definition 2 A set $V \subset X$ is called upward if $(v \in V, x \geq v) \Longrightarrow x \in V$.

Clearly V is upward if and only if $U = -V$ is downward, so all results obtained for downward sets can be easily reformulated for upward sets.

Proposition 4 *A closed upward subset V of X is proximinal.*

Proof This is an immediate consequence of Proposition 2. □

Theorem 2 *Let U be a closed upward set and $t \notin U$. Then u_0 is a solution of the problem $Pr(U, t)$ if and only if*
(i) $u_0 \leq t + r\mathbf{1}$, where $r = \min\{\lambda \geq 0: t + \lambda\mathbf{1} \in V\}$.
(ii) $p(u_0 - t) \geq p(t - u_0)$.

Proof The result can be obtained by application of Theorem 1 to the problem $Pr(-U, -t)$. □

Corollary 4 *Let $V \subset X$ be a closed upward set and $t \in X$. Then $d(t, V) = \min\{\lambda \geq 0: t + \lambda\mathbf{1} \in V\}$.*

Sets Z_+ and Z_-

Consider function s defined on X by

$$s(x) = \frac{1}{2}(p(x) - p(-x)).$$

We now indicate some properties of function s.
(1) s is homogeneous of degree one, that is, $s(\lambda x) = \lambda s(x)$ for $\lambda \in \mathbb{R}$. Indeed, we need to check that $s(-x) = -s(x)$ for all $x \in X$ and $s(\lambda x) = \lambda s(x)$ for all $x \in X$ and all $\lambda \in \mathbb{R}$. Both assertions directly follow from the definition of s.
(2) s is topical. It follows directly from the definition of s that s is increasing. We now check that $s(x + \mu\mathbf{1}) = s(x) + \mu$ for all $x \in X$ and all $\mu \in \mathbb{R}$.

Indeed,

$$s(x+\mu\mathbf{1}) = \frac{1}{2}(p(x+\mu\mathbf{1}) - (p(-x-\mu\mathbf{1}))$$
$$= \frac{1}{2}(p(x) - p(-x) + 2\mu)$$
$$= s(x) + \mu.$$

We will be interested in the level sets

$$Z_+ = \{x \in X: s(x) \geq 0\} \text{ and } Z_- = \{x \in X: s(x) \leq 0\}$$

of function s. The following holds:

$$x \in Z_+ \iff p(x) \geq p(-x) \iff p(x) = \|x\|.$$

$$x \in Z_- \iff p(x) \leq p(-x) \iff p(-x) = \|x\|.$$

Since s is homogeneous, it follows that $Z_- = -Z_+$. Let $Z_0 = \{x: s(x) = 0\}$. Then

$$Z_+ \cap Z_- = Z_0, \quad Z_- \cup Z_+ = X.$$

Since s is continuous, it follows that Z_+ and Z_- are closed subsets of X. Note that both Z_+ and Z_- are conic sets. (Recall that a set $C \subset X$ is called conic if $(x \in C, \lambda > 0) \implies \lambda x \in C$).

Since s is increasing, it follows that Z_+ is upward and Z_- is downward. Let $R = \{\lambda\mathbf{1}: \lambda \geq 0\}$ be the ray passing through $\mathbf{1}$. In view of the topicality of s,

$$Z_+ = Z_0 + R, \quad Z_- = Z_0 - R.$$

Indeed, let $x \in Z_+$; then $s(x) := \lambda \geq 0$. Let $u = x - \lambda\mathbf{1}$. Then $s(u) = 0$, hence $u \in Z_0$. We demonstrated that $x \in Z_0 + R$, so $Z_+ \subset Z_0 + R$. The opposite inclusion trivially holds. Thus, $Z_+ = Z_0 + R$. We also have $Z_- = -Z_0 - R = Z_0 - R$. We now give some examples.

Example 4 Let $X = C(Q)$ be the space of all continuous functions defined on a compact topological space Q and $p(x) = \max_{q \in Q} x(q)$. Then $s(x) = \max_{q \in Q} x(q) + \min_{q \in Q} x(q)$; therefore $Z_0 = \{x \in C(Q): \max_{q \in Q} x(q) = -\min_{q \in Q} x(q)\}$. Thus $x \in Z_0$ if and only if there exist points $q_+, q_- \in Q$ such that $|x(q_+)| = |x(q_-)| = \|x\|$ and $x(q_+) > 0$, $x(q_-) < 0$. Further, $x \in Z_+$ if and only if $\|x\| = \max_{q \in Q} x(q) > -\min_{q \in Q} x(q)$ and $x \in Z_-$ if and only if $\|x\| = \max_{q \in Q}(-x(q)) > -\min_{q \in Q}(-x(q)) = \max_{q \in Q} x(q)$.

Let Q consist of two points. Then $C(Q)$ coincides with $\mathbb{R}^2$ and $s(x) = x_1 + x_2$, that is, s is a linear function. If Q contains more than two points, then s is not linear.

Example 5 Let $X = \mathbb{R} \times Y$, where Y is a normed space (Example 2). Let $x = (c, y)$; then $p(x) = c + \|y\|$. Hence

$$s(x) = \frac{1}{2}[(c + \|y\|) - (-c + \|-y\|)] = c,$$

so s is linear. The following holds:

$$Z_0 = \{(c, y): c = 0\}, \quad Z_+ = \{(c, y): c \geq 0\},$$
$$Z_- = \{(c, y): c \leq 0\}.$$

Example 6 Let $X = l^1$ (see Example 3). Then $s(x) = x_1$ and

$$Z_0 = \{x = (x_i) \in l^1: x_1 = 0\},$$
$$Z_+ = \{x = (x_i) \in l^1: x_1 \geq 0\},$$
$$Z_- = \{x = (x_i) \in l^1: x_1 \leq 0\}.$$

Downward Hull and Upward Hull

Let U be a subset of X. The intersection U_* of all downward sets that contain U is called the *downward* hull of U. Since the intersection of an arbitrary family of downward sets is downward, it follows that U_* is downward. Clearly U_* is the least (by inclusion) downward set, which contains U. The intersection U^* of all upward sets containing U is called the *upward* hull of U. The set U^* is upward and is the least (by inclusion) upward set containing U.

Proposition 5 ([15], Proposition 3) *Let $U \subset X$. Then*

$$U_* = U - K := \{u - v: u \in U, v \in K\},$$
$$U^* = U + K := \{u + v: u \in U, v \in K\}.$$

We need the following result:

Proposition 6 *Consider a closed subset U of X.*
(1) *Let $t \in X$ be an element such that $t - U \subset Z_+$. Then $d(t, U) = d(t, U_*)$.*
(2) *Let $t \in X$ be an element such that $t - U \subset Z_-$. Then $d(t, U) = d(t, U^*)$.*

Proof We shall prove only the first part of the proposition. The second part can be proved in a similar way. Let $r = d(t, U_*)$. Since $U \subset U_*$, it follows that $r \leq d(t, U)$, so we need only check the reverse inequality. Let $u_* \in U_*$ be arbitrary. Then, by Proposition 5, there exist $u \in U$ and $v \in K$ such that $u_* = u - v$. Hence

$$t - u_* = t - u + v = x - u \text{ with } x := t + v \geq t.$$

By hypothesis, $t - u \in Z_+$. Since $x \geq t$ and Z_+ is upward, it follows that $x - u \in Z_+$. Since $\|z\| = p(z)$ for all $z \in Z_+$ and p is increasing, we have

$$\|t - u_*\| = \|x - u\| = p(x - u) \geq p(t - u) = \|t - u\|.$$

Thus for each $u_* \in U_*$ there exists $u \in U$ such that $\|t - u_*\| \geq \|t - u\|$. This means that $r := d(t, U_*) \geq d(t, U)$. We proved that $d(t, U) = r$. □

Proposition 7
(1) Let $t \in X$ be an element such that $t - U \subset Z_+$ and let U_* be a closed set. Then (U, t) is a proximinal pair.
(2) Let $t \in X$ be an element such that $t - U \subset Z_-$ and let U^* be a closed set. Then (U, t) is a proximinal pair.

Proof We shall prove only the first part of the proposition. Since U_* is a closed downward set in X, it follows, by Proposition 3, that the least element u_0 of the set $P_{U_*}(t)$ exists and $u_0 = t - r\mathbf{1}$, where $r = d(t, U_*)$. In view of Proposition 6, $r = d(t, U)$. Since $u_0 \in U_*$, by Proposition 5, there exist $u \in U$ and $v \in K$ such that $u_0 = t - r\mathbf{1} = u - v$. Then $t - u = r\mathbf{1} - v$ and

$$p(t - u) = p(r\mathbf{1} - v) \leq p(r\mathbf{1}) = r.$$

Since, by hypothesis, $t - u \in Z_+$, it follows that $\|t - u\| = p(t - u) \leq r$. On the other hand, $\|t - u\| \geq d(t, U) = r$. Hence $\|t - u\| = r$, and so $u \in P_U(t)$, which completes the proof. □

Remark 2 Let $U \subset X$ be a closed set. Assume that there exists a set $V \subset X$ such that $V \subset U \subset V_*$ and V_* is closed. Then $U_* = V_*$; hence U_* is closed. In particular, U_* is closed if there exists a compact set V such that $V \subset U \subset V_*$.

Proposition 7 can be used for the search of a metric projection of an element t onto a set U such that $t - U \subset Z_+$ and U_* is closed. In particular, we can give the following necessary and sufficient conditions for a solution of the problem $Pr(U, t)$ for these sets.

Theorem 3
(1) Let $t - U \subset Z_+$ and U_* is closed. Then $u_0 \in U$ is a solution of $Pr(U, t)$ if and only if
  (i) $u_0 \geq t - r\mathbf{1}$ where $r = \min\{\lambda \geq 0 : t - \lambda\mathbf{1} \in U - K\}$.
  (ii) $p(t - u_0) \geq p(u_0 - t)$;
(2) Let $t - U \subset Z_-$ and U^* is closed. Then $u_0 \in U$ is a solution of $Pr(U, t)$ if and only if
  (i') $u_0 \leq t + r\mathbf{1}$ where $r = \min\{\lambda \geq 0 : t + \lambda\mathbf{1} \in U + K\}$.
  (ii') $p(u_0 - t) \geq p(t - u_0)$.

Proof We again prove only the first part of the theorem. Due to Proposition 6, we get $d(t, U) = d(t, U_*) = r$. Since U_* is closed and downward, it follows (Proposition 3) that $\bar{u} := t - r\mathbf{1} \in P_{U_*}(t)$. Let $u_0 \geq \bar{u}$ and $u_0 \in U$. Then $u_0 \in U_*$ and in view of Proposition 6, it holds:

$$d(u_0, U) = d(u_0, U_*) = r$$
$$= \min\{\lambda \geq 0 : t - \lambda\mathbf{1} \in U_*\}.$$

Applying Theorem 1 we conclude that u_0 is a best approximation of t by U_*. Since $u_0 \in U$, it follows that u_0 is a best approximation of t by U.

Consider now a best approximation u_0 of t by U. Applying again Proposition 6 we deduce that $\|t - u_0\| = d(t, U) = d(t, U_*) = r$. Theorem 1 demonstrates that both (i) and (ii) hold. □

Metric Projection onto a Closed Set

Downward and upward sets can be used for examination of best approximations by arbitrary closed sets (it is assumed that a metric projection exists).

We start with the following assertion.

Proposition 8 *Let U be a closed subset of X and $t \in X$. Consider the following sets:*

$$U_t^+ = U \cap (t - Z_+), \qquad U_t^- = U \cap (t - Z_-). \quad (10)$$

Then
(1) $t - U_t^+ \subset Z_+$, $t - U_t^- \subset Z_-$.
(2) $U_t^+ \cup U_t^- = U$.

(3) $U_t^+ \cap U_t^- = U \cap (t - Z_0)$, where $Z_0 = \{x \in X : s(x) = 0\}$.
(4) U_t^+ and U_t^- are closed.
(5) If U is downward, then U_t^+ is downward; if U is upward, then U_t^- is upward.

Proof
(1) It is easy to check
$$(t - U) \cap Z_+ = t - [U \cap (t - Z_+)] = t - U_t^+.$$
Hence $t - U_t^+ \subset Z_+$. A similar argument shows that $t - U_t^- \subset Z_-$.
(2) The following holds:
$$U_t^+ \cup U_t^- = [(t - Z_+) \cap U] \cup [(t - Z_-) \cap U]$$
$$= [(t - Z_+) \cup (t - Z_-)] \cap U$$
$$= [t - (Z_+ \cup Z_-)] \cap U.$$
Since $Z_+ \cup Z_- = X$, it follows that $U_t^+ \cup U_t^- = U$.
(3) The following holds:
$$U_t^+ \cap U_t^- = [U \cap [(t - Z_+)]] \cap [U \cap (t - Z_-)]$$
$$= U \cap [(t - Z_+) \cap (t - Z_-)]$$
$$= U \cap [t - (Z_+ \cap Z_-)].$$
Since $Z_+ \cap Z_- = Z_0$, the result follows.
(4) This is clear.
(5) It follows from the fact that $t - Z_+$ is downward and $t - Z_-$ is upward. $\square$

Consider a fixed proximinal pair (U, t). Let U_t^+ and U_t^- be the sets defined by (10). Since $U_t^+ \cup U_t^- = U$, it follows that
$$\inf_{u \in U} \|t - u\| = \min(\inf_{u^+ \in U_t^+} \|t - u^+\|, \inf_{u^- \in U_t^-} \|t - u^-\|). \tag{11}$$

It follows from (11) that at least one of the pairs (U_t^+, t) and (U_t^-, t) is proximinal and a metric projection of t onto U coincides with a metric projection onto at least one of the sets U_t^+ or U_t^-. Let
$$r_+ = \inf_{u \in U_t^+} \|t - u\|,$$
$$r_- = \inf_{u \in U_t^-} \|t - u\|, \tag{12}$$
$$r = \inf_{u \in U} \|t - u\| = \min(r_+, r_-).$$

For examination of metric projections of t onto U we need to find numbers r_+ and r_-. The number r_+ can be found by solving a one-dimensional optimization problem of the form (9); r_- can be found by solving a similar problem.

If $r_+ < r_-$, then a metric projection of t onto U coincides with a metric projection of t onto U_t^+. Since $t - U_t^+ \subset Z_+$, we can use the results of this section for analyzing the problem $Pr(U, t)$ and its solution. In particular, if the downward hull $(U_t^+)_*$ of the set U_t^+ is closed, we can assert that the set $P_U(t)$ coincides with the set $P_{U_t^+}(t)$. Using Theorem 3 we can give necessary and sufficient conditions for the global minimum in this case in terms of the set U_t^+. They can be expressed in the following form:
$$P_U(t) = P_{U_t^+}(t) = \{u \in U_t^+ : u \geq t - r_+\mathbf{1},$$
$$p(t - u) \geq p(u - t)\}.$$

If $r_- < r_+$, then a metric projection of t onto U coincides with a metric projection of t onto U_t^-. If the set $(U_t^-)^*$ is closed, we can assert that
$$P_U(t) = P_{U_t^-}(t) = \{u \in U_t^- : u \leq t + r\mathbf{1},$$
$$p(u - t) \geq p(t - u)\}.$$

If $r_- = r_+$, then we can use both sets U_t^+ and U_t^-.

We assume in the rest of this section that both pairs (U_t^+, t), (U_t^-, t) are proximinal. In particular, these pairs are proximinal for arbitrary t, if U is a locally compact set.

We are now interested in metric projections u of t onto U such that $s(u - t) = 0$. We introduce the following definition.

Definition 3 A pair (U, t) with $U \subset X$, $t \in X$ is called *strongly proximinal* if $s(u - t) = 0$ for each metric projection u of t onto U.

Recall that $s(u - t) = 0$ if and only if $u - t \in Z_+ \cap Z_-$.

Proposition 9 *The following assertions (i) and (ii) are equivalent:*
(i) (U, t) *is a strongly proximinal pair;*
(ii) $P_U(t) = P_{U_t^+}(t) \cap P_{U_t^-}(t)$.

Proof
(i) $\Longrightarrow$ (ii). Let $u \in P_U(t)$. Since $u - t \in Z_- = -Z_+$ and $u \in U$, it follows that $u \in U \cap (t - Z_+) = U_t^+$.

Then $\|t-u\| = \min_{u' \in U} \|t-u'\| \leq \min_{u' \in U_t^+} \|t-u'\|$. Since $u \in U_t^+$, we conclude that the equality $\|t-u\| = \min_{u' \in U_t^+} \|t-u'\|$ holds. Thus $u \in P_{U_t^+}(t)$. A similar argument shows that $u \in P_{U_t^-}(t)$. Let $u \in P_{U_t^+}(t) \cap P_{U_t^-}(t)$. Then

$$\|u-t\| = d(t, U_t^+) = d(t, U_t^-).$$

Combining the equality $U = U_t^+ \cup U_t^-$ with (11), we get $\|u-t\| = \min_{u' \in U} \|u'-t\|$, and hence $u \in P_U(t)$.

(ii) $\Longrightarrow$ (i). Since (ii) holds, it follows that

$$P_U(t) = P_{U_t^+}(t) \cap P_{U_t^-}(t)$$
$$= \{u \in U_t^+ : t - r\mathbf{1} \leq u\}$$
$$\cap \{u \in U_t^- : u \leq t + r\mathbf{1}\}$$
$$= \{u \in U_t^+ \cap U_t^- : t - r\mathbf{1} \leq u \leq t + r\mathbf{1}\}.$$

Applying Proposition 8 (3), we conclude that

$$P_U(t) = \{u \in U \cap (t - Z_0) : t - r\mathbf{1} \leq u \leq t + r\mathbf{1}\}$$
$$= U \cap (t - Z_0) \cap B(t, r).$$

Since $P_U(t) = U \cap B(t, r)$ (by definition), it follows that $P_U(t) \subset t - Z_0$, that is, the pair (U, t) is strongly proximinal. $\square$

Let (U, t) be a proximinal pair. We are interested in a description of conditions that guarantee that $\tilde{v} := t - \tilde{u}$, where $\tilde{u}$ is a metric projection of t onto U, belongs to $Z_+ \cap Z_- = Z_0$. First, we give the following definition:

Definition 4 We say that a set $U \subset X$ is weakly K-open if for each $u \in U$ there exists an element $q \in \text{int } K$ such that $u + \delta q \in U$ for all δ with a small enough $|\delta|$.

Proposition 10 *Assume that (U, t) is a proximinal pair such that the set U is weakly K-open. Let $\tilde{u} \in P_U(t)$. Then $\tilde{v} := t - \tilde{u} \in Z_0$.*

Proof Let $\tilde{v} \notin Z_0$; then $\tilde{v} \notin (Z_+ \cap Z_-)$. Assume for the sake of definiteness that $\tilde{v} \in Z^+$, that is, $\|\tilde{v}\| = p(\tilde{v}) > p(-\tilde{v})$. Since U is weakly K-open and $\tilde{u} \in U$, it follows that there exists $q \in \text{int } K$ such that $\tilde{u} + \delta q \in U$ for all small enough $\delta > 0$. Then:

$$p(\tilde{v}) > p(\tilde{v} - \delta q) \geq p(-\tilde{v} + \delta q) = p(-(\tilde{v} - \delta q)).$$

Hence $\|\tilde{v} - \delta q\| = p(\tilde{v} - \delta q) < p(\tilde{v}) = \|\tilde{v}\|$. Let $\bar{u} = \tilde{u} + \delta q$. Because U is weakly K-open, we conclude that $\bar{u} \in U$ for all small enough $\delta > 0$. Since $\tilde{v} - \delta q = t - \tilde{u} - \delta q = t - \bar{u}$, we obtain

$$\min_{u \in U} \|t - u\| \leq \|t - \bar{u}\| = \|\tilde{v} - \delta q\| < \|\tilde{v}\| = \|t - \tilde{u}\|.$$

This is a contradiction because $\tilde{u} \in P_U(t)$. $\square$

Example 7 Let $U' \subset X$ be a locally compact set and $q \in \text{int } K$. Consider the set

$$U = U' + \{\lambda q : \lambda \in \mathbb{R}\} = \{u' + \lambda q : u' \in U', \lambda \in \mathbb{R}\}.$$

Clearly U is a locally compact set and U is weakly K-open. Then for each $t \in X$ the pair (U, t) is strongly proximinal.

Best Approximation in a Class of Normed Spaces with Star-Shaped Cones

The theory of best approximation by elements of convex sets in normed linear spaces is well developed and has found many applications [1,2,4,5,10,16,17,18,19,20]. However, convexity is sometimes a restrictive assumption, and therefore the problem arises of how to examine best approximation by not necessarily convex sets. Special tools for this are needed.

The aim of the present article is to develop a theory of best approximation by elements of closed sets in a class of normed spaces with star-shaped cones (see [9]). A star-shaped cone K in a normed space X generates a relation $\leq_K$ on X, which is an order relation if and only if K is convex. It can be shown that each star-shaped cone K, such that the interior of the kernel K is not empty, can be represented as the union of closed solid convex pointed cones K_i ($i \in I$, where I is an index set) such that the interior of the cone $K_* := \cap_{i \in I} K_i$ is not empty. A point $\mathbf{1} \in \text{int } K_*$ generates the norm $\|\cdot\|_*$ on X, where $\|x\|_* = \inf\{\lambda > 0 : x \leq_{K_*} \lambda \mathbf{1}, -x \leq_{K_*} \lambda \mathbf{1}\}$, and we assume that X is equipped with this norm. In the special case $I = \{1\}$ (that is, K is a closed convex solid pointed cone) the class of spaces under consideration contains such Banach lattices as the space $L^\infty(S, \Sigma, \mu)$ of all essentially bounded functions defined on a measure space (S, Σ, μ) and the space $C(Q)$ of all continuous functions defined on a compact topological space Q.

Now, let X be a normed space and $U \subset X$. The set $\text{kern } U$ consisting of all $u \in U$ such that

$(x \in U, 0 \leq \alpha \leq 1) \implies u + \alpha(x - u) \in U$ is called the convex kernel of U. A nonempty set U is called star-shaped if kern U is not empty. It is known (see, for example, [12]) that kern U is convex for an arbitrary star-shaped set U. If U is closed, then kern U is also closed. Indeed, let $u_k \in$ kern U, $k = 1, \ldots$ and $u_k \to u$. For each $k = 1, 2, \ldots$, $x \in U$ and $\alpha \in [0, 1]$, we have $u_k + \alpha(x - u_k) \in U$, and so $u + \alpha(x - u) \in U$. This means that $u \in$ kern U.

We need the following statement.

Proposition 11 *Let $U \subset X$ be a set and let $u \in U$. Then the following assertions are equivalent:*
(i) There exists $\varepsilon > 0$, an index set I, and a family of convex sets $(U_i)_{i \in I}$ such that

$$U = \bigcup_{i \in I} U_i \quad \text{and} \quad U_i \supset B(u, \varepsilon) \ (i \in I). \tag{13}$$

(ii) U is a star-shaped set and $u \in$ int kern U.

Proof
(i) $\implies$ (ii). Let $z \in B(u, \varepsilon)$ and let $x \in U$, $\alpha \in [0, 1]$. It follows from (13) that there exists $i \in I$ such that $x \in U_i$. Since U_i is convex and $z \in B(u, \varepsilon) \subset U_i$, we conclude that $z + \alpha(x - z) \in U_i \subset U$. Hence, $z \in$ kern U for each $z \in B(u, \varepsilon)$, and so $B(u, \varepsilon) \subset$ kern U.

(ii) $\implies$ (i). Let $I = U$. Since $u \in$ int kern U, it follows that there exists $\varepsilon > 0$ such that $B(u, \varepsilon) \subset$ kern U. Let $x \in U$ and $U_x = \text{co}(x \cup B(u, \varepsilon))$. Then the set U_x is convex and closed and $x \in U_x$. Hence, $U \subset \bigcup_{x \in U} U_x$. Applying the definition of the convex kernel we conclude that $U_x \subset U$. Hence, $\bigcup_{x \in U} U_x \subset U$. $\square$

If $0 \in$ kern U, then the Minkowski gauge μ_U of U can be defined as follows:

$$\mu_U(x) = \inf\{\lambda > 0 : x \in \lambda U\}. \tag{14}$$

(It is assumed that $\inf \emptyset = 0$.)

Let $u \in$ kern U. Then, $0 \in$ kern $(U - u)$, and so we can consider the Minkowski gauge μ_{U-u} of the set $U - u$.

Theorem 4 *Let $u \in$ int kern U. Then the Minkowski gauge μ_{U-u} of the set $U - u$ is Lipschitz.*

Theorem 4 has been proved in [11] (Theorem 5.2) for finite-dimensional spaces. The proof from [11] holds for an arbitrary normed space and we omit it.

In the sequel, we shall study star-shaped cones. Recall that a set $K \subset X$ is called a cone (or conic set) if $(\lambda > 0, x \in K) \implies \lambda x \in K$. Let K be a star-shaped cone and $K_* =$ kern K. Then, K_* is also a cone. Indeed, let $u \in K_*$, $\lambda > 0$ and $x \in K$. Let $x' = x/\lambda$. Then, $x' \in K$, and so $u + \alpha(x' - u) \in K$ for all $\alpha \in [0, 1]$. We have $\lambda u + \alpha(\lambda x' - \lambda u) = \lambda u + \alpha(x - \lambda u) \in K$. Since x is an arbitrary element of K, it follows that $\lambda u \in$ kern $K = K_*$. We now give an example.

Example 7 Let X coincide with the space $C(Q)$ of all continuous functions defined on a compact metric space Q and $K = \{x \in C(Q) : \max_{q \in Q} x(q) \geq 0\}$. Clearly K is a nonconvex cone. It is easy to check that K is a star-shaped cone and kern $K = K_+$, where

$$K_+ = \{x \in C(Q) : x(q) \geq 0 \text{ for all } q \in Q\}$$
$$= \{x \in C(Q) : \min_{x \in Q} x(q) \geq 0\}.$$

Indeed, let $u \in K_+$. Consider a point $x \in K$. Then there exists a point $q' \in Q$ such that $x(q') \geq 0$. Since $u(q) \geq 0$ for all $q \in Q$, it follows that $\alpha u(q') + (1 - \alpha)x(q') \geq 0$ for all $\alpha \in [0, 1]$. Therefore, $\alpha u + (1 - \alpha)x \in K$. We proved that $K_+ \subset$ kern K. Now, consider $u \notin K_+$. Then there exists a point q' such that $u(q') < 0$. Since u is continuous, we can find an open set $G \subset Q$ such that $u(q) < 0$ for $q \in G$. Let $x \in K$ be a function such that $x(q) < 0$ for all $q \notin G$ (such a function exists). Since the set $Q \setminus G$ is compact, it follows that $\max_{q \notin G} x(q) < 0$; hence $\alpha x(q) + (1 - \alpha)u(q) < 0$ for all $q \in Q$ and small enough $\alpha > 0$. Therefore $\alpha x + (1 - \alpha)u \notin K$ for these numbers α. The equality kern $K = K_+$ has been proved. Note that int kern $K \neq \emptyset$.

The following statement plays an important role in this paper.

Theorem 5 *Let $K \subset X$ be a closed cone and let $u \in K$. Then the following assertions are equivalent:*
(i) There exists $\varepsilon > 0$, an index set I and a family of closed convex cones $(K_i)_{i \in I}$ such that

$$K = \bigcup_{i \in I} K_i \quad \text{and} \quad K_i \supset B(u, \varepsilon) \ (i \in I). \tag{15}$$

(ii) K is a star-shaped cone and $u \in$ int kern K.

Proof
(i) $\implies$ (ii). It follows from Proposition 11 that K is a star-shaped set and $u \in$ int kern K. Since K_i is a cone for each $i \in I$, it follows that K is a cone.

(ii) $\Longrightarrow$ (i). In view of Proposition 11, there exists a family of convex sets U_i, $(i \in I)$ such that $U_i \supset B(u,\varepsilon)$ and $K = \bigcup_{i \in I} U_i$. Let K_i be the closed conic hull of U_i: $K_i = \text{cl} \bigcup_{\lambda > 0} \lambda U_i$. Then $K = \bigcup_{i \in I} K_i$. $\square$

Remark 3

(1) Let K be a closed star-shaped cone with int kern $K \neq \emptyset$. Then the set $K_* = \text{kern}\, K$ is a closed solid convex cone. (Recall that a convex cone K is called solid if int $K \neq \emptyset$.)

(2) Note that in Theorem 5, the family $(K_i)_{i \in I}$ can be chosen such that each K_i is a closed solid pointed convex cone. Indeed, if $u \in \text{int kern}\, K$, then $u \neq 0$ and a neighborhood $B(u,\varepsilon) \subset \text{kern}\, K$ can be chosen in such a way that $0 \notin B(u,\varepsilon)$. Then the closed conic hull $K_i = \text{cl} \bigcup_{\lambda > 0} \lambda U_i$ is a closed solid pointed convex cone.

Let K be a star-shaped cone and $K = \bigcup_{i \in I} K_i$, where K_i is a convex cone and $K_* = \bigcap_{i \in I} K_i$. Then kern $K \supset K_*$. Indeed, let $u \in K_*$ and $x \in K$. Then there exists $j \in I$ such that $x \in K_j$. The inclusion $u \in \bigcap_{i \in I} K_i$ implies that $u \in K_j$. Since K_j is a convex cone, it follows that $\alpha x + (1-\alpha) u \in K_j$ for all $\alpha \in (0,1)$. This means that $u \in \text{kern}\, K$.

Let K be a closed star-shaped cone and $u \in \text{int kern}\, K$. Consider the function

$$p_{u,K}(x) = \inf\{\lambda \in \mathbb{R} : \lambda u - x \in K\}. \quad (16)$$

Functions (16) are well known if K is a convex cone. These functions have been defined and studied in [12] for the so-called strongly star-shaped cones (see [11] for the definition of strongly star-shaped sets). Each star-shaped set U with int kern $U \neq \emptyset$ is strongly star-shaped. (It was shown in [11] for finite-dimensional space; however, the same argument is valid for arbitrary normed spaces.) It was shown [12] that $p_{u,K}$ is a finite positively homogeneous function of the first degree and the infimum in (16) is attained, so $p_{u,K}(x) u - x \in \text{int}\, K$. The following equality holds:

$$p_{u,K}(x - \gamma u) = \mu_{K-u}(\gamma u - x), \quad (17)$$

where μ_{K-u} is the Minkowski gauge of $K - u$. In view of Theorem 4, the function μ_{K-u} is Lipschitz, therefore $p_{u,K}$ is also Lipschitz. If K is a convex cone, then $p_{u,K}$ is a sublinear function. This function is also increasing in the sense of the order relation induced by the convex cone K. The following assertion holds (see [12]).

Proposition 12 *Let K be a star-shaped cone and $u \in \text{int kern}\, U$. Then:*

$$p_{u,K}(x + \lambda u) = p_{u,K}(x) + \lambda, \quad x \in X,\ \lambda \in \mathbb{R} \quad (18)$$

and

$$\{x : p_{u,K}(x) \leq \lambda\} = \lambda u - K, \quad \lambda \in \mathbb{R}. \quad (19)$$

We also need the following assertion.

Proposition 13 *Let $(K_i)_{i \in I}$ be a family of closed star-shaped cones such that $\bigcap_{i \in I} \text{int kern}\, K_i \neq \emptyset$. Let $u \in \bigcap_{i \in I} \text{int kern}\, K_i$. Let $K = \bigcup_{i \in I} K_i$ and $K_* = \bigcap_{i \in I} K_i$. Then*

$$p_{u,K}(x) = \inf_{i \in I} p_{u,K_i}(x),$$

$$p_{u,K_*}(x) = \sup_{i \in I} p_{u,K_i}(x), \quad (x \in X).$$

Proof Let L be a cone such that $u \in \text{int kern}\, L$. For each $x \in X$ consider the set $\Lambda_x(L) = \{\lambda \in \mathbb{R} : \lambda u - x \in L\}$. It was proved in [12], Proposition 1, that this set is a closed segment of the form $[\lambda_x, +\infty)$, where $\lambda_x = p_{u,L}(x)$. We have

$$\Lambda_{x,K} = \{\lambda \in \mathbb{R} : \lambda u \in x + \bigcup_{i \in I} K_i\}$$

$$= \{\lambda \in \mathbb{R} : \lambda u \in \bigcup_{i \in I}(x + K_i)\}$$

$$= \bigcup_{i \in I}\{\lambda \in \mathbb{R} : \lambda u \in x + K_i\} = \bigcup_{i \in I} \Lambda_{x,K_i}.$$

Hence

$$p_{u,K}(x) = \inf \Lambda_{x,K} = \inf \bigcup_{i \in I} \Lambda_{x,K_i}$$

$$= \inf_{i \in I} \inf \Lambda_{x,K_i} = \inf_{i \in I} p_{u,K_i}(x).$$

The second part of the proposition can be proved by a similar argument. $\square$

Let K be a closed star-shaped cone with int kern $K \neq \emptyset$. Then K can be represented as the union of a family of closed convex cones $(K_i)_{i \in I}$. One such family has been described in the proofs of Proposition 11 and Theorem 5: $I = K$, $K_i = \text{cl cone co}\{i \cup B(u,\varepsilon)\}$, where $u \in \text{int kern } K$ and $\varepsilon > 0$ so small such that $B(u, \varepsilon) \subset \text{kern } K$. This family is very large; often we can find a much simpler presentation. For example, assume that a cone K is given as the union of a family of closed convex cones $(K_i)_{i \in I}$ such that the cone $\bigcap_{i \in I} K_i$ has a nonempty interior. Then this cone is contained in kern K; we can use the given cones K_i in such a case. We always assume that cones K_i are pointed for all $i \in I$, that is, $K_i \cap (-K_i) = \{0\}$.

An arbitrary star-shaped cone K induces a relation $\geq_K$ on X, where $x \leq_K y$ means that $y - x \in K$. This relation is a preorder relation if and only if K is a convex set. Although $\geq_K$ is not necessarily an order relation, we will say that x is greater than or equal to y in the sense of K if $x \geq_K y$. We say that x is greater than y and write $x >_K y$ if $x - y \in K \setminus \{0\}$. Let $K = \bigcup_{i \in I} K_i$, where K_i is a convex cone. The cone K_i induces the order relation $\geq_{K_i}$. The relation $\geq_K$, which is induced by cone K, can be represented in the following form:

$$x \geq_K y \quad \text{if and only if there exists} \quad i \in I$$
$$\text{such that } x \geq_{K_i} y. \quad (20)$$

In the rest of this article, we assume that X is equipped with a closed star-shaped cone K with int kern $K \neq \emptyset$. We also assume that a family $(K_i)_{i \in I}$ of closed solid convex pointed cones K_i is given such that $K = \bigcup_{i \in I} K_i$ and $K_* = \bigcap_{i \in I} K_i$ has a nonempty interior. Let an element $\mathbf{1} \in \text{int } K_*$ be fixed. It is clear that $\mathbf{1} \in \text{int } K_i$ for all $i \in I$. We will also use the following notations:

$$p_{\mathbf{1},K} = p, \quad p_{\mathbf{1},K_i} = p_i, \quad p_{\mathbf{1},K_*} = p_*. \quad (21)$$

It follows from Proposition 13 that

$$p(x) = \inf_{i \in I} p_i(x), \quad p_*(x) = \sup_{i \in I} p_i(x). \quad (22)$$

A function $f: X \to \mathbb{R}$ is called plus-homogeneous (with respect to $\mathbf{1}$) if

$$f(x + \lambda \mathbf{1}) = f(x) + \lambda \quad \text{for all } x \in X \text{ and } \lambda \in \mathbb{R}.$$

(The term *plus homogeneous* was coined in [13].) It follows from (18) that p_i ($i \in I$), p and p_* are plus-homogeneous functions.

Let

$$B_i = \{x \in X: \mathbf{1} \geq_{K_i} x \geq_{K_i} -\mathbf{1}\} \quad i \in I. \quad (23)$$

Since K_i is a closed solid convex pointed cone, it is easy to check that B_i ($i \in I$) can be considered as the unit ball of the norm $\|\cdot\|_i$ defined on X by

$$\|x\|_i := \max(p_i(x), p_i(-x)) \quad x \in X. \quad (24)$$

Let

$$\|x\|_* = \sup_{i \in I} \|x\|_i \quad (x \in X; \, i \in I). \quad (25)$$

We now show that $\|x\|_* < +\infty$ for each $x \neq 0$. Indeed, since $\mathbf{1} \in \text{int } K_* \subset \text{int } K_i$, it follows that there exists $\varepsilon > 0$ such that $\mathbf{1} + \varepsilon \tilde{B} \subset K_i$ for all $i \in I$, where $\tilde{B} = \{x \in X: \|x\| \leq 1\}$ is the closed unit ball with respect to the initial norm $\|\cdot\|$ of the normed space X. Let $x \neq 0$. Then $x' = (\varepsilon/\|x\|)x \in \varepsilon \tilde{B}$; hence $\mathbf{1} - x' \in K_i$. This implies that

$$p_i(x') = \inf\{\lambda \in \mathbb{R}: \lambda \mathbf{1} - x' \in K_i\} \leq 1.$$

Since p_i is a positively homogeneous function, it follows that

$$p_i(x) = p_i\left(\frac{\|x\|}{\varepsilon} x'\right)$$
$$= \frac{\|x\|}{\varepsilon} p_i(x') \leq \frac{\|x\|}{\varepsilon}.$$

The same argument demonstrates that $p_i(-x) \leq \|x\|/\varepsilon$. Hence

$$\|x\|_* = \sup_{i \in I} \|x\|_i$$
$$= \sup_{i \in I} \max(p(x_i), p(-x_i)) \leq \frac{\|x\|}{\varepsilon} < +\infty.$$

Clearly $\|\cdot\|_*$ is a norm on X. It is easy to see that

$$\|x\|_* = \max(p_*(x), p_*(-x)) \quad x \in X. \quad (26)$$

Due to (23), we have

$$B_i(x, r) := \{y \in X: \|y - x\|_i \leq r\}$$
$$= \{y \in X: x + r\mathbf{1} \geq_{K_i} y \geq_{K_i} x - r\mathbf{1}\}, \quad (27)$$

where $x \in X$, $i \in I$ and $r > 0$. Let $x \in X$ and $r > 0$. Consider the closed ball $B(x, r)$ with center x and radius r with respect to $\|\cdot\|_*$:

$$B(x, r) := \{y \in X: \|y - x\|_* \leq r\}$$
$$= \{y \in X: x + r\mathbf{1} \geq_{K_*} y \geq_{K_*} x - r\mathbf{1}\}. \quad (28)$$

It follows from (20), (27), and (28) that

$$B(x,r) = \bigcap_{i \in I} B_i(x,r), \qquad (29)$$

and

$$B(x,r) \subseteq \{y \in X : x + r\mathbf{1} \geq_K y \geq_K x - r\mathbf{1}\}. \qquad (30)$$

We now present an example.

Example 8 Let $X = \mathbb{R}^2$. Consider the cones

$$A = \{(x,y) \in X : x \geq 0 \text{ and } y \geq 2x\},$$

$$B = \left\{(x,y) \in X : x \leq 0 \text{ and } y \geq \frac{1}{2}x\right\},$$

$$C = \left\{(x,y) \in X : x \geq 0 \text{ and } y \geq -\frac{1}{2}x\right\},$$

$$D = \{(x,y) \in X : x \leq 0 \text{ and } y \geq -2x\}.$$

Set $K_1 = A \cup B$, $K_2 = C \cup D$, $K = K_1 \cup K_2$, and $K_* := K_1 \cap K_2 = A \cup D$. It is easy to check that K is not a convex set while K_1, K_2 and K_* are convex sets. We also have:

$$p_*(x) = \max(y-2x, y+2x) \text{ for all } x = (x,y) \in X,$$

$$\|x\|_* = |y| + 2|x| \quad \text{for all } x = (x,y) \in X.$$

Example 9 Let X be a normed space with a norm $\|\cdot\|$. Let $Y = X \times \mathbb{R}$ and $K := \text{epi}\|\cdot\| \subset Y$ be the epigraph of $\|\cdot\|$. (Recall that $\text{epi}\|\cdot\| = \{(x,\lambda) \in Y : \lambda \geq \|x\|\}$.) Then K is a convex closed cone and $(0,1) \in \text{int}\, K$. Assume now that X is equipped with two equivalent norms $\|\cdot\|_1$ and $\|\cdot\|_2$. Let $K_i = \text{epi}\|\cdot\|_i$, $i = 1, 2$, and $K = K_1 \cup K_2$. If there exist $x' \in X$ and $x'' \in X$ such that $\|x'\|_1 < \|x'\|_2$ and $\|x''\|_1 > \|x''\|_2$, then K is not convex. Clearly K is a pointed cone. The set $\text{int}\, K$ contains $(0,1)$; hence it is nonempty. Clearly $K \setminus \{0\}$ is contained in the open half-space $\{(x,\lambda) : \lambda > 0\}$. Cone K is star-shaped. It can be proved that $\ker K = K_1 \cap K_2$.

In the remainder of the article, we consider a normed space X with a closed star-shaped cone K such that $\text{int}\ker K$ is not empty. Assume that K is given as $K = \bigcup_{i \in I} K_i$, where

- I is an arbitrary index set;
- K_i, $(i \in I)$ is a closed solid convex pointed cone;
- The interior $\text{int}\, K_*$ of the cone $K_* = \bigcap_{i \in I} K_i$ is nonempty.

In the sequel, assume that the norm $\|\cdot\|$ of X coincides with the norm $\|\cdot\|_*$ defined by (26).

Characterization of Best Approximations

Let $\varphi : X \times X \longrightarrow \mathbb{R}$ be a function defined by

$$\varphi(x,y) := \sup\{\lambda \in \mathbb{R} : x + y \geq_K \lambda\mathbf{1}\} \quad (x, y \in X). \qquad (31)$$

Since $\mathbf{1} \in \text{int}\, K_*$, it follows that the set $\{\lambda \in \mathbb{R} : x + y \geq_K \lambda\mathbf{1}\}$ is nonempty and bounded from above (by the number $\|x + y\|_*$). Clearly this set is closed. It follows from the definition of φ that the function φ has the following properties:

$$-\infty < \varphi(x,y) \leq \|x + y\|_* \quad \text{for each } x, y \in X, \qquad (32)$$

$$x + y \geq_K \varphi(x,y)\mathbf{1} \quad \text{for all } x, y \in X, \qquad (33)$$

$$\varphi(x,y) = \varphi(y,x) \quad \text{for all } x, y \in X, \qquad (34)$$

$$\varphi(x,-x) = \sup\{\lambda \in \mathbb{R} : 0 = x - x \geq_K \lambda\mathbf{1}\} \\ = 0 \quad \text{for all } x \in X, \qquad (35)$$

$$\varphi(x, y + \lambda\mathbf{1}) = \varphi(x,y) + \lambda \quad \text{for all } x, y \in X \\ \text{and } \lambda \in \mathbb{R}, \qquad (36)$$

$$\varphi(x + \lambda\mathbf{1}, y) = \varphi(x,y) + \lambda \quad \text{for all } x, y \in X \\ \text{and } \lambda \in \mathbb{R}, \qquad (37)$$

$$\varphi(\gamma x, \gamma y) = \gamma \varphi(x,y) \quad \text{for all } x, y \in X \\ \text{and } \gamma > 0. \qquad (38)$$

Proposition 14 *Let φ be the function defined by (31). Then*

$$\varphi(x,y) = -p(-x-y), \quad (x, y \in X), \qquad (39)$$

and hence

$$\varphi(x,y) = \sup_{i \in I}[-p_i(-x-y)] \quad (x, y \in X). \qquad (40)$$

Proof For each $x, y \in X$, we have

$$-\varphi(-x,-y) = -\sup\{\lambda \in \mathbb{R} : -(x+y) \geq_K \lambda\mathbf{1}\}$$
$$= \inf\{-\lambda \in \mathbb{R} : -(x+y) \geq_K \lambda\mathbf{1}\}$$
$$= \inf\{\lambda' \in \mathbb{R} : -(x+y) \geq_K -\lambda'\mathbf{1}\}$$
$$= \inf\{\lambda' \in \mathbb{R} : \lambda'\mathbf{1} \geq_K x+y\}$$
$$= p(x+y).$$

Hence $\varphi(x,y) = -p(-x-y)$. In view of (21), we get (40). □

Now, consider $x, y \in X$. We define the functions $\varphi_x \colon X \longrightarrow \mathbb{R}$ and $\varphi_y \colon X \longrightarrow \mathbb{R}$ by

$$\varphi_x(t) = \varphi(x,t) \quad t \in X \tag{41}$$

and

$$\varphi_y(t) = \varphi(t,y) \quad t \in X. \tag{42}$$

Note that φ_x and φ_y are nonincreasing functions with respect to the relation generated by K on X. We have the following result:

Corollary 5 *Let φ be the function defined by (31). Then φ is Lipschitz continuous.*

Proof This is an immediate consequence of Lipschitz continuity of p and Proposition 14. □

Corollary 6 *For each $x, y \in X$, the functions defined by (41) and (42) are Lipschitz continuous.*

Proof It follows from Corollary 5. □

Proposition 15 *Let φ be the function defined by (31) and set*

$$\Lambda(y,\alpha) = \{x \in X \colon \varphi(x,y) \geq \alpha\} \quad (y \in X; \alpha \in \mathbb{R}).$$

Then, $\Lambda(y,\alpha) = K + \alpha\mathbf{1} - y$ for all $y \in X$ and all $\alpha \in \mathbb{R}$.

Proof Fix $y \in X$ and $\alpha \in \mathbb{R}$. Then

$$x \in \Lambda(y,\alpha) \iff \varphi(x,y) \geq \alpha.$$

Due to Proposition 14, this happens if and only if $-p(-x-y) \geq \alpha$, and hence by Proposition 12, if and only if $-x - y \in -\alpha\mathbf{1} - K$. This is equivalent to $x \in K + \alpha\mathbf{1} - y$, which completes the proof. □

Corollary 7 *Under the hypotheses of Proposition 15, we have*

$$\varphi(x,y) \geq \alpha \quad \text{if and only if} \quad x + y \geq_K \alpha\mathbf{1}$$
$$(x, y \in X; \alpha \in \mathbb{R}).$$

Lemma 2 *Let W be a closed downward subset of X, $y_0 \in \mathrm{bd}\, W$ and φ be the function defined by (31). Then*

$$\varphi(w, -y_0) \leq 0 = \varphi(y_0, -y_0) \quad \forall w \in W. \tag{43}$$

Proof The proof is similar to the proof of Lemma 4.3 in [7]. □

For $x \in X$ and a nonempty subset W of X, we will use the following notations:

$$d^i(x, W) := \inf_{w \in W} \|x - w\|_i \quad i \in I$$

and

$$P_W^i(x) = \{w \in W \colon \|x - w\|_i = d^i(x, W)\} \quad i \in I.$$

Lemma 3 *Let W be a closed downward subset of X, $x \in X \setminus W$, $r > 0$, and $i \in I$. Then $r = d^i(x, W)$ if and only if $x - r\mathbf{1} \in W$ and $p_i(x - w - r\mathbf{1})) \geq 0$ for all $w \in W$.*

Proof Let $r = d^i(x, W)$. In a manner analogous to the proof of Proposition 3, one can prove that $x - r\mathbf{1} \in P_W^i(x) \subset W$. Since $P_W^i(x) \subseteq \mathrm{bd}\, W$, it follows from Lemma 2 and Proposition 14 that $p_i(x - w - r\mathbf{1}) \geq 0$ for all $w \in W$. Conversely, suppose that $x - r\mathbf{1} \in W$ and $p_i(x - w - r\mathbf{1}) \geq 0$ for all $w \in W$. Let $w \in W$ be arbitrary. Since p_i is plus-homogeneous and $p_i(x - w - r\mathbf{1}) = p_i(x - w) - r$, it follows from (24) that

$$\|x - w\|_i \geq p_i(x - w) \geq r.$$

Since $\|x - (x - r\mathbf{1})\|_i = r$ and $x - r\mathbf{1} \in W$, we conclude that $r = d^i(x, W)$. □

Lemma 4 *Let W be a closed downward subset of X, $x \in X \setminus W$, and $r > 0$. Then $r = d(x, W)$ if and only if $x - r\mathbf{1} \in W$ and for some $i \in I$, $p_i(x - w - r\mathbf{1}) \geq 0$ for all $w \in W$.*

Proof Let $r = d(x, W)$. By Proposition 3 we have $x - r\mathbf{1} \in P_W(x) \subseteq \mathrm{bd}\, W$. Then it follows from Lemma 3 that $\varphi(w, r\mathbf{1} - x) \leq 0$ for all $w \in W$. In view of (40), we get $p_i(x - w - r\mathbf{1}) \geq 0$ for all $w \in W$ and all $i \in I$. Conversely, suppose that $x - r\mathbf{1} \in W$ and for some $i \in I$, $p_i(x - w - r\mathbf{1}) \geq 0$ for all $w \in W$. Consider $w \in W$. Since p_i is plus-homogeneous and $p_i(x - w - r\mathbf{1}) = p_i(x - w) - r$, it follows from (24) and (25) that

$$\|x - w\|_* \geq \|x - w\|_i \geq p_i(x - w) \geq r.$$

Since $r = \|x - (x - r\mathbf{1})\|_*$ and $x - r\mathbf{1} \in W$, one thus has $r = d(x, W)$. □

The following result is an immediate consequence of Lemmas 3 and 4.

Corollary 8 *Let W be a closed downward subset of X, $x \in X \setminus W$. Then*

$$d(x, W) = d^i(x, W) \quad \text{for all} \quad i \in I. \tag{44}$$

Corollary 9 *Let W be a closed downward subset of X, $x \in X \setminus W$, and $w_0 \in W$. Then, $w_0 \in P_W(x)$ if and only if $w_0 \in P_W^i(x)$ for each $i \in I$.*

Proof Let $w_0 \in P_W(x)$. Then $\|x - w_0\|_* = d(x, W)$. In view of (25) and (44), we have $\|x - w_0\|_i = d^i(x, W)$ for each $i \in I$. Therefore, $w_0 \in P_W^i(x)$ for each $i \in I$. Conversely, let $w_0 \in P_W^i(x)$ for each $i \in I$. Then $\|x - w_0\|_i = d^i(x, W)$ for each $i \in I$. Hence, by (44), we get $\|x - w_0\|_* = \max_{i \in I} \|x - w_0\|_i = d(x, W)$, that is, $w_0 \in P_W(x)$. □

Theorem 6 *Let W be a closed downward subset of X, $x_0 \in X \setminus W$, $y_0 \in W$, and $r_0 := \|x_0 - y_0\|_*$. Assume that φ is the function defined by (31). Then the following assertions are equivalent:*
(1) $y_0 \in P_W(x_0)$.
(2) There exists $l \in X$ such that

$$\varphi(w, l) \leq 0 \leq \varphi(y, l), \quad \forall w \in W, \ y \in B(x_0, r_0). \tag{45}$$

Moreover, if (45) holds with $l = -y_0$, then $y_0 = w_0 = \min P_W(x_0)$, where $w_0 = x_0 - r\mathbf{1}$ is the least element of the set $P_W(x_0)$ and $r := d(x_0, W)$.

Proof
(1) $\Longrightarrow$ (2). Suppose that $y_0 \in P_W(x_0)$. Then $r_0 = \|x_0 - y_0\|_* = d(x_0, W) = r$. Since W is a closed downward subset of X, it follows from Proposition 3 that the least element $w_0 = x_0 - r_0\mathbf{1}$ of the set $P_W(x_0)$ exists. Let $l = -w_0$ and $y \in B(x_0, r_0)$ be arbitrary. Then, by (30), we have $y \geq_K -l$ or $y + l \geq_K 0$. It follows from Corollary 7 that $\varphi(y, l) \geq 0$. On the other hand, since $w_0 \in P_W(x_0)$, it follows that $w_0 \in \text{bd } W$. Hence, by Lemma 2 we have $\varphi(w, l) \leq 0$ for all $w \in W$.
(2) $\Longrightarrow$ (1). Assume that (2) holds. By (28) it is clear that $x_0 - r_0\mathbf{1} \in B(x_0, r_0)$. Therefore, by (45) we have $\varphi(x_0 - r_0\mathbf{1}, l) \geq 0$. Due to Corollary 7, we get $x_0 - r_0\mathbf{1} + l \geq_K 0$, and so $l - r_0\mathbf{1} \geq_K -x_0$. Hence there exists $j \in I$ such that

$$l - r_0\mathbf{1} \geq_{K_j} -x_0. \tag{46}$$

Now, let $w \in W$ be arbitrary. Since p_j is topical and (21), (39), and (45) hold, it follows from (46) that

$$p_j(x_0 - w) \geq p_j(r_0\mathbf{1} - l - w) = p_j(-l - w) + r_0$$
$$\geq p(-l - w) + r_0$$
$$= -\varphi(w, l) + r_0$$
$$\geq 0 + r_0 = r_0.$$

Then, by (24) and (25), we have

$$r_0 \leq p_j(x_0 - w) \leq \|x_0 - w\|_j$$
$$\leq \|x_0 - w\|_* \quad \text{for all } w \in W.$$

Thus $\|x_0 - y_0\|_* = d(x_0, W)$. Consequently, $y_0 \in P_W(x_0)$. Finally, suppose that (45) holds with $l = -y_0$. Then, by the implication (2) $\Longrightarrow$ (1), we have $y_0 \in P_W(x_0)$, and so $r_0 = \|x_0 - y_0\|_* = d(x_0, W)$ and $y_0 \geq_K w_0$, where $w_0 = x_0 - r\mathbf{1}$ is the least element of the set $P_W(x_0)$ and $r := d(x_0, W)$. Now, let $w \in P_W(x_0)$ be arbitrary. Then $\|x_0 - w\|_* = d(x_0, W) = r_0$, that is, $w \in B(x_0, r_0)$. It follows from (45) that $\varphi(w, -y_0) \geq 0$. In view of Corollary 7, we have $w - y_0 \geq_K 0$, and so $w \geq_K y_0$. This means that $y_0 = \min P_W(x_0) = w_0$. This completes the proof. □

Strictly Downward Sets and Their Best Approximation Properties

We start with the following definitions, which were introduced in [7] for downward subsets of a Banach lattice.

Definition 5 A downward subset W of X is called strictly downward if for each boundary point w_0 of W the inequality $w >_K w_0$ implies $w \notin W$.

Definition 6 Let W be a downward subset of X. We say that W is strictly downward at a point $w' \in \text{bd } W$ if for all $w_0 \in \text{bd } W$ with $w' \geq_K w_0$ the inequality $w >_K w_0$ implies $w \notin W$.

The following lemmas have been proved in [7]; however, those proofs hold for the case under consideration.

Lemma 5 Let $f\colon X \longrightarrow \mathbb{R}$ be a continuous strictly increasing function. Then all nonempty level sets $S_c(f)$ ($c \in \mathbb{R}$) of f are strictly downward.

Lemma 6 Let W be a closed downward subset of X. Then W is strictly downward at $w' \in \operatorname{bd} W$ if and only if
(i) $w >_K w' \Longrightarrow w \notin W$;
(ii) $(w' \geq_K w_0, w_0 \in \operatorname{bd} W) \Longrightarrow w_0 = w'$.

Lemma 7 Let W be a closed downward subset of X. Then W is strictly downward if and only if W is strictly downward at each of its boundary points.

Lemma 8 Let φ be the function defined by (31) and W be a closed downward subset of X that is strictly downward at a point $w' \in \operatorname{bd} W$. Then there exists unique $l \in X$ such that
$$\varphi(w, l) \leq 0 = \varphi(w', l), \quad \forall\, w \in W.$$

Theorem 7 Let φ be the function defined by (31). Then for a closed downward subset W of X the following assertions are equivalent:
(1) W is strictly downward.
(2) For each $w_0 \in \operatorname{bd} W$ there exists unique $l \in X$ such that
$$\varphi(w, l) \leq 0 = \varphi(w_0, l) \quad \forall\, w \in W.$$

Proof The implication (1) $\Longrightarrow$ (2) follows from Lemma 8. We now prove the implication (2) $\Longrightarrow$ (1). Assume that for each $w_0 \in \operatorname{bd} W$ there exists unique $l \in X$ such that
$$\varphi(w, l) \leq 0 = \varphi(w_0, l) \quad \forall\, w \in W.$$

Let $w_0 \in \operatorname{bd} W$ and $y \in X$ with $y >_K w_0$. Assume that $y \in W$. We claim that $y + \lambda \mathbf{1} \notin W$ for all $\lambda > 0$. Suppose that there exists $\lambda_0 > 0$ such that $y + \lambda_0 \mathbf{1} \in W$. Since $y + \lambda_0 \mathbf{1} >_K w_0 + \lambda_0 \mathbf{1}$ and W is a downward set, we have $w_0 + \lambda_0 \mathbf{1} \in W$. In view of Corollary 1, it contradicts with $w_0 \in \operatorname{bd} W$, and so the claim is true. Then, by Corollary 1, we have $y \in \operatorname{bd} W$. Let $l = -y$. It follows from Lemma 2 that
$$\varphi(w, l) \leq 0 = \varphi(y, l) \quad \forall\, w \in W. \tag{47}$$

On the other hand, applying Lemma 2 to the point w_0 we have for $l' = -w_0$:
$$\varphi(w, l') \leq 0 = \varphi(w_0, l') \quad \forall\, w \in W. \tag{48}$$

Since $y >_{K_i} w_0$ for some $i \in I$ and p_i is increasing, it follows from (21), (39), and (48) that $0 = p_i(-w_0 - l') \geq p_i(-y - l') \geq p(-y - l') = -\varphi(y, l') \geq 0$. This, together with (48), implies that
$$\varphi(w, l') \leq 0 = \varphi(y, l') \quad \forall\, w \in W. \tag{49}$$

Since $w_0 \neq y$, it follows that $l' \neq l$. Hence (47) and (49) contradict the uniqueness of l. We have demonstrated that the assumption $y \in W$ leads to a contradiction. Thus $y \notin W$. This means that W is strictly downward. $\square$

Corollary 10 Let $f\colon X \longrightarrow \mathbb{R}$ be a continuous strictly increasing function and φ be the function defined by (31). Then for each $x \in X$ there exists unique $l = -x$ such that
$$\varphi(w, l) \leq 0 = \varphi(x, l) \quad \forall\, w \in S_c(f),$$
where $c = f(x)$.

Proof This is an immediate consequence of Lemma 5 and Theorem 7. $\square$

Definition 7 Let W be a downward subset of X. A point $w' \in \operatorname{bd} W$ is said to be a Chebyshev point if for each $w_0 \in \operatorname{bd} W$ with $w' \geq_K w_0$ and for each $x_0 \notin W$ such that $w_0 \in P_W(x_0)$ it follows that $P_W(x_0) = \{w_0\}$, that is, the best approximation of x_0 is unique.

Definition 7 was introduced in [7] for a downward subset of a Banach lattice.

Definition 8 Let W be a downward subset of X. A point $w' \in \operatorname{bd} W$ is said to be a Chebyshev point of W with respect to each K_i ($i \in I$) if for each $w_0 \in \operatorname{bd} W$ with $w' \geq_K w_0$ and for each $x_0 \notin W$ such that $w_0 \in P^i{}_W(x_0)$ for each $i \in I$ it follows that $P^i{}_W(x_0) = \{w_0\}$ for each $i \in I$.

Remark 4 In view of Corollary 8, we have that Definitions 7 and 8 are equivalent.

Theorem 8 Let W be a closed downward subset of X and $w' \in \operatorname{bd} W$. If w' is a Chebyshev point of W with respect to each K_i ($i \in I$), then W is a strictly downward set at w'.

Proof Suppose that w' is a Chebyshev point of W with respect to each K_i ($i \in I$). Assume, if possible, that W is not strictly downward at w'. Then we can find $w_0 \in \text{bd}\,W$ and $w \in W$ such that $w' \geq_K w_0$ and $w >_K w_0$. Let $r \geq \|w - w_0\|_* > 0$. It follows from (27) that

$$r\mathbf{1} \geq_{K_i} w - w_0 \quad \forall\, i \in I.$$

Thus, $w_0 + r\mathbf{1} \geq_{K_i} w$ for all $i \in I$. Set $x_0 = w_0 + r\mathbf{1} \in X$. Since $w_0 \in \text{bd}\,W$, by Lemma 6 we have $\varphi(y, -w_0) \leq 0$ for all $y \in W$. Also, $x_0 - r\mathbf{1} = w_0 \in W$. Thus, by (21), Proposition 14, and Lemma 4 we get $r = d(x_0, W)$. Since $\|x_0 - w_0\|_i = \|r\mathbf{1}\|_i = r$ for all $i \in I$, it follows from (25) that $\|x_0 - w_0\|_* = r$, and hence $w_0 \in P_W(x_0)$. In view of Corollary 9, we obtain $w_0 \in P_W^i(x_0)$ for all $i \in I$.

On the other hand, we have $x_0 = w_0 + r\mathbf{1} \geq_{K_i} w$ for all $i \in I$. Since $w >_K w_0$, we conclude that there exists $j \in I$ such that $w >_{K_j} w_0$. It follows that $r\mathbf{1} = x_0 - w_0 >_{K_j} x_0 - w \geq_{K_j} 0$. Hence

$$\|x_0 - w\|_j \leq \|r\mathbf{1}\|_j = r = d^j(x_0, W) \leq \|x_0 - w\|_j.$$

Thus $\|x_0 - w\|_j = d^j(x_0, W)$, and so $w \in P^j{}_W(x_0)$ with $w \neq w_0$. Whence there exist a point $x_0 \in X \setminus W$ and a point $w_0 \in \text{bd}\,W$ with $w' \geq_K w_0$ such that $w_0 \in P^i{}_W(x_0)$ for each $i \in I$ and $P^j{}_W(x_0)$ contains at least one point different from w_0. This is a contradiction because w' is a Chebyshev point of W with respect to each K_i ($i \in I$), which completes the proof. □

Proposition 16 *Let W be a closed downward subset of X and $w' \in \text{bd}\,W$. If W is a strictly downward set at w', then w' is a Chebyshev point of W.*

Proof The proof is similar to that of Theorem 4.2 (the implication (2) $\Longrightarrow$ (1)) in [7]. □

Corollary 11 *Let $f: X \longrightarrow \mathbb{R}$ be a continuous strictly increasing function. Then $S_c(f)(c \in \mathbb{R})$ is a Chebyshev subset of X.*

Proof This is an immediate consequence of Lemma 5 and Proposition 16. □

References

1. Chui CK, Deutsch F, Ward JD (1990) Constrained best approximation in Hilbert space. Constr Approx 6:35–64
2. Chui CK, Deutsch F, Ward JD (1992) Constrained best approximation in Hilbert space II. J Approx Theory 71:213–238
3. Deutch F (2000) Best approximation in inner product spaces. Springer, New York
4. Deutsch F, Li W, Ward JD (1997) A dual approach to constrained interpolation from a convex subset of a Hilbert space. J Approx Theory 90:385–414
5. Deutsch F, Li W, Ward JD (2000) Best approximation from the intersection of a closed convex set and a polyhedron in Hilbert space, weak Slater conditions, and the strong conical hull intersection property. SIAM J Optim 10:252–268
6. Martinez-Legaz J-E, Rubinov AM, Singer I (2002) Downward sets and their separation and approximation properties. J Global Optim 23:111–137
7. Mohebi H, Rubinov AM (2006) Best approximation by downward sets with applications. J Anal Theory Appl 22(1):1–22
8. Mohebi H, Rubinov AM (2006) Metric projection onto a closed set: necessary and sufficent conditions for the global minimum. J Math Oper Res 31(1):124–132
9. Mohebi H, Sadeghi H, Rubinov AM (2006) Best approximation in a class of normed spaces with star-shaped cones. J Numer Funct Anal Optim 27(3–4):411–436
10. Mulansky B, Neamtu M (1998) Interpolation and approximation from convex sets. J Approx Theory 92:82–100
11. Rubinov AM (2000) Abstract convex analysis and global optimization. Kluwer, Boston Dordrecht London
12. Rubinov AM, Gasimov RN (2004) Scalarization and nonlinear scalar duality for vector optimization with preferences that are not necessarily a pre-order relation. J Glob Optim 29:455–477
13. Rubinov AM, Singer I (2001) Topical and sub-topical functions, downward sets and abstract convexity. Optimization 50:307–351
14. Rubinov AM, Singer I (2000) Best approximation by normal and co-normal sets. J Approx Theory 107:212–243
15. Singer I (1997) Abstract convex analysis. Wiley-Interscience, New York
16. Singer I (1970) Best approximation in normed linear spaces by elements of linear subspaces. Springer, New York
17. Jeyakumar V, Mohebi H (2005) A global approach to nonlinearly constrained best approximation. J Numer Funct Anal Optim 26(2):205–227
18. Jeyakumar V, Mohebi H (2005) Limiting and ε-subgradient characterizations of constrained best approximation. J Approx Theory 135:145–159
19. Vlasov LP (1967) Chebyshev sets and approximatively convex sets. Math Notes 2:600–605
20. Vlasov LP (1973) Approximative properties of sets in normed linear spaces. Russ Math Surv 28:1–66
21. Vulikh BZ (1967) Introduction to the theory of partially ordered vector spaces. Wolters-Noordhoff, Groningen

Bilevel Fractional Programming

HERMINIA I. CALVETE, CARMEN GALÉ
Dpto. de Métodos Estadísticos,
Universidad de Zaragoza, Zaragoza, Spain

MSC2000: 90C32, 90C26

Article Outline

Keywords
Introduction
Formulation
Theoretical Results
Algorithms
References

Keywords

Fractional bilevel programming; Hierarchical optimization; Nonconvex optimization

Introduction

Fractional bilevel programming (FBP), a class of bilevel programming [6,10], has been proposed as a generalization of standard fractional programming [9] for dealing with hierarchical systems with two decision levels. FBP problems assume that the objective functions of both levels are ratios of functions and the common constraint region to both levels is a nonempty and compact polyhedron.

Formulation

Using the common notation in bilevel programming, the FBP problem [1] can be formulated as:

$$\min_{x_1, x_2} f_1(x_1, x_2) = \frac{h_1(x_1, x_2)}{g_1(x_1, x_2)},$$

where x_2 solves

$$\min_{x_2} f_2(x_1, x_2) = \frac{h_2(x_1, x_2)}{g_2(x_1, x_2)}$$

s.t. $(x_1, x_2) \in S$,

where $x_1 \in \mathbb{R}^{n_1}$ and $x_2 \in \mathbb{R}^{n_2}$ are the variables controlled by the upper level and the lower level decision maker, respectively; h_i and g_i are continuous functions, h_i are nonnegative and concave and g_i are positive and convex on S; and $S = \{(x_1, x_2) : A_1 x_1 + A_2 x_2 \leq b, x_1 \geq 0, x_2 \geq 0\}$, which is assumed to be nonempty and bounded.

Let S_1 be the projection of S on $\mathbb{R}^{n_1}$. For each $\tilde{x}_1 \in S_1$ provided by the upper level decision maker, the lower level one solves the fractional problem:

$$\min_{x_2} f_2(\tilde{x}_1, x_2) = \frac{h_2(\tilde{x}_1, x_2)}{g_2(\tilde{x}_1, x_2)}$$

s.t. $A_2 x_2 \leq b - A_1 \tilde{x}_1$

$x_2 \geq 0$.

Let $M(\tilde{x}_1)$ denote the set of optimal solutions to this problem. In order to ensure that the FBP problem is well posed it is also assumed that $M(\tilde{x}_1)$ is a singleton for all $\tilde{x}_1 \in S_1$.

The feasible region of the upper level decision maker, also called the inducible region (IR), is implicitly defined by the lower level decision maker:

$$IR = \{(\tilde{x}_1, \tilde{x}_2) : \tilde{x}_1 \geq 0, \tilde{x}_2 = \operatorname{argmin} \{f_2(\tilde{x}_1, x_2) : A_1 \tilde{x}_1 + A_2 x_2 \leq b, x_2 \geq 0\}\} .$$

Therefore, the FBP problem can be stated as:

$$\min_{x_1, x_2} f_1(x_1, x_2) = \frac{h_1(x_1, x_2)}{g_1(x_1, x_2)}$$

s.t. $(x_1, x_2) \in IR$.

Theoretical Results

The FBP problem is a nonconvex optimization problem but, taking into account the quasiconcavity of f_2 and the properties of polyhedra, in [1] it was proved that the inducible region is formed by the connected union of faces of the polyhedron S.

One of the main features of FBP problems is that, even with the more complex objective functions, they retain the most important property related to the optimal solution of linear bilevel programming problems. That is, there is an extreme point of S which solves the FBP problem [1]. This result is a consequence of the properties of IR as well as of the fact of f_1 being quasiconcave. The same conclusion is also obtained when both level objective functions are defined as the minimum of a finite number of functions which are ratios with the previously stated conditions or, in general, if they are quasiconcave.

Under the additional assumption that the upper level objective function is explicitly quasimonotonic, another geometrical property of the optimal solution of the FBP problem can be obtained by introducing the concept of boundary feasible extreme point. According to [7], a point $(x_1, x_2) \in$ IR is a boundary feasible extreme point if there exists an edge E of S such that (x_1, x_2) is an extreme point of E, and the other extreme point of E is not an element of IR.

Let us consider the relaxed problem:

$$\min_{x_1, x_2} f_1(x_1, x_2) = \frac{h_1(x_1, x_2)}{g_1(x_1, x_2)}, \quad (1)$$
$$\text{s.t.} \quad (x_1, x_2) \in S .$$

Since f_1 is a quasiconcave function and S is a nonempty and compact polyhedron, an extreme point of S exists which solves (1). Obviously, if an optimal solution of (1) is a point of IR, then it is an optimal solution to the FBP problem. However, in general, this will not be true, since both decision makers usually have conflicting objectives.

Hence, if f_1 is explicitly quasimonotonic and there exists an extreme point of S not in IR that is an optimal solution of the relaxed problem (1), then a boundary feasible extreme point exists that solves the FBP problem [3].

Although FBP problems retain some important properties of linear bilevel problems, it is worth pointing out at this time some differences related to the existence of multiple optima when solving the lower level problem for given $x_1 \in S_1$. Different approaches have been proposed in the literature to make sure that the bilevel problem is well posed [6]. The most common one is to assume that $M(x_1)$ is single-valued for all $x_1 \in S_1$. Other approaches give rules for selecting $x_2 \in M(x_1)$ in order to be able to evaluate the upper level objective function $f_1(x_1, x_2)$. The optimistic approach assumes that the upper level decision maker has the right to influence the lower level decision maker so that the latter selects x_2 to provide the best value of f_1. On the contrary, the pessimistic approach assumes that the lower level decision maker always selects x_2 which gives the worst value of f_1.

It is well-known [8] that, under the optimistic approach, at least one optimal solution of the linear bilevel problem is obtained at an extreme point of the polyhedron defined by the common constraints. However, in [3] an example of the FBP problem is proposed in which $M(\tilde{x}_1)$ is not single-valued for given $\tilde{x}_1 \in S_1$ and this assertion is not true. Firstly, IR no longer consists of the union of faces of the polyhedron S. Secondly, if the pessimistic approach is used, then an optimal solution to the example does not exist. Finally, if the optimistic approach is taken the optimal solution to the example is not an extreme point of the polyhedron S.

Algorithms

Bearing in mind that there is an extreme point of S which solves the FBP problem, an enumerative algorithm can be devised which examines the set of extreme points of S in order to identify the best one regarding f_1, which is a point of IR. The bottleneck of the algorithm would be the generally large number of extreme points of a polyhedron together with the process of checking if an extreme point of S is a point of IR or not.

In the particular case in which f_1 is linear and f_2 is linear fractional (LLFBP problem), in [2] an enumerative algorithm has been proposed which finds a global optimum in a finite number of stages by examining implicitly only bases of the matrix A_2. This algorithm connects the points of IR with the bases of A_2, by applying the parametric approach to solve the fractional problem of the lower level. One of the main advantages of the procedure is that only linear problems have to be solved.

When f_1 is linear fractional and f_2 is linear (LFLBP problem), the algorithm developed in [2] combines local search in order to find an extreme point of IR with a better value of f_1 than any of its adjacent extreme points in IR and a penalty method when looking for another point of IR from which a new local search can start.

The Kth-best algorithm has been proposed in [3] to globally solve the FBP problem when both objective functions are linear fractional (LFBP). It essentially asserts that the best (in terms of the upper level objective function) of the extreme points of S which is a point of IR is an optimal solution to the problem. Moreover, the search for this point can be made sequentially by computing adjacent extreme points to the incumbent extreme point.

Finally, recently two genetic algorithms have been proposed [4,5] which allow us to solve LLFBP, LFBP and LFLBP problems. Both algorithms provide excellent results in terms of both accuracy of the solution and time invested, proving that they are effective and useful approaches for solving those problems. Both algorithms associate chromosomes with extreme points of S. The fitness of a chromosome evaluates its quality and penalizes it if the associated extreme point is not in IR. The algorithms mainly differ in the procedure of checking if an extreme point is in IR. When f_2 is linear, all lower level problems have the same dual feasible region, so it is possible to prove several properties which simplify the process.

References

1. Calvete HI, Galé C (1998) On the quasiconcave bilevel programming problem. J Optim Appl 98(3):613–622
2. Calvete HI, Galé C (1999) The bilevel linear/linear fractional programming problem. Eur J Oper Res 114(1):188–197
3. Calvete HI, Galé C (2004) Solving linear fractional bilevel programs. Oper Res Lett 32(2):143–151
4. Calvete HI, Galé C, Mateo PM (2007) A genetic algorithm for solving linear fractional bilevel problems. To appear in Annals Oper Res
5. Calvete HI, Galé C, Mateo PM (2008) A new approach for solving linear bilevel problems using genetic algorithms. Eur J Oper Res 188(1):14–28
6. Dempe S (2003) Annotated bibliography on bilevel programming and mathematical programs with equilibrium constraints. Optimization 52:333–359
7. Liu YH, Hart SM (1994) Characterizing an optimal solution to the linear bilevel programming problem. Eur J Oper Res 73(1):164–166
8. Savard G (1989) Contribution à la programmation mathématique à deux niveaux. PhD thesis, Ecole Polytechnique de Montréal, Université de Montréal, Montréal, Canada
9. Schaible S (1995) Fractional programming. In: Horst R, Pardalos PM (eds) Handbook of global optimization. Kluwer, Dordrecht, pp 495–608
10. Vicente LN, Calamai PH (1994) Bilevel and multilevel programming: a bibliography review. J Glob Optim 5:291–306

Bilevel Linear Programming

JONATHAN F. BARD
University Texas, Austin, USA

MSC2000: 49-01, 49K10, 49M37, 90-01, 91B52, 90C05, 90C27

Article Outline

Keywords
Definitions
Theoretical Properties
Algorithmic Approaches
See also
References

Keywords

Bilevel linear programming; Hierarchical optimization; Stackelberg game; Multiple objectives; Complementarity

Many hierarchical optimization problems involving two or more decision makers can be modeled as a multilevel mathematical program. The two-level structure is commonly known as a *Stackelberg game* where a leader and a follower try to minimize their individual objective functions $F(x, y)$ and $f(x, y)$, respectively, subject to a series of interdependent constraints [2,9]. Play is defined as sequential and the mood as noncooperative. The decision variables are partitioned between the players in such a way that neither can dominate the other. The leader goes first and through his choice of $x \in \mathbf{R}^n$ is able to influence but not control the actions of the follower. This is achieved by reducing the set of feasible choices available to the latter. Subsequently, the follower reacts to the leader's decision by choosing a $y \in \mathbf{R}^m$ in an effort to minimizes his costs. In so doing, he indirectly affects the leader's solution space and outcome.

Two basic assumptions underlying the Stackelberg game are that full information is available to the players and that cooperation is prohibited. This precludes the use of correlated strategies and side payments. The vast majority of research on this problem has centered on the linear case known as the linear *bilevel program* (BLP) [3,6]. Relevant notation, the basic model, and a discussion of its theoretical properties follow.

For $x \in X \subset \mathbf{R}^n$, $y \in Y \subset \mathbf{R}^m$, $F: X \times Y \to \mathbf{R}^1$, and $f: X \times Y \to \mathbf{R}^1$, the linear bilevel programming problem can be written as follows:

$$\min_{x \in X} \quad F(x, y) = c_1 x + d_1 y, \qquad (1)$$

s.t. $\quad A_1 x + B_1 y \leq b_1,$ (2)

$$\min_{y \in Y} f(x, y) = c_2 x + d_2 y,$$ (3)

s.t. $\quad A_2 x + B_2 y \leq b_2,$ (4)

where $c_1, c_2 \in \mathbf{R}^n$, $d_1, d_2 \in \mathbf{R}^m$, $b_1 \in \mathbf{R}^p$, $b_2 \in \mathbf{R}^q$, $A_1 \in \mathbf{R}^{p \times n}$, $B_1 \in \mathbf{R}^{p \times m}$, $A_2 \in \mathbf{R}^{q \times n}$, $B_2 \in \mathbf{R}^{q \times m}$. The sets X and Y place additional restrictions on the variables, such as upper and lower bounds or integrality requirements. Of course, once the leader selects an x, the first term in the follower's objective function becomes a constant and can be removed from the problem. In this case, we replace $f(x, y)$ with $f(y)$.

The sequential nature of the decisions in (1)–(4) implies that y can be viewed as a function of x; i. e., $y = y(x)$. For convenience, this dependence will not be written explicitly.

Definitions

a) Constraint region of the linear BLP:

$$S = \{(x, y) : \quad x \in X, y \in Y, \\ A_1 x + B_1 y \leq b_1, A_2 x + B_2 y \leq b_2\}.$$

b) Feasible set for follower for each fixed $x \in X$:

$$S(x) = \{y \in Y : A_2 x + B_2 y \leq b_2\}.$$

c) Projection of S onto the leader's decision space:

$$S(X) = \{x \in X : \quad \exists y \in Y, \\ A_1 x + B_1 y \leq b_1, A_2 x + B_2 y \leq b_2\}.$$

d) Follower's *rational reaction set* for $x \in S(X)$:

$$P(x) = \{y \in Y : \\ y \in \arg\min \{f(x, \widehat{y}) : \widehat{y} \in S(x)\}\}.$$

e) *Inducible region*:

$$\mathrm{IR} = \{(x, y) : (x, y) \in S, y \in P(x)\}.$$

To ensure that (1)–(4) is well posed it is common to assume that S is nonempty and compact, and that for all decisions taken by the leader, the follower has some room to respond; i. e., $P(x) \neq \emptyset$. The rational reaction set $P(x)$ defines the response while the inducible region IR represents the set over which the leader may optimize. Thus in terms of the above notation, the BLP can be written as

$$\min \{F(x, y) : (x, y) \in \mathrm{IR}\}.$$ (5)

Even with the stated assumptions, problem (5) may not have a solution. In particular, if $P(x)$ is not single-valued for all permissible x, the leader may not achieve his minimum payoff over IR. To avoid this situation in the development of algorithms, it is usually assumed that $P(x)$ is a point-to-point map. Because a simple check is available to see whether the solution to (1)–(4) is unique (see [2]) this assumption does not appear to be unduly restrictive.

It should be mentioned that in practice the leader will incur some cost in determining the decision space $S(X)$ over which he may operate. For example, when BLP is used as a model for a decentralized firm with headquarters representing the leader and the divisions representing the follower, coordination of lower level activities by headquarters requires detailed knowledge of production capacities, technological capabilities, and routine operating procedures. Up-to-date information in these areas is not likely to be available to corporate planners without constant monitoring and oversight.

Theoretical Properties

The linear bilevel program was first shown to be *NP-hard* by R.G. Jeroslow [7] using satisfiability arguments common in computer science. The complexity of the problem is further elaborated in ▶ Bilevel linear programming: Complexity, equivalence to minmax, concave programs. Issues related to the geometry of the solution space are now discussed. The main result is that when the linear BLP is written as a standard mathematical program (5), the corresponding constraint set or inducible region is comprised of connected faces of S and that a solution occurs at a vertex (see [1] or [8] for the proofs). For ease of presentation, it will be assumed that $P(x)$ is single-valued and bounded, S is bounded and nonempty, and that $Y = \{y : y \geq 0\}$.

Theorem 1 *The inducible region can be written equivalently as a piecewise linear equality constraint comprised of supporting hyperplanes of S.*

A straightforward corollary of this theorem is that the linear BLP is equivalent to minimizing F over a feasible region comprised of a piecewise linear equality constraint. In general, because a linear function $F = c_1 x + d_1 y$ is being minimized over IR, and because F is bounded below on S by, say, $\min\{c_1 x + d_1 y: (x, y) \in \text{IR}\}$, it can also be concluded that the solution to the linear BLP occurs at a vertex of IR. An alternative proof of this result was given by W.F. Bialas and M.H. Karwan [4] who noted that (5) could be written equivalently as

$$\min \{c_1 x + d_1 y: (x, y) \in \text{co IR}\},$$

where co IR is the convex hull of the inducible region. Of course, co IR is not the same as IR, but the next theorem states their relationship with respect to BLP solutions.

Theorem 2 *The solution (x^*, y^*) of the linear BLP occurs at a vertex of S.*

In general, at the solution (x^*, y^*) the hyperplane $\{(x, y): c_1 x + d_1 y = c_1 x^* + d_1 y^*\}$ will not be a support of the set S. Furthermore, a by-product of the proof of Theorem 2 is that any vertex of IR is also a vertex of S, implying that IR consists of faces of S. Comparable results were derived by Bialas and Karwan who began by showing that any point in S that strictly contributes in any convex combination of points in S to form a point in IR must also be in IR. This leads to the fact that if x is an extreme point of IR, then it is an extreme point of S. A final observation about the solution of the linear BLP can be inferred from this last assertion. Because the inducible region is not in general convex, the set of optimal solutions to (1)–(4) when not single-valued is not necessarily convex.

In searching for a way to solve the linear BLP, it would be helpful to have an explicit representation of IR rather than the implicit representation given by Definition e). This can be achieved by replacing the follower's problem (3)-(4) with his Kuhn–Tucker conditions and appending the resultant system to the leader's problem. Letting $u \in \mathbf{R}^q$ and $v \in \mathbf{R}^m$ be the dual variables associated with constraints (4) and $y \geq 0$, respectively, leads to the proposition that a necessary condition for (x^*, y^*) to solve the linear BLP is that there exists (row) vectors u^* and v^* such that (x^*, y^*, u^*, v^*) solves:

$$\min \quad c_1 x + d_1 y, \tag{6}$$
$$\text{s.t.} \quad A_1 x + B_1 y \leq b_1, \tag{7}$$
$$u B_2 - v = -d_2, \tag{8}$$
$$u(b_2 - A_2 x - B_2 y) + v y = 0, \tag{9}$$
$$A_2 x + B_2 y \leq b_2, \tag{10}$$
$$x \geq 0, \quad y \geq 0, \quad u \geq 0, \quad v \geq 0. \tag{11}$$

This formulation has played a key role in the development of algorithms. One advantage that it offers is that it allows for a more robust model to be solved without introducing any new computational difficulties. In particular, by replacing the follower's objective function (3) with a quadratic form

$$f(x, y) = c_2 x + d_2 y + x^\top Q_1 y + \frac{1}{2} y^\top Q_2 y, \tag{12}$$

where Q_1 is an $n \times m$ matrix and Q_2 is an $m \times m$ symmetric positive semidefinite matrix, the only thing that changes in (6)–(11) is constraint (8). The new constraint remains linear but now includes all problem variables; i. e.,

$$x^\top Q_1 + y^\top Q_2 + u B_2 - v = -d_2. \tag{13}$$

From a conceptual point of view, (6)–(11) is a standard mathematical program and should be relatively easy to solve because all but one constraint is linear. Nevertheless, virtually all commercial nonlinear codes find complementarity terms like (9) notoriously difficult to handle so some ingenuity is required to maintain feasibility and guarantee global optimality.

Algorithmic Approaches

There have been nearly two dozen algorithms proposed for solving the linear BLP since the field caught the attention of researchers in the mid-1970s. Many of these are of academic interest only because they are either impractical to implement or highly inefficient. In general, there are three different approaches to solving (1)–(4) that can be considered workable. The first makes use of Theorem 2 and involves some form of vertex enumeration in the context of the simplex method. W. Candler and R. Townsely [5] were the first to develop an algorithm that was globally optimal. Their scheme repeatedly solves two linear programs, one for the leader in all of the x variables and a subset of the y variables associated with an optimal basis to the follower's problem, and the other for the follower with all the x variables fixed. In a systematic way they explore optimal bases of the follower's problem for x fixed and then return to the leader's problem with the corresponding basic y variables. By focusing on the reduced cost coefficients of the y variables not in an optimal basis of the follower's problem, they are able to provide a monotonic decrease in the number of follower bases that have to be examined. Bialas and Karwan [4] offered a different approach that systematically explores vertices beginning with the basis associated with the optimal solution to the linear program created by removing (3). This is known as the *high point problem*.

The second and most popular method for solving the linear BLP is known as the *Kuhn–Tucker approach* and concentrates on (6)–(11). The fundamental idea is to use a *branch and bound strategy* to deal with the complementarity constraint (9). Omitting or relaxing this constraint leaves a standard linear program which is easy to solve. The various methods proposed employ different techniques for assuring that complementarity is ultimately satisfied (e. g., see [3,6]).

The third method is based on some form of penalty approach. E. Aiyoshi and K. Shimizu (see [8, Chap. 15]) addressed the general BLP by first converting the follower's problem to an unconstrained mathematical program using a barrier method. The corresponding stationarity conditions are then appended to the leader's problem which is solved repeatedly for decreasing values of the barrier parameter. To guarantee convergence the follower's objective function must be strictly convex. This rules out the linear case, at least in theory. A different approach using an exterior penalty method was proposed by Shimizu and M. Lu [8] that simply requires convexity of all the functions to guarantee global convergence.

In the approach of D.J. White and G. Anandalingam [10], the gap between the primal and dual solutions of the follower's problem for x fixed is used as a penalty term in the leader's problem. Although this results in a nonlinear objective function, it can be decomposed to provide a set of linear programs conditioned on either the decision variables (x, y) or the dual variables (u, v) of the follower's problem. They show that an exact penalty function exists that yields the global solution.

Related theory and algorithmic details are highlighted in [8, Chap. 16], along with presentations of several vertex enumeration and Kuhn–Tucker-based implementations.

See also

- ▶ Bilevel Fractional Programming
- ▶ Bilevel Linear Programming: Complexity, Equivalence to Minmax, Concave Programs
- ▶ Bilevel Optimization: Feasibility Test and Flexibility Index
- ▶ Bilevel Programming
- ▶ Bilevel Programming: Applications
- ▶ Bilevel Programming: Applications in Engineering
- ▶ Bilevel Programming: Implicit Function Approach
- ▶ Bilevel Programming: Introduction, History and Overview
- ▶ Bilevel Programming in Management
- ▶ Bilevel Programming: Optimality Conditions and Duality
- ▶ Multilevel Methods for Optimal Design
- ▶ Multilevel Optimization in Mechanics
- ▶ Stochastic Bilevel Programs

References

1. Bard JF (1984) Optimality conditions for the bilevel programming problem. Naval Res Logist Quart 31:13–26
2. Bard JF, Falk JE (1982) An explicit solution to the multi-level programming problem. Comput Oper Res 9(1):77–100
3. Bard JF, Moore JT (1990) A branch and bound algorithm for the bilevel programming problem. SIAM J Sci Statist Comput 11(2):281–292
4. Bialas WF, Karwan MH (1984) Two-level linear programming. Managem Sci 30(8):1004–1020
5. Candler W, Townsely R (1982) A linear two-level programming problem. Comput Oper Res 9(1):59–76
6. Hansen P, Jaumard B, Savard G (1992) New branch-and-bound rules for linear bilevel programming. SIAM J Sci Statist Comput 13(1):1194–1217
7. Jeroslow RG (1985) The polynomial hierarchy and a simple model for competitive analysis. Math Program 32:146–164
8. Shimizu K, Ishizuka Y, Bard JF (1997) Nondifferentiable and two-level mathematical programming. Kluwer, Dordrecht
9. Simaan M (1977) Stackelberg optimization of two-level systems. IEEE Trans Syst, Man Cybern SMC-7(4):554–556
10. White DJ, Anandalingam G (1993) A penalty function for solving bi-level linear programs. J Global Optim 3:397–419

Bilevel Linear Programming: Complexity, Equivalence to Minmax, Concave Programs

JONATHAN F. BARD
University Texas, Austin, USA

MSC2000: 49-01, 49K45, 49N10, 90-01, 91B52, 90C20, 90C27

Article Outline

Keywords
Related Optimization Problems
Complexity of the Linear BLPP Problem
See also
References

Keywords

Bilevel linear programming; Hierarchical optimization; Stackelberg game; Computational complexity; Concave programming; Minmax problem; Bilinear programming

A sequential optimization problem in which independent decision makers act in a noncooperative manner to minimize their individual costs, may be categorized as a *Stackelberg game*. The bilevel programming problem (BLPP) is a static, open loop version of this game where the leader controls the decision variables $x \in X \subseteq \mathbf{R}^n$, while the follower separately controls the decision variables $y \in Y \subseteq \mathbf{R}^m$ (e. g., see [3,9]).

In the model, it is common to assume that the leader goes first and chooses an x to minimize his objective function $F(x, y)$. The follower then reacts by selecting a y to minimize his individual objective function $f(x, y)$ without regard to the impact this choice has on the leader. Here, $F: X \times Y \to \mathbf{R}^1$ and $f: X \times Y \to \mathbf{R}^1$. The focus of this article is on the linear case introduced in ▶ Bilevel linear programming and given by:

$$\min_{x \in X} \ F(x, y) = c_1 x + d_1 y, \tag{1}$$
$$\text{s.t.} \ A_1 x + B_1 y \le b_1, \tag{2}$$
$$\min_{y \in Y} \ f(x, y) = c_2 x + d_2 y, \tag{3}$$
$$\text{s.t.} \ A_2 x + B_2 y \le b_2, \tag{4}$$

where $c_1, c_2 \in \mathbf{R}^n$, $d_1, d_2 \in \mathbf{R}^m$, $b_1 \in \mathbf{R}^p$, $b_2 \in \mathbf{R}^q$, $A_1 \in \mathbf{R}^{p \times n}$, $B_1 \in \mathbf{R}^{p \times m}$, $A_2 \in \mathbf{R}^{q \times n}$, $B_2 \in \mathbf{R}^{q \times m}$. The sets X and Y place additional restrictions on the variables, such as upper and lower bounds. Note that it is always possible to drop components separable in x from the follower's objective function (3).

Out of practical considerations, it is further supposed that the feasible region given by (2), (4), X and Y is nonempty and compact, and that for each decision taken by the leader, the follower has some room to respond. The rational reaction set, $P(x)$, defines these responses while the inducible region, IR, represents the set over which the leader may optimize. These terms are defined precisely in ▶ Bilevel linear programming. In the play, y is restricted to $P(x)$.

Given these assumptions, the BLPP may still not have a well-defined solution. In particular, difficulties may arise when $P(x)$ is multivalued and discontinuous. This is illustrated by way of example in [2,3].

Related Optimization Problems

The linear *minmax problem* (LMMP) is a special case of (1)–(4) obtained by omitting constraint (2) and setting $c_2 = -c_1$, $d_2 = -d_1$. It is often written compactly without the subscripts as

$$\min_{x \in X} \max_{y \in Y} \{cx + dy : Ax + By = b\} \tag{5}$$

or equivalently as

$$\min_{x \in X} \left(cx + \max_{y \in S(x)} dy \right), \tag{6}$$

where $S(x) = \{y \in Y : By \le b - Ax\}$. Several restrictive versions of (5) where, for example, X and Y are polyhedral sets and $Ax + By \le b$ is absent, as well as related optimality conditions are discussed in [8]. Although important in its own right, the LMMP plays a key role in determining the computational complexity of the linear BLPP. This is shown presently.

Consider now the inner maximization problem in (6) with $Y = \{y \ge 0\}$. Its dual is: $\min\{u^\top(b - Ax) : u \in U\}$, where u is a q-dimensional decision vector and $U = \{u : u^\top B \ge d, u \ge 0\}$. Note that the dual objective function is parameterized with respect to the vector x. Replacing the inner maximization problem with its dual leads to a second representation of (5):

$$\min_{x \in X, u \in U} (cx - u^\top Ax + u^\top b), \quad (7)$$

which is known as a disjoint *bilinear programming problem*. The theoretical properties of (7) along with its relationship to other optimization problems are highlighted in [1].

A more general version of a bilinear programming problem can be obtained directly from the linear BLPP. To see this, it is necessary to examine the Kuhn–Tucker formulation of the latter given by (6)–(11) in ▶ Bilevel linear programming. Placing the complementarity constraint in the objective function as a penalty term gives the following bilinear programming problem:

$$\begin{cases} \min & c_1 x + d_1 y \\ & + M[u^\top(b_2 - A_2 x - B_2 y) + v^\top y], \\ \text{s.t.} & A_1 x + B_1 y \leq b_1, \\ & u^\top B_2 - v^\top = -d_2, \\ & A_2 x + B_2 y \leq b_2, \\ & x \geq 0, \quad y \geq 0, \quad u \geq 0, \quad v \geq 0, \end{cases} \quad (8)$$

where M is a sufficiently large constant. In [10] it is shown that a finite M exists for the solution of (8) to be a solution of (1)–(4), and that (8) is a concave program; that is, its objective function is concave. This point is further elaborated in the next section.

Complexity of the Linear BLPP Problem

(1)–(4) can be classified as *NP*-hard which loosely means that no *polynomial time algorithm* exists for solving it unless $P = NP$. To substantiate this claim, it is necessary to demonstrate that through a polynomial transformation, some known *NP*-hard problem can be reduced to a linear BLPP. This will be done below constructively by showing that the problem of minimizing a strictly concave quadratic function over a polyhedron (see [5]) is equivalent to solving a linear minmax problem (cf. [4]). For an alternative proof based on satisfiability arguments from computer science see [7].

Theorem 1 *The linear minmax problem is NP-hard.*

To begin, let x be an n-dimensional vector of decision variables, and $c \in \mathbf{R}^n$, $b \in \mathbf{R}^q$, $A \in \mathbf{R}^{q \times n}$, $D \in \mathbf{R}^{n \times n}$ be constant arrays. For A of full row rank and D positive definite, it will be shown that the following minimization problem can be transformed into a LMMP:

$$\begin{cases} \theta^* = \min_{x} & cx - \tfrac{1}{2} x^\top D x, \\ \text{s.t.} & Ax \leq b, \end{cases} \quad (9)$$

where it is assumed that the feasible region in (9) is bounded and contains all nonnegativity constraints on the variables. The core argument centers on the fact that the Kuhn–Tucker conditions associated with the concave program (9) must necessarily be satisfied at optimality. These conditions may be stated as follows:

$$Ax \leq b, \quad (10)$$

$$x^\top D - u^\top A = c, \quad (11)$$

$$u^\top (b - Ax) = 0, \quad (12)$$

$$u \geq 0, \quad (13)$$

where u is a q-dimensional vector of dual variables. Now, multiplying (11) on the right by $x/2$, adding $cx/2$ to both sides of the equation, and rearranging gives

$$\tfrac{1}{2}(cx - u^\top Ax) = cx - \tfrac{1}{2} x^\top D x. \quad (14)$$

From (12) we observe that $u^\top b = u^\top Ax$, so (14) becomes

$$\tfrac{1}{2}(cx - u^\top b) = cx - \tfrac{1}{2} x^\top D x. \quad (15)$$

Replacing the objective function in (9) with the left-hand side of (15), and appending the Kuhn–Tucker conditions to (9) results in

$$\begin{cases} \theta^* = \min_{x,u} & cx - u^\top b, \\ \text{s.t.} & Ax \leq b, \\ & x^\top D - u^\top A = c, \\ & u^\top (b - Ax) = 0, \\ & u \geq 0, \end{cases} \quad (16)$$

which is an alternative representation of (9). Thus a quadratic objective function in (9) has been traded for a complementarity constraint in (16).

Turning attention to this term, let z be a q-dimensional nonnegative vector and note that $u^\top (b - Ax)$

can be replaced by $z_i = \min[u_i, (b - Ax)_i]$, $i = 1, \ldots, m$, where $(b - Ax)_i$ is the ith component of $b - Ax$, as long as $\sum_i z_i = 0$. The aim is to show that the following linear minmax problem is equivalent to (16):

$$\begin{cases} \theta^0 = \min_{x,u} & cx - u^\top b + \sum_{i=1}^{q} M z_i, \\ \text{s.t.} & Ax \leq b, \\ & x^\top D - u^\top A = cu \geq 0, \\ \max_z & \sum_{i=1}^{q} M z_i, \\ \text{s.t.} & z_i \leq u_i, \quad i = 1, \ldots, q, \\ & z_i \leq (b - Ax)_i, \quad i = 1, \ldots, q, \\ & z_i \geq 0, \quad i = 1, \ldots, q, \end{cases} \quad (17)$$

where M in the objective functions of problem (17) is a sufficiently large constant whose value must be determined.

Before proceeding, observe that an optimal solution to (16), call it (x^*, u^*), is feasible to (17) and yields the same value for the first objective function in (17). This follows because $\sum_i z_i^* = 0$, where $z_i^* = z_i^*(x^*, u^*)$. It must now be shown that (x^*, u^*, z^*) also solves (17). Assume the contrary; i.e., there exists a vector (x_0, u_0, z_0) in the inducible region of (17) such that $\theta^0 < \theta^*$ and $\sum_i z_i^0 > 0$. (Of course, if $\sum_i z_i^0 = 0$ and $\theta^0 < \theta^*$ this would contradict the optimality of (x^*, u^*).)

To exhibit a contradiction an appropriate value of M is needed. Accordingly, let S be the polyhedron defined by all the constraints in (17) and let

$$\theta^+ = \min\{cx - u^\top b : (x, u, z) \in S\}. \quad (18)$$

Evidently, because S is compact, θ^+ in (18) is finite. Compactness follows from the assumption that $\{x: Ax \leq b\}$ is bounded, and the fact that A has full row rank which implies that u is bounded in the second constraint in (17). Now define:

$$M > \frac{\theta^* - \theta^+}{\sum_i z_i^0 - \epsilon} \geq 0,$$

where ϵ is any value in $(0, \sum_i z_i^0)$. This leads to the following series of inequalities:

$$\theta^0 = cx^0 - b^\top u^0 + M \sum_i z_i^0 < cx^* - b^\top u^* = \theta^*$$

or

$$(cx^* - b^\top u^*) - (cx^0 - b^\top u^0) > M \sum_i z_i^0. \quad (19)$$

But from the definition of M along with (19), one has

$$M \sum_i z_i^0 - M\epsilon > \theta^* - \theta^+$$

$$\geq (cx^* - b^\top u^*) - (cx^0 - b^\top u^0) > M \sum_i z_i^0$$

which implies that the open interval $(0, \sum_i z_i^0)$ does not exist so $\sum_i z_i^0 = 0$, the desired contradiction.

Similar arguments can be used to show the reverse; therefore, if (x^*, u^*) solves (16), it also solves (17) and vice versa. Finally, note that the transformation from (9) to (17) is polynomial because it only involves the addition of $2q$ variables and $2q + n$ constraints to the formulation. The statement of the theorem follows from these developments. A straightforward corollary is that the linear BLPP is *NP*-hard.

In describing the size of a problem instance, I, it is common to reference two variables:

1) its Length[I], which is an integer corresponding to the number of symbols required to describe I under some reasonable encoding scheme, and
2) its Max[I], also an integer, corresponding to the magnitude of the largest number in I.

When a problem is said to be solvable in polynomial time, it means that an algorithm exists that will return an optimal solution in an amount of time that is a polynomial function of the Length[I]. A closely related concept is that of a *pseudopolynomial time algorithm* whose time complexity is bounded above by a polynomial function of the two variables Length[I] and Max[I]. By definition, any polynomial time algorithm is also a pseudopolynomial time algorithm because it runs in time bounded by a polynomial in Length[I]. The reverse is not true.

The theory of *NP*-completeness states that *NP*-hard problems are not solvable with polynomial time algorithms unless $P = NP$; however, a certain subclass may be solvable with pseudopolynomial time algorithms. Problems that do not yield to pseudopolynomial time algorithms are classified as *NP*-hard in the strong sense.

The linear BLPP falls into this category. The proof in [6], once again, is actually a corollary to the following theorem.

Theorem 2 *The linear minmax problem is strongly NP-hard.*

The proof is based on the notion of a *kernel K* of a graph $G = (V, E)$ which is a vertex set that is stable (no two vertices of K are adjacent) and absorbing (any vertex not in K is adjacent to a vertex of K). It is shown that the strongly NP-hard problem of determining whether or not G has a kernel (see [5]) is equivalent to determining whether or not a particular LMMP has an optimal objective function value of zero.

See also

- ▶ Bilevel Fractional Programming
- ▶ Bilevel Linear Programming
- ▶ Bilevel Optimization: Feasibility Test and Flexibility Index
- ▶ Bilevel Programming
- ▶ Bilevel Programming: Applications
- ▶ Bilevel Programming: Applications in Engineering
- ▶ Bilevel Programming: Implicit Function Approach
- ▶ Bilevel Programming: Introduction, History and Overview
- ▶ Bilevel Programming in Management
- ▶ Bilevel Programming: Optimality Conditions and Duality
- ▶ Concave Programming
- ▶ Minimax: Directional Differentiability
- ▶ Minimax Theorems
- ▶ Minimum Concave Transportation Problems
- ▶ Multilevel Methods for Optimal Design
- ▶ Multilevel Optimization in Mechanics
- ▶ Nondifferentiable Optimization: Minimax Problems
- ▶ Stochastic Bilevel Programs
- ▶ Stochastic Programming: Minimax Approach
- ▶ Stochastic Quasigradient Methods in Minimax Problems

References

1. Audet C, Hansen P, Jaumard B, Savard G (1996) On the linear maxmin and related programming problems. GERAD - École des Hautes Études Commerciales, Montreal, Working paper G-96-15
2. Bard JF (1991) Some properties of the bilevel programming problem. J Optim Th Appl 68(2):371–378
3. Bard JF, Falk JE (1982) An explicit solution to the multi-level programming problem. Comput Oper Res 9(1):77–100
4. Ben-Ayed O, Blair CE (1990) Computational difficulties of bilevel linear programming. Oper Res 38(1):556–560
5. Garey MR, Johnson DS (1979) Computers and intractability: A guide to the theory of NP-completeness. Freeman, New York
6. Hansen P, Jaumard B, Savard G (1992) New branch-and-bound rules for linear bilevel programming. SIAM J Sci Statist Comput 13(1):1194–1217
7. Jeroslow RG (1985) The polynomial hierarchy and a simple model for competitive analysis. Math Program 32:146–164
8. Shimizu K, Ishizuka Y, Bard JF (1997) Nondifferentiable and two-level mathematical programming. Kluwer, Dordrecht
9. Simaan M (1977) Stackelberg optimization of two-level systems. IEEE Trans Syst, Man Cybern SMC-7(4):554–556
10. White DJ, Anandalingam G (1993) A penalty function approach for solving bi-level linear programs. J Global Optim 3:397–419

Bilevel Optimization: Feasibility Test and Flexibility Index

MARIANTHI IERAPETRITOU
Department Chemical and Biochemical Engineering, Rutgers University, Piscataway, USA

MSC2000: 90C26

Article Outline

Keywords
Problem Statement
Local Optimization Framework
Design Optimization
Bilevel Optimization
Global Optimization Framework
Feasibility Test
Flexibility Index
Illustrative Example
Conclusions
See also
References

Keywords

Bilevel optimization; Uncertainty; Flexibility

Production systems typically involve significant *uncertainty* in their operation due to either external or internal resources. Variability of process parameters during operation and plant model mismatch (both parametric and structural) could give rise to suboptimality and even infeasibility of the deterministic solutions. Consequently, plant flexibility has been recognized to represent one of the important components in the operability of the production processes.

In a broad sense the area covers
- a *feasibility test* that requires constraint satisfaction over a specified space of uncertain parameters;
- a *flexibility index* associated with a given design that represents a quantitative measure of the range of uncertainty space that satisfies the feasibility requirement; and
- the integration of design and operations where trade-offs between design cost and plant flexibility are considered.

K.P. Halemane and I.E. Grossmann [21] proposed a feasibility measure for a given design based on the worst points for feasible operation, which can be mathematically formulated as a *max-min-max optimization problem* as will be discussed in detail in the next section.

Different approaches exist in the literature that quantify the flexibility for a given design involve the deterministic measures such as the resilience index, RI, proposed in [38], the flexibility index proposed in [41,42] and the stochastic measures such as the design reliability proposed in [27] and the stochastic flexibility index proposed in [37] and [40].

The incorporation of uncertainty into design optimization problems transforms the deterministic process models to stochastic/parametric problems, the solution of which requires the application of specialized optimization techniques. The consideration of the feasibility objective within the design optimization can be targeted towards the following two design capabilities. The first one concerns the design with *fixed degree of flexibility* that has the capability to cope with a finite number of different operating conditions ([19,20,32,34,40]). The second one considers the design optimization with *optimal degree of flexibility* that can be achieved by the trade-off of the cost of the plant and its flexibility ([22,33,35,36]). In the next section the feasibility test and the flexibility index problem will be considered in detail.

Problem Statement

The design problem can be described by a set of equality constraints I and inequality constraints J, representing plant operation and design specifications:

$$h_i(d, z, x, \theta) = 0, \quad i \in I,$$
$$g_j(d, z, x, \theta) \leq 0, \quad j \in J, \quad (1)$$

where d corresponds to the vector of design variables, z the vector of control variables, x the state variables and θ the vector of uncertain parameters. As has been shown in [21] for a specific design d, given this set of constraints, the design feasibility test problem can be formulated as the max-min-max problem:

$$\chi(d) = \max_{\theta \in T} \min_{z} \max_{\substack{j \in J, \\ i \in I}} \left\{ \begin{array}{l} h_i(d, z, x, \theta) = 0; \\ g_j(d, z, x, \theta) \leq 0 \end{array} \right\}, \quad (2)$$

where the function $\chi(d)$ represents a feasibility measure for design d. If $\chi(d) \leq 0$, design d is feasible for all $\theta \in T$, whereas if $\chi(d) > 0$, the design cannot operate for at least some values of $\theta \in T$. The above max-min-max problem defines a nondifferentiable global optimization problem which however can be reformulated as the following two-level optimization problem:

$$\begin{cases} \chi(d) = \max_{\theta \in T} \psi(d, \theta) \\ \text{s.t.} \quad \psi(d, \theta) \leq 0 \\ \psi(d, \theta) = \min_{z, u} u \\ \text{s.t.} \quad h_i(d, z, x, \theta) = 0, \quad i \in I, \\ \quad\quad g_j(d, z, x, \theta) \leq u, \quad j \in J, \end{cases} \quad (3)$$

where the function $\psi(d, \theta) = 0$ defines the boundary of the feasible region in the space of the uncertain parameters θ.

Plant feasibility can be quantified by determining the flexibility index of the design. Following the definition of flexibility index as proposed in [41], this metric expresses the largest scaled deviation δ of any expected deviations $\Delta \theta^+, \Delta \theta^-$, that the design can handle. The mathematical formulation for the evaluation of design's flexibility is the following:

$$\begin{cases} F = \max \delta \\ \text{s.t.} \quad \chi(d) \\ \qquad = \max_{\theta \in T} \min_z \max \begin{Bmatrix} h_i(d,z,x,\theta) = 0; \\ g_j(d,z,x,\theta) \le 0 \end{Bmatrix}_{i \in I}^{j \in J,} \\ T(\delta) = \left\{ \theta : \begin{matrix} \theta^N - \delta \Delta \theta^- \le \theta \\ \le \theta^N + \delta \Delta \theta^+ \end{matrix} \right\}, \\ \delta \ge 0. \end{cases} \quad (4)$$

The design flexibility index problem can be reformulated to represent the determination of the largest hyperrectangle that can be inscribed within the feasible region of the design [41]. Following this idea, the mathematical formulation of the flexibility problem has the following form:

$$\begin{cases} F = \min \delta \\ \text{s.t.} \quad \psi(d,\theta) = 0, \\ \qquad \psi(d,\theta) = \min_z u, \\ \qquad h_i(d,z,x,\theta) = 0, \quad i \in I, \\ \qquad g_j(d,z,x,\theta) \le u, \quad j \in J, \\ T(\delta) = \left\{ \theta : \begin{matrix} \theta^N - \delta \Delta \theta^- \le \theta \\ \le \theta^N - \delta \Delta \theta^+ \end{matrix} \right\}, \\ \delta \ge 0. \end{cases} \quad (5)$$

Local Optimization Framework

For the case where the constraints are jointly 1-D quasiconvex in θ and quasiconvex in z it was proven [41] that the point θ_c that defines the solution to (3) lies at one of the vertices of the parameter set T. Based on this assumption, the critical uncertain parameter points correspond to the vertices and the feasibility test problem is reformulated in the following manner:

$$\chi(d) = \max_{k \in V} \psi(d, \theta^k), \quad (6)$$

where $\psi(d, \theta^k)$ is the evaluation of the function $\psi(d, \theta)$ at the parameter vertex θ^k and V is the index set for the 2^{n_p} vertices for the n_p uncertain parameters θ. In a similar fashion for the flexibility index, problem (4) is reformulated in the following way:

$$F = \min_{k \in V} \delta^k, \quad (7)$$

where δ^k is the maximum deviation along each vertex direction $\Delta \theta^k$, $k \in V$, and is determined by the following problem:

$$\begin{cases} \delta^k = \max_{\delta, z} \delta \\ \text{s.t.} \quad g_j(d,z,x,\theta) \le 0, \quad j \in J, \\ \qquad h_i(d,z,x,\theta) = 0, \quad i \in I, \\ \qquad \theta = \theta^N + \Delta \theta^k, \\ \qquad \delta \ge 0. \end{cases} \quad (8)$$

Based on the above problem reformulations, a direct search method was proposed [21] that explicitly enumerate all the parameter set vertices. To avoid the explicit vertex enumeration, proposed two algorithms were proposed [41,42]: a heuristic vertex search and an implicit enumeration scheme. These algorithms however, rely on the assumption that the critical points correspond to the vertices of the parameter set T which is valid only for the type of constraints assumed above. To circumvent this limitation, a solution approach was proposed based on the following ideas [18]:

a) They replace the inner optimization problem:

$$\begin{cases} \psi(d,\theta) = \min_{z,u} u \\ h_i(d,z,x,\theta) = 0, \quad \forall i \in I, \\ g_j(d,z,x,\theta) \le u, \quad \forall j \in J, \end{cases}$$

by the *Karush–Kuhn–Tucker optimality conditions* (KKT):

$$\sum_{j \in J} \lambda_j = 1,$$

$$\sum_{j \in J} \lambda_j \frac{\partial g_j}{\partial z} + \sum_{i \in I} \mu_i \frac{\partial h_i}{\partial z} = 0,$$

$$\sum_{j \in J} \lambda_j \frac{\partial g_j}{\partial x} + \sum_{i \in I} \mu_i \frac{\partial h_i}{\partial x} = 0,$$

$$\lambda_j s_j = 0, \quad j \in J,$$

$$s_j = u - g_j(d,z,x,\theta), \quad j \in J,$$

$$\lambda_j, s_j \ge 0, \quad j \in J,$$

where s_j are the slack variables of constraints j, λ_j, μ_i are the Lagrange multipliers for inequality and equality constraints, respectively.

b) For the inner problem the following property holds that if each square submatrix of dimension ($n_z \times$

n_z), where n_z is the number of control variables, of the partial derivatives of the constraints g_j, $\forall j \in J$ with respect to the control variables z is of full rank, then the number of the *active constraints* is equal to $n_z + 1$.

c) They utilize the discrete nature of the selection of the active constraints by introducing a set of binary variables y_j to express if constraint g_j is active. In particular:

$$\lambda_j - y_j \leq 0, \quad j \in J,$$
$$s_j - U(1 - y_j) \leq 0, \quad j \in J,$$
$$\sum_{j \in J} y_j = n_z + 1,$$
$$\delta \geq 0,$$
$$y_j = 0, 1, \quad \lambda_j, s_j \geq 0, \quad j \in J,$$

where U represents an upper bound to the slack variables s_j. Note that if $y_j = 1$, then $\lambda_j \geq 0$, $s_j = 0$ which indicates that the constraint j is active, on the other hand if $y_j = 0$, then $\lambda_j = 0$, $s_j \geq 0$ which indicates that the constraint j is inactive.

Based on these ideas, the feasibility test problem can be reformulated in the following way:

$$\begin{cases} \chi(d) = \max u \\ \text{s.t.} \quad h_i(d, z, x, \theta) = 0, \\ \quad g_j(d, z, x, \theta) + s_j - u = 0, \\ \quad \sum_{j \in J} \lambda_j = 1, \\ \quad \sum_{j \in J} \lambda_j \frac{\partial g_j}{\partial z} + \sum_{i \in I} \mu_i \frac{\partial h_i}{\partial z} = 0, \\ \quad \sum_{j \in J} \lambda_j \frac{\partial g_j}{\partial x} + \sum_{i \in I} \mu_i \frac{\partial h_i}{\partial x} = 0, \\ \quad \lambda_j - y_j \leq 0, \quad j \in J, \\ \quad s_j - U(1 - y_j) \leq 0, \quad j \in J, \\ \quad \sum_{j \in J} y_j = n_z, +1 \\ \quad \theta^L \leq \theta \leq \theta^U, \\ \quad \delta \geq 0, \\ \quad y_j = 0, 1, \quad \lambda_j, s_j \geq 0, \quad j \in J, \end{cases}$$

which corresponds to a *mixed integer optimization problem* either linear or nonlinear depending on the nature of the constraints. In a similar way the flexibility index problem takes the following form:

$$\begin{cases} F = \min \delta \\ \text{s.t.} \quad h_i(d, z, x, \theta) = 0, \\ \quad g_j(d, z, x, \theta) + s_j - u = 0, \\ \quad u = 0, \\ \quad \sum_{j \in J} \lambda_j = 1, \\ \quad \sum_{j \in J} \lambda_j \frac{\partial g_j}{\partial z} + \sum_{i \in I} \mu_i \frac{\partial h_i}{\partial z} = 0, \\ \quad \sum_{j \in J} \lambda_j \frac{\partial g_j}{\partial x} + \sum_{i \in I} \mu_i \frac{\partial h_i}{\partial x} = 0, \\ \quad \lambda_j - y_j \leq 0, \quad j \in J, \\ \quad s_j - U(1 - y_j) \leq 0, \quad j \in J, \\ \quad \sum_{j \in J} y_j = n_z + 1, \\ \quad \theta^L \leq \theta \leq \theta^U, \\ \quad \delta \geq 0, \\ \quad y_j = 0, 1, \quad \lambda_j, s_j \geq 0, \quad j \in J. \end{cases}$$

Grossmann and C.A. Floudas [18] proposed the active set strategy for the solution of the above reformulated problems based on the property that for any combination of $n_z + 1$ binary variables that is selected (i. e., for a given set of active constraints), all the other variables can be determined as a function of θ. They proposed a procedure of systematically identifying the potential candidates for the active sets based on the signs of the gradients $\nabla_z g_j(d, z, x, \theta)$. The algorithm for the feasibility test problem involves the following steps:

a) For every potential active set determine the value u^k, $k = 1, \ldots, n_{AS}$, through the solution of the following nonlinear programming problem:

$$\begin{cases} u^k = \max u \\ \text{s.t.} \quad h_i(d, z, x, \theta) = 0, \\ \quad g_j(d, z, x, \theta) - u = 0, \quad j \in AS(k), \\ \quad \theta^L \leq \theta \leq \theta^U, \\ \quad \delta \geq 0, \\ \quad y_j = 0, 1, \quad \lambda_j, s_j \geq 0, \quad j \in J, \end{cases}$$

or if the active set AS(k) involves lower and upper bound constraints u^k is given by:

$$u^k = \frac{1}{\alpha_k} \left(\sum_{j(l) \in AS(k)} a_{j(l)} - \sum_{j(u) \in AS(k)} b_{j(u)} \right),$$

where $j(l)$, $j(u)$ are the indices that correspond to those pairs of constraints representing the lower and upper bounds on the same function, and α_k is the total number of this type of constraints.

b) The solution for the feasibility test problem is given by:

$$\chi(d) = \max_{k \in AS(k)} u^k.$$

A similar algorithm was proposed for the solution of the flexibility index problem. Under the conditions that the functions $\psi(d, \theta)$ are quasiconcave in z and θ and strictly quasiconvex in z for fixed θ, the approach guarantees global optimality.

For the linear case where the feasibility function problem has the following form:

$$\begin{cases} \psi(d, \theta) = \min_{z,u} u \\ f_j(d, z, \theta) = \\ \quad = \beta_{1j} d + \beta_{2j}(\theta) z - b_{2j}(\theta) \leq u, \\ \forall j \in J, \end{cases}$$

where f_j are the inequality constraints after the elimination of the state variables. For this case, analytical expressions have been derived [32,33] the function $\psi^k(d, \theta)$ for a given active set k:

$$\psi^k(d, \theta) = \sum_{j \in J_{AS}^k} \lambda_j^k (\beta_{1j} d + \beta_{2j}(\theta) z - b_{2j}(\theta)).$$

A branch and bound approach based on the evaluation of upper and lower bounds of function $\chi(d)$ was proposed in [30]. Although the suggested bounding problems are simpler than the original feasibility test problem they correspond to bilevel optimization problems where global optimality cannot be guaranteed using local optimization methods, (see [29]).

Design Optimization

As mentioned above, the incorporation of the feasibility objectives within the design optimization framework can be targeted towards the design with fixed degree of flexibility that is able to accommodate a finite number of changing operational conditions and the design with optimal degree of flexibility determined by proper balance of economic optimality and plant feasibility. The design optimization for fixed degree of flexibility was considered in [20], which presents a general formulation for designing multipurpose plants considering a deterministic multiperiod model of the following form:

$$\begin{cases} \min \quad C^0(d) + \sum_{i=1}^N C^i(d, z^i, x^i, t^i) \\ \text{s.t.} \quad h^i(d, z^i, x^i, t^i) = 0, \\ \quad g^i(d, z^i, x^i, t^i) \leq 0, \\ \quad r(d, z^1, \ldots, z^N, x^1, \ldots, x^N, \\ \quad t^1, \ldots, t^N) \leq 0, \end{cases}$$

where d is the vector of design variables; z^i is the vector of control variables in period i; x^i is the vector of state variables in period i; t^i is the length of time for period i; h^i is the vector of equalities for period i; g^i is the vector of inequalities for period i; r is the vector of inequalities that involve variables of all periods; N is the number of periods where different operating conditions are considered. This problem formulation exhibits a block diagonal structure which has been exploited for computational efficiency. See [19] for the *projection-restriction strategy*, which is an iterative scheme between economic optimization and design feasibility. Based of the flexibility analysis and the assumption that the critical points lie on the vertices of the uncertain parameter set T, Halemane and Grossmann [21] proposed an iterative algorithm to solve the problem of design considering specific range of uncertainty. See [31], for a nested solution procedure combining *generalized Benders decomposition* and *outer approximation* algorithms to address the problem of multiperiod design of heat integrated distillation sequences. See [44] for an outer approximation based decomposition method for the solution of multiperiod design problems.

For design optimization with optimal degree of flexibility, E.N. Pistikopoulos and Grossmann proposed [33] an iterative scheme in order to construct the trade-

off curves relating retrofit cost and expected revenue to flexibility. In their later work, [34,35], they extend their approach to nonlinear systems. Briefly, their iterative scheme consists of two phases: first the trade-off curve of retrofit cost and flexibility is determined, from this curve a number of designs are obtained for which at the second phase the expected revenue is evaluated by employing a modified Cartesian integration method. See [40], for a solution method based on generalized decomposition for the maximization of the plant flexibility subject to cost constraint. Recently (1996), a decomposition based approach for the simultaneous optimization of design economics and plant feasibility was proposed [22]. The main ideas of the proposed approach are:
- the utilization of a modified Benders decomposition scheme where the design variables correspond to the complicating variables;
- the use of a numerical integration formula for the approximation of the multiple integral of expected revenue; and
- the determination of the unknown integration points as part of the optimization procedure through the solution of a series of feasibility subproblems.

The same approach has been employed for the solution of planning and capacity expansion problems, [24], for the design of batch plants where additional properties simplified the solution procedure, [25]. The limitation of the later approach however, is that it cannot guarantee global optimality for the general case.

Bilevel Optimization

It should be noted that the feasibility test problem and the flexibility index problem correspond to bilevel optimization problems where the inner level consists of the evaluation of the function $\psi(d, \theta)$ that defines the boundary of the feasible region. Approaches that exist in the literature to deal with the solution of the bilevel optimization problem for the linear case involve the *enumeration techniques* that based on the fact that the solution must occur at an extreme point of the feasible set as the methods proposed in [13,14], and [12], *reformulation techniques* based on the transformation of the original problem to a single optimization problem by employing the optimality KKT conditions of the inner level problem as the methods proposed in [11], based on branch and bound principles [17], based on mixed integer programming [26], based on parametric complementarity pivoting [8], based on local optimization approaches for nonlinear programming such as penalty and barrier function methods, and global optimization techniques based on the reformulation of the complementarity slackness constraint to a separable quadratic reverse convex inequality constraint ([6]), or the restatement of the original problem as a reverse convex program ([43]). Recently (1996), [45], a global optimization framework was proposed based on the reformulation of the bilevel linear problem utilizing the KKT optimality conditions of the inner level and the primal-dual global optimization approach proposed in [15,16].

For the nonlinear case, local optimization techniques has been proposed based on the one-dimensional search algorithm [10], and penalty function methods as the approach proposed in [5]. Recently (1998) a general global optimization approach was proposed [23] for the solution of the feasibility test and flexibility index problem based on a utilization of a branch and bound framework and the ideas of the *deterministic global optimization algorithm* αBB, [1,3,4,9]. Although the proposed approach was applied to design feasibility/flexibility problems, it can be extended to general nonlinear bilevel problems. In the next section the main ideas and basic steps of the later approach for the solution of the feasibility test and flexibility index problems.

Global Optimization Framework

The basic idea of the proposed framework that leads to the determination of the global optimal solution for both the feasibility test and the flexibility index problem is to generate a *relaxation/enlargement of the feasible region* based on the convexification of the original problem constraints. Since the enlarged feasible region involves more feasible points than the original feasible region, the resulting feasibility test and flexibility index problem will provide lower bounds to the global solutions. Based on this relaxation idea, the proposed approach involves the following key steps:
a) Since the constraints $g_j(d, z, x, \theta)$, $h_i(d, z, x, \theta)$ are nonconvex functions, the *Karush–Kuhn–Tucker*

optimality conditions (KKT) of the inner problem that correspond to the optimization of the $\psi(d, \theta)$ function, are not necessary and sufficient to guarantee global optimality of the feasibility test and the flexibility index problems. Hence, the first step of the proposed framework involves the convexification of the constraints $g_j(d, z, x, \theta)$, $h_i(d, z, x, \theta)$ of the original problem. For the convexified problem and assuming that the constraint qualification holds, the KKT optimality conditions are necessary and sufficient, [18], and therefore we maintain the equivalence of the transformed single stage optimization problem. The solution of the single stage problem provides a lower bound of the design flexibility function $\chi(d)$ and the flexibility index of the design, F.

b) An upper bound to the design flexibility function $\chi(d)$ and flexibility index F is determined through a feasible solution of the original MINLP formulation obtained by substituting the inner problem by the KKT optimality conditions.

c) The next step after establishing an upper and a lower bound on the global solution, is to refine them. This is accomplished by successfully partitioning the initial region of the uncertain and control variables into smaller ones. The partitioning strategy involves the successive subdivision of a hyperrectangle into two subrectangles by halving on the middle point of the longest side of the initial rectangle (bisection). In each iteration the lower bound of the feasibility test and the flexibility index problem is the minimum over all the minima found in every subrectangle composing the initial rectangle. Consequently, a nondecreasing sequence of lower bounds is generated by halving the subrectangle that is responsible for the infimum over the minima obtained at each iteration. An nonincreasing sequence of upper bounds is derived by solving the nonconvex MINLP single optimization problem obtained after the substitution of the inner problem by the KKT optimality conditions, and selecting as an upper bound the minimum over all previously determined upper bounds. If at any iteration the solution of the convexified MINLP in any subrectangle is found to be greater than the upper bound, this subrectangle is fathomed since the global solution cannot be found inside it.

Feasibility Test

The procedure for the global optimization of the design feasibility problem involves the following steps:

1) Consider the whole uncertainty space. Set the lower bound LB = $-\infty$, $K = 1$ and select a tolerance ϵ.
2) Evaluate the valid underestimators of the original constraints $g_j(d, z, x, \theta)$, $h_i(d, z, x, \theta)$ utilizing the basic principles of the deterministic global optimization algorithm αBB, [1,3,4,9].
3) Considering the convexified constraints substitute the inner optimization problem by the necessary and sufficient KKT optimality conditions.
4) Solve the resulting MINLP formulation to global optimality using the deterministic global optimization algorithm SMIN-αBB, or GMIN-αBB, [2]. If the obtained solution is greater than the current LB, update the LB.
5) Substitute the inner optimization problem of the original problem (i.e., without convexifying the constraints) Solve the resulting problem using a local MINLP optimizer (e.g. DICOPT, [46], MINOPT, [39]). Set the upper bound UB equal to the obtained solution.
6) Check for convergence. If UB $-$ LB $\leq \epsilon$, STOP, otherwise continue to step 7).
7) Apply one of the branching criteria to partition the initial domain into two subdomains to be considered at the next iteration. Once the branching variable is selected the subdivision is performed by halving on the middle point of the longest side of the initial rectangle (bisection). The selection of a branching variable can be made following different branching rules. Since the aim of the branching step is the generation of problems with tighter lower bounds, the control variables, u, that participate in nonconvex terms and the uncertain parameters, θ, are involved in the set of candidate branching variables. The control variable or uncertain parameter that is selected for branching, correspond to the least-reduced axis, that is, the largest

$$r_i = \frac{x_i^U - x_i^L}{x_{i,0}^U - x_{i,0}^L}.$$

Note, that alternative branching strategies may be applied as described in [1,3].

Flexibility Index

The procedure for the global optimization of the flexibility index problem (5), involves the following steps:
1) Consider the whole uncertainty space. Set the lower bound LB = $-\epsilon$, $K = 1$ and select a tolerance ϵ.
2) Substitute the inner optimization problem by the KKT optimality conditions and construct the following single stage MINLP optimization problem. Solve the resulting problem using a local MINLP optimizer (e. g., DICOPT, [46], MINOPT, [39]). Set the upper bound UB equal to the obtained solution.
3) Determine the valid underestimators of the original constraints $g_j(d, z, x, \theta)$, $h_i(d, z, x, \theta)$ utilizing the basic principles of the deterministic global optimization algorithm αBB, [1,3,4,9].
4) Considering the convexified constraints substitute the inner optimization problem by the necessary and sufficient KKT optimality conditions.
5) Solve the resulting MINLP formulation to global optimality using the deterministic global optimization algorithm SMIN-αBB, or GMIN-αBB, [2]. If the obtained solution is greater than the current LB, update the LB.
6) Check for convergence. If UB $-$ LB $\leq \epsilon$, STOP, otherwise continue to step 7).
7) Apply one of the branching criteria to partition the initial domain into two subrectangles to be considered at the next iteration.

Illustrative Example

In this section an example of a heat exchanger network is considered to illustrate the steps of the approaches presented in the previous sections for the feasibility test and flexibility index problems.

The heat exchanger network given in [18] is considered here as shown in Fig. 1. The uncertain parameter is the heat flowrate of stream H1 which has a nominal value of 1kW/K and an expected deviation of +0.8kW/K. The following inequalities determine the feasible operation of this network as they formulated after the elimination of the state variables:

$$f_1 = -25F_{H1} + Q_c - 0.5Q_cF_{H1} + 10 \leq 0,$$
$$f_2 = -190F_{H1} + Q_c + 10 \leq 0,$$
$$f_3 = -270F_{H1} + Q_c + 250 \leq 0,$$
$$f_4 = 260F_{H1} - Q_c - 250 \leq 0.$$

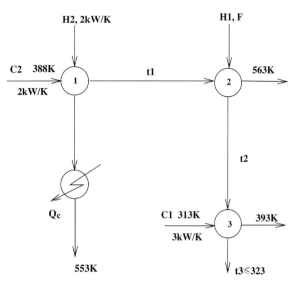

Bilevel Optimization: Feasibility Test and Flexibility Index, Figure 1
Heat exchanger network

Bilevel Optimization: Feasibility Test and Flexibility Index, Figure 2
Feasible region

The feasible region of the network is illustrated in Fig. 2. Note that the feasible region consists of the two disconnected domains which are highlighted in black.

First the feasibility test problem is solved. Constraint f_1 corresponds to the only nonconvex constraint involving the bilinear term Q_cF_{H1}. By introducing a new variable w for the bilinear term Q_cF_{H1} and introducing the four linear inequality constraints (f_5–

f_8) that define its convex envelope, [4,7,28], we have:

$$f_1 = -25F_{H1} + Q_c - 0.5w + 10 \leq 0,$$
$$f_2 = -190F_{H1} + Q_c + 10 \leq 0,$$
$$f_3 = -270F_{H1} + Q_c + 250 \leq 0,$$
$$f_4 = 260F_{H1} - Q_c - 250 \leq 0,$$
$$f_5 = 10F_{H1} + Q_c - w - 10 \leq 0,$$
$$f_6 = 236F_{H1} + 1.8Q_c - w - 424.8 \leq 0,$$
$$f_7 = -10F_{H1} - 1.8Q_c + w + 18 \leq 0,$$
$$f_8 = -236F_{H1} - Q_c + w + 236 \leq 0.$$

Considering this set of linear constraints and substituting the inner optimization problem of the feasibility test problem by the necessary and sufficient KKT optimality conditions the following MILP optimization formulation is obtained:

$$\begin{cases} \chi(d) = \max u \\ \text{s.t.} \quad f_1 = -25F_{H1} + Q_c \\ \qquad\quad +10 - 0.5w + s_1 = u, \\ \quad f_2 = -190F_{H1} + 10 + Q_c + s_2 = u, \\ \quad f_3 = -270F_{H1} + 250 + Q_c + s_3 = u \\ \quad f_4 = 260F_{H1} - 250 - Q_c + s_4 = u, \\ \quad f_5 = 10F_{H1} + Q_c - w - 10 + s_5 = u, \\ \quad f_6 = 236F_{H1} + 1.8Q_c \\ \qquad\quad -w - 424.8 + s_6 = u, \\ \quad f_7 = -10F_{H1} - 1.8Q_c \\ \qquad\quad +w + 18 + s_7 = u, \\ \quad f_8 = -236F_{H1} - Q_c \\ \qquad\quad +w + 236 + s_8 - u = u, \\ \sum_{j=1}^{8} \lambda_j = 1, \\ \lambda_1 + \lambda_2 + \lambda_3 - \lambda_4 + \lambda_5 \\ \quad +1.8\lambda_6 - 1.8\lambda_7 - \lambda_8 = 0, \\ -0.5\lambda_1 - \lambda_5 - \lambda_6 + \lambda_7 + \lambda_8 = 0, \\ \lambda_j - y_j \leq 0, \quad j = 1, \ldots, 8, \\ s_j - U(1 - y_j) \leq 0, \quad j = 1, \ldots, 8, \\ \sum_{j=1}^{8} y_j = 3, \\ F_{H1}^N - \delta \Delta F_{H1}^- \leq F_{H1} \leq F_{H1}^N + \delta \Delta F_{H1}^+, \\ \delta \geq 0, \\ y_j = 0, 1, \quad \lambda_j, s_j \geq 0, \quad j = 1, \ldots, 8. \end{cases}$$

Note that due to the introduction of an additional control variable w, the number of active constraints is increased to three. The solution of the above MILP optimization using GAMS/CPLEX is found to be equal to 0. Since this value corresponds to the lower bound of network feasibility $\chi(d)$, this result suggests that the network is not feasible within the whole range of uncertainty, $F_{H1} \in (1, 1.8)$, and no further steps are required.

The plant flexibility is then determined. First the inner feasibility problem is substituted by the KKT optimality conditions. The resulting nonconvex MINLP problem is solved using the local MINLP solver MINOPT, [39]. The solution provides an upper bound of the heat exchanger network flexibility index of 0.148,

$$\begin{cases} F = \min \delta \\ \text{s.t.} \quad f_1 = -25F_{H1} + Q_c \\ \qquad\quad +10 - 0.5Q_c F_{H1} + s_1 = 0, \\ \quad f_2 = -190F_{H1} + 10 + Q_c + s_2 = 0, \\ \quad f_3 = -270F_{H1} + 250 + Q_c + s_3 = 0, \\ \quad f_4 = 260F_{H1} - 250 - Q_c + s_4 = 0, \\ \sum_{j=1}^{4} \lambda_j = 1, \\ -0.5F_{H1}\lambda_1 + \lambda_2 + \lambda_3 - \lambda_4 = 0, \\ \lambda_j - y_j \leq 0, \quad j = 1, \ldots, 4, \\ s_j - U(1 - y_j) \leq 0, \quad j = 1, \ldots, 4, \\ \sum_{j=1}^{4} y_j = 2, \\ F_{H1}^N - \delta \Delta F_{H1}^- \leq F_{H1} \leq F_{H1}^N + \delta \Delta F_{H1}^+, \\ \delta \geq 0, \\ y_j = 0, 1, \quad \lambda_j, s_j \geq 0, \quad j = 1, \ldots, 4. \end{cases}$$

Note that this formulation corresponds to a nonconvex MINLP due to the bilinear term ($Q_c F_{H1}$) in constraint f_1 and the bilinear term ($F_{H1}\lambda_1$) in the gradient KKT constraint.

In step 3), the original constraints are convexified using the αBB resulting in the set of linear constraints $f_1, \ldots, f_8$ as presented above in the solution of the feasibility test problem. In step 4), the KKT optimality conditions are written considering the new set of linear constraints leading to the formulation of the following

MILP problem:

$$\begin{cases} F = \min \delta \\ \text{s.t.} \ f_1 = -25F_{H1} + Q_c \\ \qquad + 10 - 0.5w + s_1 = 0, \\ f_2 = -190F_{H1} + 10 + Q_c + s_2 = 0, \\ f_3 = -270F_{H1} + 250 + Q_c + s_3 = 0, \\ f_4 = 260F_{H1} - 250 - Q_c + s_4 = 0, \\ f_5 = 10F_{H1} + Q_c - w - 10 + s_5 = 0, \\ f_6 = 236F_{H1} + 1.8Q_c \\ \qquad - w - 424.8 + s_6 = 0, \\ f_7 = -10F_{H1} - 1.8Q_c \\ \qquad + w + 18 + s_7 = 0, \\ f_8 = -236F_{H1} - Q_c + w + 236 + s_8 = 0, \\ \sum_{j=1}^{8} \lambda_j = 1, \\ \lambda_1 + \lambda_2 + \lambda_3 - \lambda_4 + \lambda_5 \\ \qquad + 1.8\lambda_6 - 1.8\lambda_7 - \lambda_8 = 0, \\ -0.5\lambda_1 - \lambda_5 - \lambda_6 + \lambda_7 + \lambda_8 = 0, \\ \lambda_j - y_j \leq 0, \quad j = 1, \ldots, 8, \\ s_j - U(1 - y_j) \leq 0, \quad j = 1, \ldots, 8 \\ \sum_{j=1}^{8} y_j = 3, \\ F_{H1}^N - \delta \Delta F_{H1}^- \leq F_{H1} \leq F_{H1}^N + \delta \Delta F_{H1}^+, \\ \delta \geq 0, \\ y_j = 0, 1, \quad \lambda_j, s_j \geq 0, \quad j = 1, \ldots, 8. \end{cases}$$

The solution of this MILP problem using GAMS/CPLEX results in the network flexibility of 0.06 that provides a valid lower bound to the flexibility index problem. Hence, at the end of the first iteration we have an upper bound of 0.148 and a lower bound of 0.06 for the flexibility index problem. In step 7), since only one control variable is involved in the description of the problem, this corresponds to the branching variable resulting in the following subrectangles to be considered at the next iteration: subrectangle 1 described by $10 \leq Q_c \leq 123$ and subrectangle 1 described by $123 \leq Q_c \leq 236$. Steps 2) through 6) are then performed for each one of these subrectangles. For subrectangle 1, the resulting upper bounding MINLP gives a value of 0.148 the same as the lower bounding MILP in this region. Subrectangle 2, on the other hand results in a lower bound of 0.8138 which is larger than the current upper bound of 0.148 and consequently this region is fathomed and convergence is achieved to the global solution of network flexibility of 0.148.

Conclusions

The incorporation of uncertainty in the design stages is recognized to be one of the most important problems in the plant design analysis. Having efficient ways to test future plant feasibility and furthermore to quantify the capability of a plant to accommodate future variations of the operability parameters could lead to more efficient, economic and more flexible plants. Much of the work that appear in the literature to address the above problems was briefly presented in this paper. A general global optimization framework proposed in [23], was presented in more detail. Finally, an example problem was included to illustrate the main ideas of this framework.

See also

- ▶ Bilevel Fractional Programming
- ▶ Bilevel Linear Programming
- ▶ Bilevel Linear Programming: Complexity, Equivalence to Minmax, Concave Programs
- ▶ Bilevel Programming
- ▶ Bilevel Programming: Applications
- ▶ Bilevel Programming: Applications in Engineering
- ▶ Bilevel Programming: Implicit Function Approach
- ▶ Bilevel Programming: Introduction, History and Overview
- ▶ Bilevel Programming in Management
- ▶ Bilevel Programming: Optimality Conditions and Duality
- ▶ Minimax: Directional Differentiability
- ▶ Minimax Theorems
- ▶ Multilevel Methods for Optimal Design
- ▶ Multilevel Optimization in Mechanics
- ▶ Nondifferentiable Optimization: Minimax Problems
- ▶ Stochastic Bilevel Programs
- ▶ Stochastic Programming: Minimax Approach
- ▶ Stochastic Quasigradient Methods in Minimax Problems

References

1. Adjiman CS, Androulakis IP, Floudas CA (1997) A global optimization method αBB, for general twice-differentiable constrained NLPs - II. Implementation and computational results. Comput Chem Eng 22:1137–1158
2. Adjiman CS, Androulakis IP, Floudas CA (1997) Global optimization of MINLP problems in process synthesis. Comput Chem Eng 21:S445–S450
3. Adjiman CS, Dallwig S, Floudas CA, Neumaier A (1997) A global optimization method αBB, for general twice-differentiable constrained NLPs - I. Theoretical advances. Comput Chem Eng 22:1137–1158
4. Adjiman CS, Floudas CA (1996) Rigorous convex underestimates for general twice-differentiable problems. J Global Optim 9:23–40
5. Aiyoshi E, Shimizu K (1984) A solution method for the static constraint Stackelberg problem via penalty method. IEEE Trans Autom Control 29:1111
6. Al-Khayyal FA, Falk JE (1983) Jointly constrained biconvex programming. Math Oper Res 8:273–286
7. Al-Khayyal F, Horst R, Pardalos PM (1992) Global optimization on concave functions subject to quadratic constraints: An application in nonlinear bilevel programming. Ann Oper Res 34:125
8. Anandalingam G, White DJ (1990) A solution method for the linear static Stackelberg problem using penalty method. IEEE Trans Autom Control 35:1170
9. Androulakis IP, Maranas CD, Floudas CA (1995) αBB: A global optimization method for general constrained nonconvex problems. J Global Optim 7:337–363
10. Bard JF (1984) A solution method for the static constraint Stackelberg problem via penalty method. Naval Res Logist Quart 31:13
11. Bard JF, Falk JE (1982) An explicit solution to the multi-level programming problem. Comput Oper Res 9:77
12. Bard JF, Moore JT (1990) A branch and bound algorithm for the bilevel programming problem. SIAM J Sci Statist Comput 11:281
13. Bialas WF, Karwan MH (1984) Two-level linear programming. Managem Sci 30:1004
14. Candler W, Townsley R (1982) A linear two-level programming problem. Comput Oper Res 9:59
15. Floudas CA, Visweswaran V (1990) A global optimization algorithm (GOP) for certain classes of nonconvex NLPs - I. Theory. Comput Chem Eng 14:1397
16. Floudas CA, Visweswaran V (1993) Primal-relaxed dual global optimization approach. JOTA 78:187
17. Fortuny-Amat J, McCarl B (1981) A representation and economic interpretation of a two-level programming problem. J Oper Res Soc 32:783
18. Grossmann IE, Floudas CA (1987) Active constraint strategy for flexibility analysis in chemical processes. Comput Chem Eng 11:675–693
19. Grossmann IE, Halemane KP (1982) Decomposition strategy for designing flexible chemical plants. AIChE J 28:686–694
20. Grossmann IE, Sargent RWH (1978) Optimum design of chemical plants with uncertain parameters. AIChE J 24:1021
21. Halemane KP, Grossmann IE (1983) Optimal process design under uncertainty. AIChE J 29:425–433
22. Ierapetritou MG, Acevedo J, Pistikopoulos EN (1996) An optimization approach for process engineering problems under uncertainty. Comput Chem Eng 20:703–709
23. Ierapetritou MG, Floudas CA (1998) Global optimization in design under uncertainty: Feasibility test and flexibility index problems. Manuscript in preparation
24. Ierapetritou MG, Pistikopoulos EN (1994) A novel optimization approach of stochastic planning models. Industr Eng Chem Res 18:163–189
25. Ierapetritou MG, Pistikopoulos EN (1996) Batch plant design and operations under uncertainty. Industr Eng Chem Res 35:772–787
26. Judice JJ, Faustino AM (1992) A sequential LCP method for bilevel linear programming. Ann Oper Res 89:34
27. Kubic WL, Stein FP (1988) A theory of design reliability using probability and fuzzy sets. AIChE J 34:583
28. McCormick GP (1975) Computability of global solutions to factorable nonconvex programs: Part I - Convex underestimation problems. Math Program 10:147–175
29. Migdalas A, Pardalos PM, Varbrand P (1998) Multilevel optimization: Algorithms and applications. Kluwer, Dordrecht
30. Ostrovsky GM, Volin YM, Barit EI, Senyavin MM (1994) Flexibility analysis and optimization of chemical plants with uncertain parameters. Comput Chem Eng 18:755–767
31. Paules GE, Floudas CA (1992) Stochastic programming in process synthesis: A two-stage model with MINLP recourse for multiperiod heat-integrated distillation sequences. Comput Chem Eng 16:189–210
32. Pistikopoulos EN, Grossmann IE (1988) Optimal retrofit design for improving process flexibility in linear systems. Comput Chem Eng 12:719–731
33. Pistikopoulos EN, Grossmann IE (1988) Stochastic optimization of flexibility in retrofit design of linear systems. Comput Chem Eng 12:1215–1227
34. Pistikopoulos EN, Grossmann IE (1989) Optimal retrofit design for improving process flexibility in nonlinear systems: -I. Fixed degree of flexibility. Comput Chem Eng 13:1003–1016
35. Pistikopoulos EN, Grossmann IE (1989) Optimal retrofit design for improving process flexibility in nonlinear systems: -II. Optimal degree of flexibility. Comput Chem Eng 13:1087–1096
36. Pistikopoulos EN, Ierapetritou MG (1995) A novel approach for optimal process design under uncertainty. Comput Chem Eng 19:1089–1110

37. Pistikopoulos EN, Mazzuchi TA (1990) A novel flexibility analysis approach for processes with stochastic parameters. Comput Chem Eng 14:991–1000
38. Saboo AK, Morari M, Woodcock DC (1983) Design of resilient processing plants – VIII. A resilience index for heat exchanger networks. Chem Eng Sci 40:1553–1565
39. Schweiger CS, Floudas CA (1996) MINOPT: A software package for mixed-integer nonlinear optimization, user's guide. Manual Princeton Univ, Computer-Aided Systems Lab, Dept Chem Engin Jan
40. Straub DA, Grossmann IE (1993) Design optimization of stochastic flexibility. Comput Chem Eng 17:339
41. Swaney RE, Grossmann IE (1985) An index for operational flexibility in chemical process design - Part I: Formulation and theory. AIChE J 26:139
42. Swaney RE, Grossmann IE (1985) An index for operational flexibility in chemical process design – Part II: Computational algorithms. AIChE J 31:631
43. Tuy H, Migdalas A, Varbrand P (1993) A quasiconcave minimization method for solving linear two-level programs. J Global Optim 4:243
44. Varvarezos DK, Grossmann IE, Biegler LT (1992) An outer-approximation for multiperiod design optimization. Industr Eng Chem Res 31:1466
45. Visweswaran V, Floudas CA, Ierapetritou MG, Pistikopoulos EN (1996) A decomposition-based global optimization approach for solving bilevel linear and quadratic programs. State of the Art in Global Optimization:139
46. Viwanathan J, Grossmann IE (April 1990) DICOPT ++: A program for mixed integer nonlinear optimization, user's guide. Manual Engin Design Res Center Carnegie-Mellon Univ

Bilevel Programming

PATRICE MARCOTTE[1], GILLES SAVARD[2]
[1] University Montréal, Montréal, Canada
[2] École Polytechnique, Montréal, Canada

MSC2000: 49M37, 90C26, 91A10

Article Outline

Keywords
See also
References

Keywords

Nonconvex optimization; Equilibrium; Game theory

Let us consider a sequential game where the first player ('leader') incorporates into his optimization process the optimal reaction vector y of the second player ('follower') to the leader's decision vector x. This situation is described mathematically by the *bilevel program*

$$\text{BLP} \begin{cases} \min_{x,y} & f(x,y) \\ \text{s.t.} & (x,y) \in X \\ & y \in \operatorname*{argmin}_{y' \in Y(x)} g(x, y'), \end{cases}$$

where it is understood that the leader is requested to select a vector x such that the parameterized set $Y(x)$ is nonempty.

This formulation is extremely general in that it subsumes linear *zero-one optimization*, *quadratic concave programming*, disjoint *bilinear programming*, *nonlinear complementarity*, etc. If one denotes by $y(x)$ the set of optimal answers to a given leader vector x, the above bilevel program can be recast as the 'standard' mathematical program

$$\begin{cases} \min_{x,y} & f(x,y) \\ \text{s.t.} & (x,y) \in X \\ & y \in y(x). \end{cases}$$

The *induced region* of a bilevel program is defined as the feasible set of the above program. This set is usually nonconvex and might be disconnected. It is implicit that, whenever $y(x)$ is not a singleton, the leader is free to select that element $y \in y(x)$ that suits him best. This interpretation is legitimate in the case where side payments are allowed, i. e., the leader can bias the follower's objective in his favor. On the other hand, the behavior of a risk-averse leader which seeks to minimize, over the feasible set X, the objective

$$\max_{y \in y(x)} f(x,y).$$

has been considered in [4].

The algorithmic difficulty of bilevel programming stems mainly from the fact that the set $y(x)$ is ill-behaved, and usually not available in closed form. To gain some insight into this difficulty, let us consider the 'simple' situation where f, g are affine, the constraint $(x, y) \in X$ is absent and $Y(x) = \{y: Ax + By \geq b\}$ is a convex polyhedron. It is easy to show that, as in linear programming, bounded and feasible linear bilevel programs admit extremal solutions, hence the linear BLP lies in the class NP of problems polynomially solvable by a nondeterministic algorithm. Unfortunately, as shown in [2] and [3], the linear BLP is also strongly NP-hard. Moreover, its optimal solution(s) need not even be efficient ('Pareto optimal'). This is one of the features that distinguish bilevel programming from bicriterion optimization. Indeed consider the linear BLP illustrated in the figure below, where the arrows denote the players' respective steepest descent directions:

$$\begin{cases} \min_{x,y} & \frac{1}{2}x + y \\ \text{s.t.} & y \in \operatorname*{argmin}_{y'} -y' \\ & x + y' \leq 1 \\ & x, y' \geq 0. \end{cases}$$

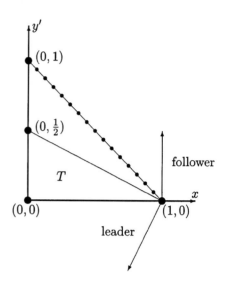

Bilevel Programming, Figure 1

The induced region of this problem reduces to the dotted line segment of Fig. 1. Optimizing over this line segment yields the solution $(x, y) = (1, 0)$, which is strictly dominated by all points inside the triangle T with vertices $(0, 1/2)$, $(0, 0)$ and $(1, 0)$. Since the set of efficient points is the segment $[(0, 0), (0, 1)]$, the only rational point that is also Pareto optimal is $(0, 1)$, which is actually the worst possible outcome for the leader. Note that, in the case where the functions f, g are affine and the sets X, $Y(x)$ are polyhedral, the induced region is in general a nonconvex piecewise linear variety that contains several local minima.

Assume now that the following conditions are satisfied:
- $Y(x) = \{y : h_i(x, y) \leq 0, 1 \leq i \leq n\}$;
- the functions g and h_i are continuously differentiable and convex;
- the set $Y(x)$ is regular for every x, i.e., some constraint qualification holds.

Then one can substitute for the follower's program its Kuhn–Tucker conditions, yielding the equivalent single-level program

$$\begin{cases} \min_{(x,y) \in X} & f(x, y) \\ \text{s.t.} & \nabla_y g(x, y) + \sum_{1 \leq i \leq n} \lambda_i \nabla_y h_i(x, y) = 0 \\ & \lambda_i h_i(x, y) = 0, \quad 1 \leq i \leq n, \\ & \lambda_i \geq 0, \quad 1 \leq i \leq n. \end{cases}$$

The complementarity constraints make this single-level problem difficult both theoretically (the constraint set is almost never regular) and algorithmically.

A useful variant of BLP occurs when $y(x)$ corresponds to the solution of an equilibrium system parameterized in x. If this system is modeled by means of a *variational inequality*, one obtains a generalized Kuhn–Tucker formulation where the gradient $\nabla_y g(x, y)$ is replaced by a function $F(x, y)$. (See [5].) In both cases, the complementarity constraint can be incorporated in the leader's objective as a penalty term $M \sum_{1 \leq i \leq n} \lambda_i h_i(x, y)$, thus greatly simplifying the constraint set. (It even becomes polyhedral in the linear case.) Under suitable assumptions, and for large but finite values of the penalty multiplier M, the penalized problem is equivalent to the original bilevel problem.

See also

- ▶ Bilevel Fractional Programming
- ▶ Bilevel Linear Programming
- ▶ Bilevel Linear Programming: Complexity, Equivalence to Minmax, Concave Programs
- ▶ Bilevel Optimization: Feasibility Test and Flexibility Index
- ▶ Bilevel Programming: Applications
- ▶ Bilevel Programming: Applications in Engineering
- ▶ Bilevel Programming: Implicit Function Approach

- Bilevel Programming: Introduction, History and Overview
- Bilevel Programming in Management
- Bilevel Programming: Optimality Conditions and Duality
- Multilevel Methods for Optimal Design
- Multilevel Optimization in Mechanics
- Stochastic Bilevel Programs

References

1. Bard JF (1998) Practical bilevel optimization: Algorithms and applications. Kluwer, Dordrecht
2. Hansen P, Jaumard B, Savard G (1992) New branch-and-bound rules for linear bilevel programming. SIAM J Sci Statist Comput 13:1194–1217
3. Jeroslow RG (1985) The polynomial hierarchy and a simple model for competitive analysis. Math Program 32:146–164
4. Loridan P, Morgan J (1996) Weak via strong Stackelberg problem: new results. J Global Optim 8:263–287 ISSN: 0925-5001
5. Luo Z-Q, Pang J-S, Ralph D (1996) Mathematical programs with equilibrium constraints. Cambridge Univ. Press, Cambridge
6. Shimizu K, Ishizuka Y, Bard JF (1997) Nondifferentiable and two-level programming. Kluwer, Dordrecht

Bilevel Programming: Applications

PATRICE MARCOTTE[1], GILLES SAVARD[2]
[1] University Montréal, Montréal, Canada
[2] École Polytechnique, Montréal, Canada

MSC2000: 91B99, 90C90, 91A65

Article Outline

Keywords
See also
References

Keywords

Network design; Economic equilibrium; Energy; Principal agent

Bilevel programming (see ▶ Bilevel programming: Introduction, history and overview; ▶ Bilevel programming) is ideally suited to model situations where the decision maker does not have full control over all decision variables. Five such situations are described in this article.

Example 1 The first example involves the improvement of a road network through either capacity expansion, traffic signals synchronization, vehicle guidance systems, etc. While management may be assumed to control the design variables, it can only affect indirectly the travel choices of the network users. Let x denote the design vector, y the flow vector, X the set of feasible design variables and $c_i(x, y)$ the travel delay along link i. One wishes to minimize over the set X the system travel cost $\sum_i y_i c_i(x, y)$, where the vector y is required to be an equilibrium *traffic assignment* corresponding to the design vector x. Neglecting the latter equilibrium requirement could lead to suboptimal policies. However, as shown in [5] for a continuous variant of the network design problem, efficient heuristic procedures can generate near-optimal solutions at a low computational cost. Indeed it is in the interest of both the management and the network users to minimize travel delays, although the former is interested in minimizing total travel time, while the users optimize their own travel time.

Example 2 Next consider the maximization of revenues raised from tolls set on a transportation network. If tolls are set too high, traffic on the corresponding arcs will drop and revenues will be affected negatively. Conversely, low toll values will generate low revenues. One could strike the right balance by maximizing total revenue, subject to the network users y achieving an equilibrium with respect to the toll vector x. In the case where the network is uncongested, users are assigned to shortest paths linking their respective origin and destination. This yields the bilevel program with bilinear objectives

$$\begin{cases} \max_{x,y} & \sum_{i \in I_1} x_i y_i \\ \text{s.t.} & y \in \operatorname*{argmin}_{y' \in Y} \sum_{i \in I_1} (c_i + x_i) y'_i + \sum_{i \in I_2} c_i y'_i, \end{cases}$$

where I_1 represents the set of toll arcs, I_2 the set of toll-free arcs, and Y the polyhedron of demand-feasible flow vectors. In [4] it has been shown that this problem is reducible to a linear bilevel program with an economic interpretation in terms of 'second-best' choices, and can also be reformulated as a *zero-one integer program* with few binary variables. Special cases are amenable to polynomial algorithms.

Example 3 The third example is the *Stackelberg–Nash–Cournot equilibrium* studied in [8] where the leader firm maximizes its revenue $x \cdot p(x + \sum_{1 \leq i \leq n} y_i) - c(x)$ (p denotes the inverse demand function and c the leader firm's production cost), subject to the vector y being a Cournot–Nash equilibrium with respect to the shifted inverse demand function $p_x(Q) = p(x + Q)$. This model subsumes the situations of monopoly ($n = 0$) as well as that of Stackelberg equilibrium ($n = 1$). It has been extended in [7] to the case of multiple leaders, but does not fit any more the framework of bilevel programming.

Example 4 A fourth example is provided by the energy sector, which is characterized by an extensive use of large scale techno-economic models describing specific subsectors or markets: gas and electricity subsectors, industrial and residential markets, etc. In this respect, it provides a rich source for bilevel models. A bilevel program arises when a utility, in its strategic planning process, takes explicitly into account the rational reaction of its competitors or customers to its own investment schedule. This approach has been applied to assess the impact of new demand management technologies for reducing power usage [3]. Another bilevel model arises when a utility is legally bound to buy any energy surplus from 'qualified small producers' at marginal cost. For example, a study of the impact of cogeneration in the pulp and paper industry on the electricity market has been conducted in [2].

Example 5 Finally we mention that bilevel programming subsumes the principal/agent paradigm of economics (see [1]), where the principal (leader) subcontracts a job to an agent (follower). The principal rewards the agent according to the quality of the final outcome ω, which may be random, while the agent maximizes its own objective, which is a function of the effort level y and the expected reward $x(\omega(y))$. The lower the effort level y, the lower the (expected) quality $\omega(y)$ of the finished job. Assuming that the agent accepts to perform the job only if his utility is larger than some 'reservation level' $g_{\min}$, one derives the bilevel program:

$$\begin{cases} \max_{x(\cdot), y} & f(x(\omega(y)), \omega(y)) \\ \text{s.t.} & g(x(\omega(y)), y) \geq g_{\min} \\ & y \in \operatorname*{argmax}_{y' \in Y} g(x(\omega(y')), y'), \end{cases}$$

where the leader's decision variable x is a function defined over the set Y of possible effort levels. Whenever the set Y is not finite, this yields an infinite-dimensional optimization problem. The situation becomes all the more complex when the output ω is a random variable of the agent's effort y.

See also

- Bilevel Fractional Programming
- Bilevel Linear Programming
- Bilevel Linear Programming: Complexity, Equivalence to Minmax, Concave Programs
- Bilevel Optimization: Feasibility Test and Flexibility Index
- Bilevel Programming
- Bilevel Programming: Applications in Engineering
- Bilevel Programming: Implicit Function Approach
- Bilevel Programming: Introduction, History and Overview
- Bilevel Programming in Management
- Bilevel Programming: Optimality Conditions and Duality
- Multilevel Methods for Optimal Design
- Multilevel Optimization in Mechanics
- Stochastic Bilevel Programs

References

1. van Ackere A (1993) The principal/agent paradigm: its relevance to various functional fields. Europ J Oper Res 70:83–103
2. Haurie A, Loulou R, Savard G (1992) A two player game model of power cogeneration in New England. IEEE Trans Autom Control 37:1451–1456
3. Hobbs BF, Nelson SK (1992) A non-linear bilevel model for analysis of electric utility demand-side planning issues. Ann Oper Res 34:255–274
4. Labbé M, Marcotte P, Savard G (1998) A bilevel model of taxation and its application to optimal highway pricing. Managem Sci 44:1608–1622
5. Marcotte P (1986) Network design with congestion effects: A case of bilevel programming. Math Program 34:142–162

6. Migdalas A, Pardalos PM, Värbrand P (1998) Multilevel optimization: Algorithms and applications. Kluwer, Dordrecht
7. Sherali HD (1984) A multiple leader Stackelberg model and analysis. Oper Res 32:390–404
8. Sherali HD, Soyster AL, Murphy FH (1983) Stackelberg–Nash–Cournot equilibria: characterizations and computations. Oper Res 31:253–276

Bilevel Programming: Applications in Engineering

ZEYNEP H. GÜMÜŞ[1,2], KEMAL SAHIN[3], AMY CIRIC[4]
[1] Department of Physiology and Biophysics, Weill Medical College, Cornell University, New York, USA
[2] The HRH Prince Alwaleed Bin Talal Bin Abdulaziz Alsaud Institute for Computational Biomedicine, Weill Medical College, Cornell University, New York, USA
[3] Huntsman (Germany) GmbH, Deggendorf, Germany
[4] Department of Chemical and Materials Engineering, University of Dayton, Dayton, USA

Article Outline

Keywords and Phrases
Introduction
   Bilevel Programming in Traffic Management
   Bilevel Programming in Chemical Process Synthesis
   Bilevel Programming in Metabolic Engineering
Conclusions
References

Keywords and Phrases

Bilevel programming; Traffic control; Phase equilibrium; Metabolic engineering; Design

Introduction

Bilevel programming problems (BLPP) are encountered when one optimization problem is embedded within another one as a constraint. BLPPs arise in many areas of engineering, where hierarchical decision models are often encountered. Almost all areas of engineering can provide some examples in which two decision models interact and the outcome of one decision influences another; applications can be found in areas as diverse as traffic control and reactive distillation.

The general BLPP formulation is as follows:

Outer optimization problem
$$\begin{cases} \min_{x} F(\mathbf{x}, \mathbf{y}) \\ \text{s.t. } \mathbf{G}(\mathbf{x}, \mathbf{y}) \geq 0 \\ \mathbf{H}(\mathbf{x}, \mathbf{y}) = 0 \\ \text{inner optimization problem} \begin{cases} \min_{y} f(\mathbf{x}, \mathbf{y}) \\ \text{s.t. } \mathbf{g}(\mathbf{x}, \mathbf{y}) \geq 0 \\ \mathbf{h}(\mathbf{x}, \mathbf{y}) = 0 \end{cases} \\ \mathbf{x} \in X \subset \Re^{n_1}, \quad \mathbf{y} \in Y \subset \Re^{n_2} \end{cases}$$

where

$f, F: \Re^{n_1} \times \Re^{n_2} \to \Re$,
$\mathbf{g} = [g_1, ..., g_J]: \Re^{n_1} \times \Re^{n_2} \to \Re^J$,
$\mathbf{G} = [G_1, ..., G_{J'}]: \Re^{n_1} \times \Re^{n_2} \to \Re^{J'}$,
$\mathbf{h} = [h_1, ..., h_I]: \Re^{n_1} \times \Re^{n_2} \to \Re^I$,
$\mathbf{H} = [H_1, ..., H_{I'}]: \Re^{n_1} \times \Re^{n_2} \to \Re^{I'}$.

The outer optimization problem, which minimizes $F(\mathbf{x}, \mathbf{y})$, is constrained by inequality constraints $\mathbf{G}$, equality constraints $\mathbf{H}$, and the inner optimization problem. This inner optimization minimizes its objective function by varying $\mathbf{y}$, while subject to its own inner constraints $\mathbf{g}$ and $\mathbf{h}$. The inner variables $\mathbf{y}$ may also appear in the outer constraints and objective function, and the inner constraints and objective function may be parameterized by $\mathbf{x}$. Novel global optimization strategies exist to solve the BLPP with twice continuously differentiable nonconvex nonlinear [18] and mixed-integer nonlinear constraints [19].

This article explores a diverse sampling of bilevel programming including examples from civil engineering traffic management, chemical engineering process design and metabolic engineering.

Bilevel Programming in Traffic Management

As urban populations increase and cities expand, traffic and its related problems effect the everyday life of all commuters.

Traffic problems follow the hierarchical structure of BLPPs. Each individual commuter travels upon a network of roads that is created and organized by a central regulatory agency. This agency plans the layout and carrying capacity of highways and streets, chooses where to place on-off ramps that connect limited access highways to local streets, decides where to install traffic lights, and sets their signaling rate.

As the regulatory agency plans and manages this network of roads, it must accommodate the traffic pattern formed by the individual decisions of the thousands of travelers who use the network each day. During each trip, each traveler takes a path that she believes will minimize her travel time, based on previous experience and ongoing traffic reports. Beckmann et al. [2] have shown that when this information is perfect and all travelers have access to it, the cumulative effect is to minimize the total time spent by all drivers on all roads in the network:

$$\min_v \sum_a \int_0^{v_a} t_a(x)\,dx.$$

This behavior by the travelling public creates bilevel programming problems in traffic management. When a regulatory agency tries to set policies that minimize gas consumption, the travel time of all drivers, or some other objective, its options are constrained by the response of the travelling public.

One application of BLPPs in traffic is signal optimization, where the objective is to minimize travel time or gasoline consumption by varying the length of green lights and the cycle time of traffic lights [9]:

$$\min \sum_a t_a v_l(t,s)$$
$$\text{s.t } \min_v \sum_a \int_0^{v_a} t_a(x)\,dx.$$

In this problem, the outer objective sums over all costs based on the signaling policies.

This bilevel programming problem is used to plan road improvements, where central planning agency minimizes the cost of construction and similar activities, subject to the inner optimization problem that predicts the behavior of traffic on the road [1,9,21]. It is also used to optimize the flow of traffic onto limited access highways. The outer problem minimizes the total travel time of all travelers by optimizing traffic light lengths and other controls at the on- and off-ramps of the highway, while the inner optimization problem predicts the behavior of traffic on the road [24,25,26].

Bilevel Programming in Chemical Process Synthesis

Chemical process synthesis by optimization techniques is a vast area that includes plant design, the synthesis of reactor networks, separation systems, heat exchanger networks, and utility plants, and the planning of batch and multiperiod operations [3,10,11,12].

Inner Problems that Minimize the Gibbs Free Energy Many chemical engineering design problems involve distillation columns, liquid-liquid extractors and decanters, and reactors; modeling these unit operations usually requires modeling the chemical equilibria and phase equilibria (vapor-liquid equilibrium, liquid-liquid equilibrium, and vapor-liquid-liquid equilibrium) occurring within them. When the number of phases is known in advance, phase and chemical equilibrium can be modeled with a set of algebraic equations. When, however, the number of phases is not known *a priori*, these algebraic equations cannot be used; in these problems, the number of phases, phase equilibrium, and in some problems chemical equilibrium can be predicted by minimizing the Gibbs free energy. Maximizing the profit, minimizing the cost, or optimizing some other measure of a chemical process that contains a unit operation with an unknown number of phases is a bilevel programming problem:

$$\max F(\mathbf{x}, n_{ik})$$
$$\text{s.t } \mathbf{G}(\mathbf{x}, n_{ik}) \geq 0$$
$$\mathbf{H}(\mathbf{x}, n_{ik}) = 0$$
$$\min_{n_{ik}} \sum_i \sum_k n_{ik} \mu_{ik}$$
$$\text{s.t. } \sum_{ik} a_{ij} n_{ik} = b_j, \quad \forall j \in E$$
$$n_{ik} \geq \delta, \quad i=1,\ldots,\text{NC}; \quad k=1,\ldots,\text{NP}.$$

Here, the outer problem maximizes the profit F. Design specifications are captured by inequality constraints $\mathbf{G}$, while equality constraints $\mathbf{H}$ are the mass and energy balances. The inner optimization minimizes the Gibbs free energy, equal to the summation of $n_{ik}\mu_{ik}$, the moles of species i and phase k multiplied by the corresponding chemical potential. This inner problem is constrained by mass balances assuring that the total number of atoms of element j is constant regardless of the phase or chemical distribution, and that the total number of moles of species i in phase k is positive.

Clark and Westerberg [7,8] used this strategy to optimize a reactor making aniline from nitrobenzene and hydrogen. The reaction also produces water, which may form a two-liquid phase mixture with nitrobenzene and aniline, depending upon the relative amounts of nitrobenzene, aniline, and water. The outer problem optimized the reactor temperature and pressure, while the inner problem found the simultaneous phase and chemical equilibrium by minimizing the Gibbs Free Energy.

Gümüş and Ciric [17] used bilevel programming to optimize a reactive distillation column that produces aniline from nitrobenzene and water. The outer problem minimizes cost by varying the number of trays, reflux and reboil ratios, and feed tray locations. A series of inner optimization problems predict the phase and chemical equilibrium in the condenser and on each tray in the column.

Bilevel Programming and Simultaneous Design and Control Bilevel programming has also been used to integrate the design of a chemical process with the synthesis of its control scheme [4]. The outer optimization problem maximizes the annual profit $D(z)$ minus the cost of off-spec product formed during process upsets, while an inner optimization problem simultaneously predicts the amount of off-spec product formed during process upsets and finds the settings of a model predictive controller that minimize this amount. The model is:

$$\max D(z) - \kappa \sum_l CO_l^P \{z, p; u_l(t), x_l(t), y_l(t), p_c(t)\} - C_H$$

$$\text{s.t } f(z, p) = 0$$
$$h(z, p) = 0$$
$$g(z, p) \geq 0$$
$$g_{h,l}^p(z, p, u_l(t), x_l(t), y_l(t), p_c(t)) \geq 0$$

$$\min_{n_{ik}} CO_l \{z, p; u_l(t), x_l(t), y_l(t), p_c(t)\}$$

$$\text{s.t. } \dot{x} = f(z, p, u_l(t), x_l(t), y_l(t), p_c(t))$$
$$x(t=0) = x_d$$
$$y(t=0) = y_d$$
$$u(t=0) = u_d$$
$$g_{h,l}(z, p, u_l(t), x_l(t), y_l(t), p_c(t)) \geq 0$$
$$h(z, p, u_l(t), x_l(t), y_l(t), p_c(t)) = 0$$
$$u^L \leq u(t) \leq u^H.$$

In this formulation, the cost of the fluctuations around the steady state, denoted by subscript d, will increase the cost of off spec production, CO_l^P. In the inner optimization, the actions u of a model predictive controller are based on the disturbance l.

Bilevel Programming and Design Under Uncertainty
In the planning stage of a design, the range of uncertain parameters that the design can tolerate for feasible operation should be determined. The design under parametric uncertainty problem can be described by a set of equality constraints I and inequality constraints J representing plant operation and design specifications:

$$h_i(\mathbf{d}, \mathbf{z}, \mathbf{x}, \boldsymbol{\theta}) = 0, \quad \forall i \in I$$
$$g_j(\mathbf{d}, \mathbf{z}, \mathbf{x}, \boldsymbol{\theta}) \geq 0, \quad \forall j \in J$$

where $\mathbf{z}$ is the vector of control variables, $\mathbf{x}$ is the vector of state variables and $\boldsymbol{\theta}$ is the vector of uncertain parameters. Feasibility concerns are incorporated into the design step by quantifying design feasibility and flexibility with the feasibility test and flexibility index measures. These measures are characterized by max-min-max formulations [20] that are further reformulated in the BLPP form [13,16,23]. For a specific design $\mathbf{d}$, the BLPP feasibility test problem is of the form:

$$\max_{\boldsymbol{\theta} \in T} \psi(\mathbf{d}, \boldsymbol{\theta})$$
$$\text{s.t } \psi(\mathbf{d}, \boldsymbol{\theta}) \leq 0$$
$$\psi(\mathbf{d}, \boldsymbol{\theta}) = \min_{z, u} u$$

$$\text{s.t. } h_i(\mathbf{d}, \mathbf{z}, \mathbf{x}, \boldsymbol{\theta}) = 0, \quad \forall i \in I$$
$$u - g_j(\mathbf{d}, \mathbf{z}, \mathbf{x}, \boldsymbol{\theta}) \geq 0, \quad \forall j \in J$$
$$T = \{\boldsymbol{\theta} | \boldsymbol{\theta}^L \leq \boldsymbol{\theta} \leq \boldsymbol{\theta}^U\}$$

where the function $\psi(\mathbf{d}, \boldsymbol{\theta})$ represents a feasibility measure for design $\mathbf{d}$. The boundary of the feasible region in the space of the uncertain variables is at $\psi(\mathbf{d}, \boldsymbol{\theta}) = 0$. If $\psi(\mathbf{d}, \boldsymbol{\theta}) \geq 0$, the design can not operate at least for some values of $\boldsymbol{\theta}$ in T, and the BLPP is infeasible.

For a specific design $\mathbf{d}$, the design flexibility test problem is also formulated as a BLPP:

$$\min_{\boldsymbol{\theta} \in T} \delta$$
$$\text{s.t } \psi(\mathbf{d}, \boldsymbol{\theta}) = 0$$
$$\psi(\mathbf{d}, \boldsymbol{\theta}) = \min_{z} u$$
$$\text{s.t. } h_i(\mathbf{d}, \mathbf{z}, \mathbf{x}, \boldsymbol{\theta}) = 0, \quad \forall i \in I$$
$$g_j(\mathbf{d}, \mathbf{z}, \mathbf{x}, \boldsymbol{\theta}) \leq u, \quad \forall j \in J$$
$$T(\delta) = \{\boldsymbol{\theta} | \boldsymbol{\theta}^N - \delta \Delta \boldsymbol{\theta}^- \leq \boldsymbol{\theta} \leq \boldsymbol{\theta}^N + \delta \Delta \boldsymbol{\theta}^+\}$$
$$\delta \geq 0$$

where δ is the largest scaled deviation of any expected deviations $\Delta \boldsymbol{\theta}^-$ and $\Delta \boldsymbol{\theta}^+$ the design can handle [13,16,23]. Higher δ signifies more flexible design towards parametric variations.

Bilevel Programming in Metabolic Engineering

Metabolic engineering involves optimization of genetic and regulatory processes within cells to increase overproduction of desired metabolites or proteins. These changes can have major effects on cell growth if the desired overproduction competes with growth resources, so the cell will redistribute the metabolic fluxes to maximize its growth rate. Metabolic flux distributions can be optimized utilizing in-silico genome scale metabolic network maps to develop overproduction strategies. Several different problems in this research area have recently been formulated as BLPPs. These involve the (i) determination of optimal gene knockouts, (ii) identification of stable steady state solutions and (iii) dynamic gene expression control strategies, all to achieve maximum product yield.

Gene Knockout Strategies Gene deletion strategies to increase the overproduction of a desired product can be straightforward and involve competing reaction pathways; however, many others can be complex and non-intuitive. Burgard et al. [5] introduced a BLPP formulation to address the optimal manipulation of gene knockout strategies to maximize overproduction, subject to maximizing cell's growth objective at the inner level. The inner problem is parameterized with gene knockout strategies that are chosen by the outer problem and constrained by metabolic flux balances and fixed substrate. This BLPP model for a steady-state metabolic network of N metabolites and M metabolic reactions fueled by a glucose substrate is formulated as:

$$\max_{y_j} v_{\text{chemical}}$$
$$\text{s.t } \max_{v_j} v_{\text{biomass}}$$
$$\text{s.t. } \sum_{j=1}^{M} S_{ij} v_j = 0, \quad \forall i \in N$$
$$v_{\text{pts}} + v_{\text{glk}} - v_{\text{glc\_uptake}} = 0$$
$$v_{\text{atp}} - v_{\text{atp\_main}} \geq 0$$
$$v_{\text{biomass}} - v_{\text{biomass}}^{\text{target}} \geq 0$$
$$v_j^{\min} \cdot y_j \leq v_j \leq v_j^{\max} \cdot y_j, \quad \forall j \in M$$
$$y_j = \{0, 1\}, \quad \forall j \in M$$
$$\sum_{j \in M} (1 - y_j) \leq K$$

where v_{chemical} is the flux of the desired product, v_{biomass} is biomass formation, S_{ij} is the stoichiometric constant for metabolite i in reaction j, v_j is the flux of reaction j, v_{pts} and v_{glk} respectively represent the uptake of glucose through the phosphotransferase system and glucokinase, $V_{\text{glk\_uptake}}$ is the basis glucose uptake scenario, $v_{\text{atp\_main}}$ is the non-growth associated ATP maintenance requirement, K is the number of allowable knockouts, and $v_{\text{biomass}}^{\text{target}}$ is a minimum level of biomass production. The BLPP can be modified further to include additional bounds on O_2, CO_2 and NH_3 transport rates and secretion pathways for key metabolites in the inner problem [22].

Stable Metabolic Networks Stability considerations of a redesigned metabolic network can be addressed within a BLPP framework, such that the new system is stable around a neighborhood of the new steady state.

Here, the outer problem maximum product flux objective is subject to flux balances and an inner stability objective [6].

Temporal Flux Control Gene expression can be controlled dynamically using the BLPP structure to optimize the temporal flux profile of a key reaction, such that at the end of a batch, the total product formation is maximized [14]. In the outer problem, a flux in a specific reaction known to have an impact on the product formation is varied with time to maximize the total product formation at the end of a batch. The inner problem maximizes cellular growth at each sampling time over the batch period by optimizing the remaining fluxes. The BLPP can be modified to determine the optimal regulation time of the specific flux from an initial to a final value. Glycerol and ethanol production in *E. coli* have been studied using the BLPP formulation [14].

Gadkar et al. [15] coupled this BLPP model with control algorithms to determine genetic manipulation strategies in bioprocess applications. They introduced three alternative BLPP models to maximize ethanol production in anaerobic batch fermentation of *E. coli*, optimizing ethanol production, batch time and multibatch scheduling in the presence of parametric uncertainty and measurement noise. These include (i) optimizing growth regulation time and batch duration time by penalizing for longer batch times in the outer objective (ii) scheduling multiple batch runs to address inhibition due to product accumulation in the reactor, optimizing the number of batch runs, batch duration times, glucose allocation per run and the manipulated flux regulation time, and (iii) optimizing genetic alterations in the presence of growth inhibition and parametric uncertainty in the inhibition constant.

Conclusions

The hierarchical structure of many engineering problems lends themselves to bilevel programming formulations, where an inner optimization problem constrains a larger, 'outer' optimization problem. Applications in civil engineering design include traffic control, where an inner optimization problem predicting driver's behavior constrains an outer optimization problem that identifies the optimal control strategies. In chemical engineering, BLPPs are used to identify processes that are both economically optimal – maximizing revenue or minimizing cost – and simultaneously ensure that multiphase equilibrium is satisfied by determining the global minimum of Gibbs Free energy. Other applications include the combined optimization of a chemical process and it's controllers and chemical process design under parametric uncertainty, to ensure operational feasibility and flexibility. Alternative BLPP formulations have been introduced in modeling metabolic engineering systems. These address the maximization of product yield by determining optimal gene knockouts, identifying stable steady state solutions and dynamic gene expression control strategies. Metabolic engineering area is a recent and growing application field for BLPP.

References

1. Ben-Ayed O, Boyce DE, Blair III. CE (1988) A general bilevel linear programming formulations of the network Design problem. Transpn Res B 22B:311–318
2. Beckmann MJ, McGuire CB, Winston C (1956) Studies in the economics of transportation. Yale University Press, New Haven, CT
3. Biegler LT, Grossmann IE, Westerberg AW (1997) Systematic methods of chemical process design. Prentice-Hall, New Jersey
4. Brengel DD, Seider WD (1992) Coordinated design and control optimization of nonlinear processes. Comput Chem Eng 16:861–886
5. Burgard AP, Pharkya P, Maranas C (2003) OptKnock: A bilevel programming framework for identifying gene knockout strategies for microbial strain optimization. Biotech Bioeng 84:647–657
6. Chang YJ, Sahinidis NV (2005) Optimization of metabolic pathways under stability considerations. Comput Chem Eng 29:467–479
7. Clark PA, Westerberg A (1990) Bilevel programming for chemical process design – I. Fundamentals and algorithms. Comput Chem Eng 14:87–97
8. Clark PA (1990) Bilevel programming for chemical process design – II. Performance study for nondegenerate problems. Comput Chem Eng 14:99–109
9. Fisk CS (1984) Game theory andr transportation systems modeling. Transp Res-B 18B:301–313
10. Floudas CA (1995) Nonlinear and mixed-integer optimization. Oxford University Press, USA
11. Floudas CA, Pardalos PM, Adjiman CS, Esposito WR, Gümüs ZH, Harding ST, Klepeis JL, Meyer CA, Schweiger CA (1999) Handbook of test problems in local and global optimization. Kluwer, Netherlands

12. Floudas CA (2000) Deterministic global optimization: theory, methods and applications. Kluwer, The Netherlands
13. Floudas CA, Gümüş ZH, Ierapetritou MG (2001) Global optimization in design under uncertainty: feasibility test and flexibility index problems. Ind Eng Chem Res 40:4267–4282
14. Gadkar KG, Doyle III. FJ, Edwards JS, Mahadevan R (2005) Estimating optimal profiles of genetic alterations using constraint-based models. Biotech Bioeng 89:243–251
15. Gadkar KG, Mahadevan R, Doyle III. FJ (2006) Optimal genetic manipulations in batch bioreactor control. Automatica 42:1723–1733
16. Grossmann IE, Floudas CA (1987) Active constraint strategy for flexible analysis in chemical processes. Comput Chem Eng 11:675–693
17. Gümüş ZH, Ciric AR (1997) Reactive distillation column design with vapor/liquid/liquid equilibria. Comput Chem Eng 21:983–988
18. Gümüş ZH, Floudas CA (2001) Global optimization of nonlinear bilevel programming problems. J Glob Optim 2:1–31
19. Gümüş ZH, Floudas CA (2005) Global optimization of mixed-integer bilevel programming problems. Comput Man Sci 2:181–212
20. Halemane KP, Grossmann IE (1983) Optimal process design under uncertainty. AIChE J 29:425–433
21. LeBlanc LJ, Boyce DE (1986) A bilevel programming algorithm for exact solution of the network design problem with user optimal flows. Transp Res-B 20B:259–265
22. Pharkya P, Burgard AP, Maranas C (2003) Exploring the overproduction of amino acids using the bilevel optimization framework optknock. Biotech Bioeng 84:887–899
23. Swaney RE, Grossmann IE (1985) An index for operational flexibility in chemical process design. Part I: Formulation and theory. AIChE J 31:621–630
24. Yang H, Yagar S, Iida Y, Asakura I (1994) An algorithm for the inflow control problem on urban freeway networks with user-optimal flows. Transp Res-B 28B:123–139
25. Yang H, Yagar S (1994) Traffic assignment and traffic control in general freeway-arterial corridor systems. Transp Res-B 28B:463–486
26. Yang H, Yagar S (1995) Traffic assignment and signal control in saturated roadnetworks. Transp Res-A 29A:125–139

Bilevel Programming Framework for Enterprise-Wide Process Networks Under Uncertainty

EFSTRATIOS N. PISTIKOPOULOS, NUNO P. FAÍSCA, PEDRO M. SARAIVA, BERÇ RUSTEM
Centre for Process Systems Engineering, Imperial College London, London, UK

Article Outline

Introduction
Formulation
  Bilevel Programming
  Bilevel Programming with Multi-Followers
Applications
Cases
  Global Optimum of a Bilevel Programming Problem
  Bilevel Programming Problem
  Bilevel Programming Problem with Multi-Followers
  Bilevel Programming with Uncertainty
References

Introduction

Optimisation of enterprise-wide process networks has attracted considerable attention in recent years; since it represents substantial economic savings there has been a growing concern to plan efficiently the operations within the complexity of decision networks. Often, in such complex networks, an hierarchy of decisions has to be followed and compromises made between identities with equivalent authority. For instance, numerous investigations have been done in the optimisation of supply chains, Fig. 1, and in the plant selection problem, Fig. 2. A detailed study of hierarchical decisions can be found in [13,14,26].

Formulation

The general multilevel decentralised optimisation problem can be described as follows:

$$\min_{x, y_1^i, y_2^k, \ldots, y_m^l} f_1(x, y_1^i, y_2^k, \ldots, y_m^l), \quad (1st\ level)$$

$$\text{s.t.}\ g_1(x, y_1^i, y_2^k, \ldots, y_m^l) \leq 0, \quad (1)$$

where $\left[y_1^i, y_2^k, \ldots, y_m^l\right]$ solve,

$$\ldots, \min_{y_1^i, y_2^k, \ldots, y_m^l} f_2^i(x, y_1^i, y_2^k, \ldots, y_m^l), \ldots \quad (2nd\ level)$$

$$\text{s.t.}\ g_2^i(x, y_1^i, y_2^k, \ldots, y_m^l), \leq 0,$$

where $\left[y_2^k, \ldots, y_m^l\right]$ solve,

$$\vdots$$

$$\ldots, \min_{y_m^l} f_m^l(x, y_1^i, y_2^k, \ldots, y_m^l), \ldots \quad (mth\ level)$$

$$\text{s.t.}\ g_m^l(x, y_1^i, y_2^k, \ldots, y_m^l) \leq 0,$$

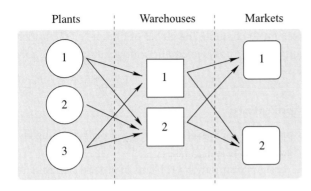

Bilevel Programming Framework for Enterprise-Wide Process Networks Under Uncertainty, Figure 1
Supply chain planning example

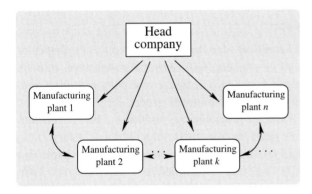

Bilevel Programming Framework for Enterprise-Wide Process Networks Under Uncertainty, Figure 2
Hierarchical decision planning example

where, f are real convex functions, g are vectorial real functions defining convex sets and x, y are sets of variables belonging to the group of real numbers; $i \in \{1, 2, \ldots, I\}$, $k \in \{1, 2, \ldots, K\}$, $l \in \{1, 2, \ldots, L\}$, implying that (*2nd level*) has I optimisation subproblems, (*3rd level*) K optimisation subproblems and (*mth level*) has L optimisation subproblems, respectively. For the sake of simplicity and without loss of generality, we analyse the relations in Problem (1) using two particular classes of multilevel programming problems: the bilevel programming problem, which organises vertically in two levels, and the bilevel programming problem with multi-followers, which is similar to bilevel programming but with several subproblems at the second level.

Bilevel Programming

Bilevel programming problems (BLPP) involve a hierarchy of two optimisation problems, of the following form [6,17,20,25,32]:

$$
\begin{aligned}
&\min_{x,y} \; F(x, y), \\
&\text{s.t.} \; G(x, y) \leq 0, \\
&\quad x \in X, \\
&\quad y \in \arg\min\{f(x, y): g(x, y) \leq 0, y \in Y\},
\end{aligned} \quad (2)
$$

where $X \subseteq \mathbb{R}^{nx}$ and $Y \subseteq \mathbb{R}^{ny}$ are both compact convex sets; F and f are real functions: $\mathbb{R}^{(nx+ny)} \to \mathbb{R}$; G and g are vectorial real functions, $G: \mathbb{R}^{(nx+ny)} \to \mathbb{R}^{nu}$ and $g: \mathbb{R}^{(nx+ny)} \to \mathbb{R}^{nl}$; $nx, ny \in \mathbb{N}$ and $nu, nl \in \mathbb{N} \cup \{0\}$. The following definitions are associated to Problem (2):

- Relaxed feasible set (or constrained region),

$$
\Omega = \{x \in X, y \in Y: G(x, y) \leq 0, g(x, y) \leq 0\}; \quad (3)
$$

- Lower level feasible set,

$$
C(x) = \{y \in Y: g(x, y) \leq 0\}; \quad (4)
$$

- Follower's rational reaction set,

$$
M(x) = \{y \in Y: y \in \arg\min\{f(x, y): y \in C(x)\}\}; \quad (5)
$$

- Inducible region,

$$
IR = \{x \in X, y \in Y: (x, y) \in \Omega, y \in M(x)\}. \quad (6)
$$

Note the parametric nature of the rational reaction set, (5), which reflects the dependence of the decisions taken at the upper levels on the decisions taken at the lower levels. This, in fact, is evidence that in bilevel programming problems the relations between the levels differ from the well-known Stackelberg game, where the decisions made by the followers don't affect the decision already taken by the leader [32].

Bilevel Programming with Multi-Followers

Bilevel programming problems with multi-followers involve two optimisation levels with several optimisation subproblems at the lower (2nd level):

$$\begin{aligned}
\min_{x, y_1, y_2, \ldots, y_m} \quad & F(x, y_1, y_2, \ldots, y_m), \quad (1st\ level) \\
\text{s.t.} \quad & G(x, y_1, y_2, \ldots, y_m) \leq 0, \\
& x \in X, \\
& y_i \in \arg\min\{f_i(x, y_1, y_2, \ldots, y_m) : \\
& g_i(x, y_1, y_2, \ldots, y_m) \leq 0, y_i \in Y_i\}, \\
& \quad (2nd\ level) \\
& i \in \{1, 2, \ldots, m\},
\end{aligned} \quad (7)$$

with the following definitions:
- Feasible set for the *ith* follower,

$$\begin{aligned}
\Omega_i(x, y_1, y_2, \ldots, y_{i-1}, y_{i+1}, \ldots, y_m) \\
= \{y_i \in Y_i : g_i(x, x, y_1, y_2, \ldots, y_m) \leq 0\},
\end{aligned} \quad (8)$$

- Rational reaction set for the *ith* follower,

$$\begin{aligned}
\phi_i(x, y_1, y_2, \ldots, y_{i-1}, y_{i+1}, \ldots, y_m) = \{y_i \in Y_i : \\
y_i \in \arg\min\{f_i(x, y_1, y_2, \ldots, y_m) : y_i \in \Omega_i(x)\}\}.
\end{aligned} \quad (9)$$

Since one assumption is that followers may exchange information, conflicts naturally occur. The Nash equilibrium is often a preferred strategy to coordinate such decentralised systems [24]. Consequently, the optimisation subproblems positioned in the lower level reach a Nash equilibrium point, $(x, y_1^*, y_2^*, \ldots, y_m^*)$ [2]:

$$\begin{cases}
f_1(x, y_1^*, y_2^*, \ldots, y_m^*) \leq f_1(x, y_1, y_2^*, \ldots, y_m^*), \\
\forall y_1 \in Y_1, \\
f_2(x, y_1^*, y_2^*, \ldots, y_m^*) \leq f_2(x, y_1^*, y_2, \ldots, y_m^*), \\
\forall y_2 \in Y_2, \\
\vdots \\
f_m(x, y_1^*, y_2^*, \ldots, y_m^*) \leq f_m(x, y_1^*, y_2^*, \ldots, y_m), \\
\forall y_m \in Y_m.
\end{cases} \quad (10)$$

Once more observe the parametric nature of the followers' rational reaction set, (9). In this case, however, each rational reaction set is a function of both the upper level decision variables and the decision variables of the other subproblems located in the same hierarchical level. Additionally, the priority remains to solve the leader's objective function to global optimality. Thus, we aim to compute the global optimum for the leader and the best possible equilibrium solution for the followers.

Applications

Applications of bilevel and multilevel programming include:
1. design optimisation problems in process systems engineering [4,5];
2. design of transportation networks [23];
3. agricultural planning [19];
4. management of multi-divisional firms [27] and
5. hierarchical decision-making structures [19].

Cases

Theoretical developments. Recently, Pistikopoulos and co-workers [9,10] have proposed novel solution algorithms which open the possibility of using a general framework to address general classes of bilevel and multilevel programming problems. These algorithms are based on parametric programming theory [1,11] and use of the Basic Sensitivity Theorem [15,16]. This approach can be classified as a *Reformulation Technique* [33] since the bilevel problem is transformed into a number of quadratic or linear problems. The main idea is to divide the follower's feasible area into different rational reaction sets, and search for the global optimum of a simple quadratic (or linear) programming problem in each area.

Global Optimum of a Bilevel Programming Problem

While for an optimal control problem (one-player problem) there is a well-defined concept for optimality, the same is not always true for multi-person games [2]. In the case of bilevel programming, [7,17,18,29,32,33] interpret the optimisation problem as a leader's problem, F, and search for the global minimum of F. The solution point obtained for the follower's problem, f, will respect the stationary (KKT) conditions and hence it can be any stationary point. Obviously, this solution strategy is acceptable when the player in the upper level of the hierarchy is in the most "powerful" position, and the other levels just react to the decision of their leader. Such an approach is sensible in many engineering applications of bilevel programming (for instance, see [4,5]). It is also a valid strategy for the cases of decentralised manufacturing and financial structures when the leader has a full insight and control of the overall objectives and strategy of the corporation, while the follower does not.

However, this is not always the case. For example, using the *feedback Stackelberg solution*, where at every level of play a Stackelberg equilibrium point is searched, the commitment of the leader for his/her decision increases with the number of players involved. [3] present an example where the sacrifice of the leader's objective on behalf of the followers results in a better solution for both levels. Similar solution strategies have also been studied [3,22,28,30].

Theorem 1 [32] *If for each $x \in X$, f and g are twice continuously differentiable functions for every $y \in C(x)$, f is strictly convex for every $y \in C(x)$ and $C(x)$ is a convex and compact set, then $M(\cdot)$ is a real-valued function, continuous and closed.* □

If Theorem 1 applies and assuming that $M(x)$ is non-empty, then $M(x)$ will have only one element, which is $y(x)$. Thus, (2) can be reformulated as:

$$\min_{x,y} \ F(x, y(x))$$
$$\text{s.t.} \ G(x, y(x)) \leq 0$$
$$x \in C_{rf} \qquad (11)$$
$$C_{rf} = \{x \in X \colon \exists y \in Y, g(x,y) \leq 0\}.$$

Considering that f is a convex real function, the function $y(x)$ can be computed as a linear conditional function based on parametric programming theory, as follows [9]:

$$y(x) = \begin{cases} m^1 + n^1 x, & \text{if } H^1 x \leq h^1 \\ m^2 + n^2 x, & \text{if } H^2 x \leq h^2 \\ \vdots \\ m^k + n^k x, & \text{if } H^k x \leq h^k \\ \vdots \\ m^K + n^K x, & \text{if } H^K x \leq h^K \end{cases} \qquad (12)$$

where, n^k, m^k and h^k are real vectors and H^k is a real matrix.

Theorem 2 [32] *If the assumptions of Theorem 1 hold, F is a real continuous function, X and the set defined by* $G(x,y)$ *are compact, and if* $\{\exists x \in X \colon G(x, y(x)) \leq 0\}$, *then there is a global solution for Problem (2).* □

Since an explicit expression for y can be computed, if the assumptions of Theorem 2 hold, and the two players have convex functions to optimise, then the global optimum for Problem (2) can be obtained via the parametric programming approach. The advantage of using this approach is that the final solution will consider the possibility of existence of other global minima, which could correspond to better solutions for the follower. Moreover, the parametric nature of the leader's problem is preserved.

Regarding computational complexity, a number of authors have shown that bilevel programming problems are $\mathcal{NP}$-Hard [8,21]. Furthermore, [31] proved that even checking for a local optimum is an $\mathcal{NP}$-Hard problem.

The objective of this section is to describe a parametric programming framework which can solve different classes of multilevel programming problems to global optimality. We describe the fundamental developments for the quadratic bilevel programming case, and how the theory unfolds to address the existence of RHS uncertainty.

Bilevel Programming Problem

Consider the following general quadratic BLLP:

$$\min_{x,y} \ F(x, y) = L_1 + L_2 x + L_3 y + \frac{1}{2} x^T L_4 x + y^T L_5 x$$
$$+ \frac{1}{2} y^T L_6 y ,$$
$$\text{s.t.} \ G_1 x + G_2 y + G_3 \leq 0 ,$$
$$\min_y \ f(x, y) = l_1 + l_2 x + l_3 y + \frac{1}{2} x^T l_4 x + y^T l_5 x$$
$$+ \frac{1}{2} y^T l_6 y ,$$
$$\text{s.t.} \ g_1 x + g_2 y + g_3 \leq 0 ,$$
$$\qquad (13)$$

where x and y are the optimisation variables, $x \in X \subseteq \mathbb{R}^{nx}$ and $y \in Y \subseteq \mathbb{R}^{ny}$. $[L_2]_{1 \times nx}$, $[L_3]_{1 \times ny}$, $[L_4]_{nx \times nx}$, $[L_5]_{ny \times nx}$, $[L_6]_{ny \times ny}$, $[l_2]_{1 \times nx}$, $[l_3]_{1 \times ny}$, $[l_4]_{nx \times nx}$, $[l_5]_{ny \times nx}$ and $[l_6]_{ny \times ny}$ are matrices defined in the real space. The matrices $[G_1]_{nu \times nx}$, $[G_2]_{nu \times ny}$, $[G_3]_{nu \times 1}$, $[g_1]_{nu \times nx}$, $[g_2]_{nu \times ny}$, $[g_3]_{nu \times 1}$ correspond to the constraints, also defined in the real space.

By focusing attention on the follower's optimisation problem, considering x as a parameter vector and operating a variable change $(z = y + l_6^{-1} l_5 x)$, it can be rewritten as the following mp-QP problem:

$$\min_z f'(x,z) = l'_1 + l'_2 x + \frac{1}{2} x^T l'_4 x + \{l'_3 z + \frac{1}{2} z^T l'_6 z\},$$
$$\text{s.t. } g'_2 z \leq g'_3 + g'_1 x, \tag{14}$$

where: $l'_1 = l_1$; $l'_2 = l_2 - l_3 l_6^{-1} l_5$; $l'_3 = l_3$; $l'_4 = l_4 - l_5^T l_6^{-1} l_5$; $l'_5 = 0$; $l'_6 = l_6$; $g'_1 = -(g_1 - g_2 l_6^{-1} l_5)$; $g'_2 = g_2$; $g'_3 = -g_3$. The mp-QP problem can be solved by applying the algorithm of [9]. As a result, a set of rational reaction sets (5) is obtained for different regions of x:

$$z^k = m^k + n^k x; \quad H^k x \leq h^k, \quad k = 1, 2, \ldots, K. \tag{15}$$

Incorporating the expressions (15) into Problem (13) results in the following K quadratic problems:

$$\min_x F'(x) = L'^k_1 + L'^k_2 x + \frac{1}{2} x^T L'^k_4 x, \tag{16}$$
$$\text{s.t. } G'^k_1 x \leq G'^k_3,$$

with:

$$L'^k_1 = L_1 + L_3 m^k + \frac{1}{2} m^{k^T} L_6 m^k;$$

$$L'^k_2 = L_2 + L_3 n^k - L_3 l_6^{-1} l_5 + m^{k^T}$$
$$\cdot L_5 + m^{k^T} L_6 n^k - m^{k^T} L_6 l_6^{-1} l_5;$$

$$L'^k_4 = L_4 + 2 n^{k^T} L_5 - 2 l_5^T l_6^{-1} L_5 + n^{k^T}$$
$$\cdot L_6 n^k - 2 n^{k^T} L_6 l_6^{-1} l_5 + l_5^T l_6^{-1} L_6 l_6^{-1} l_5;$$

$$G'_1 = G_1 + G_2 n^k - G_2 l_6^{-1} l_5;$$
$$G'_3 = -(G_3 + G_2 m^k);$$
$$G'^k_1 = [G'_1 | H^k]^T_{(nx) \times (nu + n_{hk})};$$
$$G'^k_3 = [G'_3 | h^k]^T_{(1) \times (nu + n_{hk})}.$$

Clearly, the solution of the BLLP Problem (13) is the minimum along the K solutions of Problem (16).

Remark 1 The artificial variable, z, introduced in Problem (14) is only necessary if $l_5 \neq \underline{0}$. In all other cases the multi-parametric problem can be easily formulated through algebraic manipulations.

Remark 2 When one of the matrices l'_6, L'^k_4 is null the optimisation problem where these are involved becomes linear. In particular, if $l'_6 = \underline{0}$, Problem (14) is transformed into an mp-LP; on the other hand, if $L'^k_4 = \underline{0}$, Problem (16) becomes an LP problem. In both cases, the solution procedure is not affected, due to the fact that the Basic Sensitivity Theorem [15,16] also applies to the mp-LP problem.

Remark 3 The expression for the artificial variable introduced, z, is only valid when l_6 is symmetric. If not, with the following transformation:

$$\bar{l}_6 = \left\{ \frac{l_6 + l_6^T}{2} \right\},$$

the resulting matrix is non-singular. If the resulting matrix is singular, the expression for the artificial variable should be given by:

$$z = y + Ax,$$

where A should satisfy:

$$\left\{ A \in \mathbb{R}^{nx \times nx} : l_5 - (\frac{1}{2} l_6 + \frac{1}{2} l_6^T) A = 0 \right\}.$$

In this case, several solutions for the system above can exist. However, as long as the bilinear terms are eliminated in Problem (14) any solution can be selected.

Remark 4 This technique is not valid when at the same time:
1. f is a pure quadratic cost function,
2. f involves bilinear terms and
3. matrix $\bar{l}_6$ is singular.

Observing Formulation (16) we can conclude that the parametric programming approach, Alg. 1, transforms the original quadratic bilevel programming problem into simple quadratic problems, for which a global optimum can be reached.

Bilevel Programming Problem with Multi-Followers

Consider the bilevel programming problem with multi-followers, and assume quadratic objective functions, linear constraints and two followers:

$$\min_{x,y_1,y_2} f_1 = L_1^1 +$$
$$+ L_2^1 \cdot x + L_3^1 \cdot y_1 + L_4^1 \cdot y_2 + \quad (1st\ level)$$
$$+ \frac{1}{2} x^T \cdot L_5^1 \cdot x + \frac{1}{2} y_1^T \cdot L_6^1 \cdot y_1$$
$$+ \frac{1}{2} y_2^T \cdot L_7^1 \cdot y_2 + x^T \cdot L_8^1 \cdot y_1$$
$$+ y_2^T \cdot L_9^1 \cdot x + y_2^T \cdot L_{10}^1 \cdot y_1,$$

s.t. $\Big|\ G_1^1 \cdot x + G_2^1 \cdot y_1 + G_3^1 \cdot y_2 \le 0,$

(2nd level)

$$\min_{y_1} f_2 = L_1^2 + \qquad Follower\ 1$$
$$+ L_2^2 \cdot x + L_3^2 \cdot y_1 + L_4^2 \cdot y_2$$
$$+ \frac{1}{2} x^T \cdot L_5^2 \cdot x + \frac{1}{2} y_1^T \cdot L_6^2 \cdot y_1$$
$$+ \frac{1}{2} y_2^T \cdot L_7^2 \cdot y_2 + x^T \cdot L_8^2 \cdot y_1$$
$$+ y_2^T \cdot L_9^2 \cdot x + y_2^T \cdot L_{10}^2 \cdot y_1,$$
s.t. $G_1^2 \cdot x + G_2^2 \cdot y_1 + G_3^2 \cdot y_2 \le 0,$

$$\min_{y_2} f_3 = L_1^3 + \qquad Follower\ 2$$
$$+ L_2^3 \cdot x + L_3^3 \cdot y_1 + L_4^3 \cdot y_2$$
$$+ \frac{1}{2} x^T \cdot L_5^3 \cdot x + \frac{1}{2} y_1^T \cdot L_6^3 \cdot y_1$$
$$+ \frac{1}{2} y_2^T \cdot L_7^3 \cdot y_2 + x^T \cdot L_8^3 \cdot y_1$$
$$+ y_2^T \cdot L_9^3 \cdot x + y_2^T \cdot L_{10}^3 \cdot y_1,$$
s.t. $G_1^3 \cdot x + G_2^3 \cdot y_1 + G_3^3 \cdot y_2 \le 0.$

(17)

The difference between Problem (17) and Problem (13) is the existence of two optimisation subproblems in a single level. Accordingly, the concept of Nash equilibrium is introduced.

As in the bilevel programming case, each optimisation subproblem in (2nd level) is recast as a multi-parametric programming problem. In this problem, the parameters are all the variables from the optimisation problem at (1st level) as well as the optimisation variables of the other subproblems at the same level, *Follower* 1 or *Follower* 2 in this case (17). Thus, defining vectors, $[\omega^2]^T = [x|y_2]$ and $[\omega^3]^T = [x|y_1]$, we

Algorithm – Parametric Programming Algorithm for BLPP

1. Recast the inner problem as a multi-parametric programming problem, with the leader's variables being the parameters (14);
2. Solve the resulting problem using the suitable multi-parametric programming algorithm;
3. Substitute each of the K solutions in the leader's problem, and formulate the K one-level optimisation problems;
4. Compare the K optimum points and select the best one.

Bilevel Programming Framework for Enterprise-Wide Process Networks Under Uncertainty, Algorithm 1
Parametric Programming algorithm for a BLPP

rewrite the (2nd level) optimisation subproblems as,

$$\min_{y_1} f_2(y_1, \omega^2) = L_1^2 + L_2^{2*} \cdot \omega^2 + L_3^2 \cdot y_1$$
$$+ \frac{1}{2} \omega^{2T} \cdot L_5^{2*} \cdot \omega^2 + \frac{1}{2} y_1^T \cdot L_6^2 \cdot y_1$$
$$+ y_1^T \cdot L_8^{2*} \cdot \omega^2,$$
s.t. $G_1^{2*} \cdot \omega^2 + G_2^2 \cdot y_1 \le 0,$

(18)

and,

$$\min_{y_2} f_3(y_2, \omega^3) = L_1^3 + L_2^{3*} \cdot \omega^3 + L_4^3 \cdot y_2$$
$$+ \frac{1}{2} \omega^{3T} \cdot L_5^{3*} \cdot \omega^{2a} + \frac{1}{2} y_2^T \cdot L_7^3 \cdot y_2$$
$$+ y_1^T \cdot L_9^{3*} \cdot \omega^3,$$
s.t. $G_1^{3*} \cdot \omega^3 + G_3^3 \cdot y_2 \le 0,$

(19)

where ω^2 and ω^3 are the vectors of parameters. The bilinearities can be circumvented by using a similar strategy to the one used in the bilevel case. By using a multi-parametric programming algorithm [9], problems (18) and (19) result in the following parametric expressions:

$$\begin{cases} y_1 = \phi_1(x, y_2) & \rightarrow \text{rational reaction set follower 1}, \\ y_2 = \phi_2(x, y_1) & \rightarrow \text{rational reaction set follower 2}, \end{cases}$$

(20)

Algorithm

1. Recast each of the subproblems in the lower level as a multi-parametric programming problem, with the variables out of their control being the parameters (18–19);
2. Solve the resulting problems using the suitable multi-parametric programming algorithm;
3. Compute a Nash equilibrium point by direct comparison of the rational reaction sets (21);
4. Substitute each of the K solutions in the leader's problem, and formulate the K one level optimisation problems;
5. Compare the K optima points and select the best one.

Bilevel Programming Framework for Enterprise-Wide Process Networks Under Uncertainty, Algorithm 2
Parametric programming algorithm for bilevel programming problems with multi-followers

which are then used to compute the Nash equilibrium (x, y_1^*, y_2^*):

$$\begin{cases} f_1(x, y_1^*, y_2^*) \leq f_1(x, y_1, y_2^*), & \forall y_1 \in Y_1, \\ f_2(x, y_1^*, y_2^*) \leq f_2(x, y_1^*, y_2), & \forall y_2 \in Y_2, \end{cases} \quad (21)$$

easily computed by direct comparison [24]:

$$\phi_1'(x, y_1) = \phi_2(x, y_1), \rightarrow y_1 = \phi_2^*(x), \quad (22a)$$

$$\phi_1(x, y_2) = \phi_2'(x, y_2), \rightarrow y_2 = \phi_1^*(x). \quad (22a)$$

Finally, substituting the expressions in (22) in the leader's optimisation problem, (1st level), we end up with a single-level convex optimisation problem, involving only the leader's optimisation variables, as follows:

$$\min_x f_1^*(x, y_1(x, y_2^*(x)), y_2(x, y_1^*(x))),$$

$$\text{s.t. } G_1(x, y_1(x, y_2^*), y_2(x, y_1^*)) \leq 0, \quad x \in C_{rf},$$

$$C_{rf} = \{x \in X : \exists_{y_1, y_2} \in Y, Z, G_2(x, y_1, y_2) \leq 0,$$

$$G_3(x, y_1, y_2) \leq 0\}.$$

$$(23)$$

The algorithm is summarised in Alg. 2.

Bilevel Programming with Uncertainty

[12] highlighted the importance of considering uncertainty/risk (e. g. prices, technological attributes, etc.) in the solution of decentralised decision makers. A comprehensive analysis of linear bilevel programming problems can be found in [27], where uncertainty is considered unstructured, taking any value between its bounds. Here it is extended to the quadratic case. We address the following quadratic BLPP with uncertainty, θ:

$$\min_{x, y} F(x, y, \theta) = L_1 + L_2 x + L_3 y + \frac{1}{2} x^T L_4 x$$

$$+ y^T L_5 x + \frac{1}{2} y^T L_6 y$$

$$\text{s.t. } G_1 x + G_2 y + G_3 \leq G_4 \theta \quad (24)$$

$$\min_y f(x, y, \theta) = l_1 + l_2 x + l_3 y$$

$$\frac{1}{2} x^T l_4 x + y^T l_5 x + \frac{1}{2} y^T l_6 y$$

$$\text{s.t. } g_1 x + g_2 y + g_3 \leq g_4 \theta,$$

The steps for solving (24) are as follows:

1. Recast the inner problem as an mp-QP, with parameters being both x and θ. The solution obtained is similar to (15):

$$z^k = m^k + n_b^k x + \bar{n}_c^k \theta; \quad H^k x + \bar{H}^k \theta \leq h^k,$$

$$k = 1, 2, \ldots, K.$$

$$(25)$$

2. Incorporate expressions (25) in (24) to formulate K mp-QPs, with parameters being the uncertainty θ:

$$\min_x F'(x, \theta) = \bar{L}_1'^k + \bar{L}_2'^k x + \frac{1}{2} x^T \bar{L}_4'^k x$$

$$\text{s.t. } \bar{G}_1'^k x \leq \bar{G}_3'^k + \bar{G}_4'^k \theta, \quad (26)$$

where $\bar{L}_1'^k, \bar{L}_2'^k, \bar{L}_4'^k, \bar{G}_1'^k, \bar{G}_2'^k, \bar{G}_4'^k$ are appropriate matrices derived by algebraic manipulations.

References

1. Acevedo J, Pistikopoulos EN (1997) A multiparametric programming approach for linear process engineering problems under uncertainty. Ind Eng Chem Res 36:717–728
2. Başar T, Olsder GJ (1982) Dynamic Noncooperative Game Theory. Academic Press, London
3. Cao D, Chen M (2006) Capacitated plant selection in a decentralized manufacturing environment: a bilevel optimization approach. Eur J Oper Res 169(1):97–110

4. Clark PA (1990) Bilevel programming for steady-state chemical process design – ii. performance study for non-degenerate problems. Comput Chem Eng 14(1):99–109
5. Clark PA, Westerberg AW (1990) Bilevel programming for steady-state chemical process design – i. fundamentals and algorithms. Comput Chem Eng 14(1):87–97
6. Dempe S (2003) Annotated bibliography on bilevel programming and mathematical programs with equilibrium constraints. Optimization 52(3):33–359
7. Dempe S, Kalashnikov V, Ríos-Mercado RZ (2005) Discrete bilevel programming: Application to a natural gas cash-out problem. Eur J Oper Res 166:469–488
8. Deng X (1998) Complexity issues in bilevel linear programming. In: Multilevel optimization: algorithms and applications. Kluwer, Dordrecht, pp 149–164
9. Dua V, Bozinis A, Pistikopoulos EN (2002) A multiparametric programming approach for mixed-integer quadratic engineering problems. Comput Chem Eng 26:715–733
10. Dua V, Pistikopoulos EN (2000) An algorithm for the solution of multiparametric mixed integer linear programming problems. Ann Oper Res 99:123–139
11. Dua V (2000) Parametric programming techniques for process engineering problems under uncertainty. PhD thesis, Department of Chemical Engineering and Chemical Technology Imperial College of Science, Technology and Medicine London, London
12. Evans GW (1984) An overview of thecniques for solving multiobjective mathematical programs. Manag Sci 30(11):1268–1282
13. Faísca NP, Dua V, Saraiva PM, Rustem B, Pistikopoulos EN (2007) Parametric global optimisation for bilevel programming. J Glob Optim 38(4):609–623
14. Faísca NP, Saraiva PM, Rustem B, Pistikopoulos EN (2007) A multi-parametric programming approach for multi-level hierarchical and decentralised optimisation problems. Comput Manag Sci (in press)
15. Fiacco AV (1976) Sensitivity analysis for nonlinear programming using penalty methods. Math Program 10:287–311
16. Fiacco AV (1983) Introduction to sensitivity and stability analysis in nonlinear programming. Academic Press, New York
17. Floudas CA (2000) Deterministic global optimization. Kluwer, Dordrecht
18. Floudas CA, Pardalos PM, Adjiman CS, Esposito WR, Gümüş ZH, Harding ST, Klepeis JL, Meyer CA, Schweiger CA (1999) Handbook of test problems in local and global optimization. Kluwer, Dordrecht
19. Fortuny-Amat J, McCarl B (1981) A representation and economic interpretation of a two-level programming problem. J Oper Res Soc 32(9):783–792
20. Gümüş ZH, Floudas CA (2001) Global optimization of nonlinear bilevel programming problems. J Glob Optim 20(1):1–31
21. Hansen P, Jaumard B, Savard G (1992) New brach-and-bound rules for linear bilevel programming. SIAM J Sci Stat Comput 13:1194–1217
22. Lai Y (1996) Hierarchical optimization: a satisfactory solution. Fuzzy Sets Syst 77:321–335
23. LeBlanc LJ, Boyce DE (1985) A bilevel programming algorithm for exact solution of network design problem with user-optimal flows. Transp Res B Methodol 20:259–265
24. Liu B (1998) Stackelberg-nash equilibrium for multilevel programming with multiple followers using genetic algorithms. Comput Math Appl 36(7):79–89
25. Migdalas A, Pardalos PM, Varbrand P (1997) Multilevel optimization: algorithm and applications. Kluwer, Dordrecht
26. Ryu J, Dua V, Pistikopoulos EN (2004) A bilevel programming framework for enterprise-wide process networks under uncertainty. Comput Chem Eng 28:1121–1129
27. Ryu J-H (2003) Design and operation of enterprise-wide process networks under uncertainty. PhD thesis, Department of Chemical Engineering and Chemical Technology Imperial College of Science, Technology and Medicine London, London
28. Shih H, Lai Y, Lee ES (1996) Fuzzy approach for multi-level programming problems. Comput Oper Res 23(1):73–91
29. Shimizu K, Ishizuka Y, Bard JF (1997) Nondifferentiable and two-level mathematical programming. Kluwer, Boston
30. Tabucanon MT (1988) Multiple Criteria Decision Making in Industry. Elsevier, Amsterdam
31. Vicente LN, Savard G, Júdice J (1994) Descent approaches for quadratic bilevel programming. J Optim Theor Appl 81:379–399
32. Vicente L (1992) Bilevel programming. Master's thesis, Department of Mathematics, University of Coimbra, Coimbra
33. Visweswaran V, Floudas MG, Ierapetritou CA, Pistikopoulos EN (1996) A decomposition-based global optimization approach for solving bilevel linear and quadratic programs. In: State of the art in global optimization. Kluwer, Dordrecht, pp 139–162

Bilevel Programming: Global Optimization

VISWANATHAN VISWESWARAN
SCA Technologies LLC, Pittsburgh, USA

MSC2000: 90C90, 90C30

Article Outline

Keywords
Definitions
  Complexity
  Multiple Solutions to the Follower's Problem

Solution Methods
    Enumeration Methods
    Complementary Pivot Methods
    Branch and Bound Methods
Computational Results and Test Problems
See also
References

Keywords

Bilevel programming; Global optimization; Stackelberg game

A large number of mathematical programming problems have optimization problems in their constraints. Arising from the areas of game theory and multicriteria decision making, these *bilevel programming problems* (BPP) take the form:

$$\begin{cases} \min_{x} & F(x, y) \\ \text{s.t.} & G(x, y) \leq 0 \\ & y = \begin{cases} \text{Arg}\min_{y} & f(x, y) \\ \text{s.t.} & g(x, y) \leq 0 \end{cases} \end{cases} \quad (1)$$

where $x \in \mathbf{R}^{n_1}$, $y \in \mathbf{R}^{n_2}$ and the functions $F(x, y)$, $f(x, y)$, $G(x, y)$ and $g(x, y)$ are continuous and twice differentiable. It is generally assumed that these functions are convex; the case of nonconvex functions has not been considered in the literature so far (as of 2000).

Bilevel programming has its origins in *Stackelberg game theory*, in particular from models of two-person nonzero-sum games. In these games, two players make alternate moves in a pre-established order. The first player (the leader) selects a move, x, that optimizes his own cost function. The second player (the follower) then has to make a move y that is constrained by the prior decision of the leader. The follower has access only to his own cost function, while the leader is aware of both his own as well as the follower's cost function, and can thus foresee the reaction of the follower to any move that the leader makes. If the cost functions of the two players are identical (called the *cooperative case*), then the two constraint sets can be merged and the problem can be solved as a single level optimization problem. If the cost functions are exactly opposite (that is, $f(x, y) = -F(x, y)$), then there can be neither cooperation or compromise. The most interesting (and normally studied) case is when the two objectives are neither identical nor opposite.

BPP also arises in *hierarchical decision making*. For example, a central planning office might decide upon national budgets which act as constraints for local governments and businesses. Other applications include long-range planning problems followed by short term scheduling in the chemical process industries and energy planning of businesses constrained by national government policy. A detailed list of references for applications of BPP can be found in [14]. See [13] for a full review of algorithms and applications of bilevel and multilevel programming.

Definitions

The following definitions will be used in the sequel. The *relaxed constraint region* for the BPP is defined as

$$S = \{(x, y): G(x, y) \leq 0, g(x, y) \leq 0\}.$$

The follower's *feasible region* for a fixed x, $\sigma(x)$, is defined as

$$\sigma(x) = \{y: g(x, y) \leq 0\}.$$

This set is parametric in x, and represents the allowable choices for the follower. The *rational reaction set* $M(x)$ is defined as

$$M(x) = \{y: y \in \text{Arg}\min\{f(x, y): y \in \sigma(x)\}\}.$$

Finally, the *inducible region* for the problem is

$$IR = \{(x, y): y \in M(x), (x, y) \in S\}.$$

The inducible region IR (which represents the follower's feasible region) is in general nonconvex. In terms of the bimatrix or Stackelberg games, IR represents 'equilibrium' points, that is, the set of compromise solutions between the leader and the follower. In the presence of first level constraints (1), IR may be empty, which implies that the BPP has no solution. However, it can be shown that the IR is compact and the BPP has a solution, if the following conditions are met [7]:

a) $F(x, y), f(x, y), G(x, y)$ and $g(x, y)$ are continuous and twice differentiable;
b) $f(\cdot, y)$ is strictly convex in y;
c) $\sigma(x)$ is a compact convex set; and
d) $F(x, y)$ and $G(x, y)$ are convex in x and y.

Note that the solution to the BPP need not be individually optimal for each of the leader's and follower's objective function (that is, it need not be an *efficient solution*).

The specific instance of BPP when all the functions involved are linear has received the most interest. The *linear bilevel programming problem* (BLPP) can be written as

$$\begin{cases} \min_{x} \ F_L(x, y) = c_1^\top x + d_1^\top y \\ \text{s.t.} \quad A_1 x + B_1 y \leq b_1 \\ y = \begin{cases} \text{Arg}\min_{y} \ c_2^\top x + d_2^\top y \\ \text{s.t.} \quad g(x, y) \\ \qquad = A_2 x + B_2 y \leq b_2. \end{cases} \end{cases} \quad (2)$$

Complexity

Because of the nonconvexity of the induced region IR, BPP can be a hard problem to solve. It is generally known that even the linear problem, BLPP, is *NP*-hard. This has been shown by reducing the problem to a knapsack optimization problem [3], the standard KERNEL problem [11], and by reduction to a problem of minimizing a convex quadratic function over a polyhedron [1]. In fact, even checking for local optimality in BLPP is *NP*-hard [15].

Multiple Solutions to the Follower's Problem

In the absence of dual degeneracy, the follower's subproblem has a single solution for every x. However, if the follower's subproblem has multiple solutions for any x, then the overall BPP may not be well-defined. In this case, we need further assumptions about the cooperativeness of the follower with respect to the leader. Alternately, the follower's objective function can be modified as

$$f(x, y) = f(x, y) + \epsilon F(x, y);$$

in effect, allowing the leader to 'kick back' a small portion of its earnings to ensure that the follower selects a suitable solution.

Solution Methods

From the 1980s onwards, many approaches have been proposed for the solution of BPP. These can be classified as enumerative, complementary pivot, branch and bound, descent and penalty function methods. The last two categories of methods are only useful in finding stationary points and local minima, and will not be discussed here. The vast majority of the approaches address the linear case, BLPP. Some of the global optimization methods are discussed below.

Enumeration Methods

The linear BLPP is equivalent to maximizing the linear function $F_L(x, y)$ over a piecewise linear constraint region composed of the edges and hyperplanes of S, the feasible region. It can be shown that the global optimum to BLPP occurs at a vertex of S. This suggests an extreme point search procedure for solving BLPP. One such procedure is the *Bialas–Karwan Kth-best algorithm* [4]. The basic idea is to find an 'ordered' set of extreme points to the relaxed problem

$$\begin{cases} \min_{x} \ F_L(x, y) = c_1^\top x + d_1^\top y \\ \text{s.t.} \quad A_1 x + B_1 y \leq b_1 \\ \qquad A_2 x + B_2 y \leq b_2. \end{cases}$$

The algorithm has the following steps:

0	Solve the relaxed problem. Let the solution be (x^1, y^1). Set $k = 1$.
1	Solve the inner problem with $x = x^k$. If y^k is in the solution set to the inner problem, then STOP.
2	Locate all adjacent extreme points (x_i, y_i) such that $$c_1^\top x_i + d_1^\top y_i \leq c_1^\top x^k + d_1^\top y^k \quad \forall i.$$ Choose the adjacent extreme point j that minimizes $c_1^\top x_j + d_1^\top y_j$. Set $k = k + 1$, $(x^{k+1}, y^{k+1}) = (x_j, y_j)$. Go to Step 1.

Since each successive pair of points tested in this algorithm is adjacent, it can be efficiently implemented using the dual simplex method.

Complementary Pivot Methods

Under proper regularity conditions, the inner problem to the BPP can be replaced by its Karush–Kuhn–Tucker optimality conditions. For the case of the BLPP, this results in the following single-level optimization problem (KKT):

$$\begin{cases} \min_x & F_L(x,y) = c_1^\top x + d_1^\top y \\ \text{s.t.} & A_1 x + B_1 y \le b_1 \\ & A_2 x + B_2 y \le b_2 \\ & \mu(A_2 x + B_2 y - b_2) = 0 \\ & d_2 + A_2^\top \mu = 0 \\ & \mu \ge 0. \end{cases} \quad (3)$$

The problem KKT has a linear complementarity pivot formulation. As such, it can be solved using a complementary pivoting method. Consider the following parametric formulation LCP(λ):

$$c_1^\top x + d_1^\top y \le \lambda,$$
$$A_1 x + B_1 y \le b_1,$$
$$A_2 x + B_2 y \le b_2,$$
$$\mu(A_2 x + B_2 y - b_2) = 0,$$
$$d_2 + A_2^\top \mu = 0,$$
$$\mu \ge 0.$$

The global minimization of BLPP then corresponds to the identification of the minimum value of λ such that LCP(λ) has a solution. The following method can be used to solve LCP(λ):

0	Solve LCP(λ) without the first parametric constraint. Let (x^0, y^0) be the solution to this problem, with $\lambda_0 = c_1^\top x^0 + d_1^\top y^0$.
1	Solve LCP(λ^k). If LCP(λ^k) has no solution, go to Step 3. Otherwise, let (x^k, y^k) be the solution.
2	Set $$\lambda^{k+1} = c_1^\top x^k + d_1^\top y^k - \gamma \left| c_1^\top x^k + d_1^\top y^k \right|,$$ where γ is a small positive number. Set $k = k+1$, go to Step 1.
3	If $k = 0$, then BLPP has no solution. Otherwise, x^k, y^k is an ϵ-global optimum to BLPP, where $\epsilon = \gamma \left| c_1^\top x^k + d_1^\top y^k \right|$.

The key to this algorithm is the ability to efficiently solve LCP(λ^k) in Step 1. J. Judice and A. Faustino [12] have proposed a hybrid enumerative method which works by branching on the complementarity conditions $\mu(A_2 x + B_2 y - b_2) = 0$. Numerous heuristics can be used in each node of the resulting branch and bound tree, in order to reduce the search for a complementary solution.

Branch and Bound Methods

These methods work by identifying the set of inner-level constraints that are active at the optimal solution. The simplest method, due to J. Fortuny-Amat and B. McCarl [10], works by converting the KKT complementarity conditions in (3) to

$$\mu(A_2 x + B_2 y - b_2) = 0, \mu_i \le M\alpha_i,$$
$$A_2 x + B_2 y - b_2 \ge M(1-\alpha_i),$$
$$\alpha_i = 0-1, \quad \forall i,$$

where M is a large constant. The variable α_i is equal to 1 if inner level constraint i is active at the optimal solution, and zero otherwise. This converts the one-level problem to a mixed integer linear program (MILP), which can be solved with commercial MILP codes. However, this requires the addition of $2 \cdot m$ constraints and m variables, where m is the number of inner-level constraints.

Note that at the optimal solution, at least one of the inner problem constraints must be active, that is,

$$\sum_{i=1}^m \alpha_i \ge 1. \quad (4)$$

Moreover, it can be shown that the following conditions must hold [11]:

$$\sum_{\{i:\, B_{2_{ij}}>0\}} \alpha_i \ge \quad \text{if } d_j < 0, \quad (5)$$

$$\sum_{\{i:\, B_{2_{ij}}<0\}} \alpha_i \ge 1 \quad \text{if } d_j > 0, \quad (6)$$

for $j = 1, \ldots, n_2$. It is possible to use (4)–(6) as branching criteria in a branch and bound tree. Each of these conditions, when tight, can be used to eliminate a variable from the inner constraints. By combining these conditions with the use of linear relaxations to obtain lower bounds, a branch and bound algorithm can be developed to solve the BLPP [11].

An alternate method to the use of binary variables is to establish a one-to-one correspondence between each α_i and each μ_i, as follows:

$$\frac{1}{M}\alpha_i \leq \mu_i \leq M\alpha_i,$$

where M is a suitably large number. This ensures that if $\alpha_i = 0$, then $\mu_i = 0$, while if $\alpha_i = 1$, $\mu_i \geq (1/M)$ implying an inactive constraint. With this approach, BLPP can be transformed to:

$$\begin{cases} \min_{x} & F_L(x,y) = c_1^\top x + d_1^\top y \\ \text{s.t.} & A_1 x + B_1 y \leq b_1 \\ & A_2 x + B_2 y \leq b_2 \\ & \alpha(A_2 x + B_2 y - b_2) = 0 \\ & d_2 + A_2^\top \mu = 0 \\ & \mu_i \leq M\alpha_i \\ & \alpha_i \leq M\mu_i \\ & \mu \geq 0, \quad \alpha = \{0,1\}. \end{cases}$$

By partitioning the variables into $\overline{x} = (x,y)$ and $\overline{y} = (\mu, \alpha)$, it can be seen that this problem is of the form

$$\begin{cases} \min_{\overline{x},\overline{y}} & \overline{f}(\overline{x},\overline{y}) \\ \text{s.t.} & \overline{g}(\overline{x},\overline{y}) \leq 0 \\ & \overline{h}(\overline{x},\overline{y}) = 0, \end{cases}$$

where $\overline{f}(\overline{x},\overline{y})$, $\overline{g}(\overline{x},\overline{y})$ and $\overline{h}(\overline{x},\overline{y})$ are bilinear functions. Thus, the *GOP algorithm* of [8,9] can be applied to solve this problem. The algorithm works by solving a set of primal and relaxed dual problems that bound the global solution. The primal problem is

$$\begin{cases} \min_{\overline{x}} & \overline{f}(\overline{x},\overline{y^k}) \\ \text{s.t.} & \overline{g}(\overline{x},\overline{y^k}) \leq 0 \\ & \overline{h}(\overline{x},\overline{y^k}) = 0, \end{cases}$$

where y^k is a fixed number. Because this problem is linear, it can be solved for its global solution, and yields an upper bound on the global solution. It also provides multipliers for the constraints, μ^k and λ^k, which can be used to construct a Lagrange function of the form

$$L(\overline{x},\overline{y},\mu^k,\lambda^k) = \overline{f}(\overline{x},\overline{y^k}) + \mu^k \overline{g}(\overline{x},\overline{y^k}) + \lambda^k \overline{h}(\overline{x},\overline{y^k}).$$

It is then possible to solve a dual problem

$$\begin{cases} \min_{\overline{y}} & u \\ \text{s.t.} & u \geq L(\overline{x},\overline{y},\mu^k,\lambda^k), \end{cases}$$

which provides a lower bound on the global solution. The dual problem is actually solved by partitioning the $\overline{y}$-space using the gradients of L and solving a relaxed dual subproblem in each region. In [16] it has been shown that for the bilevel problems, only one dual subproblem needs to be solved at each iteration. This approach can also be used when the inner problem objective function is quadratic.

Another approach, proposed in [2], can also be used when the inner level problem has a convex quadratic objective function. The basic idea is to first solve the one-level linear problem by dropping the complementarity conditions. At each iteration, a check is made to see if the complementarity condition is satisfied. If it is, the corresponding solution is in the inducible region IR, and hence a candidate solution for BPP. If not, a branch and bound scheme is used to implicitly examine all combinations of complementary slackness.

Let $W_1 = \{i: \mu_i = 0\}$, $W_2 = \{i: g_i = 0\}$, $W_3 = \{i: i \notin W_1 \cup W_2\}$.

0	Set $k = 0$, $W_1 = W_2 = \emptyset$, $W_3 = i$, $\overline{F} = \infty$.
1	Set $\mu_i = 0$, $i \in W_1$, $g_i = 0$, $i \in W_2$. Solve the relaxed system. Let (x^k, y^k, μ^k) be the solution. If no solution exists, or if $F(x^k, y^k) \geq \overline{F}$, go to Step 4.
2	If $\mu_i g_i = 0$, $\forall i$, go to Step 3. Otherwise select i such that $\mu_i g_i$ is maximal, say $\hat{i}$. Let $W_1 = W_1 \cup \hat{i}$, $W_3 = W_3 \cup \hat{i}$, and go to Step 1.
3	Update $\overline{F} = F(x^k, y^k)$.
4	If all nodes in the three have been exhausted, go to Step 5. Else, branch to the newest unfathomed node, say j, and set $W_1 = W_1 \cup j$, $W_2 = W_2 \cup j$. Go to Step 1.
5	If $\overline{F} = \infty$, no solution exists to BPP. Otherwise, the point corresponding to $\overline{F}$ is the optimum.

Computational Results and Test Problems

The difficulty of solving bilevel problems depends on a number of factors, including the number of inner versus outer level variables, degree of cooperation between the leader and follower objective functions, number of inner level constraints and the density of the constraints. Computational results have been reported by many authors, including [2,11,12] and [16]. Generally, these have so far been limited to problems involving up to 100 inner level variables and constraints. See [5,6] for methods for automatically generating linear and quadratic bilevel problems which can be used to test any of these and other algorithms for bilevel programming.

See also

- Bilevel Fractional Programming
- Bilevel Linear Programming
- Bilevel Linear Programming: Complexity, Equivalence to Minmax, Concave Programs
- Bilevel Optimization: Feasibility Test and Flexibility Index
- Bilevel Programming
- Bilevel Programming: Applications
- Bilevel Programming: Applications in Engineering
- Bilevel Programming: Implicit Function Approach
- Bilevel Programming: Introduction, History and Overview
- Bilevel Programming: Optimality Conditions and Duality
- Multilevel Methods for Optimal Design
- Multilevel Optimization in Mechanics
- Stochastic Bilevel Programs

References

1. Bard JF (1991) Some properties of the bilevel programming problem. J Optim Th Appl 68:371–378
2. Bard JF, Moore J (1990) A branch and bound algorithm for the bilevel programming problem. SIAM J Sci Statist Comput 11:281–292
3. Ben-Ayed O, Blair C (1990) Computational difficulties of bilevel linear programming. Oper Res 38:556–560
4. Bialas W, Karwan M (1984) Two-level linear programming. Managem Sci 30:1004–1020
5. Calamai P, Vicente L (1993) Generating linear and linear-quadratic bilevel programming problems. SIAM J Sci Statist Comput 14:770–782
6. Calamai P, Vicente L (1994) Generating quadratic bilevel programming problems. ACM Trans Math Softw 20:103–119
7. Edmunds T, Bard J (1991) Algorithms for nonlinear bilevel mathematical programming. IEEE Trans Syst, Man Cybern 21:83–89
8. Floudas CA, Visweswaran V (1990) A global optimization algorithm (GOP) for certain classes of nonconvex NLPs: I. theory. Comput Chem Eng 14:1397
9. Floudas CA, Visweswaran V (1993) A primal-relaxed dual global optimization approach. J Optim Th Appl 78(2):187
10. Fortuny-Amat J, McCarl B (1981) A representation and economic interpretation of a two-level programming problem. J Oper Res Soc 32:783–792
11. Hansen P, Jaumard B, Savard G (1992) New branching and bounding rules for linear bilevel programming. SIAM J Sci Statist Comput 13:1194–1217
12. Júdice J, Faustino A (1992) A sequential LCP method for bilevel linear programming. Ann Oper Res 34:89–106
13. Migdalas A, Pardalos PM, Värbrand P (1998) Multilevel optimization: Algorithms and applications. Kluwer, Dordrecht
14. Vicente LN, Calamai PH (1994) Bilevel and multilevel programming: A bibliography review. J Global Optim 5:291–306
15. Vicente L, Savard G, Júdice J (1994) Descent approaches for quadratic bilevel programming. J Optim Th Appl 81:379–399
16. Visweswaran V, Floudas CA, Ierapetritou MG, Pistikopoulos EN (1996) A decomposition-based global optimization approach for solving bilevel linear and quadratic programs. In: Floudas CA, Pardalos PM (eds) State of the Art in Global Optimization. Kluwer, Dordrecht, pp 139–162

Bilevel Programming: Implicit Function Approach
BP

STEPHAN DEMPE
Freiberg University Mining and Technol., Freiberg, Germany

MSC2000: 90C26, 90C31, 91A65

Article Outline

Keywords
Reformulation as a One-Level Problem
Properties of the Solution Function
Optimality Conditions
    Conditions Using the Directional Derivative of the Solution Function
    Conditions Using the Generalized Jacobian of the Solution Function

Solution Algorithms
    Descent Algorithms
    Bundle Algorithms
See also
References

Keywords

Bilevel programming problem; Stackelberg game; Implicit function approach; Strongly stable solution; Piecewise continuously differentiable function; Necessary optimality conditions; Sufficient optimality conditions; Solution algorithms

The bilevel programming problem is a hierarchical problem in the sense that its constraints are defined in part by a second parametric optimization problem. Let $\Psi(x)$ be the solution set of this second problem (the so-called lower level problem):

$$\Psi(x) := \operatorname*{Argmin}_{y} \{f(x, y):\ g(x, y) \leq 0\}, \quad (1)$$

where $f, g_i \in C^2(\mathbf{R}^n \times \mathbf{R}^m, \mathbf{R})$, $i = 1, \ldots, p$. Then, the *bilevel programming problem* is defined as

$$"\min_{x}"\{F(x, y):\ y \in \Psi(x), x \in X\} \quad (2)$$

with $F \in C^1(\mathbf{R}^n \times \mathbf{R}^m, \mathbf{R})$ and $X \subseteq \mathbf{R}^n$ is closed. Problem (2) is also called the *upper level problem*. The inclusion of equality constraints in the problem (1) is possible without difficulties. If inequalities and/or equations in both x and y appear in the problem (2), this problem becomes even more difficult since these constraints restrict the set $\Psi(x)$ after a solution y out of it has been chosen. This can make the selection of $y \in \Psi(x)$ a posteriori infeasible [6].

The bilevel programming problem can easily be interpreted in terms of *Stackelberg games* which are a special case of them widely used in economics. In Stackelberg games the inclusion of lower level constraints $g(x, y) \leq 0$ is replaced by $y \in Y$ where $Y \subseteq \mathbf{R}^m$ is a fixed closed set. Consider two decision makers which select their actions in an hierarchical manner. First the leader chooses $x \in X$ and announces his selection to the follower. Knowing the selection x the follower computes his response $y(x)$ on it by solving the problem (1). Now, the leader is able to evaluate the value of his initial choice by computing $F(x, y(x))$. Having full knowledge about the follower's responses $y(x)$ for all $x \in X$ the leader's task is it to minimize the function $G(x) := F(x, y(x))$ over the set X, i. e.to solve problem (2).

The bilevel programming problem has a large number of applications e. g.in economics, natural sciences, technology (cf. [17,25] and the references therein).

The quotation marks in (2) have been used to indicate that, due to minimization only with respect to x in the upper level problem (2), this problem is not well defined in the case that the lower level problem (1) has not a uniquely determined optimal solution for all values of x [6]. Minimization only with respect to x in (2) takes place in many applications of bilevel programming, e. g.in the cases when the lower level problem represents the reactions of the nature on the leader's actions. If $\Psi(x)$ does not reduce to a singleton for all parameter values $x \in X$, either an optimistic or a pessimistic approach has to be used to obtain a well defined auxiliary problem.

In the optimistic case, problem (2) is replaced by

$$\min_{x,y}\{F(x, y):\ y \in \Psi(x), x \in X\} \quad (3)$$

[6,11], where minimization is taken with respect to both x and y. The use of (3) instead of (2) means that the leader is able to influence the choice of the follower. If the leader is not able to force the follower to take that solution $y \in \Psi(x)$ which is the best possible for him, he has to bound the damage resulting from an unwelcome choice of the follower. Hence, the leader has to take the worst solution in $\Psi(x)$ into account for computing his decision. This leads to the auxiliary problem in the pessimistic case:

$$\min_{x}\left\{\max_{y}\{F(x, y):\ y \in \Psi(x)\}:\ x \in X\right\} \quad (4)$$

[15,16].

In the sequel it is assumed that the lower level problem (1) has a unique (global) optimal solution $y(x)$ for all $x \in X$. This is guaranteed to be true at least if the assumptions C), SCQ), and SSOC) below are satisfied. Then, the *implicit function approach to bilevel programming* can be used which means that problem (2) (and equivalently (3)) is replaced by

$$\min_{x}\{G(x) := F(x, y(x)):\ x \in X\}. \quad (5)$$

C) The functions $f(x, \cdot), g_i(x, \cdot): \mathbf{R}^m \to \mathbf{R}$ are convex in y for each $x \in X$.

SCQ) For each $x \in X$ there exists a point $\widetilde{y}(x)$ such that $g(x, \widetilde{y}(x)) < 0$.

For convex problems, *Slater's condition* SCQ) implies that a feasible point $\overline{y}(x)$ to (1) is optimal if and only if the *Karush–Kuhn–Tucker conditions* for this problem are valid: There exists a point $\lambda \in \Lambda(x, \overline{y}(x))$, where

$$\Lambda(x, \overline{y}(x)) = \{\lambda \geq 0 : \nabla_y L(x, \overline{y}(x)) = 0, \lambda^\top g(x, \overline{y}(x)) = 0\} \quad (6)$$

with $L(x, y) = f(x, y) + \lambda^\top g(x, y)$ denoting the Lagrange function of the problem (1).

Reformulation as a One-Level Problem

There are several methods to reformulate (3) as an equivalent one-level problem.

The first possibility consists in replacing the lower level problem (1) by its Karush–Kuhn–Tucker conditions (6):

$$\min_{x,y} \left\{ F(x, y) : \begin{array}{l} \nabla_y L(x, y) = 0, \\ \lambda^\top g(x, y) = 0, \\ g(x, y) \leq 0, \\ \lambda \geq 0, x \in X \end{array} \right\}. \quad (7)$$

This is an optimization problem with constraints given in part by a parametric complementarity condition.

A second possibility is to use a variational inequality describing the set $\Psi(x)$. Let assumption C) be satisfied. Then, the problem (3) is equivalent to

$$\min_{x,y} \left\{ F(x, y) : \begin{array}{l} g(x, y) \leq 0, x \in X, \\ \nabla f(x, y)(z - y) \geq 0 \\ \forall z : g(x, z) \leq 0 \end{array} \right\}. \quad (8)$$

Both approaches (7) and (8) lead to a so-called *mathematical program with equilibrium constraints* (MPEC) [17].

SSOC) For each $x \in X$, for each $y \in \Psi(x)$, for all $\lambda \in \Lambda(x, y)$ and for all $d \neq 0$ satisfying

$$\nabla_y g_i(x, y) d = 0 \text{ for all } i: \lambda_i > 0,$$

the following inequality holds:

$$d^\top \nabla^2_{yy} L(x, y, \lambda) d > 0.$$

If at an optimal solution $y(\overline{x})$ of the convex problem (1) at $x = \overline{x}$ the assumptions SCQ) and SSOC) are satisfied, then $y(\overline{x})$ is a *strongly stable optimal solution* in the sense of M. Kojima [13]. This means that there exists an open neighborhood U of $\overline{x}$ and a uniquely determined continuous function $y: U \to \mathbf{R}^m$ such that $y(x)$ is the uniquely determined optimal solution of (1) for all $x \in U$. Hence, for convex problems (1), the assumptions SCQ) and SSOC) imply that there is a uniquely determined implicit function $y(x)$ describing the unique optimal solution of the problem (1) for all $x \in X$. This function can be inserted into the problem (2) which results in the third equivalent one-level problem (5). Problem (5) consists in minimizing the implicitly determined, generally nonsmooth, nonconvex objective function $F(x, y(x))$ on the set X. It has an optimal solution if the set X is compact or the function $F(\cdot, \cdot)$ satisfies some coercivity assumption [11].

Under suitable assumptions, the parametric complementarity problem as well as the parametric variational inequality describing the constraints in a mathematical program with equilibrium constraints also possess a uniquely determined continuous solution function [17]. Then, the implicit function approach can also be used to investigate MPECs.

Properties of the Solution Function

For the investigation of bilevel programming problems via (5) the knowledge of properties of the solution function $y: X \to \mathbf{R}^m$ is needed. If the assumptions C), SCQ), and SSOC) are satisfied, this function is continuous [13], upper Lipschitz continuous [22], Hölder continuous with exponent 1/2 [9] and directionally differentiable [3,24]. Let $\overline{z} = (\overline{x}, y(\overline{x}))$, $\overline{I} := \{j : g_j(\overline{z}) = 0\}$, $J(\lambda) := \{j : \lambda_j > 0\}$. The directional derivative

$$y'(\overline{x}; r) = \lim_{t \to 0+} t^{-1}[y(\overline{x} + tr) - y(\overline{x})]$$

of the function $y(\cdot)$ at a point $\overline{x}$ can be computed as the unique optimal solution $y'(\overline{x}; r)$ of the convex quadratic problem

$$\frac{1}{2} d^\top \nabla^2_{yy} L(\overline{z}, \overline{\lambda}) d + d^\top \nabla^2_{xy} L(\overline{z}, \overline{\lambda}) r \to \min_d,$$
$$\nabla_x g_i(\overline{z}) r + \nabla_y g_i(\overline{z}) d = 0, \quad \forall i \in J(\overline{\lambda}), \quad (9)$$
$$\nabla_x g_i(\overline{z}) r + \nabla_y g_i(\overline{z}) d \leq 0, \quad \forall i \in \overline{I} \setminus J(\overline{\lambda}),$$

for some suitably chosen Lagrange multiplier

$$\bar{\lambda} \in \underset{\lambda}{\mathrm{Argmax}} \{\nabla_x L(\bar{z}, \lambda) r : \lambda \in \Lambda(\bar{z})\} \qquad (10)$$

[3]. The correct choice of $\bar{\lambda}$ is a rather difficult task since it possibly belongs to the relative interior of some facet of the polyhedral set $\Lambda(\bar{z})$ [3]. For making the application of these properties of the solution function easier, a further assumption is used:

CR) For each pair $(\bar{x}, \bar{y}), \bar{x} \in X, \bar{y} \in \Psi(\bar{x})$, there is an open neighborhood $V \subseteq \mathbf{R}^n \times \mathbf{R}^m$ of $(\bar{x}, \bar{y})$ such that, for all $I \subseteq \bar{I}$, the family of gradients $\{\nabla_y g_i(x, y) : i \in I\}$ has constant rank on V.

If the assumptions C), SCQ), SSOC), and CR) are satisfied, the function $y : X \to \mathbf{R}^m$ is a *piecewise continuously differentiable function* [21], i.e. it is continuous and there exist an open neighborhood U of $\bar{x}$ and a finite number of continuously differentiable functions y^i $U \to \mathbf{R}^m$, $i = 1, \ldots, k$, such that $y(\cdot)$ is a selection of the y^i:

$$y(x) \in \{y^i(x) : i = 1, \ldots, k\}, \quad \forall\, x \in U.$$

The functions $y^i : U \to \mathbf{R}^m$ describe locally optimal solutions of auxiliary problems

$$\min_y \{f(x, y) : g_j(x, y) = 0, j \in I_i\},$$

where the sets I_i, $i = 1, \ldots, k$, satisfy the following two conditions:

- there exists a vertex $\bar{\lambda} \in \Lambda(\bar{x}, y(\bar{x}))$ such that $J(\bar{\lambda}) \subseteq I_i \subseteq \bar{I}$; and
- the gradients $\{\nabla_y g_j(\bar{x}, y(\bar{x})) : j \in I_i\}$ are linearly independent [14].

Let $IS(\bar{x})$ denote the family of all sets I_i having these two properties. Then, k is the cardinality of $IS(\bar{x})$. The functions $y^i : U \to \mathbf{R}^m$ are continuously differentiable at $\bar{x}$ [7]. For the computation of the Jacobian of the function $y^i(\cdot)$ at $x = \bar{x}$ the unique solution of a system of linear equations is to be computed.

Moreover, the directional derivative $y'(\bar{x}; r)$ is equal to the unique optimal solution of the quadratic problem (9) for each optimal solution $\bar{\lambda}$ of the linear problem (10) [21]. For fixed $\bar{x}$, it is a continuous, piecewise linear function of the direction r. The quadratic problem (9) has an optimal solution if and only if $\bar{\lambda}$ solves the linear problem (10). Hence, for computing a linear approximation of the function $y : X \to \mathbf{R}^m$ it is sufficient to solve the parametric quadratic optimization problems (9) for all vertices $\bar{\lambda} \in \Lambda(\bar{x}, y(\bar{x}))$.

Piecewise continuously differentiable functions are locally Lipschitz continuous [10]. The generalized Jacobian [1] of the function $y(\cdot)$ satisfies

$$\partial y(\bar{x}) \subseteq \mathrm{conv}\, \{\nabla y^i(\bar{x}) : i = 1, \ldots, k\} \qquad (11)$$

[14]. Let $g_I(z) = (g_i(z))_{i \in I}$. If the assumption

FRR) For each $x \in X$, for each vertex $\bar{\lambda} \in \Lambda(\bar{z})$ with $\bar{z} = (\bar{x}, y(\bar{x}))$, the matrix

$$\begin{pmatrix} \nabla^2_{yy} L(\bar{z}, \bar{\lambda}) & \nabla^\top_y g_{J(\bar{\lambda})}(\bar{z}) & \nabla^2_{xy} L(\bar{z}, \bar{\lambda}) \\ \nabla_y g_{\bar{I}}(\bar{z}) & 0 & \nabla_x g_{\bar{I}}(\bar{z}) \end{pmatrix}$$

has full row rank

is added to C), SCQ), SSOC), and CR), then equality holds in (11) [5].

Optimality Conditions

Even under very restrictive assumptions, problem (5) is a nondifferentiable, nonconvex optimization problem. For the derivation of necessary and sufficient optimality conditions, various approaches of nondifferentiable optimization can be used.

Conditions Using the Directional Derivative of the Solution Function

Let $X = \{x : h_k(x) \leq 0, k \in K\}$, where $h_k \in C^1(\mathbf{R}^n, \mathbf{R})$, $k \in K$ and K is a finite set. Generalizations of the following results to larger classes of constraint sets are obvious. Let $\bar{x} \in X$, $y(\bar{x}) \in \Psi(\bar{x})$, $\bar{z} = (\bar{x}, y(\bar{x}))$. Let the assumptions C), SCQ), SSOC), and CR) as well as

MFCQ) There exists a direction d such that
$\nabla h_k(\bar{x}) d < 0$ for all $k \in \bar{K} := \{l : h_l(\bar{x}) = 0\}$

be valid. Then, if $\bar{x}$ is a locally optimal solution of the problem (5) (and thus of the bilevel problem (2)), there cannot exist a feasible direction of descent, i.e.

$$\nabla_x F(\bar{z}) r + \nabla_y F(\bar{z}) y'(\bar{x}; r) \geq 0 \qquad (12)$$

for all directions r satisfying $\nabla h_k(\bar{x}) \leq 0, k \in \bar{K}$. By use of the above approach for computing the directional derivative of the solution function $y(\cdot)$, the verification of this *necessary optimality condition* can be done by solving a bilevel optimization problem of minimizing the function (12) subject to the condition that $y'(\bar{x}; r)$

is an optimal solution of the problem (9). By replacing problem (9) with its Karush–Kuhn–Tucker conditions and applying an active index set strategy the following condition is obtained: If $\bar{x}$ is a locally optimal solution of the problem (2) then

$$v := \min\{\varphi(\bar{x}, I): I \in IS(\bar{x})\} \geq 0, \qquad (13)$$

where $\varphi(\bar{x}, I)$ denotes the optimal objective function value of the problem

$$\nabla_x F(\bar{z})r + \nabla_y F(\bar{z})d \to \min_{d,r,\alpha},$$
$$\nabla_x h_k(\bar{x})r \leq 0, \quad k \in \bar{K},$$
$$\nabla^2_{xy} L(\bar{z}, \bar{\lambda})r + \nabla^2_{yy} L(\bar{z}, \bar{\lambda})d + \nabla^\top_y g_I(\bar{z})\alpha = 0,$$
$$\nabla_x g_i(\bar{z})r + \nabla_y g_i(\bar{z})d = 0, \quad i \in I,$$
$$\nabla_x g_i(\bar{z})r + \nabla_y g_i(\bar{z})d \leq 0, \quad i \in \bar{I} \setminus I,$$
$$\alpha_i \geq 0, \quad i \in I \setminus J(\bar{\lambda}), \quad \|r\| = 1,$$

and $\bar{\lambda}$ is the unique vertex of $\Lambda(\bar{z})$ with $J(\bar{\lambda}) \subseteq I$ [2]. Problem (13) is a combinatorial optimization problem and can be solved by enumeration algorithms.

In [2] a more general necessary optimality condition is given even without assuming CR). Then, the directional derivative of the solution function is in general discontinuous with respect to perturbations of the direction and is to be replaced by the contingent derivative of the solution function.

In [17] it is shown that nonexistence of directions of descent in the tangent cone to the feasible set is also a necessary optimality condition for MPECs. In general, this tangent cone is not convex. Using a so-called basic constraint qualification it is shown that it is equal to the union of a finite number of polyhedral cones. The resulting condition is similar to (13). Dualizing this condition, some kind of a Karush–Kuhn–Tucker condition for MPECs is obtained.

It is also possible to obtain a *sufficient optimality condition* by use of the directional derivative. Namely, if for the optimal function value in (13) the strict inequality $v > 0$ holds then, for each $c \in (0, v)$, there exists $\varepsilon > 0$ such that

$$F(x, y(x)) \geq F(\bar{x}, y(\bar{x})) + c \|x - \bar{x}\|$$

for all x satisfying $h(x) \leq 0$ and $\|x - \bar{x}\| \leq \varepsilon$ [2]. Necessary and sufficient optimality conditions of second order based on the implicit function approach (applied to the more general MPEC formulation) are given in [17].

Conditions Using the Generalized Jacobian of the Solution Function

By [1], the generalized differential of the function $G(x) := F(x, y(x))$ is equal to

$$\partial G(\bar{x}) = \text{conv}\{\nabla_x F(\bar{z}) + \nabla_y F(\bar{z})\omega: \omega \in \partial y(\bar{x})\}, \qquad (14)$$

provided that the conditions C), SCQ), SSOC), and CR) are satisfied. Hence, the application of the necessary optimality conditions from Lipschitz optimization to problem (5) leads to necessary optimality conditions for the bilevel problem (2). Thus, if $\bar{x}$ is a locally optimal solution of the problem (2) and the assumptions C), SCQ), SSOC), CR), and MFCQ) are satisfied, then there exist Lagrange multipliers $\gamma_i \geq 0$, $i \in \bar{K}$, such that

$$0 \in \partial G(\bar{x}) + \sum_{i \in \bar{K}} \gamma_i \{\nabla h_i(\bar{x})\}.$$

This is an obvious generalization of the necessary optimality condition given in [4], where no upper level constraints in (2) appeared, and is also a special case of the results in [19], where the general constraint set $x \in X$ in the upper level problem (2) together with more restrictive assumptions for the lower level problem are used. For the use of this necessary optimality condition in computations the explicit description of the generalized Jacobian in (11) (with equality instead of inclusion) is needed.

Solution Algorithms

The implicit function approach leads to the problem (5) of minimizing a nondifferentiable, nonconvex, implicitly determined function on a fixed set. Any algorithm solving nonsmooth optimization problems can be applied to this problem. Due to the structure of (5) the computation of function values and derivative information for the objective function is expensive. Two types of algorithms are proposed: descent and bundle algorithms. The convergence proofs show that the algorithms converge to points where the above optimality conditions are satisfied, i.e. to solutions where no descent direction exists respectively to Clarke stationary points.

Descent Algorithms

Let

$$X = \{x: h_k(x) \leq 0, \ k \in K\}.$$

Descent algorithms are iterative methods which compute a sequence of feasible points $\{x^i\}_{i \in \mathbb{N}}$ by $x^{i+1} = x^i + t_i r^i$, $\forall i$, where r^i is a feasible direction of descent and t_i is a stepsize. For bilevel problems a feasible direction of descent is obtained by minimizing the function (12)

$$\nabla_x F(\bar{z})r + \nabla_y F(\bar{z}) y'(\bar{x}; r)$$

subject to r being an inner direction of the cone of feasible directions to X:

$$\min_{\alpha, r} \{\alpha: \nabla_x F(\bar{z})r + \nabla_y F(\bar{z}) y'(\bar{x}; r) \leq \alpha,$$
$$\nabla h_i(\bar{x}) r \leq \alpha, \quad i \in \overline{K}, \quad \|r\| \leq 1\}.$$

Inserting the Karush–Kuhn–Tucker conditions of the quadratic optimization problem (9) for the computation of $y'(\bar{x}; r)$ and again using an active set strategy this problem is converted into an equivalent combinatorial optimization problem. For the computation of a stepsize, e. g., Armijo's rule can be applied. Such an algorithm is described in [6,8,17]. In [6] it is also investigated how this idea can be generalized to the case when the lower level problem (1) is not assumed to have a uniquely determined optimal solution for all values of the parameter. In [17] this approach is applied to the more general MPEC.

Bundle Algorithms

Let $X = \mathbf{R}^n$. Different constraint sets can be treated by use of approaches in [12]. As in descent algorithms, in *bundle algorithms* for minimizing Lipschitz nonconvex functions a sequence of iterates $\{x^i\}_{i \in \mathbb{N}}$ with $x^{i+1} = x^i + t_i r^i$, $\forall i$, is computed. For computing a direction a model of the function to be minimized is used. In the paper [23], the following bundle algorithm has been proposed. Let two sequences of points $\{x^i\}_{i=1}^{k}, \{z^i\}_{i=1}^{k}$ have already been computed. Then, for minimizing a nonconvex function $G(x)$, this model has the form

$$\max_{1 \leq i \leq k} \{v(z^i)^\top d - \alpha_{k,i}\} + \frac{u^k d^\top d}{2}, \tag{15}$$

where

$$\alpha_{k,i} = \max \left\{ G(x^k) - v(z^i)^\top (x^k - z^i) - G(z^i), \right.$$
$$\left. c_0 \left\| x^k - z^i \right\| \right\},$$

$v(z^i)$ is a subgradient of the function $G(x)$ at $x = z^i$ and u^k is a weight. If the direction computed by minimizing the model function (15) realizes a sufficient decrease, a serious step is made (i. e. $t_k = 1$ is used). Otherwise, either a short step (which means that t_k is computed according to a stepsize rule) or a null step (only the model is updated by computing a new subgradient) is made. For updating the model (15), in each iteration of the bundle algorithm a subgradient of the objective function is needed. For its computation formula (14) can be used.

The bundle algorithm is applied to problem (5) in [4,18,20]. In [4], the lower level problem is not assumed to have a uniquely determined optimal solution for all parameter values. The Lipschitz optimization problem (5) is obtained via a regularization approach in the lower level problem (1).

Numerical experience for solving bilevel problems (in the formulation (2) as well as in the more general MPEC formulation) with the bundle algorithm is reported in [18,20].

See also

- ▶ Bilevel Fractional Programming
- ▶ Bilevel Linear Programming
- ▶ Bilevel Linear Programming: Complexity, Equivalence to Minmax, Concave Programs
- ▶ Bilevel Optimization: Feasibility Test and Flexibility Index
- ▶ Bilevel Programming
- ▶ Bilevel Programming: Applications
- ▶ Bilevel Programming: Applications in Engineering
- ▶ Bilevel Programming: Introduction, History and Overview
- ▶ Bilevel Programming in Management
- ▶ Bilevel Programming: Optimality Conditions and Duality
- ▶ Multilevel Methods for Optimal Design
- ▶ Multilevel Optimization in Mechanics
- ▶ Stochastic Bilevel Programs

References

1. Clarke FH (1983) Optimization and nonsmooth analysis. Wiley, New York
2. Dempe S (1992) A necessary and a sufficient optimality condition for bilevel programming problems. Optim 25:341–354
3. Dempe S (1993) Directional differentiability of optimal solutions under Slater's condition. Math Program 59:49–69
4. Dempe S (1997) An implicit function approach to bilevel programming problems. In: Migdalas A, Pardalos PM, Värbrand P (eds) Multilevel Optimization: Algorithms, Complexity and Applications. Kluwer, Dordrecht
5. Dempe S, Pallaschke D (1997) Quasidifferentiability of optimal solutions in parametric nonlinear optimization. Optim 40:1–24
6. Dempe S, Schmidt H (1996) On an algorithm solving two-level programming problems with nonunique lower level solutions. Comput Optim Appl 6:227–249
7. Fiacco AV, McCormic GP (1968) Nonlinear programming: Sequential unconstrained minimization techniques. Wiley, New York
8. Gauvin J, Savard G (1994) The steepest descent direction for the nonlinear bilevel programming problem. Oper Res Lett 15:265–272
9. Gfrerer H (1987) Hölder continuity of solutions of perturbed optimization problems under Mangasarian-Fromowitz constraint qualification. In: Guddat J et al (eds) Parametric Optimization and Related Topics. Akad Verlag, Berlin, pp 113–127
10. Hager WW (1979) Lipschitz continuity for constrained processes. SIAM J Control Optim 17:321–328
11. Harker PT, Pang J-S (1988) Existence of optimal solutions to mathematical programs with equilibrium constraints. Oper Res Lett 7:61–64
12. Kiwiel KC (1985) Methods of descent for nondifferentiable optimization. Springer, Berlin
13. Kojima M (1980) Strongly stable stationary solutions in nonlinear programs. In: Robinson SM (ed) Analysis and Computation of Fixed Points. Acad Press, New York pp 93–138
14. Kummer B (1988) Newton's method for non-differentiable functions. Adv Math Optim, In: Math Res, vol 45. Akad Verlag, Berlin
15. Loridan P, Morgan J (1989) ϵ-regularized two-level optimization problems: Approximation and existence results. In: Optimization: Fifth French-German Conf (Varez), In: Lecture Notes Math, vol 1405. Springer, Berlin, pp 99–113
16. Lucchetti R, Mignanego F, Pieri G (1987) Existence theorem of equilibrium points in Stackelberg games with constraints. Optim 18:857–866
17. Luo Z-Q, Pang J-S, Ralph D (1996) Mathematical programs with equilibrium constraints. Cambridge Univ Press, Cambridge
18. Outrata J (1990) On the numerical solution of a class of Stackelberg problems. ZOR: Methods and Models of Oper Res 34:255–277
19. Outrata JV (1993) Necessary optimality conditions for Stackelberg problems. J Optim Th Appl 76:305–320
20. Outrata J, Zowe J (1995) A numerical approach to optimization problems with variational inequality constraints. Math Program 68:105–130
21. Ralph D, Dempe S (1995) Directional derivatives of the solution of a parametric nonlinear program. Math Program 70:159–172
22. Robinson SM (1982) Generalized equations and their solutions, Part II: Applications to nonlinear programming. Math Program Stud 19:200–221
23. Schramm H, Zowe J (1992) A version of the bundle idea for minimizing a nonsmooth function: conceptual idea, convergence analysis, numerical results. SIAM J Optim 2:121–152
24. Shapiro A (1988) Sensitivity analysis of nonlinear programs and differentiability properties of metric projections. SIAM J Control Optim 26:628–645
25. Vicente LN, Calamai PH (1994) Bilevel and multilevel programming: A bibliography review. J Global Optim 5(3)

Bilevel Programming: Introduction, History and Overview
BP

Luis N. Vicente
Department Mat., University de Coimbra, Coimbra, Portugal

MSC2000: 90C26, 90C30, 90C31

Article Outline

Keywords
See also
References

Keywords

Bilevel programming; Multilevel programming; Hierarchical optimization; Nondifferentiable optimization; Game theory; Stackelberg problems

The *bilevel programming* (BP) problem is a hierarchical optimization problem where a subset of the variables is constrained to be a solution of a given optimization problem parameterized by the remaining variables. The BP problem is a multilevel programming problem with two levels. The hierarchical optimization structure appears naturally in many applications when lower level actions depend on upper level decisions. The applications of bilevel and multilevel programming include *transportation* (taxation, network design, trip demand estimation), *management* (coordination of multidivisional firms, network facility location, credit allocation), *planning* (agricultural policies, electric utility), and *optimal design*.

In mathematical terms, the BP problem consists of finding a solution to the upper level problem

$$\begin{cases} \min_{x,y} & F(x, y) \\ \text{s.t.} & g(x, y) \leq 0, \end{cases}$$

where y, for each value of x, is the solution of the lower level problem:

$$\begin{cases} \min_{y} & f(x, y) \\ \text{s.t.} & h(x, y) \leq 0, \end{cases}$$

with $x \in \mathbf{R}^{nx}$, $y \in \mathbf{R}^{ny}$, $F, f : \mathbf{R}^{nx+ny} \to \mathbf{R}$, $g : \mathbf{R}^{nx+ny} \to \mathbf{R}^{nu}$, and $h : \mathbf{R}^{nx+ny} \to \mathbf{R}^{nl}$ (nx, ny, nu, and nl are positive integers). The lower level problem is also referred as the follower's problem or the inner problem. In a similar way, the upper level problem is also called the leader's problem or the outer problem. One could generalize the BP problem in different ways. For instance, if either x or y or both are restricted to take integer values we would obtain an integer BP problem [22]. Or, if we replace the lower level problem by a variational inequality we would get a generalized BP problem [15].

For each value of the upper level variables x, the lower level constraints $h(x, y) \leq 0$ define the constraint set $\Omega(x)$ of the lower level problem:

$$\Omega(x) = \{y : h(x, y) \leq 0\}.$$

Then, the set $M(x)$ of solutions for the lower level problem is given by minimizing the lower level function $f(x, y)$ for all values in $\Omega(x)$ of the lower level variables y:

$$M(x) = \{y : y \in \text{argmin}\{f(x, y) : y \in \Omega(x)\}\}.$$

Given these definitions the BP problem can be reformulated as:

$$\begin{cases} \min_{x,y} & F(x, y) \\ \text{s.t.} & g(x, y) \leq 0, \\ & y \in M(x). \end{cases}$$

The feasible set

$$\{(x, y) : g(x, y) \leq 0, \ y \in M(x)\}$$

of the BP problem is called the induced or inducible region. The induced region is usually nonconvex and, in the presence of upper level constraints, can be disconnected or even empty. In fact, consider the following BP problem

$$\begin{cases} \min_{x,y} & x - 2y \\ \text{s.t.} & -x + 3y - 4 \leq 0, \end{cases}$$

where y, for each value of x, is the solution of:

$$\begin{cases} \min_{y} & x + y \\ \text{s.t.} & x - y \leq 0, \\ & -x - y \leq 0. \end{cases}$$

For this problem we have:

$$\Omega(x) = \{y : y \geq |x|\}$$

and

$$M(x) = |x|.$$

Thus, the induced region is given by:

$$\{(x, y) : -x + 3y - 4 \leq 0, \ y \in M(x)\}$$
$$= \{(x, y) : y = -x, \ -1 \leq x \leq 0\}$$
$$\cup \{(x, y) : y = x, \ 0 \leq x \leq 2\},$$

which is nonconvex but connected. If the upper level constraints were changed to

$$-x + 3y - 4 \leq 0,$$
$$-y + \frac{1}{2} \leq 0,$$

then the induced region would become

$$\left\{(x, y) : y = -x, \ -1 \leq x \leq -\frac{1}{2}\right\}$$
$$\cup \left\{(x, y) : y = x, \ \frac{1}{2} \leq x \leq 2\right\},$$

which would be a disconnected set. In either case the BP problem has two local minimizers $(-1, 1)$ and $(2, 2)$ and one global minimizer $(-1, 1)$.

This simple example illustrates many features of bilevel programming like the nonconvexity and the disconnectedness of the induced region and the existence of different local minimizers. In this example the induced region is compact. In fact, compactness of the induced region is important for the existence of a global minimizer and can be guaranteed under appropriate conditions [9].

The original formulation for bilevel programming appeared in 1973, in a paper authored by J. Bracken and J. McGill [5], although it was W. Candler and R. Norton [7] who first used the designation bilevel and multilevel programming. However, it was not until the early 1980s that these problems started receiving the attention they deserve. Motivated by the game theory of H. Stackelberg [20], several authors studied bilevel programming intensively and contributed to its proliferation in the mathematical programming community.

The theory of bilevel programming focuses on forms of optimality conditions and complexity results. A number of authors ([8,16], just to cite a few) have established original forms of optimality conditions for bilevel programming by either considering reformulations of the BP problem or by making use of nondifferentiable optimization concepts or even by appealing to the geometry of the induced region. The complexity of the problem has been addressed by a number of authors. It has been proved that even the linear BP problem, where all the involved functions are affine, is a strongly NP-hard problem [10]. It is not hard to construct a linear BP problem where the number of local minima grows exponentially with the number of variables [6]. Other theoretical results of interest have been established connecting bilevel programming to other fields in mathematical programming. For instance, one can show that minimax problems and linear, integer, bilinear and quadratic programming problems are special cases of BP. Other classes of problems different from but related to BP are multi-objective optimization problems and static Stackelberg problems. See [21] for references in these topics.

Many researchers have designed algorithms for the solution of the BP problem. One class of techniques consists of extreme point algorithms and has been mostly applied to the linear BP problem because for this problem, if there is a solution, then there is at least one global minimizer that is an extreme point of Ω [17]. Two other classes of algorithms are branch and bound algorithms and complementarity pivot algorithms that have in common the fact that exploit the complementarity part of the necessary optimality conditions of the lower level problem (assumed convex in y so that the necessary optimality conditions, under an appropriate constraint qualification, are also sufficient). These two classes of algorithms have been applied mostly to the case where the upper level is linear and the lower level is linear or convex quadratic (see for instance [10] and [12]) and, as the extreme point algorithms, find a global minimizer of the BP problem. On the other hand, the algorithms designed to solve nonlinear forms of BP appeal to descent directions (see, among others [14] and [18]) and penalty functions (for instance [1]) and are expected to find a local minimizer.

For additional material about bilevel programming, see the books [3,19], the survey papers [2,4,11,13,23], and the bibliography review [21].

See also

- ▶ Bilevel Fractional Programming
- ▶ Bilevel Linear Programming
- ▶ Bilevel Linear Programming: Complexity, Equivalence to Minmax, Concave Programs
- ▶ Bilevel Optimization: Feasibility Test and Flexibility Index
- ▶ Bilevel Programming
- ▶ Bilevel Programming: Applications
- ▶ Bilevel Programming: Applications in Engineering
- ▶ Bilevel Programming: Implicit Function Approach
- ▶ Bilevel Programming in Management
- ▶ Bilevel Programming: Optimality Conditions and Duality
- ▶ Multilevel Methods for Optimal Design
- ▶ Multilevel Optimization in Mechanics
- ▶ Stochastic Bilevel Programs

References

1. Aiyoshi E, Shimizu K (1984) A solution method for the static constrained Stackelberg problem via penalty method. IEEE Trans Autom Control 29:1111–1114

2. Anandalingam G, Friesz T (1992) Hierarchical optimization: An introduction. Ann Oper Res 34:1–11
3. Bard JF (1998) Practical bilevel optimization: Algorithms and applications. Kluwer, Dordrecht
4. Ben-Ayed O (1993) Bilevel linear programming. Comput Oper Res 20:485–501
5. Bracken J, McGill J (1973) Mathematical programs with optimization problems in the constraints. Oper Res 21:37–44
6. Calamai P, Vicente LN (1993) Generating linear and linear-quadratic bilevel programming problems. SIAM J Sci Statist Comput 14:770–782
7. Candler W, Norton R (1977) Multilevel programming. Techn Report World Bank Developm Res Center, Washington DC 20
8. Dempe S (1992) A necessary and a sufficient optimality condition for bilevel programming problems. Optim 25:341–354
9. Edmunds T, Bard J (1991) Algorithms for nonlinear bilevel mathematical programming. IEEE Trans Syst, Man Cybern 21:83–89
10. Hansen P, Jaumard B, Savard G (1992) New branch-and-bound rules for linear bilevel programming. SIAM J Sci Statist Comput 13:1194–1217
11. Hsu S, Wen U (1989) A review of linear bilevel programming problems. Proc Nat Sci Council Report China, Part A: Physical Sci Eng 13:53–61
12. Júdice J, Faustino A (1994) The linear-quadratic bilevel programming problem. INFOR 32:87–98
13. Kolstad C (1985) A review of the literature on bi-level mathematical programming. Techn Report Los Alamos Nat Lab LA-10284-MS/US-32
14. Kolstad C, Lasdon L (1990) Derivative evaluation and computational experience with large bilevel mathematical programs. J Optim Th Appl 65:485–499
15. Marcotte P, Zhu D (1996) Exact and inexact penalty methods for the generalized bilevel programming problem. Math Program 74:142–157
16. Outrata J (1994) On optimization problems with variational inequality constraints. SIAM J Optim 4:340–357
17. Savard G (1989) Contributions à la programmation mathématique à deux niveaux. PhD Thesis Ecole Polytechn. Univ. Montréal
18. Savard G, Gauvin J (1994) The steepest descent direction for the nonlinear bilevel programming problem. Oper Res Lett 15:275–282
19. Shimizu K, Ishizuka Y, Bard JF (1997) Nondifferentiable and two-level mathematical programming. Kluwer, Dordrecht
20. Stackelberg H (1952) The theory of the market economy. Oxford Univ. Press, Oxford
21. Vicente LN, Calamai PH (1994) Bilevel and multilevel programming: A bibliography review. Jogo 5:291–306
22. Vicente LN, Savard G, Júdice J (1996) The discrete linear bilevel programming problem. J Optim Th Appl 89:597–614
23. Wen U, Hsu S (1991) Linear bi-level programming problems: A review. J Oper Res Soc 42:125–133

Bilevel Programming in Management

JONATHAN F. BARD
University Texas, Austin, USA

MSC2000: 90-01, 91B52, 91B74, 91B32, 90B30, 90B50

Article Outline

Keywords
Multilevel Model
Applications
Solutions
See also
References

Keywords

Bilevel programming; Hierarchical optimization; Stackelberg game; Production planning; Government regulation; Applications; Management

Decision-making in large, hierarchical organizations rarely proceeds form a single point of view. Two of the most prominent aspects of such organizations are specialization closely followed by coordination. The former arises from a practical need to isolate individual jobs or operations and to assign them to specialized units. This leads to departmentalization; however, to accomplish the overall task, the specialized units must be coordinated. The related process divides itself naturally into two parts:

i) the establishment of individual goals and operating rules for each unit; and
ii) the enforcement of these rules within the work environment.

The first deals with the selection of appropriate divisional or lower level performance criteria and, more generally, the selection of the modes of coordination and control. The second relates to the choice of coordination inputs.

An important control variable in the theory of departmentalization is the degree of self-containment of the organization units. A unit is self-contained to the extent and degree that the conditions of carrying out its activities are independent of what is done elsewhere in the system. The corporate or higher level unit is then faced with the coordination problem of favorably resolving the divisional unit interactions. Mathematical programming has often been used as the basis for modeling these interactions with decomposition techniques providing solutions to problems of large scale (see, e. g., [9]). The central idea underlying decomposition techniques is very simple and can be envisioned as the following algorithmic process: top management, with its set of goals, asks each division of the company to calculate and submit an optimal production plan as though it were operating in isolation. Once the plans are submitted, they are modified with the overall benefit of the company in mind. Marginal profit figures are used to successively reformulate the divisional plans at each stage in the algorithm. An output plan ultimately emerges which is optimal for the company as a whole and which therefore represents the solution to the original programming problem.

Although this procedure attempts to mimic corporate behavior it fails on two counts. The first relates to the assumption that it is possible to derive a single objective or *utility function* which adequately captures the goals of both top management and each subordinate division. The second stems from lack of communications among the components of the organization; at an intermediary stage of the calculations there is no guarantee that each division's plan will satisfy the corporate constraints. In particular, if the production of some output by division k imposes burdens on other divisions by using up a scarce company resource, or by causing an upward shift in the cost functions pertaining to some other company operation, division k's calculation is likely to lead it to overproduce this item from the point of view of the company because the costs to other divisions will not enter its accounts. This is the classical problem of external diseconomies. Similarly, if one of division k's outputs yields external economies where a rise in its production increases the profitability of other divisions, division k may (considering just its own gains in its calculations) not produce enough of this product to maximize the company's profits as a whole. This may result in a final solution that does not realistically reflect the production plan that probably would have been achieved had each division been given the degree of autonomy it exercises in practice.

Another way of treating the multilevel nature of the resource allocation problem is through goal programming. T. Ruefli [11] was the first to apply this technique by proposing a generalized goal decomposition model. Others expanded on his work developing models capable of representing a wide range of operational characteristics including informational autonomy, interdependent strategies, and *bounded rationality* or individual goals. Combinations of these models have been used to solve problems related to government regulation, distribution, and control [9]. In [3], J.F. Bard presents an approach that derives from the complementary strategies of two-stage optimization [7], ▶ Bilevel linear programming, [10] and *equilibrium analysis* [13]. Decision-making between levels is assumed to proceed sequentially but with some amount of independence to account for the divergence of corporate and subordinate objectives. At the divisional level each unit simultaneously attempts to maximize its own production function and, in so doing, produces a balance of opposing forces. An example based on an integrated paper company operating three divisions is given to illustrate the differences between centralized and decentralized control. The corporate unit has little direct control over divisional schedules but may set internal transfer prices which affect production capacity and profits.

Multilevel Model

A distinguishing characteristic of multilevel systems is that the decision maker at one level may be able to influence the behavior of a decision maker at another level but not completely control his actions. In addition, the objective functions of each unit may, in part, be determined by variables controlled by other units operating at parallel or subordinate levels. For example, policies affected by corporate management relating to resource allocation and benefits may curtail the set of strategies available to divisional management. In turn, polices adopted at the lower levels affecting productivity and marketing may play a role in determining overall profitability and growth. W.F. Bialas and M.H. Karwan [7] have noted the following common features of multilevel organizations:

1) interactive decision-making units exist within a predominately hierarchical structure;
2) each subordinate level executes its policies after, and in view of, decisions made at a superordinate level;
3) these extramural effects enter a decision maker's problem through his objective function and feasible strategy set.

The need for specialization and decentralization has traditionally been met by the establishment of profit centers. In this context, divisions or departments are viewed as more or less independent units charged with the responsibility of operating in the best possible manner so as to maximize profit under the given constraints imposed by top management. The problem of decentralization is essentially how to design and impose constraints on the department units so that the well-being of the overall corporation is assured. The traditional way to coordinate decentralized organizations is by means of the pricing mechanism; coordination is designed by analogy with the operation of a free market or competitive economy. Exchange of products between departments is allowed and internal prices are specified for the exchange commodities. The problem of effective decentralization reduces to the selection of the internal prices.

The framework presented in this article is an extension of the bilevel programming problem introduced in ▶ Bilevel linear programming, and embodies a corporate management unit at the higher level and M divisions or subordinate units at the lower level. The latter may be viewed as either separate operating divisions of an organization or coequal departments within a firm, such as production, finance, and sales. This structure can be extended beyond two levels (e. g., see [4,9]) with the realization that attending behavioral and operational relationships become much more difficult to conceptualize and describe.

To formulate the problem mathematically, suppose the higher level decision maker wishes to maximize his objective function F and each of the M divisions wishes to maximize its own objective function f^i. Control of the decision variables is partitioned among the units such that the higher level decision maker may select a vector $x^0 \in S^0 \subset \mathbf{R}^{n_0}$ and each lower level decision maker may select a vector $x^i \in S^i \subset \mathbf{R}^{n_i}$ $i = 1, \ldots, M$. Letting $x \equiv (x^1, \ldots, x^M)$ and $n = \sum_{i=1}^{M} n^i$, in the most general case we have $F, f_1, \ldots, f_M : \mathbf{R}^n \to \mathbf{R}^1$. It shall be assumed that the corporate unit has the first choice and selects a strategy $x^0 \in S^0$, followed by the M subordinate units who select their strategies $x^i \in S^i$, simultaneously. In addition, the choice made at the higher level may affect the set of feasible strategies available at the lower level, while each lower-level decision maker may influence the choices available to his peers. The strategies sets will be given explicitly by

$$S^0 = \{x^0 : g(x^0, \overline{x}^0) \leq 0\},$$
$$S^i = \{x^i : g(x^i, \overline{x}^i) \leq 0\}, i = 1, \ldots, M,$$

where $\overline{x}^i \equiv (x^0, x^1, \ldots, x^{i-1}, x^{i+1}, \ldots, x^M)$ and $g^i : \mathbf{R}^n \to \mathbf{R}^{m_i}$, $i = 1, \ldots, M$.

To assure that the problem is well posed, it is common to assume that all functions are twice continuously differentiable and that the sets S^i, $i = 0, \ldots, M$, are nonempty and compact; i. e., the ith unit always has some recourse. The bilevel multidivisional programming problem (BMPP) can now be defined:

$$\max_{x^0} \quad F(x^0, \overline{x}^0), \tag{1}$$

$$\text{s.t.} \quad g^0(x^0, \overline{x}^0) \leq 0, \tag{2}$$

$$\max_{x^i} \quad f^i(x^i, \overline{x}^i), \quad i = 1, \ldots, M, \tag{3}$$

$$\text{s.t.} \quad g^i(x^i, \overline{x}^i) \leq 0, \tag{4}$$

When $M = 0$, problem (1)–(4) reduces to a standard mathematical program; when $M = 1$ a *bilevel program* results; when (1) is removed an equilibrium programming problem remains [13]. A solution to the latter is often taken as an equilibrium point; call it $x_E = (x^i, \overline{x}^i)$, where x^i solves subproblem (3)–(4), $i = 1, \ldots, M$, for $\overline{x}^i$ given. Thus, x_E represents a point of stability. No incentive exists at x_E for any of the divisions to deviate from $\overline{x}^i$ because each has optimized its individual objective function. For the linear BMPP, results similar to those presented for the linear bilevel programming problem in ▶ Bilevel linear programming hold (see [3]).

Applications

Most applications of bilevel programming, including bilevel multidivisional programming, that have appeared in the literature have dealt with central economic planning at the regional or national level. In this context, the government is considered the leader and controls a set of policy variables such as tax rates, subsidies, import quotas, and price supports (e.g., see [5] and accompanying papers). The particular industry targeted for regulation is viewed as the follower. In most cases, the follower tries to maximize net income subject to the prevailing technological, economic, and governmental constraints. Possible leader objectives include maximizing employment, maximizing production of a given product, or minimizing the use of certain resources.

The early work of W. Candler and R. Norton [8], focusing on agricultural development in northern Mexico, illustrates how bilevel programming can be used to analyze the dynamics of a regulated economy. Similarly, J. Fortuny-Amat and B. McCarl [10] present a regional model that pits fertilizer suppliers against local farm communities, while E. Aiyoshi and K. Shimizu [1] and Bard [3] discuss resource allocation in a decentralized firm. In the case of the latter, a central unit supplies resources to its manufacturing facilities which make decisions concerning production mix and output. Organizational procedures and conflicting objectives over efficiency, quality and performance lead to a hierarchical formulation. In a work related to the original Stackelberg model of a single leader-follower oligopolistic market in which a few firms supply a homogeneous product, H.D. Sherali [12] presents an extension to N leader firms and discusses issues related to the existence, uniqueness, and derivation of equilibrium solutions. His analysis provides sufficient conditions for some useful convexity and differentiability properties of the followers' reaction curves.

In a recent study [5], the French government has used bilevel programming to examine the economics of promoting biofuel production from farm crops within the petro-chemical industry. The stumbling block to this policy is that industry's costs for producing fuels from hydrocarbon-based raw materials is significantly less than it is for producing biofuels. Without incentives in the form of tax credits, industry will not buy farm output for conversion. The problem faced by the government is to determine the level of tax credits for each final product or biofuel that industry can produce while minimizing public outlays. A secondary objective is to realize some predefined level of land usage for nonfood crops. Industry is assumed to be neutral in this scenario and will produce any biofuel that is profitable. In the model, the agricultural sector is represented by a subset of farms in an agriculturally intensive region of France and is a profit maximizer. It will use the land available for nonfood crops only as long as the revenue generated from this activity exceeds the difference between the set-aside payments now received directly from the government and the maintenance costs incurred under the current support program. The resultant bilevel model contains 3628 variables and 3230 constraints at the lower level, and 8 variables and 10 constraints at the upper level. Both objective functions are quadratic and all constraints are linear.

In an earlier effort, G. Anandalingam and V. Apprey [2] investigated the problem of conflict resolution by postulating the existence of an arbitrator who acts as the leader in a Stackelberg game. They presented models for different configurations of the resulting multilevel linear programs and proposed a series of solution algorithms. The models were illustrated with an application involving a water conflict problem between India and Bangladesh; it is shown that both parties could gain by the arbitration of an international agency such as the United Nations.

Recently, researchers have tried to apply bilevel models to the *network design problem* arising in transportation and telecommunications systems. In the accompanying formulation, a central planner controls investment costs at the system level, while operational costs depend on traffic flows which are determined by the individual user's route selection. Because users are assumed to make decisions so as to maximize their individual utility functions, their choices do not necessarily coincide (and may, in fact, conflict) with the choices that are optimal for the system. Nevertheless, the central planner can influence the users' choices by improving some links to make them relatively more attractive than the others. In deciding on these improvements, the central planner tries to influence the users' preferences in such a way that total costs are minimized. The partition of the control variables between the upper and lower levels naturally leads to a bilevel formulation.

A conceptual framework for the optimization of Tunisia's inter-regional highways was proposed in [6]. The accompanying formulation included 2683 variables (2571 at the lower level) and 820 constraints (all at the lower level); the follower's problem was divided into two separate subproblems as a direct consequence of the bilevel approach. The first centered on the user-optimized flow requirement (user-equilibrium) and the second on the nonconvex improvement functions. Because none of the standard algorithmic approaches could handle problems of this size, a specialized algorithm was devised to deal with each of the two lower-level problems separately. At each iteration, the algorithm tries to find a better compromise with the user, while including the smallest possible number of nonconvex improvement functions to get the exact solution with the minimum computational effort. Despite the large number of variables and constraints, optimality was achieved.

Solutions

An assessment of existing algorithms for solving various classes of bilevel programs indicates that exact solutions can only be guaranteed for problem instances with up to a few hundred variables and constraints, and then only for the linear case. When nonlinear (nonconvex) functions are included in the model, virtually all algorithms stumble in the presence of more than a handful of variables and constraints. The ability of those working in the field to formulate problems far outstrips the capacity of current techniques to solve them optimally.

When faced with the problem of actually having to provide solutions to large scale formulations, researchers have inevitably fallen back on heuristics and ad hoc procedures. Simulated annealing, tabu search and genetic algorithm-based approaches are examples of the more formal techniques adapted, at least for the linear case. In many instances, code developers were able to demonstrate global optimality by comparing results with exact methods. The conclusion that can be drawn from these observations and related experience is that the need for efficient algorithms remains undiminished. This is the primary reason why realistic applications continue to lag behind theory and the development new codes.

See also

- ▶ Bilevel Fractional Programming
- ▶ Bilevel Linear Programming
- ▶ Bilevel Linear Programming: Complexity, Equivalence to Minmax, Concave Programs
- ▶ Bilevel Optimization: Feasibility Test and Flexibility Index
- ▶ Bilevel Programming
- ▶ Bilevel Programming: Applications
- ▶ Bilevel Programming: Applications in Engineering
- ▶ Bilevel Programming: Implicit Function Approach
- ▶ Bilevel Programming: Introduction, History and Overview
- ▶ Bilevel Programming: Optimality Conditions and Duality
- ▶ Multilevel Methods for Optimal Design
- ▶ Multilevel Optimization in Mechanics
- ▶ Stochastic Bilevel Programs

References

1. Aiyoshi E, Shimizu K (1981) Hierarchical decentralized systems and its new solution by a barrier method. IEEE Trans Syst, Man Cybern SMC-11(6):444–449
2. Anandalingam G, Apprey V (1991) Multi-level programming and conflict resolution. Europ J Oper Res 51:233–247
3. Bard JF (1983) Coordination of a multidivisional organization through two levels of management. OMEGA Internat J Management Sci 11(1):457–468
4. Bard JF (1985) Geometric and algorithmic developments for a hierarchical planning problem. Europ J Oper Res 19:372–383
5. Bard JF, Plummer J, Sourie JC (1997) Determining tax credits for converting nonfood crops to biofuels: An application of bilevel programming. In: Migdalas A, Pardalos PM, Varbrand P (eds) Multilevel Optimization: Algorithms and Applications. Kluwer, Dordrecht, pp 23–50
6. Ben-Ayed O, Blair CE, Boyce DE, Leblanc LJ (1992) Construction of a real-world bilevel programming model of the highway network design problem. Ann Oper Res 34(1–4):219–254
7. Bialas WF, Karwan MH (1984) Two-level linear programming. Managem Sci 30(8):1004–1020
8. Candler W, Norton R (1977) Multi-level programming and development policy. Working Paper 258, World Bank, Washington, DC
9. Dirickx YMI, Jennergren LP (1979) Systems analysis with multilevel methods with applications to economics and management. Wiley, New York
10. Fortuny-Amat J, McCarl B (1981) A representation and economic interpretation of a two-level programming problem. J Oper Res Soc 32(9):783–792
11. Ruefli T (1971) A generalized goal decomposition model. Managem Sci 17(9):B505–B518
12. Sherali HD (1984) A multiple leader Stackelberg model and analysis. Oper Res 3(2):390–404
13. Zangwill WI, Garcia CB (1981) Pathways to solutions, fixed points, and equilibra. Prentice-Hall, Englewood Cliffs, NJ

Bilevel Programming: Optimality Conditions and Duality

SANJO ZLOBEC
Department Math. Statist., McGill University,
Montreal, Canada

MSC2000: 90C25, 90C29, 90C30, 90C31

Article Outline

Keywords
Basic Difficulties
Optimality
Parametric Approach To Optimality
Duality
See also
References

Keywords

Bilevel programming; Optimality conditions; Duality; Stability; Parametric programming

The *bilevel programming problem* (abbreviation: BPP) is a mathematical program in two variables x and θ, where $x = x°(\theta)$ is an optimal solution of another program. Specifically, BPP can be formulated in terms of two ordered objective functions φ and Ψ as follows:

$$\begin{cases} \min_{(x,\theta)} & \varphi(x,\theta) \\ \text{s.t.} & f^i(x,\theta) \leq 0, \quad i \in P, \end{cases} \quad (1)$$

where $x = x°(\theta)$ is an optimal solution of the program

$$\begin{cases} \min_{(x)} & \Psi(x,\theta) \\ \text{s.t.} & g^j(x,\theta) \leq 0, \quad j \in Q. \end{cases} \quad (2)$$

Here the functions $\varphi, \Psi, f^i, g^j : \mathbf{R}^n \times \mathbf{R}^m \to \mathbf{R}$, $i \in P$, $j \in Q$, are assumed to be continuous; $x \in \mathbf{R}^n$, $\theta \in \mathbf{R}^m$; P, Q are finite index sets. Program (1) is often called the *upper (first level, outer, leader's)* problem; then (2) is the *lower (second level, inner, follower's)* problem. Many mathematical programs, such as minimax problems, linear integer, bilinear and quadratic programs, can be stated as special cases of bilevel programs. In view of the so-called *Reduction Ansatz*, developed in [18,44], *semi-infinite programs* can be considered as special cases of bilevel programs. For stability and deformations of these see, e. g., [20,21]. Problems appearing in such seemingly unrelated areas as best approximation problems and *data envelopment analysis* can be viewed as bilevel programs. In the former, one is often interested in finding a least-norm solution in the set of all best approximate solutions, while, in the latter, one wants to rank, or decrease the number of, efficient decision making units by a 'post-optimality analysis'. For history of *bilevel programs*, reviews of numerical methods and applications, especially for connections with *von Stackelberg games* of market economy see, e. g., [14,22,30,39]. In this contribution we will focus only on optimality conditions and duality.

Basic Difficulties

The study of bilevel programming problems requires some familiarity with point-to-set topology; see, e. g., [1,2,6,15]. Since the lower level *optimal solution mapping* $x° : \theta x°(\theta)$ is a *point-to-set mapping* (rather than a vector function), the optimal value function of the BPP may be discontinuous. This is illustrated by the following example:

Example 1 Consider the bilevel program with the upper level objective $\varphi(x,\theta) = -x_1/\theta$, the lower level objective $\Psi(x,\theta) = -x_1 - x_2$, and the lower level feasible set determined by $x_1 + \theta x_2 \leq 1$, $x_1 \geq 0$, $x_2 \geq 0$. The lower level optimal solutions $x = x°(\theta)$ are the segment $\{x_1 + x_2 = 1, x_1 \geq 0, x_2 \geq 0\}$, for $\theta = 1$, and the singleton $[0, 1/\theta]$, when $0 < \theta < 1$. The corresponding upper level optimal solutions, i. e., the BPP optimal solutions, are the points $[1, 0]$ and $[0, 1/\theta]$, respectively. Here the corresponding optimal value of the BPP jumps from -1 to 0, as θ assumes the value 1.

Note that the lower level feasible set mapping, in Example 1, is lower semicontinuous (open) at $\theta = 1$. Hence we conclude that discontinuity of the optimal value can occur even if the lower level model is stable.

The fact that the set of optimal solutions is generally discontinuous in a stable situation is well known in linear programming. It may manifest itself in a chaotic behavior of the optimal solutions, but not the optimal value, when the program is solved by computer repeatedly with small perturbations of data; see ▶ Nondifferentiable optimization: Parametric programming. The topological loss of continuity is generally unrelated to the conditioning, which describes numerical sensitivity of the solutions relative to roundoff errors. In particular, a linear program with an *ill-conditioned coefficient matrix* can be stable.

Another difficulty results from the fact that the optimal solutions mapping $x^\circ : \theta \ x^\circ(\theta)$ is not generally closed. Hence a BPP may not have an optimal solution even if the feasible set of the lower program is compact:

Example 2 Consider the bilinear BPP:

$$\min x + \theta,$$

where $x = x^\circ(\theta)$ solves

$$\begin{cases} \min & -x, \\ \text{s.t.} & x\theta = 0, \\ & 0 \le x \le 1, \quad 0 \le \theta \le 1. \end{cases}$$

Here the optimal solutions mapping is the function $x^\circ(\theta) = 0$, if $\theta > 0$, and $x^\circ(0) = 1$, if $\theta = 0$. The feasible set of the lower level problem is a unit square in the (θ, x)-plane, while the feasible set of the BPP is a disjoint noncompact set consisting of the segment $0 < \theta \le 1$ and the point $[0, 1]$. Since the origin is not a feasible point, the BPP does not have a solution. Note that the function $x^\circ(\theta)$ is not continuous here because the lower level feasible set mapping is not lower semicontinuous at the origin, i. e., the lower level problem is unstable.

Optimality

A popular approach to the study of optimality in BPP is to reduce the program to a one-level program. This can be done as follows: Denote the optimal value of the lower level program (2) by $\Psi^\circ(\theta)$ and introduce the new constraint $f^\circ(x, \theta) = \Psi(x, \theta) - \Psi^\circ(\theta)$. Now the BPP can be reformulated as

$$\begin{cases} \min_{(x,\theta)} & \varphi(x, \theta) \\ \text{s.t.} & f^i(x, \theta) \le 0, \\ & i \in R = \{0\} \cup P. \end{cases} \quad (3)$$

Difficulties with this formulation generally include discontinuity of the leading constraint f° and the lack of classical constraint qualifications. The latter can be handled in convex case using the results on optimality conditions from, e. g., [5,15,47]. One of the first attempts to formulate optimality conditions for bilevel programming problems, using (3), was made in [2]. However a counterexample to these conditions was given in [4,12,17], also see [10]. The one-level approach leads, under assumptions that guarantee Lipschitz continuity of the optimal value function, to necessary *conditions of the Fritz John type*. Under a *partial calmness condition*, and a *constraint qualification* for the lower level problem, one obtains *conditions of the Karush–Kuhn–Tucker type*. The concept of partial calmness is equivalent to the 'exact penalization' and it is satisfied, in particular, for the minimax problem and if the lower level problem is linear. This approach in a nonsmooth framework is used in, e. g., [11] and [46]. The relationship between the BPP and an associated *exact penalty function* was explored also in [7] to derive other types of necessary and sufficient optimality conditions. Other approaches to optimality conditions, that use nonsmooth analysis, include [13,19,32]. Another approach to reducing the BPP to a single-level program is to replace the lower level problem by an optimality condition. This is usually done in formulations of numerical methods; see, e. g., [42]. There are also approaches that use the specific geometry of BPP. One of these applies properties of the steepest descent directions to BPP and it yields a necessary condition for optimality, see [33]. Adaptations of the well-known first and second order optimality conditions of mathematical programming to BPP appeared in [40]. Checking local optimality for linear BPP is *NP*-hard; see [41]. Examples of linear BPPs with an exponential number of local minima can be generated by a technique proposed in [9].

Many authors have studied links between two-objective and bilevel programming, looking for conditions that guarantee that the optimal solution of a given BPP be Pareto optimal for both upper and lower level objective functions, and vice versa; e. g., [28,29,30,37]. The idea is to find an optimal solution of the BPP by solving a bi-objective program. It was shown in [43] that an optimal solution in linear BPP may not be a *Pareto optimum* for the objective function of the outer program and the optimal value function of the lower program, contrary to a claim made in [38]. The authors of [43] also give a sufficient condition for the implication to hold. If an optimal solution exists, in the linear BPP case with a compact feasible set at the lower level, then at least one optimal solution is assumed at a vertex of this set, see [3]. Necessary conditions for optimality can also be stated using marginal value formulas for optimal value functions. However, these formulas can not assume a usual constraint qualification in order to be applied to the formulation (3). One such formula in parametric convex programming is given in [48] and, under slightly different assumptions, in [49]. In the latter, it is used in the context of data envelopment analysis to rank efficiently administered university libraries by their radii of rigidity. Existence of optimal solutions is studied in [16,23,24]; constraints in [24] are defined by an implicit variational problem. Both, existence and stability of solutions and approximate solutions are studied in [27]. Optimality conditions are important for checking optimality, formulation of duality theories, and for numerical methods.

Parametric Approach To Optimality

A parametric approach to characterizing global and local optimal solutions in convex BPP can be described as follows: Denote, for every θ, the optimal value of (3) by

$$\varphi^\circ(\theta) = \begin{cases} \min_{(x)} & \varphi(x, \theta) \\ \text{s.t.} & f^i(x, \theta) \leq 0, \quad i \in R = \{0\} \cup P. \end{cases}$$

Also, denote the feasible set in the x variable by $F(\theta) = \{x : f^i(x, \theta) \leq 0, i \in R\}$, and the feasible set in the θ variable by

$$F = \{\theta \in R^m : F(\theta) \neq \emptyset\}.$$

A parametric formulation of the BPP is

$$\begin{cases} \min & \varphi^\circ(\theta) \\ \text{s.t.} & \theta \in F. \end{cases} \quad (4)$$

Here we optimize the optimal value of the outer problem over the feasible set in the variable θ, considered as a 'parameter'. The problem of the form (4) is a basic problem of *parametric programming*, e. g., ▶ Nondifferentiable optimization: Parametric programming.

It has been extensively studied in the literature from both the theoretical and the numerical side. In particular, various optimality conditions have been formulated for it, e. g., in the context of *input optimization*; see [48]. The key observation in the parametric approach is that, under the assumption that the feasible set of the lower program is compact, every θ^* that globally solves the parametric program (4), with the corresponding optimal solution x^* of the program (3), is a global optimal solution of the bilevel program, and vice versa. However, under the compactness assumption, both sets can be empty (as demonstrated by Example 2). A necessary and sufficient condition for global optimality in convex BPP can be given over a *'region of cooperation'* in terms of the existence of a saddle point; see [15]: Given a candidate for global optimality θ^* and the set of all optimal solutions at the lower level $\{x^\circ(\theta)\}$, $\theta \in F$. Denote by $K(\theta^*)$ the region in the θ-space, where the minimal index set of active constraints $R^=(\theta) = \{i \in R : x \in \{x^\circ(\theta)\} \Rightarrow f^i(x, \theta) = 0\}$ does not strictly increase, i. e., $K(\theta^*) = \{\theta \in F : R^=(\theta) \subset R^=(\theta^*)\}$. Then the region of cooperation at θ^* is defined as the set $\{(\theta, x)\} : \theta \in K(\theta^*), x \in F(\theta)\}$. One can characterize global optimality on the entire feasible set for linear BPP, and also for convex BPP provided that the constraints are *'LFS functions'*, e. g. [35,48]. These functions form a large class of convex functions that includes all linear and polyhedral functions. Characterizations of global optimality are simplified under the so-called *sandwich condition*. This is a two-sided global inclusion involving the set of optimal solutions of the inner program, e. g., [15]. Characterizations of *locally* optimal parameters θ^* for convex (4) require lower semicontinuity of the optimal solutions mapping x°. The results apply to the convex BPP with the additional assumption that the corresponding optimal solution $x^* \in \{x^\circ(\theta^*)\}$ is unique; see, e. g., [15]. The uniqueness assumption in the characterization of local optimality cannot be replaced by the requirement that the set $\{x^\circ(\theta^*)\}$ be compact. The following example illustrates a situation where a local optimum of the BPP can not be recovered by the parametric approach.

Example 3 Consider the program min $\varphi(x, \theta) = x\theta^2$, where x solves min $\Psi(x, \theta) = 0$, subject to $-1 \leq x, \theta \leq 1$. Here $x^* = 1$, $\theta^* = 0$ is a local minimum of the bilevel program. But $\varphi^\circ(\theta) = -\theta^2$ and $\theta^* = 0$ is not its local minimum; in fact, it is an isolated global maximum!

Duality

Duality theories for bilevel programming problems can be formulated by adjusting the duality theories of mathematical programming (see, e.g., [34]) to the single-objective model (3). Let us outline how this works using a parametric approach; we follow the ideas from [15]. Instead of a single 'dual' one obtains a collection of several 'subduals', each closely related to the original (primal) program. The number of these subduals is cardinality of the set

$$\Pi = \{\Omega \subset R \colon \Omega = R^=(\theta) \text{ for some } \theta \in F\}.$$

First, with each $\Omega \subset \Pi$, one associates the feasible subregion $\mathbf{S}_\Omega = \{\theta \in F : R^=(\theta) = \Omega\}$, the Lagrangian $L_\Omega(x, \theta; u) = \varphi(x, \theta) + \sum_{i \in R \setminus \Omega} u_i f^i(x, \theta)$, and the point-to-set mapping $F_\Omega : F \to R^n$ defined by $F_\Omega(\theta) = \{x : f^i(x, \theta) \leq 0, i \in \Omega\}$. The corresponding *subdual function* is

$$\Phi_\Omega(u) = \inf\{L_\Omega(x, \theta; u) \colon \theta \in \mathbf{S}_\Omega, x \in F_\Omega(\theta)\}$$

and the subdual (D, Ω) is defined as

$$\sup\left\{\Phi_\Omega(u) \colon u \in [\mathbf{S}_\Omega \to R_+^{\text{card } R \setminus \Omega}]\right\}. \quad (5)$$

Here u belongs to the set of all nonnegative vector functions defined on $\mathbf{S}_\Omega$. The duality results, stated for *partly convex programs* in, e.g., [47] can be reformulated for the outer convex model and hence BPP. In particular, if, for some $\Omega \subset \Pi$, $u^* \in [\mathbf{S}_\Omega \to R_+^{\text{card } R \setminus \Omega}]$, and an optimal solution x^* of the inner program for some fixed $\theta^* \in \mathbf{S}_\Omega$, one has $\Phi_\Omega(u^*) = \varphi(x^*, \theta^*)$, then u^* solves the subdual (5) and θ^* solves (4) on $\mathbf{S}_\Omega$.

If optimization of the optimal value function in (4) is performed from some fixed 'initial' θ, but using only parameter paths that preserve continuity of the optimal solutions mapping of the lower problem, then we talk about *stable BPP*. This approach, in the convex case, guarantees that the optimal solutions mapping in BPP is closed and that the optimal value function is continuous, thus removing the two basic difficulties mentioned in Section 1. However, the optimal solutions now depend on the initial choice of the parameter and on a particular class of stable paths used. *Stable parametric programming* has been studied in [48], stable BPP is mentioned (but not studied) in [15]; see [36].

See also

- Bilevel Fractional Programming
- Bilevel Linear Programming
- Bilevel Linear Programming: Complexity, Equivalence to Minmax, Concave Programs
- Bilevel Optimization: Feasibility Test and Flexibility Index
- Bilevel Programming
- Bilevel Programming: Applications
- Bilevel Programming: Applications in Engineering
- Bilevel Programming: Implicit Function Approach
- Bilevel Programming: Introduction, History and Overview
- Bilevel Programming in Management
- Multilevel Methods for Optimal Design
- Multilevel Optimization in Mechanics
- Stochastic Bilevel Programs

References

1. Bank B, Guddat J, Klatte D, Kummer B, Tammer K (1982) Nonlinear parametric optimization. Akad Verlag, Berlin
2. Bard J (1984) Optimality conditions for the bilevel programming problem. Naval Res Logist Quart 31:13–26
3. Bard J (1988) Convex two-level optimization. Math Program 40:15–27
4. Ben-Ayed O, Blair C (1990) Computational difficulties of bilevel linear programming. Oper Res 38:556–560
5. Ben-Israel A, Ben-Tal A, Zlobec S (1981) Optimality in nonlinear programming: A feasible directions approach. Wiley/Interscience, New York
6. Berge C (1963) Topological spaces. Oliver and Boyd, Edinburgh
7. Bi Z, Calamai P (1991) Optimality conditions for a class of bilevel programming problems. Techn Report Dept Systems Design Engin, Univ Waterloo, no. 191-O-191291
8. Bracken J, Falk J, McGill J (1974) Equivalence of two mathematical programs with optimization problems in the constraints. Oper Res 22:1102–1104
9. Calamai P, Vicente LN (1993) Generating linear and linear-quadratic bilevel programming problems. SIAM J Sci Statist Comput 14:770–782
10. Candler W (1988) A linear bilevel programming algorithm: A comment. Comput Oper Res 15:297–298
11. Chen Y, Florian M (1995) The nonlinear bilevel programming problem: formulations, regularity and optimality conditions. Optim 32:193–209
12. Clarke P, Westerberg A (1988) A note on the optimality conditions for the bilevel programming problem. Naval Res Logist Quart 35:413–418

13. Dempe S (1992) A necessary and sufficient optimality condition for bilevel programming problems. Optim 25:341–354
14. Dempe S (1997) On the leader's dilemma and a new idea for attacking bilevel programming problems. Preprint Techn Univ Chemnitz
15. Floudas CA, Zlobec S (1998) Optimality and duality in parametric convex lexicographic programming. In: Pardalos PM Migdalas A and Värbrand P (eds) Multilevel Optimization: Algorithms and Applications. Kluwer, Dordrecht, pp 359–379
16. Harker P, Pang J-S (1988) Existence of optimal solutions to mathematical programs with equilibrium constraints. Oper Res Lett 7:61–64
17. Haurie A, Savard G, White D (1990) A note on: An efficient point algorithm for a linear two-stage optimization problem. Oper Res 38:553–555
18. Hettich R, Jongen HTh (1978) Semi-infinite programming: conditions of optimality and applications. In: Stoer J (ed) Optimization Techniques, Part 2. Lecture Notes Control Inform Sci. Springer, Berlin, pp 1–11
19. Ishizuka Y (1988) Optimality conditions for quasi-differentiable programs with applications to two-level optimization. SIAM J Control Optim 26:1388–1398
20. Jongen HTh, Rückmann J-J, Stein O (1998) Generalized semi-infinite optimization: A first order optimality condition and examples. Math Program 83:145–158
21. Jongen HTh, Rückmann J-J (1998) On stability and deformation in semi-infinite optimization. In: Reemtsen R, Rückmann J-J (eds) Semi-Infinite Programming. Kluwer, Dordrecht, pp 29–67
22. Kolstad CD (Oct. 1985) A review of the literature on bilevel mathematical programming. Techn Report Los Alamos Nat Lab no. LA-10284-MS, UC-32
23. Lignola MB, Morgan J (1995) Topological existence and stability for Stackelberg problems. J Optim Th Appl 84:145–169
24. Lignola MB, Morgan J (1998) Existence of solutions to generalized bilevel programing problem. In: Migdalas A, Pardalos PM, Värbrand P (eds) Multilevel Optimization: Algorithms and Applications. Kluwer, Dordrecht, pp 315–332
25. Liu Y, Hart S (1994) Characterizing an optimal solution to the linear bilevel programming problem. Europ J Oper Res 166:164–166
26. Loridan P, Morgan J (1989) New results on approximate solutions in two-level optimization. Optim 20:819–836
27. Mallozzi L, Morgan J (1995) Weak Stackelberg problem and mixed solutions under data perturbations. Optim 32:269–290
28. Marcotte P, Savard G (1991) A note on Pareto optimality of solutions to the linear bilevel programming problem. Comput Oper Res 18:355–359
29. Migdalas A (1995) When is a Stackelberg equilibrium Pareto optimum? In: Pardalos PM, Siskos Y, Zopounidis C (eds) Advances in Multicriteria Analysis. Kluwer, Dordrecht, pp 175–181
30. Migdalas A, Pardalos PM (1996) Editorial: Hierarchical and bilevel programming. J Global Optim 8:209–215
31. Migdalas A Pardalos PM, Värbrand P (eds) (1998) Multilevel optimization: Algorithms and applications. Kluwer, Dordrecht, pp 29–67
32. Outrata J (1993) Necessary optimality conditions for Stackelberg problems. J Optim Th Appl 76:305–320
33. Savard G, Gauvin J (1994) The steepest descent direction for the nonlinear bilevel programming problem. Oper Res Lett 15:275–282
34. Tammer K, Rückmann J-J (1990) Relations between the Karush–Kuhn–Tucker points of a nonlinear optimization problem and of a generalized Lagrange dual. In: Sebastian H-J, Tammer K (eds) System Modelling and Optimization. Lecture Notes Control Inform Sci. Springer, Berlin
35. Trujillo-Cortez R (1997) LFS functions in stable bilevel programming. PhD Thesis Dept Math and Statist, McGill Univ
36. Trujillo-Cortez R (2000) Stable bilevel programming and applications. PhD Thesis McGill Univ, in preparation
37. Tuy H (1998) Bilevel linear programming, multiobjective programming, and monotonic reverse convex programming. In: Migdalas A Pardalos PM, Värbrand P (eds) Multilevel Optimization: Algorithms and Applications. Kluwer, Dordrecht, 295–314
38. Ünlü G (1987) A linear bilevel programming algorithm based on bicriteria programming. Comput Oper Res 14:173–179
39. Vicente LN, Calamai PH (1994) Bilevel and multilevel programming: A bibliography review. J Global Optim 5:291–306
40. Vicente LN, Calamai PH (1995) Geometry and local optimality conditions for bilevel programs with quadratic strictly convex lower levels. In: Dhu D-Z, Pardalos PM (eds) Minimax and Applications. Kluwer, Dordrecht, pp 141–151
41. Vicente LN, Savard G, Judice J (1994) Descent approaches for quadratic bilevel programming. J Optim Th Appl 81:379–399
42. Visweswaran V, Floudas CA, Ierapetritou MG, Pistikopoulos EN (1996) A decomposition-based global optimization approach for solving bilevel linear and quadratic programs. In: Floudas CA, Pardalos PM (eds) State of the Art in Global Optimization. Kluwer, Dordrecht, 139–162
43. Wen U-P, Hsu S-T (1989) A note on a linear bilevel programming algorithm based on bicriteria programming. Comput Oper Res 16:79–83
44. Wetterling W (1970) Definitheitsbedingungen für relative Extrema bei Optimierungs- und Approximationsaufgaben. Numerische Math 15:122–136
45. Ye J (1995) Necessary conditions for bilevel dynamic optimization problems. SIAM J Control Optim 33:1208–1223
46. Ye JJ, Zhu DL (1995) Optimality conditions for bilevel programming problems. Optim 33:9–27

47. Zlobec S (1996) Lagrange duality in partly convex programming. In: Floudas CA, Pardalos PM (eds) State of the Art in Global Optimization. Kluwer, Dordrecht, pp 1–18
48. Zlobec S (1998) Stable parametric programming. Optim 45:387–416
49. Zlobec S (2000) Parametric programming: An illustrative mini-encyclopedia. Math Commun 5:1–39

Bilinear Programming

ARTYOM G. NAHAPETYAN
Center for Applied Optimization,
Industrial and Systems Engineering Department,
University of Florida, Gainesville, USA

Article Outline

Keywords
Introduction
Formulation
  Equivalence to Other Problems
  Properties of a Solution
Methods
References

Keywords

Congestion toll pricing; Dynamic toll set; Toll pricing framework

Introduction

A function $f(x, y)$ is called bilinear if it reduces to a linear one by fixing the vector x or y to a particular value. In general, a bilinear function can be represented as follows:

$$f(x, y) = a^T x + x^T Q y + b^T y,$$

Where $a, x \in \mathbb{R}^n$, $b, y \in \mathbb{R}^m$, and Q is a matrix of dimension $n \times m$. It is easy to see that bilinear functions compose a subclass of quadratic functions. We refer to optimization problems with bilinear objective and/or constraints as bilinear problems, and they can be viewed as a subclass of quadratic programming.

Bilinear programming has various applications in constrained bimatrix games, Markovian assignment and complementarity problems. Many 0–1 integer programs can be formulated as bilinear problems. An extensive discussion of different applications can be found in [5]. Concave piecewise linear network flow problems, fixed charge network flow problems, and multi-item dynamic pricing problems, which are very common in the supply chain management, can be also solved using bilinear formulations (see, e. g., [7,8,9]). It should be noted that more general convex/non-convex optimization problems can be reduced to a bilinear problem as well, and different reduction techniques can be found in [1,2,10].

Formulation

Despite a variety of different bilinear problems, most of the practical problems involve a bilinear objective function and linear constraints, and theoretical results are derived for those cases. In our discussion we consider the following bilinear problem, which we refer to as BP.

$$\min_{x \in X, y \in Y} f(x, y) = a^T x + x^T Q y + b^T y,$$

where X and Y are nonempty polyhedra. The BP formulation is also known as a bilinear problem with a *disjoint* feasible region because the feasibility of x (y) is independent form the choice of the vector y (x).

Equivalence to Other Problems

Below we discuss some theoretical results, which reveal the equivalence between bilinear problems and some of concave minimization problems.

Let $V(x)$ and $V(y)$ denote the set of vertices of X and Y, respectively, and $g(x) = \min_{y \in Y} f(x, y) = a^T x + \min_{y \in Y} \{x^T Q y + b^T y\}$. Note that $\min_{y \in Y} f(x, y)$ is a linear programm. Because the solution of a linear problem attains on a vertex of the feasible region, $g(x) = \min_{y \in Y} f(x, y) = \min_{y \in V(Y)} f(x, y)$. Using those notations, the BP problem can be restated as

$$\min_{x \in X, y \in Y} f(x, y) = \min_{x \in X} \{\min_{y \in Y} f(x, y)\}$$
$$= \min_{x \in X} \{\min_{y \in V(Y)} f(x, y)\} = \min_{x \in X} g(x). \quad (1)$$

Observe that the set of vertices of Y is finite, and for each $y \in Y$, $f(x, y)$ is a linear function of x; therefore, function $g(x)$ is a piecewise linear concave function of x. From the later it follows that BP is equivalent to a piecewise linear concave minimization problem with linear constraints.

It also can be shown that any concave minimization problem with a piecewise linear separable objective function can be reduced to a bilinear problem. To establish this relationship consider the following optimization problem:

$$\min_{x \in X} \sum_i \phi_i(x_i), \qquad (2)$$

where X is an arbitrary nonempty set of feasible vectors, and $\phi_i(x_i)$ is a concave piecewise linear function of only one component x_i, i. e.,

$$\phi_i(x_i) = \begin{cases} c_i^1 x_i + s_i^1 (= \phi_i^1(x_i)) & x_i \in [\lambda_i^0, \lambda_i^1) \\ c_i^2 x_i + s_i^2 (= \phi_i^2(x_i)) & x_i \in [\lambda_i^1, \lambda_i^2) \\ \ldots & \ldots \\ c_i^{n_i} x_i + s_i^{n_i} (= \phi_i^{n_i}(x_i)) & x_i \in [\lambda_i^{n_i-1}, \lambda_i^{n_i}] \end{cases}$$

with $c_i^1 > c_i^2 > \ldots > c_i^{n_i}$. Let $K_i = \{1, 2, \ldots, n_i\}$. Because of the concavity of $\phi_i(x_i)$, the function can be written in the following alternative form

$$\phi_i(x_i) = \min_{k \in K_i}\{\phi_i^k(x_i)\} = \min_{k \in K_i}\{c_i^k x_i + s_i^k\}. \qquad (3)$$

Construct the following bilinear problem:

$$\min_{x \in X, y \in Y} f(x, y) = \sum_i \sum_{k \in K_i} \phi_i^k(x_i) y_i^k$$
$$= \sum_i \sum_{k \in K_i} (c_i^k x_i + s_i^k) y_i^k \qquad (4)$$

where $Y = [0, 1]^{\sum_i |K_i|}$. The proof of the following theorem follows directly from Equation (3), and for details we refer to the paper [7].

Theorem 1 *If (x^*, y^*) is a solution of the problem (4) then x^* is a solution of the problem (2).*

Observe that X is not required to be a polytop. If X is a polytop then the structure of the problem (4) is similar to BP.

Furthermore, it can be shown that any quadratic concave minimization problem can be reduced to a bilinear problem. Specifically, consider the following optimization problem:

$$\min_{x \in X} \phi(x) = 2a^T x + x^T Q x, \qquad (5)$$

where Q is a symmetric negative semi-definite matrix. Construct the following bilinear problem

$$\min_{x \in X, y \in Y} f(x, y) = a^T x + a^T y + x^T Q y, \qquad (6)$$

where $Y = X$.

Theorem 2 *(see [4]) If x^* is a solution of the problem (5) then (x^*, x^*) is a solution of the problem (6). If $(\hat{x}, \hat{y})$ is a solution of the problem (6) then $\hat{x}$ and $\hat{y}$ solve the problem (5).*

Properties of a Solution

In the previous section we have shown that BP is equivalent to a piecewise linear concave minimization problem. On the other hand it is well known that a concave minimization problem over a polytop attains its solution on a vertex (see, for instance, [3]). The following theorem follows from this observation.

Theorem 3 *(see [4] and [3]) If X and Y are bounded then there is an optimal solution of BP, (x^*, y^*), such that $x^* \in V(X)$ and $y^* \in V(Y)$.*

Let (x^*, y^*) denote a solution of BP. By fixing the vector x to the value of the vector x^*, the BP problem reduces to a linear one, and y^* should be a solution of the resulting problem. From the symmetry of the problem, a similar result holds by fixing the vector y to the value of the vector y^*. The following theorem is a necessary optimality condition, and it is a direct consequence of the above discussion.

Theorem 4 *(see [4] and [3]) If (x^*, y^*) is a solution of the BP problem, then*

$$\min_{x \in X} f(x, y^*) = f(x^*, y^*) = \min_{y \in Y} f(x^*, y) \qquad (7)$$

However, (7) is not a sufficient condition. In fact it can only guarantee a local optimality of (x^*, y^*) under some additional requirements. In particular, y^* has to be the unique solution of $\min_{y \in Y} f(x^*, y)$ problem. From the later it follows that $f(x^*, y^*) < f(x^*, y)$, $\forall y \in V(Y)$, $y \neq y^*$. Because of the continuity of the function $f(x, y)$, for any $y \in V(y)$, $y \neq y^*$, $f(x^*, y^*) < f(x, y)$ in a small neighborhood U_y of the point x^*. Let $U = \bigcap_{y \in V(Y), y \neq y^*} U_y$. Then $f(x^*, y^*) < f(x, y)$, $\forall x \in U$, $y \in V(Y)$, $y \neq y^*$. At last observe that Y is a polytop, and any point of the set can be expressed through a convex combination of its vertices. From the later it follows that $f(x^*, y^*) \leq f(x, y)$, $\forall x \in U, y \in Y$, which completes the proof of the following theorem.

Theorem 5 *If (x^*, y^*) satisfies the condition (7) and y^* is the unique solution of the problem $\min_{y \in Y} f(x^*, y)$ then (x^*, y^*) is a local optimum of BP.*

Recall that BP is equivalent to a piecewise concave minimization problem. Under the assumptions of the theorem, it is easy to show that x^* is a local minimum of the function $g(x)$ as well (see [4]).

Methods

In this section we discuss methods to find a solution of a bilinear problem. Because BP is equivalent to a piecewise linear concave minimization problem, any solution algorithm for the later can be used to solve the former. In particular, one can employ a cutting plain algorithms developed for those problems. However, the symmetric structure of the BP problem allows constructing more efficient cuts. In the paper [6], the author discusses an algorithm, which converges to a solution that satisfies condition (7), and then proposes a cutting plain algorithm to find the global minimum of the problem.

Assume that X and Y are bounded. Algorithm 1, which is also known as the "mountain climbing" procedure, starts from an initial feasible vector y^0 and iteratively solves two linear problems. The first LP is obtained by fixing the vector y to the value of the vector y^{m-1}. The solution of the problem is used to fix the value of the vector x and construct the second LP. If $f(x^m, y^{m-1}) \neq f(x^m, y^m)$, then we continue solving the linear problems by fixing the vector y to the value of y^m. If the stopping criteria is satisfied, then it is easy to show that the vector (x^m, y^m) satisfies the condition (7). In addition, observe that $V(X)$ and $V(Y)$ are finite. From the later and the fact that $f(x^m, y^{m-1}) \geq f(x^m, y^m)$ it follows that the algorithm converges in a finite number of iterations.

Let (x^*, y^*) denote the solution obtained by the Algorithm 1. Assuming that the vertex x^* is not degenerate, denote by D the set of directions d_j along the ages emanating from the point x^*. Recall that $g(x) = \min_{y \in Y} f(x, y)$ is a concave function. To con-

Step 1: Let $y^0 \in Y$ denote an initial feasible solution, and $m \leftarrow 1$.
Step 2: Let $x^m = \text{argmin}_{x \in X}\{f(x, y^{m-1})\}$, and $y^m = \text{argmin}_{y \in Y}\{f(x^m, y)\}$.
Step 3: If $f(x^m, y^{m-1}) = f(x^m, y^m)$ then stop. Otherwise, $m \leftarrow m + 1$ and go to Step 2.

Bilinear Programming, Algorithm 1
Mountain Climbing Procedure

struct a valid cut, for each direction d_j find the maximum value of θ_j such that $g(x^* + \theta_j d_j) \geq f(x^*, y^*) - \varepsilon$, i. e.,

$$\theta_j = \text{argmax}\{\theta_j | g(x^* + \theta_j d_j) \geq f(x^*, y^*) - \varepsilon\},$$

where ε is a small positive number. Let $C = (d_1, \ldots, d_n)$,

$$\Delta_x^1 = \left\{ x \Big| \left(\frac{1}{\theta_1}, \ldots, \frac{1}{\theta_n}\right)^T C^{-1}(x - x^*) \geq 1 \right\},$$

and $X_1 = X \cap \Delta_x^1$. If $X_1 = \emptyset$ then

$$\min_{x \in X, y \in Y} f(x, y) \geq f(x^*, y^*) - \varepsilon,$$

and (x^*, y^*) is a global ε-optimum of the problem. If $X_1 \neq \emptyset$ then one can replace X by the set X_1, i. e., consider the optimization problem

$$\min_{x \in X_1, y \in Y} f(x, y),$$

and run Algorithm 1 to find a better solution. However, because of the symmetric structure of the problem, a similar procedure can be applied to construct

Step 1: Apply Algorithm 1 to find a vector (x^*, y^*) that satisfies the relationship (7).
Step 2: Based on the solution (x^*, y^*), compute the appropriate cuts and construct the sets X_1 and Y_1.
Step 3: If $X_1 = \emptyset$ or $Y_1 = \emptyset$, then stop; (x^*, y^*) is a global ε-optimal solution. Otherwise, $X \leftarrow X_1$, $Y \leftarrow Y_1$, and go to Step 1.

Bilinear Programming, Algorithm 2
Cutting Plane Algorithm

a cut for the set Y. Let Δ_y^1 denote the corresponding half-space, and $Y_1 = Y \cap \Delta_y^1$. By updating both sets, i. e., considering the optimization problem

$$\min_{x \in X_1, y \in Y_1} f(x, y),$$

the cutting plane algorithm (see Algorithm 2) might find a global solution of the problem using less number of iterations.

References

1. Floudas CA, Visweswaran V (1990) A global optimization algorithm (GOP) for certain classes of nonconvex NLPs: I. Theory. Comput Chem Eng 14:1397–1417
2. Floudas CA, Visweswaran V (1993) A primal-relaxed dual global optimization approach. J Optim Theor Appl 78:187–225
3. Horst R, Pardalos P, Thoai N (2000) Introduction to global optimization, 2nd edn. Springer, Boston
4. Horst R, Tuy H (1996) Global optimization, 3rd edn. Springer, New York
5. Konno H (1971) A bilinear programming: Part II. Applications of bilinear programming. Technical Report 71-10, Operations Research House, Stanford University, Stanford, CA
6. Konno H (1976) A cutting plane algorithm for solving bilinear programs. Math Program 11:14–27
7. Nahapetyan A, Pardalos P (2007) A bilinear relaxation based algorithm for concave piecewise linear network flow problems. J Ind Manag Optim 3:71–85
8. Nahapetyan A, Pardalos P (2008) Adaptive dynamic cost updating procedure for solving fixed charge network flow problems. Comput Optim Appl 39:37–50. doi:10.1007/s10589-007-9060-x
9. Nahapetyan A, Pardalos P (2008) A Bilinear Reduction Based Algorithm for Solving Capacitated Multi-Item Dynamic Pricing Problems. J Comput Oper Res 35:1601–1612. doi:10.1016/j.cor.2006.09.003
10. Visweswaran V, Floudas CA (1990) Global Optimization Algorithm (GOP) for Certain Classes of Nonconvex NLPs: II. Applications of Theory and Test Problems. Comput Chem Eng 14:1417–1434

Bilinear Programming: Applications in the Supply Chain Management

ARTYOM G. NAHAPETYAN
Center for Applied Optimization,
Industrial and Systems Engineering Department,
University of Florida, Gainesville, USA

Article Outline

Introduction
Formulation
   Concave Piecewise Linear Network Flow Problem
   Fixed Charge Network Flow Problem
   Capacitated Multi-Item Dynamic Pricing Problem
Methods
References

Introduction

Many problems in the supply chain management can be formulated as a network flow problem with specified arc cost functions. Let $G(N, A)$ represent a network where N and A are the sets of nodes and arcs, respectively, and $f_a(x_a)$ denotes an arc cost function. In the network, there are supply and demand nodes, and the main objective of the problem is to minimize the total cost by satisfying the demand from the available supply. In addition, one can assume that the arc flows are bounded, which corresponds to the cases where a shipment along an arc should not exceed a specified capacity. The mathematical formulation of the problem can be stated as

$$\min_x f(x) = \sum_{a \in A} f_a(x_a) \tag{1}$$

$$\text{s.t.} \quad Bx = b \tag{2}$$

$$x_a \in [0, \lambda_a] \quad \forall a \in A \tag{3}$$

where B is the node-arc incident matrix of network G, and b is a supply/demand vector. In the next section, we discuss two formulations where $f_a(x_a)$ is either a concave piecewise linear or fixed charge function of the arc flow. The concave piecewise linear functions are typically used in the cases where merchandisers encourage to buy more products by offering discounts in the unit price for large orders. In [6], the authors showed that the problem in these settings is NP hard. Some heuristic procedures to solve the problem are discussed in [10] and [13]. The fixed charge functions are used in the cases where regardless the quantity of the shipment it is required to pay a fixed cost to ship along an arc. The fixed cost might be the cost of renting a truck, ship, airplane, or train to transport goods between nodes of the network. The problem can be modeled as a 0–1 mixed integer linear program and most solution approaches utilize branch-and-bound techniques to find an exact solution (see [1,2,5,7,16]). Some heuristic procedures are discussed in [3,4,8,9,12,14]. In this article we show that both problems are equivalent to a bilinear problem with a disjoint feasible region.

In addition to choosing a proper production level, sometimes managers have to make pricing decisions as well. In particular, one can assume that the satisfied demand is a function of the price, i. e., lower prices generate an additional demand. Such functional relationship between the prices and the satisfied demand is commonly used by economists. However, because of the production capacity restrictions, fixed costs related to the production process, seasonality and other factors, often it is not feasible to satisfy the optimal level of demand, and managers should consider optimal production and inventory levels in combination with pricing decisions to maximize the net profit during a specified time period. One of such problems and an equivalent bilinear formulation are discussed in the next section as well.

In addition to the bilinear formulations of the supply chain problems, in Sect. "Methods" we explore the structure of the bilinear problems and discuss difficulties in applying the standard computational methods. Despite the intricacy, the section proposes some heuristic methods to find a near optimum solution to the problems. The solution obtained by a heuristic procedure can also be used to expedite exact algorithms.

Formulation

Concave Piecewise Linear Network Flow Problem

In the problem (1)–(3), assume that $f_a(x_a)$ is a piecewise linear concave function, i. e.,

$$f_a(x_a) = \begin{cases} c_a^1 x_a + s_a^1 (= f_a^1(x_a)) & x_a \in [0, \xi_a^1) \\ c_a^2 x_a + s_a^2 (= f_a^2(x_a)) & x_a \in [\xi_a^1, \xi_a^2) \\ \ldots & \ldots \\ c_a^{n_a} x_a + s_a^{n_a} (= f_a^{n_a}(x_a)) & x_a \in [\xi_a^{n_a-1}, \lambda_a] \end{cases}$$

with $c_a^1 > c_a^2 > \ldots > c_a^{n_a}$. Let $K_a = \{1, 2, \ldots, n_a\}$. Because of the concavity of $f_a(x_a)$, it can be written in the following alternative form

$$f_a(x_a) = \min_{k \in K_a} \{f_a^k(x_a)\} = \min_{k \in K_a} \{c_a^k x_a + s_a^k\}. \quad (4)$$

By introducing additional variables $y_a^k \in [0, 1], k \in K_a$, construct the following bilinear problem.

$$\min_{x,y} g(x, y) = \sum_{a \in A} \left[\sum_{k \in K_a} c_a^k y_a^k \right] x_a + \sum_{a \in A} \sum_{k \in K_a} s_a^k y_a^k$$

$$= \sum_{a \in A} \sum_{k \in K_a} f_a^k(x_a) y_a^k \quad (5)$$

s.t. $Bx = b$ (6)

$$\sum_{k \in K_a} y_a^k = 1 \quad \forall a \in A \quad (7)$$

$x_a \in [0, \lambda_a], y_a^k \geq 0 \quad \forall a \in A \text{ and } k \in K_a$ (8)

In [13], the authors show that at any local minima of the bilinear problem, $(\hat{x}, \hat{y})$, $\hat{y}$ is either binary vector or can be used to construct a binary vector with the same objective function value. Although the vector $\hat{y}$ may have a fractional components, the authors note that in practical problems it is highly unlikely. The proof of the theorem below follows directly from (4). Details on the proof as well as transformation of the problem (1)–(3) into (5)–(8) can be found in [13].

Theorem 1 *If (x^*, y^*) is a global optima of the problem (5)–(8) then x^* is a solution of the problem (1)–(3).*

According to the theorem, the concave piecewise linear network flow problem is equivalent to a bilinear problem in a sense that the solution of the later is a solution of the former. It is important to notice that the problem (5)–(8) does not have binary variables, i. e., all variables are continuous. However, at optimum y^* is a binary vector, which makes sure that in the objective only one linear piece is employed.

Fixed Charge Network Flow Problem

In the case of the fixed charge network flow problem, we assume that the function $f_a(x_a)$ has the following structure.

$$f_a(x_a) = \begin{cases} c_a x_a + s_a & x_a \in (0, \lambda_a] \\ 0 & x_a = 0 \end{cases},$$

Observe that the function is discontinuous at the origin and linear on the interval $(0, \lambda_a]$.

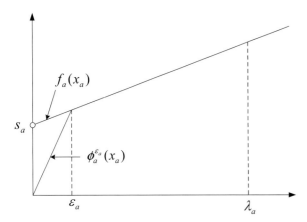

Bilinear Programming: Applications in the Supply Chain Management, Figure 1
Approximation of function $f_a(x_a)$

Let $\varepsilon_a \in (0, \lambda_a]$, and define

$$\phi_a^{\varepsilon_a}(x_a) = \begin{cases} c_a x_a + s_a & x_a \in [\varepsilon_a, \lambda_a] \\ c_a^{\varepsilon_a} x_a & x_a \in [0, \varepsilon_a) \end{cases}$$

where $c_a^{\varepsilon_a} = c_a + s_a/\varepsilon_a$. It is easy to see that $\phi_a^{\varepsilon_a}(x_a) = f_a(x_a)$, $\forall x_a \in \{0\} \bigcup [\varepsilon_a, \lambda_a]$ and $\phi_a^{\varepsilon_a}(x_a) < f_a(x_a)$, $\forall x_a \in (0, \varepsilon_a)$, i.e., $\phi_a^{\varepsilon_a}(x_a)$ approximates the function $f_a(x_a)$ from below. (see Fig. 1). Let us construct the following concave two-piece linear network flow problem.

$$\min_x \phi^{\varepsilon}(x) = \sum_{a \in A} \phi_a^{\varepsilon_a}(x_a) \quad (9)$$

$$\text{s.t.} \quad Bx = b, \quad (10)$$

$$x_a \in [0, \lambda_a], \quad \forall a \in A, \quad (11)$$

where ε denotes the vector of ε_a. Function $\phi^{\varepsilon}(x)$ as well as the problem (9)–(11) depends on the value of the vector ε. In the paper [14], the authors show that for any value of $\varepsilon_a \in (0, \lambda_a]$, a global solution of the problem (9)–(11) provides a lower bound for the fixed charge network flow problem, i.e., $\phi^{\varepsilon}(x^{\varepsilon}) \leq f(x^*)$, where x^{ε} and x^* denote the solutions of the corresponding problems.

Theorem 2 (see [14]) *For all ε such that $\varepsilon_a \in (0, \lambda_a]$ for all $a \in A$, $\phi^{\varepsilon}(x^{\varepsilon}) \leq f(x^*)$.*

Furthermore, by choosing a sufficiently small value for ε_a one can ensure that both problems have the same solution. Let $\delta = \min\{x_a^v | x^v \in V(x), a \in A, x_a^v > 0\}$, where $V(x)$ denotes the set of vertices of the polyhedra (10)–(11). Observe that δ is the minimum among all positive components of all vectors $x^v \in V(x)$; therefore, $\delta > 0$.

Theorem 3 (see [14]) *For all ε such that $\varepsilon_a \in (0, \delta]$ for all $a \in A$, $\phi^{\varepsilon}(x^{\varepsilon}) = f(x^*)$.*

Theorem 3 proves the equivalence between the fixed charge network flow problem and the concave two-piece linear network flow problem (9)–(11) in a sense that the solution of the later is a solution of the former. As we have seen in the previous section, concave piecewise linear network flow problems are equivalent to bilinear problems. In particular, problem (9)–(11) is equivalent to the following bilinear problem.

$$\min_{x,y} \sum_{a \in A} [c_a x_a + s_a] y_a + c_a^{\varepsilon_a} x_a [1 - y_a] \quad (12)$$

$$\text{s.t.} \quad Bx = b, \quad (13)$$

$$x_a \geq 0, \quad \text{and} \quad y_a \in [0, 1], \quad \forall a \in A, \quad (14)$$

where $\varepsilon_a \in (0, \delta]$.

Capacitated Multi-Item Dynamic Pricing Problem

In the problem, we assume that a company during a discrete time period Δ is able to produce different commodities from a set P. In addition, we assume that at each point of time $j \in \Delta$ and for each product $p \in P$ a functional relationship $f_{(p,j)}(d_{(p,j)})$ between the satisfied demand and the price is given, i.e., in order to satisfy the demand $d_{(p,j)}$ of the product p, the price of the product at time j should be equal to $f_{(p,j)}(d_{(p,j)})$. As a result, the revenue generated from the sales of the product p at time j is $g_{(p,j)}(d_{(p,j)}) = f_{(p,j)}(d_{(p,j)})d_{(p,j)}$. Although we do not specify the function $f_{(p,j)}(d_{(p,j)})$, it should ensure that $g_{(p,j)}(d_{(p,j)})$ is a concave function (see Fig. 2a).

Because of the concavity of $g_{(p,j)}(d_{(p,j)})$, there exists a point $\tilde{d}_{(p,j)}$, such that the function reaches its maximum, and producing and selling more than $\tilde{d}_{(p,j)}$ is not profitable. Therefore, without lost of generality, we can assume that $d_{(p,j)} \in [0, \tilde{d}_{(p,j)}]$. According to the definition of $g_{(p,j)}(d_{(p,j)})$, it is a concave monotone function on the interval $[0, \tilde{d}_{(p,j)}]$. To avoid nonlinearity in the objective, one can approximate it by a concave piecewise linear function. Doing so, divide

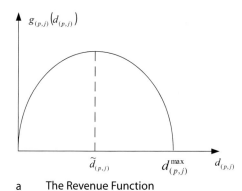
a The Revenue Function

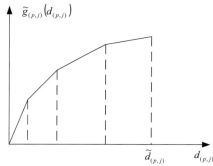
b Approximation of the Revenue Function

Bilinear Programming: Applications in the Supply Chain Management, Figure 2
The revenue function and its approximation

$[0, \tilde{d}_{(p,j)}]$ into intervals of equal length, and let $d^k_{(p,j)}$, $k \in \{1, \ldots, N\} \bigcup \{0\} = K \bigcup \{0\}$, denote the end points of the intervals. Then the approximation can be defined as

$$\tilde{g}_{(p,j)}(\lambda_{(p,j)}) = \sum_{k=1}^{N} g^k_{(p,j)} \lambda^k_{(p,j)},$$

where $g^k_{(p,j)} = g_{(p,j)}(d^k_{(p,j)}) = f_{(p,j)}(d^k_{(p,j)}) d^k_{(p,j)}$, $\sum_{k=0}^{N} \lambda^k_{(p,j)} = 1$, and $\lambda^k_{(p,j)} \geq 0, \forall p \in P, j \in \Delta$ (see Fig. 2b).

Let $x^k_{(p,i,j)}$ denote the amount of product p that is produced at time i and sold at time j using the unit price $g^k_{(p,j)}/d^k_{(p,j)} = f^k_{(p,j)} = f_{(p,j)}(d^k_{(p,j)})$. In addition, let $y_{(p,i)}$ denote a binary variable, which equals one if $\sum_k \sum_j x^k_{(p,i,j)} > 0$ and zero otherwise. Costs associated with the production process include inventory costs $c^{in}_{(p,i,j)}$, production costs $c^{pr}_{(p,i)}$, and setup costs $c^{st}_{(p,i)}$. At last, let C_i represent the production capacity at time i, which is "shared" by all products. Using those definitions, one can construct a linear mixed integer formulation of the problem. Below we provide a simplified formulation of the problem, where the variables $\lambda_{(p,j)}$ are eliminated from the formulation. For the details on the mathematical formulation of the problem and its simplification we refer to [15].

$$\max_{x,y} \sum_{p \in P} \sum_{i \in \Delta} \left[\sum_{j \in \Delta | i \leq j} \sum_{k \in K} q^k_{(p,i,j)} x^k_{(p,i,j)} - c^{st}_{(p,i)} y_{(p,i)} \right] \tag{15}$$

$$\sum_{p \in P} \sum_{j \in \Delta | i \leq j} \sum_{k \in K} x^k_{(p,i,j)} \leq C_i, \forall i \in \Delta, \tag{16}$$

$$\sum_{j \in \Delta | i \leq j} \sum_{k \in K} x^k_{(p,i,j)} \leq C_i y_{(p,i)},$$
$$\forall p \in P \text{ and } i \in \Delta, \tag{17}$$

$$\sum_{k \in K} \sum_{i \in \Delta | i \leq j} \frac{x^k_{(p,i,j)}}{d^k_{(p,j)}} \leq 1,$$
$$\forall p \in P \text{ and } j \in \Delta, \tag{18}$$

$$x^k_{(p,i,j)} \geq 0, \quad y_{(p,i)} \in \{0, 1\},$$
$$\forall p \in P, i, j \in \Delta \text{ and } k \in K, \tag{19}$$

where $q^k_{(p,i,j)} = f^k_{(p,j)} - c^{in}_{(p,i,j)} - c^{pr}_{(p,i)}$.

Let $X = \{x | x \geq 0 \text{ and } x^k_{(p,i,j)} \text{ be feasible to (16) and (18)}\}$, and $Y = [0, 1]^{|P||\Delta|}$. Consider the following bilinear problem.

$$\max_{x \in X, y \in Y} \varphi(x, y) =$$

$$\sum_{p \in P} \sum_{i \in \Delta} \left[\sum_{j \in \Delta | i \leq j} \sum_{k \in K} q^k_{(p,i,j)} x^k_{(p,i,j)} - c^{st}_{(p,i)} \right] y_{(p,i)}$$

$$\tag{20}$$

Theorem 4 (see [15]) *A global maximum of the bilinear problem (20) is a solution or can be transformed into a solution of the problem (15)–(19).*

Methods

In the previous section, we have discussed several problems arising in the supply chain management. To solve the bilinear formulations of the problems, one can employ techniques applicable for general bilinear problems. In particular, a cutting plain algorithm proposed by Konno can be applied to find a global solution of the problems. In addition, he proposes an iterative procedure, which converges to a local minimum of the problem in a finite number of iterations. For details on the procedure, which is also known as "mountain climbing" procedure (MCP), and the cutting plain algorithm we refer to the paper [11] or Bilinear Programming section of this encyclopedia.

Below, we discuss problem specific difficulties of applying the above mentioned algorithms and some effective heuristic procedures, which are able to provide a near optimum solution using negligible computer resources. The MCP, which is used by the heuristics to find a local minimum/maximum of the problems, is very fast due to a special structure of both LP problems employed by the procedure. However, to obtain a high quality solution, in some problems it is necessary to solve a sequence of approximate problems. The bilinear formulations of the supply chain problems typically have many local minima. Therefore, cutting plain algorithms may require many cuts to converge. By combining the heuristic procedures with the cutting plain algorithm, one can reduce the number of cuts by generating deep cuts.

One of the main properties of a bilinear problem with a disjoint feasible region is that by fixing vectors x or y to a particular value, the problem reduces to a linear one. The "mountain climbing" procedure employs this property and iteratively solves two linear problems by fixing the corresponding vectors to the solution of the corresponding linear programs. In the case of concave piecewise linear network flow problem, given the vector $\hat{x}$, the problem (5)–(8) can be decomposed into $|A|$ problems,

$$\min_{\{y_a^k | k \in K_a\}} \sum_{k \in K_a} [c_a^k \hat{x}_a + s_a^k] y_a^k$$

$$\text{s.t.} \quad \sum_{k \in K_a} y_a^k = 1, \quad y_a^k \geq 0 \quad \forall k \in K_a.$$

Furthermore, it can be shown that a solution of the problem is a binary vector, which has to satisfy the inequality

$$\sum_{k \in K_a} \xi_a^{k-1} y_a^k \leq \hat{x}_a \leq \sum_{k \in K_a} \xi_a^k y_a^k.$$

As a result, one can employ a search technique by assigning $y_a^k = 1$ if $\xi_a^{k-1} \leq \hat{x}_a \leq \xi_a^k$ and $y_a^k = 0$, $\forall k \in K_a, k \neq \hat{k}$. On the other hand, by fixing the vector y to the value of the constructed vector $\hat{y}$, the problem (5)–(8) reduces to the following network flow problem.

$$\min_x \sum_{a \in A} \left[\sum_{k \in K_a} c_a^k \hat{y}_a^k \right] x_a$$

$$\text{s.t.} \quad Bx = b, \quad x_a \geq 0, \quad \forall a \in A$$

Observe that $\sum_{k \in K_a} c_a^k \hat{y}_a^k = c_a^{\hat{k}}$, and different vectors $\hat{y}$ change the cost vector in the problem.

Although the MCP converges to a local minimum, it can provide a near optimum solution for the problem (5)–(8) if the initial vector $\hat{y}$ is such that $\hat{y}_a^{n_a} = 1$ and $\hat{y}_a^k = 0$, $\forall k \in K_a, k \neq n_a$. The effectiveness of the procedure is partially due to the fact that in the supply chain problems $f_a(x_a)$ is an increasing function. In addition, the procedure requires less computer resources to converge because both linear problems are relatively easy to solve. A detailed description of the procedure, properties of the linear problems, and computational experiments can be found in [13].

In the case of fixed charge network flow problems, it is not obvious how to choose the vector ε. Theorem 3 guarantees the equivalence between the fixed charge network flow problem and the bilinear problem (12)–(14) if $\varepsilon_a \in (0, \delta]$. However, according to the definition, it is necessary to find all vertices of the feasible region to compute the value of δ, which is computationally expensive. Even if the correct value of δ is known, typically it is a very small number. As a result, the value of ε_a is close to zero, and $c_a^{\varepsilon_a}$ is very large compared to the value of c_a. The later creates some difficulties for finding a global solution of the bilinear problem. In particular, the MCP may converge to a local minimum, which is far from being a global solution.

To overcome those difficulties, [14] proposes a procedure where it gradually decreases the value of ε (see Algorithm 1). The algorithm starts from an initial value for the vector ε, i.e., $\varepsilon_a = \lambda_a$. After constructing the corresponding bilinear problem, it employs the MCP to find a local minimum of the problem. If the stopping criteria is not satisfied, the value of ε is updated, i.e., $\varepsilon_a = \alpha \varepsilon_a$ where $\alpha \in (0, 1)$, and the algorithm again solves the updated bilinear problem using the current solution as an initial vector for the MCP.

The choice of α has a direct influence on the CPU time of the algorithm and the quality of the solution. Specifically, if the value of α is closer to one, then due to the fact that ε decreases slowly, the algorithm requires many iterations to stop. On the other hand, if the values of the parameter is closer to zero, it may worsen the quality of the solution. A proper choice of the parameter depends on the problem, and it should be chosen by trials and errors. In the paper [14], the authors test the algorithm on various randomly generated test problems and found satisfactory to choose $\alpha = 0.5$.

As for the stopping criteria, it is possible to show that the solution of the final bilinear problem is the solution of the fixed charge network flow problem if on Step 2 one is able to find a global solution of the corresponding bilinear problems. For details on the numerical experiments, stopping criteria and other properties of the algorithm, we refer to [14].

In the problems with pricing decisions, one may also experience some difficulties to employ the MCP for finding a near optimum solution. To explore the properties of the problem, consider the following two linear problems, which are constructed from the problem (20) by fixing either vector x or y to the value of the vector $\hat{x}$ or $\hat{y}$, respectively.

LP_1:

$$\max_{y \in Y} \sum_{p \in P} \sum_{i \in \Delta} \left[\sum_{j \in \Delta | i \leq j} \sum_{k \in K} q^k_{(p,i,j)} \hat{x}^k_{(p,i,j)} - c^{st}_{(p,i)} \right] y_{(p,i)}$$

LP_2:

$$\max_{x \in X} \sum_{p \in P} \sum_{i \in \Delta} \sum_{j \in \Delta | i \leq j} \sum_{k \in K} \left[q^k_{(p,i,j)} \hat{y}_{(p,i)} \right] x^k_{(p,i,j)}.$$

The MCP solves iteratively LP_1 and LP_2 problems, where the solution of the first problem is used to fix the corresponding vector in the second problem. However, if one of the components of the vector y equals to zero during one of the iterations, e.g., $\hat{y}_{(p,i)} = 0$, then in the second problem coefficients of the corresponding variables $x^k_{(p,i,j)}$ are equal to zero as well. As a result, changes in the values of those variables do not have any influence on the objective function value. Furthermore, because the products "share" the capacity and other products may have positive coefficients in the objective,

Step 1: Let $\varepsilon_a \leftarrow \lambda_a$, $x^0_a \leftarrow 0$, $y^0_a \leftarrow 0$, and $m \leftarrow 1$.

Step 2: Find a local minimum of the problem (12)-(14) using the MCP. Let (x^m, y^m) denote the solution found by the algorithm.

Step 3: If $\exists a \in A$ such that $x^m_a \in (0, \varepsilon^m_a)$ then $\varepsilon_a \leftarrow \alpha \varepsilon_a$, $m \leftarrow m + 1$, and go to step 2. Otherwise, stop.

Bilinear Programming: Applications in the Supply Chain Management, Algorithm 1

it is likely that at optimum of LP_2, $\hat{x}^k_{(p,i,j)} = 0$, $\forall j \in \Delta$, $k \in K$. From the later, it follows that $\hat{y}_{(p,i)} = 0$ during the next iteration, and one concludes that if some products are eliminated from the problem during the iterative process, the MCP does not consider them again. Therefore, it is likely that the solution returned by the algorithm is far from being a global one. To avoid zero coefficients in the objective of LP_2, [15] proposes an approximation to the problem (20), which can be used in the MCP to find a near optimum solution.

To construct the approximate problem, let

$$\varphi^1_{(p,i)}(x_{(p,i)}) = \sum_{j \in \Delta | i \leq j} \sum_{k \in K} q^k_{(p,i,j)} x^k_{(p,i,j)} - c^{st}_{(p,i)},$$

and

$$\varphi^2_{(p,i)}(x_{(p,i)}) = \frac{\varepsilon_{(p,i)}}{\varepsilon_{(p,i)} + c^{st}_{(p,i)}} \sum_{j \in \Delta | i \leq j} \sum_{k \in K} q^k_{(p,i,j)} x^k_{(p,i,j)},$$

Step 1: Let $\varepsilon_{(p,i)}$ be a sufficiently large number, $y^0_{(p,i)} = 1$, $\forall p \in P$, $i \in \Delta$, and $m \leftarrow 0$.

Step 2: Construct the approximation problem (21), and find a local maximum of the problem using the MSP. Let (x^{m+1}, y^{m+1}) denote the solution returned by the algorithm.

Step 3: If $\exists p \in P$ and $i \in \Delta$ such that $\sum_{j \in \Delta | i \leq j} \sum_{k \in K} q^k_{(p,i,j)} x^{(m+1)k}_{(p,i,j)} - c^{st}_{(p,i)} \leq \varepsilon^m_{(p,i)}$ and $\sum_{j \in \Delta | i \leq j} \sum_{k \in K} x^{(m+1)k}_{(p,i,j)} > 0$ then $\varepsilon \leftarrow \alpha \varepsilon$, $m \leftarrow m + 1$ and go to Step 2. Otherwise, stop.

Bilinear Programming: Applications in the Supply Chain Management, Algorithm 2

where $\varepsilon_{(p,i)} > 0$, and $x_{(p,i)}$ is the vector of $x^k_{(p,i,j)}$. Using those functions, construct the following bilinear problem

$$\max_{x \in X, y \in Y} \varphi^\varepsilon(x, y) = \sum_{p \in P} \sum_{i \in \Delta} \left[\varphi^1_{(p,i)}(x_{(p,i)}) y_{(p,i)} + \varphi^2_{(p,i)}(x_{(p,i)})(1 - y_{(p,i)}) \right], \quad (21)$$

where the feasible region is the same as in the problem (20). The authors show that $\varphi^\varepsilon(x, y)$ approximates the function $\varphi(x, y)$ from above.

Theorem 5 (see [15]) *There exists a sufficiently small $\varepsilon > 0$ such that a solution of the problem (20) is a solution of the problem (21).*

Algorithm 2 starts from a sufficiently large value of $\varepsilon_{(p,i)}$ and finds a local maximum of the corresponding bilinear problem (21) using the MCP. If the stopping criteria is not satisfied then it updates the value of ε to $\alpha\varepsilon$, updates the bilinear problem (21), and employs the MCP to find a better solution. Similar to the fixed charge network flow problem, the choice of α has a direct influence on the CPU time of the algorithm and the quality of the returned solution. The running time of the algorithm and the quality of the solution for the different values of α are studied in [15].

In addition to α, one has to find a proper initial value for the parameter $\varepsilon_{(p,i)}$. Ideally, it should be equal to the maximum profit that can be generated by producing only product p at time i. However, it requires solving a linear problem for each pair $(p, i) \in P \times \Delta$, which is computationally expensive. On the other hand, it is not necessary to find an exact solution of those LPs, and one might consider a heuristic procedure which provides a quality solution within a reasonable time. One of such procedures is discussed in [15].

References

1. Barr R, Glover F, Klingman D (1981) A New Optimization Method for Large Scale Fixed Charge Transportation Problems. Oper Res 29:448–463
2. Cabot A, Erenguc S (1984) Some Branch-and-Bound Procedures for Fixed-Cost Transportation Problems. Nav Res Logist Q 31:145–154
3. Cooper L, Drebes C (1967) An Approximate Solution Method for the Fixed Charge Problem. Nav Res Logist Q 14:101–113
4. Diaby M (1991) Successive Linear Approximation Procedure for Generalized Fixed-Charge Transportation Problem. J Oper Res Soc 42:991–1001
5. Gray P (1971) Exact Solution for the Fixed-Charge Transportation Problem. Oper Res 19:1529–1538
6. Guisewite G, Pardalos P (1990) Minimum concave-cost network flow problems: applications, complexity, and algorithms. Ann Oper Res 25:75–100
7. Kennington J, Unger V (1976) A New Branch-and-Bound Algorithm for the Fixed Charge Transportation Problem. Manag Sci 22:1116–1126
8. Khang D, Fujiwara O (1991) Approximate Solution of Capacitated Fixed-Charge Minimum Cost Network Flow Problems. Netw 21:689–704
9. Kim D, Pardalos P (1999) A Solution Approach to the Fixed Charge Network Flow Problem Using a Dynamic Slope Scaling Procedure. Oper Res Lett 24:195–203
10. Kim D, Pardalos P (2000) Dynamic Slope Scaling and Trust Interval Techniques for Solving Concave Piecewise Linear Network Flow Problems. Netw 35:216–222
11. Konno H (1976) A Cutting Plane Algorithm for Solving Bilinear Programs. Math Program 11:14–27
12. Kuhn H, Baumol W (1962) An Approximate Algorithm for the Fixed Charge Transportation Problem. Nav Res Logist Q 9:1–15
13. Nahapetyan A, Pardalos P (2007) A Bilinear Relaxation Based Algorithm for Concave Piecewise Linear Network Flow Problems. J Ind Manag Optim 3:71–85
14. Nahapetyan A, Pardalos P (2008) Adaptive Dynamic Cost Updating Procedure for Solving Fixed Charge Network Flow Problems. Comput Optim Appl 39:37–50. doi:10.1007/s10589-007-9060-x
15. Nahapetyan A, Pardalos P (2008) A Bilinear Reduction Based Algorithm for Solving Capacitated Multi-Item Dynamic Pricing Problems. Comput Oper Res J 35:1601–1612. doi:10.1016/j.cor.2006.09.003
16. Palekar U, Karwan M, Zionts S (1990) A Branch-and-Bound Method for Fixed Charge Transportation Problem. Manag Sci 36:1092–1105

Bi-Objective Assignment Problem

JACQUES TEGHEM
Lab. Math. & Operational Research Fac.,
Polytechn. Mons, Mons, Belgium

MSC2000: 90C35, 90C10

Article Outline

Keywords
Direct Methods

Two-Phase Methods
　First Step
　Second Step
Heuristic Methods
　Preliminaries
　Determination of $PE(\lambda^{(l)})$, $l = 1, \ldots, L$
　Generation of $\widehat{E(P)}$
　Concluding Remarks
See also
References

Keywords

Multi-objective programming; Combinatorial optimization; Assignment

Until recently (1998), *multi-objective combinatorial optimization* (MOCO) did not receive much attention in spite of its potential applications. The reason is probably due to specific difficulties of MOCO models as pointed out in ▶ Multi-objective combinatorial optimization. Here we consider a particular bi-objective MOCO problem, the *assignment problem* (AP). This is a basic well-known combinatorial optimization problem, important for applications and as a subproblem of more complicated ones, like the transportation problem, distribution problem or traveling salesman problem. Moreover, its mathematical structure is very simple and there exist efficient polynomial algorithms to solve it in the single objective case, like the *Hungarian method*. In a bi-objective framework, the assignment problem can be formulated as:

$$(P)\begin{cases} 'min' \quad z_k(X) = \sum_{i=1}^{n}\sum_{j=1}^{n} c_{ij}^{(k)} x_{ij}, \\ \qquad k = 1, 2, \\ \sum_{j=1}^{n} x_{ij} = 1, \quad i = 1, \ldots, n, \\ \sum_{i=1}^{n} x_{ij} = 1, \quad j = 1, \ldots, n, \\ x_{ij} \in \{0, 1\} \end{cases}$$

where c_{ij}^k are nonnegative integers and $X = (x_{11}, \ldots, x_{nn})$. Our aim is to generate the set of efficient solutions $E(P)$. It is important to stress that the distinction between the *supported efficient solutions* (belonging to $SE(P)$), i. e. those which are optimal solutions of the single objective problem obtained by a linear aggregation of the objectives, and the *nonsupported efficient solutions* (belonging to $NSE(P) = E(P)\setminus SE(P)$) (see ▶ Multi-objective integer linear programming) is still necessary even if the constraints of the problem satisfy the so-called 'totally unimodular' or 'integrality' property: when this property is verified, the integrality constraints of the single objective problem can be relaxed without any deterioration of the objective function, i. e. the optimal values of the variables are integer even if only the linear relaxation of the problem is solved. It is well known that the single objective assignment problem satisfies this integrality property, and thus this is true for the problem (see ▶ Multi-objective combinatorial optimization):

$$(P_\lambda)\begin{cases} \min \quad z_\lambda(X) = \lambda_1 z_1(X) + \lambda_2 z_2(X) \\ \sum_{j=1}^{n} x_{ij} = 1, \quad i = 1, \ldots, n, \\ \sum_{i=1}^{n} x_{ij} = 1, \quad j = 1, \ldots, n, \\ x_{ij} \in \{0, 1\} \\ \lambda_1 \geq 0, \quad \lambda_2 \geq 0. \end{cases}$$

Nevertheless, in the multi-objective framework, there exist nonsupported efficient solutions, as indicated by the following didactic example:

$$C^{(1)} = \begin{pmatrix} 5 & 1 & 4 & 7 \\ 6 & 2 & 2 & 6 \\ 2 & 8 & 4 & 4 \\ 3 & 5 & 7 & 1 \end{pmatrix},$$

$$C^{(2)} = \begin{pmatrix} 3 & 6 & 4 & 2 \\ 1 & 3 & 8 & 3 \\ 5 & 2 & 2 & 3 \\ 4 & 2 & 3 & 5 \end{pmatrix}.$$

The values of the feasible solutions are represented in the objective space in Fig. 1

There are four supported efficient solutions, corresponding to points Z_1, Z_2, Z_3 and Z_4; two nonsupported efficient solutions corresponding to points Z_5 and Z_6; the eighteen other solutions are nonefficient.

Remark 1 In [7], D.J. White analyzes a particular case of problem (P) corresponding to

$$c_{ij}^{(k)} = c_{ij}\delta_{jk}$$

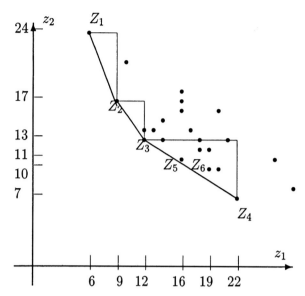

Bi-Objective Assignment Problem, Figure 1
The feasible points in the (z_1, z_2)-space for the didactic example

where

$$\delta_{jk} = \begin{cases} 1 & \text{if } j = k, \\ 0 & \text{if } j \neq k. \end{cases}$$

For this particular problem, he proves that $E(P) = SE(P)$.

We consider the problem to generate $E(P)$ and (see ▶ Multi-objective combinatorial optimization) we can distinguish three methodologies: direct methods; two-phase methods and heuristic methods.

Direct Methods

In [1], the authors propose a theoretical enumerative procedure to generate $E(P)$ in the order of increasing values of z_1: at each step they consider the admissible edges incident at the current basis and among the set of possible new bases, they selected the one with the best value of z_1: they affirm that this basis corresponds to a new efficient solution. As proved by the example described above, this procedure appears false: for instance from point $Z_5 = (16, 11)$, corresponding to the solution $x_{14} = x_{22} = x_{33} = x_{41} = 1$, it is impossible to obtain by an unique change of basis the following point $Z_6 = (19, 10)$, corresponding to the solution $x_{13} = x_{21} = x_{34} = x_{42} = 1$. Moreover the real difficulties induced by the high degeneracy of the assignment problem are not taken into account in [1].

Two-Phase Methods

The principle of this approach, and the first phase designed to generate $SE(P)$, are described in ▶ Multi-objective combinatorial optimization; by complementary, we analyse here the second phase [3].

The purpose is to examine each triangle $\Delta Z_r Z_s$ determined by two successive solutions X^r and X^s of $SE(P)$ (see Fig. 2) and to determine the possible nonsupported solutions whose image lies inside this triangle. We note that

$$z_\lambda(X) = \lambda_1 z_1(X) + \lambda_2 z_2(X)$$

with $\lambda_1 = z_{2r} - z_{2s}$ and $\lambda_2 = z_{1s} - z_{1r}$ and $c_{ij}^{(\lambda)} = \lambda_1 c_{ij}^{(1)} + \lambda_2 c_{ij}^{(2)}$.

In the first phase, the objective function $z_\lambda(X)$ has been optimized by the Hungarian method giving

- $\widetilde{z}_\lambda = \lambda_1 z_{1r} + \lambda_2 z_{2r} = \lambda_1 z_{1s} + \lambda_2 z_{2s}$, the optimal value of $z_\lambda(X)$;
- the optimal value of the reduced cost $\bar{c}_{ij}^{(\lambda)} = c_{ij}^{(\lambda)} - (u_i + v_j)$, where u_i and v_j are the dual variables associated respectively to constraints i and j of problem (P_λ).

At optimality, we have $\bar{c}_{ij}^{(\lambda)} \geq 0$ and $\widetilde{x}_{ij} = 1 \Rightarrow \bar{c}_{ij}^{(\lambda)} = 0$.

First Step

We consider $L = \{x_{ij} : \bar{c}_{ij}^{(\lambda)} > 0\}$. To generate nonsupported efficient solution in triangle $\Delta Z_r Z_s$, each variable $x_{ij} \in L$ is candidate to be fixed to 1. Nevertheless, a variable can be eliminated if we are sure that the reoptimization of problem $(P\lambda)$ will provide a dominated point in the objective space. If $x_{ij} \in L$ is set to 1, a lower bound l_{ij} of the increase of $\widetilde{z}_\lambda$ is given by

$$l_{ij} = \bar{c}_{ij}^{(\lambda)} + \min\left(\bar{c}_{i_r j_r}^{(\lambda)}; \min_{k \neq j} \bar{c}_{i_r k}^{(\lambda)} + \min_{k \neq i} \bar{c}_{k j_r}^{(\lambda)}; \right.$$
$$\left. \bar{c}_{i_s j_s}^{(\lambda)}; \min_{k \neq j} \bar{c}_{i_s k}^{(\lambda)} + \min_{k \neq i} \bar{c}_{k j_s}^{(\lambda)} \right),$$

where the indices i_r and j_r (i_s and j_s) are such that in the solution X^r (respectively, X^s) we have

$$x_{i_r j} = x_{i j_r} = 1, \qquad (x_{i_s j} = x_{i j_s} = 1).$$

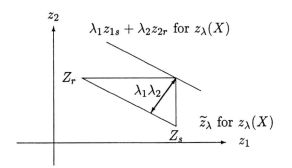

Bi-Objective Assignment Problem, Figure 2
Test 1

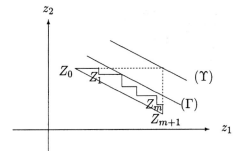

Bi-Objective Assignment Problem, Figure 3
Test 2

Effectively, to re-optimize problem (P_λ) with $x_{ij} = 1$, in regard with its optimal solution X^r (respectively, X^s), it is necessary to determine, at least, a new assignment in the line i_r (respectively, i_s) and in the column j_r (respectively, j_s). But clearly, to be inside the triangle $\triangle Z_r Z_s$, we must have (see Fig. 2)

$$\widetilde{z}_\lambda + l_{ij} < \lambda_1 z_{1s} + \lambda_2 z_{2r}.$$

Consequently, we obtain the following fathoming test:
- (Test 1): $x_{ij} \in L$ can be eliminated if $\widetilde{z}_\lambda + l_{ij} \geq \lambda_1 z_{1s} + \lambda_2 z_{2r}$ or, equivalently, if $l_{ij} \geq \lambda_1 \lambda_2$.

So in this first step, the lower bound l_{ij} is determined for all $x_{ij} \in L$; the list is ordered by increasing values of l_{ij}.

Only the variables not eliminated by test 1 are kept. Problem (P_λ) is re-optimized successively for each noneliminated variable; let us note that only one iteration of the Hungarian method is needed. After the optimization, the solution is eliminated if its image in the objective space is located outside the triangle $\triangle Z_r Z_s$. Otherwise, a nondominated solution is obtained and put in a list NS_{rs}; at this time, the second step is applied.

Second Step

When nondominated points $Z_1, \ldots, Z_m \in NS_{rs}$ are found inside the triangle $\triangle Z_r Z_s$, then test 1 can be improved. Effectively (see Fig. 3), in this test the value

$$\lambda_1 z_{1s} + \lambda_2 z_{2r}$$

can be replaced by the lower value

$$(\Gamma) \equiv \max_{i=0,\ldots,m} \left(\lambda_1 z_{1,i+1} + \lambda_2 z_{2,i} \right),$$

where $Z_o \equiv Z_r$, $Z_m + 1 \equiv Z_s$, with $\Upsilon \equiv \lambda_1 z_{1, m+1} + \lambda_2 z_{2, 0}$.

The new value corresponds to an updated upper bound of $z_\lambda(X)$ for nondominated points. More variables of L can be eliminated with the new test
- (Test 2): $x_{ij} \in L$ can be eliminated if

$$\widetilde{z}_\lambda + l_{ij} \geq \max_{i=o,\ldots,m} \left(\lambda_1 z_{1,i+1} + \lambda_2 z_{2,i} \right).$$

Each time a new nondominated point is obtained, the list NS_{rs} and the test 2 are updated. The procedure stops when all the $x_{ij} \in L$ have been either eliminated or analyzed. At this moment the list NS_{rs} contains the nonsupported solutions corresponding to the triangle $\triangle Z_r Z_s$.

When each triangle have been examined

$$NSE(P) = \cup_{rs} NS_{rs}.$$

Numerical results are given in [3].

Heuristic Methods

As described in ▶ Multi-objective combinatorial optimization, the MOSA method is an adaptation of the simulated annealing heuristic procedure to a multi-objective framework. Its aim is to generate a good approximation, denoted $\widehat{E(P)}$, of $E(P)$ and the procedure is valid for any number $K \geq 2$ of objectives. Similarly to a single objective heuristic in which a potentially optimal solution emerges, in the MOSA method the set $\widehat{E(P)}$ will contain potentially efficient solutions.

Preliminaries

- A wide diversified set of weights is considered: different weight vectors $\lambda^{(l)}$, $l \in L$, are generated where $\lambda^{(l)} = (\lambda_k^{(l)})_{k=1,\ldots,K}$ with $\lambda_k^{(l)} > 0$, $\forall k$ and

$$\sum_{k=1}^{K} \lambda_k^{(l)} = 1, \quad \forall l \in L.$$

- A scalarizing function $s(z, \lambda)$ is chosen, the effect of this choice on the procedure is small due to the stochastic character of the method. The weighted sum is very well known and it is the easiest scalarizing function:

$$s(z, \lambda) = \sum_{k=1}^{K} \lambda_k z_k.$$

- The three classic parameters of a simulated annealing procedure are initialized
  - T_0: initial temperature (or alternatively an initial acceptance probability P_0);
  - α (< 1): the cooling factor;
  - N_{step}: the length of temperature step in the cooling schedule;
  and the two stopping criteria are fixed:
  - T_{stop}: the final temperature;
  - N_{stop}: the maximum number of iterations without improvement
- A neighborhood $V(X)$ of feasible solutions in the vicinity of X is defined. This definition is problem dependent. It is particularly easy to define $V(X)$ in the case of the assignment problem: if X is characterized by $x_{ij_i} = 1$, $i = 1, \ldots, n$, then $V(X)$ contains all the solutions Y satisfying

$$y_{ij_i} = 1, \quad i \in \{1, \ldots, n\} \setminus \{a, b\},$$
$$y_{aj_b} = y_{bj_a} = 1,$$

where a, b are chosen randomly in $\{1, \ldots, n\}$.

Determination of $PE(\lambda^{(l)})$, $l = 1, \ldots, L$

For each $l \in L$ the following procedure is applied to determine a list $PE(\lambda^{(l)})$ of potentially efficient solutions.
a) (Initialization):
  - Draw at random an initial solution X_0.
  - Evaluate $z_k(X_0)$, $\forall k$.
  - $PE(\lambda^{(l)}) = \{X_0\}$; $N_c = n = 0$.
b) (Iteration n):
  - Draw at random a solution $Y \in V(X_n)$
  - evaluate $z_k(Y)$ and determine

$$\Delta z_k = z_k(Y) - z_k(X_n), \quad \forall k.$$

  - Calculate

$$\Delta s = s(z(Y), \lambda) - s(z(X_n), \lambda).$$

If $\Delta s \leq 0$, we accept the new solution:

$$X_{n+1} \leftarrow Y \qquad N_c = 0.$$

Else we accept the new solution with a certain probability $p = \exp(-\Delta s/T_n)$:

$$X_{n+1} \begin{cases} \overset{p}{\leftarrow} Y, & N_c = 0, \\ \overset{1-p}{\leftarrow} X_n, & N_c = N_c + 1. \end{cases}$$

  - If necessary, update the list $PE(\lambda^{(l)})$ in regard to the solution Y.
  - $n \leftarrow n + 1$
    IF $n(\mod N_{\text{step}}) = 0$
    THEN $T_n = \alpha T_{n-1}$; ELSE $T_n = T_{n-1}$.
    IF $N_c = N_{\text{stop}}$ OR $T < T_{\text{stop}}$
    THEN stop ELSE iterate.

Generation of $\widehat{E(P)}$

Because of the use of a scalarizing function, a given set of weights $\lambda^{(l)}$ induces a privileged direction on the efficient frontier. The procedure generates only a good subset of potentially efficient solutions in that direction. Nevertheless, it is possible to obtain solutions which are not in this direction, because of the large exploration of D at high temperature; these solutions are often dominated by some solutions generated with other weight sets.

To obtain a good approximation $\widehat{E(P)}$ to $E(P)$ it is thus necessary to filter the set

$$\cup_{l=1}^{|L|} PE(\lambda^{(l)})$$

by pairwise comparisons to remove the dominated solutions. This filtering procedure is denoted by $\wedge$ such that

$$\widehat{E(P)} = \wedge_{l=1}^{|L|} PE(\lambda^{(l)}).$$

A great number of experiments is required to determine the number L of set of weights sufficient to give a good approximation of the whole efficient frontier.

Concluding Remarks

Details and numerical results are given in [3] and [5].

Let us add that it is easy to adapt the MOSA method in an interactive way [2]; a special real case study of an assignment problem is treated in this manner in [6].

See also

- ▶ Assignment and Matching
- ▶ Assignment Methods in Clustering
- ▶ Communication Network Assignment Problem
- ▶ Decision Support Systems with Multiple Criteria
- ▶ Estimating Data for Multicriteria Decision Making Problems: Optimization Techniques
- ▶ Financial Applications of Multicriteria Analysis
- ▶ Frequency Assignment Problem
- ▶ Fuzzy Multi-Objective Linear Programming
- ▶ Maximum Partition Matching
- ▶ Multicriteria Sorting Methods
- ▶ Multi-Objective Combinatorial Optimization
- ▶ Multi-Objective Integer Linear Programming
- ▶ Multi-Objective Optimization and Decision Support Systems
- ▶ Multi-Objective Optimization: Interaction of Design and Control
- ▶ Multi-Objective Optimization: Interactive Methods for Preference Value Functions
- ▶ Multi-Objective Optimization: Lagrange Duality
- ▶ Multi-Objective Optimization: Pareto Optimal Solutions, Properties
- ▶ Multiple Objective Programming Support
- ▶ Outranking Methods
- ▶ Portfolio Selection and Multicriteria Analysis
- ▶ Preference Disaggregation
- ▶ Preference Disaggregation Approach: Basic Features, Examples From Financial Decision Making
- ▶ Preference Modeling
- ▶ Quadratic Assignment Problem

References

1. Malhotra R, Bhatia HL, Puri MC (1982) Bicriteria assignment problem. Oper Res 19(2):84–96
2. Teghem J, Tuyttens D, Ulungu EL (2000) An interactive heuristic method for multi-objective combinatorial optimization. Comput Oper Res 27:621–624
3. Tuyttens D, Teghem J, Fortemps Ph, Van Nieuwenhuyse K (1997) Performance of the MOSA method for the bicriteria assignment problem. Techn Report Fac Polytechn Mons (to appear in J. Heuristics)
4. Ulungu EL, Teghem J (1994) Multi-objective combinatorial optimization problems: A survey. J Multi-Criteria Decision Anal 3:83–104
5. Ulungu EL, Teghem J, Fortemps Ph, Tuyttens D (1999) MOSA method: A tool for solving MOCO problems. J Multi-Criteria Decision Anal 8:221–236
6. Ulungu EL, Teghem J, Ost Ch (1998) Efficiency of interactive multi-objective simulated annealing through a case study. J Oper Res Soc 49:1044–1050
7. White DJ (1984) A special multi-objective assignment problem. J Oper Res Soc 35(8):759–767

Biquadratic Assignment Problem
BiQAP

LEONIDAS PITSOULIS
Princeton University, Princeton, USA

MSC2000: 90C27, 90C11, 90C08

Article Outline

Keywords
See also
References

Keywords

Optimization

The *biquadratic assignment problem* was first introduced by R.E. Burkard, E. Çela and B. Klinz [2], as a nonlinear assignment problem that has applications in very large scale integrated (VLSI) circuit design. Given two fourth-dimensional arrays $A = (a_{ijkl})$ and $B = (b_{mpst})$ with n^4 elements each, the nonlinear integer programming formulation of the BiQAP is

$$\begin{cases} \min & \sum_{i,j,k,l} \sum_{m,p,s,t} a_{ijkl} b_{mpst} x_{im} x_{jp} x_{ks} x_{lt} \\ \text{s.t.} & \sum_{i=1}^{n} x_{ij} = 1, \quad j = 1, \dots, n, \\ & \sum_{j=1}^{n} x_{ij} = 1, \quad i = 1, \dots, n, \\ & x_{ij} \in \{0,1\}, \quad i,j = 1, \dots, n. \end{cases}$$

The BiQAP is a generalization of the quadratic assignment problem (cf. ▶ Quadratic assignment problem) (QAP), where the objective function is a fourth degree multivariable polynomial and the feasible domain is the assignment polytope as in the QAP. An equivalent formulation of the BiQAP using permutations is the following:

$$\min_{\phi \in S_n} \sum_{i=1}^{n} \sum_{j=1}^{n} \sum_{k=1}^{n} \sum_{l=1}^{n} a_{ijkl} b_{\phi(i)\phi(j)\phi(k)\phi(l)},$$

where S_n denotes the set of all permutations of the integer set $N = \{1, \ldots, n\}$.

Burkard, Çela and Klinz [2] showed that the BiQAP is *NP*-hard. They computed lower bounds for BiQAP derived from lower bounds of the QAP. The computational results showed that these bounds are weak and deteriorate as the dimension of the problem increases. This observation suggests that branch and bound methods (cf. also ▶ Integer programming: Branch and bound methods) will only be effective on very small instances. For larger instances, efficient heuristics, that find good-quality approximate solutions, are needed.

Burkard and Çela [1] developed several heuristics for the BiQAP, in particular deterministic improvement methods and variants of simulated annealing and tabu search. Computational experiments on test problems with known optimal solutions [1], suggest that one version of simulated annealing is best among those tested. T. Mavridou, P.M. Pardalos, L.S. Pitsoulis, and M.G.C. Resende develop a GRASP heuristic for solving the BiQAP in [3], which finds the optimal solution for all the test problems presented in [1].

See also

- ▶ Feedback Set Problems
- ▶ Generalized Assignment Problem
- ▶ Graph Coloring
- ▶ Graph Planarization
- ▶ Greedy Randomized Adaptive Search Procedures
- ▶ Quadratic Assignment Problem

References

1. Burkard RE, Çela E (1995) Heuristics for biquadratic assignment problems and their computational comparison. Europ J Oper Res 83:283–300
2. Burkard RE, Çela E, Klinz B (1994) On the biquadratic assignment problem. In: Pardalos PM, Wolkowicz H (eds) Quadratic assignment and related problems. DIMACS Amer Math Soc, Providence, RI, pp 117–146
3. Mavridou T, Pardalos PM, Pitsoulis LS, Resende MGC (1998) A GRASP for the biquadratic assignment problem. Europ J Oper Res 105(3):613–621

Bisection Global Optimization Methods

GRAHAM WOOD
Institute Information Sci. and Technol. College of Sci., Massey University, Palmerston North, New Zealand

MSC2000: 90C30, 65K05

Article Outline

Keywords
See also
References

Keywords

Lipschitz continuity; Bisection; Multidimensional bisection; Bracket; Simplex; Epigraph; System; Reduction; Elimination; Linear convergence; Tiling

The centuries-old method of bisection can be generalized to provide a global optimization algorithm for Lipschitz continuous functions. Full details of the algorithm, acceleration methods and its performance can be found in [1,6,7]. (Recall that $f : \mathbf{R}^n \to \mathbf{R}$ is *Lipschitz continuous* if there is an $M \geq 0$ such that $|f(x) - f(y)| \leq M \| x - y \|$ for all $x, y \in \mathbf{R}^n$. We then term M a *Lipschitz constant* of f.)

The familiar *bisection method* enables us to find a point of interest on the line by first bracketing the point in an interval, and then successively halving the interval. It is used in this way, for example, to find the root of a continuous function or to show that a bounded sequence always has a limit point. The bisection method is simple and convergence is assured and linear.

The bisection method can also be used (although we never think of it in this role) to find the minimum of a semi-infinite interval $[m, \infty)$, as illustrated in the left-hand side of Table 1. Given an initial interval bracket around m we examine the midpoint: if the midpoint is

Bisection Global Optimization Methods, Table 1
A comparison of the bisection method and the generalization to higher dimensions

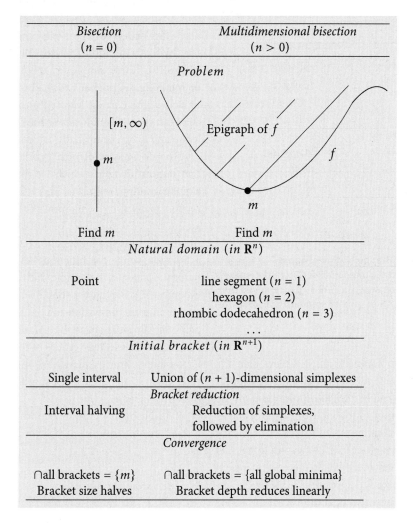

in $[m, \infty)$ then we retain the lower interval whereas if the midpoint is not in $[m, \infty)$ we retain the upper interval. It is this idea that has been generalized to higher dimensions to give the algorithm, detailed here, that has been termed in the literature *multidimensional bisection*.

It can be shown (see [7]) that the analogue in $\mathbf{R}^{n+1}$ of an upper semi-infinite interval in $\mathbf{R}$ is the *epigraph* (everything above and including the graph) of a Lipschitz continuous function. Multidimensional bisection finds the set of global minima of a Lipschitz continuous function f of n variables over a compact domain, in a manner analogous to the bisection method. At any stage in the iteration the bracket is a union of similar simplexes in $\mathbf{R}^{n+1}$, with the initial bracket a single *simplex*. (A simplex is a convex hull of affinely independent points, so a triangle, a tetrahedron and so on.) In the raw version of the algorithm the depth of the bracket decreases linearly and the infinite intersection of all brackets is the set of global minima of the graph of the function.

The algorithm works thanks to two simple facts and a very convenient piece of geometry. First, however, we note a property of a Lipschitz continuous function with Lipschitz constant M: if $x \in \mathbf{R}^n$ lies in the domain of the function and (x, y) (with $y \in \mathbf{R}$) lies in the epigraph of the function, then $(x, y) + C$ lies in the epigraph, where C is an upright spherically based cone of slope M, with apex at the origin.

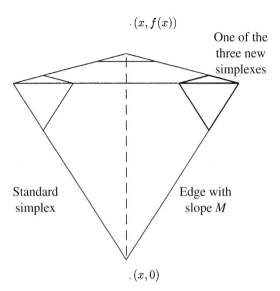

Bisection Global Optimization Methods, Figure 1
A standard simplex and the three smaller standard simplexes resulting from reduction; when $(x, f(x)) - \Delta$ is removed from the standard simplex three similar standard simplexes remain

Now for the two simple facts: if we evaluate the function f at any point in the domain, then no point higher than $(x, f(x))$ can be the global minimum on the graph of f and no point in the interior of a $(x, f(x)) - C$ can be the global minimum. Informally, this means that every evaluation of f lets us slice away an upper half space and an upside down ice-cream cone, with apex at $(x, f(x))$, from the space $\mathbf{R}^{n+1}$; we are sure the global optima are not there. These two operations coalesce in the familiar bisection method.

Now for the convenient geometry, which comes to light as soon as we attempt to generalise the bisection method. Spherically based cones are ideal to use, but hard to keep track of efficiently [3], so we use a simplicial approximation to the spherical base of the cone to make the bookkeeping easy. Such a simplex-based cone, Δ, has a cap which we call a *standard simplex*; one is shown as the large simplex in Fig. 1, for the case when $n = 2$. It fits snugly inside C, so the sloping edges have slope M. If we know that the global optimum lies in this simplex bracket and evaluate f at x, then we can remove $(x, f(x)) - \Delta$ from the space. Conveniently, this leaves three similar standard simplexes whose union must contain the global minima, as shown in Fig. 1. This process is termed *reduction* of the simplex.

What does a typical iteration of the algorithm do? At the start of each iteration the global minima are held in a *multidimensional bracket*, a union of similar standard simplexes. We denote this set of simplexes, or *system*, by $\mathcal{S}$. An iteration consists of reducing some (possibly all) of these simplexes, followed by *elimination*, or retaining the portions of the bracket at the level of, or below, the current lowest function evaluation. For this reason an iteration can be thought of informally as 'chop and drop', or formally as 'reduce and eliminate'.

How do we start off? The algorithm operates on certain *natural domains* which we must assume contain a global minimizer (just as we begin in the familiar bisection method by containing the point of interest in an interval). For functions of one variable a natural domain is an interval, for functions of two variables it is a hexagon, while for functions of three variables the natural domain is a rhombic dodecahedron (the honeycomb cell). For higher dimensions the pattern continues; in each dimension the natural domains are capable of tiling the space. By means of $n + 1$ function evaluations at selected vertices of the natural domain it is possible to bracket the global optima over the natural domain in an initial single standard simplex, termed the *initial system*.

In brief, given a Lipschitz continuous function f on a standard domain, the algorithm can be summarised as:

1. Set $i = 0$ and form the initial system S_0.
2. Form S_{i+1}, by applying reduction and then elimination to the system S_i.
3. If a stopping criterion is satisfied (such as that the variation of the system is less than a preassigned amount), then stop. Otherwise, increment i and return to Step 2.

Multidimensional bisection

By the *variation of the system* is meant the height from top to bottom of the current set of simplexes. The following example illustrates the course of a run of multidimensional bisection.

Take $f(x_1, x_2) = -e^{-x_1^2} \sin x_1 + |x_2|$, which has a global minimum on its graph at $(0.653273, 0, -0.396653)$. There are also local minima along the

Bisection Global Optimization Methods, Table 2
Example of a run of multidimensional bisection. Note how the number of simplexes in the system decreases in the 8th iteration; this corresponds to the elimination of simplexes around local, and nonglobal, minima

Iter	Simpl. in the system	Variat.	Best point to date
0	1	33.300	(10.000, −10.000, 10.000)
1	3	20.000	(10.000, 6.667, 6.667)
2	9	9.892	(1.340, 1.667, 1.505)
5	108	1.959	(25.637, 0.185, 0.185)
7	264	0.504	(0.839, 0.074, −0.294)
8	39	0.257	(0.649, −0.036, −0.361)
15	369	0.007	(0.669, 0.000, −0.396)
18	924	0.001	(0.653, 0.000, −0.397)
19	1287	0.000	(0.651, 0.000, −0.397)

x_1-axis. We use as our standard domain the regular hexagon with center at (10, 10) and radius 20, and use $M = 1$. Table 2 provides snapshots of the progress of the algorithm to convergence; it stops when the variation is less than 0.001. We carry the best point to date, shown in the final column of the table.

In this example we reduced all simplexes in the system at each iteration. This ensures that the infinite intersection of the brackets is the set of global minima. In [6] it is shown that, under certain conditions, the optimal one-step strategy is to reduce only the deepest simplex in each iteration. With this reduction and $n = 1$ multidimensional bisection is precisely the *Piyavskii–Shubert algorithm* [4,5].

Raw multidimensional bisection can require a large number of function evaluations, but can be economical with computer time (see [2]). As described so far, the method does not use the full power of the spherical cone, rather a simplicial approximation, and this approximation rapidly worsens as the dimension increases. Fortunately, much of the spherical power can be utilized very simply, by raising the function evaluation to an effective height. This is trivial to implement and has been called *spherical reduction* [6]. Reduction, as described so far, removes material only from a single simplex, whose apex determines the evaluation point. Simplexes overlap when $n \geq 2$, and it is possible to remove material from many simplexes rather than just one. This is harder to implement, but has been carried out in [1] where it is termed *complete reduction*. The algorithm operates more efficiently when such improved reduction methods are used.

Multidimensional bisection collapses to bisection with $n = 0$ when we use a primitive reduction process, one which depends only on whether the point in $\mathbf{R}^{n+1}$ considered lies in the epigraph of f; this is described in [7]. A summary comparison of bisection and multidimensional bisection is given in Table 1.

See also

▶ αBB Algorithm

References

1. Baoping Zhang, Wood GR, Baritompa WP (1993) Multidimensional bisection: the performance and the context. J Global Optim 3:337–358
2. Horst R, Pardalos PM (eds) (1995) Handbook of Global Optimization. Kluwer, Dordrecht
3. Mladineo RG (1986) An algorithm for finding the global maximum of a multimodal, multivariate function. Math Program 34:188–200
4. Piyavskii SA (1972) An algorithm for finding the absolute extremum of a function. USSR Comput Math Math Phys 12:57–67
5. Shubert BO (1972) A sequential method seeking the global maximum of a function. SIAM J Numer Anal 9:379–388
6. Wood GR (1991) Multidimensional bisection and global optimisation. Comput Math Appl 21:161–172
7. Wood GR (1992) The bisection method in higher dimensions. Math Program 55:319–337

Boolean and Fuzzy Relations

LADISLAV J. KOHOUT
Department Computer Sci., Florida State University, Tallahassee, USA

MSC2000: 03E72, 03B52, 47S40, 68T27, 68T35, 68Uxx, 91B06, 90Bxx, 91Axx, 92C60

Article Outline

Keywords
Boolean Relations
  Propositional Form
  Heterogeneous and Homogeneous Relations

The Satisfaction Set
The Extensionality Convention
The Digraph Representation
Foresets and Aftersets of Relations
Matrix Representation
Operations and Inclusions in $\mathcal{R}(A \leadsto B)$
Unary Operations
Binary Operations on Successive Relations
Matrix Formulation of the Binary Operations
Non-Associative Products of Relations
Characterization of Special Properties
of Relations Between Two Sets
Relations on a Single Set: Special Properties
Partitions IN and ON a Set
Tolerances and Overlapping Classes
Hierarchies in and on a Set:
Local and Global Orders and Pre-orders
Fuzzy Relations
Definitions
Operations and Inclusion on $\mathcal{R}_F (X \leadsto Y)$
Fuzzy Relations with Min, Max Connectives
Fuzzy Relations Based on Łukasiewicz Connectives
Fuzzy Relations With t-Norms and Co-Norms
Products: $\mathcal{R}_F(X \leadsto Y) \times \mathcal{R}_F(Y \leadsto Z) \to \mathcal{R}_F(X \leadsto Z)$
N-ary Relations
Special Properties of Fuzzy Relations
Alpha-cuts of Fuzzy Relations
Fuzzy Partitions, Fuzzy Clusters
and Fuzzy Hierarchies
Closures and Interiors with Special Properties
Applications of Relational Methods
in Engineering, Medicine and Science
Brief Review of Theoretical Development
Basic Books and Bibliographies
See also
References

Keywords

Fuzzy relations; Local relational properties; Closures; Interiors; Pre-order; Tolerances; Equivalences; BK-products; Relational compositions; Nonassociative products; Generalized morphism; Universal properties of relations; n-ary relation; Scientific applications; Medicine; Psychology; Engineering applications; Artificial intelligence; Value analysis; Decision theory

The conventional nonfuzzy relations using the classical two-valued Boolean logic connectives for defining their operations will be called crisp. The extensions that replace the 2-valued Boolean logic connectives by many-valued logic connectives will be called fuzzy. A unified approach of relations is provided here, so that the Boolean (crisp, nonfuzzy) relations and sets are just special cases of fuzzy relational structures. The first part of this entry on nonfuzzy relations can be used as reference independently, without any knowledge of fuzzy sets. The second part on fuzzy structures, however, refers frequently to the first part. This is so because most formulas in the matrix notation carry over to the many-valued logics based extensions.

In order to make this material useful not only theoretically but also in practical applications, we have paid special attention to the form in which the material is presented. There are seven distinguishing features of our approach that facilitate the unification of crisp and fuzzy relations and enhance their practical applicability:

1) Relations in their predicate forms are distinguished from their satisfaction sets.
2) Foresets and aftersets of relations are used in addition to relational predicates.
3) Relational properties are not only global but also local (important for applications).
4) Nonassociative *BK-products* are introduced and used both in definitions of relational properties and in computations.
5) The unified treatment of computational algorithms by means of matrix notation is used which is equally applicable to both crisp and fuzzy relations.
6) The theory unifying crisp and fuzzy relations makes it possible to represent a whole *finite nested family* of crisp relations with special properties as a *single* cutworthy fuzzy relation for the purpose of computation. After completing the computations, the resulting fuzzy relation is again converted by α-cuts to a nested family of crisp relations, thus increasing the computing performance considerably.
7) *Homomorphisms* between relations are extended from mappings used in the literature to general relations. This yields *generalized morphisms* important for practical solving of relational inequalities and equations.

These features were first introduced in 1977 by W. Bandler and L.J. Kohout [1] and extensively developed over the years both in theory and practical applications [7,30,52].

Boolean Relations

Propositional Form

A *binary relation* (from A to B) is given by an open predicate __P__ with two empty slots; when the first is filled with the name a of an element of A and the second with the name b of an element of B, there results a proposition, which is either true or false. If aPb is true, we write aR_Pb and say that 'a is R_P-related to b'. If $a\,P\,b$ is false, we write $a \neg R_P b$ and say that 'a is not R_P-related to b', etc. When it is unnecessary to emphasize the propositional form the subscript is dropped in R_P, writing: R, $a\,Rb$, $a \neg Rb$, respectively.

Heterogeneous and Homogeneous Relations

The lattice of all binary (two-place, 2-argument) relations from A to B is denoted by $\mathcal{R}(A \rightsquigarrow B)$. Relations of this kind are usually called *heterogeneous*. Nothing forbids the set B to be the same as A, in which case we speak of relations 'within a set' or 'in a set', or 'on a set', and call these *homogeneous*.

Relations from A to B can always be considered as relations within $A \cup B$, but so 'homogenized' relations may lose some valuable properties (discussed below), when so viewed. For this reason, we do not attempt to assimilate relations between distinct sets to those within a set.

The Satisfaction Set

The *satisfaction set* or *representative set* or *extension set* of a relation $R \in \mathcal{R}(A \rightarrow B)$ is the set of all those pairs $(a, b) \in A \times B$ for which it holds:

$$R_S = \{(a, b) \in A \times B : aRb\}.$$

Clearly R_S is a subset of the Cartesian product $A \times B$. Knowing R_P, we know R_S; knowing R_S, we know everything about R_P except the wording of its 'name' __P__.

The Extensionality Convention

This convention says that, regardless of their propositional wordings, two relations should be regarded as the same if they hold, or fail to hold between exactly the same pairs: $R_S = R_S' \Rightarrow R_P = R_P'$. In the set theory, this appears as the *axiom of extensionality*. This convention is not universally convenient; it is perhaps partly responsible for delays in the application of relation theory in the engineering, social and economical sciences and elsewhere.

Once the extensionality convention has been adopted, it becomes a matter of indifference, or mere convenience, whether a relation is given by an open predicate or by the specification of its satisfaction set. There is a one-to-one correspondence between the subsets R_S of $A \times B$ and the (distinguishable) relations R_P in $\mathcal{R}(A \rightarrow B)$. Since R_S and R_P now uniquely determine each other, the current fashion for set-theoretical parsimony suggests that they be identified. This view is common in the literature, which often *defines* relations as being satisfaction sets. We, however, maintain the distinction in principle.

Example of the failure of the extensionality convention

> $R_\geq, Q_> \in \mathcal{R}(A \rightsquigarrow B); A = \{1, 6, 8\}, B = \{0, 5, 7\}$.
> Predicates:
> $P1 := '\_\_ \geq \_\_'$ ('__ is greater than or equal to __')
> $P2 := '\_\_ > \_\_'$ ('__ is greater than __')
> Relations in their Predicate Form:
> $R_\geq = \{1 \geq 0, 8 \geq 0, 8 \geq 5, 8 \geq 7, 6 \geq 0, 6 \geq 5\}$
> $Q_> = \{1 > 0, 8 > 0, 8 > 5, 8 > 7, 6 > 0, 6 > 5\}$
> The Satisfaction Sets:
> $R_S = Q_S = \{(1, 0), (8, 0), (8, 5), (8, 7), (6, 0), (6, 5)\}$.
> By the *extensionality convention*:
> $R_S = Q_S \Rightarrow R_\geq = Q_>$.
> So, R should be the same relation as Q. This is not the case, because the predicates are not equivalent:
> $(\forall x)\, x\, P1\, x$ is true, but $(\forall x)\, x\, P2\, x$ is false.
> Hence the extensionality convention fails for these relations.

The Digraph Representation

When $B = A$, so that we are dealing with a relation within a set, we may use the *digraph* R_D to represent it; in which an arrow goes from a to a' if and only if $a\,R\,a'$. Any relation within a finite or countably infinite set can, in principle, be shown in a digraph; conversely, every digraph (with unlabelled arrows) represents a relation in the set of its vertices. Interesting properties of relations are often derived from digraphical considerations; there is a whole literature on digraphs.

Foresets and Aftersets of Relations

These are defined for any relation R from A to B.
- The *afterset* of $a \in A$ is

$$aR = \{b \in B: aRb\}.$$

- The *foreset* of $b \in B$ is

$$Rb = \{a \in A: aRb\}.$$

Mnemonically and semantically, an afterset consists of all those elements which can correctly be written after a given element, a foreset of those which can correctly be written before it. An afterset or foreset may well be empty.

Clearly, $b \in aR$ if and only if $a \in Rb$. A relation is completely known if all its foresets or all its aftersets are known.

Matrix Representation

Very important computationally and even conceptually, as well as being a useful visual aid, is the *incidence matrix* R_M of a relation R. This arises from a table in which the row-headings are the elements of A and the column-headings are the elements of B, so that the cells represent $A \times B$. In the (a,b)-cell is entered 1 if aRb, and 0 if $a \neg Rb$. For visual purposes it is better to suppress the 0s, but they should be understood to be there for computational purposes.

	b_1	b_2	b_3
a_1	1		
a_2	1	1	1
a_3	1	1	
a_4			

$a_1 R = \{b_1\}$
$a_2 R = \{b_1, b_2, b_3\}$
$a_3 R = \{b_1, b_2\}$
$a_1 R = \emptyset$

Example: The matrix representation R_M and the afterset representation of a relation R

Clearly there is a one-to-one correspondence (bijection) between distinct tables and distinct relations, and, as soon as there has been agreement on the names and ordering of the row and column headings, between either of these and distinct matrices of size $|A| \times |B|$ with entries from $\{0, 1\}$.

Furthermore, the afterset $a_i R$ is in one-to-one correspondence with the nonzero entries of the ith row of R_M; the foreset Rb_j is in one-to-one correspondence with the nonzero entries of the jth column of R_M.

Operations and Inclusions in $\mathcal{R}(A \leadsto B)$

There are a considerable number of natural and important operations. We begin with unary operations and then proceed to several kinds of binary ones.

Unary Operations

The *negated* or *complementary relation* of $R \in \mathcal{R}(A \to B)$ is $\neg R \in \mathcal{R}(A \to B)$ given by $a \neg R b$ if and only if it is not the case that aRb.

The *converse* or *transposed relation* of $R \in \mathcal{R}(A \to B)$ is $R^\mathsf{T} \in \mathcal{R}(B \to A)$ given by

$$bR^\mathsf{T} a \quad \Leftrightarrow \quad aRb.$$

(It is also called the *inverse* and is therefore often written R^{-1}. In no algebraic sense it is an inverse, in general.)

Both operators $^\mathsf{T}$ and $\neg$ are *involutory*, that is, when applied twice they give the original object: $(R^\mathsf{T})^\mathsf{T} = R$, $\neg(\neg R) = R$. They commute with each other: $\neg(R^\mathsf{T}) = (\neg R)^\mathsf{T}$, so that the parentheses may be omitted safely. One can write: $\neg R^\mathsf{T}$.

Definition 1 (Binary operators and a binary relation on $\mathcal{R}(A \to B)$)
- The *intersection* or *meet* or *AND-ing*:

$$a(R \sqcap R')b \quad \Leftrightarrow \quad aRb \text{ and } aR'b.$$

- The *union* or *join* or *OR-ing*:

$$a(R \sqcup R')b \quad \Leftrightarrow \quad aRb \text{ or } aR'b.$$

- A relation R 'is contained in' (is a *subrelation* of) a relation R', and R' 'contains' (is a *superrelation* of) R, $R \sqsubseteq R'$:

$$R \sqsubseteq R' \quad \Leftrightarrow \quad (\forall a)(\forall b)(aRb \to aR'b)$$
$$\Leftrightarrow \quad R \sqcap R' = R \quad \Leftrightarrow \quad R \sqcup R' = R',$$

where $\to$ is the Boolean *implication operator*.

Definition 2 The *relative complement* of R with respect to R', or *difference* between R' and R, is $R' \setminus R$, given by:

$$a(R' \setminus R)b \quad \Leftrightarrow \quad aR'b \text{ but } a\neg Rb,$$

that is, by $R' \setminus R = R' \sqcap \neg R$.

Binary Operations on Successive Relations

Definition 3 (Circle and square products) Where $R \in \mathcal{R}(A \to B)$ and $S \in \mathcal{R}(B \to C)$, the following compositions give a relation in $\mathcal{R}(A \to C)$:
- The *circle product* or *round composition* is $\circ$, given by $aR \circ Sc \Leftrightarrow aR \cap Sc \neq \emptyset$.
- The *square composition* or *square product* is $\square$, given by $aR \square Sc \Leftrightarrow aR = Sc$.

The circle product is the usual one, to be found throughout the literature going back at least to the nineteenth century. The square product is a more recent (1977) innovation. The $\square$ product belongs to the family of products sometimes called *BK-products*. Further interesting kinds of BK-products and their uses are discussed in the sequel.

Proposition 4 (*Properties of $\square$-product*)
1) $(R \square S) \sqcap (R' \square S) \sqsubseteq (R \sqcap R') \square S \sqsubseteq (R \sqcup R') \square S \sqsubseteq (R \square S) \sqcup (R' \square S)$;
2) $(R \square S)^{-1} = S^{-1} \square R^{-1}$;
3) $R \square S = \neg R \square \neg S$;
4) the square product is not associative.

Matrix Formulation of the Binary Operations

All of the *binary operations on relations* have a convenient formulation in matrix terms – using the matrix operations given in Proposition 6. The matrix operations use in their definitions standard Boolean logic connectives for *crisp relations*. By replacing these by the connectives of suitable many-valued logics, all the formulas easily generalize to fuzzy relations. Thus matrix formulation of binary operations and compositions unifies computationally crisp and fuzzy relations.

Definition 5 The *Boolean connectives* $\wedge, \vee, \leftrightarrow$, on the set $\mathcal{B}_2 = \{0, 1\}$ are given by:

$\wedge$	0	1
0	0	0
1	0	1

$\vee$	0	1
0	0	1
1	1	1

$\leftrightarrow$	0	1
0	1	0
1	0	1

For a pair (x_1, x_2) of elements from $\mathcal{B}_2$, we infix the operators: $x_1 \wedge x_2$, etc., while for a list $(x_k)_{k=1,\dots,n}$ or $(x_k)_{k \in K}$ of elements from $\mathcal{B}_2$, we write $\bigwedge_{k=1}^n x_k$ or $\bigwedge_{k \in K} x_k$ or simply $\bigwedge_k x_k$. (Note that K can be denumerably infinite, or even greater, without spoiling the definition; no convergence problems are involved.)

Proposition 6 (*Matrix notation*)
1) $(R \sqcap S)_{ij} = R_{ij} \wedge S_{ij}$;
2) $(R \sqcup S)_{ij} = R_{ij} \vee S_{ij}$;
3) $(R \circ S)_{ij} = \bigvee_k (R_{ik} \wedge S_{kj})$;
4) $(R \bullet S)_{ik} = \bigwedge_j (R_{ij} \vee S_{jk})$;
5) $(R \square S)_{ik} = \bigwedge_j (R_{ij} \equiv S_{jk})$;
6) $(R_1 \times R_2)_{i_1 i_2 j_1 j_2} = (R_1)_{i_1 j_1} \wedge (R_2)_{i_2 j_2}$.

Non-Associative Products of Relations

Definition 7 (Triangle products)
- *Subproduct* $\triangleleft$: $x(R \triangleleft S) z \Leftrightarrow xR \subseteq Sz$;
- *Superproduct* $\triangleright$: $x(R \triangleright S) z \Leftrightarrow xR \supseteq Sz$.

The matrix formulation of $\triangleleft$ and $\triangleright$ products uses the Boolean connectives $\to, \leftarrow, \oplus$ on the set $\mathcal{B}_2 = \{0, 1\}$ given by

$\to$	0	1
0	1	1
1	0	1

$\leftarrow$	0	1
0	1	0
1	1	1

$\oplus$	0	1
0	0	1
1	1	0

Proposition 8 (Logic notation for $\triangleleft$ and $\triangleright$)
- $(R \triangleleft S)_{ik} = \bigwedge_j (R_{ij} \to S_{jk})$;
- $(R \triangleright S)_{ik} = \bigwedge_j (R_{ij} \leftarrow S_{jk})$.

Only the conventional $\circ$-product is associative. The $\square$-product is not associative [2].

Proposition 9 *The following mixed pseudo-associativities hold for the triangle products, with $Q \in \mathcal{B}(W \rightsquigarrow X)$ and the triple products in $\mathcal{B}(W \rightsquigarrow Z)$:*
- $Q \triangleleft (R \triangleright S) = (Q \triangleleft R) \triangleright S$;
- $Q \triangleleft (R \triangleleft S) = (Q \circ R) \triangleleft S$;
- $Q \triangleright (R \triangleright S) = Q \triangleright (R \circ S)$.

Characterization of Special Properties of Relations Between Two Sets

Definition 10 (Special properties of a heterogeneous relation $R \in \mathcal{R}(X \rightsquigarrow Y)$):
- R is *covering* if and only if $(\forall x) \in X \, (\exists y) \in Y$ such that xRy.
- R is *onto* if and only if $(\forall y) \in Y \, (\exists x) \in X$ such that xRy.
- R is *univalent* if and only if $(\forall x) \in X$, if xRy and xRy' then $y = y'$.
- R is *separating* if and only if $(\forall y) \in Y$, if xRy and $x'Ry$ then $x = x'$.

Composed properties can be defined by combining these four basic properties. Well-known is the combination 'covering' and 'univalent' which defines *functional*. Other frequently used combination is 'onto' and 'separating'.

The self-inverse circle product is very useful in the characterization of *special properties of relations* between two distinct sets. Using the product, one can characterize these properties in purely relational way, without directly referring to individual elements of the relations involved.

Proposition 11 (*Special properties of a heterogeneous relation* $R \in \mathcal{R}(X \leadsto Y)$):
- R is covering if and only if $E_X \sqsubseteq R \circ R^{-1}$.
- R is univalent if and only if $R^{-1} \circ R \sqsubseteq E_Y$.
- R is onto if and only if (for all) $E_Y \sqsubseteq R^{-1} \circ R$.
- R is separating if and only if $R \circ R^{-1} \sqsubseteq E_X$.

Here E_X and E_Y are the left and right identities, respectively.

Relations on a Single Set: Special Properties

The *self-inverse products* are a fertile source of relations on the single set X. There are certain well-known special properties which a relation may possess (or may lack), of which the most important are reflexivity, symmetry, antisymmetry, strict antisymmetry, and transitivity, together with their combinations, forming *preorders* (reflexive and transitive) (*partial*) *orders* (reflexive, antisymmetric and transitive), *equivalences* (reflexive, transitive and symmetric).

Definition 12 (**Special properties of binary relations from X to X**)
- *Covering*: every x_i is related by R to something $\Leftrightarrow \forall i \in I \ \exists j \in I$ such that $R_{ij} = 1$.
- *Locally reflexive*: if x_i is related to anything, or if anything is related to x_i, then x_i is related to itself $\Leftrightarrow \forall i \in I \ R_{ii} = \max_j (R_{ij}, R_{ji})$.
- *Reflexive*: covering and locally reflexive $\Leftrightarrow \forall i \in I \ R_{ii} = 1$.
- *Transitive* $\forall i, j, k \in I \ (x_i R x_j$ and $x_j R x_k \Rightarrow x_i R x_k) \Leftrightarrow R^2 \sqsubseteq R$.
- *Symmetric*: $(x_i R x_j \Rightarrow x_j R x_i) \Leftrightarrow R^T = R$.
- *Antisymmetric*: $(x_i R x_j$ and $x_j R x_i \Rightarrow x_i = x_j) \Leftrightarrow$ if $i \neq j$ then $\min(R_{ij}, R_{ji}) = 0$.
- *Strictly antisymmetric*: never both $x_i R x_j$ and $x_j R x_i \Leftrightarrow \forall i, j \in I \ \min(R_{ij}, R_{ji}) = 0$.

Most of the properties listed above are common in the literature. Local reflexivity is worthwhile exception. It appeared in [1] and was generalized to fuzzy relations in [4], leading to new computational algorithms for both crisp and fuzzy relations [4,10]. Unfortunately, it is absent from the textbooks, yet it is extremely important in applications of relational methods to analysis of the real life data (see the notion of participant in the next two sections).

Partitions IN and ON a Set

A *partition on a set* X is a division of X into nonoverlapping (and nonempty) subsets called blocks. A *partition in a set* X is a partition on the subset of X [17,18] called the *subset of participants*.

There is a one-to-one correspondence between partitions in X and *local equivalences* (i. e. locally reflexive, symmetric and transitive relations) in $\mathcal{R}(X \leadsto B)$. The partitions in X (so also the local equivalences in $\mathcal{R}(X \leadsto B)$) form a lattice with '__is-finer-than__' as its ordering relation. This whole subject is coextensive with *classification* or *taxonomy*, i. e., very extensive indeed. Furthermore, classification is the first step in abstraction, one of the fundamental processes in human thought.

Tolerances and Overlapping Classes

Some tests for tolerance and equivalence are as follows:
- $R \circ R^T$ is always symmetric and locally reflexive.
- $R \circ R^T$ is a tolerance if and only if R is covering.
- $R \square R^T$ is always a (local) tolerance.
- $R \square R^T \sqsubseteq R$ if and only if R is reflexive.
- $E \sqsubseteq R \sqsubseteq R \square R^T$ if and only if R is an equivalence.
- $R \square R^T = R$ if and only if R is an equivalence.
- $R \square R^T \sqsubseteq R \circ R^T$ if and only if R is covering.

It is not always the case that one manages, or even attempts, to classify participants into nonoverlapping blocks. *Local tolerance relations* (i. e. locally reflexive and symmetric) lead to classes which may well overlap, where one participant may belong to more than one class. The classic case, giving its name to this kind of relation, is '__is-within-one milimeter-of__'. This is quite a different model from the severe partitions [80], and has been for a long time unduly neglected both in theory and applications, even when the data mutely favor it.

Hierarchies in and on a Set: Local and Global Orders and Pre-orders

An example of a *hierarchy in a finite set* X is displayed in Fig. 1. In such a hierarchy, there is a finite number of levels and there is no ambiguity in the assignment of a level to an element. The elements which appear eventually in the hierarchy are the *participants*; those which do not are *nonparticipants*; if all of X participates, then the hierarchy is on X.

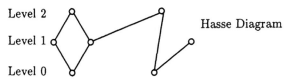

Hasse Diagram

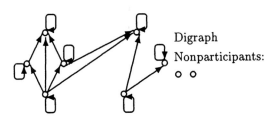

Digraph
Nonparticipants:

Every *local order* (i. e. locally reflexive, transitive and antisymmetric relation) from a finite set to itself establishes a hierarchy in that set, that is, can be used as the 'precedes' relation in the hierarchy. Conversely, given any hierarchy, its '__precedes__' is a local order. The hierarchy is on X exactly when the local order is the global one.

The picture of the hierarchy is called its *Hasse diagram*. It can always be obtained from the digraph of the local-order relation by the suppression of loops and of those arrows which directly connect nodes between which there is also a longer path.

The formulas of Theorem 13 can be used for fast computational testing of the listed properties.

Theorem 13 *The following conditions universally characterize the transitivity, reflexivity and pre-order on $R \in \mathcal{R}(X \leadsto X)$:*
- *R is transitive if and only if $R \sqsubseteq R \triangleright R^{-1}$.*
- *R is reflexive if and only if $R \triangleright R^{-1} \sqsubseteq R$.*
- *R is a pre-order if and only if $R = R \triangleright R^{-1}$.*

More complex relational structures are investigated by theories of homomorphisms, which can be further generalized [6].

Definition 14 Let F, R, G, S be heterogeneous relations between the sets A, B, C, D such that $R \in \mathcal{R}(A \leadsto B)$. The conditions that (for all $a \in A$, $b \in B$, $c \in C$, $d \in D$) the expression $(aFc \wedge aRb \wedge bGd) \rightarrow cSd$ we denote by $FRG{:}S$. We say that $FRG{:}S$ is *forward compatible*, or, equivalently, that F, G are *generalized morphisms*.

The following *Bandler–Kohout compatibility theorem* holds, [6]:

Theorem 15 (Generalized morphisms)
- *$FRG : S$ are forward compatible if and only if $F^\mathsf{T} \circ R \circ G \sqsubseteq S$.*
- *Formulas for computing the explicit compatibility criteria for F and G are: $FRG : S$ are forward-compatible if and only if $F \sqsubseteq R \triangleleft (G \triangleleft S^\mathsf{T})$.*

The R's of forward compatibility constitute a lower ideal. Similarly, the *backward compatibility* given by $F \circ S \circ G^\mathsf{T} \sqsubseteq R$ gives a generalized proteromorphism. It constitutes an upper ideal or filter: $FRG : S$ are backward compatible if and only if $F \circ S \circ G^\mathsf{T} \sqsubseteq R$ if and only if $S \sqsubseteq F^\mathsf{T} \triangleleft R \triangleright G$.

$FRG : S$ are *both-way compatible* if they are both forward and backward compatible. The conventional *homomorphism* is a special case of both-way compatibility, where F and G are not general relations but just many-to-one mappings.

The generalized morphisms of Bandler and Kohout [6] are relevant not only theoretically, but have also an important practical use in solving systems of inequalities and equations on systems of relations.

For partial homomorphisms the situation becomes more complicated. In partial structures the conventional homomorphism splits into mutually related *weak*, *strong* and *very strong* kinds of homomorphism [5].

Fuzzy Relations

Mathematical relations can contribute to investigation of properties of a large variety of structures in sciences and engineering. The power of *relational analysis* stems from the elegant algebraic structure of relational systems that is supplemented by the computational power of *relational matrix notation*. This power is further enhanced by many-valued logic based (fuzzy) extensions of the relational calculus.

As often in mathematics, where terms are used inclusively, the crisp (nonfuzzy) sets and relations are merely special cases of fuzzy sets and relations, in which the actual degrees happen to be the extreme ones. On the theoretical side, *fuzzy relations* are extensions of standard nonfuzzy (crisp) relations. By replacing the usual Boolean algebra by many-valued logic algebras, one obtains extensions that contain the classical relational theory as a special case.

Definitions

A *fuzzy set* is one to which any element may belong to various degrees, rather than either not at all (degree 0) or utterly (degree 1). Similarly, a *fuzzy relation* is one which may hold between two elements to any degree between 0 and 1 inclusive. The sentence $x_i R y_j$ takes its value $\delta \, (x_i \, R \, y_j) = R_{ij}$, from the interval $[0, 1]$ of real numbers. In early papers on fuzzy relations $\mu_R(x_i, y_j)$ was usually written instead of R_{ij}.

The matrix notation used in the previous sections for nonfuzzy (crisp) relations is directly applicable to the fuzzy case. Thus, all the definitions of operations, compositions and products can be directly extended to the fuzzy case.

Operations and Inclusion on $\mathcal{R}_F \, (X \rightsquigarrow Y)$

Fuzzy Relations with Min, Max Connectives

This has been the most common extension of relations to the fuzzy realm. Boolean $\wedge$ and $\vee$ are replaced by many-valued connectives min, max in all crisp definitions.

In matrix terms, this yields the following intersection and union operations:

$(R \sqcap S)_{ij} = \min(R_{ij}, S_{ij})$,

$(R \sqcup S)_{ij} = \max(R_{ij}, S_{ij})$.

(In older μ-notation, $\mu_{(R \sqcap S)}(x_i, x_j) = \min(\mu_R(x_i, y_j), \mu_S(x_i, y_j))$, etc.)

The negation of R is given by $(\neg R)_{ij} = 1 - R_{ij}$. The converse of R is given by $(R^T)_{ij} = R_{ji}$.

Fuzzy Relations Based on Łukasiewicz Connectives

When the *bold* (Łukasiewicz) connectives $x \vee y = \min(1, x + y)$, $x \wedge y = \max(0, x + y - 1)$ are used to define $\sqcup, \sqcap$ operations, this is an instance of relations in *MV-algebras*.

Fuzzy Relations With *t*-Norms and Co-Norms

Fuzzy logics can be further generalized. $\wedge$ and $\vee$ are obtained by replacing min and max by a *t-norm* and a *t-conorm*, respectively. A *t*-norm is an operation $* : [0, 1]^2 \rightarrow [0, 1]$ which is commutative, associative, nondecreasing in both arguments and having 1 as the unit element and 0 as the zero element. Taking a continuous *t*-norm, by *residuation* we obtain a *many-valued logic implication* $\rightarrow$. Using $\{ \wedge, \vee, *, \rightarrow \}$ one can define families of deductive systems for fuzzy logics called *BL-logics* [31]. In relational systems using BL-logics, one can define again various *t*-norm based relational properties [53,83], *BK-products* and *generalized morphisms of relations* [47].

Definition 16 (*Inclusion of relations*) A relation R is 'contained in' or is a *subrelation* of a relation S, written $R \sqsubseteq S$, if and only if $(\forall i)(\forall j) \, R_{ij} \leq S_{ij}$.

This definition guarantees that R is a subrelation of R' if and only if every R_α is a subrelation of its corresponding R'_α. (This convenient meta-property is called *cutworthiness*, see Theorem 17 below.)

Products: $\mathcal{R}_F(X \rightsquigarrow Y) \times \mathcal{R}_F(Y \rightsquigarrow Z) \rightarrow \mathcal{R}_F(X \rightsquigarrow Z)$

For fuzzy relations, there are two versions of products: harsh and mean [3,52]. Most conveniently, again, in matrix terms *harsh products* syntactically correspond to matrix formulas for the crisp relations. The *fuzzy relational products* are obtained by replacing the Boolean logic connectives AND, OR, both implications and the equivalence of crisp products by connectives of some many-valued logic chosen according to the properties of the products required. Thus the $\circ$-product and $\Box$-product are given exactly as in Proposition 6 above by formulas 3) and 5), respectively; for triangle products as given in Proposition 8 above. For the MVL implication operators most often used to define *fuzzy triangle products*, see ▶ Checklist paradigm semantics for fuzzy logics, Table 1, or [8]. The details of choice of the appropriate many-valued connectives are discussed in [3,7,8,40,43,52].

Given the general formula $(R@S)_{ik} := \# \, (R_{ij} * S_{jk})$ for a relational product, a *mean product* is obtained by

Boolean and Fuzzy Relations, Table 1
Closures and an interior

1.	The locally reflexive closure of R: locref clo $R = R \sqcup E_R$.
2.	The symmetric closure of R: sym clo $R = R \sqcup R^T$.
3.	*symmetric interior* of R: sym int $R = R \sqcap R^T$.
4.	The transitive closure of R: tra clo $R = R \sqcap R^2 \sqcap \cdots = \sqcap_{k \in \mathbb{Z}^+} R^k$.
5.	The *local tolerance closure* of R: loctol clo $R =$ locref clo (sym clo R).
6.	The *local pre-order closure* of R: locpre clo $R =$ locref clo (tra clo R) = tra clo (locref clo R).
7.	The *local equivalence closure* of R: locequ clo $R =$ tra clo (sym clo (locref clo R)).
8.	The *reflexive closure* of R: ref clo $R = R \sqcup E_X$.
9.	The *tolerance closure* of R: tol clo $R =$ ref clo (sym clo R).
10.	The *pre-order closure* of R: pre clo $R =$ ref clo (tra clo R).
11.	The *equivalence closure* of R: equ clo $R =$ tra clo (tol clo R).

replacing the outer connective # by $\sum$ and normalizing the resulting product appropriately. In more concrete terms, in order to obtain the mean products, the outer connectives $\bigvee_j$ in $\circ$ and $\bigwedge_j$ in $\square$, $\triangleleft$, $\triangleright$ are replaced by $1/n) \sum_j$ [3].

N-ary Relations

An *n-ary relation* R is an open sentence with n slots; when these are filled in order by the names of elements from sets $X_1, \ldots, X_n$, there results a proposition that is either true or false if the relation is crisp, or is judged to hold to a certain degree if the relation is fuzzy. This 'intensional' definition is matched by the satisfaction set R_S of R, which is a fuzzy subset the n-tuple of $X_1, \ldots, X_n$, and can be used, if desired as its extensional definition. The matrix notation works equally well for n-ary relations and all the types of the BK-products are also defined. For details see [9].

Special Properties of Fuzzy Relations

The *special properties of crisp relations* can be generalized to fuzzy relations exactly as they stand in Definition 12, using in each case the second of the two given definitions. It is perhaps worthwhile spelling out the requirements for transitivity in more detail:

$$R^2 \sqsubseteq R \Leftrightarrow (\forall i, k) \max_i (\min(R_{ij}, R_{jk})) \leq R_{ik}.$$

Useful references provide further pointers to the literature: general [43] on fuzzy partitions [14,69], fuzzy similarities [69], tolerances [34,75,85].

Alpha-cuts of Fuzzy Relations

It is often convenient to study fuzzy relations through their α-cuts; for any α in the half-open interval $[0, 1]$, the α-cut of a fuzzy relation R is the crisp relation R_α given by

$$(R_\alpha)_{ij} = \begin{cases} 1 & \text{if } R_{ij} \geq \alpha, \\ 0 & \text{otherwise.} \end{cases}$$

Compatibility of families of crisp relations with their fuzzy counterpart (the original relation on which the α-cuts have been performed) is guaranteed by the following theorem on cutworthy properties [10]:

Theorem 17 *It is true of each simple property P (given in Definition 12) and every compound property P (listed in Table 1), that every α-cut of a fuzzy relation R possesses P in the crisp sense, if and only if R itself possesses in the fuzzy sense. (Such properties are called cutworthy.)*

Fuzzy Partitions, Fuzzy Clusters and Fuzzy Hierarchies

Via their α-cuts, fuzzy local and global equivalences provide precisely the nested families of partitions in and on a set which are required by the theory and for the applications in taxonomy envisaged in [17,18]. Fuzzy local and global tolerances similarly provide families of tolerance classes for the cluster type of classification which allows overlaps. Fuzzy local and global orders furnish nested families of hierarchies in and on a set, with their accompanying families of Hasse diagrams.

The importance of fuzzy extensions cannot be overestimated. Thus, one may identify approximate similarities in data, approximate equivalences and orders. Such approximations are paramount in many applications, in situations when only incomplete, partial information about the domain of scientific or technological application is available.

Closures and interiors of relations play an important role in design of fast fuzzy relational closure algorithms [4,9,10,11] for computing such approximations.

Theorem 17 and other theorems on commuting of cuts with closures [11,42] guarantee their correctness.

Closures and Interiors with Special Properties

For certain properties P which a fuzzy relation may have or lack, there always exists a well-defined P-closure of R, namely the least inclusive relation V which contains R and has the property P. Also, for some properties P, the P-interior of R is the most inclusive relation Q contained in R and possessing P. Clearly, where the P-closure exists, R itself possesses P if and only if R is equal to P-clo(R), and the same for interiors.

Certain closures use the local equality E_R of R, given by $(E_R)_{ii} = \max_i(\max(R_{ij}, R_{ji}))$, $(E_R)_{ij} = 0$ if $j \neq i$. Others use the equality on X given by $(E_X)_{ii} = 1$, $(E_X)_{ij} = 0$ for $j \neq i$.

Important closures and one important interior are given in Table 1. See [4,10] for further details.

Applications of Relational Methods in Engineering, Medicine and Science

Relational properties are important for obtaining knowledge about characteristics and interactions of various parts of a relational model used in real life applications. Identification of composite properties of mathematical relations, such as local or global pre-orders, orders, tolerances or equivalences, plays an important role in *evaluation of empirical data*, (e. g. medical data, commercial data etc. or data for technological forecasting) and building and evaluating *relational models* based on such data [48,49].

The local and global properties detect important semantic distinctions between various concepts captured by relational structures. For example the interactions between technological parts, processes etc., or relationships of *cognitive constructs* elicited experimentally [37,39,41,55]. Capturing both, local and global properties is important for distinguishing participants from nonparticipants in a relational structure. This distinction is crucial for obtaining a nondistorted picture of reality.

In the general terms, the abstract theoretical tools supporting identification and representation of relational properties are fuzzy closures and interiors [4,10]. Having such means for *testing relational properties* opens the avenue to linking the empirical structures that can be observed and captured by fuzzy relations with their abstract, symbolic representations that have well defined mathematical properties.

This opens many possibilities for computer experimentation with empirically identified logical, say, predicate structures. These techniques found practical use in directing *resolution based theorem prover* strategy [56], relation-based inference in *medical diagnosis* [48,58] and at extracting predicate structures of 'train of thought' from questionnaires presented to people by means of Kelly's *repertory grids*. BK-relational products and fast fuzzy relational algorithms based on fuzzy closures and interiors have been essential for computational progress of in this field and for *optimization of computational performance*. See the survey in [52] with a list of 50 selected references on the mathematical theory and applications of BK-products in various fields of science and engineering. Further extensions or modifications of BK-products have been suggested in [19,20,21,30].

Applications of relational theories, computations and modeling include the areas of medicine [48,59], psychology [49], cognitive studies [36,38], nuclear engineering [84], industrial engineering and management [25,46], architecture and urban studies [65,66] value analysis in business and manufacturing [60] information retrieval [51,54], computer security [45,50] databases, theoretical computer science [13,68,71], software engineering [78], automated reasoning [56], and logic [12,28,63]. Particularly important for software engineering is the contribution of C.A.R. Hoare and He Jifeng [33] who use the crisp triangle BK-superproduct for software specification, calling the crisp ◁ products in fact '*weak prespecifications*'.

Relational equations [22] play an important role in applications [70] in general, and also in AI and applications of causal reasoning [24]; fuzzy inequalities in mathematical programming [72]. Applications in game theory of crisp relations is well established [78,79].

Brief Review of Theoretical Development

Binary (two place) relations were first perceived in their abstract mathematical form by Galen of Pergamon in the 2nd century AD [57]. After a long gap, first systematic development of the calculus of relations (concerned with the study of logical operations on binary relations) was initiated by A. DeMorgan, C.S. Pierce and E. Schröder [9,64]. Significant investigation into the logic of relations was the 1900 paper of B. Russell [76] and axiomatization of the relational calculus in 1941 by A. Tarski [64,81]. Extensibility of Tarski's axioms to the fuzzy domain has been investigated by Kohout [44].

Later algebraic advances in relational calculus [9] stem jointly from the elegant work of J. Riguet (1948) [74], less widely known but important work of O. Borůvka (1939) [15,16,17,18] and the stimulus of fuzzy set theory of L.A. Zadeh (1965) [35,85,86], and include a sharpened perception of special properties and the construction of new kinds of relational products [3], together with the extension of the theory from Boolean to multiple-valued logic based relations [2,9]. The triangle subproduct $R \triangleleft S$, the triangle superproduct $R \triangleright S$, and square product $R \square S$ were introduced in their general form defined below by Bandler and Kohout in 1977, and are referred to as the *BK-products* in the literature [19,20,30]. The square product, however, stems from Riguet (1948) [74], needing only to be made explicit [1,9]. E. Sanchez independently defined an α-compostition [77] which is in fact $\triangleleft$ using Heyting–Gödel implication. The special instances of the triangle BK-products were more recently rediscovered and described in 1986 by J.P. Doignon, B. Monjardet, M. Roubens, and P. Vincke [23,26] calling these 'traces of relations'. Hence, a 'trace-of-relation' is a BK-triangle superproduct in which $\rightarrow$ is the residuum of a commutative $\wedge$. The crisp square product was also independently introduced in 1986 by R. Berghammer, G. Schmidt and H. Zierer [13] as a generalization of Riguet's 'noyau' [74].

On the other hand, advances in abstract relational algebras stems from the work of Tarski [81] and his school [32,64,67,82]. Tarski's axiomatization [81] of homogeneous relational calculus takes relations and operations over relations as the primitives. It applies only to homogeneous relations as it has only one constant entity, the identity relation E. For heterogeneous relations, taking e. g. U_{XY} as the universal relation we have a finite number of separate identity relations (constants) i. e. E_{XY}, E_{YZ}, ..., etc. [4,10]. Therefore viewed syntactically through the logic axioms, the axiomatization of heterogeneous relations (containing a whole family of universal relations) would be a many-sorted theory [30], each universal relation belonging to a different sort.

Tarski's axioms of homogeneous relations

$$\begin{aligned}
&R \circ E = E \circ R; (R \circ S)^T = S^T \circ R^T; \\
&(R^T)^T = R; (\neg R)^T = \neg (R^T); \\
&(R \sqcup S)^T = R^T \sqcup S^T; \\
&(R \circ S) \circ T = R \circ (S \circ T); \\
&(R \sqcup S) \circ T = (R \circ T) \sqcup (S \circ T); \\
&R \circ (S \sqcup T) = (R \circ S) \sqcup (R \circ T); \\
&(R^T \circ \neg (R \circ S)) \sqcup \neg S = \neg S.
\end{aligned}$$

Taking the axioms on their own opens the way to abstract relational algebras (RA) with new problems at hand. Tarski and his school have investigated the interrelationship of various generalizations of associative RAs in a purely abstract way. In some of these generalizations, the axiom of associativity for relational composition is dropped. This leads from representable (RRA) to semi-associative (SA), weakly associative (WA) and nonassociative (NA) relational algebras. In 1982 R.D. Maddux [62] gave the following result:

$$RRA \subset RA \subset SA \subset WA \subset NA.$$

All these generalizations deal only with one relational composition. The equations for pseudo-associativities given above (Proposition 9) and the *nonassociativity* of the square product (Proposition 4) show that there exist nonassociative representations of relational algebras (RA) in the relational calculus. Theorem 15 and Proposition 8 show that the interplay of several relational compositions is essentially involved in the computationally more powerful formulas of the relational calculus. The Tarskian RA axiomatizations, however, do not express fully the richness of the calculus of binary relations and the mutual interplay of associative $\circ$, pseudo-associative $\triangleright$, $\triangleleft$ and nonassociative $\square$ products. Considerable scope for further research into new axiomatizations still remains. Our results based on nonassociative BK-products of Bandler and Kohout that historically precede abstract nonassociative generalizations in relational algebras of Maddux show that the nonassociative products have representations and that these representations offer various computational advantages. There is also a link of RA with projective geometries [61].

Basic Books and Bibliographies

The best general books on theory of crisp relations and applications are [78] and [80]. In fuzzy field, there is no general book available at present. There are, however, extant some more specialized monographs: on solving fuzzy relations equations [27], on preference modeling and multicriteria decision making [39], on representation of cognitive maps by relations [39] and on crisp and fuzzy BK-products of relations [53]. One can also find some specialized monographs on logic foundations and relational algebras: [32,82]. All these books also contain important list of references. The most important bibliography of selected references on the topic related to *fuzzy sets* and relations is contained in [43]. The early years of fuzzy sets (1965-1975) are covered very comprehensively in the critical survey and annotated bibliography [29]. Many-valued logic connectives form an important foundation for fuzzy sets and relations. The book of N. Rescher [73] still remains the best comprehensive survey that is also accessible to a non-logician. It contains almost complete bibliography of many-valued logics from the end of the 19th century to 1968.

See also

- ▶ Alternative Set Theory
- ▶ Checklist Paradigm Semantics for Fuzzy Logics
- ▶ Finite Complete Systems of Many-valued Logic Algebras
- ▶ Inference of Monotone Boolean Functions
- ▶ Optimization in Boolean Classification Problems
- ▶ Optimization in Classifying Text Documents

References

1. Bandler W, Kohout LJ (1977) Mathematical relations, their products and generalized morphisms. Techn Report Man-Machine Systems Lab Dept Electrical Engin Univ Essex, Colchester, Essex, UK, EES-MMS-REL 77-3. Reprinted as Chap. 2 of Kohout LJ, Bandler W (eds) Survey of Fuzzy and Crisp Relations, Lect Notes in Fuzzy Mathematics and Computer Sci, Creighton Univ Omaha
2. Bandler W, Kohout LJ (1980) Fuzzy relational products as a tool for analysis and synthesis of the behaviour of complex natural and artificial systems. In: Wang PP, Chang SK (eds) Fuzzy Sets: Theory and Appl. to Policy Analysis and Information Systems. Plenum, New York, pp 341–367
3. Bandler W, Kohout LJ (1980 1981) Semantics of implication operators and fuzzy relational products. Internat J Man-Machine Studies 12:89–116 Reprinted in: In: Mamdani EH, Gaines BR (eds) Fuzzy Reasoning and its Applications. Acad. Press, New York, 219–246
4. Bandler W, Kohout LJ (1982) Fast fuzzy relational algorithms. In: Ballester A, Cardús D, Trillas E (eds) Proc. Second Internat. Conf. Math. at the Service of Man (Las Palmas, Canary Islands, Spain, 28 June-3 July), Univ. Politechnica de Las Palmas, pp 123–131
5. Bandler W, Kohout LJ (1986) On new types of homomorphisms and congruences for partial algebraic structures and n-ary relations. Internat J General Syst 12:149–157
6. Bandler W, Kohout LJ (1986) On the general theory of relational morphisms. Internat J General Syst 13:47–66
7. Bandler W, Kohout LJ (1986) A survey of fuzzy relational products in their applicability to medicine and clinical psychology. In: Kohout LJ, Bandler W (eds) Knowledge Representation in Medicine and Clinical Behavioural Sci. Abacus Book. Gordon and Breach, New York, pp 107–118
8. Bandler W, Kohout LJ (1987) Fuzzy implication operators. In: Singh MG (ed) Systems and Control Encyclopedia. Pergamon, Oxford, pp 1806–1810
9. Bandler W, Kohout LJ (1987) Relations, mathematical. In: Singh MG (ed) Systems and Control Encyclopedia. Pergamon, Oxford, pp 4000–4008
10. Bandler W, Kohout LJ (1988) Special properties, closures and interiors of crisp and fuzzy relations. Fuzzy Sets and Systems 26(3) (June):317–332
11. Bandler W, Kohout LJ (1993) Cuts commute with closures. In: Lowen B, Roubens M (eds) Fuzzy Logic: State of the Art. Kluwer, Dordrecht, pp 161–167
12. Benthem J van (1994) General dynamic logic. In: Gabbay DM (ed) What is a Logical System. Oxford Univ. Press, Oxford pp 107–139
13. Berghammer R, Schmidt G (1989/90) Symmetric quotients and domain constructions. Inform Process Lett 33:163–168
14. Bezdek JC, Harris JD (1979) Convex decompositions of fuzzy partitions. J Math Anal Appl 67:490–512
15. Borůvka O (1939) Teorie grupoidu (Gruppoidtheorie, I. Teil). Publ Fac Sci Univ Masaryk, Brno, Czechoslovakia 275:1–17, In Czech, German summary
16. Borůvka O (1941) Über Ketten von Faktoroiden. MATH-A 118:41–64

17. Borůvka O (1945) Théorie des décompositions dans un ensemble. Publ Fac Sci Univ Masaryk, Brno, Czechoslovakia:278 1–37 (In Czech, French summary)
18. Borůvka O (1974) Foundations of the theory of groupoids and groups. VEB Deutsch. Verlag Wissenschaft., Berlin, Also published as Halsted Press book by Wiley, 1976
19. DeBaets B, Kerre E (1993) Fuzzy relational compositions. Fuzzy Sets and Systems 60(1):109–120
20. DeBaets B, Kerre E (1993) A revision of Bandler–Kohout composition of relations. Math Pannonica 4:59–78
21. DeBaets B, Kerre E (1994) The cutting of compositions. Fuzzy Sets and Systems 62(3):295–310
22. DiNola A, Pedrycz W, Sanchez E (1989) Fuzzy relation equations and their applications to knowledge engionneering. Kluwer, Dordrecht
23. Doignon JP, Monjardet B, Roubens M, Vincke P (1986) Biorders families, valued relations and preference modelling. J Math Psych 30:435–480
24. Dubois D, Prade H (1995) Fuzzy relation equations and causal reasoning. Fuzzy Sets and Systems 75(2):119–134
25. Dubrosky B, Kohout LJ, Walker RM, Kim E, Wang HP (1997) Use of fuzzy relations for advanced technological cost modeling and affordability decisions. In: 35th AIAA Aerospace Sci. Meeting and Exhibit (Reno, Nevada, January 6-9, 1997), Amer. Inst. Aeronautics and Astronautics, Reston, VA, 1–12, Paper AIAA 97-0079
26. Fodor JC (1992) Traces of fuzzy binary relations. Fuzzy Sets and Systems 50(3):331–341
27. Fodor J, Roubens M (1994) Fuzzy preference modelling and multicriteria decision support. Kluwer, Dordrecht
28. Gabbay DM (1994) What is a logical system? In: Gabbay DM (ed) What is a Logical System? Oxford Univ. Press, Oxford, pp 179–216
29. Gaines BR, Kohout LJ (1977) The fuzzy decade: A bibliography of fuzzy systems and closely related topics. Internat J Man-Machine Studies 9:1–68 (A critical survey with bibliography.) Reprinted in: Gupta MM, Saridis GN, Gaines BR (eds) (1988) Fuzzy Automata and Decision Processes. Elsevier/North-Holland, Amsterdam, pp 403–490
30. Hájek P (1996) A remark on Bandler–Kohout products of relations. Internat J General Syst 25(2):165–166
31. Hájek P (1998) Metamathematics of fuzzy logic. Kluwer, Dordrecht
32. Henkin L, Monk JD, Tarski A (1985) Cylindric algebras, vol II. North-Holland, Amsterdam
33. Hoare JAR, Jifeng He (1986) The weakest prespecification I-II. Fundam Inform 9:51–84; 217–251
34. Höhle U (1988) Quotients with respect to similarity relations. Fuzzy Sets and Systems 27(1):31–44
35. Höhle U, Klement EP (1995) Non-classical logics and their applications to fuzzy subsets: A handbook of mathematical foundations of fuzzy sets. Kluwer, Dordrecht
36. Juliano BA (1993) A fuzzy logic approach to cognitive diagnosis. PhD Thesis, Dept. Comput. Sci., Florida State Univ., Tallahassee, Fl
37. Juliano BA (1996) Towards a meaningful fuzzy analysis of urbanistic data. Inform Sci 94(1–4):191–212
38. Juliano BA, Bandler W (1989) A theoretical framework for modeling chains-of-thought: Automating fault detection and error diagnosis in scientific problem solving. In: Fishman MB (ed) Proc. Second Florida Artificial Intelligence Res. Symp., Florida AI Res. Soc., FLAIRS, pp 118–122
39. Juliano B, Bandler W (1996) Tracing chains-of-thought: Fuzzy methods in cognitive diagnosis. Physica Verlag, Heidelberg
40. Kandel A (1986) Fuzzy mathematical techniques with applications. Addison-Wesley, Reading, MA
41. Kim E, Kohout LJ, Dubrosky B, Bandler W (1996) Use of fuzzy relations for affordability decisions in high technology. In: Adey RA, Rzevski G, Sunol AK (eds) Applications of Artificial Intelligence in Engineering XI. Computational Mechanics Publ., Billerica, MA
42. Kitainik L (1992) For closeable and cutworthy properties, closures always commute with cuts. In: Proc. IEEE Internat. Conf. Fuzzy Systems, IEEE, New York, pp 703–704
43. Klir GJ, Yuan B (1995) Fuzzy sets and fuzzy logic: Theory and applications. Prentice-Hall, Englewood Cliffs, NJ
44. Kohout LJ (2000) Extension of Tarski's axioms of relations to t-norm fuzzy logics. In: Wang PP (ed) Proc. 5th Joint Conf. Information Sciences, I Assoc. Intelligent Machinery, Durham44–47
45. Kohout LJ (1990) A perspective on intelligent systems: A framework for analysis and design. Chapman and Hall and v. Nostrand, London–New York
46. Kohout LJ (1997) Fuzzy relations and their products. In: Wang P (ed) Proc. 3rd Joint Conf. Inform. Sci. JCIS'97, Duke Univ., March,), Keynote Speech VIII: Prof. W. Bandler Memorial Lecture; to appear in: Inform. Sci.
47. Kohout LJ (1998 1999) Generalized morphisms in BL-logics. In: Logic Colloquium: The 1998 ASL Europ. Summer Meeting, Prague, August 9-15 1998, Assoc. Symbolic Logic (Extended abstract presenting the main mathematical theorems). Reprinted in: Bull Symbolic Logic (1999) 5(1):116–117
48. Kohout LJ, Anderson J, Bandler W, et al. (1992) Knowledge-based systems for multiple environments. Ashgate Publ. (Gower), Aldershot, Hampshire, UK
49. Kohout LJ, Bandler W (eds) (1986) Knowledge representation in medicine and clinical behavioural science. Abacus Book. Gordon and Breach, New York
50. Kohout LJ, Bandler W (1987) Computer security systems: Fuzzy logics. In: Singh MG (ed) Systems and Control Encyclopedia. Pergamon, Oxford
51. Kohout LJ, Bandler W (1987) The use of fuzzy information retrieval techniques in construction of multi-centre knowledge-based systems. In: Bouchon B, Yager RR (eds) Uncertainty in Knowledge-Based Systems. Lecture Notes Computer Sci. Springer, Berlin, pp 257–264
52. Kohout LJ, Bandler W (1992) Fuzzy relational products in knowledge engineering. In: Novák V, et al (eds) Fuzzy Ap-

proach to Reasoning and Decision Making. Academia and Univ. Press, Prague, pp 51–66
53. Kohout LJ, Bandler W (1999) A survey of fuzzy and crisp relations. Lecture Notes Fuzzy Math and Computer Sci Creighton Univ., Omaha, NE
54. Kohout LJ, Keravnou E, Bandler W (1984) Automatic documentary information retrieval by means of fuzzy relational products. In: Zadeh LA, Gaines BR, Zimmermann H-J (eds) Fuzzy Sets in Decision Analysis. North-Holland, Amsterdam, pp 383–404
55. Kohout LJ, Kim E (1997) The role of semiotic descriptors in relational representation of fuzzy granular structures. In: Albus J (ed) ISAS '97 Intelligent Systems and Semiotics: A Learning Perspective. Special Publ. Nat. Inst. Standards and Techn., US Dept. Commerce, Washington, DC, pp 31–36
56. Kohout LJ, Kim Yong-Gi (1993) Generating control strategies for resolution-based theorem provers by means of fuzzy relational products and relational closures. In: Lowen B, Roubens M (eds) Fuzzy Logic: State of the Art. Kluwer, Dordrecht, pp 181–192
57. Kohout LJ, Stabile I (1992) Logic of relations of Galen. In: Svoboda V (ed) Proc. Internat. Symp. Logica '92, Inst. Philosophy Acad. Sci. Czech Republic, Prague, pp 144–158
58. Kohout LJ, Stabile I, Bandler W, Anderson J (1995) CLINAID: Medical knowledge-based system based on fuzzy relational structures. In: Cohen M, Hudson D (eds) Comparative Approaches in Medical Reasoning. World Sci., Singapore, pp 1–25
59. Kohout LJ, Stabile I, Kalantar H, San-Andres M, Anderson J (1995) Parallel interval-based reasoning in medical knowledge-based system Clinaid. Reliable Computing (A special issue on parallel systems) 1(2):109–140
60. Kohout LJ, Zenz G (1997) Activity structures and triangle BK-products of fuzzy relations – a useful modelling and computational tool in value analysis studies. In: Mesiar R, et al (eds) Proc. IFSA 1997 (The World Congress of Internat. Fuzzy Systems Assoc., Prague), IV, Academia, Prague, pp 211–216
61. Lyndon RC (1961) Relation algebras and projective geometries. Michigan Math J 8:21–28
62. Maddux RD (1982) Some varieties containing relation algebras. Trans Amer Math Soc 272(2):501–526
63. Maddux RD (1983) A sequent calculus for relation algebras. Ann Pure Appl Logic 25:73–101
64. Maddux RD (1991) The origin of relation algebras in the development and axiomatization of the calculus of relations. Studia Logica 50(3–4):421–455
65. Mancini V, Bandler W (1988) Congruence of structures in urban knowledge representation. In: Bouchon B, Saita L, Yager R (eds) Uncertainty and Intelligent Systems. Lecture Notes Computer Sci. Springer, Berlin, pp 219–225
66. Mancini V, Bandler W (1992) Design for designing: Fuzzy relational environmental design assistant (FREDA). In: Kandel A (ed) Fuzzy Expert Systems. Addison-Wesley, Reading, MA, pp 195–202
67. McKinsey JCC (1940) Postulates for the calculus of binary relations. J Symbolic Logic 5:85–97
68. Nemeti I (1991) Algebraization of quantifier logics, an introductory overview. Studia Logica 50(3–4):485–569
69. Ovchinikov S (1991) Similarity relations, fuzzy partitions, and fuzzy ordering. Fuzzy Sets and Systems 40(1):107–126
70. Pedrycz W (1991) Processing in relational structures: Fuzzy relational equations. Fuzzy Sets and Systems 40(1):77–106
71. Pratt V (1991) Dynamic algebras: examples, constructions, applications. Studia Logica 50(3–4):571–605
72. Ramik J, Rommelfanger H (1996) Fuzzy mathematical programming based on some new inequality relations. Fuzzy Sets and Systems 81(1):77–87
73. Rescher N (1969) Many-valued logic. McGraw-Hill, New York
74. Riguet J (1948) Relations binaires, fermetures, correspondences de Galois. Bull Soc Math France 76:114–155
75. Rundensteiner E, Bandler W, Kohout L, Hawkes LW (1987) An investigation of fuzzy nearness measure. In: Proc. Second IFSA Congress, Internat. Fuzzy Systems Assoc., pp 362–365
76. Russell B (1900/1) The logic of relations: with some applications to the theory of series. Rivisita di Mat 7:115–148, English translation (revised by Lord Russell) in: Marsh RC (ed) (1956) Logic and Knowledge – Essays 1901-1950. Allen–Unwin, London, 1–38 (in French)
77. Sanchez E (1988) Solutions in composite fuzzy relation equations. In: Gupta MM, Saridis GN, Gaines BR (eds) Fuzzy Automata and Decision Processes. Elsevier and Univ. Press, Amsterdam, pp 221–234
78. Schmidt G, Ströhlein T (1993) Relations and graphs: Discrete mathematics for computer scientists. Springer, Berlin
79. Schmidt G, Ströohlein T (1985) On kernels of graphs and solutions of games: A synopsis based on relations and fixpoints. SIAM J Alg Discrete Meth 6:54–65
80. Schreider JuA (1975) Equality, resemblance, and order. MIR, Moscow
81. Tarski A (1941) Calculus of relations. J Symbolic Logic 6(3):73–89
82. Tarski A, Givant S (1987) A formalization of set theory without variables. Colloq Publ, vol 41. Amer. Math. Soc., Providence, RI
83. Valverde L (1985) On the structure of F-indistinguishability operators. Fuzzy Sets and Systems 17:313–328
84. Walle B Van der, DeBaets B, Kerre EE (1995) Fuzzy multicriteria analysis of cutting techniques in a nuclear reactor dismantling project. Fuzzy Sets and Systems 74(1):115–126
85. Zadeh LA (1987) Fuzzy sets: Selected papers I. In: Yager R et al (eds) Wiley, New York
86. Zadeh LA (1996) Fuzzy sets: Selected papers II. In: Klir G, Yuan B (eds) World Sci., Singapore

Bottleneck Steiner Tree Problems

BSTP

ALEXANDER ZELIKOVSKY
Georgia State University, Atlanta, USA

MSC2000: 05C05, 05C85, 68Q25, 90B80

Article Outline

Keywords
See also
References

Keywords

Bottleneck Steiner trees; Facility location; Geometric algorithms; Minmax multicenter; Approximation algorithms

A *bottleneck Steiner tree* (or a *min-max Steiner tree*) is a Steiner tree (cf. ▶ Steiner tree problems) in which the maximum edge weight is minimized. Several *multifacility location* and *VLSI routing* problems ask for bottleneck Steiner trees.

Consider the problem of choosing locations for a number of hospitals serving homes where the goal is to minimize maximum weighted distance to any home from the hospital that serves it and between hospitals. The solution is a tree which spans all hospitals and connects each home to the closest hospital. This tree can be seen as a Steiner tree where the homes are terminals and hospitals are Steiner points (cf. ▶ Steiner tree problems). Unlike the classical Steiner tree problem where the total length of Steiner tree is minimized, in this problem it is necessary to minimize maximum edge weight.

The other instance of the bottleneck Steiner tree problem occurs in electronic physical design automation where nets are routed subject to delay minimization [2,3]. The terminals of a net are interconnected possibly through intermediate nodes (Steiner points) and for electrical reasons one would like to minimize maximum distance between each pair of interconnected points.

The most popular versions of the bottleneck Steiner tree problem in the literature are geometric. Note that if the number of Steiner points is not bounded, then any edge can be subdivided into infinitely small segments and the resulting maximum edge length becomes zero. Therefore, any meaningful formulation should bound the number of Steiner points. One such formulation is suggested in [9].

Problem 1 Given a set of n points in the plane (called terminals), find a bottleneck Steiner tree spanning all terminals such that degree of any Steiner point is at least 3.

Instead of introducing constraints, one can minimize the number of Steiner points. The following formulation has been proved to be *NP*-hard [15] and approximation algorithms have been suggested in [11,14].

Problem 2 Given a set of n terminals in the plane and $\lambda > 0$, find a Steiner tree spanning n terminals with the minimum number of Steiner points such that every edge is not longer than λ.

Sometimes the bottleneck Steiner tree has predefined *topology*, i. e. the unweighted tree consisting of edges between terminals and Steiner points [4,5,10]. Then it is necessary to find the optimal positions of all Steiner points. Since the number of different topologies for a given set of terminals grows exponentially, fixing the topology greatly reduces the complexity of the bottleneck Steiner tree problem.

Problem 3 Find a bottleneck Steiner tree with a given topology T which spans a set of n terminals in the plane. The first algorithms for the Euclidean case of Problem 3 are based on nonlinear optimization [7] and [13]. For a given $\lambda > 0$, the algorithm from [15] finds whether a Steiner tree ST with the maximum edge weight λ exists as follows.

The topology T is first transformed into a forest by removing edges between terminals, if any such edge has length more than λ, then ST does not exist. Each connected component T is processed separately. The following regions are computed in bottom-up fashion:

i) the region of the plane $R(s)$ where a Steiner point s can be placed; and
ii) the region $R^+(s)$ where the Steiner point adjacent to s can be placed which is the area within distance at most λ from $R(s)$.

If a Steiner point p is adjacent to nodes $s_1, \ldots, s_k$ in T_i, then $R(s) = R^+(s_1) \cap \cdots \cap R^+(s_k)$. The number $a(s)$

of arcs bounding $R(s)$ may be as high as the number of leaves in T_i. In order to keep this number low, the tree K can be decomposed in $O(\log n)$ levels such that in total there will be only $O(n)$ arcs in all regions. Thus the runtime of the algorithm is $O(n \log n)$ [15].

When the distance between points is rectilinear, several efficient algorithms are suggested for Problem 3 [4,9,10]. The algorithm above can be adjusted for the rectilinear plane: the regions $R(s)$ are rectangles. The fastest known algorithm solves Problem 3 in time $O(n^2)$ [9].

Each bottleneck Steiner problems can be generalized to arbitrary weights on edges and formulated for weighted graphs [6].

Problem 4 Given a graph $G = (V, E, w)$ with nonnegative weight w on edges, and a set of terminals $S \subset V$, find a Steiner tree spanning S with the smallest maximum edge weight.

Problem 4 can be solved efficiently in the optimal time $O(|E|)$ time [6]. Unfortunately, the above formulation does not bound the number of Steiner points. To bound the number of Steiner points it is necessary to take in account that unlike the classical Steiner tree problem in graphs (cf. ▶ Steiner tree problems), an edge cannot be replaced with a shortest path without affecting the bottleneck objective. The following graph-theoretical generalization of Problem 1 considered in [1,9] has been proved to be *NP*-hard.

Problem 5 Given a complete graph $G = (V, E, w)$ with nonnegative weight w on edges, and a set of terminals $S \subset V$, find a Steiner tree spanning S with the smallest maximum edge weight such that each Steiner point has degree at least 3.

Similarly to the classical Steiner tree problem, if no Steiner points are allowed, the minimum spanning tree (cf. also ▶ Capacitated minimum spanning trees) is the optimal solution for Problems 1 and 5. Therefore, similarly to the *Steiner ratio*, it is valid to consider the *bottleneck Steiner ratio* $\rho_B(n)$. The bottleneck Steiner ratio is defined as the supremum over all instances with n terminals of the ratio of the maximum edge weight of the minimum spanning tree over the maximum edge weight of the bottleneck Steiner tree. It has been proved that $\rho_B(n) = 2 \lfloor \log_2 n \rfloor - \delta$, where δ is either 0 or 1 depending on whether mantissa of $\log_2 n$ is greater than $\log_2 3/2$ [9].

The approximation complexity of the Problem 5 is higher than for the classical Steiner tree problem: even $(2 - \epsilon)$-approximation is *NP*-hard for any $\epsilon > 0$ [1]. On the other hand, the best known approximation algorithm for Problem 5 has *approximation ratio* $\log_2 n$ [1]. The algorithm looks for an approximate bottleneck Steiner tree in the collection C of edges between all pairs of terminals and minimum bottleneck Steiner trees for all triples of terminals. Using Lovasz' algorithm [12] it is possible to find out whether such a collection contains a valid Steiner tree, i.e. a Steiner tree with all Steiner points of degree at least three. The algorithm finds the smallest λ such that C still contains valid Steiner tree if all edges of weight more than λ are removed. It has been shown that $\lambda \leq M \cdot \log_2 n$, where M is the maximum edge weight of the optimal bottleneck Steiner tree.

See also

▶ Capacitated Minimum Spanning Trees
▶ Directed Tree Networks
▶ Minimax Game Tree Searching
▶ Shortest Path Tree Algorithms
▶ Steiner Tree Problems

References

1. Berman P, Zelikovsky A (2000) On the approximation of power-p and bottleneck Steiner trees. In: Adv. in Steiner Trees. Kluwer, Dordrecht, pp 117–135
2. Boese KD, Kahng AB, McCoy BA, Robins G (1995) Near-optimal critical sink routing tree constructions. IEEE Trans Computer-Aided Design Integr Circuits and Syst 14:1417–11436
3. Chiang C, Sarrafzadeh M, Wong CK (1990) Global routing based on Steiner min-max trees. IEEE Trans Computer-Aided Design Integr Circuits and Syst 9:1318–1325
4. Dearing PM, Francis RL (1974) A network flow solution to a multifacility location problem involving rectilinear distances. Transport Sci 8:126–141
5. Drezner Z, Wesolowsky GO (1978) A new method for the multifacility minimax location problem. J Oper Res Soc 29:1095–1101
6. Duin CW, Volgenant A (1997) The partial sum criterion for Steiner trees in graphs and shortest paths. Europ J Oper Res 97:172–182
7. Elzinga J, Hearn D, Randolph WD (1976) Minimax multifacility location with Euclidean distances. Transport Sci 10:321–336

8. Erkut E, Francis RL, Tamir A (1992) Distance-constrained multifacility minimax location problems on tree networks. Networks 22(1):37–54
9. Ganley JL, Salowe JS (1996) Optimal and approximate bottleneck Steiner trees. Oper Res Lett 19:217–224
10. Ichimori T (1996) A shortest pathe approach to a multifacility minimax location problem with rectilinear distances. J Res Soc Japan 19:217–224
11. Lin G-H, Hue G (1999) Steiner tree problem with minimum number of Steiner points and bounded edge-length. Inform Process Lett 69:53–57
12. Lovasz L, Plummer MD (1986) Matching theory. Elsevier, Amsterdam
13. Love RF, Weselowsky GO, Kraemer SA (1997) A multifacility minimax location method for Euclidean distances. Internat J Production Res 97:172–182
14. Mandoiu II, Zelikovsky AZ (2000) A note on the MST heuristic for bounded edge-length Steiner tress with minimum number of Steiner points. Inform Process Lett 75:165–167
15. Sarrafzadeh M, Wong CK (1992) Bottleneck Steiner trees in the plane. IEEE Trans Comput 41:370–374

Boundary Condition Iteration BCI

REIN LUUS
Department Chemical Engineering,
University Toronto, Toronto, Canada

MSC2000: 93-XX

Article Outline

Keywords
Illustration of the Boundary Condition Iteration
 Procedure
Sensitivity Information Without Evaluating
 the Transition Matrix
See also
References

Keywords

Optimal control; Boundary condition iteration; BCI; Control vector iteration; Pontryagin's maximum principle; Iterative dynamic programming; IDP

In solving *optimal control* problems involving nonlinear differential equations, some iterative procedure must be used to obtain the optimal control policy. From *Pontryagin's maximum principle* it is known that the minimum of the performance index corresponds to the minimum of the Hamiltonian. Obtaining the minimum value for the Hamiltonian usually involves some iterative procedure. Here we outline a procedure that uses the necessary condition for optimality, but the boundary conditions are relaxed. In essence we have the optimal control policy at each iteration to a wrong problem. Iterations are performed, so that in the limit the boundary conditions, as specified for the optimal control problem, are satisfied. Such a procedure is called *approximation to the problem* or *boundary condition iteration method* (BCI). Many papers have been written about the method. As was pointed out in [1], the method is fundamentally very simple and computationally attractive for some optimal control problems. In [3] some evaluations and comparisons of different approaches were carried out, but the conclusions were not very definitive [5]. Although for *control vector iteration* (CVI) many papers are written to describe and evaluate different approaches with widely different optimal control problems, see for example [14], for BCI such comparisons are much more limited and there is sometimes the feeling that the method works well only if the answer is already known. However, BCI is a useful procedure for determining the optimal control policy for many problems, and it is unwise to dispatch it prematurely.

To illustrate the boundary condition iteration procedure, let us consider the *optimal control problem*, where the system is described by the differential equation

$$\|x\|\frac{d\mathbf{x}}{dt} = \mathbf{f}(\mathbf{x}, \mathbf{u}), \quad \text{with } \mathbf{x}(0) \text{ given}, \quad (1)$$

where $\mathbf{x}$ is an n-dimensional state vector and $\mathbf{u}$ is an r-dimensional control vector. The optimal control problem is to determine the control $\mathbf{u}$ in the time interval $0 \le t < t_f$, so that the performance index

$$I = \int_0^{t_f} \psi(\mathbf{x}, \mathbf{u}) \, dt \quad (2)$$

is minimized. We consider the case where the final time t_f is given and there are no constraints on the control or the state variables. According to Pontryagin's maximum principle, the minimum value of the performance index in (2) is obtained by minimizing the *Hamiltonian*

$$H = \psi + \mathbf{z}^\top \mathbf{f}. \quad (3)$$

The adjoint variable $\mathbf{z}$ is defined by

$$\frac{d\mathbf{z}}{dt} = -\frac{\partial H}{\partial \mathbf{x}}, \quad \text{with } \mathbf{z}(t_f) = \mathbf{0}, \quad (4)$$

which may be written as

$$\frac{d\mathbf{z}}{dt} = -\frac{\partial \mathbf{f}^T}{\partial \mathbf{x}}\mathbf{z} - \frac{\partial \psi}{\partial \mathbf{x}}, \quad \text{with } \mathbf{z}(t_f) = \mathbf{0}. \quad (5)$$

The necessary condition for the minimum of the Hamiltonian is

$$\frac{\partial H}{\partial \mathbf{u}} = \mathbf{0}. \quad (6)$$

Let us assume that (6) can be solved explicitly for the control vector

$$\mathbf{u} = \mathbf{g}(\mathbf{x}, \mathbf{z}). \quad (7)$$

If we now substitute (7) into (1) and (5), and integrate these equations simultaneously backward from $t = t_f$ to $t = 0$ with some value assumed for $\mathbf{x}(t_f)$, we have the optimal control policy for a wrong problem, because there is no assurance that upon backward integration the given value of the initial state $\mathbf{x}(0)$ will be obtained. Therefore it is necessary to adjust the guessed value for the final state, until finally an appropriate value for $\mathbf{x}(t_f)$ is found. For this reason the method is called the boundary condition iteration method (BCI).

In order to find how to adjust the final value of the state, based on the deviation obtained from the given initial state, we need to find the mathematical relationship to establish the effect of the change in the final state on the change in initial state. Many papers have been written in this area. The development of the necessary *sensitivity* equations is presented very nicely in [1]. In essence, the sensitivity information can be obtained by getting the *transition matrix* for the linearized state equation. Linearization of (1) gives

$$\frac{d\delta\mathbf{x}}{dt} = \left(\frac{\partial \mathbf{f}^T}{\partial \mathbf{x}}\right)^T \delta\mathbf{x} + \left(\frac{\partial \mathbf{f}^T}{\partial \mathbf{u}}\right)^T \delta\mathbf{u}. \quad (8)$$

The transition matrix Φ is thus obtained from solving

$$\frac{d\Phi}{dt} = \left(\frac{\partial \mathbf{f}^T}{\partial \mathbf{x}}\right)^T \Phi, \quad \text{with } \Phi(t_f) = \mathbf{I}, \quad (9)$$

where $\mathbf{I}$ is the $(n \times n)$ identity matrix.

Suppose at iteration j the use of $\mathbf{x}^{(j)}(t_f)$ gives the initial state $\mathbf{x}^{(j)}(0)$ which is different from the given initial state $\mathbf{x}(0)$. Then a new choice will be made at iteration $(j + 1)$ through the use of

$$\mathbf{x}^{(j+1)}(t_f) = \mathbf{x}^{(j)}(t_f) + \epsilon \Phi(0)(\mathbf{x}^{(j)}(0) - \mathbf{x}(0)), \quad (10)$$

where a stabilizing parameter ϵ is introduced to avoid overstepping. A convenient way of measuring the deviation from the given initial state is to define the error as the Euclidean norm

$$e^{(j)} = \left\|\mathbf{x}^{(j)}(0) - \mathbf{x}(0)\right\|. \quad (11)$$

Once the error is sufficiently small, say less than 10^{-6}, then the iteration procedure can be stopped.

The algorithm for boundary condition iteration may thus be presented as follows:

- Choose an initial value for the final state $\mathbf{x}^{(1)}(t_f)$ and a value for ϵ; set the iteration index j to 1.
- Integrate (1), (2), (5) and (9) backwards from $t = t_f$ to $t = 0$, using for control (7). (2) is not needed for the algorithm, but it will give the performance index.
- Evaluate the error in the initial state from (11), and if it is less than the specified value, end the iteration.
- Increment the iteration index j by one. Choose a new value for the final state $\mathbf{x}^{(j)}(t_f)$ from (10) and go to step 2.

The procedure is therefore straightforward, since the equations are all integrated in the same direction. Furthermore, there is no need to store any variables over the trajectory. There is the added advantage that the control appears as a continuous variable, and therefore the accuracy of results will not depend on the size of the integration time step. Theoretically the results should be as good as can be obtained by the second variation method in control vector iteration. It is important to realize, however, that the Hamiltonian must be well behaved, so that (7) can be obtained analytically. The only drawback is the potential instability since the state equation and the sensitivity equation are integrated backwards, and problems may arise if the final time t_f is too large. For many problems in chemical engineering the BCI method can be easily applied as is shown in the following example.

Illustration of the Boundary Condition Iteration Procedure

Let us consider the nonlinear continuous stirred tank reactor that has been used for optimal control studies in [4, pp. 308–318], and which was shown in [13] to exhibit multiplicity of solutions. The system is described by the two equations

$$\frac{dx_1}{dt} = -2(x_1 + 0.25)$$
$$+ (x_2 + 0.5)\exp\left(\frac{25x_1}{x_1 + 2}\right) - u(x_1 + 0.25), \quad (12)$$

$$\frac{dx_2}{dt} = 0.5 - x_2 - (x_2 + 0.5)\exp\left(\frac{25x_1}{x_1 + 2}\right), \quad (13)$$

with the initial state $x_1(0) = 0.09$ and $x_2(0) = 0.09$. The control u is a scalar quantity related to the valve opening of the coolant. The state variables x_1 and x_2 represent deviations from the steady state of dimensionless temperature and concentration, respectively. The performance index to be minimized is

$$I = \int_0^{t_f} (x_1^2 + x_2^2 + 0.1u^2)\, dt, \quad (14)$$

where the final time $t_f = 0.78$. The Hamiltonian is

$$H = z_1(-2(x_1 + 0.25) + R - u(x_1 + 0.25))$$
$$+ z_2(0.5 - x_2 - R) + x_1^2 + x_2^2 + 0.1u^2, \quad (15)$$

where $R = (x_2 + 0.5)\exp(25 x_1/(x_1 + 2))$. The adjoint equations are

$$\frac{dz_1}{dt} = (u + 2)z_1 - 2x_1 + 50R\frac{(z_2 - z_1)}{(x_1 + 2)^2}, \quad (16)$$

$$\frac{dz_2}{dt} = -2x_2 + \frac{(z_2 - z_1)}{(x_2 + 0.5)}R + z_2. \quad (17)$$

The gradient of the Hamiltonian is

$$\frac{\partial H}{\partial u} = 0.2u - (x_1 + 0.25)z_1, \quad (18)$$

so the optimal control is given by

$$u = 5(x_1 + 0.25)z_1. \quad (19)$$

The equations for the transition matrix are:

$$\frac{d\Phi_{11}}{dt} = \frac{\partial f_1}{\partial x_1}\Phi_{11} + \frac{\partial f_1}{\partial x_2}\Phi_{21},$$

$$\frac{d\Phi_{12}}{dt} = \frac{\partial f_1}{\partial x_1}\Phi_{12} + \frac{\partial f_1}{\partial x_2}\Phi_{22},$$

$$\frac{d\Phi_{21}}{dt} = \frac{\partial f_2}{\partial x_1}\Phi_{11} + \frac{\partial f_2}{\partial x_2}\Phi_{21},$$

$$\frac{d\Phi_{22}}{dt} = \frac{\partial f_2}{\partial x_1}\Phi_{12} + \frac{\partial f_2}{\partial x_2}\Phi_{22}$$

where

$$\frac{\partial f_1}{\partial x_1} = -2 + \frac{50R}{(x_1 + 2)^2} - u,$$

$$\frac{\partial f_1}{\partial x_2} = \frac{R}{(x_2 + 0.5)},$$

$$\frac{\partial f_2}{\partial x_1} = -\frac{50R}{(x_1 + 2)^2},$$

$$\frac{\partial f_2}{\partial x_2} = -1 + \frac{R}{(x_2 + 0.5)}.$$

The adjustment of the final state is carried out by the following two equations:

$$x_1^{(j+1)}(t_f) = x_1^{(j)}(t_f) + \epsilon\left[\Phi_{11}(0)(x_1^{(j)}(0) - x_1(0))\right.$$
$$\left. + \Phi_{12}(0)(x_2^{(j)}(0) - x_2(0))\right], \quad (20)$$

$$x_2^{(j+1)}(t_f) = x_2^{(j)}(t_f) + \epsilon\left[\Phi_{21}(0)(x_1^{(j)}(0) - x_1(0))\right.$$
$$\left. + \Phi_{22}(0)(x_2^{(j)}(0) - x_2(0))\right]. \quad (21)$$

To illustrate the computational aspects of BCI, the above algorithm was used with a Pentium-120 personal computer using WATCOM Fortran compiler version 9.5. The calculations were done in double precision. When the performance index is included, there are 9 differential equations to be integrated backwards at each iteration. Standard fourth order Runge–Kutta method was used for integration with a stepsize of 0.01. For stability, it was found that ϵ had to be taken of the order of 0.1. For all the runs, therefore, this value of ϵ was used. As is shown in Table 1, to get the error less than 10^{-6}, a large number of iterations are required, but the computation time is quite reasonable. The optimal value of the performance index is very close to the value $I = 0.133094$ reported in [13] with the second variation

Boundary Condition Iteration BCI, Table 1
Application of BCI to CSTR

Initial choice $x_1(t_f) = x_2(t_f)$	Performance index	Number of iterations	CPU time s
0.045	0.133095	2657	13.9
0.00	0.133097	2858	14.9
−0.045	0.133097	2931	15.3
0.01	0.133097	2805	14.7

method and is essentially equivalent to $I = 0.133101$ obtained in [6] by using 20 stages of piecewise linear control with iterative dynamic programming. By refining the error tolerance to $e < 10^{-8}$ required no more than an additional thousand iterations with an extra expenditure of about 6 seconds of computation time in each case. Then the final value of the performance index for each of the four different initial starting points was $I = 0.133096$.

Now that computers are very fast and their speed is rapidly being improved, and computation time is no longer prohibitively expensive, the large number of iterations required by BCI should not discourage one from using the method. Since the control policy is directly inside the integration routine, equivalent results to those obtained by second variation method can be obtained. The number of equations, however, to be integrated is quite high with a moderately high-dimensional system. If we consider a system with 10 state variables, there are 121 differential equations to be integrated simultaneously. Although computationally this does not represent a problem, the programming could be a challenge to derive and enter the equations without error. Therefore, BCI methods for which the $(n \times n)$ transition matrix is not used may find a more widespread application. One possible approach is now presented.

Sensitivity Information Without Evaluating the Transition Matrix

Suppose at iteration j we have n sets of final states $\mathbf{x}^{(j-n+1)}(t_f), \ldots, \mathbf{x}^{(j)}(t_f)$ with corresponding values for the initial state obtained by integration $\mathbf{x}^{(j-n+1)}(0), \ldots, \mathbf{x}^{(j)}(0)$. Then we can write the transformation

$$\mathbf{P} = \mathbf{AQ}, \tag{22}$$

where

$$\mathbf{P} = \left(\mathbf{x}^{(j-n+1)}(t_f) \quad \cdots \quad \mathbf{x}^{(j)}(t_f)\right), \tag{23}$$

and

$$\mathbf{Q} = \left(\mathbf{x}^{(j-n+1)}(0) \quad \cdots \quad \mathbf{x}^{(j)}(0)\right). \tag{24}$$

The transformation matrix

$$\mathbf{A} = \mathbf{PQ}^{-1} \tag{25}$$

and the next vector at t_f is chosen as

$$\mathbf{x}^{(j+1)}(t_f) = \mathbf{Ax}(0). \tag{26}$$

(1) and (5) are integrated backward to obtain $\mathbf{x}^{(j+1)}(0)$, and the matrices $\mathbf{P}$ and $\mathbf{Q}$ are updated and the procedure continued. If the initial guesses are sufficiently close to the optimal, very rapid convergence is expected.

1	Pick n sets of values for $x(t_f)$ and integrate (1) and (5) backward from $t = t_f$ to $t = 0$. using (7) for control, to give n sets of initial state vectors.
2	From these two sets of vectors form the $(n \times n)$ matrices $\mathbf{P}$ and $\mathbf{Q}$.
3	Calculate $\mathbf{A}$ from (25), and calculate a new vector $\mathbf{x}^{(j+1)}(t_f)$ from (26).
4	With the vector from Step 3 as a starting condition, integrate (1) and (5) backward to give $\mathbf{x}^{(j+1)}(0)$.
5	Use the vectors in Steps 3 and 4 to replace $\mathbf{x}^{(j-n+1)}(t_f)$ and $\mathbf{x}^{(j-n+1)}(0)$ imn matrices $\mathbf{P}$ and $\mathbf{Q}$ and continue until the error as calculated from (11) is below some tolerance, such as 10^{-8}.

Boundary Condition Iteration BCI, Algorithm

For good starting conditions, one may use *iterative dynamic programming* (IDP) [9], and pick the final states obtained after each of the first n passes. F. Hartig and F.J. Keil [2] found that in the optimization of spherical reactors, IDP provided excellent values which were refined by the use of sequential quadratic programming. For convergence here we need good starting conditions. This is now illustrated with the above example.

By using IDP, as described in [6,7,8] for piecewise linear continuous control, with 3 randomly chosen points and 10 iterations per pass for piecewise linear control with 15 time stages, the data for the first four passes in Table 2 give good starting conditions for BCI.

By using as starting conditions the final states obtained in passes 1 and 2 as given in Table 2, the convergence is very fast with the above algorithm as is shown in Table 3. Only 9 iterations are required to yield $I = 0.133096$.

As expected, if the initial set of starting points is better, then the *convergence rate* is also better as is seen in comparing Table 4 to Table 3. However, in each case the total computation time was only 0.05 seconds on a Pentium-120. Taking into account that it takes 0.77 seconds to generate the initial conditions with IDP, it is observed that the optimum is obtained in less than 1 second of computation time. Therefore, BCI is a very useful procedure if (6) can be solved explicitly for the control and the final time t_f is not too large. Simple constraints on control can be readily handled by clipping technique, as shown in [12]. Further examples with this approach are given in [10].

Boundary Condition Iteration BCI, Table 2
Results of the first four passes of IDP

Pass no.	Perf. index	$x_1(t_f)$	$x_2(t_f)$	CPU time s
1	0.1627	0.05359	−0.13101	0.39
2	0.1415	0.01940	−0.05314	0.77
3	0.1357	0.05014	−0.09241	1.16
4	0.1334	0.05670	−0.10084	1.54

Boundary Condition Iteration BCI, Table 3
Convergence with the above algorithm from the starting points obtained in passes 1 and 2 by IDP

Iteration no.	Perf. index	Error ε
1	0.014520	0.1215
2	0.031301	0.1031
3	0.129568	$0.1852 \cdot 10^{-2}$
4	0.136682	$0.2414 \cdot 10^{-2}$
5	0.135079	$0.1350 \cdot 10^{-2}$
6	0.133218	$0.8293 \cdot 10^{-4}$
7	0.133093	$0.2189 \cdot 10^{-5}$
8	0.133096	$0.1373 \cdot 10^{-6}$
9	0.133096	$0.5209 \cdot 10^{-8}$

Boundary Condition Iteration BCI, Table 4
Convergence with the above algorithm from the starting points obtained in passes 3 and 4 by IDP

Iteration no.	Perf. index	Error ε
1	0.121769	$0.7353 \cdot 10^{-2}$
2	0.135249	$0.1415 \cdot 10^{-2}$
3	0.133317	$0.1531 \cdot 10^{-3}$
4	0.133138	$0.2861 \cdot 10^{-4}$
5	0.133094	$0.1703 \cdot 10^{-5}$
6	0.133096	$0.1190 \cdot 10^{-7}$
7	0.133096	$0.5364 \cdot 10^{-10}$

See also

▶ Control Vector Iteration

References

1. Denn MM, Aris R (1965) Green's functions and optimal systems – Necessary conditions and an iterative technique. Industr Eng Chem Fundam 4:7–16
2. Hartig F, Keil FJ (1993) Large scale spherical fixed bed reactors – modelling and optimization. Industr Eng Chem Res 32:57–70
3. Jaspan RK, Coull J (1972) Trajectory optimization techniques in chemical engineering. II. Comparison of the methods. AIChE J 18:867–869
4. Lapidus L, Luus R (1967) Optimal control of engineering processes. Blaisdell, Waltham
5. Luus R (1974) BCI vs. CVI. AIChE J 20:1039–1040
6. Luus R (1993) Application of iterative dynamic programming to very high dimensional systems. Hungarian J Industr Chem 21:243–250
7. Luus R (1993) Piecewise linear continuous optimal control by iterative dynamic programming. Industr Eng Chem Res 32:859–865
8. Luus R (1996) Numerical convergence properties of iterative dynamic programming when applied to high dimensional systems. Chem Eng Res Des 74:55–62
9. Luus R (1998) Iterative dynamic programming: from curiosity to a practical optimization procedure. Control and Intelligent Systems 26:1–8
10. Luus R (2000) Iterative dynamic programming. Chapman and Hall/CRC, London
11. Luus R (2000) A new approach to boundary condition iteration in optimal control. In: Proc. IASTED Internat. Conf. Control and Applications, Cancun, Mexico, May 24–27, 2000, pp 172–176
12. Luus R (2001) Further developments in the new approach to boundary condition iteration in optimal control. Canad J Chem Eng 79:968–976

13. Luus R, Cormack DE (1972) Multiplicity of solutions resulting from the use of variational methods in optimal control problems. Canad J Chem Eng 50:309–311
14. Rao SN, Luus R (1972) Evaluation and improvement of control vector iteration procedures for optimal control. Canad J Chem Eng 50:777–784

Bounding Derivative Ranges

GEORGE F. CORLISS[1], L. B. RALL[2]
[1] Marquette University, Milwaukee, USA
[2] University Wisconsin–Madison, Madison, USA

MSC2000: 90C30, 90C26

Article Outline

Keywords
Evaluation of Functions
Monotonicity
Taylor Form
Intersection and Subinterval Adaptation
Software Availability
See also
References

Keywords

Interval arithmetic; Automatic differentiation; Taylor series

Interval arithmetic can be used to bound the range of a real function over an interval. Here, we bound the ranges of its *Taylor coefficients* (and hence derivatives) by evaluating it in an interval Taylor arithmetic. In the context of classical numerical methods, truncation errors, Lipschitz constants, or other constants related to existence or convergence assertions are often phrased in terms of bounds for certain derivatives. Hence, interval inclusions of Taylor coefficients can be used to give *guaranteed bounds* for quantities of concern to classical methods.

Evaluating the expression for a function using interval arithmetic often yields overly pessimistic bounds for its range. Our goal is to tighten bounds for the range of f and its derivatives by using a *differentiation arithmetic* for series generation. We apply *monotonicity* and *Taylor form tests* to each intermediate result of the calculation, not just to f itself. The resulting inclusions for the range of derivative values are several orders of magnitude tighter than bounds obtained from differentiation arithmetic and interval calculations alone. Tighter derivative ranges allow validated applications such as optimization, nonlinear equations, quadrature, or differential equations to use larger steps, thus improving their computational efficiency.

Consider the set of q times continuously differentiable functions on the real interval $\mathbf{x} = [\underline{x}, \overline{x}]$ denoted by $f(x) \in C^q[\mathbf{x}]$. We wish to compute a tight inclusion for

$$R(f^{(p)}; \mathbf{x}) := \left\{ f^{(p)}(x) \colon \underline{x} \leq x \leq \overline{x} \right\}, \quad (1)$$

where $p \leq q$. We assume that f is sufficiently smooth for all indicated computations, and that all necessary derivatives are computed using *automatic differentiation* (cf. [5], ▶ Automatic differentiation: Point and interval Taylor operators).

Computing an inclusion for the range of $f^{(p)}$ is a generalization of the problem of computing an inclusion for the range of f, $R(f; \mathbf{x})$. Moore's *natural interval extension* [3] gives an inclusion which is often too gross an overestimation to be practical. H. Ratschek and J. Rokne [8] gives a number of improved techniques and many references. The approach of this paper follows from two papers of L.B. Rall [6,7] and from [1]. Taken together, Rall's papers outline four approaches to computing tight inclusions of $R(f; \mathbf{x})$, which we apply to derivatives:

- monotonicity,
- *mean value* and *Taylor forms*,
- *intersection*, and
- *subinterval adaptation*.

We apply the monotonicity tests and the Taylor form to each term of the Taylor polynomial of a function. Whenever we compute more than one enclosure for a quantity, either a derivative or an intermediate value, we compute intersections of all such enclosures. We apply these tests to each intermediate result of the calculation, not just to f itself. The bounds we compute for $R(f^{(p)}; \mathbf{x})$ are often *several orders of magnitude* tighter than bounds computed from natural interval extensions. In one example, we improve the interval inclusion for $R(f^{(10)}; \mathbf{x})$ from $[-3.8E10, 7.8E10]$ (width $= 1.1E11$) to $[-2.1E03, 9.6E03]$ (width $= 1.1E04$).

This improvement by a factor of 10^7 allows a Gaussian quadrature using 5 points per panel or a 10th order ODE solver (applications for which bounds for $R(f^{(10)}; \mathbf{x})$ might be needed) to increase their stepsizes, and hence their computational efficiency, by a factor of $10^{7/10} \approx 5$.

We discuss the evaluation of a function from a *code list* representation (see also ▶ Automatic differentiation: Point and interval Taylor operators). Then we discuss how monotonicity tests and Taylor form representations can be used to give tighter bounds for $R(f^{(p)}; \mathbf{x})$.

Evaluation of Functions

Functions are expressed in most computer languages by arithmetic operations and a set Φ of *standard functions*, for example, $\Phi = \{\text{abs, arctan, cos, exp, ln, sin, sqr, sqrt}\}$. A formula (or expression) can be converted into a code list or *computational graph* $\{t_1, \ldots, t_n\}$ (cf. [5], ▶ Automatic differentiation: Point and interval Taylor operators). The value of each term t_i is the result of a unary or binary operation or function applied to constants, values of variables, or one or two previous terms of the code list. For example, the function

$$f(x) = \frac{x^4 - 10x^2 + 9}{x^3 - 4x - 5}$$

can be converted into the code list

$t_1 := \text{sqr}(x);$	$t_6 := x \cdot t_1;$
$t_2 := \text{sqr}(t_1);$	$t_7 := 4 \cdot x;$
$t_3 := 10 \cdot t_1;$	$t_8 := t_6 - t_7;$
$t_4 := t_2 - t_3;$	$t_9 := t_8 - 5;$
$t_5 := t_4 + 9;$	$t_{10} := t_5/t_9.$

Bounding Derivative Ranges, Figure 1
Code list

The final term t_n of the code list (t_{10} in this case) gives the value of $f(x)$, if defined, for a given value of the variable x. The conversion of a formula into an equivalent code list can be carried out automatically by a computer subroutine.

The code list serves equally well for various kinds of arithmetic, provided the necessary arithmetic operations and standard functions are defined for the type of elements considered. Thus, the code list in Fig. 1 can serve for the computation of $f(x)$ in real, complex, interval, or differentiation arithmetic. When x is an interval, one gets an interval inclusion $f(\mathbf{x})$ of all real values $f(x)$ for real $x \in \mathbf{x}$ [3,4].

The process of automatic differentiation to obtain derivatives or Taylor coefficients of $f(x)$ can be viewed as the evaluation of the code list for $f(x)$ using a differentiation arithmetic in which the arithmetic operations and standard functions are defined on the basis of the well-known recurrence relations for Taylor coefficients (cf. also [3,4,5], ▶ Automatic differentiation: Point and interval Taylor operators). Let $(f)_i := f^{(i)}(\check{x})/i!$ be the value of the ith Taylor coefficient of $f(x) = f(\check{x} + h)$. Then we can express a Taylor series as

$$f(x) = \sum_{i=0}^{\infty} f^{(i)}(\check{x}) \frac{h^i}{i!} = \sum_{i=0}^{\infty} (f)_i h^i,$$

and the elements of Taylor series arithmetic are vectors $f = ((f)_0, \ldots, (f)_p)$. In Taylor arithmetic, constants c have the representation $c = (c, 0, \ldots, 0)$, and $x = (x_0, 1, 0, \ldots, 0)$ represents the independent variable $x = x_0 + h$. For example, multiplication $f(x) = u(x) \cdot v(x)$ of Taylor variables is defined in terms of the Taylor coefficients of u and v by $(f)_i = \sum_{j=0}^{i} (u)_j \cdot (v)_{i-j}$, $i = 0, \ldots, p$.

Monotonicity

We extend an idea of R.E. Moore for using monotonicity [4]: we check for the monotonicity of every derivative of f and of every intermediate function t_i from the code list. If the ith derivative of f is known to be of one sign on $\mathbf{x}$ ($R(f^{(i)}; \mathbf{x}) \geq 0$ or ≤ 0), then $f^{(i-1)}$ is monotonic on the interval $\mathbf{x}$, and its range is bounded by the real values $f^{(i-1)}(\underline{x})$ and $f^{(i-1)}(\overline{x})$. This is important because the bounds of $R(f^{(i-1)}; \mathbf{x})$ by $f^{(i-1)}(\underline{x})$ and $f^{(i-1)}(\overline{x})$ may be tighter than the bounds computed by the naive interval evaluation of $f^{(i-1)}(\mathbf{x})$. Hence, in addition to the ranges $R(f^{(i)}; \mathbf{x})$, we propagate enclosures of the values at the endpoints $R(f^{(i)}; \underline{x})$ and $R(f^{(i)}; \overline{x})$ so that those values are available. (We use $R(f^{(i)}; \underline{x})$ and $R(f^{(i)}; \overline{x})$ instead of $R(f^{(i)}; \underline{x})$ and $R(f^{(i)}; \overline{x})$ to denote that $f^{(i)}$ at the endpoints is evaluated in interval arithmetic.)

Similarly, if $R(f^{(i)}; \mathbf{x}) \geq 0$ (or ≤ 0), then $f^{(i-2)}$ is convex (resp. concave), and its maximum value is $\max(f^{(i-2)}(\underline{x}), f^{(i-2)}(\overline{x}))$ (resp. minimum is $\min(f^{(i-2)}(\underline{x}), f^{(i-2)}(\overline{x}))$).

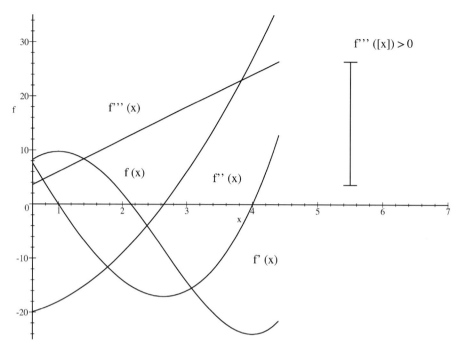

Bounding Derivative Ranges, Figure 2
$R(f^{(3)}; \mathbf{x}) \geq 0$ implies f is monotonic and f' is convex

We apply the monotonicity test to each term of each intermediate result because an intermediate result may be monotonic when f is not. Further, by proceeding with tighter inclusions for the terms of the intermediate results, we reduce subsequent over-estimations and improve our chances for validating the monotonicity of higher derivatives. If $f^{(i-1)}$ is found to be monotonic, the tightened enclosure for $R(f^{(i-1)}; \mathbf{x})$ may allow us to validate $R(f^{(i-1)}; \mathbf{x}) \geq 0$ (or ≤ 0), so we backtrack to lower terms of the series as long as we continue to find monotonicity. In the recurrence relations for divide and for all of the standard functions, the value of $f^{(i)}(x)$ depends on the value of $f^{(i-1)}(x)$. Hence, if the enclosure for $R(f^{(i-1)}; \mathbf{x})$ is tightened, we recompute the enclosure for $R(f^{(i)}; \mathbf{x})$ and all subsequent terms.

Table 1 shows (some of) the results when the monotonicity test is applied to each of the intermediate results of

$$f(x) = \frac{x^4 - 10x^2 + 9}{x^3 - 4x - 5}$$

on the interval $\mathbf{x} := [1, 2]$. Each row shows enclosures for Taylor coefficients. The row 'x' has two entries for the function x evaluates on the interval $\mathbf{x}$ and its derivative. All higher-order derivatives are zero. Similarly, rows t_4 and t_5 have five nonzero derivatives.

A few entries show where tightening occurs because of the monotonicity test. For example, at [1], the 3rd derivative of t_4 is positive. Hence, t_4 is monotonic, but that knowledge yields no tightening. Also t_4' is convex, a fact which *does* allow us to tighten the upper bound from 12 to -8. Similarly at [2], finding that t_8 is positive allows us to improve the upper bound for t_8. In this example, the monotonicity tests allow us only two relatively modest tightenings, but those two tighter values propagate through the recurrences to reduce the width of the bound finally computed for $t_{10}^{(5)}$ from about 2.3E6 to 300, an improvement of nearly a factor of 10^4.

Taylor Form

In [6], Rall proves that if $\check{x} \in x$, then

$$R(f; \mathbf{x}) \subset F_p(\mathbf{x}) := \sum_{i=0}^{p-1} (f)_i (\mathbf{x} - \check{x})^i + F^{(p)}(\mathbf{x}) \frac{(\mathbf{x} - \check{x})^p}{p!}, \quad (2)$$

Bounding Derivative Ranges, Table 1
Numerical results of applying monotonicity tests

x					
$x(\mathbf{x})$	[1, 2]	1			
$t_1 := x^2$					
$t_1(\mathbf{x})$	[1, 4]	[2, 4]	1		
$t_2 := t_1^2 = x^4$					
$t_2(\mathbf{x})$	[1, 16]	[4, 32]	[6, 24]	[4, 8]	1
$t_3 := 10 * t_1 = 10x^2$					
$t_3(\mathbf{x})$	[10, 40]	[20, 40]	10		
$t_4 := t_2 - t_3 = x^4 - 10x^2$					
$t_4(\mathbf{x})$	[−39, 6]	[−36, 12]	[−4, 14]	[4, 8][1]	1
Tightened to:					
$t_4(\mathbf{x})$	[−24, -9]	[−36, -8]	[−4, 14]	[4, 8]	1
$t_5 := t_4 + 9 = x^4 - 10x^2 + 9$					
$t_5(\mathbf{x})$	[−30, 15]	[−36, 12]	[−4, 14]	[4, 8]	1
Tightened as the result of tightening t_4:					
$t_5(\mathbf{x})$	[−15, 0]	[−36, -8]	[−4, 14]	[4, 8]	1
...					
$t_8 := t_6 - t_7 = x^3 - 4x$					
$t_8(\mathbf{x})$	[−7, 4]	[−1, 8]	[3, 6][2]	1	
Tightened to:					
$t_8(\mathbf{x})$	[−7, 0]	[−1, 8]	[3, 6]	1	
$t_9 := t_8 - 5 = x^3 - 4x - 5$					
$t_9(\mathbf{x})$	[−12, -1]	[−1, 8]	[3, 6]	1	
Tightened as the result of tightening t_8:					
$t_9(\mathbf{x})$	[−12, -5]	[−1, 8]	[3, 6]	1	
$t_{10} := t_5/t_9 = f$					
$t_{10}(\mathbf{x})$	[−15, 30]			[−10097, 20823]	
		[−132, 276]			[−87881, 181229]
			[−1160, 2392]		[−764851, 1577270]
Tightened as the result of tightening t_5 and t_9:					
$t_{10}(\mathbf{x})$	[−9.73E−0.6, 3.01]			[−9.68, 51.97]	
		[0.41, 12.01]			[−21.84, 113.67]
			[−5.21, 23.61]		[−47.58, 248.96]

and $F^{(p)}$ is an interval extension of $f^{(p)}$. The F_p given by (2) is called the (elementary) Taylor form of f of order p.

We expand the Taylor series for the function f and all intermediate functions t_i appearing in the code list at three points, $x = a := \underline{x}$, $x = c :=$ midpoint $(\mathbf{x})$, and $x = b := \overline{x}$. The series for f at $\underline{x}$ and $\overline{x}$ are already available since they were computed for the monotonicity test. The extra work required to generate the series at c is often justified because the midpoint form is much narrower than either of the endpoint forms. Let $h :=$ width $(\mathbf{x})$. We compute the Taylor form (2) for f and each t_i at the left endpoint, center, and right endpoint to all available orders and intersect. The remainder using $R(f^{(i+1)}; \mathbf{x})$ has the potential for tightening all previous terms:

$$R(f; \mathbf{x}) \subset \left(f(a) + f'(a)h[0, 1] + f''(a)\frac{h^2}{2!} * [0, 1] \right.$$
$$+ \cdots + f^{(i)}(a)\frac{h^i}{i!} * [0, 1]$$
$$\left. + R(f^{(i+1)}; \mathbf{x})\frac{h^{i+1}}{(i+1)!} * [0, 1] \right)$$

Bounding Derivative Ranges, Table 2
Numerical results of applying Taylor from tests

x
    $x(\mathbf{x})$ [1,2]   1
$t_1 = x^2$
    $t_1([a,b])$ [1, 4]   [2, 4]   1
    No tightening occurs.

...

$t_4 = t_2 - t_3 = x^4 - 10x^2$

$t_4(\mathbf{x})$	[−39, 6]	[−36, 12]	[−4, 14]	[4, 8]	1
		[−24, 12]	tightened by $f(a)$ using Fab(2)		
		[−24, -2.5]	tightened by $f(c)$ using Fab(2)		
		[−20, -7]	tightened by $f(c)$ using Fab(3)		
		[−20, -8]	tightened by $f(c)$ using Fab(4)		
	[−29, -9]	tightened by $f(a)$ using Fab(1)			
	[−27.438, -9]	tightened by $f(c)$ using Fab(1)			
	[−24, -9]	tightened by $f(b)$ using Fab(1)			

Tightened to :
    $t_4(\mathbf{x})$ [−24, -9]   [−20, -8]   [−4, 14]   [4, 8]   1

$t_5 = t_4 + 9 = x^4 - 10x^2 + 9$
    $t_5(\mathbf{x})$ [−30, 15]   [−36, 12]   [−4, 14]   [4, 8]   1
Tightened as the result of tightening t_4:
    $t_5(\mathbf{x})$ [−15, 0]   [−20, -8]   [−4, 14]   [4, 8]   1

$t_8 = t_6 - t_7 = x^3 - 4x$
    $t_8(\mathbf{x})$ [−7, 4]   [−1, 8]   [3, 6]   1
Tightened to:
    $t_8(\mathbf{x})$ [−4, 0]   [−1, 8]   [3, 6]   1

$t_{10} = t_5 / t_9$

$t_{10}(\mathbf{x})$	[−15, 30]		[−10097, 20823]	
		[−132, 276]		[−87881, 181229]
			[−1160, 2392]	[−764851, 1577270]

Tightened as the result of tightening t_5 and t_9:

$t_{10}(\mathbf{x})$	[−9.73E−06, 3.01]		[−8.57, 39.94]	
		[0.55, 8.81]		[−19.27, 87.64]
			[−4.57, 18.49]	[−42.02, 191.84]
	[−6.08E−06, 3.01] tightened by f(c) using Fab(1)			

$$\cap \left(f(c) + f'(c)h * \frac{[-1,1]}{2} + f''(c)\frac{h^2}{2!} * \frac{[-1,1]}{4} \right.$$
$$\left. + \cdots + f^{(i)}(c)\frac{h^i}{i!} * \frac{[-1,1]^i}{2^i} \right.$$
$$\left. + R(f^{(i+1)};\mathbf{x})\frac{h^{i+1}}{(i+1)!} * \frac{[-1,1]^{i+1}}{2^{i+1}} \right)$$

$$\cap \left(f(b) + f'(b)h * [-1,0] + f''(b)\frac{h^2}{2!} * [0,1] \right.$$
$$\left. + \cdots + f^{(i)}(b)\frac{h^i}{i!} * [-1,0]^i \right.$$
$$\left. + R(f^{(i+1)};\mathbf{x})\frac{h^{i+1}}{(i+1)!} * [-1,0]^{i+1} \right).$$

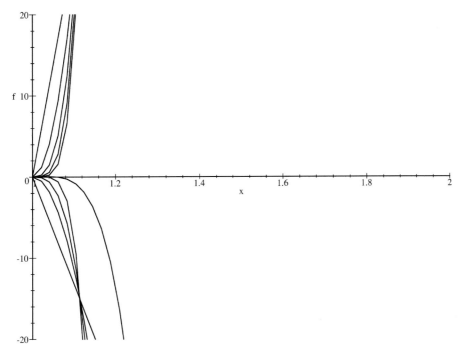

Bounding Derivative Ranges, Figure 3
Taylor polynomial enclosures for *f*, remainders from naive interval evaluation

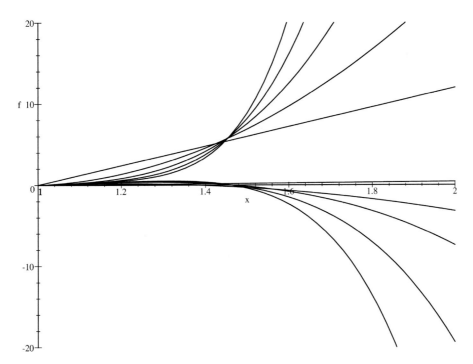

Bounding Derivative Ranges, Figure 4
Taylor polynomial enclosures for *f*, remainders tightened by Taylor form

For higher-order derivatives, $R(f^{(i)}; \mathbf{x})$ is contained in similar Taylor forms involving $f^{(i+n)}(\mathbf{x})$, for $n > 0$.

We apply the Taylor form to each intermediate result. Except for the operators +, −, *, and sqr, whenever one term is tightened, all following terms can be recomputed more tightly. This can result in an iterative process which is finite only by virtue of Moore's theorem on interval iteration [4]. In practice then, we restrict the number of times subsequent terms are recomputed starting at a given order.

Table 2 shows (some of) the results when the Taylor form is applied to each of the intermediate results of

$$f(x) = \frac{x^4 - 10x^2 + 9}{x^3 - 4x - 5}$$

on the interval $\mathbf{x} := [1, 2]$. 'Tightened by $f(a)$, $f(c)$, or $f(b)$' indicates whether the left endpoint, the midpoint, or the right endpoint expansion was used. 'Using Fab(n)' indicates that $f(i)$ was tightened using $f^{(i+n)}$.

The pattern of Table 2 is typical: Most Taylor forms give no tightening; there are many small improvements; and the compound effect of many small improvements is significant. Here we have reduced the width of the enclosure for the 6th Taylor coefficient from about $2.3E7$ to $2.3E2$. Figures 3 and 4 compare the Taylor polynomial enclosures for f resulting from naive interval evaluation of the remainders with the enclosures tightened by the Taylor form computations shown in Table 2.

For this example, the bounds achieved using the Taylor form are tighter than those achieved using the monotonicity test. For other examples, the monotonicity test performs better. Hence in practice, we apply both techniques. If the expression for f is rewritten in a mathematically equivalent form to yield tighter interval bounds for $R(f; \mathbf{x})$, the techniques of this paper can still be used profitably to tighten enclosures of higher derivatives.

Intersection and Subinterval Adaptation

The third general technique described by Rall for tightening enclosures of $R(f; \mathbf{x})$ is to intersect all enclosures for each quantity, as we have done here. That is, whatever bounds for $R(f^{(i)}; \mathbf{x})$ we compute using monotonicity or Taylor form of any degree, we intersect with the tightest bound previously computed. Each new bound may improve our lower bound, our upper bound, both, or neither. Some improvements are large. Others are so small as to seem insignificant, but even the smallest improvements may be magnified by later operations.

Rall's fourth technique is the *adaptive partitioning* of the interval $\mathbf{x}$. The over-estimation of $R(f^{(i)}; \mathbf{x})$ by naive interval evaluation decreases linearly with width ($\mathbf{x}$), while the over-estimation by the Taylor form decreases quadratically. Hence, partitioning $\mathbf{x}$ into smaller subintervals is very effective. However, we view subinterval adaptation as more effectively controlled by the application (e. g., optimization, quadrature, DE solution) than by the general-purpose interval Taylor arithmetic outlined here. Hence, we do not describe it further.

Software Availability

An implementation in Ada of interval Taylor arithmetic operators for +, −, *, /, and sqr is available at [9]. Similar implementations could be written in Fortran 90, C++, or any other language supporting operator overloading.

See also

- ▶ Automatic Differentiation: Point and Interval
- ▶ Automatic Differentiation: Point and Interval Taylor Operators
- ▶ Global Optimization: Application to Phase Equilibrium Problems
- ▶ Interval Analysis: Application to Chemical Engineering Design Problems
- ▶ Interval Analysis: Differential Equations
- ▶ Interval Analysis: Eigenvalue Bounds of Interval Matrices
- ▶ Interval Analysis: Intermediate Terms
- ▶ Interval Analysis: Nondifferentiable Problems
- ▶ Interval Analysis: Parallel Methods for Global Optimization
- ▶ Interval Analysis: Subdivision Directions in Interval Branch and Bound Methods
- ▶ Interval Analysis: Systems of Nonlinear Equations
- ▶ Interval Analysis: Unconstrained and Constrained Optimization
- ▶ Interval Analysis: Verifying Feasibility

- ▶ Interval Constraints
- ▶ Interval Fixed Point Theory
- ▶ Interval Global Optimization
- ▶ Interval Linear Systems
- ▶ Interval Newton Methods

References

1. Corliss GF, Rall LB (1991) Computing the range of derivatives. In: Kaucher E, Markov SM, Mayer G (eds) Computer Arithmetic, Scientific Computation and Mathematical Modelling. IMACS Ann Computing Appl Math. Baltzer, Basel, pp 195–212
2. Gray JH, Rall LB (1975) INTE: A UNIVAC 1108/1110 program for numerical integration with rigorous error estimation. MRC Techn Summary Report Math Res Center, Univ Wisconsin–Madison 1428
3. Moore RE (1966) Interval analysis. Prentice-Hall, Englewood Cliffs, NJ
4. Moore RE (1979) Methods and applications of interval analysis. SIAM, Philadelphia
5. Rall LB (1981) Automatic differentiation: techniques and applications. Lecture Notes Computer Sci, vol 120. Springer, Berlin
6. Rall LB (1983) Mean value and Taylor forms in interval analysis. SIAM J Math Anal 2:223–238
7. Rall LB (1986) Improved interval bounds for ranges of functions. In: Nickel KLE (ed) Interval Mathematics (Freiburg, 1985). Lecture Notes Computer Sci, vol 212. Springer, Berlin, pp 143–154
8. Ratschek H, Rokne J (eds) (1984) Computer methods for the range of functions. Horwood, Westergate
9. Website: www.mscs.mu.edu/~georgec/Pubs/eoo_da.tar.gz

Bounds and Solution Vector Estimates for Parametric NLPS

Vivek Dua, Efstratios N. Pistikopoulos
Imperial College, London, UK

MSC2000: 90C31

Article Outline

Keywords
Parametric Lower Bound
Parametric Upper Bound
See also
References

Keywords

Sensitivity analysis; Linear approximation; Parametric upper and lower bounds

In this article, we present some important theoretical results based upon which solution of parametric nonlinear programming problems can be approached. The need for these results arises from the fact that while stability, continuity and convexity properties of objective function value for linear programs are readily available [7], their counterparts in nonlinear programs are valid only for a special class of nonlinear programs. It is not surprising then that a large amount of research has been devoted towards establishing these conditions (see [1] and [3] for a comprehensive list of references). Further, due to the existence of strong duality results for linear models, parametric programming can be done by extending the simplex algorithm for linear models [6]. On the other hand, for nonlinear programs the parametric solution is given by an approximation of the optimal solution. This approximation or estimation of the optimal solution can be achieved by obtaining the optimal solution as a function of parameters. In order to derive these results we first state the following *implicit function theorem*:

Theorem 1 (see for example [3,8]) *Suppose that $\phi(x, \theta)$ is a $(r \times 1)$ vector function defined on $E^n \times E^m$, with $x \in E^n$ and $\theta \in E^m$, and $D_x \phi(x, \theta)$ and $D_\theta \phi(x, \theta)$ indicate the $(r \times n)$ and $(r \times m)$ matrix of first derivatives with respect to x and θ respectively. Suppose that $\phi: E^{m+n} \to E^n$. Let $\phi(x, \theta)$ be continuously differentiable in x and θ in an open set at (x_0, θ_0) where $\phi(x_0, \theta_0) = 0$. Suppose that $D_x \phi(x_0, \theta_0)$ has an inverse.*

Then there is a function $x(\theta)$ defined in a neighborhood of θ_0 where for each $\widehat{\theta}$ in that neighborhood $\phi[x(\widehat{\theta}), \widehat{\theta}] = 0$. Furthermore, $x(\theta)$ is a continuously differentiable function in that neighborhood and

$$\begin{aligned} x'(\theta_0) &= -D_x\phi[x(\theta_0), \theta_0]^{-1} D_\theta \phi[x(\theta_0), \theta_0] \\ &= -D_x\phi(x_0, \theta_0)^{-1} D_\theta \phi(x_0, \theta_0), \end{aligned}$$

where $x'(\theta_0)$ denotes the derivative of x evaluated at θ_0.

Consider the parametric nonlinear programming problem of the following form:

$$\begin{cases} z(\theta) = \min_x f(x,\theta) \\ \text{s.t.} \quad g_i(x,\theta) \geq 0, \quad i = 1,\ldots,p, \\ \phantom{\text{s.t.}} \quad h_j(x,\theta) = 0, \quad j = 1,\ldots,q, \\ \phantom{\text{s.t.}} \quad x \in X, \end{cases} \quad (1)$$

where f, g and h are twice continuously differentiable in x and θ. The *first order KKT conditions* for (1) are given as follows:

$$\nabla f(x,\theta) - \sum_{i=1}^{p} \lambda_i \nabla g_i(x,\theta) + \sum_{j=1}^{q} \mu_j \nabla h_j(x,\theta) = 0,$$

$$\lambda_i g_i(x,\theta) = 0, \quad i = 1,\ldots,p,$$

$$h_j(x,\theta) = 0, \quad j = 1,\ldots,q. \quad (2)$$

An application of the implicit function theorem 1 to the KKT conditions (2) results in the following *basic sensitivity theorem*:

Theorem 2 ([2,3,8]) *Let θ_0 be a vector of parameter values and (x_0, λ_0, μ_0) a KKT triple corresponding to (2), where λ_0 is nonnegative and x_0 is feasible in (1). Also assume that:*
i) *strict complementary slackness holds;*
ii) *the binding constraint gradients are linearly independent;*
iii) *the second order sufficiency conditions hold.*
Then, in neighborhood of θ_0, there exists a unique, once continuously differentiable function $[x(\theta), \lambda(\theta), \mu(\theta)]$ satisfying (2) with $[x(\theta_0), \lambda(\theta_0), \mu(\theta_0)] = (x_0, \lambda_0, \mu_0)$, where $x(\theta)$ is a unique isolated minimizer for (1), and

$$\begin{pmatrix} \frac{dx(\theta_0)}{d\theta} \\ \frac{d\lambda(\theta_0)}{d\theta} \\ \frac{d\mu(\theta_0)}{d\theta} \end{pmatrix} = -(M_0)^{-1} N_0, \quad (3)$$

where

$$M_0 = \begin{pmatrix} \nabla^2 L & -\nabla g_1 & \cdots & -\nabla g_p & \nabla h_1 & \cdots & \nabla h_q \\ \lambda_1 \nabla^\top g_1 & g_1 & & & & & \\ \vdots & & \ddots & & & & \\ \lambda_p \nabla^\top g_p & & & g_p & & & \\ \nabla^\top h_1 & & & & & & \\ \vdots & & & & & & \\ \nabla^\top h_r & & & & & & \end{pmatrix}$$

and

$$N_0 = (\nabla^2_{\theta x} L, \lambda_1 \nabla^\top_\theta g_1, \ldots, \lambda_p \nabla^\top_\theta g_p,$$
$$\nabla^\top_\theta h_1, \ldots, \nabla^\top_\theta h_q)^\top,$$
$$L(x, \lambda, \mu, \theta) = f(x, \theta)$$
$$+ \sum_{i=1}^{p} \lambda_i g_i(x,\theta) + \sum_{j=1}^{q} \mu_j h_j(x,\theta).$$

However, for a special case of (1) when the parameters are present on the right-hand side of the constraints, (1) can be rewritten in the following form:

$$\begin{cases} z(\theta) = \min_x f(x) \\ \text{s.t.} \quad g(x) \geq \theta, \\ \phantom{\text{s.t.}} \quad x \in X. \end{cases} \quad (4)$$

A simplified version of an equivalent of (3) for (4) can also be obtained (see for example [8] for details). Another important result that can be derived for (4) is that the rate of change of the optimal value function, $z(\theta)$, with change in θ is given by KKT multiplier. Thus, given an optimal solution of (4) at a fixed point in θ_0, an estimate of the optimal solution in the neighborhood of θ_0 can be obtained by using the KKT multiplier obtained at θ_0 (see [8] and [9]). For a special case of (4), when (4) is convex in x and θ is bounded between certain lower and upper bounds, say 0 and 1, one can obtain a piecewise linear approximation of the optimal value function for the whole range of θ. In order to derive these results, we first state the following properties.

Theorem 3 (*continuity property of the objective function value;* [3]) *Let*

$$z(\theta) = \inf_x \{f(x) \colon x \in X, \ g(x) \geq \theta\}.$$

Suppose
i) *X is a compact convex set in E^n,*
ii) *f and g are both continuous on $X \times E^n$, and*
iii) *each component of g is strictly concave on X for each θ.*
Then $z(\theta)$ is continuous on its effective domain.

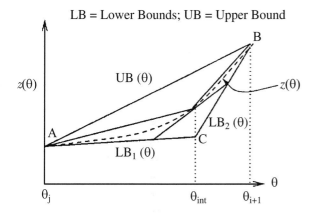

Bounds and Solution Vector Estimates for Parametric NLPS,
Figure 1
Bounds on the optimal value function $z(\theta)$

Theorem 4 (*convexity property of the solution space*; [4]) *Suppose*
i) g_i are jointly quasiconcave on X and θ, and
ii) X is convex.
Then the solution space $R(\theta) = \{x \in X: g(x) \geq \theta\}$, is convex.

Theorem 5 (*convexity property of the objective function value*; [4]) *Suppose*
i) f is convex on X, and
ii) the solution space $R(\theta)$ is essentially convex.
Then $z(\theta)$ is convex on θ.

Since $z(\theta)$ is continuous and convex under above conditions, for a given interval $[\theta_i, \theta_{i+1}]$, we can obtain [5] (see Fig. 1) parametric lower and upper bounds as follows.

Parametric Lower Bound

A linear underestimator of the convex function, $z(\theta)$, will be a global underestimator, hence lower bounds at θ_i and θ_{i+1} given by:

$$LB_i(\theta) = z^*(\theta_i) + \nabla_\theta z^*(\theta_i)(\theta - \theta_i),$$
$$LB_{i+1}(\theta) = z^*(\theta_{i+1}) + \nabla_\theta z^*(\theta_{i+1})(\theta - \theta_{i+1}),$$

where $\nabla_\theta z^*(\theta)$ is given by the Lagrange multipliers as discussed earlier, provide global underestimators to $z(\theta)$.

Parametric Upper Bound

A linear interpolation between the objective function value at the end points θ_i and θ_{i+1} given by:

$$\widehat{z}(\overline{y}, \theta) = \alpha z^*(\overline{y}, \theta_i) + (1-\alpha)z^*(\overline{y}, \theta_{i+1}), \alpha \in (0, 1),$$

gives a valid upper bound because of the convexity of the objective function.

It may be mentioned that in this simple way we can obtain a region, *ABC*, within which the value of objective function will lie. An intersection point, θ_{int}, of the two lower bounds, $LB_i(\theta)$ and $LB_{i+1}(\theta)$, is then determined. At this point the value of lower and upper bounds are compared, and if the difference is within certain tolerance, ϵ, we stop, otherwise, the interval $[\theta_i, \theta_{i+1}]$ is subdivided into two intervals $[\theta_i, \theta_{int}]$ and $[\theta_{int}, \theta_{i+1}]$. In each of these regions a similar bounding procedure is repeated until we meet the tolerance criterion.

See also

- ▶ Multiparametric Linear Programming
- ▶ Multiparametric Mixed Integer Linear Programming
- ▶ Nondifferentiable Optimization: Parametric Programming
- ▶ Parametric Global Optimization: Sensitivity
- ▶ Parametric Linear Programming: Cost Simplex Algorithm
- ▶ Parametric Mixed Integer Nonlinear Optimization
- ▶ Parametric Optimization: Embeddings, Path Following and Singularities
- ▶ Selfdual Parametric Method for Linear Programs

References

1. Bank B, Guddat J, Klatte D, Kummer B, Tammer K (1983) Nonlinear parametric optimization. Akad Verlag, Berlin
2. Fiacco AV (1976) Sensitivity analysis for nonlinear programming using penalty methods. Math Program 10:287–311
3. Fiacco AV (1983) Introduction to sensitivity and stability analysis in nonlinear programming. Acad Press, New York
4. Fiacco AV, Kyparisis J (1986) Convexity and concavity properties of the optimal value function in parametric nonlinear programming. J Optim Th Appl 48:95–126
5. Fiacco AV, Kyparisis J (1988) Computable bounds on parametric solutions of convex problems. Math Program 40:213–221

6. Gal T (1995) Postoptimal analyses, parametric programming, and related topics. De Gruyter, Berlin
7. Gal T, Nedoma J (1972) Multiparametric linear programming. Managem Sci 18:406–422
8. McCormick GP (1983) Nonlinear programming: Theory, algorithms and applications. Wiley, New York
9. Stolbjerg A, Lasdon L (1997) Nonlinear programming. In: Gal T, Greenberg HJ (eds) Advances in Sensitivity Analysis and Parametric Programming. Kluwer, Dordrecht

Branch and Price: Integer Programming with Column Generation

BP

MARTIN W. P. SAVELSBERGH
School of Industrial and Systems Engineering,
Georgia Institute Technology, Atlanta, USA

MSC2000: 68Q99

Article Outline

Keywords
Nonidentical Machines
Identical Machines
See also
References

Keywords

Optimization; Integer programming; Decomposition; Column generation

Branch and price is a generalization of *linear programming* (LP) based *branch and bound* specifically designed to handle *integer programming* (IP) formulations that contain a huge number of variables. The basic idea of branch and price is simple. Columns are left out of the *LP relaxation* because there are too many columns to handle efficiently and most of them will have their associated variable equal to zero in an optimal solution anyway. Then to check the optimality of an LP solution, a subproblem, called the *pricing problem*, is solved to try to identify columns with a profitable *reduced cost*. If such columns are found, the LP is reoptimized. Branching occurs when no profitable columns are found, but the LP solution does not satisfy the integrality conditions. Branch and price applies *column generation* at every node of the branch and bound tree.

There are several reasons for considering IP formulations with a huge number of variables.

- A compact formulation of an IP may have a weak LP relaxation. Frequently the relaxation can be tightened by a reformulation that involves a huge number of variables.
- A compact formulation of an IP may have a symmetric structure that causes branch and bound to perform poorly because the problem barely changes after branching. A reformulation with a huge number of variables can eliminate this symmetry.
- Column generation provides a *decomposition* of the problem into master and subproblems. This decomposition may have a natural interpretation in the contextual setting allowing for the incorporation of additional important constraints or nonlinear cost functions.
- A formulation with a huge number of variables may be the only choice.

At first glance, it may seem that branch and price involves nothing more than combining well-known ideas for solving linear programs by column generation with branch and bound. However, it is not that straightforward. There are fundamental difficulties in applying column generation techniques for linear programming in integer programming solution methods. These include:

- Conventional integer programming branching on variables may not be effective because fixing variables can destroy the structure of the pricing problem.
- Column generation often converges slowly and solving the LPs to optimality may become computationally prohibitive.

We illustrate the concepts of branch and price and the difficulties that may arise by means of an example.

In the *generalized assignment problem* (GAP) the objective is to find a maximum profit assignment of m tasks to n machines such that each task is assigned to precisely one machine subject to capacity restrictions on the machines. For reasons that will become apparent later, we will consider separately the two cases of nonidentical and identical machines.

Nonidentical Machines

The natural integer programming formulation of GAP is

$$\begin{cases} \max & \sum_{1\leq i\leq m}\sum_{1\leq j\leq n} p_{ij}z_{ij} \\ \text{s.t.} & \sum_{1\leq j\leq n} z_{ij} = 1, \quad i=1,\ldots,m, \\ & \sum_{1\leq i\leq m} w_{ij}z_{ij} \leq d_j, \quad j=1,\ldots,n, \\ & z_{ij} \in \{0,1\}, \\ & \quad i=1,\ldots,m, \quad j=1,\ldots,n, \end{cases}$$

where p_{ij} is the profit associated with assigning task i to machine j, w_{ij} is the amount of the capacity of machine j used by task i, d_j is the capacity of machine j, and z_{ij} is a 0–1 variable indicating whether task i is assigned to machine j.

An alternative formulation of GAP in terms of columns representing feasible assignments of tasks to machines is

$$\begin{cases} \max & \sum_{j\leq 1\leq n}\sum_{1\leq k\leq K_j} \left(\sum_{1\leq i\leq m} p_{ij}y^j_{ik}\right) \lambda^j_k \\ \text{s.t.} & \sum_{1\leq j\leq n}\sum_{1\leq k\leq K_j} y^j_{ik}\lambda^j_k = 1, \quad i=1,\ldots,m, \\ & \sum_{1\leq k\leq K_j} \lambda^j_k = 1, \quad j=1,\ldots,n, \\ & \lambda^j_k \in \{0,1\}, \\ & \quad j=1,\ldots,n, \quad k=1,\ldots,K_j, \end{cases}$$

where the first m entries of a column, given by $y^j_k = (y^j_{1k}, \ldots, y^j_{mk})$, satisfy the *knapsack constraint*

$$\sum_{1\leq i\leq m} w_{ij}x_i \leq d_j, x_i \in \{0,1\}, \quad i=1,\ldots,m,$$

and where K_j denotes the number of feasible solutions to the above knapsack constraint. The first set of constraints ensures that each task is assigned to a machine, and the second set of constraints, the convexity constraints, ensures that exactly one feasible assignment of tasks to machines is selected for each machine. This is in fact the formulation that is obtained when we apply *Dantzig–Wolfe decomposition* to the natural formulation of GAP with the assignment constraints defining the master problem and the machine capacity constraints defining the subproblems.

The reason for considering this alternative formulation of GAP is that the LP relaxation of the master problem is tighter than the LP relaxation of the natural formulation because certain fractional solutions are eliminated. Namely, all fractional solutions that are not *convex combinations* of 0–1 solutions to the knapsack constraints.

Unfortunately, the LP relaxation of the master problem cannot be solved directly due to the exponential number of columns. However, the LP relaxation of a restricted version of the master problem that considers only a subset of the columns can be solved directly using, for instance, the *simplex method*. Furthermore, if the reduced costs of all the columns that were left out are nonnegative, then the LP solution obtained is also optimal for the LP relaxation of the unrestricted master problem. To check whether there exist a column with positive reduced cost we solve the *pricing problem*

$$\max_{1\leq j\leq n} \{z(\text{KP}_j) - v_j\},$$

where v_j is the optimal *dual price* from the solution to the LP relaxation of the restricted master problem associated with the convexity constraint of machine j, and $z(\text{KP}_j)$ is the value of the optimal solution to the *knapsack problem*

$$\begin{cases} \max & \sum_{1\leq i\leq n} (p_{ij} - u_i)x_i \\ \text{s.t.} & \sum_{1\leq i\leq n} w_{ij}x_i \leq d_j \\ & x_i \in \{0,1\}, \\ & \quad i \in \{1,\ldots,n\}, \end{cases}$$

with u_i being the optimal dual price from the solution to the LP relaxation of the restricted master problem associated with the assignment constraint of task i. If the optimal value of the pricing problem is positive, we have identified a column with positive reduced cost. In that case, we add the column to the restricted master problem and reoptimize.

The LP relaxation of the master problem solved by column generation may not have an integral optimal solution and applying a standard branch and bound procedure to the master problem over the existing columns is unlikely to find an optimal, or good, or even feasible solution to the original problem. Therefore it may be necessary to generate additional columns in order to solve the linear programming relaxations of the master problem at nonroot nodes of the search tree.

Standard branching on the λ-variables creates a problem along a branch where a variable has been set to zero. Recall that y_k^j represents a particular solution to the jth knapsack problem. Thus $\lambda_k^j = 0$ means that this solution is excluded. However, it is possible (and quite likely) that the next time the knapsack problem for the jth machine is solved the optimal solution is precisely the one represented by y_k^j. In that case, it would be necessary to find the second best solution to the knapsack problem. At depth l in the branch and bound tree we may need to find the lth best solution, which is very hard. Fortunately, there is a simple remedy to this difficulty. Instead of branching on the λs in the master problem, we use a branching rule that corresponds to branching on the original variables z_{ij}. When $z_{ij} = 1$, all existing columns in the master that do not assign task i to machine j are deleted and task i is permanently assigned to machine j, i.e., variable x_i is fixed to 1 in the jth knapsack. When $z_{ij} = 0$, all existing columns in the master that assign job i to machine j are deleted and task i cannot be assigned to machine j, i.e., variable x_i is removed from the jth knapsack. Note that each of the knapsack problems contains one fewer variable after the branching has been done.

Observe that the branching scheme discussed above is specific to the GAP. This is typical of branch and price algorithms. Each problem requires its own 'problem-specific' branching scheme.

In practice, one of the computational difficulties encountered when applying branch and price is the so-called *tailing-off* effect of the column generation, i.e., the large number of iterations needed to prove the optimality of the LP solution. Potentially, this may happen at every node of the search tree. Also, the pricing problem that needs to be solved at each column generation iteration may be difficult and time consuming. Fortunately, the branch and bound framework has some inherent flexibility that can be exploited effectively in branch and price algorithms. Branch and bound is an enumeration scheme that is enhanced by fathoming based on bound comparisons. To control the size of the branch and bound tree it is best to work with strong bounds; however, the method will work with any bound. Therefore, instead of solving the linear program to optimality, i.e., generating columns as long as profitable columns exist, we can choose to prematurely end the column generation process and work with bounds on the final LP value.

Again, consider the alternative formulation of GAP. By dualizing the assignment constraints, we obtain the following *Lagrangian relaxation*, which provides an upper bound on the value of the LP for any vector u.

$$\begin{cases} \max \quad \sum_{1 \leq j \leq n} \sum_{1 \leq k \leq K_j} \left(\sum_{1 \leq i \leq m} p_{ij} y_{ik}^j \right) \lambda_k^j \\ \qquad + \sum_{1 \leq i \leq m} u_i \left(1 - \sum_{1 \leq j \leq n} \sum_{1 \leq k \leq K_j} y_{ik}^j \lambda_k^j \right) \\ \text{s.t.} \quad \sum_{1 \leq k \leq K_j} \lambda_k^j = 1, \quad j = 1, \ldots, n, \\ \qquad \lambda_k^j \in \{0, 1\}, \\ \qquad j = 1, \ldots, n, \quad ; k = 1, \ldots, K_j. \end{cases}$$

After some algebraic manipulations, we obtain

$$\begin{cases} \sum_{1 \leq i \leq m} u_i + \sum_{1 \leq j \leq n} \max \\ \qquad \sum_{1 \leq k \leq K_j} \left(\sum_{1 \leq i \leq m} (p_{ij} - u_j) y_{ik}^j \right) \lambda_k^j \\ \text{s.t.} \quad \sum_{1 \leq k \leq K_j} \lambda_k^j = 1, \quad j = 1, \ldots, n, \\ \qquad \lambda_k^j \in \{0, 1\}, \\ \qquad j = 1, \ldots, n, \quad k = 1, \ldots, K_j, \end{cases}$$

which is equivalent to

$$\sum_{1 \leq i \leq m} u_i + \sum_{1 \leq j \leq n} z(\text{KP}_j).$$

This shows that after solving the pricing problem, we have all the information necessary to compute an upper bound on the value of the final LP solution. Therefore, after every column generation iteration, we may decide to prematurely end the column generation process if the value of the LP solution to the current restricted master problem, which provides a lower bound on the final LP value, and this upper bound are sufficiently close.

Identical Machines

This is a special case of the problem with nonidentical machines and therefore the methodology described above applies. However, we need only one subproblem since all of the machines are identical, which implies that the λ_k^j can be aggregated by defining $\lambda_k = \sum_j \lambda_k^j$ and that the convexity constraints can be combined into a single constraint $\sum_{1 \leq k \leq K} \lambda_k = n$ where λ_k is restricted to be integer. In some cases the aggregated constraint will become redundant and can be deleted altogether. An example of this is when the objective is to minimize $\sum \lambda_k$, i. e., the number of machines needed to process all the tasks. Note that this special case of GAP is equivalent to a 0–1 *cutting-stock problem*.

A much more important issue here concerns symmetry, which causes branching on the original variables to perform very poorly. With identical machines, there are an exponential number of solutions that differ only by the names of the machines, i. e. by swapping the assignments of 2 machines we get 2 solutions that are the same but have different values for the variables. This statement is true for fractional as well as 0–1 solutions. The implication is that when a fractional solution is excluded at some node of the tree, it pops up again with different variable values somewhere else in the tree. In addition, the large number of alternate optima dispersed throughout the tree renders pruning by bounds nearly useless.

The remedy here is a different branching scheme that works directly on the master problem but focuses on pairs of tasks. In particular, we consider rows of the master with respect to tasks r and s. Branching is done by dividing the solution space into one set in which r and s appear together, in which case they can be combined into one task when solving the knapsack, and into another set in which they must appear separately, in which case a constraint $x_r + x_s \leq 1$ is added to the knapsack. Note that the structure of the subproblems is no longer the same on the different branches.

Most of the material presented above is based on [3], in which the term branch and price was first introduced, and [1], in which the concepts of branch and price are covered in much more detail. Another important source of information on branch and price is [4], in which various general branching schemes and bounding schemes are discussed. Routing and scheduling has been a particularly fruitful application area of branch and price, see [2] for a survey of these results.

See also

- ▶ Decomposition Techniques for MILP: Lagrangian Relaxation
- ▶ Integer Linear Complementary Problem
- ▶ Integer Programming
- ▶ Integer Programming: Algebraic Methods
- ▶ Integer Programming: Branch and Bound Methods
- ▶ Integer Programming: Branch and Cut Algorithms
- ▶ Integer Programming: Cutting Plane Algorithms
- ▶ Integer Programming Duality
- ▶ Integer Programming: Lagrangian Relaxation
- ▶ LCP: Pardalos–Rosen Mixed Integer Formulation
- ▶ Mixed Integer Classification Problems
- ▶ Multi-Objective Integer Linear Programming
- ▶ Multi-Objective Mixed Integer Programming
- ▶ Set Covering, Packing and Partitioning Problems
- ▶ Simplicial Pivoting Algorithms for Integer Programming
- ▶ Stochastic Integer Programming: Continuity, Stability, Rates of Convergence
- ▶ Stochastic Integer Programs
- ▶ Time-Dependent Traveling Salesman Problem

References

1. Barnhart C, Johnson EL, Nemhauser GL, Savelsbergh MWP, Vance PH (1998) Branch-and-price: Column generation for solving integer programs. Oper Res 46:316–329
2. Desrosiers J, Dumas Y, Solomon MM, Soumis F (1995) Time constrained routing and scheduling. In: Ball ME, Magnanti TL, Monma C, Nemhauser GL (eds) Network RoutingHandbook Oper Res and Management Sci. 8, Elsevier, Amsterdam, pp 35–140
3. Savelsbergh MWP (1997) A branch-and-price algorithm for the generalized assignment problem. Oper Res 6:831–841
4. Vanderbeck F, Wolsey LA (1996) An exact algorithm for IP column generation. Oper Res Lett 19:151–160

Branchwidth and Branch Decompositions[1]

Illya V. Hicks
Computational and Applied Mathematics,
Rice University, Houston, USA

MSC2000: 90C27, 68R10

Article Outline

Keywords
Introduction
Graph Minors Theorem
Tangles
Constructing Branch Decompositions
Branch-Decomposition-Based Algorithms
Branchwidth of Matroids
Treewidth and Tree Decompositions
References

Keywords

Branchwidth; Branch decomposition; Tangle; Graph minors theorem; Branch-decomposition-based algorithm

Introduction

Let G be a graph (or hypergraph) with node set $V(G)$ and edge set $E(G)$. Let T be a tree having $|E(G)|$ leaves in which every non-leaf node has degree 3. Let ν be a bijection (one-to-one and onto function) from the edges of G to the leaves of T. The pair (T, ν) is called a *branch decomposition* of G. Notice that removing an edge, say e, of T partitions the leaves of T and the edges of G into two subsets A_e and B_e. The *middle set* of e and of (A_e, B_e), denoted by $mid(e)$ or $mid(A_e, B_e)$, is the set $V(G[A_e]) \cap V(G[B_e])$ where $G[A_e]$ is the subgraph of G induced by A_e and similarly for $G[B_e]$. The *width* of a branch decomposition (T, ν) is the maximum order of the middle sets over all edges in T. The *branchwidth* of G, denoted by $\beta(G)$, is the minimum width over all branch decompositions of G. A branch decomposition of G is *optimal* if its width is equal to the branchwidth

---
[1]This research was partially supported by NSF grant DMI-0217265

of G. For example, Fig. 1 gives an optimal branch decomposition of an example graph where some of the middle sets of the edges of the branch decomposition are provided.

An edge e is contracted if e is deleted and the ends of e are identified into one node and a graph H is a *minor* of a graph G if H can be obtained from a subgraph of G by contracting edges. Graphs of small branchwidth are characterized by the following theorem.

Theorem 1 (Robertson and Seymour [49]) *A graph G has branchwidth:*
- *0 if and only if every component of G has ≤ 1 edge*
- *≤ 1 if and only if every component of G has ≤ 1 node of degree ≥ 2*
- *≤ 2 if and only if G has no K_4 minor.* □

Other classes of graphs with known branchwidth are grids, complete graphs, Halin graphs, and chordal graphs. The branchwidth of a $a \times b$-grid is the minimum of a and b while the branchwidth of a complete graph G with at least 3 nodes is $\lceil \frac{2}{3} |V(G)| \rceil$ [49]. Halin graphs have branchwidth 3 and the branchwidth of chordal graphs is bounded below by $\lceil \frac{2}{3} \omega(G) \rceil$ and above by $\omega(G)$ where $\omega(G)$ denotes the clique number of the graph [30].

Graph Minors Theorem

A *planar graph* is a graph that can be drawn on a sphere or plane without having edges that cross. A subdivision of a graph G is a graph obtained from G by replacing its edges by internally vertex disjoint paths. In the 1930s, Kuratowski [42] proved that a graph G is planar if and only if G does not contain a subdivision of K_5 or $K_{3,3}$. Let $\mathcal{F}$ be a class of graphs. $\mathcal{F}$ is *minor closed* when all the minors of any member of $\mathcal{F}$ also belong to $\mathcal{F}$. Given a minor closed class of graphs $\mathcal{F}$, the *obstruction set* of $\mathcal{F}$ is the set of minor minimal graphs that are not elements of $\mathcal{F}$. Clearly, any class of graphs embeddable on a given surface is a minor closed class. Erdös, also in the 1930's, posed the question of whether the obstruction set for a given surface is finite. Wagner [57] later proved that the sphere has a finite obstruction set, K_5 and $K_{3,3}$. The question of characterizing the obstruction set for surfaces other than the sphere remained open until 1979–1980 when Archdeacon [4] and Glover et al. [27] solved the case for the

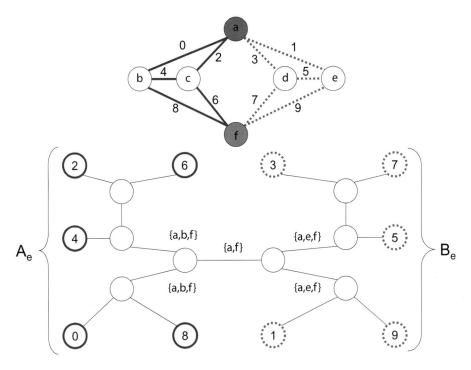

Branchwidth and Branch Decompositions, Figure 1
Example Graph G with Optimal Branch Decomposition (T, ν) with width 3

projective plane where they proved that there are 35 minor minimal "non-projective-planar" graphs. Archdeacon and Huneke [5] proved that the obstruction set for any non-orientable surface is finite and Robertson and Seymour [48] proved the case for any surface as a corollary of the Graph Minors Theorem (formerly known as Wagner's conjecture): every minor closed class has a finite obstruction set. Branch decompositions, tangles, and tree decompositions, discussed in later sections, were beneficial to the proof of the Graph Minors Theorem.

Tangles

Let G be a graph (or hypergraph) and let $k \geq 1$ be an integer. A *separation* of a graph G is a pair (G_1, G_2) of subgraphs of G with $G_1 \cup G_2 = (V(G_1) \cup V(G_2), E(G_1) \cup E(G_2)) = G$, $E(G_1) \cap E(G_2) = \emptyset$ and the *order* of this separation is defined as $|V(G_1) \cap V(G_2)|$ where $V(G_1) \cap V(G_2)$ is called the *middle set* of the separation. For a hypergraph G, define $I(G)$ to be the bipartite graph such that the nodes of $I(G)$ correspond to the nodes and edges of G and an edge ev in $I(G)$ corresponds to the edge e of G being incident with the node v in G. A hypergraph G is called *connected* if $I(G)$ is connected. Also, denote $\gamma(G)$ as the largest cardinality of a set of nodes incident to an edge of G. A *tangle* in G of order k is a set $\mathcal{T}$ of separations of G, each of order $< k$ such that:

(T1) for every separation (A, B) of G of order $< k$, one of $(A, B), (B, A)$ is an element of $\mathcal{T}$;
(T2) if $(A_1, B_1), (A_2, B_2), (A_3, B_3) \in \mathcal{T}$ then $A_1 \cup A_2 \cup A_3 \neq G$; and
(T3) if $(A, B) \in \mathcal{T}$ then $V(A) \neq V(G)$.

These are called the first, second and third tangle axioms. The *tangle number* of G, denoted by $\theta(G)$, is the maximum order of any tangle of G. Figure 2 gives an example of a tangle of order 3 for the graph in Fig. 1. Notice in Fig. 2 that the inclusion of separations of the graph of order 3 to the tangle would result in a violation of one of the tangle axioms. A tangle $\mathcal{T}$ of G with order k can be thought of as a "k-connected" component of G because some "k-connected" component of G will either be on one side or the other for any separa-

> Separation of order 0
> $(\emptyset, G)$
> Separation of order 1
> $(v, G) \; \forall v \in V(G)$
> Separation of order 2
> $(\{v, w\}, G) \; \forall v, w \in V(G)$
> $(G[e], G[E(G) \backslash e]) \; \forall e \in E(G)$
> $(G[0, 2, 4, 6, 8], G[1, 3, 5, 7, 9])$

Branchwidth and Branch Decompositions, Figure 2
Tangle of Order 3 for the Example Graph of Fig. 1

tion of $\mathcal{T}$. Robertson and Seymour [49] proved a min-max relationship between tangles and branch decompositions, given below.

Theorem 2 (Robertson and Seymour [49]) *For any hypergraph G such that $E(G) \neq \emptyset$, $\max\{\beta(G), \gamma(G)\} = \theta(G)$.* □

Related structures to tangles are respectful tangles and tangle bases. Respectful tangles of a graph G embedded on a surface Σ are tangles that are restricted according to the graph's embedding on Σ and the order of these tangles is limited by the graph's representativeness on Σ. Respectful tangles were discussed in the work of Robertson and Seymour [50] and created the foundation for the Seymour and Thomas [53] result for planar graphs. Tangle bases were introduced by Hicks [33] to assist in a branch-decomposition-based algorithm, discussed in a later section, to compute optimal branch decompositions for general graphs. Tangle bases are also restricted in the sense that the only members of a tangle basis are edges (just considering the first part of a separation) and separations which can be constructed from the union of edges. A formal definition is given below.

For an integer k and hypergraph G, a *tangle basis*, $\mathcal{B}$, of order k is a set of separations of G with order $< k$ such that:

(B1) $(G[e], G[E(G) \backslash e]) \in \mathcal{B} \forall e \in E(G)$ if $\gamma(e) < k$
(B2) if $(C, D) \in \mathcal{B}$ and $\nexists e \in E(G)$ such that $G[e] = C$, then $\exists (A_1, B_1), (A_2, B_2) \in \mathcal{B}$ such that $A_1 \cup A_2 = C$ and $B_1 \cap B_2 = D$
(B3) $\mathcal{B}$ obeys the tangle axioms T2 and T3.

> Separations of order 2
> $(G[e], G[E(G) \backslash e]) \; \forall e \in E(G)$

Branchwidth and Branch Decompositions, Figure 3
Connected Tangle Basis of Order 3 for the Graph of Fig. 1

A tangle basis, $\mathcal{B}$, in G of order k is *connected* if every separation (A, B) of $\mathcal{B}$ has A connected and define the connected tangle basis number of G, denoted by $\theta'(G)$, as the maximum order of any connected tangle basis of G. An example of a connected tangle basis for the graph in Fig. 1 is given in Fig. 3. Notice that the number of separations of the connected tangle basis of Fig. 1 is lower than the number of separations of the tangle of Fig. 1 offered by Fig. 2 but still contains the essential members of the tangle. Below is a min-max theorem relationship between tangle bases and branchwidth.

Theorem 3 (Hicks [33]) *If hypergraph G is connected such that $\beta(G) \geq \gamma(G)$, then the tangle basis number $\theta'(G)$ is equal to the $\beta(G)$.* □

Constructing Branch Decompositions

In terms of finding branch decompositions for general graphs, there is an algorithm in Robertson and Seymour [51] to approximate the branchwidth of a graph within a factor of 3. For example, the algorithm decides if a graph has branchwidth at least 10 or finds a branch decomposition with width at most 30. This algorithm has not been used in a practical implementation and its improvements by Bodlaender [8], Bodlaender and Kloks [13], and Reed [46] have not been shown to be practical either. Bodlaender and Thilikos [16] presented a tree-decomposition-based linear time algorithm for finding an optimal branch decomposition but it appears to be impractical. Tree-decomposition-based algorithms are discussed in a later section. In addition, Bodlaender and Thilikos [17] gave an algorithm to compute the optimal branch decomposition for any chordal graph with maximum clique size at most 4 but the algorithm has been only shown practical for a particular type of 3-tree.

Under practical algorithms, Kloks et al. [39] gave a polynomial time algorithm to compute the branchwidth of interval graphs, but for general graphs, one has to rely on heuristics. Cook and Seymour [20,21] gave a heuristic algorithm to produce branch decompositions that shows promise. In addition, Hicks [30,31] also found another branchwidth heuristic that was comparable to the algorithm of Cook and Seymour. Recently, Tamaki [54] has presented a linear time heuristic for constructing branch decompositions of planar graphs. This algorithm performs well when compared to the heuristics of Cook and Seymour [21] and Hicks [31]. Recently, Hicks [33] has developed a branch-decomposition-based algorithm for constructing optimal branch decompositions and it seems to be practical for sparse graphs with branchwidth at most 8.

For planar graphs, Seymour and Thomas showed that the branchwidth and an optimal branch decomposition of a graph can be computed in polynomial time. The complexity for the branchwidth is $O(n^3)$ and the complexity for computing an optimal branch decomposition is $O(n^4)$ [53]. Hicks [34,35] gave a practical implementation of these algorithms. Recently, Gu and Tamaki [28] introduced an $O(n^3)$ algorithm to compute an optimal branch decomposition of a planar graph by restricting the number of calls to the Seymour and Thomas algorithm for computing branchwidth to $O(n)$. More work in this area is encouraged to decrease the bound further.

Branch-Decomposition-Based Algorithms

Branch decompositions are of algorithmic importance for their appeal to solve intractable problems that can be modelled on graphs with bounded branchwidth. Courcelle [22] and Arnborg et al. [6] showed that several NP-complete problems can be solved in polynomial time using dynamic programming techniques on input graphs with bounded treewidth, discussed in a later section. Similar results have been obtained by Borie et al. [18]. The result is also equivalent to graphs with bounded branchwidth since the branchwidth and treewidth of a graph bound each other by constants [49]. In contrast, Seymour and Thomas [53] proved that testing if a general graph has branchwidth at most k, is NP-complete. The use of dynamic programming techniques in conjunction with a branch decomposition or a tree decomposition is referred to as a *branch-decomposition-based* or a *tree-decomposition-based* algorithm and these types of algorithms are part of the class of algorithms called *fixed parameter tractable algorithms* [1].

Some examples of branch-decomposition-based algorithms proposed in theory are Fomin and Thilikos [24] and Alekhnovich and Razborov [2]. Fomin and Thilikos used their result of improving a bound of Alon et al. [3] for the upper bound on the branchwidth of planar graphs to design a branch-decomposition-based algorithm in theory for vertex cover and dominating set for planar graphs [24]. Alekhnovich and Razborov [2] used the branchwidth of hypergraphs to design a branch-decomposition-based algorithm in theory to solve satisfiability problems.

Although theory indicates the fruitful potential of branch-decomposition-based algorithms, the number of branch-decomposition-based algorithms in the literature is exiguous. One noted exception is the work of Cook and Seymour [21] who produced the best known solutions for the 12 unsolved problems in TSPLIB95, a library of standard test instances for the TSP [47]. Hicks also presented a practical branch-decomposition-based algorithm for general minor containment [32] and constructing optimal branch decompositions [33]. One is also referred to the work of Christian [19].

Branchwidth of Matroids

Since graph theory and matroid theory have a symbiotic relationship, it is only natural that branch decompositions can be extended to matroids. In fact, branch decompositions have been used to produce a matroid analogue of the graph minors theorem [26]. A formal definition for the branchwidth of a matroid is given below.

The reader is referred to the book by Oxley [43] if not familiar with matroid theory. Let M be a matroid with finite ground set $S(M)$ and rank function ρ. The rank function of M^*, the dual of M, is denoted ρ^*.

A *separation* (A, B) of a matroid M is a pair of complementary subsets of $S(M)$ and the order of the separation, denoted $\rho(M, A, B)$, is defined to be following:

$$\rho(M, A, B) = \begin{cases} \rho(A) + \rho(B) - \rho(M) + 1 & \text{if } A \neq \emptyset \\ & \neq B, \\ 0 & \text{else}, \end{cases}$$

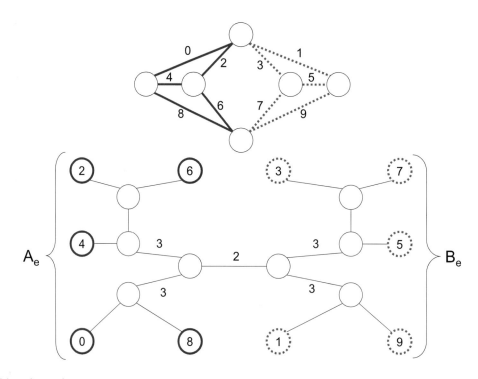

Branchwidth and Branch Decompositions, Figure 4
Example Graph G from Fig. 1 with Optimal Branch Decomposition (T, μ) with width 3 for its Cycle Matroid $M(G)$

A *branch decomposition* of a matroid M is a pair (T, μ) where T is a tree having $|S(M)|$ leaves in which every non-leaf node has degree 3 and μ is a bijection from the ground set of M to the leaves of T. Notice that removing an edge, say e, of T partitions the leaves of T and the ground set of M into two subsets A_e and B_e. The *order* of e and of (A_e, B_e), denoted $order(e)$ or $order(A_e, B_e)$, is equal to $\rho(M, A_e, B_e)$. The *width* of a branch decomposition (T, μ) is the maximum order of all edges in T. The *branchwidth* of M, denoted by $\beta(M)$, is the minimum width over all branch decompositions of M. A branch decomposition of M is *optimal* if its width is equal to the branchwidth of M. The cycle matroid of graph G, denoted $M(G)$, has $E(G)$ as its ground set and the cycles of G as the cycles of $M(G)$. For example, Fig. 4 gives an optimal branch decomposition of the cycle matroid of the example graph given Fig. 1 where all of the orders for the edges of the branch decomposition are provided.

There is also a corresponding notion of a tangle and tangle number for matroids, provided by Dharmatilake [23]. In addition, Dharmatilake gave a min-max relationship between tangles of matroids and the branchwidth of matroids, given below.

Theorem 4 (Dharmatilake [23]) *Let M be a matroid. Then $\beta(M) = \theta(M)$ if and only if M has no coloop and $\beta(M) \neq 1$.* □

It was posed by Robertson and Seymour [49] that the branchwidth of a graph and the branchwidth of the graph's cycle matroid are equivalent if the graph has a cycle of length at least 2. Recently, this conjecture was proved in the positive by Hicks and McMurray [37]. One is also referred to the work of Geelen et al. [26], Geelen et al. [25], Hall et al. [29], and Hliněný [38] for more detailed discussions on the branchwidth of matroids.

Treewidth and Tree Decompositions

This text would be remiss if a definition for treewidth and tree decompositions were not given.

A *tree decomposition* of a graph (or hypergraph) G is a pair, (T, σ), where T is a tree and for $t \in V(T), \sigma(t)$ is a subset of $V(G)$ with the following properties:

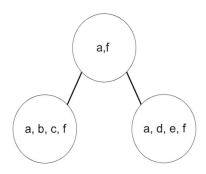

Branchwidth and Branch Decompositions, Figure 5
Optimal Tree Decomposition (T, σ) of Example Graph in Fig. 1 with width 3

- $\bigcup_{t \in V(T)} \sigma(t) = V(G)$
- $\forall e \in E(G), \exists t \in V(T)$ such that the ends of e are contained in $\sigma(t)$
- for $t, t', t'' \in V(T)$, if t' is on the path of T between t and t'' then $\sigma(t) \cap \sigma(t'') \subseteq \sigma(t')$.

The *width* of a tree decomposition is the largest value of $|\sigma(t) - 1|$ over all nodes $t \in V(T)$. The *treewidth* of a graph G, denoted by $\tau(M)$, is the minimum width over all tree decompositions of G. A tree decomposition of G is called *optimal* if its width is equal to the treewidth of G. For example, Fig. 5 gives an optimal tree decomposition of the example graph in Fig. 1. If T is restricted to be a path then (T, σ) is called a *path decomposition* and its corresponding connectivity invariant for a graph G is called the *pathwidth* of G.

The relationship between branchwidth and treewidth is characterized in the following theorem.

Theorem 5 (Robertson and Seymour [49]) *For any hypergraph* G, $max(\beta(G), \gamma(G)) \leq \tau(G) + 1 \leq max(\lfloor \frac{3}{2} \beta(G) \rfloor, \gamma(G), 1)$. □

Tree decompositions and the associated connectivity invariant, treewidth, have been extensively researched by Thomas [56], Seymour and Thomas [52], Bodlaender [8,10], Bodlaender and Kloks [12,13], Bodlaender et al. [11], Bodlaender et al. [15], Bodlaender et al. [14], Ramachandramurthi [44], Reed [45,46] and many others (see the survey papers by Bodlaender [7,9]). One is also referred to the work of Koster et al. [40], Koster et al. [41], Telle and Proskurowski [55], and Alber and Neidermeier [1] for literature related to tree decompositions and tree-decomposition-based algorithms. In addition, one is referred to Hicks et al. [36] for a more thorough survey of branch and tree decomposition techniques related to optimization.

References

1. Alber J, Niedermeier R (2002) Improved tree decomposition based algorithms for domination-like problems. In: Proceedings of the 5th Latin American Theoretical Informatics (LATIN 2002). Lecture Notes in Computer Science, vol 2286. Springer, Heidelberg, pp 613–627
2. Alekhnovich M, Razborov A (2002) Satisfiability, branchwidth and tseitin tautologies. In: 43rd Annual IEEE Symposium on Foundations of Computer Science. IEEE Computer Society, pp 593–603
3. Alon N, Seymour PD, Thomas R (1994) Planar separators. SIAM J Discret Math 7:184–193
4. Archdeacon D (1980) A Kuratowski Theorem for the Projective Plane. PhD thesis, Ohio State University
5. Archdeacon D, Huneke P (1989) A Kuratowski theorem for non-orientable surfaces. J Combin Theory Ser B 46(2):173–231
6. Arnborg S, Lagergren J, Seese D (1991) Easy problems for tree-decomposable graphs. J Algorithms 12:308–340
7. Bodlaender H (1993) A tourist guide through treewidth. Acta Cybernetica 11:1–21
8. Bodlaender H (1996) A linear time algorithm for finding tree-decompositions of small treewidth. SIAM J Comput 25:1305–1317
9. Bodlaender H (1997) Treewidth: Algorithmic techniques and results. In: Privara I, Rvzicka P (eds) Proceedings of the 22nd International Symposium on Mathematical Foundations of Computer Science, MFCS'97. Lecture Notes in Computer Science, vol 1295. Springer, Berlin, pp 29–36
10. Bodlaender H (1998) A partial k-arboretum of graphs with bounded treewidth. Theoret Comput Sci 209:1–45
11. Bodlaender H, Gilbert J, Hafsteinsson H, Kloks T (1992) Approximation treewidth, pathwidth, and minimum elimination tree height. In: Schmidt G, Berghammer R (eds) Proceedings 17th International Workshop on Graph-Theoretic Concepts in Computer Science WG 1991. Lecture Notes in Computer Science, vol 570. Springer, Berlin, pp 1–12
12. Bodlaender H, Kloks T (1992) Approximating treewidth and pathwidth of some classes of perfect graphs. In: Proceedings Third International Symposium on Algorithms and Computation, ISAAC 1992. Lecture Notes in Computer Science, vol 650. Springer, Berlin, pp 116–125
13. Bodlaender H, Kloks T (1996) Efficient and constructive algorithms for the pathwidth and treewidth of graphs. J Algorithms 21:358–402
14. Bodlaender H, Kloks T, Kratsch D, Muller H (1998) Treewidth and minimum fill-in on d-trapezoid graphs. J Graph Algorithms Appl 2(5):1–23
15. Bodlaender H, Tan R, Thilikos D, van Leeuwen J (1997) On interval routing schemes and treewidth. Inf Comput 139:92–109

16. Bodlaender H, Thilikos D (1997) Constructive linear time algorithms for branchwidth. In: Degano P, Gorrieri R, Marchetti-Spaccamela A (eds) Lecture Notes in Computer Science: Proceedings of the 24th International Colloquium on Automata, Languages, and Programming. Springer, Berlin, pp 627–637
17. Bodlaender H, Thilikos D (1999) Graphs with branchwidth at most three. J Algorithms 32:167–194
18. Borie RB, Parker RG, Tovey CA (1992) Automatic generation of linear-time algorithms from predicate calculus descriptions of problems on recursively constructed graph families. Algorithmica 7:555–581
19. Christian WA (2003) Linear-Time Algorithms for Graphs with Bounded Branchwidth. PhD thesis, Rice University
20. Cook W, Seymour PD (1994) An algorithm for the ring-router problem. Technical report, Bellcore
21. Cook W, Seymour PD (2003) Tour merging via branch-decomposition. INFORMS J Comput 15(3):233–248
22. Courcelle B (1990) The monadic second-order logic of graphs I: Recognizable sets of finite graphs. Inf Comput 85:12–75
23. Dharmatilake JS (1996) A min-max theorem using matroid separations. Contemp Math 197:333–342
24. Fomin F, Thilikos D (2003) Dominating sets in planar graphs: Branch-width and exponential speed-up. In: Proceedings of the Fourthteenth Annual ACM-SIAM Symposium on Discrete Algorithms (Baltimore, MD 2003). ACM, New York, pp 168–177
25. Geelen JF, Gerards AMH, Robertson N, Whittle GP (2003) On the excluded minors for the matroids of branch-width k. J Combin Theory Ser B 88:261–265
26. Geelen JF, Gerards AMH, Whittle G (2002) Branch width and well-quasi-ordering in matroids and graphs. J Combin Theory Ser B 84:270–290
27. Glover H, Huneke P, Wang CS (1979) 103 graphs that are irreducible for the projective plane. J Combin Theory Ser B 27:332–370
28. Gu QP, Tamaki H (2005) Optimal branch-decomposition of planar graphs in $o(n^3)$ time. In: Proceedings of the 31st International Colloquium on Automata, Languages and Programming. LNCS, vol 3580, pp 373–384
29. Hall R, Oxley J, Semple C, Whittle G (2002) On matroids of branch-width three. J Combin Theory Ser B 86:148–171
30. Hicks IV (2000) Branch Decompositions and their Applications. PhD thesis, Rice University
31. Hicks IV (2002) Branchwidth heuristics. Congressus Numerantium 159:31–50
32. Hicks IV (2004) Branch decompositions and minor containment. Networks 43(1):1–9
33. Hicks IV (2005) Graphs, branchwidth, and tangles! oh my! Networks 45:55–60
34. Hicks IV (2005) Planar branch decompositions I: The rat-catcher. INFORMS J Comput 17(4):402–412
35. Hicks IV (2005) Planar branch decompositions II: The cycle method. INFORMS J Comput 17(4):413–421
36. Hicks IV, Koster AMCA, Kolotoğlu E (2005) Branch and tree decomposition techniques for discrete optimization. In: Cole Smith J (ed) Tutorials in Operations Research 2005. INFORMS, Hanover, MD, pp 1–29
37. Hicks IV, McMurray N (2007) The branchwidth of graphs and their cycle matroids. J Combin Theory Ser B 97:681–692
38. Hliněný P (2002) On the exclued minors for matroids of branch-width three. preprint
39. Kloks T, Kratochvil J, Müller H (1999) New branchwidth territories. In: Meinel C, Tison S (eds) STAC'99, 16th Annual Symposium on Theoretical Aspects of Computer Science, Trier, Germany, March 1999 Proceedings. Springer, Berlin, pp 173–183
40. Koster A, van Hoesel S, Kolen A (2002) Solving partial constraint satisfaction problems with tree-decompositions. Networks 40:170–180
41. Koster AMCA, Bodlaender HL, van Hoesel SPM (2001) Treewidth: Computational experiments. Electr Notes Discret Math 8:54–57
42. Kuratowski K (1930) Sur le probleme des courbes gauches en topologie. Fundamenta Mathematicae 15:271–283
43. Oxley JG (1992) Matroid Theory. Oxford University Press, Oxford
44. Ramachandramurthi S (1997) The structure and number of obstructions to treewidth. SIAM J Discret Math 10:146–157
45. Reed B (1992) Finding approximate separators and computing tree width quickly. In: Proceeding of the 24th Annual Association for Computing Machinery Symposium on Theory of Computing. ACM Press, New York, pp 221–228
46. Reed B (1997) Tree width and tangles: A new connectivity measure and some applications. In: Bailey RA (ed) Survey in Combinatorics. Cambridge University Press, Cambridge, pp 87–162
47. Reinelt G (1991) TSPLIB – a traveling salesman library. ORSA J Comput 3:376–384
48. Robertson N, Seymour PD (1985) Graph minors: A survey. In: Surveys in Combinatorics, London Math Society Lecture Note Series, edition 103. Cambridge University Press, Cambridge, pp 153–171
49. Robertson N, Seymour PD (1991) Graph minors X: Obstructions to tree-decompositions. J Combin Theory Ser B 52:153–190
50. Robertson N, Seymour PD (1994) Graph minors XI: Circuits on a surface. J Combin Theory Ser B 60:72–106
51. Robertson N, Seymour PD (1995) Graph minors XIII: The disjoint paths problem. J Combin Theory Ser B 63:65–110
52. Seymour P, Thomas R (1993) Graph searching and a min-max theorem for tree-width. J Combin Theory Ser B 58:22–33
53. Seymour PD, Thomas R (1994) Call routing and the rat-catcher. Combinatorica 14(2):217–241
54. Tamaki H (2003) A linear time heuristic for the branch-decomposition of planar graphs. Technical Report MPI-I-2003-1-010, Max-Planck-Institut fuer Informatik

55. Telle JA, Proskurowski A (1997) Algorithms for vertex partitioning problems on partial *k*-trees. SIAM J Discret Math 10(4):529–550
56. Thomas R (1990) A Menger-like property of tree-width: The finite case. J Combin Theory Ser B 48:67–76
57. Wagner K (1937) Uber eine eigenschaft der ebenen komplexe. Math Annal 115:570–590

Broadcast Scheduling Problem

CLAYTON W. COMMANDER
Air Force Research Laboratory, Munitions Directorate, and Dept. of Industrial and Systems Engineering, University of Florida, Gainesville, USA

Article Outline

Synonyms
Introduction
   Organization
   Idiosyncrasies
Formulation
Methods
   Sequential Vertex Coloring
   Mixed Neural-Genetic Algorithm
   Greedy Randomized Adaptive Search Procedures (GRASP)
   Multi-start Combinatorial Algorithm
   Computational Effectiveness
Conclusion
See also
References

Synonyms

BSP; The BROADCAST SCHEDULING PROBLEM is also referred to as the TDMA MESSAGE SCHEDULING PROBLEM [6]

Introduction

Wireless mesh networks (WMNs) have become an important means of communication in recent years. In these networks, a shared radio channel is used in conjunction with a packet switching protocol to provide high-speed communication between many potentially mobile users. The stations in the network act as transmitters and receivers, and are thus capable of utilizing a multi-hop transmission procedure. The advantage of this is that several stations can be used as relays to forward messages to the intended recipient. This allows beyond line of sight communication between stations which are geographically disbursed and potentially mobile [2].

Mesh networks have increased in popularity in recent years and the number of applications is steadily increasing [25]. As mentioned in [1], WMNs allow users to integrate various networks, such as Wi-Fi, the internet and cellular systems. WMNs can also be utilized in a military setting in which tactical datalinks network various communication, intelligence, and weapon systems allowing for streamlined communication between several different entities [6]. For a survey of wireless mesh networks, the reader is referred to [1].

In WMNs, the critical problem involves efficiently utilizing the available bandwidth to provide collision free message transmissions. Unfettered transmission by the network stations over the shared channel will lead to message collisions. Therefore, some medium access control (MAC) scheme should be employed to schedule message transmissions so as to avoid message collisions. The time division multiple access (TDMA) protocol is a MAC scheme introduced by Kleinrock in 1987 which was shown to provide collision free broadcast schedules [19]. In a TDMA network, time is divided into frames with each frame consisting of a number of unit length slots in which the messages are scheduled. Stations scheduled in the same slot broadcast simultaneously. Thus, the goal is to schedule as many stations as possible in the same slot so long as there are no message collisions.

When considering the broadcast scheduling problem on TDMA networks, there are two optimization problems which must be addressed [31]. The first involves finding the minimum frame length, or the number of slots required to schedule all stations at least once. The second problem is that of maximizing the number of stations scheduled within each slot, thus maximizing the throughput. Both of these problems however, are known to be $\mathcal{NP}$-hard [2]. Therefore, efficient heuristics are typically used to quickly provide high quality solutions to real-world instances.

Organization

The organization of this article is as follows. In the following section, we formally define the problem statement and provide a mathematical programming formulation. We also examine the computational complexity the problem. In Sect. "Methods", we review several solution techniques which appear in the literature. We provide some concluding remarks in Sect. "Conclusion" and indicate directions of future research. Finally, a list of cross references is provided in Sect. "See also".

Idiosyncrasies

We will now briefly introduce some of the symbols and notations we will employ throughout this paper. Denote a graph $G = (V, E)$ as a pair consisting of a set of vertices V, and a set of edges E. All graphs in this paper are assumed to be undirected and unweighted. We use the symbol "$b := a$" to mean "the expression a defines the (new) symbol b" in the sense of King [18]. Of course, this could be conveniently extended so that a statement like "$(1 - \epsilon)/2 := 7$" means "define the symbol ϵ so that $(1 - \epsilon)/2 = 7$ holds". Finally, we will use *italics* for emphasis and SMALL CAPS for problem names. Any other locally used terms and symbols will be defined in the sections in which they appear.

Formulation

A TDMA network can be conveniently described as a graph $G = (V, E)$ where the vertex set V represents the stations and the set of edges E represents the set of communication links between adjacent stations. There are two types of message collisions which must be avoided when scheduling messages in TDMA networks. The first, called a *direct collision* occurs between *one-hop neighboring stations*, or those stations $i, j \in V$ such that $(i, j) \in E$. One-hop neighbors which broadcast during the same slot cause a direct collision. Further, if $(i, j) \notin E$, but $(i, k) \in E$ and $(j, k) \in E$, then i and j are called *two-hop neighbors*. Two-hop neighbors transmitting in the same slot cause a so-called *hidden collision* [2].

Assume that there are M slots per frame. Further, assume that packets are sent at the beginning of each time slot and are received in the same slot in which they are sent. Let $x: M \times V \mapsto \{0, 1\}$, be a surjection defined by

$$x_{mn} := \begin{cases} 1, \text{if station } n \text{ scheduled in slot } m, \\ 0, \text{otherwise}. \end{cases} \quad (1)$$

Also, let $c: E \mapsto \{0, 1\}$ return 1 if i and j are one-hop neighbors, i.e., if $(i, j) \in E$ and $i \neq j$.

Using the aforementioned definitions and assumptions, we can now formulate the BROADCAST SCHEDULING PROBLEM (BSP) on TDMA networks as the following multiobjective optimization problem:

Minimize M

Maximize $\sum_{i=1}^{M} \sum_{j=1}^{|V|} x_{ij}$

subject to:

$$\sum_{m=1}^{M} x_{mn} \geq 1, \quad \forall n \in V, \quad (2)$$

$$c_{ij} + x_{mi} + x_{mj} \leq 2, \\ \forall i, j \in V, i \neq j, m = 1, \ldots, M, \quad (3)$$

$$c_{ik}x_{mi} + c_{kj}x_{mj} \leq 1, \forall i, j, k \in V, i \neq j, j \neq k, \\ k \neq i, m = 1, \ldots, M, \quad (4)$$

$$x_{mn} \in \{0, 1\}, \quad \forall n \in V, m = 1, \ldots, M, \quad (5)$$

$$M \in \mathbb{Z}^+. \quad (6)$$

The objective provides a minimum frame length with maximum bandwidth utilization, while constraint (2) ensures that all stations broadcast at least once. Constraints (3) and (4) prevent direct and hidden collisions, respectively. Constraints (5) and (6) define the proper domain of the decision variables.

Suppose that we relax the BSP and only the consider the first objective function. This is referred to as the FRAME LENGTH MINIMIZATION PROBLEM (FLMP) and is given by the following integer program: $\min\{M: (2) - (6)\}$. Clearly any feasible solution to this problem is feasible for BSP. Now, consider a graph $G' = (V, E')$ where V follows from the original communication graph G, but whose edge set is given by $E' = E \cup \{(i, j): i, j \text{ are two-hop neighbors}\}$. Then using this augmented graph, we can formulate the following theorem due to Butenko et al. [2].

Theorem 1 *The* FRAME LENGTH MINIMIZATION PROBLEM *on* $G = (V, E)$ *is equivalent to finding an optimal coloring of the vertices of* $G'(V, E')$.

Proof Recall that in order for a message schedule to be feasible, all stations must broadcast at least once and no collisions occur, either hidden or direct. Notice now that E' contains both one-hop and two-hop neighbors, and in any feasible solution, neither of these can transmit in the same slot. Thus, there is a one-to-one correlation between time slots in G and vertex colors in G'. Hence, a minimum coloring of the vertices of G' provides the minimum required slots needed for a collision free broadcast schedule on G. □

After one has successfully solved the FLMP by solving the corresponding GRAPH COLORING PROBLEM, an optimal frame length M^* is attained. With this, the THROUGHPUT MAXIMIZATION PROBLEM (TMP) given as follows $\max\{\sum_{i=1}^{M^*}\sum_{j=1}^{|V|} x_{ij} : (2)-(6)\}$ can be solved, where M is replaced by M^* in $(2)-(6)$. A direct result of Theorem 1 is that finding an optimal frame length for a general instance of the BSP is $\mathcal{NP}$-hard [11]. The reader is referred to the paper by Butenko et al. [2] for the complete proof. Also, in [8], the TMP was also shown to be $\mathcal{NP}$-hard [8]. Thus it is unlikely that a polynomial algorithm exists for finding an optimal broadcast schedule for an instance of the BSP [11]. It is interesting to note however, that if we ignore constraint (4) which prevents two-hop neighbors from transmitting simultaneously, then the resulting problem is in $\mathcal{P}$, and a polynomial time algorithm is provided in [13].

Due to the computational complexity of the BSP, several heuristics have been applied and appear throughout the literature [2,3,6,28,31]. In the following section, we highlight several of these methods and examine their effectiveness when applied to large-scale instances.

Methods

In this section, we review many of the heuristics which have been applied to the BSP. We analyze the techniques used and compare their relative performance as reported in [6]. The particular algorithms we examine are as follows:

- Sequential vertex coloring [31];
- Mixed neural-genetic algorithm [27];
- Greedy randomized adaptive search procedures (GRASP) [2,3];
- A multi-start combinatorial algorithm [6].

We note here that none of the heuristics which we describe in this section attempt to solve the BSP by using the typical multiobjective optimization approach, in which one combines the multiple objectives into one scalar objective whose optimal value is a Pareto optimal solution to the original problem. Instead all of the methods decouple the objectives and handle each independently. This is done because for instances of the BSP, frame length minimization usually takes precedence over the utilization maximization problem [27, 28,31].

Sequential Vertex Coloring

Yeo et al. [31] propose a two-phase approach based on sequential vertex coloring (SVC). The first phase computes an approximate solution for the FLMP. Then using the computed frame length, the TMP is considered in the second phase. Specific details are as follows.

Frame Length Minimization For this phase, the FRAME LENGTH MINIMIZATION PROBLEM is considered and an approximate solution is computed by solving a graph coloring problem in the augmented graph. A sequential vertex ordering approach is used whereby the stations are first ordered in descending order of the number of one-hop and two-hop neighbors. The first vertex is colored and the list of the other $N-1$ vertices are scanned downward. The remaining vertices are colored with the smallest color which has not already been assigned to one of its one-hop neighboring station. The process is continued until all vertices have been colored.

Throughput Maximization To solve the TMP in the frame length computed in phase 1, an ordering method of the sequential vertex coloring algorithm is applied. The stations are now ordered in ascending order of the the number of one-hop and two-hop neighbors. The first ordered station is then assigned to any slots in which it can simultaneously broadcast with the previously assigned stations. This process is repeated for every station in the ordered list.

Mixed Neural-Genetic Algorithm

As with the coloring heursitic presented described above, Salcedo-Sanz et al. [27] introduced a two-phase heuristic based on combining both Hopfield neural networks [15] and genetic algorithms as in [29]. As with the vertex coloring algorithm, phase one considers the FLMP and phase two attempts to maximize the throughput.

Frame Length Minimization In order to solve the FRAME LENGTH MINIMIZATION PROBLEM, a discrete-time binary Hopfield neural network (HNN) is used. As described in [27], the HNN can be represented as a graph whose vertices are the neurons (stations) and whose edges represent the direct collisions. The neurons are updated one at a time after a randomized initialization until the system converges. For specific implementation details, the reader should see [27].

Utilization Maximization In this phase, a genetic algorithm [12] is used to maximize the throughput within the frame length that was determined in phase one. Genetic algorithms (GAs) get their names from the biological process which they mimic. Motivated by Darwin's Theory of Natural Selection [7], these algorithms evolve a *population* of solutions, called *individuals*, over several *generations* until the best solution is eventually reached. Each component of an individual is called a *allele*. Individuals in the population mate through a process called *crossover*, and new solutions having traits, i. e. alleles of both parents are produced. In successive generations, only those solutions having the best *fitness* are carried to the next generation in a process which mimics the fundamental principle of natural selection, *survival of the fittest* [12]. Again, the reader should reference [27] for implementation specific information.

Greedy Randomized Adaptive Search Procedures (GRASP)

GRASP [9] is a multi-start metaheuristic that has been used with great success to provide solutions for several difficult combinatorial optimization problems [10], including SATISFIABILITY [24], QUADRATIC ASSIGNMENT [21,23], and most recently the COOPERATIVE COMMUNICATION PROBLEM ON AD-HOC NETWORKS [4,5].

GRASP is a two-phase procedure which generates solutions through the controlled use of random sampling, greedy selection, and local search. For a given problem Π, let F be the set of feasible solutions for Π. Each solution $X \in F$ is composed of k discrete components $a_1, \ldots, a_k$. GRASP constructs a sequence $\{X\}_i$ of solutions for Π, such that each $X_i \in F$. The algorithm returns the best solution found after all iterations.

Construction Phase The construction phase for the GRASP constructs a solution iteratively from a partial broadcast schedule which is initially empty. The stations are first sorted in descending order of the number of one-hop and two-hop neighbors. Next, a so-called *Restricted Candidate List* (RCL) is created and consists of the stations which may broadcast simultaneously with the stations previously assigned to the current slot. From this RCL a station is randomly chosen and assigned. A new RCL is created and another station is randomly selected. This process continues the RCL is empty, at which time the slot number is incremented and the procedure is repeated recursively for the subgraph induced by the set of all vertices whose corresponding stations have not yet been assigned to a time slot.

Local Search The local search phase used is a swap-based procedure which is adapted from a similar method for graph coloring implemented by Laguna and Martí in [20]. First, the two slots with the fewest number of scheduled transmissions are cif stationombined and the total number of slots is now given as $k = m-1$, where m is the frame length of the schedule computed in the construction phase. Denote the new broadcast schedule as $\{x_{m',n}, m' = 1, \ldots, k, n = 1, \ldots, N\}$. Now, let the function $f(x) = \sum_{i=1}^{k} E(m'_i)$, where $E(m'_i)$ is the set of collisions in slot m'_i. $f(x)$ is then minimized by the application of a local search procedure as follows.

A colliding station in the combined slot is chosen randomly and every attempt is made to swap this station with another from the remaining $k - 1$ slots. After a swap is made, $f(x)$ is re-evaluated. If $f(x)$ has a lower value than before the swap, the swap is kept and the process repeated with the remaining colliding stations. If after every attempt to swap a colliding station the result is unimproved, a new colliding station is chosen and the swap routine is attempted. This continues until either a successful swap is made or for some specified number of iterations. If a solution is improved such that $f(x) = 0$, then the frame length has been successfully decreased by one slot. The value of k is then decremented and the process is repeated. If the procedure ends with $f(x) > 0$, then no improved solution was found.

Multi-start Combinatorial Algorithm

To our knowledge, the most recent heuristic for the BSP is a hybrid multi-start method by Commander and Pardalos [6]. This heuristic combines a graph coloring heuristic with a randomized local search to provide high-quality solutions for large-scale instances on the problem. As with the previously described method, this heuristic is also a two-phase approach. The reader should see [6] for pseudo-code and other implementation specific details.

Frame Length Minimization First a greedy randomized construction heuristic was used to determine the value for M. As a result of Theorem 1, the method is based on the construction phase of the Greedy Randomized Adaptive Search Procedure (GRASP) [26] for coloring sparse graphs proposed by Laguna and Martí in [20]. This particular method was chosen because it is able to quickly provide excellent solutions for the frame length. That being said, any other coloring heuristic would provide a value for M such as the Sequential Vertex Coloring method described above. However, the randomized approach of the selected method allows the search space to be more thoroughly investigated. This is due to the fact that different optimal colorings will yield different solutions in the second phase.

Throughput Maximization The solution from the first phase will not provide an optimal throughput in general, because each station will only be scheduled to transmit once in the frame. Therefore, a randomized local improvement method is used to schedule each station as many times as possible in the frame. This method locally optimizes each slot by considering the set of nodes which may transmit with the currently scheduled slot. A node from this set is randomly selected and the process repeats until no other stations may broadcast in the current slot. The next slot is then considered and the process is repeated until the solution is locally optimal.

Computational Effectiveness

In [6], the authors performed an extensive computational experiment comparing the effectiveness of the aforementioned heuristics. They tested all of the algorithms on a common platform and reported solutions for 63 instances ranging from 15 to 100 stations with varying densities. In addition, they implemented the integer programming model from Sect. "Formulation" using the Xpress-MP™ optimization suite from Dash Optimization [17]. Xpress-MP contains an implementation of the simplex method [14], and uses a branch and bound algorithm [30] together with advanced cutting-plane techniques [16,22].

For each instance tested, the combinatorial algorithm of [6] is superior to the other heuristics mentioned. For all 63 instances tested, the method found solutions at least as good as any of the other algorithms from the literature for all of the networks, outperforming them on 56 cases. The performance of the GRASP [2] and the Mixed Neural-Genetic Algorithm [27] were comparable, with GRASP performing slightly better on average. The weakest of the methods was the Sequential Vertex Coloring [31] algorithm. For specific numerical results, see [6].

Conclusion

In this article, we introduced the BROADCAST SCHEDULING PROBLEM on TDMA networks. The BSP is an important problem that occurs in wireless mesh networks regarding efficiently scheduling collision free broadcasts for the network stations. We formally defined the problem, examined the computational complexity, and discussed several algorithms which have been applied to the BSP, all with competitive results.

We conclude with a few words on possible directions of future research. In addition to the ones described, other metaheuristics could be considered and approximation algorithms developed. Also, a heuristic exploration of cutting plane algorithms on the IP formulation would be an interesting alternative. Another alternative would be to consider instances of the problem in which the stations are part of a mobile ad-hoc network. In this case, the topology of the network would change as the stations change position. This could potentially cause significant difficulties in determining the evolving sets of one-hop and two-hop neighbors. There is no doubt that as technology advances and research on ad-hoc networks increases, so too will applications of the BSP which will require advanced solution techniques [25].

See also

- ▶ Frequency Assignment Problem
- ▶ Genetic Algorithms
- ▶ Graph Coloring
- ▶ Greedy Randomized Adaptive Search Procedures
- ▶ Multi-objective Integer Linear Programming
- ▶ Optimization Problems in Unit-Disk Graphs
- ▶ Simulated Annealing

References

1. Akyildiz IF, Wang X, Wang W (2005) Wireless mesh networks: a survey. Comp Network 47(4):445–487
2. Commander CW, Butenko SI, Pardalos PM (2004) On the performance of heuristics for broadcast scheduling. In: Grundel D, Murphey R, Pardalos P (eds) Theory and Algorithms for Cooperative Systems. World Scientific, Singapore, pp 63–80
3. Commander CW, Butenko SI, Pardalos PM, Oliveira CAS (2004) Reactive grasp with path relinking for the broadcast scheduling problem. In: Proceedings of the 40th Annual International Telemetry Conference. pp 792–800
4. Commander CW, Festa P, Oliveira CAS, Pardalos PM, Resende MGC, Tsitselis M (2006) GRASP with path-relinking for the cooperative communication problem on ad-hoc networks. SIAM J Control Optim submitted
5. Commander CW, Oliveira CAS, Pardalos PM, Resende MGC (2005) A GRASP heuristic for the cooperative communication problem in ad hoc networks. In: Proceedings of the VI Metaheuristics International Conference. pp 225–230
6. Commander CW, Pardalos PM (2007) A combinatorial algorithm for the TDMA message scheduling problem. Comput Optim Appl to appear
7. Darwin C (1872) The Origin of Species, 6th edn. Murray, London
8. Ephremides A, Truong TV (1990) Scheduling broadcasts in multihop radio networks. IEEE Trans Commun 38(4):456–460
9. Feo TA, Resende MGC (1995) Greedy randomized adaptive search procedures. J Global Optim 6:109–133
10. Festa P, Resende MGC (2002) GRASP: An annotated bibliography. In: Ribeiro C, Hansen P (eds) Essays and surveys in metaheuristics, Kluwer, Dordrecht, pp 325–367
11. Garey MR, Johnson DS (1979) Computers and Intractability: A Guide to the Theory of NP-Completeness. Freeman WH and Company, New York
12. Goldberg DE (1989) Genetic Algorithms in Search, Optimization and Machine Learning. Kluwer, Dordrecht
13. Hajek B, Sasaki G (1988) Link scheduling in polynomial time. IEEE Trans Inf Theor 34:910–918
14. Hillier FS, Lieberman GJ (2001) Introduction to Operations Research. McGraw Hill, New York
15. Hopfield JJ, Tank DW (1982) Neural networks and physical systems with emergent collective computational abilities. In: Proceedings of the National Academy of Science, pp 2541–2554
16. Horst R, Pardalos PM, Thoai NV (1995) Introduction to Global Optimization, volume 3 of Nonconvex Optimization and its Applications. Kluwer, Dordrecht
17. Dash Optimization Inc (2003) Xpress-Optimizer Reference Manual
18. King J (1994) Three problems in search of a measure. Am Math Mon 101:609–628
19. Kleinrock L, Silvester J (1987) Spatial reuse in multihop packet radio networks. In: Proceedings of the IEEE 75
20. Laguna M, Martí R (2001) A grasp for coloring sparse graphs. Comput Optim Appl 19(2):165–178
21. Li Y, Pardalos PM, Resende MGC (1994) A greedy randomized adaptive search procedure for the quadratic assignment problem. In: Pardalos PM, Wolkowicz H (eds) Quadratic Assignment and Related Problems, volume 16 of DIMACS Series on DIscrete Mathematics and Theoretical Computer Science, pp 237–261
22. Oliveira CAS, Pardalos PM, Querido TM (2005) A combinatorial algorithm for message scheduling on controller area networks. Int J Oper Res 1(1/2):160–171
23. Oliveira CAS, Pardalos PM, Resende MGC (2003) Grasp with path-relinking for the qap. In: 5th Metaheuristics International Conference, pp 57.1–57.6
24. Resende MGC, Feo TA (1996) A grasp for satisfiability. In: Johnson DS, Trick MA (eds) Cliques, Coloring, and Satisfiability: Second DIMACS Implementation Challenges, volume 26, American Mathematical Society, Providence, RI, pp 499–520
25. Resende MGC, Pardalos PM (2006) Handbook of Optimization in Telecommunications. Springer, Berlin
26. Resende MGC, Ribeiro CC (2003) Greedy randomized adaptive search procedures. In: Glover F, Kochenberger G (eds) Handbook of Metaheuristics. Kluwer, Dordrecht, pp 219–249
27. Salcedo-Sanz S, Busoño-Calzón C, Figueiral-Vidal AR (2003) A mixed neural-genetic algorithm for the broadcast scheduling problem. IEEE Trans Wirel Commun 2(2):277–283
28. Wang G, Ansari N (1997) Optimal broadcast scheduling in packet radio networks using mean field annealing. IEEE J Sel Areas Commun 15(2):250–260
29. Watanabe Y, Mizuguchi N, and Fujii Y (1998) Solving optimization problems by using a hopfield neural network and genetic algorithm combination. Syst Comput Jpn 29(10):68–73

30. Wolsey L (1998) Integer Programming. Wiley, New York
31. Yeo J, Lee H, and Kim S (2002) An efficient broadcast scheduling algorithm for TDMA ad-hoc networks. Comput Oper Res 29:1793–1806

Broyden Family of Methods and the BFGS Update

BFM

VASSILIOS S. VASSILIADIS, RAÚL CONEJEROS
Chemical Engineering Department,
University Cambridge, Cambridge, UK

MSC2000: 90C30

Article Outline

Keywords
See also
References

Keywords

Unconstrained optimization; BFGS update; DFP update; Broyden family of methods; Rank-two updates; Quasi-Newton methods

Quasi-Newton methods attempt to update a Hessian approximation (or the inverse of it) instead of evaluating the Hessian matrix exactly at each iteration, as in the basic Newton method for *unconstrained optimization*. Consider the optimization problem:

$$\min_{\mathbf{x}} f(\mathbf{x}).$$

For this problem the Newton method requires the solution and updating iteratively of the solution point according to:

$$\mathbf{H}(x^{(k)}) \Delta x^{(k)} = -\mathbf{g}(x^{(k)}), \quad (1)$$

where $\mathbf{H}(x^{(k)})$ denotes the Hessian matrix at point $x^{(k)}$ (kth iteration of Newton's method), $\mathbf{g}(x^{(k)})$ is the gradient vector at the same point, and finally $\Delta x^{(k)}$ is the correction to the point $x^{(k)}$. The correction is applied according to:

$$x^{(k+1)} = x^{(k)} + \alpha \Delta x^{(k)}$$

where for the standard Newton method $\alpha = 1$, but otherwise in practical applications and to force theoretically 'global convergence' (not just in the neighborhood of the minimizer) one conducts a line search to estimate optimally the value of α at each iteration. Alternative algorithms use the concept of *trust regions*.

There exist symmetric updating formulae of rank-two corrections for both the inverse Hessian and the Hessian, all belonging to the broad category of *Broyden methods*. The general family updates either the Hessian ($\mathbf{H}$) or the inverse Hessian ($\mathbf{G} = \mathbf{H}^{-1}$). There are two well-known schemes, the *Davidon–Fletcher–Powell rank-two update* (DFP update), originally proposed by W.C. Davidon [3], and later by R. Fletcher and M.J.D. Powell [6], and the well-known *Broyden–Fletcher–Goldfarb–Shanno update formula* (BFGS update). This was proposed by C.G. Broyden [1,2], Fletcher [4], D. Goldfarb [7], and D.F. Shanno [9]. Both of these methods preserve positive definiteness of the updated matrices.

The definitions of $\mathbf{p}$ and $\mathbf{q}$ used below are introduced first:

$$\mathbf{p}_k = \mathbf{x}_{k+1} - \mathbf{x}_k,$$
$$\mathbf{q}_k = \mathbf{g}_{k+1} - \mathbf{g}_k.$$

The DFP updating scheme of the inverse Hessian is given by:

$$\mathbf{G}_{k+1}^{\text{DFP}} = \mathbf{G}_k + \frac{\mathbf{p}_k \mathbf{p}_k^\top}{\mathbf{p}_k^\top \mathbf{q}_k} - \frac{\mathbf{G}_k \mathbf{q}_k \mathbf{q}_k^\top \mathbf{G}_k}{\mathbf{q}_k^\top \mathbf{G}_k \mathbf{q}_k}.$$

The complementary updating formula to any updating Hessian (or inverse Hessian) scheme can be found by exchanging $\mathbf{G}$ with $\mathbf{H}$ and $\mathbf{q}$ with $\mathbf{p}$ (for example as discussed in [8]). By applying this property to the DFP update above, it is obtained:

$$\mathbf{H}_{k+1}^{\text{BFGS}} = \mathbf{H}_k + \frac{\mathbf{q}_k \mathbf{q}_k^\top}{\mathbf{q}_k^\top \mathbf{p}_k} - \frac{\mathbf{H}_k \mathbf{p}_k \mathbf{p}_k^\top \mathbf{H}_k}{\mathbf{p}_k^\top \mathbf{H}_k \mathbf{p}_k},$$

which is the BFGS updating scheme for the Hessian.

By taking the inverse of this one can obtain the inverse Hessian BFGS updating formula:

$$\mathbf{G}_{k+1}^{\text{BFGS}} = \mathbf{G}_k + \left(\frac{1 + \mathbf{q}_k^\top \mathbf{G}_k \mathbf{q}_k}{\mathbf{q}_k^\top \mathbf{p}_k} \right) \left(\frac{\mathbf{p}_k \mathbf{p}_k^\top}{\mathbf{p}_k^\top \mathbf{q}_k} \right)$$
$$- \left(\frac{\mathbf{p}_k \mathbf{q}_k^\top \mathbf{G}_k + \mathbf{G}_k \mathbf{q}_k \mathbf{p}_k^\top}{\mathbf{q}_k^\top \mathbf{p}_k} \right).$$

The general class of Broyden methods can be derived by the linear combination of the two types of updates, since they are both symmetric rank-two type corrections, being constructed from the same vectors $\mathbf{p}_k$ and $\mathbf{G}_k\mathbf{q}_k$. Thus it can be obtained that (for example see [5,8]):

$$\mathbf{G}_{k+1}^{\phi} = (1-\phi)\mathbf{G}_{k+1}^{\text{DFP}} + \phi\mathbf{G}_{k+1}^{\text{BFGS}},$$

which yields:

$$\mathbf{G}_{k+1}^{\phi} = \mathbf{G}_k + \frac{\mathbf{p}_k\mathbf{p}_k^{\top}}{\mathbf{p}_k^{\top}\mathbf{q}_k} - \frac{\mathbf{G}_k\mathbf{q}_k\mathbf{q}_k^{\top}\mathbf{G}_k}{\mathbf{q}_k^{\top}\mathbf{G}_k\mathbf{q}_k} + \phi\mathbf{v}_k\mathbf{v}_k^{\top}, \quad (2)$$

which is the general family of Broyden methods, with:

$$\mathbf{v}_k = \left(\mathbf{q}_k^{\top}\mathbf{G}_k\mathbf{q}_k\right)^{\frac{1}{2}}\left(\frac{\mathbf{p}_k}{\mathbf{p}_k^{\top}\mathbf{q}_k} - \frac{\mathbf{G}_k\mathbf{q}_k}{\mathbf{q}_k^{\top}\mathbf{G}_k\mathbf{q}_k}\right).$$

A pure Broyden method is one that uses a constant value of ϕ in all iterations. The Broyden family does not preserve positive definiteness of the updated inverse Hessian $\mathbf{G}_{k+1}^{\phi}$ for all values of ϕ.

Generally, of all these schemes the varying ϕ variant is never used nowadays (2000), with the BFGS scheme being the method of choice whenever an updating scheme is chosen. This is because computational experience has proven the BFGS to be more effective than the DFP scheme.

See also

▶ Conjugate-Gradient Methods
▶ Large Scale Unconstrained Optimization
▶ Numerical Methods for Unary Optimization
▶ Unconstrained Nonlinear Optimization: Newton–Cauchy Framework
▶ Unconstrained Optimization in Neural Network Training

References

1. Broyden CG (1970) The convergence of a class of double rank minimization algorithms. Part I. J Inst Math Appl 6:76–90
2. Broyden CG (1970) The convergence of a class of double rank minimization algorithms. Part II. J Inst Math Appl 6:222–231
3. Davidon WC (1959) Variable metric method for minimization. Techn Report AEC Res Developm ANL-5590
4. Fletcher R (1970) A new approach to variable metric algorithms. Comput J 13:317–322
5. Fletcher R (1991) Practical methods of optimization, 2nd edn. Wiley, New York
6. Fletcher R, Powell MJD (1963) A rapidly convergent descent method for minimization. Comput J 6:163–168
7. Goldfarb D (1970) A family of variable metric methods derived by variational means. Math Comput 24:23–26
8. Luenberger DG (1984) Linear and nonlinear programming, 2nd edn. Addison-Wesley, Reading, MA
9. Shanno DF (1970) Conditioning of quasi-Newton methods for function minimization. Math Comput 24:647–656

Capacitated Minimum Spanning Trees

STEFAN VOSS
Institute Wirtschaftswissenschaften,
Techn University Braunschweig,
Braunschweig, Germany

MSC2000: 90C27, 68T99

Article Outline

Keywords
Applications
Mathematical Programming Formulations
Exact Algorithms
Heuristics
    Finding Initial Feasible Solutions
    Construction Methods
    Savings Procedures
    Dual Procedures
    Additional Procedures
Improvement Procedures
    Neighborhood Definition
    Second Order Algorithms
    Computational Results
Metaheuristics
Problem Modifications and Related Problems
Conclusions
See also
References

Keywords

Telecommunication; Combinatorial optimization; Spanning tree; Capacitated minimum spanning tree problem; Terminal layout problem; Resource-constrained minimum spanning tree problem

The *capacitated minimum spanning tree problem* (CMST) or *terminal layout problem* is usually described as the problem of determining a rooted spanning tree of minimum cost in which each of the subtrees off the root node contains at most K nodes. That is, the CMST is a generalization of the well-known *minimum spanning tree problem* (MST) where the objective is to find a minimum cost tree spanning a given set of nodes such that some capacity constraints are observed.

As a graph theoretic problem we consider a connected graph $G = (V, A, b, c)$ with node set $V = \{0, \ldots, n\}$ and arc set A. Each node $i \in V$ has a nonnegative node weight b_i which may be interpreted as capacity requirement whereas a nonnegative arc weight c_{ij} represents the cost of using arc $(i, j) \in A$. Node 0 denoted as the *center node* will be the root of the tree (with $b_0 := 0$). We define a *subtree* or *component* C_i of a tree spanning V as its maximal subgraph uniquely connected to the center by arc $(0, i)$ (denoted as *central arc*). The demand of a subtree is the sum of the node weights of the included nodes. To satisfy the *capacity constraint* the demand of each subtree must not exceed a given capacity K. (Without loss of generality we may assume $b_i \leq K$ for all i.) By means of these definitions the CMST is the problem of finding a minimum cost tree spanning node set V where all subtrees satisfy the capacity constraint.

In spite of existing polynomial algorithms for the unconstrained MST the CMST has been shown to be NP-hard [32] even when all b_i-values are identical. This case of the CMST is referred to as *unit weight CMST* or *equal demand* CMST; otherwise it is called the *nonunit weight* case. Most references in the literature deal with the unit weight case with only a few exceptions treating the more general case. For a comprehensive survey of the (unit weight) CMST up to the mid-1990s see [4]. The CMST in undirected graphs requires a symmetric cost matrix. Otherwise, the direction of the arcs has to be considered, i.e., all subtrees are directed. This *capacitated minimum spanning arborescence problem* (CMDT) includes the CMST as a special case.

Motivated by the intractability of the problem both heuristic as well as exact algorithms have been developed. In the sequel various algorithmic concepts are reviewed mainly for the unit weight CMST (with special emphasis on progress made in the late nineties; for some older yet important references not given here see [4]). However, first we sketch some applications of the CMST some of which may also lead to important modifications of the problem.

Applications

The CMST has a great variety of applications especially in the field of telecommunications network design. For instance, in the design of minimum cost teleprocessing networks terminals (nodes) have to be connected to a central facility (the center node) by so-called multipoint lines (the subtrees) which have to be restricted with respect to the traffic transfered between the center and the included terminals or the number of terminals included in the line. The latter is sometimes called reliability constraint because it limits the maximal number of terminals disconnected from the central facility in the case of a single link breakdown. Although different constraints may be referred to as capacity constraints (e.g. considering arc weights instead of node weights or even nonlinear weight functions depending on the distance of a node or arc from the center) most formulations in the literature consider only one of them.

Mathematical Programming Formulations

For the CMST a great variety of formulations may be found in the literature; see, e.g., [14,19,20,21,22]. Here we restrict ourselves to the presentation of a well-known flow-based formulation. As relaxations of directed formulations may be advantageous we consider the CMDT.

Assume $b_i = 1$ for all $i = 1, \ldots, n$, and $b_0 = 0$, then the CMDT can be described as a mixed integer linear programming formulation as follows. Define $x_{ij} = 1$, if arc (i, j) is included in the solution, and $x_{ij} = 0$, otherwise. Furthermore, let y_{ij} denote the flow on arc (i, j) for all i, j, i.e., $i = 0, \ldots, n$ and $j = 1, \ldots, n$. Ensure variables x_{ij} and y_{ij} with $(i, j) \notin A$ to be equal to zero by assigning prohibitively large weights to them. The following single-commodity flow formulation gives a minimum cost directed capacitated spanning tree with center node 0 as the root:

$$(P) \begin{cases} \min \quad \sum_{i=0}^{n} \sum_{j=1}^{n} c_{ij} \cdot x_{ij} \\ \text{s.t.} \quad \sum_{i=0}^{n} x_{ij} = 1, \quad j = 1, \ldots, n, \\ \quad \sum_{i=0}^{n} y_{ij} - \sum_{i=1}^{n} y_{ji} = 1, \\ \quad j = 1, \ldots, n, \\ \quad x_{ij} \leq y_{ij} \leq (K - b_i) \cdot x_{ij} \\ \quad \text{for all } i, j, \\ \quad x_{ij} \in \{0, 1\}, \quad y_{ij} \geq 0 \quad \text{for all } i, j. \end{cases}$$

The first set of equalities ensures that exactly one arc is reaching each noncentral node. The coupling constraints in combination with the flow conservation ensure that no cycles are allowed and that the capacity constraint is satisfied in each subtree. For a formal proof of cycle prevention see [14], i.e., a tree spanning all nodes is guaranteed.

Exact Algorithms

Most exact algorithms for solving the CMST are based on the branch and bound or the branch and cut paradigm, while other approaches are usually not competitive due to time and space complexity (e.g. dynamic programming [23]).

When describing the concepts from the literature in most cases we do not report computational experiments as there is no fair comparison. When reporting problem sizes solved to optimality by a specific algorithm different authors have proposed various ways of conducting experiments (e.g. the way of data generation), i.e., comparability is not always guaranteed. Moreover, problem instances with a larger number of nodes might be easier to solve than those with a smaller number of nodes, depending on the respective values of K [20,24]. Another aspect which seems to have considerable impact on the performance of most algorithms is the location of the root. For instance, instances with the root in the center of a rectangle in the Euclidean plane may be solved very easily compared to instances with the root in the corner of the rectangle. Recently, a set of problem instances with 40 and 80 nodes became used consistently (see, e. g., [20,22]).

Branch and bound methods for the CMST can be divided into two classes. *Node oriented methods* branch by fixing nodes, *arc oriented methods* branch by including an arc (i, j) into the solution (i. e. fixing $x_{ij} = 1$) or excluding it from the solution ($x_{ij} = 0$). A node is called *established* if the path from the root to this node only consists of arcs fixed to 1 (these arcs are called established, too). Usually, only those arcs incident with exactly one established node are allowed to be fixed. If an arc is fixed to 1, it becomes established and both of its end nodes are established. Correspondingly, an arc is called *disallowed*, if $x_{ij} = 0$ is fixed.

A well-known relaxation of the CMST is the MST relaxation which can be easily solved to optimality. If the MST solution is feasible for the CMST, then it is also optimal and the respective problem can be fathomed.

In the early 1970s an arc oriented branch and bound algorithm based on the MST relaxation was very popular. Subproblems are, e. g., branched by defining the first not yet established arc of an infeasible subtree (the first counted from the center) as established or disallowed, respectively [8]. This approach may be improved by using logical tests and tighter lower bounds [10]. Let node i be established, then in a subproblem disallowing arc (i, j) all arcs (i, k) with node k being an established node of the same subtree as j may be disallowed, too, without loosing optimality. If an optimal solution is lost by disallowing these arcs, the complementary subproblem with established arc (i, j) contains another optimal solution. A dominance criterion is used to fathom some subproblems. In addition, the lower bounds are improved using a special case of the degree constrained MST considering that the degree of the center node — and hence the number of subtrees — is greater or equal to the ratio of the total demand and the capacity K of each subtree.

A. Kershenbaum and R.R. Boorstyn [27] propose two branch and bound algorithms both using last-in first-out to choose the subproblem that is next to be considered. One of the algorithms is node oriented. It starts with n subtrees and each node being 'permissible' for each subtree. A subproblem is branched by including or excluding a node from a specific subtree. Lower bounds are obtained from a partitioning algorithm. The node weights used in this algorithm are originally derived from the MST solution and then, during the branch and bound, transformed in a weight exchange process. Theoretically, these bounds are at least as good as those from the MST relaxation, in practice they are much better. With the same partitioning technique an arc oriented branch and bound algorithm similar to the one of [8] is developed.

B. Gavish [14] compares several relaxations of the CMST with respect to lower and upper bounds. Best results are obtained with a Lagrangian relaxation with an additional degree constraint combined with a subgradient optimization procedure.

Outperforming his previous methods, Gavish [15] develops a new binary programming formulation for the CMST based on an extension of the subtour elimination constraints known from the traveling salesman problem (TSP). Because of the large number of these constraints involved in the formulation an augmented Lagrangian procedure is developed where a dual ascent algorithm is used to obtain initial multipliers and a subgradient procedure to optimize them.

L. Gouveia [20] presents a flow formulation with binary variables z_{ijq} being 1 if a flow of q units goes through arc (i, j). Instead of the $O(n^2)$ constraints of the above flow formulation (P) only $O(n)$ constraints are required. The linear relaxation of the new formulation yields lower bounds as good as those produced by the original formulation. With additional constraints different Lagrangian relaxation schemes are obtained that yield some improvements on the bounds of Gavish [15], especially for problem instances with small capacity K and the center in the 'corner' of a rectangle containing the nodes.

K. Malik and G. Yu [30] present another branch and bound algorithm with Lagrangian subgradient optimization. They give a formulation for the CMST (closely related to the one of [15]) and additional tightening constraints which are added to the problem during the optimization process. Based on a multicommodity flow formulation R. Kawatra [26] uses a Lagrangian approach, too.

L. Hall [24] reports on experience with a cutting plane algorithm for instances with up to 200 nodes making clever use of polyhedral methods. Gouveia and P. Martins [22] propose a hop-indexed generalization of formulation (P). Further improvements on the lower bounds for problem instances with the root in the corner of a rectangle are obtained.

P. Toth and D. Vigo [38] provide an exact algorithm for the CMDT and numerical results are also provided for problem instances with up to 200 nodes. Their approach uses an additive lower bounding procedure combining a Lagrangian lower bound and a lower bound based on solving minimum cost flow problems.

Heuristics

Before presenting heuristics for the CMST it is useful to consider the characteristics of feasible and infeasible solutions [4]. A solution consists of a set of components $C_i = (V_i, A_i)$ with node set V_i and arc set A_i where usually C_i is a spanning tree for V_i. Each component includes only one central arc so that two different node sets V_i and V_j may have the center node as the only common node. Joining all node sets V_i would yield the entire node set V.

A component is called *feasible* if it does not violate the capacity constraint, and *infeasible*, otherwise. It is referred to as *central* if it includes the center node and *noncentral*, otherwise (i. e. a noncentral component results from a component by eliminating the central arc and the center node). Sets of components having both infeasible and noncentral components are not considered as a solution.

A solution is called *feasible*, if every component contained in the solution is central and feasible itself. It is *incomplete*, if every component is feasible but at least one is noncentral. If all components are central but at least one is not feasible then a solution is called *infeasible*.

The following special solutions of the CMST may be emphasized. The incomplete solution with $n + 1$ components $C_i = (\{i\}, \emptyset)$, $i = 0, \ldots, n$, is called an *empty tree*. All components are feasible and all except C_0 are noncentral. The feasible solution with n components $C_i = (\{0, i\}, \{(0, i)\})$, $i = 1, \ldots, n$, is called a *star*. All components are central and feasible. In the case of a sparse graph with only a subset of nodes being directly connected to the center artificial arcs with high cost values should be introduced to complete the graph. The star then might be feasible only for the modified problem.

Finding Initial Feasible Solutions

Most procedures for determining initial feasible solutions (start procedures) for the CMST may be classified as *construction procedures*, *savings procedures* or *dual procedures*.

Construction Methods

Construction methods start with an incomplete solution, usually the empty tree, and successively enlarge it until the solution is feasible. Most procedures in this category replace two components and the chosen arc that connects them by a new component. We may distinguish between arc oriented and node oriented methods.

Arc oriented (or *best arc*) procedures choose in a greedy fashion arcs which are used to join its two incident components. The procedures stop when a feasible solution is obtained. The components of the final solution generally are not built one by one but simultaneously. It is not necessary to finish one component before starting another one.

As examples one may use the basic principle of well-known MST algorithms. The *modified Kruskal algorithm* [7] in each iteration chooses a feasible arc with lowest cost and joins the two corresponding components. All arcs that have become infeasible in this step are removed from consideration for the next iterations. Correspondingly, the *modified Prim algorithm* in each iteration chooses an arc with minimal cost which is incident to the center or a central component (with not yet exhausted capacity).

Node oriented (or *best node*) procedures choose in a greedy fashion a node or component and join it to its nearest neighbor component by the best possible arc incident to the chosen component, while preserving feasibility.

An obvious idea is to cluster the nodes into groups of no more than K nodes and then to choose the arc set according to an MST for the nodes of each group and the center. Assuming that coordinates of the nodes are given this approach may be referred to as *clustering* (or *sweep*) algorithm [36].

The *Martin algorithm* [25,31] chooses the component which is most distant to the center and joins it to its nearest feasible neighbor component. If χ_i is the cost of connecting component C_i to the center, the component with maximal χ_i is chosen.

The *regret method* (or *Vogel approximation method* (VAM, [8]) computes for every component C_i a regret $r_i = a_2(i) - a_1(i)$ that has to be accepted if C_i is not joined to its nearest feasible neighbor component with cost $a_1(i)$ but to its second nearest feasible neighbor with cost $a_2(i)$. The component with maximal regret is chosen and joined to its nearest neighbor. The regrets are recomputed and the procedure continues until the solution is feasible.

Mixed procedures combine arc and node aspects. They assign a weight w_i to each node i and compute for every arc (i,j) the trade-off function value as $t_{ij} = w_i - c_{ij}$. The feasible arc with largest t_{ij} is chosen and the respective components are joined. In general, the weights have to be updated after each iteration. With an appropriate definition of the weight function and an update rule all preceding heuristics except the clustering procedures can be incorporated in this concept [29]:

- The Kruskal algorithm is obtained for $w_i = 0$ for all i. Obviously, no update is needed.
- For the Prim algorithm assign weights $w_i = 0$ to all central components and $w_i = -\infty$ to all other components. If a noncentral component is joined with a central component, the weight of the new component is set to zero.
- The Martin algorithm requires $w_i = \chi_i + a_1(i)$, and with $w_i = a_2(i) = r_i + a_1(i)$ one obtains the VAM. The weights w_i have to be recomputed if the values of $a_1(i)$ or $a_2(i)$ have changed, respectively.

Mixed VAM is a combined regret-best arc procedure [18,39]. The regret r_i is used as node weight w_i and thus, the trade-off function is $t_{ij} = r_i - c_{ij}$.

The *unified algorithm* [29] proposes a parameterization of the weight function and the trade-off function.

Savings Procedures

Savings procedures for the CMST usually start with the star. The best feasible change, i. e. the change which yields the largest savings, is performed. This is iteratively repeated until no savings can be obtained any more. The methods could easily be applied to other feasible solutions and so they could be classified as improvement procedures, too.

The *Esau–Williams algorithm* (EW, [11]) joins the two components which yield the maximal savings in cost. The savings s_{ij} of joining C_i and C_j is defined as $s_{ij} = \max\{\chi_i, \chi_j\} - c_{ij}^*$ if joining of C_i and C_j is feasible, and $s_{ij} = \infty$, otherwise, with χ_i again being the minimal cost of the connection from the center to the nodes of C_i and c_{ij}^* being the minimal cost of an arc connecting C_i and C_j. Then all savings concerning the new component have to be recomputed and again the maximal savings is chosen. The process is stopped if no more positive savings are available.

The EW is closely related to the above mentioned best node procedures. For instance, the Martin algorithm may be referred to as a less greedy version of the EW.

The EW can also be described as a special case of the unified algorithm starting with the empty tree and at each step adding a feasible arc with maximal trade-off $t_{ij} = \chi_i - c_{ij}$.

Whitney's savings heuristic [10,39] modifies the EW by allowing noncentral arcs to be deleted as well as central arcs. This leads to a possible recombination of segments of the components. Here we see again that savings algorithms are closely related to the class of improvement procedures.

The *parallel savings algorithm* (PSA) [18] computes savings like the EW. However, one iteration does not only join one pair of components but a set of pairs with maximal total savings. This set is determined by solving a maximum weight matching (maximal with respect to the savings) in an adequate graph.

To avoid the parallel construction of nearly equal sized components which cannot be joined any longer if they exceed half of the capacity, Gavish [16] proposes consideration of dummy nodes which yield high savings for any component joined with them. Thus, in this *PSA with dummy nodes* the number of joins between original components in one iteration is reduced by half the number of dummy nodes.

Dual Procedures

Dual procedures start with an infeasible low cost solution, usually the MST solution. The *violation of the con-*

straint(s) is iteratively reduced at the expense of a total cost increase until the solution becomes feasible.

The start procedure of D. Elias and M.J. Ferguson [10] examines every arc (i, j) of any infeasible component. If (i, j) is deleted the resulting noncentral component C_k is connected to another central component by arc (k, l). This arc is chosen such that the total capacity overflow is reduced. Ties are broken such that the smallest cost increase is chosen. The procedure deletes that arc (i, j) which leads to minimal total cost increase $c_{kl} - c_{ij}$ and adds (k, l) to the solution. (As a modification the arc (i, j) with minimal ratio of cost increase and capacity overflow reduction could be chosen.) Given integer cost weights, the procedure terminates with a feasible solution after a finite number of iterations, because in each iteration the total capacity overflow is reduced by at least one unit.

Given a feasible solution one can try to improve the solution by recombining segments of the components in a similar way [10]. To increase flexibility, consider a slight modification: Exchanging two arcs should be allowed even if it leads to an increase of total capacity overflow whenever the cost of the solution does not increase and the arc that is to be included never had been in the solution before.

Dual procedures may well be related to other concepts. For instance, a dual procedure may be seen as a constructive savings procedure starting within the infeasible region of the solution space. In that sense it might be related to metastrategies as, e. g., tabu search described below in the sense that it performs a recover phase within a strategic oscillation approach.

Additional Procedures

Besides classifying construction, savings or dual procedures, there are procedures using aggregation and decomposition techniques combined with dynamic programming. In addition, some heuristics which start with generating a TSP tour are not considered in that scheme.

Gouveia and J. Paixão [23] present two heuristics for the CMST which are based on problem size reduction by aggregation and decomposition techniques. In the *aggregation heuristic* the nodes are clustered using the EW — thus forming new nodes with higher and in general nonidentical weights — until the resulting aggregated problem is small enough to be solved to optimality in time limits deemed practical. The *decomposition heuristic* creates for each central arc of the MST solution a subproblem by considering only the nodes of the respective subtree. Subproblems which are small enough are solved to optimality. For the remaining subproblems the aggregation heuristic is used.

Note that the above mentioned sweep algorithm might be classified as aggregation procedure, too.

For the case of unit weights, K. Altinkemer and Gavish [2] provide a modified PSA with a worst-case error bound of $3 - 2/K$ and derive a bound of 4 for the case of nonunit weights. First, a TSP tour is constructed and then it is partitioned into feasible subtrees by adding some central arcs and removing respective noncentral arcs. Note that the (noncenter) nodes of the resulting subtrees are always connected in the same order as in the TSP tour.

In the case of unit weights a *K-iterated tour partitioning algorithm* is used: K solutions are constructed. In each solution the first subtree starts with the first (noncenter) node of the TSP-tour, the second subtree starts with node 2 (first solution), node 3 (second solution), ..., node $K + 1$ (Kth solution). Apart from the first and the last each subtree contains exactly K nodes. The best out of these K solutions is chosen.

For nonunit weights a *nearest insertion optimal partitioning algorithm* may be applied: In the nearest insertion tour the nodes are renumbered according to their position. Modified costs c_{ij}' are computed as the cost of a tree linking the center with node i, node i with node $i + 1$, etc., and node $j - 1$ with node j. If such a tree is infeasible (due to capacity), the respective cost is set to infinity. With these definitions, the shortest path with respect to c_{ij}' from the center to node n represents the optimal partitioning.

In each procedure a final step can be added: The solution is improved by computing MSTs for the derived components.

Improvement Procedures

Improvement procedures for the CMST can be classified as either *local exchange procedures* or *second order procedures*.

Neighborhood Definition

Local exchange procedures start with a feasible solution and seek to improve it by modifying the current solution in a prespecified way: Sets of arcs are included in or excluded from the solution. If more than one change of the solution is possible the best one (with respect to cost) is chosen. The procedure continues as long as improvements are possible.

Given a feasible solution, H. Frank et al. [13] examine for every node i the following exchange: Connect i to its nearest neighbor not yet connected to i and remove the arc with highest cost from the resulting cycle while still preserving feasibility. The exchange with greatest cost decrease is chosen as long as improvements are positive. The authors describe this procedure for a network design problem with variable arc capacities and cost but it can be naturally applied to the special case with only one available capacity and fixed cost for each arc as in the CMST.

Elias and Ferguson [10] try to improve the solution by recombining segments of the components, i. e., deleting an arc and reconnecting the resulting noncentral component without loosing feasibility (cf. the Whitney savings heuristic above).

The previously reported improvement procedures alter a current solution by including or excluding arcs. In contrast, a node exchange procedure transforms one feasible solution to a neighbor solution by changing the assignment of the nodes to the subtrees. Such a transformation is called *move*. Subsequently a certain number of moves is performed thus trying to find improved solutions.

Starting from the EW solution, in their CMST procedure A. Amberg et al. [4] consider two types of moves: *Shift moves* choose one node and shift it from its actual component to another one. *Exchange moves* choose two nodes belonging to different subtrees and exchange them. Both types may be simultaneously used whereas only *feasible moves* are allowed, i. e. those leading again to feasible solutions.

A modified neighborhood definition involves cutting a subtree from a given solution and to paste it within another subtree or to connect it to the root node [35]. Additional neighborhood structures are given in [1]. Contrary to the previous neighborhood structures the authors do not restrict themselves to the consideration of two subtrees to be involved in one move but into a chain of moves performed simultaneously (called cyclic exchanges and path exchanges). That is, the number of exchanges grows exponentially with the problem size. Based on a shortest path algorithm some profitable exchanges may be determined in way which may be termed *Lin–Kernighan neighborhood* or *ejection chain*.

Second Order Algorithms

Second order algorithms iteratively apply a slave procedure to different start solutions (where some arcs are fixed to be included) and/or modified cost matrices (where inhibitively high cost has been assigned to some arcs) thus forcing arcs into or out of the solution. Savings procedures as the EW or the PSA are applied as slave procedures to complete the solution. In each iteration, all possible modifications according to a given rule are checked. The best one is realized and the respective modifications are made permanent for the remaining iterations. Two important second order algorithms are inhibit and join [25].

The *inhibit procedure* examines for every arc of the current solution the effect of excluding this arc by applying the EW to a modified graph where the cost of the respective arc has been made inhibitively high. The inhibition yielding the lowest cost solution is made permanent (the arc is inhibited for the remaining iterations) and the process is repeated until no further cost reduction can be obtained. At most $O(n^2)$ iterations, each with at most $O(n)$ inhibitions, have to be considered.

The *join procedure* determines for every node i its nearest neighbor i_1 as well as the closest neighbor i_2 closer to the center than i (if different from i_1). It computes the effect on the cost of the solution if node i is directly connected to node i_1 or alternatively to node i_2 (if this is not already done in the actual solution) by applying the start procedure on a modified graph. The joining which produces the best solution is made permanent and the procedure is repeated with this solution. In each of the $O(n)$ iterations $O(n)$ joins have to be considered.

It should be noted that both procedures, inhibit and join, are already look ahead procedures (trying to overcome a shortsighted myopic behavior). Both improvement procedures can be used alone or in combination with each other performing one iteration of join after an iteration of inhibit and vice versa. Combining the procedures restricts the number of iterations to $O(n)$ (from the join procedure) yielding a complexity of $O(n^2)$ times the EW complexity.

For the improvement procedure of [28] in a first step the MST solution and the EW solution have to be determined. Then the following iteration is performed. Define T as the set of arcs which are in the MST but not in the EW solution. For every nonempty subset S of T generate a (incomplete) solution including these arcs (if this is feasible), then exclude all arcs of the remaining subset $T \setminus S$ (by modifying the respective arc costs) and complete the solution by applying the heuristic. Choose the subset S^* which yields the largest improvement and permanently include these arcs into the solution. Repeat this iteration with modified $T := T \setminus S^*$.

The *min-exchange heuristic* outlined in [17] starts with any given feasible solution and determines for every pair of components C_p and C_q the cheapest arc (i, j) connecting the two components. All arcs incident to i or j are deleted. C_p and C_q are decomposed into two noncentral single node components C_i and C_j and some remaining components. Now the noncentral components are connected with the center; hereby the minimal cost arcs are chosen. The PSA completes this modified solution. The authors propose to split all components simultaneously.

Computational Results

In the early CMST literature the EW has been found to perform best on average when compared to procedures with similar computation times. Therefore, even nowadays EW is taken as a benchmark to check the performance of other procedures. Kershenbaum and W. Chou [29] report that the unified algorithm running with 3 to 10 different parameter combinations and correspondingly multiplied computation times yields 1–5% improvement over EW. Unfortunately, no specific parameter combination produces improvements in general.

Gouveia and Paixão [23] admit the nearest insertion optimal partitioning algorithm to perform much worse than EW on average with some significant exceptions. This shows that no general dominance of EW considering single problem instances can be derived.

Gavish and Altinkemer [16,18] report for test problems with up to 400 nodes that the PSA yields improvements of 2–4% in the unit weight case, but performs poorly for nonidentical weights. In the latter case, the min-exchange heuristic applied to the PSA solution gives results comparable to those of EW [17]. Gavish [16] reports that the PSA with dummy nodes attains improvements over EW (up to 6% some cases). However, in the nonunit weight case EW performs still better. Here, the PSA with constant number of joins gives consistently better results than EW. Gouveia and Paixão [23] apply this variant of the PSA with the number of joins varying between 1 (which is in fact the EW) and 12 on unit weight test problems with up to 200 nodes: Significant improvements over EW with computation times raised by a factor of up to 250 are obtained. They also report that the (original) PSA performs best when the capacity is a power of 2 (in the unit weight case). Their aggregation heuristic on average yields a slight improvement over EW (up to 3% in some cases). In test problems, that have the center in the 'middle' of the rectangle containing the nodes, the decomposition procedure has larger computation times than the aggregation algorithm (factors of slightly more than 1 up to 3 are found) and better results, whereas in cases with the center on the 'corner' of the rectangle both methods in almost all instances have similar running times and solutions. Apart from a few cases the PSA with constant number of joins (and varied parameters) on average performs better than both procedures.

M. Karnaugh [25] tests inhibit and join on problems with up to 150 nodes. The combination of the procedures gives 2–3% improvement over EW while the running time is increased by a factor 100 (derived for the 150-node problems). Applying only inhibit performs slightly worse.

Kershenbaum et al. [28] found that inhibit, join and their own procedure yield improvements of around 2% over EW. Thus, their own procedure requiring only 2 to 3 times more computation time than EW, outperforms join.

Metaheuristics

Given a local search mechanism, a metastrategy like tabu search or simulated annealing as a *guiding process* decides which of the possible moves is chosen and forwards its decision to the *application process* which then executes the chosen move. In addition, it provides some information for the guiding process (depending on the requirements of the respective metastrategy) like the re-computed set of possible moves.

Contrary to the improvement procedures reported in the last section, the cost of a new solution may exceed the cost of the previous one. Moves leading to a cost increase are allowed in order to overcome local optima. Which of the available feasible moves should be chosen to transform the current solution? The answer to this question is not clear and various approaches may lead to good solutions. The guiding process may use, e. g., the two metastrategies simulated annealing and tabu search.

Simulated annealing (SA) randomly chooses one of the feasible moves and its change in cost is computed. If the change is a cost decrease the move is performed. Otherwise, the new solution is accepted with a certain probability. The probability function usually is logarithmic and — intending to favor good solutions — decreases with raising amount of cost increase. It decreases with the number of iterations already performed thus intensifying the search in the current area of the solution space when the execution time is growing. A parameter called *start temperature* has to be specified to adapt the probability function to the actual problem. SA does not require any additional information. If the new solution is rejected the current solution remains unchanged in this step. The next iteration tries again to alter the same solution. Simulated annealing implementations for the CMST are given in [4,6].

Tabu search (TS) examines all feasible moves. The best move — leading to the highest cost decrease or the lowest cost increase, respectively — is chosen and performed. Now suppose that a local optimum is reached. Without further instructions the procedure could permanently alternate between this local optimum and its best neighbor. For that reason a so-called *tabu list* is created: To prevent that a yet explored solution is examined again, all moves that (could) lead to such a solution are stored in the tabu list. Which moves have to be set tabu is derived from the *running list* (RL) containing all performed moves in their sequence of execution. Both lists have to be updated after each iteration.

In the literature there are several distinct ways of deriving the tabu list. They are referred to, e. g., as static tabu search STS, reverse elimination method REM and cancellation sequence method CSM (see [4] for the CMST). For STS and CSM some parameters have to be specified to adopt the methods to a specific problem and problem instances (especially problem size and scaling of cost).

The storage complexity of the application process is $O(n^2)$ and the time complexity $O(K^2)$ per iteration because of the recomputation of MSTs in the changed components. Using simulated annealing we have a time complexity of the guiding process of $O(K^2)$: To compute the probability of acceptance for the new solution two new subtrees have to be computed. This is part of the application process and need not be performed twice. Thus, additional effort only arises if a solution is rejected which does not influence the overall complexity. The storage complexity also is not raised if simulated annealing is used.

The complexity of the guiding process depends on the special tabu search method. Different tabu search implementations are described in [4,35]. Whereas [4] seem to provide better results for the benchmark instances with up to 80 nodes than [35], both seem to be outperformed by the more recent (as of 2000) algorithm in [1] based on their more powerful neighborhood structures.

Besides TS and SA additional modern heuristic search concepts have been investigated for the CMST. A neural network approach is investigated in [33]. A GRASP implementation is provided in [34]. The results for both approaches seem to be behind some of those described in the previous paragraphs.

Problem Modifications and Related Problems

Additionally to considering arc costs one may take into account unreliable arcs and node outage costs which are incurred by the user whenever a terminal node is unable to communicate with the central node, i. e., costs associated with link failures [9].

An interesting modification of the CMDT is the resource-constrained minimum spanning tree problem in directed graphs [12]. Here each node, say i, has a certain amount of scarce resources available (a capacity) which may be used to fulfill capacity requirements of all arcs leaving i. Instead of measuring capacity requirements for subgraphs off the root node here the consideration restricts to the set of incident arcs leaving a node. The current state of the art for solving this problem circumvents a branch and cut approach [12].

In practice we might be faced with the problem that a solution of the design phase need not be a tree but a forest with more than one root node. Most of the approaches developed for the CMST might be applied in a slightly modified way to this so-called multicenter CMST (see e. g. [3] for an extension of the partitioning heuristics with corresponding worst-case bounds).

When multiple centers are considered in arc oriented vehicle routing then the capacitated arc routing problem (CARP) may be transformed in a way that subproblems are successively solved as CMST. Amberg et al. [5] develop this transformation and apply their TS and SA approaches to this multiple center CARP.

Besides solving the CMST as a pure combinatorial optimization problem it may also be embedded into a problem of users with traffic requirements who have to build contracts with, e. g., a telephone company for the provision of service. This may lead to the consideration of some game-theoretic concepts associated with a cost allocation problem arising from the CMST or more general capacitated network design problems [37].

Conclusions

In this paper we have provided a survey on existing methods for solving the CMST.

With respect to considered algorithmic concepts it might be interesting to incorporate some sort of either exact or heuristic reduction techniques.

See also

- ▶ Bottleneck Steiner Tree Problems
- ▶ Directed Tree Networks
- ▶ Minimax Game Tree Searching
- ▶ Shortest Path Tree Algorithms

References

1. Ahuja RK, Orlin JB, Sharma D (1998) New neighborhood search structures for the capacitated minimum spanning tree problem. Techn Report Sloan School Management, MIT
2. Altinkemer K, Gavish B (1988) Heuristics with constant error guarantees for the design of tree networks. Managem Sci 32:331–341
3. Altinkemer K, Pirkul H (1992) Heuristics with constant error guarantees for the multi center capacitated minimum spanning tree problem. J Inform Optim Sci 13:49–71
4. Amberg A, Domschke W, Voss S (1996) Capacitated minimum spanning trees: Algorithms using intelligent search. Combin Optim: Theory and Practice 1:9–39
5. Amberg A, Domschke W, Voss S (2000) Multiple center capacitated arc routing problems: A tabu search algorithm using capacitated trees. Europ J Oper Res 124:360–376
6. Andersen K, Vidal RVV, Iversen VB (1993) Design of a teleprocessing communication network using simulated annealing. In: Vidal RVV (ed) Applied simulated annealing. Lecture Notes Economics and Math Systems. Springer, Berlin, pp 201–215
7. Boorstyn RR, Frank H (1977) Large-scale network topological optimization. IEEE Trans Communications 25:29–47
8. Chandy KM, Russell RA (1972) The design of multipoint linkages in a teleprocessing tree network. IEEE Trans Comput 21:1062–1066
9. Dutta A, Kawatra R (1994) Topological design of a centralized communication network with unreliable links and node outage costs. Europ J Oper Res 77:344–356
10. Elias D, Ferguson MJ (1974) Topological design of multipoint teleprocessing networks. IEEE Trans Communications 22:1753–1762
11. Esau LR, Williams KC (1966) On teleprocessing system design. IBM Systems J 5:142–147
12. Fischetti M, Vigo D (1997) A branch-and-cut algorithm for the resource-constrained minimum-weight arborescence problem. Networks 29:55–67
13. Frank H, Frisch IT, Slyke R Van, Chou WS (1971) Optimal design of centralized computer design network. Networks 1:43–57
14. Gavish B (1983) Formulations and algorithms for the capacitated minimal directed tree problem. J ACM 30:118–132
15. Gavish B (1985) Augmented Lagrangean based algorithms for centralized network design. IEEE Trans Communications 33:1247–1257
16. Gavish B (1991) Topological design of telecommunication networks – local access design methods. Ann Oper Res 33:17–71
17. Gavish B, Altinkemer K (1986) A parallel savings heuristic for the topological design of local access tree networks. Techn Report Graduate School Management Univ Rochester, NY
18. Gavish B, Altinkemer K (1986) Parallel savings heuristics for the topological design of local access tree networks. Proc IEEE INFOCOM 86 Conf, pp 130–139
19. Gouveia L (1993) A comparison of directed formulations for the capacitated minimal spanning tree problem. Telecommunication Systems 1:51–76

20. Gouveia L (1995) A 2n constraint formulation for the capacitated minimal spanning tree problem. Oper Res 43:130–141
21. Gouveia L, Hall L (1998) A comparative study of network flow formulations for the capacitated spanning tree problem. Working Paper Fac Ciencias Univ Lisboa 10
22. Gouveia L, Martins P (1999) The capacitated minimal spanning tree problem: An experiment with a hop-indexed model. Ann Oper Res 86:271–294
23. Gouveia L, Paixao J (1991) Dynamic programming based heuristics for the topological design of local access networks. Ann Oper Res 33:305–327
24. Hall L (1996) Experience with a cutting plane approach for the capacitated spanning tree problem. INFORMS J Comput 8:219–234
25. Karnaugh M (1976) A new class of algorithms for multipoint network optimization. IEEE Trans Communications 24:500–505
26. Kawatra R (1994) A multicommodity network flow application for the capacitated minimal spanning tree problem. Opsearch 31:296–308
27. Kershenbaum A, Boorstyn RR (1983) Centralized teleprocessing network design. Networks 13:279–293
28. Kershenbaum A, Boorstyn RR, Oppenheim R (1980) Second-order greedy algorithms for centralized teleprocessing network design. IEEE Trans Communications 28:1835–1838
29. Kershenbaum A, Chou W (1974) A unified algorithm for designing multidrop teleprocessing networks. IEEE Trans Communications 22:1762–1772
30. Malik K, Yu G (1993) A branch and bound algorithm for the capacitated minimum spanning tree problem. Networks 23:525–532
31. Martin J (1967) Design of real-time computer systems. Prentice-Hall, Englewood Cliffs, NJ
32. Papadimitriou CH (1978) The complexity of the capacitated tree problem. Networks 8:217–230
33. Patterson RA (1995) Hybrid neural networks and network design. PhD Thesis Ohio State Univ
34. Rolland E, Patterson RA, Pirkul H (1999) Memory adaptive reasoning and greedy assignment techniques for the capacitated minimum spanning tree problem. In: Voss S, Martello S, Osman IH, Roucairol C (eds) Meta-heuristics: Advances and trends in local search paradigms for optimization. Kluwer, Dordrecht, pp 487–498
35. Sharaiha YM, Gendreau M, Laporte G, Osman IH (1997) A tabu search algorithm for the capacitated shortest spanning tree problem. Networks 29:161–171
36. Sharma RL, El-Bardai MT (1970) Suboptimal communications network synthesis. Proc 1970 Internat Conf Comm (19.11-19.16.)
37. Skorin-Kapov D, Beltran HF (1994) An efficient characterization of some cost allocation solutions associated with capacitated network design problems. Telecommunication Systems 3:91–107
38. Toth P, Vigo D (1995) An exact algorithm for the capacitated shortest spanning arborescence. Ann Oper Res 61:121–141
39. Whitney VKM (1970) A study of optimal file assignment and communication network configuration in remote access computer message processing and communication systems. PhD Thesis Univ Michigan, Ann Arbor

Carathéodory, Constantine

NICOLAS HADJISAVVAS[1], PANOS M. PARDALOS[2]
[1] Department Math., University Aegean Karlovassi, Samos, Greece
[2] Center for Applied Optim. Department Industrial and Systems Engineering, University Florida, Gainesville, USA

MSC2000: 01A99

Article Outline

Keywords
See also
References

Keywords

Carathéodory; Calculus of variations; Measure theory

Constantin Carathéodory, a mathematician of Greek origin, was born in Berlin on September 13, 1873 and died on February 2, 1950, in Munich, Germany. He made important contributions to the theory of real functions, to the *calculus of variations*, and to *measure theory*.

He first studied in the Brussels' Military School, where he received a solid mathematical background. After two years as an assistant engineer with the British Asyut Dam project in Egypt, Carathéodory began his study of mathematics at the Univ. of Berlin in 1900, where he attended the courses of L. Fuchs, G. Frobenius and H. Schwarz. He was particularly influenced by Schwarz' lectures with whom he became a close friend. In 1902 he entered the Univ. of Göttingen, where he received his PhD [1] under the German mathematician H. Minkowski. In 1909 he became a full Professor in the Univ. of Hannover. In 1913 he obtained the chair held previously by F. Klein in Göttingen and in 1918 he succeeded Frobenius in the Univ. of Berlin. Then, in 1920, he accepted to help the Greek Government in creating the Univ. of Smyrna, Asia Minor, which then belonged to the Greeks. When the Turks razed Smyrna in 1922, Carathéodory managed to save the university library, which he moved to the Univ. of Athens, where he taught until 1924. He then was appointed professor of mathematics at the Univ. of Munich.

Carathéodory made important contributions to various branches of mathematics. In the calculus of variations, besides a comprehensive study of discontinuous solutions, which was contained in his PhD thesis, he also added important results linking the theory with first order partial differential equations. His work on the problems of variation of m-dimensional surfaces in an n-dimensional space marked the first far-reaching results for the general case. He also applied the calculus of variations to specific problems of mechanics and physics. He contributed important findings in his book [6]. The theory of functions and measure theory are two additional areas where the work of Carathéodory is very important. His book [3] is a classic of the field. In the theory of functions of several variables he simplified the proof of the main theorem of conformal representation of simply connected regions on the unit-radius circle. His investigations of the geometrical-set theoretic properties of boundaries resulted in his theory of boundary correspondence. Already in 1909 he published a far-reaching paper on the foundations of *thermodynamics* [2]. The paper remained unnoticed by the physicists, because it was published in a mathematical journal. Only in 1921 M. Born brought the paper to the attention of the physics community, and since then the paper and the *Carathéodory principle* became classics. He also contributed to Einstein's special theory of relativity. His published works include [4,5,7,8,9].

See also

- ▶ Carathéodory Theorem
- ▶ History of Optimization

References

1. Carathéodory C (1904) On the discontinuous solutions in the calculus of variations. PhD Thesis
2. Carathéodory C (1909) Untersuchungen über die Grundlage der Thermodynamik. Math Ann 67:355–386
3. Carathéodory C (1918) Vorlesungen über reelle Funktionen. Teubner, Leipzig, 2nd edn. 1928. Also: Chelsea Publ. 1948
4. Carathéodory C (1932) Conformal representation. Tracts in Math and Math Phys, vol 28. Cambridge Univ. Press, Cambridge
5. Carathéodory C (1935) Variationsrechnung und partielle Differentialgleichungen erster Ordnung, 2nd edn. Teubner, Leipzig, 159 p
6. Carathéodory C (1937) Geometrische Optik. Ergebnisse der Math und ihre Grenzgeb. Springer, Berlin
7. Carathéodory C (1939) Reelle Funktionen, vol I. Teubner, Leipzig
8. Carathéodory C (1950) Funktionentheorie, vol 1–2. Birkhäuser, Basel
9. Carathéodory C (1956) Mass und Integral und ihre Algebraisierung. Birkhäuser, Basel

Carathéodory Theorem

GABRIELE E. DANNINGER-UCHIDA
University Vienna, Vienna, Austria

MSC2000: 90C05

Article Outline

Keywords
See also
References

Keywords

Polytope; Convex hull; Representation

One of the basic results [3] in convexity, with many applications in different fields. In principle it states that every point in the convex hull of a set $S \subset \mathbf{R}^n$ can be represented as a convex combination of a finite number ($n + 1$) of points in the set S. See for example [1,4,6,7,9,10]. Generalizations of the theorem can be found in [2] and [5].

Theorem 1 *Let S be any subset of $\mathbf{R}^n$.*

For every $x \in \mathrm{conv}(S)$ (the convex hull of S), there exist $n + 1$ points $x_0, \ldots, x_n \in S$ such that $x \in \mathrm{conv}(x_0, \ldots, x_n)$.

Proof Since $x \in \mathrm{conv}(S)$, there exists a representation $x = \sum_{i=0}^{k} \alpha_i x_i$, $x_i \in S$ for $i = 0, \ldots, k$ and $\sum_{i=0}^{k} \alpha_i = 1$. If $k \leq n$, we are finished.

Now suppose $k > n$. Note that then $x_1 - x_0, \ldots, x_k - x_0$ are linearly dependent. There then exist scalars $\lambda_1, \ldots, \lambda_k$, not all zero, such that $\sum_{i=1}^{k} \lambda_i (x_i - x_0) = 0$.

Let now $\lambda_0 = -\sum_{i=1}^{k} \lambda_i$; it then follows that $\sum_{i=0}^{k} \lambda_i x_i = 0$ and we can find at least one $\lambda_i > 0$. So we have,

$$x = \sum_{i=0}^{k} \alpha_i x_i - \gamma \cdot 0 = \sum_{i=0}^{k} \alpha_i x_i - \gamma \sum_{i=0}^{k} \lambda_i x_i$$
$$= \sum_{i=0}^{k} (\alpha_i - \gamma \lambda_i) x_i$$

for any $\gamma \in \mathbf{R}$.

Choose γ in the following way:

$$\gamma = \min_{0 \leq i \leq k} \left\{ \frac{\alpha_i}{\lambda_i} : \lambda_i > 0 \right\} = \frac{\alpha_j}{\lambda_j}$$

for some $j \in \{0, \ldots, k\}$ so, $\alpha_i - \gamma \lambda_i \geq 0$ for all $i = 0, \ldots, k$.

Then we obtain $x = \sum_{i=0}^{k} (\alpha_i - \gamma \lambda_i) x_i$ with $\alpha_i - \gamma \lambda_i \geq 0$ for $i = 0, \ldots, k$, $\sum_{i=0}^{k} (\alpha_i - \gamma \lambda_i) = 1$ and $\alpha_j - \gamma \lambda_j = 0$.

And so x is represented as a convex combination of at most k points in S. We can now repeat these steps until $k = n$.

See also

▶ Carathéodory, Constantine
▶ Krein–Milman Theorem
▶ Linear Programming

References

1. Bazaraa MS, Sherali HD, Shetty CM (1993) Nonlinear programming. Wiley, New York
2. Bonnice W, Klee VL (1963) The generation of convex hulls. Math Ann 152:1–29
3. Carathéodory C (1911) Ueber den Variabilitätsbereich der Fourierschen Konstanten von positiven harmonischen Funktionen. Rend Circ Mat Palermo 32:193–217
4. Grünbaum B (1967) Convex polytopes. Interscience, New York
5. Reay JR (1965) Generalizations of a theorem of Carathéodory. Memoirs Amer Math Soc 54:
6. Rockafellar RT (1970) Convex analysis. Princeton Univ. Press, Princeton
7. Stoer J, Witzgall Ch (1970) Convexity and optimization in finite dimensions I. Springer, Berlin
8. Valentine FA (1964) Convex sets. McGraw-Hill, New York
9. Wets RJ-B (1976) Grundlagen Konvexer Optimierung. Lecture Notes Economics and Math Systems, vol 137. Springer, Berlin
10. Ziegler GM (1995) Lectures on polytopes. Springer, Berlin

Checklist Paradigm Semantics for Fuzzy Logics

LADISLAV J. KOHOUT
Department Computer Sci., Florida State University, Tallahassee, USA

MSC2000: 03B52, 03B50, 03C80, 62F30, 62Gxx, 68T27

Article Outline

Keywords
Why the Checklist Paradigm?
   Fuzzy Logics
   Approximate Reasoning
Many-Valued Logics in Fuzzy Sets
The Checklist Paradigm
   Mathematics of the Checklist Paradigm
   Interval Inference and The Checklist Paradigm
   Other Systems of Fuzzy Logic Connectives
      for Interval Inference
   Optimization of Interval Inference
Other Systems of Checklist Paradigm Connectives
Collapse of Intervals into Points Under
   the Additional Probabilistic Constraints
Checklist Paradigm and Generalized Quantifiers
Checklist Paradigm and Four Modes of Reasoning
Group Transformations of Logic Connectives
   and the Checklist Paradigm
   Group Transformations of Logic Connectives
     and the Checklist Paradigm
   An 8-Element Group of Logic Transformations
   A 16-Element $S_{2 \times 2 \times 2 \times 2}$ Group of Logic Transformations
Conclusion
See also
References

Keywords

Checklist paradigm; Many-valued logics; GUHA; Exploratory statistical analysis; Semantics of MVL connectives; Generalized quantifier; Observational quantifiers; Fuzzy sets; Logic of approximation; Interval computing; Approximate reasoning

Why the Checklist Paradigm?

The classical logic deals with two logical values—'truth' and 'falsity'. It can be characterised algebraically and semantically by a *Boolean algebra*. Not all issues of logic can, however, be settled by a system of classical two-valued logic. For example some modalities, such as necessity and possibility cannot, in general, be expressed in any system that admits only a finite number of logical values. Also some temporal logics [33] characterising time require an infinite number of logical values in their semantics.

Fuzzy Logics

Many-valued logic algebras are needed for developing the mathematics of *fuzzy relations* [28] and sets [18]. For example, in order to compute the degree δ to which two fuzzy sets intersect, we use the formula $\delta_{A \cap B}(x) = \delta_A(x) \wedge \delta_B(x)$, where $\wedge$ is a many-valued 'AND' connective and $\delta_A(x), \delta_B(x)$ are some logical values: either truth-values, possibilities, probabilities, etc. Depending on the *epistemological interpretation* of the logical values, we read the statement $\delta(A)(x)$ 'The degree to which it is true that $x \in A$', 'The degree to which it is possible that $x \in A$', 'The degree to which it is probable that $x \in A$', etc.

Computing the *degree of inclusion* of two sets [2] is done by the formula $\delta(A \subseteq B) = (\forall x)\delta_A(x) \rightarrow \delta_B(x)$, where x ranges over elements of the universe U from which the elements of A and B are drawn. Here $\rightarrow$ is a many-valued *implication operator*.

Approximate Reasoning

Many-valued logic systems are also required for algebraic characterization of logics of *approximate reasoning*. The *premises of an inference* (i.e. the antecedent formulas that form the arguments of the rules of approximate inference) are used by the rules to generate the succedent formulas — the *conclusion*(s).

If each of these logic formulas attains as its logic value a single value from some *lattice*, we speak of a *point-based logic* system of approximate reasoning. If the logic value is a whole interval $[\delta_k, \delta_l]$ such that $\delta_k \leq \delta_l$ it is an *interval logic*.

Hence, many-valued logics play a key role in all the areas of mathematics and logic discussed above. There is not one many-valued logic, there is an infinite number of families of logic systems of various kinds. Hence, according to the purpose of its use, one has to choose an appropriate many-valued system. But even after the choice is made, the two key questions still remain:
- Where the logic values come from?
- Is there any basic epistemic or semantic procedure by which the basic logic connectives can be meaningfully derived?

These questions are answered by the *checklist paradigm*.

Many-Valued Logics in Fuzzy Sets

The theory of fuzzy sets and relations requires a many-valued logic in which to manipulate the degrees of truth which attach to fuzzy statements. As in classical two-valued logic (in which the statements are judged to be either utterly true or utterly false), one wishes a *truth-functional* connection between the truth values assigned to 'p' and to 'q' and those to be assigned to 'p or q' and 'p and q' and 'if p then q', as well as to 'not-p' and 'not-q', that is, one wishes the evaluation of the derived formulas to depend solely on the evaluation of the original formulas, without further reference to their contents.

There are a number of such many-valued logical systems, with truth values in the closed real interval $[0, 1]$. Everyone agrees that the values assigned in the crisp 'corners', where the values $|p|$ of p and $|q|$ of q are zero (false) or one (true), must accord with the classical Boolean logic. Most agree in setting

$$|\text{not} - p| = |\neg p| = 1 - |p|$$

and the most usual 'or' and 'and' connectives are given by

$$|p \text{ or } q| = |p \vee q| = \max(p, q),$$
$$|p \text{ and } q| = |p \wedge q| = \min(p, q),$$

although other have been proposed and have something to be said for them.

Selecting max and min as the functions for computing the logical values of the *connectives* $\vee$ and $\wedge$ does not yet determine the system of many-valued logic fully. Indeed, a number of different systems employ these. Third determining factor is the choice the *impli-*

cation operator $\to$. Some frequently used $\to$ are listed below.

Checklist Paradigm Semantics for Fuzzy Logics, Table 1
Some important many-valued implication operators

No	Opr.	Definition
2.	S	Standard Strict $a \xrightarrow{2} b = \begin{cases} 1, & a \leq b \\ 0, & \text{otherwise} \end{cases}$
3.	S*	Gödel $a \xrightarrow{3} b = \begin{cases} 1, & a \leq b \\ b, & \text{otherwise} \end{cases}$
4.	G43	product ply (also: Goguen–Gaines) $a \xrightarrow{4} b = \min\left(1, \frac{b}{a}\right)$
4'.	G43'	Modified G43 $a \xrightarrow{4'} b = \min\left(1, \frac{b}{a}, \frac{1-a}{1-b}\right)$
5.	L	Lukasiewicz $a \xrightarrow{5} b = \min(1, 1 - a + b)$
5.5	KDL	Reichenbach $a \xrightarrow{5.5} b = \min(1, 1 - a + ab)$
6.	KD	Kleene–Dienes $a \xrightarrow{6} b = (1 - a) \vee b$
7.	EZ	Early Zadeh $a \xrightarrow{7} b = (a \wedge b) \vee (1 - a)$ $= (a \xrightarrow{6} b) \wedge ka$ where $ka = (1 - a) \vee a$
8.	W	Willmott $a \xrightarrow{8} b = (a \xrightarrow{7} b) \wedge kb$

Not only properties of the many-valued logic systems but also of the systems of *fuzzy sets* crucially depend on the choice of the implication operator. For example both the definition of a *fuzzy power set* (i. e. the set of all subsets) and of the *fuzzy set-inclusion operator* depend on its choice. The very first paper on fuzzy sets by L.A. Zadeh [44,45] uses max and min connectives to define the intersection ∩ and the union ∪ of two fuzzy sets. The set inclusion operator Zadeh defines by the formula

$$\mu(A \subseteq B) = 1 \quad \Leftrightarrow \quad (\forall x)\mu_A \leq \mu_B(x).$$

Using the 'Standard Strict' $\to_2$ in the formula given of the first section above we obtain

$$\delta(A \subseteq B)$$
$$= (\forall x)\delta_A(x) \to \delta_B(x)$$
$$= (\forall x)\mu_A(x) \xrightarrow{2} \mu_B(x)$$
$$= \min_{\{x \in U\}} (\mu_A(x) \xrightarrow{2} \mu_B(x)).$$

This formula is equivalent to Zadeh's early definition of fuzzy set inclusion which in fact is crisp (nonfuzzy). Power set theories with proper fuzzy set inclusion have been first investigated in [2,43] (using the *implication operators* listed in the table above).

Since 1965, when the first paper on fuzzy sets was written by Zadeh, not only max and min but also other many-valued logic connectives were used to define the union ∪ and the intersection ∩ of fuzzy sets. An important pair are the so called 'bold connectives': '$a \wedge_5 b = \max(0, a + b - 1)$' and '$a \vee_5 b = \min(1, a + b)$'. As the subscript indicates, these are related to the Łukasiewicz *implication* operator. These represent the so-called *MV algebras* which play an important role in application of fuzzy sets in quantum logics [35] and elsewhere. Both types of connectives, the pairs 'max-min' and the 'bold' connectives are special instances of the so-called triangular norms (*t-norms*) and conorms (*t-conorms*) [17,36]. These associative operations with special properties, defined on [0, 1], algebraically characterise the whole infinite family of OR-AND pairs of many-valued logic connectives and play a crucial role in the theory and applications of fuzzy sets.

The Checklist Paradigm

The *checklist paradigm* provides the mechanism by which several types of very different families of many-valued logic connectives *emerge* from some more basic considerations.

- It provides the *semantics* of systems that use single value as its logic value.
- It provides the justification for *interval logics*.
- It provides a link of *many-valued logics* connectives with *generalized quantifiers*.

Mathematics of the Checklist Paradigm

A *checklist template* Q is a finite family of properties $\langle P_1, \ldots, P_n \rangle$. With a template Q, and a given *proposition A*, one can associate a specific checklist $Q_A = \langle Q, A \rangle$. A *valuation* f_A of a checklist Q_A is a function from Q to $\{0, 1\}$.

The value a_Q of the proposition A with respect to a template Q (which is the summarised value of the valuation f_A) is given by the formula

$$a_Q = \sum_{i=1}^{n} p_i^A$$

where $n = \text{card } Q$ and $p_i^A = f_A(P_i)$.

A *fine valuation structure*, a pair of propositions A, B with respect to the template Q, is a function $f_{A,B}^Q$ from Q into $\{0, 1\}$ assigning to each attribute P_i the ordered pair of its values $\langle p^A, p^B \rangle$.

Let $\alpha_{j,k}$ be the cardinality of the set of all attributes P_i such that $f_{A,B}^Q(P_i) = \langle j, k \rangle$.

Obviously we have the following constraint on the values: $\alpha_{00} + \alpha_{01} + \alpha_{10} + \alpha_{11} = n$. Further, we define $r_0 = \alpha_{00} + \alpha_{01}$, $r_1 = \alpha_{10} + \alpha_{11}$, $c_0 = \alpha_{00} + \alpha_{10}$, $c_1 = \alpha_{01} + \alpha_{11}$.

These entities can be displayed systematically in a *contingency table*. In such a table, the inner fine-summarization structure consists of the four $\alpha_{j,k}$ appropriately arranged, and of margins c_0, c_1, r_0, r_1 (see Fig. 1).

Now let F be any logical propositional function of propositions A and B. For $i, j \in \{0, 1\}$, let $f(i, j)$ be the classical truth value of F for the pair i, j of truth values; let $u(i, j) = \alpha_{i,j}n$, the ratio of the number in the ij-cell of the constraint table, to the grand total. Then we define the (nontruth-functional) *fuzzy assessment of the truth* of the proposition $F(A, B)$ to be

$$m(F(A, B)) = \sum_{i,j} f(i, j) \cdot u_{ij}.$$

	No for B	Yes for B	Row total
No for A	α_{00}	α_{01}	r_0
Yes for A	α_{10}	α_{11}	r_1
Column Total	c_0	c_1	n

Checklist Paradigm Semantics for Fuzzy Logics, Figure 1
Checklist paradigm of the assignment of fuzzy values. Define: $a = r_1/n$; $b = c_1/n$

This assessment operator will be called the *contraction/approximation measure*.

The four interior cells $\alpha_{00}, \alpha_{01}, \alpha_{10}, \alpha_{11}$ of the constraint table constitute its *fine structure*; the margins r_0, r_1, c_0, c_1 constitute its *coarse structure* (see Fig. 1).

The fine structure gives us the appropriate fuzzy assessments for all propositional functions of A and B; the coarse structure gives us only the fuzzy assessments of A and B themselves. Our central question is:

to what extent can the fine structure be reconstructed from the coarse?

As shown elsewhere [3,5,6,8] the coarse structure imposes bounds upon the fine structure, without determining it completely. Hence, associated with the various logical connectives between propositions are their extreme values.

There are four extremes that the fine structure of the contingency table (see Fig. 1) can attain [6,8]:
i) the two *mindiag fine structures* with the diagonal values minimized ($\alpha_{00} = 0$ or $\alpha_{11} = 0$); and
ii) the two *maxdiag fine structures* with the diagonal values maximized ($\alpha_{01} = 0$ or $\alpha_{10} = 0$).

Thus we obtain the inequality restricting the possible values of $m(F)$:

$$\text{con top} \geq m(F) \geq \text{con bot},$$

where 'con' is the name of connective represented by $f(i, j)$. Choosing for the logical type of the connective 'con' the *implication* and making the assessment of the fuzzy value of the truth of a proposition by the formula $m_1(F) = 1 - u_{10}$ we obtain:

$$\min(1, 1 - a + b) \geq m_1(A \to B) \geq \max(1 - a, b).$$

We can see that the checklist paradigm generated the *Łukasiewicz implication* operator, and the *Kleene-Dienes implication* operator.

We have already noted that choosing for 'F' the connective type 'AND' [5,8]) and m_1, we obtain the bounds

$$\min(a, b) \geq m_1(\text{AND}) \geq \max(0, a + b - 1).$$

These bounds are formally identical with those of B. Schweitzer and A. Sklar [36] giving the bounds on *copulas* which play an important role in their theory of *t-norms* and *t-conorms*. Surprisingly, these checklist paradigm bounds also coincide with Novák's recent (1991) derivation [31] of bounds on fuzzy sets approximating classes of Vopěnka's alternative set theory [41,42]. E. Hisdal derives the same inequalities as the bounds on some connectives of her TEE model and comments on a possible link (cf. [16, Appendix A2]). In the context of modalities in fuzzy logics, checklist paradigm-like inequalities for $F = \{AND, OR\}$ were recently (1992) also independently discovered in [34]. Yet all these models are neither formally nor epistemologically identical. This indicates the need for a more precise meta- and metamatalogical formulation of many-valued based mathematical systems, that would include in their full definition a part formulating their 'mathematical epistemology'.

Interval Inference and The Checklist Paradigm

The checklist paradigm puts ordering on the pairs of distinct implication operators and other pairs of connectives. Hence it provides a theoretical justification of *interval-valued approximate inference*. For the m_1 contraction/approximation measure, there are 16 inequalities linking the *TOP and BOT types of connectives* [5,8], thus yielding 16 logical types of TOP-BOT pairs of connectives. Ten of these interval pairs generated by m_1 are listed in Table 2.

Other Systems of Fuzzy Logic Connectives for Interval Inference

In Boolean (crisp) logic, the values of a logical formula written in the *disjunctive normal form* (DNF) are equal to the values the formula expressed in the *conjunctive normal form* (CNF). This does not hold for every system of many-valued connectives.

I.B. Türksen [38,39,40] has shown that for $\max(a, b)$, $\min(a, b)$ and some other *t*-norm and *t*-conorm based CNFs and DNFs the inequality DNF(a CON b) $\le$ CNF(a CON b) holds for all 16 basic many-valued connectives CON.

Taking for example the max-min based CNF and DNF, the corresponding implications are given by

$$\text{CNF}(a \to b) = (\neg a \vee b),$$
$$\text{DNF}(a \to b) = (a \wedge b) \vee (\neg p \wedge q) \vee (\neg a \wedge \neg b).$$

Checklist Paradigm Semantics for Fuzzy Logics, Table 2
Two-argument interval pairs of connectives generated by m_1

Logical Type of Connective	Valuation BOTTOM$\le$TOP
AND $a\&b$	$\max(0, a+b-1)$ $\le \min(a, b)$
Nicod $a \downarrow b$	$\max(0, 1-a-b)$ $\le \min(1-a, 1-b)$
Sheffer $a\|b$	$\max(1-a, 1-b)$ $\le \min(1, 2-a-b)$
OR $a \vee b$	$\max(a, b)$ $\le \min(1, a+b)$
Nonimplication $a \not\leftarrow b$	$\max(0, b-a)$ $\le \min(1-a, b)$
Nonimplication $a \not\rightarrow b$	$\max(0, a-b)$ $\le \min(a, 1-b)$
Implication $a \leftarrow b$	$\max(a, 1-b)$ $\le \min(1, 1+a-b)$
Implication $a \to b$	$\max(1-a, b)$ $\le \min(1, 1-a+b)$
Equivalence $a \equiv b$	$\max(1-a-b, a+b-1)$ $\le$ $\min(1-a+b, 1+a-b)$
Exclusive OR $a \oplus b$	$\max(a-b, b-a)$ $\le$ $\min(2-a-b, a+b)$

For further information on other systems of connectives for *fuzzy interval inference* see [11,27].

Optimization of Interval Inference

Formulas that are equivalent logically [6,27] may not be equivalent when compared by their formula complexity. This is well-known phenomenon when expressing logical formulas in DNF or CNF normal forms [19]. The same logical function expressed in one of these forms may have more complicated expression in the other normal form. Similarly, this can be observed with other logic connectives [40]. So, transformations between logically equivalent formulas expressed by different connectives may have different formula complexity. Hence, the knowledge of such transformations is useful in optimization of interval inference.

For example, *exclusive OR*, or the *eor operator*, is conveniently defined in two ways: as '*a* without *b*' or

'b without a', or else as 'a or b but not both', thus

$$a \text{ eor } b = (a \text{ and } \neg b) \text{ or } (\neg a \text{ and } b)$$
$$= (a \text{ or } b) \text{ and } (\neg a \text{ and } b).$$

Using these definitions together with the definitions of the previous section easy calculations bring the results of the following Table [6]:

$a \equiv_{top} b$	$= (a \rightarrow_{top} b) \wedge_{top} (b \rightarrow_{top} a)$
	$= (a \wedge_{top} b) \vee_{top} (\neg a \wedge_{top} \neg b)$
$a \equiv_{bot} b$	$= (a \rightarrow_{bot} b) \wedge_{bot} (b \rightarrow_{bot} a)$
	$= (a \wedge_{bot} b) \vee_{bot} (\neg a \wedge_{bot} \neg b)$
$a \oplus_{top} b$	$= (a \wedge_{top} \neg b) \vee_{top} (\neg a \wedge_{top} b)$
	$= (a \vee_{top} b) \wedge_{top} \neg(a \wedge_{bot} b)$
$a \oplus_{bot} b$	$= (a \wedge_{bot} \neg b) \vee_{bot} (\neg a \wedge_{bot} b)$
	$= (a \vee_{bot} b) \wedge_{bot} \neg(a \wedge_{top} b)$

Formulas for equivalence (IFF) and exclusive-OR (EOR)

Other useful formulas are those that give universal bounds on classes of *fuzzy interval pairs* of formulas. If we define the *unnormalized fuzziness* of x [1] as $\phi x = \min(x, 1 - x)$, then for x in the range $[0, 1]$, ϕx is in the range $[0, 0.5]$, with value 0 if and only if x is crisp, and value .5 if and only if x is .5. The following *gap theorem* holds [8]:

Theorem 1

$$(a \wedge_{top} b) - (a \wedge_{bot} b)$$
$$= (a \vee_{top} b) - (a \vee_{bot} b)$$
$$= (a \rightarrow_{top} b) - (a \rightarrow_{bot} b)$$
$$= \min(\phi a, \phi b).$$
$$(a \equiv_{top} b) - (a \equiv_{bot} b)$$
$$= (a \oplus_{top} b) - (a \oplus_{bot} b)$$
$$= 2\min(\phi a, \phi b).$$

The width of the interval produced by an application of a pair of associated connectives (i. e. TOP and BOT connectives) characterises the margins of imprecision of an interval logic expression. Because the interval between the TOP connective and the BOT connective is directly linked to the concept of fuzziness ϕ, the margins of imprecision can be directly measured by the degree of ϕ.

Other Systems of Checklist Paradigm Connectives

Several measures other than m_1 that yield interesting results are also important. For implication again, but only the evaluation 'by performance' (that is, we are only concerned with the cases in which the evaluation of A is 1; see Fig. 1, we use $m_2 = u_{11}/u_{10} + u_{11}$) and obtain the inequality

$$\min\left(1, \frac{b}{a}\right) \geq m_2(F) \geq \max\left(0, \frac{a+b-1}{a}\right),$$

in which the left-hand side is the well-known *Goguen–Gaines implication* (cf. e. g. [3]). Still another *contracting measure* which distinguishes the proportion of satisfactions 'by performance', $u(1, 1)$, and 'by default', $u(0, 0) + u(0, 1)$. This measure given by the formula $m_3 = u_{11} \vee (u_{00} + u_{01})$ yields [3]

$$\max[\min(a, b), 1 - a] \geq m_3(F)$$
$$\geq \max(a + b - 1, 1 - a).$$

Two variations on measure m_3 have turned out to be of interest [3]. One is its lower *contrapositivization* given by the formula

$$m_4 = (u_{11} \vee (u_{00} + u_{01})) \vee (u_{00} \vee (u_{01} + u_{11}))$$

which gives the following inequality:

$$\min[\max(a + b - 1, 1 - a), \max(b, 1 - a - b)]$$
$$\leq m_4 \leq \min[\max(1 - a, b), \kappa a, \kappa b],$$

where $\kappa a = a \vee (1 - a)$.

The other arises by taking for the 'performance' part the less conservative m_2 thus obtaining the formula for $m_5 = m_2 \vee (u_{00} + u_{11})$. This yields

$$\max\left[\min\left(1, \frac{b}{a}\right), 1 - a\right] \geq m_5$$
$$\geq \max\left[\frac{a+b-1}{a}, 1 - a\right].$$

For the proofs of the results presented in this subsection and further explanation see [6, Sect. 5] (this is the first paper on the checklist paradigm, published in 1980).

Collapse of Intervals into Points Under the Additional Probabilistic Constraints

When only the row and column totals r_i, c_j of the fine structure are known (see Fig. 1), one can ask what are the expected values for the α_{ij} [3, Sect. 7].

Suppose the ways in which numbers can be distributed within the cells of the fine structure (so as to give the fixed coarse totals) constitute a hypergeometric distribution. Then the means of the distribution for each cell, give the *expected* configuration of the fine structure. The inequalities determining the interval BOTCON $\leq$ TOPCON now turn into equalities:

$$\frac{\alpha_{ij}}{\alpha_{ik}} = \frac{c_j}{c_k}; \qquad \frac{\alpha_{ij}}{\alpha_{hk}} = \frac{r_i}{r_h}.$$

Surprisingly, introducing the expected value (a probabilistic notion) this way causes the fuzzy interval to collapse into a single point: the expected value thus generating the values of the *mid connective* [3,5]. For example, the interval pair ($\rightarrow_5$, $\rightarrow_6$) generated by m_1 consisting of the Łukasiewicz implication and Kleene–Dienes implication operators collapses into the *Reichenbach implication* operator.

For further details on the *'probabilistic collapse'* of the interval pairs generated by other measures see [3,5,8,20,26].

Checklist Paradigm and Generalized Quantifiers

Some of the notions and results of the checklist paradigm are in a remarkable relation to the theory of observational generalized quantifiers, as studied by P. Hájek and T. Havránek [14] in connection with the method of *automated hypothesis formation* [12,13]. Namely, a particular type of implication operator and a particular type of implicational quantifier are mutually definable. The link is given by the contingency tables of the checklist paradigm and the statistics of the observational quantifiers [15].

Checklist Paradigm and Four Modes of Reasoning

Classical two-valued logic has presented certain modes of reasoning, of which only two concern us: *modus ponens* and *modus tollens*, respectively.

The first of these derives from the two premises 'if *a* then *b*' and '*a*', the conclusion '*b*'; the second derives from 'if *a* then *b*' and 'not *b*', the conclusion 'not *a*'. The validity of these modes is trivial.

On the other hand, there are two modes of reasoning which are classically illegitimate, although in the daily life we all use something very much like them all the time. These, so-called *plausible rules* [4,32], are shown as the central pair in [26, Fig. 1]. *Denial* derives from 'if *a* then *b*' and 'not *a*', the assertion 'not *b*', while *confirmation* derives from 'if *a* then *b*' and '*b*', the assertion '*a*'.

The reason why these errors in classical reasoning retain a strong intuitive attraction is that most human reasoning does not deal with crisp or two-valued or Boolean truth-versus-falsity, but with graded *degrees of credence*, or belief-worthiness, or whatever you like to call it. Because in multiple-valued logics the plausible modes *gain legitimacy* this intuition about human reasoning gains mathematical legitimacy. Indeed, human reasoning, 'good' human reasoning, is best modeled in multiple-valued logic which admits in addition to the modus ponens and modus tollens also the two modes of plausible reasoning.

In classical logic, an *evaluation* takes each of a given set of propositions into one of the extreme truth-values 0 (false) or 1 (true), subject to some semantic consistency rules. In multiple-valued logic, an *evaluation* is a mapping of the set of propositions into a somewhat richer set, which for present purposes may be taken to be the closed interval [0, 1] from 0 to 1, again subject to certain semantic consistency rules.

Hence, in multiple-valued logic, for any fixed choice among the distinguished implication operators, to the classically valid modes of modus ponens and modus tollens are to be added fuzzily valid modes of denial and confirmation (*modus negans* and *modus confirmans*) [4,7]. Although the out-of-bounds constraints were addressed elsewhere [4], one may wonder, what does the checklist paradigm have to offer when applied to the four plausible modes of inference.

Checklist paradigm is applicable not only to the components of the object language, such as logical operators and connectives, but also at the meta-level, thus providing an interval logic based semantics for various rules of inference. As shown below and in [8], it also provides a justification and the proofs of validity of nonclassical interval-based rules (plausible modes) of reasoning called denial and confirmation (modus negans and modus confirmans) [4,7,8,24].

As already mentioned, these do not have a nontrivial analogy in Boolean crisp logic. Thus we have the following theorems.

Theorem 2 (*checklist modus ponens*) Given $r = m(A \to B)$ and $a = m(A)$ satisfying the consistency condition $r \geq 1 - a$, the values of $b = m(B)$ are subject to

$$r - (1 - a) \leq b \leq r. \tag{1}$$

Theorem 3 (*checklist confirmation*) Given $r = m(A \to B)$ and $b = m(B)$ subject to the consistency condition $b \leq r$, the values of $a = m(A)$ are subject to

$$1 - r \leq a \leq 1 - (r - b). \tag{2}$$

Theorem 4 (*checklist modus tollens*) Given $r = m(A \to B)$ and $\neg b = m(\text{not} - B)$ satisfying the consistency condition $r \geq b$, the values of $\neg a = m(\text{not} - A)$ are subject to

$$r - b \leq \neg a \leq r. \tag{3}$$

Theorem 5 (*checklist denial*) $r = m(A \to B)$ and $\neg a = m(\text{not} - A)$ subject to the consistency condition $1 - a \leq r$, the values of $\neg b = m(\text{not} - B)$ are subject to

$$1 - r \leq \neg b \leq 2 - (r + a). \tag{4}$$

Group Transformations of Logic Connectives and the Checklist Paradigm

Let us recall that a *realization of an abstract group* is any group of concretely realizable operations which has the same algebraic structure as the given abstract group. It is well known that any abstract group can be concretely realized by a family of permutations. So a specific abstract group provides a global structural characterization of a specific family of permutations that concretely represent this abstract group. This idea can be used for global characterization of logic connectives.

The Piaget Group of Transformations

Such a global characterization of two-valued connectives of logic was first given by Piaget in the context of studies of human cognitive development. J. Piaget and his collaborators have shown that an important role in child's mental development is transition from more concrete to more abstract thinking. This transition plays a role in development of intelligence, which is viewed in the Piagetian setup as a transition from totally ambiguous and vague notions to crisp propositions in two-valued logic.

Given a family of logical connectives one can apply to them various transformations. Individual logic connectives are 2-argument logic functions. Transformations are functors that, taking one connective as the argument will produce another connective.

Let 4 transformations on basic propositional functions $f(x, y)$ of 2 arguments be given as follows:

$$I(f) = f(x, y), \quad D(f) = \neg f(\neg x, \neg y),$$
$$C(f) = f(\neg x, \neg y), \quad N(f) = \neg f(x, y).$$

In 1940, Piaget discovered experimentally a specific concrete form of such transformations. In the set of the above transformations $T_p = \{I, D, C, N\}$ these individual transformations are called *identity, dual, contradual, negation transformation*, respectively.

It has been shown that the *Piaget group of transformation* is satisfied by some many-valued logics (cf. [5,6,10,37]).

The system of connectives

$$\{\equiv_{\text{TOP}}, \oplus_{\text{BOT}}, \equiv_{\text{BOT}}, \oplus_{\text{TOP}}\}$$

obeys the Piaget group of transformations. Hence it possesses the abstract structure of the *Klein 4-element group*.

An 8-Element Group of Logic Transformations

Adding new nonsymmetrical transformations to those defined by Piaget enriches the algebraic structure of logic transformations. In 1979 L.J. Kohout and W. Bandler added the following nonsymmetric operations [22,23]:

$$LC(f) = f(\neg x, y), \quad RC(f) = f(x, \neg y),$$
$$LD(f) = \neg f(\neg x, y), \quad RD(f) = \neg f(x, \neg y)$$

to the above defined four symmetrical transformations. This yields a new 8-element group of transformations.

The abstract 8-element group $[T, *]$ that captures the structure of the above defined logic transformations is also commutative and is called the *symmetric $S_{2 \times 2 \times 2}$ group* in the standard terminology of group theory. The interval logic system based on m_1 can be characterized by such groups of transformations.

Given a set of connectives CON and a set of transformations $\mathcal{T}$, we say that $T_{\text{CON},\mathcal{T}} = \mathcal{T}(\text{CON})$ is the set of connectives generated by the application of $\mathcal{T}$ to the set CON. For example, $a \to_5 b$ will generate such a set of connectives. This generated set is a realization of $S_{2\times 2\times 2}$.

A 16-Element $S_{2\times 2\times 2\times 2}$ Group of Logic Transformations

The implication operators $a \to_5 b$ and $a \to_6 b$ yielding the measure m_1 are contrapositive. This means that their valuations satisfy the semantic equality $a \to b = \neg b \to \neg a$.

If we are interested in extending the interval logics into the domain of noncontrapositive $\to$ then the corresponding $\wedge$ is not commutative. In order to distinguish the contrapositive cases from noncontrapositive ones in a syntactically correct formal way, an additional operator is introduced.

This operator called, *commutator K*, satisfies the equality $a * b = K(b * a)$. The commutativity as well as the contrapositivity involves restrictions on transformations of connectives. In the abstract group, these restrictions are expressed abstractly as congruences [30]. It is convenient to express such restrictions equationally. For any contrapositive $\to$, the following equalities hold: $C[K(a \to b)] = K[C(a \to b)] = a \to b$. For a noncontrapositive $\to$, (1) fails, but the following equality holds:

$$(K(C(K(C(a \to b))))) = a \to b.$$

The following holds [29]:

Theorem 6 *The closed set of connectives generated by $\{\to_4, \leftarrow_4, K\}$ is a representation of the symmetric 16 element abstract group $S_{2\times 2\times 2\times 2}$.*

Conclusion

The checklist paradigm clearly demonstrates the following general meta-principle: a system of logic connectives is formed by a specific family of connectives together with some common process/structure/principles that involve the said family of connectives in some unifying way, causing these to interact.

In the checklist paradigm semantic model we use two basic unifying principles:
i) approximation (contraction) measures;
ii) transformations of logical types of connectives leading to a global characterization of logics by their groups of transformations [23].

The methods of the checklist paradigm surveyed here give the theoretical bounds on the performance of particular many-valued implication operators and other connectives by deriving these from deeper epistemological and formal assumptions. Hence, it provides a theoretical justification of interval-valued approximate inference. The checklist paradigm, together with fuzzy questionnaires and square and triangle relational products also plays an important role in the experimental identification of fuzzy membership functions and structures [21,25] (see also ▶ Boolean and fuzzy relations). The results can be extended to the groupoid-based many-valued *Pinkava algebras* (see ▶ Finite complete systems of many-valued logic algebras) that are used in the design of knowledge-based and other systems [19]. This theoretical work is supplemented by empirical studies of the adequacy of various logical connectives in practical applications of fuzzy sets and relations [9].

See also

- ▶ Alternative Set Theory
- ▶ Boolean and Fuzzy Relations
- ▶ Finite Complete Systems of Many-Valued Logic Algebras
- ▶ Inference of Monotone Boolean Functions
- ▶ Optimization in Boolean Classification Problems
- ▶ Optimization in Classifying Text Documents

References

1. Bandler W, Kohout LJ (1978) Fuzzy relational products and fuzzy implication operators. Internat. Workshop on Fuzzy Reasoning Theory and Appl. Queen Mary College Univ. London, London
2. Bandler W, Kohout LJ (1980) Fuzzy power sets and fuzzy implication operators. Fuzzy Sets and Systems 4:13–30. Reprinted in: Dubois D, Prade H, Yager R (eds) (1993) Readings in Fuzzy Sets for Intelligent Systems. Morgan Kaufmann, San MAteo, CA, pp 88–96
3. Bandler W, Kohout LJ (1980) Semantics of implication operators and fuzzy relational products. Internat J Man-Machine Studies 12:89–116 Reprinted in: Mamdani EH, Gaines BR (eds) (1981) Fuzzy Reasoning and its Applications. Acad Press, New York, pp 219–246

4. Bandler W, Kohout LJ (1984) The four modes of inference in fuzzy expert systems. In: Trappl R (ed) Cybernetics and Systems Res, vol 2. North-Holland, Amsterdam, pp 581–586
5. Bandler W, Kohout LJ (1984) Unified theory of multiple-valued logical operators in the light of the checklist paradigm. In: Proc 1984 IEEE Conf Systems, Man Cybern, IEEE, New York, pp 356–364
6. Bandler W, Kohout LJ (1985) The interrelations of the principal fuzzy logical operators. In: Gupta MM Kandel A, Bandler W, Kiszka JB (eds) Approximate Reasoning in Expert Systems. North-Holland, Amsterdam, pp 767–780
7. Bandler W, Kohout LJ (1985) Probabilistic vs. fuzzy production rules in expert systems. Internat J Man-Machine Studies 22:347–353
8. Bandler W, Kohout LJ (1986) The use of checklist paradigm in inference systems. In: Negoita CV, Prade H (eds) Fuzzy Logic in Knowledge Engineering. TÜV Rheinland, Köln, pp 95–111
9. Ben-Ahmeida B, Kohout LJ, Bandler W (1992) The use of fuzzy relational products in comparison and verification of correctness of knowledge structures. In: Kohout LJ, Anderson J, Bandler W (eds) Knowledge-Based Systems for Multiple Environments. Ashgate Publ. (Gower), Aldershot, Hampshire, UK
10. Dubois D, Prade H (1980) Fuzzy sets and systems: Theory and applications. Acad Press, New York
11. Dubois D, Prade H (1991) Fuzzy sets in approximate reasoning, Part I: Inference with possibility distributions. Fuzzy Sets and Systems 40(1):143–202
12. Hájek P (1978) Special issue on GUHA method. Internat J Man-Machine Studies 10(1):1–93 (guest editor).
13. Hájek P (1981) Special issue on GUHA method. Internat J Man-Machine Studies 15: (guest editor)
14. Hájek P, Havránek T (1978) Mechanizing hypothesis formation: Mathematical foundations of a general theory. Springer, Berlin
15. Hájek P, Kohout LJ (1996) Fuzzy implications and generalized quantifiers. Internat J Uncertainty, Fuzziness and Knowledge Based Systems 4(3):225–233
16. Hisdal E (1990) Infinite-valued logic based of two-valued logic and probability. Pt. 1.4 The TEE model for grades of membership. Inst. Informatics Univ. Oslo Oslo, Oslo
17. Klement EP, Mesiar R (1997) Triangular norms. In: Mesiar R, Riečan B (eds) Tatra Mountains, Math. Publ. (Special Issue: Fuzzy Structures – Current Trends). 13, Math Inst Slovak Acad Sci Bratislava, Bratislava, pp 169–194
18. Klir GJ, Yuan B (1995) Fuzzy sets and fuzzy logic: Theory and applications. Prentice-Hall, Englewood Cliffs, NJ
19. Kohout LJ (1990) A perspective on intelligent systems: A framework for analysis and design. Chapman and Hall and v. Nostrand, London–New York
20. Kohout LJ (1995) Epistemological aspects of many-valued logics and fuzzy structures. In: Höhle U, Klement EP (eds) Non-Classical Logics and their Applications to Fuzzy Subsets: A Handbook of Mathematical Foundations of Fuzzy Set Theory. Kluwer, Dordrecht, pp 291–339
21. Kohout LJ, Anderson J, Bandler W et al. (1992) Knowledge-based systems for multiple environments. Ashgate Publ. (Gower), Aldershot, Hampshire, UK
22. Kohout LJ, Bandler W (1979) Checklist paradigm and group transformations. Techn. Note, Dept. Electrical Engin. Univ. Essex, no. EES-MMS-ckl91.2
23. Kohout LJ, Bandler W (1992) How the checklist paradigm elucidates the semantics of fuzzy inference. Proc. IEEE Internat Conf Fuzzy Systems (1992), IEEE, New York, pp 571–578
24. Kohout LJ, Bandler W (1992) Modes of plausible reasoning viewed via the checklist paradigm. IPMU'92: Proc. Internat. Conf. Information Processing and the Management of Uncertainty in Knowlege-Based Systems, Mallorca, July 6-10 (1992), Univ. Illes Balears, Palma, Mallorca, Spain, Palma de Mallorca, pp 411–414
25. Kohout LJ, Bandler W (1992) Use of fuzzy relations in knowledge representation, acquisition and processing. In: Zadeh LA, Kacprzyk J (eds) Fuzzy Logic for the Management of Uncertainty. Wiley, New York, pp 415–435
26. Kohout LJ, Bandler W (1993) Interval-valued systems for approximate reasoning based on the checklist paradigm. In: Wang P (ed) Advances in Fuzzy Theory and Technology, vol 1, Bookwrights Press, Durham, NC, pp 167–193
27. Kohout LJ, Bandler W (1996) Fuzzy interval inference utilizing the checklist paradigm and BK-relational products. In: Kearfott RB, Kreinovich V (eds) Applications of Interval Computations. Kluwer, Dordrecht, pp 291–335
28. Kohout LJ, Bandler W (2001) A survey of fuzzy and crisp relations. Lecture Notes Fuzzy Math and Computer Sci Creighton Univ, Omaha, NE
29. Kohout LJ, Kim E (1997) Global characterization of fuzzy logic systems with para-consistent and grey set features. In: Wang P (ed) Proc 3rd Joint Conf Inform Sci. JCIS'97 (5th Internat Conf on Fuzzy Theory and Techn), 1, Duke Univ., Durham, NC, pp 238–241
30. Kohout LJ, Kim E (1997) Group transformations of systems of logic connectives. In: Proc. IEEE-FUZ'97, 1, IEEE, New York, pp 157–162
31. Novák V (1991) On the position of fuzzy sets in modelling of vague phenomena. In: Lowen R, Roubens M (eds) IFSA '91 Brussels: Artificial Intelligence, pp 165–167
32. Polya G (1954) Mathematics and plausible reasoning. Princeton Univ. Press, Princeton
33. Prior A (1962) Formal logic. Oxford Univ. Press, Oxford
34. Resconi G, Klir GJ, Clair USt (1992) Hierarchical uncertainty metatheory based upon modal logic. Internat J General Syst
35. Riečan B, Neubrunn T (1997) Integral, measure, and ordering. Kluwer, Dordrecht
36. Schweizer B, Sklar A (1983) Probabilistic metric spaces. North-Holland, Amsterdam

37. Turksen IB (1979) Containment and Klein groups of fuzzy propositions. Working Paper Dept. Industr. Engin. Univ. Toronto 79-010
38. Turksen IB (1986) Interval-valued fuzzy sets based on normal forms. Fuzzy Sets and Systems 20:191–210
39. Turksen IB (1989) Four methods of approximate reasoning with interval-valued fuzzy sets. Internat J Approximate Reasoning 3:121–142
40. Türksen IB (1995) Type I and interval-valued type II fuzzy sets and logics. In: Wang PP (ed) Advances in Fuzzy Theory and Techn, vol 3. Duke Univ, Durham, NC, pp 31–81
41. Vopěnka P (1979) Mathematics in the alternative set theory. Teubner, Leipzig
42. Vopěnka P (1991) The philosophical foundations of alternative set theory. Internat J General Syst (Special Issue of Fuzzy Sets and Systems in Czechoslovakia) 20(1):115–126
43. Willmott R (1980) Two fuzzier implication operators in the theory of fuzzy power sets. Fuzzy Sets and Systems 4(31–36):31–36
44. Zadeh LA (1965) Fuzzy sets. Inform and Control 8:338–353
45. Zadeh LA (1996) Fuzzy sets: Selected papers II. World Sci, Singapore, edited by Klir G, Yuan B

Chemical Process Planning

SHABBIR AHMED, NIKOLAOS V. SAHINIDIS
University Illinois, Urbana-Champaign, USA

MSC2000: 90C90

Article Outline

Keywords
The Long Range Planning Problem
Computational Complexity
Solution Strategies
Integer Programming Approach
Continuous Global Optimization
Approximation Schemes
Dealing with Uncertainty
Conclusion
See also
References

Keywords

Long range planning; Complexity; Mixed integer linear programming; Concave programming; Probabilistic analysis; Stochastic programming

As companies are increasingly concerned about long term stability and profitability, recent years have witnessed growing demand for long range planning tools in all sectors. The chemical process industries are no exception. New environmental regulations, rising competition, new technology, uncertainty of demand, and fluctuation of prices have all led to an increasing need for decision policies that will be 'best' over a long time horizon. Quantitative techniques have long established their importance in such decision making problems. It is therefore no surprise that there is a considerable number of papers in the optimization literature devoted to the problem of long range planning in the processing industries. The purpose of this article is to review recent advances in this area. We will describe the main modeling issues, and discuss the computational complexity, formulations and solution algorithms for this problem.

The Long Range Planning Problem

Consider a plant comprising of several processes to produce a set of chemicals for sale. Each process intakes a number of raw materials and produces a main product along with some by-products. Any of these main or by-products could be the raw materials for another process. Considering the ingredients and final product of all the processes, we have a list of chemicals consisting of all raw materials we consider purchasing from the market, all products we consider offering for sale on the market, and all possible intermediates. The plant can then be represented as a network comprising of nodes representing processes and the chemicals in the list, interconnected by arcs representing the different alternatives that are possible for processing, and purchases to and sales from different markets.

The *process planning problem* then consists of choosing among the various alternatives in such way as to maximize profit. Once we know the prices of chemicals in the various markets and the operating costs of processes, the problem is then to decide the operating level of each process and amount of each chemical to be purchased and sold to the various markets. The problem in itself grows combinatorially with the number of chemicals and processes and is further complicated once we start planning over multiple time periods.

Let us now consider the operation of the plant over a number of time periods. It is reasonable to expect that prices and demands of chemicals in various markets would fluctuate over the planning horizon. These fluctuations along with other factors, such as new environmental regulations or technology obsolescence, might necessitate the decrease or complete elimination of the production of some chemicals while requiring increase or introduction of others. Thus, we have some additional new decision variables: capacity expansion of existing processes, installation of new processes and shut down of existing processes. Moreover, due to the broadening of the planning horizon, the effect of discount factors and interest rates will become prominent in the cost and price functions. Thus, the planning objective should be to maximize the net present value instead of short term profit or revenue. This is the problem that we shall devote our attention to. The problem can be stated as follows: assuming a given network of processes and chemicals, and characterization of future demands and prices of the chemicals and operating and installation costs of the existing as well as potential new processes, we want to find an operational and capacity planning policy that would maximize the net present value.

Computational Complexity

The number of possible alternatives, regarding which processes to expand and when, increases with the number of processes and the number of time periods. Even though this increase is clearly exponential in the number of processes and time periods, it was not until recently that a formal computational complexity characterization was provided for this problem. In particular, the general long range process planning problem has been shown by S. Ahmed and N.V. Sahinidis [3] to be *NP*-hard by identifying two known *NP*-hard problems as special cases.

Consider first a single-process, multiperiod problem where the decisions consist of determining the expansion sequence to satisfy given demands over a number of time periods at a minimum cost. It can be shown that this problem is equivalent to the *NP*-hard *capacitated lot-sizing problem*, where one has to determine production lot sizes to satisfy demands at a minimum cost. Similarly, a multiple-process, single-time period problem, where the decisions are to determine which processes to install to satisfy demand at a minimum cost, can be shown to be equivalent to the *NP*-hard *knapsack problem*, where one has to select items from a set to place into a knapsack such that weight restrictions are not violated and utility is maximized.

Solution Strategies

Some of the early approaches for the long range planning problem were based on dynamic programming as described by S.M. Roberts [18]. A.S. Manne [5,15] used integer programming approaches to account for economies of scale. D.M. Himmelblau and T.C. Bickel [7] presented a nonlinear programming formulation for a hydrodesulfurization process, and I.E. Grossmann and J. Santibez [6] developed a multi period mixed integer linear programming formulation. Y. Shimizu and T. Takamatsu [22] discussed a goal programming approach where in addition to cost minimization, minimizing the number of expansions is also suggested. M. Santiago, O.A. Iglesias and C.N. Pamiagua [21] developed a method to handle nonlinear concave cost functions arising in planning models. A.G. Jimenez and D.F. Rudd [9] presented a recursive mixed integer linear programming technique and applied it to the Mexican petrochemical industry. We next describe some of the more contemporary approaches to these problems.

Integer Programming Approach

Under the assumption of linear mass balances in the processes and fixed charge cost models, Sahinidis et al. [20] developed a mixed integer linear programming (MILP) formulation of the long range process planning problem as described below.

Indices

i	For the set of *NP* processes
j	For the set of *NC* chemicals
l	For the set of *NM* markets
t	For the set of *NT* time periods

Parameters

X_{it}^L, X_{it}^U	Lower and upper bounds on the expansion of process i in period t.
a_{jlt}^L, a_{jlt}^U	Lower and upper bounds on the availability of chemical j in market l in period t.
d_{jlt}^L, d_{jlt}^U	Lower and upper bounds on the demand of chemical j in market l in period t.

$\Gamma_{jlt}, \gamma_{jlt}$ Forecasted buying and selling prices of chemical j in market l in period t.

μ_{ij}, η_{ij} Input and output proportionality constants for chemical j in process i.

α_{it}, β_{it} Variable and fixed cost for the expansion of process i at the beginning of period t.

Variables

X_{it} Capacity expansion of process i at the beginning of period t.

P_{jlt} Amount of chemical j purchased from market l at the beginning of period t.

Q_{it} Total capacity of process i in period t. The capacity of a process is expressed in terms of its main product.

S_{jlt} Units of chemical j sold to market l at the end of period t.

W_{it} Operating level of process i in period t expressed in terms of output of its main product.

y_{it} A 0–1 integer variable. If process i is expanded during period t then $y_{it} = 1$, else $y_{it} = 0$

Formulation

$$\max NPV = \sum_{t=1}^{NT} \left\{ -\sum_{i=1}^{NP}(\alpha_{it}X_{it} + \beta_{it}y_{it} + \delta_{it}W_{it}) + \sum_{j=1}^{NC}\sum_{l=1}^{NM}(\gamma_{jlt}S_{jlt} - \Gamma_{jlt}P_{jlt}) \right\} \quad (1)$$

subject to

$$y_{it}X_{it}^L \leq X_{it} \leq y_{it}X_{it}^U, \quad \forall i, t \quad (2)$$

$$Q_{it} = Q_{it-1} + X_{it}, \quad \forall i, t \quad (3)$$

$$W_{it} \leq Q_{it}, \quad \forall i, t \quad (4)$$

$$\sum_{i=1}^{NP}(\eta_{ij} - \mu_{ij})W_{it} = \sum_{l=1}^{NM}S_{jlt} - \sum_{l=1}^{NM}P_{jlt}, \quad \forall j, t \quad (5)$$

$$a_{jlt}^L \leq P_{jlt} \leq a_{jlt}^U, \quad \forall j, l, t \quad (6)$$

$$d_{jlt}^L \leq S_{jlt} \leq d_{jlt}^U, \quad \forall j, l, t \quad (7)$$

$$X_{it}, Q_{it}, W_{it} \geq 0, \quad \forall i, t \quad (8)$$

$$y_{it} \in \{0, 1\}, \quad \forall i, t. \quad (9)$$

The objective (1) in the above formulation is to maximize the difference between the sales revenues of the final products and the investment, operating, and raw material costs. Equation (2) is a constraint that bounds capacity expansions. A zero value of y_{it} forces the capacity expansion of process i at period t to zero. If the binary variable equals 1, then the capacity expansion is performed within prescribed bounds. Constraint (3) in the above formulation defines the total capacity available at period t as a sum of capacity available in period $t - 1$ and the capacity expansion at the beginning of period t. The condition that the operating level of any process cannot exceed the installed capacity is modeled by constraint (4). Equation (5) expresses mass balances for chemicals across processes and markets. Constraints (6) and (7) are bounds on the purchase and sales quantities. The nonnegativity and binary restrictions are imposed through constraints (8) and (9). Various extensions of this general model are discussed in the recent survey article [2].

Sahinidis et al. [20] developed strong bounding techniques and cutting planes to be used within a *branch and bound* framework to solve the above problem. The fact that the problem is decomposable in the number of time periods can also be exploited by using *Benders decomposition*. Further improvement of the bounding schemes are suggested by reformulating the problem to exploit lot sizing substructure in [19]. The reformulated problem results in a large number of constraints and variables. In [10], the reformulated problem is projected onto a lower-dimensional space to reduce the number of variables, and is solved using a cutting plane strategy along with branch and bound. Computational results in [10,19] and [20] suggest the following:

- Branch and bound with strong bounding techniques performs much better than Benders decomposition for large problems.
- For small sized problems, the reformulation and projection approach do not provide appreciable gains.
- For large problems, the best approach is to use a cutting plane method based on the projected model.

Continuous Global Optimization

In the MILP model, economies of scale in the investment cost functions were modeled by the introduction of a set of binary decision variables (y_{it}) to impose a fixed charge on the decision to expand in addition to the linear term for variable costs. In reality, variable costs are not directly proportional to expansion quantity. Rather, the investment cost is a concave function because of the presence of quantity discounts. Thus, a more realistic model for the investment cost would be:

$$f(X_{it}) = \begin{cases} 0 & \text{when} \quad X_{it} = 0, \\ \beta_{it} + a_{it}X_{it}^{b_{it}} & \text{when} \quad X_{it} > 0, \end{cases}$$

where $a_{it} > 0$ and $0 < b_{it} < 1$. In this formulation, the integer variables have been discarded and the linear variable cost function has been replaced by a concave function in X_{it} with coefficient a_{it} and exponent b_{it}. Note that this function is discontinuous at $X_{it} = 0$. M.L. Liu, Sahinidis and J.P. Shectman [14] present two formulations using these concave cost functions. In the fixed charge concave programming model (FCP), the linear cost relation is retained but the discrete variables are eliminated by using the following concave function:

$$f(X_{it}) = \begin{cases} 0 & \text{when} \quad X_{it} = 0, \\ \beta_{it} + \alpha_{it}X_{it} & \text{when} \quad X_{it} > 0. \end{cases}$$

In the continuous concave programming model (CCP), the discontinuity at $X_{it} = 0$ is avoided by using the following function:

$$f(X_{it}) = a_{it}X_{it}^{b_{it}}.$$

Both (FCP) and (CCP) are problems with concave objective functions to be minimized over a set of linear constraints. These can be solved by a *concave programming* method based on the branch and bound procedure. Computational experience with these models suggests that the algorithm for (FCP) outperforms the straightforward branch and bound for the MILP formulation.

Approximation Schemes

Despite the success of optimization models and algorithms in solving problems of industrial relevance, the majority of approaches in current industrial-level planning practice are still based on *heuristics* rather than integer programming techniques. However, the performance characterization of these approximate methods is based on empirical evidence and little has been done in the way of analytical investigations. Liu and Sahinidis [12] developed a simple heuristic for the process planning problem. The method is based upon solving the LP relaxation of the MILP, and then shifting capacity expansions from latter periods to earlier periods while maintaining feasibility. Worst-case bounds on the performance of this heuristic have also been developed and *probabilistic analysis* of the heuristic has shown that, under standard assumptions on the problem data, the heuristic solution converges to the optimal solution *almost surely* as the problem size increases. A modification of this heuristic for process planning problems with a restriction on the number of allowed expansions has been presented in [3]. The modified heuristic has been proven to be asymptotically optimal *in expectation*.

Dealing with Uncertainty

Uncertainty is an integral part of the long range process planning problem. In the deterministic models discussed above, it is assumed that all uncertainty has been accounted for in the estimation of the problem parameters. Stochastic models, on the other hand, provide explicit means of handling parameter uncertainties.

In process planning problems under uncertainty, the decision maker is interested in a plan that optimizes some sort of a stochastic objective. Two most common such objective functions in the literature are the expected cost/profit of the plan and the plan's flexibility.

Problems with the expected cost objective have been formulated as *two-stage stochastic linear programs* (2S-SLP). In such problems, the uncertain parameters are treated as random variables with known distributions. The desired degree of flexibility of the plan is pre-specified by identifying the probability space over which the plan is required to be feasible. The decision variables of the problem are partitioned into two sets. The *first stage* variables, which are often known as 'design' variables, have to be decided before the actual realization of the random parameters. Subsequently, once the values of the design variables have been decided and the random events presented themselves, further policy improvements can be made by deciding the values of the *second stage* variables, also known as 'control' or 'operating' variables. The choice of the design variables should be such that the first stage costs and the second stage expected costs are minimized. These problems have been solved using decomposition schemes, where the expectation functional over the uncertain parameter space has been approximated using Monte-Carlo sampling [11], successive disaggregation [4] or by Gaussian quadrature [8]. Computational results in [11] show that a combination of Benders decomposition with Monte-Carlo sampling provide optimal or excellent near-optimal solutions. Problems with up to 10 processes, 4 products, 6 chemicals, and with up to 5^{24} scenarios were solved in at most a few CPU minutes on a standard workstation.

From the flexibility objective point of view, one is interested in a plan that maximizes the range of the uncertain parameters over which the plan remains feasible. Problems of this type are typically harder to formulate and require the identification of a suitable measure of flexibility that one can optimize. Such a formulation has been presented in [24] which maximizes their *stochastic flexibility* metric [23] subject to a cost constraint.

The objectives of optimizing cost or profit and maximizing flexibility are typically conflicting. Formulations that combine the objectives by associating a retrofit cost corresponding to design flexibility have been presented in [16,17].

Liu and Sahinidis [13] applied a *fuzzy programming* approach for the problem of process planning under uncertainty. In this model, the uncertain parameters are considered to be fuzzy numbers with a known range of values, and constraints are treated as 'soft,' i. e. some violation is allowed. The degree of satisfaction of the constraints is then measured in terms of *membership functions*, and the objective is to optimize a measure of constraint satisfaction.

The standard stochastic programming formulation does not address the variability of the uncertain recourse costs across the uncertain parameter scenarios. The need for enforcing robustness of these costs is particularly important to a risk aversive planner in a high variability environment. The stochastic programming formulation of the process planning problem has been extended in [1] to account for robustness of the recourse costs through the use of an appropriate variability criterion. In particular, upper partial mean has been proposed as the measure of variability for its intuitive appeal and to avoid nonlinear formulations. These models provide the decision maker with a tool to analyze the trade-off associated with the expected profit and its variability. To overcome the difficulty associated with solving the robust models which include nonseparable terms, a heuristic procedure for the restricted recourse formulation has been developed. This method iteratively enforces recourse robustness while solving the standard stochastic program in each step. The heuristic generates similar but more conservative trade-off frontiers for the profit and its upper partial mean.

Conclusion

The purpose of this article has been to review the recent advances in the use of optimization techniques in long range chemical process planning. Considerable attention has been devoted to the mixed integer linear programming formulation of the problem and efficient solution schemes that exploit the structure of the problem have been developed. Continuous models have also been successfully solved using global optimization techniques. The combinatorial complexity of the problem has recently motivated the need for heuristics and their performance analysis. Some exciting new results have been obtained in this regard. Uncertainty of the problem parameters has been dealt with through stochastic programming and fuzzy programming models. Various different objective criteria including expected value, flexibility and variability have been considered in extensions of the two-stage stochastic programming formulation and a number of efficient algorithms have been developed for industrially relevant problems.

See also

- ▶ Extended Cutting Plane Algorithm
- ▶ Generalized Benders Decomposition
- ▶ Generalized Outer Approximation
- ▶ MINLP: Application in Facility Location-allocation
- ▶ MINLP: Applications in Blending and Pooling Problems

- MINLP: Applications in the Interaction of Design and Control
- MINLP: Branch and Bound Global Optimization Algorithm
- MINLP: Branch and Bound Methods
- MINLP: Design and Scheduling of Batch Processes
- MINLP: Generalized Cross Decomposition
- MINLP: Global Optimization with αBB
- MINLP: Heat Exchanger Network Synthesis
- MINLP: Logic-based Methods
- MINLP: Outer Approximation Algorithm
- MINLP: Reactive Distillation Column Synthesis
- Mixed Integer Linear Programming: Mass and Heat Exchanger Networks
- Mixed Integer Nonlinear Programming

References

1. Ahmed S, Sahinidis NV (1998) Robust process planning under uncertainty. Industr Engin Chem Res 37:1883–1892
2. Ahmed S, Sahinidis NV (1998) Techniques in long range planning in chemical manufacturing systems. In: Leondes CT (ed) Computer Aided and Integrated Manufacturing Systems: Techniques and Applications. Internat Ser Engin, Techn and Applied Sci. Gordon and Breach, New York
3. Ahmed S, Sahinidis NV (2000) Analytical investigations of the process planning problem. Comput Chem Eng 23:1605–1621
4. Clay RL, Grossmann IE (1996) A disaggregation algorithm for the optimization of stochastic planning models. Comput Chem Eng 21(7):751–774
5. Goreux LM, Manne AS (1973) Multi-level planning: Case studies in Mexico. North-Holland, Amsterdam
6. Grossmann IE, Santibez J (1980) Application of mixed-integer linear programming in process synthesis. Comput Chem Eng 4(205)
7. Himmelblau DM, Bickel TC (1980) Optimal expansion of a hydrodesulfurization process. Comput Chem Eng 4(101)
8. Ierapetritou MG, Pistikopoulos EN (1994) Novel optimization approach of stochastic planning models. Industr Eng Chem Res 33:1930–1942
9. Jiminez AG, Rudd DF (1987) Use of a recursive mixed-integer programming model to detect an optimal integration sequence for the mexical petrochemical industry. Comput Chem Eng 3(291)
10. Liu ML, Sahinidis NV (1995) Long range planning in the process industries: A projection approach. Comput Oper Res 23(3):237–253
11. Liu ML, Sahinidis NV (1996) Optimization in process planning under uncertainty. Industr Eng Chem Res 35:4154–4165
12. Liu M, Sahinidis NV (1997) Bridging the gap between heuristics and optimization: The capacity expansion case. AIChE J 43:2289–2299
13. Liu ML, Sahinidis NV (1997) Process planning in a fuzzy environment. Europ J Oper Res 100(1):142–169
14. Liu ML, Sahinidis NV, Shechtman JP (1996) Planning of chemical processes via global concave minimization. In: Grossmann IE (ed) Global Optimization in Engineering Design. Kluwer, Dordrecht
15. Manne AS (ed) (1967) Investments for capacity expansion. MIT, Cambridge, MA
16. Pistikopoulos EN, Grossmann IE (1988) Stochastic optimization of flexibility in retrofit design on linear systems. Comput Chem Eng 12:1215–1227
17. Pistikopoulos EN, Grossmann IE (1989) Optimal retrofit design for improving process flexibility in nonlinear systems-II. Optimal level of flexibility. Comput Chem Eng 13:1087–1096
18. Roberts SM (1964) Dynamic programming in chemical engineering and process control. Acad. Press, New York
19. Sahinidis NV, Grossmann IE (1992) Reformulation of the multiperiod MILP model for capacity expansion of chemical processes. Oper Res 40(Supp 1):S127–S144
20. Sahinidis NV, Grossmann IE, Fornari RE, Chathrathi M (1989) Optimization model for long range planning in the chemical industry. Comput Chem Eng 13(9):1049–1063
21. Santiago M, Iglesias OA, Pamiagua CN (1986) Optimal technical paths for chemical processes. Comput Chem Eng 10:421–431
22. Shimizu Y, Takamatsu T (1985) Application of mixed integer linear programming in multiterm expansion planning under multiobjectives. Comput Chem Eng 9(367)
23. Straub DA, Grossmann IE (1990) Integrated stochastic metric of flexibility for systems with discrete state and continuous parameter uncertainties. Comput Chem Eng 14:967–980
24. Straub DA, Grossmann IE (1993) Design optimization of stochastic flexibility. Comput Chem Eng 17:339–354

Cholesky Factorization

CAROLINE N. HADDAD
Department Math., State University New York, Geneseo, USA

MSC2000: 15-XX, 65-XX, 90-XX

Article Outline

Keywords
See also
References

Keywords

Symmetric; Ill-conditioned; Well-conditioned; Positive definite; Matrix decomposition or factorization; LU-decomposition; Least squares; Banded; Bandwidth; Normal equation; Pivot; Pivoting; Singular

Solving a linear system of the form $Ax = b$ where $A \in \mathbf{R}^{n \times n}$, $b \in \mathbf{R}^n$ is one of the most fundamental problems in mathematics and science. The two basic categories of numerical solutions are direct methods and iterative methods. Of the direct methods, *Gaussian elimination with backsolving* is the most commonly used technique. Straightforward Gaussian elimination uses row-operations to reduce the system to upper-triangular form before backsolving to find the solution, whereas the equivalent *LU-decomposition* initially factors A into the product of a lower- and upper-triangular matrix: $A = LU$. In the special case where A is both symmetric and positive definite, the matrix may be decomposed into $A = \widetilde{L}\widetilde{L}^\top$ where $\widetilde{L} \in \mathbb{R}^{n \times n}$ is lower-triangular. This decomposition is known as the *Cholesky factorization*, and is named for A.L. Cholesky.

The LU-decomposition of a square matrix, A, is the factorization of A into the product of a lower-triangular matrix, $L \in \mathbf{R}^{n \times n}$ and an upper-triangular matrix, $U \in \mathbf{R}^{n \times n}$. The system $Ax = (LU)x = b$ is then solved by forward solving $Ly = b$ where $y = Ux$, and then backsolving $Ux = y$. The solution can be found in roughly the same number of *floating point operations (flops)* as Gaussian elimination with backward substitution. More specifically, both methods require about $n^3/3$ multiplications/divisions and $n^3/3$ additions/subtractions for large n. The main advantage of this method is that once the matrix is factored (which requires $O(n^3)$ steps), the system can be solved repeatedly for different b, which only requires $O(n^2)$ steps. One drawback of this method is that pivoting may be required to find the decomposition.

A special class of problems arises if the matrix in the system is *positive definite*, i.e. $x^\top Ax > 0$ for $x \neq 0$. Note that if A is positive definite, but not symmetric, this implies that $1/2(A+A^\top)$, the symmetric part of A, is positive definite. The matrix A, can then be decomposed into the form $A = LDM^\top$ where L, M are lower-triangular and D is a diagonal matrix containing the *pivots* of A. If, in addition to being positive definite, A is also *symmetric*, i.e. $A = A^\top$, then L is symmetric, $M = L$, and the matrix has the special decomposition $A = LDL^\top$, where D has positive entries. Therefore $\sqrt{D}$ exists and A can be decomposed into $A = \widetilde{L}\widetilde{L}^\top$, where $\widetilde{L} = L\sqrt{D}$ and is referred to as the *Cholesky triangle*. Hence the Cholesky factorization is often referred to as the 'square-rooting method' [5]. The major advantage of this is that it requires around half the flops of the standard LU-decomposition.

The Cholesky factorization, presented below for symmetric and positive definite $A \in \mathbf{R}^{n \times n}$ in pseudocode, is taken from [2].

FOR $k = 1, \ldots, n$,
$$a_{kk} := \left(a_{kk} - \sum_{p=1}^{k-1} a_{kp}^2\right)^{1/2}$$
FOR $i = k+1, \ldots, n$
$$a_{ik} := \left(a_{ik} - \sum_{p=1}^{k-1} a_{ip}a_{kp}\right)/a_{kk}$$

Cholesky Factorization, Algorithm 1
A pseudocode for the Cholesky factorization

Example 1 Let $A = \begin{pmatrix} 4 & 2 \\ 2 & 5 \end{pmatrix}$. This matrix is both symmetric and positive definite. Therefore a Cholesky factorization exists for A (see [3] for a proof). An LU-decomposition for it is

$$A = \begin{pmatrix} 1 & 0 \\ \frac{1}{2} & 1 \end{pmatrix}\begin{pmatrix} 4 & 2 \\ 0 & 4 \end{pmatrix}.$$

Note that this is not unique, as another LU-decomposition is

$$A = \begin{pmatrix} 4 & 0 \\ 2 & 4 \end{pmatrix}\begin{pmatrix} 1 & \frac{1}{2} \\ 0 & 1 \end{pmatrix}.$$

The pivots in both cases are 4 and 4. Hence, the $LDL^\top$-decomposition is

$$A = \begin{pmatrix} 1 & 0 \\ \frac{1}{2} & 1 \end{pmatrix}\begin{pmatrix} 4 & 0 \\ 0 & 4 \end{pmatrix}\begin{pmatrix} 1 & \frac{1}{2} \\ 0 & 1 \end{pmatrix}.$$

Finally, the Cholesky factorization is

$$A = LL^\top = \begin{pmatrix} 2 & 0 \\ 1 & 2 \end{pmatrix}\begin{pmatrix} 2 & 1 \\ 0 & 2 \end{pmatrix}.$$

Additionally, this factorization is unique for symmetric, positive definite matrices. The decomposition can be performed in fixed-point with *no pivoting required* [9]. This implies that the Cholesky decomposition is *guaranteed to be stable without pivoting*.

If the matrix A is *ill-conditioned*, i.e. has *condition number* $\kappa = \|A\| \, \|A^{-1}\| \gg 1$, then the matrix may be nearly singular and the computed solution to the system, $Ax = b$, may not be sufficiently accurate. A process of iterative refinement may be used to assess the accuracy of the solution and then improve upon it when working in higher precision is not practical. See [2,9] for a more detailed explanation.

The Cholesky factorization can also be used to find the inverse and determinant of a symmetric, positive definite matrix (the LU-decomposition can be used for general $A \in \mathbf{R}^{n \times n}$). It is important for the matrix to be positive definite for a variety of reasons, for example, if A is symmetric, but not positive definite, then the stability is not guaranteed. In the case of finding the inverse of a matrix A, a poor inverse may be obtained even if A is *well-conditioned*, i.e. κ is 'close' to 1.

The efficiency of the Cholesky factorization can be further improved if the matrix is 'banded'. A matrix $A = [a_{ij}]$ is said to have *upper bandwidth* q if $a_{ij} = 0$ whenever $j > i + q$ and *lower bandwidth* p if $a_{ij} = 0$ whenever $i > j + p$. Since A is symmetric, when A has lower bandwidth p, it also has upper bandwidth p. In this case A is said to have *bandwidth* p. For example, if $p = 1$, then A is tridiagonal. The following algorithm from [2] takes advantage of the fact that A is symmetric, positive definite and has bandwidth p. It requires n square roots and

$$\frac{np^2}{2} - \frac{p^3}{3} + \frac{3}{2}(np - p^2)$$

flops or approximately $\frac{n(p^2+3p)}{2}$ flops for $p \ll n$.

```
FOR i = 1, ..., n
    FOR j = max{1, i − p}, ..., i − 1
        a_ij := (a_ij − Σ_{k=max{1,i−p}}^{j−1} a_ik a_jk) / a_jj
    a_ii := (a_ii − Σ_{k=max{1,i−p}}^{i−1} a_ik^2)^{1/2}
```

Cholesky Factorization, Algorithm 2
A pseudocode for the banded Cholesky factorization

Another important application of the Cholesky factorization is in the key role it plays in one of the most commonly used numerical techniques for solving the least squares problem (LS problem; cf. also ▶ least squares problems). The *least squares problem* is to find the 'best' solution to $Ax = b$ when the system is inconsistent for $A \in \mathbf{R}^{m \times n}$. Instead, the system $A^\mathsf{T} Ax = A^\mathsf{T} b$, more commonly known as the *normal equations*, is solved by first finding the Cholesky factorization of the symmetric matrix $A^\mathsf{T} A = \widetilde{L}\widetilde{L}^\mathsf{T}$, which is positive definite if A has rank n. Next, $\widetilde{L}y = A^\mathsf{T} b$ is forward-solved, and finally, the 'best least squares' solution, $\widehat{x}$, is found by backsolving $\widetilde{L}^\mathsf{T} x = y$. Note that $\widehat{x}$ minimizes $\|Ax - b\|_2$ and the algorithm requires $O(n^3)$ flops. For the algorithm and an analysis of the accuracy of the method, see [2].

See also

- ▶ ABS Algorithms for Linear Equations and Linear Least Squares
- ▶ Large Scale Trust Region Problems
- ▶ Large Scale Unconstrained Optimization
- ▶ Linear Programming
- ▶ Orthogonal Triangularization
- ▶ Overdetermined Systems of Linear Equations
- ▶ QR Factorization
- ▶ Solving Large Scale and Sparse Semidefinite Programs
- ▶ Symmetric Systems of Linear Equations

References

1. Burden RL, Faires JD (2005) Numerical Analysis, 8th edn. PWS/Kent, Boston
2. Golub GH, Loan CF Van (1996) Matrix Computations, 3rd edn. Johns Hopkins Univ. Press, Baltimore
3. Griffel DH (1989) Linear Algebra and its Applications, vol 2. Halsted Press, New York
4. Hager WW (1995) Applied Numerical Linear Algebra. Dept. Math. Univ. Florida, Gainesville
5. Householder AS (1964) The Theory of Matrices in Numerical Analysis. Dover, New York
6. Kincaid D, Cheney W (2002) Numerical Analysis, 3rd edn. Brooks/Cole, Pacific Grove, CA
7. Leon SJ (2005) Linear Algebra with Applications, 7th edn. Prentice-Hall, Upper Saddle River
8. Strang G (2003) Introduction to Linear Algebra, 3rd edn. Wellesley and Cambridge Press, Wellesley, MA
9. Wilkinson JH (1963) Rounding Errors in Algebraic Processes. Prentice-Hall, Upper Saddle River

Combinatorial Matrix Analysis

RICHARD A. BRUALDI
Department Math., University Wisconsin,
Madison, USA

MSC2000: 90C10, 90C09

Article Outline

Keywords
Matrix Patterns and Various Graphs
Eigenvalues and Digraphs
Sign-Nonsingular Matrices
Doubly Stochastic Matrices
See also
References

Keywords

Matrix analysis

Broadly speaking, *matrix analysis* is the study of non-algebraic properties of matrices and the analysis of matrices in order to reveal their finer properties and structure. (Here, 'algebraic' is understood at least in the classical sense of algebra.) The matrices are usually real or complex matrices. *Combinatorial matrix analysis* is the study of combinatorial properties of matrices and the analysis of matrices which takes into account combinatorial structure. Here combinatorial structure usually refers to the *zero-nonzero pattern* of a matrix, captured through the use of either the directed graph or bipartite graph associated with the nonzero entries of a matrix (or the graph of the nonzero entries in the case of a symmetric matrix), or to the *positive-negative-zero pattern* of a real matrix, captured through the use of the signed digraph or signed bipartite graph of a matrix.

This article is intended as an introduction to combinatorial matrix analysis. More detail can be found in [5] and [6], and the references contained therein. We do not discuss here the many applications of matrix theory and linear algebra to combinatorics, graphs, and discrete structures.

Matrix Patterns and Various Graphs

Let $A = [a_{ij}]$ be a matrix of order n whose entries a_{ij} are real or complex numbers. To A there corresponds a *directed graph* (or *digraph*) $D(A)$ with vertex set $V = \{1, \ldots, n\}$ and with an arc (i, j) from vertex i to vertex j if and only if $a_{ij} \neq 0$. The *bipartite graph* (or *bigraph*) $BG(A)$ of A has vertex set $\{1_r, \ldots, n_r\}$ (corresponding to the rows of A) and $\{1_c, \ldots, n_c\}$ (corresponding to the columns of A); the edges of $BG(A)$ are all pairs $\{i_r, j_c\}$ for which $a_{ij} \neq 0$. The bipartite graph of a matrix can be defined for a rectangular $m \times n$ matrix in the same way except that the vertices corresponding to the rows are $\{1_r, \ldots, m_r\}$. Both the digraph and the bigraph reveal the zero-nonzero pattern of a square matrix A.

If A is a real matrix and we want to capture the sign $(+, -, 0)$ of the entries of A, then we assign a $+$ or $-$ to each arc of $D(A)$ (to each edge of $BG(A)$) according as the corresponding entry of A is positive or negative, and in this way obtain the *signed digraph* and *signed bigraph* of A. We use the same notations $D(A)$ and $BG(A)$ for the signed versions of the digraph and bigraph of A. Thus two matrices A and B have the same *sign pattern* if and only if they have the same signed digraphs (equivalently, the same signed bigraphs).

If A is a symmetric matrix (or has a symmetric pattern in the sense that $a_{ij} \neq 0$ if and only if $a_{ji} \neq 0$), then the *graph* $G(A)$ of A has vertex set $\{1, \ldots, n\}$ with an edge $\{i, j\}$ between i and j if and only if $a_{ij} \neq 0$ (equivalently, $a_{ji} \neq 0$). Thus $G(A)$ is obtained from $D(A)$ by 'removing' the directions on arcs (this may result in two edges joining certain pairs of vertices and one edge of each such pair is removed as well). Sometimes in $D(A)$ and $G(A)$ it is convenient to ignore the arcs (i, i) and edges $\{i, i\}$ (called loops) corresponding to nonzero entries a_{ii} on the main diagonal of A.

A square matrix A of order n is *irreducible* provided there does not exist a permutation matrix P such that

$$PAP^T = \begin{bmatrix} A_1 & O \\ A_{21} & A_2 \end{bmatrix},$$

where A_1 is a square matrix of order k for some k with $0 < k < n$. (The matrix PAP^T is obtained from A by simultaneously permuting its rows and columns. The digraphs of A and PAP^T are isomorphic.) A digraph is *strongly connected* provided for each ordered pair of distinct vertices i and j there is a path from i to j.

Proposition 1 *The matrix A is irreducible if and only if the digraph $D(A)$ is strongly connected,* [5].

A square matrix can be brought to a very special form by simultaneous row and column permutations.

Theorem 2 *Let A be a matrix of order n. Then there exist a permutation matrix P and an integer $k \geq 1$ such that*

$$PAP^T = \begin{bmatrix} A_1 & O & \cdots & O \\ A_{21} & A_2 & \cdots & O \\ \vdots & \vdots & \ddots & \vdots \\ A_{k1} & A_{k2} & \cdots & A_k \end{bmatrix},$$

where $A_1, \ldots, A_k$ are square, irreducible matrices. The matrices $A_1, \ldots, A_k$ are uniquely determined to within simultaneous permutations of their rows and columns, but their order on the diagonal is not necessarily unique.

The matrices $A_1, \ldots, A_k$ in this theorem are called the *irreducible components* of A and correspond to the *strongly connected components* of the digraph $D(A)$.

Irreducible matrices have an *inductive structure* that is revealed in the next theorem [5].

Theorem 3 *Let A be an irreducible matrix of order $n \geq 2$. Then there exist a permutation matrix P and an integer $m \geq 2$ such that*

$$PAP^T = \begin{bmatrix} A_1 & O & \cdots & O & E_1 \\ E_2 & A_2 & \cdots & O & O \\ \vdots & \vdots & \cdots & \vdots & \vdots \\ O & O & \cdots & A_{m-1} & O \\ O & O & \cdots & E_m & A_m \end{bmatrix},$$

where $A_1, \ldots, A_m$ are irreducible matrices and $E_1, \ldots, E_m$ are matrices having at least one nonzero entry.

Allowing independent row and column permutations in the definition of irreducibility leads to full indecomposability. A square matrix A of order n is *fully indecomposable* provided there do not exist permutation matrices P and Q such that

$$PAQ = \begin{bmatrix} A_1 & O \\ A_{21} & A_2 \end{bmatrix},$$

where A_1 is a square matrix of order k for some k with $0 < k < n$. The matrices A and PAQ have isomorphic bigraphs.

Theorems analogous to Theorems 2 and 3 hold with independent permutations replacing simultaneous permutations and fully indecomposable replacing irreducible. The connection is provided by the fact that a square matrix A is fully indecomposable if and only if there are permutation matrices P and Q such that PAQ has a nonzero main diagonal and PAQ is irreducible [5].

Eigenvalues and Digraphs

The following theorem is the Perron–Frobenius theorem [7,8] and is one of the first instances of the influence of the digraph of a matrix on its spectral properties.

Theorem 4 (Perron–Frobenius theorem) *Let A be a matrix of order $n > 1$ each of whose entries is a nonnegative real number. Assume that A is irreducible, equivalently $D(A)$ is strongly connected. Then there is a positive number $\rho(A)$ such that*

1) *$\rho(A)$ is a simple eigenvalue of A;*
2) *every eigenvalue λ of A satisfies $|\lambda| \leq \rho(A)$;*
3) *the number of eigenvalues λ of A with $|\lambda| = \rho(A)$ equals the greatest common divisor k of the lengths of the circuits of $D(A)$, and these eigenvalues are $\rho(A)e^{2\pi ij/k}, (j = 1, \ldots, k)$.*

A more recent application of the digraph of a matrix to localization of its eigenvalues concerns a generalization [2] of the *Gershgorin theorem*. Let $A = [a_{ij}]$ be a complex matrix of order n and let

$$R_i = \sum_{j \neq i} |a_{ij}| \quad (1 \leq i \leq n).$$

Then Gershgorin's theorem asserts that the n eigenvalues of A lie in that part $\mathcal{R}$ of the complex plane determined by the union of the n closed disks

$$\{z : |z - a_{ii}| \leq R_i\}, \quad (1 \leq i \leq n).$$

If A is irreducible, then a boundary point of $\mathcal{R}$ is an eigenvalue of A only if it is a boundary point of each of the n closed disks.

By considering the *circuits*, or *directed cycles*, of the digraph of A, a better inclusion region can be obtained.

Theorem 5 *Let $A = [a_{ij}]$ be a complex matrix of order n. Then the n eigenvalues of A lie in that part $\mathcal{S}$ of the* *complex plane determined by the union of the regions*

$$S(\gamma) = \left\{ z: \prod_{\gamma} |z - a_{ii}| \leq \prod_{\gamma} R_i \right\},$$

(γ a circuit of $D(A)$),

where $\prod_\gamma$ denotes the product over all vertices i belonging to the circuit γ. If A is irreducible, then a boundary point of S is an eigenvalue of A only if it is a boundary point of each $S(\gamma)$.

Theorems 4 and 5 demonstrate how information concerning the combinatorial structure of a matrix can be used to give information on spectral properties of the matrix.

Sign-Nonsingular Matrices

It is easy to characterize real matrices $A = [a_{ij}]$ of order n whose singularity is a consequence of their zero pattern, equivalently, of their nonzero pattern or bigraph $BG(A)$. Let $Z(A)$ denote the set of all real matrices B of order n that have the same zero pattern as A, that is, satisfy $BG(B) = BG(A)$. Then the following are equivalent:
i) Each matrix $B \in Z(A)$ is singular;
ii) Each of the $n!$ terms in the standard determinant expansion of A is zero (the *standard determinant expansion of a matrix A* is

$$\det A = \sum \epsilon(i_1, \ldots, i_n) a_{1 i_1} \cdots a_{n i_n}$$

where the sum extends over each permutation $i_1 \ldots i_n$ of $\{1, \ldots, n\}$ and $\epsilon(i_1 \ldots i_n)$ is + or − depending on whether the permutation is even or odd);
iii) The bigraph $BG(A)$ does not have a perfect matching (i.e. a set of n pairwise vertex disjoint edges meeting all vertices);
iv) There is a set of fewer than n rows and columns which together contain all the nonzero entries of A;
v) There is a set of fewer that n vertices of $BG(A)$ which together meet all the edges of $BG(A)$.

Properties ii) and iii) are clearly equivalent, as are properties iv) and v). Properties i) and ii) are equivalent, since if there is a nonzero term in the standard determinant expansion of A, then by sufficiently emphasizing the entries of A in that term we obtain a nonsingular matrix. Properties iii) and iv) are equivalent by the Frobenius–König theorem [5].

Now let $Q(A)$ denote the set of all real matrices of order n that have the same sign pattern (+, −, 0) as A. $Q(A)$ consists of all real matrices of order n that have the same signed digraph (equivalently, the same signed bigraph) as A and is called the *qualitative class* of A. The matrix A is called *sign-nonsingular* provided each matrix in $Q(A)$ is nonsingular. Some equivalent characterizations of sign-nonsingularity are:
i) A is sign-nonsingular;
ii) There is a nonzero term in the standard determinant expansion of A and each such nonzero term has the same sign;
iii) $\det(A) \neq 0$ and the determinants of the matrices in $Q(A)$ all have the same sign.

The matrix

$$\begin{bmatrix} 1 & -1 & 0 \\ 1 & 1 & -1 \\ 1 & 1 & 1 \end{bmatrix}$$

is a sign-nonsingular matrix.

If a matrix $A = [a_{ij}]$ is sign-nonsingular, then so is every matrix PAQ where P and Q are permutation matrices, as is every matrix of the form DA where D is a nonsingular diagonal matrix. Also a sign-nonsingular matrix must have a nonzero term in its standard determinant expansion. Thus in dealing with sign-nonsingular matrices we may assume that each entry on the main diagonal is negative, that is, A has a *negative main diagonal*. With this normalization, we have the following theorem [1]. The *sign of a circuit*

$$\gamma: j_1 \to \cdots \to j_k \to j_1$$

of the signed digraph of A is

$$\text{sign}(\gamma) = \text{sign } a_{j_1 j_2} \cdots a_{j_{k-1} j_k} a_{j_k j_1},$$

the products of the signs of the arcs of the cycle.

Theorem 6 (Bassett–Maybee–Quirk theorem) *Let A be a real matrix of order n with a negative main diagonal. Then A is a sign-nonsingular matrix if and only if each circuit of the signed digraph of A is negative.*

Sign-nonsingularity allows one to characterize square, homogeneous systems of linear equations $Ax = 0$ for which $\widetilde{A}x = 0$ has only the zero solution (thus all solutions of $\widetilde{A}x = 0$ have the same sign pattern) for all matrices $\widetilde{A}$ with the same sign pattern as A. A more general problem is to characterize linear systems $Ax = b$ that are sign-solvable in the sense that the sign pattern of the solution is determined solely by the sign patterns of A and b. More precisely, $Ax = b$ is sign-solvable provided that for all $\widetilde{A} \in \mathcal{Q}(A)$ and all $\widetilde{b} \in \mathcal{Q}(b)$ there is a vector $\widetilde{x}$ such that $\widetilde{A}\widetilde{x} = \widetilde{b}$ and all of the vectors in

$$\{\widetilde{x}: \exists \widetilde{A} \in \mathcal{Q}(A), \widetilde{b} \in \mathcal{Q}(b) \text{ s.t. } \widetilde{A}\widetilde{x} = \widetilde{b}\}$$

have the same sign pattern.

Sign-solvable linear systems can be characterized in terms of two classes of matrices, called S^*-matrices and L-matrices. An $n \times (n+1)$ matrix B is an S^*-matrix provided each matrix of order n obtained from B by deleting a column is a sign-nonsingular matrix. Cramer's rule implies that B is an S^*-matrix if and only if there is a vector w with no zero coordinates such that the right null spaces of the matrices in $\widetilde{B} \in \mathcal{Q}(B)$ are contained in $\{0\} \cup \mathcal{Q}(w) \cup \mathcal{Q}(-w)$. The matrix

$$\begin{bmatrix} 1 & -1 & 0 & 0 \\ 1 & 1 & -1 & 0 \\ 1 & 1 & 1 & -1 \end{bmatrix}$$

is an S^*-matrix. A matrix A is an L-matrix provided every matrix in $\mathcal{Q}(A)$ has linearly independent rows. Sign-nonsingular matrices are square L-matrices. Every S^*-matrix is an L-matrix and so is any matrix obtained from an L-matrix by appending columns.

The following theorem characterizes sign-solvable linear systems [9].

Theorem 7 Let $A = [a_{ij}]$ be an $m \times n$ matrix and let b be an $m \times 1$ column vector. Let $z = (z_1, \ldots, z_n)^\mathsf{T}$ be a solution of the linear system $Ax = b$, and let

$$\beta = \{j: z_j \neq 0\},$$
$$\alpha = \{i: a_{ij} \neq 0 \text{ for some } j \in \beta\}.$$

Then $Ax = b$ is sign-solvable if and only if the matrix

$$[A[\alpha, \beta] - b[\beta]]$$

is an S^*-matrix and the matrix $A(\alpha, \beta)^\mathsf{T}$ is an L-matrix.

(Here $A[\alpha, \beta]$, respectively $A(\alpha, \beta)$, is the submatrix of A formed by the rows in α and the columns in β, respectively not in α and not in β, and $b[\alpha] = b[\alpha, \{1\}]$.)

A detailed study of sign-solvability and related issues is contained in [6].

Doubly Stochastic Matrices

A real matrix $A = [a_{ij}]$ of order n is *doubly stochastic* provided each of its entries is nonnegative, and all row and column sums equal 1:

$$a_{ij} \geq 0 \quad (i, j = 1, \ldots, n),$$

$$\sum_{j=1}^n a_{ij} = 1, \quad \sum_{i=1}^n a_{ij} = 1.$$

Doubly stochastic matrices arise quite naturally in many different contexts:

i) Let $U = [u_{ij}]$ be a real orthogonal matrix or a complex unitary matrix. Then

$$\widehat{U} = [|u_{ij}|^2] \quad (i, j = 1, \ldots, n)$$

is a doubly stochastic matrix.

ii) (*Optimal assignment problem*) Consider an assignment of n people to n positions in which the 'value' of the ith person to the jth position is $v_{ij} \geq 0$ ($i, j = 1, \ldots, n$). An *optimal assignment* is an assignment $i \to j_i$ ($i = 1, \ldots, n$) of people to positions (here $j_1 \ldots j_n$ is a permutation of $\{1, \ldots, n\}$) which maximizes the total value $\sum_{i=1}^n v_{ij_i}$. The set Ω_n of doubly stochastic matrices of order n is a convex polytope and, according to *Birkhoff's theorem* [5], the set of vertices of this polytope is the set $\mathcal{P}_n$ of permutation matrices of order n. Thus the vertices of Ω_n correspond to the $n!$ possible assignments, and the optimal assignment problem can be solved as a linear programming problem on Ω_n.

iii) Let $\mathcal{P}_n = \{P_1, \ldots, P_{n!}\}$, and let $(c_i: i = 1, \ldots, n!)$ be a probability distribution on $\mathcal{P}_n$: $c_i \geq 0$ ($i = 1, \ldots, n!$) and $\sum_{i=1}^{n!} c_i = 1$. Then the expectation of a permutation $R \in \mathcal{P}_n$ chosen at random is

$$E = E[R] = \sum_{i=1}^{n!} c_i P_i = [e_{ij}],$$

a doubly stochastic matrix. It is a consequence of Birkhoff's theorem that every doubly stochastic matrix of order n arises from a probability distribution on $\mathcal{P}_n$ in this way. The probability that a function f chosen at random according to the probabilities

$$\text{prob}(f(i) = j) = e_{ij} \quad (i, j = 1, \ldots, n)$$

is a permutation equals the *permanent* of A defined by

$$\text{per}(A) = \sum a_{1j_1} \cdots a_{nj_n},$$

where the sum extends over all permutations $j_1 \ldots j_n$ of $\{1, \ldots, n\}$.

Let $A = [a_{ij}]$ be a real, symmetric matrix (or a complex, Hermitian matrix). Then there exists a real, orthogonal matrix U such that

$$UAU^\top = \begin{bmatrix} \lambda_1 & 0 & \cdots & 0 \\ 0 & \lambda_2 & \cdots & 0 \\ \vdots & \vdots & \ddots & \vdots \\ 0 & 0 & \cdots & \lambda_n \end{bmatrix},$$

where $\lambda_1 \geq \ldots \geq \lambda_n$ are the n eigenvalues of A. Comparing diagonal entries, we get

$$(\lambda_1, \ldots, \lambda_n)^\top = S(a_{11}, \ldots, a_{nn})^\top, \quad (1)$$

where $S = \widehat{U}$ is a doubly stochastic matrix. Without loss of generality, assume that $a_{11} \geq \cdots \geq a_{nn}$. Then equation (1) implies that

$$\lambda_1 + \cdots + \lambda_i \leq a_{11} + \cdots + a_{ii} \quad (i = 1, \ldots, n), \quad (2)$$

with equality for $i = n$. When the inequalities (2), with equality for $i = n$, hold between two vectors $\lambda = (\lambda_1, \ldots, \lambda_n)$ and $\mu = (a_{11}, \ldots, a_{nn})$ (that have been arranged in nonincreasing order), then λ is said to be *majorized* by μ. A *Hardy–Littlewood–Pólya theorem* states that if λ is majorized by μ, then there exists a doubly stochastic matrix S such that $\lambda = S\mu$ [10]. Hence by Birkhoff's theorem, λ is majorized by μ if and only if λ is in the convex hull of all vectors obtained from μ by permuting its coordinates. There exist doubly stochastic matrices S of very special form such that $\lambda = S\mu$ when λ is majorized by μ [4,10].

As noted above, the vector of eigenvalues of a real, symmetric matrix is majorized by the vector of its entries on the main diagonal. Conversely, if λ and μ are two n-vectors with λ majorized by μ, then according to a theorem of A. Horn, there exists a real, symmetric matrix of order n, whose eigenvalues are given by λ and whose main diagonal entries are given by μ [10].

Let A be a doubly stochastic matrix, and let $A_1, \ldots, A_k$ ($k \geq 1$) be the fully indecomposable components of A. Since all row and column sums of A equal 1, it follows easily that up to row and column permutations A is the direct sum of its fully indecomposable components: there exist permutation matrices P and Q such that

$$PAQ = A_1 \oplus \cdots \oplus A_k,$$

where $\oplus$ denotes direct sum. The polytope Ω_n has dimension $(n-1)^2$. Each doubly stochastic matrix determines a face of Ω_n equal to the set of all doubly stochastic matrices S such that $BG(S)$ is a subgraph of $BG(A)$ (i.e. each edge of $BG(A)$ is also an edge of $BG(A)$). This face is the smallest face of Ω_n containing A, and each nonempty face of Ω_n arises in this way. Since no entry of a doubly stochastic matrix can exceed 1, the nonempty faces of Ω_n can be described as follows: Let C be a (0, 1)-matrix of order n which, up to row and column permutations, is a direct sum of fully indecomposable matrices (such matrices are said to have *total support*). Then

$$\mathcal{F}(C) = \{A : A \in \Omega_n, A \leq C \text{ entrywise}\}$$

is a face of Ω_n and its dimension equals $\sigma(C) - 2n + k$, where $\sigma(C)$ is the number of 1s of C and k is the number of its fully indecomposable components [3].

See also

- ▶ Combinatorial Optimization Algorithms in Resource Allocation Problems
- ▶ Combinatorial Optimization Games
- ▶ Evolutionary Algorithms in Combinatorial Optimization
- ▶ Fractional Combinatorial Optimization
- ▶ Multi-Objective Combinatorial Optimization
- ▶ Neural Networks for Combinatorial Optimization
- ▶ Replicator Dynamics in Combinatorial Optimization

References

1. Bassett L, Maybee J, Quirk J (1968) Qualitative economics and the scope of the correspondence principle. Econometrica 36:544–63
2. Brualdi RA (1982) Matrices, eigenvalues and directed graphs. Linear Multilinear Algebra 11:143–65
3. Brualdi RA, Gibson PM (1977) The convex polytope of doubly stochastic matrices I: Applications of the permanent function. J Combin Th A 22:194–230
4. Brualdi RA, Hwang S-G (1996) Vector majorization via Hessenberg matrices. J London Math Soc Ser 2 53:28–38
5. Brualdi RA, Ryser HJ (1991) Combinatorial matrix theory. Encycl Math Appl, vol 39. Cambridge Univ. Press, Cambridge
6. Brualdi RA, Shader BL (1995) Matrices of sign-solvable linear systems. Cambridge Tracts in Math, vol 116. Cambridge Univ. Press, Cambridge

7. Horn R, Johnson CR (1985) Matrix analysis. Cambridge Univ Press, Cambridge
8. Horn R, Johnson CR (1991) Topics in matrix analysis. Cambridge Univ Press, Cambridge
9. Klee V, Ladner R, Manber R (1984) Sign solvability revisited. Linear Alg Appl 59:131–57
10. Olkin I, Marshall AW (1979) Inequalities: Theory of majorization and its applications. Acad Press, New York

Combinatorial Optimization Algorithms in Resource Allocation Problems

NAOKI KATOH

Kyoto University, Kyoto, Japan

MSC2000: 90C09, 90C10

Article Outline

Keywords
α: Objective Functions
β: Constraints
Algorithms
Generalizations
See also
References

Keywords

Resource allocation; Combinatorial algorithm

The *resource allocation problem* seeks to find an optimal allocation of a fixed amount of resources to activities so as to minimize the cost incurred by the allocation. A simplest form of the problem is to minimize a separable convex function under a single constraint concerning the total amount of resources to be allocated. The amount of resources to be allocated to each activity is treated as a continuous or integer variable, depending on the cases. This can be viewed as a special case of the nonlinear programming problem or the nonlinear integer programming problem.

Due to its simple structure, this problem is encountered in a variety of application areas, including load distribution, production planning, computer resource allocation, queueing control, portfolio selection, and apportionment. The first explicit investigation of the resource allocation problem is due to B.O. Koopman [15] (1953), who dealt with the problem of the *optimal distribution of efforts*, which arises in the problem of searching for an object whose position is a random variable. Since then, a great number of papers have been published on resource allocation problems. Efficient algorithms have also been developed, depending on the form of objective functions and constraints or on the type of variables (i. e., continuous or integer).

See [11] for a comprehensive review of the state-of-the-art of the problems (as of 1988). After this book was published, many papers have been published on resource allocation problems. A significant progress has been made on the algorithm side. Also, new generalizations and variants of the problem have been investigated, and new application fields have been discovered. Such new progress has been reviewed in [13].

We first classify the resource allocation problems. A generic form of the resource allocation problem discussed in this article is described as follows:

$$(P) \begin{cases} \min & f(x_1, \ldots, x_n) \\ \text{s.t.} & \sum_{j=1}^{n} x_j = N, \\ & x_j \geq 0, \quad j = 1, \ldots, n. \end{cases} \quad (1)$$

That is, given one type of resource whose total amount is equal to N, we want to allocate it to n activities so that the objective value $f(x_1, \ldots, x_n)$ is minimized. The objective value may be interpreted as the cost or loss, or the profit or reward, incurred by the resulting allocation. In case of profit or reward, it is natural to maximize f, and we shall sometimes consider maximization problems. The difference between maximization and minimization is not essential because maximizing f is equal to minimizing $-f$.

Each variable x_j represents the amount of resource allocated to activity j. If it represents persons, processors or trucks, however, variable x_j becomes a discrete variable that takes nonnegative integer values, and the constraint

$$x_j : \text{integer}, \quad j = 1, \ldots, n, \quad (2)$$

is added to the constraints in (1). The resource allocation problem with this constraint is often referred to as the *discrete resource allocation problem*.

As for the objective function, it usually has some special structure according to the intended applications. Typically, the following special case, called *separable*, is often considered:

$$\sum_{j=1}^{n} f_j(x_j). \qquad (3)$$

If each f_j is convex, the objective function is called *separable convex objective function*.

Resource allocation problems are classified according to the types of objective functions, constraints and variables. We shall describe the classification scheme, and several types of problem formulations according to the classification scheme. In general, we use the notation $\alpha/\beta/\gamma$ to denote the type of a resource allocation problem. Here, α specifies the type of objective function, β the constraint type, and γ the variable type;

$$\gamma = D, \quad \gamma = C$$

denote the case of integer variable, respectively continuous variable. We shall now explain the notations for α and β.

α: Objective Functions

The objective function $f(x_1, \ldots, x_n)$ may take the following special structures:

1) *Separable* (S, for short): $\sum_{j=1}^{n} f_j(x_j)$, where each f_j is a function of one variable.
2) *Separable and convex* (SC, for short): $\sum_{j=1}^{n} f_j(x_j)$, where each f_j is a convex function of one variable. In particular, if each f_j is quadratic and convex, we denote such a subclass by SQC.
3) *Minimax*: minimize $\max_{1 \leq j \leq n} f_j(x_j)$, or *Maximin*: maximize $\min_{1 \leq j \leq n} f_j(x_j)$; here, all f_j are monotone nondecreasing in x_j.
4) *Lexicographically minimax* (Lexico-Minimax, for short): Since the objective value of Minimax is determined by the single variable x_k^* satisfying $f_k(x_k^*) = \max_j f_j(x_j^*)$, there may be many optimal solutions. To remove such ambiguity, we introduce the *lexicographical ordering for n-dimensional vectors*: Given $a = (a_1, \ldots, a_n)$ and $b = (b_1, \ldots, b_n)$, a is *lexicographically smaller* than b (or b is *lexicographically greater* than a) if $a_j = b_j$ for $j = 1, \ldots, k-1$ and $a_k < b_k$ some k. This is denoted by $a \leq_{\text{lex}} b$ or $b \geq_{\text{lex}} a$. For $a = (a_1, \ldots, a_n)$, let DEC(a) (respectively, INC(a)) denote the n-tuple of $a_j, j = 1, \ldots, n$, arranged in non-increasing order (respectively, nondecreasing order) of their values (e.g., for $a = (4, 3, 1, 5)$, we have DEC(a) = (5, 4, 3, 1) and INC(a) = (1, 3, 4, 5)). The objective of Lexico-Minimax is to find an allocation vector $x = (x_1, \ldots, x_n)$ such that DEC(x) is minimal. Notice that an optimal solution to Lexico–Minimax is also optimal to Minimax, but the converse is not generally true. This is a refined objective of Minimax. Similarly, we define Lexico-Maximin as the one that maximizes INC(x).

5) *Fair*: minimize the expression

$$g\left(\max_{1 \leq j \leq n} f_j(x_j), \min_{1 \leq j \leq n} f_j(x_j)\right),$$

where $g(u, v)$ is nondecreasing (respectively, nonincreasing) in u (respectively, v). This objective is a generalization of Minimax and Maximin.

β: Constraints

In addition to the simple first resource constraint of (1), other additional constraints are also imposed. Typical additional constraints which appeared in various resource allocation problems are as follows. We refer the case of no additional constraints as 'simple'.

1) *Lower and upper bounds* (LUB, for short): $l_j \leq x_j \leq u_j, j = 1, \ldots, n$.
2) *Generalized upper bounds* (GUB, for short): $\sum_{j \in S_i} x_j \leq b_i, i = 1, \ldots, m$, where $S_1, \ldots, S_m$ is a partition of $\{1, \ldots, n\}$.
3) *Nested constraints* (Nested, for short): $\sum_{j \in S_i} x_j \leq b_i$, $i = 1, \ldots, m$, where $S_1 \subset \ldots \subset S_m$. We can assume $b_1 \leq \ldots \leq b_m$, since if $b_i > b_{i+1}$, the ith constraint is redundant.
4) *Tree constraints* (Tree, for short): $\sum_{j \in S_i} x_j \leq b_i, i = 1, \ldots, m$, where the sets S_i are derived by some hierarchical decomposition of E into disjoint subsets.
5) *Network constraints* (Network, for short): The constraint is defined in terms of a directed network with a single source and multiple sinks. Given a directed graph $G = (V, A)$ with node set V and arc set A, let $s \in V$ be the source and $T \subseteq V$ be the set of sinks. The amount of supply from the source is $N > 0$, and

the capacity of arc (u, v) is $c(u, v)$. Denote the flow vector by $\varphi = \{\varphi(u, v) : (u, v) \in A\}$. φ is a *feasible flow* in G if it satisfies

$$0 \leq \varphi(u, v) \leq c(u, v), \quad (u, v) \in A,$$

$$\sum_{(v,w) \in A_+(v)} \varphi(v, w) - \sum_{(u,v) \in A_-(v)} \varphi(u, v) = 0, \quad (4)$$

$$v \in V - T - \{s\}, \quad (5)$$

$$\sum_{(s,v) \in A_+(s)} \varphi(s, v) - \sum_{(u,s) \in A_-(s)} \varphi(u, s) = N, \quad (6)$$

$$x_t(\varphi) \equiv \sum_{(u,t) \in A_-(t)} \varphi(u, t) - \sum_{(t,v) \in A_+(t)} \varphi(t, v) \geq 0, \quad (7)$$

$$t \in T,$$

$$\sum_t x_t(\varphi) = N. \quad (8)$$

The value $x_t(\varphi)$ denotes the amount of flow entering a sink $t \in T$. For a feasible flow φ, the vector $x_t(\varphi)x \in T\}$ is called the *feasible flow vector* with respect to φ. For instance, the problem SC/Network/C (i. e., the separable convex resource allocation problem under network constraints) is defined as follows:

$$\begin{cases} \min \quad \sum_{t \in T} f_t(x_t(\varphi)) \\ \text{s.t.} \quad (4) - (8), \end{cases} \quad (9)$$

where f_t, for each $t \in T$, is a convex function.

6) *Submodular constraints* (SM, for short): A set of feasible solutions is defined by a base polyhedron $B(r) = x \in \mathbf{R}^E : x(S) \leq r(S)$ for all $S \in \mathcal{D}, x(E) = r(E)\}$ of a submodular system $(\mathcal{D}, r)$, i. e.,

$$x \in B(r). \quad (10)$$

Here, we use the notation $E = \{1, \ldots, n\}$, and $x(S) \equiv \sum_{i \in S} x_i$ for $S \subseteq E$ and $x \in \mathbf{R}^E$. $\mathcal{D} \subseteq 2^E$ is a *distributive lattice* such that $\emptyset, E \in \mathcal{D}$, i. e., $\mathcal{D}$ is closed under union and intersection operations. Also, the function $r : \mathcal{D} \to \mathbf{Z}$ is *submodular* over $\mathcal{D}$, i. e.,

$$r(X) + r(Y) \geq r(X \cup Y) + r(X \cap Y).$$

For a *submodular system* $(\mathcal{D}, r)$,

$$P(r) = \{x \in \mathbf{R}^E : x(S) \leq r(S) \text{ for all } S \in \mathcal{D}\}$$

is called the *submodular polyhedron* of $(\mathcal{D}, r)$. Notice that the first constraint in (1) is included in the constraints of (10), as $x(E) = r(E)$ in the above definition. If we consider the case of integer variables, the constraint is defined by

$$x \in B(r) \cap \mathbf{Z}^E.$$

It is assumed, in general, that $B(r)$ of the constraint (10) is not explicitly given as an input, but is implicitly given through an *oracle* that tells the value $r(X)$ when X is given.

7) *General linear constraints* (Linear, for short): Constraints defined by a set of linear inequalities

$$\sum_{j=1}^{n} a_{ij} x_j \leq b_i, \quad i = 1, \ldots, m. \quad (11)$$

No other special assumption is imposed on the structure of the constraints. Notice that all the constraints, LUB, GUB, Nested, Tree, Network are special cases of submodular constraints (see [11]), and SM is a special case of Linear.

Algorithms

We first introduce an *incremental algorithm* for the simple resource allocation problem, SC/Simple/D. We assume that each f_j is defined over the interval $[0, N]$. Since f_j is convex, we have

$$d_j(1) \leq \cdots \leq d_j(N), \quad (12)$$

where

$$d_j(y_j) = f_j(y_j) - f(y_j - 1).$$

The incremental algorithm is a kind of *greedy algorithm*, and is also called a *marginal allocation* method. Starting with the initial solution $x = (0, \ldots, 0)$, one unit of resource is allocated at each iteration to the most favorable activity (in the sense of minimizing the increase in the current objective value) until $\sum x_j = N$ is attained.

```
Input: An instance of SC/Simple/D.
Output: An optimal solution x*.
    Let x := (0, ..., 0) and k := 0;
    WHILE k < N DO
        Find j* such that
        d_{j*}(x_{j*} + 1) = min_{1≤j≤n} d_j(x_j + 1);
        x_{j*} := x_{j*} + 1;
        k := k + 1
    END;
    Output x as x*.
```

Procedure INCREMENT

It has been shown this procedure correctly computes an optimal solution in $O(N \log n + n)$ time.

Several polynomial time algorithms have been developed for problem SC/Simple/D [3,7,14]. The fastest among them is proposed in [3]. Its running time is $O(\max\{n, n \log(N/n)\})$, but the algorithm is very complicated. All these algorithms are based on *divide-and-conquer*.

The incremental algorithm presented above also works for problem SC/SM/D. In this case, among all the elements such that $x + e(j) \in P(r)$ (i.e., feasible except for the constraint $(x + e(j))(E) = r(E)$), the x_j with the minimum increase in $f_j(x_j)$ is incremented by one. This process is repeated until $x(E) = r(E)$ is finally attained.

A polynomial time algorithm for SC/SM/D is also known [2,4]. It first solves a problem of SC/Simple/D type, which is obtained from the original problem by by considering only the simple constraint $x(E) = r(E)$ but disregarding the rest. If the obtained solution y is feasible, we are done, i.e., it is an optimal solution of the original problem. Otherwise, the problem is decomposed into subproblems using the information obtained from the vector y and the submodular constraints.

When specialized to problem SC/Network/D, the running time becomes $O(|T|(\tau(n, m, C_{\max}) + |T| \log(N|T|)))$, where $\tau(n, m, C_{\max})$ denotes the running time for the maximum flow algorithm for a graph with n vertices, m arcs and the maximum arc capacity $C_{\max}$. The direct consequence of this result is that problems SC/GUB/D, SC/Nested/D and SC/Tree/D can be solved in $O(n^2 \log(Nn))$ time. For SC/Nested/D, the running time was improved to $O(n \log n \log(N/n))$ in [8]. The idea of the improvement is based on a general and beautiful proximity theorem between integral and continuous optimal solutions for SC/SM/D and SC/SM/C.

For SQC/–/D with – equal to Simple, GUB, Nested, Tree or Network, D.S. Hochbaum and S. Hong [9] developed improved algorithms based on proximity result between SQC/–/C and SQC/–/D and efficient algorithms for SQC/–/C.

Minimax/–/D and Maximin/–/D, are equivalently transformed into problems of SC/–/D. Therefore, equally efficient algorithms can be developed for minimax and maximin problems. The transformation is done as follows: We only show this fact for the most general case, i.e., Minimax/SM/D, which are described as follows:

$$\text{MINIMAX} \begin{cases} \min \max_{j \in E} f_j(x_j), \\ \text{s.t.} \quad x \text{ is an integral base of } B(r). \end{cases}$$

Here, all f_j, $j \in E$, are assumed to be nondecreasing. When all f_j are nonincreasing, problems MINIMAX and MAXIMIN are mutually transformed into MAXIMIN and MINIMAX, respectively, by the following identities:

$$-\min_x \max_{j \in E} f_j(x_j) = \max_x \min_{j \in E} -f_j(x_j),$$

$$-\max_x \min_{j \in E} f_j(x_j) = \min_x \max_{j \in E} -f_j(x_j).$$

Define for $j \in E$,

$$g_j(x_j) = \sum_{y=0}^{x_j} f_j(y), \quad x_j = 0, 1, \ldots. \quad (13)$$

Note that

$$g_j(x_j) - g_j(x_j - 1) = f_j(x_j)$$

holds for each $x_j = 0, 1, \ldots$. From the nondecreasingness of f_j, it follows that g_j is convex over the nonnegative integers. Now consider the following problems of SC/SM/D:

$$Q_g : \min \left\{ \sum_{j \in E} g_j(x_j) : x \in B(r) \cap Z^E \right\}.$$

It is then shown that an optimal solution of problem Q_g is optimal to MINIMAX.

Generalizations

We finally note a recent development. K. Ando, S. Fujishige and T. Naitoh [1,6] considered the separable convex resource allocation problem for a *bisubmodular system* and for a *finite jump system*, whose underlying constraint can be viewed as a generalization of the submodular constraint. They developed greedy algorithms for such problems. For the case of a bisubmodular system, a polynomial time algorithm has been given in [5]. Also, Hochbaum and J.G. Shanthikumar [10] showed that, for a class of general linear constraints, efficient algorithms can be developed. The running time of their algorithm depends on the maximum absolute value of the subdeterminants, Δ, and if $\Delta = 1$ (i.e., the constraint matrix is *totally unimodular*), the running time becomes polynomial. The idea is based on the proximity result between the integral and continuous optimal solutions. When $\Delta = 1$, V.V. Karzanov and S.T. McCormick [12] proposed another polynomial time algorithm.

In addition to these efforts to generalize the constraints, new progress has recently been made towards generalizing objective functions for which efficient algorithms can still be developed. This research was done by K. Murota [16,17] who identified a subclass of nonseparable convex functions, M-convex functions, which is defined on the base polyhedron of a submodular system as follows.

A function $f : \mathbf{Z}^E \to \mathbf{R} \cup \{\infty\}$ is said to be *M-convex* if it satisfies the following property:

- (M-EXC): For any $x, y \in \text{dom } f$ and for any $i \in \text{supp}^+(x-y)$, there exists a $j \in \text{supp}^-(x-y)$ such that

$$f(x) + f(y) \geq f(x - e(i) + e(j)) + f(y + e(i) - e(j)),$$

where

$$\text{dom } f = \{x \in \mathbf{Z}^E : f(x) < +\infty\},$$
$$\text{supp}^+(x - y) = \{k \in E : x_k > y_k\},$$
$$\text{supp}^-(x - y) = \{k \in E : x_k < y_k\}.$$

The M-convex functions can enjoy nice theorems of *discrete convex analysis* in a parallel manner to the traditional convex analysis. A polynomial time algorithm has been developed for this class of problems [18].

See also

- ▶ Combinatorial Matrix Analysis
- ▶ Combinatorial Optimization Games
- ▶ Competitive Facility Location
- ▶ Evolutionary Algorithms in Combinatorial Optimization
- ▶ Facility Location with Externalities
- ▶ Facility Location Problems with Spatial Interaction
- ▶ Facility Location with Staircase Costs
- ▶ Fractional Combinatorial Optimization
- ▶ Global Optimization in Weber's Problem with Attraction and Repulsion
- ▶ MINLP: Application in Facility Location-Allocation
- ▶ Multifacility and Restricted Location Problems
- ▶ Multi-Objective Combinatorial Optimization
- ▶ Network Location: Covering Problems
- ▶ Neural Networks for Combinatorial Optimization
- ▶ Optimizing Facility Location with Rectilinear Distances
- ▶ Production-distribution System Design Problem
- ▶ Replicator Dynamics in Combinatorial Optimization
- ▶ Resource Allocation for Epidemic Control
- ▶ Simple Recourse Problem
- ▶ Single Facility Location: Circle Covering Problem
- ▶ Single Facility Location: Multi-Objective Euclidean Distance Location
- ▶ Single Facility Location: Multi-Objective Rectilinear Distance Location
- ▶ Stochastic Transportation and Location Problems
- ▶ Voronoi Diagrams in Facility Location
- ▶ Warehouse Location Problem

References

1. Ando K, Fujishige S, Naitoh T (1994) A greedy algorithm for minimizing a separable convex function over an integral bisubmodular polyhedron. J Oper Res Soc Japan 37:188–196
2. Federguruen A, Groenevelt H (1986) The greedy procedure for resource allocation problems - necessary and sufficient conditions for optimality. Oper Res 34:909–918
3. Frederickson GN, Johnson DB (1982) The complexity of selection and ranking in X +Y and matrices with sorted columns. J Comput Syst Sci 24:197–208
4. Fujishige S (1980) Lexicographically optimal base of a polymatroid with respect to a weight vector. Math Oper Res 21:186–196

5. Fujishige S (1997) A min-max theorem for bisubmodular polyhedra. SIAM J Discret Math 10:294–308
6. Fujishige S, Naitoh T (1995) A greedy algorithm for minimizing a separable convex function over a finite jump system. J Oper Res Soc Japan 38:362–375
7. Galil Z, Megiddo N (1979) A fast selection algorithm and the problem of optimum distribution of effort. J ACM 26:58–64
8. Hochbaum DS (1994) Lower and upper bounds for the allocation problem and other nonlinear optimization problems. Math Oper Res 19:390–409
9. Hochbaum DS, Hong S (1995) About strongly polynomial time algorithms for quadratic optimization over submodular constraints. Math Program 55:269–309
10. Hochbaum DS, Shanthikumar JG (1990) Nonlinear separable optimization is not much harder than linear optimization. J ACM 37:843–862
11. Ibaraki T, Katoh N (1988) Resource allocation problems: Algorithmic approaches. MIT, Cambridge, MA
12. Karzanov AV, McCormick ST (1997) Polynomial methods for separable convex optimization in totally unimodular linear spaces with applications. SIAM J Comput 26:1245–1275
13. Katoh N, Ibaraki T (1998) Resource allocation problems. In: Du D-Z, Pardalos PM (eds) Handbook Combinatorial Optim., vol 2. Kluwer, Dordrecht
14. Katoh N, Ibaraki T, Mine H (1979) A polynomial time algorithm for the resource allocation problem with a convex objective function. J Oper Res Soc 30:159–260; 449–455
15. Koopman BO (1953) The optimum distribution of effort. Oper Res 1:52–63
16. Murota K (1996) Convexity and Steinitz's exchange property. Adv Math 124:272–311
17. Murota K (1998) Discrete convex analysis. Math Program 83:313–371
18. Shioura A (1998) Minimization of an M-convex function. Discrete Appl Math 84:215–220

Combinatorial Optimization Games
CRPM

XIAOTIE DENG
City University Hong Kong, Kowloon, China

MSC2000: 91A12, 90C27, 90C60

Article Outline

Keywords
See also
References

Keywords

Combinatorial optimization; Cooperative game; Complexity; Algorithms

In combinatorial optimization games, we consider *cooperative games* for which the value of the game is obtained via a *combinatorial optimization* problem. For a cooperative game (a class of *games with side payments*), the set of participating players is denoted by N and a value $v(S)$ is achieved by each subset S of players without any help from other players (in the set $N - S$). Usually, we set $v(\emptyset) = 0$. In general, the representation of the game requires an input size exponential in the number of players. For a combinatorial optimization game, however, the value $v(S)$ is often succinctly defined as a solution to a combinatorial optimization problem for which the combinatorial structure is determined by the subset S of players. The income distributed to individual player i is represented by x_i, $1 \leq i \leq N$, and $x = (x_1, \ldots, x_N)$.

The main issue in cooperative games is how to fairly distribute the income collectively earned by the whole group of players in the game, cooperating with each other. For simplicity, let $x(s) = \sum_{i \in S} x_i$. The income vector x is called an *imputation* if $x(N) = v(N)$, and $\forall i \in N: x_i \geq v(\{i\})$ (*individual rationality*). Additional requirements may be added to ensure fairness, stability and rationality. And they lead to different sets of income vectors which are generally referred to as solution concepts. Among many of these solution concepts, the core, which consists of all the imputations satisfying the subgroup rationality condition $\forall S \subseteq N: x(s) \geq v(s)$, is naturally defined and has attracted much attention from researchers. It has also led to many fruitful results in combinatorial optimization games. Our focus in this article will be on the core. Readers interested in other solution concepts for cooperative games in general can find them in many *game theory* books and survey papers. For example, [12] gives an interesting discussion for several classical solution concepts in cooperative games and their applications to political economy.

Recently, *computational complexity* has been suggested as another metric for evaluating the rationality of these solution concepts [2]. In this argument, computational complexity is suggested as a measure of bounded rationality [13] for players not to spend *super-

polynomial time to search for the most suitable solution. For combinatorial optimization games, N. Megiddo [7] suggested that algorithms polynomial in the number of players (as good algorithms following the concept introduced by J. Edmonds [3]) be sought for solutions. As the value of any subset of players is defined as the optimal solution to a combinatorial optimization problem, the input size can often be restricted to be bounded by a polynomial in the number of players. This is usually the case for many practical collective optimization problems. The value of a subgroup of players is the optimal objective function value that this subgroup can achieve under the constraints imposed by resources controlled by players in the subgroup. Very often the collective optimization problem requires an integer solution. It is under this context the game is then referred to as a *combinatorial optimization game*.

An example to formulate a two-sided market (the assignment game) is given in [11]. The underlying structure is a bipartite graph $(V_1, V_2; E)$. One interpretation given by L.S. Shapley and M. Shubik is that V_1 is the set of sellers, and V_2 is the set of buyers. For the simplest case, each seller has an item (say a house) to sell and each buyer wants to purchase an item. The ith seller, $i \in V_1$, values its item at c_i dollars and the jth buyer values the item of the ith seller at h_{ij} dollars. Between this pair, we may define a value $v(\{i, j\}) = h_{ij} - c_i$ if $h_{ij} \geq c_i$ and set (i, j) an edge in E with weight $v(\{i, j\})$. Otherwise, there is no edge between i and j since no deal is possible if the seller values the item more than the buyer does. Considering a game with side-payment, the value $v(s)$ of a subset S of buyers and sellers is defined to be the weight of maximum matching in the bipartite graph $G[S]$ induced by the corresponding set S of vertices (an edge is in $G[S]$ if and only if its two end vertices are both in S). In a linear programming formulation, this is

$$\begin{cases} v(S) = \max & \sum_{(i,j) \in E} v(i,j) x_{ij} \\ \text{s.t.} & \sum_{i \in V_1 \cap S} x_{ij} \leq 1 \\ & \sum_{j \in V_2 \cap S} x_{ij} \leq 1 \\ & x \geq 0. \end{cases}$$

Shapley and Shubik have shown that the core for this assignment game is precisely the set of solutions for the dual program of the above linear program with $S = V_1 \cup V_2$. Such nice properties are not unusual in combinatorial optimization games. For example, the same fact is established for another game, a cost allocation game on trees, by A. Tamir. Tamir has shown that the core is exactly the set of optimal solutions to the dual program of the linear program formulation for the total cost of the cost allocation problem on trees [14].

The Shapley–Shubik model is a theoretical formulation for a *pure exchange economy*. The linear production game of G. Owen [8] applies their ideas to a production economy. In Owen's model, each player j ($j \in N$) owns a resource vector, b^j. For a subset S of players, their value is the objective function value of the optimal solution for the following linear program:

$$\begin{cases} \max & c^\top y \\ \text{s.t.} & Ay \leq \sum_{j \in S} b^j \\ & y \geq 0. \end{cases} \quad (1)$$

Thus, the value is what the subset of players can achieve in the linear production model with the resources under their control. The core for the linear production game is always nonempty [8] if all the above linear programs have finite optimum. A constructive proof presented by Owen obtains an imputation in the core from any optimal solution of the dual program

$$\begin{cases} \min & \sum_{j \in N} w^\top b^j \\ \text{s.t.} & w^\top A \geq c^\top \end{cases}$$

of the linear program for all the players

$$\begin{cases} \max & c^\top y \\ \text{s.t.} & Ay \leq \sum_{j \in N} b^j. \end{cases}$$

In fact, let w be the optimal solution for the dual program. Set $x_j = w^\top b^j, j \in N$. Then $x = (x_1, \ldots, x_N)$ is an imputation in the core. To see so, for each subset $S \subseteq N$, consider $x(s) = \sum_{i \in S} x_i$. By definition of x, we have $x(s) = \sum_{i \in S} w^\top b^i$. Let y_S^* be the optimal solution for the linear program for $v(s)$. Then, $Ay_S^* \leq \sum_{y \in S} b^j$. Therefore, $x(s) \geq w^\top A y_S^*$. On the other hand, $w^\top A \geq c^\top$. It follows that $x(s) \geq c^\top y_S^*$, which is the same as $x(s) \geq v(s)$ since $v(s) = c^\top y_S^*$.

Notice that this proof depends on the fact that, if for each $S \subseteq N$, the linear program (1) has a finite optimal value. In general, a linear program may be unbounded or infeasible. If for any $S \subseteq N$, the linear program (1) is unbounded, obviously the core does not exist. If it is infeasible, we may define $v(s) = -\infty$. This allows the extension of the above result to the case when the following conditions are satisfied:

1)
$$\begin{cases} \max & c^\top y \\ \text{s.t.} & Ay \leq \sum_{j \in N} b^j \\ & y \geq 0 \end{cases}$$

has a finite optimal value.

2) For each $S \subseteq N$, 1) has a finite optimal value or is infeasible.

However, unlike the assignment game, there may in general be imputations in the core which cannot be obtained from the dual program for Owen's linear production game [8]. In general, it is not known how to decide whether an imputation is in the core in polynomial time.

There is a weakness in applying the linear production game model to the studies of coalition optimization problems. That is, in reality, many variables are required to be of integer values. It happens that for the assignment game of Shapley and Shubik, the linear production model of Owen's always results in an integer solution. There are, however, many other situations for which the integer optimal solution cannot be obtained in the framework of the linear program.

A generalized linear production model introduced by D. Granot retains the main linear program structure of Owen's model but allows right-hand sides of the resource constraints not to be linear in the resource vectors b^j of individual players [6]. Thus, $v(S)$ is defined to be $\max\{c^\top y : Ay \leq b(S), y \geq 0\}$, where $b(S) = (b_1(s), \ldots, b_m(s))$ is a general function of S. It is shown that, if for each i, $1 \leq i \leq N$, the game consisting of player set N with value function $b_i(s)$ has a nonempty core, the generalized linear production game has a nonempty core. As the game of Owen's model, an imputation in the core is constructed from the optimal solution for the dual program and vectors in the core associated with (N, b_i) [6]. This would in general need an exponential number of function values $b(s)$ for all the subset S of N. For some collective combinatorial optimization problems, $b(S)$ is given implicitly as a solution to some optimization problem and thus the problem input size is polynomially bounded. The extended power of Granot's model can be applied to prove nonemptiness for the cores of many games beyond those of Owen's linear production game.

In particular, the generalized linear production game model is applied to show the nonemptiness of a certain minimum cost spanning tree game [6]. In this problem, we have a complete graph as the underlying structure. A cost is assigned to each edge. There is a distinguished node 0. Players are vertices $\{1, \ldots, n\}$. The cost $c(S)$ of a subset S of players is defined to be the cost of minimum spanning tree in the graph $G[S \cup 0\}]$ induced by $S \cup \{0\}$. (Notice that the cost game is different from the value game defined as above but can be handled similarly.) Even though an imputation in the core can be found in polynomial time for this game, in [4] it is shown that it is NP-hard to decide whether an imputation is not in the core.

Another way to extend Owen's model to include games of combinatorial optimization nature is to explicitly require integer solutions in the definition of the linear production model. That is, one may define game value $v(s)$ for a subset $S \subseteq N$ to be the maximum value of an integer program instead of a linear program. Therefore,

$$v(S) = \max\left\{c^\top x : Ax \leq \sum_{j \in S} b^j, x \text{ integers}\right\}.$$

For the assignment game of Shapley and Shubik and the cost allocation game on trees of Tamir, the integer program can be solved by its linear program relaxation, since there is always an integer solution for the latter. In the work of Shapley and Shubik, as well as that of Tamir, b^j is a unit vector and $b(N)$ is a vector of all ones. It is this particular structure of linear constraints that makes the core to be identified with the set of optimal solutions for the dual linear program to the linear program of the game value for the set of players [11,14]. It is no wonder this property is further exploited in [5] for a partition game, and in [1] for packing/covering games.

The *packing game*, for example, is defined for a set N of players whose game value is given by the following integer program

$$\begin{cases} \max & c^\top x \\ \text{s.t.} & x^\top A_{M,N} \leq 1^N \\ & x \in \{0,1\}^m, \end{cases}$$

where 1^N is a vector of $|N|$ ones, and $A_{M,N}$ is a 0–1 matrix of rows indexed by M and columns indexed by N. For each subset S of players, its value is given by

$$\begin{cases} \max & c^\top x \\ \text{s.t.} & x^\top A_{M,S} \leq 1^\top_{|S|}, \quad x^\top A_{M,\bar{S}} \leq 0^\top_{n-|S|}, \\ & x \in \{0,1\}^m, \end{cases}$$

where $A_{M,S}$ is the submatrix of A with row set M and column set S, $\bar{S} = N - S$ and $v(\emptyset)$ is defined to be 0.

The covering game and the partition game are defined similarly. It is a necessary and sufficient condition for the core of the packing (and covering, and partitioning) game to be nonempty that the linear relaxation of the corresponding optimization problem always has an integer optimal solution. In additional, the core, if nonempty, is exactly the set of optimal solutions to the dual program of the linear relaxation of the corresponding integer program [1,5].

These results allow for a characterization of combinatorial structures for the corresponding combinatorial optimization game to have a nonempty core. Because of the linear program characterization of the core, questions such as whether the core is empty or not, whether we can find an imputation in the core, and whether an imputation is in the core, can often be determined in polynomial time. Notice that, there are cases that the linear program may be of exponential size in the number of players, it is not immediate that all these questions can be solved in polynomial time. But even for cases when there are an exponential number of constraints, the linear program may be solvable in polynomial time [9].

First established by Shapley and Shubik for the assignment game, the connection of the core for a combinatorial optimization game with dual program of the linear program relaxation has been a successful tool in the characterization of the core, design and analysis of algorithms to find an imputation in the core and to test membership of an imputation in the core. It is expected that this approach would continously lead to fruitful results in cooperative game theory.

See also

▶ Combinatorial Matrix Analysis
▶ Evolutionary Algorithms in Combinatorial Optimization
▶ Fractional Combinatorial Optimization
▶ Multi-Objective Combinatorial Optimization
▶ Neural Networks for Combinatorial Optimization
▶ Replicator Dynamics in Combinatorial Optimization

References

1. Deng X, Ibaraki T, Nagamochi H (1999) Algorithmic aspects of the core of combinatorial optimization games. Math Oper Res 24(3):751–766
2. Deng X, Papadimitriou C (1994) On the complexity of cooperative game solution concepts. Math Oper Res 19(2):257–266
3. Edmonds J (1965) Paths, tree, and flowers. Canad J Math 17:449–469
4. Faigle U, Fekete S, Hochstättler W, Kern W (1997) On the complexity of testing membership in the core of min-cost spanning tree games. Internat J Game Theory 26:361–366
5. Faigle U, Kern W (1995) Partition games and the core of hierarchically convex cost games. Memorandum Fac Toegepaste Wiskunde Univ Twente 1269, no. June
6. Granot D (1986) A generalized linear production model: A unified model. Math Program 34:212–222
7. Megiddo N (1978) Computational complexity and the game theory approach to cost allocation for a tree. Math Oper Res 3:189–196
8. Owen G (1975) On the core of linear production games. Math Program 9:358–370
9. Schrijver A (1986) Theory of linear and integer programming. Wiley, New York
10. Shapley LS (1967) On balanced sets and cores. Naval Res Logist Quart 14:453–460
11. Shapley LS, Shubik M (1972) The assignment game. Internat J Game Theory 1:111–130
12. Shubik M (1981) Game theory models and methods in political economy. In: Arrow KJ, Intriligator MD (eds) Handbook Math. Economics, vol 1. North Holland, Amsterdam, pp 285–330
13. Simon H (1972) Theories of bounded rationality. In: Radner R (ed) Decision and Organization. North Holland, Amsterdam

14. Tamir A (1981) On the core of cost allocation games defined on location problems. Preprints, Second Internat. Conf. Locational Decisions (ISOLDE 81) (Skodsborg, Denmark), pp 387–402

Combinatorial Test Problems and Problem Generators

DON GRUNDEL[1], DAVID JEFFCOAT[2]
[1] 671 ARSS/SYEA, Eglin AFB, USA
[2] AFRL/RWGN, Eglin AFB, USA

MSC2000: 90B99, 05A99

Article Outline

Keywords
Introduction
Libraries and Generators
   Combinatorial Auctions
   Frequency Assignment Problem
   Graph Colorability
   Linear Ordering Problem
   Maximum Clique Problem
   Minimum Cut-Set
   Minimum Vertex Cover Problem
   Multidimensional Assignment Problem (MAP)
   Quadratic Assignment Problem (QAP)
   Satisfiability
   Steiner Problem in Graphs
   Traveling Salesman Problem (TSP)
   Vehicle Routing Problem
Conclusions
References

Keywords

Test problems; Problem generators

Introduction

Test problems are instances of a mathematical problem used to establish the accuracy or efficiency of a solution method. Test problems provide a common baseline against which to compare a new solution algorithm with an existing procedure. A problem generator is an algorithm to produce a test instance for a specific combinatorial problem. In this section, we provide an overview of test problems and generation methods, as well as sources of test problems for a number of well-known combinatorial problems. The design of test problems is a critical step in the design of combinatorial algorithms, and a sufficient number and variety of test problems must be available to determine the performance of a proposed algorithm across a range of problem types.

There are four basic sources of test problems:
1. Problems taken from real-world applications
2. Libraries of standard test problems
3. Test problems with parameters generated randomly from a specified probability distribution
4. Test problems generated by an algorithm designed to produce problem instances with specific characteristics: e. g., problems with a known solution.

Each of these sources has associated advantages and disadvantages. For example, problems taken from real-world applications have a degree of complexity consistent with at least some problems encountered in practice [28], and provide a context for presenting proposed solutions that promotes understanding and acceptance. However, we typically cannot find a sufficient number of such problems to constitute a satisfactory experiment.

Libraries of standard test problems can provide problems that were used by other researchers, facilitating comparisons with existing solution procedures. However, as with real-world cases, libraries may not provide a sufficient number or variety of problem instances. Procedures that randomly generate test problems can quickly provide an essentially unlimited number of problem instances, but the optimal solution to large randomly generated problems may remain unknown. An additional hazard with randomly generated problems is that such problems are sometimes artificially easy to solve [6,33].

Constructive procedures designed to generate test problems with known solutions can be very useful for evaluating an algorithm's performance, and can also provide a large number of test instances. Problem generation procedures must be carefully examined to determine the difficulty, realism, and other characteristics of the problems generated. An ideal generator would produce problems in polynomial time, with a known solution, of appropriate hardness, and with sufficient diversity [31]. Of course, it can be difficult to simultaneously meet all these requirements. For example, a trivial problem instance might be generated in polynomial time, but provide no real test for a proposed solution procedure. Problem instances should also be posed using standardized representations [10].

A good set of test problems is only one part of the evaluation of an algorithm. Barr, et al. [2] provide guidelines for designing computational experiments and for reporting results of solution algorithm performance.

The following section provides sources of standard test problems and problem generators for a number of well-known combinatorial optimization problems.

Libraries and Generators

The INFORMS OR/MS Resource Collection [16] and the OR-Library maintained by Beasley [3,4] both provide extensive collections of test data sets for a variety of operations research problems. The Zuse Institute [19] maintains a collection of various problems related to mathematical programming. A handbook of test problems [11] provides a collection of test problems from a wide variety of engineering applications. The Discrete Mathematics and Theoretical Computer Science (DIMACS) Challenges [8] encourage experimental evaluations of algorithms using standard test problems. Over the past decade, challenges have been held for TSP, cliques, coloring, and satisfiability. An overview of sources for specific combinatorial problems is provided below.

Combinatorial Auctions

This problem involves auctions in which bidders place unrestricted bids for bundles of goods. A seller faced with a set of offers for bundles of goods wishes to maximize his revenue. The Combinatorial Auction Test Suite (CATS) provides an algorithm for generating problem instances of differing levels of realism [21].

Frequency Assignment Problem

A library of frequency assignment problems in the context of wireless communication networks is available at [9]. This website includes an extensive bibliography on frequency assignment problems.

Graph Colorability

Sanchis [31] provides an algorithm for generating graph colorability problems with known solutions. This reference also provides a generator for the minimum dominating set problem.

Linear Ordering Problem

Reinelt [29] maintains a library of problems instances for the linear ordering problem, including problem data and optimal solutions. This library also includes software and data for several other discrete optimization problems. Another library is maintained by Martí [22] in which there are large randomly generated problems with best known solutions.

Maximum Clique Problem

Hasselberg, et al. [13] consider a number of interesting problems, including the maximum clique problem. They introduce different test problem generators motivated by a variety of practical applications, including coding theory and fault diagnosis.

Minimum Cut-Set

Krishnamurthy [20] provides a problem generator for partitioning heuristics, including the minimum cut-set problem. Generated instances of this problem are useful in circuit design applications.

Minimum Vertex Cover Problem

Sanchis and Jagota [32] discuss a test problem generator that builds instances of the minimum vertex cover problem. The generator provides construction parameters to control problem difficulty. Sanchis [31] provides an algorithm to generate minimum vertex cover problems that are diverse, hard and of known solution.

Multidimensional Assignment Problem (MAP)

The axial MAP is a generalization of the linear assignment problem. Grundel and Pardalos [12] provide a MAP generator that produces difficult problems with known unique optimal solutions.

Quadratic Assignment Problem (QAP)

Pardalos [25] provides a method for constructing test problems for constrained bivalent quadratic programming. This reference includes a standardized random test problem generator for the unconstrained quadratic zero-one programming problem. Yong and Pardalos [35] provide methods for generating test problems with known optimal solutions for more general cases of the QAP. Calamai, et al. [7] describe a technique for generating convex, strictly concave and indefinite QAP instances. Palubeckis [24] provides a method for generating hard rectilinear instances of the QAP with known optimal solutions. Burkard, et al. [5] give additional useful information concerning this difficult problem.

Satisfiability

Achlioptas, et al. [1] propose a generator for satisfiability problems that controls the hardness of the instances. A web page maintained by Uchida, Motoki, and Watanabe [34] is dedicated to two methods of generating instances of 3-satisfiability. A library of satisfiability problem instances and solvers is available on a Darmstadt University website [15].

Steiner Problem in Graphs

Khoury, et al. [17] use a binary-programming formulation to generate test problems with known solutions by applying the Karush-Kuhn-Tucker optimality conditions to the corresponding quadratically-constrained optimization problem. Koch, et al. [18] provide a library of Steiner tree problems with information about the origin, solvability, and other characteristics of this problem.

Traveling Salesman Problem (TSP)

Moscato [23] maintains a web site with resources for the generation of TSP instances with known optimal solutions. An approach for generating discrete instances of the symmetric TSP with known optima is provided by Pilcher and Rardin [27]. A number of libraries (e. g [4,30]) provide test cases for the TSP.

Vehicle Routing Problem

Homberger [14] provides a large set of Vehicle Routing Problems with Time Windows, including instances with up to one thousand customers.

Conclusions

Researchers need a large set of well-designed test problems to effectively compare the performance of existing solution algorithms or to evaluate a new algorithm. Although practitioners may prefer real-world problems for such tests, a sufficient number of test problems may not be available to conduct a thorough experiment. Randomly generated test problems can provide an essentially limitless supply of instances. However, random test instances may be artificially easy to solve, or, at the other extreme, may have no known solution, making it difficult to judge the performance of a new solution algorithm. Test problem generators, if properly designed, can provide a large supply of hard problem instances with known optimal solutions. Many such generators are readily available to researchers. Libraries of test problem are also available, providing a variety of problem types and sizes.

References

1. Achlioptas D, Gomes C, Kautz H, Selman B (2000) Generating satisfiable problem instances. In: Proceedings of the 17th National Conference on Artifical Intelligence, Austin, USA, 31 July–2 Aug 2000. AAAI Press, Menlo Park, USA, pp 256–261
2. Barr RS, Golden BL, Kelly JP, Resende MGC, Stewart WR (1995) Designing and Reporting on Computational Experiments with Heuristic Methods. Heuristics J 1: 9–32
3. Beasley JE (1990) OR-Library: distributing test problems by electronic mail. J Oper Res Soc 41(11):1069–1072
4. Beasley JE, OR-Library. http://people.brunel.ac.uk/~mastjjb/jeb/info.html. Accessed 7 Dec 2007
5. Burkard R, Çela E, Karisch S, Rendlqaplib F A Quadratic Assignment Problem Library. http://www.opt.math.tu-graz.ac.at/qaplib/. Accessed 7 Dec 2007
6. Burkard R, Fincke U (1985) Probabilistic asymptotic properties of some combinatorial optimization problems. Discret Appl Math 12:21–29
7. Calamai PH, Vicente LN, Júdice JJ (1993) A new technique for generating quadratic programming test problems. Math Program 61:215–231
8. Center for Discrete Mathematics & Theoretical Computer Science, Rutgers University, NJ. http://dimacs.rutgers.edu/. Accessed 7 Dec 2007
9. Eisenblätter A, Koster A. FAP web, A website about Frequency Assignment Problems. http://fap.zib.de/index.php. Accessed 7 Dec 2007
10. Fourer R, Lopes L, Martin K (2004) LPFML: A W3C XML Schema for Linear Programming. http://www.

optimization-online.org/DB_HTML/2004/02/817.html. Accessed 7 Dec 2007
11. Floudas CA, Pardalos PM, Adjiman CS, Esposito WR, Gümüs ZH, Harding ST, Klepeis JL, Meyer CA, Schweiger CA (1999) Handbook of Test Problems in Local and Global Optimization. Kluwer, Dordrecht, Netherlands
12. Grundel D, Pardalos P (2005) Test Problem Generator for the Multidimensional Assignment Problem. Comput Optim Appl 31(3)133–146
13. Hasselberg J, Pardalos PM, Vairaktarakis G (1993) Test case generators and computational results for the maximum clique problem. J Glob Optim 3:463–482
14. Homberger J Extended SOLOMON's VRPTW instances. http://www.fernuni-hagen.de/WINF/touren/menuefrm/probinst.htm. Accessed 7 Dec 2007
15. Hoos H, Stützle T (2000) SATLIB: An Online Resource for Research on SAT. In: Gent IP, v Maaren H, Walsh T (eds) SAT 2000. IOS Press, pp 283–292. http://www.satlib.org/. Accessed 7 Dec 2007
16. INFORMS® Online, OR/Resource MS Collection: Resources: Problem Instances. http://www.informs.org/Resources/Resources/Problem_Instances/. Accessed 7 Dec 2007
17. Khoury BN, Paradalos PM, Du D-Z (1993) A Test Problem Generator for the Steiner Problem in Graphs. Trans ACM Math Softw 19(4):509–522
18. Koch T, Martin A, Voß S SteinLib: An Updated Library on Steiner Tree Problems in Graphs. http://elib.zib.de/steinlib/steinlib.php. Accessed 7 Dec 2007
19. Konrad-Zuse-Zentrum für Informationstechnik Berlin (ZIB), a non-university research institute of the state Berlin (1994–2002) http://elib.zib.de/pub/Packages/mp-testdata/index.html. Accessed 7 Dec 2007
20. Krishnamurthy B (1987) Constructing test cases for partitioning heuristics. Trans IEEE Comput vol C-36, Num. 9, September 1987, pp 1112–1114
21. Leyton-Brown K, Pearson M, Shoham Y (2000) Towards a Universal Test Suite for Combinatorial Auction Algorithms. In: Proceedings of Conference ACM on Electronic Commerce, Minneapolis, USA, 17–20 Oct. Sponsored by Association for Computing Machinery (ACM) Special Interest Group on E-commerce. Test suite available via http://www.cs.ubc.ca/~kevinlb/CATS/. Accessed 7 Dec 2007
22. Martí R. Linear Ordering: Publications and Working Papers. http://www.uv.es/~rmarti/paper/lop.html. Accessed 7 Dec 2007
23. Moscato P Fractal Instances of the Traveling Salesman Problem. Densis, FEEC, Universidade UNICAMP Estadual de Campinas. http://www.ing.unlp.edu.ar/cetad/mos/FRACTAL_TSP_home.html. Accessed 7 Dec 2007
24. Palubeckis G (1999) Generating hard test instances with known optimal solution for the rectilinear quadratic assignment problem. J Glob Optim 15:127–156
25. Pardalos P (1991) Construction of test problems in quadratic bivalent programming. Trans ACM Math Softw 17(1):74–87, March 1991
26. Pardalos P, Pitsoulis L (eds) (2000) Nonlinear Assignment Problems. Algorithms and Applications. Kluwer, Dordrecht, pp 1–12
27. Pilcher M, Rardin R (1992) Partial polyhedral description and generation of discrete optimization problems with known optima. Nav Res Logist 39:839–858
28. Reilly CH (1999) Input models for synthetic optimization problems. In: Farrington PA, Nembhard HB, Sturrock DT, Evans GW (eds) Proceedings of the 1999 Winter Simulation Conference, Phoenix, USA
29. Reinelt G Linear Ordering Library (LOLIB). http://www.iwr.uni-heidelberg.de/groups/comopt/software/LOLIB/. Accessed 7 Dec 2007
30. Reinelt G. TSLIB. http://www.iwr.uni-heidelberg.de/groups/comopt/software/TSPLIB95/. Accessed 7 Dec 2007
31. Sanchis L (1995) Generating hard and diverse test sets for NP-hard graph problems. Discret Appl Math 58:35–66
32. Sanchis L, Jagota A (1996) Some experimental and theoretical results on test case generators for the maximum clique problem. INFORMS J Comput 8(2):87–102
33. Selman B, Mitchell D, Levesque H (1996) Generating Hard Satisfiability Problems. Artif Intell 81:17–29
34. Uchida T, Motoki M, Watanabe O Instance SAT Generation Page, Watanabe research group of Dept. of Math. and Computing Sciences, Tokyo Inst. of Technology. http://www.is.titech.ac.jp/~watanabe/gensat/index.html. Accessed 7 Dec 2007
35. Yong L, Pardalos PM (1992) Generating quadratic assignment test problems with known optimal permutations. Comput Optim Appl 1(2):163–184

Communication Network Assignment Problem
CAP

RAINER E. BURKARD
Technical University Graz, Graz, Austria

MSC2000: 90B80, 90C05, 90C27, 68Q25

Article Outline

Keywords
See also
References

Keywords

Optimization; Assignment problem; Communication network; Computational complexity; Exact algorithms; Heuristic approaches; Asymptotic behavior

In the *communication network assignment problem* (CAP) a system of communication centers $C_1, \ldots, C_n$ is given. The centers have to be embedded into a given (undirected) network $N = (V, E)$ with vertex set V, $|V| = n$, and edge set E. The centers exchange messages at given rates per time unit through a selected routing pattern. Let t_{ij} be the amount of messages sent from center C_i to center C_j per time unit. If there is no direct connection between C_i and C_j the messages sent from C_i to C_j pass through several intermediate centers. The messages exchanged between C_i and C_j may be sent along a single path or they may be split into several parts, each part being sent along its own path. For any fixed embedding $\mathcal{E}$ of the centers into the network and for any fixed routing pattern ρ of the messages, let $IMT_{\mathcal{E}, \rho}(C_i)$ denote the overall amount of traffic going through the center C_i as intermediate center. The goal is to find an embedding $\mathcal{E}$ of the centers into the network and a routing pattern ρ which minimizes the maximum intermediate traffic over all centers:

$$\min_{\mathcal{E}, \rho} \max \{IMT_{\mathcal{E},\rho}(C_i) \colon 1 \le i \le n\} \quad (1)$$

A typical application of the problem arises in the case of locating stations (terminals, computers) in a *local-area computer network* (LAN) as described by T.B. Boffey and J. Karkazis [1]. Usually, a given segment of the LAN serves different pairs of communicating stations. In order to prevent interference and garbled messages, only one message at a time can be sent through a given segment of the LAN. On the other hand one has to restrict the offered traffic through the same segment so as to maintain a reasonable throughput in the network. To this end it is reasonable to locate *bridges* at the endpoints of each segment. All bridges will work as intermediate centers and all stations will work as bridges. The result is that each pair of stations (or bridges) communicates through its own segment. It is reasonable to require an embedding of stations and additional bridges into the nodes of the LAN such that the intermediate traffic going through the busier station (or bridge) is minimized. Boffey and Karkazis [1] proposed and discussed also a continuous version of the problem.

A similar problem, the so-called *elevator problem* leads also to the optimization problem (1) as described by Karkazis [5]. The elevator problem arises when a single elevator has to be replaced by two elevators, each covering contiguous subsets of floors. It might be reasonable to place the connecting landing so as to minimize the traffic intensity on the busier elevator. More specifically assume that the first elevator serves floors $\{1, 2, \ldots, i\}$ and the second elevator serves floors $\{i, i + 1, \ldots, n\}$, and let t_{ij} represent the traffic intensity from floor i to floor j. Then the traffic load of the first elevator is given by $T_i^{(1)} = \sum_{k=1}^{i} \sum_{l=1}^{n} (t_{kl} + t_{lk})$ and the traffic load of the second elevator is given as $T_i^{(2)} = \sum_{k=i}^{n} \sum_{l=1}^{n} (t_{kl} + t_{lk})$. Then we want to choose i so as to minimize $\max\{T_i^{(1)}, T_i^{(2)}\}$. Obviously this problem setting can be generalized for more than two elevators.

Essentially there are two distinct models of routing patterns in (1): the *single path* model and the *fractional* model. In the single path model, for every pair of communication centers C_i and C_j, a single route in the network is selected and all t_{ij} messages are sent along this fixed route. In the fractional model, the amount t_{ij} is split into a number of positive parts and every part is sent along its own path. Most of the results available in the literature concern the CAP on *trees*. In this case, for each pair of vertices in the network there is only one path to join them and hence, both models coincide.

R.E. Burkard, E. Çela and G.J. Woeginger have proved in [3] that in general the CAP is *NP*-hard. More specifically has been shown that the CAP is *NP*-hard for networks that are i) paths; ii) stars of branch length three; iii) cycles *NP*-hardness in both models); or iv) doublestars (*NP*-hardness in the single path model). Moreover, it has been proved in [3] that the CAP is polynomially solvable in the case of stars of branch length two and in the case of doublestars in the fractional model. In the case of a star of branch length two the CAP can be formulated as a maximum weight perfect matching problem (MWPMP) if the communication center to be assigned to the central node of the star is kept fixed. Since there are only n possibilities for the selection of the center to be placed at the central node, one just has to solve n MWPMPs. In the fractional model, finding an embedding of the communication centers into the nodes of the network and a routing pattern which minimize the intermediate traffic can be done by solving a specified number of linear programs with $O(P)$ variables and $O(n^2)$ constraints each, where P is the number of pairwise disjoint paths in N. In the case of doublestars $P = O(n^2)$ and there are $O(n^2)$ programs to be solved (see [3]). This implies that in this case the CAP is polynomially solvable in the fractional model.

Some *exact algorithms* and *heuristic approaches* to solve the CAP on trees have been proposed in [3,5]. Karkazis has proposed a *branch and bound* algorithm in the case where N is a path [5], and Burkard, Çela and Woeginger [3] have proposed a branch and bound approach in the case that N is a tree. The algorithms have been tested on randomly generated trees and communication rates t_{ij}. The tests show that only small instances of the CAP of size up to 12 can be solved in reasonable time. For large instances the number of the branched nodes in the branch and bound tree explodes. In order to approximately solve larger instances of the CAP on trees Burkard, Çela and T. Dudàs proposed in [2] *simulated annealing* and *tabu search* approaches. The proposed heuristics are tested on randomly generated instances of size up to 32. The comparison of the heuristic solutions with the optimal solution produced by the branch and bound algorithm for instances of small size shows that the performance of these heuristics is quite satisfactory.

Finally, in [2] the *asymptotic behavior* of the CAP on trees has been investigated. Under natural probabilistic assumptions on the problem data the CAP on trees shows a very interesting behavior: The ratio between the maximum and the minimum values of the objective function, i. e., the ratio between the maximum and the minimum values of the intermediate traffic through the busiest center, approaches 1 with probability tending to 1 as the size of the problem tends to infinity. The proof of this fact is based on the strong relationship between the CAP-T and a special version of the *quadratic assignment problem*. It is shown that the latter fulfills the condition of a theorem of Burkard and U. Fincke [4] on the asymptotic behavior of combinatorial optimization problems. From a practical point of view the asymptotic behavior described above implies that the CAP on trees becomes trivial as its size tends the infinity: every feasible solution provides a good approximation of an optimal solution.

See also

- ▶ Assignment and Matching
- ▶ Assignment Methods in Clustering
- ▶ Auction Algorithms
- ▶ Bi-Objective Assignment Problem
- ▶ Directed Tree Networks
- ▶ Dynamic Traffic Networks
- ▶ Equilibrium Networks
- ▶ Evacuation Networks
- ▶ Frequency Assignment Problem
- ▶ Generalized Networks
- ▶ Maximum Flow Problem
- ▶ Maximum Partition Matching
- ▶ Minimum Cost Flow Problem
- ▶ Network Design Problems
- ▶ Network Location: Covering Problems
- ▶ Nonconvex Network Flow Problems
- ▶ Piecewise Linear Network Flow Problems
- ▶ Quadratic Assignment Problem
- ▶ Shortest Path Tree Algorithms
- ▶ Steiner Tree Problems
- ▶ Stochastic Network Problems: Massively Parallel Solution
- ▶ Survivable Networks
- ▶ Traffic Network Equilibrium

References

1. Boffey TB, Karkazis J (1989) Location of transfer centers on segments of a communication network with proportional traffic. J Oper Res Soc 40:729–734
2. Burkard RE, Çela E, Dudàs T (1996) A communication assignment problem on trees: heuristics and asymptotic behavior. Network optimization. In: Lecture Notes Economics and Math Systems, vol 450. Springer, Berlin, pp 127–156
3. Burkard RE, Çela E, Woeginger GJ (1995) A minimax assignment problem in treelike communication networks. Europ J Oper Res 87:670–684
4. Burkard RE, Fincke U (1985) Probabilistic asymptotic properties of some combinatorial optimization problems. Discrete Appl Math 12:21–29
5. Karkazis J (1993) A minimax assignment problem on a linear communication network. Belgian J Oper Res Statist Comput Sci 33:5–17

Competitive Facility Location

TAMMY DREZNER
California State University,
Fullerton, USA

MSC2000: 90B60, 90B80, 90B85

Article Outline

Keywords and Phrases
Introduction
The Proximity Model
The Location-Allocation Model
The Deterministic Utility Model
The Random Utility Model
Gravity Based Models
Anticipating Future Competition
Conclusions
References

Keywords and Phrases

Location; Competitive

Introduction

Facility location models deal, for the most part, with the location of plants, warehouses, distribution centers and other industrial facilities. These location models do not account for competition or for differences among facilities and therefore allocate customers to facilities by proximity. In reality, retail facilities operate in a competitive environment with an objective of profit and market share maximization. These facilities are also different from each other in their overall attractiveness to consumers. One branch of location analysis focuses on the location of retail and other commercial facilities which operate in a competitive environment, namely competitive facility location. The basic problem is the optimal location of one or more new facilities in a market area where competition already exists or will exist in the future. Assuming that profit increases when market share increases, maximizing profit is equivalent to maximizing market share. It follows, then, that the location objective is to locate the retail outlet at the location that maximizes its market share.

A unique feature of competitive facility location models is facility attractiveness (its appeal to consumers). Facilities differ in the total "bundle of benefits" they offer customers. They vary in one or more of the attributes which make up their total attractiveness to customers. Furthermore, varying importance assigned to each of these attributes by different customers will result in a selective set of consumers patronizing each. Facility attractiveness level, therefore, needs to be incorporated in the location model. Facility attractiveness needs to first be assessed using one of a variety of methods. Once attractiveness is assumed known, market share captured can be calculated. Facility attractiveness is estimated using a utility function (a composite index of attractiveness) or some other measure (floor area) serving as a surrogate for a latent attractiveness. Utility models are predicated on consumer spatial choice models as well as on the premise that facilities of the same type are not necessarily comparable.

Also unique to competitive facility location is the modeling of demand in terms of buying power. Income levels and discretionary spending become a measure of demand. For a review of competitive models see [4,15].

The underlying theme running through all competitive models is the existence of an interrelationship between four variables: buying power(demand), distance, facility attractiveness, and market share, with the first three variables being independent variables and the last the dependent variable. Buying power, or effective buying income, is known (for example, Sales and Marketing Management magazine). Distance from demand points to facilities can be measured. The most difficult link in the interrelationship between the four variables is the determination of facility attractiveness. For a discussion of the determination of facility attractiveness see [6,9]. As is mentioned above and discussed below, it is estimated using a utility function. Once buying power, distance, and attractiveness are known, market share can be calculated.

The Proximity Model

The first modern paper on competitive facility location is generally agreed to be Hotelling's paper on duopoly in a linear market [21]. Hotelling considered the location of two competing facilities on a segment (for example, two ice-cream vendors along a beach strip). The distribution of buying power along the segment is assumed uniform and customers patronize the closest facility. When one facility is located and there is no competition, all customers patronize the existing facility. However, when a competing facility is introduced and is located at a different point on the segment, the customers on one side of the midpoint between the two facilities patronize one facility and the customers on the other side of the midpoint patronize the second facility. If one facility is held fixed in place, the best location for the second is either immediately left or right of the fixed one, depending on which segment – left or right of the existing facility – is longer. In models based on Hotelling's formulation it is assumed that customers patronize the closest facility.

The Location-Allocation Model

An extension to Hotelling's approach is the location-allocation model for the selection of sites for facilities that serve a spatially dispersed population. Both the facilities' locations and the allocation of customers to them are determined simultaneously. The allocation of customers to facilities is made using Hotelling's proximity assumption – each facility attracts the consumers closest to it. The market share attracted by each facility is calculated and the best locations for the new facilities are then found. Multifacility location-allocation models analyze the system-wide interactions among all facilities. Revelle [28] introduced location-allocation models to competitive location. Goodchild [19] suggested the location-allocation market share model (MSM). A retail firm is planning to open a chain of outlets in a market in which a competing chain already exists. The entering firm's goal is to maximize the total market share captured by the entire chain. Most location-allocation solution methods rely on heuristic approaches that do not guarantee an optimal solution, rather they provide good solutions for implementation. The best locations are selected from a user-provided, prespecified set of potential sites. Typically, these problems are formulated on a network and the location solution is on a node. A book edited by Ghosh and Rushton [18] provides a collection of papers on the subject. A comprehensive review of location-allocation models can be found in [17].

The assumption that customers patronize the facility closest to them implies that the competing facilities are equally attractive. For equally attractive facilities, the plane is partitioned by a Voronoi diagram [26,27]. It is implicitly assumed that all customers located at a demand point patronize the same facility. This, in turn, implies an "all or nothing" property. The combined buying power at a demand point is assigned entirely to one facility and none is assigned to other facilities, unless two or more facilities are equidistant. A solution procedure for solving the multiple competitive facility location in the plane is proposed in [29].

The Deterministic Utility Model

When the facilities are not equally attractive, the proximity premise for allocating consumers to facilites is no longer valid. To account for variations in facility attractiveness, a deterministic utility model for competitive facility location is introduced by T. Drezner [2]. Hotelling's approach is extended by relaxing the proximity assumption. Consumers are known to make their choice of a facility based on factors other than distance alone. Therefore, it is assumed that customers base their choice of a facility on facility attractiveness which is represented by a utility function. This utility function is a composite index of facility attributes and the distance to the facility, representing the expected satisfaction from that facility (either an additive or a multiplicative utility function). It is generally agreed that customers, through a decision-making process, choose the facility with the highest utility, the facility which is expected to maximize their satisfaction. This choice is determined by some formula according to which customers evaluate alternative facilities' attributes weighted by their personal salience to arrive at an overall facility attractiveness.

A trade-off between distance and attractiveness takes place. Based on this premise the degree of expected satisfaction with each alternative as a function of the relevant characteristics of that facility is measured. It is suggested that a customer will patronize a better and farther facility as long as the extra distance to it does not exceed its attractiveness advantage. For example, paramedics transporting a motor vehicle accident victim will by-pass a nearby hospital in favor of a farther, better equipped trauma center as long as the difference in quality of care exceeds the adverse effect to the patient caused by the extra distance and time delay. A break-even distance is defined. At the break-even distance the attractiveness of two competing facilities is equal. This break-even distance, therefore, is the maximum distance that a customer will be willing to travel to a farther facility (new or existing) based on his perception of its attractiveness and advantage relative to other facilities. All customers at a demand point will patronize the new facility if it is located within the break-even distance. While customers are no longer assumed to patronize the closest facility, customers at a certain demand point are assumed to apply the same utility function, therefore, they all patronize the same facility. The "all or nothing" property is maintained in this extension.

Based on aggregated utility values for existing facilities and a utility function for a new facility, the best location is found for the new one. The optimal location for the new facility is sensitive to its attractiveness. Different attractiveness levels may yield different optimal locations.

The Random Utility Model

To address the "all or nothing" assumption of the deterministic utility model and to account for variations in individual utility functions, a random utility model is introduced by Drezner and Drezner [7]. The deterministic utility model is extended by assuming that each customer draws his utility from a random distribution of utility functions. The probability that a customer will prefer a certain facility over all other facilities is calculated by applying the multivariate normal distribution. Once the probabilities are calculated, the market share captured by a particular facility (new or existing) can be calculated as a weighted sum of the buying power at all demand points. This formulation eliminates the "all or nothing" property since a probability that a customer will patronize a particular facility can be established and is no longer either 0% or 100%. To circumvent the mathematically complicated formulation of the random utility model, Drezner et al. [14] suggested using the simpler logit model. The probability that a customer will patronize a facility as a function of the distance to that facility, can be approximated by a logit function of the distance.

Gravity Based Models

An alternative approach to the location of competing facilities, based on the gravity model, was introduced by Huff [22,23] and is extensively used by marketers. According to the gravity model two cities attract retail trade from an intermediate town in direct proportion to the populations of the two cities and in inverse proportion to the square of the distances from them to the intermediate town. Huff proposed that the probability that a consumer patronizes a retail facility is proportional to its size (floor area) and inversely proportional to a power of the distance to it. Facility size, or square footage, is a surrogate for facility attractiveness. Huff depicted equi-probability lines. A customer located on such a line between two facilities patronizes the two facilities with equal probability. These equi-probability lines divide the region into catchment areas, each dominated by a facility, in a manner similar to the Voronoi diagram [26]. These lines do not define an "all or nothing" assignment of customers to facilities, rather, at any demand point, the proportion of consumers attracted to each facility is a function of its square footage (attractiveness) and distance. The model finds the market share captured at each potential site, and thus the best location for new facilities whose individual measures of attractiveness are known.

Suppose there are k existing facilities and n demand points. The attractiveness of facility j is A_j for $j = 1, \ldots, k$, and the distance between demand point i and facility j is d_{ij}. The buying power at demand point i is b_i. Therefore, the proportion of the buying power (market share) M_j attracted by facility j is:

$$M_j = \sum_{i=1}^{n} b_i \frac{\frac{A_j}{d_{ij}^\lambda}}{\sum_{m=1}^{k} \frac{A_m}{d_{im}^\lambda}} \quad (1)$$

where λ is the power to which distances are raised.

In the original Huff formulation, facility floor area serves as a surrogate for attractiveness. A major improvement on Huff's approach was suggested by Nakanishi and Cooper [25] who introduced the multiplicative competitive interaction (MCI) model. The MCI coefficient replaces the floor area with a product of factors, each a component of attractiveness. Each factor in the product is raised to a power. Thus, the attractiveness of a facility is a composite of a set of attributes rather than the floor area alone. Nakanishi and Cooper's idea was elaborated on and applied by Jain and Mahajan [24] to food retailing using specific attractiveness attributes. Gravity based models suggest the evaluation in terms of market share of a user provided discrete set of potential sites for the location of a new facility.

Huff's and Nakanishi and Cooper's models were extended to the location of multiple facilities by [1,16].

Achabal et al. [1] extended the MCI model to the location of multiple facilities which belong to the same chain. The problem was modeled as a nonlinear integer programming problem and a random search procedure combined with an interchange heuristic was employed to identify optimal and near-optimal sets of locations. Ghosh and Craig [16] proposed a franchise distribution model. An expanding franchise seeks to maximize sales while minimizing cannibalization of franchise outlets. This model was also formulated as a nonlinear integer programming problem but included additional factors such as advertising. These two models select the best locations from a user-provided set of alternative sites as well.

Other papers [20,30] suggest variations on Huff's formulation by replacing the distance raised to a power with an exponent of the distance. This formulation accelerates the distance decay.

Finding the best location for a new facility (or multiple facilities) in a continuous space using the gravity model objective is discussed in T. Drezner [3] and Drezner and Drezner [10] for the single facility case, and in T. Drezner [5] and T. Drezner et al. [12] for the location of multiple facilities.

Finding the best location for a competing facility that minimizes the probability of not meeting a given minimum threshold of market share is discussed in [13].

All models discussed above assume that demand is distributed among the competing facilities. For nonessential services, some of the demand may not be satisfied. A model which assumes that some of the demand is lost is proposed in [11].

Anticipating Future Competition

The competitive facility location models discussed above are myopic and short-term oriented in that they attempt to find the optimal location for a new facility (facilities) by maximizing current market share against existing competition. A different approach to competitive location focuses on anticipating and preempting future competition. It is assumed that a new competing facility will enter the market at some point in the future. The competitor will establish his facility at the location which maximizes *his* market share. Therefore, one's present location decision will affect the competitor's location decision. Conversely, assuming a future competitive entry has implications for one's present location decision. The objective is to find the location that maximizes the market share captured by one's own facility *following* the competitor's entry. This problem is known in the economic literature as the Stackelberg equilibrium problem or the leader-follower problem and as the Simpson's problem in voting theory. See [8] for a review of the topic.

Conclusions

There are two main applications for competitive facility location models. The first application is the location analysis of a new facility. The best location for the new facility, based on market share maximization at that location, is found. The second application is an analysis of the impact of changes in quality in existing facilities (either own's, competitor's, or both) on the market share captured by one's facility and on its optimal location. In addition, a decision maker will be able to perform a "what-if analysis" and anticipate the impact on his facility of either competitor's improvements or of the introduction of a new facility. In this case one needs to know the overall attractiveness of the proposed new facility or the difference in overall attractiveness pre-post improvements in an existing one. Using the models, a decision maker can assess:

1. the impact on location of changes in attractiveness for his new facility;
2. the impact on market share of change in location for his new facility;
3. the impact on market share of changes in attractiveness at his existing facility(ies);
4. the impact on his facility of changes in other facilities or the introduction of a new facility.

These models afford the anticipation and analysis of the impact of likely future scenarios. In a highly competitive market such as exists domestically, and in the face of increasing global competition, the ability to optimize location in terms of market share provides a strategic advantage for decision makers.

References

1. Achabal D, Gorr WL, Mahajan V (1982) MULTILOC: A multiple store location decision model. J Retail 58:5–25

2. Drezner T (1994) Locating a single new facility among existing unequally attractive facilities. J Reg Sci 34:237–252
3. Drezner T (1994) Optimal continuous location of a retail facility, facility attractiveness, and market share: an interactive model. J Retail 70:49–64
4. Drezner T (1995) Facility Location: A survey of applications and methods, ch. Competitive facility location in the plane. Springer, New York
5. Drezner T (1998) Location of multiple retail facilities with a limited budget. J Retail Consum Serv 5:173–184
6. Drezner T (2006) Derived Attractiveness of Shopping Malls. IMA J Manag Math 4:349–358
7. Drezner T, Drezner Z (1996) Competitive facilities: market share and location with random utility. J Reg Sci 36:1–15
8. Drezner T, Drezner Z (1998) Location of retail facilities in anticipation of future competition. Locat Sci 6:155–173
9. Drezner T, Drezner Z (2002) Validating the Gravity-Based Competitive Location Model Using Inferred Attractiveness. Annals Oper Res 11:227–237
10. Drezner T, Drezner Z (2004) Finding the Optimal Solution to the Huff Based Competitive Location Model. Comput Manag Sci 1:193–208
11. Drezner T, Drezner Z (2008) Lost Demand in a Competitive Environment. J Oper Res Soc, in press
12. Drezner T, Drezner Z, Salhi S (2002) Solving the Multiple Competitive Facilities Location Problem. Eur J Oper Res 142:138–151
13. Drezner T, Drezner Z, Shiode SA (2002) Threshold Satisfying Comprtitive Location Model. J Reg Sci 42:287–299
14. Drezner T, Drezner Z, Wesolowsky GO (1998) On the logit approach to competitive facility location. J Reg Sci 38:313–327
15. Drezner Z, Drezner T (1998) Modern Methods for Business Research, ch. Applied location models. Lawrence Erlbaum Associates, Mahwah
16. Ghosh A, Craig CS (1991) FRANSYS: A franchise location model. J Retail 67:212–234
17. Ghosh A, Harche F (1993) Location-allocation models in the private sector: progress, problems, and prospects. Locat Sci 1:81–106
18. Ghosh A, Rushton G (1987) Spatial Analysis and Location Allocation Models. Van Nostrand Reinhold, New York
19. Goodchild MF (1984) ILACS: A location allocation model for retail site selection. J Retail 60:84–100
20. Hodgson JM (1981) A location-allocation model maximizing consumers' welfare. Reg Studies 15:493–506
21. Hotelling H (1929) Stability in competition. Eco J 39:41–57
22. Huff DL (1964) Defining and estimating a trade area. J Mark 28:34–38
23. Huff DL (1966) A programmed solution for approximating an optimum retail location. Land Eco 42:293–303
24. Jain AK, Mahajan V (1979) Research in Marketing, ch. Evaluating the competitive environment in retailing using multiplicative competitive interactive models. JAI Press, Greenwich
25. Nakanishi M, Cooper LG (1974) Parameter estimate for multiplicative interactive choice model: least squares approach. J Mark Res 11:303–311
26. Okabe A, Boots B, Sugihara K (1992) Spatial tessellations: concepts and applications of Voronoi diagrams. Wiley, Chichester
27. Okabe A, Suzuki A (1987) Stability of spatial competition for a large number of firms on a bounded two-dimensional space. Env Plan A 16:107–114
28. ReVelle C (1986) The maximum capture or sphere of influence problem: Hotelling revisited on a network. J Reg Sci 26:343–357
29. Suzuki A, Drezner Z, Drezner T (2007) Locating Multiple Facilities in a Planar Competitive Environment. J Oper Res Soc Japan 50:1001–1014
30. Wilson AG (1976) Theory and Practice in Regional Science, ch. Retailers' profits and consumers' welfare in a spatial interaction shopping model. Pion, London

Competitive Ratio for Portfolio Management
CRPM

XIAOTIE DENG
City University Hong Kong, Kowloon, China

MSC2000: 91B28, 68Q25

Article Outline

Keywords
See also
References

Keywords

Portfolio management; Unknown information; Algorithms; Competitive ratio

Portfolio management is a typical decision making problem under incomplete, sometimes unknown, information. Very often, a *probability distribution* is assumed for stock/bond prices in the future. In the classical work of H.M. Markowitz [9], the investors are assumed to base their decisions for portfolio management on their preference of *return and risk*. In this model, the return is specified as the expected value of the portfolio, and the risk its variance. One of the great achievements of this work is its predictive power of *diversified investment decisions*.

The assumption that future events would follow some *probability distribution* is also widely accepted for many other problems for which information on future events is uncertain. Very often, *uncertainty* is used as a synonym for probability distribution. However, a fundamental problem still remains: What decisions should we make in presence of future unknown events for which we are simply ignorant of any information? In such situations, the quality of a solution made under ignorance can only be known after future events reveal themselves. Therefore, the quality of a decision should be evaluated in comparison with the optimal available strategy we could have chosen knowing the outcome. Along this approach, the concept of *competitive ratio*, which optimizes the ratio of the outcome of a strategy under *incomplete information* and the optimal outcome under complete information, has been widely applied to solve computational problems under incomplete information, [4,7,8,11]. In particular, R. El-Yaniv, et al., applied competitive analysis to the problem of foreign currency purchase [5]. X. Deng has suggested to apply the competitive analysis to portfolio management problems [3].

Consider a maximization problem. Let $X = (x_1, \ldots, x_n)$ be the variables we have no complete information until in the future. Let $Y = (y_1, \ldots, y_m)$ be the decision variables for which we have to choose their values now. Let $A = (a_1, \ldots, a_k)$ be the variables we know of their values at the time we make decisions on Y. A decision rule is a function $S: A \rightarrow Y$. Let $v(A, Y, X)$ be the value of the *objective function*. Denote by $v_S(A, X) = v(A, S(A), X)$ be the value of the objective function achieved by the decision rule S if the future outcome is X. Let OPT $(A, X) = \max_{\text{all } Y} v(A, Y, X)$. The competitive ratio of decision rule S is

$$\min_{\text{all } X} \frac{v_S(A, X)}{\text{OPT}(A, X)}.$$

We are interested in a decision rule which achieves the optimal competitive ratio:

$$\max_{\text{all } S} \min_{\text{all } X} \frac{v_S(A, X)}{\text{OPT}(A, X)}.$$

Consider the portfolio management problem of choosing from a set of n stocks. We may scale units of stocks so that one unit of money and the current price for each stock is one. The portfolio choice decision can be represented by a vector $(x_1, \ldots, x_n)$, $1 \leq i \leq n$, $\sum_{i=1}^{n} x_i = 1$.

To illustrate the competitive analysis method, we first consider the extreme case when we know no information about future prices of the stocks. A simple strategy is to distribute the fund equally to all the stocks such that $x_1 = \cdots = x_n = 1/n$. Let $1 + c_i$ be the price of stock i at the end of the period. Therefore, in retrospective, the best strategy would be to invest all the money in the stock of the best performance: $1 + c_k = \max\{1 + c_i: 1 \leq i \leq n\}$. The income of the above strategy achieves $\sum_{i=1}^{n}(1+ c_i)/n$. Since we may assume that $1+ c_i \geq 0$, we have

$$\frac{\sum_{i=1}^{n} \frac{1+c_i}{n}}{1 + c_k} \geq \frac{1}{n}.$$

This simple strategy achieves a competitive ratio of $1/n$. On the other hand, it is natural that this strategy is optimal when we have no information whatsoever about the stocks. Consider any strategy which invests x_i in stock i ($\sum_{i=1}^{n} x_i = 1$). Its outcome will be $\sum_{i=1}^{n} x_i(1+c_i)$. Since $\sum_{i=1}^{n} x_i = 1$, there exists some j such that $x_j \leq 1/n$. In the worst case, it may happen that we have $1 + c_j = 2$ and $1 + c_i = 0$ for all other stocks. Therefore, the optimal investment will be put all the money in stock c_j. The competitive ratio of this strategy is no more than $x_j(1 + c_j)/(1 + c_j) \leq 1/n$. Therefore, the above simple strategy achieves the optimal competitive ratio when no information is available.

To illustrate this idea further, consider another case is when we have some information about future prices of the stocks. Suppose that the only information we have is that stock i will fluctuate between $[(1 - \epsilon_i), (1 + \delta_i)]$ ($-\epsilon_i \leq \delta_i$), $1 \leq i \leq n$. It is easy to see that we may normalize the value versus the risk-free rate of interest and make it as the first option so that $-\epsilon_1 = \delta_1 = 0$. E.g., we may divide outcomes of other securities by $(1 + r)$, the *riskless interest rate*.

Given a portfolio choice decision, x_i, $1 \leq i \leq n$, $\sum_{i=1} x_i = 1$. That is, one unit of investment is distributed to n options with a fraction of x_i on option i: $1 \leq i \leq n$. Let $(1 + c_i)$ be the unknown future price of option i by the projected time of sales ($-\epsilon_i \leq c_i \leq \delta_i$). In retrospect, the optimal solution would have been $\max_{j=1}^{n}(1 + c_j)$ by investing all one unit on the option achieving the optimum. For a fixed strategy of assigning $x_i: 1 \le i \le n$, its ratio versus the optimum will be

$$\frac{\sum_{i=1}^{n}(1+c_i)x_i}{\max\{1+c_i: 1 \le i \le n\}}.$$

Taking all situations into consideration, the competitive ratio of this strategy is

$$\min\left\{\frac{\sum_{i=1}^{n}(1+c_i)x_i}{\max\{1+c_j: 1 \le j \le n\}} : -\epsilon_i \le c_i \le \delta_i\right\},$$

where the minimum is taken over all the ranges of c_i: $-\epsilon_i \le c_i \le \delta_i$, $1 \le i \le n$. Suppose now that $\max_{j=1}^{n}(1+c_j)$ is achieved at some $i: 1 \le i \le n$. Then, the above ratio is at least x_i since $x_j \ge 0$ and $1 + c_j \ge 0$, for all $j: 1 \le j \le n$. If $c_i < \delta_i$, the adversary can choose a new value $c'_i = \delta_i$. In this case, the denominator $\max_{j=1}^{n}(1+c_j)$ increases by $\delta_i - c_i > 0$. The numerator $\sum_{i=1}^{n}(1+c_i)x_i$ increases by $(\delta_i - c_i) x_i$. Thereforeit is to the benefit of the adversary to choose $c_i = \delta_i$. Similarly, it is to the benefit of the adversary to set $c_j = -\epsilon_j$, for all $j \ne i$. That is, the minimum ratio is achieved at

$$\frac{(1+\delta_i)x_i + \sum_{j \ne i}(1-\epsilon_j)x_j}{1+\delta_i}$$

for some i, $1 \le i \le n$. Therefore, the adversary will choose some i such that

$$\min\left\{\frac{(1+\delta_i)x_i + \sum_{j \ne i}(1-\epsilon_j)x_j}{1+\delta_i} : 1 \le i \le n\right\}.$$

Given a portfolio decision vector x, we can search through all n possible situations to find the minimum in *polynomial time*. This allows us to evaluate the quality of portfolio choices in terms of their competitive ratios.

As a portfolio manager aiming at a solution with the best competitive ratio, its goal is to choose the decision vector x which maximizes

$$\min\left\{\frac{(1+\delta_i)x_i + \sum_{j \ne i}(1-\epsilon_j)x_j}{1+\delta_i} : 1 \le i \le n\right\}.$$

In a *linear program* formulation, this is

$$\begin{cases} \max & z \\ \text{s.t.} & \frac{(1+\delta_i)x_i + \sum_{j \ne i}(1-\epsilon_j)x_j}{1+\delta_i} \ge z, \\ & 1 \le i \le n, \\ & x_i \ge 0, \quad \sum_{i=1}^{n} x_i = 1. \end{cases}$$

Therefore, the optimal competitive ratio for the above portfolio management problem can be solved in polynomial time.

In the general case, information about future may be different for different investigators. Compare two situations where two investors each has two options, one government bond of riskless interest rate $1 + r$ and a security. One investor knows that the future price of the security will be in $[1 + \epsilon, 1 + \delta]$ with $\epsilon < \delta$ and another knows nothing about future prices of the security. The most interesting case will be $2\epsilon < r < \delta$. Apply the above analysis, the more informed investor will decide a proportion x of his fortune on riskless bond with

$$\frac{x(1+r) + (1-x)(1+\epsilon)}{1+r}$$

$$= \frac{x(1+r) + (1-x)(1+\delta)}{1+\delta}.$$

Therefore, it invests

$$x = \frac{(1+\delta)(r-\epsilon)}{(1+\delta)(r-\epsilon) + (1+r)(\delta-r)}$$

on the riskless bond and

$$1 - x = \frac{(1+r)(\delta-r)}{(1+\delta)(r-\epsilon) + (1+r)(\delta-r)}$$

on the other security. Its competitive ratio will be

$$\frac{(1+\delta)(r-\epsilon) + (1+\epsilon)(\delta-r)}{(1+\delta)(r-\epsilon) + (1+r)(\delta-r)}.$$

Applying the analysis above, we see that the person knowing nothing will invest $1/2$ for the riskless bond and $1/2$ for the other security. However, the worst situations considered by the less informed investor would not occur at all. Therefore, its competitive ratio will be the minimum of

$$\frac{(1+r) + (1+\epsilon)}{2(1+r)}$$

and

$$\frac{(1+r) + (1+\delta)}{2(1+\delta)}.$$

From the above discussion, the decision of the investor knowing nothing has a worse competitive ratio than that of the more informed one.

In comparison with the general approach of using probability distribution for events of uncertainty, the above situation shows that the *competitive analysis* method allows analysis for information asymmetry of investors. It is not easy to apply the probability method here since, in principle, the real world should not have two different probability distributions. This advantage is not only for the above case when the range of future prices is known. It can also be applied to other types of information about future.

Other decision rules based on rationality other than probability argument have also been suggested for financial problems. In particular, T.M. Cover has suggested a solution, the *universal portfolio*, which requires no information (not even probability distribution) about the future prices of the stocks under consideration [1]. In contrast to competitive analysis which evaluates a strategy with all other strategies, Cover has evaluated his solution in comparison with a class of strategies called constant rebalanced portfolio, which maintains a fixed proportion of one's fortune in each of the securities. Notice that, this would require frequent adjustment the holdings of the securities as their prices change. Surprisingly, Cover has shown his solution to approximate, under mild conditions, the best constant rebalanced portfolio (chosen after the stock outcomes are known) which out-perform any *constant rebalanced portfolio*, any single stock and index fund such as Down Jones Index Average (DJIA) [1]. However, Cover's algorithm requires higher-dimensional integration to calculate his solution and the dimension grows with the number of securities under consideration. This may make it computationally difficult to apply this method. In comparison, the competitive analysis would suggest a solution which is a constant rebalanced portfolio with the same weight for all the securities.

Dembo and King have discussed a tracking model for asset allocation which minimizes an investor's regret (defined as the difference of the solution of a strategy under incomplete information and the optimal solution) distribution in the L_2 metric [2]. In general, one may express the regret of a decision maker with strategy S as a function of $f(v_S(X), OPT(X))$, where X is the revealed future event, $v_S(X)$ is the value achieved under strategy S operating under ignorant of the future event X, and $OPT(X)$ is the optimal value achievable knowing the complete information. One such function often used is the L_∞ metric distance of these two values in the feasible space [10]. However, since the authors use the absolute difference for the basis of evaluation of strategies, probability assumption is still necessary in this model. The competitive analysis and the solution of Cover [1] base the evaluation on the ratio of the performance of a strategy with unknown information and the performance of the best solution in the class of strategies under consideration.

R.M. Hogarth and H. Kunreuther have discussed situations when financial decisions are made under ignorance. They have designed experiments to study it by evaluating human empirical judgements. However, decision making processes of economic agents are ignored in this study [6].

Some information is still available in reality, though not necessarily in the form of a well shaped probability distribution. The competitive analysis provides an approach which does not rely on probability distribution, allows for analysis under *asymmetrical information* of agents in the market, and in principle, has no difficulty to include available information in the analysis. The remaining difficulties in applying it successful to portfolio management are mainly modeling of available information and *efficient algorithms* for computational purpose.

See also

- ▶ Financial Applications of Multicriteria Analysis
- ▶ Financial Optimization
- ▶ Portfolio Selection and Multicriteria Analysis
- ▶ Robust Optimization
- ▶ Semi-Infinite Programming and Applications in Finance

References

1. Cover TM (1991) Universal portfolios. Math Finance 1:1–29
2. Dembo RS, King AJ (1992) Tracking models and the optimal regret distribution in asset allocation. Applied Stochastic Models and Data Analysis 8:151–157
3. Deng X (1996) Portfolio management with optimal regret ratio. Proc Internat Conf Management Sci, Hong Kong, pp 289–295
4. Deng X, Papadimitriou CH (1996) Competitive distributed decision-making. Algorithmica 16:133–150
5. El-Yaniv R, Fiat A, Karp RM, Turpin G (1992) Competitive analysis of financial games. In: Proc. 33rd Annual Symp. Foundations of Computer Sci., pp 327–333

6. Hogarth RM, Kunreuther H (1995) Decision making under ignorance: Arguing with yourself. J Risk and Uncertainty 10:15–36
7. Karlin AR, Manasse MS, Rudolph L, Sleator DD (1988) Competitive snoopy caching. Algorithmica 3:79–119
8. Manasse MS, McGeoch LA, Sleator DD (1990) Competitive algorithms for on-line problems. J Algorithms 11:208–230
9. Markowitz HM (1959) Portfolio selection: efficient diversification of investments. Wiley, New York
10. Savage LJ (1951) The theory of statistical decision. J Amer Statist Assoc 46:55–67
11. Sleator DD, Tarjan RE (1985) Amortized efficiency of list update and paging rules. Comm ACM 28:202–208

Complementarity Algorithms in Pattern Recognition

C. Cifarelli, Laura Di Giacomo, Giacomo Patrizi

Dipartimento di Statistica,
Probabilità e Statistiche Applicate, Università di Roma "La Sapienza", Rome, Italy

Article Outline

Introduction
Definitions
Formulation
Methods and Applications
Models
Cases
Conclusions
See also
References

Introduction

Basic to the process of human understanding and learning, the problem of recognition, which includes classification and machine learning and the more general approach of pattern recognition, consists of a set of algorithms or procedures to determine in which of a number of alternative classes an object belongs.

While recognition is a human process whose functioning is largely unknown [11], pattern recognition and classification and machine learning are algorithms or heuristic procedures with a precise functional characterization to determine as precisely as possible the class membership of an object.

The two approaches, pattern recognition on the one hand and classification and machine learning on the other, emphasize two different aspects of the learning methodology, similar to a distinction often made in numerical analysis between extrapolation and interpolation [13].

In pattern recognition, given a feature of a population, it is desired that all objects that belong to that population be recognized with an acceptable small error, since the paramount aspect of this activity is to recognize the object so as to be able to proceed accordingly. It is not of interest to diagnose a varying percentage of sick individuals, but rather it is essential to recognize correctly the pathology. Thus in pattern recognition, given an object, it is desired to determine if the object belongs to the population specified and, if so, to determine precisely to which class it belongs [5].

In classification and machine learning, a population is considered given and some objects belong to known classes, while other objects belong to as yet unknown classes, so it is desired to determine the class membership of objects that are known to belong to that population. Depending on the definition of the populations considered and the algorithms used, the classification rate may differ from one application to another. Classification and machine learning procedures are often defined in terms of heuristics, such as support vector machines with kernel methods. The kernel to be applied to a given problem cannot be determined except by trial and error, so that the existence of a suitable kernel is not guaranteed. Thus results may differ markedly from application to application [5].

Here we shall be concerned with pattern recognition problems that must consider:
- The collection of objects to examine and the training set available for learning the classes.
- The attributes that can be defined precisely on the objects in the training set and on the objects to be recognized (which may be as yet unknown).
- The precision with which the recognition is required, as well as the possible structures defined on the data sets.

The pattern recognition algorithm used to perform this will be formulated as a complementarity problem rather than an optimization algorithm as it may be considered more general, and the known differences that may exist in the attributes of the classes allows additional constraints to be defined, which permit more precise results to be obtained.

Definitions

Consider a set of objects, characterized by a set of common attributes, which have been assigned to suitable classes, so that their class labels are known. This is called a training set [5].

Definition 1 A subset of a data set is termed a training set if every entity in the training set has been assigned a class label.

Definition 2 Suppose there is a set of entities E and a set $P = \{P_1, P_2, \ldots, P_n\}$ of subsets of the set of entities, i.e. $P_j \subseteq E, j \in J = \{1, 2, \ldots, n\}$. A subset $\hat{J} \subseteq J$ forms a cover of E if $\bigcup_{j \in \hat{J}} P_j = E$. If, in addition, for every $k, j \in \hat{J}, j \neq k, P_j \cap P_k = \emptyset$, it is a partition.

Definition 3 The data set is coherent if there exists a partition that satisfies the following properties:
1. The relations defined on the training set and in particular the membership classes, defined over the data set, consist of disjoint unions of the subsets of the partition.
2. Stability: the partition is invariant to additions to the data set. This invariance should apply both to the addition of duplicate entities and to the addition of new entities obtained in the same way as the objects under consideration.
3. Extendability: if the dimension of the set of attributes is augmented, so that the basis will be composed of $p + 1$ attributes, then the partition obtained by considering the smaller set will remain valid, even for the extension, as long as this extension does not alter the relations defined on the data set.

Definition 4 A data set is linearly separable if there exist linear functions such that the entities belonging to one class can be separated from the entities belonging to the other classes. It is pairwise linearly separable if every pair of classes is linearly separable. A set is piecewise separable if every element of each class is separable from all the other elements of all the other classes.

Clearly if a set is linearly separable, it is pairwise linearly separable and piecewise separable, but the converse is not true. The following results are straightforward:

Theorem 1 *If a data set is coherent, then it is piecewise separable.*

A given class is formed from distinct subsets of the partition, so no pattern can belong to two classes. Therefore each pattern of a given class will be separable from every pattern in the other subsets of the partition. Consequently the data set is piecewise separable.

Theorem 2 *Given a data set that does not contain two identical patterns assigned to different classes, a correct classifier can be formulated that realizes the given partition on this training set.*

Corollary 1 *Given that the training set does not contain two or more identical patterns assigned to different classes, the given partition yields a completely correct classification of the patterns.*

The avoidance of the juxtaposition property, i.e. two identical patterns belong to different classes, entails that the Bayes error is zero [2].

In general this does not mean that in any given neighborhood of a pattern there cannot be other patterns of other classes, but only that they cannot lie on the same point. Thus the probability distribution of the patterns with respect to the classes may overlap, if such distributions exist, although they will exhibit discontinuities in the overlap region, so that juxtaposition is avoided.

Formulation

The classification algorithm to be formulated may be specified as a combinatorial problem in binary variables [6].

Suppose that a training set is available with n patterns, represented by appropriate feature vectors indicated by $x_i \in \mathbf{R}^p, \forall i = 1, 2, \ldots, n$ and grouped in c classes. An upper bound is selected to the number of barycentres that may result from the classification, which can be taken "ad abundantiam" as m, or on the basis of a preliminary run of some classification algorithm.

The initial barycenter matrix will be an $p \times mc$ matrix which is set to zero. The barycentres when calculated will be written in the matrix by class. Thus a barycenter of class k will occupy a column of the matrix between $(m(k-1)+1)$ and mk.

Since we are considering a training set, the feature vectors can be ordered by increasing class label. Thus the first n_1 columns of the training set matrix consists of patterns of class 1, from $n_1 + 1$ to n_2 of class 2 and in general from $n_{k-1} + 1$ to n_k of class k.

Thus consider the following inequality constrained optimization problem, from which we shall derive the non-linear complementarity specification. Let the following hold:

- $x_i \in \mathbf{R}^p$: the p-dimensional pattern vector of pattern i;
- c classes are considered, $k = 0, 1, \ldots, (c-1)$. Let the number of patterns in class c_k be indicated by n_k; then the n patterns can be subdivided by class so that $n = \sum_{k=0}^{c-1} n_k$;
- $z_j \in \{0, 1\}$, integer: $\{j = 1, 2, \ldots mc\}$ if $z_j = 1$ then the barycenter vector $j \in \{mk+1\}, \ldots, m(k+1)\}$ belonging to recognition class $c_k \in \{0, \ldots, c-1\}$,
- $y_{ij} \in \{0, 1\}$, integer: pattern i has been assigned to the barycenter j $(y_{ij} = 1)$;
- $t_j \in \mathbf{R}^p$: the sum of the elements of the vectors of the patterns assigned to the barycenter $j = \{1, 2, \ldots, mc\}$;
- M is a large scalar.

$$\text{Min } Z = \sum_{j=1}^{mc} z_j \qquad (1)$$

s.t. $\sum_{j=km+1}^{m(k+1)} y_{ij} - 1 \geq 0 \quad \forall k = 0, 1, \ldots, (c-1);$

$$\forall i = n_{k-1} + 1, \ldots, n_k \qquad (2)$$

$$-\sum_{i=1}^{n} \sum_{j=1}^{mc} y_{ij} + n \geq 0 \qquad (3)$$

$$Mz_j - \sum_{i=1}^{n} y_{ij} \geq 0 \quad \forall j = 1, 2, \ldots, mc \qquad (4)$$

$$t_j - \sum_{i=1}^{n} x_i y_{ij} \geq 0 \quad \forall j = 0, 1, \ldots, mc \qquad (5)$$

$$-\sum_{j=1}^{mc} \left(t_j - \sum_{i=1}^{n} x_i y_{ij} \right) \geq 0 \qquad (6)$$

$$\left(x_i - \frac{t_h}{\sum_{s=lm+1}^{m(l+1)} y_{sh}} \right)^{\mathrm{T}} \left(x_i - \frac{t_h}{\sum_{s=lm+1}^{m(l+1)} y_{sh}} \right)$$

$$- \sum_{j=km+1}^{m(k+1)}$$

$$\left(x_i - \frac{t_j}{\sum_{r=km+1}^{m(k+1)} y_{rj}} \right)^{\mathrm{T}} \left(x_i - \frac{t_j}{\sum_{r=km+1}^{m(k+1)} y_{rj}} \right)$$

$$\times y_{ij} \geq 0$$

$\forall i = 1, 2, \ldots, n; \quad h = 1, 2, \ldots, mc;$
$\quad k, l = 0, 1, \ldots, c-1; \qquad (7)$

$$z_j, y_{ij} \in \{0, 1\} \text{ integer}. \qquad (8)$$

The solution of this optimization problem assigns each pattern to a mean vector, called a barycenter $(z_j, j = 1, 2, \ldots, mc)$, whose values are given by the vectors $t_j \in \mathbf{R}^p$, $j = \{1, 2, \ldots, mc\}$ divided by the number of patterns assigned to that barycenter. The least number of barycentres, indicated by the objective function Eq. (1), which will satisfy the stated constraints is determined.

The n constraints Eqs. (2) and (3) state that each feature vector from a pattern in a given class must be assigned to some barycenter vector of that class. As patterns and barycentres have been ordered by class, the summation should be run over the appropriate index sets.

The mc constraints Eq. (4) impose that no pattern be assigned to a non-existing barycenter.

Instead, constraints Eqs. (5) and (6) determine the vector of the total sum element by element assigned to a barycenter. Notice that x_i is a vector, so the number of inequalities will be $2mc$ times the number of elements in the feature vector.

The last set of inequalities Eq. (7) indicates that each feature vector must be nearer to the assigned barycenter of its own class than to any other barycenter. Should the barycenter be null, this is immediately verified, while if it is non-zero, this must be imposed.

Finally, Eq. (8) indicates that the vectors $z \in R^{mc}$ and $y \in R^{nmc}$ are binary.

The solution will determine that each pattern of the training set is nearer to a barycenter of its own class than to a barycenter of another class. Each barycenter has the class label of the patterns assigned to it, which will belong by construction to a single class. This defines a partition of the pattern space.

A new pattern can be assigned to a class by determining its distance from each barycenter formed by the algorithm and then assigning the pattern to the class of the barycenter to which it is nearest.

In general, other constraints which characterize relationships between objects of different classes can be easily introduced in this specification, as well as dynamical relationships regarding the attributes of the objects.

The problem can also be formulated as a non-linear complementarity problem in binary variables, which will be solved through iterating on a set of linear complementarity problems in binary variables, by using a linear programming technique with parametric variation in one scalar variable [7] which has given good results [3].

For simplicity in the representation and analysis, write the constraints (7) as:

$$g(y, x, t)$$

$$= \left(x_i - \frac{t_h}{\sum_{s=lm+1}^{m(l+1)} y_{sh}}\right)^T \left(x_i - \frac{t_h}{\sum_{s=lm+1}^{m(l+1)} y_{sh}}\right)$$

$$- \sum_{j=km+1}^{m(k+1)} \left(x_i - \frac{t_j}{\sum_{r=km+1}^{m(k+1)} y_{rj}}\right)^T \left(x_i - \frac{t_j}{\sum_{r=km+1}^{m(k+1)} y_{rj}}\right)$$

$$\times y_{ij} \quad (9)$$

The following additional notation should be adopted to write the optimization problem (1)–(8) as a non-linear complementarity problem:
- e is an appropriate dimensional vector of ones.
- $E \in \mathbf{R}^{n \times nmc}$ is a matrix composed of mc identity matrices of dimension $n \times n$.
- $H \in \mathbf{R}^{mc \times n}$ matrix of ones.
- η is a scalar to be assigned by dichotomous search during the iterations.

The data matrix of patterns indicated as X of dimension $(p \times m \times c) \times (n \times m \times c)$ is written in diagonal block form with blocks of dimension $p \times n$ elements containing the original data matrix.

This block is repeated mc times with the first element of the block placed at the position $((j-1)p+1, (j-1)n), j = 1, 2, \ldots, mc$.

In fact the size of matrices E, H and X can be greatly reduced in applications since the patterns in the training set are ordered conformably with the barycenter vector $t = \{t_j\} \in \mathbf{R}^{pmc}$ and each class is of known cardinality.

The non-linear complementarity problem can therefore be written as:

$$\begin{pmatrix} -z \\ -y \\ 0 \\ Ey \\ -e^T y \\ Mz - Hy \\ t - Xy \\ -e^T(t - Xy) \\ g(y, x, t) \\ -e^T z \end{pmatrix} + \begin{pmatrix} e \\ e \\ 0 \\ -e \\ n \\ 0 \\ 0 \\ 0 \\ 0 \\ \eta \end{pmatrix} \geq 0 \quad (10)$$

$$\begin{pmatrix} z \\ y \\ t \\ \lambda_1^T \\ \lambda_2^T \\ \lambda_3^T \\ \lambda_4^T \\ \lambda_5^T \\ \lambda_6^T \\ \lambda_7^T \end{pmatrix} \geq 0 \quad (11)$$

$$(z^T, \; y^T \; t \; \lambda_1^T, \; \lambda_2^T, \; \lambda_3^T, \; \lambda_4^T, \; \lambda_5^T, \; \lambda_6^T, \; \lambda_7^T)$$

$$\times \left(\begin{pmatrix} -z \\ -y \\ 0 \\ Ey \\ -e^T y \\ Mz - Hy \\ t - Xy \\ -e^T(t - Xy) \\ g(y, x, t) \\ -e^T z \end{pmatrix} + \begin{pmatrix} e \\ e \\ 0 \\ -e \\ n \\ 0 \\ 0 \\ 0 \\ 0 \\ \eta \end{pmatrix} \right) = 0$$

$$(12)$$

Binary values to the z, y variables are imposed by the constraints Eq. (10) and the complementarity condition Eq. (12).

Finally, by recursing on the parameter η fewer and fewer barycentres will be created, as long as the problem remains feasible and thus ensuring a minimal solution.

Methods and Applications

The aim of this section is to describe the method to solve the non-linear complementarity problem specified in the previous section. The convergence of the non-linear complementarity problem Eqs. (10)–(12) has been given elsewhere [1].

In a small enough neighborhood, the approximation of the non-linear complementarity problem by a linear complementarity problem will be sufficiently accurate so that, instead of solving the original system, a linear complementarity system approximation can be solved, which may be thus represented:

$$\begin{pmatrix} -I & 0 & 0 & 0\,0\,0\,0\,0\,0\,0\,0 \\ 0 & -I & 0 & 0\,0\,0\,0\,0\,0\,0\,0 \\ 0 & 0 & 0 & 0\,0\,0\,0\,0\,0\,0\,0 \\ 0 & E & 0 & 0\,0\,0\,0\,0\,0\,0\,0 \\ 0 & -e^T & 0 & 0\,0\,0\,0\,0\,0\,0\,0 \\ MI & -H & 0 & 0\,0\,0\,0\,0\,0\,0\,0 \\ 0 & -X & I & 0\,0\,0\,0\,0\,0\,0\,0 \\ 0 & e^T X & -e^T & 0\,0\,0\,0\,0\,0\,0\,0 \\ 0 & \nabla g_y(\hat{t},\hat{y}) & \nabla g_t(\hat{t},\hat{y}) & 0\,0\,0\,0\,0\,0\,0\,0 \\ 0 & 0 & D & 0\,0\,0\,0\,0\,0\,0\,0 \\ -e^T & 0 & 0 & 0\,0\,0\,0\,0\,0\,0\,0 \end{pmatrix} \begin{pmatrix} z \\ y \\ t \\ \lambda_1 \\ \lambda_2 \\ \lambda_3 \\ \lambda_4 \\ \lambda_5 \\ \lambda_6 \\ \lambda_7 \\ \lambda_8 \end{pmatrix}$$

$$+ \begin{pmatrix} e \\ e \\ 0 \\ -e \\ n \\ 0 \\ 0 \\ 0 \\ -g(\hat{t},\hat{y}) + \nabla g(\hat{t},\hat{y})\hat{y} \\ -d \\ \eta \end{pmatrix} \geq 0 \qquad (13)$$

$$\begin{pmatrix} z \\ y \\ t \\ \lambda_1 \\ \lambda_2 \\ \lambda_3 \\ \lambda_4 \\ \lambda_5 \\ \lambda_6 \\ \lambda_7 \\ \lambda_8 \end{pmatrix} \geq 0 \qquad (14)$$

$$\left(z^T,\ y^T,\ t^T,\ \lambda_1^T,\ \lambda_2,\ \lambda_3^T,\ \lambda_4^T,\ \lambda_5,\ \lambda_6^T,\ \lambda_7^T,\ \lambda_8 \right) \times$$

$$\left(\begin{pmatrix} -I & 0 & 0 & 0\,0\,0\,0\,0\,0\,0\,0 \\ 0 & -I & 0 & 0\,0\,0\,0\,0\,0\,0\,0 \\ 0 & 0 & 0 & 0\,0\,0\,0\,0\,0\,0\,0 \\ 0 & E & 0 & 0\,0\,0\,0\,0\,0\,0\,0 \\ 0 & -e^T & 0 & 0\,0\,0\,0\,0\,0\,0\,0 \\ MI & -H & 0 & 0\,0\,0\,0\,0\,0\,0\,0 \\ 0 & -X & I & 0\,0\,0\,0\,0\,0\,0\,0 \\ 0 & e^T X & -e^T & 0\,0\,0\,0\,0\,0\,0\,0 \\ 0 & \nabla g_y(\hat{t},\hat{y}) & \nabla g_t(\hat{t},\hat{y}) & 0\,0\,0\,0\,0\,0\,0\,0 \\ 0 & 0 & D & 0\,0\,0\,0\,0\,0\,0\,0 \\ -e^T & 0 & 0 & 0\,0\,0\,0\,0\,0\,0\,0 \end{pmatrix} \begin{pmatrix} z \\ y \\ t \\ \lambda_1 \\ \lambda_2 \\ \lambda_3 \\ \lambda_4 \\ \lambda_5 \\ \lambda_6 \\ \lambda_7 \\ \lambda_8 \end{pmatrix} \right.$$

$$\left. + \begin{pmatrix} e \\ e \\ 0 \\ -e \\ n \\ 0 \\ 0 \\ 0 \\ -g(\hat{t},\hat{y}) + \nabla g(\hat{t},\hat{y})\hat{y} \\ -d \\ \eta \end{pmatrix} \right) = 0$$

(15)

The problem (10)–(12) is then solved by expanding the vectorial function $g(y,x,t)$ in a Taylor series around the iteration point and solving the resulting linear complementarity problem approximation (13)–(15) of the given non-linear complementarity problem within a suitable trust region. It is easy to show:

Theorem 3 *The following are equivalent:*
1. *The non-linear optimization problem defined by (1)–(8) has a solution;*

2. The non-linear complementarity problem defined by (10)–(12) has a solution;
3. The linear complementarity problem defined by (13)–(15) has a solution.

Thus the computational specification of this algorithm is:

Algorithm 1 (CASTOR)
Begin;
- **Given**: a training set $A \in R^{p \times n}$ with n patterns each with p elements belonging to c classes;
- **Construct**: the matrices $E \in \mathbf{R}^{n \times nmc}$, $H \in \mathbf{R}^{mc \times n}$, $X \in \mathbf{R}^{(pmc) \times (mnc)}$, $D \in \mathbf{R}^{pmc \times pmc}$;
- **Set** y^0, d^0, η^0;

For $k = 1, 2, \ldots$;
- **while** $z^{k+1}, y^{k+1}, t^{k+1}$ is a solution to LCP Eqs. (13)–(15) **Do**;
- **Begin**: recursion on $g(x,y,t)$
  - **while** $(z^{k+1}, y^{k+1}, t^{k+1}) \neq (z^k, y^k, t^k)$ **Do**;
  - $(z^k, y^k, t^k) \leftarrow (z^{k+1}, y^{k+1}, t^{k+1})$
  - **Determine** $\nabla g_y(x^k, y^k, t^k,)$
    *__Begin__: dichotomous search on η^k;
    $(z^{k+1}, y^{k+1}, t^{k+1}) \leftarrow LCP(z^k, y^k, t^k)$
    *__end__;
- **end**;
  the solution is (z^k, y^k, t^k)
  end;

The termination of the classification algorithm may now be proved under a consistency condition.

Theorem 4 *Given a set which does not contain two identical patterns assigned to different classes, a correct classifier will be determined by Algorithm 1.*

Models

Suppose a training set is available, defined over a suitable representation space, which is piecewise separable and coherent, what properties should such a training set have to determine a precise classification with regard to a set of data of as yet unknown classes?

The algorithm **CASTOR** (**C**omplementarity **A**lgorithm **S**ystem for **TO**tal **R**ecognition) described in the Sects. "Formulation" and "Methods" will determine a classification rule to apply, on the data set, just that partition which has been found for the training set, so that to each entity in the data set a class is assigned. If the training set forms a random sample and the data set which includes the training set is coherent, then this classification can be performed to any desired degree of accuracy by extending the size of the training sample. Sufficient conditions to ensure that these properties hold are given by selecting the data set and the verification set by non-repetitive random sampling.

Theorem 5 *Suppose that the data set is coherent; then the data set can be classified correctly.*

To avoid having to introduce distributional properties on the data set considered, the empirical risk minimization inductive principle may be applied [12]:

Definition 5 A data set is stable, according to definition 3, with respect to a partition and a population of entities if the relative frequency of misclassification is $R_{emp}(\alpha^*) \geq 0$ and

$$\lim_{n \to \infty} pr\{R_{emp}(\alpha^*) > \epsilon\} = 0, \qquad (16)$$

where α^* is the classification procedure applied, $\epsilon > 0$ for given arbitrary small value and $pr\{.\}$ is the probability of the event included in the braces.

In some diagnostic studies the set of attributes considered have no significant relationship with the outcome or the classification of the entity. Typically the classes could be eye color and the attributes the weight, height and sex of a person. Such a classification would be spurious since there is no relation between eye color and body indices.

A spurious collection of entities, in which there is no similarity relations, may occur and should be recognized. With this algorithm, this occurrence is easily determined, as very many barycentres are formed, almost one per object. Such spuriousness may arise even in the presence of some meaningful relationships in the data, which are, however, swamped by noise, and so data reduction techniques may be useful [5].

In general, by considering smaller and smaller subsets of the attribute space X, if there exists a relationship between the attributes and the classes of the entities, for certain of these subsets the frequency of the entities of a given class will increase to the upper limit of one, while in other subsets it will decrease to a lower limit of zero. Thus for a very fine subdivision of the attribute space, each subset will tend to include entities only of a given class.

Definition 6 A proper subset S_k of the attribute space X of the data set will give rise to a spurious classification if the conditional probability of a pattern belonging to a given class c is equal to its unconditional probability over the attribute space. The data set is spurious if this holds for all subsets of the attribute space X.

$$pr\{y_i = c \mid (y_i, x_i) \cap S_k\} = pr\{y_i = c \mid (y_i, x_i) \cap X\} \quad (17)$$

The following results can now be presented, which are proved elsewhere [1].

Theorem 6 *Consider a training set of n patterns randomly selected, assigned to two classes, where the unconditional probability of belonging to class one is p. Let a be a suitable large number and let ($n > a$). Let the training set form b_n barycentres. Then, under* **CASTOR**, *this training set will provide a spurious classification if*

$$\frac{b_n}{n} \geq (1-p) \quad n > a. \quad (18)$$

Theorem 7 *Let the probability of a pattern belonging to class one be p. Then the number of barycentres required to partition correctly a subset S, containing $n_s > a$ patterns, which is not spurious, formed from the* **CASTOR** *algorithm, is $b_s < n_s$, $\forall n_s > a$.*

Corollary 2 ([12]) *The Vapnik–Cervonenkis dimension (VC dimension), s(C,n), for the class of sets defined by the* **CASTOR** *algorithm, restricted to the classification of a non-spurious data set which is piecewise separable, with n_s elements and two classes, is less than 2^{n_s}, if $n_s > a$.*

Theorem 8 ([2]) *Let C be a class of decision functions and ψ_n^* a classifier restricted to the classification of a data set which is not spurious and returns a value of empirical error equal to zero based on the training sample $(z_1, z_2, \ldots, z_n)$. Thus $Inf_{\psi \in C} L(\psi) = 0$, i.e. the Bayes decision is contained in C. Then*

$$pr\{L(\psi_n^*) > \epsilon\} \leq 2s(C, 2n) 2^{\frac{-n\epsilon}{2}}. \quad (19)$$

By calculating bounds on the VC dimension, the universal consistency property can be established for this algorithm applied to the classification of a data set which is not spurious.

Corollary 3 ([5]) *A non-spurious classification problem with a piecewise separable training set is strongly universally consistent.*

Cases

To use the **CASTOR** algorithm in applications, it is necessary to determine, first, whether the data set is spurious or not, for the given problem with the specific pattern vectors adopted. The way the pattern vectors are defined based on the data available may affect strongly the results obtainable.

Further, the coherence of the data set must be tested to ensure that the patterns extracted are sufficiently rich to ensure the proper classification, stability and extendability of the data set (Definition 5). Then the algorithm can be applied, but the results will only hold if the data set, training set and verification set are random samples, taken from a known or unknown population, as otherwise the sample may not be representative of the population.

Note that with this method, if the data come from a set of unknown populations, a suitable partition of the data set will form accordingly, even though the operator may not know to which population an individual barycenter belongs. If the number of objects coming from different populations is so high with respect to the training set, then the problem may be recognized as spurious, only to signify that too many barycentres are formed with respect to the available training objects [8,9].

Consider a set of proteins randomly sampled from a population of proteins, and the set of proteins whose structure should be determined also belongs to that population, but are of unknown structure. Probability limits can be imposed on the likelihood of the structure identified being the correct one. Therefore, accurate limits on the precision of the recognition of the the new protein's structure can be specified.

Results could be obtained also by selecting "purposefully" representative proteins and subjecting these to a suitable algorithm. The results could be better on particular sets than the asymptotic mean precision measures, but generally, and, almost surely, on using new data, the results will turn out to have a greater variance and a lower mean precision, as is well known from sampling theory [4]. Thus to minimize the asymp-

Complementarity Algorithms in Pattern Recognition, Table 1
Q_3 Classification results on the Rost 126 verification set by similarity classes (15 proteins selected)

Sim. class	CASTOR	PHD	DSC	PRED	MUL	NNSSP	Zpred	CONS
0	0.82	0.74	0.73	0.72	0.68	0.78.	0.66.	0.80
1	0.96	0.75	0.77	0.64	0.64	0.70	0.76	0.76
3	1.00	0.84	0.87	0.83	0.69	0.83	0.69	0.84
5	1.00	0.81	0.83	0.75	0.76	0.68	0.73	0.77
6	1.00	1.00	1.00	1.00	1.00	1.00	1.00	1.00
7	0.98	0.67	0.69	0.61	0.59	0.66	0.55	0.68

Complementarity Algorithms in Pattern Recognition, Table 2
Q_3 estimate of the classification precision of CASTOR and other classification procedures on 56 randomly selected proteins of the Cuff 513 data set

Sim. class	CASTOR	PHD	DSC	PRED	MUL	NNSSP	Zpred	CONS
0	0.73	0.73	0.71	0.72	0.66	0.71	0.63	0.75
1	0.95	0.79	0.76	0.69	0.68	0.74	0.75	0.78
2	0.87	0.84	0.81	0.89	0.62	0.95	0.65	0.90
3	0.94	0.77	0.79	0.78	0.72	0.82	0.67	0.80
5	1.00	0.81	0.83	0.75	0.76	0.68	0.73	0.77
6	0.95	0.85	0.81	0.80	0.79	0.79	0.78	0.82
7	0.85	0.71	0.70	0.67	0.61	0.69	0.57	0.72
Mean	0.84	0.73	0.72	0.70	0.64	0.71	0.61	0.74

totic misclassification error a sample, as large as possible, drawn randomly from the given population should be used, which will then ensure that under mild conditions the properties derived above are satisfied.

Also a distinction is often introduced regarding the similarity between classes of subsets of proteins [10]. In this case, it may be considered relevant that a cover be defined on the population of proteins, so as to form eight or more subpopulations. The samples can still be drawn randomly from the relevant subpopulation and the classifier determined for each subpopulation. To determine the structure of a new protein, first the similarity in the residue chain must be determined with respect to each subpopulation and then the classifier of the subpopulation with the highest similarity coefficient is applied. Here, the proper sampling method should consist of a stratified non-repetitive random sampling design, but this would be warranted only if there are significant differences in the results for the subpopulations.

The limitations of not using stratified random data sets and using ad hoc heuristics, instead of demonstrably convergent algorithms, is well brought out in the following tables [1], where classification results are compared between the **CASTOR** algorithm and seven popular alternative procedures, for two well-known data sets the Rost 126 and the Cuff 513 data sets.

Table 1 presents the Q_3 classification results on the Rost 126 verification set for the various similarity classes which have appeared in the sample of 15 proteins and in the random verification set. The precision of the classification results found by applying the **CASTOR** algorithm dominate all other procedures, and usually by over 15%.

In Table 2 the Q_3 estimates are given for the classification precision of **CASTOR** and the other classification procedures on 56 randomly selected proteins of the Cuff 513 data set. It is seen that the precision obtained by **CASTOR** dominates all the other entries except four. Two of these entries occur for the CONS algorithm for similarity classes 0 and 2, while the two other entries which dominate the results by the **CASTOR** algorithm occur for similarity class 2, for the procedures PRED and NNSSP.

Conclusions

The experiments described above shed some light on two important aspects, which are very closely related: the sampling procedure to adopt and the classification procedure to apply. Moreover, these results show the essential non-linearity and complexity of the pattern recognition problem.

Random sampling is necessary for precise and stable estimates, with the required accuracy, obtainable in predictions, since invariably the choice of special sets in verification or in training will alter the expected prediction accuracy, as a non-random sample will contain a different distribution of classes from the one regarding the population. As the prediction precision varies with the class distribution, this will have a significant effect on recognition.

Heuristics compared to algorithms will bias the results in the same way: they will be accurate in some cases, unstable in others. Moreover, when the classification results are poor, with a heuristic the source of the problem cannot usually be determined. With an algorithm, such as the one indicated, the root of the problem will invariably be tied to one of the mild assumptions not being satisfied.

This can be checked and remedied.

See also

- ▶ Generalizations of Interior Point Methods for the Linear Complementarity Problem
- ▶ Generalized Eigenvalue Proximal Support Vector Machine Problem

References

1. Cifarelli C, Patrizi G (2007) Solving large protein secondary structure classification problems by non-linear complementarity algorithm with {0,1} variables. Optim Softw 22:25–49
2. Devroye L, Gyorfi L, Lugosi G (1996) A Probabilistic Theory of Pattern Recognition. Springer, Berlin
3. Di Giacomo L, Argento E, Patrizi G (2004) Linear complementarity methods for the solution of combinatorial problems. Submitted for publication: copy at http://banach.sta.uniroma1.it/patrizi/
4. Konjin HS (1973) Statistical Theory of Sample Design and Analysis. North Holland, Amsterdam
5. Nieddu L, Patrizi G (2000) Formal properties of pattern recognition algorithms: A review. Eur J Oper Res 120:459–495
6. Patrizi G (1979) Optimal clustering properties. Ricerca Operativa 10:41–64
7. Patrizi G (1991) The equivalence of an lcp to a parametric linear program with a scalar parameter. Eur J Oper Res 51:367–386
8. Patrizi G, Addonisio G, Giannakakis C, AO Muda, Patrizi G, Faraggiana T (2007) Diagnosis of alport syndrome by pattern recognition techniques. In: Pardalos PM, Boginski VL, Vazacopoulos A (eds) Data Mining in Biomedicine. Springer, Berlin, pp 209–230
9. Patrizi G, Patrizi G, Di Cioccio L, Bauco C (2007) Clinical analysis of the diagnostic classification of geriatric disorders. In: Pardalos PM, Boginski VL, Vazacopoulos A (eds) Data Mining in Biomedicine. Springer, Berlin, pp 231–260
10. Rost B (1999) Twilight zone of protein sequence alignment. Protein Eng 12:85–94
11. Simon HA, Newell A (1972) Human problem solving. Prentice-Hall, Englewood Cliffs, NJ
12. Vapnik VN (1998) Learning Theory. Wiley, New York
13. Watanabe S (1985) Pattern Recognition: Human and Mechanical. Wiley, New York

Complexity Classes in Optimization

H. B. HUNT III
University Albany, New York, USA

MSC2000: 90C60

Article Outline

Keywords
Time and Space Complexity of Turing Machines
Definition of the Complexity Classes
    General Comments
Efficient Reducibility and 'Hard' Problems
See also
References

Keywords

Complexity; Complexity classes

We survey a number of the basic ideas, results, and references, for the complexity classes *P*, *NP*, *CoNP*, *PSPACE*, *DEXPTIME*, *NDEXPTIME*, and *EXSPACE*, the most important complexity classes in optimization. These ideas and results include the following:

i) the time and space complexities of both deterministic and nondeterministic multitape Turing machines as formalized in [1];
ii) the complexity classes above including the known results on their intercontainments; and
iii) the concepts of *polynomial reducibility*, *F-hardness*, and *F-completeness*, for the complexity classes *F* above.

We present brief historical surveys of the results obtained in ii) and iii). We also briefly survey many of the results on the above complexity classes in the basic references [1,12,16,25,29,30,37], emphasizing results especially relevant to the area of optimization.

The following are a list of some of the basic notation and terminology used here.

Definition 1 A *finite alphabet* Σ is a finite nonempty set of characters. A set of strings over some finite alphabet is said to be a *language*.

Further, we denote 'infinitely often' by 'i. o.'.

Definition 2 By an *exponential function* in n we mean a function $f(n) = 2^{c \cdot n^r}$ where $c, r > 0$ are constants independent of n.

The languages or problems 3-SATISFIABILITY (3-SAT), 3-DIMENSIONAL MATCHING, VERTEX COVER, CLIQUE, HAMILTON CIRCUIT, and PARTITION are defined as in [12]. Thus, for example, the language 3-SAT is defined to be the set of all satisfiable CNF formulas with no more than 3 literals per clause, when suitably encoded as a language over some finite alphabet. We note that this language, its quantified variants, and its succinctly-specified variants are the languages in the literature most widely used to prove *NP*-, *PSPACE*-, *DEXPTIME*-, and *NDEXPTIME*-hardness results (Definition 8 and [12,21,22,29,30]).

Finally, we denote the linear programming, $\{0, 1\}$-integer linear programming, integer linear programming, and quadratic programming problems as defined in [12,25,30,37] by LP, $\{0, 1\}$-ILP, ILP, and QP, respectively.

Time and Space Complexity of Turing Machines

In the literature of computational complexity, the most common models of computational devices and the problems solvable by such devices are *Turing machines* (TMs) and *language recognition problems*, respectively [1,9,12,15,17,29]. Here, we only consider multitape deterministic and nondeterministic Turing machines (denoted DTMs and NDTMs, respectively) and their associated language recognition problems as described in [1]. Informally, such a Turing machine M consists of the following:

1) a finite state control together with a finite nonempty set Q of possible *states of the control*;
2) finite nonempty *tape* and *input alphabets* T and I, respectively, and distinct symbols b and $\vdash$, denoting 'blank' and 'leftmost cell of tape', such that $I \subset T$ and $b, \vdash \in T - I$;
3) a finite number $k \geq 1$ of *tapes*, each of which is infinite to the right only and is divided into individual *tape cells* such that each cell can contain exactly one symbol in T at any one time;
4) k *tape heads*, one for each tape, each head capable of scanning a single cell at any one time;
5) a *start state* $q^o \in Q$ and a set of *accepting states* $F \subset Q$; and
6) a finite set μ of *moves*, each of the form

$$(q, s_1, \ldots, s_k, r, (t_1, d_1), \ldots, (t_k, d_k))$$
$$\in Q \times T^k \times Q \times (T \times \{L, R, S\})^k.$$

M is said to be *deterministic* if, for each $k + 1$ tuple $(q, s_1, \ldots, s_k) \in Q \times T^k$, there is at most one move in μ whose initial $k + 1$ components are $q, s_1, \ldots, s_k$, respectively. Otherwise, M is said to be *nondeterministic*. A state $q \in Q$ is said to be *final* if there is no move in μ whose first component equals q. We assume that the following two restrictions hold on F and μ:

7) Each accepting state is final.
8) There are no moves

$$(q, s_1, \ldots, s_k, r, (t_1, d_1), \ldots, (t_k, d_k))$$

in μ such that letting $1 \leq i \leq k$, $s_i = \vdash$ and $t_i \neq \vdash$, $s_i \neq \vdash$ and $t_i = \vdash$, or $s_i = \vdash$ and $d_i = L$.

Let $w \in I^*$. A a *partial computation* of M on w is a finite sequence $\sigma_w = (m_1, \ldots, m_l)$ with $l \geq 0$ of moves of M such that the following hold:

1) Initially, M is in its start state q^o; the first tape of M holds the string $\vdash w$, one symbol per cell starting at its leftmost cell; the contents of the leftmost cells of each of the other tapes of M equal $\vdash$ the contents of all other cells of M equal b; and each of the tape heads of M scans the leftmost cell of its corresponding tape.

2) When initialized as in 1), M executes the move-rules $m_1, \ldots, m_l$ consecutively and in that order. The *length of the partial computation* $\sigma_w = (m_1, \ldots, m_l)$ of M on w equals l. A partial computation $\sigma_w = (m_1, \ldots, m_l)$ of M on w is said to be an *accepting computation* (respectively, a *nonaccepting computation*) of M on w if, after executing the sequence of moves σ_w, M is in an accepting state (respectively, M is in a final state that is not an accepting state).

The restrictions 7) and 8) on the sets F and μ ensure that *no* accepting or nonaccepting computation of M on w is an initial subsequence of any other partial computation of M on w and at *no* point during a partial computation of M on w does one of the tape heads of M attempt to move off the left end of its corresponding tape.

The *language accepted by a Turing machine* M, denoted by $L(M)$, is the set of all strings $w \in I^*$ such that there exists an accepting computation on M on w. The *language recognition problem* of M is the problem of verifying, given $w \in I^* \cap L(M)$, that w is, in fact an element of $L(M)$. The time and space complexities of M are defined in terms of partial computations of M on strings $w \in I^*$ as follows:

Definition 3 Let M be a deterministic Turing machine. The *time complexity* of M, denoted by $T_M(\cdot)$, is the function from $\mathbf{N}$ to $\mathbf{N} \cup \{\infty\}$ defined, for all $n \in \mathbf{N}$, by
- $T_M(n) = \max\{l \in \mathbf{N} : l$ is the length of a partial computation of M on w, where $w \in I^n\}$, if this maximum exists;
- $T_M(n) = \infty$ otherwise.

The *space complexity* of M, denoted by $S_M(\cdot)$ is the function from $\mathbf{N}$ to $\mathbf{N} \cup \{\infty\}$ defined, for all $n \in \mathbf{N}$, by
- $S_M(n) = \max\{s \in \mathbf{N} : s$ is the maximum number of tape cells scanned on any of the tapes of M during a partial computation of M on $w\}$, where $w \in I^n$, if this maximum exists;
- $S_M(n) = \infty$ otherwise.

Let M be a nondeterministic Turing machine. The *time complexity* of M, denoted by $T_M(\cdot)$, is the function from $\mathbf{N}$ to $\mathbf{N}$ defined, for all $n \in \mathbf{N}$, by
- $T_M(n) = 0$ if no string $w \in I^n$ is in $L(M)$;
- $T_M(n) = \max\{l \in \mathbf{N} : l$ is the minimum length of an accepting computation M on $w\}$, where $w \in I^n \cap L(M)$;
- $T_M(n) = \infty$ otherwise.

The *space complexity* of M, denoted by $S_M(\cdot)$ is the function from $\mathbf{N}$ to $\mathbf{N}$ defined, for all $n \in \mathbf{N}$, by
- $S_M(n) = 0$ if no string $w \in I^n$ is in $L(M)$;
- $S_M(n) = \max\{m \in \mathbf{N} \mid m$ equals the minimum of the maximum numbers of tape cells scanned on any of the tapes of M during an accepting computation of M on w, where $w \in I^n \cap L(M)\}$;
- $S_M(n) = \infty$ otherwise.

There is a fundamental difference between the definitions of time and space complexity, for DTMs and for NDTMs, respectively. For a DTM M, the functions T_M and S_M are defined in terms of the numbers of moves executed and tape cells scanned in an *arbitrary* partial computation of M on strings $w \in I^*$. In contrast, for an NDTM M, these functions are defined *only* in terms of the numbers of moves executed and tape cells scanned in minimum 'cost accepting computations' of M on strings $w \in I^* \cap L(M)$. The following are two easy implications of this difference:

Proposition 4 *If L is accepted by a DTM M with time and space complexities T_M and S_M, then L is also accepted by an NDTM M' with time and space complexities $T_{M'}$ and $S_{M'}$ such that, for all $n \in \mathbf{N}$,*

$$T_{M'}(n) \leq T_M(n) \quad \text{and} \quad S_{M'}(n) \leq S_{M'}(n).$$

Proposition 5 *If L is accepted by a DTM M with input alphabet I such that, for all $n \in \mathbf{N}$, $T_M(n) \in \mathbf{N}$ (equivalently, $T_M(n) \neq \infty$), then the language $I^* - L$ is accepted by a DTM M' such that, for all $n \in \mathbf{N}$, $T_{M'}(n) = T_M(n)$.*

Consequently, deterministic time complexity classes as defined in Definition (6) are closed under complementation. At present (1999), nondeterministic time complexity classes are not known to be closed under complementation.

Definition of the Complexity Classes

In Definition (6) we use Definition (3) to define the time and space complexity classes most relevant to optimization. Next in Theorem (7), we give several basic properties of these classes, whose proofs do not require the concept of polynomial time reducibility defined in Definition (8).

Definition 6 Let $T, S : \mathbf{N} \to \mathbf{N}$.

A Turing machine M is said to be *polynomially time-bounded*, respectively *polynomially space-bounded*, if the function $T_M(n)$, respectively $S_M(n)$, is bounded above by a polynomial function in n. M is said to be *exponentially time-bounded*, respectively *exponentially space-bounded*, if the function $T_M(n)$, respectively $S_M(n)$, is bounded above by an exponential function in n.

- $DTIME(T(n))$, respectively $NDTIME(T(n))$, is the class of all languages L for which there exist a DTM, respectively an NDTM, M and and a $c > 0$ such that

$$L = L(M)$$

and, for all $n \in \mathbf{N}$,

$$T_M(n) \leq c \cdot T(n).$$

- $DSPACE(S(n))$, respectively $NDSPACE(S(n))$, is the class of all languages L for which there exist a DTM, respectively an NDTM, M and a $c > 0$ such that

$$L = L(M)$$

and, for all $n \in \mathbf{N}$,

$$S_M(n) \leq c \cdot S(n).$$

- P, NP, PSPACE, DEXPTIME, NDEXPTIME, and EXSPACE are the classes of all languages L such that L is the language accepted by a polynomially time-bounded DTM, a polynomially time-bounded NDTM, a polynomially space-bounded TM, an exponentially time-bounded DTM, an exponentially time-bounded NDTM, and an exponentially space-bounded TM, respectively.

- CoNP, respectively CoNDEXPTIME, is the class of all languages L for which there exists an NDTM M with tape alphabet I such that $L = I^* - L(M)$ and the function $T_M(n)$ is polynomially time-bounded, respectively exponentially time-bounded.

Theorem 7

1) The following containments hold among the complexity classes defined in Definition 6:
    a) $P \subset NP \cap CoNP$.
    b) $NP, CoNP \subset PSPACE \subset DEXPTIME$.
    c) $DEXPTIME \subset NDEXPTIME \cap CoNDEXPTIME$.
    d) $NDEXPTIME, CoNDEXPTIME \subset EXSPACE$.

2)
    a) $P = NP$ if and only if $NDTIME(n) \subset P$;
    b) $P = PSPACE$ if and only if $\exists \epsilon > 0$ such that $DSPACE(n^\epsilon) \subset P$;
    c) $NP = PSPACE$ if and only if $\exists \epsilon \geq 0$ such that $DSPACE(n\epsilon) \subset NP$;
    d) for all integers $k \geq 1$, $NDTIME(n^k) \subset DSPACE(n^k)$.

3) $PSPACE = DEXPTIME$ if and only if $\exists \epsilon \geq 0$ such that $DTIME(2^{n\epsilon}) \subset PSPACE$.

4) If we restrict the classes P and NP to languages over a single letter alphabet, denoting these restrictions by P_{sla} and NP_{sla}, respectively, then
    a) $P_{sla} = NP_{sla}$ if and only if $\cup_{k \geq 1} DTIME(2^{kn}) = \cup_{k \geq 1} NDTIME(2^{kn})$;
    b) NP_{sla} is closed under complementation if and only if the class $\cup_{k \geq 1} NDTIME(2^{kn})$ is closed under complementation.

Proof The claims in 1), 2), and 3) of the theorem follow directly from Definitions 3 and 6, the discussion after Definition 3, and simple well-known arguments involving 'padding', e. g. see [4,13,17]. As an example, let $L \in DSPACE(n^l)$, where $l \geq 1$ is an integer. Let $k \geq 2$ be an integer. Let $L' = \{w \cdot \#^m : m = k \cdot l, w \in L\}$, with $\#$ a symbol not occurring in L. Then $L' \in DSPACE(n^{1/k})$; and $L' \in P$ (respectively $L' \in NP$) if and only if $L \in P$ (respectively $L \in NP$). The claims of 4) follow from simple arguments about 'tally languages', see [13].

General Comments

The Turing machine model is due to A.M. Turing [36]. Additional discussions of this model can be found in [1,9,17].

The time and space complexity of DTMs were first studied in [15], and [14], respectively.

The time-bounded complexity classes P, NP, CoNP, etc. are invariant under several otherformal computer models, including multihead multitape Turing machines, Turing machines with multidimensional tapes, and both the *RAM* and *RASP* models of [35] and [11] under the logarithmic cost function. (For a detailed discussion of this, see [1, Chap. 1].

It is widely assumed that a computational problem is 'practically computationally tractable' only if it can be solved by a deterministic polynomially time-bounded Turing machine [1,12,25,29,30,37]. The resulting importance of the class *P* was first observed by A. Cobham [7] and J. Edmonds [10].

By the well-known *Savitch theorem* [33] that NDSPACE(log n) $\subset$ DSPACE($[\log n]^2$), the classes of languages accepted by polynomially space-bounded DTMs and NDTMs are equal and the classes of languages accepted by exponentially space-bounded DTMs and NDTMs are equal. This is the reason for defining the complexity classes *PSPACE* and *EXSPACE*, rather than the classes *DPSPACE*, *NDPSPACE*, *DEXSPACE*, and *NDEXSPACE*.

Efficient Reducibility and 'Hard' Problems

Throughout this section *F* is a class of languages. Following [1,12,17], we define polynomial reducibility, *F*-hardness and *F*-completeness under such reducibilities. To do this, we must extend the definition of DTMs given above so that the resulting machines have outputs, and thus can be viewed as computing partial functions of their inputs. This is accomplished by augmenting each DTM *M* with an additional 'output' tape o_M such that the tape head of o_M can only move one cell to the right or stay stationary during any move of *M*. In addition to all usual constraints on *M*, for all inputs *w* of *M*:

1) initially, during a partial computation of *M* on *w*, all cells of o_M are blank and the tape head of o_M scans its leftmost cell;
2) the value computed of *M* on *w* is the final nonblank contents of o_M during the accepting computation or the nonaccepting computation of *M* on *w*, if such a computation exists, and is undefined otherwise.

The time complexity T_M of such augmented DTMs (henceforth, referred to simply as DTMs) is defined exactly as in Definition 3.

Definition 8 Let Σ and Δ be finite alphabets.

A function *f* from Σ^* to Δ^* is said to be *polynomial time computable* if and only if there exists a polynomially time-bounded DTM *M* with input alphabet Σ such that, for all $w \in \Sigma^*$, the value computed by *M* on input *w* is $f(w)$.

Let $L \subset \Sigma^*$ and $M \subset \Delta^*$. *L* is said to be *polynomially reducible* to *M*, denoted $L \leq_p M$, if and only if there is an $f: \Sigma^* \to \Delta^*$ such that *f* is polynomial time computable and, for all $w \in \Sigma^*: w \in L$ if and only if $f(w) \in M$.

A language *L* is said to be *F-hard* if and only if for all languages $L' \in F$, $L' \leq_p L$. *L* is said to be *F-complete* if and only if *L* is both *F*-hard and is in *F*.

Henceforth, let *F* be any of the complexity classes *NP*, *CoNP*, *PSPACE*, *DEXPTIME*, *NDEXPTIME*, *CoNDEXPTIME*, or *EXSPACE*. The following two propositions underlie most of the work on *F*-hard and *F*-complete problems in the literature on algorithmic analysis and computational complexity.

Proposition 9
1) Let Σ, Δ, and Π be finite alphabets. Let $L \subset \Sigma^*$, $M \subset \Delta^*$, and $N \subset \Pi^*$. If $L \leq_p M$ and $M \leq_p N$, then $L \leq_p N$. (Thus, polynomial reducibility is transitive.)
2) If *L* and *M* are languages such that $L \leq_p M$ and *L* is *F*-hard, then *M* is also *F*-hard.
3) *P* = *NP* if and only if some *NP*-complete language is in *P*.
4) *P* = *PSPACE* if and only if some *PSPACE*-complete language is in *P*.
5) *NP* = *PSPACE* if and only if some *PSPACE*-complete language is in *NP*.
6) *NP* = *CoNP* if and only if some *CoNP*-complete language is in *NP* if and only if some *NP*-complete language is in *CoNP*.
7) If *L* is *DEXPTIME*-, *NDEXPTIME*-, or *EXSPACE*-hard, then the recognition of *L* requires more than 2^{n^ϵ} time, 2^{n^ϵ} time, respectively 2^{n^ϵ} space, i. o. on any DTM, NDTM, respectively TM, where $\epsilon > 0$ is a constant independent of *n*.

Proof We sketch a proof: 1) follows from the fact that the polynomial time computable functions are closed under composition. The proofs of 2) through 6) follow directly from the correctness of 1) and Definitions 6 and 8. The proof of 7) follows directly from well-known 'hierarchy' theorems, for deterministic and nondeterministic time and space-bounded Turing machines [1,12,14,15,17,34].

Proposition 10 *There exists an F-complete language, for each of the complexity classes F.*

Proof It is easy to construct *F*-complete languages defined in terms of Turing machines and coded versions of their inputs, for each of the complexity classes F. For example, the language $L = \{\# \cdot M_i \cdot\# \cdot \text{code}(x_1, \ldots, x_n) \cdot \#^m : x_1 \cdot x_n$ is accepted by the one-tape NDTM M_i in time $t\}$, where $m = 3 \cdot |M_i| \cdot t$, is both an element of $NDTIME(n)$ and is NP-hard.

The first complete problems for NP and, consequently, the importance of the class NP in nonnumerical computation are due to S.A. Cook [8], and R.M. Karp [18]. Following the terminology of [12], these initial NP-complete problems include the problems

- 3-SAT,
- 3-DIMENSIONAL MATCHING,
- VERTEX COVER, CLIQUE,
- HAMILTONIAN CIRCUIT, and
- PARTITION.

The first complete problems for $PSPACE$ and ND-$EXPTIME$, are due to A.R. Meyer and L.J. Stockmeyer [24]. Subsequently, a very large number of natural computational problems have been shown to be F-hard or F-complete, for each of the complexity classes F. Many examples and historical references can be found in [1,12,16,25,29,30,37]. References [12,25,28,30,37], are especially relevant to problems in the area of optimization, including:

- LP (which is solvable deterministically in polynomial time as show initially in [19]);
- $\{0, 1\}$-ILP and ILP (the feasibility problems of which are NP-complete [5,12,17,30]); and
- QP (which is NP-complete) [12,32,37].

Reference [12] also discusses much of the early work (prior to 1979) on the complexity of approximating NP-hard optimization problems. Reference [16] consists of several separately authored chapters surveying many of the more recent results on the complexity of approximating NP-hard optimization problems. These results include the important result of [3] that unless $P = NP$, no MAX SNP-hard optimization problem [31] has a PTAS (i. e. a polynomial time approximation scheme, [12]).

Many basic polynomial time solvable and NP-hard optimization problems become $PSPACE$-hard, $DEXPTIME$-hard, $NDEXPTIME$-hard, and even $EXSPACE$-hard, when problem instances are specified succinctly by hierarchical specifications or by 1- and 2-dimensional periodic specifications, see [20,21,22,26,27,29]. References [2,6] discuss algorithms for

and the computational complexity of solving systems of multivariable polynomial equations over real closed fields. Most of these last problems are NP-hard or worse. Finally, the problems of determining the solvability of a system of multivariable polynomial equations over $\mathbf{N}$ or $\mathbf{Z}$ are recursively undecidable, by a straightforward effective reduction from *Hilbert's tenth problem* [9,23].

See also

- ▶ Complexity of Degeneracy
- ▶ Complexity of Gradients, Jacobians, and Hessians
- ▶ Complexity Theory
- ▶ Complexity Theory: Quadratic Programming
- ▶ Computational Complexity Theory
- ▶ Fractional Combinatorial Optimization
- ▶ Information-based Complexity and Information-based Optimization
- ▶ Kolmogorov Complexity
- ▶ Mixed Integer Nonlinear Programming
- ▶ NP-complete Problems and Proof Methodology
- ▶ Parallel Computing: Complexity Classes

References

1. Aho AV, Hopcroft JE, Ullman JD (1974) The design and analysis of computer algorithms. Addison-Wesley, Reading, MA
2. Arnon DS, Buchberger B (1988) Algorithms in real algebraic geometry. Acad. Press, New York
3. Arora S, Lund C, Motwani R, Sudan M, Szegedy M (1992) Proof verification and hardness of approximation problems. Proc 33rd IEEE Symp Foundations of Computer Sci., pp 14–23
4. Book RV (1974) Comparing complexity classes. JCCS 9:213–229
5. Borosh I, Treybig LB (1976) Bounds on positive integral solutions of linear Diophantine equations. In: Proc Amer Math Soc, vol 55, pp 294–304
6. Canny J (1988) Some algebraic and geometric computations in PSPACE. Proc. 20th Annual ACM Symposium on Theory of Computing, 460–467
7. Cobham A (1964) The intrinsic computational difficulty of functions. In: Bar-Hillel Y (ed) Proc. 1964 Internat. Congress for Logic, Methodology and Philosophy of Sci. North-Holland, Amsterdam, 24–30
8. Cook SA (1971) The complexity of theorem-proving procedures, Proc. Third ACM Symp. Theory of Computing, pp 151–158
9. Davis M (1958) Computability and unsolvability. McGraw-Hill, New York

10. Edmonds J (1965) Paths, trees, and flowers. Canad J Math 17:449–467
11. Elgot CC, Robinson A (1964) Random access stored program machines. J ACM 11:365–399
12. Garey MR, Johnson DS (1979) Computers and intractability. A guide to the theory of NP-completeness. Freeman, New York
13. Hartmanis J, Hunt III HB (1974) The LBA problem and its importance in the theory of computing. Complexity of Computation 1–26
14. Hartmanis J, Lewis PM, Stearns RE (1965) Classification of computations by time and memory requirements. Proc. IFIP Congress, 31–35
15. Hartmanis J, Stearns RE (1965) On the computational complexity of algorithms. Trans Amer Math Soc 117:285–306
16. Hochbaum DS (1979) Approximation algorithm for NP-hard problems. PWS, Boston, MA
17. Hopcroft JE, Ullman JD (1969) Formal languages and their relation to automata. Addison-Wesley, Reading, MA
18. Karp RM (1972) Reducibility among combinatorial problems. In: Miller RE, Thatcher JW (eds) Complexity of Computer Computations. Plenum, New York, pp 85–103
19. Khachian LG (1979) A polynomial algorithm for linear IV programming. Soviet Math Dokl 20, 191–194. (Dokl Akad Nauk USSR 224 (1979), 1093–1096)
20. Lengauer T, Wagner KW (1992) The correlation between the complexities of non-hierarchical and hierarchical versions of graph problems. JCSS 44:63–93
21. Marathe MV, Hunt III HB, Rosenkrantz DJ, Stearns RE (1998) Theory of periodically specified problems: complexity and approximability. Proc. 13th IEEE Conf. Computational Complexity
22. Marathe MV, Hunt III HB, Stearns RE, Radhakrishnan V (1997) Complexity of hierarchically and 1-dimensional periodically specified problems I: hardness results. In: Du D-Z, Gu J, Pardalos PM (eds), Satisfiability Problem Theory and Appl. DIMACS 35. Amer. Math. Soc., pp 225–259
23. Matijasevic YV (1970) Enumerable sets are Diophantine, Soviet Math Dokl 11:354–357 (Dokl Akad Nauk USSR 191 (1970), 279–282)
24. Meyer AR, Stockmeyer LJ (1972) The equivalence problem for regular expressions with squaring requires exponential times. Proc. 13th Annual Symposium on Switching and Automata Theory, pp 125–129
25. Nemhauser GL, Wolsey LA (1988) Integer and combinatorial optimization. Interscience Ser Discrete Math and Optim. Wiley, New York
26. Orlin JB (1984) Some problems on dynamic/periodic graphs. Program. Combin. Optim. Acad. Press, New York, pp 273–293
27. Orlin JB (1985 1978) The complexity of dynamic/periodic languages and optimization problems. Sloan WP, vol 1676-86. MIT, Cambridge, MA. A preliminary version of this paper appears in Proc. 13th annual ACM Symposium on Theory of Computing, 1978, pp 218–227
28. Padalos PM (1993) Complexity in numerical optimization. World Sci., Singapore
29. Papadimitriou CH (1994) Computational complexity. Addison-Wesley, Reading, MA
30. Papadimitriou CH, Steiglitz K (1982) Combinatorial optimization: Algorithms and complexity. Prentice-Hall, Englewood Cliffs, NJ
31. Papadimitriou CH, Yannakakis M (1991) Optimization, approximation, and complexity classes. JCCS 43:425–440
32. Sahni S (1974) Computationally related problems. SIAM J Comput 3:262–279
33. Savitch WJ (1970) Relationship between nondeterministic and deterministic tape complexities. JCSS 4:177–192
34. Seiferas JJ, Fischer MJ, Meyer AR (1973) Refinements of nondeterministic time and space hierarchies, Proc. 14th Annual IEEE Symp. Switching and Automata Theory, pp 130–137
35. Sheperdson JC, Sturgis HE (1963) Computability of recursive functions. JACM 10:217–255
36. Turing AM (1936/7) On computable numbers, with an application to the Entscheidungsproblem. Proc London Math Soc (2) 49:230–265
37. Vavasis SA (1991) Nonlinear optimization: Complexity issues. Oxford Sci. Publ., Oxford

Complexity of Degeneracy

KATTA G. MURTY
Department IOE, University Michigan,
Ann Arbor, USA

MSC2000: 90C60

Article Outline

Keywords
Degeneracy in Linear Programming
   Degeneracy in Standard Form Systems
   The Complexity of Checking Whether a System
     of Constraints is Degenerate
   Problems Posed By Degeneracy for the Simplex Method of LP
   Cycling
   Stalling
   Degeneracy Handling in Commercial Codes
   Effect of Degeneracy on the Optimum Face of an LP
   Effects of Degeneracy on Post-optimality Analysis in LP
   Effects of Degeneracy in Interior Point Methods for LP
   Effect of Degeneracy on Algorithms
     for Enumerating Extreme Point Solutions
   Effect of Degeneracy
     in Extreme Point Ranking Methods

Degeneracy in Nonlinear Programming
See also
References

Keywords

Degeneracy; Nondegeneracy; Near degeneracy; Active constraints; Tight constraints; Inactive constraints; Slack constraints; Basic feasible solution; Regular polyhedron; Simple polyhedron; *NP*-complete problem; Cycling; Stalling; Resolving degeneracy; Positive marginal values; Negative marginal values; Extreme point enumeration; Extreme point ranking; Assignment ranking; Segments of polyhedra; Active set methods

Degeneracy in Linear Programming

In mathematical programming, the terms *degeneracy*, and its absence, *nondegeneracy*, have arisen first in the simplex method of linear programming (LP), where they have been given precise definitions. The notions were first introduced by G.B. Dantzig in his seminal paper [7] when he invoked the nondegeneracy assumption to prove the finite convergence of the simplex method.

In the study of the simplex method for LP, degeneracy and nondegeneracy are properties defined for basic feasible (or *extreme point*) solutions of systems of linear constraints or for the systems themselves. To give the definition, let us consider the general system of linear constraints

$$A_{i\cdot}x \begin{cases} = b_i, & i = 1, \ldots, r, \\ \geq b_i, & i = r+1, \ldots, m, \end{cases} \quad (1)$$

where $A_{i\cdot} \in \mathbf{R}^n$ is the row vector of coefficients in the *i*th constraint, and we assume that the equality constraints in it are linearly independent. Let K denote its set of feasible solutions.

Given $\bar{x} \in K$, the *i*th constraint in (1) is said to be: *active* or *tight* at $\bar{x}$ if either it is an equality constraint (i.e., $i \in \{1, \ldots, r\}$), or it is an inequality constraint that holds as an equation at $\bar{x}$; *inactive* or *slack* at $\bar{x}$ otherwise. We denote the index set of active constraints at $\bar{x}$, i.e., $\{i: i \in \{1, \ldots, m\}, A_{i\cdot}\bar{x} = b_i\}$ by $I(\bar{x})$.

The feasible solution $\bar{x}$ for (1) is said to be an *extreme feasible solution* or a *BFS* (*basic feasible solution*)

if it is the unique solution of the system of equations defined by the active constraints at it in (1); i.e., $A_{i\cdot}\bar{x} = b_i$ for $i \in I(\bar{x})$.

The BFS $\bar{x}$ for (1) is said to be: a *nondegenerate BFS* if the set of active constraints at it, treated as equations, forms a square nonsingular system of equations; a *degenerate BFS* if that system has one or more redundant equations, i.e., if $|I(\bar{x})| > n$. Thus at a degenerate BFS, this system of equations formed from the active constraints is an overdetermined system of linear equations with a unique solution.

The general system (1) is said to be: a *degenerate system* if it has at least one degenerate BFS; *nondegenerate system* if all its BFSs are nondegenerate.

Degeneracy in Standard Form Systems

Before solving an LP, the simplex method transforms the constraints into a *standard form* which is

$$\begin{aligned} Ax &= b, \\ x &\geq 0, \end{aligned} \quad (2)$$

where the matrix A is of order $m \times n$, and without any loss of generality we assume that rank(A) = m. Let $A_{\cdot j}$ denote the *j*th column vector of A for $j = 1, \ldots, n$, it is the column of x_j in (2) and we assume it is $\neq 0$ for all j. Let Γ denote the set of feasible solutions of (2).

Specializing the above definitions to the standard form, we conclude that a feasible solution $\bar{x}$ of (2) is a BFS if and only if $\{A_{\cdot j}: j \text{ such that } \bar{x}_j > 0\}$ is linearly independent. The BFS $\bar{x}$ is: degenerate if $|\{j: \bar{x}_j > 0\}| < m$, nondegenerate if $|\{j: \bar{x}_j > 0\}| = m$.

So, for a nondegenerate BFS $\bar{x}$, the submatrix with column vectors $\{A_{\cdot j}: j \text{ such that } \bar{x}_j > 0\}$ is a basis for the system of equations in (2). If $\bar{x}$ is a degenerate BFS, this submatrix has to be augmented with the columns of some variables having 0 values in $\bar{x}$ in order to become a basis; usually this augmentation can be carried out in many ways. Hence, while each nondegenerate BFS is associated with a unique basis, each degenerate BFS is associated with several (usually a huge number of) bases.

System (2) is said to be: degenerate if it has at least one degenerate BFS; nondegenerate otherwise. From this we see that if system (2) is degenerate, then the right-hand side constants vector b lies in a subspace that is the linear hull of a set of $m-1$ or less column vectors of the coefficient matrix A. These observations imply the following facts.

1) Keeping the coefficient matrix A fixed in (2), but letting the right-hand side constants vector b vary over $\mathbf{R}^m$, the set of all b for which (2) is degenerate is a set of Lebesgue measure zero in $\mathbf{R}^m$.
2) If (2) is degenerate, the right-hand side constants vector b in it can be perturbed ever so slightly to make the *perturbed system* nondegenerate.

   The *perturbation technique* for resolving the problem of cycling in the simplex method caused by degeneracy is based on this fact. We will discuss more on this later.
3) If (2) has at least one nondegenerate BFS, the dimension of Γ is $d = n - m$.

   Whether (2) is degenerate or not, every nondegenerate BFS of (2) is incident to exactly $d = n - m$ edges of Γ.

   However, a degenerate BFS is usually incident to more than d (could be very large) edges of Γ, and the number of edges of Γ incident at different degenerate extreme points of Γ may be very different.
4) If (2) is nondegenerate, Γ is said to be a *regular* or *simple polyhedron* because every one of its vertices is incident to exactly d (the dimension of Γ) edges. This nice regular property may not hold for Γ if (2) is degenerate. Thus, degeneracy has the effect of making the polyhedron more complex geometrically.

The Complexity of Checking Whether a System of Constraints is Degenerate

Despite the rarity of degeneracy among all possible LP models, surprisingly, many real world LP models turn out to be degenerate. However, R. Chandrasekaran, S.N. Kabadi and K.G. Murty [4] showed that the problem of checking whether a given system of linear constraints is degenerate is *NP-complete*.

A nondegenerate BFS of (2) is said to be *nearly degenerate* if some variables have positive values which are very close to zero in it. In practice, while computing BFSs of (2), unless exact arithmetic is used, it is very hard to distinguish between degenerate and nearly degenerate BFSs because of *round-off errors* introduced in digital computation.

Problems Posed By Degeneracy for the Simplex Method of LP

Cycling

Very soon after developing the simplex method for LP, Dantzig realized that it may not lead to an optimum solution under degeneracy but instead may cycle indefinitely among a set of nonoptimal degenerate bases. The first example of *cycling* in the simplex method was constructed by A.J. Hoffman [13].

For implementing the simplex method, the user has to select two *tie breaking rules* to be used in each pivot step, one for selecting the entering nonbasic variable, and the other for selecting the dropping basic variable, among all those that tie. For cycling to occur, these tie breaking rules are very crucial.

Any technique that makes sure that the simplex method cannot cycle under degeneracy is said to *resolve degeneracy*. Quite early in the development of LP, techniques for resolving degeneracy in theory, using a virtual perturbation involving powers of an infinitesimal indeterminate, without altering the data were developed ([5,8,21]; also see [16] for an extension of this technique to the bounded variable simplex method). These fix the tie breaker for the dropping variable as one based on lexicographic ordering, but leave the entering variable choice arbitrary among those eligible. Over the years several other techniques have been developed for resolving degeneracy in theory; some, e. g., Bland's technique [3], fix the tie breakers for both the entering and dropping variables. Bland's technique and others like it, however, lead to implementations which are very slow in practice.

Computationally, it is also important that techniques for resolving degeneracy pay attention to the possible effects of round-off errors in near degenerate solutions. See [10].

It is commonly believed that the problem of cycling is not encountered in practice; however, degeneracy related problems have been discovered to contribute substantially to the difficulty in using LP based methods in scheduling and related combinatorial and integer programming problems.

Stalling

Even after resolving the problem of cycling, yet another phenomenon called *stalling* at a degenerate BFS can occur in the simplex method. Unlike cycling which is an infinite repetition of the same sequence of degenerate bases, stalling is a finite but exponentially long sequence of consecutive degenerate pivot steps at the same objective value. Examples of stalling in network flow models have been exhibited by J. Edmonds, see [6,17].

A technique is said to resolve both cycling and stalling under degeneracy, if it can be established that the total number of consecutive degenerate pivot steps in the simplex method, using this technique, is bounded above by a polynomial function of n and the size of the LP. Such techniques that fix the tie breakers for both the entering and the dropping variables have been developed in [6] for the special case of minimum cost pure network flow problems. Extending this work to the general LP model seems to be hard, as it can be shown that resolving both cycling and stalling in a general LP model is only possible if there exist tie breakers for both the entering and dropping variables which guarantee to make the simplex method a polynomial time method for LP. To establish whether such tie breakers exist has been a long standing open problem in LP theory.

Degeneracy Handling in Commercial Codes

In spite of the folklore that cycling is very unlikely to occur in practice, commercial LP codes have sought to implement *anti-cycling procedures* that involve little overhead and are effective in practice.

The lexicographic technique for resolving degeneracy is not very desirable, as it needs the explicit basis inverse in every step (most commercial codes do not compute the basis inverse explicitly, they use matrix factorizations of the basis inverse for preserving sparsity and for numerical stability).

For handling degeneracy, commercial codes normally use procedures based on perturbing the bounds on the variables. If there is no progress in the objective value after some number of iterations dependent on problem size, then the bounds on the variables in the present basic vector are enlarged (i. e., if the previous lower and upper bounds on x_j are ℓ_j and u_j, they are changed to $\ell_j - \delta_j$ and $u_j + \delta_j$, where δ_j is a small positive quantity chosen appropriately), and the application of the algorithm is continued on the perturbed problem. When the perturbed problem reaches optimality, the bounds are reset to their original values to see if the resulting basis is optimal to the original problem (this happens very often). Otherwise, the resulting basis satisfies the optimality criterion but may be infeasible. Then a Phase I procedure is used to get feasibility, this works fine in almost all cases since the optimal basis for the perturbed problem is close to one for the original problem. The dual simplex algorithm can also be used for this later part. See [11] for details.

Effect of Degeneracy on the Optimum Face of an LP

From LP theory we know that if an LP has at least one optimum nondegenerate BFS, then the dual problem has a unique optimum solution. Conversely, if the dual problem has at least one optimum nondegenerate BFS, then the primal LP optimum solution is unique.

Effects of Degeneracy on Post-optimality Analysis in LP

After having found an optimum solution for an LP, an integral part of a good report generator is *marginal analysis*.

Consider the LP in standard form: minimize $z = cx$ subject to (2), and let $z^*(b)$ denote the optimum objective value in this LP as a function of the right-hand side constants vector b while all the other data remains fixed. The *marginal value* or *shadow price* vector for this LP is defined to be $(\partial z(b)/\partial b_i : i = 1, \ldots, m)$ when it exists.

If the LP has a nondegenerate optimum BFS, then the dual optimum solution is unique, it is the vector of marginal values; and for each right hand side constant b_i there is an interval of positive length containing the present value of b_i in its interior, which is its optimality range. As b_i varies in this range while all the other data remains fixed, the dual optimum solution remains unchanged and remains as the marginal value vector.

In practical applications, the right-hand side constants vector b in the LP model for a company's operations usually contains parameters such as the limits on raw material supplies, etc. When it exists, practitioners use the marginal value vector to derive many facts of great use in planning, such as identifying which raw material supplies are critical, what the break even price is for additional supply of each raw material, etc. Marginal analysis is the process of drawing such conclusions, and practitioners rely on it heavily to provide valuable planning information.

The situation changes dramatically when all the optimum BFSs of the LP are degenerate. The dual optimum solution may not be unique, and the marginal value vector as defined above may not exist. In its place we have two-sided marginal values: a *positive* (and a *negative*) *marginal value* giving the rate of change in the optimum objective value per unit increase (decrease) in b_i. M. Akgul [1] proved the existence of these two-sided marginal values using convex analysis. Simple proofs based on parametric LP are given in [16] where it is shown that the positive (negative) marginal value with respect to b_i is

max $\{\pi_i:$ over the dual optimum face$\}$

(min $\{\pi_i:$ over the dual optimum face$\}$).

Effects of Degeneracy in Interior Point Methods for LP

Unlike the simplex method which walks along edges of the polyhedron, the paths traced by interior point methods (IPMs) are contained in the strict interior of the polyhedron. There are many different classes of IPMs based on the strategy used. At first glance, degeneracy, a concept based on properties of extreme point solutions, does not seem to be as serious a problem for IPMs as it is for simplex methods. In fact proofs of polynomiality for IPMs of the projective, path following, and affine potential reduction categories hold true without any nondegeneracy assumption.

However, degeneracy affects the convergence of the primal-dual pair in the affine scaling method. Under primal nondegeneracy, this method has been shown to be globally convergent for any steplength as long as all the iterates remain in the interior of the feasible region. But this technique breaks down when the primal nondegeneracy assumption is removed, in fact L.H. Hall and R.J. Vanderbei [12] constructed a degenerate example to show that the dual sequence cannot be convergent anymore if any fixed steplength greater than 2/3 to the boundary is taken. Thus stepsize 2/3 to the boundary is the longest stepsize for the affine scaling algorithm that guarantees convergence of the primal-dual pair in the presence of degeneracy.

Although other IPMs go through the interior of the feasible region, degeneracy still has a role to play in them. But the problems here are different from the cycling and stalling problems occurring in the simplex method. Degeneracy and redundant constraints affect the central path which most IPMs aim to follow. Numerical performance of the algorithms may suffer from numerical instability and ill-conditioning if the optimum solutions are degenerate or near degenerate. Also, generating an optimum basis from the near optimum interior solution at the termination of the IPM is strongly polynomial, but the computational effort depends on the degree of degeneracy.

Effect of Degeneracy on Algorithms for Enumerating Extreme Point Solutions

Consider the problem of *enumerating all the extreme point solutions* of a system of linear constraints, say (2). Let ℓ_0 denote the unknown number of extreme point solutions.

If (2) is nondegenerate, all its extreme point solutions can be enumerated in time $O(\ell_0 mn)$, an effort which grows linearly with ℓ_0, D. Avis and K. Fukuda [2]. If (2) is degenerate and is the system of constraints for a network linear program, J.S. Provan [19] has an algorithm for enumerating all its extreme point solutions in time polynomial in ℓ_0 and the input size.

However, it remains an open question whether there is an algorithm for enumerating all extreme point solutions in time polynomial in ℓ_0 and input size, when (2) is a general degenerate system of constraints.

Murty and S.-J. Chung [18] have shown that degenerate polyhedra have proper subsets called *segments* satisfying certain facial incidence properties. For each nondegenerate polyhedron, the only segment possible is the whole polyhedron itself. The difficulty of enumerating extreme point solutions of degenerate systems efficiently is related to the problem of recognizing whether a given segment is the whole polyhedron or a proper subset of it.

Effect of Degeneracy in Extreme Point Ranking Methods

Consider the objective function $z(x) = cx$ defined over Γ, the set of feasible solutions of (2). For simplicity assume that Γ is a convex polytope, i.e., it is bounded. An algorithm for *ranking the extreme points* of Γ in increasing order of $z(x)$ has been discussed in [15]. In each step, this algorithm carries out the operation of enumerating the adjacent extreme points of a given extreme point of Γ.

If (2) is nondegenerate, every extreme point has exactly $n - m$ adjacent extreme points, and the above operation can be carried out efficiently by pivot steps. Hence, the complexity of generating k extreme points in the ranked sequence grows linearly with k, and the ranking algorithm becomes practically effective.

If (2) is degenerate, the number of adjacent extreme points of a degenerate extreme point of Γ may be very large, and the ranking algorithm becomes almost impractical.

The *assignment problem* is a well known example of a highly degenerate problem. However all its extreme point solutions known as assignments are $0 - 1$ vectors, using this property an efficient special algorithm has been developed in [14] for ranking the assignments in increasing order of a linear objective function.

Degeneracy in Nonlinear Programming

In contrast to linear programming where the concept of degeneracy is defined purely using extreme point solutions; in nonlinear programming it is defined for any solution point.

Discussion of degeneracy arises in nonlinear programming, particularly in methods known as *active set methods*. These methods are popular for solving nonlinear programs in which the constraints are linear, say of the form (1); but also used when there are nonlinear constraints in the system. In these methods, when at a feasible point x^0, certain constraints indexed by an active set $\mathcal{A}$ are treated as equations, and the rest are temporarily disregarded, and a search direction y^0 is generated. The next point is taken ideally as the best feasible point on the half-line $\{x^0 + \lambda y^0 : \lambda \geq 0\}$. However, if one or more inequality constraints not from the set $\mathcal{A}$ are violated by $x^0 + \lambda y^0$ whenever $\lambda > 0$ and sufficiently small, then those constraints allow no progress in the search direction, and we have a degenerate situation. See [10] and [11] for a discussion of this degeneracy in active set methods, and its resolution.

There are also other generalizations of the notions of degeneracy and nondegeneracy, to systems of nonlinear constraints. In these generalizations, degeneracy is taken to mean any measure of departure of problem structure from some idealized norm. Simply put, nondegenerate means well-posed in some context, degenerate means absence of such nice structure. For a nonlinear program, nondegeneracy at a solution point has been defined variously as the satisfaction of: LICQ (linear independence constraint qualification of the binding constraint gradients), KKT first order necessary conditions for a local minimum, second order sufficient conditions for a local minimum, or the strict complementary slackness condition. Also connections between nondegeneracy and performance of algorithms has been studied, addressing the local effects of special kinds of nondegeneracy or its lack at a local minimizer. See [9].

See also

- ▶ Complexity Classes in Optimization
- ▶ Complexity of Gradients, Jacobians, and Hessians
- ▶ Complexity Theory
- ▶ Complexity Theory: Quadratic Programming
- ▶ Computational Complexity Theory
- ▶ Fractional Combinatorial Optimization
- ▶ Information-Based Complexity and Information-Based Optimization
- ▶ Kolmogorov Complexity
- ▶ Mixed Integer Nonlinear Programming
- ▶ NP-Complete Problems and Proof Methodology
- ▶ Parallel Computing: Complexity Classes

References

1. Akgul M (1984) A note on shadow prices in linear programming. J Oper Res Soc 35:425–431
2. Avis D, Fukuda K (1992) A pivoting algorithm for convex hulls and vertex enumeration of arrangements and polyhedra. Discrete Comput Geom 8:295–313
3. Bland RG (1977) New finite pivoting rules for the simplex method. Math Oper Res 2:103–107
4. Chandrasekaran R, Kabadi SN, Murty KG (1982) Some NP-complete problems in linear programming. Oper Res Lett 1:101–104
5. Charnes A (1952) Optimality and degeneracy in linear programming. Econometrica 20:160–170
6. Cunningham WH (1979) Theoretical properties of the network simplex method. Math Oper Res 4:196–208
7. Dantzig GB (1951) Maximization of a linear function of variables subject to linear inequalities. In: Koopmans TC (ed) Activity Analysis of Production and Allocation. Wiley, New York, pp 339–347

8. Dantzig GB, Orden A, Wolfe P (1955) The generalized simplex method for minimizing a linear form under linear inequality restraints. Pacific J Math 5:183–195
9. Fiacco AV, Liu J (1993) Degeneracy in nonlinear programming and the development of results motivated by its presence. In: Gal T (ed) Degeneracy in optimization problems. vol 46 of Ann Oper Res, pp 61–80
10. Fletcher R (1987) Practical methods of optimization, 2nd edn. Wiley, New York
11. Gill PE, Murray W, Saunders MA, Wright MH (1989) A practical anti-cycling procedure for linearly constrained optimization. Math Program 45:437–474
12. Hall LA, Vanderbei RJ (1993) Two-thirds is sharp for affine scaling. Oper Res Lett 13:197–201
13. Hoffman AJ (1953) Cycling in the simplex algorithm. NBS Report 2974 Washington D.C.
14. Murty KG (1968) An algorithm for ranking all the assignments in increasing order of cost. Oper Res 16:682–687
15. Murty KG (1968) Solving the fixed charge problem by ranking the extreme points. Oper Res 16:268–279
16. Murty KG (1983) Linear programming. Wiley, New York
17. Murty KG (1992) Network programming. Prentice-Hall, Englewood Cliffs, NJ
18. Murty KG, Chung S-J (1995) Segments in enumerating faces. Math Program 70:27–45
19. Provan JS (1994) Efficient enumeration of the vertices of polyhedra associated with network LPs. Math Program 63:47–64
20. Gal T (ed) (1993) Degeneracy in optimization problems. Ann Oper Res, vol 46
21. Wolfe P (1963) A technique for resolving degeneracy in linear programming. SIAM J 11:205–211

Complexity of Gradients, Jacobians, and Hessians

SA

ANDREAS GRIEWANK
Institute Sci. Comput. Department Math.,
Techn. University Dresden, Dresden, Germany

MSC2000: 65D25, 68W30

Article Outline

Keywords
'Numerical' Differentiation Methods
'Analytical' Differentiation Methods
Two-Stranded Chain Scenario
Computational Model
Indefinite Integral Scenario
Lack of Smoothness
Predictability of Complexities
Goal-Oriented Differentiation
The Computational Graph
Forward Mode
Bauer's Formula
Reverse Mode
Second Order Adjoints
Operations Counts and Overheads
Worst-Case Optimality
Expensive ≡ Redundant?
Preaccumulation and Combinatorics
Summary
See also
References

Keywords

Automatic differentiation; Divided differences; Computer algebra; Symbolic manipulation

The evaluation or approximation of derivatives is a central part of most nonlinear optimization calculations. The gradients of objectives and active constraints enter directly into they Karush–Kuhn–Tucker conditions so that inaccuracies in their evaluation limit the achievable solution accuracy. The latter depends also crucially on the conditioning of the projected Hessian of the Lagrangian. Hence accurate values of this symmetric matrix allow the design of appropriate stopping criteria including the verification of second order conditions. Second derivatives also facilitate a rapid final rate of convergence, provided the step-defining linear systems can be solved by factorization or iteration at a reasonable cost. The same observations apply to more general optimization calculations like the solution of nonlinear complementarity problems.

Whether or not the obvious benefits of evaluating first and higher derivatives accurately justify the costs incurred, does strongly depend on the suitability of the differentiation method employed for the particular problem at hand. We may distinguish five principal options for evaluating or approximating derivatives

- *symbolic differentiation*;
- *handcoded derivatives*;
- *automatic differentiation*;

- *difference quotients*;
- *secant updating*.

'Numerical' Differentiation Methods

The last two options are widely used in practical optimization, primarily because they require no extra effort whatsoever on the part of the user. Difference quotients are often called *divided differences* or *finite differences*, though the last term invites confusion with a related method for discretizing differential equations. Other popular labels are *differencing* or *numerical differentiation*, because the results are floating point numbers rather than algebraic expressions. The latter are often presumed to be the output of more symbolic methods. Even though we shall see that the distinction is not quite that easy, there is no doubting the importance of the fundamental relation

$$F'(x)\dot{x} = \frac{d}{d\alpha} F(x + \alpha \dot{x}) \Big|_{\alpha=0}$$
$$= \frac{1}{\varepsilon} \big[F(x + \varepsilon \dot{x}) - F(x) \big] + O(\varepsilon).$$

Here, the vector function $F: \mathbf{R}^n \to \mathbf{R}^m$ is assumed Lipschitz-continuously differentiable on some neighborhood of the base point $x \in \mathbf{R}^n$. In other words, the directional derivative of F along some vector $\dot{x} \in \mathbf{R}^n$ is the product of the Jacobian matrix $F'(x) \in \mathbf{R}^{m \times n}$ with the direction $\dot{x}$ and it can be approximated by a difference quotient. The quality of this approximation depends strongly on the choice of ε and one must expect a halving in the number of significant digits under the best of circumstances. Quasi-Newton, or secant methods may be viewed as an ingenious way of sequentially incorporating difference quotients into a Jacobian approximation while iterating towards the solution vector of a nonlinear system of equations. The corresponding theory of superlinear convergence is quite beautiful from a mathematical point of view, though perhaps not terribly relevant in practice for large, structured problems.

It is important to note that the quality of the approximate derivative matrices generated by quasi-Newton methods influences only the rate of convergence but not so much the solution accuracy itself. The latter depends on the accurate evaluation of residual vectors, which may be composed of gradients as is the case for the KKT conditions. The importance of accurate residual values is particularly well understood in numerical linear algebra, and replacing them with approximations of uncertain reliability is generally a dicy proposition. Fortunately, it just so happens that gradients can usually be evaluated with working precision at a moderate cost relative to that of the underlying functions. This is far from true for Jacobians and Hessians, whose cost is very hard to predict (and even define) as we shall demonstrate further below on various examples.

The relative cost of evaluating one-sided difference quotients in p directions $\dot{x}$ from the same base point x is clearly $p + 1$. Theoretically one might sometimes reduce the evaluation costs by exploiting the fact that the p points $x + \varepsilon \dot{x}$ are close to x. This proximity may arise in the topological sense that the stepsize $\varepsilon \|\dot{x}\|$ is small as well as in the structural sense that $\dot{x}$ is sparse and thus leaves many components of x unchanged. In practice such savings are rarely realized and they would certainly destroy the main advantage of differencing, namely its black box quality, which does not require any insight or access to the process by which function values are generated. Of course, there is the optimistic assumption that they vary smoothly as a function of the argument x, and usually the selection of a suitable increment ε causes enough trouble for the user and possibly even quite a few extra trial evaluations.

Hence it is indeed fair to assume that one-sided or centered differences in p directions $\dot{x}$ at a common x require $1 + p$ or $1 + 2p$ separate function evaluations but little extra storage. By letting $\dot{x}$ range over all n Cartesian basis vectors one obtains an approximate Jacobian with first or second order accuracy at the cost of $1 + n$ or $1 + 2n$ function evaluations. The number of dependent variables does not matter for differencing so that the cost of a gradient, where $m = 1$, is also $1 + n$ or $1 + 2n$ times that of the underlying scalar function. To compute the Hessian or more generally a full second derivative tensor one needs $n(n + 1)/2$ function evaluations for one-sided and twice that many for the more accurate centered differences.

Since multiple function evaluations are an 'embarrassingly' parallel task the availability of several processors can be used to achieve a nearly perfect speed up for derivative approximations by differencing [7]. In the sparse case, the number of independent variables n can be replaced in the cost ratios above by a number $p \leq n$ that represents either the maximal number of nonzeros in any row of $F'(x)$ or the usually slightly larger chromatic number of the *column incidence graph*. The latter reduction can be achieved by the by now classical grouping or coloring technique originally due to Curtis–Powell–Reid [6] and further developed by Coleman–Moré [4]. An alternative way to compress the rows of the Jacobian even further at the expense of some linear equation solving is due to Newsam–Ramsdell [12] and has recently been adopted to automatic differentiation.

'Analytical' Differentiation Methods

The first three options listed at the beginning are based on the chain rule and may therefore be combined under the label *analytical differentiation*. They all would yield exact derivative values if real arithmetic could be performed in infinite precision. Moreover, even the actual sequence of operations performed to evaluate a particular partial derivative would quite likely be the same and thus yield identical results if the same floating point arithmetic was used. Only the way in which the instruction for this floating point calculation are generated and stored differ significantly between the three approaches. Also, there may be more or less recalculation of intermediates that are common to several partial derivatives, which can have drastic effects on the computational efficiency.

The result of the second option *handcoding* may in principle be always similarly obtained by symbolic or automatic differentiation, provided the computer algebra package or the differentiation software is sufficiently smart. Hence we will discuss only the pure options one and three, which might of course also be combined by a highly sophisticated programmer or software tool.

Symbolic differentiation is usually performed in *computer algebra* packages like Maple, Mathematica and Reduce. Most users have the notion that the differentiation commands in these sophisticated systems turn formulas for functions into formulas for derivatives. Moreover there is a tendency to assume that having a 'formula' means directly expressing dependent variables as algebraic expressions of independents without allowing any named intermediates. The natural data structures for such formulas would be expression trees. There, every node has only one parent, so that the whole thing can be easily linearized and printed by enumeration in a depth first order. In reality computer algebra packages do not restrict themselves to expression trees, because for any nontrivial function the corresponding tree structure is very likely to represent an incredible amount of *redundancy*, even before any differentiation takes place.

Two-Stranded Chain Scenario

Consider for example a sequence of complex function evaluations

$$x_{k+1} + iy_{k+1} = \phi_k(x_k + iy_k) + i\psi_k(x_k + iy_k)$$

for $k = 0, \ldots, l - 1$ starting from some initial $x_0 + iy_0 \in \mathbf{C}$. Suppose all function pairs $\phi_k + i\psi_k$ are nonlinear and do not allow any algebraic simplifications. Then eliminating the intermediates x_1 and y_1 yields the formula

$$\begin{aligned} x_2 + iy_2 = {} & \phi_1(\phi_0(x_0, y_0) + i\varphi_0(x_0, y_0)) \\ & + \psi_1(\phi_0(x_0, y_0) + i\varphi_0(x_0, y_0)), \end{aligned}$$

which involves already twice as many terms as the one-level original formula. The same doubling occurs at each subsequent level so that expressing x_l and y_l directly in terms of the initial components x_0 and y_0 yields an exponentially long formula with the symbols x_0 and y_0 each occurring exactly 2^l times. In this case one could avoid the highly undesirable expression swell by merely substituting $z_k \equiv x_k + iy_k$, which turns the binary expression tree into a simple chain of the same height l.

While this example may appear rather algebraic and somewhat contrived, exactly the same effect occurs if the real pairs (x_k, y_k) specify straight lines in the plane. Specifically, one might think of light-beams being reflected in a maze of mirrors or some other optical arrangement in the plane. Each ray (x_k, y_k) that is incoming to a mirror or lens uniquely determines an outgoing ray $(x_k + 1, y_k + 1)$ via some simple algebraic relationship. Then expressing the final ray parameters (x_l, y_l) directly as functions of the initial parameters (x_0, y_0) will again yield an expression of size 2^l.

Rather than dealing with this algebraic monster one should of course keep all the intermediate pairs (x_k, y_k) with $0 \leq k \leq l$ as named variables. Along this chain one can easily propagate all information of interest, including the 2×2 Jacobian of (x_k, y_k) with respect to (x_0, y_0), at a temporal and spatial complexity of order l. To achieve this result one may employ suitable variants of computer algebra, automatic differentiation or, of course, hand-coding. Before discussing them in more detail let us discuss a general model of function and derivative evaluations.

Computational Model

All analytical differentiation methods are based on the observation that most vector functions F of practical interest are being evaluated by a sequence of assignments

$$v_i = \varphi_i(v_j)_{j<i} \quad \text{for } i = 1,\ldots,l+m. \tag{1}$$

Here, the variables v_i are real scalars and the elemental functions φ_i are either binary arithmetic operations or univariate intrinsics. Consequently, only one or two of the partial derivatives

$$c_{ij} \equiv \frac{\partial}{\partial v_j}\varphi_i(v_k)_{k<i}$$

do not vanish identically and can be evaluated at a cost comparable to that of the underlying φ_i itself.

Without loss of generality we may require that the first n variables $v_{j-n} = x_j$ with $j = 1,\ldots, n$ represent the *independent variables* and the last m variables $y_i = v_l + i$ with $i = 1,\ldots, m$ represent the *dependent variables*. Then the function $y = F(x)$ is defined by the program (1). Here, the nonnegative integer l represents the number of intermediate variables, which we expect to be much larger than both n and m for seriously nonlinear problems. We will also assume that within a small constant all elemental functions have the same complexity so that we have the approximate operations count

$$\text{OPS}(x \stackrel{\text{prog}}{\longmapsto} y) \sim l \equiv \#\text{intermediates}.$$

Throughout this article, $\sim$ means proportional with small constants that are independent of the particular problem at hand. Each intermediate variable may be viewed and thus later differentiated as a function $v_i \equiv v_i(x)$ of the independent variable vector x. As long as all intermediates v_i are stored in separate locations the memory requirement for evaluating F will also be proportional to l. This is a very unrealistic assumption as most evaluation programs involve shared allocation of intermediates. Due to space constraints we will not be able to discuss any aspects of spatial complexity in this article. For a detailed treatment of various trade-offs between space and time see [9].

The way in which the elemental partials c_{ij} are handled differs amongst various analytical differentiation methods. They are always evaluated as floating point numbers at the current argument in what is variously known as *automatic* or *algorithmic* or *computational differentiation*. The same can be assumed for hand written derivative codes unless they are programmed within a computer algebra system, where the c_{ij} can be defined and manipulated as algebraic expressions. In some cases applying the chain rule to these expressions may theoretically lead to significant simplifications and thus potentially provide the user with analytical insight. In the following section we reverse engineer one such class of examples and arrive at the tentative conclusion that the practical potential for symbolic simplifications during the differentiation process appears to be very slim indeed.

Indefinite Integral Scenario

Suppose that

$$F(x) = \int_a^x \frac{P(\tilde{x})}{Q(\tilde{x})}\, d\tilde{x}$$

for two polynomials $P(x)$ and $Q(x)$ with $\deg(Q) > \deg(P)$. Besides a rational term the symbolic expression for $F(x)$ is then likely to contain a welter of logarithms and arcus tangents, whose complexity may easily exceed that of the integrand $f(x) \equiv P(x)/Q(x)$ by orders of magnitude. Then fully symbolic differentiation will of course lead back to an algebraic expression for $f(x)$, while automatic differentiation will combine the c_{ij} in floating point arithmetic according to some variant of the chain rule and obtain 'just' a numerical value of $f(x)$ at the given point $x \in \mathbf{R}$. Moreover, due to cancellations that value may well be less accurate than that obtained by plugging the particular argument x into the formula for $f(x)$.

However, similar numerical instabilities are likely to already affect the evaluation of $F(x)$ itself. They may also show up in the form of an imaginary component when the coefficients of $P(x)$ and $Q(x)$ are real but given in floating point format. Then the roots of the denominator polynomial are already perturbed by unavoidable round-off and symbolic differentiation of the resulting expression for $F(x)$ will usually not lead back to $f(x)$ but some other rational function with a higher polynomial degree in the numerator or denominator. To avoid this effect all coefficients of $f(x)$ must be specified as algebraic numbers so that the symbolic integration can be performed exactly. This process which typically involves rational numbers with enormous coefficients and thus requires a large computational effort.

Hence on practical models one may well be better advised to evaluate $F(x)$ by a numerical quadrature yielding highly accurate results at a fraction of the computing time. Analytically differentiating a nonadaptive quadrature procedure yields the same quadrature applied to the derivatives of the integrand, namely $f'(x) = F''(x)$. Hence the resulting values are quite likely to be good approximations to the original integrand $f(x)$ and they are the exact derivatives of the approximate values computed for $F(x)$ by the quadrature.

Lack of Smoothness

Adaptive quadratures on the other hand may vary grid points and coefficient values in a nondifferentiable or even discontinuous fashion. Then derivatives of the quadrature value may well not exist in the classical sense at some critical arguments x. This difficulty is likely to arise in the form of program branches in all substantial scientific codes and there is no agreement yet on how to deal with it. In most situations one can still compute one-sided directional derivatives as well as generalized gradients and Jacobians [9]. Naturally, computing difference quotients of nonsmooth functions is also a risky proposition. Generally, optimal results in terms of accuracy and efficiency can only be expected from a derivatives code developed by a knowledgeable user, possibly with the help of program analysis and transformation tools.

Predictability of Complexities

With regards to spatial and temporal complexity the following basic distinction applies between the analytical differentiation methods sketched above. The cost of fully symbolic differentiation seems impossible to predict. It can sometimes be very low due to fortuitous cancellations but it is more likely to grow drastically with the complexity of the underlying function. In contrast the relative cost incurred by the various modes of automatic differentiation can always be a priori bounded in terms of the number of independent and dependent variables. Moreover, as we will see below these bounds can sometimes be substantially undercut for certain structured problems.

Another advantage of automatic differentiation compared to a fully symbolic approach is that restrictions and projections of Jacobians and Hessians to certain subspaces of the functions domain and range can be built into the differentiation process with corresponding savings in computational complexity. In the remainder of this article we will therefore focus on the complexity of various automatic differentiation techniques; always making sure that no other known approach is superior in terms of accuracy and complexity on general vector functions defined by a sequence of elemental assignments.

Goal-Oriented Differentiation

The two-stranded chain scenario above illustrates the crucial importance of suitable representations of the mathematical objects, whose complexity we try to quantify here. So one really has to be more specific about what one means by *computing* a function, gradient, Jacobian, Hessian, or their restriction and projection to certain subspaces. At the very least we have to distinguish the (repeated) *evaluation* in floating point arithmetic at various arguments from the *preparation* of a suitable procedure for doing so. This preparation stage comes actually first and might be considered the symbolic part of the differentiation process. It usually involves no floating point operations, except possibly the propagation and simplification of some constants. This happens for example when a source code for evaluating F is precompiled into a source code for jointly evaluating $F(x)$ and its Jacobian $F'(x)$ at a given argument x. In the remainder we will neglect the preparation effort presuming that it can be amortized over many numerical evaluations as is typically the case in iterative or time-dependent computations.

In general, it is not a priori understood that $F'(x)$ should be returned as a rectangular array of floating point numbers, especially if it is sparse or otherwise structured. Its cheapest representation is the sparse triangular matrix

$$C = C(x) \equiv (c_{ij})_{i=1-n,\ldots,l+m}^{j=1-n,\ldots,l+m}.$$

The nonzero entries in C can be obtained during the evaluation of F at a given x for little extra cost in terms of arithmetic operations so that

$$\text{OPS}\{x \mapsto C\} \sim \text{OPS}\{x \overset{\text{prog}}{\mapsto} F\}.$$

As we will see below, the nonzeros in C allow directly the calculation of the products

$$F'(x)\dot{x} \in \mathbb{R}^m \quad \text{for } \dot{x} \in \mathbb{R}^n,$$

and

$$F'(x)^\top \bar{y}^\top \in \mathbb{R}^n \quad \text{for } \bar{y}^\top \in \mathbb{R}^m,$$

using just one multiplication and addition per $c_{ij} \neq 0$. So if our goal is the iterative calculation of an approximate Newton-step using just a few matrix-vector products, we are well advised to just work with the collection of nonzero entries of C provided it can be kept in memory. If on the other hand we expect to take a large number of iterations or wish to compute a matrix factorization of the Jacobian we have to first accumulate all mn partial derivatives $\partial y_i / \partial x_j$ from the elemental partials c_{ij}. It is well understood that a subsequent inplace triangular factorization of the Jacobian $F'(x)$ yields an ideal representation if one needs to multiply itself as well as its inverse by several vectors and matrices from the left or right. Hence we have at least three possible ways in which a Jacobian can be represented and kept in storage:

- unaccumulated: computational graph;
- accumulated: rectangular array;
- factorized: two triangular arrays.

Here the arrays may be replaced by sparse matrix structures. For the time being we note that Jacobians and Hessians can be provided in various representation at various costs for various purposes. Which one is most appropriate depends strongly on the structure of the problem function $F(x)$ at hand and the final numerical

purpose of evaluating derivatives in the first place. The interpretation of C as computational graph goes back to L.V. Kantorovich and requires a little more explanation.

The Computational Graph

With respect to the precedence relation

$$j \prec i \quad \Longleftrightarrow \quad c_{ij} \neq 0 \quad \Longleftrightarrow \quad (j, i) \in E,$$

the indices $i, j \in V \equiv [1-n, \ldots, 1+m]$ form a directed graph with the edge set E. Since by assumption $j \prec i$ implies $j < i$ the graph is acyclic and the transitive closure of $\prec$ defines a partial ordering between the corresponding variables v_i and v_j. The minimal and maximal elements with respect to that order are exactly the independent and dependent variables $v_{j-n} \equiv x_j$ with $j = 1, \ldots, n$ and the $v_{m+i} \equiv y_i$ with $i = 1, \ldots, m$, respectively. For the two stranded chain scenario with $l = 3$ one obtains a computational graph of the following form:

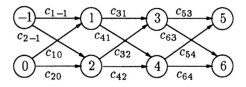

Assuming that all elemental φ_i are unary functions or binary operations we find $|E| \leq 2(l+m) \sim l$. One may always annotate the graph vertices with the elemental functions φ_i and the edges with the nonvanishing elemental partials c_{ij}. For most purposes the φ_i do not really matter and we may represent the graph (V, E) simply by the sparse matrix C.

Forward Mode

Given some vector $\dot{x} \equiv (\dot{v}_{j-n})_j = 1, \ldots, n \in \mathbb{R}^n$, there exist derivatives

$$\dot{v}_i \equiv \frac{d}{d\alpha} v_i(x + \alpha \dot{x})\Big|_{\alpha=0} \quad \text{for } 1 \leq i \leq l+m.$$

By the chain rule these $\dot{v}_i$ satisfy the recurrence

$$\dot{v}_i \equiv \sum_{j \prec i} c_{ij} \dot{v}_j \quad \text{for } i = 1, \ldots, l+m. \tag{2}$$

The resulting tangent vector $\dot{y} \equiv (\dot{v}_{l+i})_{i=1,\ldots,m}$ satisfies $\dot{y} = F'(x)\dot{x}$ and it is obtained at a cost proportional to l. Instead of propagating derivatives with respect to just one direction vector $\dot{x}$ one may amortize certain overheads by bundling p of them into a matrix $\dot{X} \in \mathbf{R}^{n \times p}$ and then computing simultaneously $\dot{Y} = F'(x)\dot{X} \in \mathbf{R}^{m \times p}$. The cost of this *vector forward* mode of automatic differentiation is given by

$$\text{OPS}\{C \xmapsto{\text{forw}} \dot{Y}\} \sim pl \sim p\,\text{OPS}\{x \xmapsto{\text{prog}} y\}. \quad (3)$$

If the columns of $\dot{X}$ are Cartesian basis vectors $e_j \in \mathbf{R}^n$ the corresponding columns of the resulting $\dot{Y}$ are the jth columns of the Jacobian. Hence by setting $\dot{X} = I$ with $p = n$ we may compute the whole Jacobian at a temporal complexity proportional to nl. Fortunately, in many applications the whole Jacobian is either not needed at all or due to its sparsity pattern it may be reconstructed from its *compression* $\dot{Y} = F'(x)\dot{X}$ for a suitable *seed matrix* $\dot{X}$. As in the case of difference quotients this matrix may be chosen according to the Curtis–Powell–Reid [6] or the Newsam–Ramsdell [12] approach with p usually close to the maximal number of nonzeros in any row of the Jacobian.

Bauer's Formula

Using the recurrence for the $\dot{v}_i$ given above one may also obtain an explicit expression for each individual partial derivative $\partial y_i / \partial x_j$. Namely, it is given by the sum over the products of all arc values c_{ij} along all paths connecting the minimal node v_{j-n} with the maximal node v_{l+i}. This formula due to F.L. Bauer [1] implies in particular that the ijth Jacobian entry vanishes identically exactly when there is no path connecting nodes $j - n$ and $l + i$ in the computational graph. In general the number of distinct paths in the graph is very large and it represents exactly the lengths of the formulas obtained if one expresses each y_i directly in terms of all x_j that it depends on. Hence we may conclude

$$\text{OPS}\{C \xmapsto{\text{bauer}} F'\} \sim \text{OPS}\{x \xmapsto{\text{formul}} y\}.$$

In the two-stranded chain scenario considered above, both operations counts would be of order 2^l, which is obviously an unacceptable effort. Fortunately, vector forward and Bauer's formula are just two special choices amongst many ways for accumulating the Jacobian $F'(x)$ from the computational graph C. The most celebrated alternative is the reverse or backward mode of automatic differentiation.

Reverse Mode

Rather than propagating directional derivatives $\dot{v}_i$ forward through the computational graph one may also propagate adjoint quantities $\bar{v}_i$ backward. To define them properly one must perturb the original evaluation loop by rounding errors δ_i so that now

$$v_i = \delta_i + \varphi_i(v_j)_{j \prec i} \quad \text{for } i = 1 - n, \ldots, l.$$

Then the resulting vector y is a function not only of x but also of the vector of small perturbations $(\delta_i)_{i=1-n,\ldots,l}$. Given any row vector of weights $\bar{y} = (\bar{v}_{l+i})_{i=1,\ldots,m}$ we obtain the sensitivities

$$\bar{v}_i \equiv \left.\frac{\partial}{\partial \delta_i}\bar{y}y\right|_{\delta_i=0} \quad \text{for } 1 - n \leq i \leq l,$$

where all other perturbations δ_j with $j \neq i$ are set to zero during the differentiation. The adjoint components $\bar{v}_{j-n} = \bar{x}_j$ form the row vector $\bar{x} = \bar{y}F'(x) \in \mathbb{R}^n$, which is simply the gradient of the linear combination $\bar{y}F(x)$. In the optimization context this scalar valued function is usually a Lagrangian, whose gradient and Hessian figure prominently in the first and second order optimality conditions. The amazing thing is that as a consequence of the chain rule such gradients can be computed at the same cost as tangents by using the backward recurrence

$$\bar{v}_j = \sum_{i \succ j} \bar{v}_i c_{ij} \quad \text{for } j = l, \ldots, 1 - n. \quad (4)$$

Just like in the forward scalar recurrence (2), each elemental partial $c_{ij} \neq 0$ occurs exactly once and we may amortize costs by bundling several $\bar{y}$ into an adjoint seed matrix $\bar{Y} \in \mathbb{R}^{q \times m}$. This vector reverse mode yields the matrix $\bar{X} = \bar{Y}F'(x) \in \mathbb{R}^{q \times n}$ at the cost

$$\text{OPS}\{C \xmapsto{\text{rev}} \bar{X}\} \sim ql \sim q\,\text{OPS}\{x \xmapsto{\text{prog}} y\}.$$

Again the whole Jacobian is obtained directly if we seed $\bar{Y} = I$ with $q = m$. Hence we find by comparison with (3) as a rule of thumb that the reverse mode is preferable if $m \ll n$, i.e., if there are not nearly as many dependents as independents. In classical NLPs we may think of m as the number of active constraints plus one, which is often much smaller then n, the number of variables. In unconstrained optimization we have $m = 1$ so that the gradient of the objective F can be computed with essentially the same effort as F itself. In the sparse case we may now employ column rather than row compression with q roughly equal to the maximal number of nonzeros in any column of $F'(x)$.

For suitable seeds $\overline{Y}$ the column compression $\overline{X} = \overline{Y} F'(x)$ allows the reconstruction of the complete Jacobian $F'(x)$. Furthermore, row and column compression can be combined yielding for example Jacobians with arrow head structure at the cost of roughly $p + q = 3$ function evaluations. In that case one may use

$$\dot{X} \equiv \begin{pmatrix} 1 & 1 & \cdots & 1 & 0 \\ 0 & 0 & \cdots & 0 & 1 \end{pmatrix}^\top \quad \text{and} \quad \overline{Y} = e_m^\top .$$

Then $\overline{X} = \overline{Y} F'(x)$ is the last row of the arrowhead matrix $F'(x)$ and the two columns of $\dot{Y} = F'(x)\dot{X}$ contain all other nonzero entries. For pure row or column compression dense rows or columns always force $p = n$ or $q = m$, respectively. Hence the combination of forward and reverse differentiation offers the potential for great savings. In either case projections and restrictions of the Jacobian to subspaces of the vector functions domain and range can be built into the differentiation process, which is part of the goal-orientation we alluded to before.

Second Order Adjoints

Rather than separately propagating some first derivatives forward, others reverse, and then combining the results to compute Jacobian matrices efficiently, one may compose these two fundamental modes to compute second derivatives like Hessians of Lagrangians. More specifically, we obtain by directional differentiation of the adjoint relation $\overline{x} = \overline{y} F'(x)$ the second order adjoint

$$\dot{\overline{x}} = \overline{y} F''(x) \dot{x} \in \mathbb{R}^n .$$

Here we have assumed that the adjoint vector $\overline{y}$ is constant. We also have taken liberties with matrix vector notation by suggesting that the $m \times n \times n$ derivative tensor $F''(x)$ can be multiplied by the row vector $\overline{y} \in \mathbb{R}^m$ from the left and the column vector $\dot{x} \in \mathbb{R}^n x \in \mathbb{R}^n$ from

the right yielding a row vector $\dot{\overline{x}}$ of dimension n. In an optimization context $\overline{y}$ should be thought of as a vector of Lagrange multipliers and $\dot{x}$ as a feasible direction. By composing the complexity bounds for the reverse and the forward mode one obtains the estimates

$$\mathrm{OPS}\{x \stackrel{\mathrm{prog}}{\longmapsto} y\} \sim \mathrm{OPS}\{x, \dot{x} \stackrel{\mathrm{forw}}{\longmapsto} \dot{y}\}$$
$$\sim \mathrm{OPS}\{x, \overline{y} \stackrel{\mathrm{rev}}{\longmapsto} \overline{x}\} \sim \mathrm{OPS}\{x, \dot{x}, \overline{y} \stackrel{\mathrm{ad}}{\longmapsto} \dot{\overline{x}}\} .$$

Here, ad represents reverse differentiation followed by forward differentiation or vise versa. The former interpretation is a little easier to implement and involves only one forward and one backward sweep through the computational graph.

Operations Counts and Overheads

From a practical point of view one would of course like to know the proportionality factors in the relations above. If one counts just multiplication operations then $\dot{y}$ and $\overline{x}$ are at worst 3 times as expensive as y, and $\dot{\overline{x}}$ is at most 9 times as expensive. A nice intuitive example is the calculation of the determinant y of a $\sqrt{n} \times \sqrt{n}$ matrix whose entries form the variable vector x. Then we have $m = 1$ and

$$\mathrm{OPS}\{x \mapsto y\} = \frac{1}{3}\sqrt{n}^3 + O(n)$$

multiplications if one uses an LU factorization. Then it can be seen that $\overline{y} = 1/y$ makes $\overline{x}$ the transpose of the inverse matrix and the resulting cost estimate of $\sqrt{n}^3 + O(n)$ multiplications conforms exactly with that for the usual substitution procedure.

However, these operations count ratios are no reliable indications of actual runtimes, which depend very strongly on the computing platform, the particular problem an hand, and the characteristics of the AD tool. Implementations of the vector forward mode like ADIFOR [3] that generate compilable source codes can easily compete with divided differences, i. e. compute p directional derivatives in the form $\dot{Y} = F'(x)\dot{X}$ at the cost of about p function evaluations. For sizeable $p \approx 10$ they are usually faster than divided differences, unless the roughly p-fold increase in storage results in too much paging onto disk. The reverse mode is an entirely different ball-game since most intermediate values v_i and some control flow hints need to be first saved and later retrieved, which can easily make the calculation of adjoints memory bound. This memory access overhead can be partially amortized in the vector reverse mode, which yields a bundle $\overline{X} = \overline{Y}F'(x)$ of q gradient vectors. For example in multicriteria optimization one may well have $q \approx 10$ objectives or soft constraints, whose gradients are needed simultaneously.

Worst-Case Optimality

Counting only multiplications we obtain for Jacobians $F' \in \mathbf{R}^{m \times n}$ the complexity bound

$$\mathrm{OPS}\{x \overset{\mathrm{ad}}{\mapsto} F'\} \leq 3 \min(n, m) \, \mathrm{OPS}\{x \overset{\mathrm{prog}}{\mapsto} y\} \,.$$

Here, n and m can be reduced to the maximal number of nonzero entries in the rows and columns of the Jacobian, respectively.

Similarly, we have for the one-sided projection of the Lagrangian Hessian

$$H(x, \overline{y}) \equiv \overline{y}F'' \equiv \sum_{i=1}^{m} \overline{y}_i \nabla^2 F_i \in \mathbf{R}^{n \times n}$$

onto the space spanned by the columns of $\dot{X}$:

$$\mathrm{OPS}\{x \overset{\mathrm{ad}}{\mapsto} H(x, \overline{y})\dot{X}\} \leq 9p \, \mathrm{OPS}\{x \overset{\mathrm{prog}}{\mapsto} y\} \,.$$

As we already discussed for indefinite integrals there are certainly functions whose derivatives can be evaluated much cheaper than they themselves for example using a computer algebra package. Note that here again we have neglected the preparation effort, which may be very substantial for symbolic differentiation. Nevertheless, the estimates given above for AD are optimal in the sense that there are vector functions F defined by evaluation procedures of the form (1), for which no differentiation process imaginable can produce the Jacobian and projected Hessian significantly cheaper than the given cost bound divided by a small constant. Here, producing these matrices is understood to mean calculating all its elements explicitly, which may or may not be actually required by the overall computation.

Consider, for example, the cubic vector function

$$F(x) = x + \frac{b(a^\top x)^3}{2} \quad \text{with } a, b \in \mathbf{R}^n \,.$$

Its Jacobian and projected Hessian are given by

$$F'(x) = I + b \left(a^\top x \right)^2 a^\top \in \mathbf{R}^{n \times n}$$

and

$$H(x, \overline{y})\dot{X} = 2a(\overline{y}b)(a^\top x)a^\top \dot{X} \in \mathbf{R}^{n \times p} \,.$$

For general a, b and $\dot{X}$, all entries of the matrices $F'(x)$ and $H(x, \overline{y})\dot{X}$ are distinct and depend nontrivially on x. Hence their explicit calculation by any method requires at least n^2 or np arithmetic operations, respectively. Since the evaluation of F itself can be performed using just $3n$ multiplications and a few additions, the operations count ratios given above cannot be improved by more than a constant. There are other, more meaningful examples [9] with the same property, namely that their Jacobians and projected Hessians are orders of magnitude more expensive than the vector function itself. At least this is true if we insist on representing them as rectangular arrays of reals. This does not contradict our earlier observation that gradients are cheap, because the components of $F(x)$ cannot be considered as independent scalar functions. Rather, their simultaneous evaluation may involve many common subexpressions, as is the case for our rank-one example. These appear to be less beneficial for the corresponding derivative evaluation, thus widening the gap between function and derivative complexities.

Expensive ≡ Redundant?

The rank-one problem and similar examples for which explicit Jacobians or Hessians appear to be expensive have a property that one might call *redundancy*. Namely, as x varies over some open neighborhood in its domain, the Jacobian $F'(x)$ stays in a lower-dimensional manifold of the linear space of all matrices with its format and sparsity pattern. In other words, the nonzero entries of the Jacobian are not truly independent of each other so that computing them all and storing them separately may be wasteful. In the rank-one example the Jacobian $F'(x)$ is dense but belongs at all x to the one-dimensional affine variety $\{I + b\alpha a^\top : \alpha \in \mathbf{R}\}$. Note that the vectors $a, b \in \mathbf{R}^n$ are assumed to be dense and constant parameter vectors of the problem at hand. Their elements all play the role of elemental partials c_{ij} with the corresponding operation φ_i being multiplications. Hence accumulating the extremely sparse triangular matrix C, which involves only $O(n)$ nonzero entries, to the dense $n \times n$ array $F'(x)$ is almost certainly a bad idea, no matter what the ultimate purpose of the calculation. In particular, if one wishes to solve linear systems in the Jacobian, the inverse formula of Sherman–Morrison–Woodbury provides a way of computing the solution of rank-one perturbations to diagonal matrices with $O(n)$ effort. This formula may be seen as a very special case of embedding linear systems in F' into a much larger and sparse linear system involving C as demonstrated in [11] and [5].

As of now, all our examples for which the array representation of Jacobians and Hessians are orders of magnitude more expensive to evaluate than the underlying vector function exhibit this redundancy property. In other words, we know of no convincing example where vectors that one may actually wish to calculate as end products are necessarily orders of magnitude more expensive than the functions themselves. Especially for large problems it seems hard to imagine that array representations of the Jacobians and Hessians themselves are really something anybody would wish to look at rather than just use as auxiliary quantities within the overall calculation.

So evaluating complete derivative arrays is a bit like fitting a handle to a wooden crate that needs to be moved about frequently. If the crate is of small weight and size this job is easily performed using a few screws. If, on the other hand, the crate is large and heavy, fitting a handle is likely to require additional bracing and other reinforcements. Moreover, this effort is completely pointless since nobody can just pick up the crate by the handle anyhow and one might as well use a fork left in the first place.

Preaccumulation and Combinatorics

The temporal complexity for both the forward and the reverse (vector) mode are proportional to the number of edges in the linearized computational graph. Hence one may try to reduce the number of edges by certain algebraic manipulations that leave the corresponding Jacobian, i.e., the linear mapping between $\dot{x}$ and $\dot{y} = F'(x)\dot{x}$ and equivalently also that between $\bar{y}$ and $\bar{x} = \bar{y}F'(x)$ unchanged. It can be easily checked that this is the case if given an index j one updates first

$$c_{ik} + = c_{ij}c_{jk}$$

either for fixed $i \succ j$ and all $k \prec j$, or for fixed $k \prec j$ and all $i \succ j$, and then sets $c_{ij} = 0$ or $c_{jk} = 0$, respectively. In other words, either the edge (j, i) or the edge (k, j) is eliminated from the graph. This leads to fill-in by the creation of new arcs, unless all updated c_{ik} were already nonzero beforehand. Eliminating all edges (k, j) with $k \prec j$ or all edges (j, i) with $i \succ j$ is equivalent and amounts to eliminating the vertex j completely from the graph. After all intermediate vertices $1 \leq j \leq l$ are eliminated in some arbitrary order, the remaining edges c_{ij} directly connect independent variables with dependent variables and are therefore entries of the Jacobian $F'(x)$. Hence, one refers to the *accumulation* of the Jacobian F' if all intermediate nodes are eliminated and to *preaccumulation* if some of them remain so that the Jacobian is represented by a simplified graph.

As we have indicated in the section on goal oriented differentiation one would have to carefully look at the problem function and the overall computational task to decide how much preaccumulation should be performed. Moreover, there are $\tilde{l}!$ different orders in which a particular set of $\tilde{l} \leq l$ intermediate nodes can be eliminated and even many more different ways of eliminating the corresponding set of edges. So far there have only been few studies of heuristic criteria for finding efficient elimination orderings down to an appropriate preaccumulation level [9].

Summary

First and second derivative vectors of the form $\dot{y} = F'(x)\dot{x}, \bar{x} = \bar{y}F'(x)$ and $\dot{\bar{x}} = \bar{y}F''(x)\dot{x}$ can be evaluated for a fixed small multiple of the temporal complexity of the underlying relation $y = F(x)$. The calculation of the gradient $\bar{x}$ and the second order adjoint $\dot{\bar{x}}$ by the basic reverse method may require storage of order $l \equiv$ #intermediates. This possibly unacceptable amount can be reduced to order $\log(l)$ at a slight increase in the operations count (see [8]).

Jacobians and one-sided projected Hessians can be composed column by column or row by row from vectors of the kind $\dot{y}, \bar{x}$ and $\dot{\bar{x}}$. For sparse derivative matrices row and/or column compression using suitable seed matrices of type CPR or NR allow a substantial reduction of the computational effort. In some cases the nonzero entries of derivative matrices may be redundant, so that their calculation should be avoided, if the overall computational goal can be reached in some other way. The attempt to evaluate derivative array with absolutely minimal effort leads to hard combinatorial problems.

See also

- ▶ Complexity Classes in Optimization
- ▶ Complexity of Degeneracy
- ▶ Complexity Theory
- ▶ Complexity Theory: Quadratic Programming
- ▶ Computational Complexity Theory
- ▶ Fractional Combinatorial Optimization
- ▶ Information-Based Complexity and Information-Based Optimization
- ▶ Kolmogorov Complexity
- ▶ Mixed Integer Nonlinear Programming
- ▶ *NP*-Complete Problems and Proof Methodology
- ▶ Parallel Computing: Complexity Classes

References

1. Bauer FL (1974) Computational graphs and rounding error. SIAM J Numer Anal 11:87–96
2. Berz M, Bischof Ch, Corliss G, Griewank A (eds) (1996) Computational differentiation: Techniques, applications, and tools. SIAM, Philadelphia
3. Bischof Ch, Carle A, Corliss G, Griewank A, Hovland P (1992) ADIFOR: Generating derivative codes from Fortran programs. Scientif Program 1:1–29
4. Coleman TF, Morée JJ (1984) Estimation of sparse Jacobian matrices and graph coloring problems. SIAM J Numer Anal 20:187–209
5. Coleman TF, Verma A (1996) Structure and efficient Jacobian calculation. In: Berz M, Bischof Ch, Corliss G, Griewank A (eds) Computational Differentiation: Techniques, Applications, and Tools. SIAM, Philadelphia, pp 149–159
6. Curtis AR, Powell MJD, Reid JK (1974) On the estimation of sparse Jacobian matrices. J Inst Math Appl 13:117–119
7. Griewank A (1991) The chain rule revisited in scientific computing, I–II. SIAM News
8. Griewank A (1992) Achieving logarithmic growth of temporal and spatial complexity in reverse automatic differentiation. Optim Methods Softw 1:35–54
9. Griewank A (2000) Evaluating derivatives, principles and techniques of algorithmic differentiation. Frontiers in Appl Math, vol 19. SIAM, Philadelphia
10. Griewank A, Corliss GF (eds) (1991) Automatic differentiation of algorithms: Theory, implementation, and application. SIAM, Philadelphia
11. Griewank A, Reese S (1991) On the calculation of Jacobian matrices by the Markowitz rule. In: Griewank A and Corliss GF (eds) Automatic Differentiation of Algorithms: Theory, Implementation, and Application. SIAM, Philadelphia, pp 126–135
12. Newsam GN, Ramsdell JD (1983) Estimation of sparse Jacobian matrices. SIAM J Alg Discrete Meth 4:404–417

Complexity and Large-Scale Least Squares Problems

JOSEF KALLRATH
GVCS, BASF Aktiengesellschaft,
Ludwigshafen, Germany

MSC2000: 93E24, 34-xx, 34Bxx, 34Lxx

Article Outline

Introduction
  A Standard Formulation for Unconstrained
    Least Squares Problem
  Solution Methods
  Explicit Versus Implicit Models
  Practical Issues of Solving Least Squares Problems
Parameter Estimation in ODE Models
  The Initial Value Problem Approach
  The Boundary Value Problem Approach
Parameter Estimation in DAE Models
Parameter Estimation in PDE Models
  Methodology
Least Squares Problems with Massive Data Sets
  The Matching Approach
Conclusions
Acknowledgments
References

Introduction

Least squares problems and solution techniques to solve them have a long history briefly addressed by Björck [4]. In this article we focus on two classes of complex least squares problems. The first one is established by models involving *differential equations*. The other class is made by least squares problems involving difficult models which need to be solved for many independent observational data sets. We call this *least squares problems with massive data sets*.

A Standard Formulation for Unconstrained Least Squares Problem

The unconstrained least squares problem can be expressed by

$$\min_{\mathbf{p}} l_2(\mathbf{p}), \quad l_2(\mathbf{p}) := \left\| \mathbf{r}_1\left[\mathbf{x}(t_1), \ldots, \mathbf{x}(t_k), \mathbf{p}\right]\right\|_2^2$$

$$= \sum_{k=1}^{N} \left[r_{1k}(\mathbf{p})\right]^2, \quad \mathbf{r}_1 \in \mathbb{R}^N. \quad (1)$$

The minimization of this functional, i. e., the minimization of the sum of weighted quadratic residuals, under the assumption that the statistical errors follow a Gaußian distribution with variances as in (4), provides a maximum likelihood estimator ([7] Chap. 7) for the unknown parameter vector $\mathbf{p}$. This objective function dates back to Gauß [14] and in the mathematical literature the problem is synonymously called least squares or ℓ_2 approximation problem.

The least squares structure (1) may arise either from a nonlinear over-determined system of equations

$$r_{1k}(\mathbf{p}) = 0, \quad k = 1, \ldots, N, \quad N > n, \quad (2)$$

or from a data fitting problem with N given data points $(t_k, \tilde{Y}_k)$ and variances σ_v, a model function $\tilde{F}(t, \mathbf{p})$, and n adjustable parameters $\mathbf{p}$:

$$r_{1k} := r_{1k}(\mathbf{p}) = Y_k - F_k(\mathbf{p}) = \sqrt{w_k}\left[\tilde{Y}_k - \tilde{F}(t_k, \mathbf{p})\right]. \quad (3)$$

The weights w_k are related to the variances σ_k by

$$w_k := \beta/\sigma_k^2. \quad (4)$$

Traditionally, the weights are scaled to a variance of unit weights. The factor β is chosen so as to make the weights come out in a convenient range. In short vector notation we get

$$\mathbf{r}_1 := \mathbf{Y} - \mathbf{F}(\mathbf{p}) = [r_{11}(\mathbf{p}), \ldots, r_{1N}(\mathbf{p})]^T,$$
$$\mathbf{F}(\mathbf{p}), \mathbf{Y} \in \mathbb{R}^N.$$

Our least squares problem requires us to provide the following input:
1. model,
2. data,
3. variances associated with the data,
4. measure of goodness of the fit, e. g., the Euclidean norm.

In many practical applications, unfortunately, less attention is paid to the variances. It is also very important to point out that the use of the Euclidean norm requires pre-information related to the problem and statistical properties of the data.

Solution Methods

Standard methods for solving linear version of (1), i. e., $\mathbf{F}(\mathbf{p}) = \mathbf{A}\mathbf{p}$, are reviewed by Björck [4]. Nonlinear methods for unconstrained least squares problems are covered in detail by Xu [35,36,37]. In addition, we mention a popular method to solve unconstrained least squares problems: the Levenberg–Marquardt algorithm proposed independently by Levenberg [21] and Marquardt [22] and sometimes also called "damped least squares". It modifies the eigenvalues of the normal equation matrix and tries to reduce the influence of eigenvectors related to small eigenvalues (cf. [8]). Damped (step-size cutting) Gauß–Newton algorithms combined with orthogonalization methods control the damping by natural level functions [6,9,10] seem to be superior to Levenberg–Marquardt type schemes and can be more easily extended to nonlinear constrained least squares problems.

Explicit Versus Implicit Models

A common basic feature and limitation of least squares methods, but seldom explicitly noted, is that they require some *explicit* model to be fitted to the data. However, not all models are explicit. For example, some pharmaceutical applications for receptor-ligand binding studies are based on specifically coupled mass equilibrium models. They are used, for instance, for the radioimmunological determination of Fenoterol or related substances, and lead to least squares problems in systems of nonlinear equations [31], in which the model function $\mathbf{F}(\mathbf{p})$ is replaced by $\mathbf{F}(t; \mathbf{p}, \mathbf{z})$ which, besides the parameter vector $\mathbf{p}$ and the time t, depends on a vector function $\mathbf{z} = \mathbf{z}(t; \mathbf{p})$ implicitly defined as the solution of the nonlinear equations

$$\mathbf{F}_2(t; \mathbf{p}, \mathbf{z}) = \mathbf{0}, \quad \mathbf{F}_2(\mathbf{p}) \in \mathbb{R}^{n_2}. \quad (5)$$

This is a special case of an implicit model. There is a much broader class of implicit models. Most models in science are based on physical, chemical and biological laws or include geometry properties, and very often lead to differential equations which may, however, not be solvable in a closed analytical form. Thus, such models do not lead to explicit functions or models we want to fit to data. We rather need to fit an implicit model (represented by a system of differential equations or another implicit model). The demand for and the applications of such techniques are widespread in science, especially in the rapidly increasing fields of nonlinear dynamics in physics and astronomy, nonlinear reaction kinetics in chemistry [5], nonlinear models in material sciences [16] and biology [2], and nonlinear systems describing ecosystems [28,29] in biology, or the environmental sciences. Therefore, it seems desirable to focus on least squares algorithms that use nonlinear equations and differential equations as constraints or side conditions to determine the solution implicitly.

Practical Issues of Solving Least Squares Problems

Solving least squares problems involves various difficulties among them to find an appropriate model, nonsmooth models with discontinuous derivatives, data quality and checking the assumption of the underlying error distribution, and dependence on initial parameter or related questions of global convergence.

Models and Model Validation   A model may be defined as an appropriate abstract representation of a real system. In the natural sciences (e. g., Physics, Astronomy, Chemistry and Biology) models are used to gain a deeper understanding of processes occurring in nature (an epistemological argument). The comparison of measurements and observations with the predictions of a model is used to determine the appropriateness and quality of the model. Sir Karl Popper [26] in his famous book *Logic of Scientific Discovery* uses the expressions *falsification* and *verification* to describe tasks that the models can be used to accomplish as an aid to scientific process. Models were used in early scientific work to explain the movements of planets. Then, later, aspects and questions of accepting and improving global and fundamental models (e. g., general relativity or quantum physics) formed part of the discussion of the philosophy of science. In science models are usually falsified, and, eventually, replaced by modified or completely different ones.

In industry, models have a rather local meaning. A special aspect of reality is to be mapped in detail. Pragmatic and commercial aspects are usually the motivation. The model maps most of the relevant features and neglect less important aspects. The purpose is to
- provide insight into the problem,
- allow numerical, *virtual* experimentation but avoid expensive and/or dangerous *real* experiments, or
- tune a model for later usage, i. e., determine, for instance, the reaction coefficients of a chemical system – once these parameters are known the dynamics of the process can be computed.

A (mathematical) model represents a *real-world problem* in the language of mathematics, i. e., by using mathematical symbols, variables (in this context: the adjustable least squares parameters), equations, inequalities, and other relations. How does one get a mathematical model for a real-world problem? To achieve that is neither easy nor unique. In some sense it is similar to solving exercises in school where problems are put in a verbal way [25]. The following points are useful to remember when trying to build a model:
- there will be no precise recipe telling the user how to build a model,
- experience and judgment are two important aspects of model building,
- there is nothing like a *correct* model,
- there is no concept of a *unique* model, as different models focusing on different aspects may be appropriate.

Industrial models are eventually validated which means that they reached a sufficient level of consensus among the community working with these models.

Statistics provide some means to discriminate models but this still is an art and does not replace the need for appropriate model validation. The basic notion is: *with a sufficient number of parameters on can fit an elefant*. This leads us to one important consequence: it seems to be necessary that one can interpret these model parameters. A reasonable model derived from the laws of science with interpretable parameters is a good candidate to become accepted. Even, if it may lead to a somewhat worse looking fits than a model with a larger number of formal parameters without interpretation.

Non-Smooth Models The algorithm reviewed by Xu [35,36,37] for solving least squares problems usually require the continuous first derivatives of the model function with respect to the parameters. We might, however, encounter models for which the first derivatives are discontinuous. Derive-free methods such as Nelder and Mead's [23] downhill Simplex method, or direction set methods; cf. ([27], p. 406) have been successfully used to solve least squares problems. The Simplex method provides the benefit of exploring parameter space and good starting values for derivative based methods. Powell's direction set method with appropriate conjugate directions preserve the derivative free nature of the method.

Global Convergence Nonlinear least squares algorithms usually converge only if the initial parameters are close to the best fit parameters. Global convergence can be established for some algorithms, i. e., they converge for all initial parameters. An essential support tool accompanying the analysis of difficult least squares problem is to visualize the data and the fits. Inappropriate or premature fits can easily be excluded. Inappropriate fits are possible because all algorithms mentioned in Sect. "Introduction", "Parameter Estimation in ODE Models", and "Parameter Estimation in DAE Models" are local algorithm. Only if the least squares problem is convex, they yield the global least squares minimum. Sometimes, it is possible to identify false local minima from the residuals.

Data and Data Quality Least squares analysis is concerned by fitting data to a model. The data are not exact but subject to unknown random errors ε_k. In ideal cases these errors follow a Gaussian normal distribution. One can test this assumption after the least squares fit by analyzing the distribution of the residuals as described in Sect. "Residual Distributions, Covariances and Parameter Uncertainties". Another important issue is whether the data are appropriate to estimate all parameters. Experimental design is the discipline which addresses this issue.

Residual Distributions, Covariances and Parameter Uncertainties Once the minimal least squares solution has been found one should at first check with the χ^2-test or Kolmogoroff–Smirnov test whether the usual assumption that the distribution really follows a Gaussian normal distribution. With the Kolmogoroff–Smirnov test (see, e. g., [24]) it is possible to check as follows whether the residuals of a least-squares solution are normally distributed around the mean value 0.

1. let $M := (x_1, x_2, \ldots, x_n)$ be a set of observations for which a given hypothesis should be tested;
2. let $G : x \in M \to \mathbb{R}, x \to G(x)$, be the corresponding cumulative distribution function;
3. for each observation $x \in M$ define $S_n(x) := k/n$, where k is the number of observations less than or equal to x;
4. determine the maximum $D := \max(G(x) - S_n(x) \mid x \in M)$;
5. D_{crit} denotes the maximum deviation allowed for a given significance level and a set of n elements. D_{crit} is tabulated in the literature, e. g., ([24], Appendix 2, p. 560); and
6. if $D < D_{\text{crit}}$, the hypothesis is accepted.

For the least squares problem formulated in Sect. "A Standard Formulation for Unconstrained Least Squares Problem" the hypothesis is "The residuals $\mathbf{x} := \mathbf{r}_1 = \mathbf{Y} - \mathbf{F}(\mathbf{p})$ are normally distributed around the mean value 0". Therefore, the cumulative distribution function $G(x)$ takes the form

$$\sqrt{2\pi} G(x) = \int_{-\infty}^{x} g(z) dz$$
$$= \int_{-\infty}^{-x_0} g(z) dz + \int_{-x_0}^{x} g(z) dz,$$
$$g(z) := e^{-\frac{1}{2}z^2}.$$

The value x_0 separates larger residuals; this is problem specific control parameter.

The derivative based least squares methods usually also give the covariance matrix from which the uncertainties of the parameter are derived; cf. [7], Chap. 7. Least squares parameter estimations without quantifying the uncertainty of the parameters are very doubtful.

Parameter Estimation in ODE Models

Consider a differential equation with independent variable t for the state variable

$$\mathbf{x}'(t) = \frac{d\mathbf{x}}{dt} = \mathbf{f}(t, \mathbf{x}, \mathbf{p}), \quad \mathbf{x} \in \mathbb{R}^{n_d}, \quad \mathbf{p} \in \mathbb{R}^{n_p} \quad (6)$$

with a right hand side depending on an unknown parameter vector $\mathbf{p}$. Additional requirements on the solution of the ODE (1) like periodicity, initial or boundary conditions or range restrictions to the parameters can be formulated in vectors $\mathbf{r}_2$ and $\mathbf{r}_3$ of (component wise) equations and inequalities

$$\begin{aligned} \mathbf{r}_2\left[\mathbf{x}(t_1), \ldots, \mathbf{x}(t_k), \mathbf{p}\right] &= 0 \quad \text{or} \\ \mathbf{r}_3\left[\mathbf{x}(t_1), \ldots, \mathbf{x}(t_k), \mathbf{p}\right] &\geq 0 \,. \end{aligned} \quad (7)$$

The multi-point boundary value problem is linked to experimental data via minimization of a least squares objective function

$$l_2(\mathbf{x}, \mathbf{p}) := \left\| \mathbf{r}_1\left[\mathbf{x}(t_1), \ldots, \mathbf{x}(t_k), \mathbf{p}\right] \right\|_2^2 \,. \quad (8)$$

In a special case of (8) the components ℓ of the vector $\mathbf{r}_1 \in \mathbb{R}^L$ are "equations of condition" and have the form

$$r_{1\ell} = \sigma_{ij}^{-1}[\eta_{ij} - g_i(\mathbf{x}(t_j), \mathbf{p})] \,,$$

$$\ell = 1, \ldots, L := \sum_{i=1}^{N_j} J_i \,. \quad (9)$$

This case leads us to the least squares function

$$l_2(\mathbf{x}, \mathbf{p}) := \sum_{j=1}^{N^D} \sum_{i=1}^{N_j} \sigma_{ij}^{-2}[\eta_{ij} - g_i(\mathbf{x}(t_j), \mathbf{p})]^2 \,. \quad (10)$$

Here, N^D denotes the number of values of the independent variable (here called time) at which observed data are available, N_j denotes the number of observables measured at time t_j and η_{ij} denotes the observed value which is compared with the value of observable i evaluated by the model where the functions $g_i(\mathbf{x}(t_j), \mathbf{p})$ relate the state variables to $\mathbf{x}$ this observable

$$\eta_{ij} = g_i(\mathbf{x}(t_j), \mathbf{p}) + \varepsilon_{ij} \,. \quad (11)$$

The numbers ε_{ij} are the measurement errors and σ_{ij}^2 are weights that have to be adequately chosen due to statistical considerations, e. g. as the variances. The unknown parameter vector $\mathbf{p}$ is determined from the measurements such that the model is optimally adjusted to the measured (observed) data. If the errors ε_{ij} are independent, normally distributed with the mean value zero and have variances σ_{ij}^2 (up to a common factor β^2), then the solution of the least squares problem is a *maximum likelihood estimate*.

The Initial Value Problem Approach

An obvious approach to estimate parameters in ODE which is also implemented in many commercial packages is the initial value problem approach. The idea is to guess parameters and initial values for the trajectories, compute a solution of an initial value problem (IVP) (6) and iterate the parameters and initial values in order to improve the fit. Characteristic features and disadvantages are discussed in, e. g., [6] or [18]. In the course of the iterative solution one has to solve a sequence of IVPs. The state variable $\mathbf{x}(t)$ is eliminated for the benefit of the unknown parameter $\mathbf{p}$ and the initial values. Note that no use is made of the measured data while solving the IVPs. They only enter in the performance criterion. Since initial guesses of the parameters may be poor, this can lead to IVPs which may be hard to solve or even have no solution at all and one can come into badly conditioned regions of the IVPs, which can lead to the loss of stability.

The Boundary Value Problem Approach

Alternatively to the IVP approach, in the "boundary value problem approach" invented by Bock [5], the inverse problem is interpreted as an over-determined, constrained, multiple-point boundary problem. This interpretation does not depend on whether the direct problem is an initial or boundary value problem. The algorithm used here consists of an adequate combination of a multiple shooting method for the discretization of the boundary value problem side condition in combination with a generalized Gauss-Newton method for the solution of the resulting structured nonlinear constrained least squares problem [5,6]. Depending on the vector of signs of the state and parameter dependent switching functions $\mathbf{Q}$ it is even possible to allow piecewise smooth right hand side functions f, i. e., differential equations with switching conditions

$$\mathbf{x}' = f(t, \mathbf{x}, \mathbf{p}; sign(\mathbf{Q}(t, \mathbf{x}, \mathbf{p}))) \,, \quad (12)$$

where the right side may change discontinuously if the vector of signs of the switching functions $\mathbf{Q}$ changes. Such discontinuities can occur, e. g. as a result of unsteady changes of physical values. The switching points are in general given by the roots of the state-dependent components of the switching functions

$$\mathbf{Q}_i(t, \mathbf{x}, \mathbf{p}) = 0 \,. \quad (13)$$

Depending on the stability behavior of the ODE and the availability of information about the process (measured data, qualitative knowledge about the problem, etc.) a grid $\mathcal{T}_m$

$$\mathcal{T}_m : \tau_1 < \tau_2 < \ldots < \tau_m, \quad \Delta\tau_j := \tau_{j+1} - \tau_j,$$
$$1 \leq j \leq m - 1, \quad (14)$$

of m multiple shooting nodes τ_j ($m - 1$ subintervals I_j) is chosen. The grid is adapted to the problem and data and is defined such that it includes the measuring interval ($[\tau_1, \tau_m] = [t_0, t_f]$). Usually, the grid points τ correspond to values of the independent variable t at which observations are available but additional grid points may be chosen for strongly nonlinear models. At each node τ_j an IVP

$$\mathbf{x}'(t) = \mathbf{f}(t, \mathbf{x}, \mathbf{p}), \quad \mathbf{x}(t = \tau_j) = \mathbf{s}_j \in \mathbb{R}^{n_d} \quad (15)$$

has to be integrated from τ_j to τ_{j+1}. The $m - 1$ vectors of (unknown) initial values $\mathbf{s}_j$ of the partial trajectories, the vector $\mathbf{s}_m$ representing the state at the end point and the parameter vector $\mathbf{p}$ are summarized in the (unknown) vector $\mathbf{z}$

$$\mathbf{z}^T := (\mathbf{s}_1^T, \ldots, \mathbf{s}_m^T, \mathbf{p}^T). \quad (16)$$

For a given guess of $\mathbf{z}$ the solutions $\mathbf{x}(t; \mathbf{s}_j, \mathbf{p})$ of the $m-1$ independent initial value problems in each sub interval I_j are computed. This leads to an (at first discontinuous) representation of $\mathbf{x}(t)$. In order to replace (6) equivalently by these $m - 1$ IVPs matching conditions

$$\mathbf{h}_j(\mathbf{s}_j, \mathbf{s}_{j+1}, p) := \mathbf{x}(\tau_{j+1}; \mathbf{s}_j, \mathbf{p}) - \mathbf{s}_{j+1} = 0,$$
$$\mathbf{h}_j : \mathbb{R}^{2n_d + n_p} \to \mathbb{R}^{n_d} \quad (17)$$

are added to the problem. (17) ensures the continuity of the final trajectory $\mathbf{x}(t)$.

Replacing $\mathbf{x}(t_i)$ and $\mathbf{p}$ in (10) by $\mathbf{z}$ the least squares problem is reformulated as a nonlinear constrained optimization problem with the structure

$$\min_{\mathbf{z}} \left\{ \frac{1}{2} \|\mathbf{F}_1(\mathbf{z})\|_2^2 \, | \, \mathbf{F}_2(\mathbf{z}) = 0 \in \mathbb{R}^{n_2}, \right.$$
$$\left. \mathbf{F}_3(\mathbf{z}) \geq 0 \in \mathbb{R}^{n_3} \right\}, \quad (18)$$

wherein n_2 denotes the number of the equality and n_3 the number of the inequality constraints. This usually large constrained structured nonlinear problem is solved by a damped generalized Gauss-Newton method [5]. If $\mathsf{J}_1(\mathbf{z}_k) := \partial_z \mathbf{F}_1(\mathbf{z}_k)$, $\mathsf{J}_2(\mathbf{z}_k) := \partial_z \mathbf{F}_2(\mathbf{z}_k)$ vis. $\mathsf{J}_3(\mathbf{z}_k) := \partial_z \mathbf{F}_3(\mathbf{z}_k)$ denote the Jacobi matrices of $\mathbf{F}_1$, $\mathbf{F}_2$ vis. $\mathbf{F}_3$, then the iteration proceeds as

$$\mathbf{z}_{k+1} = \mathbf{z}_k + \alpha_k \Delta \mathbf{z}_k \quad (19)$$

with damping constant $\alpha_k, 0 < \alpha_{min} \leq \alpha_k \leq 1$, and the increment $\Delta \mathbf{z}_k$ determined as the solution of the constrained linear problem

$$\min_{\mathbf{z}} \left\{ \frac{1}{2} \|\mathsf{J}_1(\mathbf{z}_k)'\mathbf{z}_k + \mathbf{F}_1(\mathbf{z}_k)\|_2^2 \right.$$
$$\left. \begin{array}{l} \mathsf{J}_2(\mathbf{z}_k)\Delta\mathbf{z}_k + \mathbf{F}_2(\mathbf{z}_k) = 0 \\ \mathsf{J}_3(\mathbf{z}_k)\Delta\mathbf{z}_k + \mathbf{F}_3(\mathbf{z}_k) \geq 0 \end{array} \right\}. \quad (20)$$

Global convergence can be achieved if the damping strategy is properly chosen [6].

The inequality constraints that are active in a feasible point are defined by the index set

$$\mathcal{I}(\mathbf{z}_k) := \{i | F_{3i}(\mathbf{z}_k) = 0, \quad i = 1, \ldots, n_3\}. \quad (21)$$

The inequalities which are defined by the index set $\mathcal{I}(\mathbf{z}_k)$ or their derivatives are denoted with $\hat{\mathbf{F}}_3$ or $\hat{\mathsf{J}}_3$ in the following. In addition to (21) we define

$$\mathbf{F}_c := \begin{pmatrix} \mathbf{F}_2 \\ \hat{\mathbf{F}}_3 \end{pmatrix}, \quad \mathsf{J}_c := \begin{pmatrix} \mathsf{J}_2 \\ \hat{\mathsf{J}}_3 \end{pmatrix}. \quad (22)$$

In order to derive the necessary conditions that have to be fulfilled by the solution of the problem (18) the Lagrangian

$$L(\mathbf{z}, \lambda, \mu) := \frac{1}{2}\|\mathbf{F}_1(\mathbf{z})\|_2^2 - \lambda^T \mathbf{F}_2(\mathbf{z}) - \mu^T \mathbf{F}_3(\mathbf{z}) \quad (23)$$

and the reduced Lagrangian

$$\hat{L}(\mathbf{z}, \lambda_c) := \frac{1}{2}\|\mathbf{F}_1(\mathbf{z})\|_2^2 - \lambda_c^T \mathbf{F}_c(\mathbf{z}), \quad \lambda_c := \begin{pmatrix} \lambda \\ \mu_c \end{pmatrix} \quad (24)$$

are defined. The Kuhn–Tucker-conditions, i. e. the necessary conditions of first order, are the feasibility conditions

$$\mathbf{F}_2(\mathbf{z}^*) = 0, \quad \mathbf{F}_3(\mathbf{z}^*) \geq 0 \quad (25)$$

ensuring that $\mathbf{z}^*$ is feasible, and the stationarity conditions stating that the adjoined variables λ^*, μ^* exist as solution of the stationary conditions

$$\frac{\partial L}{\partial z}(\mathbf{z}^*, \boldsymbol{\lambda}^*, \boldsymbol{\mu}^*) = \mathbf{F}_1^T(\mathbf{z}^*) \cdot \mathbf{J}(\mathbf{z}^*) - (\boldsymbol{\lambda}^*)^T \mathbf{J}_2(\mathbf{z}^*)$$
$$- (\boldsymbol{\mu}^*)^T \mathbf{J}_3(\mathbf{z}^*) = 0 \quad (26)$$

and

$$\boldsymbol{\mu}^* \geq 0, \quad i \notin \mathcal{I}(\mathbf{z}^*) \Rightarrow \mu_i^* = 0. \quad (27)$$

If $(\mathbf{z}^*, \boldsymbol{\lambda}^*, \boldsymbol{\mu}^*)$ fulfills the conditions (25), (26) and (27), it is called a Kuhn–Tucker-point and $\mathbf{z}^*$ a stationary point. The necessary condition of second order means that for all directions

$$\mathbf{s} \in T(\mathbf{x}^*) := \left\{ \mathbf{s} \neq 0 \,\middle|\, \begin{array}{l} \mathbf{J}_2(\mathbf{z}^*)\mathbf{s} = 0 \\ \mathbf{J}_3(\mathbf{z}^*)\mathbf{s} \geq 0 \end{array}, \ \mu_i \mathbf{J}_{3i}(\mathbf{z}^*)\mathbf{s} = 0 \right\}$$
(28)

the Hessian $G(\mathbf{z}^*, \boldsymbol{\lambda}^*, \boldsymbol{\mu}^*)$ of the Lagrangian is positive semi-definite:

$$\mathbf{s}^T G(\mathbf{z}^*, \boldsymbol{\lambda}^*, \boldsymbol{\mu}^*)\mathbf{s} \geq 0,$$
$$G(\mathbf{z}^*, \boldsymbol{\lambda}^*, \boldsymbol{\mu}^*) := \frac{\partial^2}{\partial \mathbf{z}^2} L(\mathbf{z}^*, \boldsymbol{\lambda}^*, \boldsymbol{\mu}^*). \quad (29)$$

As $\mu_i = 0$ for $i \notin \mathcal{I}(\mathbf{z}^*)$ it is sufficient to postulate the stationary condition for the reduced Lagrangian (24). For the linear problem (20) follows: $(\mathbf{z}^*, \boldsymbol{\lambda}^*, \boldsymbol{\mu}^*)$ is a Kuhn–Tucker-point of the nonlinear problem (18) if and only, if $(0, \boldsymbol{\lambda}^*, \boldsymbol{\mu}^*)$ is a Kuhn–Tucker-point of the linear problem. The necessary conditions for the existence of a local minimum of problem (18) are:
1. $(\mathbf{z}^*, \boldsymbol{\lambda}^*, \boldsymbol{\mu}^*)$ is a Kuhn–Tucker-point of the nonlinear problem
2. the Hessian $G(\mathbf{z}^*, \boldsymbol{\lambda}^*, -^*)$ of the Lagrangian is positive definite for all directions $\mathbf{s} \in T(\mathbf{x}^*)$, vis. $\mathbf{s}^T G(\mathbf{z}^*, \boldsymbol{\lambda}^*, \boldsymbol{\mu}^*)\mathbf{s} > 0$

If the necessary conditions for the existence of the local minimum and the condition $\mu_i \neq 0$ for $i \in \mathcal{I}(\mathbf{z}^*)$ are fulfilled, two perturbation theorems [6] can be formulated. If the sufficient conditions are fulfilled it can be shown for the neighborhood of a Kuhn–Tucker-point $(\mathbf{z}^*, \boldsymbol{\lambda}^*, \boldsymbol{\mu}^*)$ of the nonlinear problem (18) that the local convergence behavior of the inequality constrained problem corresponds to that of the equality constrained problem which represents active inequalities and equations. Under the assumption of the regularity of the Jacobians J_1 and J_c, i. e.

$$\operatorname{rank}\begin{pmatrix} J_1(\mathbf{z}_k) \\ J_c(\mathbf{z}_k) \end{pmatrix} = n_d + n_p, \quad \operatorname{rank}(J_c(\mathbf{z}_k)) = n_c,$$
(30)

a unique solution $\Delta \mathbf{z}_k$ of the linear problem (20) exists and an unique linear mapping J_k^+ can be constructed which satisfies the relation

$$\Delta \mathbf{z}_k = -J_k^+ \mathbf{F}(\mathbf{z}_k), \quad J_k^+ J_k J_k^+ = J_k^+,$$
$$J_k^T := [J_1^T(\mathbf{z}_k), J_c^T(\mathbf{z}_k)]. \quad (31)$$

The solution $\Delta \mathbf{z}_k$ of the linear problem or formally the generalized inverse J_k^+ [5] of J_k results from the Kuhn–Tucker conditions. But it should be noticed that $\mathbf{z}_k$ is not calculated from (31) because of reasons of numerical efficiency but is based on a decomposition procedure using orthogonal transformations.

By taking into consideration the special structure of the matrices J_i caused by the continuity conditions of the multiple shooting discretization (18) can be reduced by a condensation algorithm described in [5,6] to a system of lower dimension

$$\min \left\{ \frac{1}{2} \|A_1 \mathbf{x}_k + \mathbf{a}_1\|_2^2 | A_2 \mathbf{x}_k + \mathbf{a}_2 = 0, \right.$$
$$\left. A_3 \mathbf{x}_k + \mathbf{a}_3 \geq 0 \right\}, \quad (32)$$

from which $\mathbf{x}_k$ can be derived at first and at last $\Delta \mathbf{z}_k$. This is achieved by first performing a "backward recursion", the "solution of the condensed problem" and a "forward recursion" [6]. Kilian [20] has implemented an active set strategy following the description in [6] and [33] utilizing the special structure of J_2.

The details of the parameter estimation algorithms which are incorporated in the efficient software package PARFIT (a software package of stable and efficient boundary value problem methods for the identification of parameters in systems of nonlinear differential equations) are found in [6]. The damping constant α^k in the k-th iteration is computed with the help of *natural level functions* which locally approximate the distance $\|\mathbf{z}_k - \mathbf{z}^*\|$ of the solution from the Kuhn–Tucker point $\mathbf{z}^*$.

The integrator METANB (for the basic discretization see, for instance, [3]) embedded in PARFIT is also suitable for the integration of stiff differential equation systems. It allows the user to compute simultaneously the sensitivity matrixes G,

$$G(t; t_0, \mathbf{x}_0, \mathbf{p}) := \frac{\partial}{\partial \mathbf{x}_0} \mathbf{x}(t; t_0, \mathbf{x}_0, \mathbf{p}) \in \mathcal{M}(n_d, n_d) \quad (33)$$

and H,

$$H(t; t_0, \mathbf{x}_0, \mathbf{p}) := \frac{\partial}{\partial \mathbf{p}} \mathbf{x}(t; t_0, \mathbf{x}_0, \mathbf{p}) \in \mathcal{M}(n_d, n_p) \quad (34)$$

which are the most costly blocks of the Jacobians J_i via the so-called *internal numerical differentiation* as introduced by Bock [5]. This technique does not require the often cumbersome and error prone formulation of the variational differential equations

$$G' = \mathbf{f}_\mathbf{x}(t, \mathbf{x}, \mathbf{p}) \cdot G, \quad G(t_0; t_0, \mathbf{x}_0, \mathbf{p}) = \mathbb{1} \quad (35)$$

and

$$H' = \mathbf{f}_x(t, \mathbf{x}, \mathbf{p}) \cdot H + \mathbf{f}_p(t, \mathbf{x}, \mathbf{p}),$$
$$H(t_0; t_0, \mathbf{x}_0, \mathbf{p}) = 0 \quad (36)$$

by the user.

Using the multiple shooting approach described above, differential equation systems with poor stability properties and even chaotic systems can be treated [18].

Parameter Estimation in DAE Models

Another, even more complex class of problems, are parameter estimation in mechanical multibody systems, e.g., in the planar slider crank mechanisms, a simple model for a cylinder in an engine. These problems lead to boundary problems for higher index differential algebraic systems [34]. Singular controls and state constraints in optimal control also lead to this structure. Inherent to such problems are invariants that arise from index reduction but also additional physical invariants such as the total energy in conservative mechanical systems or the Hamiltonian in optimal control problems.

A typical class of DAEs in mechanical multibody systems is given by the equations of motion

$$\dot{\mathbf{x}} = \mathbf{v}$$
$$M(t, \mathbf{x})\dot{\mathbf{v}} = \mathbf{f}(t, \mathbf{x}) - \nabla_x g(t, \mathbf{x})\lambda, \quad (37)$$
$$0 = \mathbf{g}(t, \mathbf{x})$$

where $\mathbf{x} = \mathbf{x}(t)$ and $\mathbf{v} = \mathbf{v}(t)$ are the coordinates and velocities, M is the mass matrix, $\mathbf{f}$ denotes the applied forces, $\mathbf{g}$ are the holonomic constraints, and λ are the generalized constraint forces. Usually, M is symmetric and positive definite. A more general DAE system might have the structure

$$\dot{\mathbf{x}} = \mathbf{f}(t, \mathbf{x}, \mathbf{z}; \mathbf{p})$$
$$0 = \mathbf{g}(t, \mathbf{x}, \mathbf{z}; \mathbf{p}),$$

where $\mathbf{p}$ denotes some parameters and $\mathbf{z} = \mathbf{z}(t)$ is a set of algebraic variables, i.e., the differentials $\dot{\mathbf{z}}$ do not appear; in (37) λ is the algebraic variable. In addition we might have initial values $\mathbf{x}_0$ and $\mathbf{z}_0$. Obviously, some care is needed regarding the choice of $\mathbf{z}_0$ because it needs to be consistent with the constraint. In some exceptional cases (in which $Z := \nabla_z \mathbf{g}$ has full rank and can be inverted analytically) we might insert $\mathbf{z} = \mathbf{z}(t, \mathbf{x}; \mathbf{p})$ into the differential equation. DAE systems with a regular matrix Z are referred to as index-1 systems. Index-1-DAEs can be transformed into equivalent ordinary differential equations by differencing the equations w.r.t. t. At first we get the implicit system of differential equations

$$\mathbf{g}_t + X\dot{\mathbf{x}} + Z\dot{\mathbf{z}} = 0, \quad X := \nabla_x \mathbf{g}$$

which, according to the assumption of the regularity of Z, can be written as the explicit system

$$\dot{\mathbf{z}} = Z^{-1}(\mathbf{g}_t + X\mathbf{f}).$$

Many practical DAEs have index 1, e.g., in some chemical engineering problems, where algebraic equations are introduced to describe, for instance, mass balances or the equation of state. However, multibody systems such as (37) have higher indices; (37) is of index 3. The reason is, that the multiplier variables, i.e., the algebraic variables, do not occur in the algebraic constraints and it is therefore not possible to extract them directly without further differentiation. If Z does not have full rank the equations are differentiated successively, until the algebraic variables can be eliminated. The smallest number of differentiations required to transform the original DAE system to an ODE system is called the *index* of the DAE. The approach developed and described by Schulz et al. [34] is capable to handle least squares problems without special assumption to the index.

An essential problem for the design, optimization and control of chemical systems is the estimation of parameters from time-series. These problems lead to nonlinear DAEs. The parameters estimation problem leads to a non-convex optimization problem for which several local minima exist. Esposito and Floudas [13] developed two global branch-and-bound and convex-underestimator based optimization approaches to solve this problem. In the first approach, the dynamical system is converted into an algebraic system using orthogonal collocation on finite elements. In the second approach, state profiles are computed by integration. In Esposito and Floudas [12] a similar approach is used to solve optimal control problems.

Parameter Estimation in PDE Models

A very complex class of least squares problems are data fitting problems in partial differential equations based models. These include eigenvalue problems, as well as initial and boundary value problems and cover problems in atomic physics, elasticity, electromagnetic fields, fluid flow or heat transfer. Some recent problems are, for instance, in models describing the water balance and solid transport used to analyze the distributions of nutrients and pesticides [1], in the determination of diffusive constants in water absorption processes in hygroscopic liquids discussed in [15], or in multispecies reactive flows through porous media [38]. Such nonlinear multispecies transport models can be used to describe the interaction between oxygen, nitrate, organic carbon and bacteria in aquifers. They may include convective transport and diffusion/dispersion processes for the mobile parts (that is the mobile pore water) of the species. The immobile biophase represents the part where reactions caused by microbial activity take place and which is coupled to transport through mobile pore water. The microorganisms are assumed to be immobile. The model leads to partial differential algebraic equations

$$M\partial_t \mathbf{u} - \nabla(D\nabla \mathbf{u}) + q\nabla \mathbf{u} = \mathbf{f}_1(\mathbf{u}, \mathbf{v}, \mathbf{z}, \mathbf{p}),$$
$$\partial_t \mathbf{v} = \mathbf{f}_2(\mathbf{u}, \mathbf{v}, \mathbf{z}, \mathbf{p}), \quad (38)$$
$$0 = \mathbf{g}(\mathbf{u}, \mathbf{v}, \mathbf{z}, \mathbf{p}),$$

where D and q denote the hydraulic parameters of the model, $\mathbf{p}$ denotes a set of reaction parameters, $\mathbf{u}$ and $\mathbf{v}$ refer to the mobile and immobile species, and $\mathbf{z}$ is related to source and sink terms.

Methodology

To solve least squares problems based on PDE models requires sophisticated numerical techniques but also great attention with respect to the quality of data and identifiability of the parameters. To solve such problems we might use the following approaches:

1. Unstructured approach: The PDE model is, for fixed parameters $\mathbf{p}$, integrated by any appropriate method yielding estimations of the observations. The parameters are adjusted by a derivative-free optimization procedure, e.g., by the Simplex method by Nelder and Mead [23]. This approach is relatively easy to implement, it solves a sequence of direct problems, and is comparable to what in Sect. "Parameter Estimation in ODE Models" has been called the IVP approach. Arning [1] uses such an approach.

2. Structured approach (for initial value PDE problems): Within the PDE model spatial coordinates and time are discretized separately. Especially for models with only one spatial coordinate, it is advantageous to apply finite difference or finite element discretizations to the spatial coordinate. The PDE system is transformed into a system of (usually stiff) ordinary differential equations. This approach is known as the *method of lines* (see, for example, [30]). It reduces parameter estimation problems subject to time-dependent partial differential equations to parameter identification problems in systems of ordinary differential equations to be integrated w.r.t. time. Now it is possible to distinguish again between the IVP and BVP approach. Schittkowski [32] in his software package EASY-FIT applies the method of lines to PDEs with one spatial coordinate and uses several explicit and implicit integration methods to solve the ODE system. The integration results are used by an SQP optimization routine or a Gauß–Newton method to estimate the parameters. Zieße et al. 38 and Dieses et al. [11], instead, couple the method of lines (in one and two spatial coordinates) with Bock's [6] BVP approach, discretize time, for instance, by multiple shooting and use an extended version of PARFIT.

The method of lines has become one of the standard approaches for solving time-dependent PDEs with only one spatial coordinate. It is based on a partial discretization, which means that only the spatial derivative is discretized but not the time derivative. This leads to a system of N coupled ordinary differential equation, where N is the number of discretization points. Let us demonstrate the method by applying it to the diffusion equation

$$\frac{\partial}{\partial t} c(t, z) = D \frac{\partial^2}{\partial z^2} c(t, z), \quad \begin{array}{l} 0 \leq t < \infty \\ 0 \leq z \leq L \end{array} \quad (39)$$

with constant diffusion coefficient D. We discretize the spatial coordinate z according to

$$z_i = i \Delta z, \quad \Delta z := \frac{L}{N},$$

$$c_i = c_i(t) = c(t, z_i), \quad i = 0, \ldots, N. \quad (40)$$

If we choose a finite difference approximation we get

$$\frac{\partial^2}{\partial z^2} c(t, z) \approx \frac{c(t, z - \Delta z) - 2c(t, z) + c(t, z + \Delta z)}{(\Delta z)^2}$$

$$= \frac{c_{i-1} - 2c_i + c_{i+1}}{(\Delta z)^2}, \quad (41)$$

which replaces the diffusion Eq. (39) by N ordinary differential equations

$$\dot{c}_i(t) = \frac{c_{i-1} - 2c_i + c_{i+1}}{(\Delta z)^2}. \quad (42)$$

A detailed example of this method is discussed in [15]. The water transport and absorption processes within a hygroscopic liquid are described by a model containing the diffusion Eq. (39) describing the water transport within the hygroscopic liquid, a mixed Dirichlet–Neumann condition representing a flux balance equation at the surface of the liquid, and an additional integral relation describing the total amount of water in the liquid. The model included three parameters to be estimated.

The available measurement data provide the total time dependent concentration $C(t)$ of water in the liquid. A further complication was that the mathematical solution of the diffusion equation is the water concentration $c(t, z)$ in the hygroscopic liquid and it is a function of time *and* location. Therefore, in order to compare the mathematical solution with the observed data one had to integrate $c(t, z)$ over the space coordinate z, i.e., the depth of the fluid.

Least Squares Problems with Massive Data Sets

We motivate the necessity to analyze massive data sets by an example taken from astrophysics [19]. We outline the method for a huge set of millions of observed data curves in which time is the independent parameter and for each of the N, $N \simeq 10^6$, curves there is a different underlying parameter set we want to estimate by a least squares method. Note that we assume that there is a model in the sense of (6) or (10) available involving an adjustable parameter vector $\mathbf{p}$. We are further assume that we are dealing with nonlinear least squares problems which are not easy to solve. The difficulties could arise from the dependence on initial parameters, non-smoothness of the model, the number of model evalutions, or the CPU time required for one model evaluation. For each available curve we can, of course, solve this least squares problem by the techniques mentioned or discussed earlier in this article. However, the CPU time required to solve this least squares problem for several million curves is prohibitive. The archive approach described in this section is appropriate for this situation.

Examples of massive data sets subject to least squares analyses are surveys in astrophysics where millions of stars are observed over a range of time. About 50% of them are binary stars or multiple systems. The observed data could be flux of photons (just called *light* in the discipline of binary star researchers) in a certain wavelength region or radial velocity as a function of time. Thus we have to analyze millions of light and radial velocity curves. There are well validated models and methods (cf., [17] to compute such curves on well defined physical and geometrical parameters of the binary systems, e.g., the mass ratio, the ratio of their radii, their temperatures, inclination, semi-major axis and eccentricity to mention a few. Thus one is facing the problem how to analyze the surveys and to derive the stellar parameters $\mathcal{P}$ relevant to astrophysicists. In this eclipsing binary star example it suffices to consider the range $[0, P]$ for the independent parameter time because the observed curves are periodic with respect to the period P. The period could be determined a priori from a frequency analysis of the observed curve. Under certain assumptions, in eclipsing binary star analyses, time can be replaced by phase.

The critical issues are speed and stability. Speed is obviously necessary to analyze large number of data, light and radial velocity curves in the example. Stability is required to automatize the procedure. Automatization enables the user to analyze large sets of eclipsing binary data produced by surveys. Stability and automatization need to overcome the problem of initial parameters usually experienced in nonlinear least squares. There is a price to be paid in terms of accuracy. But nevertheless, such an approach will produce good approximate results and may indicate interesting eclipsing binary stars for detailed follow-up analysis.

The method we propose to solve least squares problems with massive data sets is a matching approach: match one or several curves to a large test sets of precomputed archive curves for an appropriate set of combinations of $|\mathcal{P}|$ parameters.

The Matching Approach

Let for a given binary system ℓ_{ic}^o be any observed light value for observable c, $c = 1 \ldots C$, at phase θ_i, $i = 1, \ldots, I$. Correspondingly, ℓ_{ick}^c denotes the computed light value at the same phase θ_i for the *archive* light curve k, $k = 1 \ldots K$. Note that K easily might be a large number such as 10^{10}. Each archive light curve k is computed by a certain parameter combination.

The idea of the matching approach is to pick that light curve from the archive which matches the observed curve of binary j best. The best fit solution is obtained by linear regression. The matching approach returns, for each j, the number of the archive light curve which fits best, a scaling parameter, a, and a shift parameter, b, (which might be interpreted as a constant third light) by solving the following nested minimization problem for all j, $j = 1, \ldots, N$:

$$\min_k \left\{ \min_{a_{kc}, b_{kc}} \sum_{i=1}^{I} w_i \left[\ell_{ic}^o - \left(a_{kc} \ell_{ick}^c + b_{kc} \right) \right]^2 \right\}$$

Note that the inner minimization problem requires just to solve a linear regression problem. Thus, for each k, there exists an analytic solution for the unknown parameters a_{kc} and b_{kc}. Further note that the ℓ_{ick}^c values might be obtained by interpolation. The archive light curves are generated in such a way that they have a good covering in the eclipses while a few points will do in those parts of the light curves which show only small variation with phase. Thus, there might be a non-equidistant distribution of phase grid points. A cubic interpolation will probably suffice.

Thus, the matching approach requires us to provide the following components:
1. solving linear regression problems determining a and b for all archive curves and all observed curves (the sequence of the loops is important),
2. generating the archive curves,
3. cubic interpolation in the independent time-like quantity and interpolation after the best matching solution has been found.

In the sequel we briefly comment on the last two components.

Generating and Storing the Archive Curves As the number of archive curves can easily reach 10^{10} one should carefully think about storing them. That requires also appropriate looping over the parameters $p = 1, \ldots, |\mathcal{P}|$. For the eclipsing binary example the details are given in [19]. Among the efficiency issues is the usage of non-equidistant parameter grids exploiting the sensitivity of the parameters on the model function ℓ_{ic}^c.

One might think to store the archive light curves in a type of data base. However, data base techniques become very poor when talking about 10^{10} curves. Therefore, it is probably easier to use a flat storage scheme. In the simplest case, for each k we store the physical and geometric parameters, then those parameters describing observable c, and then the values of the observable. If we use the same number of phase values for each observable and each k, we have the same amount of data to be stored.

Exploiting Interpolation Techniques Within the matching approach interpolation can be used at two places. The first occurrence is in the regression phase. The test curves in the archive are computed for a finite grid of the independent parameter time (phase in this example). The observed curves might be observed at time values not contained in the archive. We can interpolate from the archive values by linear or cubic interpolation to the observed time values. However, it may well pay out to have some careful thoughts on the generation of the time grid points.

The second occurrence is when it comes to determining the best fit. The linear regression returns that parameter set which matches the observed one best. Alternatively, we could exploit several archive points to obtain a better fit to the observed curve. Interpolation in an appropriately defined neighborhoods of the best archive solution can improve the fit of the observed curve.

Numerical Efficiency The efficiency of a least squares method could be measured by the number of function or model evaluation per unknown parameter. If we assume that for each model parameter p we generate n_p archive curves in the archive, the archive contains test curves $N_c = \prod_{p=1}^{|\mathcal{P}|} n_p$ and thus requires N_c model evaluation; n_p is the number of archive grid points of parameter p.

Conclusions

This contribution outlines how to solve ODE and PDE based least squares problems. Academic and commercial least squares solvers as well as software packages are available. Massive data sets and observations arise in data mining problems, medicine, the stock market, and surveys in astrophysics. The approach described in Sect. "The Matching Approach" has been proven efficient for surveys in astrophysics. It can also support the generation of impersonal good initial parameter estimations for further analysis. The archive approach is also suitable for parameter fitting problems with non-smooth models. Another advantage is that on the archve grid it provides the global least squares minimum.

Acknowledgments

Thanks is directed to Steffen Rebennack (University of Florida, Gainesville, FL) for a careful reading of the manuscript, and Johannes P. Schlöder (IWR, Universität Heidelberg, Germany) and Gerhard Krennrich (BASF Aktiengesellschaft, Ludwigshafen) for discussions on the subject of parameter estimation.

References

1. Arning M (1994) Lösung des Inversproblems von partiellen Differentialgleichungen beim Wassertransport im Boden. Dissertation, TU Carolo-Wilhelmina zu Braunschweig
2. Baake E, Schlöder JP (1992) Modelling the Fast Fluorescence Rise of Photosynthesis. Bull Math Biol 54:999–1021
3. Bader G, Deuflhard P (1981) A Semi-Implicit Mid-Point Rule for Stiff Systems of Ordinary Differential Equations. Preprint 114, Universität Heidelberg SFB 123, Institut für Angewandte Mathematik, Heidelberg
4. Björck A (2001) Least Squares Problems. In: Floudas CA, Pardalos P (eds) Encyclopedia of Optimization. Kluwer, Dordrecht, pp 160–170
5. Bock HG (1981) Numerical Treatment of Inverse Problems in Chemical Reaction Kinetics. In: Ebert KH, Deuflhard P, Jäger W (eds) Modelling of Chemical Reaction Systems, Series in Chemical Physics. Springer, Heidelberg, pp 102–125
6. Bock HG (1987) Randwertproblemmethoden zur Parameteridentifizierung in Systemen nichtlinearer Differentialgleichungen. Preprint 142, Universität Heidelberg SFB 123, Institut für Angewandte Mathematik, Heidelberg
7. Brand S (1976) Statistical and Computational Methods in Data Analysis, 2nd edn. North Holland, Amsterdam
8. Dennis JE, Schnabel RB (1983) Numerical Methods for Unconstrained Optimisation and Nonlinear Equations. Prentice Hall, Englewood Cliffs
9. Deuflhard P, Apostolescu V (1977) An Underrelaxed Gauss-Newton Method for Equality Constrained Nonlinear Least Squares Problems. In: Stoer J (ed) Proc. 8th Conf IFIP Würzburg Symposium on the Theory of Computing, number 23 in Springer Lecture Notes Control Inf. Sci. Springer, Heidelberg Berlin New York
10. Deuflhard P, Apostolescu V (1980) A Study of the Gauss–Newton Method for the Solution of Nonlinear Least Squares Problems. In: Frehse J, Pallaschke D, Trottenberg U (eds) Special Topics of Applied Mathematics. North-Holland, Amsterdam, pp 129–150
11. Dieses AE, Schlöder JP, Bock HG, Richter O (1999) Parameter Estimation for Nonlinear Transport and Degradation Processes of Xenobiotica in Soil. In: Neumann J (ed) Proceedings of the 2nd International Workshop on Scientific Computing in Chemical Engineering. Technical University Hamburg-Harburg (TUUH), Hamburg
12. Esposito WR, Floudas CA (2000) Deterministic Global Optimization in Nonlinear Optimal Control Problems. J Glob Optim 17:97–126
13. Esposito WR, Floudas CA (2000) Global Optimization for the Parameter Estimation of Differential-Algebraic Systems. Ind Eng Chem Res 39(5):1291–1310
14. Gauß CF (1809) Theoria Motus Corporum Coelestium in Sectionibus Conicus Solem Ambientium. Perthes, Hamburg
15. Kallrath J (1999) Least Squares Methods for Models Including Ordinary and Partial Differential Equations. In: Dvorak

R, Haupt HF, Wodnar K (eds) Modern Astrometry and Astrodynamics honouring Eichhorn Heinrich. Austrian Academy of Sciences, Vienna, pp 61–75
16. Kallrath J, Altstädt V, Schlöder JP, Bock HG (1999) Analysis of Crack Fatigue Growth Behaviour in Polymers and their Composites based on Ordinary Differential Equations Parameter Estimation. Polym Test 18:11–35
17. Kallrath J, Milone EF (1999) Eclipsing Binary Stars: Modeling and Analysis. Springer, New York
18. Kallrath J, Schlöder J, Bock HG (1993) Parameter Fitting in Chaotic Dynamical Systems. CMDA 56:353–371
19. Kallrath J, Wilson RE (2008) Eclipsing Binary Analysis via Light Curve Archives. ApJ, in preparation
20. Kilian C (1992) Numerische Behandlung von Ungleichheitsrestriktionen für Parameterschätzprobleme in Systemen gewöhnlicher Differentialgleichungen. Diploma thesis, Fachhochschule Darmstadt
21. Levenberg K (1944) A Method for the Solution of Certain Non-Linear Problems in Least Squares. Q Appl Math 2:164–168
22. Marquardt DW (1963) An Algorithm for Least Squares Estimation of Nonlinear Parameters. Appl SIAM J Math 11:431–441
23. Nelder JA, Mead R (1965) A Simplex Method for Function Minimization. Comput J 7:308–313
24. Ostle B (1963) Statistics in Research. Iowa State University Press, Ames
25. Polya G (1979) Vom Lernen und Lösen mathematischer Aufgaben. Einsicht und Entdeckung. Lernen und Lehren. Birkhäuser Verlag, Basel
26. Popper KR (1980) The Logic of Scientific Discovery, 10th edn. Hutchinson, London
27. Press WH, Flannery BP, Teukolsky SA, Vetterling WT (1992) Numerical Recipes The – Art of Scientific Computing, 2nd edn. Cambridge University Press, Cambridge
28. Richter O, Nörtersheuser P, Pestemer W (1992) Non-linear parameter Estimation in Pesticide Degradation. Sci Total Environ 123/124:435–450
29. Richter O, Söndgerath D (1990) Parameter Estimation in Ecology. VCH-Verlag, Weinheim
30. Schiesser WE (1991) The Numerical Methods of Lines. Academic Press, San Diego
31. Schittkowski K (1994) Parameter Estimation in Systems of Nonlinear Equations. Numer Math 68:129–142
32. Schittkowski K (1997) Parameter Estimation in One-Dimensional Time-Dependent Partial Differential Equations. Optim Method Softw 7:165–210
33. Schlöder JP (1988) Numerische Methoden zur Behandlung hochdimensionaler Aufgaben der Parameteridentifizierung. Preprint 187, Bonner Mathematische Schriften, Institut für Angewandte Mathematik, Bonn
34. Schulz VH, Bock HG, Steinbach MC (1998) Exploiting Invariants in the Numerical Solution of Multipoint Boundary Value Problems for DAE. J SIAM Sci Comput 19:440–467
35. Xu C (2001) Nonlinear Least Squares. In: Floudas CA, Pardalos P (eds) Encyclopedia of Optimization. Kluwer, Dordrecht, pp 75–80
36. Xu C (2001) Nonlinear Least Squares: Newton-type Methods. In: Floudas CA, Pardalos P (eds) Encyclopedia of Optimization. Kluwer, Dordrecht, pp 67–69
37. Xu C (2001) Nonlinear Least Squares: Trust Region Methods. In: Floudas CA, Pardalos P (eds) Encyclopedia of Optimization. Kluwer, Dordrecht, pp 80–86
38. Zieße MW, Bock HG, Gallitzendörfer JV, Schlöder JP (1996) Parameter Estimation in Multispecies Transport Reaction Systems Using Parallel Algorithms. In: Gottlieb J, DuChateau P (eds) Parameter Identification and Inverse Problems in Hydrology, Geology and Ecology. Kluwer, Dordrecht, pp 273–282

Complexity Theory

STEPHEN A. VAVASIS
Cornell University, Ithaca, USA

MSC2000: 90C60

Article Outline

Keywords
See also
References

Keywords

Complexity; Turing machine; NP-hard; NP-complete; Polynomial time; Information-based complexity; Real number model; Decision problem

Complexity theory poses the question: How much computing time is required to solve a problem, as a function of the size of the problem? A similar questions may be asked about other computing resources like memory space. In the context of optimization, the commonly-asked complexity question is how much computing time, as a function of m and n, is required to solve a certain class of mathematical programming problems with n variables and m constraints. This form of asymptotic complexity analysis was introduced by J. Hartmanis and R.E. Stearns [4].

Several different complexity theories have been developed to address this question. The best known complexity theory is based on Turing machines. Before defining this term, we start with a definition of 'problem'. Formally, a *problem* is a function F that takes as input an *instance* and produces as output a *result*. For example, in the context of *linear programming* the instance would be a triple $(A, \mathbf{b}, \mathbf{c})$ specifying a standard-form linear program LP:

$$\begin{cases} \min & \mathbf{c}^\top \mathbf{x} \\ \text{s.t.} & A\mathbf{x} = \mathbf{b}, \\ & \mathbf{x} \geq \mathbf{0}. \end{cases}$$

The value of $F(A, \mathbf{b}, \mathbf{c})$ is the optimal value of the LP instance, or perhaps the optimizer. The range of F must also include special output values to signify an infeasible instance, an unbounded instance, or an ill-formed instance, e.g., dimensions of A and $\mathbf{b}$ are incompatible. Thus, in the context of complexity theory, the word 'problem' refers not to a specific instance but to a class of instances.

For a Turing machine, all instances must be specified as finite-length strings of symbols where the symbols are chosen from a fixed, finite alphabet. For LP and other optimization problems, a reasonable alphabet would include the ten digits and delimiter marks like decimal points, commas, parentheses. A cardinality argument shows that this stipulation of finite string over finite alphabet precludes the consideration of problems with arbitrary real number data. Thus, Turing machine solution of linear programming is generally restricted to rational or integer data. Rational and integer data are essentially equivalent since one can transform rational to integer by multiplying by a common denominator.

A second limitation of the Turing machine model is that there is no simple way to specify a general objective function or constraint function of an optimization problem as part of the input. There is a generalization of the Turing machine definition to overcome this limitation (so-called 'oracle' Turing machines), but in this article we limit attention to conventional Turing machines. This limitation means that our Turing machine complexity analysis focuses on optimization problems with predefined classes of objective functions and constraints in which the only free parameters are numeric data, e.g., $(A, \mathbf{b}, \mathbf{c})$ in linear programming.

A *Turing machine* (TM) is a computational device equipped with an infinitely-long tape used for memory and a controller with a finite program. The tape contains an infinite number of cells numbered 0, 1, 2, ..., and each cell is capable of holding one symbol chosen from a finite alphabet. The alphabet of the tape is a superset of the alphabet used for the input. Initially the tape contains the input instance written one symbol per cell starting at the left end of the tape (cell 0). The remaining cells contain a special symbol meaning 'blank'.

The Turing machine controller has a tape head that is above one cell of the tape at any particular time. The controller is always in a *state* chosen from a finite list of states. Finally, the TM obeys a finite list of *transition rules*. Each transition rule has the form: 'if the current symbol under the head is x and the current state is y, then change the symbol to x', change the state to y' and move the tape head one cell in direction d', where d is either 'left' or 'right'. Thus, a TM is fully specified by its input alphabet, its tape alphabet, its list of states, and its list of transition rules. If, for any given combination of current symbol/current state, there is at most one applicable transition rule, the TM is said to be *deterministic* else it is said to be *nondeterministic*. In this article we consider deterministic TMs only.

An *execution* of a Turing machine consists of a sequence of *moves*. Initially, as mentioned above, the input is written on the tape, the head is at position 0, and the machine is in a specially designated state called the 'start' state. The applicable transition rule is selected and executed, meaning that cell 0 is rewritten and the head is moved. Each execution of a transition rule is called a 'move'. The Turing machine continues to make moves until it reaches another special state called the 'halt' state.

The Turing machine is said to *solve* problem F, if given an input instance x, the TM (eventually) writes $F(x)$ on its tape starting at position 0, followed by blanks, and then halts. If for some input the TM could ever execute an illegal operation, e.g., move left from cell 0, or enter a state/symbol combination before halting for which there is no applicable transition rule, then it does not solve F. Furthermore, we require that the Turing machine can correctly handle every possible finite string that can be written with the input alphabet. For incorrectly formatted strings, the Turing machine should output a special string indicating incorrect formatting.

The *running time* of a Turing machine for a given input instance is the number of moves required before it halts. The running time for the whole problem F is usually expressed as a function of the *size of the input*, that is, the number of symbols in the input string.

It can be proved using lengthy constructions that a Turing machine is capable of all the operations of an ordinary computer: it can simulate consecutively numbered memory cells each holding a separate integer or rational number that are individually addressable, it can multiply, divide, add, and subtract two such numbers, etc. For a more detailed treatment of Turing machines, see [5].

A Turing machine is said to solve problem F in *polynomial time* if its running time is no more than a polynomial function of the size of the input. Examples of optimization problems that can be solved in polynomial time include linear and convex quadratic programming.

A *decision problem* is a problem F in which the range of F consists of just two entries, 'YES' and 'NO'. Optimization problems can often be recast as decision problems. For instance, in the case of linear programming, the input instance consists of $(A, \mathbf{b}, \mathbf{c}, r)$, where r is a rational number, and the TM outputs 'YES' if the minimal solution to the LP problem is r or less, else it outputs 'NO'. For incorrectly formatted and infeasible problems, the TM also outputs 'NO'. The decision problem F partitions the input space into two sets of strings, 'YES'-instances and 'NO'-instances. A synonym for 'decision problem' is *language recognition problem*.

The set $\mathcal{P}$ is defined to be all decision problems that can be solved in polynomial time. This set includes linear programming (as recast in the previous paragraph) and many combinatorial optimization problems such as the minimum spanning tree problem and the shortest path problem (cf. also ▶ Shortest path tree algorithms).

Many interesting problems, such as *nonconvex quadratic programming* and *Boolean satisfiability*, are not known to be in $\mathcal{P}$, but are also not proven to lie outside of $\mathcal{P}$. To analyze these cases, we introduce a second complexity class called NP. A decision problem F is said to lie in NP if there exists a polynomial time 'certificate-checking' machine M outputting 'YES' or 'NO', and polynomials $p(\cdot), q(\cdot)$ with the following properties. For every 'YES'-instance x of F, there exists another string y, called the *certificate* of x, such that the size of y is no more than $p(\text{size}(x))$ such that the pair (x, y) (i.e., the string concatenation of x and y properly delimited) is a 'YES'-instance of M. On the other hand, for every 'NO'-instance x of F, and for every possible certificate y, M outputs 'NO' for the input pair (x, y). Finally, in both cases, M is required to run in time no more than $q(\text{size}(x) + \text{size}(y))$.

Notice this definition is asymmetric between 'YES'- and 'NO'-instances. Thus, it is not necessarily true that if problem F is in NP, then the problem $\overline{F}$ that results from complementing F's output (i.e., 'YES'-instances of F are 'NO'-instances of $\overline{F}$ and vice-versa) is still in NP. Indeed, the question of whether $F \in NP \Leftrightarrow \overline{F} \in NP$ is a well-known open question.

Another observation from the above definition is that $\mathcal{P} \subset NP$. In particular, if a decision problem F lies in $\mathcal{P}$, then it has polynomial time Turing machine T that distinguishes 'YES'-instances from 'NO'-instances. In this case, it is simple to design the certificate checking machine M needed for the definition of NP: in particular, M takes as input (x, y), it discards y, i.e., overwrites it with blank cells, and then switches to running T on x.

The most famous open question in complexity theory is whether this containment is actually equality, i.e., whether $\mathcal{P} = NP$. It turns out that the $\mathcal{P} = NP$ question hinges on NP-*complete problem* the prototype of which is the satisfiability problem. The *satisfiability problem* is as follows. An instance is a Boolean formula with variables $x_1, \ldots, x_m$, conjunctions, disjunctions and complement operations. For example, $x_1 \wedge (\overline{x_1} \vee \overline{x_2}) \wedge (x_2 \vee x_3)$ is a satisfiability instance. The decision problem is to determine whether there is an *assignment* of the variables, each one either 'TRUE' or 'FALSE', to make the entire formula true following the usual laws of boolean algebra. For example, the preceding formula is a 'YES'-instance because there is a satisfying assignment, namely x_1 = 'TRUE', x_2 = 'FALSE', x_3 = 'TRUE'. It is easy to see that this problem is in NP: the certificate for a 'YES'-instance is the satisfying assignment. The certificate-checking machine M substitutes the satisfying assignment into the formula and verifies that the formula evaluates as 'TRUE'. Thus, every satisfiable formula has a certificate, but every nonsatisfiable instance is rejected by M no matter what certificate is given.

S.A. Cook [2] proved that every problem in NP is polynomially transformable to satisfiability. We say that decision problem F is *polynomially transformable* to F' if there exists a Turing machine T that takes as input an instance x of F and produces as output an instance x' of F', such that the running time of T is no more than a polynomial in the size of the input, and such that x is a 'YES'-instance of F if and only if x' is a 'YES'-instance of F'.

Cook's result means that given any decision problem F in *NP*, there is a Turing machine M depending on F that takes as input an instance x of F and proceeds as follows. In polynomial time, M constructs a Boolean formula x' from x such that x' is satisfiable if and only x is a 'YES'-instance of F. The construction of M is as as follows. The Boolean formula x' simulates the action of the certificate-checking machine in on x. The actual entries of the certificate are represented by unknown Boolean variables, as are the entries on the tape of M after the first move. The formula is composed of clauses that require the Turing machine to obey all transition rules and to end up halting with 'YES' as the output.

Thus, Cook's theorem implies that if there were a polynomial time algorithm for satisfiability, then there would be a polynomial time algorithm for every other problem in *NP*. Thus, the famous open question 'is $\mathcal{P} = NP$?' is now reduced to the (apparently) simpler question 'is there a polynomial time algorithm for the satisfiability problem?'

It is not yet (1999) known whether the answer to either question in the last paragraph is 'yes'. But many in the field suspect that the answer is 'no', i. e., many believe that there is no polynomial time algorithm for satisfiability. If indeed it is proved some day that no such polynomial time algorithm exists, we would say that satisfiability is *intractable*.

A decision problem F is said to be *NP-complete* if it has these two properties, namely
1) F is in *NP*; and
2) every problem in *NP* can be polynomially transformed to F.

Cook's result can be restated as: satisfiability is *NP*-complete. Furthermore, since polynomial transformations can be composed, any problem F in *NP* to which any known *NP*-complete problem F' can be transformed must itself be *NP*-complete. After Cook's result was announced, R.M. Karp [6] showed that many well-known combinatorial problems, such as the *Hamiltonian cycle problem* ('given an undirected graph, is there a cycle containing each vertex exactly once?') and the *max-clique problem* ('given an undirected graph and an integer k, is there a set of k vertices that are all mutually connected by edges?'), are *NP*-complete. By 1979, already thousands of problems were known to be *NP*-complete and many were catalogued in [3]. A proof that a problem is *NP*-complete is regarded as strong evidence of the problem's intractability.

Although the first batch of *NP*-completeness proofs applied to combinatorial problems, many continuous optimization problems are also known to be *NP*-complete; see ▶ *NP*-complete problems and proof methodology.

A generalization of '*NP*-complete' is the notion of '*NP*-hard'. A problem F is said to be *NP-hard* if satisfiability (or any other *NP*-complete problem) can be polynomially transformed to F. Thus, an *NP*-hard problem does not necessarily lie in *NP*. Indeed, the term *NP*-hard is often used to describe problems that are not even decision problems.

The Turing machine is not the only model of complexity used in the literature. In fact, some feel that the TM is inadequate for modeling continuous optimization problems. Most continuous optimization problem are based on computation with real numbers, but true real number computation is not possible with a TM. One model of real number computation is the *information-based model*. In this model, an algorithm is composed of operations on real numbers. Operations on real numbers are often counted as cost-free in this model, and the only costly operation is the evaluation of the functions defining the objective and constraints of the optimization problem. The objective functions and constraints are considered external black-box subroutines that take as input a vector and return as output the value of the function and possibly derivative values. In information-based complexity, a parameter $\epsilon > 0$ that specifies the desired degree of accuracy in the solution is always part of the input, since the information-based model rarely permits any problem to be solved exactly. This information-based model was used to analyze the *ellipsoid method* by its inventors D.B. Yudin and A.S. Nemirovsky [7]. It has also been used to analyze complexity of *local optimization* by S.A. Vavasis [12]. This model has also been used extensively to analyze other numerical algorithms not related to optimization, e. g., quadrature and other linear problems [9].

A second model of computation is a *real number model* in which each operation is unit cost, and in which there is no concept of external black-box function evaluation. In this model it is possible to develop real number analogs of complexity classes $\mathcal{P}$ and $\mathcal{NP}$, and also a reasonable definition for *NP*-complete; see [1]. This model can be used to analyze linear programming and other problems specified by a finite number of real parameters. The complexity of linear programming problem in this model is not fully understood (see [13] and [10]).

For a more detailed look at complexity in optimization up to 1991, see [11]. For a more recent collection of papers on this topic, see [8].

See also

- ▶ Complexity Classes in Optimization
- ▶ Complexity of Degeneracy
- ▶ Complexity of Gradients, Jacobians, and Hessians
- ▶ Complexity Theory: Quadratic Programming
- ▶ Computational Complexity Theory
- ▶ Fractional Combinatorial Optimization
- ▶ Information-Based Complexity and Information-Based Optimization
- ▶ Kolmogorov Complexity
- ▶ Mixed Integer Nonlinear Programming
- ▶ *NP*-Complete Problems and Proof Methodology
- ▶ Parallel Computing: Complexity Classes

References

1. Blum L, Cucker F, Shub M, Smale S (1998) Complexity and real computation. Springer, Berlin
2. Cook SA (1971) The complexity of theorem-proving procedures. In: Proc 3rd Annual ACM Symp Theory of Computing. ACM, pp 151–158
3. Garey MR, Johnson DS (1979) Computers and intractability: A guide to the theory of NP-completeness. Freeman, New York
4. Hartmanis J, Stearns RE (1965) On the computational complexity of algorithms. Trans Amer Math Soc 117:285–306
5. Hopcroft J, Ullman J (1979) Introduction to automata theory, languages and computation. Addison-Wesley, Reading, MA
6. Karp RM (1972) Reducibility among combinatorial problems. In: Miller RE, Thatcher JW (eds) Complexity of Computer Computations. Plenum, New York, pp 85–103
7. Nemirovsky AS, Yudin DB (1983) Problem complexity and method efficiency in optimization. Wiley, New York
8. Pardalos PM (1993) Complexity in numerical optimization. World Sci, Singapore
9. Traub JF, Wasilkowski GW, Woźniakowski H (1988) Information-based complexity. Acad Press, New York
10. Traub JF, Woźniakowski H (1982) Complexity of linear programming. Oper Res Lett 1:59–62
11. Vavasis SA (1991) Nonlinear optimization: Complexity issues. Oxford Univ. Press, Oxford
12. Vavasis SA (1993) Black-box complexity of local minimization. SIAM J Optim 3:60–80
13. Vavasis SA, Ye Y (1996) A primal-dual interior point method whose running time depends only on the constraint matrix. Math Program 74:79–120

Complexity Theory: Quadratic Programming

STEPHEN A. VAVASIS
Cornell University, Ithaca, USA

MSC2000: 90C60

Article Outline

Keywords
See also
References

Keywords

Quadratic programming; Complexity; *NP*-hard; *NP*-complete; Local minimization; Trust region problem; Polynomial time; Strongly polynomial time; Knapsack problem; Simplicial constraints; Box constraints; Ellipsoid method; Interior point methods; Approximation algorithms

Nowhere in optimization is the dichotomy between convex and nonconvex programming more apparent than in complexity issues for quadratic programming. *Quadratic programming*, abbreviated QP, refers to minimizing a quadratic function $q(\mathbf{x}) = \mathbf{x}^\top H\mathbf{x}/2 + \mathbf{c}^\top \mathbf{x}$ subject to linear constraints $A\mathbf{x} \geq \mathbf{b}$. The problem is thus specified by the four-tuple $(H, A, \mathbf{b}, \mathbf{c})$ where H is a symmetric $n \times n$ matrix, A is an $m \times n$ matrix, $\mathbf{b}$ is an m-vector and $\mathbf{c}$ is an n-vector. Minimizing a quadratic function subject to convex quadratic constraints is also an interesting problem and is considered at the end of this article. The quadratic function $q(\mathbf{x})$ is said to be *convex* if the matrix H is positive semidefinite. A special case of a convex function is when $H = 0$, in which case the problem is now called *linear programming*.

Convex quadratic programming inherits all the desirable attributes of the general convex programming. In particular, there is no local minimizer other than the global minimizer(s). Furthermore, general convex programming techniques like the *ellipsoid method* and *interior point methods* can be applied.

With either the ellipsoid or interior point method, convex quadratic programming can be solved in *polynomial time*. In more detail, assume that $(H, A, \mathbf{b}, \mathbf{c})$ contain all integer data so that the problem is finitely represented for a *Turing machine* (see ▶ Complexity theory). Let L denote the length of the input data, that is, the total number of digits to write $(H, A, \mathbf{b}, \mathbf{c})$. Assume $L \geq m^2$ since H has m^2 entries. Then M.K. Kozlov et al. [9] showed that the ellipsoid algorithm of A.S. Nemirovsky and D.B. Yudin [14] can solve a convex QP instance in time $O(m^2 L)$ iterations, where each iteration requires $O(m^2)$ arithmetic operations on integers, each of which has at most $O(L)$ digits. This result built on an analogous result for the LP case by L.G. Hačijan (also spelled L.G. Khachiyan) [4]. Thus, the total running time of this algorithm is polynomial in the size of the input. Note that the the global minimizer for quadratic programming (either convex or nonconvex), if it exists, can be written down with $O(nL)$ digits, and hence computing the true global minimizer in a Turing machine setting is possible.

Later, S. Kapoor and P.M. Vaidya [6] and Y. Ye and E. Tse [29] proved that an interior point method can solve convex quadratic programming in polynomial time under similar assumptions. This result built on the earlier result for the LP case by N.K. Karmarkar [7]. The best known running time for an interior point method for convex QP is $O(m^{1/2}L)$ iterations, each iteration requiring $O(m^3)$ arithmetic operations on integers each of which has at most $O(L)$ digits and is based on work by J. Renegar [17].

The running times of both the ellipsoid and interior point algorithms are 'weakly' polynomial, meaning that the number of arithmetic operations is bounded by a polynomial in L rather than by a polynomial in m and n. In contrast, polynomial time algorithms for other problems like solving a system of linear equations or finding a minimum flow a network are *strongly polynomial time*, meaning that the number of operations is bounded by a polynomial in the combinatorial dimension of the input data. A well-known open (1999) question asks whether there is a strongly polynomial time algorithm for convex QP (or, more specifically, for LP). A strongly polynomial algorithm would involve a number of arithmetic operations polynomially bounded in m, n, in which each operation involves integers with a number of digits bounded by a polynomial in L. Some progress related to this question is as follows. If the dimension n is restricted to a small integer, then QP can be solved in time linear in m. This result is due to I. Adler and R. Shamir [1] and builds on [10] and [2]. Since the constant of proportionality (or perhaps an additive term) is exponential in n, this algorithm is not so useful except when $n \ll m$. An example would be quadratic programming arising from a geometric problem, such as finding the point in a 3D polyhedron closest to the origin.

In the case of linear programming, a modified ellipsoid algorithm has a number of operations depending only on L_A, where L_A is the number of digits in A (i. e., the number of operations no longer depends on $\mathbf{b}$ or $\mathbf{c}$), a result due to É. Tardos [19] and extended in [25]. Finally, some special cases of quadratic programming are known to be solvable in strongly polynomial time such as the *convex quadratic knapsack problem* [5] which is:

$$\begin{cases} \min & q_1(x_1) + \cdots + q_n(x_n) \\ \text{s.t.} & \mathbf{l} \leq \mathbf{x} \leq \mathbf{u}, \\ & \mathbf{b}^\top \mathbf{x} = \gamma. \end{cases} \quad (1)$$

Here $q_1, \ldots, q_n$ are convex quadratic functions of one variables (each specified by a quadratic and a linear coefficient) $\mathbf{l}, \mathbf{u}, \mathbf{b}, \gamma$ are also part of the problem data.

Nonconvex quadratic programming is much harder than convex quadratic programming. If H is not positive semidefinite, then the QP instance is said to be *nonconvex*. A special case of nonconvex problems is when H is negative semidefinite, in which case the problem is said to be *concave quadratic programming*. When H is neither positive nor negative semidefinite, the problem is *indefinite*. Nonconvex quadratic programming was shown to be *NP*-hard by S. Sahni [18]. If the problem is posed as a decision problem, then it lies in *NP* (and is therefore *NP*-complete), a result due to S.A. Vavasis [20]. (See ▶ Complexity theory or ▶ *NP*-complete problems and proof methodology for the definitions of *NP*-complete and *NP*-hard.) Even the problem of finding a *local minimizer* is known to be *NP*-hard, a result due to K.G. Murty and S.N. Kabadi [13].

Many restricted versions of the problem are still *NP*-hard. The nonconvex quadratic knapsack problem, that is, (1) with general (not necessarily convex) quadratic functions $q_1, \ldots, q_n$, is *NP*-hard [18]. QP with only *box constraints*, that is, minimize $q(\mathbf{x})$ subject to $\mathbf{l} \leq \mathbf{x} \leq \mathbf{u}$, is also *NP*-hard. Similarly, minimizing $q(\mathbf{x})$ subject to *simplicial constraints*, that is, constraints $\mathbf{x} \geq \mathbf{0}$ and $x_1 + \cdots + x_n = 1$, is *NP*-hard as proved in [15] using a theorem of T.S. Motzkin and E.G. Straus [12]. The simplicial case is interesting because minimizing either a concave or convex quadratic function on a simplex can be solved in polynomial time. P.M. Pardalos and Vavasis [16] showed that quadratic programming in which H has a single negative eigenvalue (i. e., H is 'almost' positive semidefinite) is *NP*-hard.

The hardness results have motivated a search for approximation algorithms to nonconvex quadratic programming problems. Vavasis [22,24] proposed approximation algorithms for concave and indefinite QP in which the complexity depends exponentially on the number of negative eigenvalues of H. An additional result is a fully polynomial time approximation scheme for the indefinite knapsack problem. Ye [28] gave a constant-factor polynomial time approximation scheme for indefinite qua- dratic programming with box constraints.

Because computing even a local minimum of a quadratic programming instance is hard, several researchers have looked at approximations and special cases for the local minimization problem. J.J. Moré and Vavasis [11] proved that a local minimizer for the concave knapsack problem can be found in polynomial time; this result was extended to the indefinite case in [23]. Ye [28] gave a polynomial time algorithm to find an approximate *KKT point* of general nonconvex QP.

So far we have considered only linear constraints. A *convex quadratic constraint*, also called an *ellipsoidal constraint*, is a constraint of the form $(\mathbf{x} - \mathbf{c})^\top A (\mathbf{x} - \mathbf{c}) \leq 1$, where A is a symmetric positive semidefinite matrix. The problem of minimizing a nonconvex quadratic function subject to a single ellipsoidal constraint is called the *trust region problem* and has received extensive attention in the literature because algorithms to solve this problem are often used as subroutines by general-purpose optimization algorithms. A polynomial time algorithm for the trust region problem was proposed independently by Ye [27] and Karmarkar [8]. The sense in which this algorithm is 'polynomial time' is weaker than the analogous claim for QP because in the trust region case, the optimizer $\mathbf{x}$ cannot be written in a finite number of digits even if the input data is all integer (because the solution may be irrational). But Vavasis and R. Zippel [26] showed nonetheless that this algorithm leads to a proof that the associated decision problem lies in $\mathcal{P}$. The trust region problem is thus one of the very few nonconvex optimization problem solvable in polynomial time. M. Fu, Z.-Q. Luo and Ye [3] have considered generalizing this result to more than one ellipsoidal constraint, although the results are not as strong as the single-constraint case.

All of the pre-1991 material in this article is covered in more depth by [21].

See also

- Complexity Classes in Optimization
- Complexity of Degeneracy
- Complexity of Gradients, Jacobians, and Hessians
- Complexity Theory
- Computational Complexity Theory
- Fractional Combinatorial Optimization
- Information-Based Complexity and Information-Based Optimization
- Kolmogorov Complexity
- Mixed Integer Nonlinear Programming
- NP-Complete Problems and Proof Methodology
- Parallel Computing: Complexity Classes
- Quadratic Assignment Problem
- Quadratic Fractional Programming: Dinkelbach Method
- Quadratic Knapsack
- Quadratic Programming with Bound Constraints
- Quadratic Programming Over an Ellipsoid
- Standard Quadratic Optimization Problems: Algorithms
- Standard Quadratic Optimization Problems: Applications
- Standard Quadratic Optimization Problems: Theory

References

1. Adler I, Shamir R (1993) A randomization scheme for speeding up algorithms for linear and convex quadratic programming problems with a high constraints-to-variables ratio. Math Program 61:39–52
2. Clarkson KL (1995) Las Vegas algorithms for linear and integer programming when the dimension is small. J ACM 42(2):488–499
3. Fu M, Luo Z-Q, Ye Y (1996) Approximation algorithms for quadratic programming. Working Paper Dept. Management Sci Univ Iowa
4. Hačijan LG (1979) A polynomial algorithm in linear programming. Soviet Math Dokl 20:191–194. (Dokl. Akad. Nauk SSSR 244 (1979), 1093–1096)
5. Helgason R, Kennington J, Lall H (1980) A polynomially bounded algorithm for a singly constrained quadratic program. Math Program 18:338–343
6. Kapoor S, Vaidya PM (1986) Fast algorithms for convex quadratic programming and multicommodity flows. In: Proc. 18th Annual ACM Symp. Theory of Computing. ACM, pp 147–159
7. Karmarkar N (1984) A new polynomial-time algorithm for linear programming. Combinatorica 4:373–395
8. Karmarkar N (1989) An interior-point approach to NP-complete problems. Preprint
9. Kozlov MK, Tarasov SP, Hačijan LG (1979) Polynomial solvability of convex quadratic programming. Soviet Math Dokl 20:1108–1111 (Dokl Akad Nauk SSSR 248 (1979), 1049–1051)
10. Megiddo N (1984) Linear programming in linear time when the dimension is fixed. J ACM 31:114–127
11. Moré JJ, Vavasis SA (1991) On the solution of concave knapsack problems. Math Program 49:397–411
12. Motzkin TS, Straus EG (1965) Maxima for graphs and a new proof of a theorem of Turán. Canad J Math 17:553–540
13. Murty KG, Kabadi SN (1987) Some NP-complete problems in quadratic and nonlinear programming. Math Program 39:117–123
14. Nemirovsky AS, Yudin DB (1983) Problem complexity and method efficiency in optimization. Wiley, New York
15. Pardalos PM, Han C, Ye Y (1991) Algorithms for the solution of quadratic knapsack problems. Linear Alg Appl 152:69–91
16. Pardalos PM, Vavasis SA (1991) Quadratic programming with one negative eigenvalue is NP-hard. J Global Optim 1:15–22
17. Renegar J (1988) A polynomial-time algorithm based on Newton's method for linear programming. Math Program 40:59–94
18. Sahni S (1974) Computationally related problems. SIAM J Comput 3:262–279
19. Tardos E (1986) A strongly polynomial algorithm to solve combinatorial linear programs. Oper Res 34:250–256
20. Vavasis SA (1990) Quadratic programming is in NP. Inform Process Lett 36:73–77
21. Vavasis SA (1991) Nonlinear optimization: Complexity issues. Oxford Univ. Press, Oxford
22. Vavasis SA (1992) Approximation algorithms for indefinite quadratic programming. Math Program 57:279–311
23. Vavasis SA (1992) Local minima for indefinite quadratic knapsack problems. Math Program 54:127–153
24. Vavasis SA (1992) On approximation algorithms for concave quadratic programming. In: Floudas CA, Pardalos PM (eds) Recent Advances in Global Optimization. Princeton Univ. Press, Princeton, pp 3–18
25. Vavasis SA, Ye Y (1996) A primal-dual interior point method whose running time depends only on the constraint matrix. Math Program 74:79–120
26. Vavasis SA, Zippel R (1990) Proving polynomial-time for sphere-constrained quadratic programming. Techn. Report Dept Computer Sci. Cornell Univ. 90-1182
27. Ye Y (1992) On affine scaling algorithms for nonconvex quadratic programming. Math Program 56:285–300
28. Ye Y (1997) Approximating quadratic programming with bound constraints. Working Paper Dept Management Sci Univ Iowa
29. Ye Y, Tse E (1989) An extension of Karmarkar's projective algorithm for convex quadratic programming. Math Program 44:157–179

Composite Nonsmooth Optimization
CNSO

V. Jeyakumar
School of Math., University Sydney, Sydney, Australia

MSC2000: 46A20, 90C30, 52A01

Article Outline

Keywords
Real-Valued CNSO
Extended Real-Valued CNSO
Multi-Objective CNSO
See also
References

Keywords

Nonsmooth analysis; Convex composite programming; Optimality conditions; Nonsmooth optimization; Vector optimization

By composite nonsmooth optimization (CNSO) we mean a class of optimization problems involving composite functions of the form $f(x) := g(F(x))$, where $F: \mathbf{R}^n \to \mathbf{R}^m$ is a (differentiable) smooth map and $g: \mathbf{R}^m \to \mathbf{R}$ is a nonsmooth function. The function g is often a nonsmooth convex function. Problems of CNSO occur when solving nonlinear equations $F_i(x) = 0$, $i = 1, \ldots, m$, by minimizing the norm $\| F(x) \|$. Similar problems arise when finding a feasible point of a system of nonlinear inequalities $F_i(x) \leq 0$, $i = 1, \ldots, m$, by minimizing $\| F(x)^+ \|$ where $F_i^+ = \max(F_i, 0)$. Composite functions f also appear in the form of an exact penalty function when solving a nonlinear programming problem. Another type of CNSO problem which frequently arises in (electrical) engineering [4] is to minimize the max-function $\max_i F_i(x)$, where the maximum is taken over some (finite) set. All these examples can be cast within the structure of CNSO. Moreover, CNSO provides a unified framework in which to study theoretical properties and convergence behavior of various numerical methods for constrained optimization problems. There have been many contributors to the study of CNSO problems both in finite and infinite dimensions. (See for example [2,6,7,10,11,13,15,16,19].) In this article we only discuss different forms of composite model problems in finite dimensions and provide a brief account of their first and *second order Lagrangian theory of CNSO problems*. The implications for numerical optimization are not discussed here. For details on this see, for instance, [1,6].

Real-Valued CNSO

Consider the problem

(P) $\quad \min\limits_{x \in \mathbf{R}^n} g(F(x))$.

Notably, A.D. Ioffe [7,8,9] provided the theoretical foundation for CNSO problems in the case where the function g is sublinear (convex plus positively homogeneous). Then J.V. Burke [2] extended the theory to the case where g is convex. A fundamental local dualization technique plays a significant role in the development of first – and second order *Lagrangian theory* for (P). To see the dualization result, let us define the *Lagrangian* of (P) as

$$L(x, y^*) = \langle y^*, F(x) \rangle - g^*(y^*),$$

where g^* is the Fenchel conjugate of g [14]. Let

$$L_0(z) = \{y^*: \ y^* \in \partial g(F(z)), \ y^* F'(z) = 0\}$$

and let

$$L_{\eta\epsilon}(z) = \{y^*: \ y^* \in \partial_\epsilon g(F(z)), \ \|y^* F'(z)\| \leq \eta\},$$

where $F'(z)$ is the derivative of F at z, $\partial g(y)$ is the *convex subdifferential* of g at y, $\epsilon > 0$, $\eta > 0$ and $\partial_\epsilon g(F(z))$ is the ϵ-*subdifferential* of g at $F(z)$. The set $L_0(z)$ is the set of Lagrange multipliers for (P) at z (see [2,11]). Define

$$\phi_{\eta\epsilon}(x) := \max\limits_{y^* \in L_{\eta\epsilon}(z)} L(x, y^*).$$

A general form of the *Ioffe–Burke local dualization* result [11] states that if g is a lower semicontinuous convex function and F is a locally Lipschitzian and (Gâteaux) differentiable function then the following statements are equivalent:

i) $g(F(x))$ attains a local minimum at z.
ii) $L_0(z) \neq \emptyset$ and $\phi_{\eta\epsilon}$ attains a local minimum at z, for any $\eta > 0$, $\epsilon > 0$.
iii) $L_0(z) \neq \emptyset$ and $\phi_{\eta\epsilon}$ attains a local minimum at z, for some $\eta > 0$, $\epsilon > 0$.

These conditions also provide first order Lagrangian conditions for (P). Moreover, this local *dualization* result and a generalized Taylor expansion of V. Jeyakumar and X.Q. Yang [11] yield second order optimality conditions for (P). If g is a lower semicontinuous convex function and F is a differentiable map with locally Lipschitzian derivative F' (i.e. $C^{1,1}$) then a necessary condition for $a \in \mathbf{R}^n$ to be a local minimizer of (P) is

$$\max\limits_{y^* \in L_0(a)} L^{\circ\circ}(a, y^*; u, u) \geq 0, \quad \forall u \in \overline{K(a)}.$$

On the other hand if $a \in \mathbf{R}^n$, $L_0(a) \neq \emptyset$ and

$$\max\limits_{y^* \in L_0(a)} -L^{\circ\circ}(a, y^*; u, -u) > 0, \quad \forall u \in D(a),$$

then a is a strict local minimizer of order 2 for (P), i.e., there exist $\epsilon > 0$, $\rho > 0$ such that whenever $\| x - a \| < \rho$, $f(x) \geq f(a) + \epsilon \| x - a \|^2$. Here

$$K(a) = \left\{ u \in \mathbf{R}^n : \begin{array}{l} \exists t > 0, \\ g(F(a) + tF'(a)u) \\ \leq g(F(a)) \end{array} \right\},$$

$D(a) = \{u \in \mathbf{R}^n : f'(a; u) \leq 0\}$, and the directional derivative of f at a is given by $f'(a; d) = g'(F(a); F'(a)d)$. The *generalized second order directional derivative* of L at a in the directions $(u, v) \in \mathbf{R}^n \times \mathbf{R}^n$, $L^{\infty\infty}(a; u, v)$, is defined by

$$\limsup_{y \to a,\, s \to 0} \frac{\langle \nabla L(y + su), v \rangle - \langle \nabla L(y), v \rangle}{s}.$$

Special cases of these optimality conditions under twice continuously differentiability hypothesis can be found in [2,9]. Composite problems where the map F is $C^{1,1}$, but is not necessarily twice continuously differentiable are discussed in [19].

Extended Real-Valued CNSO

A composite problem form which has greater versatility than the traditional form (P) is the following nonfinite valued problem [15,16]

$$\text{(PE)} \begin{cases} \min & g(F(x)) \\ \text{s.t.} & x \in \mathbb{R}^n, \\ & F(x) \in \text{dom}(g), \end{cases}$$

where $g: \mathbf{R}^m \to \mathbf{R} \cup \{+\infty\}$ is a convex function and $F: \mathbf{R}^n \to \mathbf{R}^m$ is a smooth map. For instance, constrained CNSO problems of the form,

$$\begin{cases} \min & g_0(F_0(x)) \\ \text{s.t.} & x \in C, \\ & g_j(F_j(x)) \leq 0, \quad j = 1, \ldots, m, \end{cases}$$

studied in [10,17], can be re-written in the form of (PE) [11]. Here C is a closed convex subset of $\mathbf{R}^n$, g_j, $j = 0, \ldots, m$, are locally Lipschitz functions and F_j, $j = 0, \ldots, m$, are differentiable functions. Optimality conditions for (PE) can be derived by reducing (PE) to a real-valued minimization problem as it was shown in [3]. This requires a regularity condition known as a constraint qualification in the nonlinear programming literature. The following regularity condition, introduced in [15] as a *basic constraint qualification*, permits one to establish a reduction theorem. If $g: \mathbf{R}^m \to \mathbf{R} \cup \{+\infty\}$ is a lower semicontinuous convex function and if $F: \mathbf{R}^n \to \mathbf{R}^m$ is locally Lipschitzian then the function $f(x) := g(F(x))$ is said to satisfy the basic constraint qualification at a point $x \in \text{dom}(f)$ if the only point $w \in N(F(x)|\text{dom}(g))$ for which $0 \in w^{\mathsf{T}} \partial F(x)$ is $w = 0$, where $N(F(x)|\text{dom}(g))$ is the normal cone to $\text{dom}(g)$ at $F(x)$ and $\partial F(x)$ is the generalized Jacobian of F at x [5]. The basic constraint qualification is equivalent to the *Mangasarian–Fromovitz constraint qualification* for the standard nonlinear programming problem with inequality and equality constraints (see [15]). The *Burke–Poliquin reduction* result gives us the following second order conditions for (PE). For problem (PE), suppose that $F(a) \in \text{dom}(g)$, g is lower semicontinuous convex and F is $C^{1,1}$. Then the following statements (i) and (ii) hold.

i) If a is a local minimizer of (PE) at which the basic constraint qualification holds, then

$$\max_{y^* \in L_0(a)} L^{\infty\infty}(a, y^*; u, u) \geq 0, \quad \forall u \in \overline{K(a)}.$$

ii) If $L_0(a) \neq \emptyset$ and

$$\max_{y^* \in L_0(a)} -L^{\infty\infty}(a, y^*; u, -u) > 0, \quad \forall u \in D(a),$$

then a is a strict local minimizer of order 2 for (PE). With the aid of a representation condition, second order conditions can also be obtained for a *global minimizer* of (PE) in the case where F is twice strictly differentiable. This was shown in [19]. The problems (PE) have also been extensively studied by R.T. Rockafellar [15,16] in the case where F is twice continuously differentiable and g is a proper convex function that is *piecewise linear quadratic* in the sense that the $\text{dom}(g)$ is expressible as the union of finitely many polyhedral sets, relative to each of which g is given by the formula that is quadratic (or affine).

Multi-Objective CNSO

Nonsmooth vector optimization problems (cf. ▶ Vector optimization) where the functions involved are compositions of convex functions and smooth functions arise in various applications. The following model problem was examined in [12]:

$$\text{(MP)} \begin{cases} V-\min & (f_1(F_1(x)), \ldots, f_p(F_p(x))) \\ \text{s.t.} & x \in C, \\ & g_j(G_j(x)) \leq 0, \\ & j = 1, \ldots, m, \end{cases}$$

where C is a convex subset of $\mathbf{R}^n$, f_i, g_j are real valued convex functions on $\mathbf{R}^n$, F_i, G_j are locally Lipschitz and differentiable functions from $\mathbf{R}^n$ into $\mathbf{R}^n$. Note here that 'V-min' stands for vector minimization. This model is broad and flexible enough to cover many common types of vector optimization problems. In particular, this model includes the penalty representation of the standard vector nonlinear programming problems, examined in [18], and many vector approximation problems. By employing the Clarke subdifferential, first order Lagrangian optimality and duality results can be discussed as it was shown in [12]. Second order optimality conditions for a special case of the problem (MP) are discussed in [19,21].

See also

- ▶ Nonconvex-Nonsmooth Calculus of Variations
- ▶ Nonsmooth and Smoothing Methods for Nonlinear Complementarity Problems and Variational Inequalities
- ▶ Solving Hemivariational Inequalities by Nonsmooth Optimization Methods

References

1. Burke JV (1985) Descent methods for composite nondifferentiable optimization problems. Math Program 33:260–279
2. Burke JV (1987) Second-order necessary and sufficient conditions for convex composite NDO. Math Program 38:287–302
3. Burke JV, Poliquin RA (1992) Optimality conditions for nonfinite valued convex composite functions. Math Program B 57:103–120
4. Charalambous C (1979) Acceleration of the least pth- algorithm for minimax optimization with engineering applications. Math Program 17:270–297
5. Clarke FH (1983) Optimization and nonsmooth analysis. Wiley, New York
6. Fletcher R (1987) Practical methods of optimization. Wiley, New York
7. Ioffe AD (1979) Necessary and sufficient conditions for a local minimum, 1: A reduction theorem and first-order conditions. SIAM J Control Optim 17:245–250
8. Ioffe AD (1979) Necessary and sufficient conditions for a local minimum, 2: Conditions of Levitin–Miljutin–Osmolovskii type. SIAM J Control Optim 17:251–265
9. Ioffe AD (1979) Necessary and sufficient conditions for a local minimum, 3: second order conditions and augmented duality. SIAM J Control Optim 17:266–288
10. Jeyakumar V (1991) Composite nonsmooth programming with Gâteaux differentiability. SIAM J Optim 1:30–41
11. Jeyakumar V, Yang XQ (1993) Convex composite multi-objective nonsmooth programming. Math Program 59:325–343
12. Jeyakumar V, Yang XQ (1995) Convex composite minimization with C1, 1 functions. J Optim Th Appl 86:631–648
13. Penot JP (1994) Optimality conditions in mathematical programming and composite optimization. Math Program 67:225–245
14. Rockafellar RT (1970) Convex analysis. Princeton Univ. Press, Princeton
15. Rockafellar RT (1988) First- and second-order epi-differentiability in nonlinear programming. Trans Amer math Soc 307:75–108
16. Rockafellar RT (1989) Second-order optimality conditions in nonlinear programming obtained by way of epi-derivatives. Math Oper Res 14:462–484
17. Studniarski M, Jeyakumar V (1995) A generalized mean-value theorem and optimality conditions in composite nonsmooth minimization. Nonlinear Anal Th Methods Appl 24:883–894
18. White DJ (1984) Multi-objective programming and penalty functions. J Optim Th Appl 43:583–599
19. Yang XQ (1994) Generalized second-order directional derivatives and optimality conditions. PhD Thesis Univ. New South Wales, Australia
20. Yang XQ (1998) Second-order global optimality conditions for convex composite optimization. Math Program 81:327–347
21. Yang XQ, Jeyakumar V (1997) First and second-order optimality conditions for convex composite multi-objective optimization. J Optim Th Appl 95:209–224

Computational Complexity Theory

HAMILTON EMMONS, SANATAN RAI
Department OR and Operations Management,
Case Western Reserve University, Cleveland, USA

MSC2000: 90C60

Article Outline

Keywords
Definitions
The Nature of the Time Complexity Function
Polynomial Versus Exponential Algorithms
Reducibility
Classification of Hard Problems
Using Reduction to Establish Complexity
See also
References

Keywords

Computational complexity; Complexity theory; Combinatorial optimization; Decision problem; Recognition problem; Time complexity function; Efficient algorithm; Polynomial algorithm; Exponential algorithm; Reducibility; Nondeterministic polynomial algorithm; *NP*-hard problem; *NP*-complete problem

Many problems that arise in operations research and related fields are *combinatorial* in nature: problems where we seek the optimum from a very large but finite number of solutions. Sometimes such problems can be solved quickly and efficiently, but often the best solution procedures available are slow and tedious. It therefore becomes important to assess how well a proposed procedure will perform.

The theory of *computational complexity* addresses this issue. Complexity theory is a comparatively young field, with seminal papers dating from 1971–1972 ([1,5]). Today, it is a wide field encompassing many subfields. For a formal treatment, see [6]. As we shall see, the theory partitions all realistic problems into two groups: the 'easy' and the 'hard' to solve, depending on how complex (hence how fast or slow) the computational procedure for that problem is. The theory defines still other classes, but all but the most artificial mathematical constructs fall into these two. Each of them can be further subdivided in various ways, but these refinements are beyond our scope. It should be noted that we have not here used the accepted terminology, which is introduced below.

Definitions

A *problem* is a well-defined question to which an unambiguous answer exists. *Solving the problem* means answering the question. The problem is stated in terms of several *parameters*, numerical quantities which are left unspecified but are understood to be predetermined. They make up the data of the problem. An *instance* of a problem gives specified values to each parameter. A *combinatorial optimization problem*, whether *maximization* or *minimization*, has for each instance a finite number of candidates from which the answer, or optimal solution, is selected. The choice is based on a real-valued objective function which assigns a value to each candidate solution. A *decision problem* or *recognition problem* has only two possible answers, YES or NO.

Example 1 For example, consider the problem of solving a given system of linear equations. Stated as a question, it becomes: 'what is the solution to $\mathbf{A}\,\mathbf{x} = \mathbf{b}$?' with parameters $m, n, a_{i,j}, b_i, x_j$ where $i = 1, \ldots, m, j = 1, \ldots, n$. An instance might be: 'What is the solution to $7x_1 - 3x_2 = 16$ and $2x_1 + 5x_2 = 9$?' with parameters $m = 2$, $n = 2$ etc.

This is neither an optimization problem nor a decision problem. An example of an optimization problem is a linear program, which asks: 'what is the greatest value of $\mathbf{cx}$ subject to $\mathbf{Ax} \leq \mathbf{b}$?' To make this a combinatorial optimization problem, we might make the variable $\mathbf{x}$ bounded and integer-valued so that the number of candidate solutions is finite. A decision problem is: 'does there exist a solution to the linear program with $\mathbf{cx} \geq k$?'

To develop *complexity theory*, it is convenient to state all problems as decision problems. An optimization (say, maximization) problem can always be replaced by a sequence of problems of determining the existence of solutions with values exceeding $k_1, k_2, \ldots$. An *algorithm* is a step-by-step procedure which provides a solution to a given problem; that is, to all instances of the problem. We are interested in how fast an algorithm is. We now introduce a measure of algorithmic speed: the time complexity function.

The Nature of the Time Complexity Function

Complexity theory does not measure the speed of an algorithm directly; that would depend on the speed of the computer being used and other extraneous factors. Rather, it considers the rate of growth of the solution time as a function of the instance size. Since different instances of the same size may require dramatically different solution times, we use the 'worst case' or longest time that any instance of that size requires. This maximal time needed to solve a problem instance, as a function of its size, is called the *time complexity function* (TCF) or simply the *complexity of the algorithm*. When we speak of the complexity of the problem, we mean the complexity of the most efficient algorithm (known or unknown) that solves it.

We need to clarify what we mean by the 'time required' and the 'size of an instance'. First, note that we always think of solving problems using a computer. Thus, an algorithm is a piece of computer code. Similarly, the *size of a problem instance* is technically the number of characters needed to specify the data, or the length of the input needed by the program. For a decision problem, an algorithm receives as input any string of characters, and produces as output either YES or NO or 'this string is not a problem instance'. An algorithm *solves the instance or string in time m* if it requires *m* basic operations to reach one of the three conclusions and stop.

In order to avoid detailed consideration of the exact input length (are binary or alphanumeric characters used? what encoding scheme is used?), as well as avoiding precise measurement of solution times, the theory requires no more than *orders of magnitude* of these measurements. Recall, we are only concerned with the rate of increase of solution time as instances grow. For example, we may ask how much longer it takes if we double the instance size. As long as we enter data consistently, an instance that is twice as big as another under one data entry scheme remains twice as big under another. (For a rigorous proof and other technical issues, see [4]). Indeed, it is customary to use as a surrogate for instance size, any number that is roughly proportional to the true value. We shall use the symbol n, $n = 1, 2, \ldots$, to represent the size of a problem instance. In summary, for a decision problem Π:

Definition 2 The *time complexity function* (TCF) of algorithm A is:

$$T_A(n) = \begin{cases} \text{maximal time for } A \\ \text{to solve any string of length } n. \end{cases}$$

In what follows, the big oh notation ($\mathcal{O}$) introduced in [3] will be used when expressing the time complexity function. We say that, for two real-valued functions f and g, '$f(n)$ is $\mathcal{O}(g(n))$', or '$f(n)$ is of the same order as $g(n)$', if $|f(n)| \leq k \cdot |g(n)|$ for all $n \geq 0$ and some $k > 0$.

Polynomial Versus Exponential Algorithms

An *efficient, polynomially bounded, polynomial time algorithm*, or simply a *polynomial algorithm*, is one which solves a problem instance in time bounded by a power of the instance size. Formally:

Definition 3 An algorithm A is *polynomial time* if there exists a polynomial p such that

$$T_A(n) \leq p(n), \quad \forall n \in \mathbb{Z}^+ \equiv \{1, 2, \ldots\}.$$

More specifically, an algorithm is *polynomial of degree c*, or *has complexity* $\mathcal{O}(n^c)$, or *runs in* $\mathcal{O}(n^c)$ *time* if, for some $k > 0$, the algorithm never takes longer than kn^c (the TCF) to solve an instance of size n.

Definition 4 The collection P comprises all problems for which a polynomial time algorithm exists.

Problems which belong to P are the ones we referred to earlier as 'easy'. All other algorithms are called *exponential time* or just *exponential*, and problems for which nothing quicker exists are 'hard'. Although not all algorithms in this class have TCF's that are technically exponential functions, we may think of a typical one as running in $\mathcal{O}(c^{p(n)})$ for some polynomial $p(n)$. Other examples of exponential rates of growth are n^n and $n!$.

The terms 'hard' and 'easy' are somewhat misleading, even though exponential TCFs clearly lead to far more rapid growth in solution times. Suppose an 'easy' problem has an algorithm with running time bounded by, say kn^5. Such a TCF may not be exponential, but it may well be considered pretty rapidly growing. Furthermore, some algorithms take a long time to solve even small problems (large k), and hence are unsatisfactory in practice even if the time grows slowly. On the other hand, an algorithm for which the TCF is exponential is not always useless in practice. Recall, the concept of the TCF is a worst case estimate, so complexity is only an upper bound on the amount of time required by an algorithm. This is a conservative measure and usually useful, but it is too pessimistic for some popular algorithms. The simplex algorithm for linear programming, for example, has a TCF that is $\mathcal{O}(2^n)$, but it has been shown that for the average case the complexity is only $\mathcal{O}(n)$. Thus, the algorithm is actually very fast for most problems encountered.

Despite these caveats, exponential algorithms generally have running times that tend to increase at an exponential rate and often seem to 'explode' when a certain problem size is exceeded. Polynomial time algorithms usually turn out to be of low degree ($\mathcal{O}(n^3)$ or better), run pretty efficiently, and are considered desirable.

Reducibility

A problem can be placed in P as soon as a polynomial time algorithm is found for it. Sometimes, rather than finding such an algorithm, we may place it in P by showing that it is 'equivalent' to another problem which is already known to be in P. We explain what we mean by equivalence between problems with the following definitions.

Definition 5 A problem Π' is *reducible* or *transformable* to a problem Π ($\Pi' \propto \Pi$) if, for any instance I' of Π', an instance I of Π can be constructed in polynomially bounded time, such that the solution to I is sufficient to find the solution to I' in polynomial time.

We call the construction of the I that corresponds to I' a *polynomial transformation* of I' into I.

Definition 6 Two problems are *equivalent* if each is reducible (or simply *reduces*) to the other.

Since reduction, and hence equivalence, are clearly transitive properties, we can define *equivalence classes of problems*, where all problems in the same equivalence class are reducible (or equivalent) to each other. Consider polynomial problems. Clearly, for two equivalent problems, if one is known to be polynomial, the other must be, too. Also, if two problems are each known to be polynomial, they are equivalent. This is because any problem $\Pi' \in P$ is reducible to any other problem $\Pi \in P$, or indeed to any $\Pi \notin P$, in the following trivial sense. For any instance I' of Π', we can pick any instance of Π, ignore its solution, and find the solution to I' directly. We conclude that P is an equivalence class.

We state a third simple result for polynomial problems as a theorem.

Theorem 7 If $\Pi \in P$, then $\Pi' \propto \Pi \Rightarrow \Pi' \in P$.

Given any instance I' of Π', one can find an instance I of Π by applying a polynomial time transformation to I'. Since $\Pi \in P$, there is a polynomial time algorithm that solves I. Hence, using the transformation followed by the algorithm, I' can be solved in polynomial time.

Classification of Hard Problems

In practice, we do not usually use reduction to show a problem is polynomial. We are more likely to start optimistically looking for an *efficient algorithm* directly, which may be easier than seeking another problem known to be polynomial, for which we can find an appropriate transformation. But suppose we cannot find either an efficient algorithm or a suitable transformation. We begin to suspect that our problem is not 'easy' (i. e., is not a member of P). How can we establish that it is in fact 'hard'? We start by defining a larger class of problems, which includes P and also all the difficult problems we ever encounter. To describe it, consider any combinatorial decision problem. For a typical instance, there may be a very large number of possible solutions which may have to be searched. Picture a candidate solution as a set of values assigned to the variables $\mathbf{x} = (x_1, \ldots, x_n)$. The question may be 'for a given vector $\mathbf{c}$ is there a solution $\mathbf{x}$ such that $\mathbf{cx} \leq B$?' and the algorithm may search the solutions until it finds one satisfying the inequality (whereupon it stops with the answer YES) or exhausts all solutions (and stops at NO).

This may well be a big job. But suppose we are told 'the answer is YES, and here is a solution $\mathbf{x}$ that satisfies the inequality'. We feel we must at least *verify* this, but that is trivial. Intuitively, even for the hardest problems, the amount of work to check that a given candidate solution confirms the answer YES should be small, even for very large instances. We will now define our 'hard' problems as those which, though hard to solve, are easy to verify, where as usual 'easy' means taking a time which grows only polynomially with instance size. To formalize this, let:

$$V_A(n) = \begin{cases} \text{maximal time for } A \\ \text{to verify that a given solution} \\ \text{establishes the answer YES} \\ \text{for any instance of length } n. \end{cases}$$

Definition 8 An algorithm $\tilde{A}$ is *nondeterministic polynomial time* if there exists a polynomial p such that for every input of length n with answer YES, $V_{\tilde{A}}(n) \leq p(n)$

Definition 9 The collection *NP* comprises all problems for which a nondeterministic polynomial algorithm exists.

It may be noted that a problem in *NP* is solvable by searching a decision tree of polynomially bounded depth, since verifying a solution is equivalent to tracing one path through the tree. From this, it is easy to see that $P \subseteq NP$. Strangely, complexity theorists have been unable to show that $P \subset NP$; it remains possible that all the problems in *NP* could actually be solved by polynomial algorithms, so that $P = NP$. However, since so many brilliant researchers have worked on so many difficult problems in *NP* for so many years without success, this is regarded as being very unlikely. Assuming $P \neq NP$, as we shall hereafter, it can be shown that the problems in *NP* include an infinite number of equivalence classes, which can be ranked in order of increasing difficulty; where an equivalence class *C* is 'more difficult' than another class C' if, for every problem $\Pi \in C$ and every $\Pi' \in C'$, $\Pi' \propto \Pi$ but $\Pi \not\propto \Pi'$. There also exist problems that cannot be compared: neither $\Pi \propto \Pi'$ nor $\Pi' \propto \Pi$.

Fortunately, however, all problems that arise naturally have always been found to lie in one of two equivalence classes: the 'easy' problems in *P*, and the 'hard' ones, which we now define.

The class of *NP-hard problem* (*NPH*) is a collection of problems with the property that every problem in *NP* can be reduced to the problems in this class. More formally,

Definition 10

$$NPH = \{\Pi : \forall \Pi' \in NP : \Pi' \propto \Pi\}.$$

Thus each problem in *NPH* is at least as hard as any problem in *NP*. We know that some problems in *NPH* are themselves in *NP*, though some are not. Those that are include the toughest problems in *NP*, and form the class of *NP-complete problem* (*NPC*). That is,

Definition 11

$$NPC = \left\{ \Pi : \begin{array}{c} (\Pi \in NP) \\ \text{and} \\ (\forall \Pi' \in NP : \Pi' \propto \Pi) \end{array} \right\}.$$

The problems in *NPC* form an equivalence class. This is so because all problems in *NP* reduce to them, hence, since they are all in *NP*, they reduce to each other. The class *NPC* includes the most difficult problems in *NP*. As we mentioned earlier, by a surprising but happy chance, all the problems we ever encounter outside the most abstract mathematics turn out to belong to either *P* or *NPC*.

Using Reduction to Establish Complexity

When tackling a new problem Π, we naturally wonder whether it belongs to *P* or *NPC*. As we said above, to show that the problem belongs to *P*, we usually try to find a polynomial time algorithm, though we could seek to reduce it to a problem known to be polynomial. If we are unable to show that the problem is in *P*, the next step generally is to attempt to show that it lies in *NPC*; if we can do so, we are justified in not developing an efficient algorithm. To do this, clearly no direct algorithmic development is possible, and only a reduction argument will do. This is based on the following theorem, which should be clear enough to require no proof.

Theorem 12 $\forall \Pi\ \Pi' \in NP : (\Pi' \in NPC)$ and $(\Pi' \propto \Pi)$ imply $\Pi \in NPC$.

Thus, we need to find a problem $\Pi' \in NPC$ and show $\Pi' \propto \Pi$, thereby demonstrating that Π is at least as hard as any problem in *NPC*. To facilitate this, we need a list of problems known to be in *NPC*. Several hundred are listed in [2] in a dozen categories such as graph theory, mathematical programming, sequencing and scheduling, number theory, etc., and more are being added all the time. Even given an ample selection, a good deal of tenacity and ingenuity are needed to pick one with appropriate similarities to ours and to fill in the precise details of the transformation.

Of course, to build up the membership in *NPC* using Theorem 12, we need other problems that have already been show to belong to that class. To begin this process, at least one problem needs to be in *NPC*. It was S.A. Cook [1] who showed that the satisfiability problem is *NP*-complete, using direct arguments that did not involve reduction. This very important result is called *Cook's theorem*. For a proof, see [2].

As a simple illustration of reduction, we show that the traveling salesperson decision problem (TSP) is in *NPC*. To do so, we first select a closely related problem, the Hamiltonian circuit problem (HCP), which we assume has already been shown to be *NP*-complete. We then find a reduction of HCP to TSP. The problems are defined as follows.

Definition 13 (*TSP, traveling salesperson problem*).
Instance:
- a positive integer n;
- a finite set $C = \{c_1, \ldots, c_n\}$ of 'cities';
- 'distances', $d_{ij} \in \mathbf{Z}^+$. $\forall i, j : c_i, c_j \in C$;
- a bound $B \in \mathbf{Z}^+$.

Question: Does there exist a *tour* (i.e., a closed path that visits every city exactly once), of length no greater than B?

Definition 14 (HCP, Hamiltonian circuit problem)
Instance: A graph $G = (V, E)$, where V is the set of m vertices, and E the set of edges.

Question: Does G contain a *Hamiltonian circuit*, i.e., a tour that traverses all vertices exactly once?

Example 15 To show: HCP $\propto$ TSP.

In TSP, we have a complete graph and seek the shortest tour, whereas in HCP, given an arbitrary graph we require any tour. Thus, given the challenge of showing that the traveling salesperson problem (or the decision version of it) is *NP*-complete, we have found a similar problem whose membership in *NPC* is already established. We may still be unable to find a polynomial transformation from HCP, in which case another problem must be sought. A transformation of Π' to Π is a way of computing each parameter of Π in terms of the parameters of Π'. In this case, the reduction is relatively simple. The parameters of HCP are:
- m = cardinality V;
- $E = \left\{ (i, j) : \begin{array}{l} G \text{ contains an arc} \\ \text{between vertices } i, j \end{array} \right\}$.

The parameters of TSP are computed as follows:
- $n = m$;
- $d_{ij} = \begin{cases} 1 & (i, j) \in E, \\ N & \text{otherwise}; \end{cases}$
- $B = m$.

Here, N can be any number larger than 1; say, 2. Clearly, the shortest possible tour in TSP has length m, and this only occurs when arcs in E are used exclusively; that is, when a tour in HCP exists.

To complete the reduction, we need to show that the transformation can be performed in polynomial time. For that, given a pair of nodes in TSP, we need to check if an arc exists in HC, and this requires time

$$\mathcal{O}\left[\frac{m(m-1)}{2}\right] = \mathcal{O}(m^2).$$

See also

- ▶ Complexity Classes in Optimization
- ▶ Complexity of Degeneracy
- ▶ Complexity of Gradients, Jacobians, and Hessians
- ▶ Complexity Theory
- ▶ Complexity Theory: Quadratic Programming
- ▶ Fractional Combinatorial Optimization
- ▶ Information-Based Complexity and Information-Based Optimization
- ▶ Kolmogorov Complexity
- ▶ Mixed Integer Nonlinear Programming
- ▶ Parallel Computing: Complexity Classes

References

1. Cook SA (1971) The complexity of theorem proving procedures. Proc. 3rd Annual ACM Symposium on Theory of Computing. ACM, New York, pp 151–158
2. Garey MR, Johnson DS (1979) Computers and intractability. Freeman, New York
3. Hardy GH, Wright EM (1979) An introduction to the theory of numbers. Clarendon Press, Oxford
4. Hopcroft JE, Ullman JD (1979) Introduction to automata theory, languages, and computation. Addison-Wesley, Reading, MA
5. Karp RM (1972) Reducibility among combinatorial problems. In: Miller RE, Thatcher JW (eds) Complexity of Computer Computations. Plenum, New York, pp 85–103
6. Papadimitriou CH (1994) Computational complexity. Addison-Wesley, Reading, MA
7. Pinedo M (1995) Scheduling: Theory, algorithms and systems. Prentice-Hall, Englewood Cliffs, NJ

Concave Programming

HAROLD P. BENSON
Department Decision and Information Sci., University Florida, Gainesville, USA

MSC2000: 90C25

Article Outline

Keywords
See also
References

Keywords

Concave programming; Multi-extremal global optimization

Concave programming constitutes one of the most fundamental and intensely-studied problem classes in deterministic nonconvex optimization. There are at least three reasons for this. First, many of the mathematical properties of concave programming are intriguingly attractive. Some are even identical to properties of linear programming. Second, concave programming has a remarkably broad range of direct and indirect applications. Third, the algorithmic ideas used in concave programming have played and continue to play an active and often fundamental role in the development of solution procedures for other types of nonconvex programming problems.

The *concave programming*, or *concave minimization*, problem (CMP) can be written

$$\begin{cases} \text{globmin} & f(x), \\ \text{s.t.} & x \in D, \end{cases}$$

where D is a nonempty, closed convex set in $\mathbf{R}^n$ and f is a real-valued, concave function defined on some open convex set A in $\mathbf{R}^n$ that contains D. The goal in CMP is to find the global minimum value that f achieves over D, and, if this value is not equal to $-\infty$, to find, if it exists, at least one point in D that achieves this value. In many applications, D is compact and A equals all of $\mathbf{R}^n$. CMP invariably contains many points in D that are local, but not global, minimizers of f over D. For this reason, CMP is an example of a *(multi-extremal) global optimization* problem [7].

The application of standard algorithms designed for solving constrained convex programming problems generally will fail to solve CMP. Even instances of CMP with relatively simple components can apparently present very significant solution challenges. For example, B. Kalantari [8] has shown that in problems involving the minimization of concave quadratic functions over rectangles in $\mathbf{R}^n$, an exponential number of extreme point local minima can exist. Additionally, P.M. Pardalos and G. Schnitger [13] have shown that minimizing a concave quadratic function over a hypercube is an *NP*-hard problem.

Although CMP is more difficult to solve than a convex programming problem, it possesses some highly interesting, special mathematical properties. A number of these properties have been exploited by researchers to create successful algorithms for solving the problem.

For instance, if D contains at least one extreme point, and CMP has at least one global optimal solution, then there must exist a global optimal solution which is an extreme point of D [14]. This is perhaps the most important and striking property of concave minimization problems. As a result of this property, just as in linear programming, if CMP has a global optimal solution, then one can confine the search for such a solution to the set of extreme points of D, provided that this set is nonempty. This property holds, as in linear programming, even when D is unbounded. A number of algorithms for CMP are based upon this property.

Another highly important property for CMP is that if D is a compact set, then CMP must have a global optimal solution which is an extreme point of D. This is perhaps the most widely-known theoretical result in concave minimization [1]. Like the property stated in the previous paragraph, it forms the basis for a number of important concave minimization algorithms.

For cases where D is a polyhedron, possibly unbounded, that contains at least one extreme point, it has been shown that either CMP has a global optimal solution which is an extreme point of D, or CMP must be unbounded and f must be unbounded from below over some extreme direction of D. Notice that the same property, remarkably, holds in the case of linear programming. This property is used by a large number of the algorithms designed to solve CMP when D is a nonempty polyhedron.

CMP displays a remarkable diversity of applications. Each application is either direct or indirect. By direct, we mean that the original model formulation takes the form of CMP immediately or, if not, with only relatively simple algebraic manipulations. The indirect applications involve problems whose direct formulations do not take the form of CMP, but existing theory can be used to reformulate these problems in the form of CMP.

Some of the oldest and most diverse direct applications of CMP belong to a class of problems called *fixed charge* problems. In these problems, the objective function f is *separable*, i. e., it is of the form

$$f(x) = \sum_{i=1}^{n} f_i(x_i) .$$

For each $i = 1, \ldots, n$, in these problems f_i is a concave function on $\{x_i \in \mathbf{R}: x_i \geq 0\}$ of the form

$$f_i(x_i) = \begin{cases} 0 & \text{if } x_i = 0, \\ c_i + w_i(x_i) & \text{if } x_i > 0, \end{cases}$$

where $c_i > 0$ is the fixed *setup cost* of undertaking activity i at a positive level, and $w_i(x_i)$ is a continuous concave function on $\{x_i \in \mathbf{R} : x_i > 0\}$ that represents the *variable cost* of undertaking the activity at level x_i.

When the functions $w_i(x_i)$, $i = 1, \ldots, n$, are linear, the classical *linear fixed charge* problem is obtained. Some of the oldest applications of concave minimization involve solving problems of this type. Among these are applications to transportation planning, site selection, production lot sizing and network design. For cases where at least one of the functions $w_i(x_i), i = 1, \ldots, n$, is piecewise linear, several types of applications have been reported. Included among these are problems involving price breaks, such as bid evaluation problems, certain inventory planning problems and various plant location problems. When $w_i(x_i)$ is a general concave function for some $i = 1, \ldots, n$, applications involving *economies of scale*, for instance, can be solved.

More recently, CMP has been directly applied to a class of problems called *multiplicative programming* problems. These are problems of the form CMP where

$$f(x) = \prod_{j=1}^{p} f_j(x) \tag{1}$$

for some set of $p \geq 2$ functions $f_j, j = 1, \ldots, p$, that are each nonnegative over D. Notice that if $f_k(\overline{x}) = 0$ for some $k \in \{1. \ldots, p\}$ and some $\overline{x} \in D$, then it is easy to see that the global minimum value for CMP is 0, and $\overline{x}$ is a global optimal solution. Therefore, it is generally assumed in multiplicative programming problems that each function f_j is positive over D. Let us make this assumption henceforth.

The objective function (1) of a multiplicative program is generally not a concave function. But, when each function $f_j, j = 1, \ldots, p$, is a concave function on $\mathbf{R}^n$, some simple transformations of (1) yield concave functions over D. For instance, if, for each $x \in D$, we define w_1 and w_2 by

$$w_1(x) = \ln f(x) = \sum_{j=1}^{p} \ln f_j(x),$$

$$w_2(x) = [f(x)]^{\frac{1}{p}},$$

respectively, then, whenever each $f_j, j = 1, \ldots, p$, is a concave function on $\mathbf{R}^n$, both w_1 and w_2 are concave functions on D [3,10]. Thus, by using w_1 or w_2, multiplicative programming problems in which $f_j, j = 1, \ldots, p$, are concave functions can be easily transformed to concave minimization problems.

Various applications of multiplicative programming problems with concave or linear functions $f_j, j = 1, \ldots, p$, in (1) have arisen, especially since the 1960s. For example, the linear case has been applied to the problem of optimizing value functions for multiple objective programming problems subject to linear or nonlinear constraints. For $p = 2$, the linear case has been used to help solve the modular design problem, to design integrated circuit chips and to select bond portfolios. The concave case has been used to analyze and solve a number of problems in microeconomics.

Subject to occasional restrictions, large classes of integer programming problems can be converted by various means into the form of CMP and solved as concave minimization problems. As a result, these integer programming problems, indirectly, are applications of concave minimization. The transformation processes used to accomplish the conversion, however, can be rather involved. They may also increase the size of the original problem [4], and they may call for choosing values for parameters that are difficult to determine [5,9].

In particular, by using a certain general transformation process, any feasible linear integer or quadratic integer programming problem over a polyhedron with nonnegative, bounded variables can be converted into the form of CMP and solved as a concave minimization problem. By using more customized conversion processes, linear zero-one programs, quadratic assignment problems, and other special integer programming problems can also be transformed to the form of CMP and solved as concave minimization problems. The specialized transformations generally take advantage of some aspect of the original integer programming problem that the general processes ignore.

There are many other indirect applications of CMP, including *d.c. optimization*, *indefinite quadratic programming*, and *bilinear programming*, for instance. For further details and additional direct and indirect applications, see [1,2,6,7,11,12].

To solve CMP, a large number of algorithms have been developed. Many of these rely on one or more of the following four approaches.

In the *cutting plane approach*, a local minimum for f over D is found. Subsequently, a hyperplane is constructed and used to cut off all points of D whose objective function values are not less than that of the local minimum. This yields a new closed convex set $D^1 \subset D$. This process is then repeated with D^1 in the role of D. By iterating this process, the portion of D remaining to be explored is progressively reduced. Termination occurs when it can be shown that $f(y) \geq f(x^k)$ for all $y \in D^k$, where D^k is the portion of D remaining at iteration k, and x^k is the local minimum found through iteration k with the smallest objective function value.

In a typical *outer approximation* approach, D is assumed to be compact. To initiate the approach, a simple bounded polyhedron P^1 containing D whose vertices can be enumerated is constructed. A vertex v^1 of P^1 of minimum objective function value among all of the vertices of P^1 is found. This gives a lower bound $f(v^1)$ for the optimal value of CMP. If $v^1 \in D$, v^1 is a global optimal solution for CMP and termination occurs. Otherwise, a new bounded polyhedron $P^2 \subset P^1$ is constructed that contains D, and its vertices are enumerated. With P^2 in the role of P^1, the process is repeated. By repeating this process, a sequence of telescoping bounded polyhedra containing D is obtained. Termination occurs in the first iteration k where the vertex v^k found that minimizes f over all of the vertices of P^k lies in D.

In *inner approximation* (*polyhedral annexation*) approaches for CMP, D is assumed to be a bounded polyhedron. Typically, at each major iteration of an inner approximation algorithm, a local minimum extreme point solution $\bar{x}$ for CMP is available. A sequence of expanding inner approximating compact polyhedra for $(D \cap G)$ is constructed via a series of subiterations, where $G = \{x \in \mathbb{R}^n : f(x) \geq f(\bar{x})\}$. During this process, either an improved local minimum extreme point $\bar{\bar{x}}$ is found, or, after k subiterations, the algorithm detects that $D \subseteq P^k$, where P^k is the current inner approximation of $(D \cap G)$. In the former case, $\bar{\bar{x}}$ replaces $\bar{x}$ and a new major iteration begins. In the latter case, since $P^k \subseteq (D \cap G)$, it follows that $D \subseteq G$, and the algorithm therefore terminates with the global optimal solution $\bar{x}$.

In the *branch and bound* approaches for CMP, D is repeatedly subdivided into finer and finer partitions. A lower bound for f over each partition element is calculated. The lowest of these lower bounds at any step k of the process gives a global lower bound LB_k for f over D. At any stage, typically some feasible solutions for CMP have been detected. A feasible solution $\bar{y}$ with the smallest f value among all feasible solutions detected through any point in the algorithm is always available. This solution is called the *incumbent solution*. When, at some step k, the inequality $LB_k \geq f(\bar{y})$ holds for the first time, the algorithm stops and returns the global optimal solution $\bar{y}$.

Details concerning these and other solution approaches can be found in [1,2,6,7].

See also

▶ Bilevel Linear Programming: Complexity, Equivalence to Minmax, Concave Programs
▶ Minimum Concave Transportation Problems

References

1. Benson HP (1995) Concave minimization: Theory, applications and algorithms. In: Horst R, Pardalos PM (eds) Handbook global optimization. Kluwer, Dordrecht, pp 43–148
2. Benson HP (1996) Deterministic algorithms for constrained concave minimization: A unified critical survey. Naval Res Logist 73:765–795
3. Benson HP, Boger GM (1997) Multiplicative programming problems: Analysis and efficient point search heuristic. J Optim Th Appl 94:487–510
4. Garfinkel RS, Nemhauser GL (1972) Integer programming. Wiley, New York
5. Giannessi F, Niccolucci F (1976) Connections between nonlinear and integer programming problems. Symp Mat Inst Naz di Alta Mat 19:161–176
6. Horst R, Pardalos PM, Thoai NV (1995) Introduction to global optimization. Kluwer, Dordrecht
7. Horst R, Tuy H (1993) Global optimization: Deterministic approaches, 2nd revised edn. Springer, Berlin
8. Kalantari B (1986) Quadratic functions with exponential number of local maxima. Oper Res Lett 5:47–49
9. Kalantari B, Rosen JB (1982) Penalty for zero-one integer equivalent problems. Math Program 24:229–232

10. Konno H, Kuno T (1995) Multiplicative programming problems. In: Horst R, Pardalos PM (eds) Handbook Global Optim. Kluwer, Dordrecht, pp 369–405
11. Pardalos PM (1994) On the passage from local to global in optimization. In: Birge JR, Murty KG (eds) Mathematical Programming: State of the Art. Braun-Brumfield, Ann Arbor, MI, pp 220–247
12. Pardalos PM, Rosen JB (1987) Constrained global optimization: Algorithms and applications. Springer, Berlin
13. Pardalos PM, Schnitger G (1988) Checking local optimality in constrained quadratic programming is NP-hard. Oper Res Lett 7:33–35
14. Rockafellar RT (1970) Convex analysis. Princeton Univ. Press, Princeton

Conjugate-Gradient Methods

J. L. Nazareth[1,2]

[1] Department Pure and Applied Math., Washington State University, Pullman, USA

[2] Department Applied Math., University Washington, Seattle, USA

MSC2000: 90C30

Article Outline

Keywords
Introduction
  Notation and Preliminaries
Linear CG Algorithms
Nonlinear CG Algorithms
Nonlinear CG-Related Algorithms
  Classical Alternatives to CG
  Nonlinear CG Variants
  Variable-Storage/Limited-Memory Algorithms
  Affine-Reduced-Hessian Algorithms
Conclusion
See also
References

Keywords

Unconstrained minimization; Conjugate gradients; Linear CG method; Nonlinear CG method; Nonlinear CG-related algorithms; Variable-storage; Limited-memory; Affine-reduced-Hessian; Three-term-recurrence

Introduction

Conjugate-gradient methods (CG methods) are used to solve large-dimensional problems that arise in computational linear algebra and computational nonlinear optimization. These two subjects share a broad common frontier, and one of the most easily traversed crossing points is via the following simple observation: the problem of solving a system of linear equations $\mathbf{A}\mathbf{x} = \mathbf{b}$ for the unknown vector $\mathbf{x} \in \mathbf{R}^n$, where $\mathbf{A}$ is a positive definite, symmetric matrix and $\mathbf{b}$ is a given vector, is mathematically equivalent to finding the minimizing point of the strictly convex quadratic function

$$q(\mathbf{x}) = -\mathbf{b}^\top \mathbf{x} + \frac{1}{2}\mathbf{x}^\top \mathbf{A}\mathbf{x}.$$

The *linear CG method* for solving the system of linear equations is able to capitalize on this equivalent optimization formulation. It was developed in the pioneering 1952 paper of M.R. Hestenes and E.L. Stiefel [11] who, in turn, cite antecedents in the contributions of several other authors (see [9]). The method fell out of favor with numerical analysts during the 1960s because it did not compete with direct methods, in particular, Gaussian elimination, but it continued to be widely used in real-world applications by specialists in other areas. Interest in CG as an iterative method, downplaying its finite-termination properties, revived in the 1970s when the solution of *large scale linear systems* was coming to the forefront of academic research.

The *nonlinear CG method* extends the linear CG approach to the problem of minimizing a smooth, nonlinear function $f(\mathbf{x})$, $\mathbf{x} \in \mathbf{R}^n$, where n can be large. It was developed in another landmark article published in 1964 by R. Fletcher and C. Reeves [8]. Optimization techniques for this class of problems, which are inherently iterative in nature, form a direction of descent at an approximation to the solution (the current iterate), and search along this direction to obtain a new iterate with an improved function value. The *Fletcher–Reeves algorithm* combined a search direction derived from the *Hestenes–Stiefel approach* with an efficient line search procedure along this direction vector adapted from the 1959 variable-metric breakthrough algorithm of W.C. Davidon [6]. The resulting CG algorithm was a marked enhancement of the classical steepest descent method of A.-L. Cauchy.

Like steepest descent, the nonlinear CG method is storage-efficient and requires only a few n-vectors of computer storage beyond that needed to specify the problem itself. During the three decades after its discovery, a large number of storage-efficient *nonlinear CG-related algorithms* were proposed. In particular, a structural connection between conjugate-gradient and variable-metric techniques for defining search direction vectors provided the springboard for effective new families of variable-storage/limited-memory algorithms and affine-reduced-Hessian algorithms that occupy a middle ground.

These three classes of algorithms, namely, linear CG, nonlinear CG and nonlinear CG-related, will be discussed in the respective Sections below.

Notation and Preliminaries

Lowercase boldface letters, e.g., $\mathbf{x}$, denote vectors, and uppercase boldface letters, e.g., $\mathbf{A}$, denote symmetric, positive definite matrices. The *residual* at $\mathbf{x}$ of the linear system $\mathbf{Ax} = \mathbf{b}$ is $\mathbf{r} = \mathbf{Ax} - \mathbf{b}$. It equals the gradient vector $\mathbf{g} = -\mathbf{b} + \mathbf{Ax}$ of the strictly convex, quadratic form

$$q(\mathbf{x}) = -\mathbf{b}^\top \mathbf{x} + \frac{1}{2}\mathbf{x}^\top \mathbf{Ax}$$

at $\mathbf{x}$. The gradient vector of q vanishes only at the unique solution $\mathbf{A}^{-1}\mathbf{b}$ of the linear system.

Linear CG Algorithms

A basic CG algorithm for solving the system of linear equations $\mathbf{Ax} = \mathbf{b}$, where $\mathbf{A}$ is a positive definite, symmetric matrix, is as follows:

	(Initialization)
0	$\mathbf{x}_1$ = arbitrary;
	$\mathbf{r}_1$ = residual of linear system at x_1;
	$\mathbf{d}_1 = -\mathbf{r}_1$;
	(Iteration i)
1	$\mathbf{x}_{i+1}$ = unique minimazing point of q on halfline through $\mathbf{x}_i$ along direction $\mathbf{d}_i$;
2	$\mathbf{r}_{i+1}$ = residual of linear system at $\mathbf{x}_{i+1}$;
3	$\beta_i = \|\mathbf{r}_{i+1}\|^2/\|\mathbf{r}_i\|^2$;
4	$\mathbf{d}_{i+1} = -\mathbf{r}_{i+1} + \beta_i \mathbf{d}_i$.

In the computational linear algebra setting, the matrix $\mathbf{A}$ is provided exogenously. The residual $\mathbf{r}_{i+1}$, at step 2, is computed as $\mathbf{Ax}_{i+1} - \mathbf{b}$ or else obtained by updating $\mathbf{r}_i$.

The direction $\mathbf{d}_i$ is always a descent direction for q at $\mathbf{x}_i$. At step 1, the minimizing point is computed as follows:

$$\alpha_i = -\frac{\mathbf{r}_i^\top \mathbf{d}_i}{\mathbf{d}_i^\top \mathbf{A}\mathbf{d}_i}; \quad \mathbf{x}_{i+1} = \mathbf{x}_i + \alpha_i \mathbf{d}_i.$$

There are numerous variants on this basic algorithm that seek to enhance convergence through problem preconditioning (transformation of variables), to improve algorithm stability and to solve related computational linear algebra problems. A contextual overview and further references can be found in [2].

Here our focus is optimization. In this setting, the residual at $\mathbf{x}_{i+1}$ is given its alternative interpretation and representation as the gradient vector $\mathbf{g}_{i+1}$ of q at $\mathbf{x}_{i+1}$, and this gradient is assumed to be provided exogenously. The minimizing point at step 1 is computed, alternatively, as follows:

$$\alpha_i = -\frac{\mathbf{g}_i^\top (\bar{\mathbf{x}}_i - \mathbf{x}_i)}{\mathbf{d}_i^\top (\bar{\mathbf{g}}_i - \mathbf{g}_i)},$$

where $\bar{\mathbf{x}}_i$ is any point on the ray through $\mathbf{x}_i$ in the direction $\mathbf{d}_i$ and $\bar{\mathbf{g}}_i$ is its corresponding gradient vector; $\mathbf{x}_{i+1} = \mathbf{x}_i + \alpha_i \mathbf{d}_i$. The expression for α_i is derived from the previous linear systems version and the relation $\mathbf{A}(\bar{\mathbf{x}}_i - \mathbf{x}_i) = \bar{\mathbf{g}}_i - \mathbf{g}_i$.

We will call the resulting optimization algorithm the *CG-standard for minimizing q*. It will provide an important guideline for defining CG algorithms in the subsequent discussion.

Nonlinear CG Algorithms

A nonlinear CG algorithm is used to find a minimizing point of the nonlinear function $f(\mathbf{x})$, $\mathbf{x} \in \mathbf{R}^n$, when n is large and/or computer storage is at a premium. A basic algorithm can be stated as follows:

	(Initialization)
0	$\mathbf{x}_1$ = arbitrary;
	$\mathbf{g}_1$ = gradient of f at $\mathbf{x}_1$;
	$\mathbf{d}_1 = -\mathbf{g}_1$;
	(Iteration i)
1	$\mathbf{x}_{i+1}$ = an improved iterate on halfline through x_i along direction $\mathbf{d}_i$;
2	$\mathbf{g}_{i+1}$ = gradient of f at $\mathbf{x}_{i+1}$;
3	$\beta_i = \|\mathbf{g}_{i+1}\|^2/\|\mathbf{g}_i\|^2$;
4	$\mathbf{d}_{i+1} = -\mathbf{g}_{i+1} + \beta_i \mathbf{d}_i$.

Note that this algorithm is closely patterned after the *CG-standard*. The improved iterate at step 1 is obtained by a *line search* procedure, which is normally based on quadratic or cubic polynomial fitting (suitably safeguarded). When $f = q$, such a line search procedure can immediately locate the minimizing point along the line of search, once it has gathered the requisite information to make an exact fit. In other words, when applied to the minimization of q, the foregoing nonlinear CG algorithm is able to replicate the CG-standard precisely. This property characterizes a nonlinear CG algorithm.

Considerable research has gone into alternative expressions for the quantity β_i. The four leading contenders are the *Fletcher–Reeves* (FR), *Hestenes–Stiefel* (HS), *Polyak–Polak–Ribière* (PPR) and *Dai–Yuan* (DY) choices (see [5,8,11,20,22]). These define β_i as follows:

$$\text{FR}: \quad \beta_i = \frac{\mathbf{g}_{i+1}^\top \mathbf{g}_{i+1}}{\mathbf{g}_i^\top \mathbf{g}_i};$$

$$\text{HS}: \quad \beta_i = \frac{\mathbf{g}_{i+1}^\top \mathbf{y}_i}{\mathbf{d}_i^\top \mathbf{y}_i}; \quad (1)$$

$$\text{PPR}: \quad \beta_i = \frac{\mathbf{g}_{i+1}^\top \mathbf{y}_i}{\mathbf{g}_i^\top \mathbf{g}_i};$$

$$\text{DY}: \quad \beta_i = \frac{\mathbf{g}_{i+1}^\top \mathbf{g}_{i+1}}{\mathbf{d}_i^\top \mathbf{y}_i}, \quad (2)$$

where $\mathbf{y}_i = \mathbf{g}_{i+1} - \mathbf{g}_i$ is the gradient change that corresponds to the step $\mathbf{s}_i = \mathbf{x}_{i+1} - \mathbf{x}_i$.

When line searches are exact and the function is quadratic, the following relations hold:

$$\mathbf{g}_{i+1}^\top \mathbf{g}_{i+1} = \mathbf{g}_{i+1}^\top \mathbf{y}_i, \quad \mathbf{g}_i^\top \mathbf{g}_i = \mathbf{d}_i^\top \mathbf{y}_i. \quad (3)$$

Thus, the values of the scalar β_i are identical for all four choices, and each of the associated algorithms becomes the CG-standard. In general, however, they are applied to nonquadratics and use inexact line searches, resulting in four distinct, nonlinear CG algorithms that can exhibit behavior very different from one another.

The following generalization yields a *two-parameter family*:

$$\beta_i = \frac{\lambda_i (\mathbf{g}_{i+1}^\top \mathbf{g}_{i+1}) + (1-\lambda_i)(\mathbf{g}_{i+1}^\top \mathbf{y}_i)}{\mu_i (\mathbf{g}_i^\top \mathbf{g}_i) + (1-\mu_i)(\mathbf{d}_i^\top \mathbf{y}_i)}, \quad (4)$$

where $\lambda_i \in [0, 1]$ and $\mu_i \in [0, 1]$. For any choice of λ_i and μ_i in these ranges, the associated algorithm reduces to the CG-standard when f is quadratic and line searches are exact, which follows from (3). If the scalars λ_i and μ_i take only their extreme values, 0 or 1, then one obtains four possible combinations corresponding to (1)–(2). The above two-parameter family of nonlinear CG algorithms, which subsumes FR, HS, PPR and DY, and its subfamilies (defined, for example, by taking $\lambda_i \equiv 1$) are currently a topic of active research.

When the line search is sufficiently accurate, a nonlinear CG algorithm always produces a direction of descent at step 4. Suitable inexact line search termination conditions, in conjunction with different choices of β_i, have been extensively studied, both theoretically and computationally. A good overview of the theory and convergence analysis of nonlinear CG algorithms can be found in [19]. The nonlinear CG algorithm based on the PPR choice for β_i [20,22] and a suitable restarting strategy [23] has emerged as the most efficient in practice. However, it is well known that no single nonlinear CG algorithm works well all the time. There is enormous variability in performance on different problems or even within different regions of the same problem.

Nonlinear CG-Related Algorithms

We informally characterize a *nonlinear CG-related algorithm* as follows:

- its computer storage requirements are 'similar' to those of an implemented nonlinear CG algorithm, for example, [23];
- its path traverses the iterates of the CG-standard when the function is a strictly convex quadratic, line searches are exact, and the same initialization is used.

In other words, it may use a few more n-vectors of computer storage than, say, the PPR nonlinear CG algorithm, and it is permitted to generate additional intermediate iterates and form search vectors in novel ways. A nonlinear CG-related algorithm does not have to imitate the 'structure' of the basic nonlinear CG algorithm of the previous Section. But the above requirement that its path must cover the iterates of the CG-standard of the first Section implies the following: if the candidate algorithm does not exhibit finite termination when applied to a quadratic q then it is not a CG-related algorithm.

Let us now briefly categorize the main lines of development.

Classical Alternatives to CG

These seek to enhance or accelerate the steepest descent algorithm of Cauchy more directly, without explicitly introducing notions of conjugacy. The year of publication of [8] was indeed a banner year for such developments. Two particularly noteworthy contributions, which coincidentally also appeared in 1964, were the *parallel-tangents* or *PARTAN* algorithm of B.V. Shah, R.J. Buehler and O. Kempthorne [24] and the *heavy ball algorithm* of B.T. Polyak [21]. For a modern description of the former, and its subsequently discovered CG-related properties, see [14]. The basic iteration of the latter algorithm is as follows:

$$\mathbf{x}_{i+1} = \mathbf{x}_i - \alpha \mathbf{g}_i + \beta(\mathbf{x}_i - \mathbf{x}_{i-1}),$$

where α is a constant positive stepsize and β is a scalar with $0 < \beta < 1$. Although, strictly speaking, it is not CG-related in this form, the algorithm has CG-like rate of convergence properties on a quadratic under optimal choices of the algorithm parameters α and β (see [3]). The algorithms in [24] and [21] both used very simple steplength techniques to move from one iterate to the next, in contrast to the nonlinear CG implementation of [8]. The introduction of the line-search technique of Davidon [6] into either of these 1964 algorithms, as in [8], could have propelled them much more into the limelight at the time. More recently, the important contribution of Yu.E. Nesterov [17] on global rate of convergence is based on an algorithm akin to PARTAN [24]. However, in general, classical alternatives to CG remain on the sidelines.

Nonlinear CG Variants

These are premised on retaining the algorithmic structure, and the conjugacy properties on quadratics, of the basic nonlinear CG algorithm of the above Section when the initial direction is not along the negative gradient and/or line searches are inexact. For instance, the *three-term-recurrence algorithm* (TTR) is able to simultaneously relax both CG-standard requirements. The overall iteration follows the basic algorithm of the previous Section, but with the computation of the search direction at steps 3 and 4 replaced as follows:

$$\beta_i = \frac{\mathbf{y}_i^\top \mathbf{y}_i}{\mathbf{y}_i^\top \mathbf{d}_i}, \quad \gamma_i = \frac{\mathbf{y}_{i-1}^\top \mathbf{y}_i}{\mathbf{y}_{i-1}^\top \mathbf{d}_{i-1}},$$

and

$$\mathbf{d}_{i+1} = -\mathbf{y}_i + \beta_i \mathbf{d}_i + \gamma_i \mathbf{d}_{i-1}.$$

Conjugacy of search directions is retained when the initial search direction is chosen to be an arbitrary direction of descent. If this initial direction is along the negative gradient and line searches are exact, then the TTR generates the same search directions and iterates as the CG-standard on a positive definite quadratic. A drawback of the TTR algorithm is that it does not guarantee a descent direction on more general functions even if line searches are exact. But, in practice, the direction almost always satisfies the condition $\mathbf{g}_{i+1}^\top \mathbf{d}_{i+1} < 0$.

For references to other nonlinear CG variants, see [10] and the survey articles in [1]. Despite theoretical advantages on quadratics, algorithms in this category, in practice, have not proved to be significantly superior to the PPR nonlinear CG algorithm of the above Section.

Variable-Storage/Limited-Memory Algorithms

These are premised on a key structural relationship between the nonlinear CG algorithm and the BFGS variable-metric algorithm, and its properties on quadratics, see [4,12,15]. The most effective CG-related algorithm, to date, is the L-BFGS algorithm of J. Nocedal [18]. This is described in more detail in ▶ Unconstrained nonlinear optimization: Newton–Cauchy framework and is not repeated here. The algorithm has the property that it produces a descent direction under weak termination conditions on the line search. It has proved to be an efficient and versatile algorithm (it can exploit additional computer storage when available) that generally outperforms the PPR algorithm in practice.

For an overview of other variable-storage algorithms that draw on the *BFGS-CG relationship*, see the survey articles in [1].

Affine-Reduced-Hessian Algorithms

These make estimates of curvature, i.e., approximations to the Hessian or its inverse, in an affine subspace usually of low dimension and defined by the most recent gradient and one or more prior steps and/or gradients. For algorithms of this type and their CG-related properties, see [7,13,16] and references cited therein.

The foregoing principle, on which such algorithms are premised, has the conceptual advantage that it provides a *true continuum* between the nonlinear CG and full-storage variable-metric (and Newton) algorithms. But, practical affine-reduced-Hessian implementations are not yet widespread.

Conclusion

CG algorithms are among the simplest and most elegant algorithms of computational nonlinear optimization. They can be surprisingly effective in practice, and thus will always have an honored place in the repertoire. Nevertheless, the subject still lacks a comprehensive underlying theory, and many interesting algorithmic issues remain to be explored.

Some references cited in the present discussion are listed below, and other key references can be traced, in turn, through their bibliographies.

See also

- ▶ Broyden Family of Methods and the BFGS Update
- ▶ Large Scale Trust Region Problems
- ▶ Large Scale Unconstrained Optimization
- ▶ Local Attractors for Gradient-Related Descent Iterations
- ▶ Nonlinear Least Squares: Newton-Type Methods
- ▶ Nonlinear Least Squares: Trust Region Methods
- ▶ Unconstrained Nonlinear Optimization: Newton–Cauchy Framework
- ▶ Unconstrained Optimization in Neural Network Training

References

1. Adams L, Nazareth JL (1996) Linear and nonlinear conjugate gradient-related methods. SIAM, Philadelphia
2. Barrett R, Berry M, Chan TF, Demmel J, Donato J, Dongarra J, Eijkhout V, Pozo R, Romine C, Vorst H van der (1993) Templates for the solution of linear systems. SIAM, Philadelphia
3. Bertsekas DP (1999) Nonlinear programming, 2nd edn. Athena Sci, Belmont, MA
4. Buckley A (1978) Extending the relationship between the conjugate gradient and BFGS algorithms. Math Program 15:343–348
5. Dai YH (1997) Analyses of nonlinear conjugate gradient methods. PhD Diss Inst Computational Math Sci/Engin Computing Chinese Acad Sci, Beijing, China
6. Davidon WC (1991) Variable metric method for minimization. SIAM J Optim 1:1–17, (Original (with a different preface): Argonne Nat Lab Report ANL-5990 (Rev, Argonne, Illinois)
7. Fenelon MC (1981) Preconditioned conjugate-gradient-type methods for large-scale unconstrained optimization. PhD Diss Stanford Univ, Stanford, CA
8. Fletcher R, Reeves C (1964) Function minimization by conjugate gradients. Computer J 7:149–154
9. Golub GH, O'Leary DP (1989) Some history of the conjugate gradient and Lanczos algorithms 1948–1976. SIAM Rev 31:50–102
10. Hestenes MR (1980) Conjugate direction methods in optimization. Appl Math, vol 12. Springer, Berlin
11. Hestenes MR, Stiefel EL (1952) Methods of conjugate gradients for solving linear systems. J Res Nat Bureau Standards (B) 49:409–436
12. Kolda TG, O'Leary DP, Nazareth JL (1998) BFGS with update skipping and varying memory. SIAM J Optim 8:1060–1083
13. Leonard MW (1995) Reduced Hessian quasi-Newton methods for optimization. PhD Diss Univ Calif, San Diego, CA
14. Luenberger DG (1984) Linear and nonlinear programming, 2nd edn. Addison-Wesley, Reading, MA
15. Nazareth JL (1979) A relationship between the BFGS and conjugate gradient algorithms and its implications for new algorithms. SIAM J Numer Anal 16:794–800
16. Nazareth JL (1986) The method of successive affine reduction for nonlinear minimization. Math Program 35:97–109
17. Nesterov YE (1983) A method of solving a convex programming problem with convergence rate $O(1/k2)$. Soviet Math Dokl 27:372–376
18. Nocedal J (1980) Updating quasi-Newton matrices with limited storage. Math Comput 35:773–782
19. Nocedal J (1992) Theory of algorithms for unconstrained optimization. Acta Numer 1:199–242
20. Polak E, Ribière G (1969) Note sur la convergence de methode de directions conjuguees. Revue Franc Inform Rech Oper 16:35–43
21. Polyak BT (1964) Some methods of speeding up the convergence of iteration methods. USSR Comput Math Math Phys 4:1–17
22. Polyak BT (1969) The conjugate gradient method in extremal problems. USSR Comput Math Math Phys 9:94–112
23. Powell MJD (1977) Restart procedures for the conjugate gradient method. Math Program 12:241–254
24. Shah BV, Buehler RJ, Kempthorne O (1964) Some algorithms for minimizing a function of several variables. J SIAM 12:74–91

Contact Map Overlap Maximization Problem, CMO

WEI XIE[1], NIKOLAOS V. SAHINIDIS[2]
[1] American Airlines Operations Research and Decision Support Group, Fort Worth, USA
[2] Department of Chemical Engineering, Carnegie Mellon University, Pittsburgh, USA

Article Outline

Introduction
Definition
Methods
    Exact algorithms
    Approximate Algorithms
Conclusions
References

Introduction

Contact map overlap maximization is a problem that arises in computational biology as an important approach to compare structural similarity of proteins. Contact map overlap was proposed in [6] as a measure for protein structural similarity, and is employed in the Critical Assessment of Techniques for Protein Structure Prediction (CASP).

Proteins consist of amino acid residues and assume specific 3-dimensional structures. Proteins of similar structures often have similar function and properties. Therefore, structure alignment provides critical insights into the relation of existing proteins, and has important applications in designing knowledge-based potential functions that are useful for protein folding prediction.

Definition

Given two proteins A and B with m and n residues, respectively, we denote the residues of A with indices i, i', and i'', and the residues of B with indices j, j' and j''. If two residues in the same protein are close in space, we say that they are *in contact*. A list of residue pairs is known as the *contact map* of a given protein. In this article, the contact map for protein A (resp. B) is denoted as E^A (resp. E^B) so that $E^A_{ii'} = 1$ (resp. $E^B_{jj'} = 1$) if residues i and i' in protein A (resp. j and j' in protein B) are in contact, and $E^A_{ii'} = 0$ (resp. $E^B_{jj'} = 0$) otherwise.

From a graph-theoretic perspective, a contact map is a node-node incident graph where nodes represent residues and edges represent contacts. The *contact map overlap* maximization problem aims at identifying an ordered residue correspondence between two contact maps so as to result in a maximum common subgraph. To solve this problem, a correspondence (alignment) must be established between the node sets (residues) of the contact maps so that the number of common contacts (edges) can be maximized. If residue i in protein A aligns with residue j in protein B, they form a *pair* (i, j). If pairs (i, j) and (i', j') result in common contacts, i.e., $E^A_{ii'} = 1$ and $E^B_{jj'} = 1$, then they form an *overlap*. If two pairs (i, j) and (i', j') form an overlap, we set $h(i, j, i', j') = 1$. Otherwise, $h(i, j, i', j')$ is set to zero.

An important requirement for structure alignment is that the relative orders of residues in the original sequences agree – a property that is known as the *non-crossing* property in the CMO literature. For two pairs (i, j) and (i', j') to be non-crossing, either $i < i'$ and $j < j'$ or $i > i'$ and $j > j'$ must hold. In this paper, non-crossing pairs are also referred to as *parallel* pairs.

For convenience of the presentation, $[i, i']$ (resp. $[j, j']$) will denote the set of residues $\{i'' : i \leq i'' \leq i'\}$ (resp. $\{j'' : j \leq j'' \leq j'\}$). The interval product $[i, i'] \times [j, j']$ therefore denotes the set of pairs $\{(i'', j'') : i \leq i'' \leq i', j \leq j'' \leq j'\}$. For any given set of residue pairs S, we will use $\mathbb{Q}(S)$ to denote the set of subsets of S that contain only parallel pairs. The objective of CMO is to identify a set of parallel pairs that maximize the resultant number of overlaps. For proteins A and B, the problem can be stated as follows:

$$\max_{Q \in \mathbb{Q}([1,m] \times [1,n])} \frac{1}{2} \sum_{(i,j) \in Q} \sum_{(i',j') \in Q} h(i, j, i', j').$$

An optimal alignment of the contact maps of the human Rap30 DNA-binding domain (1BBY) and the DNA binding domain of Escherichia coli LexA repressor (1LEA) is shown in Fig. 1 together with their 3-dimensional structures.

Methods

Goldman et al. [7] proved that CMO is APX-hard, which practically defies the existence of a polynomial time exact algorithm or even a polynomial time approximation scheme. In the remainder of this section, we

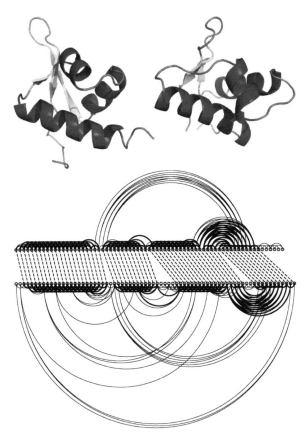

Contact Map Overlap Maximization Problem, CMO, Figure 1
An instance of CMO involving 1BBY and 1LEA

will survey both exact and approximate algorithms for this problem.

Exact algorithms

Four exact algorithms have been proposed for the CMO problem so far. All embrace the branch-and-bound framework to ensure global optimality. The branch-and-reduce algorithm [12,13] currently has an edge against all other exact algorithms both in terms of computational speed as well as its ability to solve challenging instances.

Integer Linear Program Carr et al. [5] proposed an integer linear programming formulation for the problem. This formulation was further developed in [9] and involves two sets of binary variables: x_{ij} and $y_{iji'j'}$. Variable x_{ij} equals one if pair (i,j) is chosen in the optimal alignment and zero otherwise. Variable $y_{iji'j'}$ equals one if both pairs (i,j) and (i',j') are chosen in the final solution. Otherwise, variable $y_{iji'j'}$ is set to zero. The model is as follows:

$(M1)$

$$\max \sum_{(i,j,i',j'): E^A(i,i')=1, E^B(j,j')=1, i<i', j<j'} y_{iji'j'} \quad (1)$$

$$\text{s.t.} \sum_{i': E^A(i,i')=1, i'>i} y_{iji'j'} \leq x_{ij} \quad (2)$$

$$\sum_{j': E^B(j,j')=1, j'>j} y_{iji'j'} \leq x_{ij} \quad (3)$$

$$\sum_{i: E^A(i,i')=1, i<i'} y_{iji'j'} \leq x_{i'j'} \quad (4)$$

$$\sum_{j: E^B(j,j')=1, j<j'} y_{iji'j'} \leq x_{i'j'} \quad (5)$$

$$x_{ij} + x_{i'j'} \leq 1 \quad \forall \text{ crossing pairs } (i,j) \text{ and } (i',j') \quad (6)$$

$$x_{ij} \in \{0,1\}, y_{iji'j'} \in \{0,1\}$$

The sum of common contacts in Eq. (1) constitutes the number of overlaps by definition since Eq. (6) prohibits the existence of crossing pairs in the final solution. Equations (2)–(5) ensure that the optimal solution contains at most one pair from each residue that does not cross a chosen pair. In addition to these necessary constraints, two classes of cuts, *clique-cuts* and *odd-hole-cuts*, can be optionally generated in polynomial time and append to (M1) as was shown in [9].

Lagrangian Relaxation In [3], Caprara and Lancia proposed a Lagrangian relaxation approach for the CMO problem. Their approach begins with an integer linear program formulation of CMO as is shown in model (M2) below. In this model, x_{ij} and $y_{iji'j'}$ bear same definitions as their counterparts in (M1), and $\mathcal{I}$ is the set of maximal set of crossing pairs.

$(M2)$

$$\max \sum_{(i,j,i',j')} \frac{1}{2} h_{iji'j'} y_{iji'j'} \quad (7)$$

$$\text{s.t.} \sum_{(i,j):(i,j) \in I} x_{ij} \leq 1, \quad \forall I \in \mathcal{I} \quad (8)$$

$$\sum_{(i,j)} y_{iji'j'} \leq x_{i'j'}, \quad \forall I \in \mathcal{I}, (i', j') \tag{9}$$

$$y_{iji'j'} = y_{i'j'ij}, \quad \forall i < i', j < j' \tag{10}$$

$$x_{ij}, y_{iji'j'} \in \{0, 1\} \tag{11}$$

It is easy to verify that Eq. (7) yields the number of overlaps given the validity of Eq. (10). Since any two pairs in a maximal set of crossing pairs would cross, constraint (8) enforces the non-crossing property by prohibiting any such two pairs to co-exist in the final alignment. Constraint (9) ensures that, if an arbitrary pair (i', j') is chosen, the final solution should contain no more than one pair (i, j) from any maximal crossing pair set that could form an overlap with pair (i', j').

By introducing multipliers $\lambda_{iji'j'}$ for Eq. (9), Caprara and Lancia obtained the following Lagrangian relaxation of (M2):

(LM2)

$$\min_{\lambda} \max_{i,j} \sum_{(i,j,i',j')} \frac{1}{2} h_{iji'j'} y_{iji'j'}$$
$$+ \sum_{(i,j,i',j'): i<i',j<j'} \lambda_{iji'j'} \left(y_{iji'j'} - y_{i'j'ij} \right)$$

s.t. Constraints (8), (9), and (11).

A subgradient method was used to iteratively improve the multipliers, while an $\mathcal{O}(|E^A||E^B|)$ algorithm was employed to solve (LM2) for the set of multipliers at each iteration.

Reformulation as Maximum Clique Strickland et al. [11] showed that CMO can be cast into a maximum clique problem. To this end, they considered a two-dimensional lattice whose rows correspond to the contacts in E^A and columns correspond to the contacts in E^B. Each vertex of the lattice, if chosen, contributes a unit overlap to the objective value. In addition, an edge is drawn between two vertices if the corresponding pair of overlaps are admissible to some feasible alignments. It is not difficult to see that a maximum clique in the resultant graph indeed corresponds to an optimal solution for the CMO problem. This algorithm involves a number of preprocessing routines to reduce the problem size before calling a maximum clique solver. In a more recent work [10], the authors proposed improved data structures to enhance the algorithm performance.

Combinatorial Branch-and-Reduce Xie and Sahinidis [12,13] developed a branch-and-reduce algorithm, which combines the generic branch-and-bound framework with problem-specific reduction schemes. The algorithm initializes a branch-and-bound tree with a root node where all pairs are allowed. Reduction schemes, both based on domination and the current best solution value, are used to remove inferior pairs. Lower and upper bounds on the overlaps for the current node are then computed using dynamic programming. If the lower and upper bounds agree, the search is terminated with a global optimal alignment. Otherwise, the algorithm chooses a branching pair and creates two children nodes. The branching pair is enforced in one of the node, while it is disallowed in the other node.

A key step in their algorithm is the computation of the contribution to the overlaps by a given pair on a set of pairs. Define $\mathbb{Q}(S)$ to be the set of all subsets of S that contain only parallel pairs. Then, the contribution of pair (i, j) to the objective value on the set S is defined as

$$p(i, j, S) := \max_{Q \in \mathbb{Q}(S)} \sum_{(i',j') \in Q} h(i, j, i', j').$$

In particular, let $p^+(i, j, i', j')$ and $p^-(i, j, i', j')$ denote $p(i, j, S)$ when $S = [i', m] \times [j', n]$ and when $S = [1, i'] \times [1, j']$, respectively. They proved that computing a single term of $p^+(\cdot)$ or $p^-(\cdot)$ can be accomplished in $\mathcal{O}(mn)$ time with preprocessing time and space complexity of $\mathcal{O}(m + n)$.

The upper bounding scheme is summarized in Proposition 1, where, for a node V of the search tree, $\mathbb{C}(V)$ is the set of pairs that must be included and $\mathbb{F}(V)$ is the set of pairs that have the freedom to be in the solution or not. In addition, define

$$g(i, j) := \sum_{(i',j') \in \mathbb{C}(V)} h(i, j, i', j'). \forall (i, j) \in \mathbb{F}(V)$$

Then:

Proposition 1 *Define*

$$t(i,j) := \begin{cases} g(i,j), \\ \quad \text{if } (i,j) \in \mathbb{F}(V); \; i=1 \text{ or } j=1, \\ \max\{g(i,j), g(i,j)+w(i,j)\}, \\ \quad \text{if } (i,j) \in \mathbb{F}(V); i>1 \text{ and } j>1, \\ -\infty, \quad \text{otherwise}, \end{cases}$$

where

$$\bar{h}(i,j,i',j') := \begin{cases} h(i,j,i',j'), \\ \quad \text{if } (i,j), (i',j') \in \mathbb{F}(V) \\ 0, \quad \text{otherwise}, \end{cases}$$

$$w(i,j) := \begin{cases} -\infty, \\ \quad \text{if } [1, i-1] \times [1, j-1] \cap \mathbb{F}(V) = \emptyset, \\ \max_{(i',j') \in [1, i-1] \times [1, j-1] \cap \mathbb{F}(V)} \{t(i',j') \\ \quad + \bar{h}(i',j',i,j) + \frac{1}{2}u(i',j',i,j)\}, \\ \quad \text{otherwise}, \end{cases}$$

and

$$u(i',j',i,j) := p^+(i',j',i+1,j+1) \\ + p^-(i,j,i'-1,j'-1).$$

Then

$$\frac{1}{2} \sum_{(i,j) \in \mathbb{C}(V)} \sum_{(i',j') \in \mathbb{C}(V)} h(i,j,i',j') + \left\lfloor \max_{(i,j) \in \mathbb{F}(V)} t(i,j) \right\rfloor$$

is an upper bound for the current node V.

Proposition 1 suggests an $\mathcal{O}(m^2 n^2)$ algorithm to compute the upper bound. In addition, the upper bounding scheme also provides a natural lower bound as was shown in [13].

Approximate Algorithms

This subsection outlines both approximation algorithms (i. e., with performance guarantee) and heuristics (i. e., without performance guarantee) that have been proposed in the literature for the CMO problem.

Goldman et al. [7] considered a special class of CMO instances from 2-dimensional self-avoiding walk, and proposed a 3-approximation algorithm that runs in $\mathcal{O}(n^6)$ time. Agarwal et al. [1] proposed a 6-approximation algorithm for the same class of problems, with however a better complexity of $\mathcal{O}(n^3 \log n)$. In addition, they proved the special class of CMO from 3-dimensional self-avoiding walk is MAXSNP-hard, and proposed a $\mathcal{O}(\sqrt{n})$-approximation algorithm for this class of CMO instances.

Carr et al. [4] proposed to use a memetic algorithm, which combines global search and local search, to solve general instances of CMO. In [2], Caprara et al. proposed several heuristics, including a genetic algorithm, local search, and greedy algorithms based on Lagrangian relaxation. Existing computational studies [8,13] suggest that Lagrangian-relaxation-based greedy algorithms perform the best among existing heuristics for a large set of test instances.

Conclusions

The contact map overlap maximization problem is a very important problem in computational biology. Efficient algorithms, both exact and approximate, for this problem are of great interest. Despite the inherent difficulty of this problem, many large-scale practical instances have been solved within reasonable amounts of time. Many challenging instances of the problem currently remain unsolved [13].

References

1. Agarwal PK, Mustafa NH, Wang Y (2007) Fast molecular shape matching using contact maps. J Comput Biol 14(2):131–143
2. Caprara A, Carr R, Istrail S, Lancia G, Walenz B (2004) 1001 optimal PDB structure alignments: Integer programming methods for finding the maximum contact map overlap. J Comput Biol 11(1):27–52
3. Caprara A, Lancia G (2002) Structural alignment of large-size proteins via Lagrangian relaxation. In: Proceedings of International Conference on Computational Biology (RECOMB), Washington, April 2002. pp 100–108
4. Carr B, Hart WE, Krasnogor N, Burke EK, Hirst JD, Smith JE (2002) Alignment of protein structures with a memetic evolutionary algorithm. In: Proceedings of the Genetic and Evolutionary Computation Conference, New York, July 2002. Morgan Kaufmann Publishers, pp 1027–1034
5. Carr RD, Lancia G, Istrail S (2000) Branch-and-cut algorithms for independent set problems: Integrality gap and an application to protein structural alignment, Technical report. Sandia National laboratories
6. Godzik A, Skolnick J (1994) Flexible algorithm for direct multiple alignment of protein structures and sequences. Comput Appl Biosci 10(6):587–596

7. Goldman D, Papadimitriou C, Istrail S (1999) Algorithmic aspects of protein structure similarity. In: Proceedings of the 40th Annual Symposium on Foundations of Computer Science (FOCS), New York, October 1999. IEEE Computer societypages, pp 512–522
8. Krasnogor N, Lancia G, Zemla A, Hart WE, Carr RD, Hirst JD, Burke EK (2005) A comparison of computational methods for the maximum contact map overlap of protein pairs. http://citeseer.ist.psu.edu/659931.html. Accessed 30 Jan 2008
9. Lancia G, Carr R, Walenz B, Istrail S (2001) 101 optimal PDB structure alignments: A branch-and-cut algorithm for the maximum contact map overlap problem. In: Proceedings of Annual International Conference on Computational Biology (RECOMB), Montreal, April 2001. pp 193–202
10. Melvin J, Sokol J, Tovey C (2006) Finding optimal solutions to large CMO instances. (preprint)
11. Strickland DM, Barnes E, Sokol JS (2005) Optimal protein structure alignment using maximum cliques. Oper Res 53(3):389–402
12. Xie W, Sahinidis NV (2006) A branch-and-reduce algorithm for the contact map overlap problem. In: Apostolico A, Guerra C, Istrail S, Pevzner P, Waterman M (eds) RECOMB 2006. Lect Note Comput Sci 3909:516–529
13. Xie W, Sahinidis NV (2006) A reduction-based exact algorithm for the contact map overlap problem. J Comput Biol 14(5):637–654

Continuous Approximations to Subdifferentials

ADIL BAGIROV
Centre for Informatics and Applied Optimization,
School of Information Technology
and Mathematical Sciences, University of Ballarat,
Ballarat, Australia

MSC2000: 65K05, 90C56

Article Outline

Introduction
Definitions
    Continuous Approximations
Methods
    Computation of the Continuous Approximations
    Computation of Subgradients
    Discrete Gradients
    Continuous Approximations to the Quasidifferential
Conclusions
References

Introduction

In this paper we use the following notation: $\mathbb{R}^n$ is an n-dimensional space, where the scalar product will be denoted by $\langle x, y \rangle$:

$$\langle x, y \rangle = \sum_{i=1}^{n} x_i y_i ,$$

and $\|\cdot\|$ will denote the associated norm. The gradient of a function $f : \mathbb{R}^n \to \mathbb{R}^1$ will be denoted by ∇f.

A function f is differentiable at a point $x \in \mathbb{R}^n$ with respect to a direction $g \in \mathbb{R}^n$ if the limit

$$f'(x, g) = \lim_{\alpha \downarrow 0} \frac{f(x + \alpha g) - f(x)}{\alpha}$$

exists. The closed unit ball will be denoted by B: $B = \{x \in \mathbb{R}^n : \|x\| \leq 1\}$.

Consider a function f defined on $\mathbb{R}^n$. This function is called locally Lipschitz continuous if for any bounded subset $X \subset \mathbb{R}^n$ there exists an $L > 0$ such that

$$|f(x) - f(y)| \leq L\|x - y\| \quad \forall x, y \in X .$$

If f is convex then one can define a subdifferential $\partial f(x)$ of this function at a point $x \in \mathbb{R}^n$ as follows [13]:

$$\partial f(x) = \{v \in \mathbb{R}^n : f(y) - f(x) \geq \langle v, y - x \rangle, y \in \mathbb{R}^n\} .$$

Elements $v \in \partial f(x)$ of the subdifferential $\partial f(x)$ are called subgradients of f at the point x. For a convex function f defined on $\mathbb{R}^n$ the subdifferential $\partial f(x)$ is nonempty, convex and compact at any $x \in \mathbb{R}^n$. A set-valued mapping $x \mapsto \partial f(x)$ is upper semicontinuous.

An ε-subdifferential $\partial_\varepsilon f(x)$, $\varepsilon > 0$ of the convex function f at a point $x \in \mathbb{R}^n$ is defined as [13]

$$\partial_\varepsilon f(x) = \{v \in \mathbb{R}^n : f(y) - f(x) \geq \langle v, y - x \rangle - \varepsilon, y \in \mathbb{R}^n\} .$$

Elements $v \in \partial_\varepsilon f(x)$ of the subdifferential $\partial_\varepsilon f(x)$ are called ε-subgradients of f at the point x. For a convex function f defined on $\mathbb{R}^n$ the ε-subdifferential $\partial_\varepsilon f(x)$ is nonempty, convex and compact at any $x \in \mathbb{R}^n$ and a set-valued mapping $x \mapsto \partial_\varepsilon f(x)$ is continuous in Hausdorff metric at any x.

Most efficient methods in nonsmooth optimization such as the bundle method and its variations are based on ε-subgradients of convex functions (see, for example, [10,11]).

The analysis of nonsmooth, nonconvex functions has been area of intensive research for more than three decades. Clarke [4,5] introduced the notion of generalized gradient. We define a Clarke subdifferential for locally Lipschitz continuous functions defined on $\mathbb{R}^n$. A locally Lipschitz function f is differentiable almost everywhere and one can define for it a Clarke subdifferential [5] by

$$\partial f(x) = \text{co}\left\{v \in \mathbb{R}^n : \exists (x^k \in D(f), x^k \to x, \; k \to +\infty) : v = \lim_{k \to +\infty} \nabla f(x^k)\right\},$$

where $D(f)$ denotes the set where f is differentiable, and co denotes the convex hull of a set. It is shown in [5] that the mapping $\partial f(x)$ is upper semicontinuous and bounded on bounded sets. The generalized directional derivative of f at x in the direction g is defined to be

$$f^0(x, g) = \limsup_{y \to x, \alpha \downarrow 0} \alpha^{-1}[f(y + \alpha g) - f(y)].$$

The generalized directional derivative always exists and

$$f^0(x, g) = \max\{\langle v, g \rangle : v \in \partial f(x)\}.$$

f is called a Clarke regular function on $\mathbb{R}^n$ if it is differentiable with respect to any direction $g \in \mathbb{R}^n$ and $f'(x, g) = f^0(x, g)$ for all $x, g \in \mathbb{R}^n$. For nonregular functions the Clarke subdifferential has calculus only by means of inclusions, which makes very difficult the computation of subgradients of some complex nonsmooth, nonconvex functions. The cluster function from the cluster analysis is one such example [2,3].

Demyanov and Rubinov [6,7] introduced the notion of quasidifferential. Let f be a locally Lipschitz continuous function defined on $\mathbb{R}^n$. This function is called quasidifferentiable at a point $x \in \mathbb{R}^n$ if it is directionally differentiable and there exist compact, convex sets $\underline{\partial} f(x)$ and $\overline{\partial} f(x)$ such that

$$f'(x, g) = \max\{\langle v, g \rangle, \sim v \in \underline{\partial} f(x)\}$$
$$+ \min\{\langle w, g \rangle, \sim w \in \overline{\partial} f(x)\}.$$

The pair

$$Df(x) = [\underline{\partial} f(x), \overline{\partial} f(x)]$$

is called a quasidifferential of the function f at a point x. The set $\underline{\partial} f(x)$ is said to be a subdifferential and the set $\overline{\partial} f(x)$ a superdifferential of the function f at x.

Unlike the Clarke subdifferential, the quasidifferential enjoys a full-scale calculus; however, set-valued mappings $x \mapsto \underline{\partial} f(x)$ and $x \mapsto \overline{\partial} f(x)$ need not be even upper semicontinuous.

Unfortunately, the notion of an ε-subdifferential cannot be extended for nonsmooth, nonconvex functions. Instead one can define the Goldstein ε-subdifferential [9]. However, in general the Goldstein ε-subdifferential is only upper semicontinuous.

In this paper, we consider continuous approximations to subdifferentials and quasidifferentials. We will also describe an algorithm for computation of elements of such approximations.

Definitions

Continuous Approximations

Let X be a compact subset of the space $\mathbb{R}^n$. We consider a family $C(x, \varepsilon) = C_\varepsilon(x)$ of set-valued mappings depending on a parameter $\varepsilon > 0$. For each $\varepsilon > 0$

$$C(\cdot, \varepsilon): X \to 2^{\mathbb{R}^n}.$$

We suppose that $C(x, \varepsilon)$ is a compact convex set for all $x \in X$ and $\varepsilon > 0$. It is assumed that there exists a number $K > 0$ such that

$$\sup\{\|v\|: v \in C(x, \varepsilon), x \in X, \varepsilon > 0\} \leq K. \tag{1}$$

Definition 1 The limit $C_L(x)$ of the family $\{C(x, \varepsilon)\}, \varepsilon > 0$ at a point x is defined as follows:

$$C_L(x) = \left\{v \in \mathbb{R}^n : \exists (x^k \to x, \varepsilon_k \to +0, k \to +\infty, \right.$$
$$\left. v^k \in C(x^k, \varepsilon_k)) : v = \lim_{k \to +\infty} v^k\right\}.$$

It is possible that the limit $C_L(x)$ is not convex even if all the sets $C(x, \varepsilon)$ are convex. We consider $\text{co} C_L(x)$ the convex hull of $C_L(x)$. It follows from Definition 1 and the inequality (1) that the mapping $\text{co} C_L$ has compact convex images.

Definition 2 A family $\{C(x, \varepsilon)\}$, $\varepsilon > 0$ is called a continuous approximation to a subdifferential ∂f on X if the following holds:

1. $C(x, \varepsilon)$ is a Hausdorff continuous mapping with respect to x on X for all $\varepsilon > 0$.
2. The subdifferential $\partial f(x)$ is the convex hull of the limit of the family $\{C(x, \varepsilon)\}$, $\varepsilon > 0$ on X, i.e. for all $x \in X$

$$\partial f(x) = \operatorname{co} C_L(x).$$

Some properties of the continuous approximations were studied in [1].

Such continuous approximations need not be monotonically decreasing as $\varepsilon \to +0$. Uniform and strongly continuous approximations to the subdifferential studied in [14] have such a property. Let f be a locally Lipschitz continuous function defined on an open set which contains a compact set X. We consider a family of set-valued mappings $A_\varepsilon f : \mathbb{R}^n \to 2^{\mathbb{R}^n}$, $\varepsilon > 0$. Assume that the sets $A_\varepsilon f(x)$ are nonempty and compact for all $\varepsilon > 0$ and $x \in X$. We will denote by $\partial f(x + B_\delta)$ the set $\bigcup \{\partial f(y) : y \in B_\delta(x)\}$, where $B_\delta(x) = \{y \in \mathbb{R}^n : \|x - y\| \leq \delta\}$.

Definition 3 ([14]) We say that the family $\{A_\varepsilon f(\cdot)\}_{\varepsilon > 0}$ is a uniform continuous approximation to the subdifferential ∂f on X, if the following conditions are satisfied:

1. For each given $\varepsilon > 0$, $\mu > 0$ there exists $\tau > 0$, such that for all $x \in X$

$$\partial f(x + B_\tau) \subset A_\varepsilon f(x) + B_\mu.$$

2. For each $x \in X$ and for all $0 < \varepsilon_1 < \varepsilon_2$:

$$A_{\varepsilon_1} f(x) \subset A_{\varepsilon_2} f(x).$$

3. $A_\varepsilon f(x)$ is Hausdorff continuous with respect to x on X.
4. For each $x \in X$

$$\bigcap_{\varepsilon > 0} A_\varepsilon f(x) = \partial f(x).$$

Definition 4 ([14]) We say that the family $\{A_\varepsilon f(\cdot)\}_{\varepsilon > 0}$ is a strong continuous approximation to the subdifferential ∂f on X, if $\{A_\varepsilon f(\cdot)\}_{\varepsilon > 0}$ satisfies properties 1–3 above and instead of property 4 the following is valid:

4′. For every $\gamma, \mu > 0$ there exists $\varepsilon > 0$ such that for all $x \in X$

$$\partial f(x) \subset A_\varepsilon f(x) \subset \partial f(x + B_\gamma) + B_\mu.$$

For the set-valued mapping $C(x, \varepsilon)$ we set

$$C_0(x) = \Big\{v \in \mathbb{R}^n : \exists (\varepsilon_k \to +0, k \to +\infty, \\ v^k \in C(x, \varepsilon_k)) : v = \lim_{k \to +\infty} v^k \Big\}$$

and let

$$C(x, 0) = C_0(x).$$

Theorem 1 ([1]) Let the family $\{A_\varepsilon f(\cdot)\}_{\varepsilon > 0}$ be a uniform continuous approximation to the subdifferential ∂f on a compact set X. Then $C(x, \varepsilon) = A_\varepsilon f(x)$ is a continuous approximation to the subdifferential ∂f in the sense of Definition 2.

Corollary 1 It was shown in [14] that a strong continuous approximation is a uniform continuous approximation. So a strong continuous approximation is a continuous approximation in the sense of Definition 2.

Theorem 2 ([1]) Let the family $C(x, \varepsilon)$ be a continuous approximation to the subdifferential ∂f on a compact set X and the mapping $C(x, \varepsilon)$ be continuous with respect to (x, ε), $x \in X$, $\varepsilon > 0$. Assume $\operatorname{co} C_L(x) = C_0(x)$ for all $x \in X$. Then the mapping

$$Q(x, \varepsilon) = \operatorname{co} \bigcup \{C(x, t) : 0 \leq t \leq \varepsilon\}$$

is a uniform continuous approximation to ∂f on X.

Corollary 2 Let the family $C(x, \varepsilon)$ be a continuous approximation to the subdifferential ∂f on a compact set X and the mapping $C(x, \varepsilon)$ be a continuous with respect to (x, ε), $x \in X$, $\varepsilon > 0$. Assume $\operatorname{co} C_L(x) = C_0(x)$ for all $x \in X$. Then the mapping Q is upper semicontinuous with respect to (x, ε) at the point $(x, 0)$.

One can get a chain rule for continuous approximations [1,14]. However it is not always applicable to compute elements of continuous approximations. In the next section we propose an algorithm to compute those elements.

Methods

Computation of the Continuous Approximations

We consider a locally Lipschitz continuous function f defined on $\mathbb{R}^n$ and assume that this function is semismooth and quasidifferentiable (for the definition of semismooth functions see [12]). We also assume that both sets $\underline{\partial}f(x)$ and $\overline{\partial}f(x)$ are represented as a convex hull of a finite number of points at any $x \in \mathbb{R}^n$, that is at a point $x \in \mathbb{R}^n$ there exist sets

$$A = \{a^1, \ldots, a^m\}, \ a^i \in \mathbb{R}^n, \ i = 1, \ldots, m, m \geq 1$$

and

$$B = \{b^1, \ldots, b^p\}, \ b^j \in \mathbb{R}^n, \ j = 1, \ldots, p, p \geq 1$$

such that

$$\underline{\partial}f(x) = \text{co}A, \ \overline{\partial}f(x) = \text{co}B.$$

In other words we assume that the subdifferential and the superdifferential of the function f are polytopes at any $x \in \mathbb{R}^n$. This assumption is true, for example, for functions represented as a maximum, minimum or max–min of a finite number of smooth functions.

We take a direction $g \in \mathbb{R}^n$ such that

$$g = (g_1, \ldots, g_n), \ |g_i| = 1, \ i = 1, \ldots, n$$

and consider the sequence of n vectors $e^j = e^j(\alpha)$, $j = 1, \ldots, n$ with $\alpha \in (0, 1]$:

$$e^1 = (\alpha g_1, 0, \ldots, 0),$$
$$e^2 = (\alpha g_1, \alpha^2 g_2, 0, \ldots, 0),$$
$$\ldots = \ldots\ldots\ldots,$$
$$e^n = (\alpha g_1, \alpha^2 g_2, \ldots, \alpha^n g_n).$$

We introduce the following sets:

$$\underline{R}_0 = A, \ \overline{R}_0 = B,$$
$$\underline{R}_j = \{v \in \underline{R}_{j-1} : v_j g_j = \max\{w_j g_j : w \in \underline{R}_{j-1}\}\},$$
$$\overline{R}_j = \{v \in \overline{R}_{j-1} : v_j g_j = \min\{w_j g_j : w \in \overline{R}_{j-1}\}\}.$$

It is clear that

$$\underline{R}_j \neq \emptyset, \ \forall j \in \{0, \ldots, n\}, \ \underline{R}_j \subseteq \underline{R}_{j-1},$$
$$\forall j \in \{1, \ldots, n\}$$

and

$$\overline{R}_j \neq \emptyset, \ \forall j \in \{0, \ldots, n\}, \ \overline{R}_j \subseteq \overline{R}_{j-1},$$
$$\forall j \in \{1, \ldots, n\}.$$

Moreover

$$v_r = w_r \ \forall v, w \in \underline{R}_j, \ r = 1, \ldots, j \quad (2)$$

and

$$v_r = w_r \ \forall v, w \in \overline{R}_j, \ r = 1, \ldots, j. \quad (3)$$

Consider the following two sets:

$$\underline{R}(x, e^j(\alpha)) = \left\{v \in A : \langle v, e^j \rangle = \max_{u \in A} \langle u, e^j \rangle \right\},$$
$$\overline{R}(x, e^j(\alpha)) = \left\{w \in B : \langle w, e^j \rangle = \min_{u \in B} \langle u, e^j \rangle \right\}.$$

Proposition 1 *Assume that the function f is quasidifferentiable and its subdifferential and superdifferential are polytopes at a point x. Then there exists $\alpha_0 > 0$ such that*

$$\underline{R}(x, e^j(\alpha)) \subset \underline{R}_j, \ \overline{R}(x, e^j(\alpha)) \subset \overline{R}_j, \ j = 1, \ldots, n$$

for all $\alpha \in (0, \alpha_0)$.

Corollary 3 *Assume that the function f is quasidifferentiable and its subdifferential and superdifferential are polytopes at a point x. Then there exists $\alpha_0 > 0$ such that*

$$f'(x, e^j(\alpha)) = f'(x, e^{j-1}(\alpha)) + v_j \alpha^j g_j + w_j \alpha^j g_j,$$
$$\forall v \in \underline{R}_j, \ w \in \overline{R}_j, \ j = 1, \ldots, n$$

for all $\alpha \in (0, \alpha_0]$.

Proposition 2 *Assume that the function f is quasidifferentiable and its subdifferential and superdifferential are polytopes at a point x. Then the sets $\underline{R}_n$ and $\overline{R}_n$ are singletons.*

In the next subsection we propose an algorithm to approximate subgradients. This algorithm finds a subgradient which can be represented as a sum of elements of the sets $\underline{R}_n$ and $\overline{R}_n$.

Computation of Subgradients

Let $g \in \mathbb{R}^n, |g_i| = 1, i = 1, \ldots, n$ be a given vector and $\lambda > 0, \alpha > 0$ be given numbers. We define the following points

$$x^0 = x, \ x^j = x^0 + \lambda e^j(\alpha), \ j = 1, \ldots, n.$$

It is clear that

$$x^j = x^{j-1} + (0, \ldots, 0, \lambda \alpha^j g_j, 0, \ldots, 0), \ j = 1, \ldots, n.$$

Let $v = v(\alpha, \lambda) \in \mathbb{R}^n$ be a vector with the following coordinates:

$$v_j = (\lambda \alpha^j g_j)^{-1} \left[f(x^j) - f(x^{j-1})\right], \ j = 1, \ldots, n. \quad (4)$$

For any fixed $g \in \mathbb{R}^n, |g_i| = 1, i = 1, \ldots, n$ and $\alpha > 0$ we introduce the following set:

$$V(g, \alpha) = \left\{ w \in \mathbb{R}^n : \exists (\lambda_k \to +0, k \to +\infty), \right.$$
$$\left. w = \lim_{k \to +\infty} v(\alpha, \lambda_k) \right\}.$$

Proposition 3 *Assume that f is a quasidifferentiable function and its subdifferential and superdifferential are polytopes at x. Then there exists $\alpha_0 > 0$ such that*

$$V(g, \alpha) \subset \partial f(x)$$

for all $\alpha \in (0, \alpha_0]$.

Remark 1 It follows from Proposition 3 that in order to approximate subgradients of quasidifferentiable functions one can choose a vector $g \in \mathbb{R}^n$ such that $|g_i| = 1, i = 1, \ldots, n$, sufficiently small $\alpha > 0, \lambda > 0$ and apply (4) to compute a vector $v(\alpha, \lambda)$. This vector is an approximation to a certain subgradient.

Remark 2 A class of quasidifferentiable functions presents a broad class of nonsmooth functions, including many interesting nonregular functions. Thus, the scheme proposed in this section allows one to approximate subgradients of a broad class of nonsmooth functions.

Discrete Gradients

In the previous subsection we demonstrated an algorithm for the computation of subgradients. In this subsection we consider an algorithm for the computation of subdifferentials. This algorithm is based on the notion of a discrete gradient. We start with its definition [1].

Let f be a locally Lipschitz continuous function defined on $\mathbb{R}^n$. Let

$S_1 = \{g \in \mathbb{R}^n : \|g\| = 1\}$,
$G = \{e \in \mathbb{R}^n : e = (e_1, \ldots, e_n), |e_j| = 1,$
$\quad j = 1, \ldots, n\}$,
$P = \{z(\lambda) : z(\lambda) \in \mathbb{R}^1, z(\lambda) > 0, \lambda > 0,$
$\quad \lambda^{-1} z(\lambda) \to 0, \lambda \to 0\}$.

Here S_1 is the unit sphere, G is the set of vertices of the unit hypercube in $\mathbb{R}^n$ and P is the set of univariate positive infinitesimal functions.

We take any $g \in S_1$ and define $|g_i| = \max\{|g_k|, k = 1, \ldots, n\}$. We also take any $e = (e_1, \ldots, e_n) \in G$, a positive number $\alpha \in (0, 1]$ and define the sequence of n vectors $e^j(\alpha), j = 1, \ldots, n$. Then for given $x \in \mathbb{R}^n$ and $z \in P$ we define a sequence of $n + 1$ points as follows:

$$x^0 = x + \lambda g,$$
$$x^1 = x^0 + z(\lambda) e^1(\alpha),$$
$$x^2 = x^0 + z(\lambda) e^2(\alpha),$$
$$\ldots = \ldots \ldots,$$
$$x^n = x^0 + z(\lambda) e^n(\alpha).$$

Definition 5 The discrete gradient of the function f at the point $x \in \mathbb{R}^n$ is the vector $\Gamma^i(x, g, e, z, \lambda, \alpha) = (\Gamma_1^i, \ldots, \Gamma_n^i) \in \mathbb{R}^n, g \in S_1$ with the following coordinates:

$$\Gamma_j^i = [z(\lambda) \alpha^j e_j)]^{-1} \left[f(x^j) - f(x^{j-1}) \right],$$
$$j = 1, \ldots, n, \ j \neq i,$$

$$\Gamma_i^i = (\lambda g_i)^{-1} \left[f(x + \lambda g) - f(x) - \lambda \sum_{j=1, j \neq i}^{n} \Gamma_j^i g_j \right].$$

It follows from the definition that

$$f(x + \lambda g) - f(x) = \lambda \langle \Gamma^i(x, g, e, z, \lambda, \alpha), g \rangle \quad (5)$$

for all $g \in S_1, e \in G, z \in P, \lambda > 0, \alpha > 0$.

Remark 3 One can see that the discrete gradient is defined with respect to a given direction $g \in S_1$ and in order to compute the discrete gradient $\Gamma^i(x, g, e, z, \lambda, \alpha)$ first we define a sequence of points $x^0, \ldots, x^n$ and compute the values of the function f at these points, that is we compute $n + 2$ values of this function including the point x. $n - 1$ coordinates of the discrete gradient are defined similar to those of the vector $v(\alpha, \lambda)$ from (4) and ith coordinate is defined so as to satisfy the equality (5), which can be considered as some version of the mean value theorem.

Proposition 4 *Let f be a locally Lipschitz continuous function defined on $\mathbb{R}^n$ and $L > 0$ be its Lipschitz constant. Then for any $x \in \mathbb{R}^n, g \in S_1, e \in G, \lambda > 0, z \in P, \alpha > 0$*

$$\|\Gamma^i\| \leq C(n) L, \ C(n) = (n^2 + 2n^{3/2} - 2n^{1/2})^{1/2}.$$

For a given $\alpha > 0$ we define the following set:

$$B(x, \alpha) = \{v \in \mathbb{R}^n : \exists (g \in S_1, e \in G, z_k \in P,$$
$$z_k \to +0, \lambda_k \to +0, k \to +\infty),$$

$$v = \lim_{k \to +\infty} \Gamma^i(x, g, e, z_k, \lambda_k, \alpha)\}. \quad (6)$$

Proposition 5 *Let the function f be a differentiable with respect to any direction $g \in \mathbb{R}^n$. Then for any $g \in \mathbb{R}^n$ there exists $v \in B(x, \alpha)$, $\alpha > 0$ such that*

$$f'(x, g) = \langle v, g \rangle.$$

Proposition 6 *Let the function f be a locally Lipschitz continuous, differentiable with respect to any direction $g \in \mathbb{R}^n$ and $x \in D(f)$. Then $\nabla f(x) \in B(x, \alpha)$, $\alpha > 0$.*

Proposition 7 *Assume that f is a semismooth, quasidifferentiable function and its subdifferential and superdifferential are polytopes at a point x. Then there exists $\alpha_0 > 0$ such that*

$$\text{co} B(x, \alpha) \subset \partial f(x)$$

for all $\alpha \in (0, \alpha_0]$.

Remark 4 Proposition 7 implies that discrete gradients can be applied to approximate subdifferentials of a broad class of semismooth, quasidifferentiable functions.

Remark 5 The discrete gradient contains three parameters: $\lambda > 0$, $z \in P$ and $\alpha > 0$. $z \in P$ is used to exploit semismoothness of the function f and it can be chosen sufficiently small. In general α depends on x. However if f is a semismooth quasidifferentiable function and its subdifferential and superdifferential are polytopes at any $x \in \mathbb{R}^n$ then there exist $\delta > 0$ and $\alpha_0 > 0$ such that $\alpha(x) \in (0, \alpha_0]$ for all $y \in B_\delta(x)$. The most important parameter is $\lambda > 0$. In the sequel we assume that $z \in P$ and $\alpha > 0$ are sufficiently small.

Consider the following set at a point $x \in \mathbb{R}^n$:

$$D_0(x, \lambda, z) = \text{cl co}\Big\{v \in \mathbb{R}^n : \exists (g \in S_1, e \in G) : v$$
$$= \Gamma^i(x, g, e, \lambda, z, \alpha)\Big\}.$$

Proposition 4 implies that the set $D_0(x, \lambda, z)$ is compact and convex for any $x \in \mathbb{R}^n$.

Corollary 4 *Let f be a quasidifferentiable semismooth function. Assume that its subdifferential and superdifferential are polytopes and that in the equality*

$$f(x + \lambda g) - f(x) = \lambda f'(x, g) + o(\lambda, g), \ g \in S_1$$

$\lambda^{-1} o(\lambda, g) \to 0$ *as $\lambda \to +0$ uniformly with respect to $g \in S_1$. Then for any $\delta > 0$ there exist $\lambda_0 > 0$ and $z_0 \in P$ such that*

$$D_0(x, \lambda, z) \subset \partial f(x) + B_\delta$$

for all $\lambda \in (0, \lambda_0)$ and $z \in (0, z_0)$.

Consider the continuous approximation $C(x, \varepsilon)$ to the subdifferential $\partial f(x)$. Then Corollary 4 implies that for any $\delta > 0$ there exist $\varepsilon_0 > 0, \lambda_0 > 0$ and $z_0 \in P$ such that

$$D_0(x, \lambda, z) \subset C(x, \varepsilon) + B_\delta$$

for all $\varepsilon \in (0, \varepsilon_0)$, $\lambda \in (0, \lambda_0)$ and $z \in (0, z_0)$. Thus, discrete gradients can be used to compute subsets of continuous approximations to the subdifferential in the sense of Definition 2. Consequently they can also be used to compute subsets of uniform and strongly uniform continuous approximations.

Continuous Approximations to the Quasidifferential

In this subsection we will consider continuous approximations to the Demyanov–Rubinov quasidifferential. We consider a function of the form

$$f(x) = F(x, y_1(x), \ldots, y_m(x)), \quad (7)$$

where $x \in \mathbb{R}^n$, the function F is continuously differentiable in $\mathbb{R}^{n+m}$, $y_i(x)$, $i \in I = \{1, \ldots, m\}$, are semismooth, regular functions and their subdifferentials are polytopes. It is easy to see that the function f is differentiable with respect to any direction and

$$f'(x, g) = \left\langle \frac{\partial F(x, y(x))}{\partial x}, g \right\rangle$$
$$+ \sum_{i \in I} \frac{\partial F(x, y(x))}{\partial y_i} y'_i(x, g),$$

where $y(x) = (y_1(x), \ldots, y_m(x))$, $\partial F(x, y(x))/\partial x$ is the gradient of the function F with respect to x, and $\partial F(x, y(x))/\partial y_i$ is the partial derivative of the function F with respect to y_i, $i \in I$.

Let

$$I_1(x) = \left\{i \in I : \frac{\partial F(x, y(x))}{\partial y_i} > 0\right\},$$

$$I_2(x) = \left\{i \in I : \frac{\partial F(x, y(x))}{\partial y_i} < 0\right\}.$$

Consider the mappings $B_i(x, \alpha)$ corresponding to the functions $y_i(x)$, $i \in I$. We introduce the following two sets:

$$\underline{Z}(x, \alpha) = \mathrm{co}\left\{v \in \mathbb{R}^n : v = \frac{\partial F(x, y(x))}{\partial x}\right.$$
$$+ \sum_{i \in I_1(x)} \frac{\partial F(x, y(x))}{\partial y_i} v^i,$$
$$\left. v^i \in B_i(x, \alpha), i \in I_1(x)\right\},$$

$$\overline{Z}(x, \alpha) = \mathrm{co}\left\{w \in \mathbb{R}^n : w = \sum_{i \in I_2(x)} \frac{\partial F(x, y(x))}{\partial y_i} w^i,\right.$$
$$\left. w^i \in B_i(x, \alpha), i \in I_2(x)\right\}.$$

Proposition 8 *Assume that the function F is continuously differentiable in $\mathbb{R}^{n+m}$, functions $y_i(x)$, $i \in I$, are semismooth and regular and their subdifferentials are polytopes. Then the function f is quasidifferentiable and there exists $\alpha_0 > 0$ such that*

$$\underline{Z}(x, \alpha) \subset \underline{\partial} f(x)$$

and

$$\overline{Z}(x, \alpha) \subset \overline{\partial} f(x)$$

for all $\alpha \in (0, \alpha_0)$.

Corollary 5 *Suppose we are given the function $f(x) = f_1(x) - f_2(x)$, where f_1 and f_2 are semismooth, regular functions and their subdifferentials are polytopes, and $B_1(x, \alpha)$ and $B_2(x, \alpha)$ are mappings corresponding to the functions f_1 and f_2, respectively. Then the function f is quasidifferentiable and there exists $\alpha_0 > 0$ such that*

$$B_1(x, \alpha) \subset \partial f_1(x), \ B_2(x, \alpha) \subset \partial f_2(x)$$

for all $\alpha \in (0, \alpha_0)$.

Let $D_{0i}(x, \lambda, z)$ be mappings corresponding to the functions $y_i(x)$, $i \in I$. We set

$$D_1(x, \lambda, z) = \mathrm{co}\left\{v \in \mathbb{R}^n : v = \frac{\partial F(x, y(x))}{\partial x}\right.$$
$$+ \sum_{i \in I_1(x)} \frac{\partial F(x, y(x))}{\partial y_i} v^i,$$
$$\left. v^i \in D_{0i}(x, z, \lambda), i \in I_1(x)\right\},$$

$$D_2(x, z, \lambda, \beta) = \mathrm{co}\left\{w \in \mathbb{R}^n :\right.$$
$$w = \sum_{i \in I_2(x)} \frac{\partial F(x, y(x))}{\partial y_i} w^i,$$
$$\left. w^i \in D_{0i}(x, z, \lambda), i \in I_2(x)\right\}.$$

Note that the mappings $D_1(x, \lambda, z), D_2(x, \lambda, z)$ are Hausdorff continuous with respect to x for any fixed $\lambda > 0, z \in P$.

It follows from Corollary 4 that the sets $D_1(x, \lambda, z), D_2(x, \lambda, z)$ can be used to compute subsets of continuous approximations to the subdifferential and superdifferential of the function (7).

Conclusions

In this paper we introduced continuous approximations to the subdifferential and the quasidifferential of the nonsmooth, nonconvex functions. We proposed the algorithm for their computation. This algorithm allows one to approximate subgradients a broad class of nonsmooth functions.

References

1. Bagirov AM (2003) Continuous subdifferential approximations and their applications. J Math Sci 115(5):2567–2609
2. Bagirov AM, Rubinov AM, Sukhorukova NV, Yearwood J (2003) Supervised and unsupervised data classification via nonsmooth and global optimisation. TOP: Spanish Oper Res J 11(1):1–93
3. Bagirov AM, Yearwood J (2006) A new nonsmooth optimisation algorithm for minimum sum-of-squares clustering problems. Eur J Oper Res 170(2):578–596
4. Clarke FH (1973) Necessary Conditions for Nonsmooth Problems in Optimal Control and the Calculus of Variations, PhD Thesis. University of Washington, Seattle
5. Clarke FH (1990) Optimization and Nonsmooth Analysis. Classics Appl Math 5. SIAM, Philadelphia
6. Demyanov VF, Rubinov AM (1985) Quasidifferentiable Calculus. Optimization Software, New York

7. Demyanov VF, Rubinov AM (1995) Constructive Nonsmooth Analysis. Peter Lang, Frankfurt am Main
8. Demyanov VF, Vasilyev L (1985) Nondifferentiable Optimization. Optimization Software, New York
9. Goldstein AA (1977) Optimization of Lipschitz continuous functions. Math Programm 13:14–22
10. Hiriart-Urruty JB, Lemarechal C (1993) Convex Analysis and Minimization Algorithms, II: Advanced Theory and Bundle Methods. Springer, Berlin
11. Kiwiel KC (1985) Methods of Descent for Nondifferentiable Optimization, Lecture Notes in Math., 1133. Springer, Berlin, New York
12. Mifflin R (1977) Semismooth and semiconvex functions in constrained optimization. SIAM J Control Optim 15(6): 959–972
13. Rockafellar RT (1970) Convex Analysis. Princeton University Press, Princeton
14. Xu H, Rubinov AM, Glover B (1999) Continuous approximations to generalized Jacobians. Optim 46:221–246

Continuous Global Optimization: Applications

János D. Pintér
Pintér Consulting Services, Inc.,
and Dalhousie University, Halifax, Canada

MSC2000: 90C05

Article Outline

Keywords
Nonlinear Systems and Global Optimization
The Continuous Global Optimization Model
Test Problems
Illustrative Applications
See also
References

Keywords

Nonlinear decision models; Multi-extremality; Global optimization; Scientific applications; Engineering applications; Economic applications

Nonlinear Systems and Global Optimization

Man-made systems and processes can often be modeled to reasonable accuracy by postulating the exclusive use of continuous linear functions. For instance, one may think of the simplest production and distribution models known from the OR literature. (Models with integer variables will not be discussed here, even though they can be equivalently reformulated, to fit into the present framework.) If we attempt, however, the analysis of natural — physical, chemical, biological, environmental, or even economic, financial and societal — systems and their governing processes, then nonlinear functions start to play a significant role in the quantitative description. To illustrate this point, one may think of the most prominent (basic) function forms in physics: probably polynomials, power functions, the exponential-logarithmic pair and trigonometric functions come to mind first. Clouds, water flows, rugged terrains, plants and animals — as well as many other natural objects — all possess visible nonlinearities. For sophisticated examples and general principles, one may think of discussions of nonlinear dynamics, chaos, self-organizing systems and the fractal nature of the Universe: consult, e. g., [3,5,14,19,25].

Prescriptive (control, management, optimization) models which attempt to describe and optimize the behavior of inherently nonlinear systems — as a rule — lead to nonlinear decision problems. Since nonlinear decision models frequently possess multiple local optima, the general relevance of global optimization (GO) becomes obvious. In this brief article we present a list of important and challenging GO applications. We also provide several illustrative references: these describe numerous further application areas.

The Continuous Global Optimization Model

We shall consider problems in the general form

$$\begin{cases} \min_{x \in D} \ f(x) \\ \text{s.t.} \quad D := \left\{ x : \begin{array}{l} l \leq x \leq u; \\ g(x) \leq 0, \\ j = 1, \ldots, J \end{array} \right\}. \end{cases} \quad (1)$$

In (1) we use the following notation:
- x is a vector which represents decision alternatives in $\mathbf{R}^n$;
- $f(x)$ is a continuous objective function;

- *D* is a nonempty set of feasible decisions, defined by
  - *g*(*x*), an *m*-vector of continuous constraint functions; and
  - *l*. *u*, explicit (finite, componentwise) *n*-vector bounds.

Explicit bounds on the constraint function values can also be imposed; however, such more specialized models are directly amenable to the form (1).

First of all, note that if all functions are continuous, then – by the classical theorem of Weierstrass – the optimal solution set to (1) is nonempty. At the same time, without further structural assumptions, (1) can be a very difficult global optimization problem. In other words – unless additional information is provided – there may well exist multiple (local) solutions of various quality to (1). Naturally, in most cases one would like to find the 'very best' (global) solution to the underlying decision problem, avoiding the 'traps' offered by local optima. To attain this objective, a considerable variety of GO models and solution approaches have been suggested: consult, e. g., [12].

Test Problems

Although our primary topic is real-world GO applications, one should at least mention several standardized test problem suites, since these often originate from real-world applications. For collections of (both convex and nonconvex) nonlinear programming test problems, consult, e. g., [11,18]. See [6,7,13,22] for collections of GO test problems. On the WWW, see [1,8] and [21]; especially [21] provides numerous further links and pointers, including discussions of test and real-world problems.

Illustrative Applications

Since GO problems are literally ubiquitous in scientific, engineering and economic decision making, we shall only list a number of illustrative applications. All application areas will be listed simply in alphabetical order, by information source. (The reader will notice overlaps among the problems studied in different works.)

The test problem collection [6] presents application models from the following fields:
- chemical reactor networks;
- distillation column sequencing;
- heat exchanger network synthesis;
- indefinite quadratic programming;
- mechanical design;
- general nonlinear programming;
- phase and chemical reaction equilibrium;
- pooling and blending;
- quadratically constrained problems;
- reactor-separator-recycling systems;
- VLSI design.

The volume [7] significantly expands upon the above material, adding more specific classes of nonlinear programming models, combinatorial optimization problems, and dynamic models, as well as further practical examples (see later on).

The MINPACK-2 collection presented at [1] includes models related to the following types of problems:
- brain activity;
- Chebychev quadrature;
- chemical and phase equilibria;
- coating thickness standardization;
- combustion of propane;
- control systems (analysis and design);
- database optimization;
- design with composites;
- elastic-plastic torsion;
- enzyme reaction analysis;
- exponential data fitting;
- flow in a channel;
- flow in a driven cavity;
- Gaussian data fitting;
- Ginzburg–Landau problem;
- human heart dipole;
- hydrodynamic modeling;
- incompressible elastic rods;
- isomerization of alpha-pinene;
- Lennard–Jones clusters;
- minimal surfaces;
- pressure distribution;
- Ramsey graphs;
- solid fuel ignition;
- steady-state combustion;
- swirling flow between disks;
- thermistor resistance analysis;
- thin film design;
- VLSI design.

A detailed discussion of several GO case studies and applications is presented in [22]. These problems were analyzed by using LGO, an integrated model development environment to formulate and solve GO problems; consult also [23]. The current list of LGO applications includes, for instance, the following areas:

- bio-mechanical design;
- 'black box' (closed, confidential, etc.) system design and operation;
- combination of negotiated expert opinions (forecasts, votes, assessments, etc.);
- data classification, pattern recognition;
- dynamic population and resource management;
- extremal energy (point arrangement) problems, free and surface-constrained forms;
- inverse model fitting to observation data (calibration);
- multifacility location-allocation problems;
- nonlinear approximation, nonlinear regression, and other curve/surface fitting problems;
- optimized tuning of equipment and instruments in medical research and other areas;
- reactor maintenance policy analysis;
- resource allocation (in cutting, loading, scheduling, sequencing, etc. problems);
- risk analysis and control in various environmental management contexts;
- robotics design issues;
- robust product/mixture design;
- satisfiability problems;
- statistical modeling;
- systems of nonlinear equations and inequalities;
- therapy (dosage and schedule) optimization.

The WWW site [21] discusses, inter alia, the following application areas:

- bases for finite elements;
- boundary value problems;
- chemical engineering problems;
- chemical phase equilibria;
- complete pivoting example;
- distance geometry models;
- extreme forms;
- identification of dynamical systems with matrices depending linearly on parameters;
- indefinite quadratic programming models;
- minimax problems;
- nonlinear circuits;
- optimal control problems;
- optimal design;
- parameter identification with data of bounded error;
- PDE defect bounds;
- PDE solution by least squares;
- pole assignment;
- production planning;
- propagation of discrete dynamical systems;
- protein-folding problem;
- pseudospectrum;
- quadrature formulas;
- Runge–Kutta formulas;
- spherical designs (point configurations);
- stability of parameter matrices.

The collection of test problems [7] includes models from the following application areas:

- batch plant design under uncertainty;
- chemical reactor network synthesis;
- conformational problems in clusters of atoms and molecules;
- dynamic optimization problems in parameter estimation;
- homogeneous azeotropic separation system;
- network synthesis;
- optimal control problems;
- parameter estimation and data reconcilliation;
- phase and chemical reaction equilibrium;
- pooling/blending operations;
- pump network synthesis;
- robust stability analysis;
- trim loss minimization.

The article [4] reviews several significant applications of rigorous global optimization (based on the interval branch and bound approach). These applications include:

- currency trading;
- finite element analysis (in a high-tech engineering design context);
- gene prediction in genome therapeutics;
- magnetic resonance imaging (in a medical application);
- numerical mathematics (search for an approximate greatest common divisor of given polynomials);
- parameter estimation in signal processing;
- portfolio management;

One can immediately add here the application of interval techniques to an issue of paramount significance in numerical modeling:

- solving systems of nonlinear (and linear) equations; consult, e. g., [20]

The volume [24] also covers a broad range of applications from the following areas:
- agro-ecosystem management;
- analysis of nucleid acid sequences;
- assembly line design;
- cellular mobile network design;
- chemical process optimization;
- chemical product design;
- computational modeling of atomic and molecular structures;
- controller design for motors;
- electrical engineering design;
- feeding strategies in animal husbandry;
- financial modeling;
- laser equipment design;
- mechanical engineering design;
- radiotherapy equipment calibration;
- robotics design;
- satellite data analysis (interferometry problem);
- virus structure reconstruction;
- water resource distribution systems.

As the above lists illustrate, the application potentials of global optimization are indeed most diverse.

For additional literature on real-world applications, see, e. g., the following references:
- network problems, combinatorial optimization (knapsack, traveling salesman, flow-shop problems), batch process scheduling: [17];
- GO algorithms and their applications (primarily) in chemical engineering design: [9];
- contributions on decision support systems and techniques for solving GO problems, but also on molecular structures, queueing systems, image reconstruction, location analysis and process network synthesis: [2];
- multilevel optimization algorithms and their applications: [15];
- engineering applications of the finite element method: [16];
- a variety of applications, e. g., from the fields of environmental management, geometric design, robust product design, and parameter estimation: [10].

Numerous issues of the *Journal of Global Optimization* – as well as a large number of other professional OR/MS, natural science and engineering journals – also publish articles describing interesting GO applications.

See also

▶ αBB Algorithm
▶ Continuous Global Optimization: Models, Algorithms and Software
▶ Differential Equations and Global Optimization
▶ DIRECT Global Optimization Algorithm
▶ Forecasting
▶ Global Optimization in the Analysis and Management of Environmental Systems
▶ Global Optimization Based on Statistical Models
▶ Global Optimization in Binary Star Astronomy
▶ Global Optimization Methods for Systems of Nonlinear Equations
▶ Global Optimization Using Space Filling
▶ Interval Global Optimization
▶ Mixed Integer Nonlinear Programming
▶ Topology of Global Optimization

References

1. Argonne National Lab (1993) MINPACK-2 test problem collection. Argonne National Lab., Argonne, IL, see also the accompanying notes Large-scale optimization: Model problems. BM Averick and Moré JJ www-c.mcs.anl/gov/gov/home/more/tprobs/html
2. Bomze IM, Csendes T, Horst R, Pardalos PM (eds) (1997) Developments in global optimization. Kluwer, Dordrecht
3. Casti JL (1990) Searching for certainty. Morrow, New York
4. Corliss GF, Kearfott RB (1999) Rigorous global search: Industrial applications. In: Csendes T (ed) Developments in Reliable Computing. Kluwer, Dordrecht, pp 1–16
5. Eigen M, Winkler R (1975) Das Spiel. Piper, Munich
6. Floudas CA, Pardalos PM (1990) A collection of test problems for constrained global optimization algorithms. No. 455 of Lecture Notes Economics and Math Systems. Springer, Berlin
7. Floudas CA, Pardalos PM, Adjiman C, Esposito WR, Gumus ZH, Harding ST, Klepeis JL, Meyer CA, Schweiger CA (1999) Handbook of test problems in local and global optimization. Kluwer, Dordrecht
8. Gray P, Hart WE, Painton L, Phillips C, Trahan M, Wagner J (1999) A survey of global optimization methods. Sandia National Lab, Albuquerque, NM http://www.cs.sandia.gov/opt/survey/main.html
9. Grossmann IE (ed) (1996) Global optimization in engineering design. Kluwer, Dordrecht
10. Hendrix EMT (1998) Global optimization at work. PhD Thesis LU Wageningen, The Netherlands

11. Hock W, Schittkowski K (eds) (1987) Test examples for nonlinear programming codes. Lecture Notes Economics and Math Systems. Springer, Berlin
12. Horst R, Pardalos PM (eds) (1995) Handbook of global optimization. Kluwer, Dordrecht
13. Jansson C, Knöppel O (1992) A global minimization method: The multi-dimensional case. Res Report TUHH, Hamburg-Harburg, Germany
14. Mandelbrot BB (1983) The fractal geometry of nature. Freeman, New York
15. Migdalas A, Pardalos PM, Värbrand P (eds) (1997) Multilevel optimization: Algorithms and applications. Kluwer, Dordrecht
16. Mistakidis E, Stavroulakis GE (1997) Nonconvex optimization. Algorithms, heuristics and engineering applications of the F.E.M. Kluwer, Dordrecht
17. Mockus J, Eddy W, Mockus A, Mockus L, Reklaitis G (1996) Bayesian heuristic approach to discrete and global optimization. Kluwer, Dordrecht
18. Moré JJ, Garbow BS, Hillström KE (1981) Testing unconstrained optimization software. ACM Trans Math Softw 7:17–41
19. Murray JD (1983) Mathematical biology. Springer, Berlin
20. Neumaier A (1990) Interval methods for systems of equations. Cambridge Univ. Press, Cambridge
21. Neumaier A (1999) Global optimization. http://solon.cma.univie.ac.at/~neum/glopt.html
22. Pinter JD (1996) Global optimization in action. Kluwer, Dordrecht
23. Pintér JD (1999) LGO-A model development system for continuous global optimization. User's guide. Pintér Consulting Services, Halifax, NS, Halifax, NS
24. Pintér JD (ed) (2001) Global optimization – Selected case studies. Kluwer, Dordrecht
25. Schroeder M (1991) Fractals, chaos, power laws. Freeman, New York

Continuous Global Optimization: Models, Algorithms and Software

János D. Pintér
Pintér Consulting Services, Inc.,
and Dalhousie University, Halifax, Canada

MSC2000: 90C05

Article Outline

Keywords
The Continuous Global Optimization Model
Model Types
Exact Methods
  Naive Approaches
  Complete (Enumerative) Search Strategies
  Homotopy (Parameter Continuation),
    Trajectory Methods, and Related Approaches
  Successive Approximation (Relaxation) Methods
  Branch and Bound Algorithms
  Bayesian Search (Partition) Algorithms
  Adaptive Stochastic Search Algorithms
Heuristic Strategies
  'Globalized' Extensions of Local Search Methods
  Genetic Algorithms, Evolution Strategies
  Simulated Annealing
  Tabu Search
  Approximate Convex Global Underestimation
  Continuation Methods
  Sequential Improvement of Local Optima
Global Optimization Software
Software Evaluation
  Software Applicability Range (Solvable Model Types)
  GO Methodology Applied
  Hardware and Software Requirements
  Test Results
  Additional Software Information
See also
References

Keywords

Nonlinear decision models; Continuous global optimization; Model types; Solution strategies; Software development and evaluation

The Continuous Global Optimization Model

We shall consider the continuous global optimization problem (GOP) in the general form

$$\begin{cases} \min_{x \in D} \quad f(x) \\ \text{s.t.} \quad D := \left\{ x : \begin{array}{l} l \leq x \leq u; \\ g(x) \leq 0, \\ j = 1, \ldots, J \end{array} \right\} \end{cases} \qquad (1)$$

In (1) the following assumptions are used:
- x is a vector representing decision alternatives in $\mathbf{R}^n$;
- D is a nonempty set of feasible decisions, defined by
  - l, u: explicit (finite, componentwise) n-vector bounds of x, and
  - $g(x)$ is an m-vector of continuous constraint functions defined on $[l, u]$;

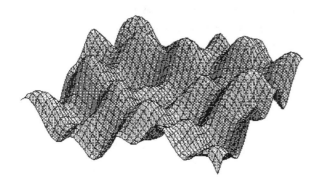

Continuous Global Optimization: Models, Algorithms and Software, Figure 1
A two-variable multi-extremal function

- $f(x)$ is a continuous objective function defined on D.

Explicit bounds on the constraint function values can also be imposed; however, such more specialized models are also directly amenable to the form (1).

Since the functions f and g are all continuous in D, the GOP (1) evidently has a nonempty globally optimal solution set X^*. At the same time, one can immediately realize that — in its full generality — instances of model (1) can pose a very significant numerical challenge. Since the usual convexity assumptions are absent, D may be disconnected and/or nonconvex, and the objective function f may also be multi-extremal. That is, the number of local (pseudo) solutions to (1) is typically unknown and it can be large; the quality of the various local and global solutions may differ significantly. To illustrate this point, see Fig. 1, which depicts a 'hilly landscape' (in fact, the surface plot of a relatively simple composition of trigonometric functions with embedded polynomial arguments, in just two variables). For instance, this function could be the objective in (1) defined on the corresponding interval feasible region $[l, u]$.

To solve the GOP (1) – in a strict mathematical sense – means to find the complete set of globally optimal solutions X^*, and the associated global optimum value $f^* = f(x^*)$, $x^* \in X^*$. In most cases, at least in the realm of continuous GO, we need to replace this 'ambitious' objective by finding a verified estimate – upper and lower bounds – of f^*, and corresponding approximation(s) of points from the set X^*. Naturally, such estimates are to be determined on the basis of a finite number of algorithmically generated sample points from D, or from the embedding interval $[l, u]$.

For reasons of better analytical and numerical tractability, usually the following additional assumptions are made:
- D is a full-dimensional subset (a 'body') in $\mathbf{R}^n$;
- X^* is at most countable;
- g (i. e., each of its component functions) and f are Lipschitz-continuous on $[l, u]$.

Observe that the first assumption makes algorithmic search possible within the set D. With respect to the second assumption, note that – in most practical contexts – the set of global optimizers consists only of a single point, or of several points. Finally, the Lipschitz assumption – i. e., that changes in function values are uniformly controlled by changes in their argument – is a sufficient condition for estimating f^* on the basis of a finite set of search points. We emphasize that the factual knowledge of the smallest suitable Lipschitz constant is not required – and in practice it is typically unknown indeed. The Lipschitz criterion is evidently met, e. g., by all continuously differentiable functions defined on $[l, u]$; however, their class is even broader.

Due to the very general model structure postulated above, classical (convexity-based) numerical approaches are, generally speaking, not directly applicable to solve GOPs: instead, truly global scope methodology is needed. In the past decades, a considerable variety of GO models and solution approaches have been proposed and analyzed. Below we shall provide a concise review, with a view towards software development. For detailed discussions, consult, e. g., the illustrative list of references.

Model Types

The most important GO model classes that have been extensively studied include the following. (Note that postulated properties of g – such as e. g., convexity – are required componentwise.)
- Bilinear and biconvex programming (f is bilinear or biconvex, D is convex).
- Combinatorial optimization (problems that have discrete decision variables in f and/or in g can be equivalently reformulated as GO problems in continuous variables).

- Concave minimization (f is concave, D is convex).
- Continuous global optimization (f and g are arbitrary continuous functions).
- Differential convex (DC) optimization (f and the components in g can all be explicitly represented, as the difference of two corresponding convex functions).
- Fractional programming (f is the ratio of two real functions, and g is convex).
- Linear and nonlinear complementarity problems (f is the scalar product of two vector functions, D is typically assumed to be convex).
- Lipschitz optimization (f and g are arbitrary Lipschitz-continuous functions).
- Minimax problems (f is some minimax objective, the maximum is considered over a discrete set or a convex set, D is convex).
- Multilevel optimization (e. g., models of noncooperative games, involving hierarchies of decision makers, the conflicting criteria are aggregated by f; D is usually assumed to be convex).
- Multi-objective programming (e. g., determination of the efficient set, when several conflicting objectives are to be optimized over the region D).
- Multiplicative programming (f is the product of several convex functions, and g is convex, or – more generally – also multiplicative).
- Network problems (f can be taken from several nonconvex function classes, and g is typically linear or convex).
- Parametric nonconvex programming (in these the feasible region D and/or the objective f may also depend also on a parameter vector).
- Quadratic optimization (f is an arbitrary – indefinite – quadratic function; g is linear or, in the more general case, is also made up by arbitrary quadratic functions).
- Reverse convex programming (at least one of the functions in g expresses a reverse convex constraint).
- Separable global optimization (f is an arbitrary nonlinear – in general, nonconvex – separable function, D is typically convex).
- Stochastic (nonconvex) models in which the functions f, g depend on random factors.
- Various other nonlinear programming problems, in absence of a verified convex structure: this broad category includes, e. g., models in which some of the functions f, g are defined by complex 'black box' computational procedures.

Note that the problem classes listed are not necessarily distinct; in fact, several of them are hierarchically contained in the more general problem types listed. For detailed descriptions of most of these model types and their connections consult, e. g., [13], with numerous further references.

Observe also that in the list presented, there are specifically structured models (such as e. g., a concave minimization problem under linear or convex constraints), as well as far more general ones (such as e. g., differential convex, Lipschitz or continuous problems). Hence, one can reasonably expect that the most suitably tailored solution approaches will also vary to a considerable extent. Very general search strategies should work for most models – albeit their efficiency might be low for specialized problems. At the same time, strictly specialized solvers may not work at all for problem classes outside of their scope.

Several of the most important GO strategies are listed below, together with additional remarks and references. Again, the items of the list are not necessarily exclusive. Most GO software implementations are based upon one of these approaches, possibly combining ideas from several strategies.

Exact Methods

Naive Approaches

These include the most well known passive (simultaneous) or direct (not fully adaptive) sequential GO strategies: uniform grid, space covering, and pure random searches. Note that such methods are obviously convergent under mild assumptions, but are – as a rule – impracticable in higher-dimensional problems. Consult corresponding chapters in [13,24,30].

Complete (Enumerative) Search Strategies

These are based upon an exhaustive (and typically streamlined) enumeration of all possible solutions. Applicable to combinatorial problems, as well as to certain 'well-structured' continuous GO problems such as, e. g., concave programming. See, e. g., [14].

Homotopy (Parameter Continuation), Trajectory Methods, and Related Approaches

These methods have the 'ambitious' objective of visiting all stationary points of the objective function: this, in turn, leads to the list of all – global as well as local – optima. This general approach includes differential equation model based, path following search strategies, as well as fixed-point methods and pivoting algorithms. See, for instance, [5] and [8].

Successive Approximation (Relaxation) Methods

The initial optimization problem is replaced by a sequence of relaxed subproblems that are easier to solve. Successive refinement of subproblems to approximate the initial problem; cutting planes and more general cuts, diverse minorant function constructions, nested optimization and decomposition strategies are also possible. Applicable to structured GO problems such as, e. g., concave minimization and DC problems [14].

Branch and Bound Algorithms

A variety of partition strategies have been proposed to solve GOPs. These are based upon adaptive partition, sampling, and subsequent lower and upper bounding procedures: these operations are applied iteratively to the collection of active (remaining 'candidate') subsets within the feasible set D. Their exhaustive search feature is similar in spirit to analogous integer programming methodology. Branch and bound subsumes many specific approaches, and allows for a range of implementations.

Branch and bound methods typically rely on some a priori structural knowledge about the problem. This information may relate, for instance to how rapidly each function can vary (e. g. the knowledge of a suitable 'overall' Lipschitz constant, for each function f and g); or to the availability of an analytic formulation – and guaranteed smoothness – of all functions (for instance, in interval arithmetic based methods).

The branch and bound methodology is applicable to broad classes of GO problems: e. g., in combinatorial optimization, concave minimization, reverse convex programs, DC programming, and Lipschitz optimization. For details, consult [12,14,15,20,24,26].

Bayesian Search (Partition) Algorithms

These methods are based upon some postulated statistical information, to enable a prior stochastic description of the function class modeled. During optimization, the problem instance characteristics are adaptively estimated and updated. Note that, typically only the corresponding one-dimensional model development is exact; furthermore, that in most practical cases 'myopic' approximate decisions govern the search procedure.

This general approach is applicable also to (merely) continuous GO problems. Theoretically, convergence to the optimal solution set is guaranteed only by generating an everywhere dense set of search points. One of the obvious challenges of using statistical methods is the choice and verification of an 'appropriate' statistical model, for the class of problems to which they are applied. Additionally, it seems to be difficult to implement rigorous and computationally efficient versions of these algorithms for higher-dimensional optimization problems. Note, however, that if one 'skips' the underlying Bayesian paradigm, then these methods can also be pragmatically viewed as adaptive partition algorithms, and – as such – they can be directly extended to higher dimensions: see [24]. For detailed expositions on Bayesian approaches, consult, e. g., [18,19,27].

Adaptive Stochastic Search Algorithms

This is another broad class of methods, based upon random sampling in the feasible set. In its basic form, it includes various random search strategies that are convergent, with probability one. Search strategy adjustments, clustering and deterministic solution refinement options, statistical stopping rules, etc. can also be added as enhancements.

The methodology is applicable to both discrete and continuous GO problems under very mild conditions. Consult, for instance, [2,24,30].

Heuristic Strategies

'Globalized' Extensions of Local Search Methods

These are partially heuristic algorithms, yet often successful in practice. The essential idea is to apply a preliminary grid search or random search based global phase, followed by applying a local (convex programming) method. For instance, random multistart performs a local search from several points selected randomly from the search domain D. Note that even such sampling is not trivial, when D has a complicated shape, as being defined, e. g., by (merely) continuous nonlinear functions.

Frequently, sophisticated algorithm enhancements are added to this basic strategy. For instance, the clustering of sample points is aimed at selecting only a single point from each sampled 'basin' of f from which then a local search method is initiated. For more details, consult, for instance, [27].

Genetic Algorithms, Evolution Strategies

These methods 'mimic' biological evolution: namely, the process of natural selection and the 'survival of the fittest' principle. An adaptive search procedure based on a 'population' of candidate solution points is used. Iterations involve a competitive selection that drops the poorer solutions. The remaining pool of candidates with higher 'fitness value' are then 'recombined' with other solutions by swapping components with another; they can also be 'mutated' by making some smaller-scale change to a candidate. The recombination and mutation moves are applied sequentially; their aim is to generate new solutions that are biased towards subsets of D in which good – although not necessarily globally optimized – solutions have already been found.

Numerous variants of this general strategy, based on diverse evolution 'game rules', can be constructed. The different types of evolutionary search methods include approaches that are aimed at continuous GOPs, and also others that are targeted towards solving combinatorial problems. The latter group is often called genetic algorithms. For details, consult, e. g., [10,17,22,29].

Simulated Annealing

These techniques are based upon the physical analogy of cooling crystal structures that spontaneously attempt to arrive at some stable (globally or locally minimal potential energy) equilibrium. This general principle is applicable to both discrete and continuous GO problems under mild structural requirements: consult, e. g., [1,22,28].

Tabu Search

In this general category of metaheuristics, the essential idea during search is to 'forbid' moves to points already visited in the (usually discrete) search neighborhood, at least for a number of upcoming steps. This way, one can temporarily accept new inferior solutions, in order to avoid (sub)paths already investigated. This approach can lead to exploring new regions of D, with the goal of finding a solution by 'globalized' search.

Tabu search has traditionally been applied to combinatorial optimization (e. g., scheduling, routing, traveling salesman) problems. The technique can be made – at least, in principle – directly applicable to continuous GOPs by a discrete approximation (encoding) of the problem, but other extensions are also possible. See [9,22,29].

Approximate Convex Global Underestimation

This heuristically attractive strategy attempts to estimate the (postulated) large scale, 'overall' convexity characteristics of the objective function f based on directed sampling in D. Applicable to smooth problems. See, e. g., [6].

Continuation Methods

These first transform the potential function into a more smooth ('simpler') function which has fewer local minimizers, and then attempt to trace the minimizers back to the original function. Again, this methodology is applicable to smooth problems. For theoretical background, see, for instance, [8].

Sequential Improvement of Local Optima

These methods usually operate on adaptively constructed auxiliary functions, to assist the search for gradually better optima. The general heuristic principle is realized by so-called tunneling, deflation, and filled function approaches; consult, for example, [16].

Global Optimization Software

In spite of significant theoretical advances in GO, software development and 'standardized' use lag behind. This can be expected due to the potential numerical difficulty of GOPs; recall Fig. 1. Even 'much simpler' problem instances – such as e. g., concave minimization, or indefinite quadratic programming – belong to the hardest (*NP*) class of mathematical programming problems.

As summarized above, there exist several broad classes of algorithmic GO approaches that possess strong theoretical convergence properties, and – at least in principle – are straightforward to implement. However, all such rigorous approaches involve a computational demand that increases exponentially as a function of problem size, even in case of the simplest GO problem instances. (Consult, for example, [13] for related discussions.) Therefore many practical GO strategies are completed by a 'traditional' local optimization phase. Global convergence, however, needs to be guaranteed by the global scope algorithm component: the latter – at least in theory – should be used in a complete, 'exhaustive' fashion. The above remarks indicate the basic inherent theoretical (and practical) difficulty of developing robust, yet efficient GO software.

Since the computational demand of rigorous strategies can be expected to be some exponential function of the problem dimensionality, GO problems in $\mathbf{R}^n$ (n being just 5, 10, 20, 50, 100, …) may have rapidly increasing – possibly straight enormous – numerical complexity. This is (and will remain) true, in spite of the fact that computational power seems to grow at an unbelievable pace: the so-called 'curse of dimensionality' is here to stay.

In 1996, a survey on continuous GO software was prepared for the newsletter of the Mathematical Programming Society [23]. Additional information has been collected from the Internet, from several GO books, and from the *Journal of Global Optimization*. Drawing on the responses of software developers and the additional information available, over 50 software products were annotated in that review. (In order to assist in obtaining further information, contact person(s), their e-mail addresses, ftp and/or WWW sites have also been listed.)

Most probably, by now the number of solvers aimed at GOPs is around one hundred (or even more). The general impression is, however, that many of these software products are still at an experimental development stage, and of dominantly 'academic' character, as opposed to 'industrial strength' tools. (Of course, it is not impossible that proprietary software products used by industry and private companies are not announced publicly.)

Below we shall list some key aspects that should be addressed by professional quality GO software development:

- well-specified hardware and software environments (supported development platforms);
- quality user guidance: clearly outlined model development procedure, sensible modeling and troubleshooting tips, user file templates, and (also) nontrivial numerical examples;
- fully functional, 'friendly' user interface;
- 'fool-proof' solver selection and execution procedures;
- good runtime communication and documentation: clear system output for all foreseeable program execution versions and situations, including proper error messages, and result file(s);
- visualization features which are especially desirable in nonlinear systems modeling, to avoid problem misrepresentation, and to assist in finding alternative models and solution procedures;
- reliable, high-quality user support;
- continuous product maintenance and development (since not only science progresses, but hardware devices, operating systems, as well as development platforms are in permanent change).

This tentative 'wish-list' of requirements indicates that although the task is not impossible, it is a challenge. As for an example, we refer to LGO – an integrated model development and solver system – that has been developed with a view towards the desiderata listed above. Details regarding LGO are described, e. g., in [24,25].

Software Evaluation

In order to obtain information regarding the scope and usability of GO software, it needs to be thoroughly tested. This is a demanding task, when done properly. Consideration needs to be given to the selection of appropriate – nontrivial and practically meaningful – examples. Computational experiments should be carefully designed; and the results should be reported in sufficient details, to assure a fair and accurate assessment. For corresponding discussions and GO (or other) test problems, consult, e. g., [3,4,7,21].

A GO software evaluation framework can be proposed, for instance, along the following guidelines.

Software Applicability Range (Solvable Model Types)

- objective function: concave, DC, Lipschitz, continuous, or some other (general or more special) function form;
- constraints: unconstrained problems, bound constrained problems, linear constraints, general nonlinear smooth constraints;
- additional information related to solvable model types and sizes, with corresponding expected runtimes (within given hardware and software environments).

GO Methodology Applied

- summary (or more detailed) description of basic principles;
- adequate list of references;

Hardware and Software Requirements

- supported hardware platforms;
- minimal hardware configuration needed;
- operating systems;
- programming languages and environments;
- compiler(s) needed;
- connectivity to other development environments;
- portability to other hardware and software platforms.

Test Results

- test problem description, mathematical and/or coded form;
- real world background information (when applicable);
- best known results, with references;
- accuracy requirements, stopping criteria;
- hardware and software environment used in testing;
- standard timing (to facilitate comparisons among different platforms);
- time and computational demand, in order to find the (estimated) global optimum;
- comparative success rate;
- information regarding the reproducibility of results.

Additional Software Information

- installation procedure;
- user interface features;
- academic and/or professional licenses; conditions of use;
- user support (manual, on-line help, example files, input and result handling, etc.);
- other points of interest.

Of course – at least from a practical point of view – the most meaningful test is to apply GO methods to problems that are of interest in the real world. For numerous existing and prospective GO applications, please consult the related articles ▶ Global Optimization in the Analysis and Management of Environmental Systems and ▶ Continuous Global Optimization: Applications.

See also

- ▶ αBB Algorithm
- ▶ Convex Envelopes in Optimization Problems
- ▶ Differential Equations and Global Optimization
- ▶ DIRECT Global Optimization Algorithm
- ▶ Global Optimization Based on Statistical Models
- ▶ Global Optimization in Batch Design Under Uncertainty
- ▶ Global Optimization in Binary Star Astronomy
- ▶ Global Optimization in Generalized Geometric Programming
- ▶ Global Optimization of Heat Exchanger Networks
- ▶ Global Optimization Methods for Systems of Nonlinear Equations
- ▶ Global Optimization in Phase and Chemical Reaction Equilibrium
- ▶ Global Optimization Using Space Filling
- ▶ Interval Global Optimization
- ▶ Large Scale Unconstrained Optimization
- ▶ MINLP: Branch and Bound Global Optimization Algorithm
- ▶ MINLP: Heat Exchanger Network Synthesis
- ▶ MINLP: Mass and Heat Exchanger Networks
- ▶ Mixed Integer Linear Programming: Heat Exchanger Network Synthesis
- ▶ Mixed Integer Linear Programming: Mass and Heat Exchanger Networks
- ▶ Modeling Languages in Optimization: A New Paradigm
- ▶ Optimization Software
- ▶ Smooth Nonlinear Nonconvex Optimization
- ▶ Topology of Global Optimization

References

1. Aarts E, Lenstra JK (eds) (1997) Local search in combinatorial optimization. Wiley, New York
2. Boender CGE, Romeijn E (1995) Stochastic methods. In: Horst R, Pardalos PM (eds) Handbook of Global of Optimization. Kluwer, Dordrecht, pp 829–869
3. Bomze IM, Csendes T, Horst R, Pardalos PM (eds) (1997) Developments in global optimization. Kluwer, Dordrecht
4. De Leone R, Murli A, Pardalos PM, Toraldo G (eds) (1998) High performance software for nonlinear optimization: Status and perspectives. Kluwer, Dordrecht
5. Diener I (1995) Trajectory methods in global optimization. In: Horst R, Pardalos PM (eds) Handbook of Global Optimization. Kluwer, Dordrecht, pp 649–668
6. Dill KA, Phillips AT, Rosen JB (1997) Molecular structure prediction by global optimization. In: Bomze IM, Csendes T, Horst R, Pardalos PM (eds) Developments in Global Optimization. Kluwer, Dordrecht, pp 217–234
7. Floudas CA, Pardalos PM, Adjiman C, Esposito WR, Gumus ZH, Harding ST, Klepeis JL, Meyer CA, Schweiger CA (1999) Handbook of test problems in local and global optimization. Kluwer, Dordrecht
8. Forster W (1995) Homotopy methods. In: Horst R, Pardalos PM (eds) Handbook of Global Optimization. Kluwer, Dordrecht, pp 669–750
9. Glover F, Laguna M (1997) Tabu search. Kluwer, Dordrecht
10. Goldberg DE (1989) Genetic algorithms in search, optimization, and machine learning. Addison-Wesley, Reading
11. Gray P, Hart WE, Painton L, Phillips C, Trahan M, Wagner J (1999) A survey of global optimization methods. Sandia National Laboratories, http://www.cs.sandia.gov/opt/survey/main.html
12. Hansen ER (1992) Global optimization using interval analysis. M. Dekker, New York
13. Horst R, Pardalos PM (eds) (1995) Handbook of global optimization. Kluwer, Dordrecht
14. Horst R, Tuy H (1996) Global optimization–deterministic approaches, 3rd edn. Springer, Berlin
15. Kearfott RB (1996) Rigorous global search: Continuous problems. Kluwer, Dordrecht
16. Levy AV, Gomez S (1984) The tunneling method applied to global optimization. Numerical Optimization. In: Boggs PT (ed). SIAM, Philadelphia, pp 213–244
17. Michalewicz Z (1996) Genetic algorithms + data structures = evolution programs, 3rd edn. Springer, Berlin
18. Mockus J (1989) Bayesian approach to global optimization. Kluwer, Dordrecht
19. Mockus J, Eddy W, Mockus A, Mockus L, Reklaitis G (1996) Bayesian heuristic approach to discrete and global optimization. Kluwer, Dordrecht
20. Neumaier A (1990) Interval methods for systems of equations. Cambridge Univ. Press, Cambridge
21. Neumaier A (1999) Global optimization. http://solon.cma.univie.ac.at/~neum/glopt.html
22. Osman IH, Kelly JP (eds) (1996) Meta-heuristics: Theory and applications. Kluwer, Dordrecht
23. Pintér JD (1996) Continuous global optimization software: A brief review. Optima 52:1–8, http://plato.la.asu.edu/gom.html
24. Pintér JD (1996) Global optimization in action. Kluwer, Dordrecht
25. Pintér JD (1998) A model development system for global optimization. In: De Leone R, Murli A, Pardalos PM, Toraldo G (eds) High Performance Software for Nonlinear Optimization: Status and Perspectives. Kluwer, Dordrecht, pp 301–314
26. Ratschek H, Rokne JG (1995) Interval methods. In: Horst R, Pardalos PM (eds) Handbook of Global Optimization. Kluwer, Dordrecht, pp 751–828
27. Törn AA, Žilinskas A (1989) Global optimization. In: Lecture Notes Computer Sci, vol 350. Springer, Berlin
28. Van Laarhoven PJM, Aarts EHL (1987) Simulated annealing: Theory and applications. Kluwer, Dordrecht
29. Voss S, Martello S, Osman IH, Roucairol C (eds) (1999) Meta-heuristics: Advances and trends in local search paradigms for optimization. Kluwer, Dordrecht
30. Zhigljavsky AA (1991) Theory of global random search. Kluwer, Dordrecht

Continuous Reformulations of Discrete-Continuous Optimization Problems

OLIVER STEIN

School of Economics and Business Engineering, University of Karlsruhe, Karlsruhe, Germany

MSC2000: 90C11, 90C10, 90C33, 90C27

Article Outline

Introduction
Definitions
Formulations
   Representing the Discrete Decisions by Approximate Continuous Variables
   Representing the Discrete Decisions by Exact Continuous Variables
   Modeling Propositional Logic Constraints with Exact Continuous Variables
   The Case of Inconsistent Equalities
Conclusions
See also
References

Introduction

Nonlinear optimization problems involving discrete decision variables, also known as generalized disjunctive programming (GDP) or mixed-integer nonlinear programming (MINLP) problems, arise frequently in applications. Examples from process engineering include the synthesis of heat exchanger or reactor networks, the optimization of separation processes, such as sequencing and tray optimization problems of distillation columns, and the optimization of entire process flowsheets [3]. The discrete decisions in these problems are usually related to the structure of the process whereas typical continuous variables are process states such as temperatures, concentrations or flows.

Connections between continuous and discrete optimization problems have been studied for several decades (see, e. g., [4]). In particular, in [9] it was observed that discrete variables can be modeled by complementarity constraints, that is, the discrete model is replaced by a continuous model. A broad survey on other approaches to model discrete decisions by continuous formulations is given in [10], including concave optimization problems and relaxation by semi-definite programming, with applications to the maximum clique problem, satisfiability, the Steiner tree problem, and minimax problems.

Extensive work has been addressed to discrete-continuous problems with linear objective function and constraints, known as mixed-integer linear programming (MILP) problems. In fact, a number of powerful algorithms have been developed which are ready to solve practically relevant, large-scale problems of this type. As soon as the objective function and the constraints comprise nonlinear terms in the continuous variables, as it is usually the case, for example, for problems in process engineering, the optimization problem is referred to as a mixed-integer nonlinear programming (MINLP) problem. Algorithms for MINLP problems are either based on branch and bound with nonlinear programming (NLP) subproblems or on decomposition methods that alternately solve NLP and MILP subproblems. These algorithms are guaranteed to locate the global optimum if the nonlinearities are *convex*.

Optimization problems involving *nonconvex* objective function and constraints are by far more difficult to solve. In [11] it is proposed to reformulate discrete-continuous optimization problems by the idea from [9], as purely continuous optimization problems with complementarity constraints. In this approach, the discrete variable set of an MINLP problem is replaced by continuous variables which are restricted to take discrete values by enforcing a special type of either non-differentiable or degenerate continuous constraints.

[15] complements this approach by purely continuous reformulations of MINLP problems with better theoretical properties, as will be explained below. As all continuous reformulation approaches inevitably lead to nonconvex optimization problems, searching for a global solution may be numerically challenging. On the other hand, these continuous reformulations yield efficient ways to locally solve MINLP problems on the basis of NLP solution methods.

Definitions

Consider a generalized disjunctive representation [12] of nonlinear optimization problems, where an objective function is minimized subject to two different types of constraints, namely global constraints that hold irrespectively of any discrete decision, and constraints contained in disjunctions that are only enforced if a corresponding Boolean variable $Y_{i,k}$ is True. The optimization problem is then formulated as follows:

$$(GDP) \min_{x,Y} \Phi(x) + \sum_{k \in K} b_k$$

s.t. $f(x) = 0,$ (1)

$g(x) \leq 0,$ (2)

$$\bigvee_{i \in D_k} \begin{bmatrix} Y_{i,k} \\ h_{i,k}(x) = 0, \\ r_{i,k}(x) \leq 0, \\ b_k = \gamma_{i,k}, \end{bmatrix}, k \in K,$$ (3)

$D_k = \{1, 2, \ldots, n_k\},$

$\Omega(Y) = \text{True}, Y_{i,k} \in \{\text{True}, \text{False}\}.$ (4)

In *GDP*, x represents a vector of continuous decision variables and $Y_{i,k}$ are Boolean variables. b_k is a scalar and $\gamma_{i,k}$ represents a fixed charge. The objective function comprises the sum of all fixed charges and a nonlinear term $\Phi(x)$. Whereas the model equations (1) and inequality constraints (2) hold irrespective of discrete choices, there are further equations and inequality constraints (3) that are contained in n_k, $k \in K$, disjunctions. Each disjunction k may consist of several terms $i \in D_k$, where the index set D_k defines the number of terms for each disjunction. Note that exactly one term $i \in D_k$ holds per disjunction, that is, $\bigvee_{i \in D_k}$ is understood as an 'exclusive or' operator. The disjunctive constraints are only enforced if the Boolean variable value $Y_{i,k}$ is True. Otherwise, if $Y_{i,k}$ is False, the corresponding constraints are removed from the optimization problem.

The Boolean variables themselves are related to each other by so called propositional logic constraints (4). These logic constraints are used to model interrelationships between disjunctive constraints. For example, assume that the first disjunctive term from disjunction $k = 1$ has to be selected ($Y_{1,1}$ = True) if another term from disjunction $k = 2$ is removed from the constraint set ($Y_{1,2}$ = False). This situation can be expressed by the implication $\neg Y_{1,2} \Rightarrow Y_{1,1}$, which can be transformed into a constraint of type (4):

$$Y_{1,2} \vee Y_{1,1} = \text{True} . \tag{5}$$

Any optimization problem in disjunctive form GDP can be posed as an equivalent MINLP problem [5] by, for example, transforming the disjunctive constraints into big-M or binary multiplication constraints and by replacing the Boolean variables $Y_{i,k}$ by binary variables $y_{i,k} \in \{0, 1\}$.

A problem reformulation based on binary multiplication is:

$$(BM) \quad \min_{x,y} \quad \Phi(x) + \sum_{k \in K} b_k$$

s.t. $f(x) = 0$,

$g(x) \leq 0$,

$$y_{i,k} \cdot h_{i,k}(x) = 0, \tag{6}$$

$$y_{i,k} \cdot r_{i,k}(x) \leq 0, \tag{7}$$

$$y_{i,k} \cdot (b_k - \gamma_{i,k}) = 0, \tag{8}$$

$$Ay \leq a, \tag{9}$$

$$\sum_{i \in D_k} y_{i,k} = 1, \tag{10}$$

$$y_{i,k} \in \{0, 1\}, \quad i \in D_k, k \in K, \tag{11}$$

where each disjunctive constraint is multiplied by a variable $y_{i,k}$. If $y_{i,k} = 0$, the corresponding constraint becomes redundant. On the other hand, a constraint contained in a disjunction is enforced with $y_{i,k} = 1$. The propositional logic constraints (4) can be modeled by the linear constraints (9) on the binary variables. Note that with these inequalities not only exclusive but also inclusive 'or'-relations can be modeled, although a binary variable itself takes only *exclusively* the values 0 or 1. In fact, for two binary variables y_1 and y_2 the inclusive relation $y_1 + y_2 \geq 1$, modeling (5), becomes exclusive under the additional relation $y_1 + y_2 \leq 1$.

It is important to note that the problem formulation BM has the drawback of being nonconvex even if the nonlinear, disjunctive constraints of the original optimization problem are convex. Thus, a problem reformulation based on binary multiplication would be employed only if the disjunctive optimization problem was nonconvex itself, as it is the case, for example, in a large portion of process engineering applications. Hence, this drawback should not be regarded as a strong limitation. Also note that the nonconvex expressions in (7) can be convexified if $r_{i,k}$ is a convex function [16]. However, in the following this assumption will not be made.

Formulations

Instead of applying an MINLP algorithm for solving the discrete-continuous optimization problem BM introduced in the previous section directly, one can reformulate the problem such that no discrete variables are present anymore. In particular, the discrete set defined in (10),(11) can be replaced by a set of restrictions involving continuous variables only, which can be used as constraints to form a purely continuous NLP. Since NLP solvers are usually designed to work with continuous variables, that is, variables from at least one-dimensional sets, the basic idea here is to increase the dimension of the constraint sets for $y_{i,k}$. Note that the discrete variables $y_{i,k}$ as defined in (11) are contained in a set of dimension zero.

For the explanation of the main ideas consider a single disjunction with $n_k = 2$ as it appears in (3). Put $y_k := y_{1,k}$ as well as $z_k := y_{2,k}$ and drop the fixed index k. This leads to a single binary decision variable

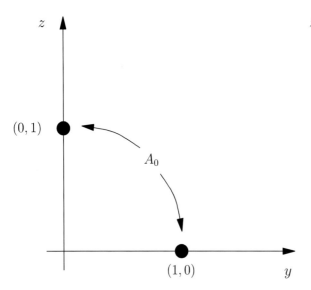

Continuous Reformulations of Discrete-Continuous Optimization Problems, Figure 1
Values of the discrete variable (y, z)

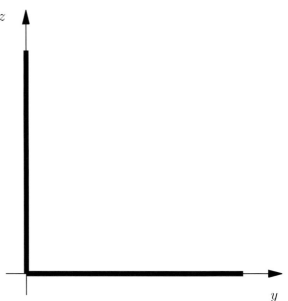

Continuous Reformulations of Discrete-Continuous Optimization Problems, Figure 2
The points which satisfy the complementarity condition (14),(15)

$y \in \{0, 1\}$ and its negation z, where the pair (y, z) can then attain exactly one of the values $(1, 0)$ and $(0, 1)$, that is,

$$(y, z) \in A_0 = \{(1, 0), (0, 1)\}, \qquad (12)$$

(cf. Fig. 1). Hence, in this case the conditions (10),(11) are replaced equivalently by (12). In the general case $n_k \geq 2$ there are several ways to use the set A_0 to reformulate (10),(11) equivalently. A first possibility is to introduce additional variables $z_{i,k} = 1 - y_{i,k}$ and replace only (11) by the conditions $(y_{i,k}, z_{i,k}) \in A_0, i \in D_k$.

Note that the constraint (10) guarantees that exactly one of the variables $y_{i,k}$, $i \in D_k$, is equal to 1, since these variables *can only take* the values 0 and 1. This restriction can be relaxed in conjunction with an alternative approach for modeling binary decision variables explained below. Having these later developments in mind, note that an alternative reformulation of (10), (11) using A_0 is

$$\left(y_{i,k}, \sum_{j \in D_k \setminus \{i\}} y_{j,k} \right) \in A_0, \quad i \in D_k, k \in K. \quad (13)$$

An advantage of the latter reformulation is that it does not increase the problem dimension by auxiliary variables $z_{i,k}$.

Representing the Discrete Decisions by Approximate Continuous Variables

There are a number of ways to describe A_0 with continuous constraints. A suggestion of [9,11] is to replace (12) with the equivalent set of constraints

$$y \cdot z = 0, \qquad (14)$$

$$y \geq 0, z \geq 0, \qquad (15)$$

$$y + z = 1. \qquad (16)$$

In fact, the constraints (14),(15) are known as a *complementarity* condition. They model a piecewise linear set with one kink at the origin in $\mathbb{R}^2$, as depicted in Fig. 2. Together with the constraint (16) one obtains exactly the set A_0 (cf. Fig. 3).

It is well-known that sets whose description contains complementarity conditions are not easy to treat numerically. In fact, the so-called Mangasarian–Fromovitz constraint qualification is violated everywhere in the feasible set as soon as a complementarity condition appears [14]. This constraint qualification, however, is known to characterize the (numerical) stability of the described set [6,13].

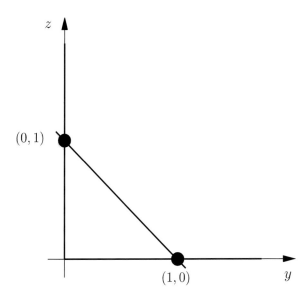

Continuous Reformulations of Discrete-Continuous Optimization Problems, Figure 3
Modeling discrete variables with a complementarity condition

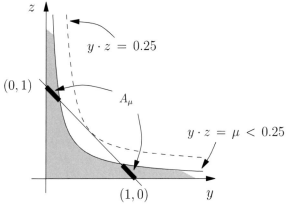

Continuous Reformulations of Discrete-Continuous Optimization Problems, Figure 4
Continuous variables for the relaxed complementarity condition

There are many suggestions on how a complementarity condition can be treated numerically, in particular in the literature on so-called mathematical programs with equilibrium constraints (MPECs) which are optimization problems with complementarity conditions in the constraints [7,8].

One approach is to use regularization techniques, for example to replace the condition (14) by its relaxation $y \cdot z \leq \mu$ with some positive parameter μ. The idea is to trace the solutions of the corresponding auxiliary problems to a solution of the original problem while driving μ to zero.

For the reformulation of binary variables this approach means that the discrete set A_0 is replaced by the one-dimensional set

$$A_\mu = \{(0, 1) + t \cdot (1, -1) | t \in [0, 0.5 - \sqrt{0.25 - \mu}] \cup [0.5 + \sqrt{0.25 - \mu}, 1]\},$$

which is disconnected for $\mu < 0.25$ as illustrated in Fig. 4.

Hence, this approach replaces discrete by continuous variables, at least via an approximation. [15] refers to the variables from the set A_μ as *approximate continuous*. In view of (13), a possible approximation of binary variables from a general disjunction is

$$\left(y_{i,k}, \sum_{j \in D_k \setminus \{i\}} y_{j,k} \right) \in A_\mu, \quad i \in D_k, \; k \in K$$

with $\mu > 0$.

Note that there are two serious drawbacks of the reformulation by a complementarity condition. First, a look at Fig. 3 shows that the kink at the origin is irrelevant for the description of A_0 because of the additional constraint (16). Thus, there is no need to use the numerically demanding complementarity condition together with (16), but any function with a smooth zero set and the correct intersection points would do. For example, one can use the constraint

$$\left(y - \frac{1}{2} \right)^2 + \left(z - \frac{1}{2} \right)^2 = \frac{1}{2},$$

which is illustrated in Fig. 5.

Here, the Mangasarian–Fromovitz constraint qualification and even the stronger linear independence constraint qualification are satisfied everywhere in the set A_0 (for background information on constraint qualifications see [1]). This can be seen as an important advantage when compared to the properties of A_0 represented by the complementarity condition.

A second drawback which the circle condition shares with the reformulation by a complementarity condition is that the variables are still contained in the

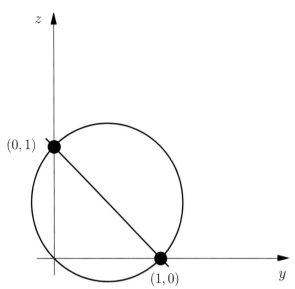
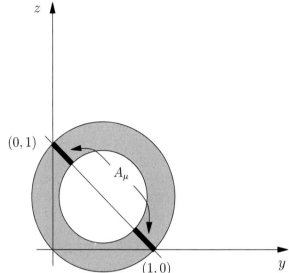

Continuous Reformulations of Discrete-Continuous Optimization Problems, Figure 5
The circle condition

Continuous Reformulations of Discrete-Continuous Optimization Problems, Figure 6
The circle relaxation

discrete set A_0. In order to obtain a one-dimensional set one can again relax the conditions that describe A_0 to obtain a set A_ν corresponding to A_μ from the MPEC relaxation above. In fact, the constraints

$$\left(y - \frac{1}{2}\right)^2 + \left(z - \frac{1}{2}\right)^2 \leq \frac{1}{2}$$

$$\left(y - \frac{1}{2}\right)^2 + \left(z - \frac{1}{2}\right)^2 \geq \left(\frac{1}{\sqrt{2}} - \nu\right)^2$$

$$y + z = 1,$$

with $\nu > 0$ describe sets A_ν (cf. Fig. 6), which correspond to the sets $A_\mu (\mu > 0)$ via a reparametrization, that is, one arrives at the same set of approximate continuous variables.

Unfortunately, for the limiting case $\nu = 0$ the circle is not described by one equality constraint but by two inequalities with gradients pointing in opposite directions, so that the Mangasarian–Fromovitz constraint qualification is then again violated in A_0.

Representing the Discrete Decisions by Exact Continuous Variables

Although both the reformulation by a complementarity condition and the reformulation by a circle condition lead to well performing numerical methods for small examples [11,15] they share two intrinsic drawbacks:

- the replacement for A_0 is one-dimensional, but only approximate,
- in the (limiting) case of an exact description for A_0, the Mangasarian–Fromovitz constraint qualification is violated,
- the one-dimensional set becomes discrete if equality constraints are inconsistent (see below).

Since these properties may affect the numerical solution of large problems, [15] proposes a different continuous reformulation of the integrality constraints with better theoretical features.

The subsequent considerations are based on an alternative model reformulation that allows to replace the discrete decision variables defined in (10) by variables $y_{i,k}$, which are *not* defined on a discrete set as, for example, A_0. In fact, this model reformulation has the property of being equivalent to the corresponding disjunctive optimization problem in conjunction with one-dimensional rather than discrete variables $y_{i,k}$. Before describing the model reformulation in detail, we focus on the variables $y_{i,k}$ and show how a continuous, one-dimensional set A_1 can be defined using appropriate constraints.

In fact, since in BM any disjunctive constraint is not only enforced by $y_{i,k} = 1$, but alternatively also by

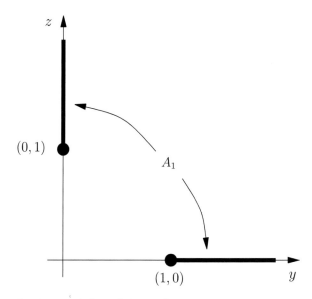

Continuous Reformulations of Discrete-Continuous Optimization Problems, Figure 7
A one-dimensional feasible set for (y, z).

$y_{i,k} \geq 1$, one may define $y_{i,k}$ as continuous variables of dimension one according to:

$$y_{i,k} \in \{0\} \cup [1, \infty).$$

Of course, now the negation of $y_{i,k}$ in general does not coincide with $1 - y_{i,k}$, as the value of $y_{i,k}$ might exceed 1. On the other hand, for the case $n_k = 2$ as above we obtain

$$(y, z) \in A_1 = ([1, \infty) \times \{0\}) \cup (\{0\} \times [1, \infty)) \quad (17)$$

(cf. Fig. 7). Hence, the negation of $y_{i,k}$ is coded by the variable $z_{i,k}$. Moreover, one can now describe the binary decisions via

$$\left(y_{i,k}, \sum_{j \in D_k \setminus \{i\}} y_{j,k} \right) \in A_1, \quad i \in D_k, \, k \in K. \quad (18)$$

The set A_1 is obviously one-dimensional and is an *exact* rather than approximate model of a discrete decision. Therefore, the variables defined by the set A_1 are referred to as *exact continuous*.

To be able to apply an NLP solution algorithm, one has to describe A_1 by continuous constraints. One possibility, of course, is to use the (degenerate) complementarity condition (14),(15) with the additional constraint $y + z \geq 1$. However, it is also possible to choose a function with an appropriate zero set, such that the linear independence constraint qualification holds everywhere in the feasible set. A function with these properties is the so-called *Fischer–Burmeister* function $\varphi_{FB}(y, z) = y + z - \sqrt{y^2 + z^2}$. This means that one can write

$$A_1 = \{ (y, z) \in \mathbb{R}^2 \mid \varphi_{FB}(y, z) = 0, \, y + z \geq 1 \}. \quad (19)$$

Equivalently, one could use a multitude of other so-called NCP-functions (for a survey see [2]). NCP-functions are used for the description of nonlinear complementarity problems. They are designed such that their zero set coincides with the set defined by (14),(15) (cf. Fig. 2). A description like (19) reveals better numerical properties than the original description via (14),(15). For example, whereas the Mangasarian–Fromovitz constraint qualification is violated everywhere in the set under a description via (14),(15), the description as the zero set of the Fischer–Burmeister function even leads to the validity of the linear independence constraint qualification everywhere in the set, except for the origin (which does not play a role here). In terms of the Fischer–Burmeister function, and using (19), the condition (18) is equivalent to

$$\varphi_{FB}\left(y_{i,k}, \sum_{j \in D_k \setminus \{i\}} y_{j,k} \right) = 0, \quad i \in D_k, \, k \in K,$$

$$\sum_{i \in D_k} y_{i,k} \geq 1, \quad k \in K.$$

Modeling Propositional Logic Constraints with Exact Continuous Variables

The question remains how logical conditions on two logical variables Y_1 and Y_2 should be modeled when (y_1, z_1) and (y_2, z_2) are not discrete but continuous as proposed in (17). This can easily be done by adding inequality constraints. In fact, $Y_1 \wedge Y_2$ is true if and only if $y_1 \geq 1$ and $y_2 \geq 1$. Moreover, $Y_1 \vee Y_2$ is true if and only if $y_1 + y_2 \geq 1$. For the negation of Y_1 one may *not* use $1 - y_1$, as y_1 might take a value strictly larger than one. On the other hand, for $n_k = 2$ the negation of Y_1 is already coded in the variable z_1. Moreover, for $n_k > 2$ the negation of $y_{i,k}$ is coded in $\sum_{j \in D_k \setminus \{i\}} y_{j,k}$, and one can proceed as above. Just like in the discrete case, inclusive as well as exclusive 'or'-relations can be modeled with exact continuous variables.

To circumvent the introduction of nonconvexity into the model by binary multiplication in *BM*, [15] presents an alternative, convex reformulation approach on the basis of tailored big-M constraints which can also be used in conjunction with exact continuous variables as defined in equation (17). A distinctive property of the binary multiplication-based model formulation *BM*, however, is the treatment of inconsistent equality constraints.

The Case of Inconsistent Equalities

In many applications, the constraints (6)–(8) in *BM* lead to implicit restrictions on the exact continuous variables. In particular, (6) and (8) have to hold simultaneously for $i \in D_k, k \in K$. In process engineering applications, the underlying equations (i. e. $h_{i,k}(x) = 0, i \in D_k$ as well as $b_k - \gamma_{i,k} = 0, i \in D_k$) are often inconsistent for fixed $k \in K$, that is, they do not admit a common solution or, put geometrically, the sets described by these equations are disjoint. Note that this is an inherent property of a GDP problem with so-called disjoint disjunctions which have non-empty intersecting feasible regions [17].

This is particularly the case, if for fixed $k \in K$ the values $\gamma_{i,k}$ are pairwise distinct for $i \in D_k$. It implies that *at most* one of the variables $y_{i,k}, i \in D_k$, is non-vanishing. In the case $n_k = 2$ with $y = y_1$ and $z = y_2$ this means that the equation $y \cdot z = 0$ holds automatically. As a consequence, the only constraint needed for the description of A_1 (cf. Fig. 7) is $y + z \geq 1$, that is, the set

$$A_2 = \{ (y, z) \in \mathbb{R}^2 \mid y + z \geq 1 \}$$

coincides with A_1 in the case of inconsistent equalities (cf. Fig. 8).

Although the pair (y, z) does not vary in the complete two-dimensional set A_2 from Fig. 8, in the restrictions one does not code the same information twice. This can be expected to lead to better numerical performance when NLP solvers are applied.

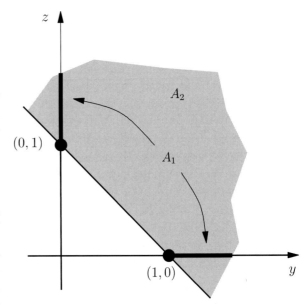

Continuous Reformulations of Discrete-Continuous Optimization Problems, Figure 8
A two-dimensional feasible set for (y, z)

neering are treated numerically, with approximate as well as exact continuous variables representing the discrete decisions. It is shown that, using these reformulations, an efficient numerical treatment of disjunctive optimization problems is possible, but one can only expect to find local solutions when using standard NLP solvers. This is due to the fact that any continuous reformulation of a disjunctive optimization problem leads to a nonconvex optimization problem. Consequently, the reformulation approaches may be combined with global optimization algorithms whenever the problem size admits to do so.

See also

▶ Disjunctive Programming
▶ Mixed Integer Programming/Constraint Programming Hybrid Methods
▶ Order Complementarity

Conclusions

In [15] several example problems involving discrete and continuous decision variables from process engi-

References

1. Bazaraa M, Sherali H, Shetty C (1993) Nonlinear Programming. Wiley, Hoboken, New Jersey

2. Chen B, Chen X, Kanzow C (2000) A penalized Fischer–Burmeister NCP-function. Math Program 88:211–216
3. Floudas CA (1995) Nonlinear and Mixed-Integer Optimization: Fundamentals and Applications. Oxford University Press, New York
4. Giannesi F, Niccolucci F (1976) Connections between nonlinear and integer programming problems. In: Institut Nazionale di Alta Mathematica (ed) Symposia Mathematica, vol XIX. Acad. Press, New York, pp 161–176
5. Grossmann IE, Hooker J (2000) Logic based approaches for mixed integer programming models and their application in process synthesis. In: Malone M, Trainham J, Carnahan B (eds) Foundations of Computer-Aided Process Design 323. AIChE Symp. Series. CACHE Publications, Austin, Texas, pp 70–83
6. Jongen HT, Weber G-W (1991) Nonlinear optimization: characterization of structural stability. J Glob Optim 1:47–64
7. Leyffer S (2006) Complementarity constraints as nonlinear equations: theory and numerical experiences. In: Dempe S, Kalashnikov V (eds) Optimization and Multivalued Mappings. Springer, Dordrecht, pp 169–208
8. Luo Z, Pang J, Ralph D (1996) Mathematical Programs with Equilibrium Constraints. Cambridge University Press, Cambridge
9. Pardalos PM (1994) The linear complementarity problem. In: Gomez S, Hennart JP (eds) Advances in Optimization and Numerical Analysis. Springer, New York, pp 39–49
10. Pardalos PM, Prokopyev OA, Busygin S (2006) Continuous approaches for solving discrete optimization problems. In: Appa G, Pitsoulis L, Williams HP (eds) Handbook on Modelling for Discrete Optimization. Springer, New York, pp 39–60
11. Raghunathan A, Biegler L (2003) Mathematical programs with equilibrium constraints (MPEC) in process engineering. Comp Chem Eng 27(10):1381–1392
12. Raman R, Grossmann IE (1994) Modelling and computational techniques for logic based integer programming. Comput Chem Eng 18(7):563–578
13. Robinson S (1976) Stability theory for systems of inequalities, part II: differentiable nonlinear systems. SIAM J Numer Anal 13:497–513
14. Scheel H, Scholtes S (2000) Mathematical programs with complementarity constraints: stationarity, optimality, and sensitivity. Math Oper Res 25:1–22
15. Stein O, Oldenburg J, Marquardt W (2004) Continuous reformulations of discrete-continuous optimization problems. Comput Chem Eng 28:1951–1966
16. Stubbs R, Mehrotra S (1999) A branch-and-cut method for 0-1 mixed convex programming. Math Program 86:515–532
17. Vecchietti A, Lee S, Grossmann IE (2003) Modeling of discrete/continuous optimization problems: characterization and formulation of disjunctions and their relaxations. Comput Chem Eng 27(3):433–448

Continuous Review Inventory Models: (Q, R) Policy

ISMAIL CAPAR[1], BURAK EKSIOGLU[2]

[1] Department of Engineering Technology and Industrial Distribution, Texas A&M University, College Station, USA

[2] Department of Industrial and Systems Engineering, Mississippi State University, Mississippi State, USA

MSC2000: 49-02, 90-02

Article Outline

Keywords
Introduction
Models
   Single-Echelon Models
   Multi-Echelon Models
Conclusions
References

Keywords

Continuous review inventory models; (Q, R) models

Introduction

Inventory control is an important issue in supply chain management. Today, many different approaches are used to solve the complicated inventory control problems. While some of the approaches use a periodic review cycle, others use methods based on continuous review of inventory. In this survey, stochastic inventory theory that is based on continuous review is analyzed.

One of the challenging tasks in continuous review inventory problems is finding the order quantity (Q) and the reorder point (R) such that the total cost is minimized and fill rate constraints are satisfied. The total cost includes ordering cost, backorder cost, and inventory holding cost. The fill rate is defined as the fraction of demand satisfied from inventory on hand. Under the continuous inventory control methodology, when the inventory position (on-hand inventory plus outstanding orders minus backorders) drops down to or below a reorder point, R, an order of size Q is placed.

Although they are all the same, there are many different representations of this inventory model such as

(Q, r) (Boyaci and Gallego [5]), (Q, R) (Hing et al. [14]), (R, Q) (Axsater [2,3] and Marklund [16]). In addition, for some of the problems it is assumed that the order quantity (nQ) is a multiple of minimum batch size, Q. Here, n is the minimum integer required to increase the inventory position to above R (Chen and Zheng[8]). In this case, the problem is formulated as a (R, nQ) type model.

Models

When the literature of (Q, R) models is investigated, some similarities and differences among the publications can easily be identified. Thus, the publications can be classified according to those similarities and differences. Two of the most distinctive attributes of (Q, R) models are as follows:

1. Type of supply chain: While some articles only consider one entity that uses a (Q, R) policy [1,5,14], others consider a multi echelon inventory system [2,3,8,16].
2. Exact evaluation or near-optimal evaluation: The (Q, R) inventory problems are not easy to solve; thus, many of the research papers give approximate solution approaches or try to find bounds on the solutions [1,2,4,5,20], only a small number of articles give the exact evaluation of the (Q, R) inventory system [3,9,21].

In the next section, the literature based on type of supply chain considered and the evaluation methods used is reviewed. First, heuristic methods are analyzed. Second, publications providing optimal methods are reviewed. In the last section, we give some concluding remarks.

Single-Echelon Models

Hing et al. [14] focus on average inventory level approximation in a (Q, R) system with backorders. They compare different approaches proposed in the literature. Their numerical analysis shows that the approximation developed by Hadley and Whitin [13], 1/2Q + safety stock, is more robust than other approximations that have been proposed so far. Then, the authors propose a new methodology based on spreadsheet optimization. Using numerical examples they show that spreadsheet optimization based approach is better than those methods proposed in the literature.

Agrawal and Seshadri [1] provide upper and lower bounds for optimal R and Q subject to fill rate constraints. Although the authors consider backorder cost, the algorithm that was developed to find bounds can be used when backorder costs are zero. Another important application of the algorithm is that it can be applied when there are no service level constraints.

Like Agrawal and Seshadri [1], Platt et al. [19] also consider fill rate constraints and propose two heuristics that can be used for (Q, R) policy models. While the first heuristic is suitable for deterministic lead time demand models, the second one assumes that demand during the lead time follows a normal distribution. Both heuristics are used to find R and Q values. The authors compare the proposed heuristics with others that have been proposed in the literature. Their analysis shows that their heuristics do not necessarily outperform the other heuristics in each problem instance.

Boyaci and Gallego [5] propose a new (Q, R) model that minimizes average holding and ordering costs subject to upper bounds on the expected and maximum waiting times for the backordered items. They provide optimality conditions and an exact algorithm for the problem. Boyaci and Gallego [5] conclude their study by performing a numerical analysis.

Gallego [12] proposes heuristics to find distribution-free bounds on the optimal cost and optimal batch size when a (Q, R) policy is used. He also shows that the heuristics work well when the demand distribution is Poisson or compound Poisson.

Bookbinder and Cakanyildirim [4] consider a (Q, R) policy where lead time is not constant. They treat lead time as a random variable and develop two probabilistic models. While in the first model the lead time is fixed, in the second model the lead time can be reduced by using an expediting factor (τ). The order quantity, reorder point, and expediting factor are the the three decision variables in the second model. The authors show that for both models the expected cost per unit time is jointly convex. They also make a sensitivity analysis with respect to cost parameters.

Ryu and Lee [20] consider the lead time as a decision variable. However, in this study the demand is constant. In their model, Ryu and Lee [20] assume that there are two suppliers for the items to be procured. They mainly consider two cases. In the first case, lead time cannot be decreased but in the second case, orders can be expedited. The authors also assume that lead-time distributions are non-identical exponential. For the first case, their objective is to determine a Q, an R, and an order-splitting proportion. In the second case, they find new values for the lead times using the order-splitting proportion. Their sensitivity analysis shows that the order-splitting proportion tends to be a half, and it is biased by the coefficient of the expediting function.

Cakanyildirim et al. [6] develop a model that considers lead-time variability. The authors assume that lead time is effected by both the lot size and the reserved capacity. The authors come up with a closed-form solution for the situation where lead time is proportional to the lot size. Cakanyildirim et al. [6] also present the effect of linear and concave lead times on the value of cost function. In the model, in addition to the order quantity and the reorder point, the reserved capacity is also a decision variable. Finally, the authors consider a case in which fixed proportion of capacity is allocated at the manufacturing facility.

Most of the articles in the literature consider lead time as a constant and focus on demand during the lead time. However, Wu and Ouyang [21] assume that lead time is a decision variable and that lead-time demand follows a normal distribution. They also assume that an arrival order may contain some defective parts and that those parts will be kept in inventory until next delivery. Moreover, they include an inspection cost for defective parts to the model. Their model is defined as (Q, R, L) inventory model where order quantity (Q), reorder point (R), and lead time (L) are decision variables. The objective is to minimize the total cost which includes ordering costs, inventory holding costs (defective and non-defective), lost sales costs, backorder costs, and inspection costs. The authors present an algorithm to find the optimal solutions for the given problem.

Duran et al. [9] present a (Q, R) policy where orders can be expedited. At the time of order release, if inventory position is less than or equal to a critical value r_e, the order is expedited at an additional cost. If the inventory level is higher then r_e and lower than or equal to the reorder point R, then order is not expedited. The aim is to find the order quantity (Q), the reorder point (R), and the expediting point r_e which minimize average cost (note that this does not include backorder costs). The authors present an optimal algorithm to obtain the Q, R, and r_e values if they are restricted to be integers.

The model proposed by Kao and Hsu [15] is different from other models reviewed in this paper because the authors discuss the order quantity and reorder point with fuzzy demand. Kao and Hsu [15] use this fuzzy demand to construct the fuzzy total inventory cost. The authors derive five pairs of simultaneous nonlinear equations to find the optimal order quantity Q and the reorder point R. The authors show that when the demand is a trapezoid fuzzy number, the equations can be reduced to a set of closed-form equations. Then, they prove that the solution to these equations give an optimal solution. Kao and Hsu [15] also present a numerical example to show that the solution methodology developed in the paper is easy to apply in practice.

Multi-Echelon Models

Moinzadeh and Lee [18] present a model to determine the batch size in a multi-echelon system with one central depot and M sites. In their problem, when the number of failed items is equal to the order quantity Q at any site, then those items are sent to the depot. If the depot has sufficient inventory on hand, it delivers the items immediately; otherwise, the items are backlogged. Although all sites use a (Q, R) policy, the depot uses a $(S-1, S)$ policy. In other words, whenever the depot receives an order of size Q, it places an order simultaneously to replenish its stock. After determining the Q and R values for each site, the authors use an approximation to estimate the total system stock and the backorder levels. The numerical results show that the (Q, R) policy is better than the $(S-1, S)$ policy for such systems.

Forsberg [10] deals with a multi-echelon inventory system with one warehouse and multiple non-identical retailers. The author assumes that the retailers face independent Poisson demand and both the warehouse and the retailers use (Q, R) policies. Forsberg [10] evaluates inventory holding and shortage costs using an exact solution approach.

Chen and Zheng [8] study a (nQ, R) policy in a multi-stage serial inventory system where *stage 1* orders from *stage 2*, *stage 2* from *stage 3*, etc., and *stage N* places orders to an outside supplier with unlimited capacity. The demand seen by *stage 1* is compound Poisson and excess demand is backlogged at every stage. The transportation lead times among stages are constant. By using a two-step approach, Chen and Zheng [8] provide near-optimal solution. In the first step, they find the lower and upper bounds on the cost function by changing the penalty cost of being short on inventory. In the second step, the authors minimize the bounds by using three different heuristic approaches. Chen and Zheng [8] also propose an optimal algorithm that requires additional computational effort.

Axsater [2] considers a two-stage inventory system with one warehouse and N non-identical retailers. He presents an exact method to evaluate inventory holding and shortage costs when there are only two retailers. He focuses on the timing of the warehouse orders for the sub-batches of Q. He identifies three possibilities and evaluates the cost for each case separately. At the end, total cost is calculated by summing the costs for the three cases. When there are more than two retailers, he extends his evaluation technique by combining the retailers into two groups, and then uses the same approach he developed for the two retailer case. The author also presents a model where the lead times are constant and all facilities use (Q, R) policies with different Q and R values. In this model, all stockouts are backordered, delayed orders are delivered on a first-come-first-served basis, and partial shipments are also allowed. In order to simplify the problem, Axsater [2] assumes that all batch sizes are multiples of the smallest batch size. In the objective function, the author only considers expected inventory holding cost and backorder cost.

Like Axsater [2], Marklund [16] also considers a two-stage supply chain with one central warehouse and an arbitrary number of non-identical retailers. Customer demands occur only at the retailers. The retailers use (Q, R) policies with different parameters, and they request products from the central warehouse whenever their inventory positions reach R or fall below R. The author proposes a new policy (Q_0, a_0) that is motivated by relating the traditional echelon stock model to the installation stock (Q, R) model where the order quantity Q is a multiple of a minimum batch size. In the article, Marklund [16] gives the detailed derivation of the exact cost function when the retailers use (Q, R) policies and the warehouse uses the new (Q_0, a_0) policy. The performance of the new policy is compared to traditional echelon stock policy and (Q, R) policy through numerical examples. Although the results show that the proposed policy outperforms the other policies in all numerical examples, the author does not guarantee that the policy will always give the best result.

Fujiwara and Sedarage [11] apply a (Q, R) policy for a multi-part assembly system under stochastic lead times. The objective of the article is to simultaneously determine the order quantity and the assembly lot size so that the average total cost per unit time is minimized. The total cost includes setup costs, inventory holding costs of parts and assembled items, and shortage costs of assembled items. The authors try to find separate reorder points, r_i, for each part and a global order quantity, Q, which will be used for all parts. Although the authors propose a global order quantity Q, they also mention that this kind of policy may not be optimal. They suggest that instead of a global Q, a common Q where all order quantities are multiples of Q might be more sensible.

Chen and Zheng [7] consider a distribution system with one warehouse and multiple retailers. The retailers' demands follow an independent compound Poisson process. It is assumed that the order quantity is a multiple of the smallest batch size. The order quantity and the reorder point are calculated by using a heuristic. The authors present an exact procedure for evaluating the performance (average cost) of the (nQ, R) policy when the demand is a Poisson process. Chen and Zheng [7] also give two approximation procedures for the case with compound Poisson processes. The approximations are based on exact formulations of the case with Poisson processes.

Axsater [3] presents an exact analysis of a two-stage inventory system with one warehouse and multiple retailers. The demand for each retailer follows an independent compound Poisson process. The retailers replenish their stock from the warehouse, and the warehouse replenishes its stock from an outside supplier. The transportation times from the warehouse to the retailers and from the outside supplier to the warehouse are constant. In addition, if there is a shortage, then additional delay may also occur since shortages and stockouts are backordered. The author emphasizes that the approach developed is not directly applicable for items with large demand. Instead, it is suitable mostly for slow-moving parts such as spare parts.

Moinzadeh [17] also considers a supply chain with one warehouse and multiple identical retailers. The author assumes that demand at the retailers is random but stationary and that each retailer places its order according to a (Q, R) policy. In addition, Moinzadeh [17] assumes that the warehouse receives online information about the demand. The author shows the effect of information sharing on order replenishment decisions of the supplier. In the article, the author first proposes a possible replenishment policy for the supplier and then provides an exact analysis for the operating measures of such systems. The author concludes the article by giving information about when information sharing is most beneficial.

Conclusions

We provide a literature review on continuous review (Q, R) inventory policies. Although we review most of the well known papers that deal with (Q, R) policy, this is not an exhaustive review of the literature. Our aim is to present the importance of the (Q, R) policy and show possible extensions of the simple (Q, R) model.

References

1. Agrawal V, Seshadri S (2000) Distribution free bounds for service constrained (Q, r)inventory systems. Nav Res Logist 47:635–656
2. Axsater S (1998) Evaluation of installation stock based (R, Q)-policies for two-level inventory systems with poisson demand. Oper Res 46(3):135–145
3. Axsater S (2000) Exact analysis of continuous review (R, Q) policies in two-echelon inventory systems with compound poisson demand. Oper Res 48(5):686–696
4. Bookbinder JH, Cakanyildirim M (1999) Random lead times and expedited orders in (Q, r) inventory systems. Eur J Oper Res 115:300–313
5. Boyaci T, Gallego G (2002) Managing waiting times of backordered demands in single-stage (Q, r) inventory systems. Nav Res Logist 49:557–573
6. Cakanyildirim M, Bookbinder JH, Gerchak Y (2000) Continuous review inventory models where ran- dom lead time depends on lot size and reserved capacity. Int J Product Econ 68:217–228
7. Chen F, Zheng YS (1997) One warehouse multi-retailer system with centralized stock information. Oper Res 45(2):275–287
8. Chen F, Zheng YS (1998) Near-optimal echelon-stock (R, nQ) policies in multistage serial systems. Oper Res 46(4):592–602
9. Duran A, Gutierrez G, Zequeira RI (2004) A continuous review inventory model with order expediting. Int J Product Econ 87:157–169
10. Forsberg R (1997) Exact evaluation of (R, Q)-policies for two-level inventory systems with Poisson demand. Eur J Oper Res 96:130–138
11. Fujiwara O, Sedarage D (1997) An optimal (Q, r) policy for a multipart assembly system under stochastic part procurement lead times. Eur J Oper Res 100:550–556
12. Gallego G (1998) New bounds and heuristics for (Q, r) policies. Manag Sci 44(2):219–233
13. Hadley G, Whitin TM (1963) Analysis of inventory systems. Prentice-Hall, Englewood Cliffs, NJ
14. Hing A, Lau L, Lau HS (2002) A comparison of different methods for estimating the average inventory level in a (Q; R) system with backorders. Int J Product Eco 79:303–316
15. Kao C, Hsu WH (2002) Lot size-reorder point inventory model with fuzzy demands. Comput Math Appl 43:1291–1302
16. Marklund J (2002) Centralized inventory control in a two-level distribution system with poisson demand. Nav Res Logist 49:798–822
17. Moinzadeh K (2002) A multi-echelon inventory system with information exchange. Manag Sci 48(3):414–426
18. Moinzadeh K, Lee HL (1986) Batch size and stocking levels in multi-echelon reparable systems. Manag Sci 32(12):1567–1581
19. Platt DE, Robinson LW, Freund RB (1997) Tractable (Q, R) heuristic models for constrained service level. Manag Sci 43(7):951–965
20. Ryu SW, Lee KK (2003) A stochastic inventory model of dual sourced supply chain with lead-time reduction. Int J Product Eco 81–82:513–524
21. Wu KS, Ouyang LY (2001) (Q, r, L) inventory model with defective item. Comput Ind Eng 39:173–185

Contraction-Mapping

C. T. Kelley

Department Math. Center for Research in Sci., North Carolina State University, Raleigh, USA

MSC2000: 65H10, 65J15

Article Outline

Keywords
Statement of the Result
Affine Problems
Nonlinear Problems
Integral Equations Example

See also
References

Keywords

Nonlinear equations; Linear equations; Integral equations; Iterative method; Contraction mapping

Statement of the Result

The method of *successive substitution*, *Richardson iteration*, or *direct iteration* seeks to find a *fixed point* of a map K, that is a point u^* such that

$$u^* = K(u^*).$$

Given an initial iterate u_0, the iteration is

$$u_{k+1} = K(u_k), \quad \text{for} \quad k \geq 0. \tag{1}$$

Let X be a Banach space and let $D \subset X$ be closed. A map $K : D \to D$ is a *contraction* if

$$\|K(u) - K(v)\| \leq \alpha \|u - v\| \tag{2}$$

for some $\alpha \in (0, 1)$ and all $u, v \in D$. The contraction mapping theorem, [3,7,13,14], states that if K is a contraction on D then
- K has a unique fixed point u^* in D, and
- for any $u_0 \in D$ the sequence $\{u_k\}$ given by (1) converges to u^*.

The message of the contraction mapping theorem is that if one wishes to use direct iteration to solve a fixed point problem, then the fixed point map K must satisfy (2) for some D and relative to some choice of norm. The choice of norm need not be made explicitly, it is determined implicitly by the K itself. However, if there is no norm for which (2) holds, then another, more robust, method, such as Newton's method with a line search, must be used, or the problem must be reformulated.

One may wonder why a Newton-like method is not always better than a direct iteration. The answer is that the cost for a single iteration is very low for Richardson iteration. So, if the equation can be set up to make the contraction constant α in (2) small, successive substitution, while taking more iterations, can be more efficient than a Newton-like iteration, which has costs in linear algebra and derivative evaluation that are not incurred by successive substitution.

Affine Problems

An affine fixed point map has the form

$$K(u) = Mu + b$$

where M is a linear operator on the space X. The fixed point equation is

$$(I - M)u = b, \tag{3}$$

where I is the identity operator. The classical stationary iterative methods in numerical linear algebra, [8,13], are typically analyzed in terms of affine fixed point problems, where M is called the iteration matrix. Multigrid methods, [2,4,5,9], are also stationary iterative methods. We give an example of how multigrid methods are used later in this article.

The contraction condition (2) holds if

$$\|M\| \leq \alpha < 1. \tag{4}$$

In (4) the norm is the operator norm on X. M may be a well defined operator on more than one space and (4) may not hold in all of them. Similarly, if X is finite dimensional and all norms are equivalent, (4) may hold in one norm and not in another. It is known, [10], that (4) holds for some norm if and only if the spectral radius of M is < 1.

When (4) does not hold it is sometimes possible to form an approximate inverse *preconditioner* P so that direct iteration can be applied to the equivalent problem

$$u = (I - P(I - M))u - Pb. \tag{5}$$

In order to apply the contraction mapping theorem and direct iteration to (5) we require that

$$\|I - P(I - M)\| \leq \alpha < 1$$

in some norm. In this case we say that P is an *approximate inverse* for $I - M$. In the final section of this article we give an example of how approximate inverses can be built for discretizations of integral operators.

Nonlinear Problems

If the nonlinear fixed point map K is sufficiently smooth, then a Newton-like method may be used to solve

$$F(u) = u - K(u) = 0.$$

The transition from a current approximation u_c of u^* to an update u_+ is

$$u_+ = u_c - P(u_c - K(u_c)), \quad (6)$$

where

$$P \approx F'(u^*)^{-1} = (I - K'(u^*))^{-1}.$$

$P = F(u_c)^{-1}$ is Newton's method and $P = F'(u_0)^{-1}$ is the chord method.

It is easy to show [7,13,14] that if u is near u^* and P is an approximate inverse for $F'(u^*)$ then the preconditioned fixed point problem

$$u = u - P(u - K(u))$$

is a contraction on a neighborhood D of u^*. This is, in fact, one way to analyze the convergence of Newton's method. In this article our focus is on preconditioners that remain constant for several iterations and do not require computation of the derivative of K.

The point to remember is that, if the goal is to transform a given fixed point map into a contraction, preconditioning of nonlinear problems can be done by the same process (formation of an approximate inverse) as for linear problems.

Integral Equations Example

We close this article with the *Atkinson–Brakhage preconditioner* for integral operators [2,4]. We will begin with the linear case, from which the nonlinear algorithm is a simple step. Let $\Omega \in \mathbf{R}^N$ be compact and let $k(x, y)$ be a continuous function on $\Omega \times \Omega$. We consider the affine fixed point problem

$$u(x) = f(x) + (\mathbf{K}u)(x) = f(x) + \int_\Omega k(x,y) u(y) \, dy,$$

where $f \in X = C(\Omega)$ is given and a solution $u^* \in X$ is sought. In this example $D = X$. We will assume that the linear operator $I - \mathbf{K}$ is nonsingular on X.

We consider a family of increasingly accurate quadrature rules, indexed with a level l, with weights $\{w_i^l\}_{i=1}^{N_l}$ and nodes $\{x_i^l\}_{i=1}^{N_l}$ that satisfy

$$\lim_{l \to \infty} \sum_{j=1}^{N_l} f(x_j^l) w_j^l = \int_\Omega f(x) \, dx$$

for all $f \in X$. The family of operators $\{\mathbf{K}_l\}$ defined by

$$\mathbf{K}_l u(x) = \sum_{j=1}^{N_l} k(x, x_j^l) u(y) w_j^l$$

converges strongly to $\mathbf{K}$, that is

$$\lim_{l \to \infty} \mathbf{K}_l u = \mathbf{K} u$$

for all $u \in X$. The family $\{\mathbf{K}_l\}$ is also *collectively compact*, [1]. This means that if $\mathbf{B}$ is a bounded subset of X, then

$$\cup_l \mathbf{K}_l(\mathbf{B})$$

is precompact in X. The direct consequences of the strong convergence and collective compactness are that $I - \mathbf{K}_l$ are nonsingular for l sufficiently large and

$$(I - \mathbf{K}_l)^{-1} \to (I - \mathbf{K})^{-1} \quad (7)$$

strongly in X. The Atkinson–Brakhage preconditioner is based on these results.

For $g \in X$ one can compute

$$v = (I - \mathbf{K}_l)^{-1} g$$

by solving the finite-dimensional linear system

$$v_i = g(x_i^l) + \sum_{j=1}^{N_l} k(x_i^l, x_j^l) v_j w_j^l \quad (8)$$

for the values $v(x_i^l) = v_i$ of v at the nodal points and then applying the *Nyström interpolation*

$$v(x) = g(x) + \sum_{j=1}^{N_l} k(x, x_j^l) v_j w_j^l = g(x) + (\mathbf{K}_l v)(x)$$

to recover $v(x)$ for all $x \in \Omega$. (8) can be solved at a cost of $O(N_l^3)$ floating point operations if direct methods for linear equations are used and for much less if iterative methods such as GMRES [15] are used. In that case, only $O(1)$ matrix-vector products are need to obtain a solution that is accurate to truncation error [6]. This is, up to a multiplicative factor, optimal. The Atkinson–Brakhage preconditioner can dramatically reduce this factor, however.

The results in [1] imply that

$$M_l = I + (I - \mathbf{K}_l)^{-1} \mathbf{K},$$

the Atkinson–Brakhage preconditioner, converges to $(I - K)^{-1}$ *in the operator norm*. Hence, for l sufficiently large (coarse mesh sufficiently fine) Richardson iteration can be applied to the system

$$u = u - M_l(I - \mathbf{K}_l)u - M_l f,$$

where $L \gg l$. Applying this idea for a sequence of grids or levels leads to the optimal form of the Atkinson–Brakhage iteration [11]. The algorithm uses a coarse mesh, which we index with $l = 0$, to build the preconditioner and then cycles through the grids sequentially until the solution at a desired fine ($l = L$) mesh is obtained. One example of this is a sequence of composite midpoint rule quadratures in which $N_{l+1} = 2N_l$. Then, [2,11], if the coarse mesh is sufficiently fine, only one Richardson iteration at each level will be needed. The cost at each level is two matrix vector products at level l and a solve at level 0.

1) Solve $u_0 - \mathbf{K}_0 u_0 = f$; set $u = u_0$.
2) For $l = 1, \ldots, L$:
   a) Compute $r = u - \mathbf{K}_l u - f$;
   b) $u = u - M_0 r$.

Nonlinear problems can be solved with exactly the same idea. We will consider the special case of *Hammerstein equations*

$$u(x) = \mathbf{K}(u)(x) = \int_\Omega k(x, y, u(y)) \, dy.$$

If we use a sequence of quadrature rules as in the linear case we can define

$$\mathbf{K}_l(u)(x) = \sum_{j=1}^{N_l} k(x, x_j^l, u(x_j^l)) w_j^l.$$

The nonlinear form of the Atkinson–Brakhage algorithm for Hammerstein equations simply uses the approximation

$$I + (I - \mathbf{K}'_0(u_0))^{-1} \mathbf{K}'(u) \approx (I - \mathbf{K}'_l(u))^{-1}$$

in a Newton-like iteration. One can see from the formal description below that little has changed from the linear case.

1) Solve $u_0 - \mathbf{K}_0 (u_0) = 0$; set $u = u_0$.
2) For $l = 1, \ldots, L$:
   a) Compute $r = u - \mathbf{K}_l(u)$;
   b) $u = u - (I + (I - \mathbf{K}'_l(u_0))^{-1} \mathbf{K}'(u)) r$.

The Atkinson–Brakhage algorithm can, under some conditions, be further improved, [12] and the number of fine mesh operator-function products per level reduced to one. There is also no need to explicitly represent the operator as an integral operator with a kernel.

See also

▶ Global Optimization Methods for Systems of Nonlinear Equations
▶ Interval Analysis: Systems of Nonlinear Equations
▶ Nonlinear Least Squares: Newton-Type Methods
▶ Nonlinear Systems of Equations: Application to the Enclosure of all Azeotropes

References

1. Anselone PM (1971) Collectively compact operator approximation theory. Prentice-Hall, Englewood Cliffs, NJ
2. Atkinson KE (1973) Iterative variants of the Nyström method for the numerical solution of integral equations. Numer Math 22:17–31
3. Banach S (1922) Sur les opérations dans les ensembles abstraits et leur applications aux équations intégrales. Fundam Math 3:133–181
4. Brakhage H (1960) Über die numerische Behandlung von Integralgleichungen nach der Quadraturformelmethode. Numer Math 2:183–196
5. Briggs W (1987) A multigrid tutorial. SIAM, Philadelphia
6. Campbell SL, Ipsen ICF, Kelley CT, Meyer CD, Xue ZQ (1996) Convergence estimates for solution of integral equations with GMRES. J Integral Eq Appl 8:19–34
7. Dennis JE, Schnabel RB (1996) Numerical methods for nonlinear equations and unconstrained optimization. Classics Appl Math, vol 16, SIAM, Philadelphia
8. Golub GH, VanLoan CG (1983) Matrix computations. Johns Hopkins Univ. Press, Baltimore, MD
9. Hackbusch W (1985) Multi-grid methods and applications. Comput Math, vol 4. Springer, Berlin
10. Isaacson E, Keller HB (1966) Analysis of numerical methods. Wiley, New York
11. Kelley CT (1990) Operator prolongation methods for nonlinear equations. In: Allgower EL, Georg K (eds) Computational Solution of Nonlinear Systems of Equations. Lect Appl Math Amer Math Soc, Providence, RI, pp 359–388
12. Kelley CT (1995) A fast multilevel algorithm for integral equations. SIAM J Numer Anal 32:501–513
13. Kelley CT (1995) Iterative methods for linear and nonlinear equations. No. in Frontiers in Appl Math, vol 16 SIAM, Philadelphia
14. Ortega JM, Rheinboldt WC (1970) Iterative solution of nonlinear equations in several variables. Acad Press, New York

15. Saad Y, Schultz MH (1986) GMRES a generalized minimal residual algorithm for solving nonsymmetric linear systems. SIAM J Sci Statist Comput 7:856–869

Control Vector Iteration CVI

REIN LUUS

Dept. Chemical Engineering, Univ. Toronto, Toronto, Canada

MSC2000: 93-XX

Article Outline

Keywords and Phrases
Optimal Control Problem
Second Variation Method
Determination of Stepping Parameter
Illustration of the First Variation Method
See also
References

Keywords and Phrases

Optimal control; Control vector iteration; Variation method; Pontryagin's maximum principle

In solving optimal control problems involving nonlinear differential equations, some iterative procedure must be used to obtain the optimal control policy. As is true with any iterative procedure, one is concerned about the convergence rate and also about the reliability of obtaining the optimal control policy. Although from Pontryagin's maximum principle it is known that the minimum of the performance index corresponds to the minimum of the Hamiltonian, to obtain the minimum value for the Hamiltonian is not always straightforward. Here we outline a procedure that changes the control policy from iteration to iteration, improving the value of the performance index at each iteration, until the improvement is less than certain amount. Then the iteration procedure is stopped and the results are analyzed. Such a procedure is called control vector iteration method (CVI), or iteration in the policy space.

Optimal Control Problem

To illustrate the procedure, let us consider the optimal control problem, where the system is described by the differential equation

$$\frac{d\mathbf{x}}{dt} = \mathbf{f}(\mathbf{x}, \mathbf{u}), \quad \text{with } \mathbf{x}(0) \text{ given}, \quad (1)$$

where $\mathbf{x}$ is an n-dimensional state vector and $\mathbf{u}$ is an r-dimensional control vector. The optimal control problem is to determine the control $\mathbf{u}$ in the time interval $0 \le t < t_f$, so that the performance index

$$I = \int_0^{t_f} \phi(\mathbf{x}, \mathbf{u}) \, dt \quad (2)$$

is minimized. We consider the case where the final time t_f is given. To carry out the minimization of the performance index in (2) subject to the constraints in (1), we consider the augmented performance index

$$J = \int_0^{t_f} \left[\phi + \mathbf{z}^T \left(\mathbf{f} - \frac{d\mathbf{x}}{dt} \right) \right] dt, \quad (3)$$

where the n-dimensional vector of Lagrange multipliers $\mathbf{z}$ is called the adjoint vector. The last term in (3) can be thought of as a penalty function to ensure that the state equation is satisfied throughout the given time interval. We introduce the Hamiltonian

$$H = \phi + \mathbf{z}^T \mathbf{f} \quad (4)$$

and use integration by parts to simplify (3) to

$$J = \int_0^{t_f} \left(H + \frac{d\mathbf{z}^T}{dt} \mathbf{x} \right) dt - \mathbf{z}(t_f) \mathbf{x}(t_f) + \mathbf{z}^T(0) \mathbf{x}(0). \quad (5)$$

The optimal control problem now reduces to the minimization of J.

To minimize J numerically, we assume that we have evaluated J at iteration j by using control policy denoted by $\mathbf{u}^{(j)}$. Now the problem is to determine the control policy $\mathbf{u}^{(j+1)}$ at the next iteration. Since the goal is to minimize J, obviously we want to make the change in J negative and numerically as large as possible. If we let $\delta \mathbf{u} = \mathbf{u}^{(j+1)} - \mathbf{u}^{(j)}$, the corresponding change in J is obtained by using Taylor series expansion up to the quadratic terms:

$$\delta J = \int_0^{t_f} \left[\left(\left(\frac{\partial H}{\partial \mathbf{x}}\right)^T + \frac{d\mathbf{z}^T}{dt} \right) \delta \mathbf{x} + \left(\frac{\partial H}{\partial \mathbf{u}}\right)^T \delta \mathbf{u} \right] dt$$
$$+ \frac{1}{2} \int_0^{t_f} \left[\delta \mathbf{x}^T \frac{\partial^2 H}{\partial \mathbf{x}^2} \delta \mathbf{x} + 2 \delta \mathbf{x}^T \frac{\partial^2 H}{\partial \mathbf{x} \partial \mathbf{u}} \delta \mathbf{u} \right.$$
$$\left. + \delta \mathbf{u}^T \frac{\partial^2 H}{\partial \mathbf{u}^2} \delta \mathbf{u} \right] dt - \mathbf{z}^T(t_f) \delta \mathbf{x}(t_f) . \quad (6)$$

The necessary condition for minimum of J is that the first integral in (6) should be zero; i. e.,

$$\frac{d\mathbf{z}}{dt} = -\frac{\partial H}{\partial \mathbf{x}}, \quad \text{with } \mathbf{z}(t_f) = \mathbf{0} . \quad (7)$$

and

$$\frac{\partial H}{\partial \mathbf{u}} = \mathbf{0} . \quad (8)$$

In control vector iteration, we relax the necessary condition in (8) and choose $\delta \mathbf{u}$ to make δJ negative and in the limit (8) is satisfied. One approach is to choose

$$\delta \mathbf{u} = -\epsilon \frac{\partial H}{\partial \mathbf{u}} , \quad (9)$$

where ϵ is a positive parameter which may vary from iteration to iteration. This method is sometimes called first variation method, since the driving force for the change in the control policy is based only on the first term of the Taylor series expansion. The negative sign in (9) is required to minimize the Hamiltonian, as is required by Pontryagin's maximum principle. Numerous papers have been written on the determination of the stepping parameter ϵ [7].

Second Variation Method

Instead of arbitrarily determining the stepping parameter ϵ, one may solve the accessory minimization problem, where $\delta \mathbf{u}$ is chosen to minimize δJ given by (6) after the requirements for the adjoint are satisfied; i. e., it is required to find $\delta \mathbf{u}$ to minimize δJ given by

$$\delta J = \int_0^{t_f} \left[\left(\frac{\partial H}{\partial \mathbf{u}}\right)^T \delta \mathbf{u} + \frac{1}{2} \delta \mathbf{x}^T \frac{\partial^2 H}{\partial \mathbf{x}^2} \delta \mathbf{x} \right.$$
$$\left. + \delta \mathbf{x}^T \frac{\partial^2 H}{\partial \mathbf{x} \partial \mathbf{u}} \delta \mathbf{u} + \frac{1}{2} \delta \mathbf{u}^T \frac{\partial^2 H}{\partial \mathbf{u}^2} \delta \mathbf{u} \right] dt , \quad (10)$$

subject to the differential equation

$$\frac{d \delta \mathbf{x}}{dt} = \left(\frac{\partial \mathbf{f}^T}{\partial \mathbf{x}}\right)^T \delta \mathbf{x} + \left(\frac{\partial \mathbf{f}^T}{\partial \mathbf{u}}\right)^T \delta \mathbf{u},$$
$$\text{with } \delta \mathbf{x}(0) = \mathbf{0} . \quad (11)$$

The solution to this accessory minimization problem is straightforward, since (11) is linear and the performance index in (10) is almost quadratic, and can be easily done, as shown in ([1], pp. 259–266) and [7]. The resulting equations, to be integrated backwards from $t = t_f$ to $t = 0$ with zero starting conditions, are

$$\frac{d\mathbf{J}}{dt} + \mathbf{J} \left(\frac{\partial \mathbf{f}^T}{\partial \mathbf{x}}\right)^T + \frac{\partial \mathbf{f}^T}{\partial \mathbf{x}} \mathbf{J} + \frac{\partial^2 H}{\partial \mathbf{x}^2}$$
$$- \mathbf{S}^T \left(\frac{\partial^2 H}{\partial \mathbf{u}^2}\right)^{-1} \mathbf{S} = \mathbf{0} , \quad (12)$$

where the $(r \times n)$-matrix $\mathbf{S} = \partial^2 H / \partial \mathbf{u} \partial \mathbf{x} + \partial \mathbf{f}^T / \partial \mathbf{x} \mathbf{J}$, and

$$\frac{d\mathbf{g}}{dt} - \mathbf{S}^T \left(\frac{\partial^2 H}{\partial \mathbf{u}^2}\right)^{-1} \left(\frac{\partial H}{\partial \mathbf{u}} + \frac{\partial \mathbf{f}^T}{\partial \mathbf{x}} \mathbf{g}\right) + \left(\frac{\partial \mathbf{f}^T}{\partial \mathbf{x}}\right) \mathbf{g} = \mathbf{0} .$$
$$(13)$$

The control policy is then updated through the equation

$$\mathbf{u}^{(j+1)} = \mathbf{u}^{(j)} - \left(\frac{\partial^2 H}{\partial \mathbf{u}^2}\right)^{-1} \left(\frac{\partial H}{\partial \mathbf{u}} + \frac{\partial \mathbf{f}^T}{\partial \mathbf{x}} \mathbf{g}\right)$$
$$- \left(\frac{\partial^2 H}{\partial \mathbf{u}^2}\right)^{-1} \mathbf{S} \left(\mathbf{x}^{(j+1)} - \mathbf{x}^{(j)}\right) . \quad (14)$$

This method of updating the control policy is called the second variation method. In (12) the $(n \times n)$-matrix $\mathbf{J}$ is symmetric, so the total number of differential equations to be integrated backwards is $n(n + 1)/2 + 2n$. However, the convergence is quadratic if the initial control policy is close to the optimum.

To obtain good starting conditions, Luus and Lapidus [6] suggested the use of first variation method for the first few iterations and then to switch over to the second variation method.

One additional feature of the second variation method is that the control policy given in (14) is a function of the present state, so that the control policy is treated as being continuous and is not restricted to being piecewise constant over an integration time step, as is the case with the first variation method. As was shown in ([1], pp. 316–317), for the linear six-plate gas absorber example, when the system equation is linear and the performance index is quadratic, the second variation method yields the optimal control policy in a single step.

Determination of Stepping Parameter

However, the large number and complexity of equations required for obtaining the control policy and the instability of the method for very complex systems led to investigating different means of obtaining faster convergence with the first variation method. The effort was directed on the best means of obtaining the stepping parameter ϵ in (9). When ϵ is too large, overstepping occurs, and if ϵ is too small, the convergence rate is very small.

Numerous papers have been written on the determination of ϵ. Several methods were compared by Rao and Luus [8] in solving typical optimal control problems. Although they suggested a means of determining the 'best' method for performance indices that are almost quadratic, it is found that a very simple scheme is quite effective for a wide variety of optimal control problems. Instead of trying to get very fast convergence and risk instability, the emphasis is placed on the robustness. The strategy is to obtain the initial value for ϵ from the magnitude of $\partial H/\partial \mathbf{u}$, and then increasing ϵ when the iteration has been successful, and reducing its value if overstepping occurs. This type of approach was used in [2] in solving the optimal control of a pyrolysis problem. When the iteration was successful, the stepping parameter was increased by 10 percent, and when overstepping resulted, the stepping parameter was reduced to half its value. The algorithm for first variation method may be presented as follows:

- Choose an initial control policy $\mathbf{u}^{(0)}$ and a value for ϵ; set the iteration index j to 0.
- Integrate (1) from $t = 0$ to $t = t_f$ and evaluate the performance index in (2). Store the values of the state vector at the end of each integration time step.
- Integrate the adjoint equation (7) from $t = t_f$ to $t = 0$, using for $\mathbf{x}$ the stored values of the state vector in Step 2. At each integration time step evaluate the gradient $\partial H/\partial \mathbf{u}$.
- Choose a new control policy

$$\mathbf{u}^{(j+1)} = \mathbf{u}^{(j)} - \epsilon \frac{\partial H}{\partial \mathbf{u}}. \tag{15}$$

- Integrate (1) from $t = 0$ to $t = t_f$ and evaluate the performance index in (2). Store the values of the state vector at the end of each integration time step. If the performance index is worse (i. e., overstepping has occurred), reduce ϵ to half its value and go to Step 4. If the performance index has been improved increase ϵ by a small factor, such as 1.10 and go to Step 3, and continue for a number of iterations, or terminate the iterations when the change in the performance index in an iteration is less than some criterion, and interpret the results.

Illustration of the First Variation Method

Let us consider the nonlinear continuous stirred tank reactor that has been used for optimal control studies in ([1], pp. 308–318) and [6], and which was shown in [4] to exhibit multiplicity of solutions. The system is described by the two equations

$$\frac{dx_1}{dt} = -2(x_1 + 0.25) \\ + (x_2 + 0.5) \exp\left(\frac{25x_1}{x_1 + 2}\right) - u(x_1 + 0.25), \tag{16}$$

$$\frac{dx_2}{dt} = 0.5 - x_2 - (x_2 + 0.5) \exp\left(\frac{25x_1}{x_1 + 2}\right), \tag{17}$$

with the initial state $x_1(0) = 0.09$ and $x_2(0) = 0.09$. The control u is a scalar quantity related to the valve opening of the coolant. The state variables x_1 and x_2 represent deviations from the steady state of dimensionless temperature and concentration, respectively. The performance index to be minimized is

$$I = \int_0^{t_f} \left(x_1^2 + x_2^2 + 0.1u^2\right) dt, \tag{18}$$

where the final time $t_f = 0.78$. The Hamiltonian is

$$H = z_1\left(-2(x_1 + 0.25) + R - u(x_1 + 0.25)\right) \\ + z_2(0.5 - x_2 - R) + x_1^2 + x_2^2 + 0.1u^2, \tag{19}$$

where $R = (x_2 + 0.5) \exp(25x_1/(x_1 + 2))$. The adjoint equations are

$$\frac{dz_1}{dt} = (u + 2)z_1 - 2x_1 + 50R\frac{(z_2 - z_1)}{(x_1 + 2)^2}, \quad (20)$$

$$\frac{dz_2}{dt} = -2x_2 + \frac{(z_2 - z_1)}{(x_2 + 0.5)}R + z_2, \quad (21)$$

and the gradient of the Hamiltonian is

$$\frac{\partial H}{\partial \mathbf{u}} = 0.2u - (x_1 + 0.25)z_1. \quad (22)$$

To illustrate the computational aspects of CVI, the above algorithm was used with a Pentium-120 personal computer using WATCOM Fortran compiler version 9.5. The calculations were done in double precision. As found in [4], convergence to the local optimum was obtained when small values for the initial control policy were used, and the global optimum was obtained when large values were used as initial policy. As is seen in Table 1, when an integration time step of 0.0065 was used (allowing 120 piecewise constant steps), in spite of the large number of iterations, the optimal control policy can be obtained in less than 2 s of computer time. The iterations were stopped when the change in the performance index from iteration to iteration was less than 10^{-6}.

The total computation time for making this run with 11 different initial control policies was 9.6 s on the Pentium-120 digital computer. When an integration time step of 0.00312 was used, the value of the performance index at the global optimum was improved to 0.133104. When a time step of 0.001 was used, giving 780 time steps, the optimal control policy yielded $I = 0.133097$. Even here the computation time for the 11 different initial conditions was only 31 s. With the use of piecewise linear control and only 20 time stages, a performance index of $I = 0.133101$ was obtained in [3] with iterative dynamic programming (IDP). To obtain this result with IDP, by using 5 randomly chosen points and 10 passes, each consisting of 20 iterations, took 13.4 s on a Pentium-120. The use of 15 time stages yielded $I = 0.133112$ and required 7.8 s. Therefore, computationally CVI is faster than IDP for this problem, but the present formulation does not allow piecewise linear control to be used in CVI.

The effect of the number of time stages for piecewise constant control is shown in Table 2, where CVI results are compared to those obtained by IDP in [5].

As can be seen, the given algorithm gives results very close to those obtained by IDP, and the deviations decrease as the number of time stages increases, because the approximations introduced during the backward integration when the stored values for the state vector are used, and in the calculation of the gradient of the Hamiltonian in CVI become negligible as the time stages become very small. As is shown in Fig. 1, when the optimal value of the performance index is plotted against $1/P^2$, the extrapolated value, as $1/P^2$ approaches zero, gives the value obtained with the second variation method.

The first variation method is easy to program and will continue to be a very useful method of determining the optimal control of nonlinear systems.

Control Vector Iteration CVI, Table 1
Application of First Variation Method to CSTR

Initial policy $u^{(0)}$	Performance index	Number of iterations	CPU time s
1.0	0.244436	16	0.16
1.2	0.244436	17	0.17
1.4	0.244436	18	0.11
1.6	0.244436	18	0.16
1.8	0.244436	19	0.22
2.0	0.133128	143	1.49
2.2	0.133128	149	1.53
2.4	0.133128	149	1.54
2.6	0.133130	133	1.43
2.8	0.133129	142	1.37
3.0	0.133130	136	1.38

Control Vector Iteration CVI, Table 2
Effect of the number of time stages P on the optimal performance index

Number of time stages P	Optimal I by CVI	Optimal I by IDP
20	0.13429	0.13416
30	0.13363	0.13357
40	0.13339	0.13336
60	0.13323	0.13321
80	0.13317	0.13316
120	0.13313	0.13313
240	0.13310	0.13310

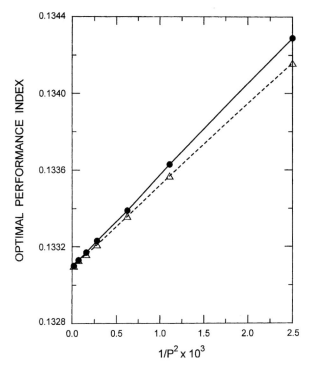

Control Vector Iteration CVI, Figure 1
Linear variation of optimal performance index with P^{-2};
•–•–• CVI, △ -- △ -- △ IDP

See also

- ▶ Boundary Condition Iteration BCI
- ▶ Duality in Optimal Control with First Order Differential Equations
- ▶ Dynamic Programming: Continuous-time Optimal Control
- ▶ Dynamic Programming and Newton's Method in Unconstrained Optimal Control
- ▶ Dynamic Programming: Optimal Control Applications
- ▶ Hamilton–Jacobi–Bellman Equation
- ▶ Infinite Horizon Control and Dynamic Games
- ▶ MINLP: Applications in the Interaction of Design and Control
- ▶ Multi-objective Optimization: Interaction of Design and Control
- ▶ Optimal Control of a Flexible Arm
- ▶ Optimization Strategies for Dynamic Systems
- ▶ Robust Control
- ▶ Robust Control: Schur Stability of Polytopes of Polynomials
- ▶ Semi-infinite Programming and Control Problems
- ▶ Sequential Quadratic Programming: Interior Point Methods for Distributed Optimal Control Problems
- ▶ Suboptimal Control

References

1. Lapidus L, Luus R (1967) Optimal control of engineering processes. Blaisdell, Waltham
2. Luus R (1978) On the optimization of oil shale pyrolysis. Chem Eng Sci 33:1403–1404
3. Luus R (1993) Application of iterative dynamic programming to very high dimensional systems. Hungarian J Industr Chem 21: 243–250
4. Luus R, Cormack DE (1972) Multiplicity of solutions resulting from the use of variational methods in optimal control problems. Canad J Chem Eng 50:309–311
5. Luus R, Galli M (1991) Multiplicity of solutions in using dynamic programming for optimal control. Hungarian J Industr Chem 19:55–62
6. Luus R, Lapidus L (1967) The control of nonlinear systems. Part II: Convergence by combined first and second variations. AIChE J 13:108–113
7. Merriam CW (1964) Optimization theory and the design of feedback control systems, McGraw-Hill, New York, pp 259–261
8. Rao SN, Luus R (1972) Evaluation and improvement of control vector iteration procedures for optimal control. Canad J Chem Eng 50:777–784

Convex Discrete Optimization

SHMUEL ONN
Technion – Israel Institute of Technology, Haifa, Israel

MSC2000: 05A, 15A, 51M, 52A, 52B, 52C, 62H, 68Q, 68R, 68U, 68W, 90B, 90C

Article Outline

Abstract
Introduction
  Limitations
  Outline and Overview of Main Results and Applications
  Terminology and Complexity
Reducing Convex to Linear Discrete Optimization
  Edge-Directions and Zonotopes
  Strongly Polynomial Reduction of Convex to Linear Discrete Optimization
  Pseudo Polynomial Reduction when Edge-Directions Are not Available

Convex Combinatorial Optimization and More
  From Membership to Linear Optimization
  Linear and Convex Combinatorial Optimization
    in Strongly Polynomial Time
  Linear and Convex Discrete Optimization
    over any Set in Pseudo Polynomial Time
  Some Applications
Linear N-fold Integer Programming
  Oriented Augmentation and Linear Optimization
  Graver Bases and Linear Integer Programming
  Graver Bases of N-fold Matrices
  Linear N-fold Integer Programming in Polynomial Time
  Some Applications
Convex Integer Programming
  Convex Integer Programming
    over Totally Unimodular Systems
  Graver Bases and Convex Integer Programming
  Convex N-fold Integer Programming in Polynomial Time
  Some Applications
Multiway Transportation Problems and Privacy
  in Statistical Databases
  Multiway Transportation Problems and Privacy
    in Statistical Databases
  The Universality Theorem
  The Complexity of the Multiway Transportation Problem
  Privacy and Entry-Uniqueness
References

Abstract

We develop an algorithmic theory of convex optimization over discrete sets. Using a combination of algebraic and geometric tools we are able to provide polynomial time algorithms for solving broad classes of convex combinatorial optimization problems and convex integer programming problems in variable dimension. We discuss some of the many applications of this theory including to quadratic programming, matroids, bin packing and cutting-stock problems, vector partitioning and clustering, multiway transportation problems, and privacy and confidential statistical data disclosure. Highlights of our work include a strongly polynomial time algorithm for convex and linear combinatorial optimization over any family presented by a membership oracle when the underlying polytope has few edge-directions; a new theory of so-termed n-fold integer programming, yielding polynomial time solution of important and natural classes of convex and linear integer programming problems in variable dimension; and a complete complexity classification of high dimensional transportation problems, with practical applications to fundamental problems in privacy and confidential statistical data disclosure.

Introduction

The general linear discrete optimization problem can be posed as follows.

LINEAR DISCRETE OPTIMIZATION. Given a set $S \subseteq \mathbb{Z}^n$ of integer points and an integer vector $w \in \mathbb{Z}^n$, find an $x \in S$ maximizing the standard inner product $wx := \sum_{i=1}^n w_i x_i$.

The algorithmic complexity of this problem, which includes *integer programming* and *combinatorial optimization* as special cases, depends on the presentation of the set S of feasible points. In integer programming, this set is presented as the set of integer points satisfying a given system of linear inequalities, which in standard form is given by

$$S = \{x \in \mathbb{N}^n : Ax = b\},$$

where $\mathbb{N}$ stands for the nonnegative integers, $A \in \mathbb{Z}^{m \times n}$ is an $m \times n$ integer matrix, and $b \in \mathbb{Z}^m$ is an integer vector. The input for the problem then consists of A, b, w. In combinatorial optimization, $S \subseteq \{0, 1\}^n$ is a set of $\{0, 1\}$-vectors, often interpreted as a family of subsets of a ground set $N := \{1, \dots, n\}$, where each $x \in S$ is the indicator of its support $\mathrm{supp}(x) \subseteq N$. The set S is presented implicitly and compactly, say as the set of indicators of subsets of edges in a graph G satisfying a given combinatorial property (such as being a matching, a forest, and so on), in which case the input is G, w. Alternatively, S is given by an oracle, such as a *membership oracle* which, queried on $x \in \{0, 1\}^n$, asserts whether or not $x \in S$, in which case the algorithmic complexity also includes a count of the number of oracle queries needed to solve the problem.

Here we study the following broad generalization of linear discrete optimization.

CONVEX DISCRETE OPTIMIZATION. Given a set $S \subseteq \mathbb{Z}^n$, vectors $w_1, \dots, w_d \in \mathbb{Z}^n$, and a convex functional $c : \mathbb{R}^d \longrightarrow \mathbb{R}$, find an $x \in S$ maximizing $c(w_1 x, \dots, w_d x)$.

This problem can be interpreted as *multi-objective* linear discrete optimization: given d linear functionals $w_1 x, \dots, w_d x$ representing the values of points $x \in S$ under d criteria, the goal is to maximize their "convex balancing" defined by $c(w_1 x, \ldots, w_d x)$. In fact, we have a hierarchy of problems of increasing generality and complexity, parameterized by the number d of linear functionals: at the bottom lies the linear discrete optimization problem, recovered as the special case of $d = 1$ and c the identity on $\mathbb{R}$; and at the top lies the problem of maximizing an arbitrary convex functional over the feasible set S, arising with $d = n$ and with $w_i = \mathbf{1}_i$ the ith standard unit vector in $\mathbb{R}^n$ for all i.

The algorithmic complexity of the convex discrete optimization problem depends on the presentation of the set S of feasible points as in the linear case, as well as on the presentation of the convex functional c. When S is presented as the set of integer points satisfying a given system of linear inequalities we also refer to the problem as *convex integer programming*, and when $S \subseteq \{0, 1\}^n$ and is presented implicitly or by an oracle we also refer to the problem as *convex combinatorial optimization*. As for the convex functional c, we will assume throughout that it is presented by a *comparison oracle* that, queried on $x, y \in \mathbb{R}^d$, asserts whether or not $c(x) \leq c(y)$. This is a very broad presentation that reveals little information on the function, making the problem, on the one hand, very expressive and applicable, but on the other hand, very hard to solve.

There is a massive body of knowledge on the complexity of linear discrete optimization – in particular (linear) integer programming [55] and (linear) combinatorial optimization [31]. The purpose of this monograph is to provide the first comprehensive unified treatment of the extended convex discrete optimization problem. The monograph follows the outline of five lectures given by the author in the Séminaire de Mathématiques Supérieures Series, Université de Montréal, during June 2006. Colorful slides of theses lectures are available online at [46] and can be used as a visual supplement to this monograph. The monograph has been written under the support of the ISF – Israel Science Foundation. The theory developed here is based on and is a culmination of several recent papers including [5,12,13,14,15,16,17,25,39,47, 48,49,50,51] written in collaboration with several colleagues – Eric Babson, Jesus De Loera, Komei Fukuda, Raymond Hemmecke, Frank Hwang, Vera Rosta, Uriel Rothblum, Leonard Schulman, Bernd Sturmfels, Rekha Thomas, and Robert Weismantel. By developing and using a combination of geometric and algebraic tools, we are able to provide polynomial time algorithms for several broad classes of convex discrete optimization problems. We also discuss in detail some of the many applications of our theory, including to quadratic programming, matroids, bin packing and cutting-stock problems, vector partitioning and clustering, multiway transportation problems, and privacy and confidential statistical data disclosure.

We hope that this monograph will, on the one hand, allow users of discrete optimization to enjoy the new powerful modelling and expressive capability of convex discrete optimization along with its broad polynomial time solvability, and on the other hand, stimulate more research on this new and fascinating class of problems, their complexity, and the study of various relaxations, bounds, and approximations for such problems.

Limitations

Convex discrete optimization is generally intractable even for small fixed d, since already for $d = 1$ it includes linear integer programming which is NP-hard. When d is a variable part of the input, even very simple special cases are NP-hard, such as the following problem, so-called *positive semi-definite quadratic binary programming*,

$$\max \{(w_1 x)^2 + \cdots + (w_n x)^2 : x \in \mathbb{N}^n,$$
$$x_i \leq 1, \quad i = 1, \ldots, n \}.$$

Therefore, throughout this monograph we will assume that d is fixed (but arbitrary).

As explained above, we also assume throughout that the convex functional c which constitutes part of the data for the convex discrete optimization problem is presented by a comparison oracle. Under such broad presentation, the problem is generally very hard. In particular, if the feasible set is $S := \{x \in \mathbb{N}^n : Ax = b\}$ and the underlying polyhedron $P := \{x \in \mathbb{R}_+^n : Ax = b\}$ is *unbounded*, then the problem is inaccessible even in one variable with no equation constraints. Indeed, consider the following family of univariate convex integer programs with convex functions parameterized by $-\infty < u \leq \infty$,

$$\max \{c_u(x) : x \in \mathbb{N}\},$$

$$c_u(x) := \begin{cases} -x, & \text{if } x < u; \\ x - 2u, & \text{if } x \geq u. \end{cases}$$

Consider any algorithm attempting to solve the problem and let u be the maximum value of x in all queries to the oracle of c. Then the algorithm can not distinguish between the problem with c_u, whose objective function is unbounded, and the problem with c_∞, whose optimal objective value is 0. Thus, convex discrete optimization (with an oracle presented functional) over an *infinite* set $S \subset \mathbb{Z}^n$ is quite hopeless. Therefore, an algorithm that solves the convex discrete optimization problem will either return an optimal solution, or assert that the problem is infeasible, or assert that the underlying polyhedron is unbounded. In fact, in most applications, such as in combinatorial optimization with $S \subseteq \{0, 1\}^n$ or integer programming with $S := \{x \in \mathbb{Z}^n : Ax = b, l \leq x \leq u\}$ and $l, u \in \mathbb{Z}^n$, the set S is finite and the problem of unboundedness does not arise.

Outline and Overview of Main Results and Applications

We now outline the structure of this monograph and provide a brief overview of what we consider to be our main results and main applications. The precise relevant definitions and statements of the theorems and corollaries mentioned here are provided in the relevant sections in the monograph body. As mentioned above, most of these results are adaptations or extensions of results from one of the papers [5,12,13,14,15,16,17,25,39, 47,48,49,50,51]. The monograph gives many more applications and results that may turn out to be useful in future development of the theory of convex discrete optimization.

The rest of the monograph consists of five sections. While the results evolve from one section to the next, it is quite easy to read the sections independently of each other (while just browsing now and then for relevant definitions and results). Specifically, Sect. "Convex Combinatorial Optimization and More" uses definitions and the main result of Sect. "Reducing Convex to Linear Discrete Optimization"; Sect. "Convex Integer Programming" uses definitions and results from Sections "Reducing Convex to Linear Discrete Optimization" and "Linear N-fold Integer Programming"; and Sect. "Multiway Transportation Problems and Privacy in Statistical Databases" uses the main results of Sections "Linear N-fold Integer Programming" and "Convex Integer Programming".

In Sect. "Reducing Convex to Linear Discrete Optimization" we show how to reduce the convex discrete optimization problem over $S \subset \mathbb{Z}^n$ to strongly polynomially many linear discrete optimization counterparts over S, provided that the convex hull conv(S) satisfies a suitable geometric condition, as follows.

Theorem 1 *For every fixed d, the convex discrete optimization problem over any finite $S \subset \mathbb{Z}^n$ presented by a linear discrete optimization oracle and endowed with a set covering all edge-directions of* conv(S)*, can be solved in strongly polynomial time.*

This result will be incorporated in the polynomial time algorithms for convex combinatorial optimization and convex integer programming to be developed in Sect. "Convex Combinatorial Optimization and More" and Sect. "Convex Integer Programming".

In Sect. "Convex Combinatorial Optimization and More" we discuss convex combinatorial optimization. The main result is that convex combinatorial optimization over a set $S \subseteq \{0, 1\}^n$ presented by a membership oracle can be solved in strongly polynomial time provided it is endowed with a set covering all edge-directions of conv(S). In particular, the standard linear combinatorial optimization problem over S can be solved in strongly polynomial time as well.

Theorem 2 *For every fixed d, the convex combinatorial optimization problem over any $S \subseteq \{0, 1\}^n$ presented by a membership oracle and endowed with a set covering all edge-directions of the polytope* conv(S)*, can be solved in strongly polynomial time.*

An important application of Theorem 2 concerns convex matroid optimization.

Corollary 1 *For every fixed d, convex combinatorial optimization over the family of bases of a matroid presented by membership oracle is strongly polynomial time solvable.*

In Sect. "Linear N-fold Integer Programming" we develop the theory of linear *n-fold integer programming*. As a consequence of this theory we are able to solve a broad class of linear integer programming problems in variable dimension in polynomial time, in contrast with the general intractability of linear integer programming. The main theorem here may seem a bit technical at a first glance, but is really very natural and has many applications discussed in detail in Sect. "Linear N-fold Integer Programming", Sect. "Convex Integer Programming" and Sect. "Multiway Transportation Problems and Privacy in Statistical Databases". To state it we need a definition. Given an $(r + s) \times t$ matrix A, let A_1 be its $r \times t$ sub-matrix consisting of the first r rows and let A_2 be its $s \times t$ sub-matrix consisting of the last s rows. We refer to A explicitly as $(r + s) \times t$ matrix, since the definition below depends also on r and s and not only on the entries of A. The *n-fold matrix* of an $(r + s) \times t$ matrix A is then defined to be the following $(r + ns) \times nt$ matrix,

$$A^{(n)} := (\mathbf{1}_n \otimes A_1) \oplus (I_n \otimes A_2)$$

$$= \begin{pmatrix} A_1 & A_1 & A_1 & \cdots & A_1 \\ A_2 & 0 & 0 & \cdots & 0 \\ 0 & A_2 & 0 & \cdots & 0 \\ \vdots & \vdots & \vdots & \ddots & \vdots \\ 0 & 0 & 0 & \cdots & A_2 \end{pmatrix}.$$

Given now any $n \in \mathbb{N}$, lower and upper bounds $l, u \in \mathbb{Z}_\infty^{nt}$ with $\mathbb{Z}_\infty := \mathbb{Z} \uplus \{\pm\infty\}$, right-hand side $b \in \mathbb{Z}^{r+ns}$, and linear functional wx with $w \in \mathbb{Z}^{nt}$, the corresponding linear n-fold integer programming problem is the following program in variable dimension nt,

$$\max \{wx \,:\, x \in \mathbb{Z}^{nt},\ A^{(n)}x = b,\ l \leq x \leq u\}\,.$$

The main theorem of Sect. "Linear N-fold Integer Programming" asserts that such integer programs are polynomial time solvable.

Theorem 3 *For every fixed $(r + s) \times t$ integer matrix A, the linear n-fold integer programming problem with any n, l, u, b, and w can be solved in polynomial time.*

Theorem 3 has very important applications to high-dimensional transportation problems which are discussed in Sect. "Three-Way Line-Sum Transportation Problems" and in more detail in Sect. "Multiway Transportation Problems and Privacy in Statistical Databases". Another major application concerns bin packing problems, where items of several types are to be packed into bins so as to maximize packing utility subject to weight constraints. This includes as a special case the classical cutting-stock problem of [27]. These are discussed in detail in Sect. "Packing Problems and Cutting-Stock".

Corollary 2 *For every fixed number t of types and type weights $v_1, \ldots, v_t$, the corresponding integer bin packing and cutting-stock problems are polynomial time solvable.*

In Sect. "Convex Integer Programming" we discuss convex integer programming, where the feasible set S is presented as the set of integer points satisfying a given system of linear inequalities. In particular, we consider convex integer programming over n-fold systems for any fixed (but arbitrary) $(r + s) \times t$ matrix A, where, given $n \in \mathbb{N}$, vectors $l, u \in \mathbb{Z}_\infty^{nt}$, $b \in \mathbb{Z}^{r+ns}$ and $w_1, \ldots, w_d \in \mathbb{Z}^{nt}$, and convex functional $c : \mathbb{R}^d \longrightarrow \mathbb{R}$, the problem is

$$\max \{c(w_1 x, \ldots, w_d x) \,:\, x \in \mathbb{Z}^{nt},$$
$$A^{(n)} x = b,\ l \leq x \leq u\}\,.$$

The main theorem of Sect. "Convex Integer Programming" is the following extension of Theorem 3, asserting that convex integer programming over n-fold systems is polynomial time solvable as well.

Theorem 4 *For every fixed d and $(r + s) \times t$ integer matrix A, convex n-fold integer programming with any $n, l, u, b, w_1, \ldots, w_d$, and c can be solved in polynomial time.*

Theorem 4 broadly extends the class of objective functions that can be efficiently maximized over n-fold systems. Thus, all applications discussed in Sect. "Some Applications" automatically extend accordingly. These include convex high-dimensional transportation problems and convex bin packing and cutting-stock problems, which are discussed in detail in Sect. "Transportation Problems and Packing Problems" and Sect. "Multiway Transportation Problems and Privacy in Statistical Databases".

Another important application of Theorem 4 concerns vector partitioning problems which have applications in many areas including load balancing, circuit layout, ranking, cluster analysis, inventory, and reliability, see e. g. [7,9,25,39,50] and the references therein. The problem is to partition n items among p players so as to maximize social utility. With each item is associated a k-dimensional vector representing its utility under k criteria. The social utility of a partition is a convex function of the sums of vectors of items that each player receives. In the constrained version of the problem, there are also restrictions on the number of items each player can receive. We have the following consequence of Theorem 4; more details on this application are in Sect. "Vector Partitioning and Clustering".

Corollary 3 *For every fixed number p of players and number k of criteria, the constrained and unconstrained vector partition problems with any item vectors, convex utility, and constraints on the number of item per player, are polynomial time solvable.*

In the last Sect. "Multiway Transportation Problems and Privacy in Statistical Databases" we discuss multiway (high-dimensional) transportation problems and secure statistical data disclosure. Multiway transportation problems form a very important class of discrete optimization problems and have been used and studied extensively in the operations research and mathematical programming literature, as well as in the statistics literature in the context of secure statistical data disclosure and management by public agencies, see e. g. [4,6, 11,18,19,42,43,53,60,62] and the references therein. The feasible points in a transportation problem are the multiway tables ("contingency tables" in statistics) such that the sums of entries over some of their lower dimensional sub-tables such as lines or planes ("margins" in statistics) are specified. We completely settle the algorithmic complexity of treating multiway tables and discuss the applications to transportation problems and secure statistical data disclosure, as follows.

In Sect. "The Universality Theorem" we show that "short" 3-way transportation problems, over $r \times c \times 3$ tables with variable number r of rows and variable number c of columns but fixed small number 3 of layers (hence "short"), are *universal* in that *every* integer programming problem is such a problem (see Sect. "The Universality Theorem" for the precise stronger statement and for more details).

Theorem 5 *Every linear integer programming problem* $\max\{cy : y \in \mathbb{N}^n : Ay = b\}$ *is polynomial time representable as a short 3-way line-sum transportation problem*

$$\max\left\{wx \ : \ x \in \mathbb{N}^{r \times c \times 3} \ : \ \sum_i x_{i,j,k} = z_{j,k}, \ \sum_j x_{i,j,k} = v_{i,k}, \ \sum_k x_{i,j,k} = u_{i,j}\right\}.$$

In Sect. "The Complexity of the Multiway Transportation Problem" we discuss k-way transportation problems of any dimension k. We provide the first polynomial time algorithm for convex and linear "long" ($k+1$)-way transportation problems, over $m_1 \times \cdots \times m_k \times n$ tables, with k and $m_1, \ldots, m_k$ fixed (but arbitrary), and variable number n of layers (hence "long"). This is best possible in view of Theorem 21. Our algorithm works for any *hierarchical collection of margins*: this captures common margin collections such as all line-sums, all plane-sums, and more generally all h-flat sums for any $0 \le h \le k$ (see Sect. "Tables and Margins" for more details). We point out that even for the very special case of linear integer transportation over $3 \times 3 \times n$ tables with specified line-sums, our polynomial time algorithm is the only one known. We prove the following statement.

Corollary 4 *For every fixed $d, k, m_1, \ldots, m_k$ and family $\mathcal{F}$ of subsets of $\{1, \ldots, k+1\}$ specifying a hierarchical collection of margins, the convex (and in particular linear) long transportation problem over $m_1 \times \cdots \times m_k \times n$ tables is polynomial time solvable.*

In our last subsection Sect. "Privacy and Entry-Uniqueness" we discuss an important application concerning privacy in statistical databases. It is a common practice in the disclosure of a multiway table containing sensitive data to release some table margins rather than the table itself. Once the margins are released, the security of any specific entry of the table is related to the set of possible values that can occur in that entry in any table having the same margins as those of the source table in the data base. In particular, if this set consists of a unique value, that of the source table, then this entry can be exposed and security can be violated. We show that for multiway tables where one category is significantly richer than the others, that is, when each sample point can take many values in one category and only few values in the other categories, it is possible to check entry-uniqueness in polynomial time, allowing disclosing agencies to make learned decisions on secure disclosure.

Corollary 5 *For every fixed $k, m_1, \ldots, m_k$ and family $\mathcal{F}$ of subsets of $\{1, \ldots, k+1\}$ specifying a hierarchical collection of margins to be disclosed, it can be decided in polynomial time whether any specified entry $x_{i_1, \ldots, i_{k+1}}$ is the same in all long $m_1 \times \cdots \times m_k \times n$ tables with the disclosed margins, and hence at risk of exposure.*

Terminology and Complexity

We use $\mathbb{R}$ for the reals, $\mathbb{R}_+$ for the nonnegative reals, $\mathbb{Z}$ for the integers, and $\mathbb{N}$ for the nonnegative integers. The sign of a real number r is denoted by sign$(r) \in \{0, -1, 1\}$ and its absolute value is denoted by $|r|$. The ith standard unit vector in $\mathbb{R}^n$ is denoted by $\mathbf{1}_i$. The *support* of $x \in \mathbb{R}^n$ is the index set supp$(x) := \{i : x_i \neq 0\}$ of nonzero entries of x. The *indicator* of a subset $I \subseteq \{1, \ldots, n\}$ is the vector $\mathbf{1}_I := \sum_{i \in I} \mathbf{1}_i$ so that supp$(\mathbf{1}_I) = I$. When several vectors are indexed by subscripts, $w_1, \ldots, w_d \in \mathbb{R}^n$, their entries are indicated by pairs of subscripts, $w_i = (w_{i,1}, \ldots, w_{i,n})$. When vectors are indexed by superscripts, $x^1, \ldots, x^k \in \mathbb{R}^n$, their entries are indicated by subscripts, $x^i = (x_1^i, \ldots, x_n^i)$. The integer lattice $\mathbb{Z}^n$ is naturally embedded in $\mathbb{R}^n$. The space $\mathbb{R}^n$ is endowed with the standard inner product which, for $w, x \in \mathbb{R}^n$, is given by $wx := \sum_{i=1}^n w_i x_i$. Vectors w in $\mathbb{R}^n$ will also be regarded as linear functionals on $\mathbb{R}^n$ via the inner product wx. Thus, we refer to elements of $\mathbb{R}^n$ as points, vectors, or linear functionals, as will be appropriate from the context. The *convex hull* of a set $S \subseteq \mathbb{R}^n$ is denoted by conv(S) and the set of *vertices* of a polyhedron $P \subseteq \mathbb{R}^n$ is denoted by vert(P). In linear discrete optimization over $S \subseteq \mathbb{Z}^n$, the *facets* of conv(S) play an important role, see Chvátal [10] and the references therein for earlier work, and Grötschel, Lovász and Schrijver [31,45] for the later culmination in the equivalence of separation and linear optimization via the ellipsoid method of Yudin and Nemirovskii [63]. As will turn out in Sect. "Reducing Convex to Linear Discrete Optimization", in convex discrete optimization over S, the *edges* of conv(S) play an important role (most significantly in a way which is *not* related to the Hirsch conjecture discussed in [41]). We therefore use extensively convex polytopes, for which we follow the terminology of [32,65].

We often assume that the feasible set $S \subseteq \mathbb{Z}^n$ is finite. We then define its *radius* to be its l_∞ radius $\rho(S) := \max\{\|x\|_\infty : x \in S\}$ where, as usual, $\|x\|_\infty := \max_{i=1}^n |x_i|$. In other words, $\rho(S)$ is the smallest $\rho \in \mathbb{N}$ such that S is contained in the cube $[-\rho, \rho]^n$.

Our algorithms are applied to rational data only, and the time complexity is as in the standard Turing machine model, see e. g. [1,26,55]. The input typically consists of rational (usually integer) numbers, vectors, matrices, and finite sets of such objects. The *binary length* of an integer number $z \in \mathbb{Z}$ is defined to be the number of bits in its binary representation, $\langle z \rangle := 1 + \lceil \log_2(|z| + 1) \rceil$ (with the extra bit for the sign). The length of a rational number presented as a fraction $r = \frac{p}{q}$ with $p, q \in \mathbb{Z}$ is $\langle r \rangle := \langle p \rangle + \langle q \rangle$. The length of an $m \times n$ matrix A (and in particular of a vector) is the sum $\langle A \rangle := \sum_{i,j} \langle a_{i,j} \rangle$ of the lengths of its entries. Note that the length of A is no smaller than the number of entries, $\langle A \rangle \geq mn$. Therefore, when A is, say, part of an input to an algorithm, with m, n variable, the length $\langle A \rangle$ already incorporates mn, and so we will typically not account additionally for m, n directly. But sometimes, especially in results related to n-fold integer programming, we will also emphasize n as part of the input length. Similarly, the length of a finite set E of numbers, vectors or matrices is the sum of lengths of its elements and hence, since $\langle E \rangle \geq |E|$, automatically accounts for its cardinality.

Some input numbers affect the running time of some algorithms through their unary presentation, resulting in so-called "pseudo polynomial" running time. The *unary length* of an integer number $z \in \mathbb{Z}$ is the number $|z| + 1$ of bits in its unary representation (again, an extra bit for the sign). The unary length of a rational number, vector, matrix, or finite set of such objects are defined again as the sums of lengths of their numerical constituents, and is again no smaller than the number of such numerical constituents.

When studying convex and linear integer programming in Sect. "Linear N-fold Integer Programming" and Sect. "Convex Integer Programming" we sometimes have lower and upper bound vectors l, u with entries in $\mathbb{Z}_\infty := \mathbb{Z} \uplus \{\pm\infty\}$. Both binary and unary lengths of a $\pm\infty$ entry are constant, say 3 by encoding $\pm\infty := \pm"00"$.

To make the input encoding precise, we introduce the following notation. In every algorithmic statement we describe explicitly the input encoding, by listing in square brackets all input objects affecting the running time. Unary encoded objects are listed directly whereas binary encoded objects are listed in terms of their length. For example, as is often the case, if the input of an algorithm consists of binary encoded vectors (linear functionals) $w_1, \ldots, w_d \in \mathbb{Z}^n$ and unary encoded integer $\rho \in \mathbb{N}$ (bounding the radius $\rho(S)$ of the feasible set) then we will indicate that the input is *encoded as* $[\rho, \langle w_1, \ldots, w_d \rangle]$.

Some of our algorithms are strongly polynomial time in the sense of [59]. For this, part of the input is regarded as "special". An algorithm is then *strongly polynomial time* if it is polynomial time in the usual Turing sense with respect to all input, and in addition, the number of arithmetic operations (additions, subtractions, multiplications, divisions, and comparisons) it performs is polynomial in the special part of the input. To make this precise, we extend our input encoding notation above by splitting the square bracketed expression indicating the input encoding into a "left" side and a "right" side, separated by semicolon, where the entire input is described on the right and the special part of the input on the left. For example, Theorem 1, asserting that the algorithm underlying it is strongly polynomial with data *encoded as* $[n, |E|; \langle \rho(S), w_1, \ldots, w_d, E \rangle]$, where $\rho(S) \in \mathbb{N}$, $w_1, \ldots, w_d \in \mathbb{Z}^n$ and $E \subset \mathbb{Z}^n$, means that the running time is polynomial in the binary length of $\rho(S)$, $w_1, \ldots, w_d$, and E, and the number of arithmetic operations is polynomial in n and the cardinality $|E|$, which constitute the special part of the input.

Often, as in [31], part of the input is presented by oracles. Then the running time and the number of arithmetic operations count also the number of oracle queries. An oracle algorithm is *polynomial time* if its running time, including the number of oracle queries, and the manipulations of numbers, some of which are answers to oracle queries, is polynomial in the length of the input encoding. An oracle algorithm is *strongly polynomial time* (with specified input encoding as above), if it is polynomial time in the entire input (on the "right"), and in addition, the number of arithmetic operations it performs (including oracle queries) is polynomial in the special part of the input (on the "left").

Reducing Convex to Linear Discrete Optimization

In this section we show that when suitable auxiliary geometric information about the convex hull conv(S) of a finite set $S \subseteq \mathbb{Z}^n$ is available, the convex discrete optimization problem over S can be reduced to the solution of strongly polynomially many linear discrete optimization counterparts over S. This result will be incorporated into the polynomial time algorithms developed in Sect. "Convex Combinatorial Optimization and More" and Sect. "Convex Integer Programming" for convex combinatorial optimization and convex integer programming respectively. In Sect. "Edge-Directions and Zonotopes" we provide some preliminaries on edge-directions and zonotopes. In Sect. "Strongly Polynomial Reduction of Convex to Linear Discrete Optimization" we prove the reduction which is the main result of this section. In Sect. "Pseudo Polynomial Reduction when Edge-Directions are not Available" we prove a pseudo polynomial reduction for any finite set.

Edge-Directions and Zonotopes

We begin with some terminology and facts that play an important role in the sequel. A *direction* of an edge (1-dimensional face) $e = [u, v]$ of a polytope P is any nonzero scalar multiple of $u - v$. A set of vectors E *covers all edge-directions of* P if it contains a direction of each edge of P. The *normal cone* of a polytope $P \subset \mathbb{R}^n$ at its face F is the (relatively open) cone C_P^F of those linear functionals $h \in \mathbb{R}^n$ which are maximized over P precisely at points of F. A polytope Z is a *refinement* of a polytope P if the normal cone of every vertex of Z is contained in the normal cone of some vertex of P. If Z refines P then, moreover, the closure of each normal cone of P is the union of closures of normal cones of Z. The *zonotope* generated by a set of vectors $E = \{e_1, \ldots, e_m\}$ in $\mathbb{R}^d$ is the following polytope, which is the projection by E of the cube $[-1, 1]^m$ into $\mathbb{R}^d$,

$$\begin{aligned}Z &:= \mathrm{zone}(E) \\ &:= \mathrm{conv}\left\{\sum_{i=1}^m \lambda_i e_i : \lambda_i = \pm 1\right\} \subset \mathbb{R}^d.\end{aligned}$$

The following fact goes back to Minkowski, see [32].

Lemma 1 *Let P be a polytope and let E be a finite set that covers all edge-directions of P. Then the zonotope $Z := \mathrm{zone}(E)$ generated by E is a refinement of P.*

Proof Consider any vertex u of Z. Then $u = \sum_{e \in E} \lambda_e e$ for suitable $\lambda_e = \pm 1$. Thus, the normal cone C_Z^u consists of those h satisfying $h\lambda_e e > 0$ for all e. Pick any $\hat{h} \in C_Z^u$ and let v be a vertex of P at which $\hat{h}$ is maximized over P. Consider any edge $[v, w]$ of P. Then $v - w = \alpha_e e$ for some scalar $\alpha_e \neq 0$ and some $e \in E$, and $0 \leq \hat{h}(v - w) = \hat{h}\alpha_e e$, implying $\alpha_e \lambda_e > 0$.

It follows that every $h \in C_Z^u$ satisfies $h(v - w) > 0$ for every edge of P containing v. Therefore h is maximized over P uniquely at v and hence is in the cone C_P^v of P at v. This shows $C_Z^u \subseteq C_P^v$. Since u was arbitrary, it follows that the normal cone of every vertex of Z is contained in the normal cone of some vertex of P. □

The next lemma provides bounds on the number of vertices of any zonotope and on the algorithmic complexity of constructing its vertices, each vertex along with a linear functional maximized over the zonotope uniquely at that vertex. The bound on the number of vertices has been rediscovered many times over the years. An early reference is [33], stated in the dual form of 2-partitions. A more general treatment is [64]. Recent extensions to p-partitions for any p are in [3,39], and to Minkowski sums of arbitrary polytopes are in [29]. Interestingly, already in [33], back in 1967, the question was raised about the algorithmic complexity of the problem; this is now settled in [20,21] (the latter reference correcting the former). We state the precise bounds on the number of vertices and arithmetic complexity, but will need later only that for any fixed d the bounds are polynomial in the number of generators. Therefore, below we only outline a proof that the bounds are polynomial. Complete details are in the above references.

Lemma 2 *The number of vertices of any zonotope $Z :=$ zone(E) generated by a set E of m vectors in $\mathbb{R}^d$ is at most $2 \sum_{k=0}^{d-1} \binom{m-1}{k}$. For every fixed d, there is a strongly polynomial time algorithm that, given $E \subset \mathbb{Z}^d$, encoded as $[m := |E|; \langle E \rangle]$, outputs every vertex v of $Z := $ zone(E) along with a linear functional $h_v \in \mathbb{Z}^d$ maximized over Z uniquely at v, using $O(m^{d-1})$ arithmetics operations for $d \geq 3$ and $O(m^d)$ for $d \leq 2$.*

Proof We only outline a proof that, for every fixed d, the polynomial bounds $O(m^{d-1})$ on the number of vertices and $O(m^d)$ on the arithmetic complexity hold. We assume that E linearly spans $\mathbb{R}^d$ (else the dimension can be reduced) and is generic, that is, no d points of E lie on a linear hyperplane (one containing the origin). In particular, $0 \notin E$. The same bound for arbitrary E then follows using a perturbation argument (cf. [39]).

Each oriented linear hyperplane $H = \{x \in \mathbb{R}^d : hx = 0\}$ with $h \in \mathbb{R}^d$ nonzero induces a partition of E by $E = H^- \uplus H^0 \uplus H^+$, with $H^- := \{e \in E : he < 0\}$, $E^0 := E \cap H$, and $H^+ := \{e \in E : he > 0\}$. The vertices of $Z = $ zone(E) are in bijection with ordered 2-partitions of E induced by such hyperplanes that avoid E. Indeed, if $E = H^- \uplus H^+$ then the linear functional $h_v := h$ defining H is maximized over Z uniquely at the vertex $v := \sum\{e : e \in H^+\} - \sum\{e : e \in H^-\}$ of Z.

We now show how to enumerate all such 2-partitions and hence vertices of Z. Let M be any of the $\binom{m}{d-1}$ subsets of E of size $d-1$. Since E is generic, M is linearly independent and spans a unique linear hyperplane lin(M). Let $\hat{H} = \{x \in \mathbb{R}^d : \hat{h}x = 0\}$ be one of the two orientations of the hyperplane lin(M). Note that $\hat{H}^0 = M$. Finally, let L be any of the 2^{d-1} subsets of M. Since M is linearly independent, there is a $g \in \mathbb{R}^d$ which linearly separates L from $M \setminus L$, namely, satisfies $gx < 0$ for all $x \in L$ and $gx > 0$ for all $x \in M \setminus L$. Furthermore, there is a sufficiently small $\epsilon > 0$ such that the oriented hyperplane $H := \{x \in \mathbb{R}^d : hx = 0\}$ defined by $h := \hat{h} + \epsilon g$ avoids E and the 2-partition induced by H satisfies $H^- = \hat{H}^- \uplus L$ and $H^+ = \hat{H}^+ \uplus (M \setminus L)$. The corresponding vertex of Z is $v := \sum\{e : e \in H^+\} - \sum\{e : e \in H^-\}$ and the corresponding linear functional which is maximized over Z uniquely at v is $h_v := h = \hat{h} + \epsilon g$.

We claim that any ordered 2-partition arises that way from some M, some orientation $\hat{H}$ of lin(M), and some L. Indeed, consider any oriented linear hyperplane $\tilde{H}$ avoiding E. It can be perturbed to a suitable oriented $\hat{H}$ that touches precisely $d-1$ points of E. Put $M := \hat{H}^0$ so that $\hat{H}$ coincides with one of the two orientations of the hyperplane lin(M) spanned by M, and put $L := \tilde{H}^- \cap M$. Let H be an oriented hyperplane obtained from M, $\hat{H}$ and L by the above procedure. Then the ordered 2-partition $E = H^- \uplus H^+$ induced by H coincides with the ordered 2-partition $E = \tilde{H}^- \uplus \tilde{H}^+$ induced by $\tilde{H}$.

Since there are $\binom{m}{d-1}$ many $(d-1)$-subsets $M \subseteq E$, two orientations $\hat{H}$ of lin(M), and 2^{d-1} subsets $L \subseteq M$, and d is fixed, the total number of 2-partitions and hence also the total number of vertices of Z obey the upper bound $2^d \binom{m}{d-1} = O(m^{d-1})$. Furthermore, for each choice of M, $\hat{H}$ and L, the linear functional $\hat{h}$ defining $\hat{H}$, as well as g, ϵ, $h_v = h = \hat{h} + \epsilon g$, and the vertex $v = \sum\{e : e \in H^+\} - \sum\{e : e \in H^-\}$ of Z at which h_v is uniquely maximized over Z, can all be computed using $O(m)$ arithmetic operations. This shows the claimed bound $O(m^d)$ on the arithmetic complexity. □

We conclude with a simple fact about edge-directions of projections of polytopes.

Lemma 3 *If E covers all edge-directions of a polytope P, and $Q := \omega(P)$ is the image of P under a linear map $\omega: \mathbb{R}^n \longrightarrow \mathbb{R}^d$, then $\omega(E)$ covers all edge-directions of Q.*

Proof Let f be a direction of an edge $[x, y]$ of Q. Consider the face $F := \omega^{-1}([x, y])$ of P. Let V be the set of vertices of F and let $U = \{u \in V : \omega(u) = x\}$. Then for some $u \in U$ and $v \in V \setminus U$, there must be an edge $[u, v]$ of F, and hence of P. Then $\omega(v) \in (x, y]$ hence $\omega(v) = x + \alpha f$ for some $\alpha \neq 0$. Therefore, with $e := \frac{1}{\alpha}(v - u)$, a direction of the edge $[u, v]$ of P, we find that $f = \frac{1}{\alpha}(\omega(v) - \omega(u)) = \omega(e) \in \omega(E)$. □

Strongly Polynomial Reduction of Convex to Linear Discrete Optimization

A *linear discrete optimization oracle* for a set $S \subseteq \mathbb{Z}^n$ is one that, queried on $w \in \mathbb{Z}^n$, either returns an optimal solution to the linear discrete optimization problem over S, that is, an $x^* \in S$ satisfying $wx^* = \max\{wx : x \in S\}$, or asserts that none exists, that is, either the problem is infeasible or the objective function is unbounded. We now show that a set E covering all edge-directions of the polytope conv(S) underlying a convex discrete optimization problem over a finite set $S \subset \mathbb{Z}^n$ allows to solve it by solving polynomially many linear discrete optimization counterparts over S. The following theorem extends and unifies the corresponding reductions in [49] and [12] for convex combinatorial optimization and convex integer programming respectively. Recall from Sect. "Terminology and Complexity" that the *radius* of a finite set $S \subset \mathbb{Z}^n$ is defined to be $\rho(S) := \max\{|x_i| : x \in S, i = 1, \ldots, n\}$.

Theorem 6 *For every fixed d there is a strongly polynomial time algorithm that, given finite set $S \subset \mathbb{Z}^n$ presented by a linear discrete optimization oracle, integer vectors $w_1, \ldots, w_d \in \mathbb{Z}^n$, set $E \subset \mathbb{Z}^n$ covering all edge-directions of conv(S), and convex functional $c: \mathbb{R}^d \longrightarrow \mathbb{R}$ presented by a comparison oracle, encoded as $[n, |E|; \langle \rho(S), w_1, \ldots, w_d, E \rangle]$, solves the convex discrete optimization problem*

$$\max\{c(w_1 x, \ldots, w_d x) : x \in S\}.$$

Proof First, query the linear discrete optimization oracle presenting S on the trivial linear functional $w = 0$. If the oracle asserts that there is no optimal solution then S is empty so terminate the algorithm asserting that no optimal solution exists to the convex discrete optimization problem either. So assume the problem is feasible. Let $P := \text{conv}(S) \subset \mathbb{R}^n$ and $Q := \{(w_1 x, \ldots, w_d x) : x \in P\} \subset \mathbb{R}^d$. Then Q is a projection of P, and hence by Lemma 3 the projection $D := \{(w_1 e, \ldots, w_d e) : e \in E\}$ of the set E is a set covering all edge-directions of Q. Let $Z := \text{zone}(D) \subset \mathbb{R}^d$ be the zonotope generated by D. Since d is fixed, by Lemma 2 we can produce in strongly polynomial time all vertices of Z, every vertex v along with a linear functional $h_v \in \mathbb{Z}^d$ maximized over Z uniquely at v. For each of these polynomially many h_v, repeat the following procedure. Define a vector $g_v \in \mathbb{Z}^n$ by $g_{v,j} := \sum_{i=1}^{d} w_{i,j} h_{v,i}$ for $j = 1, \ldots, n$. Now query the linear discrete optimization oracle presenting S on the linear functional $w := g_v \in \mathbb{Z}^n$. Let $x_v \in S$ be the optimal solution obtained from the oracle, and let $z_v := (w_1 x_v, \ldots, w_d x_v) \in Q$ be its projection. Since $P = \text{conv}(S)$, we have that x_v is also a maximizer of g_v over P. Since for every $x \in P$ and its projection $z := (w_1 x, \ldots, w_d x) \in Q$ we have $h_v z = g_v x$, we conclude that z_v is a maximizer of h_v over Q. Now we claim that each vertex u of Q equals some z_v. Indeed, since Z is a refinement of Q by Lemma 1, it follows that there is some vertex v of Z such that h_v is maximized over Q uniquely at u, and therefore $u = z_v$. Since $c(w_1 x, \ldots, w_d x)$ is convex on $\mathbb{R}^n$ and c is convex on $\mathbb{R}^d$, we find that

$$\max_{x \in S} c(w_1 x, \ldots, w_d x)$$
$$= \max_{x \in P} c(w_1 x, \ldots, w_d x)$$
$$= \max_{z \in Q} c(z)$$
$$= \max\{c(u) : u \text{ vertex of } Q\}$$
$$= \max\{c(z_v) : v \text{ vertex of } Z\}.$$

Using the comparison oracle of c, find a vertex v of Z attaining maximum value $c(z_v)$, and output $x_v \in S$, an optimal solution to the convex discrete optimization problem. □

Pseudo Polynomial Reduction when Edge-Directions Are not Available

Theorem 6 reduces convex discrete optimization to polynomially many linear discrete optimization counterparts when a set covering all edge-directions of the underlying polytope is available. However, often such a set is not available (see e.g. [8] for the important case of bipartite matching). We now show how to reduce convex discrete optimization to many linear discrete optimization counterparts when a set covering all edge-directions is not offhand available. In the absence of such a set, the problem is much harder, and the algorithm below is polynomially bounded only in the unary length of the radius $\rho(S)$ and of the linear functionals $w_1, \ldots, w_d$, rather than in their binary length $\langle \rho(S), w_1, \ldots, w_d \rangle$ as in the algorithm of Theorem 6. Moreover, an upper bound $\rho \geq \rho(S)$ on the radius of S is required to be given explicitly in advance as part of the input.

Theorem 7 *For every fixed d there is a polynomial time algorithm that, given finite set $S \subseteq \mathbb{Z}^n$ presented by a linear discrete optimization oracle, integer $\rho \geq \rho(S)$, vectors $w_1, \ldots, w_d \in \mathbb{Z}^n$, and convex functional $c: \mathbb{R}^d \longrightarrow \mathbb{R}$ presented by a comparison oracle, encoded as $[\rho, w_1, \ldots, w_d]$, solves the convex discrete optimization problem*

$$\max \{c(w_1 x, \ldots, w_d x) : x \in S\}.$$

Proof Let $P := \text{conv}(S) \subset \mathbb{R}^n$, let $T := \{(w_1 x, \ldots, w_d x) : x \in S\}$ be the projection of S by $w_1, \ldots, w_d$, and let $Q := \text{conv}(T) \subset \mathbb{R}^d$ be the corresponding projection of P. Let $r := n\rho \max_{i=1}^{d} \|w_i\|_\infty$ and let $G := \{-r, \ldots, -1, 0, 1, \ldots, r\}^d$. Then $T \subseteq G$ and the number $(2r+1)^d$ of points of G is polynomially bounded in the input as encoded.

Let $D := \{u - v : u, v \in G, u \neq v\}$ be the set of differences of pairs of distinct point of G. It covers all edge-directions of Q since $\text{vert}(Q) \subseteq T \subseteq G$. Moreover, the number of points of D is less than $(2r+1)^{2d}$ and hence polynomial in the input. Now invoke the algorithm of Theorem 6: while the algorithm requires a set E covering all edge-directions of P, it needs E only to compute a set D covering all edge-directions of the projection Q (see proof of Theorem 6, which here is computed directly). □

Convex Combinatorial Optimization and More

In this section we discuss convex combinatorial optimization. The main result is that convex combinatorial optimization over a set $S \subseteq \{0, 1\}^n$ presented by a membership oracle can be solved in strongly polynomial time provided it is endowed with a set covering all edge-directions of $\text{conv}(S)$. In particular, the standard linear combinatorial optimization problem over S can be solved in strongly polynomial time as well. In Sect. "From Membership to Linear Optimization" we provide some preparatory statements involving various oracle presentation of the feasible set S. In Sect. "Linear and Convex Combinatorial Optimization in Strongly Polynomial Time" we combine these preparatory statements with Theorem 6 and prove the main result of this section. An extension to arbitrary finite sets $S \subset \mathbb{Z}^n$ endowed with edge-directions is established in Sect. "Linear and Convex Discrete Optimization over any Set in Pseudo Polynomial Time". We conclude with some applications in Sect. "Some Applications".

As noted in the introduction, when S is contained in $\{0, 1\}^n$ we refer to discrete optimization over S also as *combinatorial optimization* over S, to emphasize that S typically represents a family $\mathcal{F} \subseteq 2^N$ of subsets of a ground set $N := \{1, \ldots, n\}$ possessing some combinatorial property of interest (for instance, the family of bases of a matroid over N, see Sect. "Matroids and Maximum Norm Spanning Trees". The convex combinatorial optimization problem then also has the following interpretation (taken in [47,49]). We are given a weighting $\omega : N \longrightarrow \mathbb{Z}^d$ of elements of the ground set by d-dimensional integer vectors. We interpret the weight vector $\omega(j) \in \mathbb{Z}^d$ of element j as representing its value under d criteria (e.g., if N is the set of edges in a network then such criteria may include profit, reliability, flow velocity, etc.). The weight of a subset $F \subseteq N$ is the sum $\omega(F) := \sum_{j \in F} \omega(j)$ of weights of its elements, representing the total value of F under the d criteria. Now, given a convex functional $c: \mathbb{R}^d \longrightarrow \mathbb{R}$, the objective function value of $F \subseteq N$ is the "convex balancing" $c(\omega(F))$ of the values of the weight vector of F. The convex combinatorial optimization problem is to find a family member $F \in \mathcal{F}$ maximizing $c(\omega(F))$. The usual linear combinatorial optimization problem over $\mathcal{F}$ is the special case of $d = 1$ and c the identity on

$\mathbb{R}$. To cast a problem of that form in our usual setup just let $S := \{\mathbf{1}_F : F \in \mathcal{F}\} \subseteq \{0,1\}^n$ be the set of indicators of members of $\mathcal{F}$ and define weight vectors $w_1, \ldots, w_d \in \mathbb{Z}^n$ by $w_{i,j} := \omega(j)_i$ for $i = 1, \ldots, d$ and $j = 1, \ldots, n$.

From Membership to Linear Optimization

A *membership oracle* for a set $S \subseteq \mathbb{Z}^n$ is one that, queried on $x \in \mathbb{Z}^n$, asserts whether or not $x \in S$. An *augmentation oracle* for S is one that, queried on $x \in S$ and $w \in \mathbb{Z}^n$, either returns an $\hat{x} \in S$ with $w\hat{x} > wx$, i.e. a better point of S, or asserts that none exists, i.e. x is optimal for the linear discrete optimization problem over S.

A membership oracle presentation of S is very broad and available in all reasonable applications, but reveals little information on S, making it hard to use. However, as we now show, the edge-directions of conv(S) allow to convert membership to augmentation.

Lemma 4 *There is a strongly polynomial time algorithm that, given set $S \subseteq \{0,1\}^n$ presented by a membership oracle, $x \in S$, $w \in \mathbb{Z}^n$, and set $E \subset \mathbb{Z}^n$ covering all edge-directions of the polytope conv(S), encoded as $[n, |E|; \langle x, w, E \rangle]$, either returns a better point $\hat{x} \in S$, that is, one satisfying $w\hat{x} > wx$, or asserts that none exists.*

Proof Each edge of $P := \mathrm{conv}(S)$ is the difference of two $\{0,1\}$-vectors. Therefore, each edge direction of P is, up to scaling, a $\{-1,0,1\}$-vector. Thus, scaling $e := \frac{1}{\|e\|_\infty} e$ and $e := -e$ if necessary, we may and will assume that $e \in \{-1,0,1\}^n$ and $we \geq 0$ for all $e \in E$. Now, using the membership oracle, check if there is an $e \in E$ such that $x + e \in S$ and $we > 0$. If there is such an e then output $\hat{x} := x + e$ which is a better point, whereas if there is no such e then terminate asserting that no better point exists.

Clearly, if the algorithm outputs an $\hat{x}$ then it is indeed a better point. Conversely, suppose x is not a maximizer of w over S. Since $S \subseteq \{0,1\}^n$, the point x is a vertex of P. Since x is not a maximizer of w, there is an edge $[x, \hat{x}]$ of P with $\hat{x}$ a vertex satisfying $w\hat{x} > wx$. But then $e := \hat{x} - x$ is the one $\{-1,0,1\}$ edge-direction of $[x, \hat{x}]$ with $we \geq 0$ and hence $e \in E$. Thus, the algorithm will find and output $\hat{x} = x + e$ as it should. □

An augmentation oracle presentation of a finite S allows to solve the linear discrete optimization problem $\max\{wx : x \in S\}$ over S by starting from any feasible $x \in S$ and repeatedly augmenting it until an optimal solution $x^* \in S$ is reached. The next lemma bounds the running time needed to reach optimality using this procedure. While the running time is polynomial in the binary length of the linear functional w and the initial point x, it is more sensitive to the radius $\rho(S)$ of the feasible set S, and is polynomial only in its unary length. The lemma is an adaptation of a result of [30,57] (stated therein for $\{0,1\}$-sets), which makes use of bit-scaling ideas going back to [23].

Lemma 5 *There is a polynomial time algorithm that, given finite set $S \subset \mathbb{Z}^n$ presented by an augmentation oracle, $x \in S$, and $w \in \mathbb{Z}^n$, encoded as $[\rho(S), \langle x, w \rangle]$, provides an optimal solution $x^* \in S$ to the linear discrete optimization problem $\max\{wz : z \in S\}$.*

Proof Let $k := \max_{j=1}^n \lceil \log_2(|w_j| + 1) \rceil$ and note that $k \leq \langle w \rangle$. For $i = 0, \ldots, k$ define a linear functional $u_i = (u_{i,1}, \ldots, u_{i,n}) \in \mathbb{Z}^n$ by $u_{i,j} := \mathrm{sign}(w_j) \lfloor 2^{i-k} |w_j| \rfloor$ for $j = 1, \ldots, n$. Then $u_0 = 0$, $u_k = w$, and $u_i - 2u_{i-1} \in \{-1, 0, 1\}^n$ for all $i = 1, \ldots, k$.

We now describe how to construct a sequence of points $y_0, y_1, \ldots, y_k \in S$ such that y_i is an optimal solution to $\max\{u_i y : y \in S\}$ for all i. First note that all points of S are optimal for $u_0 = 0$ and hence we can take $y_0 := x$ to be the point of S given as part of the input. We now explain how to determine y_i from y_{i-1} for $i = 1, \ldots, k$. Suppose y_{i-1} has been determined. Set $\tilde{y} := y_{i-1}$. Query the augmentation oracle on $\tilde{y} \in S$ and u_i; if the oracle returns a better point $\hat{y}$ then set $\tilde{y} := \hat{y}$ and repeat, whereas if it asserts that there is no better point then the optimal solution for u_i is read off to be $y_i := \tilde{y}$. We now bound the number of calls to the oracle. Each time the oracle is queried on $\tilde{y}$ and u_i and returns a better point $\hat{y}$, the improvement is by at least one, i.e. $u_i(\hat{y} - \tilde{y}) \geq 1$; this is so because u_i, $\tilde{y}$ and $\hat{y}$ are integer. Thus, the number of necessary augmentations from y_{i-1} to y_i is at most the total improvement, which we claim satisfies

$$u_i(y_i - y_{i-1}) = (u_i - 2u_{i-1})(y_i - y_{i-1})$$
$$+ 2u_{i-1}(y_i - y_{i-1}) \leq 2n\rho + 0 = 2n\rho,$$

where $\rho := \rho(S)$. Indeed, $u_i - 2u_{i-1} \in \{-1, 0, 1\}^n$ and $y_i, y_{i-1} \in S \subset [-\rho, \rho]^n$ imply $(u_i - 2u_{i-1})(y_i - $

$y_{i-1}) \leq 2n\rho$; and y_{i-1} optimal for u_{i-1} gives $u_{i-1}(y_i - y_{i-1}) \leq 0$.

Thus, after a total number of at most $2n\rho k$ calls to the oracle we obtain y_k which is optimal for u_k. Since $w = u_k$ we can output $x^* := y_k$ as the desired optimal solution to the linear discrete optimization problem. Clearly the number $2n\rho k$ of calls to the oracle, as well as the number of arithmetic operations and binary length of numbers occurring during the algorithm, are polynomial in $\rho(S)$, $\langle x, w \rangle$. This completes the proof. □

We conclude this preparatory subsection by recording the following result of [24] which incorporates the heavy simultaneous Diophantine approximation of [44].

Proposition 1 *There is a strongly polynomial time algorithm that, given $w \in \mathbb{Z}^n$, encoded as $[n; \langle w \rangle]$, produces $\hat{w} \in \mathbb{Z}^n$, whose binary length $\langle \hat{w} \rangle$ is polynomially bounded in n and independent of w, and with $\text{sign}(\hat{w}z) = \text{sign}(wz)$ for every $z \in \{-1, 0, 1\}^n$.*

Linear and Convex Combinatorial Optimization in Strongly Polynomial Time

Combining the preparatory statements of Sect. "From Membership to Linear Optimization" with Theorem 6, we can now solve the convex combinatorial optimization over a set $S \subseteq \{0, 1\}^n$ presented by a membership oracle and endowed with a set covering all edge-directions of conv(S) in strongly polynomial time. We start with the special case of linear combinatorial optimization.

Theorem 8 *There is a strongly polynomial time algorithm that, given set $S \subseteq \{0, 1\}^n$ presented by a membership oracle, $x \in S$, $w \in \mathbb{Z}^n$, and set $E \subset \mathbb{Z}^n$ covering all edge-directions of the polytope conv(S), encoded as $[n, |E|; \langle x, w, E \rangle]$, provides an optimal solution $x^* \in S$ to the linear combinatorial optimization problem* $\max\{wz : z \in S\}$.

Proof First, an augmentation oracle for S can be simulated using the membership oracle, in strongly polynomial time, by applying the algorithm of Lemma 4.

Next, using the simulated augmentation oracle for S, we can now do linear optimization over S in strongly polynomial time as follows. First, apply to w the algorithm of Proposition 1 and obtain $\hat{w} \in \mathbb{Z}^n$ whose binary length $\langle \hat{w} \rangle$ is polynomially bounded in n, which satisfies $\text{sign}(\hat{w}z) = \text{sign}(wz)$ for every $z \in \{-1, 0, 1\}^n$. Since $S \subseteq \{0, 1\}^n$, it is finite and has radius $\rho(S) = 1$. Now apply the algorithm of Lemma 5 to S, x and $\hat{w}$, and obtain a maximizer x^* of $\hat{w}$ over S. For every $y \in \{0, 1\}^n$ we then have $x^* - y \in \{-1, 0, 1\}^n$ and hence $\text{sign}(w(x^* - y)) = \text{sign}(\hat{w}(x^* - y))$. So x^* is also a maximizer of w over S and hence an optimal solution to the given linear combinatorial optimization problem. Now, $\rho(S) = 1$, $\langle \hat{w} \rangle$ is polynomial in n, and $x \in \{0, 1\}^n$ and hence $\langle x \rangle$ is linear in n. Thus, the entire length of the input $[\rho(S), \langle x, \hat{w} \rangle]$ to the polynomial-time algorithm of Lemma 5 is polynomial in n, and so its running time is in fact strongly polynomial on that input. □

Combining Theorems 6 and 8 we recover at once the following result of [49].

Theorem 9 *For every fixed d there is a strongly polynomial time algorithm that, given set $S \subseteq \{0, 1\}^n$ presented by a membership oracle, $x \in S$, vectors $w_1, \ldots, w_d \in \mathbb{Z}^n$, set $E \subset \mathbb{Z}^n$ covering all edge-directions of the polytope conv(S), and convex functional $c : \mathbb{R}^d \longrightarrow \mathbb{R}$ presented by a comparison oracle, encoded as $[n, |E|; \langle x, w_1, \ldots, w_d, E \rangle]$, provides an optimal solution $x^* \in S$ to the convex combinatorial optimization problem*

$$\max\{c(w_1 z, \ldots, w_d z) : z \in S\}.$$

Proof Since S is nonempty, a linear discrete optimization oracle for S can be simulated in strongly polynomial time by the algorithm of Theorem 8. Using this simulated oracle, we can apply the algorithm of Theorem 6 and solve the given convex combinatorial optimization problem in strongly polynomial time. □

Linear and Convex Discrete Optimization over any Set in Pseudo Polynomial Time

In Sect. "Linear and Convex Combinatorial Optimization in Strongly Polynomial Time" above we developed strongly polynomial time algorithms for linear and convex discrete optimization over $\{0, 1\}$-sets. We now provide extensions of these algorithms to arbitrary finite sets $S \subset \mathbb{Z}^n$. As can be expected, the algorithms become slower.

We start by recording the following fundamental result of Khachiyan [40] asserting that linear programming is polynomial time solvable via the ellipsoid method [63]. This result will be used below as well as several more times later in the monograph.

Proposition 2 *There is a polynomial time algorithm that, given $A \in \mathbb{Z}^{m \times n}$, $b \in \mathbb{Z}^m$, and $w \in \mathbb{Z}^n$, encoded as $[\langle A, b, w \rangle]$, either asserts that $P := \{x \in \mathbb{R}^n : Ax \leq b\}$ is empty, or asserts that the linear functional wx is unbounded over P, or provides a vertex $v \in \text{vert}(P)$ which is an optimal solution to the linear program $\max\{wx : x \in P\}$.*

The following analog of Lemma 4 shows how to covert membership to augmentation in polynomial time, albeit, no longer in strongly polynomial time. Here, both the given initial point x and the returned better point $\hat{x}$ if any, are *vertices* of conv(S).

Lemma 6 *There is a polynomial time algorithm that, given finite set $S \subset \mathbb{Z}^n$ presented by a membership oracle, vertex x of the polytope conv(S), $w \in \mathbb{Z}^n$, and set $E \subset \mathbb{Z}^n$ covering all edge-directions of conv(S), encoded as $[\rho(S), \langle x, w, E \rangle]$, either returns a better vertex $\hat{x}$ of conv(S), that is, one satisfying $w\hat{x} > wx$, or asserts that none exists.*

Proof Dividing each vector $e \in E$ by the greatest common divisor of its entries and setting $e := -e$ if necessary, we can and will assume that each e is *primitive*, that is, its entries are relatively prime integers, and $we \geq 0$. Using the membership oracle, construct the subset $F \subseteq E$ of those $e \in E$ for which $x + re \in S$ for some $r \in \{1, \ldots, 2\rho(S)\}$. Let $G \subseteq F$ be the subset of those $f \in F$ for which $wf > 0$. If G is empty then terminate asserting that there is no better vertex. Otherwise, consider the convex cone cone(F) generated by F. It is clear that x is incident on an edge of conv(S) in direction f if and only if f is an extreme ray of cone(F). Moreover, since $G = \{f \in F : wf > 0\}$ is nonempty, there must be an extreme ray of cone(F) which lies in G. Now $f \in F$ is an extreme ray of cone(F) if and only if there do not exist nonnegative λ_e, $e \in F \setminus \{f\}$, such that $f = \sum_{e \neq f} \lambda_e e$; this can be checked in polynomial time using linear programming. Applying this procedure to each $f \in G$, identify an extreme ray $g \in G$. Now, using the membership oracle, determine the largest $r \in \{1, \ldots, 2\rho(S)\}$ for which $x + rg \in S$. Output $\hat{x} := x + rg$ which is a better vertex of conv(S). □

We now prove the extensions of Theorems 8 and 9 to arbitrary, not necessarily $\{0, 1\}$-valued, finite sets.

While the running time remains polynomial in the binary length of the weights $w_1, \ldots, w_d$ and the set of edge-directions E, it is more sensitive to the radius $\rho(S)$ of the feasible set S, and is polynomial only in its unary length. Here, the initial feasible point and the optimal solution output by the algorithms are vertices of conv(S). Again, we start with the special case of linear combinatorial optimization.

Theorem 10 *There is a polynomial time algorithm that, given finite $S \subset \mathbb{Z}^n$ presented by a membership oracle, vertex x of the polytope conv(S), $w \in \mathbb{Z}^n$, and set $E \subset \mathbb{Z}^n$ covering all edge-directions of conv(S), encoded as $[\rho(S), \langle x, w, E \rangle]$, provides an optimal solution $x^* \in S$ to the linear discrete optimization problem $\max\{wz : z \in S\}$.*

Proof Apply the algorithm of Lemma 5 to the given data. Consider any query $x' \in S$, $w' \in \mathbb{Z}^n$ made by that algorithm to an augmentation oracle for S. To answer it, apply the algorithm of Lemma 6 to x' and w'. Since the first query made by the algorithm of Lemma 5 is on the given input vertex $x' := x$, and any consequent query is on a point $x' := \hat{x}$ which was the reply of the augmentation oracle to the previous query (see proof of Lemma 5), we see that the algorithm of Lemma 6 will always be asked on a vertex of S and reply with another. Thus, the algorithm of Lemma 6 can answer all augmentation queries and enables the polynomial time solution of the given problem. □

Theorem 11 *For every fixed d there is a polynomial time algorithm that, given finite set $S \subseteq \mathbb{Z}^n$ presented by membership oracle, vertex x of conv(S), vectors $w_1, \ldots, w_d \in \mathbb{Z}^n$, set $E \subset \mathbb{Z}^n$ covering all edge-directions of the polytope conv(S), and convex functional $c : \mathbb{R}^d \longrightarrow \mathbb{R}$ presented by a comparison oracle, encoded as $[\rho(S), \langle x, w_1, \ldots, w_d, E \rangle]$, provides an optimal solution $x^* \in S$ to the convex combinatorial optimization problem*

$$\max\{c(w_1 z, \ldots, w_d z) : z \in S\}.$$

Proof Since S is nonempty, a linear discrete optimization oracle for S can be simulated in polynomial time by the algorithm of Theorem 10. Using this simulated oracle, we can apply the algorithm of Theorem 6 and solve the given problem in polynomial time. □

Some Applications

Positive Semidefinite Quadratic Binary Programming The quadratic binary programming problem is the following: given an $n \times n$ matrix M, find a vector $x \in \{0, 1\}^n$ maximizing the quadratic form $x^T M x$ induced by M. We consider here the instance where M is positive semidefinite, in which case it can be assumed to be presented as $M = W^T W$ with W a given $d \times n$ matrix. Already this restricted version is very broad: if the rank d of W and M is variable then, as mentioned in the introduction, the problem is NP-hard. We now show that, for fixed d, Theorem 9 implies at once that the problem is strongly polynomial time solvable (see also [2]).

Corollary 6 *For every fixed d there is a strongly polynomial time algorithm that given $W \in \mathbb{Z}^{d \times n}$, encoded as $[n; \langle W \rangle]$, finds $x^* \in \{0, 1\}^n$ maximizing the form $x^T W^T W x$.*

Proof Let $S := \{0, 1\}^n$ and let $E := \{\mathbf{1}_1, \ldots, \mathbf{1}_n\}$ be the set of unit vectors in $\mathbb{R}^n$. Then $P := \mathrm{conv}(S)$ is just the n-cube $[0, 1]^n$ and hence E covers all edge-directions of P. A membership oracle for S is easily and efficiently realizable and $x := 0 \in S$ is an initial point. Also, $|E|$ and $\langle E \rangle$ are polynomial in n, and E is easily and efficiently computable.

Now, for $i = 1, \ldots, d$ define $w_i \in \mathbb{Z}^n$ to be the ith row of the matrix W, that is, $w_{i,j} := W_{i,j}$ for all i, j. Finally, let $c \colon \mathbb{R}^d \longrightarrow \mathbb{R}$ be the squared l_2 norm given by $c(y) := \|y\|_2^2 := \sum_{i=1}^d y_i^2$, and note that the comparison of $c(y)$ and $c(z)$ can be done for $y, z \in \mathbb{Z}^d$ in time polynomial in $\langle y, z \rangle$ using a constant number of arithmetic operations, providing a strongly polynomial time realization of a comparison oracle for c.

This translates the given quadratic programming problem into a convex combinatorial optimization problem over S, which can be solved in strongly polynomial time by applying the algorithm of Theorem 9 to $S, x = 0, w_1, \ldots, w_d, E$, and c. □

Matroids and Maximum Norm Spanning Trees Optimization problems over matroids form a fundamental class of combinatorial optimization problems. Here we discuss matroid bases, but everything works for independent sets as well. Recall that a family $\mathcal{B}$ of subsets of $\{1, \ldots, n\}$ is the family of *bases* of a *matroid* if all members of $\mathcal{B}$ have the same cardinality, called the *rank* of the matroid, and for every $B, B' \in \mathcal{B}$ and $i \in B \setminus B'$ there is a $j \in B'$ such that $B \setminus \{i\} \cup \{j\} \in \mathcal{B}$. Useful models include the *graphic matroid* of a graph G with edge set $\{1, \ldots, n\}$ and $\mathcal{B}$ the family of spanning forests of G, and the *linear matroid* of an $m \times n$ matrix A with $\mathcal{B}$ the family of sets of indices of maximal linearly independent subsets of columns of A.

It is well known that linear combinatorial optimization over matroids can be solved by the fast greedy algorithm [22]. We now show that, as a consequence of Theorem 9, convex combinatorial optimization over a matroid presented by a membership oracle can be solved in strongly polynomial time as well (see also [34,47]). We state the result for bases, but the analogous statement for independent sets hold as well. We say that $S \subseteq \{0, 1\}^n$ is the *set of bases of a matroid* if it is the set of indicators of the family $\mathcal{B}$ of bases of some matroid, in which case we call $\mathrm{conv}(S)$ the *matroid base polytope*.

Corollary 7 *For every fixed d there is a strongly polynomial time algorithm that, given set $S \subseteq \{0, 1\}^n$ of bases of a matroid presented by a membership oracle, $x \in S$, $w_1, \ldots, w_d \in \mathbb{Z}^n$, and convex functional $c \colon \mathbb{R}^d \longrightarrow \mathbb{R}$ presented by a comparison oracle, encoded as $[n; \langle x, w_1, \ldots, w_d \rangle]$, solves the convex matroid optimization problem*

$$\max\{c(w_1 z, \ldots, w_d z) : z \in S\}.$$

Proof Let $E := \{\mathbf{1}_i - \mathbf{1}_j : 1 \leq i < j \leq n\}$ be the set of differences of pairs of unit vectors in $\mathbb{R}^n$. We claim that E covers all edge-directions of the matroid base polytope $P := \mathrm{conv}(S)$. Consider any edge $e = [y, y']$ of P with $y, y' \in S$ and let $B := \mathrm{supp}(y)$ and $B' := \mathrm{supp}(y')$ be the corresponding bases. Let $h \in \mathbb{R}^n$ be a linear functional uniquely maximized over P at e. If $B \setminus B' = \{i\}$ is a singleton then $B' \setminus B = \{j\}$ is a singleton as well in which case $y - y' = \mathbf{1}_i - \mathbf{1}_j$ and we are done. Suppose then, indirectly, that it is not, and pick an element i in the symmetric difference $B \triangle B' := (B \setminus B') \cup (B' \setminus B)$ of minimum value h_i. Without loss of generality assume $i \in B \setminus B'$. Then there is a $j \in B' \setminus B$ such that $B'' := B \setminus \{i\} \cup \{j\}$ is also a basis. Let $y'' \in S$ be the indicator of B''. Now $|B \triangle B' q| > 2$ implies that B'' is neither B nor B'. By the choice of i we have $h y'' = h y - h_i + h_j \geq h y$. So y'' is also a maximizer of h over P and hence $y'' \in e$. But no $\{0, 1\}$-vector is a convex combination of others, a contradiction.

Now, $|E| = \binom{n}{2}$ and $E \subset \{-1, 0, 1\}^n$ imply that $|E|$ and $\langle E \rangle$ are polynomial in n. Moreover, E can be easily computed in strongly polynomial time. Therefore, applying the algorithm of Theorem 9 to the given data and the set E, the convex discrete optimization problem over S can be solved in strongly polynomial time. □

One important application of Corollary 7 is a polynomial time algorithm for computing the *universal Gröbner basis* of any system of polynomials having a finite set of common zeros in fixed (but arbitrary) number of variables, as well as the construction of the *state polyhedron* of any member of the *Hilbert scheme*, see [5,51]. Other important applications are in the field of *algebraic statistics* [52], in particular for *optimal experimental design*. These applications are beyond our scope here and will be discussed elsewhere.

Here is another concrete example of a convex matroid optimization application.

Example 1 (MAXIMUM NORM SPANNING TREE). Fix any positive integer d. Let $\|\cdot\|_p : \mathbb{R}^d \longrightarrow \mathbb{R}$ be the l_p norm given by $\|x\|_p := (\sum_{i=1}^d |x_i|^p)^{\frac{1}{p}}$ for $1 \leq p < \infty$ and $\|x\|_\infty := \max_{i=1}^d |x_i|$. Let G be a connected graph with edge set $N := \{1, \ldots, n\}$. For $j = 1, \ldots, n$ let $u_j \in \mathbb{Z}^d$ be a weight vector representing the values of edge j under some d criteria. The weight of a subset $T \subseteq N$ is the sum $\sum_{j \in T} u_j$ representing the total values of T under the d criteria. The problem is to find a spanning tree T of G whose weight has maximum l_p norm, that is, a spanning tree T maximizing $\|\sum_{j \in T} u_j\|_p$.

Define $w_1, \ldots, w_d \in \mathbb{Z}^n$ by $w_{i,j} := u_{j,i}$ for $i = 1, \ldots, d, j = 1, \ldots, n$. Let $S \subseteq \{0, 1\}^n$ be the set of indicators of spanning trees of G. Then, in time polynomial in n, a membership oracle for S is realizable, and an initial $x \in S$ is obtainable as the indicator of any greedily constructible spanning tree T. Finally, define the convex functional $c := \|\cdot\|_p$. Then for most common values $p = 1, 2, \infty$, and in fact for any $p \in \mathbb{N}$, the comparison of $c(y)$ and $c(z)$ can be done for $y, z \in \mathbb{Z}^d$ in time polynomial in $\langle y, z, p \rangle$ by computing and comparing the integer valued pth powers $\|y\|_p^p$ and $\|z\|_p^p$. Thus, by Corollary 7, this problem is solvable in time polynomial in $\langle u_1, \ldots, u_n, p \rangle$.

Linear N-fold Integer Programming

In this section we develop a theory of linear *n-fold integer programming*, which leads to the polynomial time solution of broad classes of linear integer programming problems in variable dimension. This will be extended to convex n-fold integer programming in Sect. "Convex Integer Programming".

In Sect. "Oriented Augmentation and Linear Optimization" we describe an adaptation of a result of [56] involving an oriented version of the augmentation oracle of Sect. "From Membership to Linear Optimization". In Sect. "Graver Bases and Linear Integer Programming" we discuss Graver bases and their application to linear integer programming. In Sect. "Graver Bases of N-fold Matrices" we show that Graver bases of n-fold matrices can be computed efficiently. In Sect. "Linear N-fold Integer Programming in Polynomial Time" we combine the preparatory statements from Sect. "Oriented Augmentation and Linear Optimization", Sect. "Graver Bases and Linear Integer Programming", and Sect. "Graver Bases of N-fold Matrices", and prove the main result of this section, asserting that linear n-fold integer programming is polynomial time solvable. We conclude with some applications in Sect. "Some Applications".

Here and in Sect. "Convex Integer Programming" we concentrate on discrete optimization problems over a set S presented as the set of integer points satisfying an explicitly given system of linear inequalities. Without loss of generality we may and will assume that S is given either in standard form $S := \{x \in \mathbb{N}^n : Ax = b\}$ where $A \in \mathbb{Z}^{m \times n}$ and $b \in \mathbb{Z}^m$, or in the form

$$S := \{x \in \mathbb{Z}^n : Ax = b, \, l \leq x \leq u\}$$

where $l, u \in \mathbb{Z}_\infty^n$ and $\mathbb{Z}_\infty = \mathbb{Z} \uplus \{\pm\infty\}$, where some of the variables are bounded below or above and some are unbounded. Thus, S is no longer presented by an oracle, but by the explicit data A, b and possibly l, u. In this setup we refer to discrete optimization over S also as *integer programming* over S. As usual, an algorithm solving the problem must either provide an $x \in S$ maximizing wx over S, or assert that none exists (either because S is empty or because the objective function is unbounded over the underlying polyhedron). We will sometimes assume that an initial point $x \in S$ is given, in which case b will be computed as $b := Ax$ and not be part of the input.

Oriented Augmentation and Linear Optimization

We have seen in Sect. "From Membership to Linear Optimization" that an augmentation oracle presentation of a finite set $S \subset \mathbb{Z}^n$ enables to solve the linear discrete optimization problem over S. However, the running time of the algorithm of Lemma 5 which demonstrated this, was polynomial in the unary length of the radius $\rho(S)$ of the feasible set rather than in its binary length.

In this subsection we discuss a recent result of [56] and show that, when S is presented by a suitable stronger oriented version of the augmentation oracle, the linear optimization problem can be solved by a much faster algorithm, whose running time is in fact polynomial in the binary length $\langle \rho(S) \rangle$. The key idea behind this algorithm is that it gives preference to augmentations along interior points of conv(S) staying far off its boundary. It is inspired by and extends the combinatorial interior point algorithm of [61].

For any vector $g \in \mathbb{R}^n$, let $g^+, g^- \in \mathbb{R}^n_+$ denote its *positive* and *negative* parts, defined by $g_j^+ := \max\{g_j, 0\}$ and $g_j^- := -\min\{g_j, 0\}$ for $j = 1, \ldots, n$. Note that both g^+, g^- are nonnegative, supp(g) = supp(g^+) $\uplus$ supp(g^-), and $g = g^+ - g^-$.

An *oriented augmentation oracle* for a set $S \subset \mathbb{Z}^n$ is one that, queried on $x \in S$ and $w_+, w_- \in \mathbb{Z}^n$, either returns an *augmenting vector* $g \in \mathbb{Z}^n$, defined to be one satisfying $x + g \in S$ and $w_+ g^+ - w_- g^- > 0$, or asserts that none exists.

Note that this oracle involves *two* linear functionals $w_+, w_- \in \mathbb{Z}^n$ rather than one (w_+, w_- are two distinct independent vectors and *not* the positive and negative parts of one vector). The conditions on an augmenting vector g indicate that it is a feasible direction and has positive value under the nonlinear objective function determined by w_+, w_-. Note that this oracle is indeed stronger than the augmentation oracle of Sect. "From Membership to Linear Optimization": to answer a query $x \in S$, $w \in \mathbb{Z}^n$ to the latter, set $w_+ := w_- := w$, thereby obtaining $w_+ g^+ - w_- g^- = wg$ for all g, and query the former on x, w_+, w_-; if it replies with an augmenting vector g then reply with the better point $\hat{x} := x + g$, whereas if it asserts that no g exists then assert that no better point exists.

The following lemma is an adaptation of the result of [56] concerning sets of the form $S := \{x \in \mathbb{Z}^n : Ax = b, 0 \leq x \leq u\}$ of nonnegative integer points satisfying equations and upper bounds. However, the pair A, b is neither explicitly needed nor does it affect the running time of the algorithm underlying the lemma. It suffices that S is of that form. Moreover, an arbitrary lower bound vector l rather than 0 can be included. So it suffices to assume that S coincides with the intersection of its affine hull and the set of integer points in a box, that is, $S = \text{aff}(S) \cap \{x \in \mathbb{Z}^n : l \leq x \leq u\}$ where $l, u \in \mathbb{Z}^n$. We now describe and prove the algorithm of [56] adjusted to any lower and upper bounds l, u.

Lemma 7 *There is a polynomial time algorithm that, given vectors $l, u \in \mathbb{Z}^n$, set $S \subset \mathbb{Z}^n$ satisfying $S = \text{aff}(S) \cap \{z \in \mathbb{Z}^n : l \leq z \leq u\}$ and presented by an oriented augmentation oracle, $x \in S$, and $w \in \mathbb{Z}^n$, encoded as $[\langle l, u, x, w \rangle]$, provides an optimal solution $x^* \in S$ to the linear discrete optimization problem $\max\{wz : z \in S\}$.*

Proof We start with some strengthening adjustments to the oriented augmentation oracle. Let $\rho := \max\{\|l\|_\infty, \|u\|_\infty\}$ be an upper bound on the radius of S. Then any augmenting vector g obtained from the oriented augmentation oracle when queried on $y \in S$ and $w_+, w_- \in \mathbb{Z}^n$, can be made in polynomial time to be *exhaustive*, that is, to satisfy $y + 2g \notin S$ (which means that no longer augmenting step in direction g can be taken). Indeed, using binary search, find the largest $r \in \{1, \ldots, 2\rho\}$ for which $l \leq y + rg \leq u$; then $S = \text{aff}(S) \cap \{z \in \mathbb{Z}^n : l \leq z \leq u\}$ implies $y + rg \in S$ and hence we can replace $g := rg$. So from here on we will assume that if there is an augmenting vector then the oracle returns an exhaustive one. Second, let $\mathbb{R}_\infty := \mathbb{R} \uplus \{\pm \infty\}$ and for any vector $v \in \mathbb{R}^n$ let $v^{-1} \in \mathbb{R}_\infty^n$ denote its entry-wise reciprocal defined by $v_i^{-1} := \frac{1}{v_i}$ if $v_i \neq 0$ and $v_i^{-1} := \infty$ if $v_i = 0$. For any $y \in S$, the vectors $(y - l)^{-1}$ and $(u - y)^{-1}$ are the reciprocals of the "entry-wise distance" of y from the given lower and upper bounds. The algorithm will query the oracle on triples y, w_+, w_- with $w_+ := w - \mu(u - y)^{-1}$ and $w_- := w + \mu(y - l)^{-1}$ where μ is a suitable positive scalar and w is the input linear functional. The fact that such w_+, w_- may have infinite entries does not cause any problem: indeed, if g is an augmenting vector then $y + g \in S$ implies that $g_i^+ = 0$ whenever $y_i = u_i$

and $g_i^- = 0$ whenever $l_i = y_i$, so each infinite entry in w_+ or w_- occurring in the expression $w_+g^+ - w_-g^-$ is multiplied by 0 and hence zeroed out.

The algorithm proceeds in phases. Each phase i starts with a feasible point $y_{i-1} \in S$ and performs repeated augmentations using the oriented augmentation oracle, terminating with a new feasible point $y_i \in S$ when no further augmentations are possible. The queries to the oracle make use of a positive scalar parameters μ_i fixed throughout the phase. The first phase ($i=1$) starts with the input point $y_0 := x$ and sets $\mu_1 := \rho \|w\|_\infty$. Each further phase $i \geq 2$ starts with the point y_{i-1} obtained from the previous phase and sets the parameter value $\mu_i := \frac{1}{2}\mu_{i-1}$ to be half its value in the previous phase. The algorithm terminates at the end of the first phase i for which $\mu_i < \frac{1}{n}$, and outputs $x^* := y_i$. Thus, the number of phases is at most $\lceil \log_2(2n\rho\|w\|_\infty) \rceil$ and hence polynomial in $\langle l, u, w \rangle$.

We now describe the ith phase which determines y_i from y_{i-1}. Set $\mu_i := \frac{1}{2}\mu_{i-1}$ and $\hat{y} := y_{i-1}$. Iterate the following: query the strengthened oriented augmentation oracle on $\hat{y}$, $w_+ := w - \mu_i(u - \hat{y})^{-1}$, and $w_- := w + \mu_i(\hat{y} - l)^{-1}$; if the oracle returns an exhaustive augmenting vector g then set $\hat{y} := \hat{y} + g$ and repeat, whereas if it asserts that there is no augmenting vector then set $y_i := \hat{y}$ and complete the phase. If $\mu_i \geq \frac{1}{n}$ then proceed to the $(i+1)$th phase, else output $x^* := y_i$ and terminate the algorithm.

It remains to show that the output of the algorithm is indeed an optimal solution and that the number of iterations (and hence calls to the oracle) in each phase is polynomial in the input. For this we need the following facts, the easy proofs of which are omitted:

1. For every feasible $y \in S$ and direction g with $y + g \in S$ also feasible, we have

$$(u - y)^{-1}g^+ + (y - l)^{-1}g^- \leq n.$$

2. For every $y \in S$ and direction g with $y + g \in S$ but $y + 2g \notin S$, we have

$$(u - y)^{-1}g^+ + (y - l)^{-1}g^- > \frac{1}{2}.$$

3. For every feasible $y \in S$, direction g with $y + g \in S$ also feasible, and $\mu > 0$, setting $w_+ := w - \mu(u - y)^{-1}$ and $w_- := w + \mu(y - l)^{-1}$ we have

$$w_+g^+ - w_-g^- = wg - \mu\left((u-y)^{-1}g^+ + (y-l)^{-1}g^-\right).$$

Now, consider the last phase i with $\mu_i < \frac{1}{n}$, let $x^* := y_i := \hat{y}$ be the output of the algorithm at the end of this phase, and let $\hat{x} \in S$ be any optimal solution. Now, the phase is completed when the oracle, queried on the triple $\hat{y}$, $w_+ = w - \mu_i(u - \hat{y})^{-1}$, and $w_- = w + \mu_i(\hat{y} - l)^{-1}$, asserts that there is no augmenting vector. In particular, setting $g := \hat{x} - \hat{y}$, we find $w_+g^+ - w_-g^- \leq 0$ and hence, by facts 1 and 3 above,

$$w\hat{x} - wx^* = wg \leq \mu_i\left((u - \hat{y})^{-1}g^+ + (\hat{y} - l)^{-1}g^-\right) < \frac{1}{n} \cdot n = 1.$$

Since $w\hat{x}$ and wx^* are integer, this implies that in fact $w\hat{x} - wx^* \leq 0$ and hence the output x^* of the algorithm is indeed an optimal solution to the given optimization problem.

Next we bound the number of iterations in each phase i starting from $y_{i-1} \in S$. Let again $\hat{x} \in S$ be any optimal solution. Consider any iteration in that phase, where the oracle is queried on $\hat{y}$, $w_+ = w - \mu_i(u - \hat{y})^{-1}$, and $w_- = w + \mu_i(\hat{y} - l)^{-1}$, and returns an exhaustive augmenting vector g. We will now show that

$$w(\hat{y} + g) - w\hat{y} \geq \frac{1}{4n}(w\hat{x} - wy_{i-1}), \quad (1)$$

that is, the increment in the objective value from $\hat{y}$ to the augmented point $\hat{y} + g$ is at least $\frac{1}{4n}$ times the difference between the optimal objective value $w\hat{x}$ and the objective value wy_{i-1} of the point y_{i-1} at the beginning of phase i. This shows that at most $4n$ such increments (and hence iterations) can occur in the phase before it is completed.

To establish (1), we show that $wg \geq \frac{1}{2}\mu_i$ and $w\hat{x} - wy_{i-1} \leq 2n\mu_i$. For the first inequality, note that g is an exhaustive augmenting vector and so $w_+g^+ - w_-g^- > 0$ and $\hat{y} + 2g \notin S$ and hence, by facts 2 and 3, $wg > \mu_i((u - \hat{y})^{-1}g^+ + (\hat{y} - l)^{-1}g^-) > \frac{1}{2}\mu_i$. We proceed with the second inequality. If $i = 1$ (first phase) then this indeed holds since $w\hat{x} - wy_0 \leq 2n\rho\|w\|_\infty = 2n\mu_1$. If $i \geq 2$, let $\tilde{w}_+ := w - \mu_{i-1}(u - y_{i-1})^{-1}$ and $\tilde{w}_- :=$

$w + \mu_{i-1}(y_{i-1} - l)^{-1}$. The $(i-1)$th phase was completed when the oracle, queried on the triple $y_{i-1}, \tilde{w}_+$, and $\tilde{w}_-$, asserted that there is no augmenting vector. In particular, for $\tilde{g} := \hat{x} - y_{i-1}$, we find $\tilde{w}_+ \tilde{g}^+ - \tilde{w}_- \tilde{g}^- \leq 0$ and so, by facts 1 and 3,

$$w\hat{x} - wy_{i-1}$$
$$= w\tilde{g} \leq \mu_{i-1}\left((u - y_{i-1})^{-1}\tilde{g}^+ + (y_{i-1} - l)^{-1}\tilde{g}^-\right)$$
$$\leq \mu_{i-1} n = 2n\mu_i \,.$$
$\square$

Graver Bases and Linear Integer Programming

We now come to the definition of a fundamental object introduced by Graver in [28]. The *Graver basis* of an integer matrix A is a canonical finite set $G(A)$ that can be defined as follows. Define a partial order $\sqsubseteq$ on $\mathbb{Z}^n$ which extends the coordinate-wise order $\leq$ on $\mathbb{N}^n$ as follows: for two vectors $u, v \in \mathbb{Z}^n$ put $u \sqsubseteq v$ and say that u is *conformal* to v if $|u_i| \leq |v_i|$ and $u_i v_i \geq 0$ for $i = 1, \ldots, n$, that is, u and v lie in the same orthant of $\mathbb{R}^n$ and each component of u is bounded by the corresponding component of v in absolute value. It is not hard to see that $\sqsubseteq$ is a *well* partial ordering (this is basically Dickson's lemma) and hence every subset of $\mathbb{Z}^n$ has finitely-many $\sqsubseteq$-minimal elements. Let $\mathcal{L}(A) := \{x \in \mathbb{Z}^n : Ax = 0\}$ be the lattice of linear integer dependencies on A. The *Graver basis* of A is defined to be the set $G(A)$ of all $\sqsubseteq$-minimal vectors in $\mathcal{L}(A) \setminus \{0\}$.

Note that if A is an $m \times n$ matrix then its Graver basis consist of vectors in $\mathbb{Z}^n$. We sometimes write $G(A)$ as a suitable $|G(A)| \times n$ matrix whose rows are the Graver basis elements. The Graver basis is centrally symmetric ($g \in G(A)$ implies $-g \in G(A)$); thus, when listing a Graver basis we will typically give one of each antipodal pair and prefix the set (or matrix) by $\pm$. Any element of the Graver basis is primitive (its entries are relatively prime integers). Every circuit of A (nonzero primitive minimal support element of $\mathcal{L}(A)$) is in $G(A)$; in fact, if A is totally unimodular then $G(A)$ coincides with the set of circuits (see Sect. "Convex Integer Programming over Totally Unimodular Systems" in the sequel for more details on this). However, in general $G(A)$ is much larger. For more details on Graver bases and their connection to Gröbner bases see Sturmfels [58] and for the currently fastest procedure for computing them see [35,36].

Here is a quick simple example; we will see more structured and complex examples later on. Consider the 1×3 matrix $A := (1, 2, 1)$. Then its Graver basis can be shown to be the set $G(A) = \pm\{(2, -1, 0), (0, -1, 2), (1, 0, -1), (1, -1, 1)\}$. The first three elements (and their antipodes) are the circuits of A; already in this small example non-circuits appear as well: the fourth element (and its antipode) is a primitive linear integer dependency whose support is not minimal.

We now show that when we do have access to the Graver basis, it can be used to solve linear integer programming. We will extend this in Sect. "Convex Integer Programming", where we show that the Graver basis enables to solve convex integer programming as well. In Sect. "Graver Bases of N-fold Matrices" we will show that there are important classes of matrices for which the Graver basis is indeed accessible.

First, we need a simple property of Graver bases. A finite sum $u := \sum_i v_i$ of vectors $v_i \in \mathbb{R}^n$ is *conformal* if each summand is conformal to the sum, that is, $v_i \sqsubseteq u$ for all i.

Lemma 8 *Let A be any integer matrix. Then any $h \in \mathcal{L}(A) \setminus \{0\}$ can be written as a conformal sum $h := \sum g_i$ of (not necessarily distinct) Graver basis elements $g_i \in G(A)$.*

Proof By induction on the well partial order $\sqsubseteq$. Recall that $G(A)$ is the set of $\sqsubseteq$-minimal elements in $\mathcal{L}(A) \setminus \{0\}$. Consider any $h \in \mathcal{L}(A) \setminus \{0\}$. If it is $\sqsubseteq$-minimal then $h \in G(A)$ and we are done. Otherwise, there is a $h' \in G(A)$ such that $h' \sqsubset h$. Set $h'' := h - h'$. Then $h'' \in \mathcal{L}(A) \setminus \{0\}$ and $h'' \sqsubset h$, so by induction there is a conformal sum $h'' = \sum_i g_i$ with $g_i \in G(A)$ for all i. Now $h = h' + \sum_i g_i$ is the desired conformal sum of h. $\square$

The next lemma shows the usefulness of Graver bases for oriented augmentation.

Lemma 9 *Let A be an $m \times n$ integer matrix with Graver basis $G(A)$ and let $l, u \in \mathbb{Z}_\infty^n$, $w_+, w_- \in \mathbb{Z}^n$, and $b \in \mathbb{Z}^m$. Suppose $x \in T := \{y \in \mathbb{Z}^n : Ay = b, l \leq y \leq u\}$. Then for every $g \in \mathbb{Z}^n$ which satisfies $x + g \in T$ and $w_+ g^+ - w_- g^- > 0$ there exists an element $\hat{g} \in G(A)$ with $\hat{g} \sqsubseteq g$ which also satisfies $x + \hat{g} \in T$ and $w_+ \hat{g}^+ - w_- \hat{g}^- > 0$.*

Proof Suppose $g \in \mathbb{Z}^n$ satisfies the requirements. Then $Ag = A(x + g) - Ax = b - b = 0$ since $x, x + g \in$

T. Thus, $g \in \mathcal{L}(A) \setminus \{0\}$ and hence, by Lemma 8, there is a conformal sum $g = \sum_i h_i$ with $h_i \in \mathcal{G}(A)$ for all i. Now, $h_i \sqsubseteq g$ is equivalent to $h_i^+ \leq g^+$ and $h_i^- \leq g^-$, so the conformal sum $g = \sum_i h_i$ gives corresponding sums of the positive and negative parts $g^+ = \sum_i h_i^+$ and $g^- = \sum_i h_i^-$. Therefore we obtain

$$0 < w_+ g^+ - w_- g^- = w_+ \sum_i h_i^+ - w_- \sum_i h_i^-$$
$$= \sum_i (w_+ h_i^+ - w_- h_i^-)$$

which implies that there is some h_i in this sum with $w_+ h_i^+ - w_- h_i^- > 0$. Now, $h_i \in \mathcal{G}(A)$ implies $A(x + h_i) = Ax = b$. Also, $l \leq x, x + g \leq u$ and $h_i \sqsubseteq g$ imply that $l \leq x + h_i \leq u$. So $x + h_i \in T$. Therefore the vector $\hat{g} := h_i$ satisfies the claim. □

We can now show that the Graver basis enables to solve linear integer programming in polynomial time provided an initial feasible point is available.

Theorem 12 *There is a polynomial time algorithm that, given $A \in \mathbb{Z}^{m \times n}$, its Graver basis $\mathcal{G}(A)$, $l, u \in \mathbb{Z}_\infty^n$, $x, w \in \mathbb{Z}^n$ with $l \leq x \leq u$, encoded as $[\langle A, \mathcal{G}(A), l, u, x, w\rangle]$, solves the linear integer program $\max\{wz : z \in \mathbb{Z}^n, Az = b, l \leq z \leq u\}$ with $b := Ax$.*

Proof First, note that the objective function of the integer program is unbounded if and only if the objective function of its relaxation $\max\{wy : y \in \mathbb{R}^n, Ay = b, l \leq y \leq u\}$ is unbounded, which can be checked in polynomial time using linear programming. If it is unbounded then assert that there is no optimal solution and terminate the algorithm.

Assume then that the objective is bounded. Then, since the program is feasible, it has an optimal solution. Furthermore, (as basically follows from Cramer's rule, see e.g. [13, Theorem 17.1]) it has an optimal x^* satisfying $|x_j^*| \leq \rho$ for all j, where ρ is an easily computable integer upper bound whose binary length $\langle\rho\rangle$ is polynomially bounded in $\langle A, l, u, x\rangle$. For instance, $\rho := (n+1)(n+1)! r^{n+1}$ will do, with r the maximum among $\max_i |\sum_j A_{i,j} x_j|$, $\max_{i,j} |A_{i,j}|$, $\max\{|l_j| : |l_j| < \infty\}$, and $\max\{|u_j| : |u_j| < \infty\}$.

Let $T := \{y \in \mathbb{Z}^n : Ay = b, l \leq y \leq u\}$ and $S := T \cap [-\rho, \rho]^n$. Then our linear integer programming problem now reduces to linear discrete optimization over S. Now, an oriented augmentation oracle for S can be simulated in polynomial time using the given Graver basis $\mathcal{G}(A)$ as follows: given a query $y \in S$ and $w_+, w_- \in \mathbb{Z}^n$, search for $g \in \mathcal{G}(A)$ which satisfies $w_+ g^+ - w_- g^- > 0$ and $y + g \in S$; if there is such a g then return it as an augmenting vector, whereas if there is no such g then assert that no augmenting vector exists. Clearly, if this simulated oracle returns a vector g then it is an augmenting vector. On the other hand, if there exists an augmenting vector g then $y + g \in S \subseteq T$ and $w_+ g^+ - w_- g^- > 0$ imply by Lemma 9 that there is also a $\hat{g} \in \mathcal{G}(A)$ with $\hat{g} \sqsubseteq g$ such that $w_+ \hat{g}^+ - w_- \hat{g}^- > 0$ and $y + \hat{g} \in T$. Since $y, y + g \in S$ and $\hat{g} \sqsubseteq g$, we find that $y + \hat{g} \in S$ as well. Therefore the Graver basis contains an augmenting vector and hence the simulated oracle will find and output one.

Define $\hat{l}, \hat{u} \in \mathbb{Z}^n$ by $\hat{l}_j := \max(l_j, -\rho)$, $\hat{u}_j := \min(u_j, \rho)$, $j = 1, \ldots, n$. Then it is easy to see that $S = \mathrm{aff}(S) \cap \{y \in \mathbb{Z}^n : \hat{l} \leq y \leq \hat{u}\}$. Now apply the algorithm of Lemma 7 to $\hat{l}, \hat{u}, S, x$, and w, using the above simulated oriented augmentation oracle for S, and obtain in polynomial time a vector $x^* \in S$ which is optimal to the linear discrete optimization problem over S and hence to the given linear integer program. □

As a special case of Theorem 12 we recover the following result of [55] concerning linear integer programming in standard form when the Graver basis is available.

Theorem 13 *There is a polynomial time algorithm that, given matrix $A \in \mathbb{Z}^{m \times n}$, its Graver basis $\mathcal{G}(A)$, $x \in \mathbb{N}^n$, and $w \in \mathbb{Z}^n$, encoded as $[\langle A, \mathcal{G}(A), x, w\rangle]$, solves the linear integer programming problem $\max\{wz : z \in \mathbb{N}^n, Az = b\}$ where $b := Ax$.*

Graver Bases of N-fold Matrices

As mentioned above, the Graver basis $\mathcal{G}(A)$ of an integer matrix A contains all circuits of A and typically many more elements. While the number of circuits is already typically exponential and can be as large as $\binom{n}{m+1}$, the number of Graver basis elements is usually even larger and depends also on the entries of A and not only on its dimensions m, n. So unfortunately it is typically very hard to compute $\mathcal{G}(A)$. However, we now show that for the important and useful broad class of n-fold matrices, the Graver basis is better behaved and can be computed in polynomial time. Recall the following definition from the introduction. Given an $(r + s) \times t$

matrix A, let A_1 be its $r \times t$ sub-matrix consisting of the first r rows and let A_2 be its $s \times t$ sub-matrix consisting of the last s rows. We refer to A explicitly as $(r+s) \times t$ matrix, since the definition below depends also on r and s and not only on the entries of A. The *n-fold matrix* of an $(r+s) \times t$ matrix A is then defined to be the following $(r+ns) \times nt$ matrix,

$$A^{(n)} := (\mathbf{1}_n \otimes A_1) \oplus (I_n \otimes A_2)$$

$$= \begin{pmatrix} A_1 & A_1 & A_1 & \cdots & A_1 \\ A_2 & 0 & 0 & \cdots & 0 \\ 0 & A_2 & 0 & \cdots & 0 \\ \vdots & \vdots & \vdots & \ddots & \vdots \\ 0 & 0 & 0 & \cdots & A_2 \end{pmatrix}.$$

We now discuss a recent result of [54], which originates in [4], and its extension in [38], on the stabilization of Graver bases of n-fold matrices. Consider vectors $x = (x^1, \ldots, x^n)$ with $x^k \in \mathbb{Z}^t$ for $k = 1, \ldots, n$. The *type* of x is the number $|\{k : x^k \neq 0\}|$ of nonzero components $x^k \in \mathbb{Z}^t$ of x. The *Graver complexity* of an $(r+s) \times t$ matrix, denoted $c(A)$, is defined to be the smallest $c \in \mathbb{N} \uplus \{\infty\}$ such that for all n, the Graver basis of $A^{(n)}$ consists of vectors of type at most $c(A)$. We provide the proof of the following result of [38,54] stating that the Graver complexity is always finite.

Lemma 10 *The Graver complexity $c(A)$ of any $(r+s) \times t$ integer matrix A is finite.*

Proof Call an element $x = (x^1, \ldots, x^n)$ in the Graver basis of some $A^{(n)}$ *pure* if $x^i \in \mathcal{G}(A_2)$ for all i. Note that the type of a pure $x \in \mathcal{G}(A^{(n)})$ is n. First, we claim that if there is an element of type m in some $\mathcal{G}(A^{(l)})$ then for some $n \geq m$ there is a pure element in $\mathcal{G}(A^{(n)})$, and so it will suffice to bound the type of pure elements. Suppose there is an element of type m in some $\mathcal{G}(A^{(l)})$. Then its restriction to its m nonzero components is an element $x = (x^1, \ldots, x^m)$ in $\mathcal{G}(A^{(m)})$. Let $x^i = \sum_{j=1}^{k_i} g_{i,j}$ be a conformal decomposition of x^i with $g_{i,j} \in \mathcal{G}(A_2)$ for all i, j, and let $n := k_1 + \cdots + k_m \geq m$. Then $g := (g_{1,1}, \ldots, g_{m,k_m})$ is in $\mathcal{G}(A^{(n)})$, else there would be $\hat{g} \sqsubseteq g$ in $\mathcal{G}(A^{(n)})$ in which case the nonzero $\hat{x}$ with $\hat{x}^i := \sum_{j=1}^{k_i} \hat{g}_{i,j}$ for all i would satisfy $\hat{x} \sqsubseteq x$ and $\hat{x} \in \mathcal{L}(A^{(m)})$, contradicting $x \in \mathcal{G}(A^{(m)})$. Thus g is a pure element of type $n \geq m$, proving the claim.

We proceed to bound the type of pure elements. Let $\mathcal{G}(A_2) = \{g_1, \ldots, g_m\}$ be the Graver basis of A_2 and let G_2 be the $t \times m$ matrix whose columns are the g_i. Suppose $x = (x^1, \ldots, x^n) \in \mathcal{G}(A^{(n)})$ is pure for some n. Let $v \in \mathbb{N}^m$ be the vector with $v_i := |\{k : x^k = g_i\}|$ counting the number of g_i components of x for each i. Then $\sum_{i=1}^m v_i$ is equal to the type n of x. Next, note that $A_1 G_2 v = A_1(\sum_{k=1}^n x^k) = 0$ and hence $v \in \mathcal{L}(A_1 G_2)$. We claim that, moreover, $v \in \mathcal{G}(A_1 G_2)$. Suppose indirectly not. Then there is $\hat{v} \in \mathcal{G}(A_1 G_2)$ with $\hat{v} \sqsubseteq v$, and it is easy to obtain a nonzero $\hat{x} \sqsubseteq x$ from x by zeroing out some components so that $\hat{v}_i = |\{k : \hat{x}^k = g_i\}|$ for all i. Then $A_1(\sum_{k=1}^n \hat{x}^k) = A_1 G_2 \hat{v} = 0$ and hence $\hat{x} \in \mathcal{L}(A^{(n)})$, contradicting $x \in \mathcal{G}(A^{(n)})$.

So the type of any pure element, and hence the Graver complexity of A, is at most the largest value $\sum_{i=1}^m v_i$ of any nonnegative element v of the Graver basis $\mathcal{G}(A_1 G_2)$. □

Using Lemma 10 we now show how to compute $\mathcal{G}(A^{(n)})$ in polynomial time.

Theorem 14 *For every fixed $(r+s) \times t$ integer matrix A there is a strongly polynomial time algorithm that, given $n \in \mathbb{N}$, encoded as $[n; n]$, computes the Graver basis $\mathcal{G}(A^{(n)})$ of the n-fold matrix $A^{(n)}$. In particular, the cardinality $|\mathcal{G}(A^{(n)})|$ and binary length $\langle \mathcal{G}(A^{(n)}) \rangle$ of the Graver basis of the n-fold matrix are polynomially bounded in n.*

Proof Let $c := c(A)$ be the Graver complexity of A and consider any $n \geq c$. We show that the Graver basis of $A^{(n)}$ is the union of $\binom{n}{c}$ suitably embedded copies of the Graver basis of $A^{(c)}$. For every c indices $1 \leq k_1 < \cdots < k_c \leq n$ define a map $\phi_{k_1, \ldots, k_c}$ from $\mathbb{Z}^{ct}$ to $\mathbb{Z}^{nt}$ sending $x = (x^1, \ldots, x^c)$ to $y = (y^1, \ldots, y^n)$ with $y^{k_i} := x^i$ for $i = 1, \ldots, c$ and $y^k := 0$ for $k \notin \{k_1, \ldots, k_c\}$. We claim that $\mathcal{G}(A^{(n)})$ is the union of the images of $\mathcal{G}(A^{(c)})$ under the $\binom{n}{c}$ maps $\phi_{k_1, \ldots, k_c}$ for all $1 \leq k_1 < \cdots < k_c \leq n$, that is,

$$\mathcal{G}(A^{(n)}) = \bigcup_{1 \leq k_1 < \cdots < k_c \leq n} \phi_{k_1, \ldots, k_c}(\mathcal{G}(A^{(c)})). \quad (2)$$

If $x = (x^1, \ldots, x^c) \in \mathcal{G}(A^{(c)})$ then x is a $\sqsubseteq$-minimal nonzero element of $\mathcal{L}(A^{(c)})$, implying that $\phi_{k_1, \ldots, k_c}(x)$ is a $\sqsubseteq$-minimal nonzero element of $\mathcal{L}(A^{(n)})$ and therefore we have $\phi_{k_1, \ldots, k_c}(x) \in \mathcal{G}(A^{(n)})$. So the right-hand side of (2) is contained in the left-hand side. Conversely, consider any $y \in \mathcal{G}(A^{(n)})$. Then, by Lemma 10, the type of y is at most c, so there are indices $1 \leq k_1 < \cdots < k_c \leq n$

such that all nonzero components of y are among those of the reduced vector $x := (y^{k_1}, \ldots, y^{k_c})$ and therefore $y = \phi_{k_1,\ldots,k_c}(x)$. Now, $y \in G(A^{(n)})$ implies that y is a $\sqsubseteq$-minimal nonzero element of $\mathcal{L}(A^{(n)})$ and hence x is a $\sqsubseteq$-minimal nonzero element of $\mathcal{L}(A^{(c)})$. Therefore $x \in G(A^{(c)})$ and $y \in \phi_{k_1,\ldots,k_g}(G(A^{(c)}))$. So the left-hand side of (2) is contained in the right-hand side.

Since A is fixed we have that $c = c(A)$ and $G(A^{(c)})$ are constant. Then (2) implies that $|G(A^{(n)})| \leq \binom{n}{c}|G(A^{(c)})| = O(n^c)$. Moreover, every element of $G(A^{(n)})$ is an nt-dimensional vector $\phi_{k_1,\ldots,k_c}(x)$ obtained by appending zero components to some $x \in G(A^{(c)})$ and hence has linear binary length $O(n)$. So the binary length of the entire Graver basis $G(A^{(n)})$ is $O(n^{c+1})$. Thus, the $\binom{n}{c} = O(n^c)$ images $\phi_{k_1,\ldots,k_c}(G(A^{(c)}))$ and their union $G(A^{(n)})$ can be computed in strongly polynomial time, as claimed. □

Example 2 Consider the $(2+1) \times 2$ matrix A with $A_1 := I_2$ the 2×2 identity and $A_2 := (1,1)$. Then $G(A_2) = \pm(1,-1)$ and $G(A_1 G_2) = \pm(1,1)$ from which the Graver complexity of A can be concluded to be $c(A) = 2$ (see the proof of Lemma 10). The 2-fold matrix of A and its Graver basis, consisting of two antipodal vectors only, are

$$A^{(2)} = \begin{pmatrix} 1 & 0 & 1 & 0 \\ 0 & 1 & 0 & 1 \\ 1 & 1 & 0 & 0 \\ 0 & 0 & 1 & 1 \end{pmatrix},$$

$$G(A^{(2)}) = \pm \begin{pmatrix} 1 & -1 & -1 & 1 \end{pmatrix}.$$

By Theorem 14, the Graver basis of the 4-fold matrix $A^{(4)}$ is computed to be the union of the images of the $6 = \binom{4}{2}$ maps $\phi_{k_1,k_2}: \mathbb{Z}^{2 \cdot 2} \longrightarrow \mathbb{Z}^{4 \cdot 2}$ for $1 \leq k_1 < k_2 \leq 4$, getting

$$A^{(4)} = \begin{pmatrix} 1 & 0 & 1 & 0 & 1 & 0 & 1 & 0 \\ 0 & 1 & 0 & 1 & 0 & 1 & 0 & 1 \\ 1 & 1 & 0 & 0 & 0 & 0 & 0 & 0 \\ 0 & 0 & 1 & 1 & 0 & 0 & 0 & 0 \\ 0 & 0 & 0 & 0 & 1 & 1 & 0 & 0 \\ 0 & 0 & 0 & 0 & 0 & 0 & 1 & 1 \end{pmatrix},$$

$$G(A^{(4)}) = \pm \begin{pmatrix} 1 & -1 & -1 & 1 & 0 & 0 & 0 & 0 \\ 1 & -1 & 0 & 0 & -1 & 1 & 0 & 0 \\ 1 & -1 & 0 & 0 & 0 & 0 & -1 & 1 \\ 0 & 0 & 1 & -1 & -1 & 1 & 0 & 0 \\ 0 & 0 & 1 & -1 & 0 & 0 & -1 & 1 \\ 0 & 0 & 0 & 0 & 1 & -1 & -1 & 1 \end{pmatrix}.$$

Linear N-fold Integer Programming in Polynomial Time

We now proceed to provide a polynomial time algorithm for linear integer programming over n-fold matrices. First, combining the results of Sect. "Graver Bases and Linear Integer Programming" and Sect. "Graver Bases of N-fold Matrices", we get at once the following polynomial time algorithm for converting any feasible point to an optimal one.

Lemma 11 *For every fixed $(r+s) \times t$ integer matrix A there is a polynomial time algorithm that, given $n \in \mathbb{N}$, $l, u \in \mathbb{Z}_\infty^{nt}$, $x, w \in \mathbb{Z}^{nt}$ satisfying $l \leq x \leq u$, encoded as $[\langle l, u, x, w \rangle]$, solves the linear n-fold integer programming problem with $b := A^{(n)}x$,*

$$\max\{wz : z \in \mathbb{Z}^{nt}, A^{(n)}z = b, l \leq z \leq u\}.$$

Proof First, apply the polynomial time algorithm of Theorem 14 and compute the Graver basis $G(A^{(n)})$ of the n-fold matrix $A^{(n)}$. Then apply the polynomial time algorithm of Theorem 12 to the data $A^{(n)}$, $G(A^{(n)})$, l, u, x and w. □

Next we show that an initial feasible point can also be found in polynomial time.

Lemma 12 *For every fixed $(r+s) \times t$ integer matrix A there is a polynomial time algorithm that, given $n \in \mathbb{N}$, $l, u \in \mathbb{Z}_\infty^{nt}$, and $b \in \mathbb{Z}^{r+ns}$, encoded as $[\langle l, u, b \rangle]$, either finds an $x \in \mathbb{Z}^{nt}$ satisfying $l \leq x \leq u$ and $A^{(n)}x = b$ or asserts that none exists.*

Proof If $l \not\leq u$ then assert that there is no feasible point and terminate the algorithm. Assume then that $l \leq u$ and determine some $x \in \mathbb{Z}^{nt}$ with $l \leq x \leq u$ and $\langle x \rangle \leq \langle l, u \rangle$. Now, introduce $n(2r+2s)$ auxiliary variables to the given n-fold integer program and denote by $\hat{x}$ the resulting vector of $n(t+2r+2s)$ variables. Suitably extend the lower and upper bound vectors to $\hat{l}, \hat{u}$ by setting $\hat{l}_j := 0$ and $\hat{u}_j := \infty$ for each auxiliary variable $\hat{x}_j$. Consider the auxiliary integer program of finding an integer vector $\hat{x}$ that minimizes the sum of auxiliary variables subject to the lower and upper bounds $\hat{l} \leq \hat{x} \leq \hat{u}$ and the following system of equations, with I_r and I_s the

$r \times r$ and $s \times s$ identity matrices,

$$\begin{pmatrix} A_1 & I_r & -I_r & 0 & 0 & A_1 & I_r & -I_r & 0 \\ A_2 & 0 & 0 & I_s & -I_s & 0 & 0 & 0 & 0 \\ 0 & 0 & 0 & 0 & 0 & A_2 & 0 & 0 & I_s \\ \vdots & \vdots & \vdots & \vdots & \vdots & \vdots & \vdots & \vdots & \vdots \\ 0 & 0 & 0 & 0 & 0 & 0 & 0 & 0 & 0 \end{pmatrix}$$

$$\left. \begin{pmatrix} 0 & \cdots & A_1 & I_r & -I_r & 0 & 0 \\ 0 & \cdots & 0 & 0 & 0 & 0 & 0 \\ -I_s & \cdots & 0 & 0 & 0 & 0 & 0 \\ \vdots & \ddots & \vdots & \vdots & \vdots & \vdots & \vdots \\ 0 & \cdots & A_2 & 0 & 0 & I_s & -I_s \end{pmatrix} \right) \hat{x} = b.$$

This is again an n-fold integer program, with an $(r+s) \times (t+2r+2s)$ matrix $\hat{A}$, where $\hat{A}_1 = (A_1, I_r, -I_r, 0, 0)$ and $\hat{A}_2 = (A_2, 0, 0, I_s, -I_s)$. Since A is fixed, so is $\hat{A}$. It is now easy to extend the vector $x \in \mathbb{Z}^{nt}$ determined above to a feasible point $\hat{x}$ of the auxiliary program. Indeed, put $\hat{b} := b - A^{(n)}x \in \mathbb{Z}^{r+ns}$; now, for $i = 1, \ldots, r+ns$, simply choose an auxiliary variable $\hat{x}_j$ appearing only in the ith equation, whose coefficient equals the sign $\text{sign}(\hat{b}_i)$ of the corresponding entry of $\hat{b}$, and set $\hat{x}_j := |\hat{b}_i|$. Define $\hat{w} \in \mathbb{Z}^{n(t+2r+2s)}$ by setting $\hat{w} := 0$ for each original variable and $\hat{w} := -1$ for each auxiliary variable, so that maximizing $\hat{w}\hat{x}$ is equivalent to minimizing the sum of auxiliary variables. Now solve the auxiliary linear integer program in polynomial time by applying the algorithm of Lemma 11 corresponding to $\hat{A}$ to the data n, $\hat{l}$, $\hat{u}$, $\hat{x}$, and $\hat{w}$. Since the auxiliary objective $\hat{w}\hat{x}$ is bounded above by zero, the algorithm will output an optimal solution $\hat{x}^*$. If the optimal objective value is negative, then the original n-fold program is infeasible, whereas if the optimal value is zero, then the restriction of $\hat{x}^*$ to the original variables is a feasible point x^* of the original integer program. □

Combining Lemmas 11 and 12 we get at once the main result of this section.

Theorem 15 *For every fixed $(r+s) \times t$ integer matrix A there is a polynomial time algorithm that, given n, lower and upper bounds $l, u \in \mathbb{Z}_\infty^{nt}$, $w \in \mathbb{Z}^{nt}$, and $b \in \mathbb{Z}^{r+ns}$, encoded as $[\langle l, u, w, b \rangle]$, solves the following linear n-fold integer programming problem,*

$$\max \{ wx \,:\, x \in \mathbb{Z}^{nt},\, A^{(n)} x = b,\, l \leq x \leq u \}.$$

Again, as a special case of Theorem 15 we recover the following result of [13] concerning linear integer programming in standard form over n-fold matrices.

Theorem 16 *For every fixed $(r+s) \times t$ integer matrix A there is a polynomial time algorithm that, given n, linear functional $w \in \mathbb{Z}^{nt}$, and right-hand side $b \in \mathbb{Z}^{r+ns}$, encoded as $[\langle w, b \rangle]$, solves the following linear n-fold integer program in standard form,*

$$\max \{ wx \,:\, x \in \mathbb{N}^{nt},\, A^{(n)} x = b \}.$$

Some Applications

Three-Way Line-Sum Transportation Problems
Transportation problems form a very important class of discrete optimization problems studied extensively in the operations research and mathematical programming literature, see e. g. [6,42,43,53,60,62] and the references therein. We will discuss this class of problem and its applications to secure statistical data disclosure in more detail in Sect. "Multiway Transportation Problems and Privacy in Statistical Databases".

It is well known that 2-way transportation problems are polynomial time solvable, since they can be encoded as linear integer programs over totally unimodular systems. However, already 3-way transportation problem are much more complicated. Consider the following 3-way transportation problem over $p \times q \times n$ tables with all line-sums fixed,

$$\max \Big\{ wx \,:\, x \in \mathbb{N}^{p \times q \times n},\, \sum_i x_{i,j,k} = z_{j,k},\, \sum_j x_{i,j,k} = v_{i,k},\, \sum_k x_{i,j,k} = u_{i,j} \Big\}.$$

The data for the problem consist of given integer numbers (lines-sums) $u_{i,j}$, $v_{i,k}$, $z_{j,k}$ for $i = 1, \ldots, p$, $j = 1, \ldots, q$, $k = 1, \ldots, n$, and a linear functional given by a $p \times q \times n$ integer array w representing the transportation profit per unit on each cell. The problem is to find a transportation, that is, a $p \times q \times n$ nonnegative integer table x satisfying the line sum constraints, which attains maximum profit $wx = \sum_{i=1}^p \sum_{j=1}^q \sum_{k=1}^n w_{i,j,k} x_{i,j,k}$.

When at least two of the table sides, say p, q, are variable part of the input, and even when the third side is fixed and as small as $n = 3$, this problem is already *universal* for integer programming in a very strong sense [14,16], and in particular is NP-hard [15]; this will be discussed in detail and proved in Sect. "Multiway Transportation Problems and Privacy in Statistical Databases". We now show that in contrast, when two sides, say p, q, are fixed (but arbitrary), and one side n is variable, then the 3-way transportation problem over such *long* tables is an n-fold integer programming problem and therefore, as a consequence of Theorem 16, can be solved is polynomial time.

Corollary 8 *For every fixed p and q there is a polynomial time algorithm that, given n, integer profit array $w \in \mathbb{Z}^{p \times q \times n}$, and line-sums $u \in \mathbb{Z}^{p \times q}$, $v \in \mathbb{Z}^{p \times n}$ and $z \in \mathbb{Z}^{q \times n}$, encoded as $[\langle w, u, v, z \rangle]$, solves the integer 3-way line-sum transportation problem*

$$\max\left\{ wx \ : \ x \in \mathbb{N}^{p \times q \times n}, \ \sum_i x_{i,j,k} = z_{j,k}, \right.$$

$$\left. \sum_j x_{i,j,k} = v_{i,k}, \ \sum_k x_{i,j,k} = u_{i,j} \right\}.$$

Proof Re-index $p \times q \times n$ arrays as $x = (x^1, \ldots, x^n)$ with each component indexed as $x^k := (x^k_{i,j}) := (x_{1,1,k}, \ldots, x_{p,q,k})$ suitably indexed as a pq vector representing the kth layer of x. Put $r := t := pq$ and $s := p + q$, and let A be the $(r + s) \times t$ matrix with $A_1 := I_{pq}$ the $pq \times pq$ identity and with A_2 the $(p+q) \times pq$ matrix of equations of the usual 2-way transportation problem for $p \times q$ arrays. Re-arrange the given line-sums in a vector $b := (b^0, b^1, \ldots, b^n) \in \mathbb{Z}^{r+ns}$ with $b^0 := (u_{i,j})$ and $b^k := ((v_{i,k}), (z_{j,k}))$ for $k = 1, \ldots, n$.

This translates the given 3-way transportation problem into an n-fold integer programming problem in standard form,

$$\max\{wx \ : \ x \in \mathbb{N}^{nt}, \ A^{(n)}x = b\},$$

where the equations $A_1(\sum_{k=1}^n x^k) = b^0$ represent the constraints $\sum_k x_{i,j,k} = u_{i,j}$ of all line-sums where summation over layers occurs, and the equations $A_2 x^k = b^k$ for $k = 1, \ldots, n$ represent the constraints $\sum_i x_{i,j,k} = z_{j,k}$ and $\sum_j x_{i,j,k} = v_{i,k}$ of all line-sums where summations are within a single layer at a time.

Using the algorithm of Theorem 16, this n-fold integer program, and hence the given 3-way transportation problem, can be solved in polynomial time. □

Example 3 We demonstrate the encoding of the $p \times q \times n$ transportation problem as an n-fold integer program as in the proof of Corollary 8 for $p = q = 3$ (smallest case where the problem is genuinely 3-dimensional). Here we put $r := t := 9$, $s := 6$, write

$$x^k := (x_{1,1,k}, x_{1,2,k}, x_{1,3,k}, x_{2,1,k}, x_{2,2,k}, x_{2,3,k},$$
$$x_{3,1,k}, x_{3,2,k}, x_{3,3,k}), \quad k = 1, \ldots, n,$$

and let the $(9 + 6) \times 9$ matrix A consist of $A_1 = I_9$ the 9×9 identity matrix and

$$A_2 := \begin{pmatrix} 1 & 1 & 1 & 0 & 0 & 0 & 0 & 0 & 0 \\ 0 & 0 & 0 & 1 & 1 & 1 & 0 & 0 & 0 \\ 0 & 0 & 0 & 0 & 0 & 0 & 1 & 1 & 1 \\ 1 & 0 & 0 & 1 & 0 & 0 & 1 & 0 & 0 \\ 0 & 1 & 0 & 0 & 1 & 0 & 0 & 1 & 0 \\ 0 & 0 & 1 & 0 & 0 & 1 & 0 & 0 & 1 \end{pmatrix}.$$

Then the corresponding n-fold integer program encodes the $3 \times 3 \times n$ transportation problem as desired. Already for this case, of $3 \times 3 \times n$ tables, the only known polynomial time algorithm for the transportation problem is the one underlying Corollary 8.

Corollary 8 has a very broad generalization to multiway transportation problems over long k-way tables of any dimension k; this will be discussed in detail in Sect. "Multiway Transportation Problems and Privacy in Statistical Databases".

Packing Problems and Cutting-Stock We consider the following rather general class of packing problems which concern maximum utility packing of many items of several types in various bins subject to weight constraints. More precisely, the data is as follows. There are t types of items. Each item of type j has integer weight v_j. There are n_j items of type j to be packed. There are n bins. The weight capacity of bin k is an integer u_k. Finally, there is a utility matrix $w \in \mathbb{Z}^{t \times n}$ where $w_{j,k}$ is the utility of packing one item of type j in bin k. The problem is to find a feasible packing of maximum total utility. By incrementing the number t of types by 1 and suitably augmenting the data, we may assume that the last type t represents "slack items" which occupy the unused capacity in each bin, where the weight of each slack item is 1, the utility of packing any slack item in any bin is 0, and the number of slack items is the total residual weight capacity $n_t := \sum_{k=1}^n u_k - \sum_{j=1}^{t-1} n_j v_j$. Let $x \in \mathbb{N}^{t \times n}$ be a variable matrix where $x_{j,k}$ represents the number of items of type j to be packed in bin k. Then the packing problem becomes the following linear integer program,

$$\max\left\{wx\ :\ x\in\mathbb{N}^{t\times n},\ \sum_j v_j x_{j,k} = u_k,\ \sum_k x_{j,k} = n_j\right\}.$$

We now show that this is in fact an n-fold integer programming problem and therefore, as a consequence of Theorem 16, can be solved is polynomial time. While the number t of types and type weights v_j are fixed, which is natural in many bin packing applications, the numbers n_j of items of each type and the bin capacities u_k may be very large.

Corollary 9 *For every fixed number t of types and integer type weights $v_1,\ldots,v_t$, there is a polynomial time algorithm that, given n bins, integer item numbers $n_1,\ldots,n_t$, integer bin capacities $u_1,\ldots,u_n$, and $t\times n$ integer utility matrix w, encoded as $[\langle n_1,\ldots,n_t, u_1,\ldots,u_n,w\rangle]$, solves the following integer bin packing problem,*

$$\max\left\{wx\ :\ x\in\mathbb{N}^{t\times n},\ \sum_j v_j x_{j,k} = u_k,\ \sum_k x_{j,k} = n_j\right\}.$$

Proof Re-index the variable matrix as $x = (x^1,\ldots,x^n)$ with $x^k := (x_1^k,\ldots,x_t^k)$ where x_j^k represents the number of items of type j to be packed in bin k for all j and k. Let A be the $(t+1)\times t$ matrix with $A_1 := I_t$ the $t\times t$ identity and with $A_2 := (v_1,\ldots,v_t)$ a single row. Re-arrange the given item numbers and bin capacities in a vector $b := (b^0, b^1,\ldots,b^n) \in \mathbb{Z}^{t+n}$ with $b^0 := (n_1,\ldots,n_t)$ and $b^k := u_k$ for all k. This translates the bin packing problem into an n-fold integer programming problem in standard form,

$$\max\left\{wx\ :\ x\in\mathbb{N}^{nt},\ A^{(n)}x = b\right\},$$

where the equations $A_1(\sum_{k=1}^n x^k) = b^0$ represent the constraints $\sum_k x_{j,k} = n_j$ assuring that all items of each type are packed, and the equations $A_2 x^k = b^k$ for $k=1,\ldots,n$ represent the constraints $\sum_j v_j x_{j,k} = u_k$ assuring that the weight capacity of each bin is not exceeded (in fact, the slack items make sure each bin is perfectly packed).

Using the algorithm of Theorem 16, this n-fold integer program, and hence the given integer bin packing problem, can be solved in polynomial time. □

Example 5 (cutting-stock problem). This is a classical manufacturing problem [27], where the usual setup is as follows: a manufacturer produces rolls of material (such as scotch-tape or band-aid) in one of t different widths $v_1,\ldots,v_t$. The rolls are cut out from standard rolls of common large width u. Given orders by customers for n_j rolls of width v_j, the problem facing the manufacturer is to meet the orders using the smallest possible number of standard rolls. This can be cast as a bin packing problem as follows. Rolls of width v_j become items of type j to be packed. Standard rolls become identical bins, of capacity $u_k := u$ each, where the number of bins is set to be $n := \sum_{j=1}^t \lceil n_j/\lfloor u/v_j\rfloor\rceil$ which is sufficient to accommodate all orders. The utility of each roll of width v_j is set to be its width negated $w_{j,k} := -v_j$ regardless of the standard roll k from which it is cut (paying for the width it takes). Introduce a new roll width $v_0 := 1$, where rolls of that width represent "slack rolls" which occupy the unused width of each standard roll, with utility $w_{0,k} := -1$ regardless of the standard roll k from which it is cut (paying for the unused width it represents), with the number of slack rolls set to be the total residual width $n_0 := nu - \sum_{j=1}^t n_j v_j$. Then the cutting-stock problem becomes a bin packing problem and therefore, by Corollary 9, for every fixed t and fixed roll widths $v_1,\ldots,v_t$, it is solvable in time polynomial in $\sum_{j=1}^t \lceil n_j/\lfloor u/v_j\rfloor\rceil$ and $\langle n_1,\ldots,n_t, u\rangle$.

One common approach to the cutting-stock problem uses so-called *cutting patterns*, which are feasible solutions of the knapsack problem $\{y\in\mathbb{N}^t : \sum_{j=1}^t v_j y_j \leq u\}$. This is useful when the common width u of the standard rolls is of the same order of magnitude as the demand role widths v_j. However, when u is much larger than the v_j, the number of cutting patterns becomes prohibitively large to handle. But then the values $\lfloor u/v_j\rfloor$ are large and hence $n := \sum_{j=1}^t \lceil n_j/\lfloor u/v_j\rfloor\rceil$ is small, in which case the solution through the algorithm of Corollary 9 becomes particularly appealing.

Convex Integer Programming

In this section we discuss convex integer programming. In particular, we extend the theory of Sect. "Linear N-fold Integer Programming" and show that convex n-fold integer programming is polynomial time solvable as well. In Sect. "Convex Integer Programming over Totally Unimodular Systems" we discuss convex integer programming over totally unimodular matrices. In Sect. "Graver Bases and Convex Integer Programming" we show the applicability of Graver bases to convex integer programming. In Sect. "Convex N-fold Integer Programming in Polynomial Time" we combine Theorem 6, the results of Sect. "Linear N-fold Integer Programming", and the preparatory facts from Sect. "Graver Bases and Convex Integer Programming", and prove the main result of this section, asserting that convex n-fold integer programming is polynomial time solvable. We conclude with some applications in Sect. "Some Applications".

As in Sect. "Linear N-fold Integer Programming", the feasible set S is presented as the set of integer points satisfying an explicitly given system of linear inequalities, given in one of the forms

$$S := \{x \in \mathbb{N}^n : Ax = b\} \quad \text{or}$$
$$S := \{x \in \mathbb{Z}^n : Ax = b, \ l \leq x \leq u\},$$

with matrix $A \in \mathbb{Z}^{m \times n}$, right-hand side $b \in \mathbb{Z}^m$, and lower and upper bounds $l, u \in \mathbb{Z}_\infty^n$.

As demonstrated in Sect. "Limitations", if the polyhedron $P := \{x \in \mathbb{R}^n : Ax = b, \ l \leq x \leq u\}$ is unbounded then the convex integer programming problem with an oracle presented convex functional is rather hopeless. Therefore, an algorithm that solves the convex integer programming problem should either return an optimal solution, or assert that the program is infeasible, or assert that the underlying polyhedron is unbounded.

Nonetheless, we do allow the lower and upper bounds l, u to lie in $\mathbb{Z}_\infty^n$ rather than $\mathbb{Z}^n$, since often the polyhedron is bounded even though the variables are not bounded explicitly (for instance, if each variable is bounded below only, and appears in some equation all of whose coefficients are positive). This results in broader formulation flexibility. Furthermore, in the next subsections we prove auxiliary lemmas asserting that certain sets cover all edge-directions of relevant polyhedra, which do hold also in the unbounded case. So we now extend the notion of edge-directions, defined in Sect. "Edge-Directions and Zonotopes" for polytopes, to polyhedra. A *direction* of an edge (1-dimensional face) e of a polyhedron P is any nonzero scalar multiple of $y - x$ where x, y are any two distinct points in e. As before, a set *covers all edge-directions of* P if it contains a direction of each edge of P.

Convex Integer Programming over Totally Unimodular Systems

A matrix A is *totally unimodular* if the determinant of every square submatrix of A lies in $\{-1, 0, 1\}$. Such matrices arise naturally in network flows, ordinary (2-way) transportation problems, and many other situations. A fundamental result in integer programming [37] asserts that polyhedra defined by totally unimodular matrices are integer. More precisely, if A is an $m \times n$ totally unimodular matrix, $l, u \in \mathbb{Z}_\infty^n$, and $b \in \mathbb{Z}^m$, then

$$P_I := \operatorname{conv}\{x \in \mathbb{Z}^n : Ax = b, \ l \leq x \leq u\}$$
$$= \{x \in \mathbb{R}^n : Ax = b, \ l \leq x \leq u\} := P,$$

that is, the underlying polyhedron P coincides with its integer hull P_I. This has two consequences useful in facilitating the solution of the corresponding convex integer programming problem via the algorithm of Theorem 6. First, the corresponding linear integer programming problem can be solved by linear programming over P in polynomial time. Second, a set covering all edge-directions of the implicitly given integer hull P_I, which is typically very hard to determine, is obtained here as a set covering all edge-directions of P which is explicitly given and hence easier to determine.

We now describe a well known property of polyhedra of the above form. A *circuit* of a matrix $A \in \mathbb{Z}^{m \times n}$ is a nonzero primitive minimal support element of $\mathcal{L}(A)$. So a circuit is a nonzero $c \in \mathbb{Z}^n$ satisfying $Ac = 0$, whose entries are relatively prime integers, such that no nonzero c' with $Ac' = 0$ has support strictly contained in the support of c.

Lemma 13 *For every $A \in \mathbb{Z}^{m \times n}$, $l, u \in \mathbb{Z}_\infty^n$, and $b \in \mathbb{Z}^m$, the set of circuits of A covers all edge-directions of the polyhedron $P := \{x \in \mathbb{R}^n : Ax = b, \ l \leq x \leq u\}$.*

Proof Consider any edge e of P. Pick two distinct points $x, y \in e$ and set $g := y - x$. Then $Ag = 0$

and therefore, as can be easily proved by induction on $|\mathrm{supp}(g)|$, there is a finite decomposition $g = \sum_i \alpha_i c_i$ with α_i positive real number and c_i circuit of A such that $\alpha_i c_i \sqsubseteq g$ for all i, where $\sqsubseteq$ is the natural extension from $\mathbb{Z}^n$ to $\mathbb{R}^n$ of the partial order defined in Sect. "Graver Bases and Linear Integer Programming". We claim that $x + \alpha_i c_i \in P$ for all i. Indeed, c_i being a circuit implies $A(x + \alpha_i c_i) = Ax = b$; and $l \leq x, x + g \leq u$ and $\alpha_i c_i \sqsubseteq g$ imply $l \leq x + \alpha_i c_i \leq u$.

Now let $w \in \mathbb{R}^n$ be a linear functional uniquely maximized over P at the edge e. Then $w\alpha_i c_i = w(x + \alpha_i c_i) - wx \leq 0$ for all i. But $\sum(w\alpha_i c_i) = wg = wy - wx = 0$, implying that in fact $w\alpha_i c_i = 0$ and hence $x + \alpha_i c_i \in e$ for all i. This implies that each c_i is a direction of e (in fact, all c_i are the same and g is a multiple of some circuit). □

Combining Theorem 6 and Lemma 13 we obtain the following statement.

Theorem 17 *For every fixed d there is a polynomial time algorithm that, given $m \times n$ totally unimodular matrix A, set $C \subset \mathbb{Z}^n$ containing all circuits of A, vectors $l, u \in \mathbb{Z}_\infty^n$, $b \in \mathbb{Z}^m$, and $w_1, \ldots, w_d \in \mathbb{Z}^n$, and convex $c : \mathbb{R}^d \longrightarrow \mathbb{R}$ presented by a comparison oracle, encoded as $[\langle A, C, l, u, b, w_1, \ldots, w_d \rangle]$, solves the convex integer program*

$$\max \{c(w_1 x, \ldots, w_d x) : x \in \mathbb{Z}^n,$$
$$Ax = b, \ l \leq x \leq u\}.$$

Proof First, check in polynomial time using linear programming whether the objective function of any of the following $2n$ linear programs is unbounded,

$$\max \{\pm y_i : y \in P\}, \ i = 1, \ldots, n,$$
$$P := \{y \in \mathbb{R}^n : Ay = b, \ l \leq y \leq u\}.$$

If any is unbounded then terminate, asserting that P is unbounded. Otherwise, let ρ be the least integer upper bound on the absolute value of all optimal objective values. Then $P \subseteq [-\rho, \rho]^n$ and $S := \{y \in \mathbb{Z}^n : Ay = b, \ l \leq y \leq u\} \subset P$ is finite of radius $\rho(S) \leq \rho$. In fact, since A is totally unimodular, $P_I = P = \mathrm{conv}(S)$ and hence $\rho(S) = \rho$. Moreover, by Cramer's rule, $\langle \rho \rangle$ is polynomially bounded in $\langle A, l, u, x \rangle$.

Now, since A is totally unimodular, using linear programming over $P_I = P$ we can simulate in polyno-

mial time a linear discrete optimization oracle for S. By Lemma 13, the given set C, which contains all circuits of A, also covers all edge-directions of $\mathrm{conv}(S) = P_I = P$. Therefore we can apply the algorithm of Theorem 6 and solve the given convex n-fold integer programming problem in polynomial time. □

While the number of circuits of an $m \times n$ matrix A can be as large as $2\binom{n}{m+1}$ and hence exponential in general, it is nonetheless relatively small in that it is bounded in terms of m and n only and is independent of the matrix A itself. Furthermore, it may happen that the number of circuits is much smaller than the upper bound $2\binom{n}{m+1}$. Also, if in a class of matrices, m grows slowly in terms of n, say $m = O(\log n)$, then this bound is subexponential. In such situations, the above theorem may provide a good strategy for solving convex integer programming over totally unimodular systems.

Graver Bases and Convex Integer Programming

We now extend the statements of Sect. "Convex Integer Programming over Totally Unimodular Systems" about totally unimodular matrices to arbitrary integer matrices. The next lemma shows that the Graver basis of any integer matrix covers all edge-directions of the integer hulls of polyhedra defined by that matrix.

Lemma 14 *For every $A \in \mathbb{Z}^{m \times n}$, $l, u \in \mathbb{Z}_\infty^n$, and $b \in \mathbb{Z}^m$, the Graver basis $\mathcal{G}(A)$ of A covers all edge-directions of the polyhedron $P_I := \mathrm{conv}\{x \in \mathbb{Z}^n : Ax = b, \ l \leq x \leq u\}$.*

Proof Consider any edge e of P_I and pick two distinct points $x, y \in e \cap \mathbb{Z}^n$. Then $g := y - x$ is in $\mathcal{L}(A) \setminus \{0\}$. Therefore, by Lemma 8, there is a conformal sum $g = \sum_i h_i$ with $h_i \in \mathcal{G}(A)$ for all i. We claim that $x + h_i \in P_I$ for all i. Indeed, first note that $h_i \in \mathcal{G}(A) \subset \mathcal{L}(A)$ implies $Ah_i = 0$ and hence $A(x + h_i) = Ax = b$; and second note that $l \leq x, x + g \leq u$ and $h_i \sqsubseteq g$ imply that $l \leq x + h_i \leq u$.

Now let $w \in \mathbb{Z}^n$ be a linear functional uniquely maximized over P_I at the edge e. Then $wh_i = w(x + h_i) - wx \leq 0$ for all i. But $\sum(wh_i) = wg = wy - wx = 0$, implying that in fact $wh_i = 0$ and hence $x + h_i \in e$ for all i. Therefore each h_i is a direction of e (in fact, all h_i are the same and g is a multiple of some Graver basis element). □

Combining Theorems 6 and 12 and Lemma 14 we obtain the following statement.

Theorem 18 *For every fixed d there is a polynomial time algorithm that, given integer $m \times n$ matrix A, its Graver basis $G(A)$, $l, u \in \mathbb{Z}_\infty^n$, $x \in \mathbb{Z}^n$ with $l \leq x \leq u$, $w_1, \ldots, w_d \in \mathbb{Z}^n$, and convex $c : \mathbb{R}^d \longrightarrow \mathbb{R}$ presented by a comparison oracle, encoded as $[\langle A, G(A), l, u, x, w_1, \ldots, w_d \rangle]$, solves the convex integer program with $b := Ax$,*

$$\max \{c(w_1 z, \ldots, w_d z) \,:\, z \in \mathbb{Z}^n,$$
$$Az = b, \; l \leq z \leq u\} \,.$$

Proof First, check in polynomial time using linear programming whether the objective function of any of the following $2n$ linear programs is unbounded,

$$\max \{\pm y_i \,:\, y \in P\}, \; i = 1, \ldots, n,$$
$$P := \{y \in \mathbb{R}^n \,:\, Ay = b, \; l \leq y \leq u\} \,.$$

If any is unbounded then terminate, asserting that P is unbounded. Otherwise, let ρ be the least integer upper bound on the absolute value of all optimal objective values. Then $P \subseteq [-\rho, \rho]^n$ and $S := \{y \in \mathbb{Z}^n \,:\, Ay = b, \; l \leq y \leq u\} \subset P$ is finite of radius $\rho(S) \leq \rho$. Moreover, by Cramer's rule, $\langle \rho \rangle$ is polynomially bounded in $\langle A, l, u, x \rangle$.

Using the given Graver basis and applying the algorithm of Theorem 12 we can simulate in polynomial time a linear discrete optimization oracle for S. Furthermore, by Lemma 14, the given Graver basis covers all edge-directions of the integer hull $P_I := \operatorname{conv}\{y \in \mathbb{Z}^n \,:\, Ay = b, l \leq y \leq u\} = \operatorname{conv}(S)$. Therefore we can apply the algorithm of Theorem 6 and solve the given convex program in polynomial time. □

Convex N-fold Integer Programming in Polynomial Time

We now extend the result of Theorem 15 and show that convex integer programming problems over n-fold systems can be solved in polynomial time as well. As explained in the beginning of this section, the algorithm either returns an optimal solution, or asserts that the program is infeasible, or asserts that the underlying polyhedron is unbounded.

Theorem 19 *For every fixed d and fixed $(r + s) \times t$ integer matrix A there is a polynomial time algorithm that, given n, lower and upper bounds $l, u \in \mathbb{Z}_\infty^{nt}$, $w_1, \ldots, w_d \in \mathbb{Z}^{nt}$, $b \in \mathbb{Z}^{r+ns}$, and convex functional $c : \mathbb{R}^d \longrightarrow \mathbb{R}$ presented by a comparison oracle, encoded as $[\langle l, u, w_1, \ldots, w_d, b \rangle]$, solves the convex n-fold integer programming problem*

$$\max \{c(w_1 x, \ldots, w_d x) \,:\, x \in \mathbb{Z}^{nt},$$
$$A^{(n)} x = b, \; l \leq x \leq u\} \,.$$

Proof First, check in polynomial time using linear programming whether the objective function of any of the following $2nt$ linear programs is unbounded,

$$\max \{\pm y_i \,:\, y \in P\}, \; i = 1, \ldots, nt,$$
$$P := \{y \in \mathbb{R}^{nt} \,:\, A^{(n)} y = b, \; l \leq y \leq u\} \,.$$

If any is unbounded then terminate, asserting that P is unbounded. Otherwise, let ρ be the least integer upper bound on the absolute value of all optimal objective values. Then $P \subseteq [-\rho, \rho]^{nt}$ and $S := \{y \in \mathbb{Z}^{nt} \,:\, A^{(n)} y = b, \; l \leq y \leq u\} \subset P$ is finite of radius $\rho(S) \leq \rho$. Moreover, by Cramer's rule, $\langle \rho \rangle$ is polynomially bounded in n and $\langle l, u, b \rangle$.

Using the algorithm of Theorem 15 we can simulate in polynomial time a linear discrete optimization oracle for S. Also, using the algorithm of Theorem 14 we can compute in polynomial time the Graver basis $G(A^{(n)})$ which, by Lemma 14, covers all edge-directions of $P_I := \operatorname{conv}\{y \in \mathbb{Z}^{nt} \,:\, A^{(n)} y = b, l \leq y \leq u\} = \operatorname{conv}(S)$. Therefore we can apply the algorithm of Theorem 6 and solve the given convex n-fold integer programming problem in polynomial time. □

Again, as a special case of Theorem 19 we recover the following result of [12] concerning convex integer programming in standard form over n-fold matrices.

Theorem 20 *For every fixed d and fixed $(r + s) \times t$ integer matrix A there is a polynomial time algorithm that, given n, linear functionals $w_1, \ldots, w_d \in \mathbb{Z}^{nt}$, right-hand side $b \in \mathbb{Z}^{r+ns}$, and convex functional $c : \mathbb{R}^d \longrightarrow \mathbb{R}$ presented by a comparison oracle, encoded as $[\langle w_1, \ldots, w_d, b \rangle]$, solves the convex n-fold integer program in standard form*

$$\max \{c(w_1 x, \ldots, w_d x) \,:\, x \in \mathbb{N}^{nt}, \; A^{(n)} x = b\} \,.$$

Some Applications

Transportation Problems and Packing Problems
Theorems 19 and 20 generalize Theorems 15 and 16 by broadly extending the class of objective functions that can be maximized in polynomial time over n-fold systems. Therefore all applications discussed in Sect. "Some Applications" automatically extend accordingly.

First, we have the following analog of Corollary 8 for the *convex integer transportation problem* over long 3-way tables. This has a very broad further generalization to multiway transportation problems over long k-way tables of any dimension k, see Sect. "Multiway Transportation Problems and Privacy in Statistical Databases".

Corollary 10 *For every fixed d, p, q there is a polynomial time algorithm that, given n, arrays $w_1, \ldots, w_d \in \mathbb{Z}^{p \times q \times n}$, line-sums $u \in \mathbb{Z}^{p \times q}$, $v \in \mathbb{Z}^{p \times n}$ and $z \in \mathbb{Z}^{q \times n}$, and convex functional $c: \mathbb{R}^d \longrightarrow \mathbb{R}$ presented by a comparison oracle, encoded as $[\langle w_1, \ldots, w_d, u, v, z \rangle]$, solves the convex integer 3-way line-sum transportation problem*

$$\max \left\{ c(w_1 x, \ldots, w_d x) \ : \ x \in \mathbb{N}^{p \times q \times n}, \right.$$

$$\sum_i x_{i,j,k} = z_{j,k}, \quad \sum_j x_{i,j,k} = v_{i,k},$$

$$\left. \sum_k x_{i,j,k} = u_{i,j} \right\}.$$

Second, we have the following analog of Corollary 9 for convex bin packing.

Corollary 11 *For every fixed d, number of types t, and type weights $v_1, \ldots, v_t \in \mathbb{Z}$, there is a polynomial time algorithm that, given n bins, item numbers $n_1, \ldots, n_t \in \mathbb{Z}$, bin capacities $u_1, \ldots, u_n \in \mathbb{Z}$, utility matrices $w_1, \ldots, w_d \in \mathbb{Z}^{t \times n}$, and convex functional $c: \mathbb{R}^d \longrightarrow \mathbb{R}$ presented by a comparison oracle, encoded as $[\langle n_1, \ldots, n_t, u_1, \ldots, u_n, w_1, \ldots, w_d \rangle]$, solves the convex integer bin packing problem,*

$$\max \left\{ c(w_1 x, \ldots, w_d x) \ : \ x \in \mathbb{N}^{t \times n}, \right.$$

$$\left. \sum_j v_j x_{j,k} = u_k, \quad \sum_k x_{j,k} = n_j \right\}.$$

Vector Partitioning and Clustering The vector partition problem concerns the partitioning of n items among p players to maximize social value subject to constraints on the number of items each player can receive. More precisely, the data is as follows. With each item i is associated a vector $v_i \in \mathbb{Z}^k$ representing its utility under k criteria. The utility of player h under ordered partition $\pi = (\pi_1, \ldots, \pi_p)$ of the set of items $\{1, \ldots, n\}$ is the sum $v_h^\pi := \sum_{i \in \pi_h} v_i$ of utility vectors of items assigned to h under π. The social value of π is the balancing $c(v_{1,1}^\pi, \ldots, v_{1,k}^\pi, \ldots, v_{p,1}^\pi, \ldots, v_{p,k}^\pi)$ of the player utilities, where c is a convex functional on $\mathbb{R}^{pk}$. In the constrained version, the partition must be of a given *shape*, i.e. the number $|\pi_h|$ of items that player h gets is required to be a given number λ_h (with $\sum \lambda_h = n$). In the unconstrained version, there is no restriction on the number of items per player.

Vector partition problems have applications in diverse areas such as load balancing, circuit layout, ranking, cluster analysis, inventory, and reliability, see e. g. [7,9,25,39,50] and the references therein. Here is a typical example.

Example 6 (minimal variance clustering). This problem has numerous applications in the analysis of statistical data: given n observed points $v_1, \ldots, v_n$ in k-space, group them into p clusters $\pi_1, \ldots, \pi_p$ that minimize the sum of cluster variances given by

$$\sum_{h=1}^p \frac{1}{|\pi_h|} \sum_{i \in \pi_h} \left\| v_i - \left(\frac{1}{|\pi_h|} \sum_{i \in \pi_h} v_i \right) \right\|^2.$$

Consider instances where there are $n = pm$ points and the desired clustering is balanced, that is, the clusters should have equal size m. Suitable manipulation of the sum of variances expression above shows that the problem is equivalent to a constrained vector partition problem, where $\lambda_h = m$ for all h, and where the convex functional $c: \mathbb{R}^{pk} \longrightarrow \mathbb{R}$ (to be maximized) is the Euclidean norm squared, given by

$$c(z) = \|z\|^2 = \sum_{h=1}^p \sum_{i=1}^k |z_{h,i}|^2.$$

If either the number of criteria k or the number of players p is variable, the partition problem is intractable since it instantly captures NP-hard problems [39].

When both k, p are fixed, both the constrained and unconstrained versions of the vector partition problem are polynomial time solvable [39,50]. We now show that vector partition problems (either constrained or unconstrained) are in fact convex n-fold integer programming problems and therefore, as a consequence of Theorem 20, can be solved is polynomial time.

Corollary 12 *For every fixed number p of players and number k of criteria, there is a polynomial time algorithm that, given n, item vectors $v_1, \ldots, v_n \in \mathbb{Z}^k$, $\lambda_1, \ldots, \lambda_p \in \mathbb{N}$, and convex functional $c: \mathbb{R}^{pk} \longrightarrow \mathbb{R}$ presented by a comparison oracle, encoded as $[\langle v_1, \ldots, v_n, \lambda_1, \ldots, \lambda_p \rangle]$, solves the constrained and unconstrained partitioning problems.*

Proof There is an obvious one-to-one correspondence between partitions and matrices $x \in \{0,1\}^{p \times n}$ with all column-sums equal to one, where partition π corresponds to the matrix x with $x_{h,i} = 1$ if $i \in \pi_h$ and $x_{h,i} = 0$ otherwise. Let $d := pk$ and define d matrices $w_{h,j} \in \mathbb{Z}^{p \times n}$ by setting $(w_{h,j})_{h,i} := v_{i,j}$ for all $h = 1, \ldots, p$, $i = 1, \ldots, n$ and $j = 1, \ldots, k$, and setting all other entries to zero. Then for any partition π and its corresponding matrix x we have $v_{h,j}^\pi = w_{h,j} x$ for all $h = 1, \ldots, p$ and $j = 1, \ldots, k$. Therefore, the unconstrained vector partition problem is the convex integer program

$$\max \left\{ c(w_{1,1} x, \ldots, w_{p,k} x) \,:\, x \in \mathbb{N}^{p \times n}, \; \sum_h x_{h,i} = 1 \right\} .$$

Suitably arranging the variables in a vector, this becomes a convex n-fold integer program with a $(0+1) \times p$ defining matrix A, where A_1 is empty and $A_2 := (1, \ldots, 1)$.

Similarly, the constrained vector partition problem is the convex integer program

$$\max \left\{ c(w_{1,1} x, \ldots, w_{p,k} x) \,:\, x \in \mathbb{N}^{p \times n}, \; \sum_h x_{h,i} = 1, \; \sum_i x_{h,i} = \lambda_h \right\} .$$

This again is a convex n-fold integer program, now with a $(p+1) \times p$ defining matrix A, where now $A_1 := I_p$ is the $p \times p$ identity matrix and $A_2 := (1, \ldots, 1)$ as before.

Using the algorithm of Theorem 20, this convex n-fold integer program, and hence the given vector partition problem, can be solved in polynomial time. □

Multiway Transportation Problems and Privacy in Statistical Databases

Transportation problems form a very important class of discrete optimization problems. The feasible points in a transportation problem are the multiway tables ("contingency tables" in statistics) such that the sums of entries over some of their lower dimensional subtables such as lines or planes ("margins" in statistics) are specified. Transportation problems and their corresponding transportation polytopes have been used and studied extensively in the operations research and mathematical programming literature, as well as in the statistics literature in the context of secure statistical data disclosure and management by public agencies, see [4,6,11,18,19,42,43,53,60,62] and references therein.

In this section we completely settle the algorithmic complexity of treating multiway tables and discuss the applications to transportation problems and secure statistical data disclosure, as follows. After introducing some terminology in Sect. "Tables and Margins", we go on to describe, in Sect. "The Universality Theorem", a universality result that shows that "short" 3-way $r \times c \times 3$ tables, with variable number r of rows and variable number c of columns but fixed small number 3 of layers (hence "short"), are *universal* in a very strong sense. In Sect. "The Complexity of the Multiway Transportation Problem" we discuss the general multiway transportation problem. Using the results of Sect. "The Universality Theorem" and the results on linear and convex n-fold integer programming from Sect. "Linear N-fold Integer Programming" and Sect. "Convex Integer Programming", we show that the transportation problem is intractable for short 3-way $r \times c \times 3$ tables but polynomial time treatable for "long" $(k+1)$-way $m_1 \times \cdots \times m_k \times n$ tables, with k and the sides $m_1, \ldots, m_k$ fixed (but arbitrary), and the number n of layers variable (hence "long"). In Sect. "Privacy and Entry-Uniqueness" we turn to discuss data privacy and security and consider the central problem of detecting entry uniqueness in tables with disclosed margins. We show that as a consequence of the results of Sect. "The Universality Theorem" and Sect. "The Complexity of the Multiway Transportation Problem", and in analogy to the complexity of the transportation problem established in Sect. "The Complexity of the Multiway Transportation Problem", the entry uniqueness problem is intractable for short 3-way $r \times c \times 3$ tables but polynomial time decidable for long $(k+1)$-way $m_1 \times \cdots \times m_k \times n$ tables.

Tables and Margins

We start with some terminology on tables, margins and transportation polytopes. A *k-way table* is an $m_1 \times \cdots \times m_k$ array $x = (x_{i_1,\ldots,i_k})$ of nonnegative integers. A *k-way transportation polytope* (or simply *k-way polytope* for brevity) is the set of all $m_1 \times \cdots \times m_k$ nonnegative arrays $x = (x_{i_1,\ldots,i_k})$ such that the sums of the entries over some of their lower dimensional sub-arrays (margins) are specified. More precisely, for any tuple $(i_1,\ldots,i_k)$ with $i_j \in \{1,\ldots,m_j\} \cup \{+\}$, the corresponding *margin* $x_{i_1,\ldots,i_k}$ is the sum of entries of x over all coordinates j with $i_j = +$. The *support* of $(i_1,\ldots,i_k)$ and of $x_{i_1,\ldots,i_k}$ is the set $\mathrm{supp}(i_1,\ldots,i_k) := \{j : i_j \neq +\}$ of non-summed coordinates. For instance, if x is a $4 \times 5 \times 3 \times 2$ array then it has 12 margins with support $F = \{1,3\}$ such as $x_{3,+,2,+} = \sum_{i_2=1}^{5} \sum_{i_4=1}^{2} x_{3,i_2,2,i_4}$. A collection of margins is *hierarchical* if, for some family $\mathcal{F}$ of subsets of $\{1,\ldots,k\}$, it consists of all margins $u_{i_1,\ldots,i_k}$ with support in $\mathcal{F}$. In particular, for any $0 \leq h \leq k$, the collection of all *h-margins* of *k*-tables is the hierarchical collection with $\mathcal{F}$ the family of all *h*-subsets of $\{1,\ldots,k\}$. Given a hierarchical collection of margins $u_{i_1,\ldots,i_k}$ supported on a family $\mathcal{F}$ of subsets of $\{1,\ldots,k\}$, the corresponding *k-way polytope* is the set of nonnegative arrays with these margins,

$$T_\mathcal{F} := \{ x \in \mathbb{R}_+^{m_1 \times \cdots \times m_k} : x_{i_1,\ldots,i_k} = u_{i_1,\ldots,i_k},$$
$$\mathrm{supp}(i_1,\ldots,i_k) \in \mathcal{F} \}.$$

The integer points in this polytope are precisely the *k*-way tables with the given margins.

The Universality Theorem

We now describe the following *universality* result of [14,16] which shows that, quite remarkably, *any* rational polytope is a short 3-way $r \times c \times 3$ polytope with all line-sums specified. (In the terminology of Sect. "Tables and Margins" this is the $r \times c \times 3$ polytope $T_\mathcal{F}$ of all 2-margins fixed, supported on the family $\mathcal{F} = \{\{1,2\},\{1,3\},\{2,3\}\}$.) By saying that a polytope $P \subset \mathbb{R}^p$ is *representable* as a polytope $Q \subset \mathbb{R}^q$ we mean in the strong sense that there is an injection $\sigma : \{1,\ldots,p\} \longrightarrow \{1,\ldots,q\}$ such that the coordinate-erasing projection

$$\pi : \mathbb{R}^q \longrightarrow \mathbb{R}^p : x = (x_1,\ldots,x_q)$$
$$\mapsto \pi(x) = (x_{\sigma(1)},\ldots,x_{\sigma(p)})$$

provides a bijection between Q and P and between the sets of integer points $Q \cap \mathbb{Z}^q$ and $P \cap \mathbb{Z}^p$. In particular, if P is representable as Q then P and Q are isomorphic in any reasonable sense: they are linearly equivalent and hence all linear programming related problems over the two are polynomial time equivalent; they are combinatorially equivalent and hence they have the same face numbers and facial structure; and they are integer equivalent and therefore all integer programming and integer counting related problems over the two are polynomial time equivalent as well.

We provide only an outline of the proof of the following statement; complete details and more consequences of this theorem can be found in [14,16].

Theorem 21 *There is a polynomial time algorithm that, given $A \in \mathbb{Z}^{m \times n}$ and $b \in \mathbb{Z}^m$, encoded as $[\langle A, b \rangle]$, produces r, c and line-sums $u \in \mathbb{Z}^{r \times c}$, $v \in \mathbb{Z}^{r \times 3}$ and $z \in \mathbb{Z}^{c \times 3}$ such that the polytope $P := \{y \in \mathbb{R}_+^n : Ay = b\}$ is representable as the 3-way polytope*

$$T := \left\{ x \in \mathbb{R}_+^{r \times c \times 3} : \sum_i x_{i,j,k} = z_{j,k}, \right.$$
$$\left. \sum_j x_{i,j,k} = v_{i,k}, \sum_k x_{i,j,k} = u_{i,j} \right\}.$$

Proof The construction proving the theorem consists of three polynomial time steps, each representing a polytope of a given format as a polytope of another given format.

First, we show that any $P := \{y \geq 0 : Ay = b\}$ with A, b integer can be represented in polynomial time as $Q := \{x \geq 0 : Cx = d\}$ with C matrix all entries of which are in $\{-1, 0, 1, 2\}$. This reduction of coefficients will enable the rest of the steps to run in polynomial time. For each variable y_j let $k_j := \max\{\lfloor \log_2 |a_{i,j}| \rfloor :$

$i = 1, \ldots m\}$ be the maximum number of bits in the binary representation of the absolute value of any entry $a_{i,j}$ of A. Introduce variables $x_{j,0}, \ldots, x_{j,k_j}$, and relate them by the equations $2x_{j,i} - x_{j,i+1} = 0$. The representing injection σ is defined by $\sigma(j) := (j, 0)$, embedding y_j as $x_{j,0}$. Consider any term $a_{i,j} y_j$ of the original system. Using the binary expansion $|a_{i,j}| = \sum_{s=0}^{k_j} t_s 2^s$ with all $t_s \in \{0, 1\}$, we rewrite this term as $\pm \sum_{s=0}^{k_j} t_s x_{j,s}$. It is not hard to verify that this represents P as Q with defining $\{-1, 0, 1, 2\}$-matrix.

Second, we show that any $Q := \{y \geq 0 : Ay = b\}$ with A, b integer can be represented as a face F of a 3-way polytope with all plane-sums fixed, that is, a face of a 3-way polytope $T_{\mathcal{F}}$ of all 1-margins fixed, supported on the family $\mathcal{F} = \{\{1\}, \{2\}, \{3\}\}$.

Since Q is a polytope and hence bounded, we can compute (using Cramer's rule) an integer upper bound U on the value of any coordinate y_j of any $y \in Q$. Note also that a face of a 3-way polytope $T_{\mathcal{F}}$ is the set of all $x = (x_{i,j,k})$ with some entries forced to zero; these entries are termed "forbidden", and the other entries are termed "enabled".

For each variable y_j, let r_j be the largest between the sum of positive coefficients of y_j and the sum of absolute values of negative coefficients of y_j over all equations,

$$r_j := \max\left(\sum_k \{a_{k,j} : a_{k,j} > 0\}, \sum_k \{|a_{k,j}| : a_{k,j} < 0\}\right).$$

Assume that A is of size $m \times n$. Let $r := \sum_{j=1}^n r_j$, $R := \{1, \ldots, r\}$, $h := m + 1$ and $H := \{1, \ldots, h\}$. We now describe how to construct vectors $u, v \in \mathbb{Z}^r$, $z \in \mathbb{Z}^h$, and a set $E \subset R \times R \times H$ of triples – the enabled, non-forbidden, entries – such that the polytope Q is represented as the face F of the corresponding 3-way polytope of $r \times r \times h$ arrays with plane-sums u, v, z and only entries indexed by E enabled,

$$F := \left\{ x \in \mathbb{R}_+^{r \times r \times h} : x_{i,j,k} = 0 \right.$$

for all $(i, j, k) \notin E$,

and $\sum_{i,j} x_{i,j,k} = z_k$, $\sum_{i,k} x_{i,j,k} = v_j$,

$$\left. \sum_{j,k} x_{i,j,k} = u_i \right\}.$$

We also indicate the injection $\sigma : \{1, \ldots, n\} \longrightarrow R \times R \times H$ giving the desired embedding of coordinates y_j as coordinates $x_{i,j,k}$ and the representation of Q as F.

Roughly, each equation $k = 1, \ldots, m$ is encoded in a "horizontal plane" $R \times R \times \{k\}$ (the last plane $R \times R \times \{h\}$ is included for consistency with its entries being "slacks"); and each variable y_j, $j = 1, \ldots, n$ is encoded in a "vertical box" $R_j \times R_j \times H$, where $R = \biguplus_{j=1}^n R_j$ is the natural partition of R with $|R_j| = r_j$ for all $j = 1, \ldots, n$, that is, with $R_j := \{1 + \sum_{l < j} r_l, \ldots, \sum_{l \leq j} r_l\}$.

Now, all "vertical" plane-sums are set to the same value U, that is, $u_j := v_j := U$ for $j = 1, \ldots, r$. All entries not in the union $\biguplus_{j=1}^n R_j \times R_j \times H$ of the variable boxes will be forbidden. We now describe the enabled entries in the boxes; for simplicity we discuss the box $R_1 \times R_1 \times H$, the others being similar. We distinguish between the two cases $r_1 = 1$ and $r_1 \geq 2$. In the first case, $R_1 = \{1\}$; the box, which is just the single line $\{1\} \times \{1\} \times H$, will have exactly two enabled entries $(1, 1, k^+), (1, 1, k^-)$ for suitable k^+, k^- to be defined later. We set $\sigma(1) := (1, 1, k^+)$, namely embed $y_1 = x_{1,1,k^+}$. We define the *complement* of the variable y_1 to be $\bar{y}_1 := U - y_1$ (and likewise for the other variables). The vertical sums u, v then force $\bar{y}_1 = U - y_1 = U - x_{1,1,k^+} = x_{1,1,k^-}$, so the complement of y_1 is also embedded. Next, consider the case $r_1 \geq 2$. For each $s = 1, \ldots, r_1$, the line $\{s\} \times \{s\} \times H$ (respectively, $\{s\} \times \{1 + (s \mod r_1)\} \times H$) will contain one enabled entry $(s, s, k^+(s))$ (respectively, $(s, 1 + (s \mod r_1), k^-(s))$). All other entries of $R_1 \times R_1 \times H$ will be forbidden. Again, we set $\sigma(1) := (1, 1, k^+(1))$, namely embed $y_1 = x_{1,1,k^+(1)}$; it is then not hard to see that, again, the vertical sums u, v force $x_{s,s,k^+(s)} = x_{1,1,k^+(1)} = y_1$ and $x_{s,1+(s \mod r_1),k^-(s)} = U - x_{1,1,k^+(1)} = \bar{y}_1$ for each $s = 1, \ldots, r_1$. Therefore, both y_1 and $\bar{y}_1$ are each embedded in r_1 distinct entries.

We now encode the equations by defining the horizontal plane-sums z and the indices $k^+(s), k^-(s)$ above as follows. For $k = 1, \ldots, m$, consider the kth equation $\sum_j a_{k,j} y_j = b_k$. Define the index sets $J^+ := \{j : a_{k,j} > 0\}$ and $J^- := \{j : a_{k,j} < 0\}$, and set $z_k := b_k + U \cdot \sum_{j \in J^-} |a_{k,j}|$. The last coordinate of z is set for consistency with u, v to be $z_h = z_{m+1} := r \cdot U - \sum_{k=1}^m z_k$. Now, with $\bar{y}_j := U - y_j$ the complement of variable y_j as above, the kth equation can be rewritten as

$$\sum_{j \in J^+} a_{k,j} y_j + \sum_{j \in J^-} |a_{k,j}| \bar{y}_j$$

$$= \sum_{j=1}^{n} a_{k,j} y_j + U \cdot \sum_{j \in J^-} |a_{k,j}|$$

$$= b_k + U \cdot \sum_{j \in J^-} |a_{k,j}| = z_k .$$

To encode this equation, we simply "pull down" to the corresponding kth horizontal plane as many copies of each variable y_j or $\bar{y}_j$ by suitably setting $k^+(s) := k$ or $k^-(s) := k$. By the choice of r_j there are sufficiently many, possibly with a few redundant copies which are absorbed in the last hyperplane by setting $k^+(s) := m+1$ or $k^-(s) := m+1$. This completes the encoding and provides the desired representation.

Third, we show that any 3-way polytope with plane-sums fixed and entry bounds,

$$F := \left\{ y \in \mathbb{R}_+^{l \times m \times n} : \sum_{i,j} y_{i,j,k} = c_k , \right.$$

$$\sum_{i,k} y_{i,j,k} = b_j ,$$

$$\sum_{j,k} y_{i,j,k} = a_i ,$$

$$\left. y_{i,j,k} \le e_{i,j,k} \right\} ,$$

can be represented as a 3-way polytope with line-sums fixed (and no entry bounds),

$$T := \left\{ x \in \mathbb{R}_+^{r \times c \times 3} : \sum_{I} x_{I,J,K} = z_{J,K} , \right.$$

$$\left. \sum_{J} x_{I,J,K} = v_{I,K} , \sum_{K} x_{I,J,K} = u_{I,J} \right\} .$$

In particular, this implies that any face F of a 3-way polytope with plane-sums fixed can be represented as a 3-way polytope T with line-sums fixed: forbidden entries are encoded by setting a "forbidding" upper-bound $e_{i,j,k} := 0$ on all forbidden entries $(i, j, k) \notin E$ and an "enabling" upper-bound $e_{i,j,k} := U$ on all enabled entries $(i, j, k) \in E$. We describe the presentation, but omit the proof that it is indeed valid; further details on this step can be found in [14,15,16]. We give explicit formulas for $u_{I,J}, v_{I,K}, z_{J,K}$ in terms of a_i, b_j, c_k and $e_{i,j,k}$ as follows. Put $r := l \cdot m$ and $c := n + l + m$. The first index I of each entry $x_{I,J,K}$ will be a pair $I = (i, j)$ in the r-set

$$\{(1, 1), \dots, (1, m), (2, 1), \dots,$$
$$(2, m), \dots, (l, 1), \dots, (l, m)\} .$$

The second index J of each entry $x_{I,J,K}$ will be a pair $J = (s, t)$ in the c-set

$$\{(1, 1), \dots, (1, n), (2, 1), \dots, (2, l), (3, 1), \dots, (3, m)\} .$$

The last index K will simply range in the 3-set $\{1, 2, 3\}$. We represent F as T via the injection σ given explicitly by $\sigma(i, j, k) := ((i, j), (1, k), 1)$, embedding each variable $y_{i,j,k}$ as the entry $x_{(i,j),(1,k),1}$. Let U now denote the minimal between the two values $\max\{a_1, \dots, a_l\}$ and $\max\{b_1, \dots, b_m\}$. The line-sums (2-margins) are set to be

$$u_{(i,j),(1,t)} = e_{i,j,t} ,$$

$$u_{(i,j),(2,t)} = \begin{cases} U & \text{if } t = i, \\ 0 & \text{otherwise.} \end{cases} ,$$

$$u_{(i,j),(3,t)} = \begin{cases} U & \text{if } t = j, \\ 0 & \text{otherwise.} \end{cases} ,$$

$$v_{(i,j),t} = \begin{cases} U & \text{if } t = 1, \\ e_{i,j,+} & \text{if } t = 2, \\ U & \text{if } t = 3. \end{cases} ,$$

$$z_{(i,j),1} = \begin{cases} c_j & \text{if } i = 1, \\ m \cdot U - a_j & \text{if } i = 2, \\ 0 & \text{if } i = 3. \end{cases}$$

$$z_{(i,j),2} = \begin{cases} e_{+,+,j} - c_j & \text{if } i = 1, \\ 0 & \text{if } i = 2, \\ b_j & \text{if } i = 3. \end{cases} ,$$

$$z_{(i,j),3} = \begin{cases} 0 & \text{if } i = 1, \\ a_j & \text{if } i = 2, \\ l \cdot U - b_j & \text{if } i = 3. \end{cases} .$$

Applying the first step to the given rational polytope P, applying the second step to the resulting Q, and applying the third step to the resulting F, we get in polynomial time a 3-way $r \times c \times 3$ polytope T of all line-sums fixed representing P as claimed. □

The Complexity of the Multiway Transportation Problem

We are now finally in position to settle the complexity of the general multiway transportation problem. The data for the problem consists of: positive integers k (table dimension) and $m_1, \ldots, m_k$ (table sides); family $\mathcal{F}$ of subsets of $\{1, \ldots, k\}$ (supporting the hierarchical collection of margins to be fixed); integer values $u_{i_1,\ldots,i_k}$ for all margins supported on $\mathcal{F}$; and integer "profit" $m_1 \times \cdots \times m_k$ array w. The transportation problem is to find an $m_1 \times \cdots \times m_k$ table having the given margins and attaining maximum profit, or assert than none exists. Equivalently, it is the linear integer programming problem of maximizing the linear functional defined by w over the transportation polytope $T_\mathcal{F}$,

$$\max\{wx : x \in \mathbb{N}^{m_1 \times \cdots \times m_k} : x_{i_1,\ldots,i_k} = u_{i_1,\ldots,i_k},$$
$$\text{supp}(i_1,\ldots,i_k) \in \mathcal{F}\}.$$

The following result of [15] is an immediate consequence of Theorem 21. It asserts that if two sides of the table are variable part of the input then the transportation problem is intractable already for short 3-way tables with $\mathcal{F} = \{\{1, 2\}, \{1, 3\}, \{2, 3\}\}$ supporting all 2-margins (line-sums). This result can be easily extended to k-way tables of any dimension $k \geq 3$ and $\mathcal{F}$ the collection of all h-subsets of $\{1, \ldots, k\}$ for any $1 < h < k$ as long as two sides of the table are variable; we omit the proof of this extended result.

Corollary 13 *It is NP-complete to decide, given r, c, and line-sums $u \in \mathbb{Z}^{r \times c}$, $v \in \mathbb{Z}^{r \times 3}$, and $z \in \mathbb{Z}^{c \times 3}$, encoded as $[\langle u, v, z \rangle]$, if the following set of tables is nonempty,*

$$S := \left\{ x \in \mathbb{N}^{r \times c \times 3} : \sum_i x_{i,j,k} = z_{j,k}, \right.$$
$$\left. \sum_j x_{i,j,k} = v_{i,k}, \sum_k x_{i,j,k} = u_{i,j} \right\}.$$

Proof The integer programming feasibility problem is to decide, given $A \in \mathbb{Z}^{m \times n}$ and $b \in \mathbb{Z}^m$, if $\{y \in \mathbb{N}^n : Ay = b\}$ is nonempty. Given such A and b, the polynomial time algorithm of Theorem 21 produces r, c and $u \in \mathbb{Z}^{r \times c}$, $v \in \mathbb{Z}^{r \times 3}$, and $z \in \mathbb{Z}^{c \times 3}$, such that $\{y \in \mathbb{N}^n : Ay = b\}$ is nonempty if and only if the set S above is nonempty. This reduces integer programming feasibility to short 3-way line-sum transportation feasibility. Since the former is NP-complete (see e.g. [55]), so turns out to be the latter. □

We now show that in contrast, when all sides but one are fixed (but arbitrary), and one side n is variable, then the corresponding long k-way transportation problem for any hierarchical collection of margins is an n-fold integer programming problem and therefore, as a consequence of Theorem 16, can be solved is polynomial time. This extends Corollary 8 established in Sect. "Three-Way Line-Sum Transportation Problems" for 3-way line-sum transportation.

Corollary 14 *For every fixed k, table sides $m_1, \ldots, m_k$, and family $\mathcal{F}$ of subsets of $\{1, \ldots, k+1\}$, there is a polynomial time algorithm that, given n, integer values $u = (u_{i_1,\ldots,i_{k+1}})$ for all margins supported on $\mathcal{F}$, and integer $m_1 \times \cdots \times m_k \times n$ array w, encoded as $[\langle u, w \rangle]$, solves the linear integer multiway transportation problem*

$$\max\{wx : x \in \mathbb{N}^{m_1 \times \cdots \times m_k \times n},$$
$$x_{i_1,\ldots,i_{k+1}} = u_{i_1,\ldots,i_{k+1}}, \text{supp}(i_1,\ldots,i_{k+1}) \in \mathcal{F}\}.$$

Proof Re-index the arrays as $x = (x^1, \ldots, x^n)$ with each $x^j = (x_{i_1,\ldots,i_k,j})$ a suitably indexed $m_1 m_2 \cdots m_k$ vector representing the jth layer of x. Then the transportation problem can be encoded as an n-fold integer programming problem in standard form,

$$\max\{wx : x \in \mathbb{N}^{nt}, A^{(n)}x = b\},$$

with an $(r + s) \times t$ defining matrix A where $t := m_1 m_2 \cdots m_k$ and r, s, A_1 and A_2 are determined from $\mathcal{F}$, and with right-hand side $b := (b^0, b^1, \ldots, b^n) \in \mathbb{Z}^{r+ns}$ determined from the margins $u = (u_{i_1,\ldots,i_{k+1}})$, in such a way that the equations $A_1(\sum_{j=1}^n x^j) = b^0$ represent the constraints of all margins $x_{i_1,\ldots,i_k,+}$ (where summation over layers occurs), whereas the equations $A_2 x^j = b^j$ for $j = 1, \ldots, n$ represent the constraints of all margins $x_{i_1,\ldots,i_k,j}$ with $j \neq +$ (where summations are within a single layer at a time).

Using the algorithm of Theorem 16, this n-fold integer program, and hence the given multiway transportation problem, can be solved in polynomial time. □

The proof of Corollary 14 shows that the set of feasible points of any long k-way transportation problem, with all sides but one fixed and one side n variable, for any hierarchical collection of margins, is an n-fold integer programming problem. Therefore, as a consequence of Theorem 20, we also have the following extension of Corollary 14 for the convex integer multiway transportation problem over long k-way tables.

Corollary 15 *For every fixed d, k, table sides $m_1, \ldots, m_k$, and family $\mathcal{F}$ of subsets of $\{1, \ldots, k+1\}$, there is a polynomial time algorithm that, given n, integer values $u = (u_{i_1,\ldots,i_{k+1}})$ for all margins supported on $\mathcal{F}$, integer $m_1 \times \cdots \times m_k \times n$ arrays $w_1, \ldots, w_d$, and convex functional $c: \mathbb{R}^d \longrightarrow \mathbb{R}$ presented by a comparison oracle, encoded as $[\langle u, w_1, \ldots, w_d \rangle]$, solves the convex integer multiway transportation problem*

$$\max \{ c(w_1 x, \ldots, w_d x) : x \in \mathbb{N}^{m_1 \times \cdots \times m_k \times n},$$
$$x_{i_1,\ldots,i_{k+1}} = u_{i_1,\ldots,i_{k+1}}, \; \mathrm{supp}(i_1, \ldots, i_{k+1}) \in \mathcal{F} \}.$$

Privacy and Entry-Uniqueness

A common practice in the disclosure of a multiway table containing sensitive data is to release some of the table margins rather than the table itself, see e. g. [11,18,19] and the references therein. Once the margins are released, the security of any specific entry of the table is related to the set of possible values that can occur in that entry in any table having the same margins as those of the source table in the data base. In particular, if this set consists of a unique value, that of the source table, then this entry can be exposed and privacy can be violated. This raises the following fundamental *entry-uniqueness problem*: given a consistent disclosed (hierarchical) collection of margin values, and a specific entry index, is the value that can occur in that entry in any table having these margins unique? We now describe the results of [48] that settle the complexity of this problem, and interpret the consequences for secure statistical data disclosure.

First, we show that if two sides of the table are variable part of the input then the entry-uniqueness problem is intractable already for short 3-way tables with all 2-margins (line-sums) disclosed (corresponding to $\mathcal{F} = \{\{1,2\},\{1,3\},\{2,3\}\}$). This can be easily extended to k-way tables of any dimension $k \geq 3$ and $\mathcal{F}$ the collection of all h-subsets of $\{1,\ldots,k\}$ for any $1 < h < k$ as long as two sides of the table are variable; we omit the proof of this extended result. While this result indicates that the disclosing agency may not be able to check for uniqueness, in this situation, some consolation is in that an adversary will be computationally unable to identify and retrieve a unique entry either.

Corollary 16 *It is coNP-complete to decide, given r, c, and line-sums $u \in \mathbb{Z}^{r \times c}$, $v \in \mathbb{Z}^{r \times 3}$, $z \in \mathbb{Z}^{c \times 3}$, encoded as $[\langle u, v, z \rangle]$, if the entry $x_{1,1,1}$ is the same in all tables in*

$$\left\{ x \in \mathbb{N}^{r \times c \times 3} : \sum_i x_{i,j,k} = z_{j,k}, \right.$$
$$\left. \sum_j x_{i,j,k} = v_{i,k}, \sum_k x_{i,j,k} = u_{i,j} \right\}.$$

Proof The subset-sum problem, well known to be NP-complete, is the following: given positive integers $a_0, a_1, \ldots, a_m$, decide if there is an $I \subseteq \{1,\ldots,m\}$ with $a_0 = \sum_{i \in I} a_i$. We reduce the complement of subset-sum to entry-uniqueness. Given $a_0, a_1, \ldots, a_m$, consider the polytope in $2(m+1)$ variables $y_0, y_1, \ldots, y_m, z_0, z_1, \ldots, z_m$,

$$P := \left\{ (y,z) \in \mathbb{R}_+^{2(m+1)} : a_0 y_0 - \sum_{i=1}^m a_i y_i = 0, \right.$$
$$\left. y_i + z_i = 1, \; i = 0, 1 \ldots, m \right\}.$$

First, note that it always has one integer point with $y_0 = 0$, given by $y_i = 0$ and $z_i = 1$ for all i. Second, note that it has an integer point with $y_0 \neq 0$ if and only if there is an $I \subseteq \{1,\ldots,m\}$ with $a_0 = \sum_{i \in I} a_i$, given by $y_0 = 1$, $y_i = 1$ for $i \in I$, $y_i = 0$ for $i \in \{1,\ldots,m\} \setminus I$, and $z_i = 1 - y_i$ for all i. Lifting P to a suitable $r \times c \times 3$ line-sum polytope T with the coordinate y_0 embedded in the entry $x_{1,1,1}$ using Theorem 21, we find that T has a table with $x_{1,1,1} = 0$, and this value is unique among the tables in T if and only if there is *no* solution to the subset-sum problem with $a_0, a_1, \ldots, a_m$. □

Next we show that, in contrast, when all table sides but one are fixed (but arbitrary), and one side n is variable, then, as a consequence of Corollary 14, the corresponding long k-way entry-uniqueness problem for any hierarchical collection of margins can be solved is polynomial time. In this situation, the algorithm of Corollary 17 below allows disclosing agencies to efficiently check possible collections of margins before disclosure: if an entry value is not unique then disclosure may be assumed secure, whereas if the value is unique then disclosure may be risky and fewer margins should be released. Note that this situation, of long multiway tables, where one category is significantly richer than the others, that is, when each sample point can take many values in one category and only few values in the other categories, occurs often in practical applications, e. g., when one category is the individuals age and the other categories are binary ("yes-no"). In such situations, our polynomial time algorithm below allows disclosing agencies to check entry-uniqueness and make learned decisions on secure disclosure.

Corollary 17 *For every fixed k, table sides $m_1, \ldots, m_k$, and family $\mathcal{F}$ of subsets of $\{1, \ldots, k+1\}$, there is a polynomial time algorithm that, given n, integer values $u = (u_{j_1,\ldots,j_{k+1}})$ for all margins supported on $\mathcal{F}$, and entry index $(i_1, \ldots, i_{k+1})$, encoded as $[n, \langle u \rangle]$, decides if the entry $x_{i_1,\ldots,i_{k+1}}$ is the same in all tables in the set*

$$\{x \in \mathbb{N}^{m_1 \times \cdots \times m_k \times n} : x_{j_1,\ldots,j_{k+1}} = u_{j_1,\ldots,j_{k+1}},$$
$$\mathrm{supp}(j_1, \ldots, j_{k+1}) \in \mathcal{F}\}.$$

Proof By Corollary 14 we can solve in polynomial time both transportation problems

$$l := \min\{x_{i_1,\ldots,i_{k+1}} : x \in \mathbb{N}^{m_1 \times \cdots \times m_k \times n},$$
$$x \in T_{\mathcal{F}}\},$$

$$u := \max\{x_{i_1,\ldots,i_{k+1}} : x \in \mathbb{N}^{m_1 \times \cdots \times m_k \times n},$$
$$x \in T_{\mathcal{F}}\},$$

over the corresponding k-way transportation polytope

$$T_{\mathcal{F}} := \{x \in \mathbb{R}_+^{m_1 \times \cdots \times m_k \times n} : x_{j_1,\ldots,j_{k+1}} = u_{j_1,\ldots,j_{k+1}},$$
$$\mathrm{supp}(j_1, \ldots, j_{k+1}) \in \mathcal{F}\}.$$

Clearly, entry $x_{i_1,\ldots,i_{k+1}}$ has the same value in all tables with the given (disclosed) margins if and only if $l = u$, completing the description of the algorithm and the proof. □

References

1. Aho AV, Hopcroft JE, Ullman JD (1975) The Design and Analysis of Computer Algorithms. Addison-Wesley, Reading
2. Allemand K, Fukuda K, Liebling TM, Steiner E (2001) A polynomial case of unconstrained zero-one quadratic optimization. Math Prog Ser A 91:49–52
3. Alon N, Onn S (1999) Separable partitions. Discret Appl Math 91:39–51
4. Aoki S, Takemura A (2003) Minimal basis for connected Markov chain over $3 \times 3 \times K$ contingency tables with fixed two-dimensional marginals. Austr New Zeal J Stat 45:229–249
5. Babson E, Onn S, Thomas R (2003) The Hilbert zonotope and a polynomial time algorithm for universal Gröbner bases. Adv Appl Math 30:529–544
6. Balinski ML, Rispoli FJ (1993) Signature classes of transportation polytopes. Math Prog Ser A 60:127–144
7. Barnes ER, Hoffman AJ, Rothblum UG (1992) Optimal partitions having disjoint convex and conic hulls. Math Prog 54:69–86
8. Berstein Y, Onn S: Nonlinear bipartite matching. Disc Optim (to appear)
9. Boros E, Hammer PL (1989) On clustering problems with connected optima in Euclidean spaces. Discret Math 75:81–88
10. Chvátal V (1973) Edmonds polytopes and a hierarchy of combinatorial problems. Discret Math 4:305–337
11. Cox LH (2003) On properties of multi-dimensional statistical tables. J Stat Plan Infer 117:251–273
12. De Loera J, Hemmecke R, Onn S, Rothblum UG, Weismantel R: Integer convex maximization via Graver bases. E-print: arXiv:math.CO/0609019. (submitted)
13. De Loera J, Hemmecke R, Onn S, Weismantel R: N-fold integer programming. Disc Optim (to appear)
14. De Loera J, Onn S (2004) All rational polytopes are transportation polytopes and all polytopal integer sets are contingency tables. In: Proc IPCO 10 – Symp on Integer Programming and Combinatoral Optimization, Columbia University, New York. Lec Not Comp Sci. Springer, 3064, pp 338–351
15. De Loera J, Onn S (2004) The complexity of three-way statistical tables. SIAM J Comput 33:819–836
16. De Loera J, Onn S (2006) All linear and integer programs are slim 3-way transportation programs. SIAM J Optim 17:806–821
17. De Loera J, Onn S (2006) Markov bases of three-way tables are arbitrarily complicated. J Symb Comput 41:173–181
18. Domingo-Ferrer J, Torra V (eds) (2004) Privacy in Statistical Databases. Proc. PSD 2004 – Int Symp Privacy in Statistical Databases, Barcelona, Spain. Lec Not Comp Sci. Springer, 3050
19. Doyle P, Lane J, Theeuwes J, Zayatz L (eds) (2001) Confidentiality, Disclosure and Data Access: Theory and Practical Applications for Statistical Agencies. North-Holland, Amsterdam
20. Edelsbrunner H, O'Rourke J, Seidel R (1986) Constructing arrangements of lines and hyperplanes with applications. SIAM J Comput 15:341–363
21. Edelsbrunner H, Seidel R, Sharir M (1991) On the zone theorem for hyperplane arrangements. In: New Results and

Trends in Computer Science. Lec Not Comp Sci. Springer, 555, pp 108–123
22. Edmonds J (1971) Matroids and the greedy algorithm. Math Prog 1:127–136
23. Edmonds J, Karp RM (1972) Theoretical improvements in algorithmic efficiency of network flow problems. J Ass Comput Mach 19:248–264
24. Frank A, Tardos E (1987) An application of simultaneous Diophantine approximation in combinatorial optimization. Combinatorica 7:49–65
25. Fukuda K, Onn S, Rosta V (2003) An adaptive algorithm for vector partitioning. J Global Optim 25:305–319
26. Garey MR, Johnson DS (1979) Computers and Intractability. Freeman, San Francisco
27. Gilmore PC, Gomory RE (1961) A linear programming approach to the cutting-stock problem. Oper Res 9:849–859
28. Graver JE (1975) On the foundation of linear and integer programming I. Math Prog 9:207–226
29. Gritzmann P, Sturmfels B (1993) Minkowski addition of polytopes: complexity and applications to Gröbner bases. SIAM J Discret Math 6:246–269
30. Grötschel M, Lovász L (1995) Combinatorial optimization. In: Handbook of Combinatorics. North-Holland, Amsterdam, pp 1541–1597
31. Grötschel M, Lovász L, Schrijver A (1993) Geometric Algorithms and Combinatorial Optimization, 2nd edn. Springer, Berlin
32. Grünbaum B (2003) Convex Polytopes, 2nd edn. Springer, New York
33. Harding EF (1967) The number of partitions of a set of n points in k dimensions induced by hyperplanes. Proc Edinburgh Math Soc 15:285–289
34. Hassin R, Tamir A (1989) Maximizing classes of two-parameter objectives over matroids. Math Oper Res 14:362–375
35. Hemmecke R (2003) On the positive sum property and the computation of Graver test sets. Math Prog 96:247–269
36. Hemmecke R, Hemmecke R, Malkin P (2005) 4ti2 Version 1.2–Computation of Hilbert bases, Graver bases, toric Gröbner bases, and more. http://www.4ti2.de/. Accessed Sept 2005
37. Hoffman AJ, Kruskal JB (1956) Integral boundary points of convex polyhedra. In: Linear inequalities and Related Systems, Ann Math Stud 38. Princeton University Press, Princeton, pp 223–246
38. Hoşten S, Sullivant S (2007) A finiteness theorem for Markov bases of hierarchical models. J Comb Theory Ser A 114:311–321
39. Hwang FK, Onn S, Rothblum UG (1999) A polynomial time algorithm for shaped partition problems. SIAM J Optim 10:70–81
40. Khachiyan LG (1979) A polynomial algorithm in linear programming. Sov Math Dok 20:191–194
41. Klee V, Kleinschmidt P (1987) The d-step conjecture and its relatives. Math Oper Res 12:718–755
42. Klee V, Witzgall C (1968) Facets and vertices of transportation polytopes. In: Mathematics of the Decision Sciences, Part I, Stanford, CA, 1967. AMS, Providence, pp 257–282
43. Kleinschmidt P, Lee CW, Schannath H (1987) Transportation problems which can be solved by the use of Hirsch-paths for the dual problems. Math Prog 37:153–168
44. Lenstra AK, Lenstra HW Jr, Lovász L (1982) Factoring polynomials with rational coefficients. Math Ann 261:515–534
45. Lovàsz L (1986) An Algorithmic Theory of Numbers, Graphs, and Convexity. CBMS-NSF Ser App Math, SIAM 50:iv+91
46. Onn S (2006) Convex discrete optimization. Lecture Series, Le Séminaire de Mathématiques Supérieures, Combinatorial Optimization: Methods and Applications, Université de Montréal, Canada, June 2006. http://ie.technion.ac.il/~onn/Talks/Lecture_Series.pdf and at http://www.dms.umontreal.ca/sms/ONN_Lecture_Series.pdf. Accessed 19–30 June 2006
47. Onn S (2003) Convex matroid optimization. SIAM J Discret Math 17:249–253
48. Onn S (2006) Entry uniqueness in margined tables. In: Proc. PSD 2006 – Symp. on Privacy in Statistical Databases, Rome, Italy. Lec Not Comp Sci. Springer, 4302, pp 94–101
49. Onn S, Rothblum UG (2004) Convex combinatorial optimization. Disc Comp Geom 32:549–566
50. Onn S, Schulman LJ (2001) The vector partition problem for convex objective functions. Math Oper Res 26:583–590
51. Onn S, Sturmfels B (1999) Cutting Corners. Adv Appl Math 23:29–48
52. Pistone G, Riccomagno EM, Wynn HP (2001) Algebraic Statistics. Chapman and Hall, London
53. Queyranne M, Spieksma FCR (1997) Approximation algorithms for multi-index transportation problems with decomposable costs. Disc Appl Math 76:239–253
54. Santos F, Sturmfels B (2003) Higher Lawrence configurations. J Comb Theory Ser A 103:151–164
55. Schrijver A (1986) Theory of Linear and Integer Programming. Wiley, New York
56. Schulz A, Weismantel R (2002) The complexity of generic primal algorithms for solving general integral programs. Math Oper Res 27:681–692
57. Schulz A, Weismantel R, Ziegler GM (1995) (0, 1)-integer programming: optimization and augmentation are equivalent. In: Proc 3rd Ann Euro Symp Alg. Lec Not Comp Sci. Springer, 979, pp 473–483
58. Sturmfels B (1996) Gröbner Bases and Convex Polytopes. Univ Lec Ser 8. AMS, Providence
59. Tardos E (1986) A strongly polynomial algorithm to solve combinatorial linear programs. Oper Res 34:250–256
60. Vlach M (1986) Conditions for the existence of solutions of the three-dimensional planar transportation problem. Discret Appl Math 13:61–78
61. Wallacher C, Zimmermann U (1992) A combinatorial interior point method for network flow problems. Math Prog 56:321–335

62. Yemelichev VA, Kovalev MM, Kravtsov MK (1984) Polytopes, Graphs and Optimisation. Cambridge University Press, Cambridge
63. Yudin DB, Nemirovskii AS (1977) Informational complexity and efficient methods for the solution of convex extremal problems. Matekon 13:25–45
64. Zaslavsky T (1975) Facing up to arrangements: face count formulas for partitions of space by hyperplanes. Memoirs Amer Math Soc 154:vii+102
65. Ziegler GM (1995) Lectures on Polytopes. Springer, New York

Convex Envelopes in Optimization Problems

YASUTOSHI YAJIMA
Tokyo Institute Technol., Tokyo, Japan

MSC2000: 90C26

Article Outline

Keywords
See also
References

Keywords

Convex underestimator; Nonconvex optimization

Let $f: S \to \mathbf{R}$ be a *lower semicontinuous function*, where $S \subseteq \mathbf{R}^n$ is a nonempty convex subset. The convex envelope taken over S is a function $f_S: S \to \mathbf{R}$ such that
- f_S is a convex function defined over the set S;
- $f_S(x) \leq f(x)$ for all $x \in S$;
- if h is any other convex function such that $h(x) \leq f(x)$ for all $x \in S$, then $h(x) \leq f_S(x)$ for all $x \in S$.

In other words, f_S is the pointwise supremum among any *convex underestimators* of f over S, and is uniquely determined. The following demonstrates the most fundamental properties shown by [3,6]. Suppose that the minimum of f over S exists. Then,

$$\min\{f(x): x \in S\} = \min\{f_S(x): x \in S\}$$

and

$$\{x^*: f(x^*) \leq f(x), \forall x \in S\} \subseteq \{x^*: f_S(x^*) \leq f_S(x), \forall x \in S\}.$$

The properties indicate that an optimal solution of a nonconvex minimization problem could be obtained by minimizing the associated convex envelope. In general, however, finding the convex envelope is at least as difficult as solving the original one.

Several practical results have been proposed for special classes of objective functions and constraints. Suppose that the function f is concave and S is a polytope with vertices $v^0, \ldots, v^K$. Then, the convex envelope f_S over S can be expressed as:

$$\begin{cases} f_S(x) = \min & \sum_{i=0}^{K} \alpha_i f(v^i) \\ \text{s.t.} & \sum_{i=0}^{K} \alpha_i v^i = x, \\ & \sum_{i=0}^{K} \alpha_i = 1, \\ & \alpha_i \geq 0, \quad i = 0, \ldots, K. \end{cases}$$

Especially, if S is an n-simplex with vertices $v^0, \ldots, v^n$, f_S is the affine function

$$f_S(x) = a^\top x + b,$$

which is uniquely determined by solving the following linear system

$$a^\top v^i + b = f(v^i), \quad i = 0, \ldots, n.$$

The properties above have been used to solve concave minimization problems with linear constraints [4,6].

The following property shown in [1,5] is frequently used in the literature. For each $i = 1, \ldots, p$, let $f^i: S_i \to \mathbf{R}$ be a continuous function, where $S_i \subseteq \mathbf{R}^{n_i}$, and let $n = n_1 + \cdots + n_p$. If

$$f(x) = \sum_{i=1}^{p} f^i(x^i)$$

and

$$S = S_1 \times \cdots \times S_p,$$

where $x^i \in \mathbf{R}^{n_i}$, $i = 1, \ldots, p$, and $x = (x^1, \ldots, x^p) \in \mathbf{R}^n$, then the convex envelope $f_S(x)$ can be expressed as:

$$f_S(x) = \sum_{i=1}^{p} f_S^i(x^i).$$

In particular, let $f(x) = \sum_{i=1}^{n} f_i(x_i)$ be a separable function, where $x = (x_1, \ldots, x_n) \in \mathbf{R}^n$, and let $f_i(x_i)$ be concave for each $i = 1, \ldots, n$. Then the convex envelope of $f(x)$ over the rectangle $R = \{x \in \mathbf{R}^n: a_i \leq x_i \leq b_i, i = 1, \ldots, n\}$ can be the affine function, which is given by the sum of the linear functions below:

$$f_R(x) = \sum_{i=1}^{n} l_i(x_i),$$

where $l_i(x_i)$ meets $f_i(x_i)$ at both ends of the interval $a_i \leq x_i \leq b_i$ for each $i = 1, \ldots, n$. B. Kalantari and J.B. Rosen [7] show an algorithm for the global minimization of a quadratic concave function over a polytope. They exploit convex envelopes of separable functions over rectangles to generate lower bounds in a *branch and bound scheme*.

Also, convex envelopes of *bilinear functions* over rectangles have been proposed in [2]. Consider the following rectangles:

$$\Omega_i = \left\{ (x_i, y_i): \begin{array}{l} l_i \leq x_i \leq L_i, \\ m_i \leq y_i \leq M_i \end{array} \right\},$$

$$i = 1, \ldots, n,$$

and let

$$f^i(x_i, y_i) = x_i y_i, \quad i = 1, \ldots, n,$$

be bilinear functions with two variables. It has been shown that for each $i = 1, \ldots, n$, the convex envelope of $f^i(x_i, y_i)$ over Ω_i is expressed as:

$$f^i_{\Omega_i}(x_i, y_i) = \max\{m_i x_i + l_i y_i - l_i m_i, M_i x_i + L_i y_i - L_i M_i\}.$$

Moreover, it can be verified that $f^i_{\Omega_i}(x_i, y_i)$ agrees with $f^i(x_i, y_i)$ at the four extreme points of Ω_i. Thus, the convex envelope of the general bilinear function

$$f(x, y) = x^\top y = \sum_{i=1}^{n} f^i(x_i, y_i),$$

where $x^\top = (x_1, \ldots, x_n)$ and $y^\top = (y_1, \ldots, y_n)$ over $\Omega = \Omega_1 \times \cdots \times \Omega_n$ can be expressed as

$$f_\Omega(x, y) = \sum_{i=1}^{n} f^i_{\Omega_i}(x_i, y_i).$$

Another characterization of convex envelopes of bilinear functions over a special type of polytope, which includes a rectangle as a special case, is derived in [8].

See also

▶ αBB Algorithm
▶ Global Optimization in Generalized Geometric Programming
▶ MINLP: Global Optimization with αBB

References

1. Al-Khayyal FA (1990) Jointly constrained bilinear programs and related problems: An overview. Comput Math Appl 19(11):53–62
2. Al-Khayyal FA, Falk JE (1983) Jointly constrained biconvex programming. Math Oper Res 8(2):273–286
3. Bazaraa MS, Sherali HD, Shetty CM (1993) Nonlinear programming: Theory and algorithms. Wiley, New York
4. Benson HP, Sayin S (1994) A finite concave minimization algorithm using branch and bound and neighbor generation. J Global Optim 5(1):1–14
5. Falk JE (1969) Lagrange multipliers and nonconvex programs. SIAM J Control 7:534–545
6. Falk JE, Hoffman K (1976) A successive underestimation method for concave minimization problems. Math Oper Res 1(3):251–259
7. Kalantari B, Rosen JB (1987) An algorithm for global minimization of linearly constrained concave quadratic functions. Math Oper Res 12(3):544–561
8. Sherali HD, Alameddine A (1990) An explicit characterization of the convex envelope of a bivariate bilinear function over special polytopes. Ann Oper Res 25(1–4):197–209

Convexifiable Functions, Characterization of[1]

SANJO ZLOBEC
Department of Mathematics and Statistics,
McGill University, Montreal, Canada

MSC2000: 90C25, 90C26, 90C30, 90C31, 25A15, 34A05

Article Outline

Introduction
Definitions
Characterizations of a Convexifiable Function
Canonical Form of Smooth Programs
Other Applications
Conclusions
References

---

[1] Research partly supported by NSERC of Canada.

Introduction

A twice continuously differentiable function in several variables, when considered on a compact convex set C, becomes convex if an appropriate convex quadratic is added to it, e. g. [2]. Equivalently, a twice continuously differentiable function is the difference of a convex function and a convex quadratic on C. This decomposition is valid also for smooth functions with Lipschitz derivatives [8]. Here we recall three conditions that are both necessary and sufficient for the decomposition [9,10]. We also list several implications of the convexification in optimization and applied mathematics [10,11]. A different notion of convexification is studied in, e. g. [6]; see also [3,5].

Definitions

Definition 1. ([7,10]) Given a continuous function $f: \mathbb{R}^n \to \mathbb{R}$ on a compact convex set C of the Euclidean space $\mathbb{R}^n$, consider $\phi: \mathbb{R}^{n+1} \to \mathbb{R}$ defined by $\phi(x, \alpha) = f(x) - 1/2\alpha x^T x$ where x^T is the transpose of x. If $\phi(x, \alpha)$ is convex on C for some $\alpha = \alpha^*$, then $\phi(x, \alpha)$ is said to be a *convexification* of f and α^* is its *convexifier* on C. Function f is *convexifiable* if it has a convexification.

Observation. If α^* is a convexifier of f on a compact convex set C, then so is every $\alpha < \alpha^*$.

Illustration 1. Consider $f(t) = \cos t$ on, say, $-\pi \leq t \leq 2\pi$. This function is convexifiable, its convexifier is any $\alpha \leq -1$. For, e. g., $\alpha^* = -2$, its convexification is $\phi(t, -2) = \cos t + t^2$. Note that $f(t)$ is the difference of (strictly) convex $\phi(t, \alpha) = \cos t - 1/2\alpha t^2$ and (strictly) convex quadratic $-1/2\alpha x^T x$ for every *sufficiently small* α. The graphs of $f(t)$ and its convexification $\phi(t, -2)$ are depicted in Fig. 1.

Characterizations of a Convexifiable Function

One can characterize convexifiable functions using the fact that a continuous $f: \mathbb{R}^n \to \mathbb{R}$ is convex if, and only if, f is mid-point convex, i. e., $f((x+y)/2) \leq 1/2(f(x) + f(y))$, $x, y \in C$, e. g. [4]. Denote the norm of $u \in \mathbb{R}^n$ by $||u|| = (u^T u)^{1/2}$. With a continuous $f: \mathbb{R}^n \to \mathbb{R}$ one can associate $\Psi: \mathbb{R}^n \times \mathbb{R}^n \to \mathbb{R}$:

Definition 2. ([10]) Consider a continuous $f: \mathbb{R}^n \to \mathbb{R}$ on a compact convex set C in $\mathbb{R}^n$. The mid-point acceleration function of f on C is the function

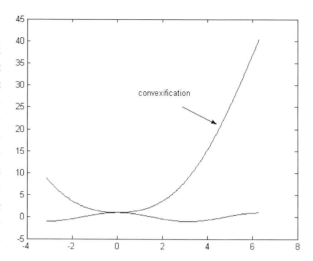

Convexifiable Functions, Characterization of, Figure 1
Function $f(t) = \cos t$ and its convexification

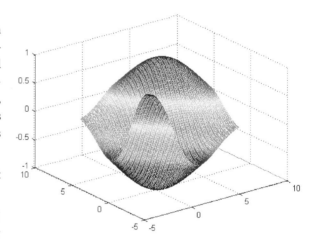

Convexifiable Functions, Characterization of, Figure 2
Mid-point acceleration function of $f(t) = \cos t$

$$\Psi(x, y) = (4/||x - y||^2)[f(x) + f(y) - 2f((x + y)/2)], \quad x, y \in C, x \neq y.$$

Function Ψ describes a mid-point "displacement of the displacement" (i. e., the "acceleration") of f at x between x and y along the direction $y - x$. The graph of Ψ for the scalar function $f(t) = \cos t$ is depicted in Fig. 2.

Using Ψ one can characterize a convexifiable function:

Theorem 1. ([10]) *Consider a continuous $f: \mathbb{R}^n \to \mathbb{R}$ on a compact convex set C in $\mathbb{R}^n$. Function f is convexifi-*

able on C if, and only if, its mid-point acceleration function Ψ is bounded from below on C.

For scalar functions one can also use a determinant:

Theorem 2. ([9] Determinant Characterization of Scalar Convexifiable Function) *A continuous scalar function $f\colon \mathbb{R} \to \mathbb{R}$ is convexifiable on a compact convex interval I if, and only if, there exists a number α such that for every three points $s < t < \xi$ in I*

$$\det \begin{pmatrix} 1 & 1 & 1 \\ s & t & \xi \\ f(s) & f(t) & f(\xi) \end{pmatrix} \geq \frac{1}{2}\alpha(s-t)(t-\xi)(\xi-s).$$

Illustration 2. Function $f(t) = -|t|^{3/2}$ on $C = [-1, 1]$ is continuously differentiable but it is not convexifiable. Indeed, for $s = 0, \Psi(0, t) = 2^{5/2}(1 - 2^{1/2})/t^{-1/2} \to -\infty$ as $t > 0, t \to 0$. Also, using Theorem 2 at $s = -\epsilon, t = 0, \xi = \varepsilon > 0$, we find that there is no α such that $\alpha \leq -2/\epsilon$ as $\varepsilon \to 0$. Function $g(t) = -|t|$ is not convexifiable around the origin.

Scalar convexifiable functions can be represented explicitly on a compact interval I:

Theorem 3. ([9] Explicit Representation of Scalar Convexifiable Functions) *A continuous scalar function $f\colon I \to \mathbb{R}$ is convexifiable if, and only if, there exists a number α such that*

$$f(t) = f(c) + \frac{1}{2}\alpha(t^2 - c^2) + \int_c^t g(\xi, \alpha) \, d\xi.$$

Here $c, t \in I, c < t$, and $g = g(\cdot, \alpha)\colon I \to \mathbb{R}$ is a nondecreasing right-continuous function.

An implication of this result is that every smooth function with a Lipschitz derivative, in particular every analytic function and every trajectory of an object governed by Newton's Second Law, is of this form.

Two important classes of functions are convexifiable and a convexifier α can be given explicitly. First, if f is twice continuously differentiable, then the second derivative of f at x is represented by the Hessian matrix $H(x) = (\partial^2 f(x)/\partial x_i \partial x_j), i, j = 1, \ldots, n$. This is a symmetric matrix with real eigenvalues. Denote its smallest eigenvalue at x by $\lambda(x)$ and its "globally" smallest eigenvalue over a compact convex set C by

$$\lambda^* = \min_{x \in C} \lambda(x).$$

Corollary 1. ([7]) *A twice continuously differentiable function $f\colon \mathbb{R}^n \to \mathbb{R}$ is convexifiable on a compact convex set C in $\mathbb{R}^n$ and $\alpha = \lambda^*$ is a convexifier.*

Suppose that f is a continuously differentiable (smooth) function with the derivative satisfying the Lipschitz property, i. e., $\|\nabla f(x) - \nabla f(y)\| \leq L\|x - y\|$ for every $x, y \in C$ and some constant L. Here $\nabla f(u)$ is the (Frechet) derivative of f at u. We represent the derivative of f at x by a column n-tuple gradient $\nabla f(x) = (\partial f(x)/\partial x_i)$.

Corollary 2. ([8]) *A continuously differentiable function $f\colon \mathbb{R}^n \to \mathbb{R}$, with the derivative having the Lipschitz property with a constant L on a compact convex set C in $\mathbb{R}^n$, is convexifiable on C and $\alpha = -L$ is a convexifier.*

A Lipschitz function may not be convexifiable. For example, $f(t) = t^2 \sin(1/t)$ for $t \neq 0$ and $f(0) = 0$ is a Lipschitz function and it is also differentiable (not continuously differentiable). Its derivative is uniformly bounded, but the function is not convexifiable, e. g. [7,11].

Canonical Form of Smooth Programs

Every mathematical program (NP)

$$\operatorname{Min} f(x), f^i(x) \leq 0, \quad i \in P = \{1, \ldots, m\}, x \in C,$$

where the functions $f, f^i\colon \mathbb{R}^n \to \mathbb{R}, i \in P$ are continuous and convexifiable on a compact convex set C can be reduced to a canonical form. First one considers some convexifications of these functions: $\phi(x, \alpha) = f(x) - 1/2\alpha x^T x$ and $\phi^i(x, \alpha_i) = f^i(x) - 1/2\alpha_i x^T x$, where α, α_i are, respectively, arbitrary convexifiers of $f, f^i, i \in P$. Then one associates with (NP) the following program with partly linear convexifications

$$(\text{LF}; \theta, \varepsilon):$$

$$\operatorname{Min}_{(x,\theta)} \phi(x, \alpha) + \frac{1}{2}\alpha x^T \theta,$$

$$\phi^i(x, \alpha_i) + \frac{1}{2}\alpha_i x^T \theta \leq 0, \quad i \in P,$$

$$x \in C, \|x - \theta\| \leq \varepsilon.$$

Here $\varepsilon \geq 0$ is a scalar parameter. This parameter was fixed at zero value in [2]. For the sake of "numerical stability" it was extended to $\varepsilon \geq 0$ in [7]. If the norm is chosen to be uniform, i. e., $\|u\|_\infty = \max_{i=1,\ldots,n} |u_i|$

then (LF; θ, ε) is a convex program in x for every fixed (θ, ε) and linear in (θ, ε) for every x. Such programs are called partly linear-convex. Theory of optimality and stability for such programs and related models is well studied, e. g. [1,8].

Remark Since one can construct the program (LF; θ, ε) for every (NP) with convexifiable functions, we refer to (LF; θ, ε) as the *parametric Liu–Floudas canonical form* of (NP).

Let us relate an optimal solution of (NP) to optimal solutions $x^0(\varepsilon), \theta^0(\varepsilon)$ of (LF; θ, ε).

Theorem 4. ([8,9]) *Consider (NP) with a unique optimal solution, where all functions are assumed to be convexifiable, and its partly linear-convex program (LF; θ, ε). Then a feasible x^* is an optimal solution of (NP) if, and only if, $x^* = \lim_{\epsilon \to 0} x^0(\varepsilon)$ and $\theta^* = \lim_{\varepsilon \to 0} \theta^0(\varepsilon)$, with $x^* = \theta^*$. Moreover, the feasible set mapping of (LF; θ, ε) is lower semi-continuous at θ^* and $\varepsilon^* = 0$, relative to all feasible perturbations of (θ, ε).*

Other Applications

There are many other areas of applications of convexifiable functions:

(i) Every convexifiable function f is the difference of a convex function $\phi(x, \alpha)$ and a convex quadratic $1/2\alpha x^T x$ for every sufficiently small α on a compact convex set. Hence it follows that the results for convex functions can be applied to $\phi(x, \alpha)$. With minor adjustments, pertaining to the quadratic term, such results can be extended to convexifiable (generally nonconvex) functions. Here is an illustration of how this works for the mean value. The result is well known for convex functions (the case $\alpha = 0$).

Theorem 5. ([9]) *Consider a continuous scalar convexifiable function $f : \mathbb{R} \to \mathbb{R}$ on an open interval (a, b) with a convexifier α. Then*

$$1/(d-c) \int_c^d f(\xi) d\xi \leq \frac{1}{2}[f(c) + f(d)]$$

$$- \frac{1}{12}\alpha(d-c)^2, \quad \text{for every} \quad a < c < d < b.$$

A composite version of this result follows.

Theorem 6. ([9]) *Integral Mean-Value for Composite Convexifiable Function) Let $f : (a, b) \to \mathbb{R}$ be continous and convexifiable with a convexifier α and let $g : [c, d] \to (a, b)$ be continuous. Then*

$$f\left(1/(d-c) \cdot \int_c^d g(t) dt\right) \leq 1/(d-c)$$

$$\cdot \int_c^d (f \circ g)(t) dt + \frac{1}{2}\alpha \cdot R(c, d; g)$$

where

$$R(c, d; g) = [1/(d-c) \cdot \int_c^d g(t) dt]^2 - 1/(d-c)$$

$$\cdot \int_c^d [g(t)]^2 dt.$$

Remarks If $f: (a, b) \to \mathbb{R}$ and $g: [c, d] \to (a, b)$ are continuous on (a, b), then, in Theorem 5 and 6, one can specify $\alpha = 0$, if f is convex on (a, b). Also $\alpha = \lambda^* = \min_{t \in I} f''(t)$, if f is twice continuously differentiable or analytic on (a, b), and $\alpha = -L$ a negative Lipschitz constant of the derivative of f on (a, b), if f is continuously differentiable.

(ii) Convexification of differential equations: A solution $y(t)$ of an ordinary differential equation of second or higher order, over a compact interval, is continuously differentiable with a Lipschitz derivative. Such y is convexifiable. Using $y(t) = \phi(t, \alpha) + 1/2\alpha t^2$, the problems in differential equations can be "convexified", i. e., transformed to equivalent differential equations with convex solutions $\phi(t, \alpha)$. After back-substitution, the true solution $y(t)$ is recovered. In particular, the problems of theoretical mechanics based on the Second Newton Law, can be "convexified", e. g. [9,11].

(iii) Convexification in linear algebra: Some of the basic eigenvalue inequalities for symmetric matrices follow from inequalities for convex functions, e. g. [4]. Using a convexification, one can extend these results to non-convexity. For example, after finding a convexifier of the product function $f(x) = x_1 x_2 \cdots x_n$, the following follows:

Corollary 3. (Bounds for the Determinant of an Arbitrary Symmetric Matrix [11]) *Let $A = (a_{ij})$ be an*

$n \times n$ real symmetric matrix where $n \geq 2$ and let ρ be its spectral radius. Then

$$\prod_{i=1,\ldots,n} a_{ii} - (n-1)\rho^{n-2} \sum_{i,j=1,\ldots,n; i<j} a_{ij}^2 \leq \det A$$
$$\leq \prod_{i=1,\ldots,n} a_{ii} + (n-1)\rho^{n-2} \sum_{i,j=1,\ldots,n; i<j} a_{ij}^2.$$

If the left hand-side in Corollary 3 is positive, then the matrix A is non-singular.

From the two-sided Jensen inequalities, obtained by convexification, new estimates follow for the absolute value of the inner product function in arbitrary inner product spaces. In some situations the new estimates are sharper than the Cauchy–Schwarz inequality. Many results for convexifiable functions, including the canonical form (LF; θ, ε), can be extended to non-smooth Lipschitz functions. This can be done using the fact that every Lipschitz function, when considered on a compact convex set, is only a linear function away from the set of all coordinate-wise monotone functions [11].

Conclusions

A necessary and sufficient condition for convexifiability on a compact convex set is given using the mid-point acceleration function. For scalar functions there are also characterizations given in terms of determinants and integrals. In particular, every smooth function with a Lipschitz derivative is convexifiable. Such function is only a convex quadratic away from the set of all convex functions. Using this "closeness", many results for convex functions can be extended to smooth (generally non-convex) functions with Lipschitz derivatives. On the other hand, mathematical programs with convexifiable functions can be reduced to the parametric Liu-Floudas partly linear-convex canonical form.

References

1. Floudas CA (2000) Deterministic Global Optimization. Kluwer, Dordrecht
2. Liu WB, Floudas CA (1993) A remark on the GOP algorithm for global optimization. J Global Optim 3:519–521
3. Pardalos PM (1988) Enumerative techniques in global optimization. Oper Res Spektr 10:29–35
4. Roberts AW, Varberg DE (1973) Convex Functions. Academic Press, New York
5. Vial JP (1983) Strong and weak convexity of sets and functions. Math Oper Res 8(2):231–259
6. Wu ZY, Lee HWJ, Yang XM (2005) A class of convexification and concavification methods for non-monotone optimization problems. Optimization 54:605–625
7. Zlobec S (2003) Estimating convexifiers in continuous optimization. Math Commun 8:129–137
8. Zlobec S (2005) On the Liu-Floudas convexification of smooth programs. J Glob Optim 32:401–407
9. Zlobec S (2005) Convexifiable functions in integral calculus. Glasnik Matematički 40(60):241–247, Erratum (2006) Ibid 41(61):187–188
10. Zlobec S (2006) Characterization of convexifiable functions. Optimization 55:251–262
11. Zlobec S (2008) On two simple decompositions of continuous functions. Optimization 57(2):249–261

Convex Max-Functions

CLAUDIA SAGASTIZÁBAL
IMPA, Jardim Botânico, Brazil

MSC2000: 49K35, 49M27, 65K10, 90C25

Article Outline

Keywords
Synonyms
Examples of the Problem
Continuity and Optimality Conditions
Algorithms of Minimization
    Nonlinear Programming
    Nonsmooth Optimization
    Other Methods of Resolution
See also
References

Keywords

Minimax problem; Convex optimization; Max-function

Synonyms

Minimax

Examples of the Problem

In a classic nonlinear program (NLP) a smooth objective function is minimized on a feasible set defined by finitely many smooth constraints. However, many optimization problems have an objective function that is not smooth but is of the max type, that is, defined as

$$f(x) := \max\{f_\ell(x) : \ell \in \mathcal{L}\}, \tag{1}$$

where the functions $f_\ell : \mathbf{R}^n \to \mathbf{R}$ are themselves smooth.

When $\mathcal{L}$ is finite, the minimization of f is called a *finite minimax problem*. In a general *minimax problem*, the indices ℓ can range over an infinite compact set $\mathcal{L}$, see [6]. Those f_ℓ's realizing the maximum in (1) are called *active functions*. The corresponding indices form the *active index set*, defined as $\mathcal{L}_a(x) := \{l \in \mathcal{L} : f(x) = f_l(x)\}$.

There are numerous examples of optimization problems dealing with the minimization of convex max-functions:

- When processing empirical data $\{(t_\ell, p_\ell) : \ell = 1, \ldots, m\}$, consider the problem of selecting the coefficients x of a polynomial $p_x(t) = \sum_{j=0}^n x_{j+1} t^j$ that fits well the data. The quality of the approximation can be measured through the deviation $f_\ell(x) := |p_\ell - p_x(t_\ell)|$, defined for all ℓ. Depending on the nature of the problem, it can be interesting to minimize either the sum of squared deviations or the maximum deviation. The first case is a *least squares problem*, while the second is known as *Chebyshev best approximation*, and is a particular instance of (1), with index set $\ell = \{\ell : \ell = 1, \ldots, m\}$.
- A more general case is finding the best approximation of a continuous function φ_0 on a compact interval $\mathcal{L}$. Instead of n powers t^j, n linearly independent functions $\varphi_j : \mathcal{L} \to \mathbf{R}$ are given. To find the linear combination $\sum_j x_j \varphi_j$ which best approximates φ_0 in the max-norm comes to solve (1), with $f_\ell(x) = |\sum_j x_j \varphi_j(\ell) - \varphi_0(\ell)|$. Because the problem has infinitely many constraints, it is also an example of semi-infinite programming.
- A basic problem in structural optimization, see [2], is to find the stiffest structure of a given volume that is able to carry loads varying on a given set. Optimization can be performed through the variation of sizing variables, like the thickness of bars in a truss; shape variables, as the splines defining the boundary of the body; or even the distribution and properties of the composite material used to make the structure itself. After discretization by finite elements, the design problem has an objective function as in (1). The f_ℓ's therein have usually the form $f_\ell(x) = (1/2) x^\mathsf{T} A_\ell x - b^\mathsf{T} x$, with A_ℓ denoting the stiffness matrix of the ℓth element of the grid.
- Some large scale mixed integer problems can be solved by *decomposition techniques* using Lagrangian relaxation. The idea is to coordinate, by means of a master program, the iterative resolution of problems of smaller dimension or complexity, called local problems. When applying price decomposition, the master is led to maximize a dual function, say $\theta(\lambda)$, on the space of multipliers Λ, using some iterative method. In production planning problems, the iterate λ^k sent to the local solvers can be interpreted as prices paid by the master. Let $\ell = \prod_i \mathcal{L}^i$ denote the (decomposed) primal space and let $L(x, \lambda) = \sum_i L^i(x^i, \lambda)$ be the (decomposed) Lagrangian of the original problem. Each local unit i decides its corresponding optimal level of production by solving $\min_{x^i \in \mathcal{L}^i} L^i(x^i, \lambda^k)$. Having those local optimal levels, the master adjusts prices by updating λ^k in order to maximize the dual function θ, defined as the pointwise minimum of the Lagrangian: $\theta(\lambda) = \min_{x \in \mathcal{L}} L(x, \lambda) = \sum_i \min_{x^i \in \mathcal{L}^i} L^i(x^i, \lambda)$. An equivalent problem for the master is to minimize $f(\lambda) := -\theta(\lambda)$. This last formulation has an objective f that is a max-function as in (1), letting $f_\ell(\lambda) = -\sum_i L^i(x_\ell^i, \lambda)$, for each $x_\ell \in \mathcal{L}$.
- Solving a nonlinear program using *exact penalties* leads to the iterative minimization of penalized objective which are max-functions, with the max-operation involving the constraints.
- In game theory, consider a zero sum game, with two players whose strategy is to optimize their individual choice against the worst possible selection by the other player. The first player can choose his action over n possible moves, with probability distribution $x = (x_1, \ldots, x_n)$. To every possible move $j = 1, \ldots, m$ of the second player, corresponds a loss $a_{i,j}$ player I pays to player II. Let ℓ be a continuous index counting elements in the set of probability distributions of the second player: $\mathcal{L} = \{z \in \mathbf{R}^m : \sum_{j=1}^m z_j = 1, z_j \geq 0\}$. Calling $A = [a_{i,j}]$ the $n \times m$ loss-matrix, the average amount to be paid by player I is $f_\ell(x) = x^\mathsf{T} A z_\ell$. It follows that the first player needs to solve a problem like (1). The mini-maximization of the bivariate function $F(x, z) = x^\mathsf{T} A z$ is also called a *saddle-point problem*.

Continuity and Optimality Conditions

When taking the pointwise maximum in (1), some properties of the functions f_ℓ are transmitted to f. Such is the case of continuity and convexity, but not of differentiability.

More precisely, f is continuous when both the f_ℓ-s and its gradients ∇f_ℓ are continuous. When the underlying functions f_ℓ are convex, so is f.

Convexity implies that the max-function f is differentiable almost everywhere. Nevertheless, at those points where more than one underlying function f_l realizes the maximum, the gradient fails to exist. Typically, such is the case at $\bar{x}$, a minimizer of f. For instance, suppose that $\mathcal{L}_a(\bar{x}) = \{1, 2\}$, i.e., there are two active functions at $\bar{x}$: $f(\bar{x}) = f_1(\bar{x}) = f_2(\bar{x})$. Clearly, for $\nabla f(\bar{x})$ to exist, the unlikely equality $\nabla f(\bar{x}) = \nabla f_1(\bar{x}) = \nabla f_2(\bar{x})$ needs to hold. Moreover, the optimality condition for minimizing f on $\mathbf{R}^n$ would require all the involved gradients to be null.

Rather than a single gradient, it is possible to define a whole set of *subgradients* by making convex combinations of $\nabla f_1(\bar{x})$ and $\nabla f_2(\bar{x})$. For an arbitrary convex max-function f, at any given x, the set of subgradients is the so-called *subdifferential* of f at x. Its expression is given by the formula

$$\partial f(x) = \left\{ \sum_{l \in \mathcal{L}_a(x)} \alpha_l \nabla f_l(x) \colon \alpha \in \Delta \right\}, \quad (2)$$

where Δ is the unit simplex

$$\Delta := \left\{ \alpha \in \mathbb{R}^{|\mathcal{L}_a(x)|} \colon \sum_{l \in \mathcal{L}_a(x)} \alpha_l = 1, \alpha_l \geq 0 \right\}. \quad (3)$$

When $\mathcal{L}$ in (1) is an infinite compact set, (2) still holds, provided the application $\ell \mapsto f_\ell(x)$ is upper semicontinuous for each x, see [8, Chap. VI, §4.4].

Consider the constrained problem

$$\min_{x \in \Omega} f(x), \quad (4)$$

where $\Omega \subset \mathbf{R}^n$ is a closed convex set and f is the function defined in (1). Assume the index set $\mathcal{L}$ is infinite and suppose (4) has a solution $\bar{x}$ such that $f(\bar{x}) = \max\{f_\ell(\bar{x}) \colon \ell \in \mathcal{L}\}$. Then it can be proved (see [6, Chap. VI, Thm. 3.3]) that (4) is equivalent to the finite minimax

$$\min_{x \in \Omega} \max_{i=1,\ldots,r} \{f_{\ell_i}(x) \colon \ell_i \in \mathcal{L}\}, \quad (5)$$

with $r \leq n + 1$. The set $\{\ell_1, \ldots, \ell_r\}$ is called an *extremal basis* of (4).

When Ω satisfies a constraint qualification condition of Slater type, see, for instance, [7, Chap. III], a necessary optimality condition (OC) characterizing a minimizer $\bar{x}$ of (4) is

$$0 \in \partial f(\bar{x}) + N_\Omega(\bar{x}), \quad (6)$$

where N is the normal cone of convex analysis. Because f is convex, the optimality condition is also sufficient.

The optimality condition (6) can be further specified when Ω is represented by a set of convex inequalities:

$$\Omega := \{x \in \mathbb{R}^n \colon c_j(x) \leq 0, j \in \mathcal{J}\}. \quad (7)$$

Observe that Ω may contain an infinite number of constraints c_j, assumed to be smooth and convex. Using (1) together with (7), the following characterization of $\bar{x}$ results:

Lemma 1 *There exist $r \leq n + 1$ and $s \leq n$ such that for the index sets $\mathcal{L}_a(\bar{x}) := \{l_1, \ldots, l_r\} \subset \mathcal{L}$ and $\{j_1, \ldots, j_s\} \subset \mathcal{J}$ it holds*

$$\sum_{i=1}^{r} \alpha_i \nabla f_{l_i}(\bar{x}) + \sum_{i=1}^{s} \mu_i \nabla c_{j_i}(\bar{x}) = 0, \quad (8)$$

where the multipliers α and μ are positive and α is an element of the simplex Δ in $\mathbb{R}^{|\mathcal{L}_a(\bar{x})|}$ from (3).

This characterization ensures the existence of an extremal basis of (4) near $\bar{x}$.

Algorithms of Minimization

Depending on the nature of the problem, several approaches have been proposed to solve (4):

- Reformulation as a NLP.
- Minimization of the nonsmooth max-function.
- Determination of a saddle point.
- Search of an extremal basis.

For example, the amount of available information can determine the method of resolution: if for any given x, all the active indices in (1) are known, then the full subdifferential (2) is available and a nonlinear programming technique can be applied.

Nonlinear Programming

An important feature of this approach is that smooth NLP techniques have a superlinear rate of convergence. The essential idea is first to write (4) as an NLP with an additional variable:

$$\begin{cases} \min\limits_{\substack{r\in\mathbb{R}\\x\in\Omega}} & r \\ \text{s.t.} & r \geq f_\ell(x), \quad \ell \in \mathcal{L}, \end{cases} \quad (9)$$

and then solve the associated optimality conditions by using a Newton-like method, such as sequential quadratic programming (cf. also ▶ Successive quadratic programming) or interior point schemes, see [4, Parts III–IV] and [3, §§4.3–4.4].

Nonsmooth Optimization

Sometimes the explicit knowledge of all the active constraints in (9) can be difficult, if not impossible, to obtain; such is the case for structural optimization problems.

On the other hand, it is often possible to obtain a single subgradient almost for free when computing $f(x)$. Indeed, suppose that just one active index l in $\ell_a(x)$ is known: $f(x) = f_l(x)$. Then, because of (2), $\nabla f_l(x) \in \partial f(x)$.

Algorithms from nonsmooth optimization, such as bundle methods [8, Chaps. XIV–XV], are designed to minimize a general convex function, possibly nondifferentiable, with the information furnished by an oracle that gives $f(x)$ and only one subgradient at x. Nonsmooth optimization techniques, specialized to a max-function like f in (1), have been successfully used in [12] and [9] to solve general minimax problems.

Although bundle methods are essentially first order methods, in recent years some proposals have been given that aim at obtaining better than linear convergence. They consist of a combination of bundle, proximal and quasi-Newton techniques [5,10,11].

Other Methods of Resolution

V.F. Demyanov and V.N. Malozemov treat (4) in an indirect way, by solving an infinite sequence of finite minimax problems. Keeping (5) in mind, the idea is to asymptotically identify an extremal basis by making successive approximations on a finite grid of the index set $\mathcal{L}$.

In game theory, rather than solving (4) by some 'mini-maximization' procedure, it can be more convenient to find a *saddle point*. That is, an equilibrium point satisfying $\min_x \max_{z_\ell} F(x, z_\ell) = \max_{z_\ell} \min_x F(x, z_\ell)$, with $F(x, z_\ell) := f_\ell(x)$. The determination of saddle points of F can be performed taking advantage of the extra structure of the problem. Some popular methods are Arrow–Hurwicz's and Uzawa's, see [1].

See also

▶ Lagrangian Multipliers Methods for Convex Programming

References

1. Arrow KJ, Hurwicz L, Uzawa H (1958) Studies in linear and nonlinear programming. Stanford Univ Press, Palo Alto, CA
2. Bendsøe MP (1995) Optimization of structural topology, shape and material. Springer, Berlin
3. Bertsekas DP (1995) Nonlinear programming. Athena Sci, Belmont, MA
4. Bonnans JF, Gilbert JCh, Lemaréchal C, Sagastizábal C (1997) Optimisation numérique: Aspects théoriques et pratiques. Springer, Berlin
5. Chen X, Fukushima M (1999) Proximal quasi-Newton methods for nondifferentiable convex optimization. Math Program 85(2):313–334
6. Dem'yanov VF, Malozemov VN (1974) Introduction to minimax. Wiley, New York
7. Hiriart-Urruty J-B (1996) L'optimisation. Que sais-je?, vol 3184. Press. Univ. France, Paris
8. Hiriart-Urruty J-B, Lemaréchal C (1993) Convex analysis and minimization algorithms. Grundl Math Wiss, vol 305-306. Springer, Berlin
9. Kiwiel KC (1987) A direct method of linearization for continuous minimax problems. J Optim Th Appl 55:271–287
10. Lemaréchal C, Sagastizábal C (1997) Variable metric bundle methods: from conceptual to implementable forms. Math Program 76:393–410
11. Mifflin R, Sun DF, Qi LQ (1998) Quasi-Newton bundle-type methods for nondifferentiable convex optimization. SIAM J Optim 8(2):583–603
12. Panin VM (1981) Linearization method for continuous minimax problems. Kibernetika 2:75–78

Convex-Simplex Algorithm

SIRIPHONG LAWPHONGPANICH
Naval Postgraduate School, Monterey, USA

MSC2000: 90C30

Article Outline

Keywords
See also
References

Keywords

Basic component; Convex-simplex algorithm; Extreme point; Feasible direction methods; First order Taylor series expansion; Generalized networks; Generalized upper bounding structure; Golden section method; Improving feasible direction; Inexact line search technique; Karush–Kuhn–Tucker conditions; Linearly independent; Line search problem; Linear program; Nonbasic component; Nonlinear network flow problems; Nonlinear programming; Nonsingular; Polyhedron; Pseudoconvex function; Reduced gradient algorithm; Second order approximation; Simplex algorithm; Superbasic variables

W.I. Zangwill [9] first proposed the *convex-simplex algorithm* (CSA) for the following problem:

$$\min_{x \in S} f(x), \quad (1)$$

where $f(x)$ is a *pseudoconvex function* on $\mathbf{R}^n$. The set S is a nonempty *polyhedron*, i.e., $S = \{x \in \mathbf{R}^n: Ax = b, x \geq 0\}$, A is a $m \times n$ matrix, and b is a vector in $\mathbf{R}^m$. For simplicity, S is assumed to be bounded also.

CSA belongs to a class of algorithms called *feasible direction methods*. Given an initial feasible solution, algorithms in this class solve problem (1) by iteratively generating an improving feasible direction that leads to another feasible solution with an improved objective value. The name 'convex-simplex' is to indicate that the algorithm generates improving feasible directions in manner similar to the simplex algorithm for linear programs. When $f(x)$ is linear, the algorithm is identical to the simplex algorithm. In general, a vector d is an *improving feasible direction* at a point, x, feasible to problem (1) if the following (sufficient) conditions hold
a) $\nabla f(x)^\top d < 0$,
b) $Ad = 0$, and
c) $d_j \geq 0$ if $x_j = 0$.

It follows from the first order Taylor series expansion of $f(x)$ that

$$f(x + \lambda d) = f(x) + \lambda \nabla f(x)^\top d + \lambda \|d\| \alpha(x; \lambda d),$$

where $\lim_{\lambda \to 0} \alpha(x; \lambda d) = 0$. Via the above expansion, condition a) implies that $f(x + \lambda d) < f(x)$ for a sufficiently small $\lambda > 0$, i.e., d leads to an improvement in the objective function. The remaining two conditions guarantee that d can produce a point in S. In particular, condition b) yields the following:

$$A(x + \lambda d) = Ax + \lambda Ad = Ax = b.$$

This shows that $x + \lambda d$ is always feasible with respect to the equality constraint. Next, each component of $x + \lambda d$ can be written as

$$x_i + \lambda d_i = \begin{cases} x_i + \lambda d_i & \text{if } x_i > 0, \\ \lambda d_i & \text{if } x_i = 0. \end{cases}$$

When λ is a sufficiently small positive number, $x_i + \lambda d_i$ remains nonnegative in the first case. For the second case, it follows directly from condition c) that $\lambda d_i \geq 0$ for all $\lambda > 0$. Thus, $x + \lambda d \in S$ when λ is sufficiently small.

To describe how CSA generates an improving feasible direction, let a_j denote the jth column of A. Also, assume that every m columns of A are linearly independent and every extreme point of S has m strictly positive components. Under these assumptions, every feasible solution has at least m positive components and at most $(n - m)$ zero components. Given a feasible solution x, let $I(x)$ be the set of indices for the m largest components of x. Then, A can be partitioned into $[B, N]$, where $B = [a_j : j \in I(x)]$ and $N = [a_j : j \notin I(x)]$. Similarly, $x^\top$ can be partitioned into $[x_B^\top, x_N^\top]$ where $x_B^\top$, the *basic component*, corresponds to components of x belonging to $I(x)$, and $x_N^\top$, the *nonbasic component*, corresponds to components not in $I(x)$. By the above assumptions, $x_B^\top > 0$ and B is nonsingular.

Partitioning the direction vector, $d^\top$, into its basic and nonbasic components, i.e., $[d_B^\top, d_N^\top]$, produces the following sequence of relationships:

$$Ad = 0,$$
$$Bd_B + Nd_N = 0,$$
$$d_B = -B^{-1} N d_N.$$

The last equality yields the following:

$$\nabla f(x)^\top d = \nabla_B f(x)^\top d_B + \nabla_N f(x)^\top d_N$$
$$= [\nabla_N f(x)^\top - \nabla_B f(x)^\top B^{-1} N] d_N$$
$$= r_N^\top d_N = \sum_{j \notin I(x)} r_j d_j,$$

where $r_N^\top \equiv \nabla_N f(x)^\top - \nabla_B f(x)^\top B^{-1} N$. In order for d to be an improving direction, $\nabla f(x)^\top d < 0$. There are several approaches to make this inner product negative. Each approach generates a different algorithm and all of which can be viewed as an extension or variant of the *reduced gradient algorithm* first proposed by P. Wolfe [8].

Like the simplex algorithm, CSA allows only one nonbasic component of d to be nonzero. In particular, let

$$j^+ = \underset{j \notin I(x)}{\mathrm{argmax}} \{-r_j : r_j < 0\}$$

and

$$j^- = \underset{j \notin I(x)}{\mathrm{argmax}} \{x_j r_j : x_j > 0, r_j > 0\}.$$

Then, CSA chooses d_N as follows. If $-r_{j^+} \geq x_{j^-} r_{j^-}$, then $d_{j^+} = 1$ and $d_j = 0$ for the remaining nonbasic components. This makes $\nabla f(x)^\top d = r_{j^+} < 0$. Otherwise, $d_{j^-} = -1$ instead and $\nabla f(x)^\top d = -r_{j^-} < 0$. Given d_N, the basic component can be computed using the relationship $d_B = -B^{-1} N d_N$.

When $r_N \geq 0$ and $r_N^\top x_N = 0$, the indices j^+ and j^- are undefined in the above construction. When this occurs, x is globally optimal and d_N is usually set to zero to indicate that there is no improving feasible direction. To demonstrate, it is sufficient to show that there exist vectors $\mu^\top = [\mu_B^\top, \mu_N^\top] \geq 0$ and ν (unrestricted) satisfying the following equations:

$$\nabla_B f(x) + B^\top \nu - \mu_B = 0,$$
$$\nabla_N f(x) + N^\top \nu - \mu_N = 0,$$
$$\mu_B^\top x_B = 0,$$
$$\mu_B^\top x_N = 0.$$

These equations are known as the *Karush–Kuhn–Tucker conditions* (see, e. g., [2]) and they are sufficient optimality conditions for problem (1). Letting $\mu_B = 0$, $\mu_N = \nabla_N f(x) - N(B^\top)^{-1} \nabla_N f(x)$, and $\nu = -(B^\top)^{-1} \nabla_B f(x)$ satisfies the first three conditions. Since $\mu_N = r_N$, the above assumptions concerning r_N imply that $\mu_N \geq 0$ and $\mu_N^\top x_N = 0$. Thus, the Karush–Kuhn–Tucker conditions hold and x must be globally optimal.

When $d_N \neq 0$, a better feasible point can be obtained from a solution to the following problem, typically called the *line search problem*:

$$\min_{0 \leq \lambda \leq \lambda_{\max}} f(x + \lambda d),$$

where $\lambda_{\max} = \min_j \{-x_j/d_j : d_j < 0\}$. This prevents components of $x + \lambda d$ from being negative. (If S is unbounded, then every component of d may be nonnegative and $\lambda_{\max} = \infty$.) Algorithms such as the bisection search, the golden section method, and an inexact line search technique (e. g., the *Armijo rule*, [1]) can efficiently solve the line search problem.

To summarize, CSA can be stated as follows:

0	Select $x^1 \in S$ and set $k = 1$.
1	Identify $I(x^k)$ and form the submatrices B and N. Compute $$r_N^\top = \nabla_N f(x^k)^\top - \nabla_B f(x^k)^\top B^{-1} N.$$
2	IF $r_N \geq 0$ and $r_N^\top x_N^\top = 0$, THEN stop and x^k is an optimal solution. ELSE, let $$j^+ = \underset{j \notin I(x^k)}{\mathrm{argmax}}\{-r_j : r_j < 0\},$$ $$j^- = \underset{j \notin I(x^k)}{\mathrm{argmax}}\{x_j^k r_j : x_j^k > 0, r_j > 0\}.$$ Set $d_N^k = 0$. IF $-r_{j^+} \geq x_{j^-}^k r_{j^-}$, THEN set $d_{j^+}^k = 1$. ELSE, set $d_{j^-}^k = -1$ instead. Set $d_B^k = B^{-1} N d_N^k$. Go to Step 3.
3	Set $\lambda_{\max} = \min_j\{-x_j^k/d_j^k : d_j^k < 0\}$ and compute $$\lambda^k = \underset{0 \leq \lambda \leq \lambda_{\max}}{\mathrm{argmin}} f(x^k + \lambda d^k).$$ Then, set $x^{k+1} = x^k + \lambda^k d^k$ and $k = k + 1$. Return to Step 1.

The convex-simplex algorithm (CSA)

The convergence proof for CSA is the same as that of the reduced gradient algorithm and follows standard arguments in nonlinear programming. Although CSA behaves like the simplex algorithm, CSA converges slowly when compared to other algorithms. This is due in part to the restriction that only one nonbasic component of the improving feasible direction can be nonzero. To accelerate CSA, B.A. Murtagh and M.A.

Saunder [5] used a second order approximation for the objective function and allowed several nonbasic components to be nonzero. The latter is often referred to as *superbasic variables*.

For other developments, S. Nguyen [6] and R.V. Helgason and J.L. Kennington [4] specialized CSA to nonlinear network flow problems. D.P. Rutenberg [7] (see also [3]) demonstrated that special techniques for solving linear programs with generalized network and generalized upper bounding structure also extend to CSA.

See also

- ▶ Lemke Method
- ▶ Linear Complementarity Problem
- ▶ Linear Programming
- ▶ Parametric Linear Programming: Cost Simplex Algorithm
- ▶ Sequential Simplex Method

References

1. Armijo L (1966) Minimization of functions having Lipschitz continuous first-partial derivatives. Pacific J Math 16:1–3
2. Bazaraa MS, Sherali HD, Shetty CM (1993) Nonlinear programming: Theory and algorithm. Wiley, New York
3. Hsia WS (1974) On Rutenberg's decomposition method. Managem Sci 21:10–12
4. Kennington JL, Helgason RV (1980) Algorithms for network programming. Wiley, New York
5. Murtagh BA, Saunder MA (1978) Large-scale linearly constrained optimization. Math Program 14:14–72
6. Nguyen S (1974) An algorithm for the traffic assignment problem. Transport Sci 8:203–216
7. Rutenberg DP (1970) Generalized networks, generalized upper bounding, and decomposition of the convex simplex method. Managem Sci 16:388–401
8. Wolfe P (1963) Methods of nonlinear programming. In: Graves RL, Wolfe P (eds) Recent Advances in Mathematical Programming. McGraw-Hill, New York
9. Zangwill WI (1967) The convex simplex method. Managem Sci 14:221–283

Copositive Optimization

IMMANUEL M. BOMZE
University of Vienna, Vienna, Austria

MSC2000: 90C20, 90C22, 90C26

Article Outline

References

Copositive optimization (or copositive programming, coined in [4]) is a special case of conic optimization that consists in extremizing a linear function over a (convex) cone subject to additional (inhomogeneous) linear (inequality or equality) constraints.

It is well known that the simplest class of hard problems in continuous optimization is that of quadratic optimization problems [20] – to extremize a (possibly indefinite) quadratic form $x^\top Q x$ over a polyhedron $\{x \in \mathbb{R}^n_+ : Ax = b\}$. Note that a linear term in the objective function can be removed by an affine transformation of the polyhedron. The number of local, non-global solutions to this problem may be exponential in the number of variables and/or constraints.

This class has a close connection to copositive optimization. The idea here is to linearize the quadratic form

$$x^\top Q x = \text{trace}(x^\top Q x) = \text{trace}(Q, xx^\top) = \langle Q, xx^\top \rangle$$

by introducing the new symmetric matrix variable $X = xx^\top$ and Frobenius duality $\langle X, Y \rangle = \text{trace}(X, Y)$. If $Ax \in \mathbb{R}^m_+$ for all $x \in \mathbb{R}^n_+$ and $b \in \mathbb{R}^m_+$, then the linear constraints can be squared, to arrive in a similar way at constraints of the form $\langle A_i, X \rangle = b_i^2$.

Now the set of all these X generated by feasible x is nonconvex since $\text{rank}(xx^\top) = 1$. The convex hull

$$\mathcal{K} = \text{conv}\left\{xx^\top : x \in \mathbb{R}^n_+\right\},$$

results in a convex matrix cone known as the cone of completely positive matrices since [14]; see [1]. Note that a similar construction dropping nonnegativity constraints leads to

$$\mathcal{P} = \text{conv}\left\{xx^\top : x \in \mathbb{R}^n\right\},$$

the cone of positive-semidefinite matrices, the basic set in semidefinite optimization (or semidefinite programming, SDP).

The first account of copositive optimization goes back to [4], who established a copositive representation of a subclass of particular interest, namely, in standard quadratic optimization (StQP). Here the feasible polyhedron is the standard simplex $\Delta = \{x \in \mathbb{R}^n_+ :$

$\sum_i x_i = 1\}$: this subclass is also NP-hard from the worst-case complexity but allows for a polynomial-time approximation scheme [3]. There can be up to $\approx 2^n/(1.25\sqrt{n})$ local nonglobal solutions. Now, with J the $n \times n$ all-ones matrix, we have

$$\min\{x^\top Q x : x \in \Delta\}$$
$$= \min\{\langle Q, X \rangle : \langle J, X \rangle = 1, X \in \mathcal{K}\}.$$

Note that the right-hand problem is convex, so there are no more local, nonglobal solutions. In addition, the objective function is now linear, and there is just one linear equality constraint. The complexity has been completely pushed into the feasibility condition $X \in \mathcal{K}$, which also shows that there are indeed convex minimization problems that cannot be solved easily.

The duality theory for conic optimization problems requires the dual cone $\mathcal{K}^*$ of $\mathcal{K}$ w.r.t. the Frobenius inner product $\langle \ldots, \rangle$, which is

$$\mathcal{K}^* = \{S \text{ symmetric } n \times n : \langle S, X \rangle \geq 0$$
$$\text{for all } X \in \mathcal{K}\}.$$

Here it can easily be shown that $\mathcal{K}^*$ coincides with the cone of copositive matrices, which justifies the terminology:

$$\mathcal{K}^* = \{S \text{ symmetric } n \times n : x^\top S x \geq 0 \text{ if } x \in \mathbb{R}_+^n\},$$

i. e., a matrix S is copositive [18] (most probably abbreviating "*conditionally positive*-semidefinite"), if S generates a quadratic form $x^\top S x$ taking no negative values over the positive orthant. The dual of the special program over $\mathcal{K}$ above is then

$$\max\{y : S = Q - yJ \in \mathcal{K}^*\},$$

a linear objective in just one variable y with the innocent-looking feasibility constraint $S \in \mathcal{K}^*$. This shows that checking membership of $\mathcal{K}^*$ (and, similarly, of $\mathcal{K}$) is already NP-hard, and there are many approaches to algorithmic copositivity detection; for recent developments see, e. g., [7] and references therein.

More generally, a typical primal-dual pair in copositive optimization (COP) is of the following form:

$$\inf\{\langle C, X \rangle : \langle A_i, X \rangle = b_i, \quad i = 1: m, X \in \mathcal{K}\}$$
$$\geq \sup\left\{\sum_i b_i y_i : y \in \mathbb{R}^m, S = C - y_i A_i \in \mathcal{K}^*\right\}.$$

The inequality above is just standard weak duality, but observe we have to use inf and sup since – as in general conic optimization – there may be problems with the attainability of either or both problems above, and likewise there could be a (finite or infinite) positive duality gap without any further conditions such as strict feasibility (Slater's condition). For the above representation of standard quadratic optimization problems, this is not the case:

$$\min\{\langle Q, X \rangle : \langle J, X \rangle = 1, \quad X \in \mathcal{K}\}$$
$$= \max\{y : S = Q - yJ \in \mathcal{K}^*\}.$$

But for a similar class arising in many applications, the multi-standard quadratic optimization problems [6], dual attainability is not guaranteed while the duality gap is zero – an intermediate form between weak and strong duality [25].

Recently Burer [8] showed a more general result: any mixed-binary quadratic optimization problem

$$\min\left\{\frac{1}{2}x^\top Q x + c^\top x : Ax = b, x \in \mathbb{R}_+^n, \right.$$
$$\left. x_j \in \{0, 1\}, \text{ all } j \in B\right\}$$

can (under mild conditions) be represented as COP:

$$\min\left\{\frac{1}{2}\langle \hat{Q}, \hat{X} \rangle : \mathcal{A}(\hat{X}) = \hat{b}, X \in \mathcal{K}\right\},$$

where $\hat{X}$ and $\hat{Q}$ are $(n+1) \times (n+1)$ matrices, and the size of $(\mathcal{A}, \hat{b})$ is polynomial in the size of (A, b).

Denote by $\mathcal{N} = \{N \text{ symmetric } n \times n : N_{ij} \geq 0 \text{ for all } i, j = 1: n\}$ the cone of nonnegative matrices. Then evidently

$$\mathcal{K} \subseteq \mathcal{P} \cap \mathcal{N} \subset \mathcal{P} + \mathcal{N} \subseteq \mathcal{K}^*,$$

which also shows that $\mathcal{K}$ never can be self-dual (note $(\mathcal{P} \cap \mathcal{N})^* = \mathcal{P}^* + \mathcal{N}^* = \mathcal{P} + \mathcal{N}$), unlike $\mathcal{P} = \mathcal{P}^*$ and $\mathcal{N} = \mathcal{N}^*$. For $n \geq 5$, A. Horn noted that the leftmost and the rightmost inclusion above is strict [10,14], so the middle sets $\mathcal{P} \cap \mathcal{N}$ and $\mathcal{P} + \mathcal{N}$ can only be used as tractable approximations for the intractable cones $\mathcal{K}$ and $\mathcal{K}^*$, respectively.

Copositive approximation hierarchies [3,15,21,22] start with $\mathcal{K}^{(0)} = \mathcal{P} + \mathcal{N}$ and consist of an increasing sequence $\mathcal{K}^{(r)}$ of cones satisfying $\cup_{r \geq 0} \mathcal{K}^{(r)} = \text{int } \mathcal{K}^*$, the cone of strictly copositive matrices, i. e., those that generate quadratic forms strictly positive over Δ. For instance, a higher-order approximation due to [21] squares the variables to get rid of sign constraints: $S \in \mathcal{K}^*$ if and only if $y^\top S y \geq 0$ for all y s.t. $y_i = x_i^2$, some $x \in \mathbb{R}^n$, and this is guaranteed if the n-variable polynomial of degree $2(r+2)$ in x,

$$p_S^{(r)}(x) = \Big(\sum x_i^2\Big)^r y^\top S y = \Big(\sum x_i^2\Big)^r \sum_{j,k} S_{jk} x_j^2 x_k^2,$$

is nonnegative for all $x \in \mathbb{R}^n$. But this holds in particular if
(a) $p_S^{(r)}$ has no negative coefficients; or if
(b) $p_S^{(r)}$ is a sum-of-squares (s.o.s.):

$$p_S^{(r)}(x) = \sum_i [f_i(x)]^2, \quad f_i \text{ some polynomials.}$$

This gives the approximation cones

$$C^{(r)} = \{S \text{ symmetric } n \times n \colon S \text{ satisfies (a)}\}$$

and

$$\mathcal{K}^{(r)} = \{S \text{ symmetric } n \times n \colon S \text{ satisfies (b)}\}.$$

While $C^{(r)}$ can be described by linear constraints on the entries of S, leading to LP formulations, the cones $\mathcal{K}^{(r)}$ need for their description linear matrix inequalities (LMIs), leading to SDP formulations. However, for large r both are also intractable as they generate problems on matrices of order $\mathcal{O}(n^{r+1} \times n^{r+1})$, see [3].

Copositive optimization has been receiving increasing attention also because many NP-hard combinatorial problems have a representation in this domain; we start with the historically first such representation, the maximum (weight) clique problem, which amounts to finding a largest (or heaviest) clique in an undirected graph G (with weights on the vertices). Using an StQP formulation going back to [19] and applying some regularization [2], the following copositive formulation was introduced in [4]:

$$1/\omega(G) = \min \{x^\top Q_G x \colon x \in \Delta\}$$
$$= \min \{\langle Q_G, X\rangle \colon \langle J, X\rangle = 1, X \in \mathcal{K}\},$$

where Q_G is a matrix derived from the adjacency matrix of G (and the weights). Taking the inverse $t = 1/y$ in the dual of the last problem above, we also arrive at the formulation of [9] (for the complementary graph):

$$\omega(G) = \min \{t \colon tQ_G - J \in \mathcal{K}^*\}.$$

Here $\omega(G)$ is the clique number of G, i.e., the size (weight) of a maximum (weight) clique in G. Replacing $\mathcal{K}^*$ with its zero-order approximation, we get a strengthening $\theta'(G)$ of the well-known Lovász bound $\theta(G)$ [16,17,26]:

$$\theta'(G) = \min \{t \colon tQ_G - J \in \mathcal{P} + \mathcal{N}\} \geq \omega(G),$$

while shrinking further the feasible set to $\mathcal{P}$, we finally arrive at the Lovász number $\theta(G)$ which – as $\theta'(G)$ – can be computed in polynomial time:

$$\theta(G) = \min \{t \colon tQ_G - J \in \mathcal{P}\} \geq \omega(G).$$

Strong duality yields, as above,

$$1/\theta'(G) = \min \{\langle Q_G, X\rangle \colon \langle J, X\rangle = 1, X \in \mathcal{P} \cap \mathcal{N}\},$$

and a recent improvement over $\theta'(G)$ adding a single valid linear cut motivated by the COP representation is

$$1/\theta^C(G) = \min \{\langle Q_G, X\rangle \colon \langle J, X\rangle = 1,$$
$$\langle C, X\rangle \geq 0, X \in \mathcal{P} \cap \mathcal{N}\} \geq 1/\theta'(G),$$

where $C \in \mathcal{K}^*$ is arbitrary: indeed, for any $X \in \mathcal{K}$ we then have $\langle C, X\rangle \geq 0$. See [5] for appropriate choices of C and results, and [9,13,15,22] for higher-order approximation alternatives, with a particular emphasis on SDP-based bounds on the clique number. Similar copositivity optimization approaches, among many others, were employed to obtain bounds on the (fractional) chromatic number of a graph [11,12], and graph partitioning and quadratic assignment problems [23,24].

References

1. Berman A, Shaked-Monderer N (2003) Completely positive matrices. World Scientific, Singapore
2. Bomze IM (1998) On standard quadratic optimization problems. J Glob Optim 13:369–387
3. Bomze IM, de Klerk E (2002) Solving standard quadratic optimization problems via linear, semidefinite and copositive programming. J Glob Optim 24:163–185
4. Bomze IM, Dür M, de Klerk E, Quist A, Roos C, Terlaky T (2000) On copositive programming and standard quadratic optimization problems. J Glob Optim 18:301–320
5. Bomze IM, Frommlet F, Locatelli M (2007) The first cut is the cheapest: improving SDP bounds for the clique number via copositivity. Technical Report TR-ISDS 2007-05,

University of Vienna, submitted. Available at http://www.optimization-online.org/DB_HTML/2006/08/1443.html. Accessed Nov 2007
6. Bomze IM, Schachinger W (2007) Multi-standard quadratic optimization problems. Technical Report TR-ISDS 2007-12, University of Vienna, submitted. Available at http://www.optimization-online.org/DB_HTML/2007/11/1843.html. Accessed Mar 2008
7. Bundfuss S, Dür M (2008) Algorithmic copositivity detection by simplicial partition. Linear Algebr Appl 428:1511–1523
8. Burer S (2006) On the copositive representation of binary and continuous nonconvex quadratic programs. Manuscript, Department of Management Sciences, University of Iowa, October (2006). Available at http://www.optimization-online.org/DB_HTML/2006/10/1501.html. Accessed Nov 2007
9. de Klerk E, Pasechnik DV (2002) Approximation of the stability number of a graph via copositive programming. SIAM J Optim 12:875–892
10. Diananda PH (1967) On non-negative forms in real variables some or all of which are non-negative. Proc Cambridge Philos Soc 58:17–25
11. Dukanović I, Rendl F (2007) Semidefinite programming relaxations for graph coloring and maximal clique problems. Math Programm 109:345–365
12. Gvozdenović N, Laurent M (2008) The operator Ψ for the chromatic number of a graph. To appear in SIAM J Optim. Available at http://www.optimization-online.org/DB_HTML/2007/02/1592.html. Accessed Mar 2008
13. Gvozdenović N, Laurent M (2006) Semidefinite bounds for the stability number of a graph via sums of squares of polynomials. Lecture Notes in Computer Science, vol 3509/2005, pp 136–151. Integer Programming and Combinatorial Optimization: 11th International IPCO Conference
14. Hall M, Newman M (1963) Copositive and completely positive quadratic forms. Proc Cambridge Philos Soc 59:329–339
15. Lasserre JB (2001) Global optimization with polynomials and the problem of moments. SIAM J Optim 11:796–817
16. Lovász L (1979) On the Shannon capacity of a graph. IEEE Trans Inf Theory IT 25:1–7
17. McEliece RJ, Rodemich ER, Rumsey HC (1978) The Lovász' bound and some generalizations. J Combinat Inf Syst Sci 3:134–152
18. Motzkin TS (1952) Copositive quadratic forms. Natl Bur Stand Rep 1818 11–22
19. Motzkin TS, Straus EG (1965) Maxima for graphs and a new proof of a theorem of Turán. Can J Math 17:533–540
20. Murty KG, Kabadi SN (1987) Some NP-complete problems in quadratic and linear programming. Math Programm 39:117–129
21. Parrilo P (2003) Semidefinite programming relaxations for semi-algebraic problems. Math Programm 696B:293–320
22. Peña J, Vera J, Zuluaga L (2007) Computing the stability number of a graph via linear and semidefinite programming. SIAM J Optim 18:87–105
23. Povh J, Rendl F (2006) Copositive and semidefinite relaxations of the quadratic assignment problem. Univ. of Klagenfurt, submitted. Available at http://www.optimization-online.org/DB_HTML/2006/10/1502.html. Accessed Nov 2007
24. Povh J, Rendl F (2007) A copositive programming approach to graph partitioning. SIAM J Optim 18:223–241
25. Schachinger W, Bomze IM (2007) A conic duality Frank–Wolfe type theorem via exact penalization in quadratic optimization. Technical Report 2006-10, ISDS, University of Vienna. to appear in Math Oper Res. Available at http://www.optimization-online.org/DB_HTML/2007/02/1596.html. Accessed Nov 2007
26. Schrijver A (1979) A comparison of the Delsarte and Lovasz bounds. IEEE Trans Inf Theory IT 25:425–429

Copositive Programming

STANISLAV BUSYGIN

Department of Industrial and Systems Engineering, University of Florida, Gainesville, USA

MSC2000: 90C25, 90C22

Article Outline

Introduction
Applications
  Complexity of Copositive Programming
Models
  Approximating C_n with Linear Matrix Inequalities
References

Introduction

Let us denote the set of $n \times n$ real symmetric matrices by

$$S_n = \{X \in \mathbb{R}^{n \times n}, X = X^T\}.$$

We will be considering the following subsets of S_n:

- The $n \times n$ symmetric *positive semidefinite* matrices

$$S_n^+ = \{X \in S_n, y^T X y \geq 0 \, \forall y \in \mathbb{R}^n\};$$

- The $n \times n$ symmetric *copositive* matrices

$$C_n = \{X \in S_n, y^T X y \geq 0 \, \forall y \in \mathbb{R}^n, y \geq 0\};$$

- The $n \times n$ symmetric *completely positive* matrices

$$C_n^* = \{X = \sum_{i=1}^{k} y_i y_i^T, y_i \in \mathbb{R}^n, y_i \geq 0 \, (i = 1, \ldots, k)\};$$

- The $n \times n$ symmetric *nonnegative* matrices

$$\mathcal{N}_n = \{X \in S_n, X \geq 0\}.$$

It is easy to see that all these sets are *convex cones* (that is, if X and Y belong to one of these sets, then cX, for any $c \geq 0$, and $\alpha X + (1-\alpha)Y$, for any $0 \leq \alpha \leq 1$, also do so).

We will denote the Euclidean inner product of $A \in S_n$ and $B \in S_n$ by

$$A \bullet B = \sum_{i=1}^{n}\sum_{j=1}^{n} a_{ij}b_{ij} = \text{trace}(AB).$$

For an arbitrary cone of matrices $\mathcal{K}$, we define its *dual cone* $\mathcal{K}^*$ as

$$\mathcal{K}^* = \{Y | X \bullet Y \geq 0, \forall X \in \mathcal{K}\}.$$

The cone of positive semidefinite matrices S_n^+ is dual to itself, and it is easy to see that the cone of copositive matrices C_n and the cone of completely positive matrices C_n^* are dual to each other (generally, $\mathcal{K}^{**} = \mathcal{K}$).

For a given cone of $n \times n$ matrices $\mathcal{K}_n$, and its dual cone $\mathcal{K}_n^*$, we define a pair of conic linear programs called, correspondingly, *primal* and *dual*:

$$p^* = \inf_X \{C \bullet X : A_i \bullet X = b_i \, (i = 1, \ldots, m),$$

$$X \in \mathcal{K}_n\}; \quad (1)$$

$$d^* = \sup_y \{b^T y : C - \sum_{i=1}^{m} y_i A_i + S = C, S \in \mathcal{K}_n^*\}. \quad (2)$$

When $\mathcal{K}_n = \mathcal{N}_n$, we refer to *linear programming*, when $\mathcal{K}_n = S_n^+$, we refer to *semidefinite programming*, and when $\mathcal{K}_n = C_n$, we refer to *copositive programming*.

The *conic duality theorem* (see, e. g., [9]) establishes the duality relation between (1) and (2):

Theorem 1 (Conic Duality Theorem) *If there exists an interior feasible solution $X^0 \in int(\mathcal{K})$ of (1) and a feasible solution of (2), then $p^* = d^*$ and the supremum in (2) is attained. Similarly, if there exist feasible y^0 and S^0 for (2), where $S^0 \in int(\mathcal{K}^*)$, and a feasible solution for (1), then $p^* = d^*$ and the infimum in (1) is attained.*

It is well-known that optimization over the cones S_n^+ and $\mathcal{N}_n$ can be done in polynomial time (in sense of computing ε-optimal solution), but copositive programming is NP-hard as we will see below.

Applications

Copositive matrices have a great variety of applications in mathematics and, especially, in optimization. They play an essential part in characterization of local solutions of constrained optimization problems [5], including the linear complementarity problem. In [1,2,8] the authors used copositivity to improve convex relaxation bounds for quadratic programming problems. Generally, convex relaxations are the underlying basis of many crucial results in robustness analysis. For example, copositive matrices have been used in the stability analysis of piecewise linear control systems (in context of using piecewise quadratic Lyapunov functions [4]).

Complexity of Copositive Programming

Copositive programming can be easily shown to be NP-hard by reduction from the maximum independent set problem. In [3], the authors established the following theorem:

Theorem 2 *Let $G(V, E)$ be a given graph with $V = \{1 \ldots n\}$. Then the maximum independent set size of G is the optimum value of the following program:*

$$\alpha(G) = \max \, O_n \bullet X \quad (3)$$

subject to

$$\begin{cases} x_{ij} = 0, \, (i,j) \in E \\ \text{trace}(X) = 1 \\ X \in C_n^* \end{cases} \quad (4)$$

where O_n is the all-one $n \times n$ matrix.

Proof. Extreme rays of C_n^* are rank-one matrices of the form xx^T for nonnegative $x \in \mathbb{R}^n$. Then, considering the convex cone

$$C_G = \{X \in C_n^* : x_{ij} = 0, (i,j) \in E\},$$

we can conclude that its extreme rays are of the form xx^T, where the nonnegative $x \in \mathbb{R}^n$ supports an independent set (i.e., the set $\{i : x_i > 0\}$ is independent). Therefore, the extreme points of the set defined by (4) are given by the intersection of the extreme rays of C_G with the hyperplane $O_n \bullet X = 1$.

Since the optimum value of a linear function over a convex set is attained at an extreme point, there is an optimum solution of the form

$$X^* = x^* x^{*T}, \ x^* \in \mathbb{R}^n, \ x^* \geq 0, \|x^*\| = 1,$$

and where x^* supports an independent set. Therefore, we can reformulate the program (3) as

$$\max_x \left(\sum_{i=1}^n x_i\right)^2, \ \|x\| = 1,$$

$$x \geq 0, \ x_i x_j > 0 \Rightarrow (i, j) \notin E.$$

Then, it is easy to see that the maximum is attained when x supports a maximum independent set and all $x_i > 0$ are equal to $1/\alpha(G)$. This provides the optimum value to the program (3) equal to $\alpha(G)$. QED.

Since $X \in C_n^*$ is always nonnegative, we can reduce the set of constraints $x_{ij} = 0, (i, j) \in E$ in (4) to a single constraint $A \bullet X = 0$. Thus, the following copositive program is dual to (3), (4):

$$\alpha(G) = \min_{\lambda, y \in \mathbb{R}} \{\lambda : \lambda I + yA - O_n = Q, \ Q \in C_n\}.$$

Therefore, the maximum independent set problem is reducible to copositive programming. See also [1,2] for reduction of the standard quadratic optimization problem to copositive programming.

Furthermore, it can be shown that checking if a given matrix is not copositve is NP-complete [5] and, hence, checking matrix copositivity is co-NP-complete.

Models

Approximating C_n with Linear Matrix Inequalities

While, in general, there is no polynomial-time verifiable certificate of copositivity, unless co-NP $=$ NP, in many cases it is still possible to show by a short argument that a matrix is copositive. For instance, if the matrix M can be represented as sum of a positive semidefinite matrix $S \in S_n^+$ and a nonnegative matrix $N \in \mathcal{N}_n$, then it follows that $M \in C_n$. Hence, we can obtain a semidefinite relaxation of a copositive program over M introducing the linear matrix constraints:

$$\begin{cases} M = S + N \\ N \geq 0, \ S \in S_n^+ \end{cases}$$

Parrilo showed in [6] that using sufficiently large systems of linear matrix inequalities, one can approximate the copositive cone C_n to any desired accuracy.

Obviously, copositivity of the matrix M is equivalent to (global) nonnegativity of the fourth-degree form:

$$P(x) = (x \circ x)^T M (x \circ x)$$

$$= \sum_{i=1}^n \sum_{j=1}^n M_{ij} x_i^2 x_j^2 \geq 0, \ x \in \mathbb{R}^n, \quad (5)$$

where "$\circ$" denotes the componentwise (Hadamard) product. It is shown in [6] that the mentioned decomposition into positive semidefinite and nonnegative matrices exists if and only if $P(x)$ can be represented as sum of squares. Higher-order sufficient conditions for copositivity proposed by Parrilo in [6] correspond to checking whether the polynomial

$$P^{(r)}(x) = \left(\sum_{i=1}^n x_i^2\right) P(x) \quad (6)$$

has a sum-of-squares decomposition (or – a weaker condition – whether $P^{(r)}(x)$ has only nonnegative coefficients). These conditions can be expressed via linear matrix inequalities over $n^r \times n^r$ symmetric matrices. In particular, for $r = 1$, Parrilo showed that the existence of a sum-of-squares decomposition of $P^{(1)}(x)$ is equivalent to feasibility of the following system (see also [3]):

$$\begin{cases} M - M^{(i)} \in S_n^+, & i = 1, \ldots, n, \\ M_{ii}^{(i)} = 0, & i = 1, \ldots, n, \\ M_{jj}^{(i)} + 2M_{ij}^{(j)} = 0, & i \neq j, \\ M_{jk}^{(i)} + M_{ik}^{(j)} + M_{ij}^{(k)} \geq 0, & i < j < k, \end{cases}$$

where $M^{(i)}$ ($i = 1, \ldots, n$) are symmetric matrices.

With sufficiently large r, the convergence to the copositivity constraint on M is guaranteed by the famous theorem of Pólya [7]:

Theorem 3 (Pólya) *Let f be a homogeneous polynomial which is positive on the simplex*

$$\Delta = \left\{ x \in \mathbb{R}^n : \sum_{i=1}^{n} x_i = 1, \ x \geq 0 \right\}.$$

Then, for a sufficiently large N, all the coefficients of the polynomial

$$\left(\sum_{i=1}^{n} x_i \right)^N f(x)$$

are positive.

References

1. Bomze IM, de Klerk E (2002) Solving standard quadratic optimization problems via linear, semidefinite and copositive programming. J Glob Optim 24:163–185
2. Bomze IM, Dür M, de Klerk E, Roos C, Quist AJ, Terlaky T (2000) On copositive programming and standard quadratic optimization problems. J Glob Optim 18:301–320
3. de Klerk E, Pasechnik D (2001) Approximation of the stability number of a graph via copositive programming. SIAM J Optim 12(4):875–892
4. Johansson M (1999) Piecewise linear control systems. PhD thesis, Lund Institute of Technology, Lund
5. Murty KG, Kabadi SN (1987) Some *NP*-complete problems in quadratic and nonlinear programming. Math Prog 39:117–129
6. Parrilo PA (2000) Structured semidefinite programs and semialgebraic geometry methods in robustness and optimization. PhD thesis, California Institute of Technology, Pasadena, CA, http://www.mit.edu/~parrilo/pubs
7. Pólya G (1928) Über positive Darstellung von Polynomen, Vierteljschr. Naturforsch Ges Zürich 73:141–145 (Collected Papers, vol 2, MIT Press, Cambridge, MA, London, 1974, 309–313)
8. Quist AJ, de Klerk E, Roos C, Terlaky T (1998) Copositive relaxation for general quadratic programming. Optim Method Softw 9:185–208
9. Renegar J (2001) A Mathematical View of Interior-Point Methods in Convex Optimization. SIAM, Philadelphia

Cost Approximation Algorithms
CA Algorithms

MICHAEL PATRIKSSON
Department Math., Chalmers University Technol., Göteborg, Sweden

MSC2000: 90C30

Article Outline

Keywords
Instances of the CA Algorithm
   Linearization Methods
   Regularization, Splitting and Proximal Point Methods
   Perturbation Methods
Variational Inequality Problems
Descent Properties
   Optimization
   Variational Inequality Problems
Steplength Rules
Convergence Properties
Decomposition CA Algorithms
   Sequential Decomposition
   Synchronized Parallel Decomposition
   Asynchronous Parallel Decomposition
See also
References

Keywords

Linearization methods; Gradient projection; Quasi-Newton; Frank–Wolfe; Sequential quadratic programming; Operator splitting; Proximal point; Levenberg–Marquardt; Auxiliary problem principle; Subgradient optimization; Variational inequality problem; Merit function; Cartesian product; Gauss–Seidel; Jacobi; Asynchronous computation

The notion of cost approximation (CA) was created in the thesis [39], to describe the construction of the subproblem of a class of iterative methods in mathematical programming. In order to explain the notion of CA, we will consider the following conceptual problem (the full generality of the algorithm is explained in detail in [45] and in [37,38,40,42,43,44,46]):

$$\begin{cases} \min & T(x) := f(x) + u(x), \\ \text{s.t.} & x \in X, \end{cases} \quad (1)$$

where $X \subseteq \mathbf{R}^n$ is nonempty, closed and convex, $u: \mathbf{R}^n \to \mathbf{R} \cup \{+\infty\}$ is lower semicontinuous (l.s.c.), proper and convex, and $f: \mathbf{R}^n \to \mathbf{R} \cup \{+\infty\}$ is continuously differentiable (for short, in C^1) on dom $u \cap X$, where dom denotes 'effective domain'. This problem is general enough to cover convex optimization ($f = 0$), unconstrained optimization ($f = 0$ and $X = \mathbf{R}^n$), and differentiable constrained optimization ($u = 0$). We note that if int (dom u) $\cap$ X is nonempty, then any locally optimal solution x^* satisfies the inclusion

$$-\nabla f(x^*) \in \partial u(x^*) + N_X(x^*), \qquad (2)$$

where N_X is the normal cone operator for X and ∂u is the subdifferential mapping for u. Equivalently, by the definitions of these two operators,

$$\nabla f(x^*)^\top (x - x^*) + u(x) - u(x^*) \geq 0, \quad x \in X.$$

The CA algorithm was devised in order to place a number of existing algorithms for (1) in a common framework, thereby facilitating comparisons, for example, between their convergence properties. In short, the method works iteratively as follows. Note that, from (2), we seek a zero of the mapping $[\nabla f + \partial u + N_X]$. Given an iterate, $x^t \in$ dom $u \cap X$, this mapping is approximated by a monotone mapping, constructed so that a zero of which is easier to find. Such a point, y^t, is then utilized in the search for a new iterate, x^{t+1}, having the property that the value of some merit function for (1) is reduced sufficiently, for example through a line search in T along the direction of $d^t := y^t - x^t$.

Instances of the CA Algorithm

To obtain a monotone approximating mapping, we introduce a monotone mapping Φ^t: dom $u \cap X \to \mathbf{R}^n$, which replaces the (possibly nonmonotone) mapping ∇f; by subtracting off the error at x^t, $[\Phi^t - \nabla f](x^t)$, from Φ^t, so that the resulting mapping becomes $[\Phi^t + \partial u + N_X] + [\nabla f - \Phi^t](x^t)$, the CA subproblem becomes the inclusion

$$[\Phi^t + \partial u + N_X](y^t) + [\nabla f - \Phi^t](x^t) \ni 0^n. \qquad (3)$$

We immediately reach an interesting fixed-point characterization of the solutions to (2):

Theorem 1 *(Fixed-point, [45]) The point x^t solves (2) if and only if $y^t = x^t$ solves (3).*

This result is a natural starting point for devising stopping criteria for an algorithm.

Assume now that $\Phi^t \equiv \nabla \varphi^t$ for a convex function φ^t. We may then derive the inclusion equivalently as follows. At x^t, we replace f with the function φ^t, and subtract off the linearization of the error at x^t; the subproblem objective function then becomes

$$T_{\varphi^t}(y) := \varphi^t(y) + u(y) + [\nabla f(x^t) - \nabla \varphi^t(x^t)]^\top y.$$

It is straightforward to establish that (3) is the optimality conditions for the convex problem of minimizing T_{φ^t} over X.

Linearization Methods

Our first example instances utilize Taylor expansions of f to construct the approximations.

Let $u = 0$ and $X = \mathbf{R}^n$. Let $\Phi^t(y) := (1/\gamma_t) Q^t y$, where $\gamma_t > 0$ and Q^t is a symmetric and positive definite mapping in $\mathbf{R}^{n \times n}$. The inclusion (3) reduces to

$$\nabla f(x^t) + \frac{1}{\gamma_t} Q^t (y^t - x^t) = 0^n,$$

that is, $y^t = x^t - \gamma_t (Q^t)^{-1} \nabla f(x^t)$. The direction of $y^t - x^t$, $d^t := -\gamma_t (Q^t)^{-1} \nabla f(x^t)$, is the search direction of the class of *deflected gradient methods*, which includes the *steepest descent method* ($Q^t := I^n$, the identity matrix) and *quasi-Newton methods* (Q^t equals (an approximation of) $\nabla^2 f(x^t)$, if positive definite). (See further [5,35,47,50].)

In the presence of constraints, this choice of Φ^t leads to $y^t = P_X^{Q^t}[x^t - \gamma_t (Q^t)^{-1} \nabla f(x^t)]$, where $P_X^{Q^t}[\cdot]$ denotes the projection onto X with respect to the norm $\|z\|_{Q^t} := \sqrt{z^\top Q^t z}$. Among the algorithms in this class we find the *gradient projection algorithm* ($Q^t := I^n$) and *Newton's method* ($Q^t := \nabla^2 f(x^t)$, $\gamma_t := 1$). (See [5,19,27,50].)

A first order Taylor expansion of f is obtained from choosing $\varphi^t(y) := 0$; this results in $T_{\varphi^t}(y) = \nabla f(x^t)^\top y$ (if $u = 0$ is still assumed), which is the subproblem objective in the Frank–Wolfe algorithm ([5,17]; cf. also ▶ Frank–Wolfe algorithm).

We next provide the first example of the very useful fact that the result of the cost approximation (in the above examples a linearization), leads to different approximations of the original problem, and ultimately to different algorithms, depending on which *representation* of the problem to one applies the cost approximation.

Consider the problem

$$\begin{cases} \min & f(x) \\ \text{s.t.} & g_i(x) = 0, \quad i = 1, \ldots, \ell, \end{cases} \qquad (4)$$

where f and g_i, $i = 1, \ldots, \ell$, are functions in C^2. We may associate this problem with its first order optimality conditions, which in this special case is

$$F(x^*, \lambda^*) := \begin{pmatrix} \nabla_x L(x^*, \lambda^*) \\ -\nabla_\lambda L(x^*, \lambda^*) \end{pmatrix} = \begin{pmatrix} 0^n \\ 0^\ell \end{pmatrix}, \quad (5)$$

where $\lambda \in \mathbf{R}^\ell$ is the vector of Lagrange multipliers for the constraints in (4), and $L(x, \lambda) := f(x) + \lambda^\mathsf{T} g(x)$ is the associated Lagrangian function. We consider using Newton's method for this system, and therefore introduce a (primal-dual) mapping $\Phi: \mathbf{R}^{2(n+\ell)} \to \mathbf{R}^{n+\ell}$ of the form

$$\Phi((y, p), (x, \lambda)) := \nabla F(x, \lambda) \begin{pmatrix} y \\ p \end{pmatrix}$$
$$= \begin{pmatrix} \nabla_x^2 L(x, \lambda) & \nabla g(x)^\mathsf{T} \\ -\nabla g(x) & 0 \end{pmatrix} \begin{pmatrix} y \\ p \end{pmatrix}.$$

The resulting CA subproblem in (y, p) can be written as the following linear system:

$$\nabla_x^2 L(x, \lambda)(y - x) + \nabla g(x)^\mathsf{T} p = 0^n,$$
$$\nabla g(x)(y - x) = -g(x);$$

this system constitutes the first order optimality conditions for (e. g., [4, Sec. 10.4])

$$\begin{cases} \min & f(x) + \nabla f(x)^\mathsf{T}(y - x) \\ & + \tfrac{1}{2}(y - x)^\mathsf{T} \nabla_x^2 L(x, \lambda)(y - x) \\ \text{s.t.} & g(x) + \nabla g(x)^\mathsf{T}(y - x) = 0^\ell, \end{cases}$$

where we have added some fixed terms in the objective function for clarity. This is the generic subproblem of *sequential quadratic programming* (SQP) methods for the solution of (4); see, for example, [5,16].

Regularization, Splitting and Proximal Point Methods

We assume now that $f := f_1 + f_2$, where f_1 is convex on dom $u \cap X$, and rewrite the cost mapping as

$$[\nabla f + \partial u + N_X] = [\nabla \varphi^t + \nabla f_1 + \partial u + N_X] \\ - [\nabla \varphi^t - \nabla f_2].$$

The CA subproblem is, as usual, derived by fixing the second term at x^t; the difference to the original setup is that we have here performed an *operator splitting* in the mapping ∇f to keep an additive part from being approximated. (Note that such a splitting can always be found by first choosing f_1 as a convex function, and then define $f_2 := f - f_1$. Note also that we can derive this subproblem from the original derivation by simply redefining $\varphi^t := \varphi^t + f_1$.) We shall proceed to derive a few algorithms from the literature.

Consider choosing $\varphi^t(y) = 1/(2\gamma_t) \| y - x^t \|^2$, $\gamma_t > 0$. If $f_2 = 0$, then we obtain the subproblem objective $T_{\varphi^t}(y) = T(y) + 1/(2\gamma_t) \| y - x^t \|^2$, which is the subproblem in the *proximal point algorithm* (e. g., [32,33,34,51,52]). This is the most classical algorithm among the regularization methods. More general choices of strictly convex functions φ^t are of course possible, leading for example to the class of regularization methods based on Bregman functions ([9,14,22]) and ψ-divergence functions ([23,54]). If, on the other hand, $f_1 = 0$, then we obtain the gradient projection algorithm if also $u = 0$.

We can also construct algorithms in between these two extremes, yielding a true operator splitting. If both f_1 and f_2 are nonzero, choosing $\varphi^t = 0$ defines a *partial linearization* ([25]) of the original objective, wherein only f_2 is linearized. Letting $x = (x_1^\mathsf{T}, x_2^\mathsf{T})^\mathsf{T}$, the choice $\varphi^t(y) = 1/(2\gamma_t) \| y_1 - x_1^t \|^2$ leads to the *partial proximal point algorithm* ([7,20]); choosing $\varphi^t(y) = f(y_1, x_2^t)$ leads to a linearization of f in the variables x_2.

Several well-known methods can be derived either directly as CA algorithms, or as *inexact proximal point algorithms*. For example, the *Levenberg–Marquardt algorithm* ([5,49]), which is a Newton-like algorithm wherein a scaled diagonal matrix is added to the Hessian matrix in order to make the resulting matrix positive definite, is the result of solving the proximal point subproblem with one iteration of a Newton algorithm. Further, the *extra-gradient algorithm* of [24] is the result of instead applying one iteration of the gradient projection algorithm to the proximal point subproblem.

The perhaps most well-known splitting algorithm is otherwise the class of *matrix splitting methods in quadratic programming* (e. g., [28,29,35,36]). In a quadratic programming problem, we have

$$f(x) = \frac{1}{2} x^\mathsf{T} A x + q^\mathsf{T} x,$$

where $A \in \mathbf{R}^{n \times n}$. A splitting (A_1^t, A_2^t) of this matrix is one for which $A = A_1^t + A_2^t$, and it is further termed regular if $A_1^t - A_2^t$ is positive definite. Matrix splitting methods correspond to choosing

$$f_1(x) = \frac{1}{2} x^\top A_1^t x,$$

and results in the CA subproblem mapping $y \mapsto A_1^t y + [A_2^t x^t + q]$, which obviously is monotone whenever A_1^t was chosen positive semidefinite.

Due to the fact that proximal point and splitting methods have dual interpretations as augmented Lagrangian algorithms ([51]), a large class of multiplier methods is included among the CA algorithms. See [45, Chapt. 3.2–3.3] for more details.

Perturbation Methods

All the above algorithms assume that
i) the mappings ∂u and N_X are left intact; and
ii) the CA subproblem has the fixed-point property of Theorem 1.

We here relax these assumptions, and are then able to derive subgradient algorithms as well as perturbed CA algorithms which include both regularization algorithms and exterior penalty algorithms.

Let $[\Phi^t + N_X] + [\nabla f + \partial u + \Phi^t]$ represent the original mapping, having moved ∂u to the second term. Then by letting any element $\xi_u(x^t) \in \partial u(x^t)$ represent this point-to-set mapping at x^t, we reach the subproblem mapping of the *auxiliary problem principle* of [12]. Further letting $\Phi^t(y) = (1/\gamma_t)[y - x^t]$ yields the subproblem in the classical *subgradient optimization* scheme ([48,53]), where, assuming further that $f = 0$, $y^t := P_X[x^t - \gamma_t \xi_u(x^t)]$. (Typically, $\ell_t := 1$ is taken.)

Let again $[\Phi^t + \partial u + N_X] + [\nabla f + \Phi^t]$ represent the original problem mapping, but further let u be replaced by an *epiconvergent sequence* $\{u^t\}$ of l.s.c., proper and convex functions. An example of an epiconvergent sequence of convex functions is provided by convex exterior penalty functions. In this way, we can construct CA algorithm that approximate the objective function and simultaneously replace some of the constraints of the problem with exterior penalties. See [3,13] for example methods of this type.

One important class of regularization methods takes as the subproblem mapping $[\Phi^t + \nabla f + \partial u + N_X]$, where Φ^t is usually taken to be strongly monotone (cf. (12)). This subproblem mapping evidently does not have the fixed-point property, as it is not identical to the original one at x^t unless $\Phi^t(x^t) = 0^n$ holds. In order to ensure convergence, we must therefore force the sequence $\{\Phi^t\}$ of mappings to tend to zero; this is typically done by constructing the sequence as $\Phi^t := (1/\gamma_t)\Phi$ for a fixed mapping Φ and for a sequence of $\gamma_t > 0$ constructed so that $\{\gamma_t\} \to \infty$ holds. For this class of algorithms, F. Browder [10] has established convergence to a unique limit point x^* which satisfies $-\Phi(x^*) \in N_{X^*}(x^*)$, where X^* is the solution set of (2). The origin of this class of methods is the work of A.N. Tikhonov [55] for *ill-posed problems*, that is, problems with multiple solutions. The classical regularization mapping is the scaled identity mapping, $\Phi^t(y) := (1/\gamma_t)[y]$, which leads to least squares (least norm) solutions. See further [49,56].

Variational Inequality Problems

Consider the following extension of (2):

$$-F(x^*) \in \partial u(x^*) + N_X(x^*), \qquad (6)$$

where $F: X \to \mathbf{R}^n$ is a continuous mapping on X. When $F = \nabla f$ we have the situation in (2), and also in the case when $F(x, y) = (\nabla_x \Pi(x, y)^\top, -\nabla_y \Pi(x, y)^\top)^\top$ holds for some saddle function Π on some convex product set $X \times Y$ (cf. (5)), the *variational inequality problem* (6) has a direct interpretation as the necessary optimality conditions for an optimization problem. In other cases, however, a merit function (or, objective function), for the problem (6) is not immediately available. We will derive a class of suitable merit functions below.

Given the convex function $\varphi: \text{dom } u \cap X \to \mathbf{R}$ in C^1 on dom $u \cap X$, we introduce the function

$$\psi(x) := \sup_{y \in X} L(y, x), \quad x \in \text{dom } u \cap X, \qquad (7)$$

where

$$L(y, x) := u(x) - u(y) + \varphi(x) - \varphi(y)$$
$$+ [F(x) - \nabla \varphi(x)]^\top (x - y). \qquad (8)$$

We introduce the optimization problem

$$\min_{x \in X} \psi(x). \qquad (9)$$

Theorem 2 *(Gap function, [45])* For any $x \in X$, $\psi(x) \geq 0$ holds. Further, $\psi(x) = 0$ if and only if x solves (6). Hence, the solution set of (6) (if nonempty) is identical to that of the optimization problem (9), and the optimal value is zero.

The Theorem shows that the CA subproblem defines an auxiliary function ψ which measures the violation of (6), and which can be used (directly or indirectly) as a merit function in an algorithm.

To immediately illustrate the possible use of this result, let us consider the extension of Newton's method to the solution of (6). Let $x \in \text{dom } u \cap X$, and consider the following cost approximating mapping: $y \mapsto \Phi(y, x) := \nabla F(x)(y - x)$. The CA subproblem then is the linearized variational inequality problem of finding $y \in \text{dom } u \cap X$ such that

$$[F(x) + \nabla F(x)^\top (y-x)]^\top (z-y) + u(z) - u(y) \geq 0,$$
$$\forall z \in X. \quad (10)$$

Assuming that x is not a solution to (6), we are interested in utilizing the direction $d := y - x$ in a line search based on a merit function. We will utilize the *primal gap function* ([2,62]) for this purpose, which corresponds to the choice $\varphi := 0$ in the definition of ψ. We denote the primal gap function by ψ_P. Let w be an arbitrary solution to its inner problem, that is, $\psi_P(x) = u(x) - u(w) + F(x)^\top (x - w)$. The steplength is chosen such that the value of ψ_P decreases sufficiently; to show that this is possible, we use Danskin's theorem and the variational inequality (10) with $z = w$ to obtain (the maximum is taken over all w defining $\psi_P(x)$)

$$\psi_P'(x; d)$$
$$:= \max_w \left\{ [F(x) + \nabla F(x)^\top (x - w)]^\top d + u'(x; d) \right\}$$
$$\leq -\psi_P(x) - d^\top \nabla F(x)^\top d,$$

which shows that d defines a direction of descent with respect to the merit function ψ_P at all points outside the solution set, whenever F is monotone and in C^1 on dom $u \cap X$. (See also [30] for convergence rate results.) So, if Newton's method is supplied with a line search with respect to the primal gap function, it is globally convergent for the solution of variational inequality problems.

The merit function ψ and the optimization problem (9) cover several examples previously considered for the solution of (6).

The primal gap function, as typically all other gap functions, is nonconvex, and further also nondifferentiable in general. In order to utilize methods from differentiable optimization, we consider letting φ be strictly convex, whence the solution y^t to the inner problem (7) is unique. Under the additional assumption that dom $u \cap X$ is bounded and that u is in C^1 on this set, ψ is in C^1 on dom $u \cap X$. Among the known differentiable gap functions that are covered by this class of merit functions we find those of [1,18,26,40], and [31,59,60,61].

Descent Properties

Optimization

Assume that x^t is not a solution to (2). We are interested in the conditions under which the direction of $d^t := y^t - x^t$ provides a descent direction for the merit function T. Let $d^t := \bar{y}^t - x^t$, where $\bar{y}^t$ is a possibly inexact solution to (3). Then, if $\Phi^t = \nabla \varphi^t$, the requirement is that

$$T_{\varphi^t}(\bar{y}^t) < T_{\varphi^t}(x^t), \quad (11)$$

that is, any improvement in the value of the subproblem objective over that at the current iterate is enough to provide a descent direction. To establish this result, one simply utilizes the convexity of φ^t and u and the formula for the directional derivative of T in the direction of d^t (see [45, Prop. 2.14.b]). We further note that (11) is possible to satisfy if and only if x^t is not a solution to (2); this result is in fact a special case of Theorem 1.

If Φ^t has stronger monotonicity properties, descent is also obtained when Φ^t is not necessarily a gradient mapping, and, further, if it is Lipschitz continuous then we can establish measures of the steepness of the search directions, extending the gradient relatedness conditions of unconstrained optimization. Let Φ^t be *strongly monotone* on dom $u \cap X$, that is, for $x, y \in \text{dom } u \cap X$,

$$[\Phi^t(x) - \Phi^t(y)]^\top (x - y) \geq m_{\Phi^t} \|x - y\|^2, \quad (12)$$

for some $m_{\Phi^t} > 0$. This can be used to establish that

$$T'(x^t; d^t) \leq -m_{\Phi^t} \|d^t\|^2.$$

If y^t is not an exact solution to (3), in the sense that for a vector $\bar{y}^t$, we satisfy a perturbation of (3) where its right-hand side 0^n is replaced by $r^t \neq 0^n$, then $d^t := \bar{y}^t - x^t$ is a descent direction for T at x^t if $\|r^t\| < m_{\Phi^t} \|d^t\|$.

Variational Inequality Problems

The requirements for obtaining a descent direction in the problem (6) are necessarily much stronger than in the problem (2), the reason being the much more complex form that the merit functions for (6) takes. (For example, the directional derivative of T at x in any direction d depends only on those quantities, while the directional derivative of ψ depends also on the argument y which defines its value at x.) Typically, monotonicity of the mapping F is required, as is evidenced in the above example of the Newton method. If further a differentiable merit function is used, the requirements are slightly strengthened, as the following example result shows.

Theorem 3 *(Descent properties, [45,60]) Assume that X is bounded, u is finite on X and F is monotone and in C^1 on X. Let $\varphi\colon X \times X \to R$ be a continuously differentiable function on $X \times X$ of the form $\varphi(y, x)$, strictly convex in y for each $x \in X$. Let $\alpha > 0$. Let $x \in X$, y be the unique vector in X satisfying*

$$\psi_\alpha(x) := \max_{y \in X} L_\alpha(y, x),$$

where

$$L_\alpha(y, x) := u(x) - u(y)$$
$$+ \frac{1}{\alpha}[\varphi(x, x) - \varphi(y, x)]$$
$$+ \left[F(x) - \frac{1}{\alpha}\nabla_y \varphi(x, x)\right]^\top (x - y).$$

Then, with $d := y - x$, either d satisfies

$$\psi'_\alpha(x; d) \leq -\gamma \psi_\alpha(x), \quad \gamma \in (0, 1),$$

or

$$\psi_\alpha(x) \leq -\frac{1}{\alpha(1-\gamma)}(\varphi(y, x) + \nabla_x \varphi(y, x)^\top d).$$

A descent algorithm is devised from this result as follows. For a given $x \in X$ and choice of $\alpha > 0$, the CA subproblem is solved with the scaled cost approximating, continuous and iteration-dependent function φ. If the resulting direction does not have the descent property, then the value of α is increased and the CA subproblem rescaled and resolved. Theorem 3 shows that a sufficient increase in the value of α will produce a descent direction unless x solves (6).

Steplength Rules

In order to establish convergence of the algorithm, the steplength taken in the direction of d^t must be such that the value of the merit function decreases sufficiently. An exact line search obviously works, but we will introduce simpler steplength rules that do not require a one-dimensional minimization to be performed.

The first is the *Armijo rule*. We assume temporarily that $u = 0$. Let $\alpha, \beta \in (0, 1)$, and $\ell := \beta^{\bar{i}}$, where $\bar{i}$ is the smallest nonnegative integer i such that

$$f(x^t + \beta^i d^t) - f(x^t) \leq \alpha \beta^i \nabla f(x^t)^\top d^t. \tag{13}$$

There exists a finite integer such that (13) is satisfied for any search direction $\bar{d}^t := \bar{y}^t - x^t$ satisfying (11), by the descent property and Taylor's formula (see [45, Lemma 2.24.b]).

In the case where $u \neq 0$, however, the situation becomes quite different, since $T := f + u$ is nondifferentiable. Simply replacing $\nabla f(x^t)^\top d^t$ with $T'(x^t; d^t)$ does not work. We can however use an overestimate of the predicted decrease $T'(x^t; d^t)$. Let $\alpha, \beta \in (0, 1)$, and $\ell := \beta^{\bar{i}}$, where $\bar{i}$ is the smallest nonnegative integer i such that

$$T(x^t + \beta^i d^t) - T(x^t)$$
$$\leq \alpha \beta^i [\nabla \varphi^t(x^t) - \nabla \varphi^t(y^t)]^\top d^t,$$

where now y^t necessarily is an exact solution to (3), and φ^t must further be strictly convex. We note that $T'(x^t; d^t) \leq [\nabla \varphi^t(x^t) - \nabla \varphi^t(y^t)]^\top d^t$ indeed holds, with equality in the case where $u = 0$ and $X = \mathbf{R}^n$ (see [45, Remark 2.28]).

To develop still simpler steplength rules, we further assume that ∇f is Lipschitz continuous, that is, that for $x, y \in \mathrm{dom}\, u \cap X$,

$$\|\nabla f(x) - \nabla f(y)\| \leq M_{\nabla f} \|x - y\|,$$

for some $M_\nabla f > 0$. The Lipschitz continuity assumption implies that for every $\ell \in [0, 1]$,

$$T(x^t + \ell d^t) - T(x^t)$$
$$\leq \ell \left[\nabla \varphi^t(x^t) - \nabla \varphi^t(y^t)\right]^\top d^t$$
$$+ \frac{M_{\nabla f}}{2} \ell^2 \|d^t\|^2;$$

adding a strong convexity assumption on φ^t yields that

$$T(x^t + \ell d^t) - T(x^t) \leq \ell \left(-m_{\varphi^t} + \frac{M_{\nabla f} \ell}{2}\right) \|d^t\|^2.$$

This inequality can be used to validate the *relaxation step*, which takes

$$\ell_t \in \left(0, \frac{2m_{\varphi^t}}{M_{\nabla f}}\right) \cap [0, 1], \quad (14)$$

and the *divergent series steplength rule*,

$$[0, 1] \supset \{\ell_t\} \to 0, \quad \sum_{t=0}^{\infty} \ell_t = \infty. \quad (15)$$

In the case of (14), descent is guaranteed in each step, while in the case of (15), descent is guaranteed after a finite number of iterations.

Convergence Properties

Convergence of the CA algorithm can be established under many combinations of
i) the properties of the original problem mappings;
ii) the choice of forms and convexity properties of the cost approximating mappings;
iii) the choice of accuracy in the computations of the CA subproblem solutions;
iv) the choice of merit function; and
v) the choice of steplength rule.

A subset of the possible results is found in [45, Chapt. 5-9]. Evident from these results is that convergence relies on reaching a critical mass in the properties of the problem and algorithm, and that, given that this critical mass is reached, there is a very large freedom-of-choice how this mass is distributed. So, for example, weaker properties in the monotonicity of the subproblem must be compensated both by stronger coercivity conditions on the merit function and by the use of more accurate subproblem solutions and steplength rules.

Decomposition CA Algorithms

Assume that dom $u \cap X$ is a *Cartesian product set*, that is, for some finite index set $\mathcal{C}$ and positive integers n_i with $\sum_{i \in \mathcal{C}} n_i = n$,

$$X = \prod_{i \in \mathcal{C}} X_i, \quad X_i \subseteq \mathbb{R}^{n_i};$$

$$u(x) = \sum_{i \in \mathcal{C}} u_i(x_i), \quad u_i \colon \mathbb{R}^{n_i} \to \mathbb{R} \cup \{+\infty\}.$$

Such problems arise in applications of equilibrium programming, for example in traffic ([41]) and Nash equilibrium problems ([21]); of course, box constrained and unconstrained problems fit into this framework as well.

The main advantage of this problem structure is that one can devise several decomposition versions of the CA algorithm, wherein components of the original problem are updated upon in parallel or sequentially, independently of each other. With the right computer environment at hand, this can mean a dramatic increase in computing efficiency. We will look at three computing models for decomposition CA algorithm, and compare their convergence characteristics. In all three cases, decomposition is achieved by choosing the cost approximating mapping separable with respect to the partition of $\mathbf{R}^n$ defined by $\mathcal{C}$:

$$\Phi(x)^\top = [\Phi_1(x_1)^\top, \ldots, \Phi_{|\mathcal{C}|}(x_{|\mathcal{C}|})^\top]. \quad (16)$$

The individual subproblems, given x, then are to find y_i, $i \in \mathcal{C}$, such that

$$\Phi_i(y_i) + \partial u_i(y_i) + N_{X_i}(y_i) + F_i(x) - \Phi_i(x_i)$$
$$\ni 0^{n_i};$$

if $\Phi_i \equiv \nabla \varphi_i$ for some convex function $\varphi_i \colon \text{dom } u_i \cap X_i \to \mathbf{R}$ in C^1 on dom $u_i \cap X_i$, then this is the optimality conditions for

$$\min_{y_i \in X_i} T_{\varphi_i}(y_i)$$
$$:= \varphi_i(y_i) + u_i(y_i) + [F_i(x) - \nabla \varphi_i(x_i)]^\top y_i.$$

Sequential Decomposition

The *sequential CA algorithm* proceeds as follows. Given an iterate $x^t \in \text{dom } u \cap X$ at iteration t, choose an index $i_t \in \mathcal{C}$ and a cost approximating mapping $\Phi_{i_t}^t$, and solve the problem of finding $y_{i_t}^t \in \mathbf{R}^{n_{i_t}}$ such that $(i = i_t)$

$$\Phi_i^t(y_i^t) + \partial u_i(y_i^t) + N_{X_i}(y_i^t) + F_i(x^t) - \Phi_i^t(x_i^t)$$
$$\ni 0^{n_i}.$$

Let $y_j^t := x_j^t$ for all $j \in \mathcal{C} \setminus \{i_t\}$ and $d^t := y^t - x^t$. The next iterate, x^{t+1}, is then defined by $x^{t+1} := x^t + \ell_t d^t$, that is,

$$x_j^{t+1} := \begin{cases} x_j^t + \ell_t(y_j^t - x_j^t), & j = i_t, \\ x_j^t, & j \neq i_t, \end{cases}$$

for some value of ℓ_t such that $x_{i_t}^t + \ell_t(y_{i_t}^t - x_{i_t}^t) \in \text{dom } u_{i_t} \cap X_{i_t}$, and the value of a merit function ψ is reduced sufficiently.

Assume that F is the gradient of a function $f: \text{dom } u \cap X \to \mathbf{R}$. Let the sequence $\{i_t\}$ be chosen according to the *cyclic rule*, that is, in iteration t,

$$i_t := t \pmod{|\mathcal{C}|} + 1.$$

Choose the cost approximating mapping ($i = i_t$)

$$y_i \mapsto \Phi_i^t(y_i) := \nabla_i f(x_{\neq i}^t, y_i),$$
$$y_i \in \text{dom } u_i \cap X_i.$$

Note that this mapping is monotone whenever f is convex in x_i. Since $\Phi_i^t(x_i^t) = \nabla_i f(x^t)$, the CA subproblem is equivalent (under this convexity assumption) to finding

$$y_i^t \in \arg\min_{y_i \in X_i}\{f(x_{\neq i}^t, y_i) + u_i(y_i)\}.$$

An exact line search would produce $\ell_t := 1$, since y_i^t minimizes $f(x_{\neq i}, \cdot) + u_i$ over $\text{dom } u_i \cap X_i$ (the remaining components of x kept fixed), and so $x_i^{t+1} := y_i^t$. The iteration described is that of the classic *Gauss–Seidel algorithm* ([35]) (also known as the *relaxation algorithm*, the *coordinate descent method*, and the *method of successive displacements*), originally proposed for the solution of unconstrained problems. The Gauss–Seidel algorithm is hence a special case of the sequential CA algorithm.

In order to compare the three decomposition approaches, we last provide the steplength requirement in the relaxation steplength rule (cf. (14)). The following interval is valid under the assumptions that for each $i \in \mathcal{C}$, $\nabla_i f$ is Lipschitz continuous on $\text{dom } u_i \cap X_i$ and each mapping Φ_i^t is strongly monotone:

$$\ell_{i,t} \in \left(0, \frac{2m_{\Phi_i^t}}{M_{\nabla_i f}}\right) \cap [0,1]. \tag{17}$$

Synchronized Parallel Decomposition

The synchronized parallel CA algorithm is identical to the original scheme, where the CA subproblems are constructed to match the separability structure in the constraints.

We presume the existence of a multiprocessor powerful enough to solve the $|\mathcal{C}|$ CA subproblems in parallel. (If fewer than $|\mathcal{C}|$ processors are available, then either some of the subproblems are solved in sequence or, if possible, the number of components is decreased; in either case, the convergence analysis will be the same, with the exception that the value of $|\mathcal{C}|$ may change.)

In the sequential decomposition CA algorithm, the steplengths are chosen individually for the different variable components, whereas the original CA algorithm uses a uniform steplength, ℓ_t. If the relative scaling of the variable components is poor, in the sense that F or u changes disproportionally to unit changes in the different variables x_i, $i \in \mathcal{C}$, then this ill-conditioning may result in a poor performance of the parallel algorithm. Being forced to use the same steplength in all the components can also have an unwanted effect due to the fact that the values of some variable components are close to their optimal ones while others may be far from optimal, in which case one might for example wish to use longer steps for the latter components. These two factors lead us to introduce the possibility to scale the component directions in the synchronized parallel CA algorithm. We stress that such effects cannot in general be accommodated into the original algorithm through a scaling of the mappings Φ_i^t. The scaling factors $s_{i,t}$ introduced are assumed to satisfy

$$0 < \underline{s}_i \leq s_{i,t} \leq 1, \quad i \in \mathcal{C}.$$

Note that the upper bound of one is without any loss of generality.

Assume that F is the gradient of a function $f: \text{dom } u \cap X \to \mathbf{R}$. In the parallel algorithm, choose the cost approximating mapping of the form (16), where for each $i \in \mathcal{C}$,

$$y_i \mapsto \Phi_i^t(y_i) := \nabla_i f(x_{\neq i}^t, y_i),$$
$$y_i \in \text{dom } u_i \cap X_i.$$

This mapping is monotone on $\text{dom } u \cap X$ whenever f is convex in each component x_i. Since $\Phi_i^t(x_i^t) = \nabla_i f(x^t)$, $i \in \mathcal{C}$, it follows that the CA subproblem is equivalent (under the above convexity assumption on f) to finding

$$y_i^t \in \arg\min_{y_i \in X_i}\{f(x_{\neq i}^t, y_i) + u_i(y_i)\}.$$

Choosing $\ell_t := 1$ and $s_{i,t} := 1$, $i \in \mathcal{C}$, yields $x^{t+1} := y^t$, and the resulting iteration is that of the *Jacobi algorithm* [8,35] (also known as the *method of simultaneous displacements*). The Jacobi algorithm, which was originally proposed for the solution of systems of equations, is therefore a parallel CA algorithm where the cost approximating mapping is (18) and unit steps are taken.

The admissible step in component i is $\ell s_{i,t} \in [0, 1]$, where

$$\ell \in \left(0, \min_{i \in \mathcal{C}} \left\{ \frac{2m_{\Phi_i^t}}{s_{i,t} M_{\nabla f}} \right\} \right). \qquad (18)$$

The maximal step is clearly smaller than in the sequential approach. To this conclusion contributes both the minimum operation and that $M_{\nabla_i f} \leq M_{\nabla f}$; both of these requirements are introduced here because the update is made over all variable components simultaneously. (An intuitive explanation is that the sequential algorithm utilizes more recent information when it constructs the subproblems.) One may therefore expect the sequential algorithm to converge to a solution with a given accuracy in less iterations, although the parallel algorithm may be more efficient in terms of solution time; the scaling introduced by $s_{i,t}$ may also improve the performance of the parallel algorithm to some extent.

Although the parallel version of the algorithm may speed-up the practical convergence rate compared to the sequential one, the need for synchronization in carrying out the updating step will generally deteriorate performance, since faster processors must wait for slower ones. In the next section, we therefore introduce an asynchronous version of the parallel algorithm, in which processors do not wait to receive the latest information available.

Asynchronous Parallel Decomposition

In the algorithms considered in this Section, the synchronization step among the processors is removed. Because the speed of computations and communications can vary among the processors, and communication delays can be substantial, processors will perform the calculations out of phase with each other. Thus, the advantage of reduced synchronization is paid for by an increase in interprocessor communications, the use of outdated information, and a more difficult convergence detection (see [8]). (Certainly, the convergence analysis also becomes more complicated.) Recent numerical experiments indicate, however, that the introduction of such *asynchronous computations* can substantially enhance the efficiency of parallel iterative methods (e. g., [6,11,15]).

The model of partial asynchronism that we use is as follows. For each processor (or, variable component) $i \in \mathcal{C}$, we introduce

a) initial conditions, $x_i(t) := x_i^0 \in X_i$, for all $t \leq 0$;
b) a set $\mathcal{T}^i$ of times at which x_i is updated; and
c) a variable $\tau_j^i(t)$ for each $j \in \mathcal{C}$ and $t \in \mathcal{T}^i$, denoting the time at which the value of x_j used by processor i at time t is generated by processor j, satisfying $0 \leq \tau_j^i(t) \leq t$ for all $j \in \mathcal{C}$ and $t \geq 0$.

We note that the sequential CA algorithm and the synchronized parallel CA algorithm can both be expressed as asynchronous algorithms: the cyclic sequential algorithm model is obtained from the choices $\mathcal{T}^i := \cup_{k \geq 0} \{|\mathcal{C}| \, k + i - 1\}$ and $\tau_j^i(t) := t$, while the synchronous parallel model is obtained by choosing $\mathcal{T}^i := \{1, 2, \dots\}$ and $\tau_j^i(t) := t$, for all i, j and t.

The communication delay from processor j to processor i at time t is $t - \tau_j^i(t)$. The convergence of the *partially asynchronous parallel decomposition CA algorithm* is based on the assumption that this delay is upper bounded: there exists a positive integer P such that

i) for every $i \in \mathcal{C}$ and $t \geq 0$, at least one element of $\{t, \dots, t + P - 1\}$ belongs to $\mathcal{T}^i$;
ii) $0 \leq t - \tau_j^i(t) \leq P - 1$ holds for all $i, j \in \mathcal{C}$ and all $t \geq 0$; and
iii) $\tau_i^i(t) = t$ holds for all $i \in \mathcal{C}$ and all $t \geq 0$.

In short, parts i) and ii) of the assumption state that no processor waits for an arbitrarily long time to compute a subproblem solution or to receive a message from another processor. (Note that a synchronized model satisfies $P = 1$.) Part iii) of the assumption states that processor i always uses the most recent value of its own component x_i of x, and is in [58] referred to as a *computational nonredundancy condition*. This condition holds in general when no variable component is updated simultaneously by more than one processor, as, for example, in message passing systems. For further discussions on the assumptions, we refer the reader to [8,57]; we only remark that they are easily enforced in practical implementations.

The iterate $x(t)$ is defined by the vector of $x_i(t)$, $i \in \mathcal{C}$. At a given time t, processor i has knowledge of a possibly outdated version of $x(t)$; we let

$$x^i(t)^\top := \left[x_1(\tau_1^i(t))^\top, \dots, x_{|\mathcal{C}|}(\tau_{|\mathcal{C}|}^i(t))^\top \right]$$

denote this vector. (Note that iii) above implies the relation $x_i^i(t) := x_i(\tau_i^i(t)) = x_i(t)$.)

To describe the (partially) asynchronous parallel CA algorithm, processor i updates $x_i(t)$ according to

$$x_i(t+1) := x_i(t) + \ell s_i(y_i(t) - x_i^i(t)),$$

$$t \in \mathcal{T}^i,$$

where $y_i(t)$ solves the CA subproblem defined at $x^i(t)$, and $s_i \in (0, 1]$ is a scaling parameter. (We define $d_i(t) := y_i(t) - x_i^i(t)$ to be zero at each $t \notin \mathcal{T}^i$.)

The admissible steplength for $i \in \mathcal{C}$ is $\ell s_i \in [0, 1]$, where

$$\ell \in \left(0, \frac{2 \min_{i \in \mathcal{C}}\{\frac{m\Phi_i}{s_i}\}}{M_{\nabla f}[1 + (|\mathcal{C}| + 1)P]}\right). \quad (19)$$

If further for some $M \geq 0$ and every $i \in \mathcal{C}$, all vectors x, y in dom u $\cap$ X with $x_i = y_i$ satisfy

$$\|\nabla_i f(x) - \nabla_i f(y)\| \leq M \|x - y\|, \quad (20)$$

then, in the above result, the steplength restrictions are adjusted to

$$\ell \in \left(0, \frac{2 \min_{i \in \mathcal{C}}\{\frac{m\Phi_i}{s_i}\}}{M_{\nabla f} + (|\mathcal{C}| + 1)MP}\right).$$

(We interpret the property (20) as a quantitative measure of the coupling between the variables.)

Most important to note is that the upper bound on ℓ is (essentially) inversely proportional to the maximal allowed asynchronism P; this is very intuitive, since if processors take longer steps then they should exchange information more often. Conversely, the more outdated the information is, the less reliable it is, hence the shorter step.

The relations among the steplengths in the three approaches (cf. (17), (18), and (19)) quantify the intuitive result that utilizing an increasing degree of parallelism and asynchronism results in a decreasing quality of the step directions, due to the usage of more outdated information; subsequently, smaller steplengths must be used. More detailed discussions about this topic is found in [45, Sect. 8.7.2].

See also

▶ Dynamic Traffic Networks

References

1. Auchmuty G (1989) Variational principles for variational inequalities. Numer Funct Anal Optim 10:863–874
2. Auslender A (1976) Optimisation: Méthodes numériques. Masson, Paris
3. Auslender A, Crouzeix JP, Fedit P (1987) Penalty-proximal methods in convex programming. J Optim Th Appl 55:1–21
4. Bazaraa MS, Sherali HD, Shetty CM (1993) Nonlinear programming: Theory and algorithms, 2nd edn. Wiley, New York
5. Bertsekas DP (1999) Nonlinear programming, 2nd edn. Athena Sci., Belmont, MA
6. Bertsekas DP, Castanon DA (1991) Parallel synchronous and asynchronous implementations of the auction algorithm. Parallel Comput 17:707–732
7. Bertsekas DP, Tseng P (1994) Partial proximal minimization algorithms for convex programming. SIAM J Optim 4:551–572
8. Bertsekas DP, Tsitsiklis JN (1989) Parallel and distributed computation: Numerical methods. Prentice-Hall, Englewood Cliffs, NJ
9. Bregman LM (1966) A relaxation method of finding a common point of convex sets and its application to problems of optimization. Soviet Math Dokl 7:1578–1581
10. Browder FE (1966) Existence and approximation of solutions of nonlinear variational inequalities. Proc Nat Acad Sci USA 56:1080–1086
11. Chajakis ED, Zenios SA (1991) Synchronous and asynchronous implementations of relaxation algorithms for nonlinear network optimization. Parallel Comput 17:873–894
12. Cohen G (1978) Optimization by decomposition and coordination: A unified approach. IEEE Trans Autom Control AC-23:222–232
13. Cominetti R (1997) Coupling the proximal point algorithm with approximation methods. J Optim Th Appl 95:581–600
14. Eckstein J (1993) Nonlinear proximal point algorithms using Bregman functions, with applications to convex programming. Math Oper Res 18:202–226
15. El Baz D (1989) A computational experience with distributed asynchronous iterative methods for convex network flow problems. Proc. 28th IEEE Conf. Decision and Control, pp 590–591
16. Fletcher R (1987) Practical methods of optimization, 2nd edn. Wiley, New York
17. Frank M, Wolfe P (1956) An algorithm for quadratic programming. Naval Res Logist Quart 3:95–110
18. Fukushima M (1992) Equivalent differentiable optimization problems and descent methods for asymmetric variational inequality problems. Math Program 53:99–110
19. Goldstein AA (1964) Convex programming in Hilbert space. Bull Amer Math Soc 70:709–710

20. Ha CD (1990) A generalization of the proximal point algorithm. SIAM J Control Optim 28:503–512
21. Harker PT, Pang J-S (1990) Finite-dimensional variational inequality and nonlinear complementarity problems: A survey of theory, algorithms and applications. Math Program 48:161–220
22. Kiwiel KC (1998) Generalized Bregman projections in convex feasibility problems. J Optim Th Appl 96:139–157
23. Kiwiel KC (1998) Subgradient method with entropic projections for convex nondifferentiable minimization. J Optim Th Appl 96:159–173
24. Korpelevich GM (1977) The extragradient method for finding saddle points and other problems. Matekon 13:35–49
25. Larsson T, Migdalas A (1990) An algorithm for nonlinear programs over Cartesian product sets. Optim 21:535–542
26. Larsson T, Patriksson M (1994) A class of gap functions for variational inequalities. Math Program 64:53–79
27. Levitin ES, Polyak BT (1966) Constrained minimization methods. USSR Comput Math Math Phys 6:1–50
28. Luo Z-Q, Tseng P (1991) On the convergence of a matrix splitting algorithm for the symmetric monotone linear complementarity problem. SIAM J Control Optim 29:1037–1060
29. Mangasarian OL (1991) Convergence of iterates of an inexact matrix splitting algorithm for the symmetric monotone linear complementarity problem. SIAM J Optim 1:114–122
30. Marcotte P, Dussault J-P (1989) A sequential linear programming algorithm for solving monotone variational inequalities. SIAM J Control Optim 27:1260–1278
31. Marcotte P, Zhu D (1995) Global convergence of descent processes for solving non strictly monotone variational inequalities. Comput Optim Appl 4:127–138
32. Martinet B (1972) Détermination approchée d'un point fixe d'une application pseudo-contractante. CR Hebdom Séances de l'Acad Sci (Paris), Sér A 274:163–165
33. Minty GJ (1962) Monotone (nonlinear) operators in Hilbert space. Duke Math J 29:341–346
34. Moreau J-J (1965) Proximité et dualité dans un espace Hilbertien. Bull Soc Math France 93:273–299
35. Ortega JM, Rheinboldt WC (1970) Iterative solution of nonlinear equations in several variables. Acad. Press, New York
36. Pang J-S (1982) On the convergence of a basic iterative method for the implicit complementarity problem. J Optim Th Appl 37:149–162
37. Patriksson M (1993) Partial linearization methods in nonlinear programming. J Optim Th Appl 78:227–246
38. Patriksson M (1993) A unified description of iterative algorithms for traffic equilibria. Europ J Oper Res 71:154–176
39. Patriksson M (1993) A unified framework of descent algorithms for nonlinear programs and variational inequalities. PhD Thesis Dept. Math. Linköping Inst. Techn.
40. Patriksson M (1994) On the convergence of descent methods for monotone variational inequalities. Oper Res Lett 16:265–269
41. Patriksson M (1994) The traffic assignment problem – Models and methods. Topics in Transportation. VSP, Utrecht
42. Patriksson M (1997) Merit functions and descent algorithms for a class of variational inequality problems. Optim 41:37–55
43. Patriksson M (1998) Cost approximation: A unified framework of descent algorithms for nonlinear programs. SIAM J Optim 8:561–582
44. Patriksson M (1998) Decomposition methods for differentiable optimization problems over Cartesian product sets. Comput Optim Appl 9:5–42
45. Patriksson M (1998) Nonlinear programming and variational inequality problems: A unified approach. Applied Optim, vol 23. Kluwer, Dordrecht
46. Patriksson M (1999) Cost approximation algorithms with nonmonotone line searches for a general class of nonlinear programs. Optim 44:199–217
47. Polyak BT (1963) Gradient methods for the minimisation of functionals. USSR Comput Math Math Phys 3:864–878
48. Polyak BT (1967) A general method of solving extremum problems. Soviet Math Dokl 8:593–597
49. Polyak BT (1987) Introduction to optimization. Optim. Software, New York
50. Pshenichny BN, Danilin Yu M. (1978) Numerical methods in extremal problems. MIR, Moscow
51. Rockafellar RT (1976) Augmented Lagrangians and applications of the proximal point algorithm in convex programming. Math Oper Res 1:97–116
52. Rockafellar RT (1976) Monotone operators and the proximal point algorithm. SIAM J Control Optim 14:877–898
53. Shor NZ (1985) Minimization methods for non-differentiable functions. Springer, Berlin
54. Teboulle M (1997) Convergence of proximal-like algorithms. SIAM J Optim 7:1069–1083
55. Tikhonov AN (1963) Solution of incorrectly formulated problems and the regularization method. Soviet Math Dokl 4:1035–1038
56. Tikhonov AN, Arsenin VYa (1977) Solutions of ill-posed problems. Wiley, New York (translated from Russian)
57. Tseng P, Bertsekas DP, Tsitsiklis JN (1990) Partially asynchronous, parallel algorithms for network flow and other problems. SIAM J Control Optim 28:678–710
58. Üresin A, Dubois M (1992) Asynchronous iterative algorithms: Models and convergence. In: Kronsjö L, Shumsheruddin D (eds) Advances in Parallel Algorithms. Blackwell, Oxford, pp 302–342
59. Wu JH, Florian M, Marcotte P (1993) A general descent framework for the monotone variational inequality problem. Math Program 61:281–300
60. Zhu DL, Marcotte P (1993) Modified descent methods for solving the monotone variational inequality problem. Oper Res Lett 14:111–120
61. Zhu DL, Marcotte P (1994) An extended descent framework for variational inequalities. J Optim Th Appl 80:349–366

62. Zuhovickiĭ SI, Polyak RA, Primak ME (1969) Two methods of search for equilibrium points of n-person concave games. Soviet Math Dokl 10:279–282

Credit Rating and Optimization Methods

CONSTANTIN ZOPOUNIDIS, MICHAEL DOUMPOS
Department of Production Engineering and Management, Financial Engineering Laboratory, Technical University of Crete, Chania, Greece

MSC2000: 91B28 90C90 90C05 90C20 90C30

Article Outline

Synonyms
Introduction/Background
Definitions
Formulation
Methods/Applications
   Logistic Regression
   Neural Networks
   Support Vector Machines
   Multicriteria Value Models
      and Linear Programming Techniques
   Evolutionary Optimization
Conclusions
References

Synonyms

Credit scoring; Credit granting; Financial risk management; Optimization

Introduction/Background

Financial risk management has evolved over the past two decades in terms of both its theory and its practices. Economic uncertainties, changes in the business environment and the introduction of new complex financial products (e. g., financial derivatives) led financial institutions and regulatory authorities to the development of a new framework for financial risk management, focusing mainly on the capital adequacy of banks and credit institutions.

Banks and other financial institutions are exposed to many different forms of financial risks. Usually these are categorized as [14]:

- Market risk that arises from the changes in the prices of financial securities and currencies.
- Credit risk originating from the inability of firms and individuals to meet their debt obligations to their creditors.
- Liquidity risk that arises when a transaction cannot be conducted at the existing market prices or when early liquidation is required in order to meet payments obligations.
- Operational risk that originate from human and technical errors or accidents.
- Legal risk which is due to legislative restrictions on financial transactions.

Among these types of risk, credit risk is considered as the primary financial risk in the banking system and exists in virtually all income-producing activities [7]. How a bank selects and manages its credit risk is critically important to its performance over time.

In this context credit risk management defines the whole range of activities that are implemented in order to measure, monitor and minimize credit risk. Credit risk management has evolved dramatically over the last 20 years. Among others, some factors that have increased the importance of credit risk management include [2]: (i) the worldwide increase in the number of bankruptcies, (ii) the trend towards disintermediation by the highest quality and largest borrowers, (iii) the increased competition among credit institutions, (iv) the declining value of real assets and collateral in many markets, and (v) the growth of new financial instruments with inherent default risk exposure, such as credit derivatives.

Early credit risk management was primarily based on empirical evaluation systems of the creditworthiness of a client. CAMEL has been the most widely used system in this context, which is based on the empirical combination of several factors related to capital, assets, management, earnings and liquidity.

It was soon realized, however, that such empirical systems cannot provide a solid and objective basis for credit risk management. This led to an outgrowth of studies from academics and practitioners on the development of new credit risk assessment systems. These efforts were also motivated by the changing regulatory framework that now requires banks to implement specific methodologies for managing and monitoring their credit portfolios [4].

The existing practices are based on sophisticated statistical and optimization methods, which are used to develop a complete framework for measuring and monitoring credit risk. Credit rating models are in the core of this framework and are used to assess the creditworthiness of firms and individuals. The following sections describe the functionality of credit rating systems and the type of optimization methods that are used in some popular techniques for developing rating systems.

Definitions

As already noted, credit risk is defined as the likelihood that an obligor (firm or individual) will be unable or unwilling to fulfill debt obligations towards the creditors. In such a case, the creditors will suffer losses that have to be measured as accurately as possible.

The expected loss L_{it} over a period t from granting credit to a given obligor i can be measured as follows:

$$L_{it} = PD_{it} LGD_i EAD_i$$

where PD_{it} is the probability of default for the obligor i in the time period t, LGD_i is the percentage of exposure the bank might lose in case the borrower defaults and EAD_i is the amount outstanding in case the borrower defaults. The time period t is usually taken equal to one year.

In the new regulatory framework default is considered to have occurred with regard to a particular obligor when one or more of the following events has taken place [4,11]:

- it is determined that the obligor is unlikely to pay its debt obligations in full;
- a credit loss event associated with any obligation of the obligor;
- the obligor is past due more than 90 days on any credit obligation; or
- the obligor has filed for bankruptcy or similar protection from creditors.

The aim of credit rating models is to assess the probability of default for an obligor, whereas other models are used to estimate LGD and EAD. Rating systems measure credit risk and differentiate individual credits and groups of credits by the risk they pose. This allows bank management and examiners to monitor changes and trends in risk levels thus promoting safety and soundness in the credit granting process. Credit rating systems are also used for credit approval and underwriting, loan pricing, relationship management and credit administration, allowance for loan and lease losses and capital adequacy, credit portfolio management and reporting [7].

Formulation

Generally, a credit rating model can be considered as a mapping function $f : \mathbb{R}^n \longrightarrow G$ that estimates the probability of default of an obligor described by a vector $x \in \mathbb{R}^n$ of input features and maps the result to a set G of risk categories. The feature vector x represents all the relevant information that describes the obligor, including financial and nonfinancial data.

The development of a rating model is based on the process of Fig. 1.

The process begins with the collection of appropriate data regarding known cases in default and nondefault cases. These data can be taken from the historical database of a bank, or from external resources. At this data selection stage, some preprocessing of the data is necessary in order to transform the obtained data into useful features, to clean out the data from possible outliers and to select the appropriate set of features for the analysis. These steps lead to the final data $\{x_i, y_i\}_{i=1}^m$, where x_i is the input feature vector for obligor i, y_i in the known status of the obligor (e. g. $y_i = -1$ for cases

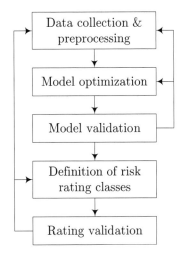

Credit Rating and Optimization Methods, Figure 1
The process for developing credit rating models

in default and $y_i = 1$ for nondefault cases), and m in the number of observations in the data set. These data, which are used for model development, are usually referred to as *training data*.

The second stage involves the optimization process, which refers to the identification of the model's parameters that best fit the training data. In the simplest case, the model can be expressed as a linear function of the form:

$$f(x) = x\beta + \beta_0$$

where $\beta \in \mathbb{R}^n$ is the vector with the coefficients of the selected features in the model and β_0 is a constant term. Other types of nonlinear models are also applicable.

In the above linear case, the objective of the optimization process is to identify the optimal parameter vector $\alpha = (\beta, \beta_0)$ that best fit the training data. This can be expressed as an optimization problem of the following general form:

$$\min_{\alpha \in S} \mathcal{L}(\alpha, X) \quad (1)$$

where S is a set of constraints that define the feasible (acceptable) values for the parameter vector α, X is the training data set and $\mathcal{L}$ is a loss function measuring the differences between the model's output and the given classification of the training observations.

The result of the model optimization process are validated using another sample of obligors with known status. This is referred to as the *validation sample*. Typically it consists of cases different than the ones of the training sample and for a future time period. The optimal model is applied to these new observations and its predictive ability is measured. If this is acceptable, then the model's outputs are used to define a set of risk rating classes (usually 10 classes are used). Each rating class is associated with a probability of default and it includes borrowers with similar credit risk levels. The defined rating needs also to be validated in terms of its stability over time, the distribution of the borrowers in the rating groups, and the consistency between the estimated probabilities of default in each group and the empirical ones which are taken from the population of rated borrowers.

Methods/Applications

The optimization problem (1) is expressed in different forms depending on the method used to develop the rating model. The characteristics of some popular methods are outlined below.

Logistic Regression

Logistic regression is the most widely used method in financial decision-making problems, with numerous applications in credit risk rating. Logistic regression assumes that the log of the probability odds is a linear function:

$$\log \frac{p}{1-p} = \beta_0 + x\beta$$

where $p = \Pr(1 \mid x)$ is the probability that an obligor x is a member of class 1, which is then expressed as

$$p = \left[1 + \exp^{-(\beta_0 + x\beta)}\right]^{-1}$$

The parameters of the model (constant term β_0 and coefficient vector β) are estimated to maximize the conditional likelihood of the classification given the training data. This is expressed as

$$\max_{\beta_0, \beta \in \mathbb{R}} \prod_{i=1}^{m} \Pr(y_i \mid x_i)$$

which can be equivalently written as

$$\max_{\beta_0, \beta \in \mathbb{R}} \sum_{i=1}^{m} \left[\frac{y_i + 1}{2} \ln(p_i) + \frac{1 - y_i}{2} \ln(1 - p_i)\right]$$

where $y_i = 1$ if obligor i is in the nondefault group and $y_i = -1$ otherwise.

Nonlinear optimization techniques such as the Newton algorithm are used to perform this optimization.

Logistic regression has been widely applied in credit risk rating both by academics and by practitioners [1]. Its advantages are mainly related to its simplicity and transparency: it provides direct estimates of the probabilities of default as well as estimates for the significance of the predictor variables and it is computationally feasible even for large data sets.

Neural Networks

Neural networks is a popular methodology for developing decision-making models in complex domains. A neural network is a network of parallel processing units (neurons) organized into layers. A typical structure of a neural network (Fig. 2) includes the following structural elements:

1. An input layer consisting of a set of nodes (processing units – neurons); one for each input to the network.
2. An output layer consisting of one or more nodes depending on the form of the desired output of the network. In classification problems, the number of nodes of the output layer is determined in accordance with the the number of groups.
3. A series of intermediate layers referred to as hidden layers. The nodes of each hidden layer are fully connected with the nodes of the subsequent and the proceeding layer.

Each connection between two nodes of the network is assigned a weight representing the strength of the connection. On the basis of the connections' weights, the input to each node is determined as the weighted average of the outputs from all the incoming connections. Thus, the input in_{ir} to node i of the hidden layer r is defined as follows:

$$in_{ir} = \sum_{j=0}^{r-1} \sum_{k=1}^{n_j} w_{ik}^j o_{ik} + \phi_{ir}$$

where n_j is the number of nodes at the hidden layer j, w_{ik} is the weight of the connection between node i at layer r and node k at layer j, o_{kj} is the output of node k at layer j and ϕ_{ir} an bias term.

The output of each node is specified through a transformation function. The most common form of this function is the logistic function:

$$o_{ir} = (1 + \exp^{-in_{ij}})^{-1}$$

The determination of the optimal neural network model requires the estimation of the connection weights and the bias terms of the nodes. The most widely used network training methodology is the backpropagation approach [18]. Nonlinear optimization techniques are used for this purpose [10,13,16].

Neural networks have become increasingly popular in recent years for the development of credit rating models [3]. Their main advantages include their ability to model complex nonlinear relationships in credit data, but they have also been criticized for their lack of transparency, the difficulty of specifying a proper architecture and the increased computational resources that are needed for large data sets.

Support Vector Machines

Support vector machines (SVMs) have become an increasingly popular nonparametric methodology for developing classification models. In a dichotomous classification setting, SVMs can be used to develop a linear decision function $f(x) = \text{sgn}(x\beta + \beta_0)$.

The optimal decision function f should maximize the margin induced in the separation of the classes [24], which is defined as $2/\|\beta\|$. Thus, the estimation of the optimal model is expressed as a quadratic programming problem of the following from:

$$\min \tfrac{1}{2} \beta^\top \beta + C e^\top d$$
$$\text{subject to } Y(X\beta + e\beta_0) + d \geq e \qquad (2)$$
$$\beta, \beta_0 \in \mathbb{R}, d \geq 0$$

where X is an $m \times n$ matrix with the training data, Y is an $m \times m$ matrix such that $Y_{ii} = y_i$ and $Y_{ij} = 0$ for all $i \neq j$, d is $m \times 1$ vector with nonnegative error (slack) variables defined such that $d_i > 0$ iff $y_i(x_i\beta + \beta_0) < 1$, e is a $m \times 1$ vector of ones, and $C > 0$ is a user-defined constant representing the trade-off between the two con-

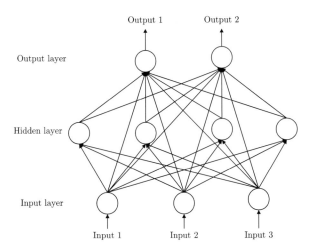

Credit Rating and Optimization Methods, Figure 2
A typical architecture of a neural network

flicting objectives (maximization of the separating margin and minimization of the training errors).

SVMs can also be used to develop nonlinear models. This is achieved by mapping the problem data to a higher-dimensional space H (feature space) through a transformation of the form $x_i x_j^\top = \phi(x_i)\phi^\top(x_j)$. The mapping function ϕ is implicitly defined through a symmetric positive definite kernel function $K(x_i, x_j) = \phi(x_i)\phi^\top(x_j)$ [22]. The representation of the data using the kernel function enables the development of a linear model in the feature space H.

For large training sets several computational procedures have been proposed to enable the fast training of SVM models. Most of these procedures are based on a decomposition scheme. The optimization problem (2) is decomposed into smaller subproblems taking advantage of the sparse nature of SVM models, since only a small part of the data (the support vectors), contribute to the final form of the model. A review of the algorithms for training SVMs can be found in [6].

SVMs seem to be a promising methodology for developing credit rating models. The algorithmic optimization advances enable their application to large credit data sets and they provide a unified framework for developing both linear and nonlinear models. Recent application of SVMs in credit rating can be found in [9,12,21].

Multicriteria Value Models and Linear Programming Techniques

The aforementioned classification methods assume that the groups are defined in a nominal way (i.e., the grouping provides a simple description of the cases). However, in credit risk modeling the groups are defined in an ordinal way, in the sense that an obligor classified in a low risk group is preferred to an obligor classified in a high risk group (in terms of its probability of default). Multicriteria methods are well-suited to the study of ordinal classification problems [26].

A typical multicriteria method that is well-suited for the development of credit rating models is the UTADIS method. The method leads to the development of an multiattribute additive value function:

$$V(x) = \sum_{j=1}^{n} w_j v_j(x_j)$$

where w_j is the weight of attribute j, and $v_j(x_j)$ is the corresponding marginal value function. Each marginal value function provides a monotone mapping of the performance of the obligors on the corresponding attribute in a scale between 0 (high risk) and 1 (low risk). According to [15], such an additive value function model is well-suited for credit scoring and is widely used by banks in their internal rating systems.

Using a piece-wise linear modeling approach, the estimation of the value function is performed based on a set of training data using linear programming techniques. For a two-class problem, the general form of the linear programming formulation is as follows [8]:

$$\min d_1 + d_2 + \cdots + d_m$$
$$\text{subject to: } y_i[V(x_i) - \beta] + d_i \geq \delta, \quad 1, 2, \ldots, m$$
$$w_1 + w_2 + \cdots + w_n = 1$$
$$w_j, d_i, \beta \geq 0$$

where β is a value threshold that distinguishes the two classes, δ is a small positive user-defined constant and $d_i = \max\{0, \delta - y_i[V(x_i) - \beta]\}$ denotes the classification error for obligor i.

Extensions of this framework and alternative linear programming formulations with applications to credit risk rating have been presented by [5,17,19]. The main advantages of these methodologies involve their computational efficiency and the simplicity and transparency of the resulting models.

Evolutionary Optimization

Evolutionary algorithms (EA) are stochastic search and optimization heuristics inspired from the theory of natural evolution. In an EA, different possible solutions of an optimization problem constitute the individuals of a population. The quality of each individual is assessed with a fitness (objective) function. Better solutions are assigned higher fitness values than worse performing solutions. The key idea of EAs is that the optimal solution can be found if an initial population is evolved using a set of stochastic genetic operators, similar to the "survival of the fittest" mechanism of natural evolution. The fitness values of the individuals in a population are used to define how they will be propagated to subsequent generations of populations. Most EAs include operators that select individuals for reproduction, produce new individuals based on those selected, and determine the composition of the population at the subsequent generation.

Well-known EAs and similar metaheuristic techniques include, among others, genetic algorithms, genetic programming, tabu search, simulated annealing, ant colony optimization and particle swarm optimization. EAs have been used to facilitate the development of credit rating systems addressing some important issues such as feature selection, rule extraction, neural network development, etc. Some recent applications can be found in the works of Varetto [20], Salcedo-Sanza et al. [25] and Tsakonas et al. [23].

Conclusions

Credit rating systems are in the core of the new regulatory framework for the supervision of financial institutions. Such systems support the credit granting process and enable the measurement and monitoring of credit risk exposure.

The increasing volume of credit data which are available for developing rating systems highlight the importance of implementing efficient optimization techniques for the construction of rating models. The existing optimization methods used in this field, are mainly based on nonlinear optimization, linear programming and evolutionary algorithms.

Future research is expected to take advantage of the advances in computer science, algorithmic developments regarding new forms of decision models, the analysis of the combination of different models, the comparative investigation on the performance of the existing methods and the implementation into decision support system that can be used by credit analysts in their daily practice.

References

1. Altman EI, Avery R, Eisenbeis R, Stinkey J (1981) Application of Classification Techniques in Business, Banking and Finance. JAI Press, Greenwich
2. Altman EI, Saunders A (1998) Credit risk measurement: Developments over the last 20 years. J Banking Finance 21:1721–1742
3. Atiya AF (2001) Bankruptcy prediction for credit risk using neural networks: A survey and new results. IEEE Trans Neural Netw 12:929–935
4. Basel Committee on Banking Supervision (2004) International Convergence of Capital Measurement and Capital Standards: A Revised Framework. Bank for International Settlements, Basel, Switzerland
5. Bugera V, Konno H, Uryasev S (2002) Credit cards scoring with quadratic utility functions. J Multi-Criteria Decis Anal 11:197–211
6. Campbell C (2002) Kernel methods: A survey of current techniques. Neurocomput 48:63–84
7. Comptroller of the Currency Administrator of National Banks (2001) Rating Credit Risk: Comptrollers Handbook. Comptroller of the Currency Administrator of National Banks, Washington, DC
8. Doumpos M, Zopounidis C (2002) Multicriteria Decision Aid Classification Methods. Kluwer, Dordrecht
9. Friedman C (2002) CreditModel technical white paper, Techical Report. Standard and Poor's, New York
10. Hagan MT, Menhaj M (1994) Training feedforward networks with the Marquardt algorithm. IEEE Trans Neural Netw 5:989–993
11. Hayden E (2003) Are credit scoring models sensitive with respect to default definitions? Evidence from the Austrian market. EFMA 2003 Helsinki Meetings (Available at SSRN: http://ssrn.com/abstract=407709)
12. Huang Z, Chen H, Hsu CJ, Chen WH, Wu S (2004) Credit rating analysis with support vector machines and neural networks: A market comparative study. Decis Support Syst 37:543–558
13. Hung MS, Denton JW (1993) Training neural networks with the GRG2 nonlinear optimizer. Eur J Oper Res 69:83–91
14. Jorion P (2000) Value at Risk: The New Benchmark for Managing Financial Risk, 2nd edn. McGraw-Hill, New York
15. Krahnen JP, Weber M (2001) Generally accepted rating principles: A primer. J Banking Finance 25:3–23
16. Moller MF (1993) A scaled conjugate gradient algorithm for fast supervised learning. Neural Netw 6:525–533
17. Mou TY, Zhou ZF, Shi Y (2006) Credit risk evaluation based on LINMAP. Lecture Notes Comput Sci 3994:452–459
18. Rumelhart DE, Hinton GE, Williams RJ (1986) Learning internal representation by error propagation. In: Rumelhart DE, Williams JL (eds) Parallel Distributed Processing: Explorations in the Microstructure of Cognition. MIT Press, Cambridge, pp 318–362
19. Ryu YU, Yue WT (2005) Firm bankruptcy prediction: Experimental comparison of isotonic separation and other classification approaches. IEEE Trans Syst, Man Cybern – Part A 35:727–737
20. Salcedo-Sanza S, Fernández-Villacañasa JL, Segovia-Vargas MJ, Bousoño-Calzón C (2005) Genetic programming for the prediction of insolvency in non-life insurance companies. Comput Oper Res 32:749–765
21. Schebesch KB, Stecking R (2005) Support vector machines for classifying and describing credit applicants: detecting typical and critical regions. J Oper Res Soc 56:1082–1088

22. Schölkopf B, Smola A (2002) Learning with Kernels: Support Vector Machines, Regularization, Optimization and Beyond. MIT Press, Cambridge
23. Tsakonas A, Dounias G, Doumpos M, Zopounidis C (2006) Bankruptcy prediction with neural logic networks by means of grammar-guided genetic programming. Expert Syst Appl 30:449–461
24. Vapnik VN (1998) Statistical Learning Theory. Wiley, New York
25. Varetto F (1998) Genetic algorithms applications in the analysis of insolvency risk. J Banking Finance 22:1421–1439
26. Zopounidis C, Doumpos M (2002) Multicriteria classification and sorting methods: A literature review. Eur J Oper Res 138:229–246

Criss-Cross Pivoting Rules

TAMÁS TERLAKY
Department Comput. & Software,
McMaster University, Hamilton, Canada

MSC2000: 90C05, 90C33, 90C20, 05B35, 65K05

Article Outline

Keywords
Synonyms
Introduction
   Ziont's Criss-Cross Method
The Least-Index Criss-Cross Method
   Other Interpretations
   Recursive Interpretation
   Lexicographically Increasing List
   Other Finite Criss-Cross Methods
   First-in Last-out Rule (FILO)
   Most Often Selected Variable Rule
   Exponential and Average Behavior
   Best-Case Analysis of Admissible Pivot Methods
Generalizations
   Fractional Linear Optimization
   Linear Complementarity Problems
   Convex Quadratic Optimization
   Oriented Matroids
See also
References

Keywords

Pivot rules; Criss-cross method; Cycling; Recursion; Linear optimization; Oriented matroids

Synonyms

Criss-cross

Introduction

From the early days of linear optimization (LO) (or linear programming), many people have been looking for a *pivot algorithm* that avoids the *two-phase procedure* needed in the simplex method when solving the general LO problem in standard primal form

$$\min \{c^\top x : Ax = b, x \geq 0\},$$

and its dual

$$\max \{b^\top y : A^\top y \leq c\}.$$

Such a method was assumed to rely on the intrinsic symmetry behind the *primal and dual problems* (i. e. it hoped to be *selfdual*), and it should be able to start with any *basic solution*.

There were several attempts made to relax the feasibility requirement in the simplex method. It is important to mention Dantzig's [7] parametric selfdual simplex algorithm. This algorithm can be interpreted as *Lemke's algorithm* [22] for the corresponding linear complementarity problem (cf. ▶ Linear complementarity problem) [23]. In the 1960s people realized that pivot sequences through possibly infeasible basic solutions might result in significantly shorter paths to the optimum. Moreover a selfdual one phase procedure was expected to make linear programming more easily accessible for broader public. Probably these advantages stimulated the introduction of the *criss-cross method* by S. Ziont [39,40].

Ziont's Criss-Cross Method

Assuming that the reader is familiar with both the primal and dual simplex methods, Ziont's criss-cross method can easily be explained.

- It can be initialized by any, possibly both primal and dual infeasible *basis*.

  If the basis is optimal, we are done.

  If the basis is not *optimal*, then there are some primal or dual infeasible variables. One might choose any of these.

  It is advised to choose once a primal and then a dual infeasible variable, if possible.

- If the selected variable is dual infeasible, then it enters the basis and the leaving variable is chosen among the primal feasible variables in such a way that primal feasibility of the currently primal feasible variables is preserved.

    If no such basis exchange is possible another infeasible variable is selected.

- If the selected variable is primal infeasible, then it leaves the basis and the entering variable is chosen among the dual feasible variables in such a way that dual feasibility of the currently dual feasible variables is preserved.

    If no such basis exchange is possible another infeasible variable is selected.

If the current basis is infeasible, but none of the infeasible variables allows a pivot fulfilling the above requirements then it is proved that the LO problem has no optimal solution.

Once a primal or dual feasible solution is reached then Ziont's method reduces to the primal or dual simplex method, respectively.

One attractive character of Ziont's criss-cross method is primal-dual symmetry (selfduality), and this alone differentiates itself from the simplex method. However it is not clear how one can design a finite version (i.e. a finite pivot rule) of this method. Both lexicographic perturbation and minimal index resolution seem not to be sufficient to prove finiteness in the general case when the initial basis is both primal and dual infeasible. Nevertheless, although not finite, this is the first published criss-cross method in the literature.

The other thread, that lead to finite criss-cross methods, was the intellectual effort to find finite, other than the lexicographic rule [4,8], variants of the simplex method. These efforts were also stimulated by studying the combinatorial structures behind linear programming. From the early 1970s in several branches of the optimization theory, finitely convergent algorithms were published. In particular A.W. Tucker [32] introduced the *consistent labeling* technique in the Ford–Fulkerson maximal flow algorithm; pivot selection rules based on least-index ordering, such as the Bard-type scheme for the *P*-matrix linear complementarity problem (K.G. Murty, [24]) and the celebrated least-index rule in linear and oriented matroid programming (R.G. Bland, [2]). A thorough survey of pivot algorithms can be found in [29].

It is remarkable that almost at the same time, in different parts of the world (China, Hungary, USA) essentially the same result was obtained independently by approaching the problem from quite different directions.

Below we will refer to the standard simplex (basis) tableau. A tableau is called *terminal* if it gives a primal and dual optimal solution or evidence of primal or dual infeasibility/inconsistency of the problem. Terminal tableaus have the following sign structure.

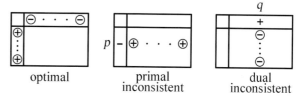

Terminal tableaus

The pivot operations at all known pivot methods, including all variants of the primal and dual simplex method and Ziont's criss-cross method have the following properties. When a primal infeasible variable is selected to leave the basis, the entering variable is selected so that after the pivot both variables involved in the pivot will be primal feasible. Analogously, when a dual infeasible variable is selected to enter the basis, then the leaving variable is selected in such a way that after the pivot both variables involved in the pivot will be dual feasible. Such pivots are called *admissible*. The sign structure of tableaus at an admissible pivot of 'type I' and 'type II' are demonstrated by the following figure.

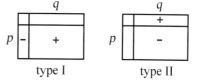

Admissible pivot situations

Observe that, while dual(primal) simplex pivots preserve dual(primal) feasibility of the basic solution, admissible pivots do not in general. Admissible pivots extract the greedy nature of pivot selection, i.e. 'repair primal/dual infeasibility' of the pivot variables.

The Least-Index Criss-Cross Method

The first finite criss-cross algorithm, which we call the *least-index criss-cross method*, was discovered independently by Y.Y. Chang [26], T. Terlaky [26,27,28] and Zh. Wang [34]; further, a strongly related general recursion by D. Jensen [18]. Chang presented the algorithm for positive semidefinite linear complementarity problems, Terlaky for linear optimization, for oriented matroids and with coauthors for QP, LCP and for oriented matroid LCP [9,16,19], while Wang primarily for the case of oriented matroids.

The least-index criss-cross method is perhaps the simplest finite pivoting method to LO problems. This criss-cross method is a purely combinatorial pivoting method, it uses admissible pivots and traverses through different (possibly both primal and dual infeasible) bases until the associated basic solution is optimal, or an evidence of primal or dual infeasibility is found.

To ease the understanding a figure is included that shows the scheme of the least-index criss-cross method.

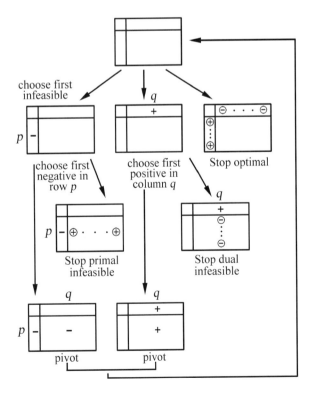

Scheme of the least-index criss-cross method

Observe the simplicity of the algorithm:
- It can be initiated with any basis.
- No two phases are needed.
- No ratio test is used to preserve feasibility, only the signs of components in a basis tableau and a prefixed ordering of variables determine the pivot selection.

Several finiteness proofs for the least-index criss-cross method can be found in the literature. The proofs are quite elementary, they are based on the orthogonality of the primal and dual spaces [14,26,28,29,34]; on recursive argumentation [11,18,33] or on lexicographically increasing lists [11,14].

0	Let an ordering of the variables be fixed. Let $T(B)$ be an arbitrary basis tableau (it can be neither primal nor dual feasible);
1	Let r be the minimal i such that either x_i is primal infeasible or x_i has a negative reduced cost. IF there is no r, THEN stop; the first terminal tableau is obtained, thus $T(B)$ is optimal.
2	IF x_r is primal infeasible THEN let $p := r$; $q := \min\{\ell : t_{p\ell} < 0\}$. IF there is no q, THEN stop; the second terminal tableau is obtained, thus the primal problem is infeasible. Go to Step 3. IF x_r is dual infeasible, THEN let $q := r$: $p := \min\{\ell : t_{\ell q} > 0\}$. IF there is no q, THEN stop: the third terminal tableau is obtained, thus the dual problem is infeasible. Go to Step 3.
3	Pivot on (p, q). Go to Step 1.

The least-index criss-cross rule

One of the most important consequences of the finiteness of the least-index criss-cross method is the *strong duality theorem* of linear optimization. This gives probably the simplest algorithmic proof of this fundamental result:

Theorem 1 (*Strong duality theorem*) *Exactly one of the following two cases occurs.*
- *At least one of the primal problem and the dual problem is infeasible.*
- *Both problems have an optimal solution and the optimal objective values are equal.*

Other Interpretations

The least-index criss-cross method can be interpreted as a *recursive algorithm*. This recursive interpretation and the finiteness proof based on it can be derived from the results in [2,3,18] and can be found in [33].

Recursive Interpretation

As performing the least-index criss-cross method at each pivot one can make a note of the larger of the two indices $r = \max\{p, q\}$ that entered or left the basis. In this list, an index must be followed by another larger one before the same index occurs anew.

The recursive interpretation is becoming apparent when one notes that it is based on the observation that the size of the solved subproblem (the subproblem for which a terminal tableau is obtained) is monotonically increasing.

The third interpretation is based on the proof technique developed by J. Edmonds and K. Fukuda [9] and adapted by Fukuda and T. Matsui [11] to the case of the least-index criss-cross method.

Lexicographically Increasing List

Let u be an binary vector with appropriate dimension, set initially to be the zero vector. In applying the algorithm let $r = \max\{p, q\}$ be the larger of the two indices p that entered or q that left the basis.

At each pivot update u as follows: let $u_r = 1$ and $u_i = 0$, $\forall i < r$. The remaining components of u stay unchanged. Then at each step of the least-index criss-cross method the vector u strictly increases lexicographically, thus the method terminates in a finite number of steps.

Other Finite Criss-Cross Methods

Both the recursive interpretation and the hidden flexibility of pivot selection in the least-index criss-cross method make it possible to develop other finite variants. Such finite criss-cross methods, which do not rely on a fixed minimal index ordering, were developed on the basis of the finite simplex rules presented by S. Zhang [37]. These finite criss-cross rules [38] are as follows.

First-in Last-out Rule (FILO)

First, choose a primal or dual infeasible variable that has changed its basis-nonbasis status most recently.

Then choose a variable in the selected row or column so that the pivot entry fulfills the sign requirement of the admissible pivot selection and which has changed its basis/nonbasis status most recently.

When more than one candidates occur with the same pivot age then one break tie as you like (e. g. randomly).

This rule can easily be realized by assigning an 'age' vector u to the vector of the variables and using a pivot counter k. Initially we set $k = 0$ and $u = 0$. Increase k by one at each pivot and we set the pivot coordinates of u equal to k. Then the pivot selections are made by choosing the variable with the highest possible u_i value satisfying the sign requirements.

Most Often Selected Variable Rule

First, choose a primal or dual infeasible variable that has changed its basis-nonbasis status most frequently.

Then choose a variable in the selected row or column so that the pivot entry fulfills the sign requirement of the admissible pivot selection and which has changed its basis/nonbasis status most frequently.

When more than one candidates occur with the same pivot age then one break tie as you like (e. g. randomly).

The most often selected rule can also be realized by assigning another 'age' vector u to the vector of the variables. Initially we set $u = 0$. At each pivot we increase the pivot-variable components of u by one. Then the pivot selections are made by choosing the variable with the highest possible u_i value satisfying the sign requirement.

Exponential and Average Behavior

The worst-case exponential behavior of the least-index criss-cross method was studied by C. Roos [25]. Roos' exponential example is a variant of the cube of V. Klee and G.J. Minty [21]. In this example the starting solution is the origin defined by a feasible basis, the variables are ordered so that the least-index criss-cross method follows a simplex path, i. e. without making any ratio test feasibility of the starting basis is preserved. Another exponential example was presented by Fukuda and Namiki [12] for linear complementarity problems.

Contrary to the clear result on the worst-case behavior, to date not much is known about the expected or average number of pivot steps required by finite criss-cross methods.

Best-Case Analysis of Admissible Pivot Methods

As it was discussed above, and it is the case for many simplex algorithms, the least-index criss-cross method is not a polynomial time algorithm. A question naturally arises: whether there exists a polynomial criss-cross method? Unfortunately no answer to this question is available at this moment. However some weaker variants of this question can be answered positively. The problem is stated as follows: An arbitrary basis is given. What is the shortest admissible pivot path from this given basis to an optimal basis?

For *nondegenerate problems*, [10] shows the existence of such an admissible pivot sequence of length at most m. The nondegeneracy assumption is removed in [15]. This result solves a relaxation of the d-step conjecture.

Observe, that we do not know of any such result for feasibility preserving, i. e. simplex algorithms. In fact, the maximum length of feasibility-preserving pivot sequences between two feasible bases is not known to be bounded by a polynomial in the size of the given LO problem.

Generalizations

Finite criss-cross methods were generalized to solve fractional linear optimization problems, to large classes of linear complementarity problems (LCPs; cf. ▶ Linear complementarity problem) and to oriented matroid programming problems (OMPs).

Fractional Linear Optimization

Fractional linear or, as it is frequently referred to, *hyperbolic programming*, can be reformulated as a linear optimization problem. Thus it goes without surprise that the least-index criss-cross method is generalized to this class of optimization problems as well [17].

Linear Complementarity Problems

The largest solvable class of LCPs is the class of LCPs with a *sufficient matrix* [5,6]. The LCP least-index criss-cross method is a proper generalization of the LO criss-cross method. When the LCP arises from a LO problem, the LO criss-cross method is obtained.

Convex Quadratic Optimization

Convex quadratic optimization problems give an LCP with a bisymmetric coefficient matrix. Because a bisymmetric matrix is semidefinite and semidefinite matrices form a subclass of sufficient matrices, one obtain a finite criss-cross algorithm for convex quadratic optimization problems as well. Such criss-cross algorithms were published e. g. in [20]. The least-index criss-cross method is extremely simple for the *P-matrix* LCP. Starting from an arbitrary complementary basis, here the least-indexed infeasible variable leaves the basis and it is replaced by its complementary pair. This algorithm was originally proposed in [24], and studied in [12]. The general case of sufficient LCPs was treated in [4,13,16].

Oriented Matroids

The intense research in the 1970s on *oriented matroids* and oriented matroid programming [2,9] gave a new insight in pivot algorithms. It became clear that although the simplex method has rich combinatorial structures, some essential results like the finiteness of Bland's least-index simplex rule [2] does not hold in the oriented matroid context. Edmonds and Fukuda [9] showed that it might cycle in the oriented matroid case due to the possibility of nondegenerate cycling which is impossible in the linear case.

The predecessors of finite criss-cross rules are: Bland's recursive algorithm [2,3], the Edmonds–Fukuda algorithm [9], its variants and generalizations [1,35,36,37]. All these are variants of the simplex method in the linear case, i. e. they preserve the feasibility of the basis, but not in the oriented matroid case. In the case of oriented matroid programming only Todd's finite lexicographic method [30,31] preserves feasibility of the basis and therefore yields a finite simplex algorithm for oriented matroids.

The least-index criss-cross method is a finite criss-cross method for oriented matroids [28,34]. A general recursive scheme of finite criss-cross methods is given in [18]. Finite criss-cross rules are also presented for oriented matroid quadratic programming and for oriented matroid linear complementarity problems [13,19].

See also

- ▶ Least-Index Anticycling Rules
- ▶ Lexicographic Pivoting Rules
- ▶ Linear Programming
- ▶ Pivoting Algorithms for Linear Programming Generating Two Paths
- ▶ Principal Pivoting Methods for Linear Complementarity Problems
- ▶ Probabilistic Analysis of Simplex Algorithms
- ▶ Simplicial Pivoting Algorithms for Integer Programming

References

1. Bjorner A, Vergnas M LAS, Sturmfels B, White N, Ziegler G (1993) Oriented matroids. Cambridge Univ Press, Cambridge
2. Bland RG (1977) A combinatorial abstraction of linear programming. J Combin Th B 23:33–57
3. Bland RG (1977) New finite pivoting rules for the simplex method. Math Oper Res 2:103–107
4. Chang YY (1979) Least index resolution of degeneracy in linear complementarity problems. Techn Report Dept Oper Res Stanford Univ 14
5. Cottle R, Pang JS, Stone RE (1992) The linear complementarity problem. Acad Press, New York
6. Cottle RW, Pang J-S, Venkateswaran V (1987) Sufficient matrices and the linear complementarity problem. Linear Alg Appl 114/115:235–249
7. Dantzig GB (1963) Linear programming and extensions. Princeton Univ Press, Princeton
8. Dantzig GB, Orden A, Wolfe P (1955) Notes on linear programming: Part I – The generalized simplex method for minimizing a linear form under linear inequality restrictions. Pacific J Math 5(2):183–195
9. Fukuda K (1982) Oriented matroid programming. PhD Thesis Waterloo Univ
10. Fukuda K, Luethi H-J, Namiki M (1997) The existence of a short sequence of admissible pivots to an optimal basis in LP and LCP. ITOR 4:273–284
11. Fukuda K, Matsui T (1991) On the finiteness of the criss-cross method. Europ J Oper Res 52:119–124
12. Fukuda K, Namiki M (1994) On extremal behaviors of Murty's least index method. Math Program 64:365–370
13. Fukuda K, Terlaky T (1992) Linear complementarity and oriented matroids. J Oper Res Soc Japan 35:45–61
14. Fukuda K, Terlaky T (1997) Criss-cross methods: A fresh view on pivot algorithms. Math Program (B) In: Lectures on Math Program, vol 79. ISMP97, Lausanne, pp 369–396
15. Fukuda K, Terlaky T (1999) On the existence of short admissible pivot sequences for feasibility and linear optimization problems. Techn Report Swiss Federal Inst Technol
16. Hertog D Den, Roos C, Terlaky T (1993) The linear complementarity problem, sufficient matrices and the criss-cross method. Linear Alg Appl 187:1–14
17. Illés T, Szirmai Á, Terlaky T (1999) A finite criss-cross method for hyperbolic programming. Europ J Oper Res 114:198–214
18. Jensen D (1985) Coloring and duality: Combinatorial augmentation methods. PhD Thesis School OR and IE, Cornell Univ
19. Klafszky E, Terlaky T (1989) Some generalizations of the criss-cross method for the linear complementarity problem of oriented matroids. Combinatorica 9:189–198
20. Klafszky E, Terlaky T (1992) Some generalizations of the criss-cross method for quadratic programming. Math Oper Statist Ser Optim 24:127–139
21. Klee V, Minty GJ (1972) How good is the simplex algorithm? In: Shisha O (ed) Inequalities-III. Acad Press, New York, pp 1159–175
22. Lemke CE (1968) On complementary pivot theory. In: Dantzig GB, Veinott AF (eds) Mathematics of the Decision Sci Part I. Lect Appl Math 11. Amer Math Soc. Providence, RI, pp 95–114
23. Lustig I (1987) The equivalence of Dantzig's self-dual parametric algorithm for linear programs to Lemke's algorithm for linear complementarity problems applied to linear programming. SOL Techn Report Dept Oper Res Stanford Univ 87(4)
24. Murty KG (1974) A note on a Bard type scheme for solving the complementarity problem. Opsearch 11(2–3):123–130
25. Roos C (1990) An exponential example for Terlaky's pivoting rule for the criss-cross simplex method. Math Program 46:78–94
26. Terlaky T (1984) Egy új, véges criss-cross módszer lineáris programozási feladatok megoldására. Alkalmazott Mat Lapok 10:289–296 English title: A new, finite criss-cross method for solving linear programming problems. (In Hungarian)
27. Terlaky T (1985) A convergent criss-cross method. Math Oper Statist Ser Optim 16(5):683–690
28. Terlaky T (1987) A finite criss-cross method for oriented matroids. J Combin Th B 42(3):319–327

29. Terlaky T, Zhang S (1993) Pivot rules for linear programming: A survey on recent theoretical developments. Ann Oper Res 46:203–233
30. Todd MJ (1984) Complementarity in oriented matroids. SIAM J Alg Discrete Meth 5:467–485
31. Todd MJ (1985) Linear and quadratic programming in oriented matroids. J Combin Th B 39:105–133
32. Tucker A (1977) A note on convergence of the Ford–Fulkerson flow algorithm. Math Oper Res 2(2):143–144
33. Valiaho H (1992) A new proof of the finiteness of the criss-cross method. Math Oper Statist Ser Optim 25:391–400
34. Wang Zh (1985) A conformal elimination free algorithm for oriented matroid programming. Chinese Ann Math 8(B1)
35. Wang Zh (1991) A modified version of the Edmonds–Fukuda algorithm for LP in the general form. Asia–Pacific J Oper Res 8(1)
36. Wang Zh (1992) A general deterministic pivot method for oriented matroid programming. Chinese Ann Math B 13(2)
37. Zhang S (1991) On anti-cycling pivoting rules for the simplex method. Oper Res Lett 10:189–192
38. Zhang S (1999) New variants of finite criss-cross pivot algorithms for linear programming. Europ J Oper Res 116:607–614
39. Zionts S (1969) The criss-cross method for solving linear programming problems. Managem Sci 15(7):426–445
40. Zionts S (1972) Some empirical test of the criss-cross method. Managem Sci 19:406–410

Cutting Plane Methods for Global Optimization

HOANG TUY
Institute Math., Hanoi, Vietnam

MSC2000: 90C26

Article Outline

Keywords
Outer Approximation
Inner Approximation
Concavity Cut
Nonlinear Cuts
References

Keywords

Cutting plane method; Outer approximation; Inner approximation; Polyhedral annexation; Concavity cut; Intersection cut; Convexity cut; Nonlinear cut; Polyblock approximation; Monotonic optimization

In solving global and combinatorial optimization problems cuts are used as a device to discard portions of the feasible set where it is known that no optimal solution can be found. Specifically, given the optimization problem

$$\min \{f(x) \colon x \in D \subset \mathbb{R}^n\}, \qquad (1)$$

if x^0 is an unfit solution and there exists a function $l(x)$ satisfying $l(x^0) > 0$, while $l(x) \leq 0$ for every optimal solution x, then by adding the inequality $l(x) \leq 0$ to the constraint set we exclude x^0 without excluding any optimal solution. The inequality $l(x) \leq 0$ is called a *valid cut*, or briefly, a *cut*. Most often the function $l(x)$ is affine: the cut is then said to be linear, and the hyperplane $l(x) = 0$ is called a cutting plane. However, nonlinear cuts have proved to be useful, too, for a wide class of problems.

Cuts may be employed in different contexts: outer and inner approximation (conjunctive cuts), branch and bound (disjunctive cuts), or in combined form.

Outer Approximation

Let $\Omega \subset \mathbf{R}^n$ be the set of optimal solutions of problem (2). Suppose there exists a family $\mathcal{P}$ of polytopes $P \supset \Omega$ such that for each $P \in \mathcal{P}$ a *distinguished point* $z(P) \in P$ (conceived of as some approximate solution) can be defined satisfying the following conditions:

A1) $z(P)$ always exists (unless $\Omega = \emptyset$) and can be computed by an efficient procedure;

A2) given any $P \in \mathcal{P}$ and the associated distinguished point $z = z(P)$, we can recognize when $z \in \Omega$ and if $z \notin \Omega$, we can construct an affine function $l(x)$ such that $P' = P \cap \{x \colon l(x) \leq 0\} \in \mathcal{P}$, and $l(z) > 0$, while $l(x) \leq 0$, $\forall x \in \Omega$, i.e. $\Omega \subset P' \subset P \setminus \{z\}$.

Under these conditions, one can attempt to solve problem (2) by the following *outer approximation method* (*OA method*) [8]:

Prototype OA (outer approximation) procedure

0	Start with an initial polytope $P_1 \in \mathcal{P}$. Set $k = 1$.
1	Compute the distinguished point $z^k = z(P_k)$ (by A1)). If $z(P_k)$ does not exist, terminate: the problem is infeasible. If $z(P_k) \in \Omega$, terminate. Otherwise, continue.
2	Using A2), construct an affine function $l_k(x)$ such that $P_{k+1} = P_k \cap \{x \colon l_k(x) \leq 0\} \in \mathcal{P}$ and $l_k(x)$ strictly separates z^k from Ω, i.e. satisfies $$l_k(z^k) > 0, \quad l_k(x) \leq 0 \; \forall x \in \Omega. \qquad (2)$$ Set $k \leftarrow k + 1$ and return to Step 1.

The algorithm is said to be *convergent* if it is either finite or generates an infinite sequence $\{z^k\}$ every cluster point of which is an optimal solution of problem (2).

Usually the distinguished point z^k is defined as a vertex of the polytope P_k satisfying some criterion (e. g., minimizing a given concave function). In these cases, the implementation of the above algorithm requires a procedure for computing, at iteration k, the vertex set V_k of the current polytope P_k. At the beginning, V_1 is supposed to be known, while P_{k+1} is obtained from P_k simply by adding one more linear constraint $l_k(x) \leq 0$. Using this information V_{k+1} can be derived from V_k by an on-line vertex enumeration procedure [1].

Example 1 (Concave minimization.) Consider the problem (1) where $f(x)$ is concave and D is a convex compact set with int $D \neq \emptyset$.

Assume that D is defined by a convex inequality $g(x) \leq 0$ and let $w \in$ int D. Take $\mathcal{P}$ to be the collection of all polytopes containing D. For every $P \in \mathcal{P}$ define $z := z(P)$ to be a minimizer of $f(x)$ over the vertex set V of P (hence, by concavity of $f(x)$, a minimizer of $f(x)$ over P). Clearly, if $z \in D$, it solves the problem. Otherwise, the line segment joining z to w meets the boundary of D at a unique point y and the affine function $l(x) = \langle p, x-y \rangle + g(y)$ with $p \in \partial g(y)$ strictly separates D from z (indeed, $l(z) = g(z) > 0$ while $l(x) \leq g(x) - g(z) + g(z) \leq 0$ for all $x \in D$. Obviously $P' = P \cap \{x : l(x) \leq 0\} \in \mathcal{P}$, so Assumptions A1) and A2) are fulfilled and the OA algorithm can be applied. The convergence of the algorithm is easy to establish.

Example 2 (Reverse convex programming.) Consider the problem (1) where $f(x) = \langle c, x \rangle$, while $D = \{x \in \mathbf{R}^n : h(x) \leq 0 \leq g(x)\}$ with $g(x), h(x)$ continuous convex functions. Assume that the problem is *stable*, i. e. that $D = \text{cl}(\text{int } D)$, so a feasible solution $\overline{x} \in D$ is optimal if and only if

$$\{x \in D: \ \langle c, x - \overline{x} \rangle \leq 0\} \subset \{x: \ g(x) \leq 0\}. \quad (3)$$

Also for simplicity assume a point w is available satisfying $\max\{h(w), g(w)\} < 0$ and $\langle c, w \rangle < \min\{\langle c, x \rangle : h(x) \leq 0 \leq g(x)\}$ (the latter assumption amounts to assuming that the constraint $g(x) \geq 0$ is essential).

Let Ω be the set of optimal solutions, $\mathcal{P}$ the collection of all polytopes containing Ω. For every $P \in \mathcal{P}$ let $z = z(P)$ be a maximizer of $g(x)$ over the vertex set V of the polyhedron $P \cap \}x : \langle c, x \rangle \leq \gamma\}$, where γ is the value of the objective function at the best feasible solution currently available (set $\gamma = +\infty$ if no feasible solution is known yet). By (3), if $g(z) \leq 0$, then γ is the optimal value (for $\gamma < +\infty$), or the problem is infeasible (for $\gamma = +\infty$). Otherwise, $g(z) > 0$, and we can construct an affine function $l(x)$ strictly separating z from Ω as follows. Since $\max\{h(w), g(w)\} < 0$ while $\max\{h(z), g(z)\} > 0$ the line segment joining z, w meets the surface $\max\{h(x), g(x)\} = 0$ at a unique point y.

1) If $g(y) = 0$ (while $h(y) \leq 0$), then y is a feasible solution and since $y = \lambda w + (1-\lambda) z$ for some $\lambda \in (0, 1)$ we must have $\langle c, y \rangle = \lambda \langle c, w \rangle + (1-\lambda)\langle c, z \rangle < \gamma$, so the cut $l(x) = \langle c, x-y \rangle \leq 0$ strictly separates z from Ω.

2) If $h(y) = 0$, then the cut $l(x) = \langle p, x-y \rangle + h(y) \leq 0$, where $p \in \partial h(y)$, strictly separates z from Ω (indeed, $l(x) \leq h(x) - h(y) + h(y) = h(x) \leq 0$ for all $x \in \Omega$ while $l(z) > 0$ because $l(w) < 0, l(y) = 0$).

Thus assumptions A1), A2) are satisfied, and again the OA algorithm can be applied. The convergence of the OA algorithm for this problem is established by a more elaborate argument than for the concave minimization problem (see [3,8]).

Various variants of OA method have been developed for a wide class of optimization problems, since any optimization problem described by means of differences of convex functions can be reduced to a reverse convex program of the above form [3]. However, a difficulty with this method when solving large scale problems is that the size of the vertex set V_k of P_k may grow exponentially with k, creating serious storage problems and making the computation of V_k almost impracticable.

Inner Approximation

Consider the concave minimization problem under linear constraints, i. e. the problem (2) when $f(x)$ is a concave function and D is a polytope in $\mathbf{R}^n$.

Without loss of generality we may assume that 0 is a vertex of D. For any real number $\gamma \leq f(0)$, the set $C_\gamma = \{x \in \mathbf{R}^n\}\{f(x) \geq \gamma\}$ is convex and $0 \in D \cap C_\gamma$. Of course, $D \subset C_\gamma$ if and only if $f(x) \geq \gamma$ for all $x \in D$.

The idea of the *inner approximation* method (IA method), also called the *polyhedral annexation* method (or PA method)[3], is to construct a sequence of expanding polytopes $P_1 \subset P_2 \subset \cdots$ together with a nonincreasing sequence of real numbers $\gamma_1 \geq \gamma_2 \geq \cdots$, such that $\gamma_k \in f(D)$, $P_k \subset C_{\gamma k}$, $k = 1, 2, \ldots$, and eventually $D \subset P_h$ for some h: then $\gamma_h \leq f(x)$ for all $x \in D$, i.e. γ_h will be the optimal value.

For every set $P \subset \mathbf{R}^n$ let P° be the *polar* of P, i.e. $P^\circ = \{y \in \mathbf{R}^n : \langle y, x \rangle \leq 1, \forall x \in P\}$. As is well known P° is a closed convex set containing 0 (in fact a polyhedron if P is a polyhedron), and $P \subset Q$ only if $P^\circ \supset Q^\circ$; moreover, if C is a closed convex set containing 0, then $(C^\circ)^\circ = C$. Therefore, setting $S_k = (P^k)^\circ$, the IA method amounts to constructing a sequence of nested polyhedra $S_1 \supset \cdots \supset S_h$ satisfying $S_k^\circ \subset C_{\gamma k}$, $k = 1, \ldots, h$ and $S_h \subset D^\circ$. The key point in this scheme is: Given $\gamma_k \in f(D)$ and a polyhedron S_k such that $S_k^\circ \subset C_{\gamma k}$, check whether $S_k \subset D^\circ$ and if there is $y^k \in S_k \setminus D^\circ$, then construct a cut $l_k(y) \leq 1$ to exclude y^k and to form a smaller polyhedron S_{k+1} such that $S_{k+1}^\circ \subset C_{\gamma k+1}$ for some $\gamma_{k+1} \in f(D)$ satisfying $\gamma_{k+1} \leq \gamma_k$.

To deal with this point, define $s(y) = \max\{\langle y, x \rangle : x \in D\}$. Since $y \in D^\circ$ whenever $s(y) \leq 1$ we will have $S_k \subset D^\circ$ whenever

$$\max\{s(y): y \in S_k\} \leq 1. \qquad (4)$$

But clearly the function $s(y)$ is convex as the pointwise maximum of a family of linear functions. Therefore, denoting the vertex set and the extreme direction set of S_k by V_k, U_k, respectively, we will have (4) (i.e. $S_k \subset D^\circ$) whenever

$$\begin{cases} \max\{s(y): y \in V_k\} \leq 1, \\ \max\{s(y): y \in U_k\} \leq 0. \end{cases} \qquad (5)$$

Thus, checking the inclusion $S_k \subset D^\circ$ amounts to checking (5), a condition that fails to hold in either of the following cases:

$$s(y^k) > 1 \quad \text{for some } y^k \in V_k \qquad (6)$$

$$s(y^k) > 0 \quad \text{for some } y^k \in U_k. \qquad (7)$$

In each case, it can be verified that if x^k maximizes $\langle y^k, x \rangle$ over D, and $\gamma_{k+1} = \min\{\gamma_k, f(x^k)\}$ while

$$\theta_k = \sup\left\{\theta: f(\theta x^k) \geq \gamma_{k+1}\right\},$$

then $S_{k+1} = S_k \cap \{y : \langle x^k, y \rangle \leq 1/\theta_k\}$ satisfies

$$P_{k+1} := S_{k+1}^\circ = \operatorname{conv}(P_k \cup \{\theta_k x^k\}) \subset C_{\gamma_{k+1}}.$$

In the case (6), S_{k+1} no longer contains y^k while in the case (7), y^k is no longer an extreme direction of S_{k+1}. In this sense, the cut $\langle x^k, y \rangle \leq 1/\theta_k$ excludes y^k. We can thus state the following algorithm.

IA Algorithm (for concave minimization)

0	By translating if necessary, make sure that 0 is a vertex of D. Let $\bar{x}^1$ be the best basic feasible solution available, $\gamma_1 = f(\bar{x}^1)$. Take a simplex $P_1 \subset C_{\gamma_1}$ and let $S_1 = P_1^\circ$, V_1 = vertex set of S_1, U_1 = extreme direction set of S_1. Set $k = 1$.
1	Compute $s(y)$ for every new $y \in (V_k \cup U_k) \setminus \{0\}$. If (5) holds, then terminate: $S_k \subset D^\circ$ so $\bar{x}^k$ is a global optimal solution.
2	If (6) or (7) holds, then let $$x^k \in \arg\max\{\langle y^k, x \rangle : x \in D\}.$$ Update the current best feasible solution by comparing x^k and $\bar{x}^k$. Set $\gamma_{k+1} = f(\bar{x}^{k+1})$.
3	Compute $\theta_k = \max\{\theta \geq 1 : f(\theta x^k) \geq \gamma_{k+1}\}$ and let $$S_{k+1} = S_k \cap \left\{y : \langle x^k, y \rangle \leq \tfrac{1}{\theta_k}\right\}.$$ From V_k and U_k derive the vertex set V_{k+1} and the extreme direction set U_{k+1} of S_{k+1}. Set $k \leftarrow k+1$ and go to Step 1.

It can be shown that the IA algorithm is finite [3]. Though this algorithm can be interpreted as dual to the OA algorithm, its advantage over the OA method is that it can be started at any vertex of D, so that each time the set V_k has reached a certain critical size, it can be stopped and 'restarted' at a new vertex of D, using the last obtained best value of $f(x)$ as the initial γ_1. In that way the set V_k can be kept within manageable size.

Note that if D is contained in a cone M and $P_1 = \{x \in M : \langle v^1, x \rangle \leq 1\} \subset C_{\gamma_1}$, then it can be shown that (7) automatically holds, and only (6) must be checked [6].

Concavity Cut

The cuts mentioned above are used to separate an unfit solution from some convex set containing at least one optimal solution. They were first introduced in convex programming [2,4]. Another type of cuts originally devised for concave minimization [7] is the following.

Suppose that a feasible solution $\bar{x}$ has already been known with $f(\bar{x}) = \gamma$ and we would like to check whether there exists a better feasible solution. One way to do that is to take a vertex x^0 of D with $f(x^0) > \gamma$ and to construct a cone M, as small as possible, vertexed at x^0, containing D and having exactly n edges. Since x^0 is interior to the convex set $C_\gamma = \{x : f(x) \geq \gamma\}$, each

ith edge of M, for $i = 1, \ldots, n$, meets the boundary of C_γ at a uniquely defined point y^i (assuming that C_γ is bounded). Through these n points $y^1, \ldots, y^n$ (which are affinely independent) one can draw a unique hyperplane, of equation $\pi(x - x^0) = 1$ such that $\pi(y^i - x^0) = 1$ $(i = 1, \ldots, n)$, hence $\pi = e^\top U^{-1}$, where U is the matrix of columns $y^1 - x^0, \ldots, y^n - x^0$ and e denotes a vector of n ones. Since the linear inequality

$$e^\top U^{-1}(x - x^0) \geq 1 \qquad (8)$$

excludes x^0 without excluding any feasible solution x better than $\bar{x}$, this inequality defines a valid cut. In particular, if it so happens that the whole polytope D is cut off, i. e. if

$$D \subset \{x: \ e^\top U^{-1}(x - x^0) \leq 1\}, \qquad (9)$$

then $\bar{x}$ is a global optimal solution.

This cut is often referred to as a γ-valid *concavity cut* for (f, D) at x^0 [3]. Its construction requires the availability of a cone $M \supset S$ vertexed at x^0 and having exactly n edges. In particular, if the vertex x^0 of D has exactly n neighboring vertices then M can be taken to be the cone generated by the n halflines from x^0 through each of these neighbors of x^0. Note, however, that the definition of the concavity cut can be extended so that its construction is possible even when the cone M has more than n edges (as e. g., when x^0 is a degenerated vertex of D).

Condition (9), sufficient for optimality, suggests a cutting method for solving the linearly constrained concave minimization problem by using concavity cuts to iteratively reduce the feasible polyhedron. Unfortunately, experience has shown that concavity cuts, when applied repeatedly, tend to become shallower and shallower. Though these cuts can be significantly strengthened by exploiting additional structure of the problem (e. g., in concave quadratic minimization, bilinear programming [5] and also in low rank nonconvex problems [6]), pure cutting methods are often outperformed by *branch and cut* methods where cutting is combined with successive partition of the space [8].

Concavity cuts have also been used in combinatorial optimization ('intersection cuts', or in a slightly extended form, 'convexity cuts').

Nonlinear Cuts

In many problems, *nonlinear cuts* arise in a quite natural way.

For example, consider the following problem of *monotonic optimization* [10]:

$$\max\{f(x): \ g(x) \leq 1, \ h(x) \geq 1, \ x \in \mathbf{R}^n_+\}, \qquad (10)$$

where f, g, h are continuous increasing functions on $\mathbf{R}^n_+$ (a function $f(x)$ is said to be *increasing* on $\mathbf{R}^n_+$ if $0 \leq x \leq x' \Rightarrow f(x) \leq f(x')$; the notation $x \leq x'$ means $x_i \leq x'_i$ for all i while $x < x'$ means $x_i < x'_i$ for all i). As argued in [10], a very broad class of optimization problems can be cast in the form (10). Define $G = \{x \in \mathbf{R}^n_+ g(x) \leq 1\}$, $H = \{x \in \mathbf{R}^n_+ : h(x) \geq 1\}$, so that the problem is to maximize $f(x)$ over the feasible set $G \cap H$. Clearly

$$0 \leq x \leq x' \in G \quad \Rightarrow \quad x \in G, \qquad (11)$$

$$0 \leq x \leq x' \notin H \quad \Rightarrow \quad x \notin H. \qquad (12)$$

Assume that $g(0) < 1$ and $0 < a \leq x \leq b$ for all $x \in G \cap H$ (so $0 \in \mathrm{int}\, G$, $b \in H$). From (11) it follows that if $z \in \mathbf{R}^n_+ \setminus G$ and $\pi(z)$ is the last point of G on the halfline from 0 through z, then the cone $K\pi(z)\} = \{x \in \mathbf{R}^n_+ : x > \pi(z)\}$ separates z from G, i. e. $G \cap K_\pi(z) = \emptyset$, while $z \in K_\pi(z)$.

A set of the form $P = \bigcup_{y \in V}\{x: 0 \leq y\}$, where V is a finite subset of $\mathbf{R}^n_+$, is called a *polyblock* of vertex set V [9]. A vertex v is said to be *improper* if $v \leq v'$ for some $v' \in V \setminus \{v\}$. Of course, improper vertices can be dropped without changing P. Also if $P \supset G \cap H$ then the polyblock of vertex set $V' = V \cap H$ still contains $G \cap H$ because $v \notin H$ implies that $[0, v] \cap H = \emptyset$. With these properties in mind we can now describe the *polyblock approximation* procedure for solving (10).

Start with the polyblock $P_1 = [0, b] \supset G \cap H$ and its vertex set $V_1 = \{b\} \subset H$. At iteration k we have a polyblock $P_k \supset G \cap H$ with vertex set $V_k \subset H$. Let $y^k \in \arg\max\{f(x) : x \in V_k\}$. Clearly y^k maximizes $f(x)$ over P_k, and $y^k \in H$, so if $y^k \in G$ then y^k is an optimal solution. If $y^k \notin G$ then the point $x^k = \pi(y^k)$ determines a cone K_{x^k} such that the set $P_{k+1} = P_k \setminus K_{x^k}$ excludes y^k but still contains $G \cap H$. It turns out that P_{k+1} is a polyblock whose vertex set V_{k+1} is obtained from V_k by adding n points $v^{k,1}, \ldots, v^{k,n}$ (which are the n vertices of the hyperrectangle $[x^k, y^k]$ adjacent to y^k) and then dropping all those which do not belong to H. With this polyblock P_{k+1}, we pass to iteration $k+1$.

In that way we generate a nested sequence of polyblocks $P_1 \supset P_2 \supset \cdots \supset G \cap H$. It can be proved that either y^k is an optimal solution at some iteration k or $f(y^k) \searrow \gamma := \max\{f(x) : x \in G \cap H\}$.

A similar method can be developed for solving the problem

$$\min \{f(x): \ g(x) \leq 1, h(x) \geq 1, x \in \mathbb{R}_+^n\}$$

by interchanging the roles of g, h and a, b. In contrast with what happens in OA methods, the vertex set V_k of the polyblock P_k in the polyblock approximation algorithm is extremely easy to determine. Furthermore this method admits restarts, which provide a way to prevent stall and overcome storage difficulties when solving large scale problems [10].

References

1. Chen P, Hansen P, Jaumard B (1991) On-line and off-line vertex enumeration by adjacent lists. Oper Res Lett 10:403–409
2. Cheney EW, Goldstein AA (1959) Newton's method for convex programming and Tchebycheff approximation. Numerische Math 1:253–268
3. Horst R, Tuy H (1996) Global optimization: deterministic approaches, 3rd edn. Springer, Berlin
4. Kelley JE (1960) The cutting plane method for solving convex programs. J SIAM 8:703–712
5. Konno H (1976) A cutting plane algorithm for solving bilinear programs. Math Program 11:14–27
6. Konno H, Thach PT, Tuy H (1997) Optimization on low rank nonconvex structures. Kluwer, Dordrecht
7. Tuy H (1964) Concave programming under linear constraints. Soviet Math 5:1437–1440
8. Tuy H (1998) Convex analysis and global optimization. Kluwer, Dordrecht
9. Tuy H (1999) Normal sets, polyblocks and monotonic optimization. Vietnam J Math 27(4):277–300
10. Tuy H (2000) Monotonic optimization: Problems and solution approaches. SIAM J Optim 11(2):464–494

Cutting-Stock Problem

ANDRÁS PRÉKOPA[1], CSABA I. FÁBIÁN[2]
[1] RUTCOR, Rutgers Center for Operations Research, Piscataway, USA
[2] Eötvös Loránd University, Budapest, Hungary

MSC2000: 90B90, 90C59

Article Outline

Keywords
See also
References

Keywords

Cutting-stock problem; Cutting pattern; Column generation; Knapsack problem

A company that produces large rolls of paper, textile, steel, etc., usually faces the problem of how to cut the large rolls into smaller rolls, called finished rolls, in such a way that the demands for all finished rolls be satisfied. Any large roll is cut according to some cutting pattern and the problem is to find the cutting patterns to be used and to how many large rolls they should be applied. We assume, for the sake of simplicity, that each large roll has width W, an integer multiple of some unit and the finished roll widths are also specified by some integers $w_1, \ldots, w_m$. Let a_{ij} designate the number of rolls of width w_i produced by the use of the jth pattern, $i = 1, \ldots, m, j = 1, \ldots, n$. Let further b_i designate the demand for roll i, $i = 1, \ldots, m$, and $c_j = 1, j = 1, \ldots, n$. If $A = (a_{ij})$, $\mathbf{b} = (b_1, \ldots, b_m)^\mathsf{T}$, $\mathbf{c} = (c_1, \ldots, c_n)^\mathsf{T}$, then the problem is:

$$\begin{cases} \min & \mathbf{c}^\mathsf{T} \mathbf{x} \\ \text{s.t.} & A\mathbf{x} = \mathbf{b} \\ & \mathbf{x} \geq \mathbf{0}. \end{cases}$$

Here x_j means the number of jth cutting patterns to be used and as such, an all integer solution would be required to the problem. However, one is usually satisfied with an optimal solution of the above problem without the integrality restriction and, having that, a simple round-off procedure provides us with the solution to the problem.

In the above problem, however, the matrix A is huge, therefore we do not, and in most cases cannot, create it, by enumerating the cutting patterns. P.C. Gilmore and R.E. Gomory [3,4] resolved this difficulty by an ingenious column generation technique. It works in such a way that we generate column j, in the course of the simplex algorithm, whenever needed. Assume that B is the current basis and designate by π the corresponding dual vector, i. e., the solution of the linear equation: $\pi^\top B = c_B^\top$. Now, if $\mathbf{a} = (a_1, \ldots, a_m)^\top \in \mathbb{Z}_+^m$ satisfies the inequality $\mathbf{w}^\top \mathbf{a} \leq W$, then, by definition, $\mathbf{a}$ represents a cutting pattern, a column of the matrix A. Since the cutting-stock problem is a minimization problem, the basis B is optimal if $\pi^\top \mathbf{a} \leq 1$ for any $\mathbf{a}$ that satisfies $\mathbf{w}^\top \mathbf{a} \leq W$. We can check it by solving the linear program:

$$\begin{cases} \min & \pi^\top \mathbf{a} \\ \text{s.t.} & \mathbf{w}^\top \mathbf{a} \leq W \\ & \mathbf{a} \in \mathbb{Z}_+^m. \end{cases}$$

If the optimum value is greater than 1, then the optimal $\mathbf{a}$ vector may enter the basis, otherwise B is an optimal basis and $\mathbf{x}_B$ is an optimal solution to the problem. The problem to find the vector $\mathbf{a}$ is a knapsack problem for which efficient solution methods exist.

In practice, however, frequently more complicated cutting-stock problems come up, due to special customer requirements depending on quality and other characteristics. In addition, we frequently need to include set up costs, capacity constraints and costs due to delay in manufacturing. These lead to the development of special algorithms as described in [1,4,5,6,7]. Recently Cs.I. Fábián [2] formulated stochastic variants of the cutting-stock problem, for use in fiber manufacturing.

See also

▶ Integer Programming

References

1. Dyckhoff H, Kruse HJ, Abel D, Gal T (1985) Trim loss and related problems. OMEGA Internat J Management Sci 13: 59–72
2. Fábián CsI (1998) Stochastic programming model for optical fiber manufacturing. RUTCOR Res Report 34–98
3. Gilmore PC, Gomory RE (1961) A linear programming approach to the cutting stock problem. Oper Res 9:849–859
4. Gilmore PC, Gomory RE (1963) A linear programming approach to the cutting stock problem, Part II. Oper Res 11:863–888
5. Gilmore PC, Gomory RE (1965) Multistage cutting stock problems of two and more dimensions. Oper Res 13:94–120
6. Johnson MP, Rennick C, Zak E (1998) Skiving addition to the cutting stock problem in the paper industry. SIAM Rev 39(3):472–483
7. Nickels W (1988) A knowledge-based system for integrated solving cutting stock problems and production control in the paper industry. In: Mitra G (ed) Mathematical Models for Decision Support. Springer, Berlin

Cyclic Coordinate Method
CCM

Vassilios S. Vassiliadis, Raúl Conejeros
Chemical Engineering Department,
University Cambridge, Cambridge, UK

MSC2000: 90C30

Article Outline

Keywords
See also
References

Keywords

Cyclic coordinate search; Line search methods; Pattern search; Aitken double sweep method; Gauss–Southwell method; Nondifferentiable optimization

Often the solution of multivariable optimization problems it is desired to be done with a *gradient-free algorithm*. This may be the case when gradient evaluations are difficult, or in fact gradients of the underlying optimization method do not exist. Such a method that offers this feature is the method of the cyclic coordinate search and its variants.

The minimization problem considered is:

$$\min_{\mathbf{x}} f(\mathbf{x}).$$

The method in its basic form uses the coordinate axes as the search directions. In particular, the search directions $\mathbf{d}^{(1)}, \ldots, \mathbf{d}^{(n)}$, where the $\mathbf{d}^{(i)}$ are vectors of zeros, except for a 1 in the ith position. Therefore along each search direction $\mathbf{d}^{(i)}$ the corresponding variable x_i is changed only, with all remaining variables being kept constant to their previous values.

It is assumed here that the minimization is carried out in order over all variables with indices $1, \ldots, n$ at each iteration of the algorithm. However there are variants. The first of these is the *Aitken double sweep method*, which processes first the variables in the order mentioned above, and then in the second sweep returns in reverse order, that is $n-1, \ldots, 1$. The second variant is termed the *Gauss–Southwell method* [2], according to which the component (variable) with largest partial derivative magnitude in the gradient vector is selected for line searching. The latter requires the availability of first derivatives of the objective function.

The algorithm of the cyclic coordinate method can be summarized as follows:

---

1. Initialization
Select a tolerance $\epsilon > 0$, to be used in the termination criterion of the algorithm. Select an initial point $\mathbf{x}^{(0)}$ and initialize by setting $\mathbf{z}^{(1)} = \mathbf{x}^{(0)}$. Set $k = 0$ and $i = 1$.

---

2. Main iteration
Let α_i^* (scalar variable) be the optional solution to the line search problem of minimizing $f(\mathbf{z}^{(i)} + \alpha \mathbf{d}_i)$. Set $\mathbf{z}^{(i+1)} = \mathbf{z}^{(i)} + \alpha_i^* \mathbf{d}^{(i)}$. If $j < n$, then increase i to $i+1$ and repeat step 2. Otherwise, if $j = n$, then go to step 3.

---

3. Termination check
Set $\mathbf{x}^{k+1} = \mathbf{z}^{(n)}$. If the termination criterion is satisfied, for example $\|\mathbf{x}^{(k+1)} - \mathbf{x}^{(k)}\| \leq \epsilon$, then stop. Else, set $\mathbf{z}^{(1)} = \mathbf{x}^{(k+1)}$. Increase k to $k+1$, set $i = 1$ and repeat step 2.

The steps above outline the basic cyclic coordinate method, the Aitken and Gauss–Southwell variants can be easily included by modifying the main algorithm.

In terms of convergence rate comparisons, D.G. Luenberger [3] remarks that such comparisons are not easy. However, an interesting analysis presented there indicates that roughly $n-1$ coordinate searches can be as effective as a single gradient search. Unless the variables are practically uncoupled from one another then coordinate search seems to require approximately n line searches to bring about the same effect as one step of steepest descent.

It can generally be proved that the cyclic coordinate method, when applied to a differentiable function, will converge to a stationary point [1,3]. However, when differentiability is not present then the method can stall at a suboptimal point. Interestingly there are ways to overcome such difficulties, such as by applying at every pth iteration (a heuristic number, user specified) the search direction $\mathbf{x}^{(k+1)} - \mathbf{x}^{(k)}$. This is even applied in practice for differentiable functions, as it is found to be helpful in accelerating convergence. These modifications are referred to as *acceleration steps* or *pattern searches*.

See also

▶ Powell Method
▶ Rosenbrock Method
▶ Sequential Simplex Method

References

1. Bazaraa MS, Sherali HD, Shetty CM (1993) Nonlinear programming, theory and algorithms. Wiley, New York
2. Forsythe GE (1960) Finite difference methods for partial differential equations. Wiley, New York
3. Luenberger DG (1984) Linear and nonlinear programming, 2nd edn. Addison-Wesley, Reading, MA

Printing: Krips bv, Meppel, The Netherlands
Binding: Stürtz, Würzburg, Germany